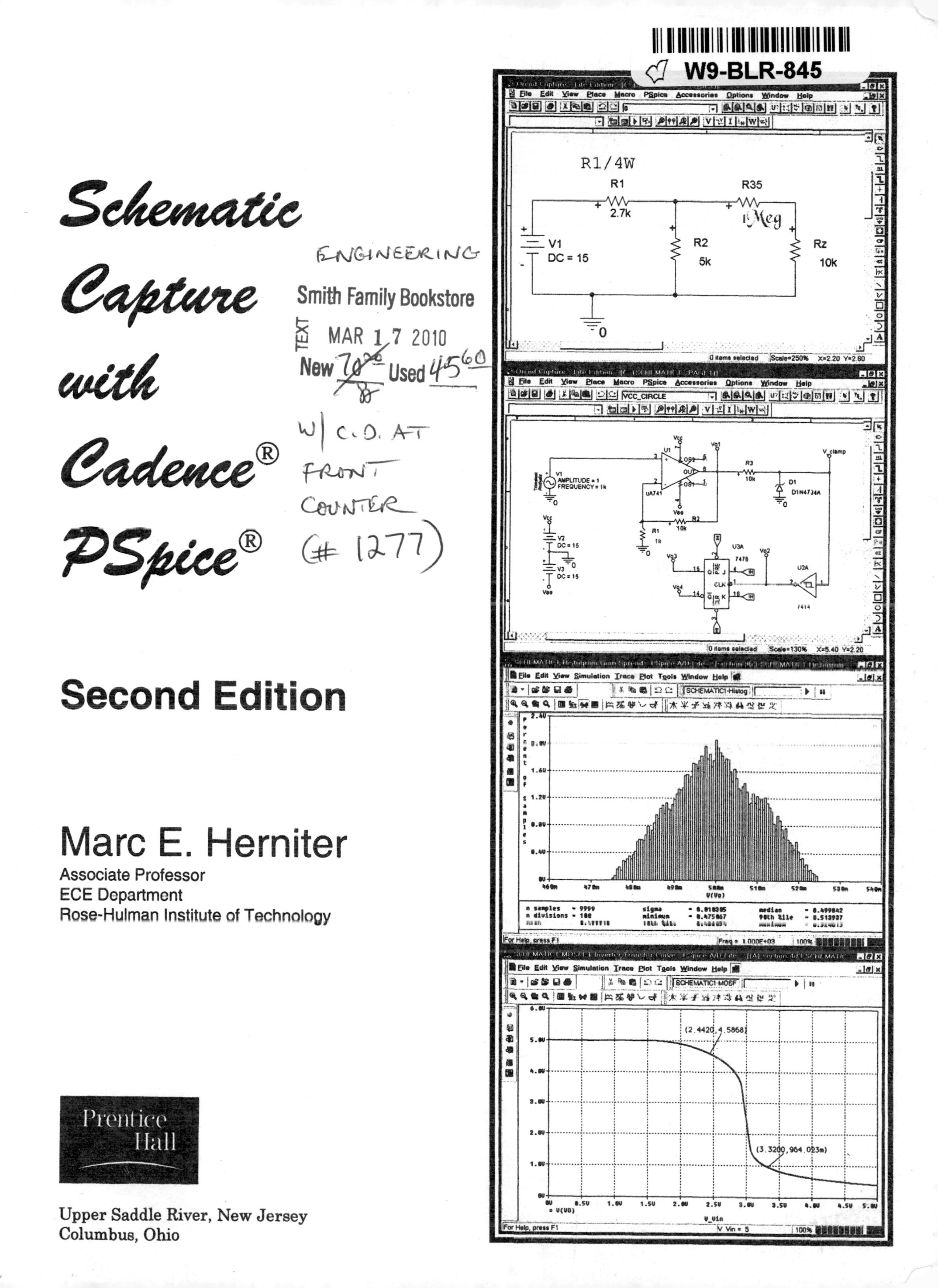
W9-BLR-845
Schematic
Capture
with
Cadence®
PSpice®
ENGINEERING
Smith Family Bookstore
TEXT MAR 17 2010
New 70.75 88 Used 45.60
W/ C.D. AT
FRONT
COUNTER
(# 1277)
Second Edition
Marc E. Herniter
Associate Professor
ECE Department
Rose-Hulman Institute of Technology
Prentice Hall
Upper Saddle River, New Jersey
Columbus, Ohio
R1/4W
R1
2.7k
R35
1Meg
V1
DC = 15
R2
5k
Rz
10k
0
(2.4420,4.5868)
(3.3200,964.023m)

**Library of Congress Cataloging-in-Publication Data**
Herniter, Marc E.
Schematic capture with Cadence PSpice / Marc E. Herniter.—2nd ed.
p. cm.
ISBN 0-13-048400-8
1. Electronics—Notation—Data processing. 2. Electronics—Charts, diagrams, etc. 3. Electronic drafting. 4. PSpice. I. Title
TK7866 .H438 2003
621.3815—dc21

2002019751

Editor in Chief: Stephen Helba
Acquisitions Editor: Dennis Williams
Production Editor: Stephen C. Robb
Design Coordinator: Karrie Converse-Jones
Cover Designer: Jeff Vanik
Production Manager: Pat Tonneman
Marketing Manager: David Gesell

The book was set in Times New Roman by Marc E. Herniter. It was printed and bound by Courier Kendallville, Inc.

Pearson Education Ltd.
Pearson Education of Australia Pty. Limited
Pearson Education Singapore Pte. Ltd.
Pearson Education North Asia Ltd.
Pearson Education Canada Ltd.
Pearson Educación de Mexico, S.A. de C.V.
Pearson Education–Japan
Pearson Education Malaysia Pte. Ltd.
Pearson Education, *Upper Saddle River, New Jersey*

---



10 9 8 7 6 5

ISBN 0-13-048400-8

# Warning and Disclaimer

This book is designed to provide tutorial information about the Cadence® set of computer programs. Every effort has been made to make this book as complete and accurate as possible, but no warranty or fitness is implied.

The information is provided on an "as is" basis. The author and Pearson Education shall have neither liability nor responsibility to any person or entity with respect to any loss or damages arising from the information contained in this book or from the use of the disks and programs that may accompany it.

# Trademarks

Adobe® and Acrobat® are registered trademarks of Adobe Systems Incorporated.

Cadence®, Designlab®, MicroSim®, Monte Carlo®, Orcad®, Orcad Capture®, Orcad Capture CIS®, Orcad Express®, Orcad Express CIS®, Orcad Layout®, Orcad PSpice®, Orcad PSpice A/D®, Parts®, Probe®, PSpice®, Schematics®, Stimulus Editor®, and StmEd® are registered trademarks of Cadence Design Systems, Inc.

IBM® is a registered trademark of International Business Machines Corporation.

Microsoft®, MS®, and MS-DOS® are registered trademarks and Windows™ and Win32s™ are trademarks of Microsoft Corporation.

Sun® is a registered trademark and Open Windows™ is a trademark of Sun Microsystems Incorporated.

To my wife, Corena,
my daughters, Katarina Alexis Sierra, Laina Calysta Dine'h, Darrian Iliana Francheska,
my parents, Mom and Dad,
my cats, Zak, Chester, Tipper, Cajsa, Kennedy, Blaise, and Sedona,
my dogs, Samantha, Sage, and Wolfee

# Preface

This manual is designed to show students how to use the PSpice circuit simulation program from Orcad with the schematic capture front end, Capture. It is a collection of examples that show students how to create a circuit, how to run the different analyses, and how to obtain the results from those analyses. This manual does not attempt to teach students circuit theory or electronics; that task is left for the main text. Instead, the manual takes the approach of showing students how to simulate many circuits found throughout the engineering curriculum. An example is the DC circuit shown below.

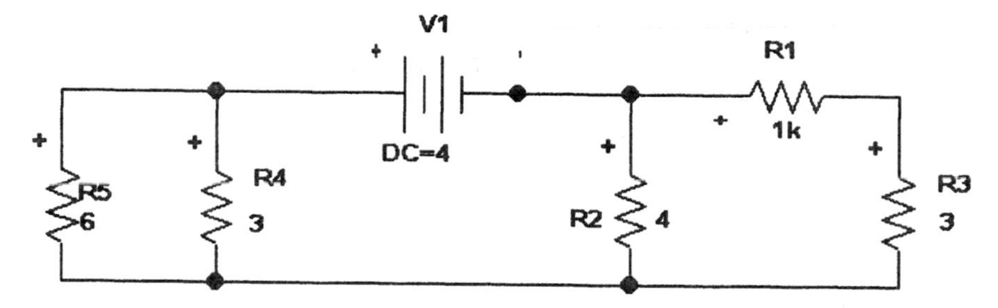


It is assumed that the student has been given enough information to completely analyze the circuit. This manual assumes that the student wishes to check his or her answers (or intuition) with this program. The student would construct the circuit as shown in Part 1 and then run either the node voltage analysis in Part 3 or the DC Sweep in Part 4. This circuit is different from the circuits in Parts 3 and 4, but the procedure given in those parts can be applied to the circuit.

This manual was designed to be used by students for their entire educational career and beyond. Since the parts are arranged by analysis type, they contain a range of examples from circuits covered in first-semester circuit theory courses to senior-level amplifier and switching circuits. Sections that are too advanced for beginning students may be skipped without loss of continuity. All parts contain both simple circuits and advanced circuits to illustrate the analysis types. Sections do not have to be covered sequentially. Individual examples can be identified that apply to specific courses. However, the following sequence is suggested for first-time users. All beginning users should follow Parts 1 and 2 completely to learn how to draw, print, and save schematics, and how to use Probe, the graphical post-processor, for viewing results. All students should follow some of the examples in Parts 3 and 4 that are relevant to the course and also cover a few of the examples that may apply to earlier courses (if any). The early examples in these parts have the most step-by-step detail of how to use the software.

This manual contains examples that apply to courses throughout the engineering curriculum. Introductory circuits classes usually cover DC circuits, AC circuits with phasors, and transient circuits with a single capacitor or inductor and a switch. Examples are given to cover these types of problems. After reviewing the examples in this manual, a student should be able to simulate similar problems. A typical first electronics course may cover transistor biasing, amplifier gain, and amplifier frequency response. Examples of these analyses are also given. Higher-level electronics courses would cover Monte Carlo analysis, Worst Case analysis, and distortion. Examples of these types of analyses are included.

Exercises are given at the end of each section. These exercises specify a circuit and give the simulation results. The students are encouraged to work these problems to see if they can obtain the same simulation results. The exercises are intended to give students practice in using the software, not to teach them circuits. Since this software covers such a wide variety of courses, problems are not given. These problems are best left to the instructor or main text. Using PSpice on problems specific to the class material is far more instructional than using it on problems designed to teach PSpice. My philosophy is that PSpice should be used only to verify one's own calculations or intuition. In my classes I assign problems that are worked by hand calculation, simulated with PSpice, and then tested in the lab. The students then compare the measured results to the hand calculations and PSpice simulations. Without hand calculation, it is impossible to know if the PSpice simulations are correct.

The book is written as if the instructor were giving a class demonstration on how to use the software. Intermediate windows are shown, and all mouse selections are specified. When I first started teaching the schematic capture version, I brought the students into the computer lab. I gave a lecture using an LCD projection screen and an overhead projector. The students could see the screens projected by the overhead and could follow along using their own computers. It required too much lecture time to cover the wide scope of the Orcad Lite software, so I wrote this manual using the philosophy that the screen captures presented in this manual would be the same as if I were presenting the software in a lecture.

The main advantage of the schematic capture front end of PSpice is its ease of use. At first this may not be apparent. When I first started to use Capture, I tested it with a simple three- or four-node circuit. Since I was familiar with writing netlists, it was far easier to write a simple four-line netlist than to search through the many menus of Capture and create a schematic. As I became more proficient at using the program and remembering the standard parts, I could create a schematic faster than I could type a netlist. The schematic version becomes much easier when you use parts with which you are not familiar, such as an exponential or pulsed voltage source. How many of us can remember the order of the parameters in a pulsed voltage source? Usually you have to look them up in the manual. With Capture, a manual is not necessary. Suppose that you get a part called **Vpulse**. The parameters of the source are listed in the part's attributes, and the order is not needed since Capture takes care of the order automatically. Another example would be an operational amplifier. If you were describing the circuit using a netlist, you would first have to find the order and number of calling nodes of the op-amp subcircuit. To figure out the calling nodes, you have to look at the library listing that contains the netlist of the op-amp subcircuit. Since the MS-DOS operating system did not allow multitasking, this usually involved exiting PSpice and listing the library. In the schematic capture version, you only need to get the op-amp part. All nodes are shown on the schematic, and the correct calling order is not needed. This makes the schematic capture version far simpler to use.

Another major advantage of the schematic capture version is that students find drawing circuits much more interesting than writing netlists. Students tend to explore the schematic capture front end much more than a text-based shell. Since all of the analyses and parts are available on-line as windows, graphics, or help files, students tend to explore the abilities of the program and they don't have to dig through a manual. In the text-based version of PSpice, students would first come to the instructor rather than look through a manual.

There are many other advantages to using the schematic capture version. The enormous popularity of the Microsoft Windows operating system should attest to the ease of use of a graphics-based interface compared to a text-based interface. One such advantage is automatic documentation. When you simulate a circuit, you automatically have a circuit schematic. This schematic can be incorporated into lab notebooks and reports. In a corporate environment, documentation of this type is extremely important. With the Windows operating system, the schematics can be incorporated into written documents using screen captures. This manual is an example of what can be accomplished.

The version of Orcad Lite described in this manual is Version 9.2 displayed using Windows 2000. The parts libraries described in this manual have been changed slightly from Orcad's distribution libraries. New parts were added to make the program simpler for beginning students to use. Please note that the libraries contained in this manual are slightly different from the factory distribution libraries.

**Software Included with the Manual**

The software is provided on a CD-ROM. The following items are included:

- Orcad Lite Version 9.2 – This is the version of Capture and PSpice that was used to compose this manual. The software is included courtesy of Cadence Design Systems, Inc..
- Adobe Acrobat Reader Version 4.05.
- All circuit files used in this book.

**Software Updates and Manual Supplements**

Periodic updates of the software can be obtained from Orcad's web site at www.orcad.com. Updates to the libraries and supplements to the text can be obtained from the author's web site at http://www.rose-hulman.edu/~herniter/.

**PSpice Mailing List**

This mailing list will only be used to:

- Notify you when new libraries are available for download.
- Notify you when new sections have been added to the this text. Usually these sections are available for download.

To subscribe to the mailing list, send an email to majordomo@seawolf.rose-hulman.edu. In the body of the email place the line "subscribe pspice" (without the quotes).

### Comments and Suggestions

The author would appreciate any comments or suggestions on this manual. Comments and suggestions from students are especially welcome. Please feel free to contact the author using any of the methods listed below:


- **E-mail:** Marc.Herniter@rose-hulman.edu.
- **Phone:** (812) 877-8512
- **FAX:** (253) 369-9536
- **Mail:** Rose-Hulman Institute of Technology, CM123, 5500 Wabash Avenue, Terre Haute, IN, 47803-3999.


### Acknowledgments

I would like to thank my students at Rose-Hulman Institute of Technology for giving me continued inspiration to improve this manual. Without their constant curiosity, this book would not be necessary. I am extremely grateful to Cadence Design Systems, Inc., for allowing us to distribute the Lite version of Capture and PSpice with this manual. I would like to thank Phil Kilcoin of Cadence Design Systems, Inc., for his help with this text and on future projects, and Tim Christensen, also of Cadence Design Systems, Inc., for answering my many questions. Finally, I would like to express my deepest appreciation to my wife, Corena, who now sits in front of her computer twenty-four hours a day.

# Before You Begin

**Software Limitations**

- Capture Lite and Capture CIS Lite Edition
    - Save ability for designs or libraries falls under the following limitations:
        - 60 instances maximum. For example, you can place 1 part 60 times, 60 parts once each, or any combination in between. Note: some parts contain more than 1 "instance," thus lowering the actual number of parts that can be saved.
        - 15 parts per library. You cannot modify or save a part if its library contains more than 15 parts.
    - No support for EDIF export.
    - Limited set of part libraries (about 25%) is included. All additional libraries can be found on our FTP site or activeparts.com.

- PSpice A/D Lite Edition
    - Circuit simulation is limited to circuits with up to:
        - 64 nodes
        - 10 transistors
        - 65 digital primitive devices
        - 10000 digital transitions
        - 10 transmission lines in total (ideal or non-ideal)
        - 4 pairwise coupled transmission lines
    - Device characterization (Model Editor) is limited to diodes.
    - Stimulus generation is limited to sine waves (analog) and clocks (digital).
    - Includes a sample library of 39 analog and 134 digital parts.
    - Creation of CSDF format data files is not supported.
    - In the Simulation Manager, only one simulation may be running or paused at a time.
    - Modifying runtime settings, with the Edit Runtime Settings dialog in the Simulation menu, is not supported. (The simulation terminates rather than pausing and allowing recovery from convergence failure.)
    - Display of simulation data is limited to simulations done with the Lite version of the simulator.

- PSpice Optimizer has the following limitations:
    - Requires PSpice A/D Lite Edition.
    - Limited to one goal, one parameter, and one constraint.

- Layout Lite Edition
    - No limitations are put on the ability to create, save, and use new library parts.
    - A single footprint library, demo.llb, which contains basic resistor, capacitor, diode, transistor patterns, plus selected DIP, SOIC, and PLCC patterns, is provided.
    - Designs containing a maximum of 100 connections can be saved.
    - Designs containing a maximum of 15 components can be saved.
    - Competitive PCB translators are not included (PCAD, PADS, Protel, Cadstar, Tango).
    - Autobackup is disabled.

**Minimum System Requirements**

- Pentium 90 MHz PC

- 32 MB RAM
- Hard disk space:
    - Capture CIS 89 MB
    - Layout Plus 66 MB
    - PSpice A/D 46 MB
- 800x600, 256 color VGA display
- Microsoft Windows 95/98, or Windows NT 4.0 Service Pack 3
- 4x CD-ROM drive
- 16-bit audio (recommended)
- Capture CIS requires three additional pieces of software that Orcad provides as part of the install if the user does not have them:
    - Microsoft Data Access Pack (ODBC)
    - DCOM95 (only needed with Microsoft Windows 95a)
    - Internet Explorer 4.0

**General Conventions**

- This manual assumes that you have a two- or three-button mouse. The words ***LEFT*** and ***RIGHT*** refer to the left and right mouse buttons.
- All text highlighted in bold refers to menu selections. Examples would be **File** and **Run.**
- All text in capital letters refers to keyboard selections. For example, press the **ENTER** key.
- ***All text in this font refers to text you will see on the computer screen. This applies to all text except menu selections.***
- **All text in this font refers to text you will type into the program.**
- The word "select" is to be interpreted as "click the left mouse button on."

**Keyboard Conventions**

Throughout the manual many keyboard sequences are given as shortcuts for making menu selections. The explanation of these sequences will be given later. It is important to know the conventions used to specify the sequences.

- Many control key sequences will be specified. For example, **CTRL-R** means hold down the "Ctrl" key and press the "R" key simultaneously. **CTRL-A** means hold down the "Ctrl" key and press the "A" key simultaneously. Not all keyboards are the same. Some keyboards may have a key labeled "Control" rather than "Ctrl."
- The keyboard sequence **ALT-TAB** in Microsoft Windows is used to toggle the active window. **ALT-TAB** means hold down the "Alt" key and press the "Tab" key simultaneously.

**Sign Conventions**

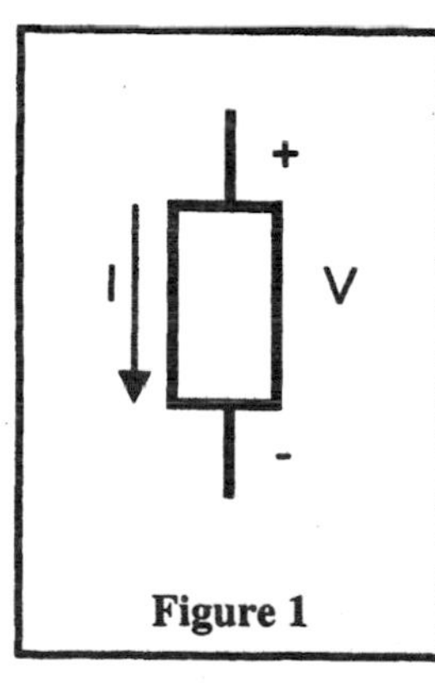

Figure 1

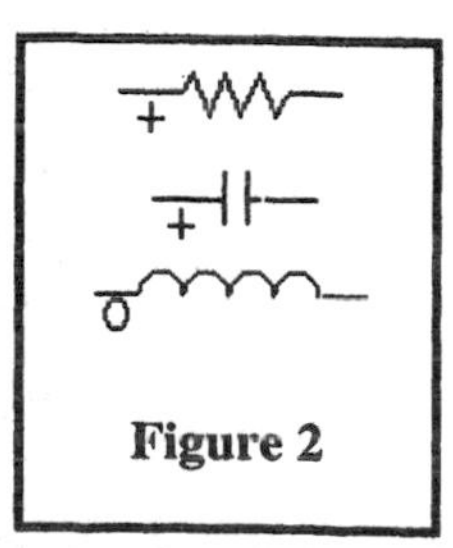

Figure 2

In circuit analysis it is important to know the polarity of voltage and direction of positive current flow. For resistors, capacitors, and inductors, the person doing the analysis usually assigns voltage and current references. The standard convention is that current is always positive entering the positive voltage terminal. This convention is shown in Figure 1. With this convention, if we know which terminal is marked positive for voltage, we know the direction for positive current flow. PSpice follows this convention and always assigns one of the terminals as the positive terminal. Knowing the polarity that PSpice assigns to components is not necessary until you ask for the current through a device or specify an initial condition. If we were interested in the current through a resistor, R1, for example, we would specify this current as I(R1). PSpice will give us the current through R1. The question is, which direction through R1 does PSpice interpret as positive? To solve this problem, a positive sign has been added to the resistor and capacitor graphics, and a dot has been added to the inductor graphic, as shown in Figure 2.

Also note that the dot in the inductor symbol is usually used to indicate polarity for mutual inductance. It is standard not to have a dot in the inductor symbol unless there is mutual coupling between inductors. However, to indicate polarity for a single inductor, we would have had to add a plus sign to the inductor symbol. The plus sign together with a dot could be confusing when talking about mutual inductance. **Thus, the dot is always specified in the inductor symbol. For single inductors, the dot means the same thing as the plus sign in the capacitor and resistor graphics. When mutual inductance is specified, the dot refers to the dot convention in mutual inductance.** The dot does not imply mutual inductance. To specify mutual coupling we must specify two parts, the inductor (L) and the coupling (K).

### Nomenclature Used in This Manual

This manual uses many terms associated with Windows. Some of the terms are shown here:

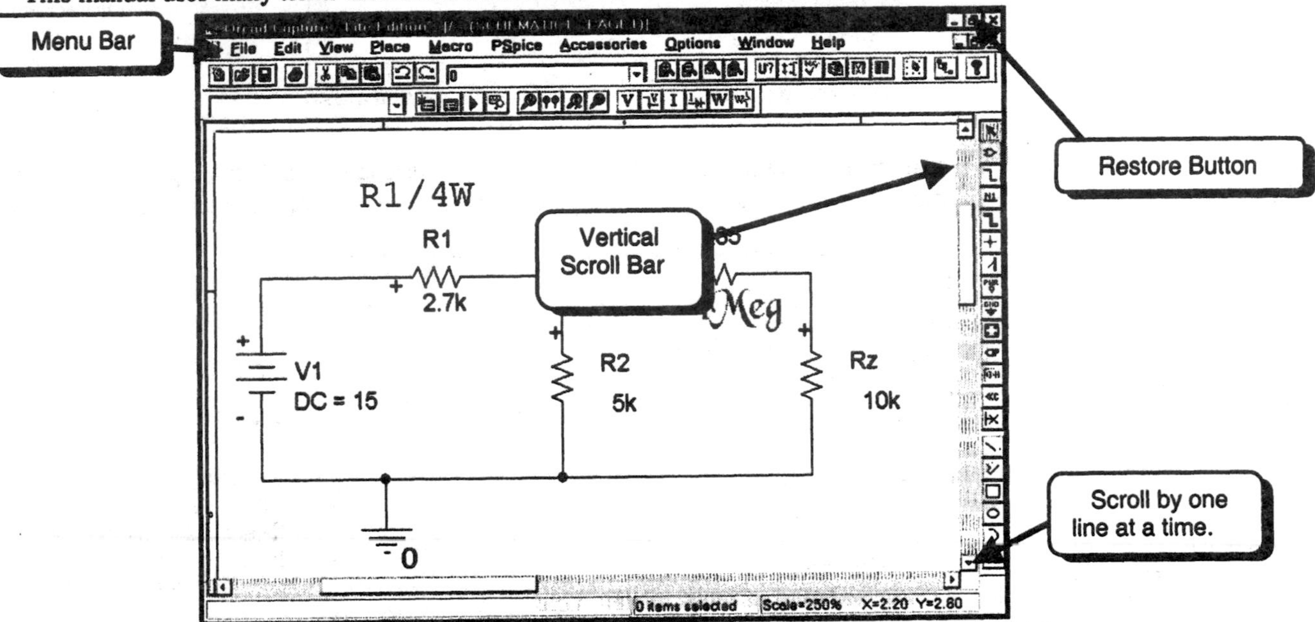


Note: The restore button shown above changes depending on the size of the window. The graphic is the restore button and means restore the window to its last dimensions, which were not full screen. The graphic is the maximize button and means expand a window to occupy the entire screen.

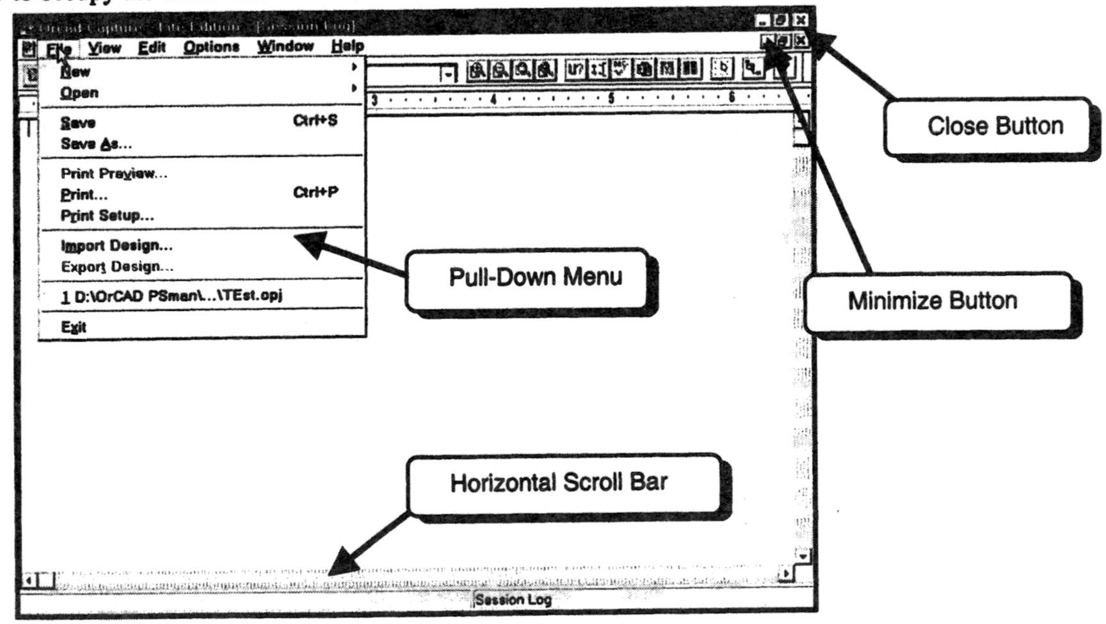

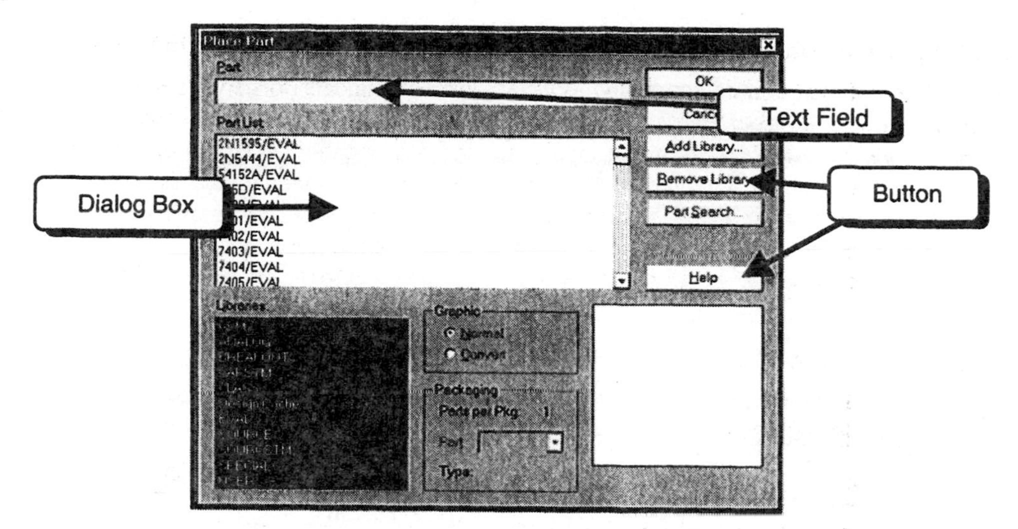


## Part Tables

All circuit examples in the manual are accompanied by part tables. These tables contain the graphic symbols of the parts in the circuit, give descriptions of the parts, and specify the part names. An example of a part table is shown below:

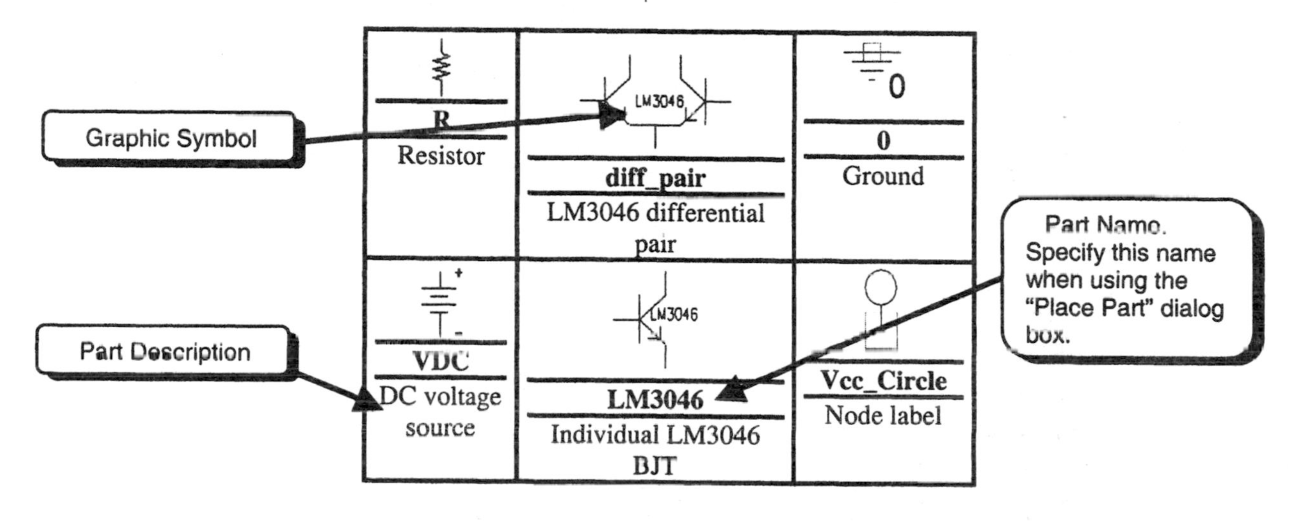

# Contents

# PART 1
# Editing a Basic Schematic with Orcad Capture

This part assumes that the Microsoft Windows operating system and the Orcad Lite Software are already installed on your computer. If the Orcad Lite software has not been installed, refer to Appendix A for installation instructions. **If you do not follow the installation instructions in Appendix A, the libraries specific to this manual will not be installed correctly.** If Windows has not been installed, refer to the Windows operating system documentation for instructions. The portions of the Orcad Lite software we will be demonstrating in this manual are:

- Capture – The schematic capture front end that has replaced MicroSim Schematics.
- PSpice – The mixed signal simulation tool. For those readers familiar with MicroSim Design Center, this tool is relatively unchanged.
- Probe – The graphical post-processor for viewing the results generated by PSpice. For those readers familiar with MicroSim Design Center, this tool is relatively unchanged as well.

This part covers creating a circuit using Orcad Capture. Part 2 covers the basics of using Probe. The remaining parts cover the simulation of specific circuits using the suite of programs, Capture, PSpice, and Probe.

## 1.A. Starting Orcad Capture

If Capture was installed properly on your computer, it can be easily started from the Windows ***Start*** menu. However, depending on how the Windows desktop is configured, the ***Start*** menu may appear differently. Usually, the ***Start*** menu is displayed at the bottom of the desktop:

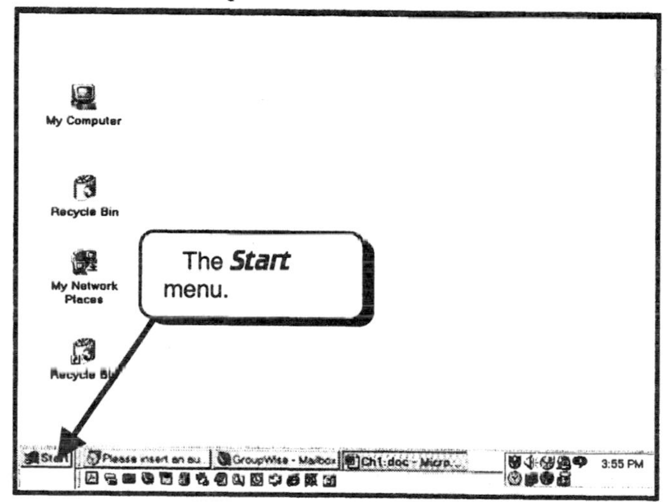


At this point, you could click the ***Start*** button and continue:

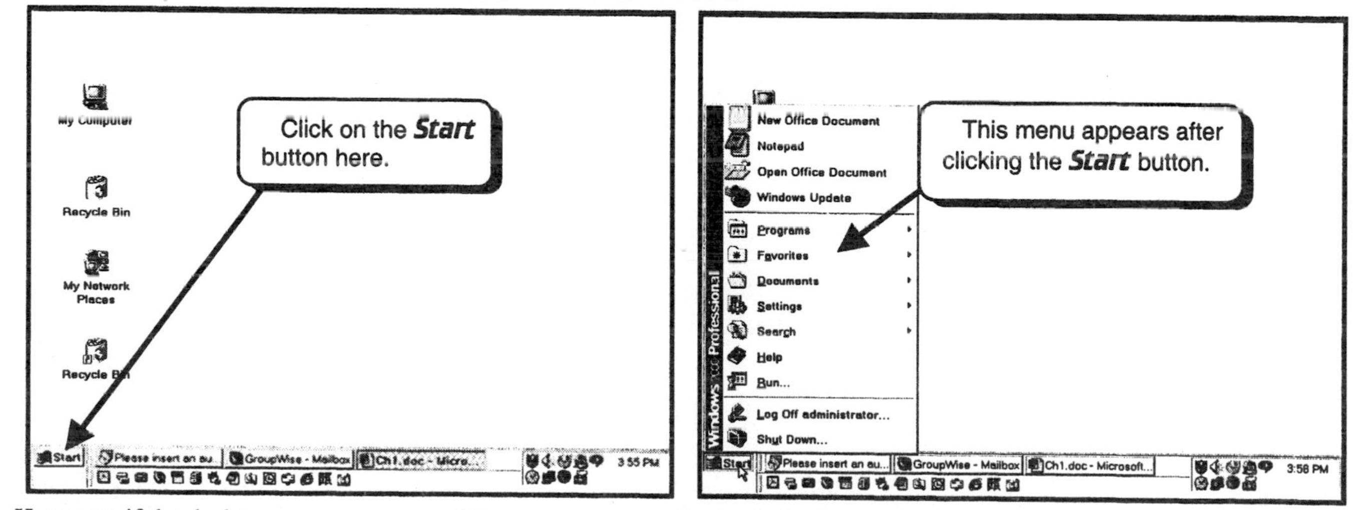


However, if the desktop appears as one of the screen captures shown below

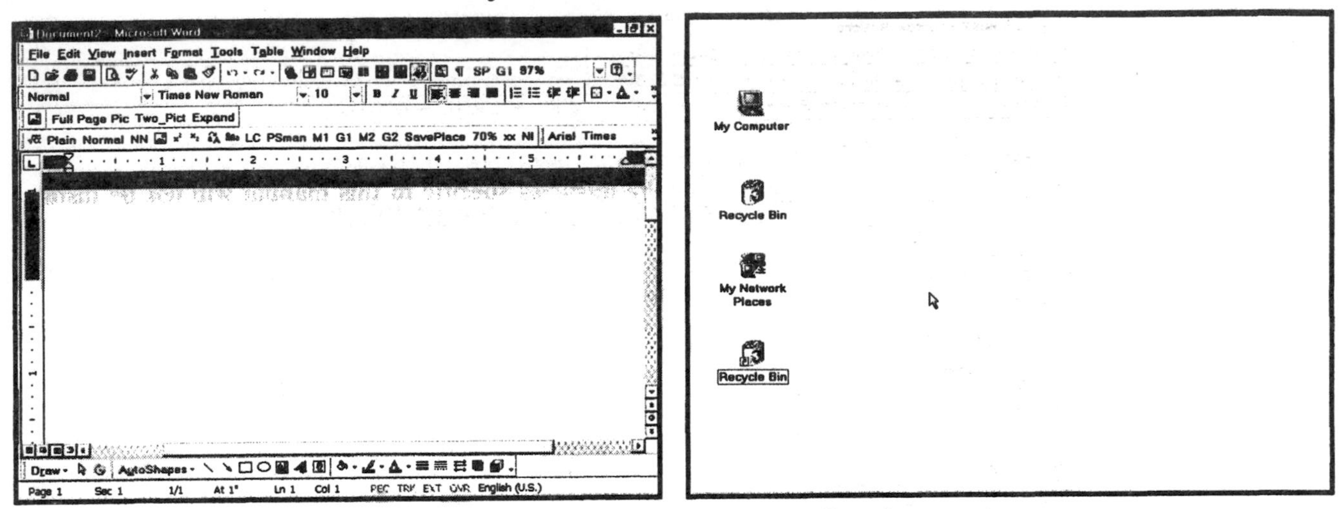

then the ***Start*** menu is hidden from view. There are two ways to make the ***Start*** menu appear. The first way is to bring the mouse pointer down to the bottom of the screen. After a moment, the ***Start*** menu should appear:

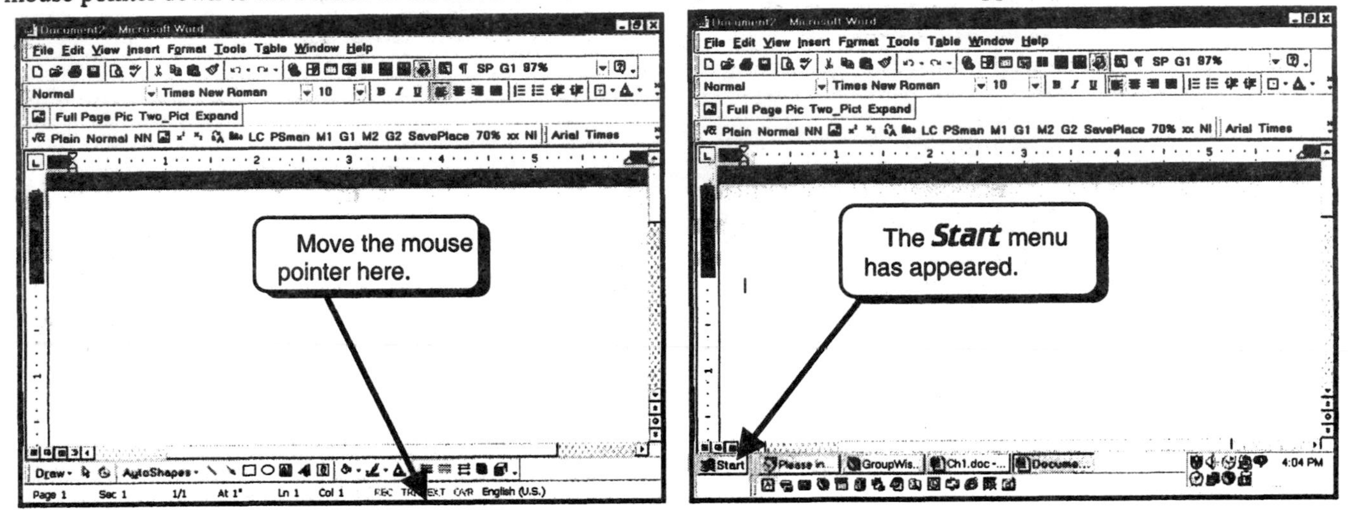


This method works if the ***Start*** menu is configured so that it is always on top and auto-hide mode is selected. At this point, you could click the ***Start*** button [Start] and continue:

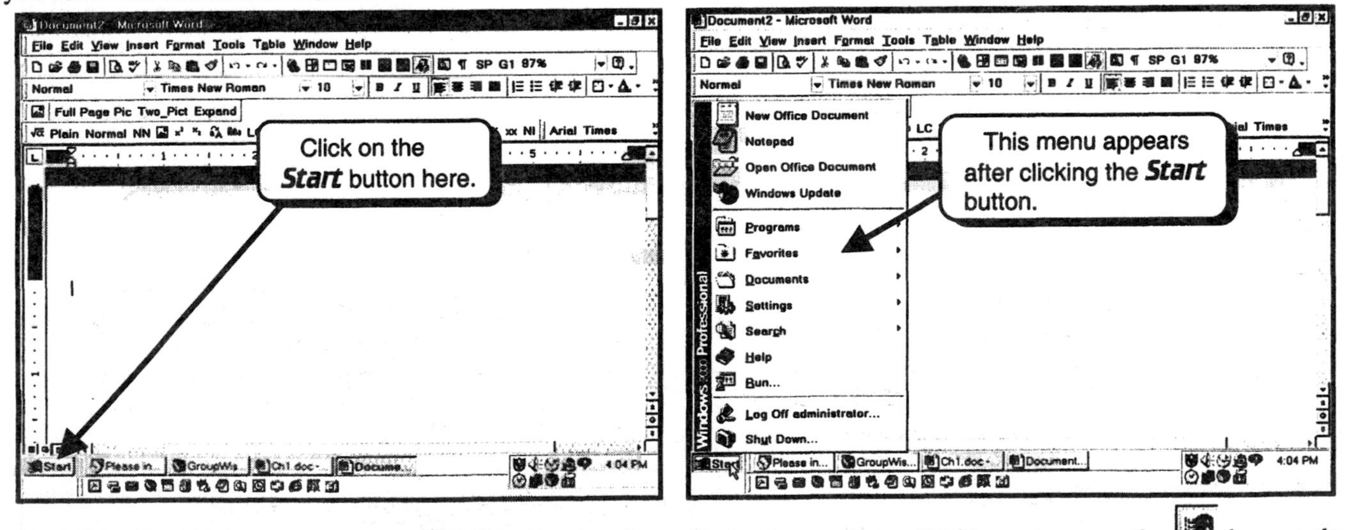


However, if you still cannot see the ***Start*** menu, you have three more options. (1) You can press the [Windows] key on the keyboard or type **CTRL-ESC**. These keys will bring the ***Start*** menu to the top and also select the ***Start*** button. (2) If you have an older keyboard and do not have the [Windows] key, you can move the mouse to the bottom of the screen and drag the ***Start*** menu up. For this example, we will press the [Windows] key:

We now have the ***Start*** menu displayed. Select **Programs**:

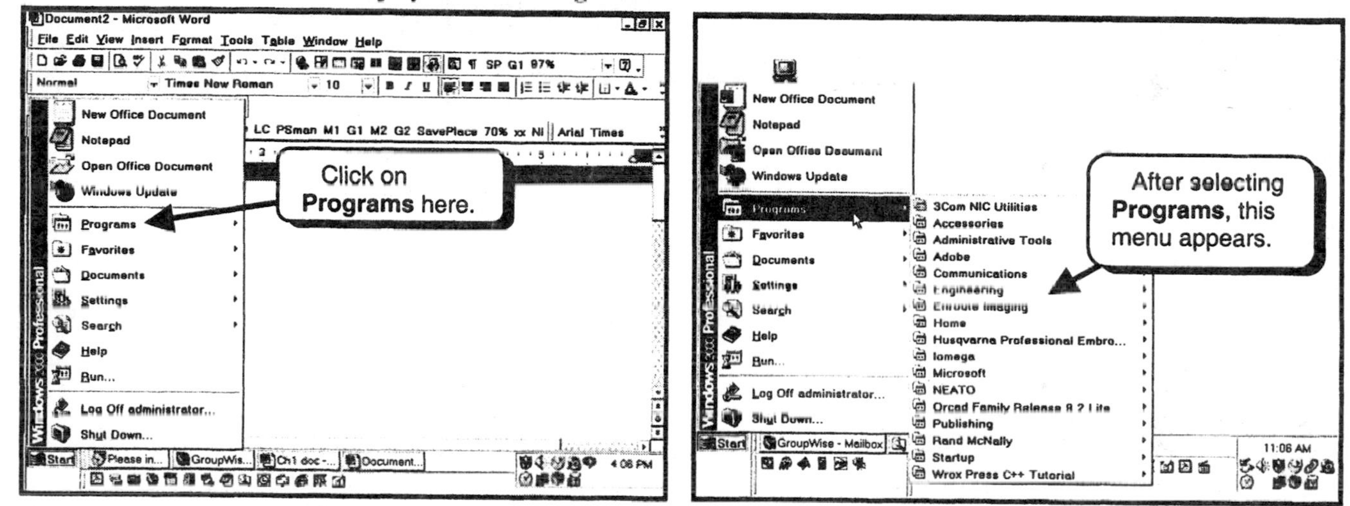


After selecting **Programs**, the programs and program groups for your computer appear. The Orcad programs are contained in the group **Orcad Family Release 9.2 Lite**. Click the ***LEFT*** mouse button on the text **Orcad Family Release 9.2 Lite**. This will display the programs contained in this group:

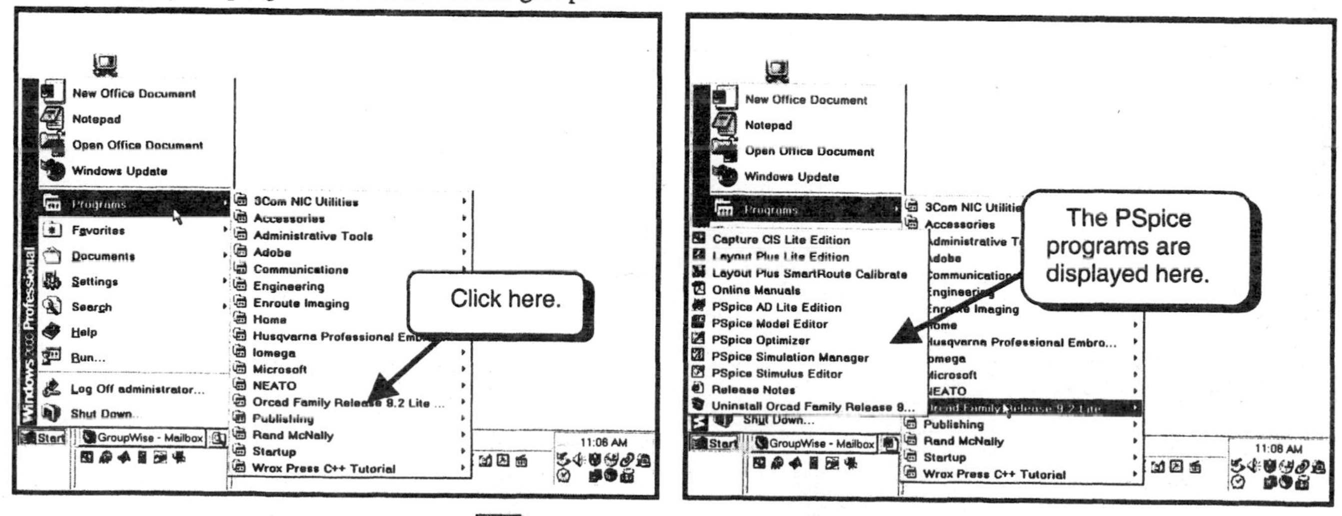


Click the ***LEFT*** mouse button on the item **Capture CIS Lite Edition** to run the schematic capture circuit simulation package.

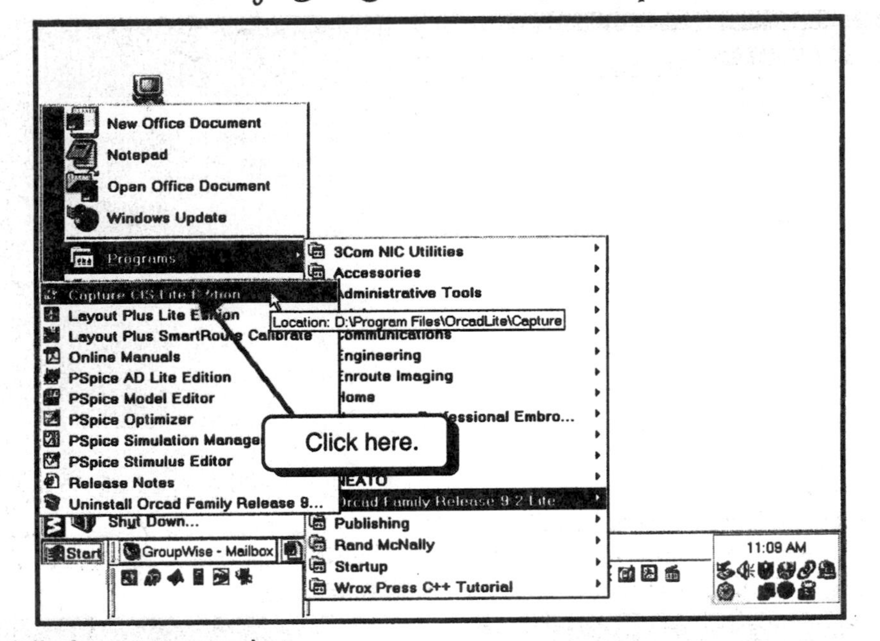


Capture should run and display an empty project:

We must now create a new project. Select **File** and then **New** from the Capture menus:

The items listed in the cascaded menu are the types of objects we can create using Capture as the front end. For this text, we are concerned only with the **Project** selection, which is used to draw schematics and simulate circuits, and the **Library** selection, which is used to create part libraries. Select **Project** to create a new project:

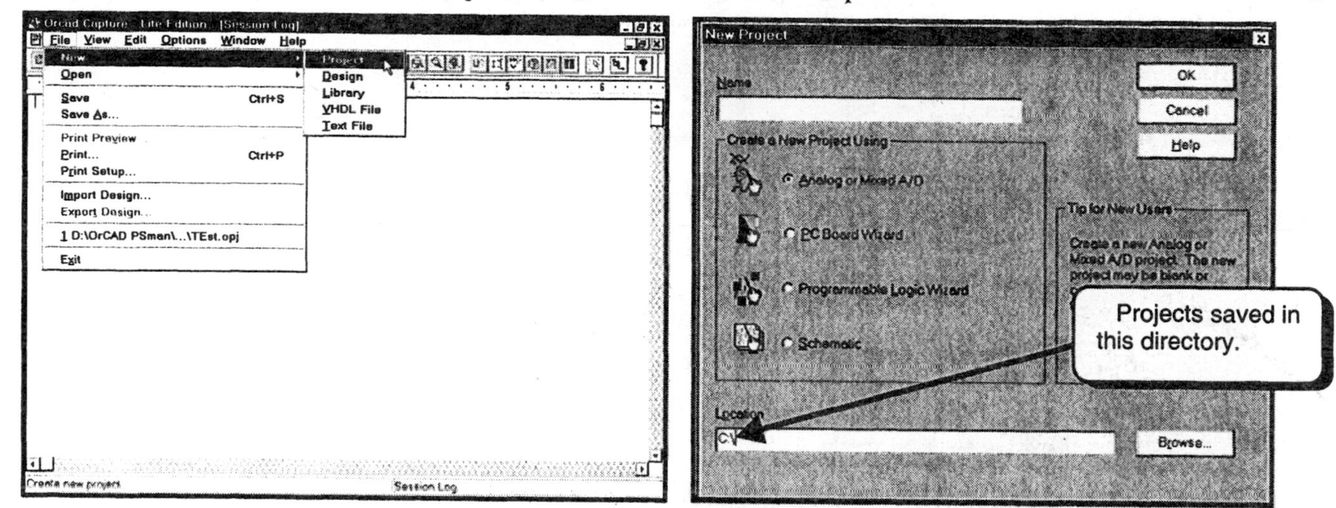


During installation, we have set the default directory for saving files to C:\. You should change this directory to one in which you would like your circuit files to be saved. For this book, I am saving all files in the directory D:\Orcad PSman\Circuit Files. Click on the ***Browse*** button to specify a different directory:

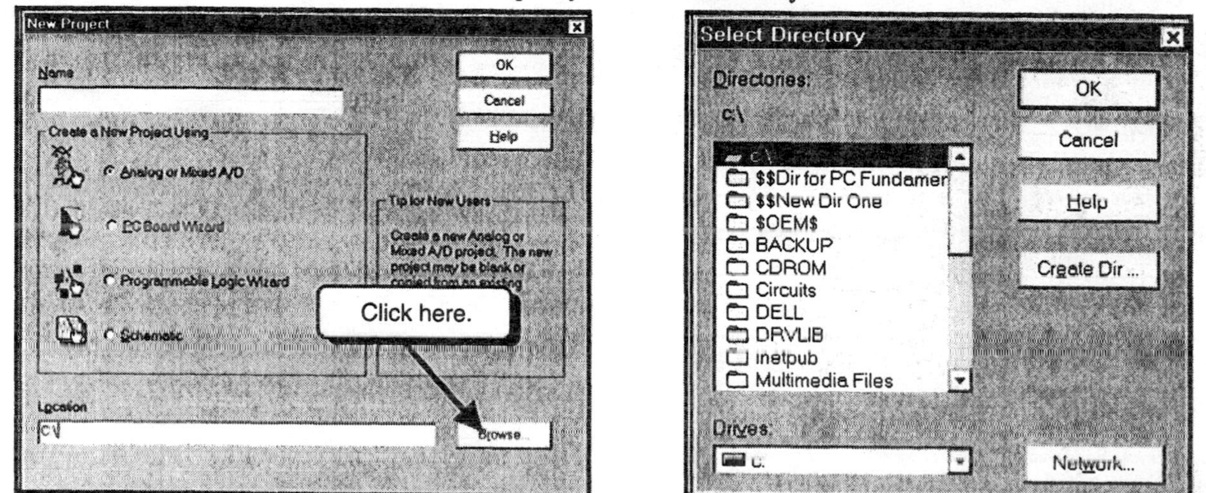


You can use this dialog box to change the current directory and to create a new directory. Select a directory in which to save your circuit files and then click the ***OK*** button:

The ***Name*** field specifies the name of the project. Enter the name **`Part 1 Schematic`**:

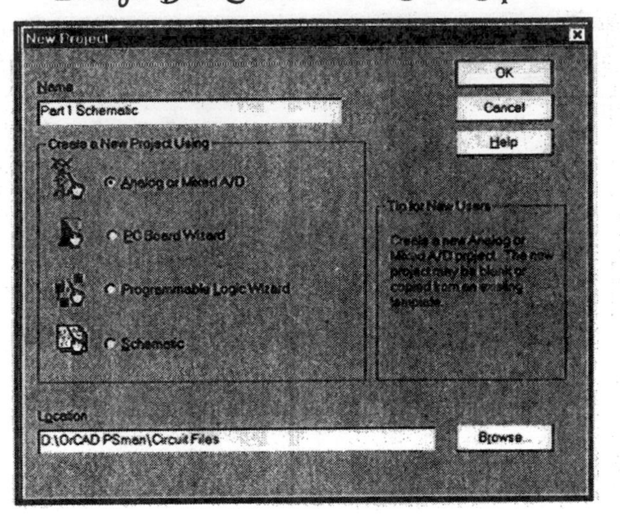


There are four types of projects you can create:

- **Analog or Mixed A/D** – Allows us to draw a circuit with Capture and then simulate the circuit with PSpice. This is the selection we will choose.
- **PC Board Wizard** – Allows us to draw a circuit with Capture and then create a PC board layout with Layout Plus.
- **Programmable Logic Wizard** – Allows us to use Capture to design a CPLD or an FPGA.
- **Schematic** – Allows us to create a schematic using Capture.

To simulate circuits, select option ***Analog or Mixed A/D*** and click the ***OK*** button:

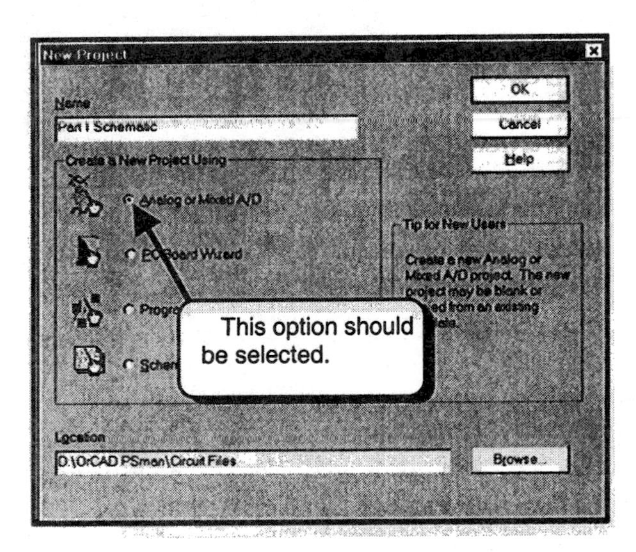


This dialog box allows us to select a project template. A hierarchical design is one with many pages, and blocks that connect one page to another. We will choose to start with a ***blank project***:

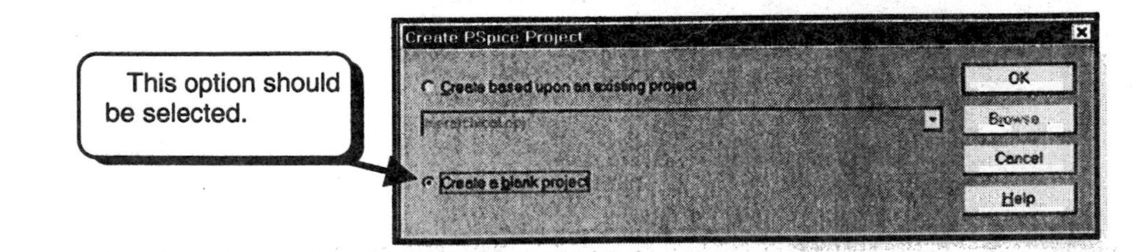


Click the ***OK*** button. You will be presented with an empty schematic page:

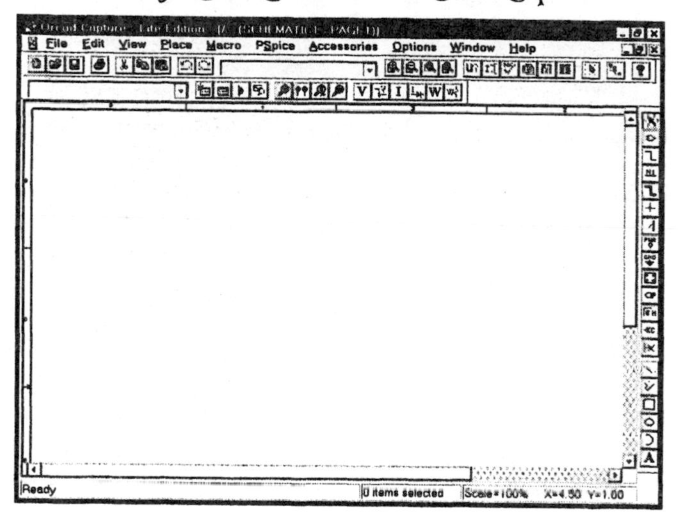

## 1.B. Placing Parts

We will now create a schematic. The first part we will get is an independent voltage source. All parts are retrieved using the procedure outlined below. First, click the ***LEFT*** mouse button on the **Place** menu selection:

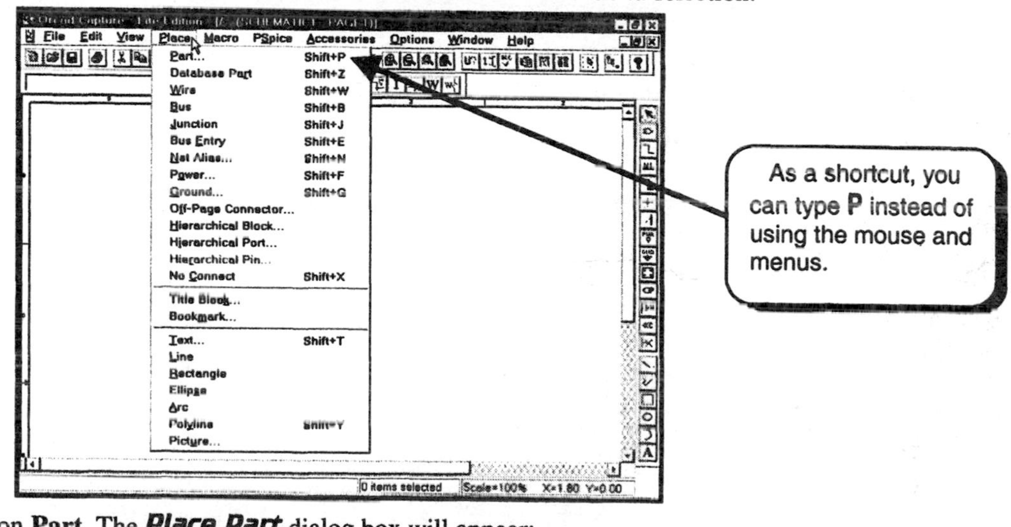

As a shortcut, you can type **P** instead of using the mouse and menus.


Click the ***LEFT*** mouse button on **Part**. The ***Place Part*** dialog box will appear:

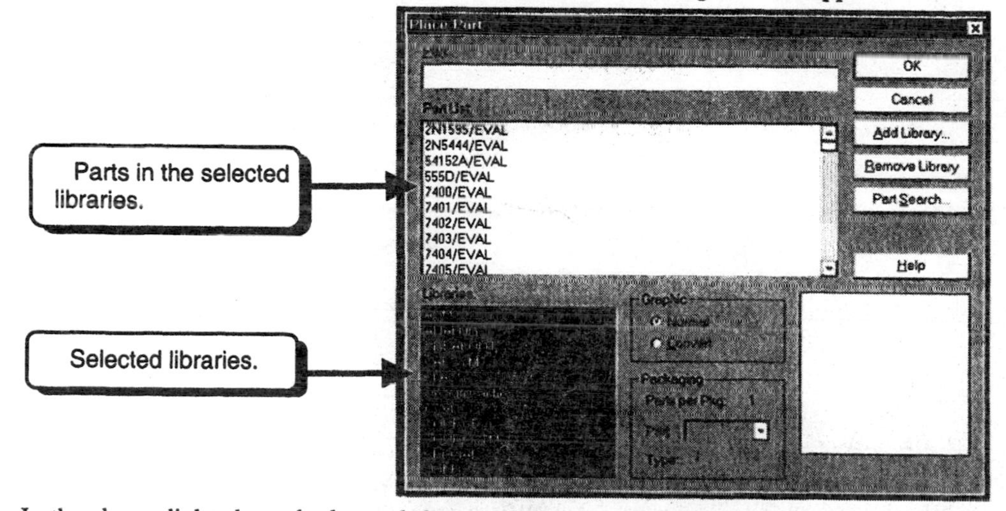


In the above dialog box, the lower left pane shows that all configured libraries are selected. The left-center pane displays the parts contained in the selected libraries. Since all libraries are selected, the left-center pane will display all parts available to us.

If you know the name of the part you want to place, you can type the part name in the box. Most independent voltage sources start with the letter v. Type the letter **v**. The left-center pane will display the parts that begin with the letter v:

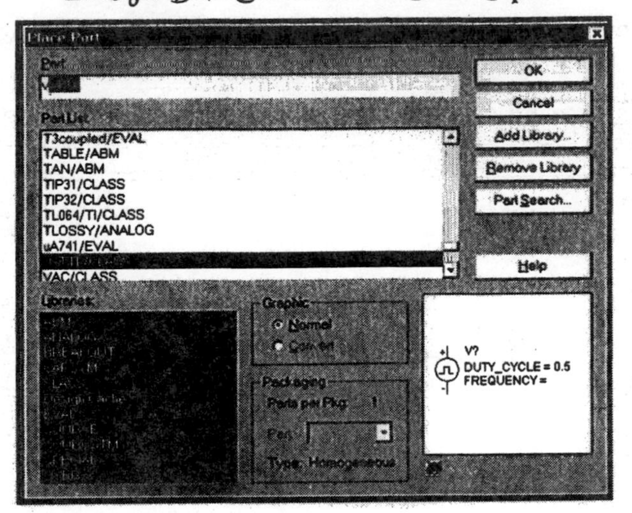


The information displayed in the highlighted line of the left-center pane is the name of the part and the library in which it is located. For example, the text ***V_TTL/CLASS*** indicates that the part name is ***V_TTL*** and the part is contained in library class.olb. The line ***TAN/ABM*** indicates that part ***TAN*** is contained in library ***ABM***.olb.

The presently selected part is ***V_TTL***, which was created for this textbook to generate a TTL-compatible waveform (0 to 5 V) with a specified duty cycle and frequency. We wish to select a generic DC voltage supply. Type the text **dc** to complete the typing of "vdc" and select the part:

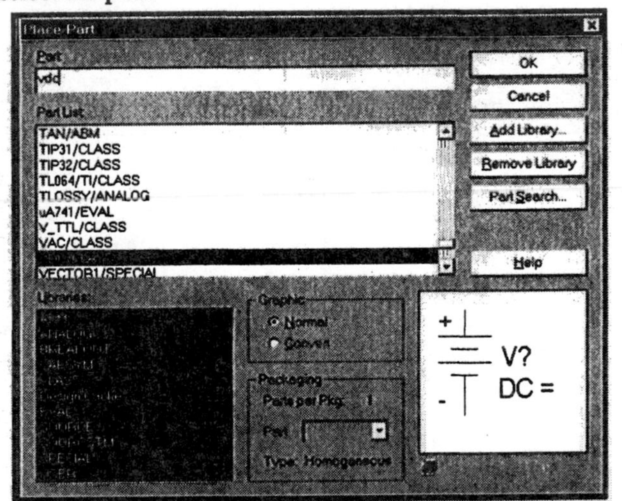


To accept the part and place it in your circuit, click the ***OK*** button. When you click the button, the program will return to the schematic with the graphic for the part attached to the mouse pointer:

The graphic moves with the mouse. Move the graphic to the location where you want to put the voltage source. Click the ***LEFT*** mouse button **once** to place the part, and then move the mouse pointer. **Note on your screen and in the figure below**

**that when you place a part, a second part appears attached to the mouse pointer. This is the auto repeat function for placing parts.**

If you want to add another voltage source to your schematic, you can move the mouse to where you want to place the source and click the ***LEFT*** mouse button. We do not need another DC source at this time, so press the **ESC** key to make the second source disappear.

To make the schematic more readable, we will zoom in on the DC source. Select **View** and then **Zoom** from the Capture menus:

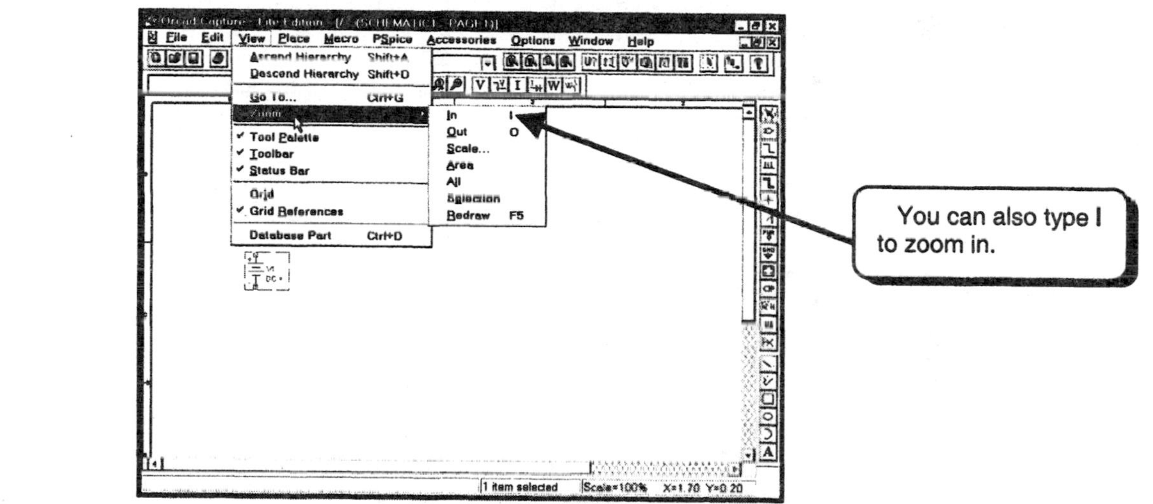


To zoom in, click the ***LEFT*** mouse button on the **In** menu selection:

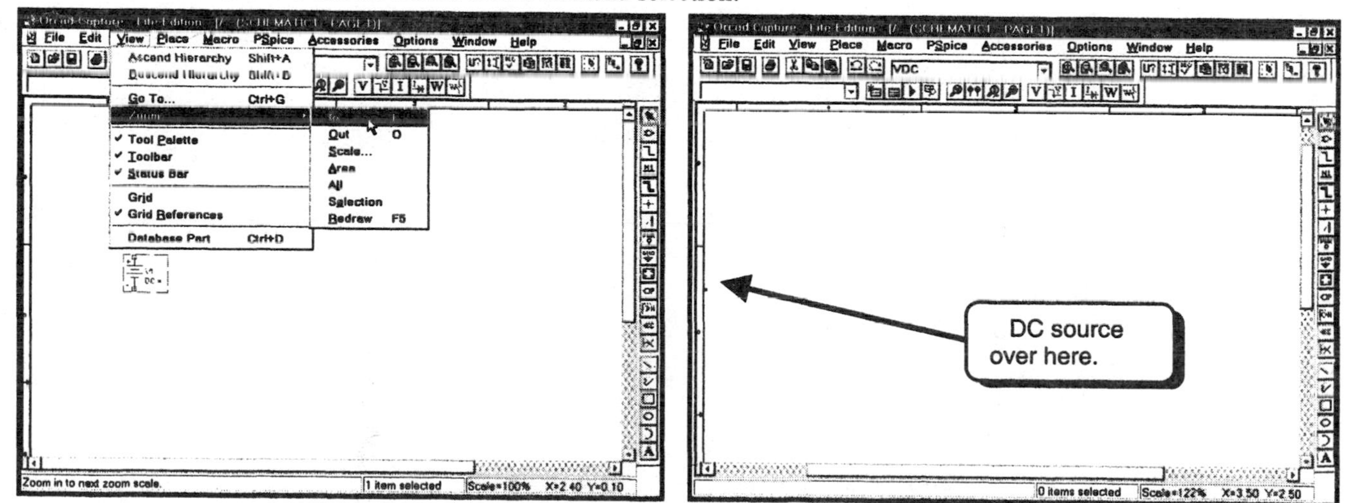


Capture zoomed in on the schematic, but the DC source is no longer visible. Click the ***LEFT*** mouse button as shown to scroll the window to the left:

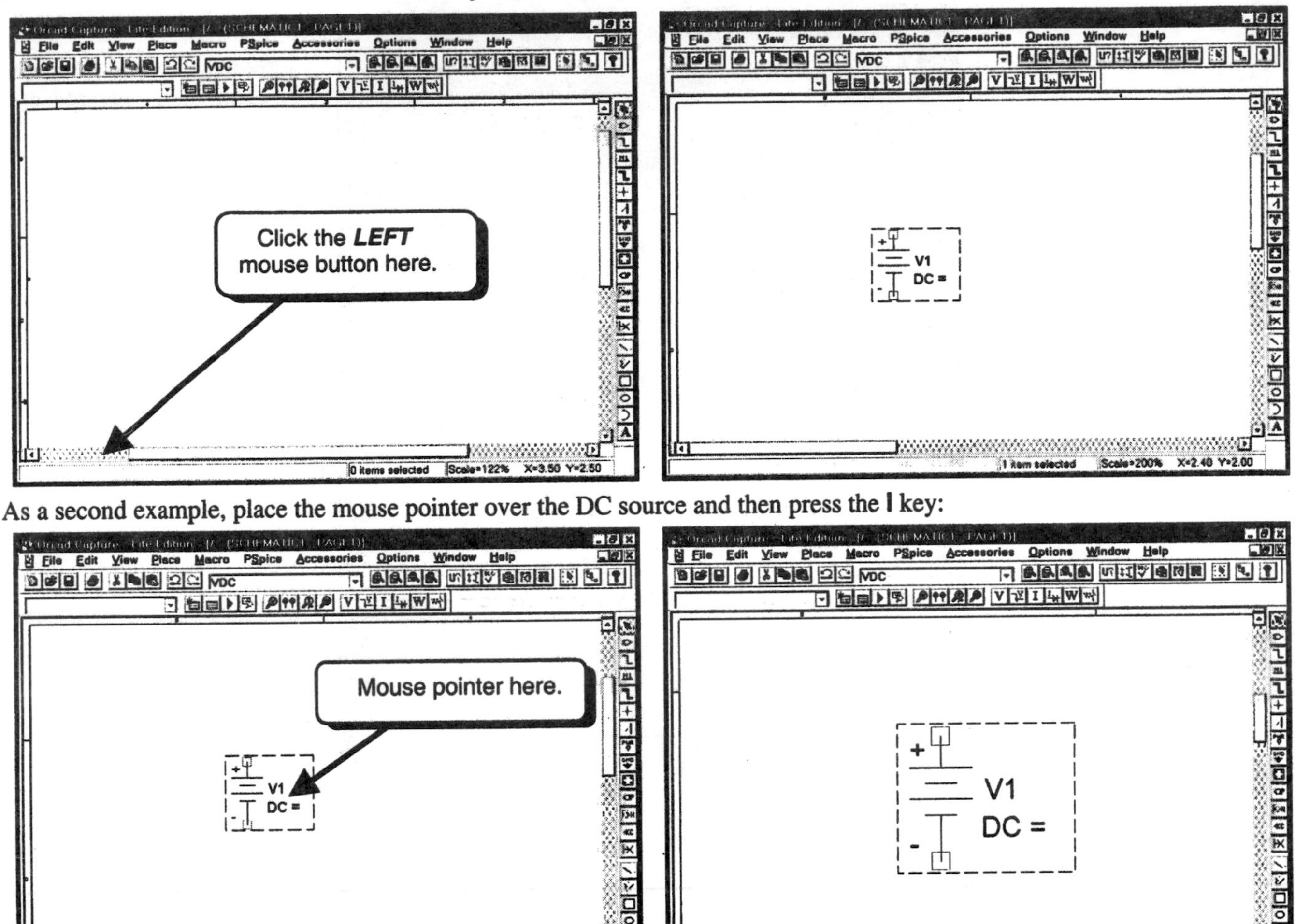


As a second example, place the mouse pointer over the DC source and then press the **I** key:

We see that the Capture zooms in around the mouse pointer. Unless you cannot remember the keyboard shortcuts, the easiest way to zoom in on an object is to place the mouse pointer at the object and then press the **I** key. (To zoom out, press the **O** key.)

We will now edit the properties of the source to make it a 12 VDC source. There are two ways to change the properties of a part. The first way we will look at is editing the individual properties that are displayed on the screen. We will first change the voltage of the DC source. To do this, double-click the *LEFT* mouse button on the ***DC=*** text next to the DC source. The dialog box below will appear:

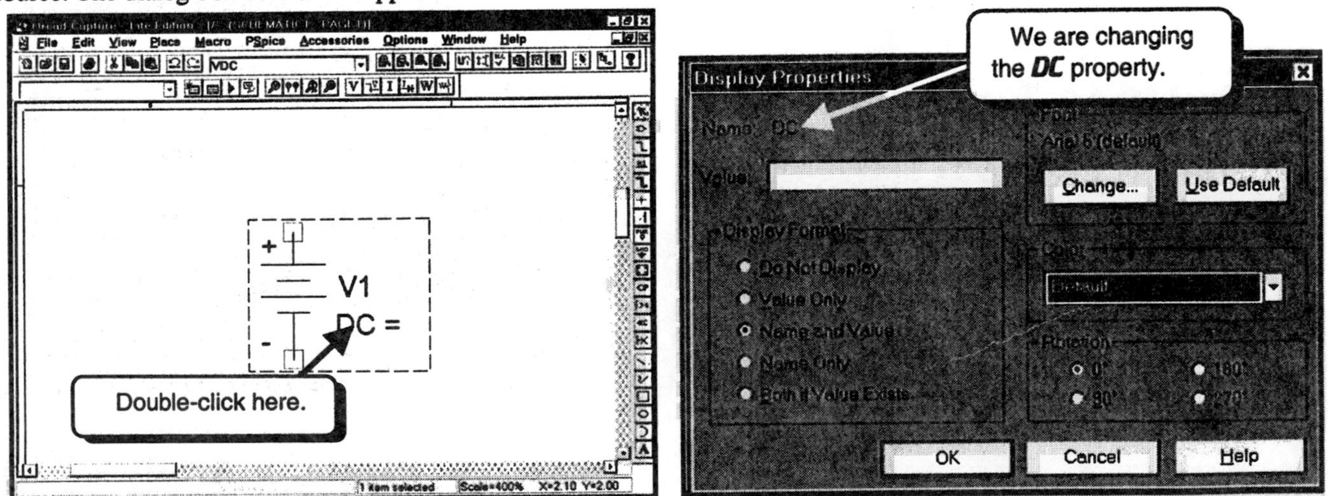


Note that the dialog box says that the property we are changing is the ***DC*** property. To change the property, type in the desired value. We will create a 12-volt source, so type in **12**:

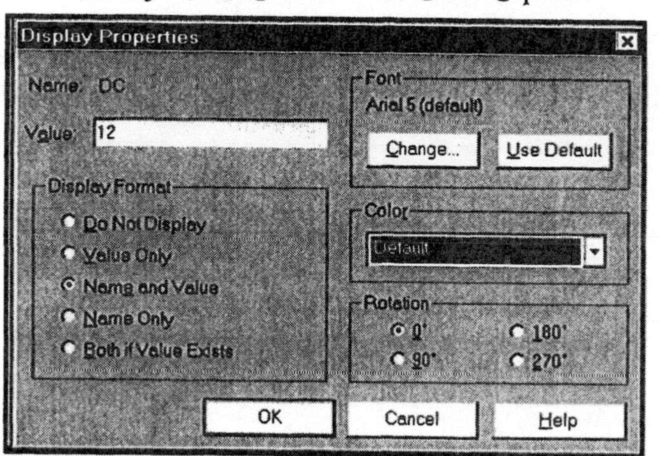


Click the ***OK*** button to change the property. The part will appear in the schematic with the line ***DC=12*** displayed in the schematic:

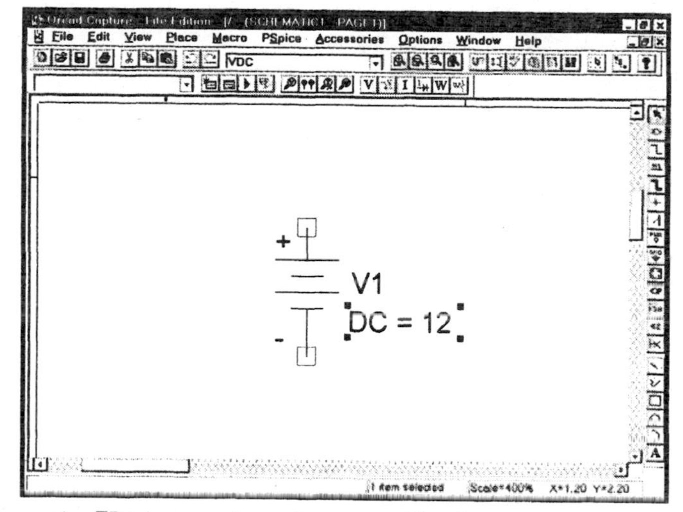


If your text appears garbled, press the **F5** key to redraw the screen. To change the name of the voltage source, double-click the ***LEFT*** mouse button on the text ***V1***.* The dialog box below will appear:

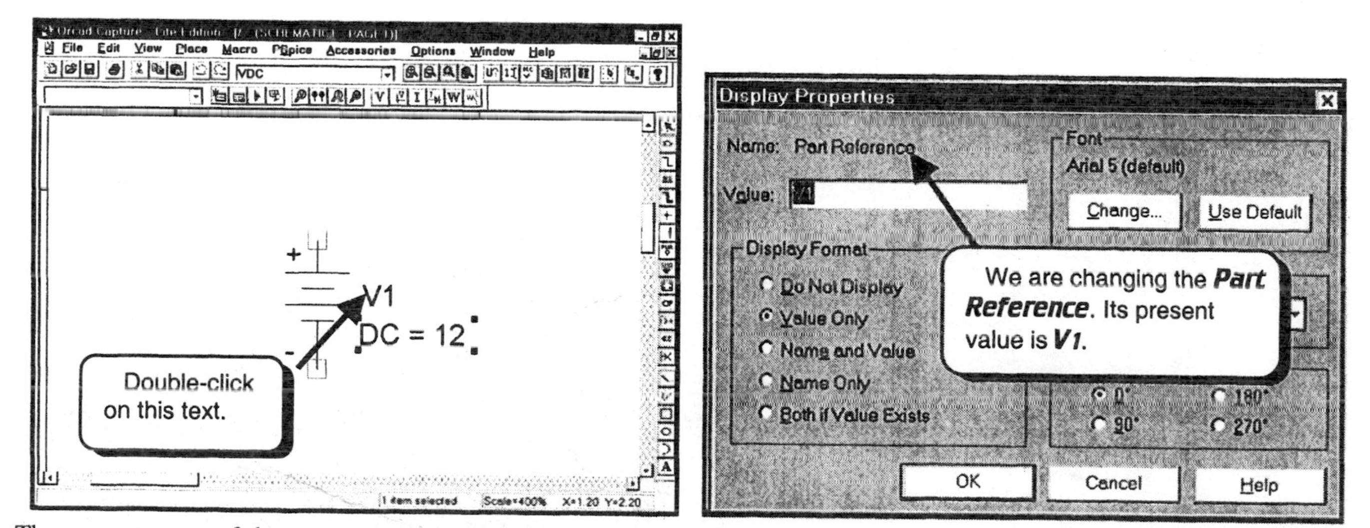


The present name of the source appears highlighted. To change it, type the desired name of the source. We will change the name of the source to Vx, so type the text **Vx**:

*The name of the source may vary depending on how many sources you have placed in your schematic.

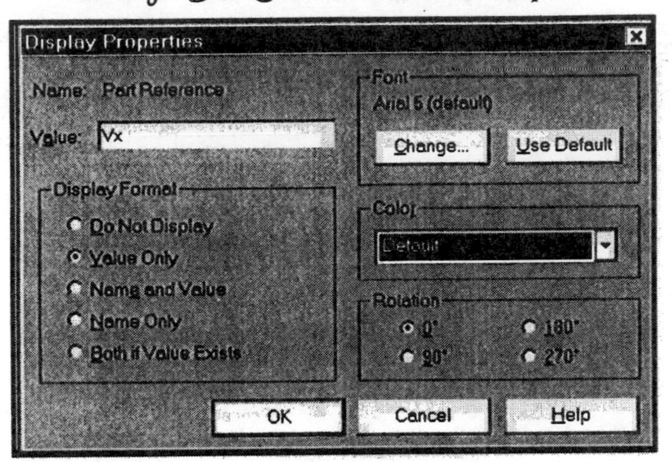


Click the ***OK*** button. The updated part will appear as shown below:

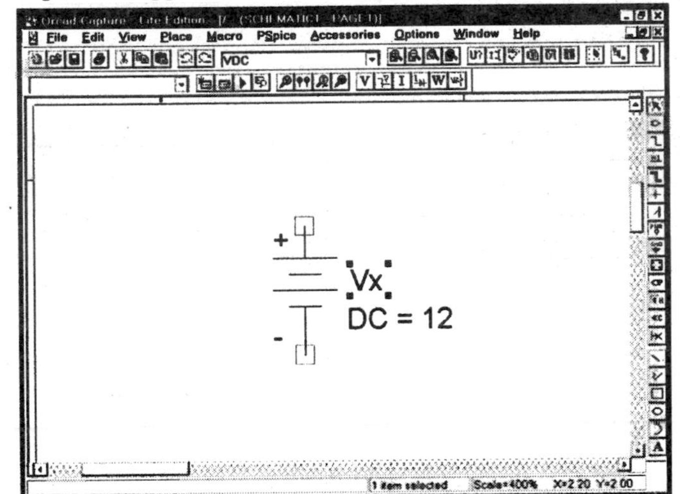


The part now has the desired properties. To illustrate the second method of changing a part's properties, we will change the source voltage to 15 volts and the name back to V1. To change the properties, click the ***LEFT*** mouse button once on the DC source graphic, ⊣|⊢. The graphic will be highlighted in pink when the part is selected. When the part is highlighted, click the ***LEFT*** mouse button on the **Edit** menu selection:

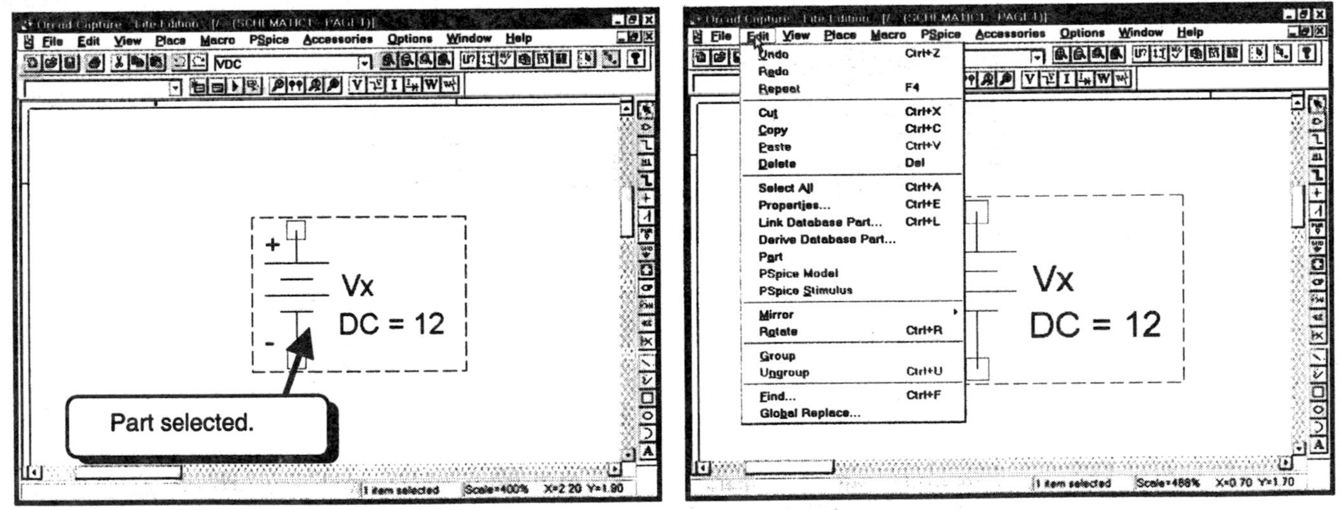


Click the ***LEFT*** mouse button on the **Properties** menu selection:*

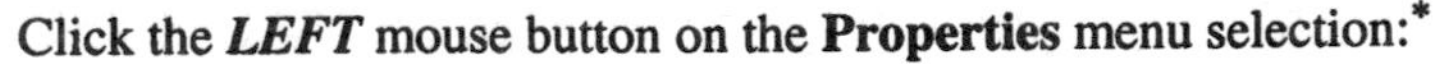

*This dialog box can also be obtained by double-clicking the ***LEFT*** mouse button on the DC voltage source graphic, ⊣|⊢. Note also that if the text **Properties** in the menu appears as **Properties**, you have not properly selected the graphic. Click the ***LEFT*** mouse button on the graphic again until it becomes highlighted in red. When the graphic is highlighted in red, you may edit its properties.

This window displays a spreadsheet of all of the part's properties.* Note that there are more properties than appear on the schematic. Not all properties of a part are displayed on the schematic. To change the voltage of the source, click the ***LEFT*** mouse button on the text ***12*** in the ***DC*** column to select the cell:

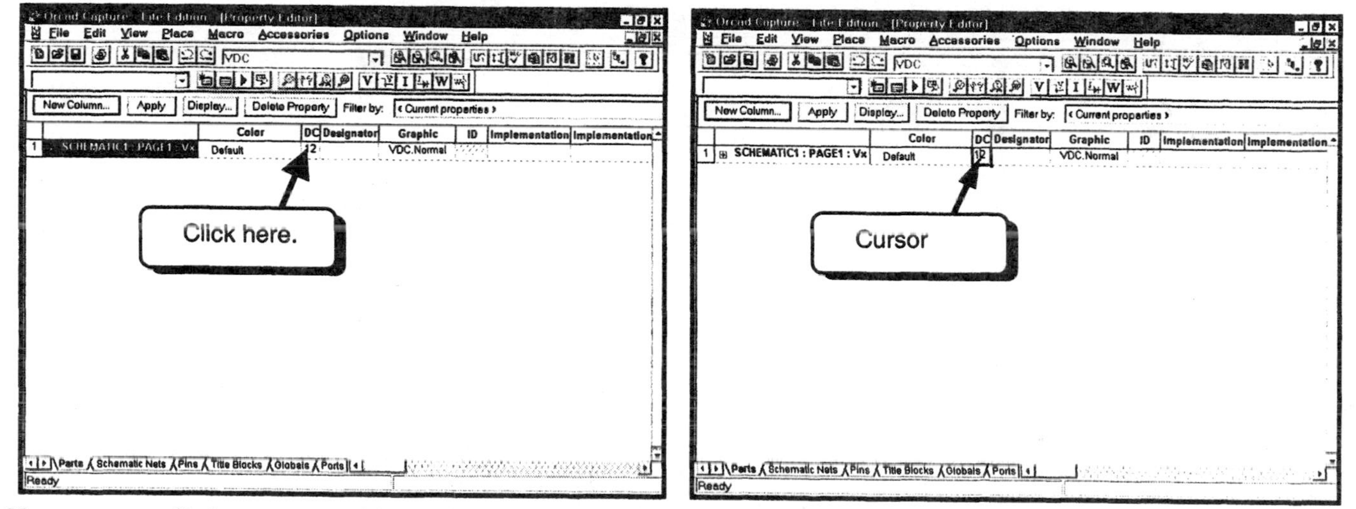


You can now edit the contents of the cell. Use the **BACKSPACE** and **DELETE** keys to clear out the cell and then enter the text **15** to make a 15-volt DC source:

Next, we will attempt to change the name of the source. The property that specifies the name of the source is not displayed in the window. Click the ***LEFT*** mouse button as shown below to scroll the spreadsheet to the right:

* You can also obtain the properties spreadsheet by clicking the ***LEFT*** mouse button on a part to select it. Once the part is selected, click the ***RIGHT*** mouse button on the part and then select **Edit Properties** from the menu that appears.

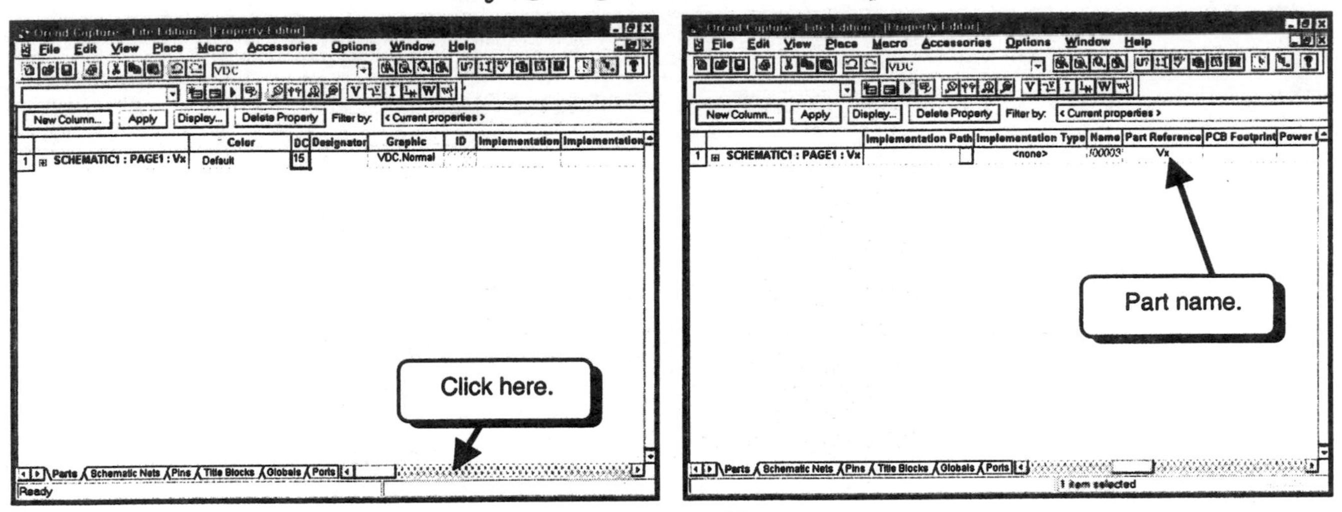


We can now see the ***Part Reference*** property. Change the value ***VX*** to **V1**:

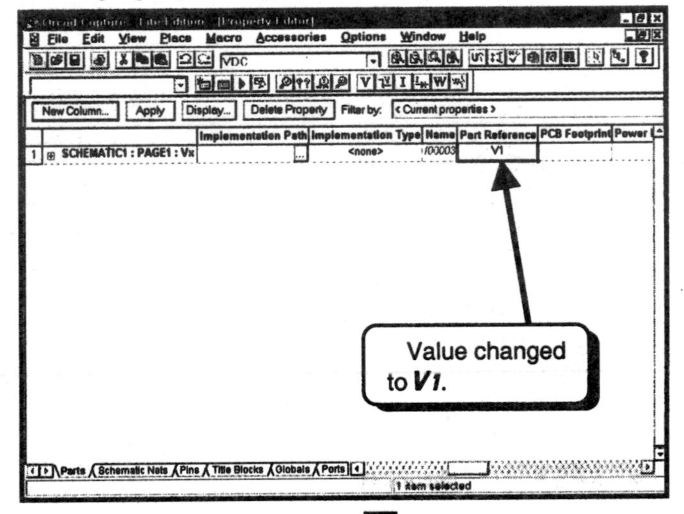


To return to the schematic, click the ***LEFT*** mouse button on the ☒ as shown below or type **CTRL-F4**:

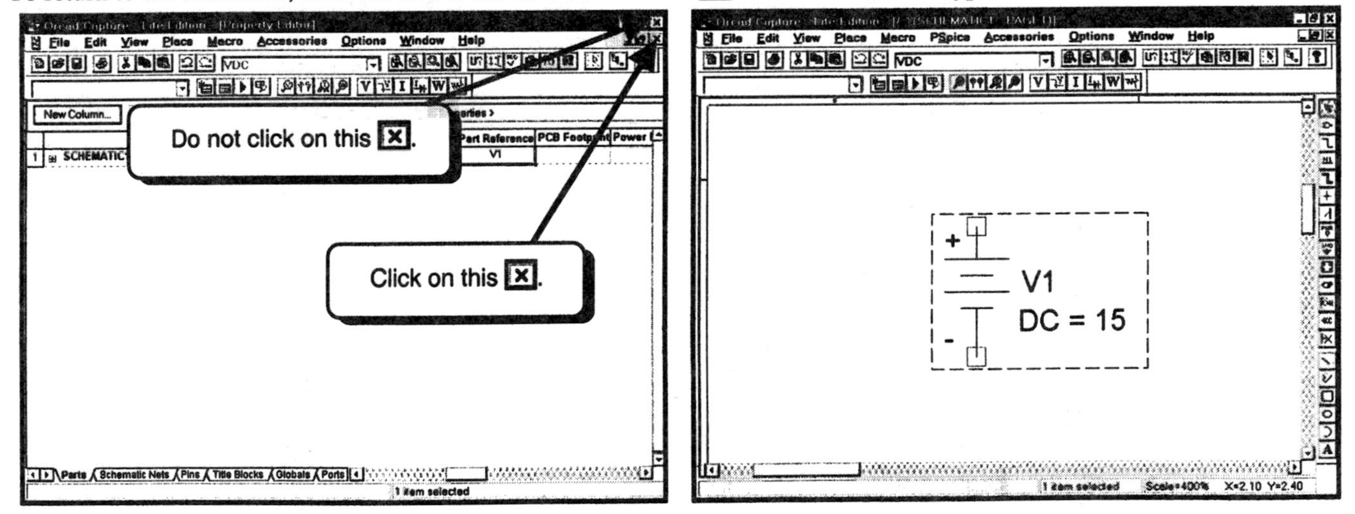


We see that both the line ***DC=15*** and the name of the source, ***V1***, have changed in the schematic. For practice, change the name of the source back to Vx.

We now need to place some resistors in the circuit. The part name for a resistor is R. We could choose the **Place** and then select **Part** to get the resistor. Instead, we will type **p**. The ***Place Part*** dialog box will appear:

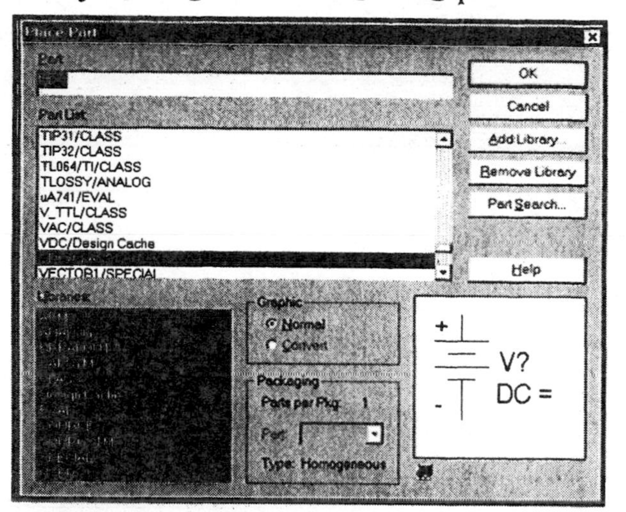


Since we know that a resistor part is called "R" we will just type in **r**:

Place Part
Part
Part List:
R1gauss/CLASS
R1pcnt/CLASS
R5gauss/CLASS
R5pcnt/CLASS
R_var/ANALOG
RAM8Kx1break/BREAKOUT
RAM8Kx8break/BREAKOUT
Rbreak/BREAKOUT
Libraries:
OK
Cancel
Add Library...
Remove Library
Part Search...
Help
Graphic
Normal
Convert
Packaging
Parts per Pkg: 1
Part:
Type: Homogeneous
R?
1k

Click the ***OK*** button to accept the part. The resistor graphic will appear attached to the mouse pointer:

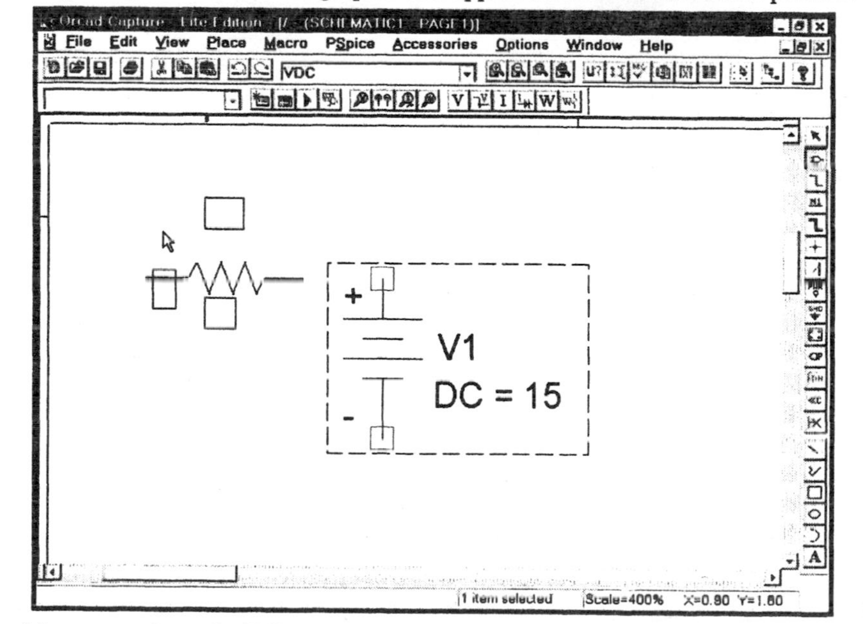


The schematic is zoomed in too much on the DC source to place any resistors. To zoom out, press the **O** key:

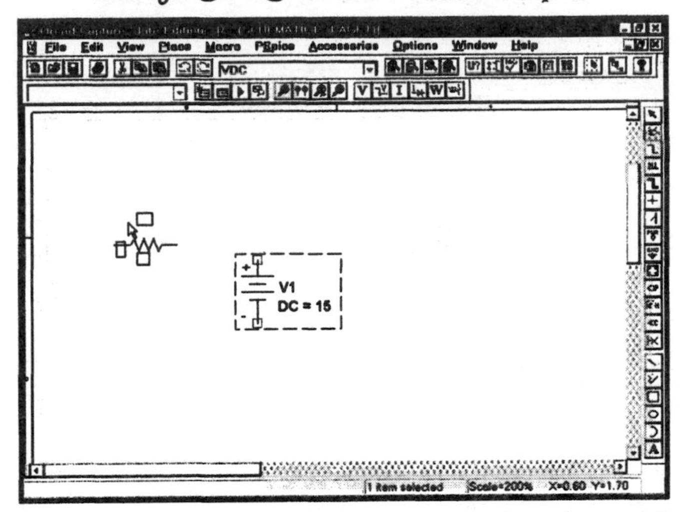


We now have a better view of the schematic page. Note that we can zoom in and out while still placing parts and that the resistor moves with the mouse. Zoom out until you see a screen close to the one shown below:

Move the resistor to where you want to place it and click the ***LEFT*** mouse button. The resistor is placed, and a second resistor appears attached to the mouse pointer:

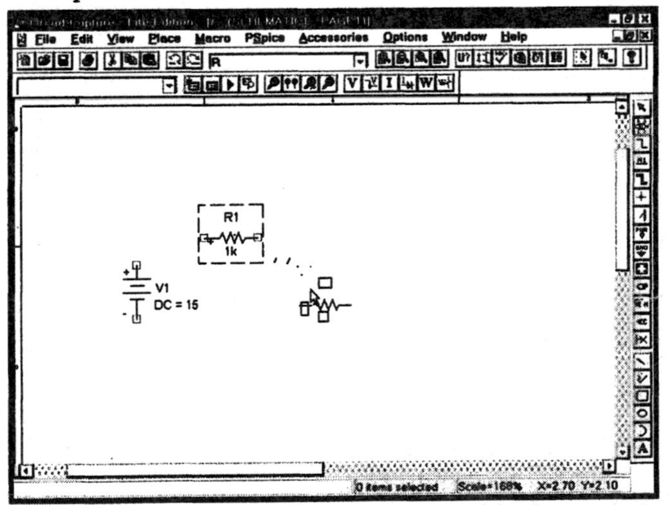


We now wish to add a second resistor to the schematic. The next resistor we want to place will be vertically oriented. Currently the resistor attached to the mouse is horizontal. To rotate the part, type **R**. The part will rotate while attached to the mouse. Orient the part vertically as shown in the screen below.

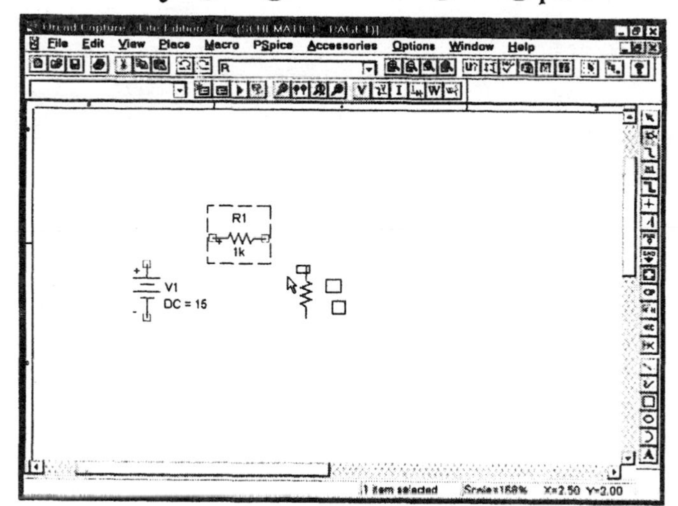

Move the mouse to place the part in the location shown below. To place the part click the ***LEFT*** mouse button. When you click the ***LEFT*** mouse button, the part is placed, and a third resistor appears attached to the mouse pointer:

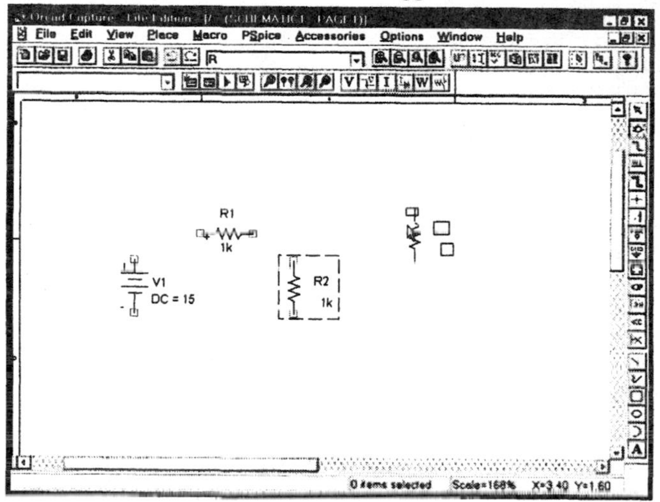

Add two more resistors as shown in the figure below:

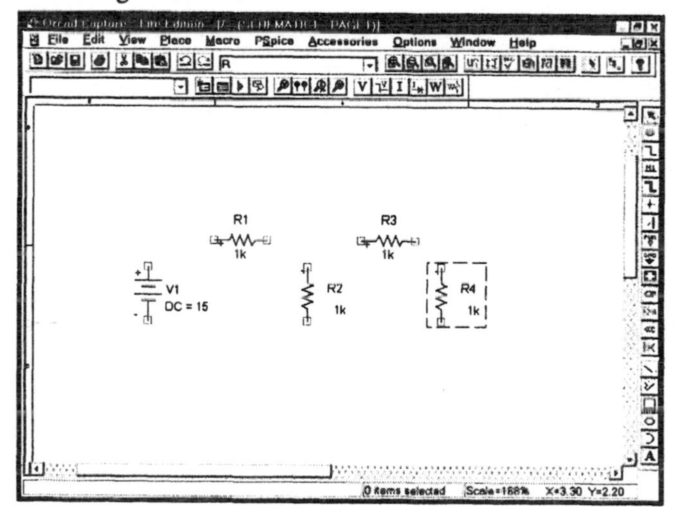

Press the **R** key to rotate the part when necessary and click the ***LEFT*** mouse button to place the part. When you have placed all four resistors, press the **ESC** key to stop placing resistors. When you press the **ESC** key, the resistor graphic at the mouse pointer disappears.

Before we continue, we would like to zoom in on the circuit and make it as large as possible while still displaying the entire circuit. Select **View**, **Zoom**, and then **Area** from the menus. The mouse pointer will be replaced by a magnifying glass:

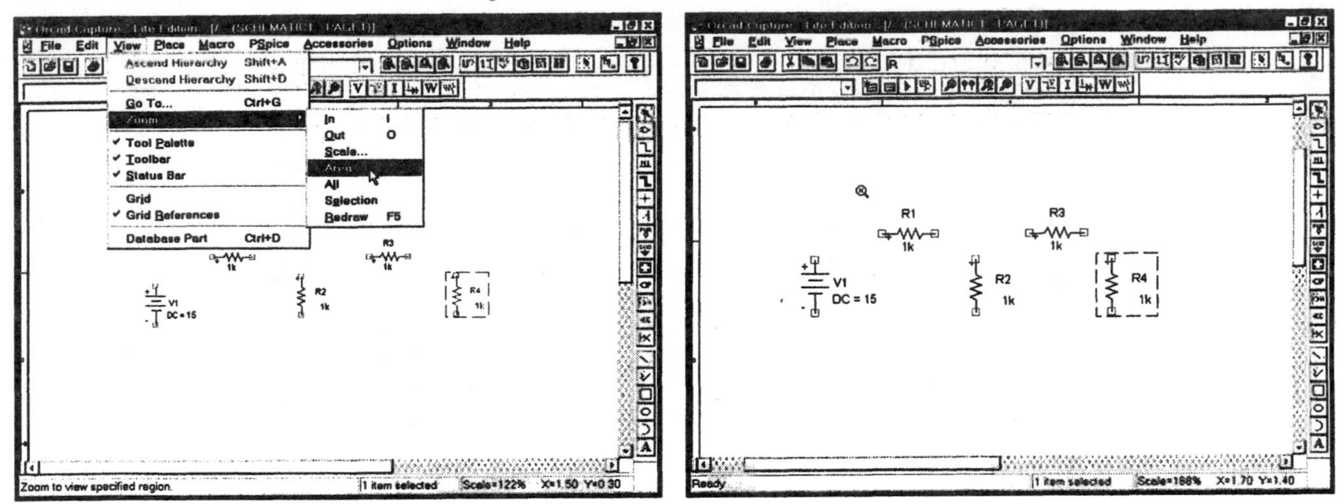


Place the mouse pointer near the upper left corner of the area in which you would like to zoom and then click and **HOLD** the ***LEFT*** mouse button. When you move the mouse away, a zoom rectangle will be shown:

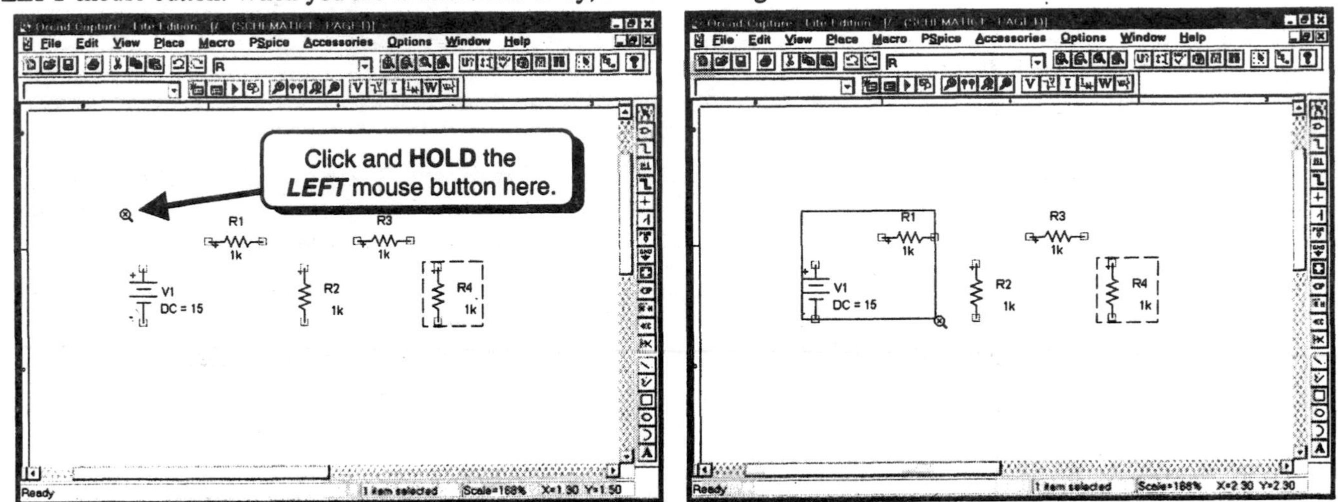


Drag the mouse to create a zoom rectangle that encloses the entire circuit:

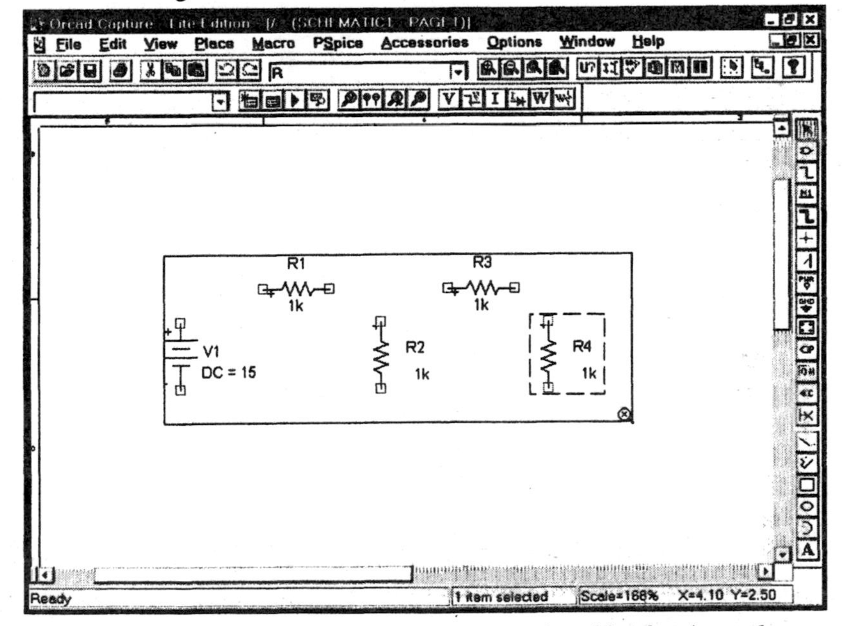


When you release the mouse button, Capture will zoom in on the area enclosed by the rectangle:

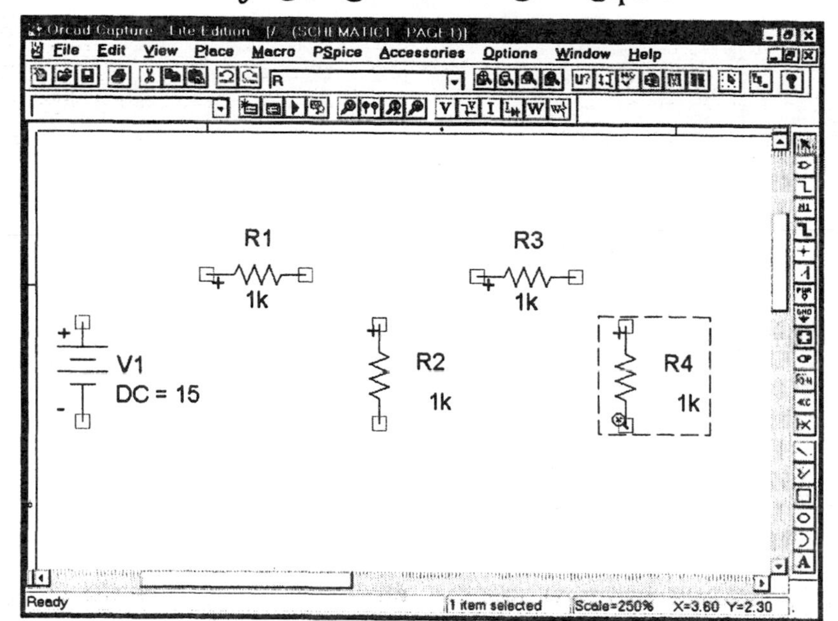

To end zoom mode, press the **ESC** key.

# 1.C. Correcting Mistakes

At this point you may have some mistakes in your schematic. You may have clicked the ***LEFT*** mouse button one too many times and placed too many resistors on your schematic, or you may have placed resistors too close to each other. To move parts follow the procedure below. For the moment we will assume that you wish to move a resistor.

1. Click the ***LEFT*** mouse button on the resistor graphic, , you wish to move. When the resistor graphic is highlighted in pink, it has been selected. It may take several tries to highlight the resistor.
2. When the graphic is highlighted in pink, drag the resistor graphic to the desired spot in the schematic.

If you need to delete a part, follow Step 1 above. When the appropriate part is selected, press the **DELETE** key.

# 1.D. Changing Properties

There are several ways to change the properties of parts. Some have already been illustrated previously, but we will now go over all of the different methods. The first thing we need to do is to select a part. We will select resistor R1. Place the mouse pointer to the left of and above ***R1***. Press and hold the ***LEFT*** mouse button and move the mouse down and to the right. Notice that a box is drawn. Move the mouse so that the box touches ***R1***:

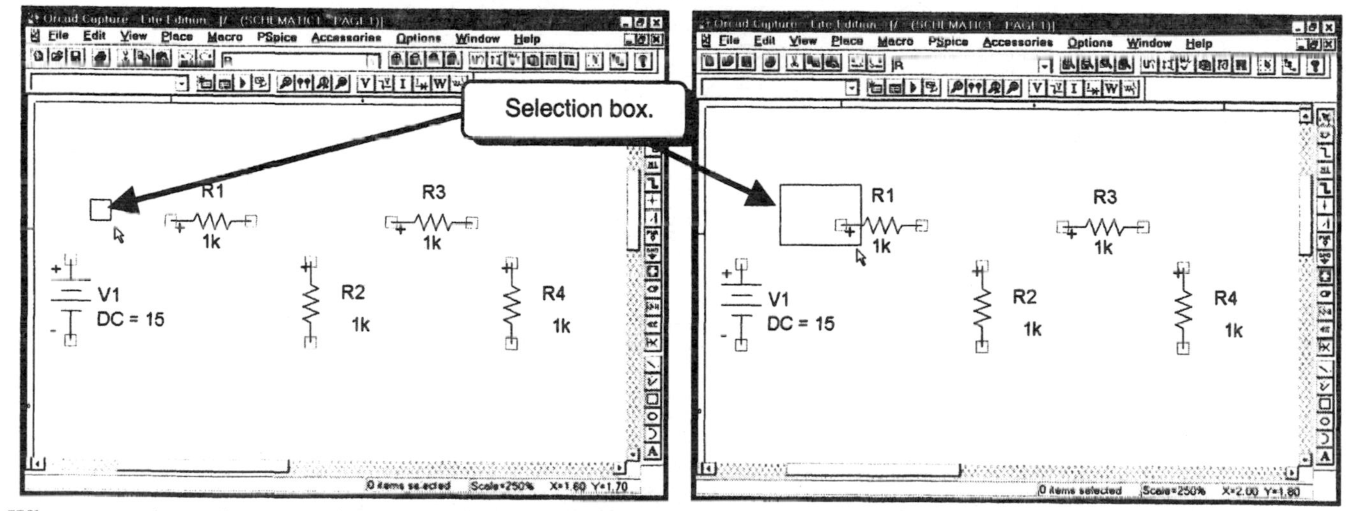


When you release the mouse button, everything touching the box will be selected. In the screen capture above, the box is touching the ***R1*** graphic, so it will be selected:

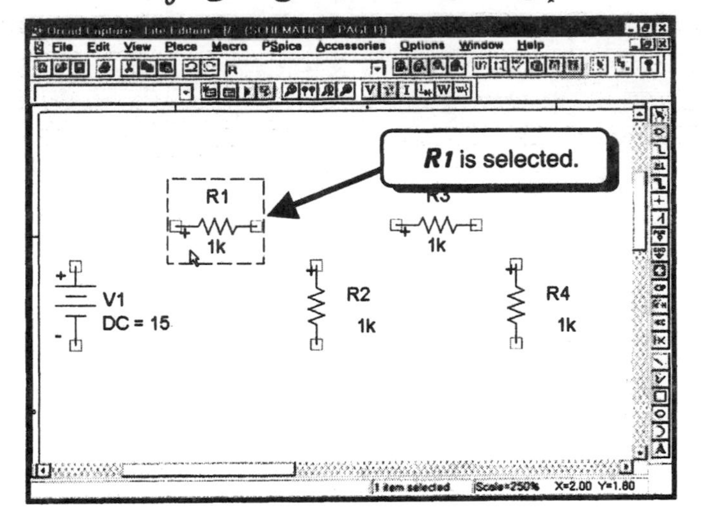


Resistor R1 is highlighted in pink and enclosed in a dashed box, indicating that it is selected. To edit all of the selected item's properties at the same time, click the ***LEFT*** mouse button on the **Edit** menu selection and then select **Properties**:

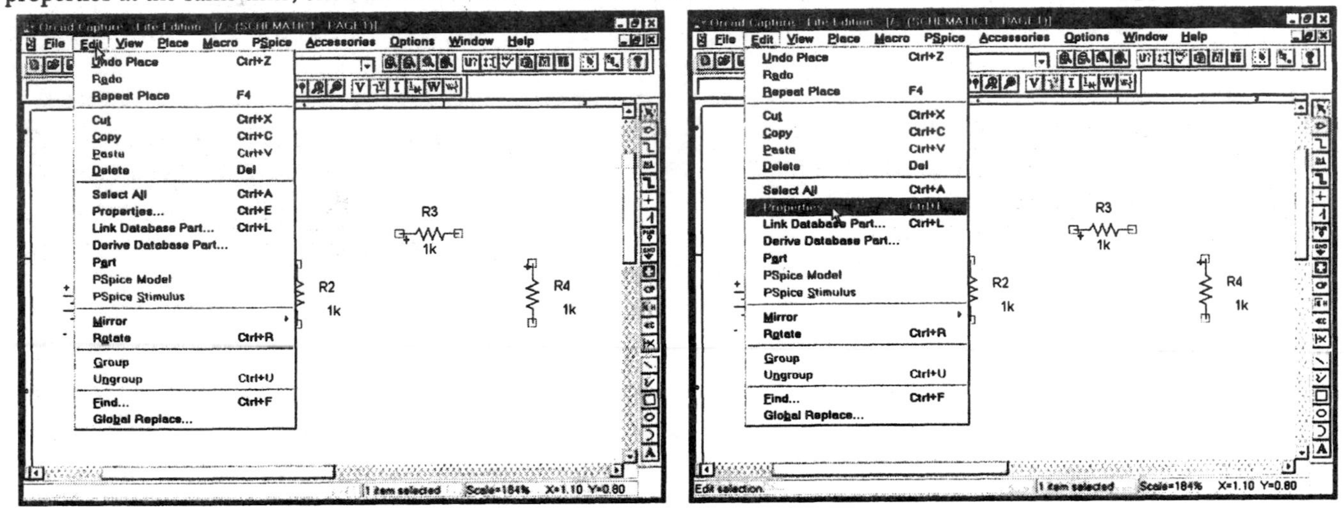

The properties spreadsheet for ***R1*** will appear:

This screen shows all the properties of R1. Presently, none of the properties displayed in the window are of interest to us. Click the ***LEFT*** mouse button on the horizontal scroll bar as shown below to scroll the spreadsheet right:

Click here.

The ***Part Reference*** property is the name of the resistor, ***R1*** in this case. The ***PCB Footprint*** property is the graphic that will be used if you use this schematic to create a PC board. If we scroll the window to the right, we will see more properties. Two screen captures are shown below to display all of the properties:

The value of the ***PSpice Template*** property generates the PSpice netlist line for the resistor when you create a netlist for the schematic. The ***TC*** property is a temperature coefficient for the resistor. Its default is zero (no temperature dependence). The property ***Value*** is the value of the resistor in ohms. The Source Library (not shown in the screen captures above) is the name of the .olb file in which the part is located.

As you scroll through the properties, you will notice that two properties appear to have the same value, ***Reference*** and ***Part Reference***. Both have a value of R1 for this part. The difference between the two properties appears when you use packages that have multiple parts, such as NAND gates and quad op-amps. For a NAND gate, the package would have a reference such as U1. Each gate within the package has its own designation, such as A, B, C, or D. An individual gate in the package would have a unique part reference, such as U1A, U1B, and so on. The ***Reference*** property is the package reference, such as U1, while the ***Part Reference*** property is the reference to an individual gate in the package, such as U1A. The ***Part Reference*** is used by PSpice and the ***Reference*** is used by the layout program.

We will now change a few of the properties. To change the resistance value, click the ***LEFT*** mouse button on the text ***1k***. The cell will be selected and the cursor will appear in the cell:

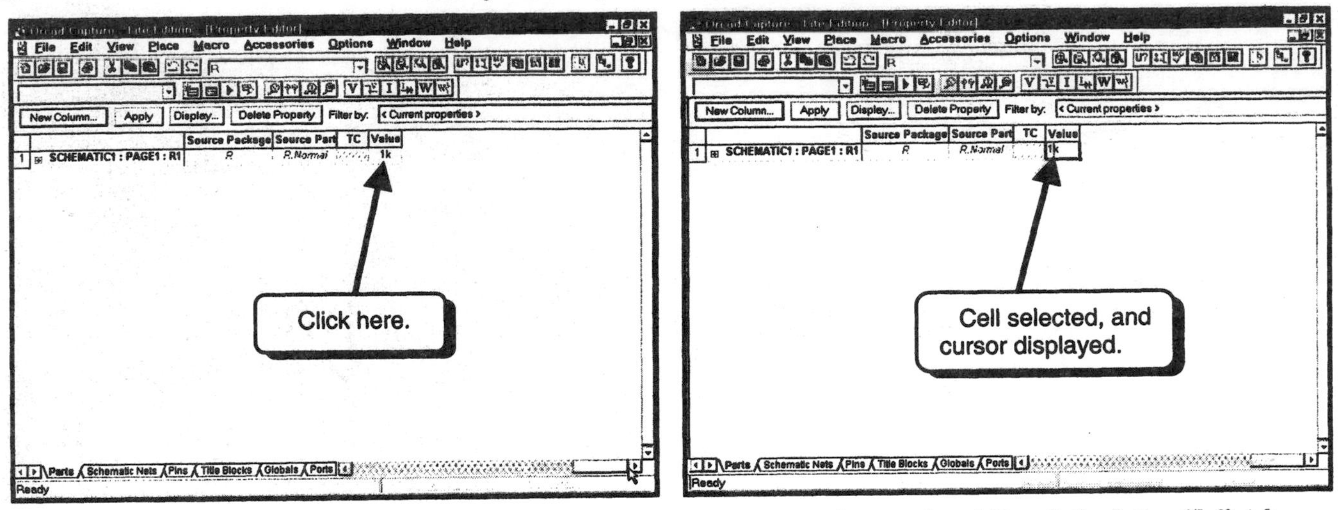


Change the value from ***1k*** to **2.7k. There must not be a space between the number 2.7 and the letter "k."** After you change the value to 2.7k, click the ***Apply*** button:

Value changed to ***2.7k.***

If you remember, the footprint of the resistor was not shown on the schematic. This is because the ***PCB Footprint*** property is not being displayed. All properties can either be displayed or not. We will illustrate how to display the ***PCB Footprint*** property. Scroll the properties window to the left to display the ***PCB Footprint*** property and then click the ***LEFT*** mouse button on the ***PCB Footprint*** column to select the column:

Click here to select the column.

Click here to scroll the window left.

Column selected.

Click the ***LEFT*** mouse button on the ***Display*** button. The dialog box shown below will appear:

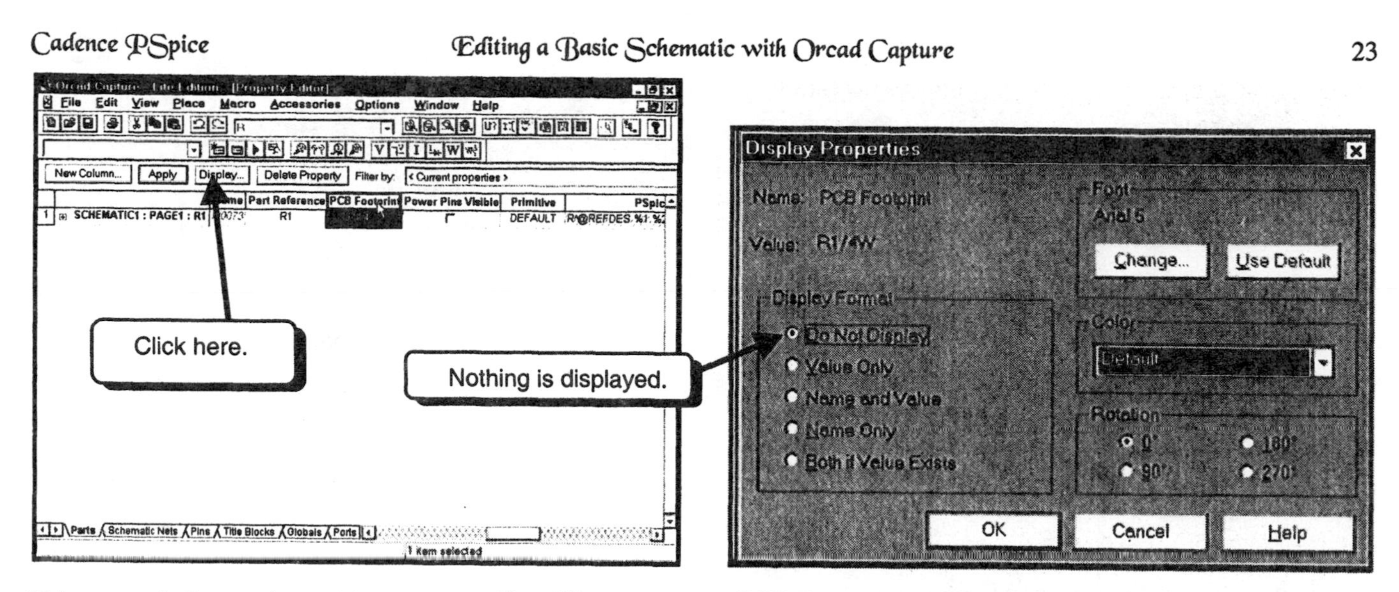


This screen indicates that neither the value ***R1/4W*** nor the name ***PCB Footprint*** will be displayed on the schematic. There are several display options. Selecting the ***Value Only*** option will display the value of the property. In this case, the text ***R1/4W*** will be displayed. Selecting ***Name Only*** will display the name of the property. In this case, the text ***PCB Footprint*** will be displayed. Selecting the option ***Name and Value*** will display the name of the property and its value. In this case the text ***PCB Footprint=R1/4W*** will be displayed. Selecting ***Both If Value Exists*** will display the property name and the value on the schematic if a value is assigned to the property. Presently, ***Do Not Display*** is selected, meaning that nothing related to this property will be displayed on the schematic. To select another display option, click the ***LEFT*** mouse button on the circle ○ next to the option you wish to select. The circle should fill with a dot ◉, indicating that the option has been selected. I will display only the value of the property:

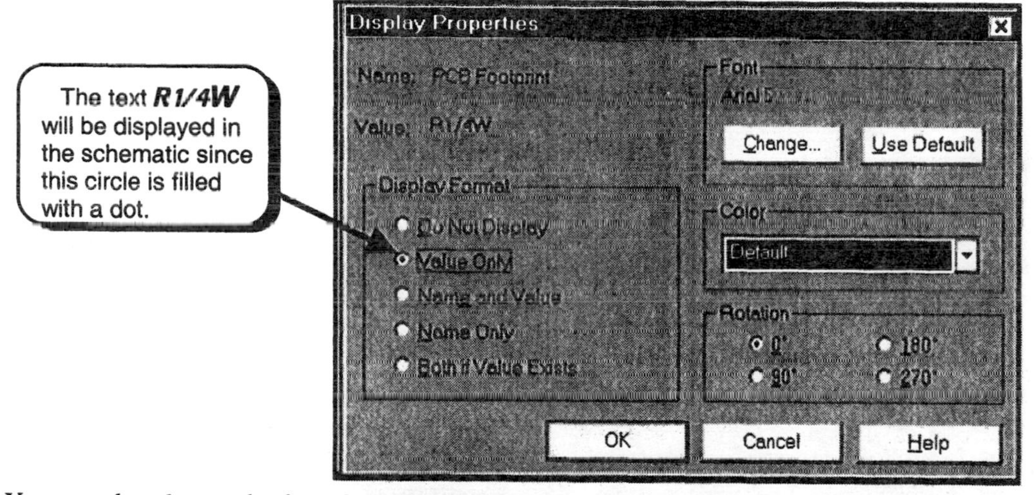


You can also change the font, font size, and the color of the font in which the property value is displayed. I will select a 20 point Courier New font and choose a color of red. You can choose different properties if you wish.

To accept the changes click the ***LEFT*** mouse button on the ***OK*** button. You will return to the properties spreadsheet:

Click the ***LEFT*** mouse button on the lower ☒ in the upper right corner of the spreadsheet window to close the window as shown below. You will return to the schematic with all of the properties updated:

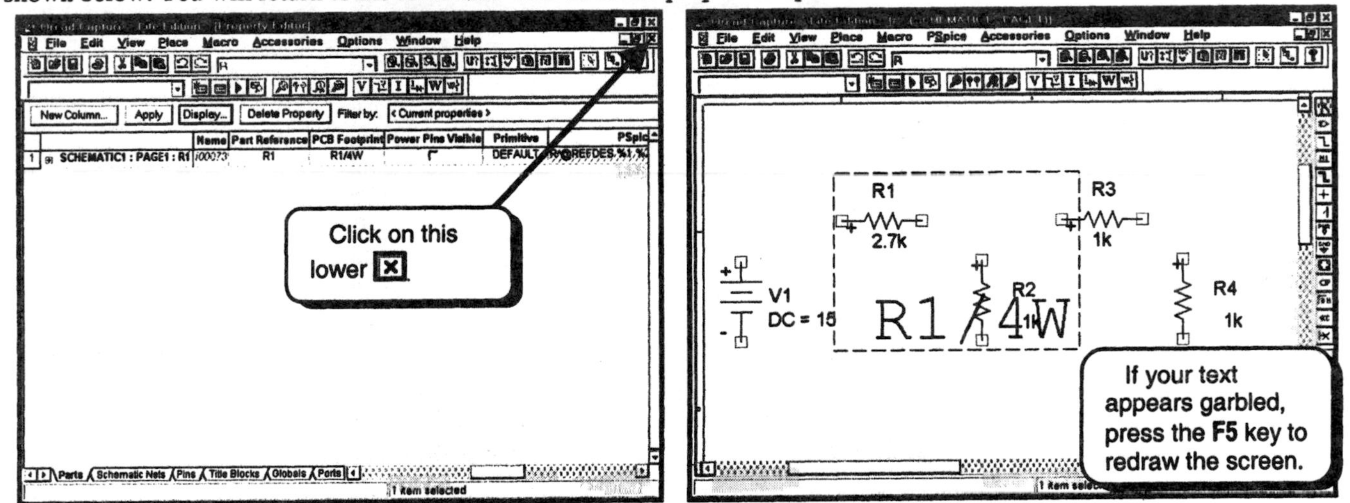


The text ***R1/4W*** is a bit too large. Repeat the previous procedure to change the font size to 9 points:

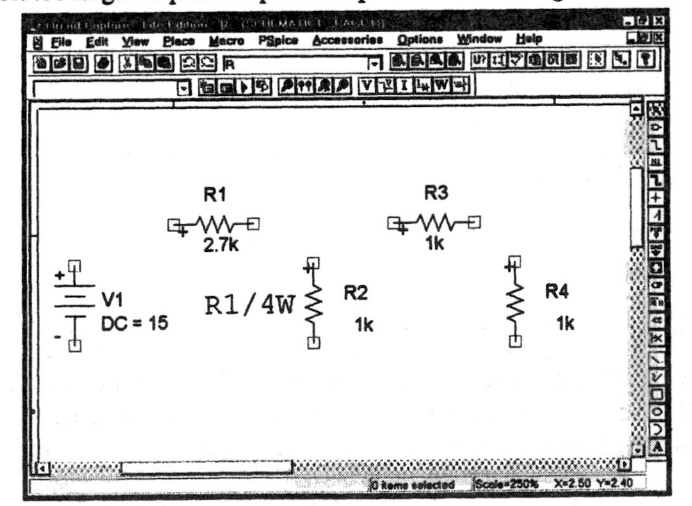


On my display, the text for ***R1/4W*** was placed in a poor location and we will move it. Click the ***LEFT*** mouse button on the text ***R1/4W*** to select it. It should turn pink and display pink handles:

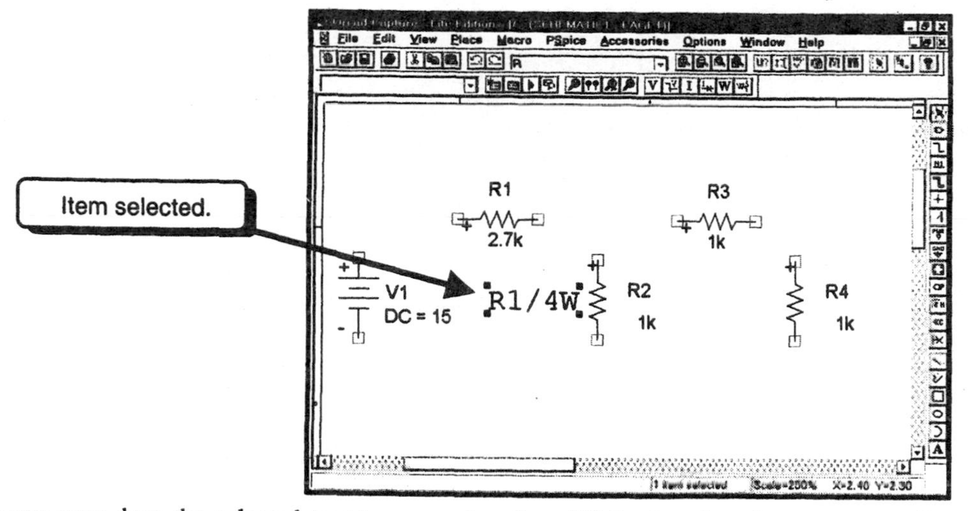


You can now drag the selected text to a new location. While dragging the text, an outline of the text will move with the mouse:

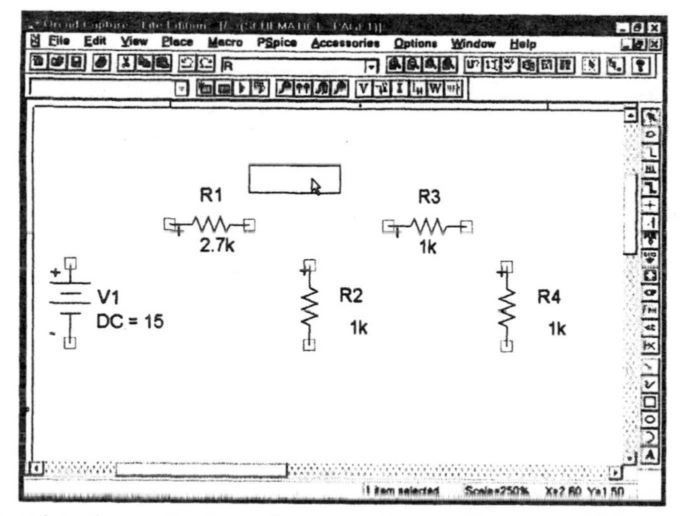


Move the outline to a convenient location and release the mouse button. The text will be moved to the new location of the box:

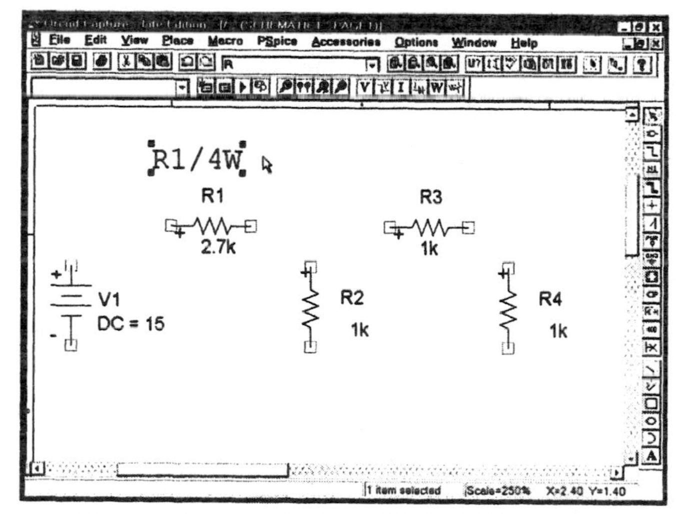


The second way to edit a part's properties is to double-click the ***LEFT*** mouse button on the graphic symbol for that part. To edit the properties of ***R2,*** double-click the ***LEFT*** mouse button on the ***R2*** resistor graphic, ⊢ΛΛΛ–. Make sure that you click on the center of the graphic. If you double-click fast enough, the properties spreadsheet will appear:

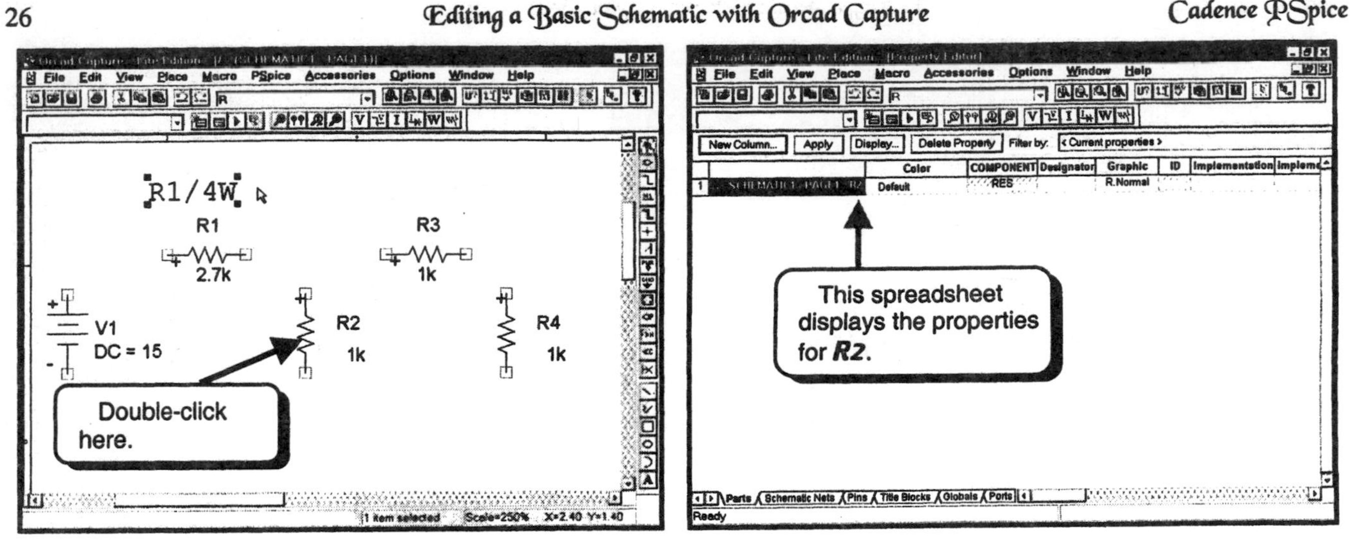


If you double-click at the wrong place, you might get the properties of another item, such as the properties of a pin as shown below:

This spreadsheet displays the properties for ***R2*** pin ***2***.

If you get the properties of the wrong item, close the spreadsheet by clicking on the lower [X] and try again. Keep trying until you see the spreadsheet below:

This spreadsheet displays the properties for ***R2***.

Edit the properties to change the value to **5k** and then close the spreadsheet by typing **CTRL-F4**.

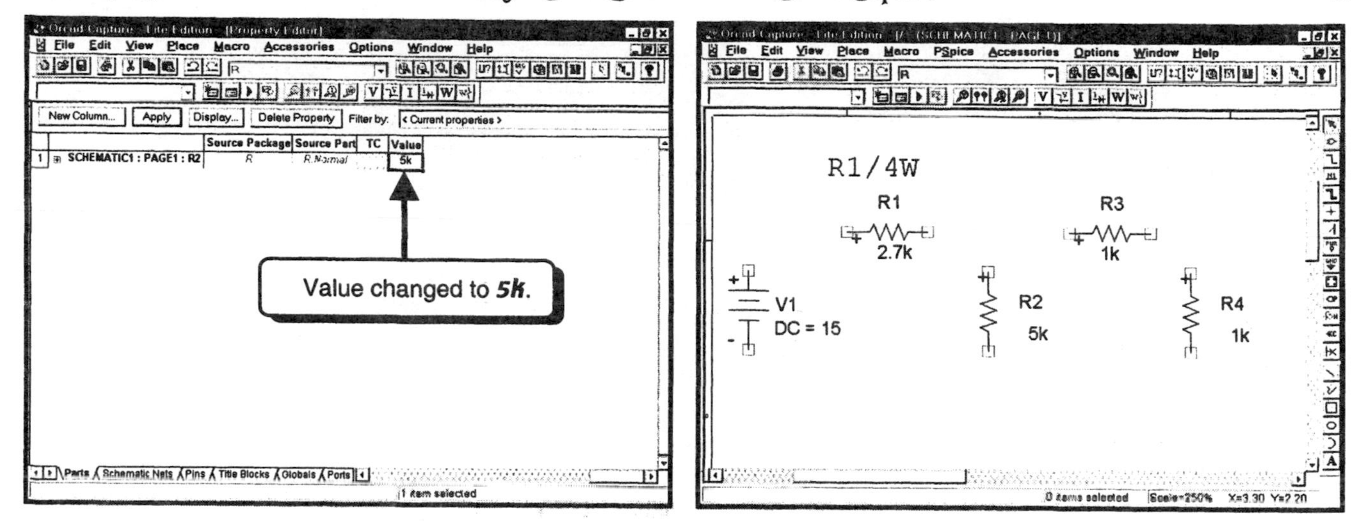


Next, we would like to change the name of ***R3*** to R35. Double-click the ***LEFT*** mouse button on the text ***R3***. If you double-click fast enough, the dialog box below will appear:

Display Properties

Name: Part Reference

Value: R3

Font: Arial 5 (default) — Change... — Use Default

Display Format: Do Not Display / Value Only / Name and Value / Name Only / Both if Value Exists

Color: Default

Rotation: 0° / 90° / 180° / 270°

OK — Cancel — Help

Change the name to R35 and click the ***OK*** button. The name will be changed in the schematic.

Display Properties

Name: Part Reference

Value: R35

Font: Arial 5 (default) — Change... — Use Default

Display Format: Do Not Display / Value Only / Name and Value / Name Only / Both if Value Exists

Color: Default

Rotation: 0° / 90° / 180° / 270°

OK — Cancel — Help

Orcad Capture - Lite Edition - [/ - (SCHEMATIC1 : PAGE1)]

File Edit View Place Macro PSpice Accessories Options Window Help

R1/4W — R1 2.7k — R35 1k — V1 DC = 15 — R2 5k — R4 1k

0 items selected Scale=250% X=3.60 Y=1.40

To change the value of ***R35***, double-click on the text ***1k***. A dialog box will appear:

The dialog box indicates that the current value is ***1k***. To change the value, type in the new value, **1Meg**, for example:

***1Meg*** stands for $1 \times 10^6$. There must not be any spaces between the ***1*** and the ***Meg***.* You can also change the font, font size, and font color with this dialog box:

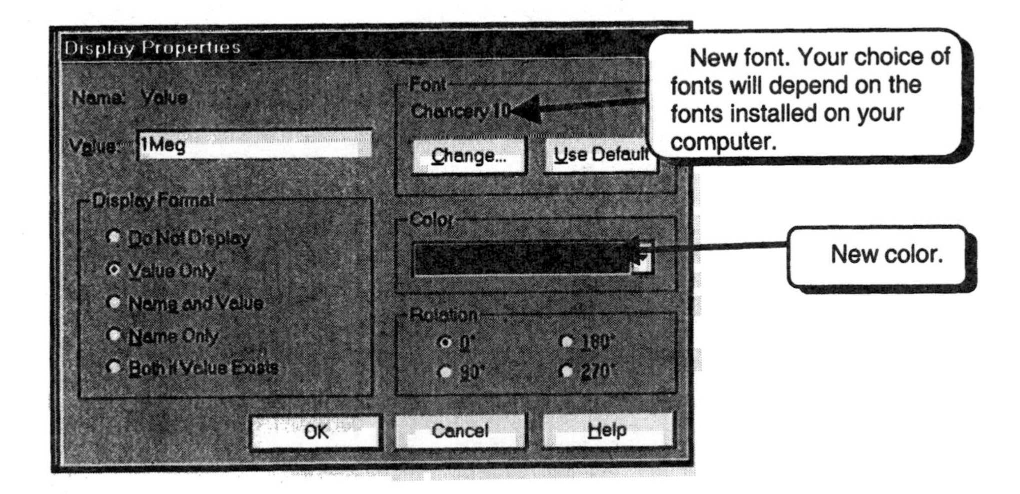


Click the ***OK*** button to accept the changes:

*In PSpice the multipliers m and M both refer to "milli." Thus the numbers 1m and 1M both equal 0.001.

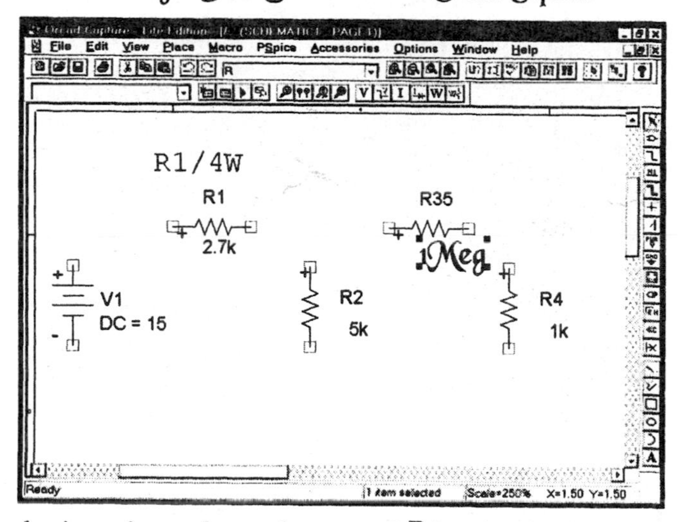


Using one of the methods given above, change the name of ***R4*** to Rz, and the value to 10k:

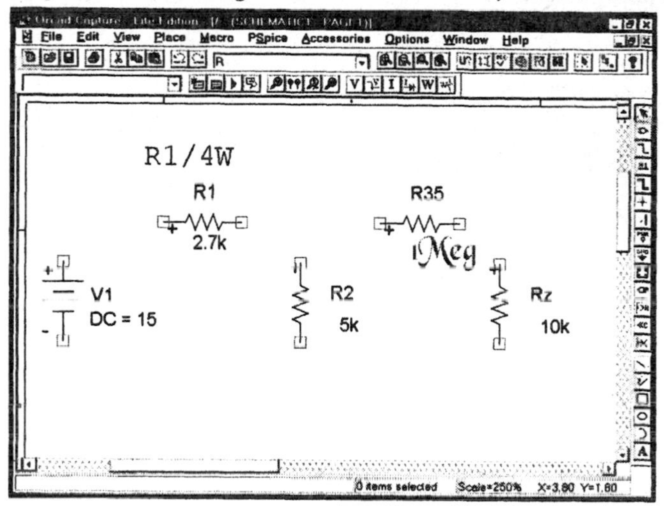


We are now finished placing parts and changing properties. Using the method given on pages 24–25 to move text items, move all the necessary text to make the schematic more readable. When you are done, you should have a schematic that looks something like the previous screen.

At this point your schematic may have bits of garbage floating around due to the editing. To get a fresh copy of the screen, select **View** and then **Zoom** from the menus:

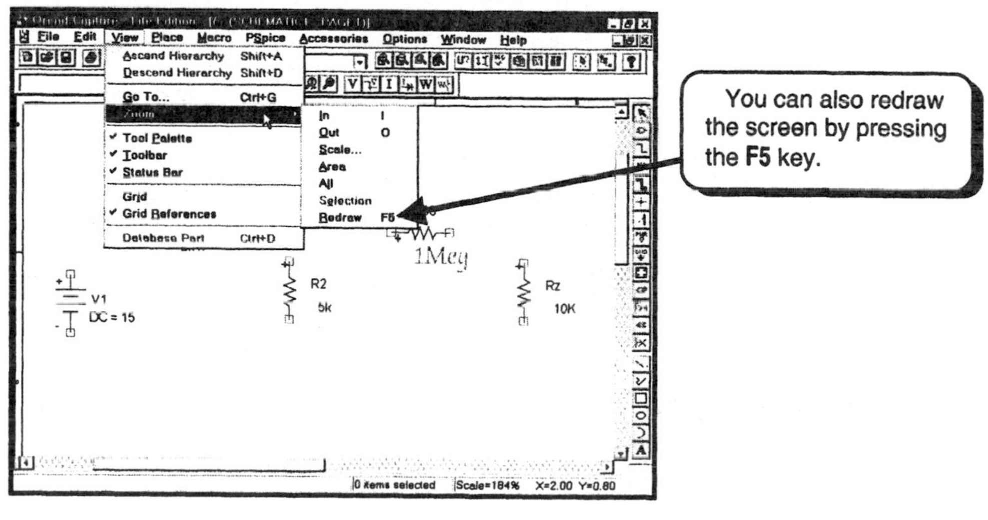


Click the ***LEFT*** mouse button on **Redraw**. The screen will be cleared and redrawn. You can also redraw the screen by pressing the **F5** key.

# 1.E. Wiring Components

We now must wire the resistors together. Click the ***LEFT*** mouse button on the **Place** menu selection:

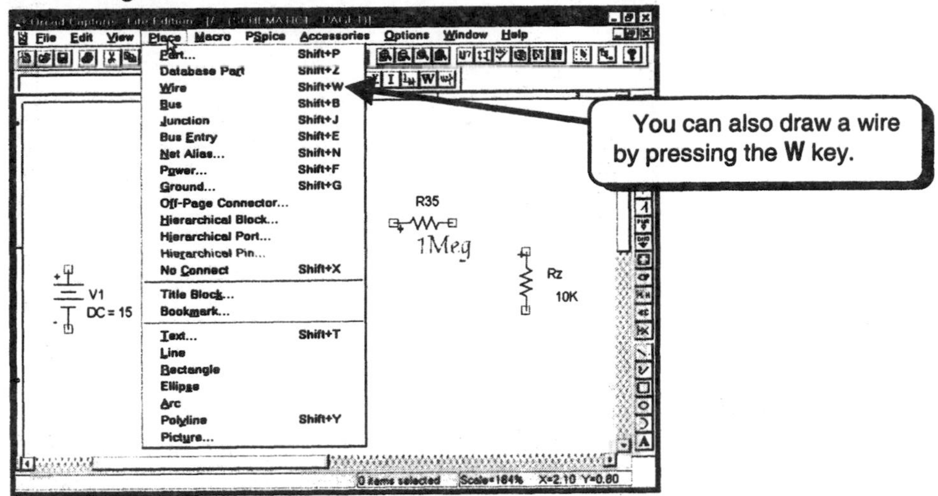


Click the ***LEFT*** mouse button on the **Wire** menu selection. Crosshairs will replace the mouse pointer. Move the crosshairs so that they point toward the top of the positive (+) terminal of the DC voltage source:

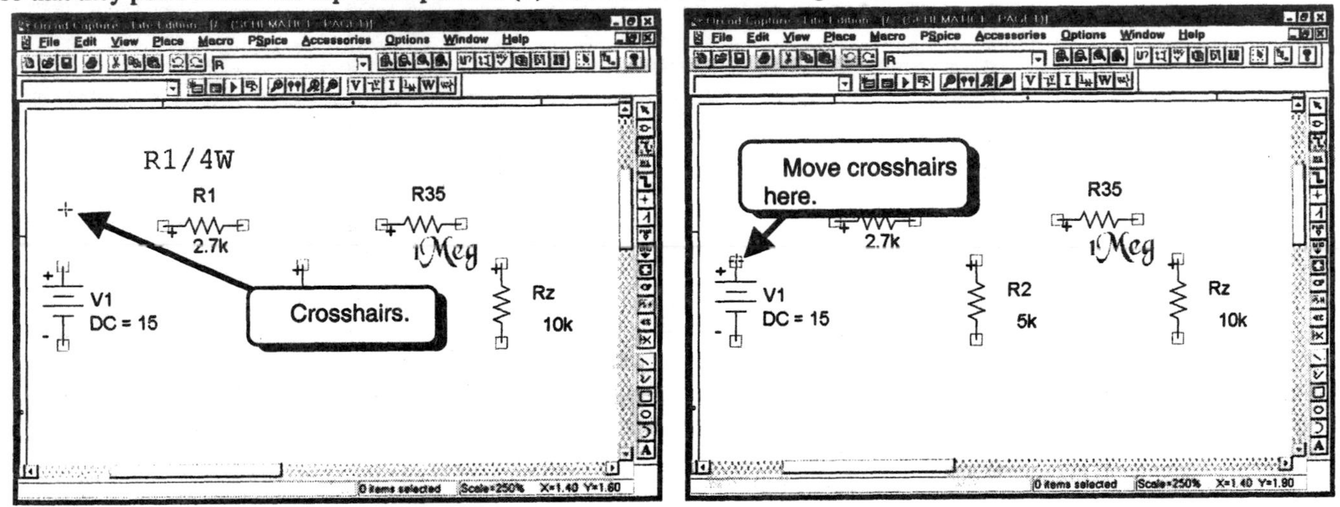


To start drawing a wire, click the ***LEFT*** mouse button on top of the positive terminal and then move the crosshairs away:

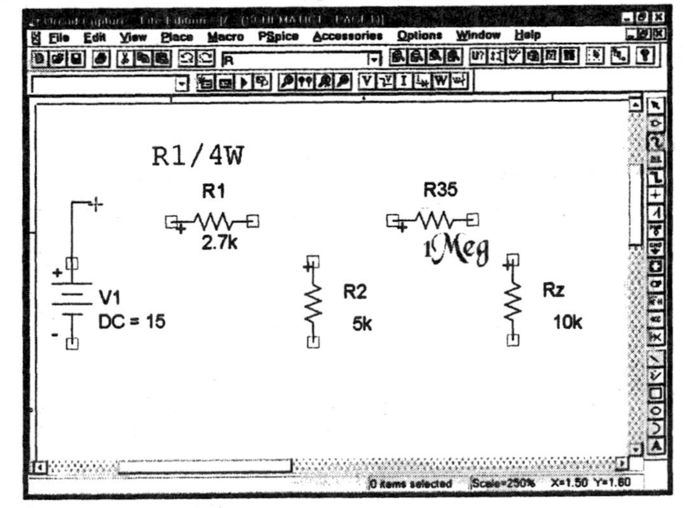


**If you missed the positive terminal, press the ESC key and start over.** Next, move the crosshairs to point toward the left terminal of the ***2.7k*** resistor. A red dot should appear, indicating that you are at a pin, and will make a connection if you click the ***LEFT*** mouse button:

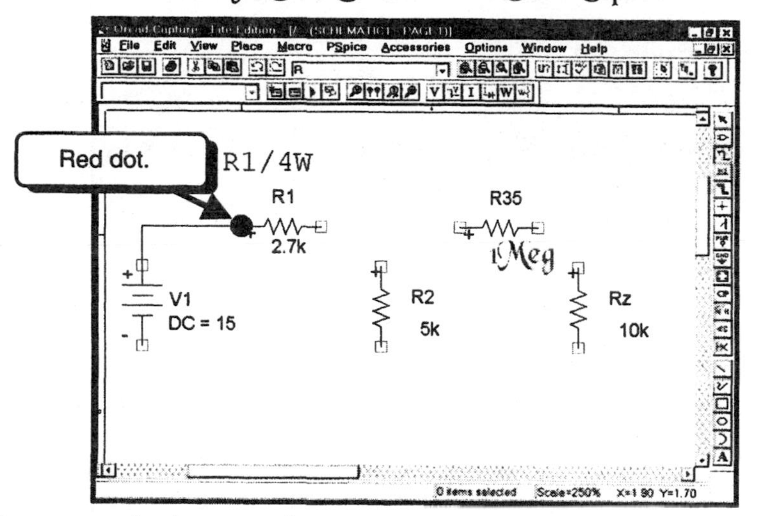


Click the ***LEFT*** mouse button to make the connection:

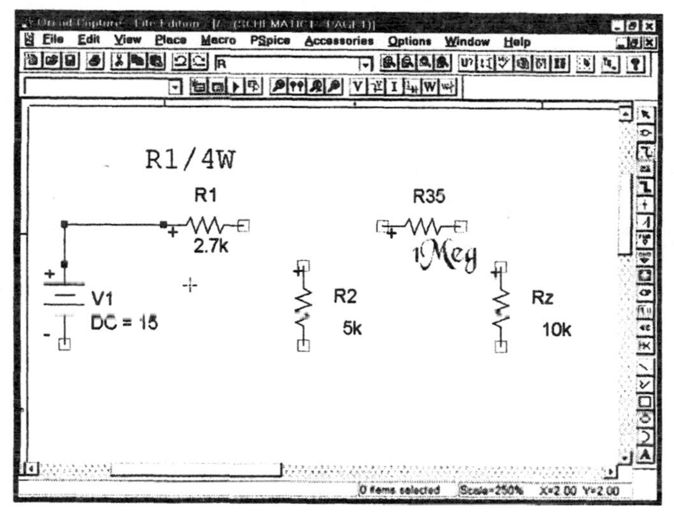


The schematic shows that the voltage source and the ***2.7k*** resistor are now wired together. The schematic also shows that the crosshairs are still on the screen, indicating that we are still drawing wires. If your cursor is not shown as crosshairs, you can start drawing wires using three methods:

1. Select **Place** and then **Wire** from the Capture menus.
2. Press the **W** key. This is the keyboard shortcut for selecting **Place** and then **Wire** from the menus.
3. Click the ***LEFT*** mouse button on the draw wires icon, ⌐L.

   Move the crosshairs to the right terminal of the ***2.7k*** resistor:

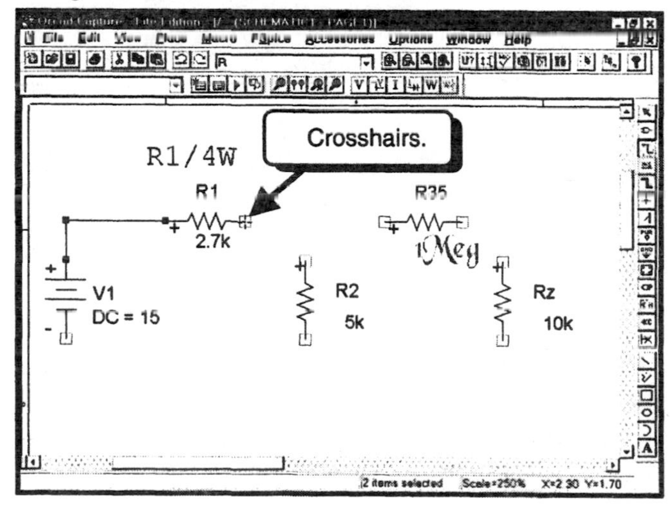

Click the ***LEFT*** mouse button and then move the mouse. You should have a wire connected to the right terminal of the ***2.7k*** resistor:

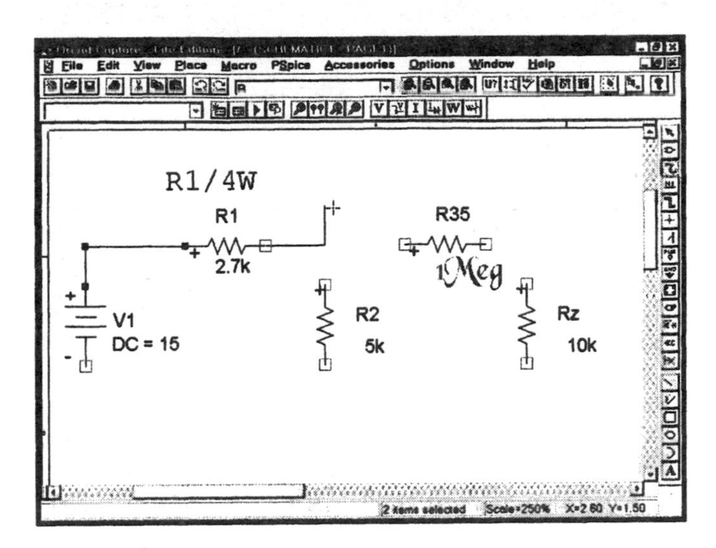


Move the crosshairs to the left terminal of ***R35*** and click the ***LEFT*** mouse button to make the connection:

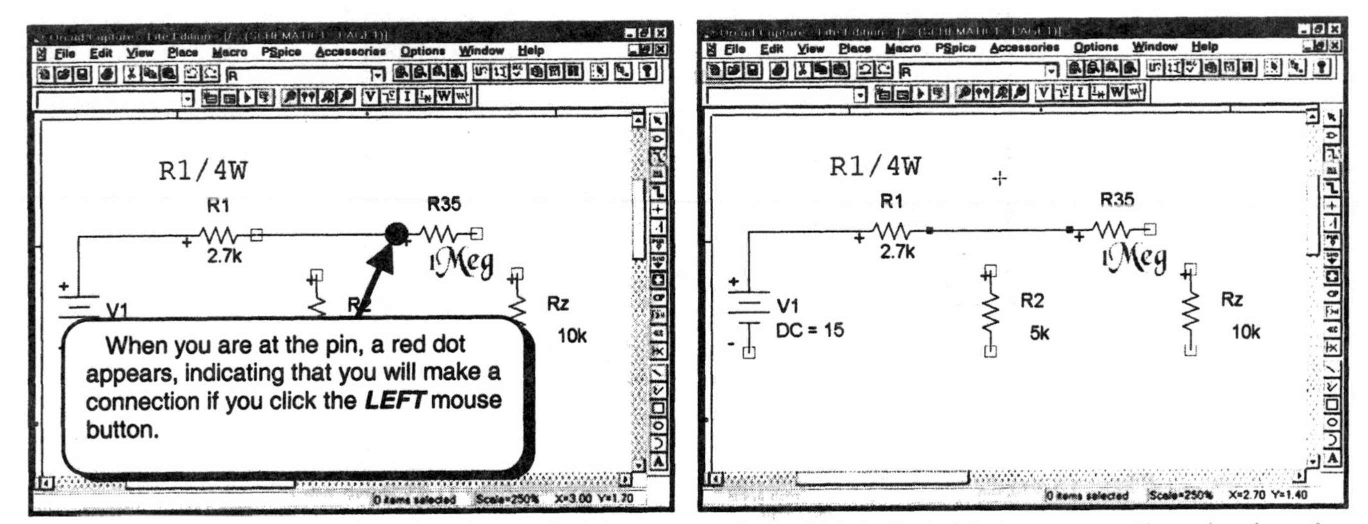


Note that the crosshairs are still displayed instead of the mouse pointer. This indicates that we can continue drawing wires. Move the crosshairs to point at the top pin of ***R2***:

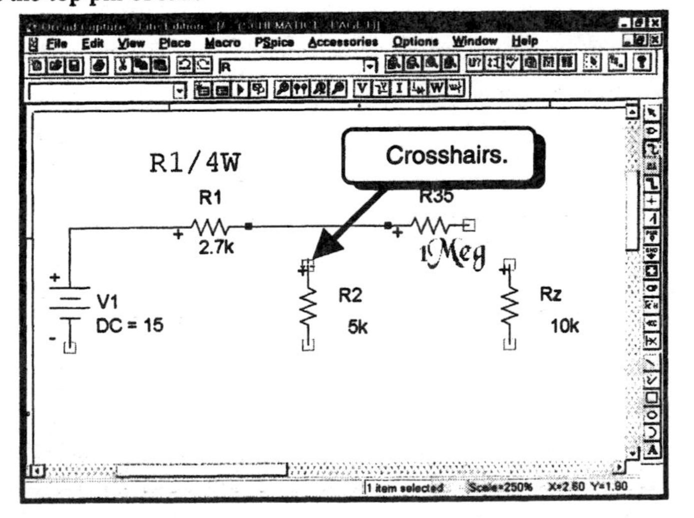

Click the ***LEFT*** mouse button to start drawing a wire and then move the crosshairs up. A wire should join with the top pin of ***R2***:

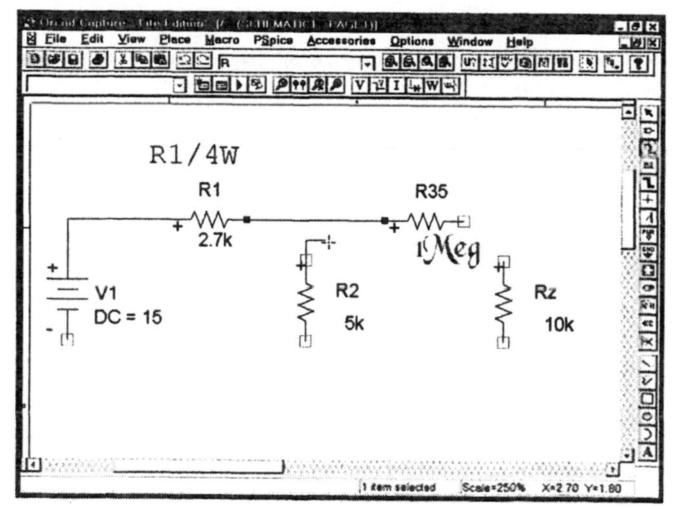


Move the crosshairs to the wire connecting R1 and R35. A red dot should appear, indicating that you will make a connection if you click the ***LEFT*** mouse button.

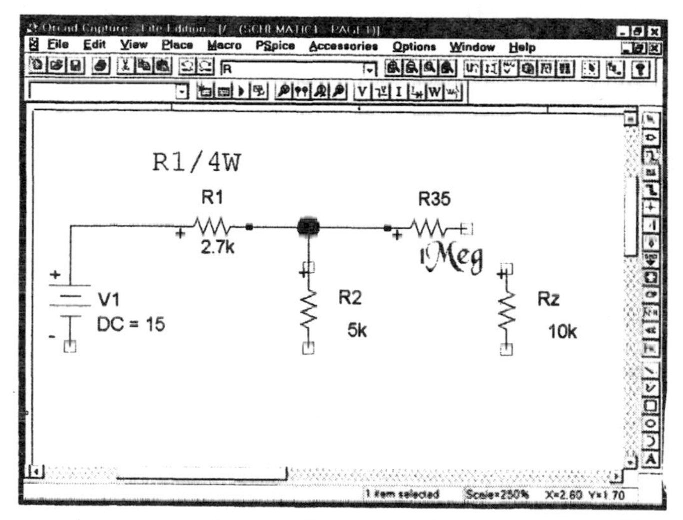


Click the ***LEFT*** mouse button to make a connection and then move the mouse away. You should now have the schematic shown below:

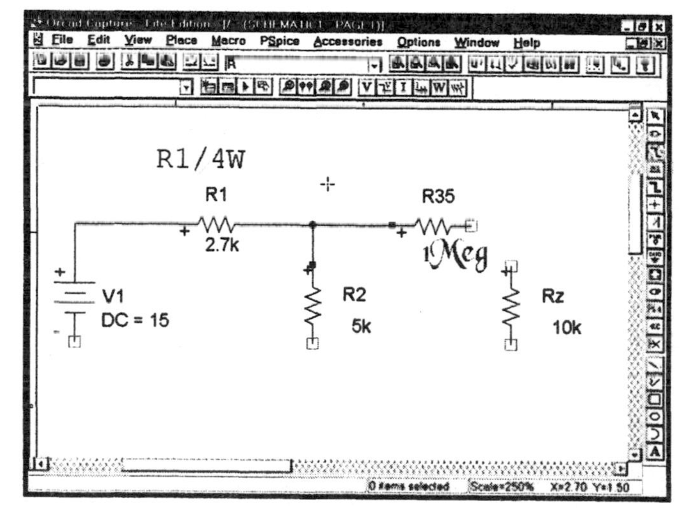


Continue wiring until you obtain the circuit shown below:

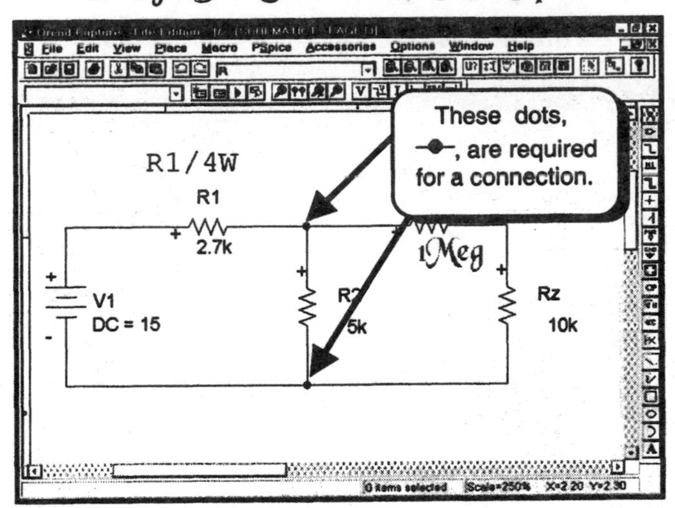


To stop drawing wires, press the **ESC** key.

Note in the circuit above that some of the connections have a dot, ┯. A dot indicates a connection. It is not necessary to have a dot present when a wire joins a pin. Dots are always drawn when wires meet in a "T," ┯. If two wires cross and do not display a dot, then the wires are not connected. If two wires cross and display a dot, then the wires are connected.

# 1.F. Correcting Wiring Mistakes

If you made a mistake drawing a wire, you can use the following procedure to delete unwanted wires, as well as components.

1. Make sure you are not in "wire" mode. If you are in "wire" mode, the mouse pointer is displayed as crosshairs. To terminate "wire" mode, press the **ESC** key.
2. Move the mouse pointer to the segment of wire or the part that you wish to remove.
3. Click the ***LEFT*** mouse button on the wire or part you wish to remove. This will select the wire or part. When the wire or part has been selected, it will turn pink.
4. Press the **DELETE** key to delete the selected wire or part.

# 1.G. Grounding Your Circuit

To run a circuit on PSpice, you must have at least one ground connection in your circuit. To ground your circuit, you must place a symbol called "0" (the number zero). There are three ways to place ground symbols:

1. Select **Place** and then **Ground** from the menus.
2. Press the **G** key.
3. Click the ***LEFT*** mouse button on the ground button, [GND], displayed in the toolbar on the right side of the Capture window (button shown below).

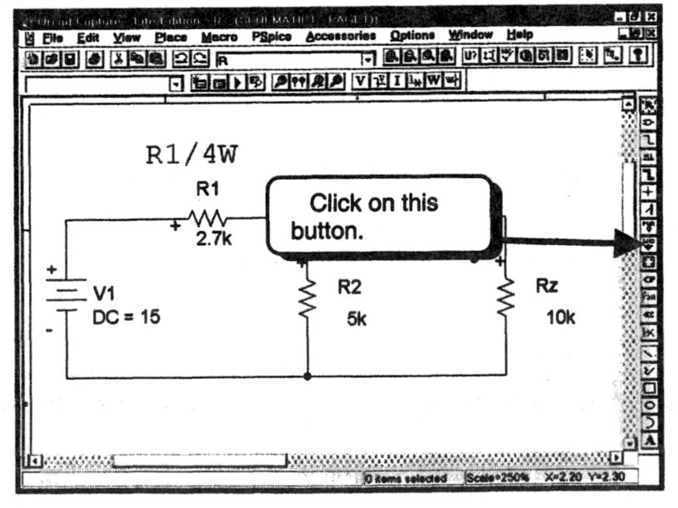

When you use any of these methods, the ***Place Ground*** dialog box will be displayed:

Capture has many ground symbols available for schematic drawing purposes. **Only the *0/SOURCE* ground can be used for PSpice simulations.** Select the ***0/SOURCE*** part:

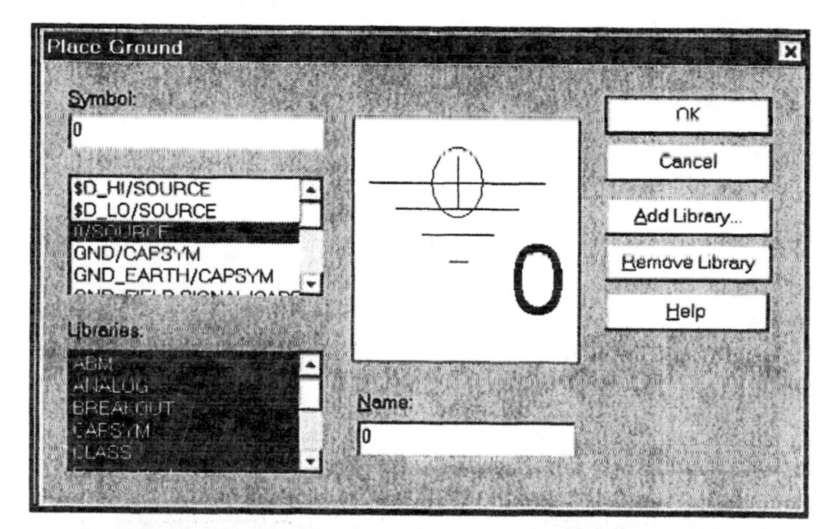


Click the ***OK*** button to accept the part. The ground symbol will appear attached to the mouse pointer:

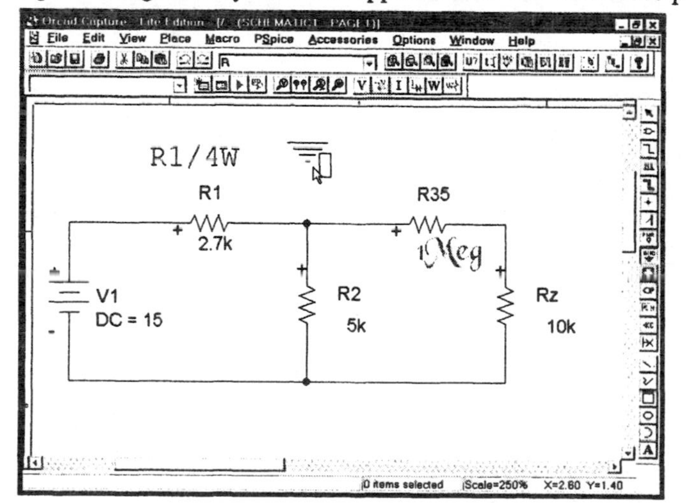


Note that the ground symbol moves with the mouse. Move the ground symbol as shown below:

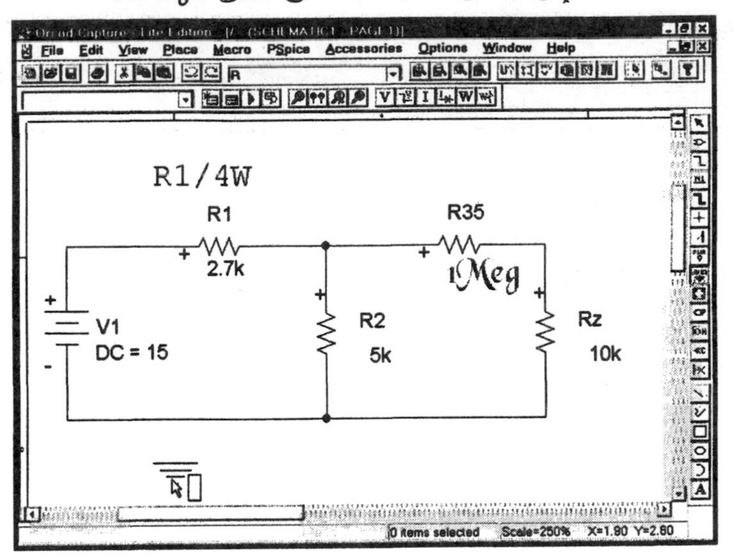


To place the ground symbol, click the ***LEFT*** mouse button. The ground symbol is placed on the schematic and, when you move the mouse, a second ground symbol appears attached to the mouse pointer:

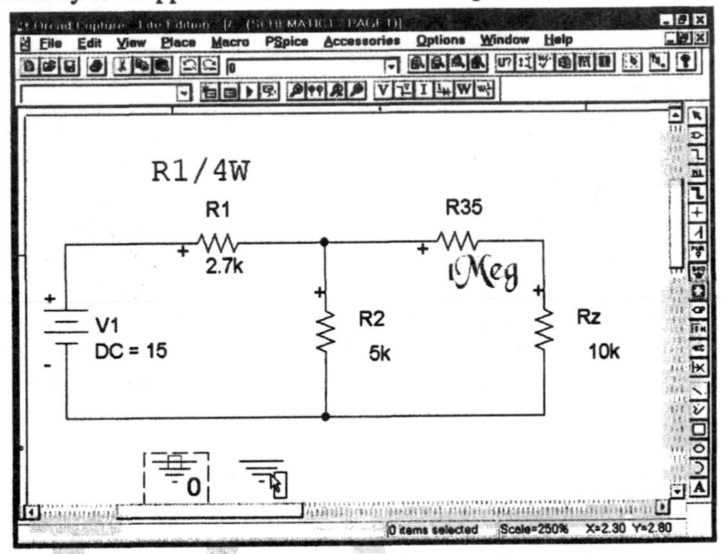


Since we do not need any more ground symbols, press the **ESC** key to terminate drawing ground symbols.

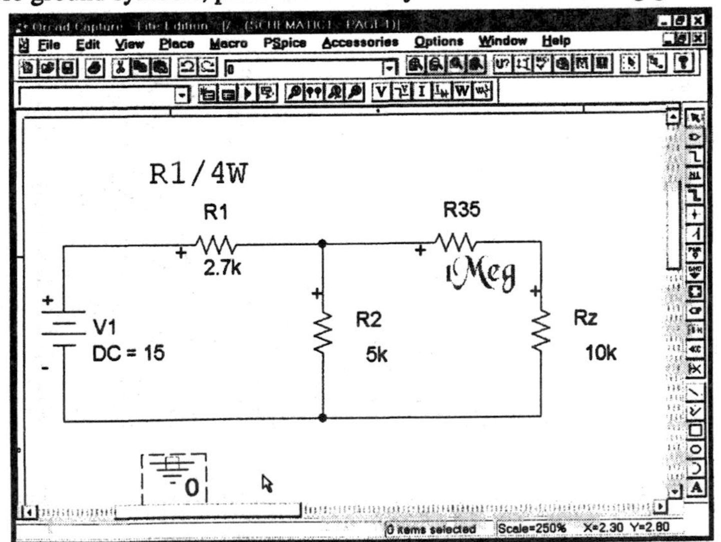


You must now connect the ground symbol to the circuit by connecting a wire from the ground to the circuit. Press the **W** key to start drawing wires and wire the circuit as shown below. To stop drawing wires, press the **ESC** key.

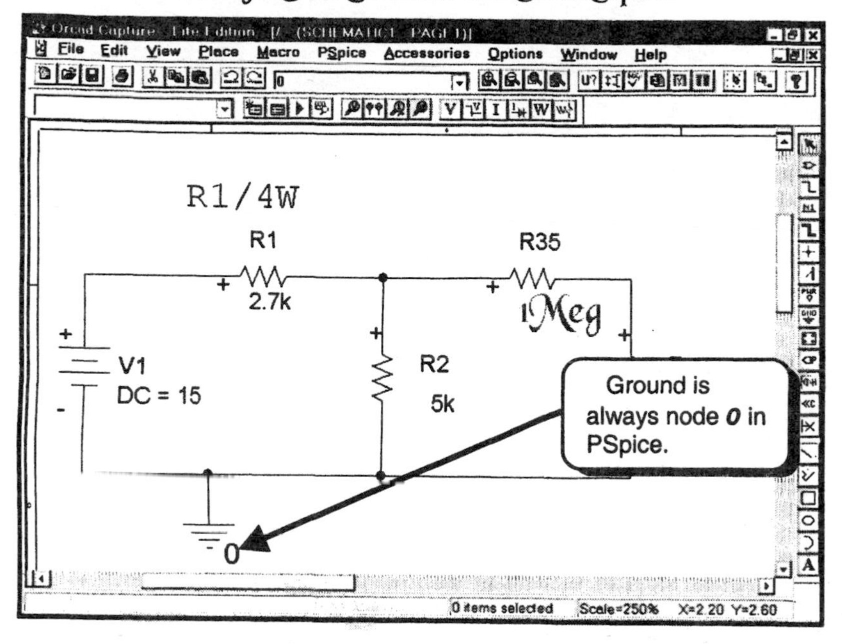


# 1.H. Labeling Nodes

Capture arbitrarily labels nodes with names such as N01015 or N01985, and users do not know the names unless they double-click on a wire to view the name. This is normally not a problem because we are usually not concerned with probing all of the nodes of a circuit. However, there are usually a few nodes that we would like to look at, and if we do not know the names of those nodes, viewing the voltages at that nodes is more difficult. Capture provides three methods for labeling a node: placing a net alias on a wire, connecting a power connector to a node, or connecting an offpage connector to a node. All methods have the same effect of naming a node. A net alias places a text label next to a wire, while power and offpage connectors place a graphic on your schematic and then you can label the connector. Note that if you give two nodes the same name, those two nodes will be connected together. This is a convenient way of connecting parts, but if you make a mistake and accidentally give two nodes the same label, those nodes will be connected together and your circuit will not operate properly.

We will first label a node using a net alias. Select **Place** and then **Net Alias** from the menus:

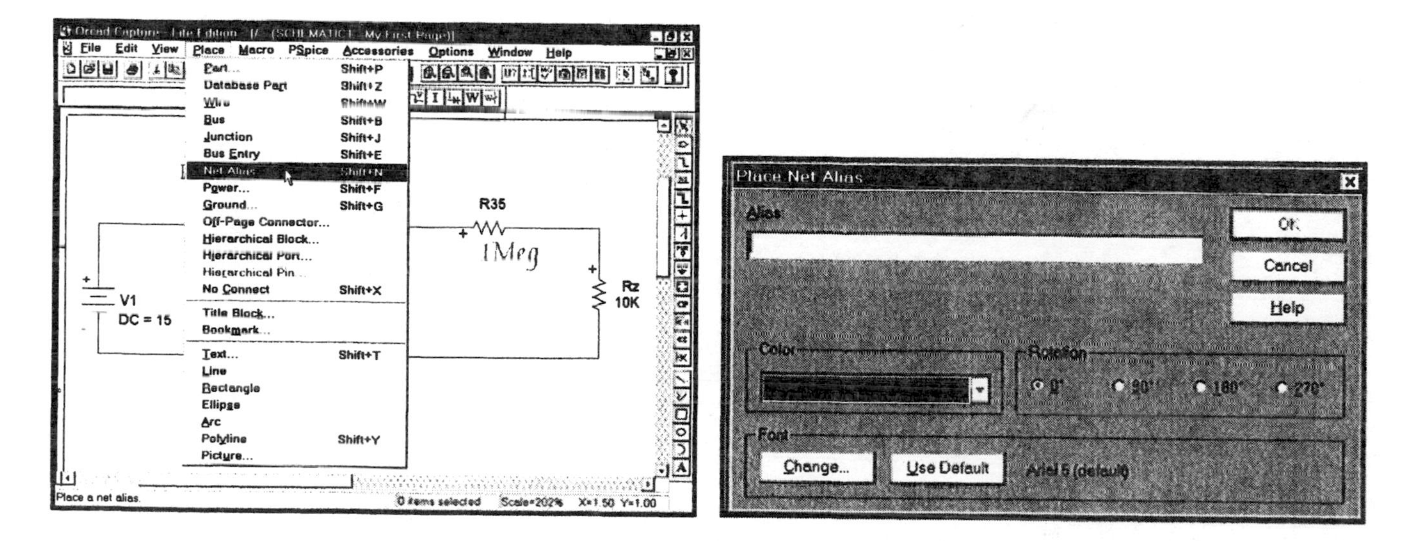


Enter a name for the alias and then click the ***OK*** button. An outline for the alias will be attached to the mouse pointer:

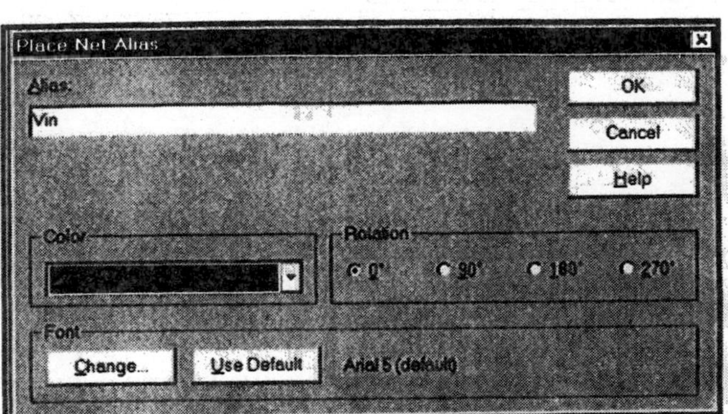


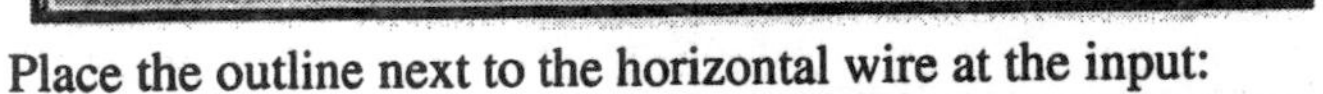

Place the outline next to the horizontal wire at the input:

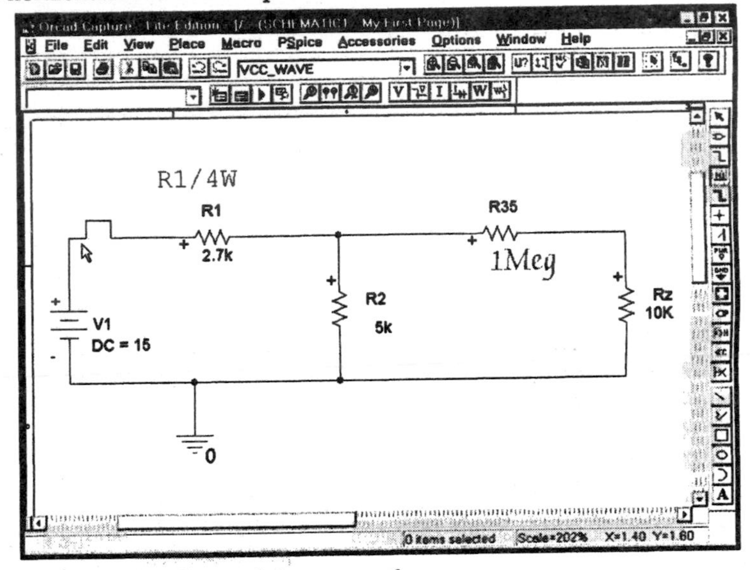


Click the ***LEFT*** mouse button to place the alias and then move the mouse away:

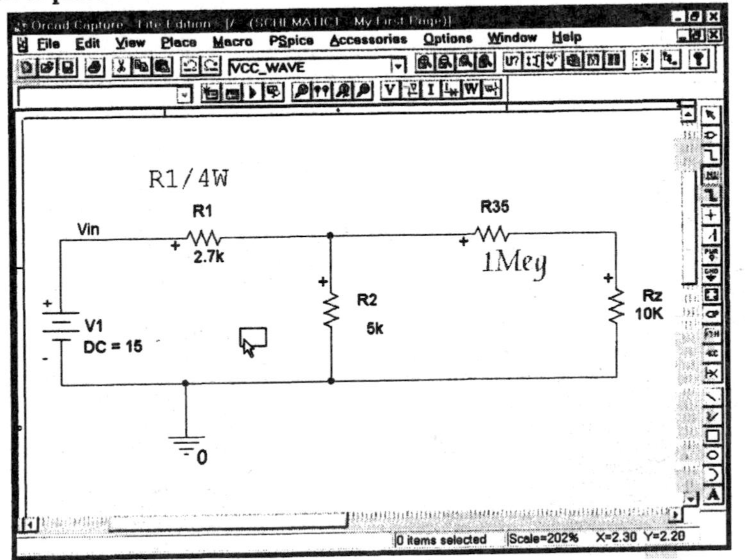


Press the **ESC** key to stop placing net aliases. This node is now named ***VIn***.

If you double-click the ***LEFT*** mouse button on the wire between V1 and R1, the name of that node will be listed:

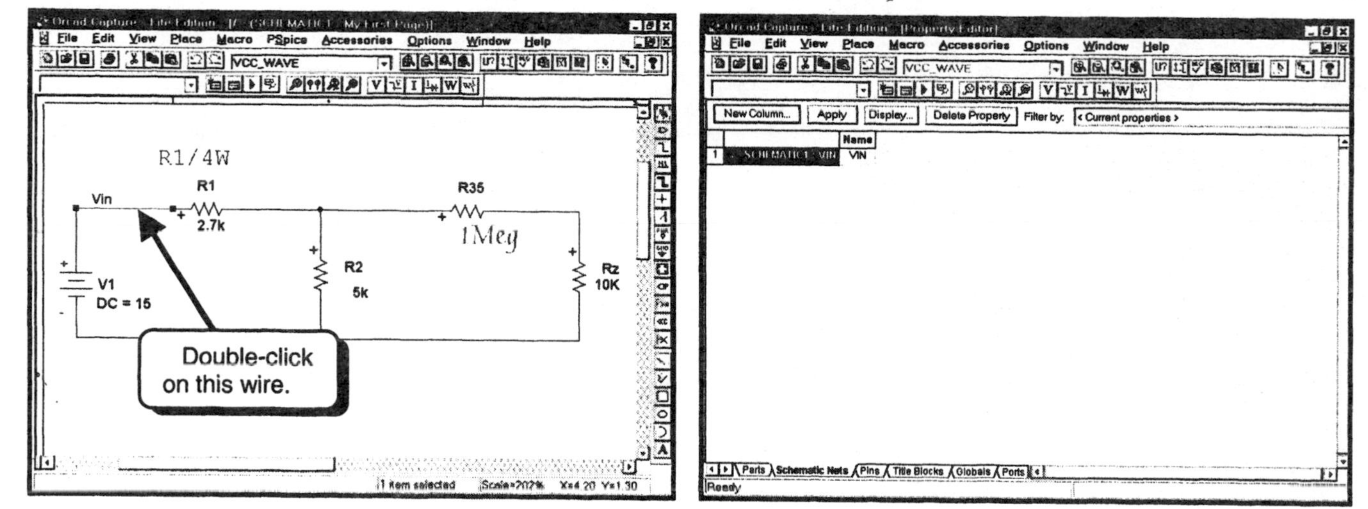


We see that the name of the wire is now ***Vin***. Type **CTRL-F4** to close the spreadsheet.

Capture provides seven different graphics symbols that we can use to label nodes. The names of the parts and their symbols are shown in **Table 1-1**.

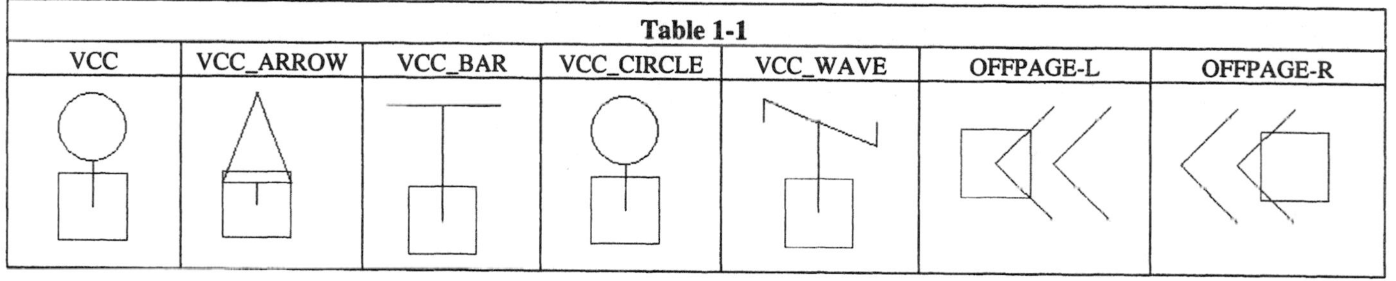

**Table 1-1**

| VCC | VCC_ARROW | VCC_BAR | VCC_CIRCLE | VCC_WAVE | OFFPAGE-L | OFFPAGE-R |
|---|---|---|---|---|---|---|

To place a power connector, select **Place** and then **Power** from the menus:

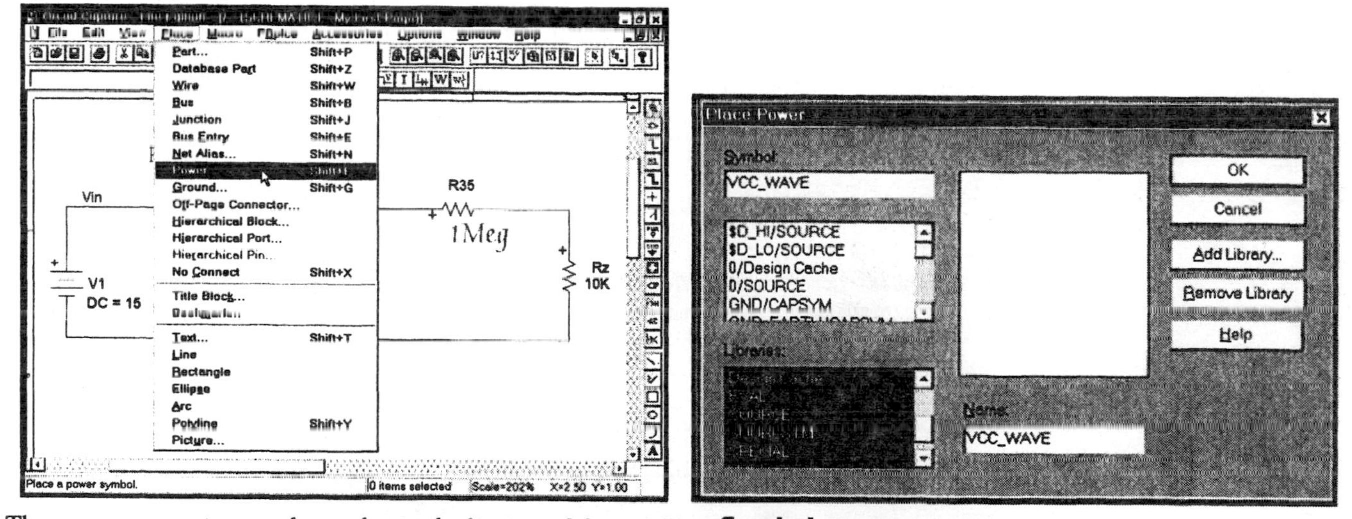


The power connectors are located near the bottom of the scrolling ***Symbol*** window pane:

Click the ***LEFT*** mouse button on one of the power connectors, ***VCC_BAR*** for example, and then click the ***OK*** button. The graphic for the power connector will become attached to the mouse pointer.

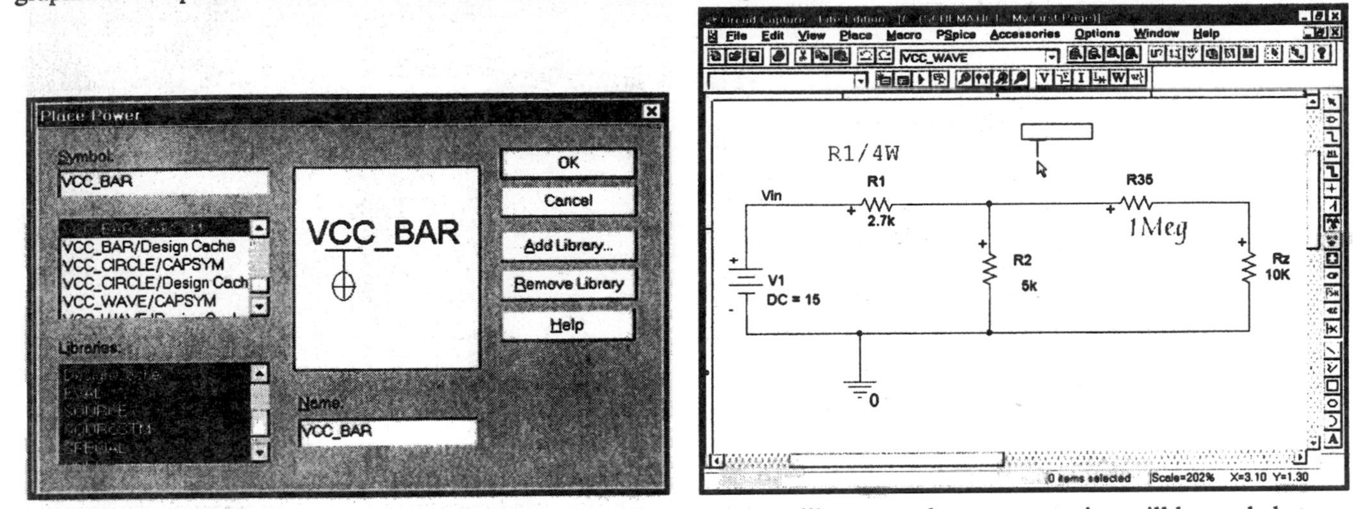


Place the power connector next to a wire at the center node. A red dot will appear where a connection will be made between the connector and the wire:

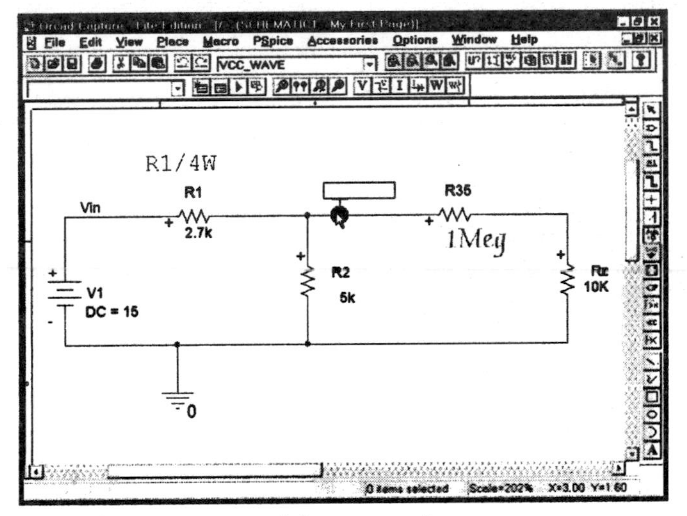


Click the ***LEFT*** mouse button to place the connector, and then move the mouse away:

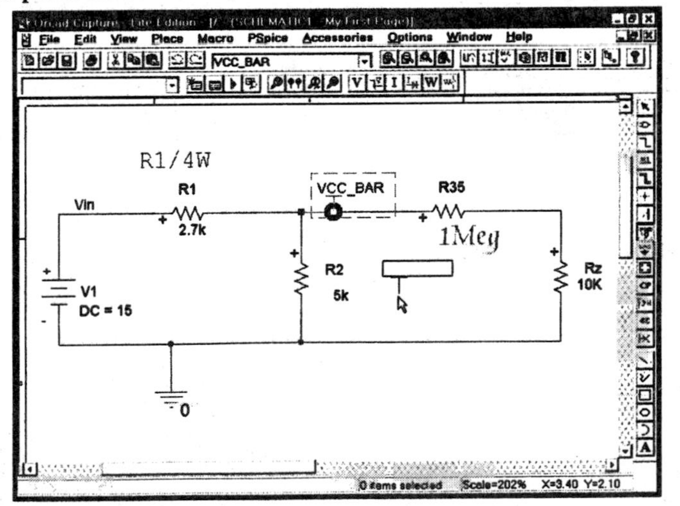


A second power connector is now connected to the mouse pointer. Press the **ESC** key to stop placing power connectors.

We must now change the label. Double-click on the text ***VCC_BAR*** to change it:

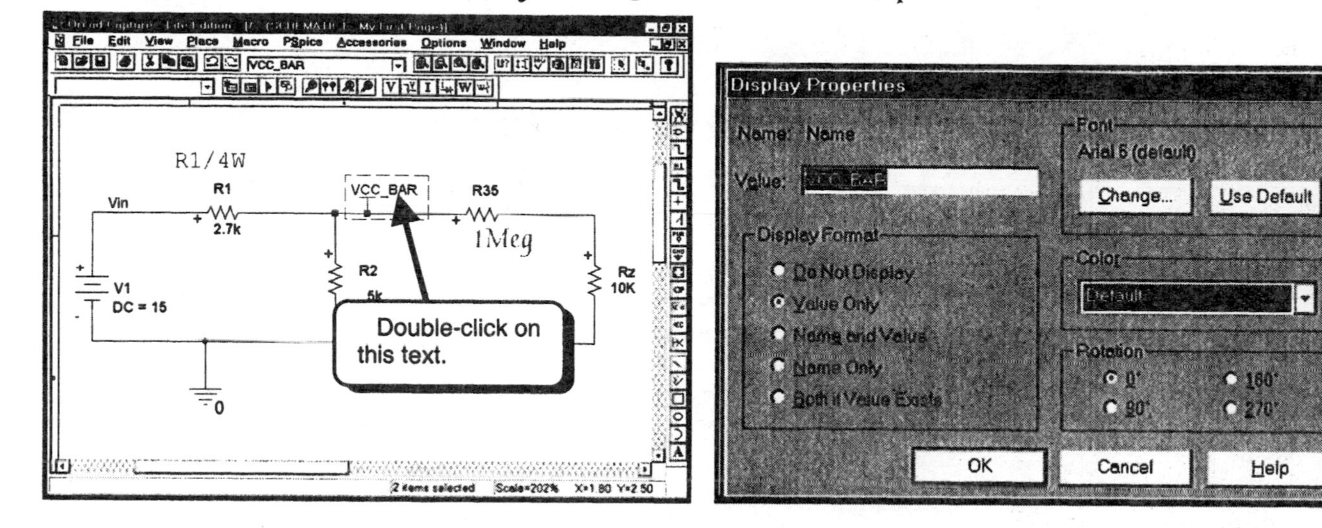


Enter a new name for the label and click the ***OK*** button:

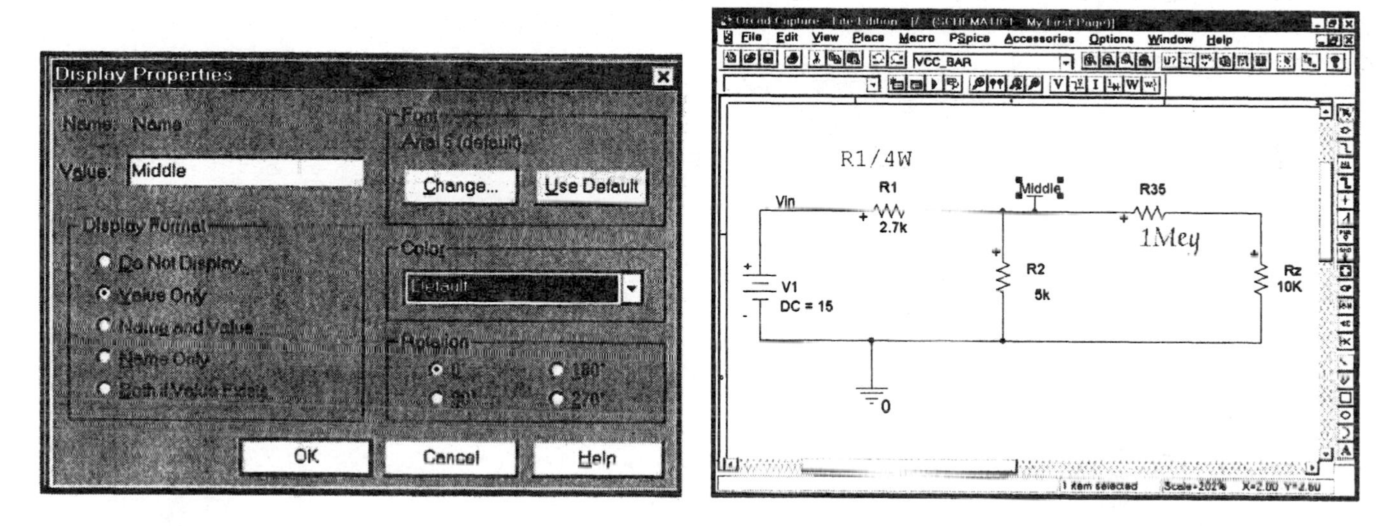

The center node is now labeled ***Middle***.

To place an offpage connector, select **Place** and then **Off-Page Connector** from the menus:

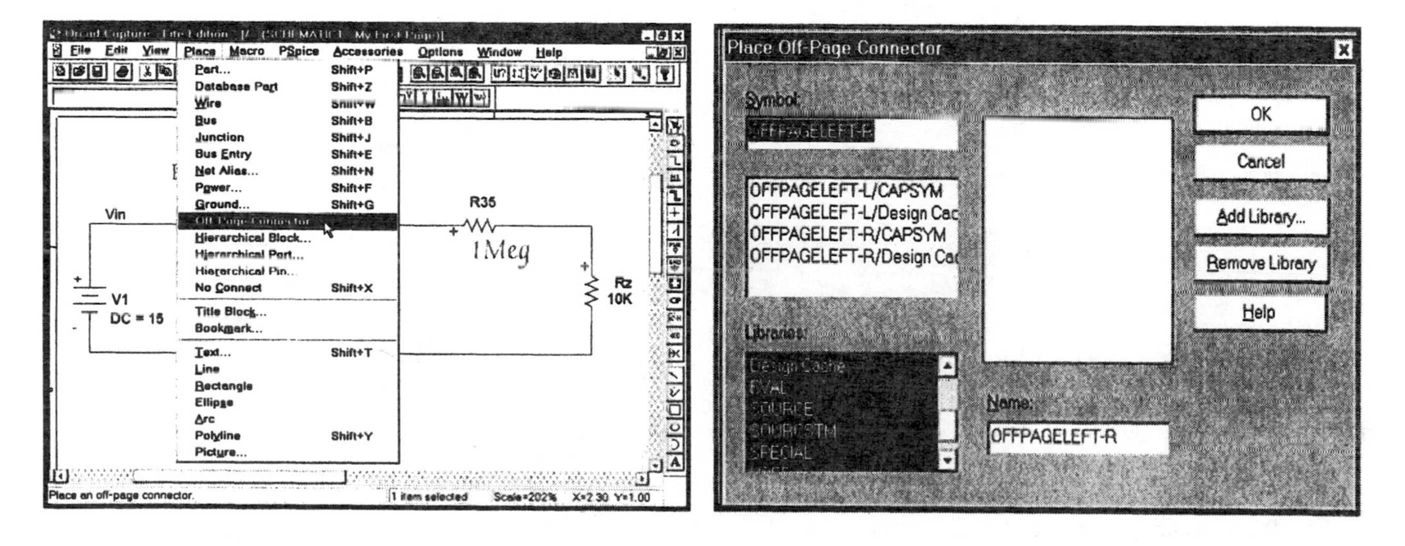


Select a connector and then place the connector as shown:

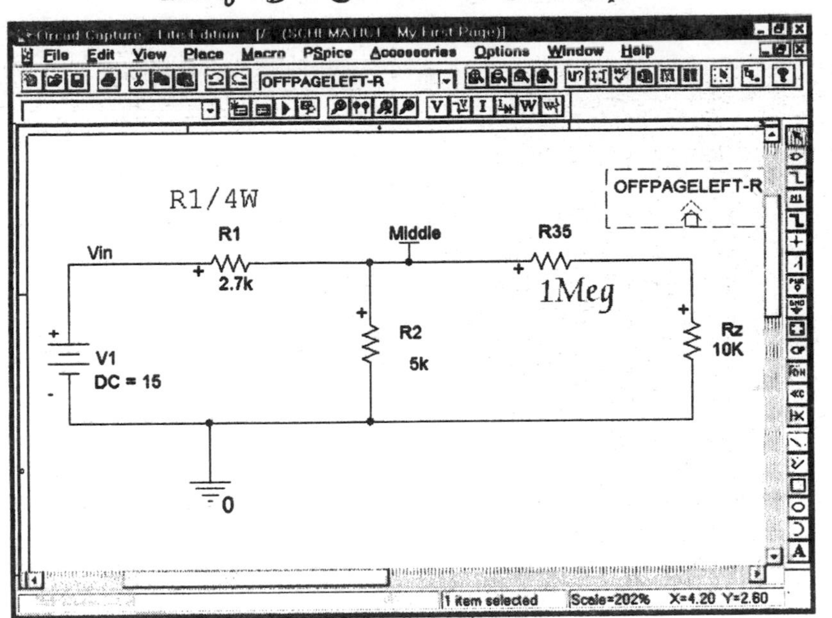


Wire the connector to your circuit:

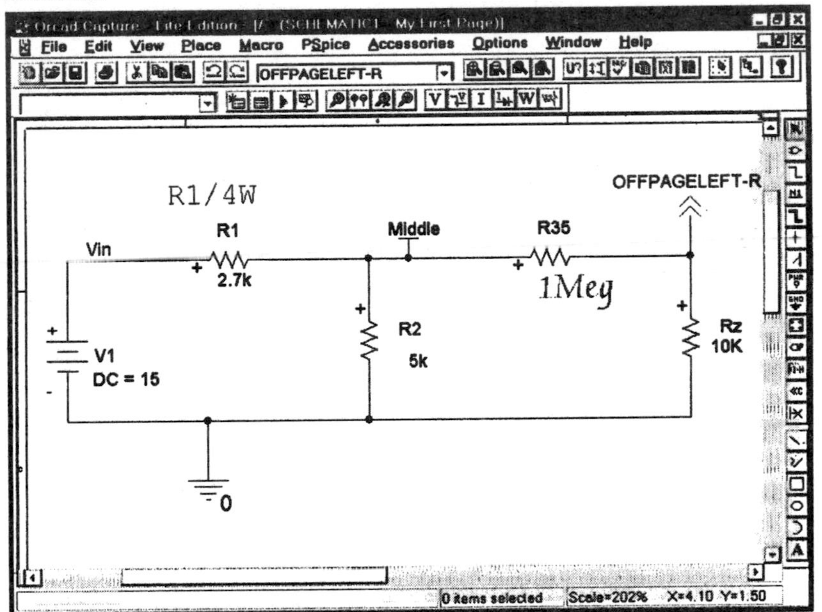


To change the label, double-click on the text ***OFFPAGELEFT-R***:

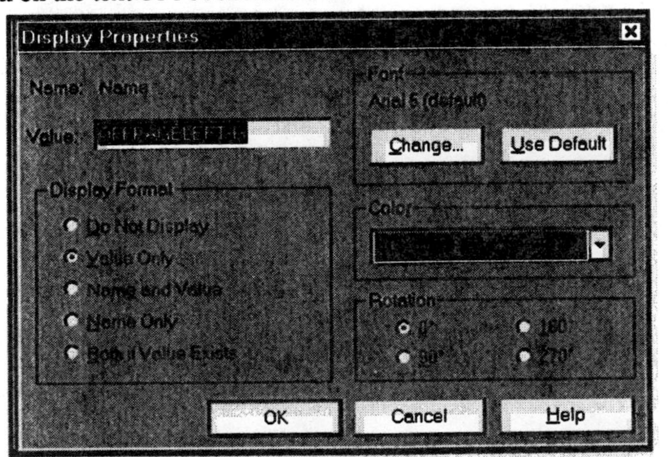


Type the text **Vo** to change the label to Vo and then click the ***OK*** button:

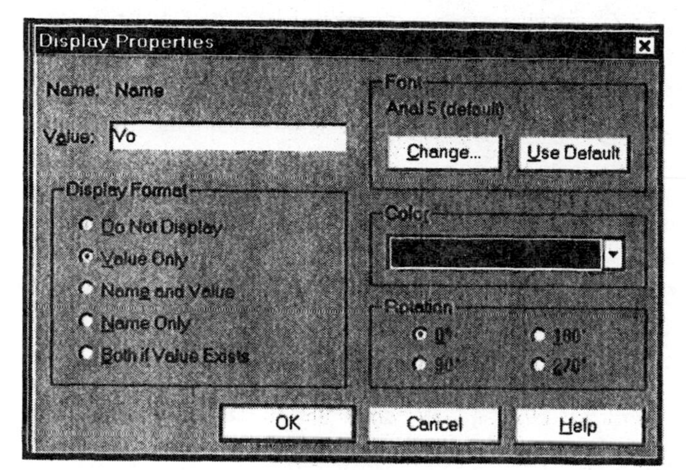


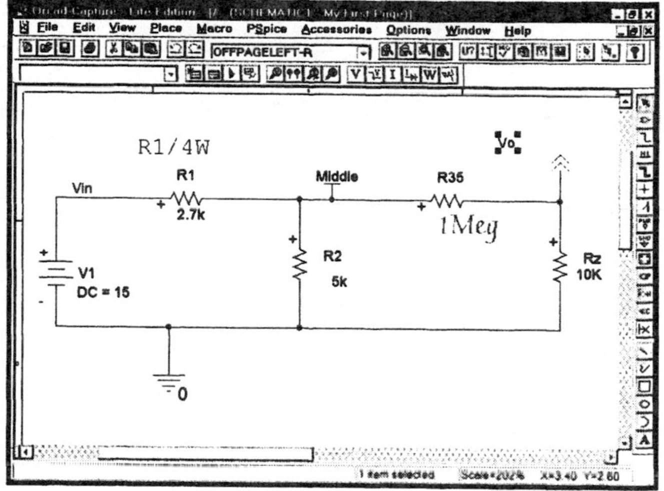


You will need to drag the text ***Vo*** closer to the offpage connector:

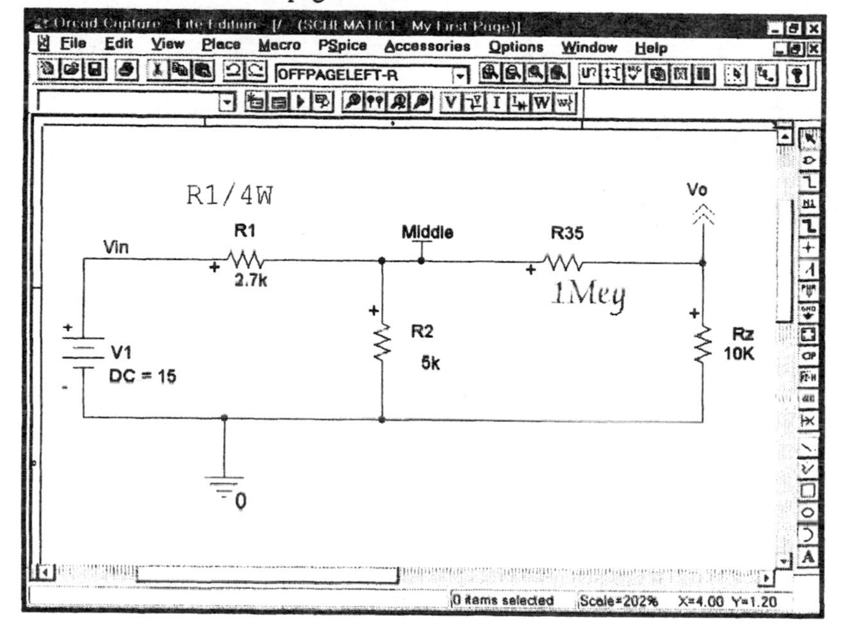


# 1.I. Saving Your Schematic

Now that we have finished the schematic, we need to save it. Select **File** and then **Save** from the menus. The **File** pull-down menu will appear:

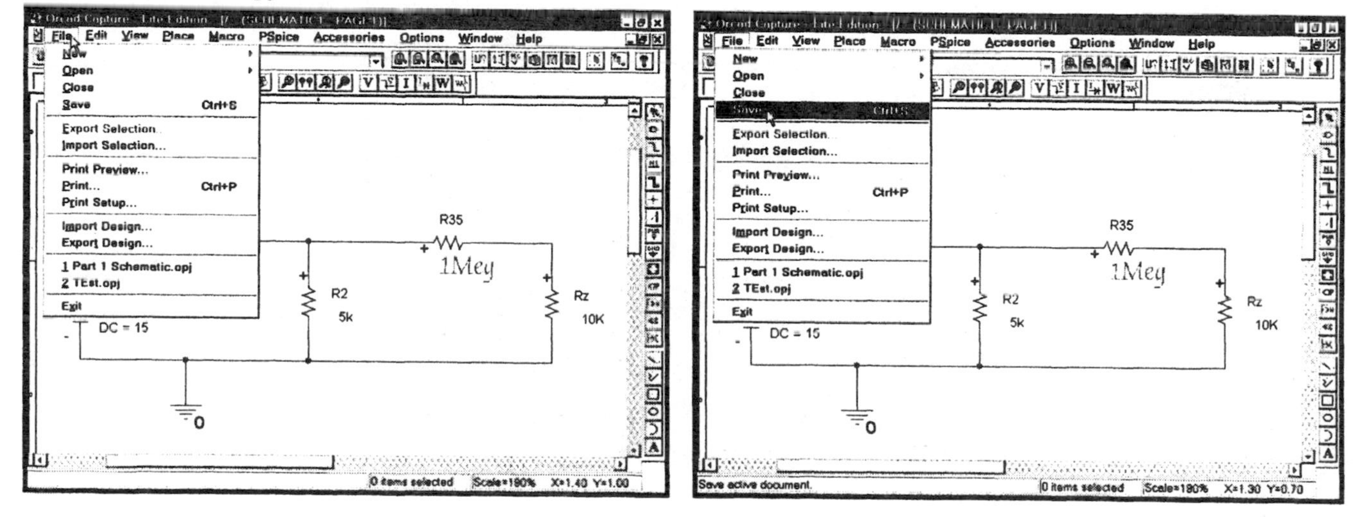

When we first created the project, we specified a name and directory location for the project, so we do not need to specify them here.

# 1.J. Editing the Title Block

The title block contains information about you, your company, and the schematic. We will be using the title block to identify the student, class, and university where the schematic was created. There are four ways we can modify the title block. The first way shows how to change the title block for the current schematic. This method must be repeated each time you create a new schematic. The second way shows you how to tell Capture information about yourself that is used in every schematic you create. This could be information such as your name, company, address, and phone number. The third method shows how to select a different title block from a list of available title blocks already created. The fourth method is to create a new title block and save it in a symbol library.

## 1.J.1. Modifying the Title Block in the Current Schematic

The first thing we want to do is place your name and class information on your schematic. In the bottom right corner of your schematic is a title block. To see this block we must scroll the view of the page to the bottom right corner. To do this, click the ***LEFT*** mouse button on the vertical and horizontal scroll bars as shown below:

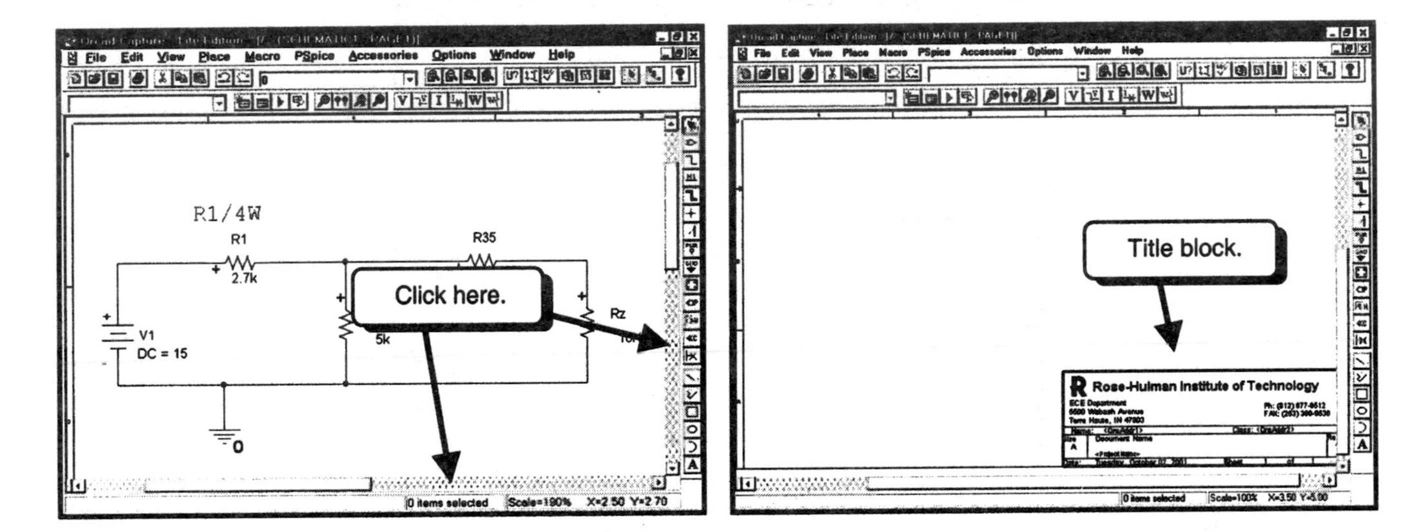


This is a title block that I created for this text and my students. In Section 1.J.4 we will show you how to customize this title block for your needs. There are three items that we can change: **<OrgAddr1>**, **<OrgAddr2>**, **<Revision>**, and ***<Project Name>***. To edit an item, double-click the ***LEFT*** mouse button on the item. For example, double-click the ***LEFT*** mouse button on the text **<OrgAddr1>**

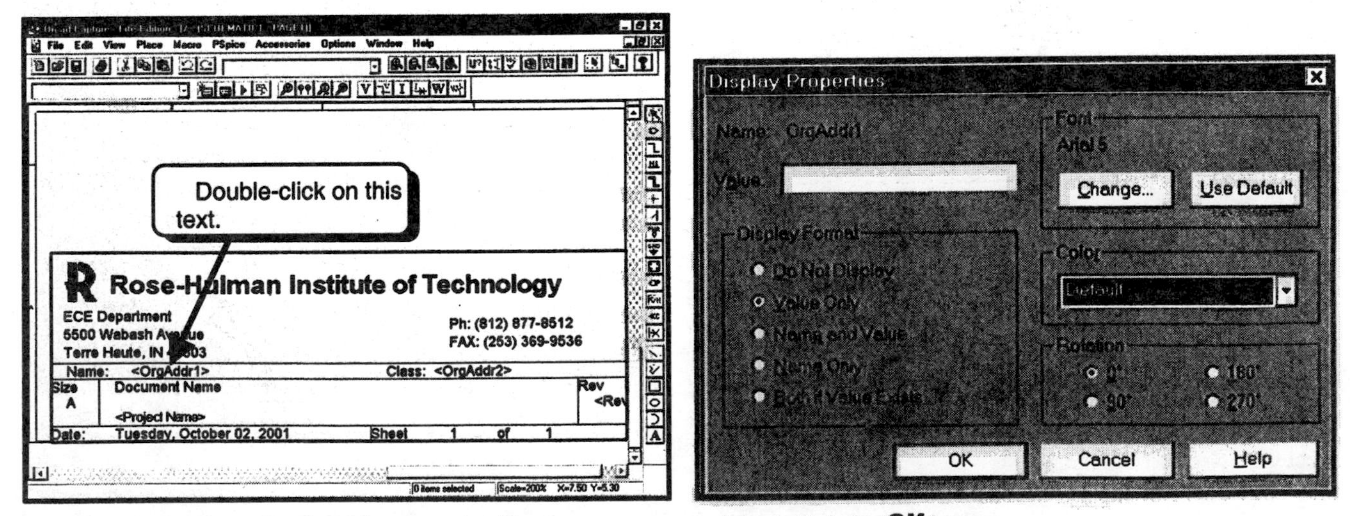


You can now enter a value for this parameter. Enter your name and then click the ***OK*** button:

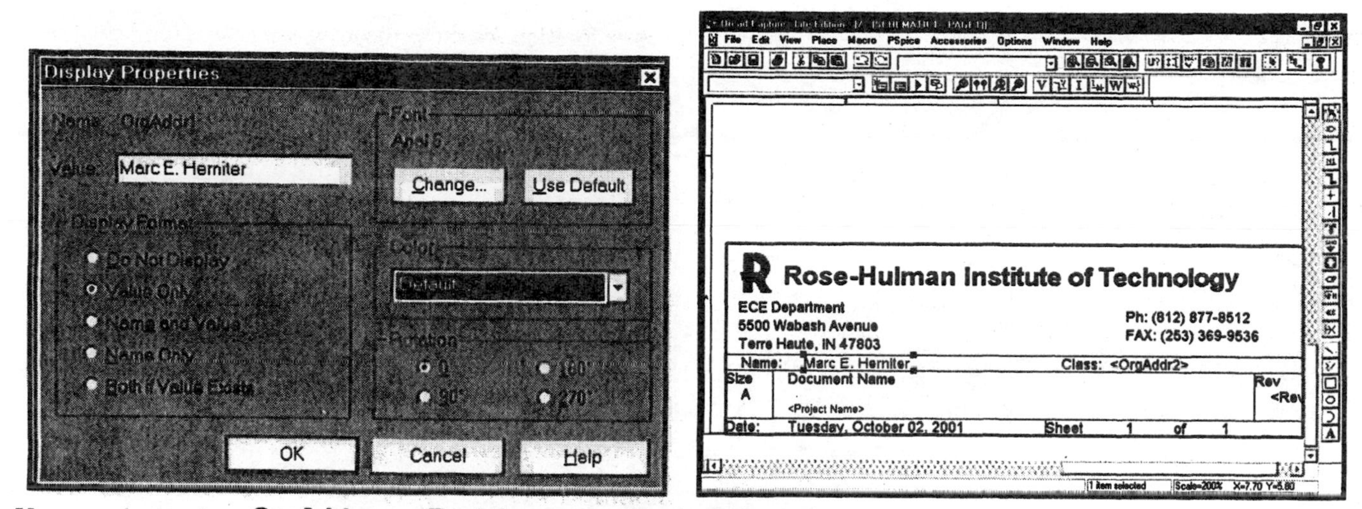


You can change the ***<OrgAddr2>***, ***<Revision>***, and ***<Project Name>*** items using the same method:

The other items such as the date and page size are set automatically.

## 1.J.2. Specifying Commonly Used Information for All Title Blocks

If we use the method in Section 1.J.1, we will have to enter information, such as our name and class, in every schematic we create. This can be avoided by specifying this information in Capture. Select **Options** and then **Design Template** from the Capture menus:

Select the ***Title Block*** tab:

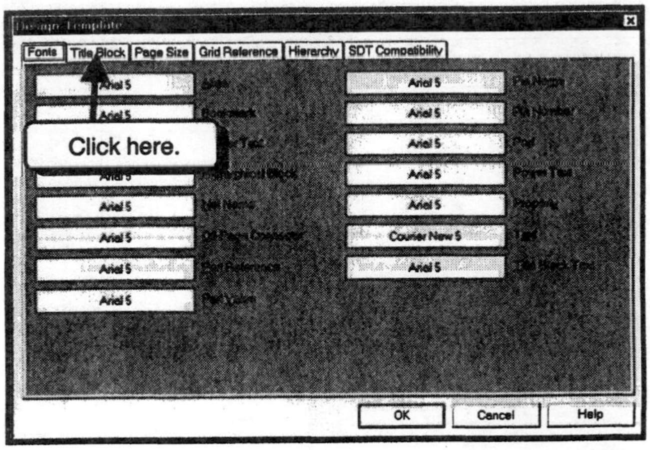

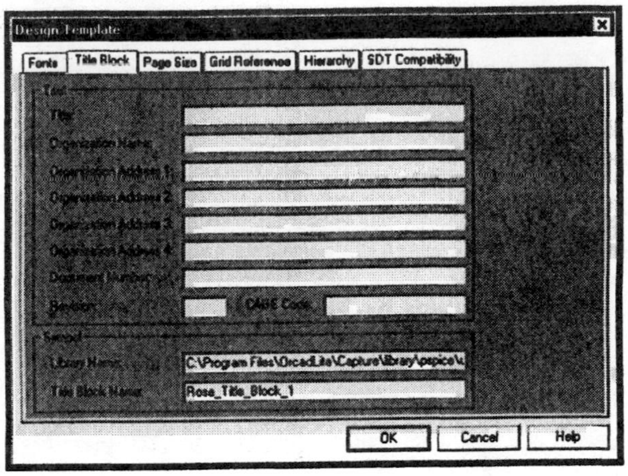

The information here corresponds to the information displayed in the title block. Field <OrgAddr1> in the title block corresponds to the information you enter in ***Organization Address 1***, and field <OrgAddr2> in the title block corresponds to the information you enter in ***Organization Address 2***. The title block we are using only has these two fields. When we look at other title blocks, they will have more fields that are associated with the information in this title block.

For a school, the commonly used information is your name and the name of your class. For the Rose title block, this information goes in fields ***Organization Address 1*** and ***Organization Address 2***. You can also specify a value for the ***Revision***. You can add information in the other fields, but it will not be used by the currently displayed title block:

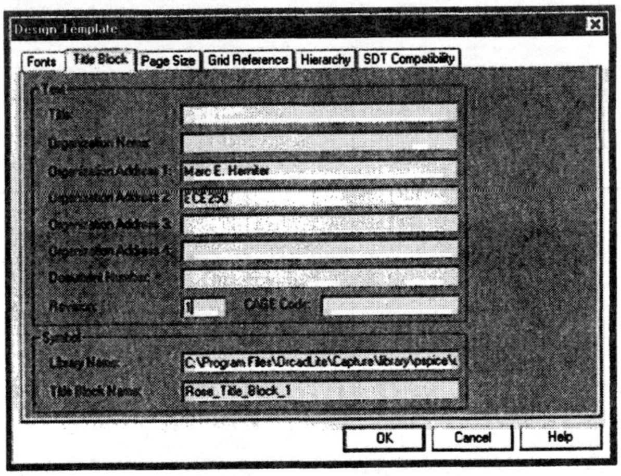

Click the ***OK*** button when you are finished. You will return to the schematic:

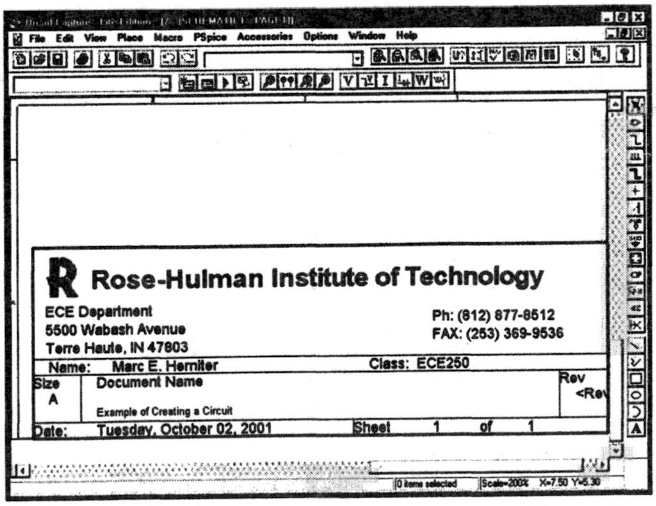

To see how the information is used, we will create a new project. We will save the present project, close it, and then create a new project. Select **File** and then **Save** to save any changes we made to the schematic:

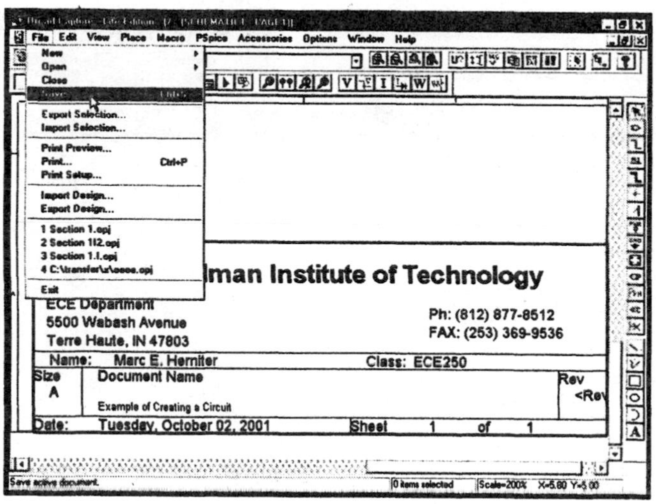

File Edit View Place Macro PSpice Accessories Options Window Help

New
Open
Close
Save Ctrl+S
Export Selection...
Import Selection...
Print Preview...
Print... Ctrl+P
Print Setup...
Import Design...
Export Design...
1 Section 1.opj
2 Section 112.opj
3 Section 1.1.opj
4 C:\transfer\x\eeee.opj
Exit

...man Institute of Technology
ECE Department
5500 Wabash Avenue
Terre Haute, IN 47803
Ph: (812) 877-8512
FAX: (253) 369-9536
Name: Marc E. Herniter Class: ECE250
Size A Document Name Rev
Example of Creating a Circuit
Date: Tuesday, October 02, 2001 Sheet 1 of 1

Save active document. 0 items selected Scale=200% X=5.80 Y=5.00

Select **File** and then **Close** to close the schematic page:

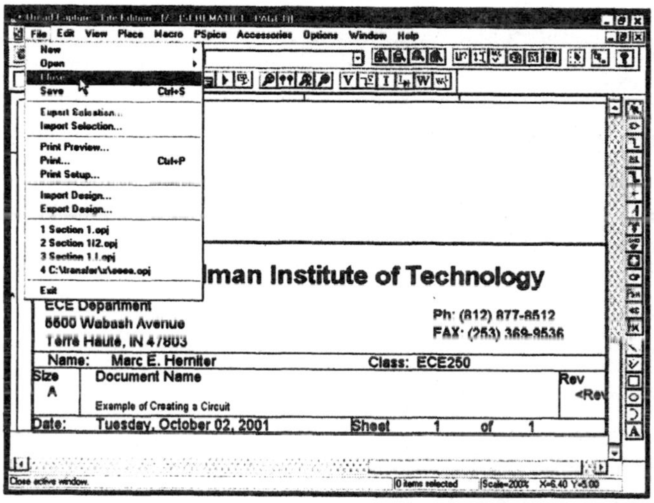

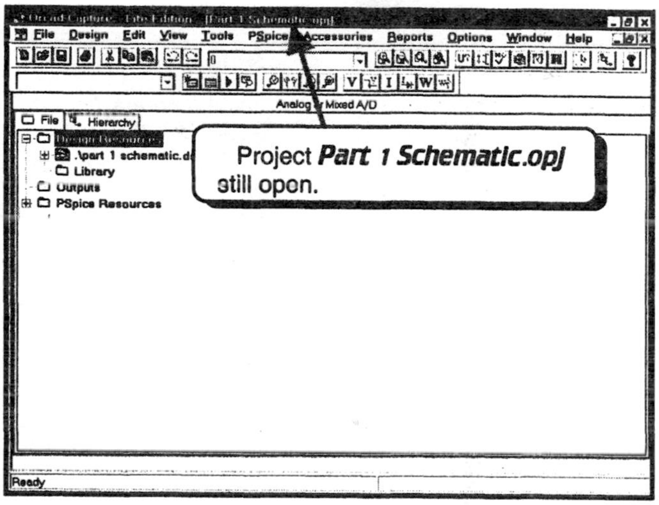

File Edit View Place Macro PSpice Accessories Options Window Help

New
Open
Close
Save Ctrl+S
Export Selection...
Import Selection...
Print Preview...
Print... Ctrl+P
Print Setup...
Import Design...
Export Design...
1 Section 1.opj
2 Section 112.opj
3 Section 1.1.opj
4 C:\transfer\x\eeee.opj
Exit

...man Institute of Technology
ECE Department
5500 Wabash Avenue
Terre Haute, IN 47803
Ph: (812) 877-8512
FAX: (253) 369-9536
Name: Marc E. Herniter Class: ECE250
Size A Document Name Rev
Example of Creating a Circuit
Date: Tuesday, October 02, 2001 Sheet 1 of 1

Close active window. 0 items selected Scale=200% X=6.40 Y=5.00

File Design Edit View Tools PSpice Accessories Reports Options Window Help
Analog or Mixed A/D
File Hierarchy
Design Resources
.\part 1 schematic.dsn
Library
Outputs
PSpice Resources
Ready

Project *Part 1 Schematic.opj* still open.

Note that the schematic page was closed, but the project is still open. To close the project, select **File** and then **Close Project** from the menus:

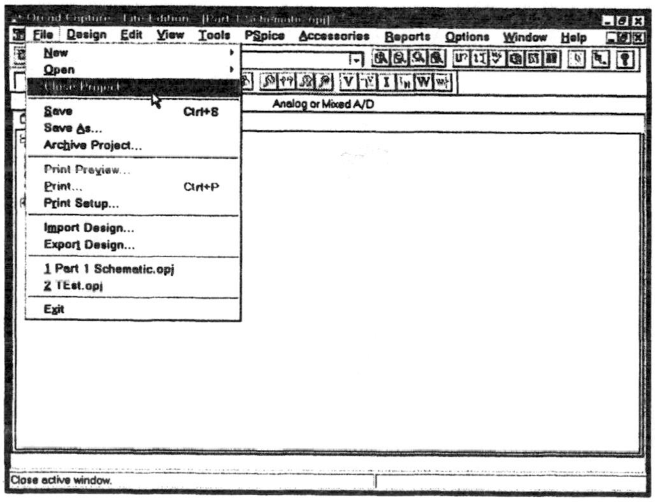

File Design Edit View Tools PSpice Accessories Reports Options Window Help
New
Open
Close Project
Save Ctrl+S
Save As...
Archive Project...
Print Preview...
Print... Ctrl+P
Print Setup...
Import Design...
Export Design...
1 Part 1 Schematic.opj
2 TEst.opj
Exit
Analog or Mixed A/D
Close active window.

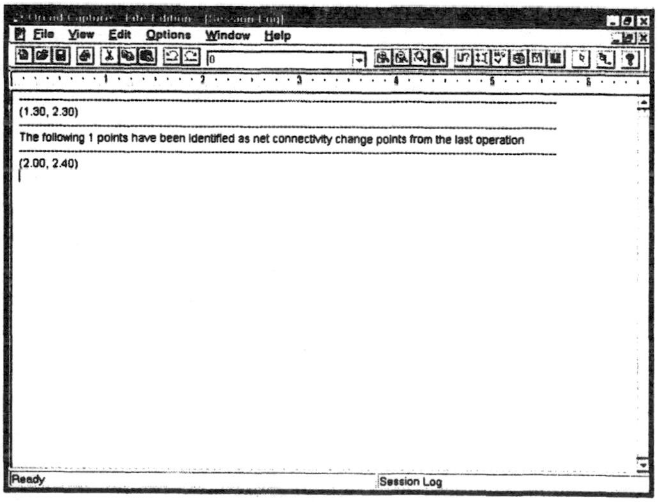

File View Edit Options Window Help
(1.30, 2.30)
The following 1 points have been identified as net connectivity change points from the last operation
(2.00, 2.40)
Ready Session Log

There are now no projects open. The text shown is from the ***Session Log***. This log displays all informational notes and error messages for some of the activities you perform using Capture.

We will now create a new project. Select **File, New,** and then **Project** from the menus:

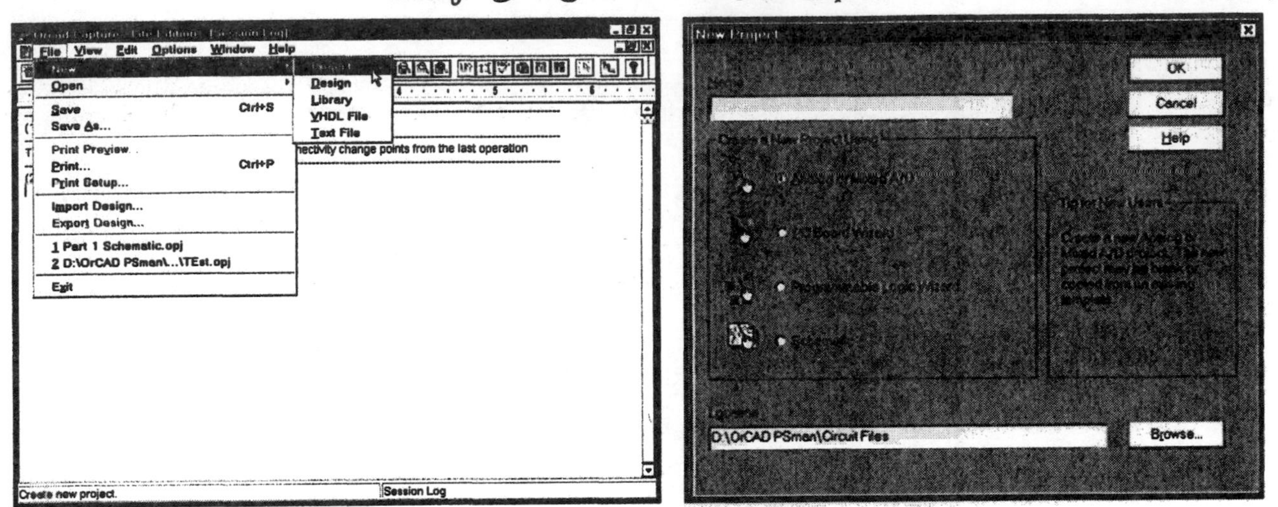

Note that the ***Location*** or directory in which the project will be created is the same directory we specified on page 6 when we created the first project. This location will always be the location you specified for the last project. This is the directory I wish to use so I will not change it. Enter a name for the new project and make sure that the ***Analog or Mixed A/D*** option is selected. I chose the name **Section 1I2**:

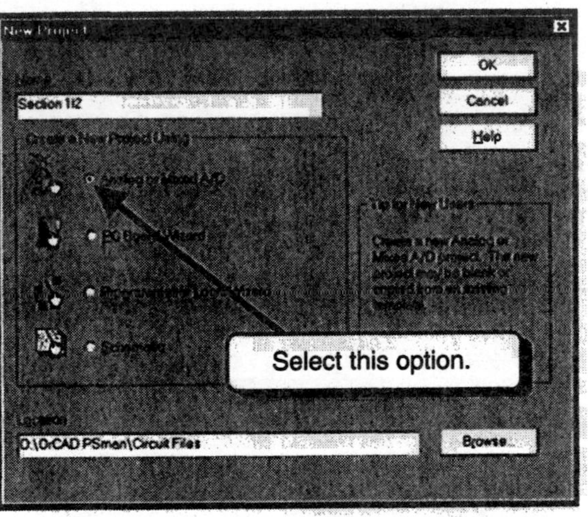


Click the ***OK*** button:

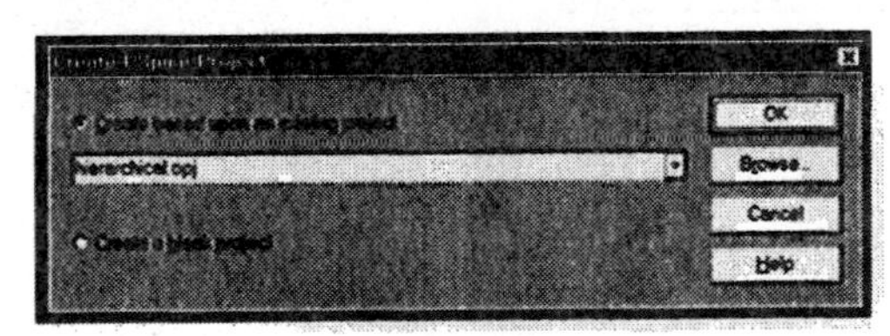

Select to ***Create a blank project*** and then click the ***OK*** button:

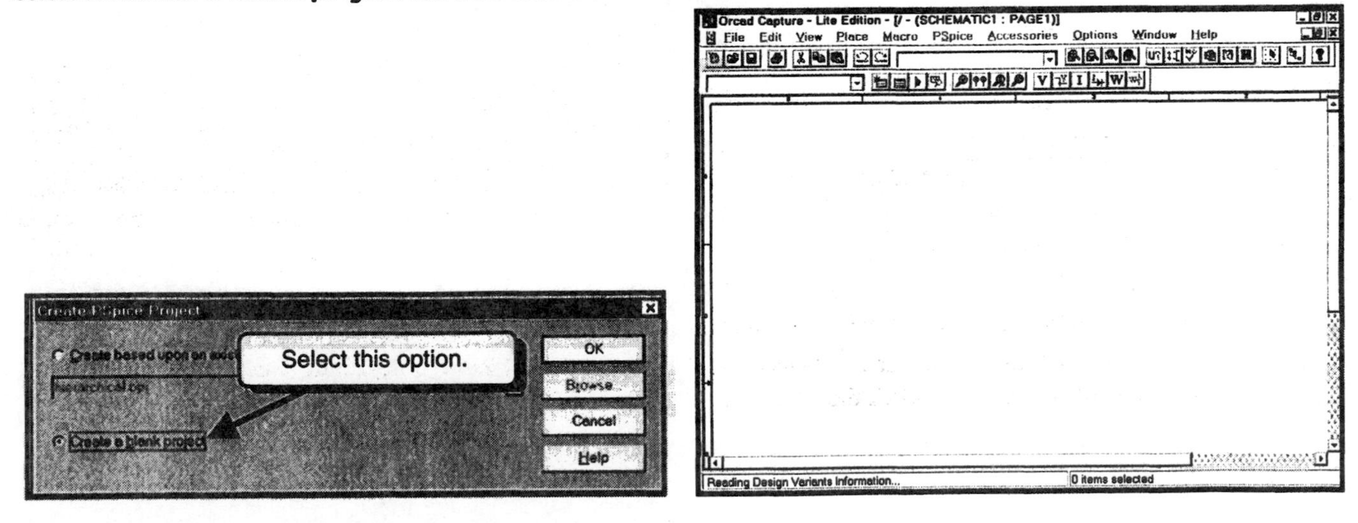

A blank schematic is displayed. The title block is in the lower right corner. Use the scroll bars to scroll the page to the lower right corner of the page. I will zoom in on my schematic to see the title block more clearly. (To zoom in, place the mouse pointer over the title block and press the **I** key.)

We see that my name and class are entered automatically, but the **<Project Name>** has to be specified. This is O.K. since the project name will probably be different for each schematic we create.

## 1.J.3. Selecting a Different Title Block

There are several predefined title blocks from which you can choose. In this section we will show how to choose a different title block. In the next section, we will show how to modify an existing title block to create your own custom title block. We will first look at the available title blocks. We will assume that you still have the project open from the previous example. This is a blank schematic, and we can place title blocks in it without worrying about cluttering up a useful schematic.

To place a title block, select **Place** and then **Title Block** from the menus:

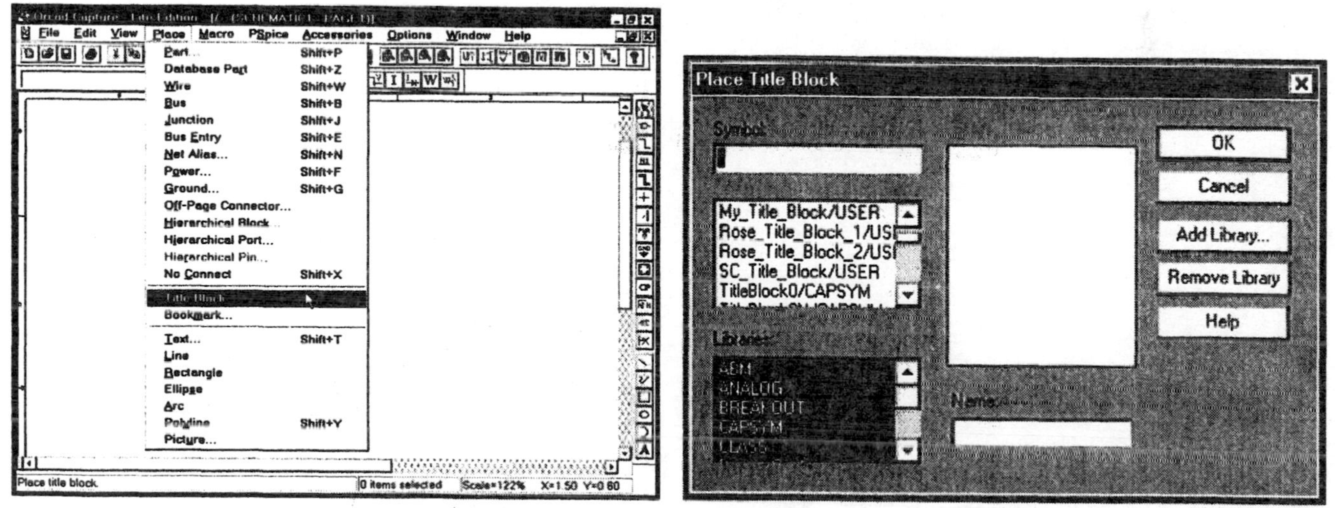


This dialog box lists all of the predefined title blocks and the library in which they reside. If you installed the demo software in the default directory, then the two libraries are:

- d:\Program Files\OrcadLite\Capture\library\pspice\user.olb
- d:\Program Files\OrcadLite\Capture\library\capsym.olb

where d: is the drive on which the demo software is installed. The block named Rose_Title_Block_1 is the default title block for this text. To view another title block, click the ***LEFT*** mouse button on its name. For example, select ***SC_Title_Block***:

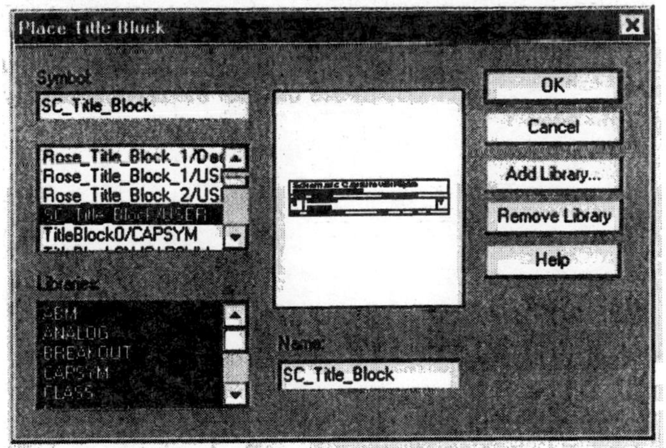

A picture of the title block is displayed in the dialog box. The picture of the title block is a bit too small to see all of the details, so we will place it in the schematic. Click the ***OK*** button to place the block in the schematic page. The outline of the title block will appear attached to the mouse pointer:

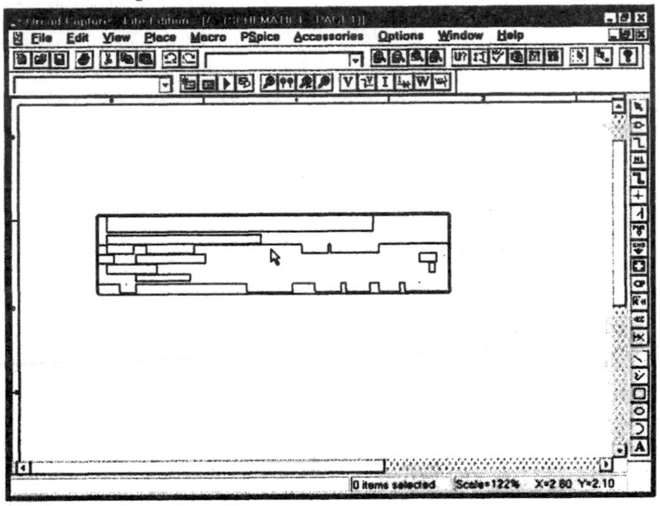

Move the mouse to locate the outline in a convenient location and then click the ***LEFT*** mouse button to place the block. Next, press the **ESC** key to terminate placing blocks.

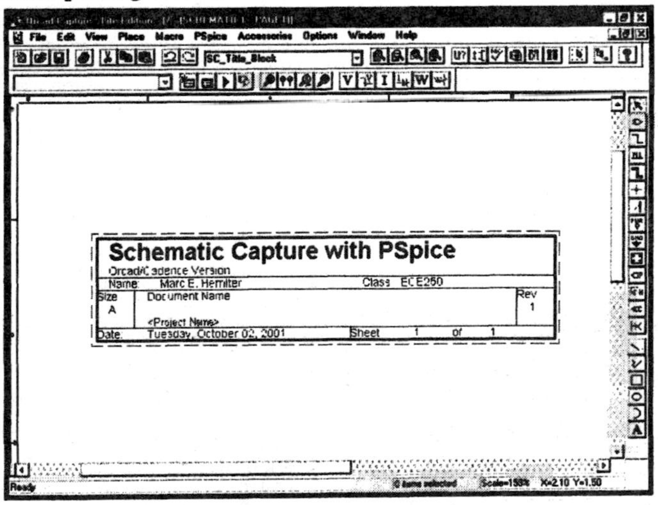

We can now see the block more clearly.

Continue to place different title blocks until you find a title block that you like. To place a title block, select **Place** and then **Title Block**:

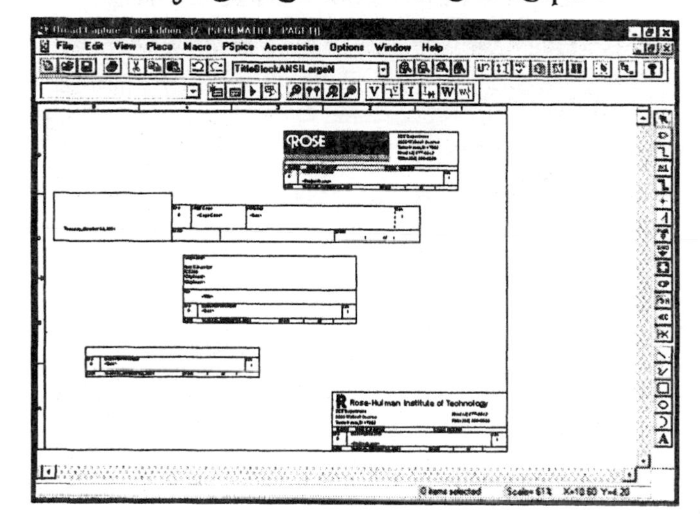

I like the block named TitleBlock5 located in the Capsym.olb library, and I would like to use it as my default title block in all future schematics. You can choose a different title block. If we zoom in on this title block, we will see that it has five properties we can specify. The properties are ***<OrgName>***, ***<OrgAddr1>***, ***<OrgAddr2>***, ***<OrgAddr3>***, and ***<OrgAddr4>***. ***<OrgAddr1>*** is set to my name and ***<OrgAddr2>*** is set to ***ECE250*** from the last example. When you choose a title block, note down its name and library. For example, SC_Title_Block/USER specifies that the name of the title block is SC_Title_Block and that it is in the library named USER. TitleBlock5/CAPSYM specifies that the name of the title block is TitleBlock5 and that it is in the library named CAPSYM.

Now that we have found the title block we wish to use, we want to specify it as our default title block for all schematics. Select **Options** and then **Design Template** from the Capture menus:

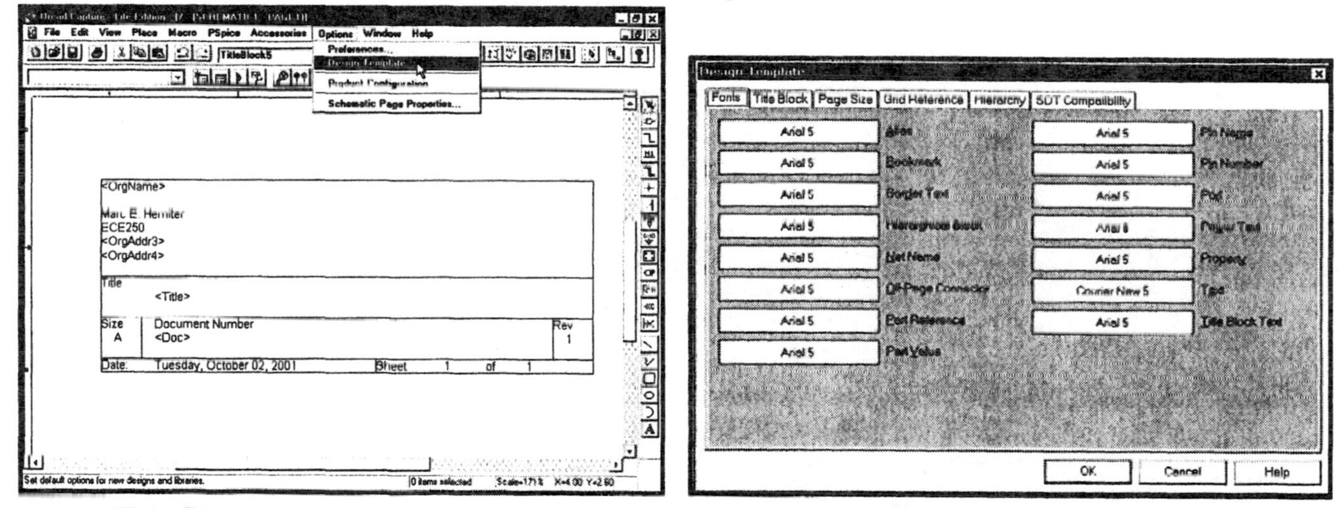

Select the ***Title Block*** tab:

Click here.

Title block location.

Title block name.

The last two fields in the dialog box specify the ***Library Name*** and the ***Title Block Name***. Change these to the appropriate values for your selected title block. The information for the title block I have chosen is:

- d:\Program Files\OrcadLite\Capture\library\capsym.olb
- TitleBlock5

where d: is the drive on which the demo software is installed. Your directory may be different depending on the title block you chose and whether or not you installed the software in the default directory. I have entered this information in the screen capture shown below:

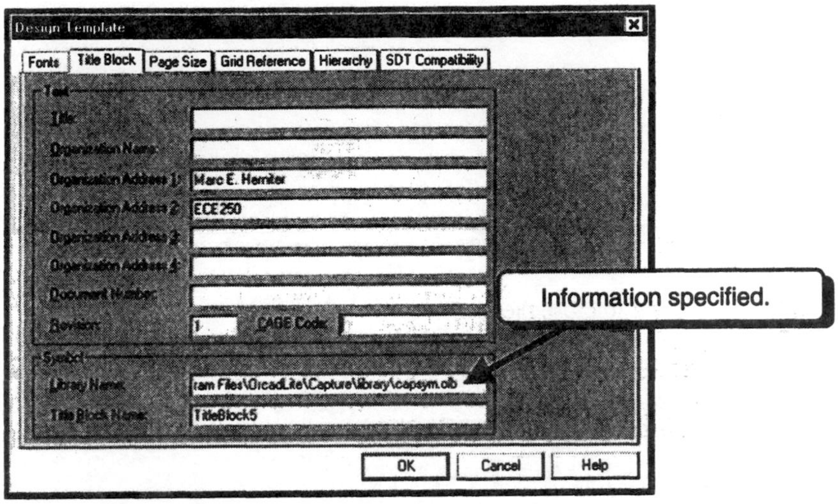

The title block I have chosen has 5 fields that take information from this title block. The fields are ***<OrgName>***, ***<OrgAddr1>***, ***<OrgAddr2>***, ***<OrgAddr3>***, and ***<OrgAddr4>***. I will enter this information for these fields:

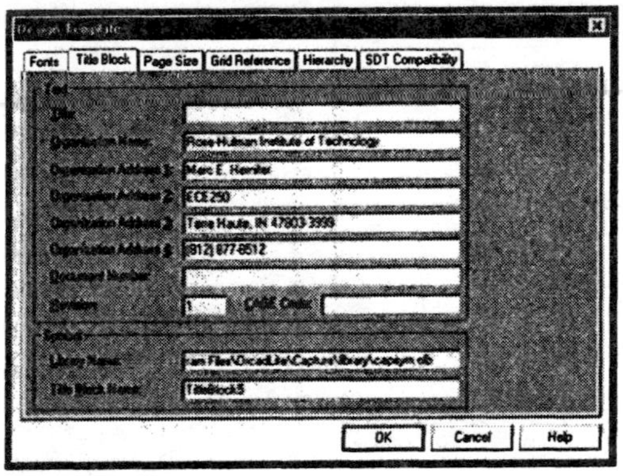

Click the ***OK*** button when you are finished. You will return to the schematic page:

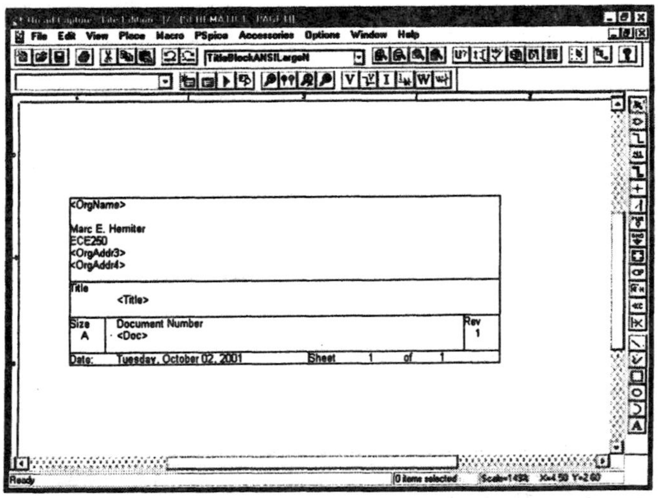

We will not see any changes until we create a new schematic. We will close this project and then create a new project. Select **File** and then **Close** from the menus. You can choose to save or not save the changes.

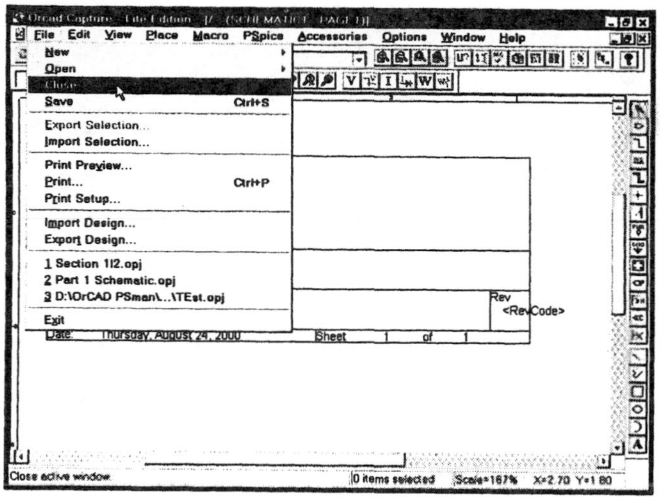

Orcad Capture - Lite Edition - [/ - (SCHEMATIC1 : PAGE1)]
File Edit View Place Macro PSpice Accessories Options Window Help
New
Open
Close
Save Ctrl+S
Export Selection...
Import Selection...
Print Preview...
Print... Ctrl+P
Print Setup...
Import Design...
Export Design...
1 Section 1l2.opj
2 Part 1 Schematic.opj
3 D:\OrCAD PSman\...\TEst.opj
Exit
Rev
<RevCode>
Date: Thursday, August 24, 2000 Sheet 1 of 1
Close active window 0 items selected Scale=167% X=2.70 Y=1.80

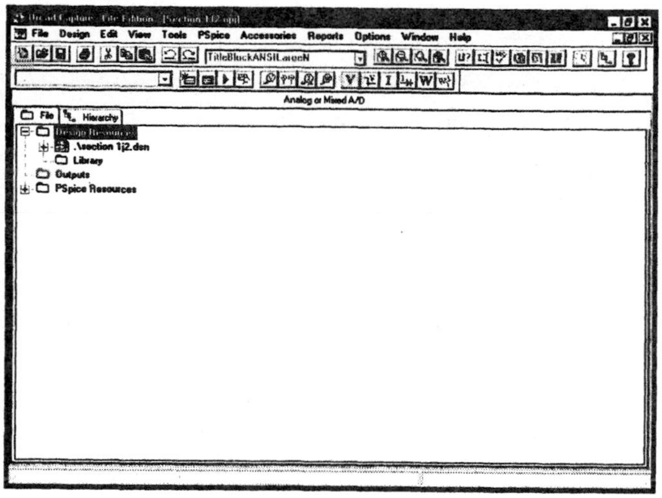

Orcad Capture - Lite Edition - [Section 1l2.opj]
File Design Edit View Tools PSpice Accessories Reports Options Window Help
TitleBlockANSILarge...
Analog or Mixed A/D
File Hierarchy
Design Resources
.\section 1l2.dsn
Library
Outputs
PSpice Resources

Select **File** and then **Close Project** to close the project.

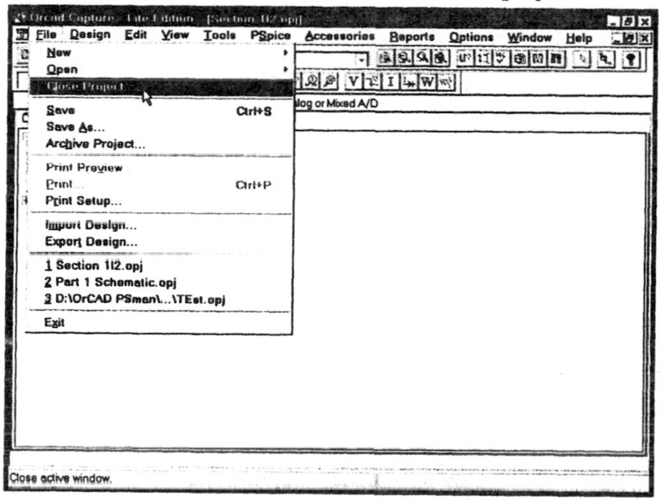

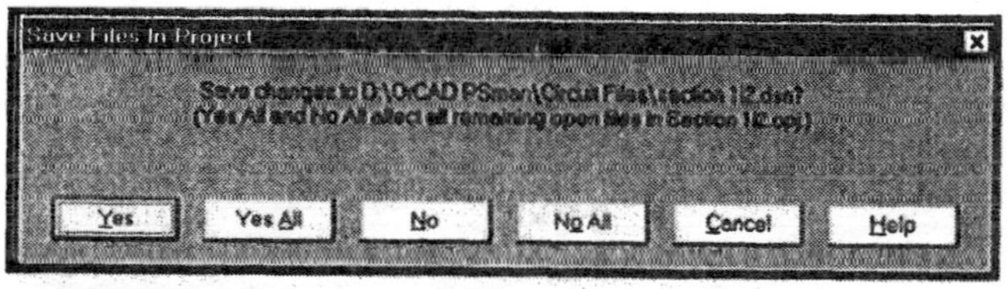

Orcad Capture - Lite Edition - [Section 1l2.opj]
File Design Edit View Tools PSpice Accessories Reports Options Window Help
New
Open
Close Project
Save Ctrl+S
Save As...
Archive Project...
Print Preview
Print... Ctrl+P
Print Setup...
Import Design...
Export Design...
1 Section 1l2.opj
2 Part 1 Schematic.opj
3 D:\OrCAD PSman\...\TEst.opj
Exit
Analog or Mixed A/D
Close active window.

Save Files In Project
Save changes to D:\OrCAD PSman\Circuit Files\section 1l2.dsn?
(Yes All and No All affect all remaining open files in Section 1l2.opj.)
Yes Yes All No No All Cancel Help

You can choose to either save or not save the changes to the last project.

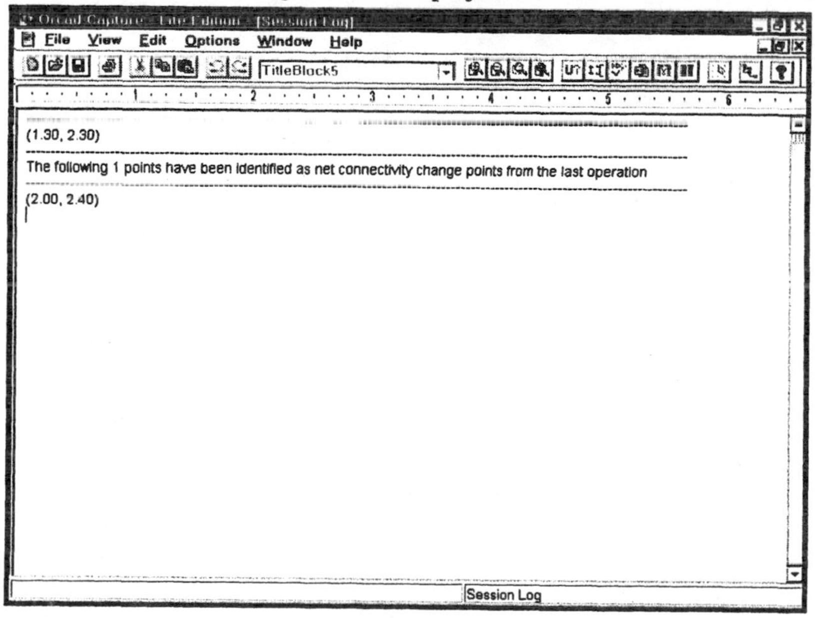

Orcad Capture - Lite Edition - [Session Log]
File View Edit Options Window Help
TitleBlock5
(1.30, 2.30)
The following 1 points have been identified as net connectivity change points from the last operation
(2.00, 2.40)
Session Log

To create a new project, select **File**, **New**, and then **Project**:

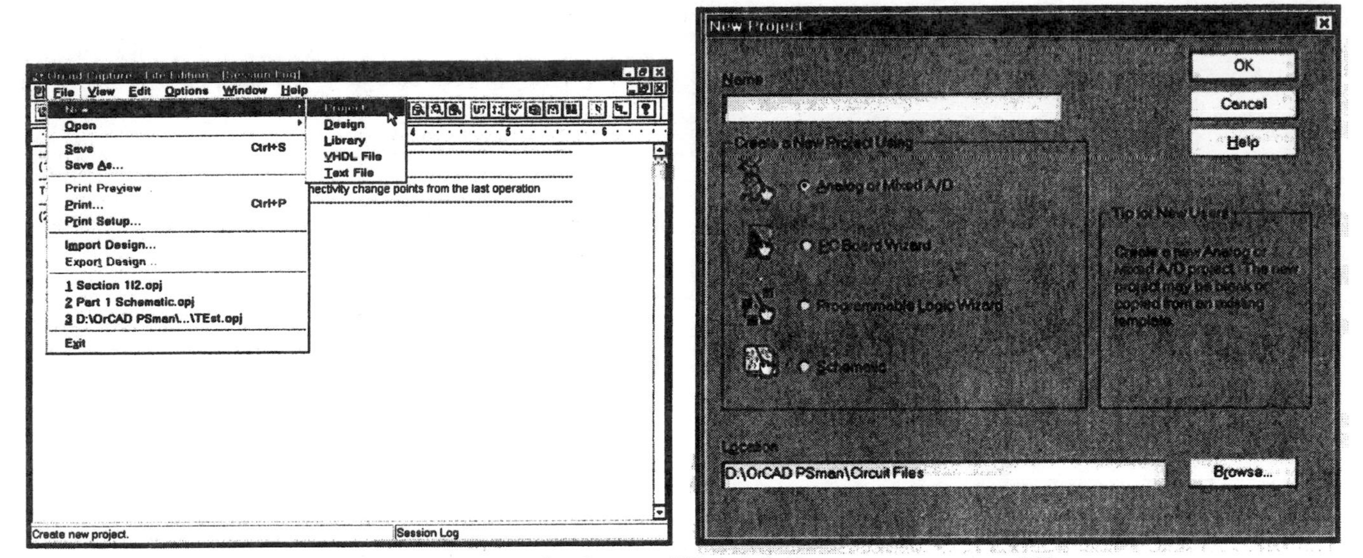


Specify a name for the project, select the option ***Analog or Mixed A/D***, and then click the ***OK*** button:

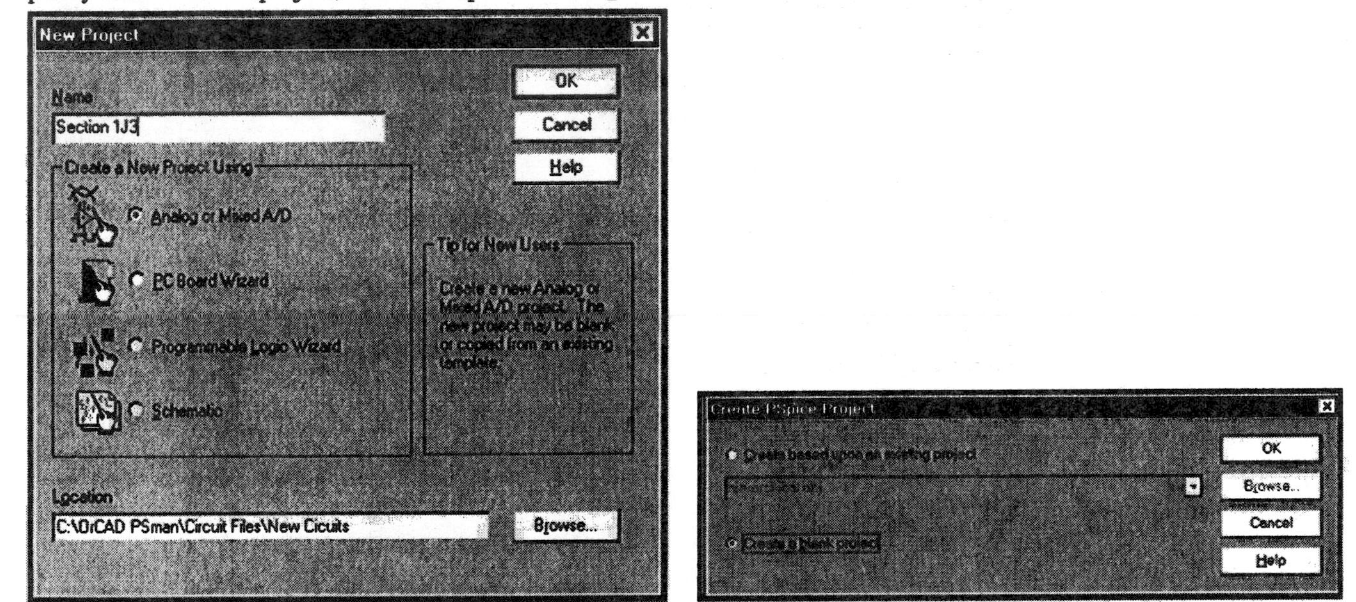


Select the option to ***Create a blank project*** and then click the ***OK*** button. If you scroll the window to the bottom right of the page, you should see the new title block with the specified information. I have zoomed in on the title block to see the information more clearly. (To zoom in, place the mouse pointer over the title block and type **I**.)

Rose-Hulman Institute of Technology
Marc E. Herniter
ECE250
Terre Haute, IN 47803-3999
(812) 877-8512
Title
<Title>
Size A
Document Number <Doc>
Rev 1
Date: Tuesday, October 02, 2001
Sheet 1 of 1

## 1.J.4. Creating Your Own Title Block

For our last example, we will modify the Rose_Title_Block_1 to match your company or school name and address. Follow the procedure starting on page 52 to close any open projects. You should have the window shown below:



The Rose_Title_Block_1 is in library user.olb. This library is small enough so that you can modify the items in the library with the demo version of the software. You cannot modify the title blocks in the CAPSYM.olb library with the Lite vérsion, but you can copy them to the user.olb library and then modify them. Here, we will modify the Rose_Title_Block_1 since it is already in the user.olb library.

We must first open the library. Select **File** and then **Open** from the menus:



Select **Library**:

Select ***USER.OLB*** and then click the ***Open*** button:

This library has only three parts,* ***Rose_Title_Block_1***, ***Rose_Title_Block_2***, and ***SC_Title_Block***.

We will first duplicate the title block. Click the ***LEFT*** mouse button on the text ***Rose_Title_Block_1*** to select it:

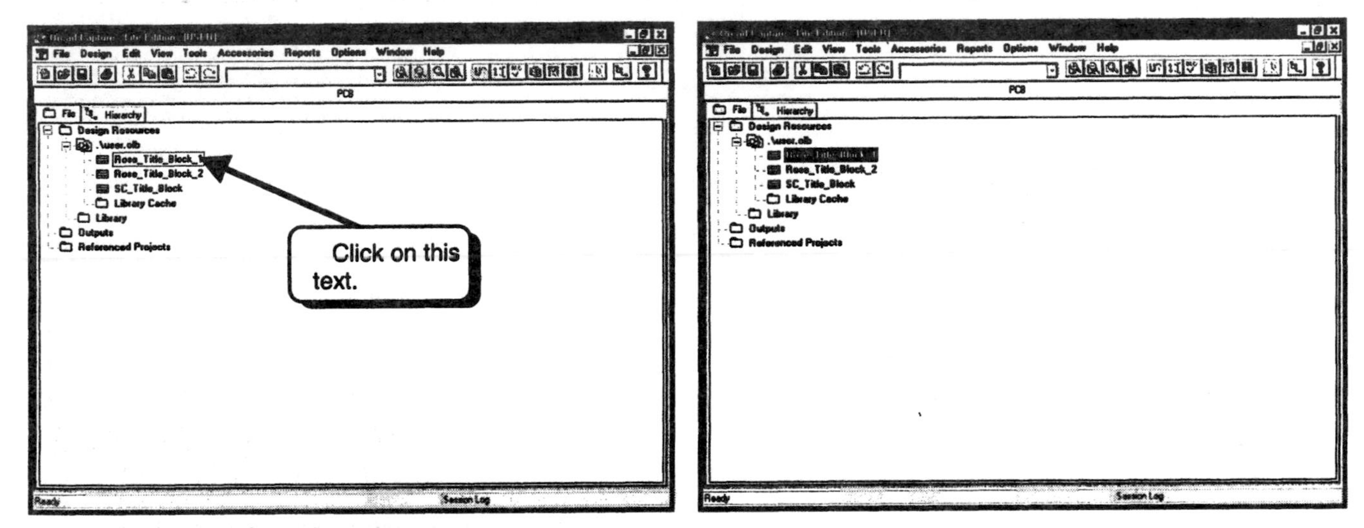


Select **Edit** and then **Copy** from the menus:

---

* If your library already contains the My_Title_Block part, you should delete it before continuing or use a different name for your title block. Click the ***LEFT*** mouse button on the text ***My_Title_Block*** to select the part and then press the **DELETE** key.

Before we can paste the title block to create a new part, we must rename the old part. Click the ***RIGHT*** mouse button on the text ***Rose_Title_Block_1***. A menu will appear:





Select **Rename**:





Enter a name for the title block and click the ***OK*** button:



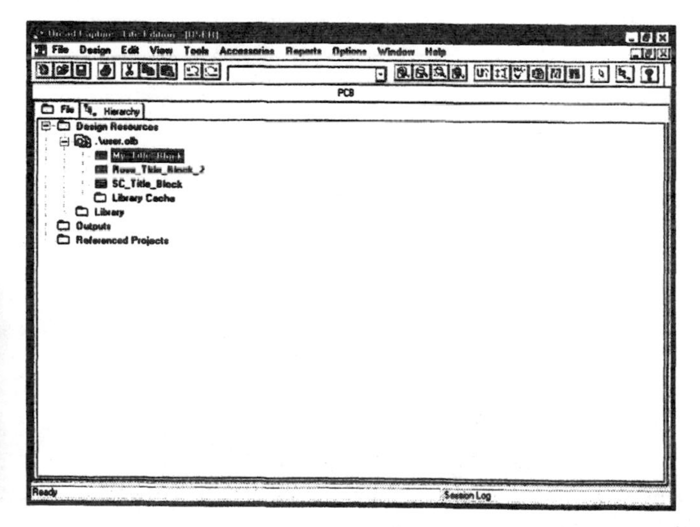

We now need to paste the old title block back into the library. Click the ***LEFT*** mouse button on the text **\user.olb** to select the library:

File Design Edit View Tools Accessories Reports Options Window Help

Design Resources

\user.olb

My_Title_Block

Rose_Title_Block_2

SC_Title_Block

Library Cache

Library

Outputs

Referenced Projects

Click the ***LEFT*** mouse button on this text.

Ready Session Log

Library selected.

Select **Edit** and then **Paste** from the menus:

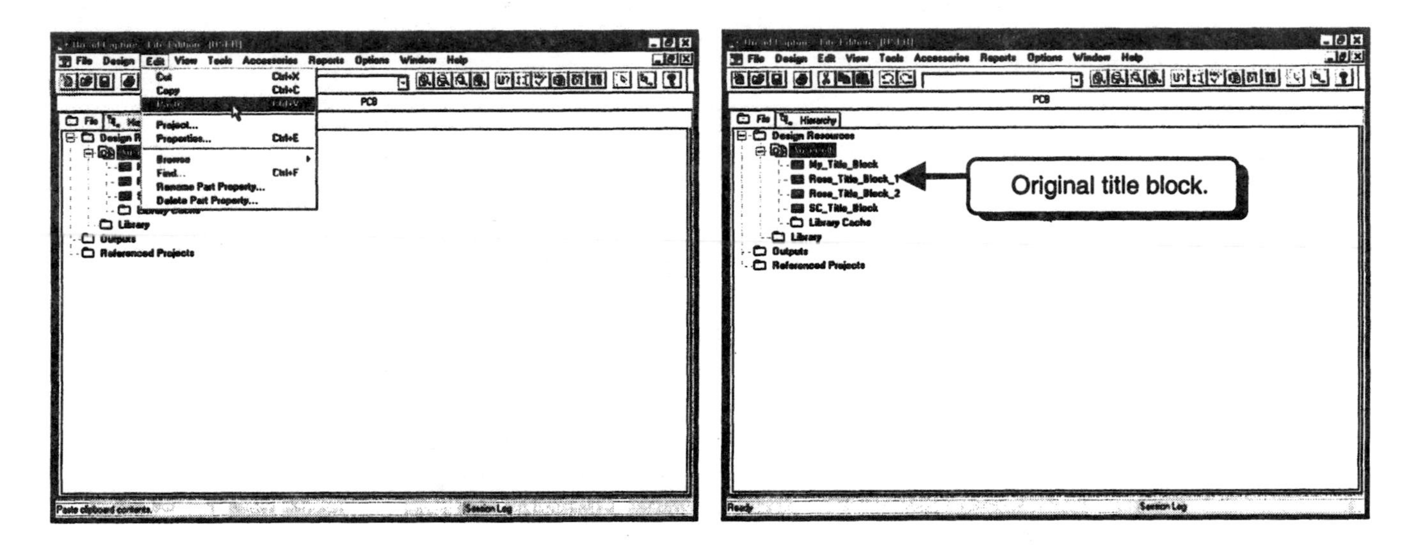


The original title block is now back in the library.

We will now edit the title block named ***My_Title_Block***. Click the ***RIGHT*** mouse button on the text ***My_Title_Block***:

File Design Edit View Tools Accessories Reports Options Window Help

PCB

Design Resources

\user.olb

My_Title_Block

Rose_Title_Block_1

Rose_Title_Block_2

SC_Title_Block

Library Cache

Library

Outputs

Referenced Projects

Click the ***RIGHT*** mouse button on this text.

Edit Symbol

Rename

Properties

Part manager

Ready Session Log

Select **Edit Symbol**:

File Design Edit View Tools Accessories Reports Options Window Help
PCB
File Hierarchy
Design Resources
.\user.olb
My Title Block
Rose_Title_Block
Rose_Title_Block
SC_Title_Block
Library Cache
Edit Symbol
Rename
Properties
Part manager
Library
Outputs
Referenced Projects
Edit selection
Session Log

File Edit View Place Options Window Help
Rose-Hulman Institute of Technology
ECE Department
5500 Wabash Avenue
Terre Haute, IN 47803
Ph: (812) 877-8512
FAX: (253) 369-9536
Name: <OrgAddr1> Class: <OrgAddr2>
Size A Document Name <Project Name> Rev ?
Date: <Page Modify Date> Sheet 1 of 1
Ready

You can now change text, add pictures, and draw objects. The main menu for adding items is the **Place** menu:

File Edit View Place Options Window Help
IEEE Symbol...
Text... Shift+T
Line
Rectangle
Ellipse
Arc
Polyline Shift+Y
Picture...
Rose-Hulman Institute of Technology
ECE Department
5500 Wabash Avenue
Terre Haute, IN 47803
Ph: (812) 877-8512
FAX: (253) 369-9536
Name: <OrgAddr1> Class: <OrgAddr2>
Size A Document Name <Project Name> Rev ?
Date: <Page Modify Date> Sheet 1 of 1

We will not demonstrate these functions here. Press the **ESC** key to return to the symbol. To change any piece of text, double-click the ***LEFT*** mouse button on the text. For example, I will change the name of the school:

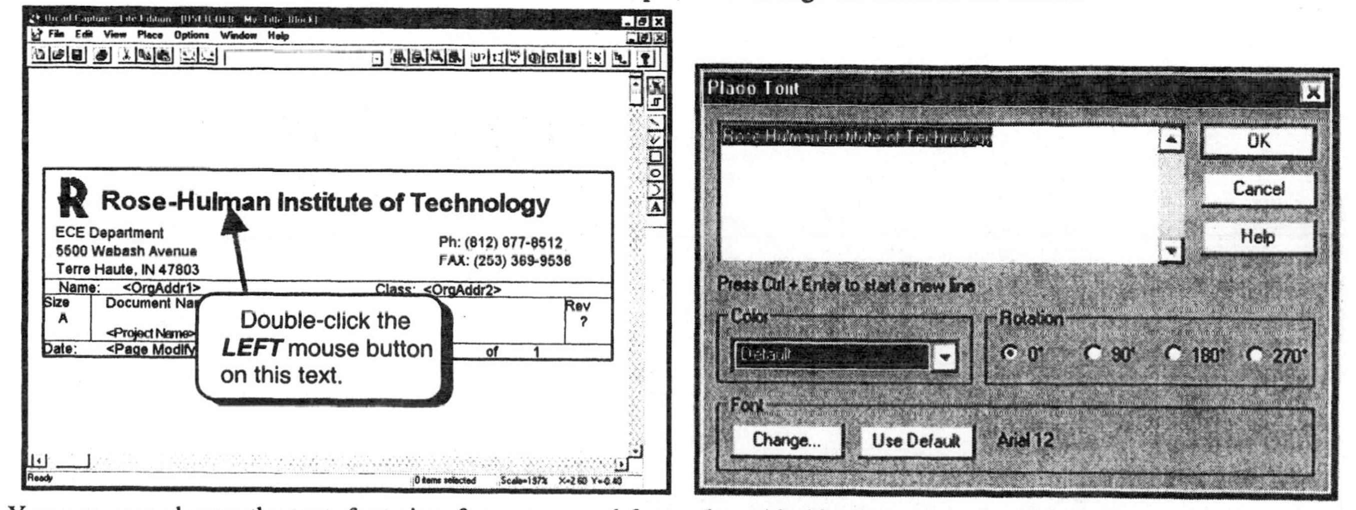


You can now change the text, font size, font name, and font color with this dialog box. I will just change the name of the school and click the ***OK*** button:

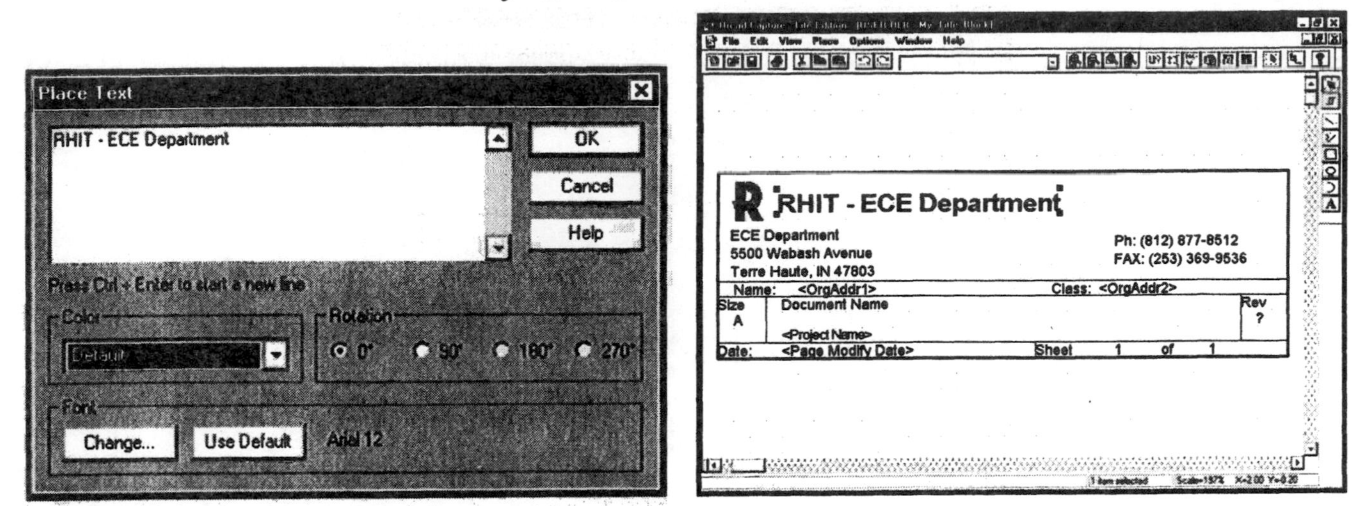


To change the address information, double-click the ***LEFT*** mouse button on the address text:

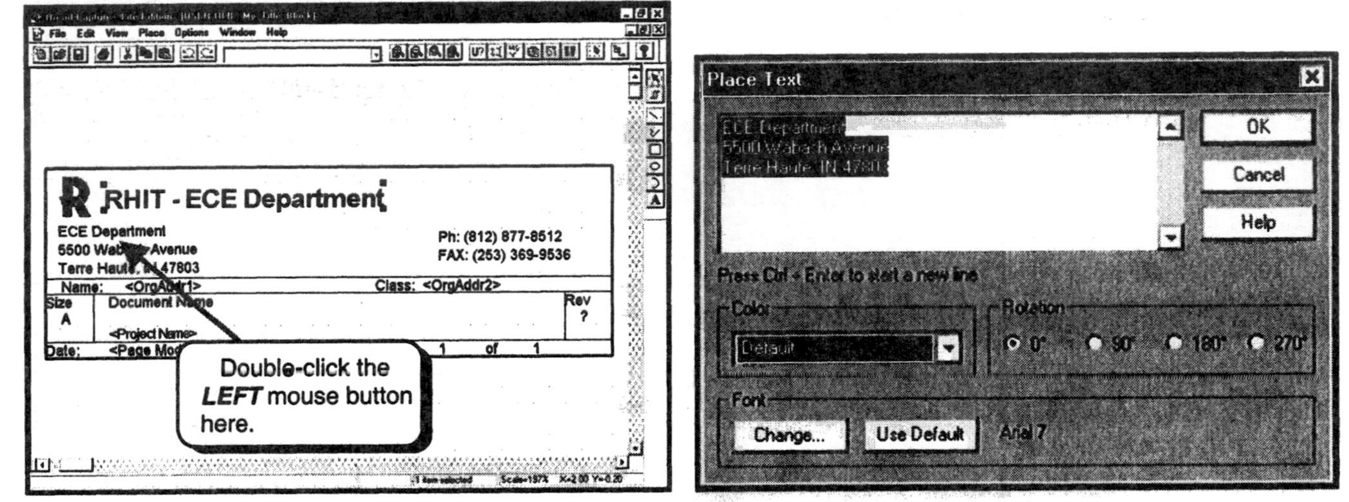


You can now change the address information. **Note that to create a new line of text in this dialog box, you must type CTRL-ENTER**. Just pressing the **ENTER** key will select the ***OK*** button and return you to the symbol. Enter the new information and click the ***OK*** button:

The text in the above dialog box was created by typing `My Dept` **CTRL-ENTER** `My Street` **CTRL-ENTER** `My Town, State, Zip`.

When you are finished making changes, select **File** and then **Save** from the menus:

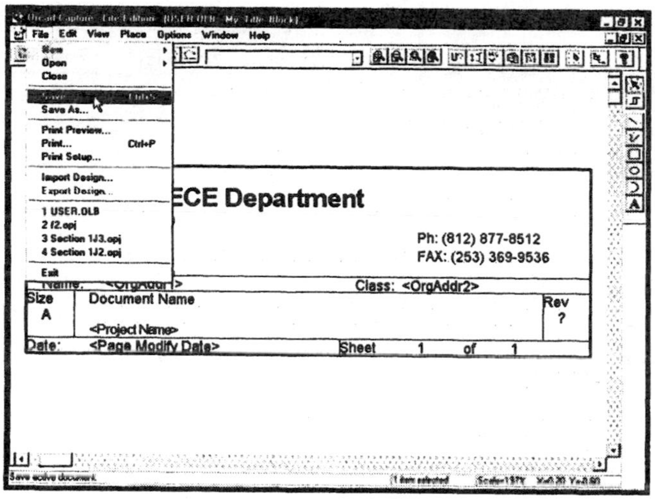

Orcad Capture - Lite Edition - [USER.OLB - My Title Block]
File Edit View Place Options Window Help
New
Open
Close
Save Ctrl+S
Save As...
Print Preview...
Print... Ctrl+P
Print Setup...
Import Design...
Export Design...
1 USER.OLB
2 f2.opj
3 Section 1J3.opj
4 Section 1J2.opj
Exit
ECE Department
Ph: (812) 877-8512
FAX: (253) 369-9536
Name: <OrgAddr1>
Class: <OrgAddr2>
Size A
Document Name
<Project Name>
Rev ?
Date: <Page Modify Date>
Sheet 1 of 1
Save active document.
1 item selected Scale=197% X=0.20 Y=0.60

Select **File** and then **Close** to return to the library:

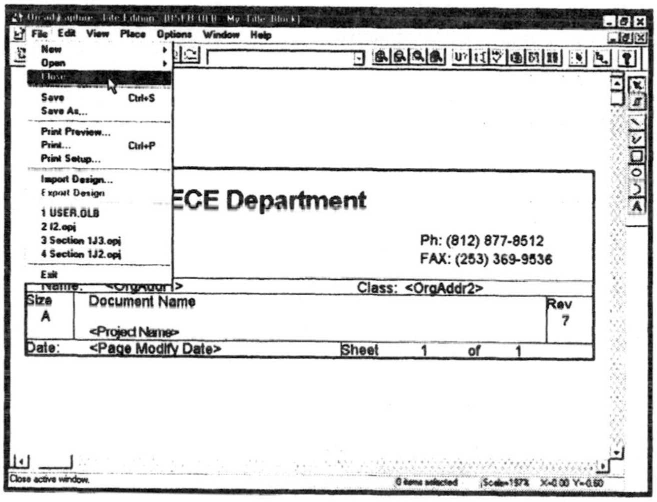

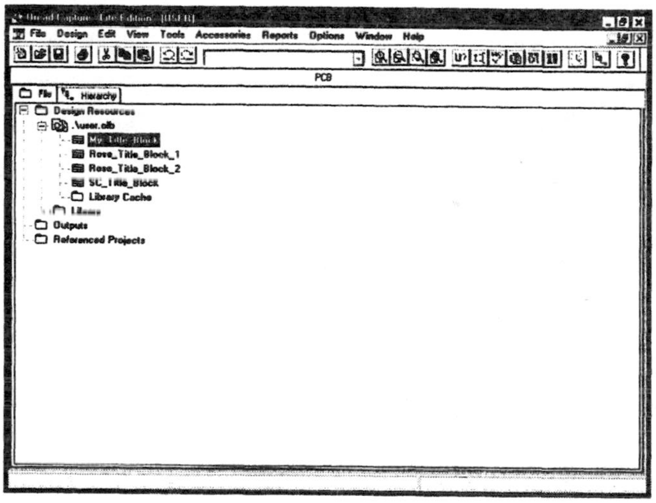

Orcad Capture - Lite Edition - [USER.OLB - My Title Block]
File Edit View Place Options Window Help
New
Open
Close
Save Ctrl+S
Save As...
Print Preview...
Print... Ctrl+P
Print Setup...
Import Design...
Export Design...
1 USER.OLB
2 f2.opj
3 Section 1J3.opj
4 Section 1J2.opj
Exit
ECE Department
Ph: (812) 877-8512
FAX: (253) 369-9536
Class: <OrgAddr2>
Size A
Document Name
<Project Name>
Rev 7
Date: <Page Modify Date>
Sheet 1 of 1
Close active window.
0 items selected Scale=197% X=0.00 Y=0.60

Orcad Capture - Lite Edition - [USER.OLB]
File Design Edit View Tools Accessories Reports Options Window Help
PCB
File Hierarchy
Design Resources
.\user.olb
My Title Block
Rose_Title_Block_1
Rose_Title_Block_2
SC_Title_Block
Library Cache
Library
Outputs
Referenced Projects

We now need to specify our new title block as the default. Select **Options** and then **Design Template** from the menus:

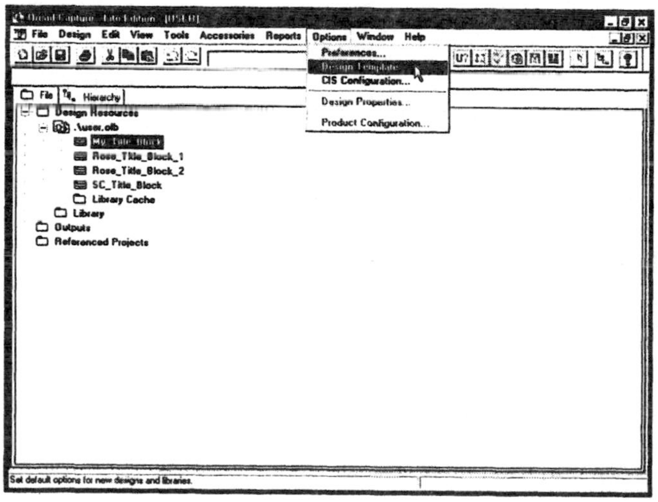

Orcad Capture - Lite Edition - [USER.OLB]
File Design Edit View Tools Accessories Reports Options Window Help
Preferences...
Design Template...
CIS Configuration...
Design Properties...
Product Configuration...
File Hierarchy
Design Resources
.\user.olb
My Title Block
Rose_Title_Block_1
Rose_Title_Block_2
SC_Title_Block
Library Cache
Library
Outputs
Referenced Projects
Set default options for new designs and libraries.

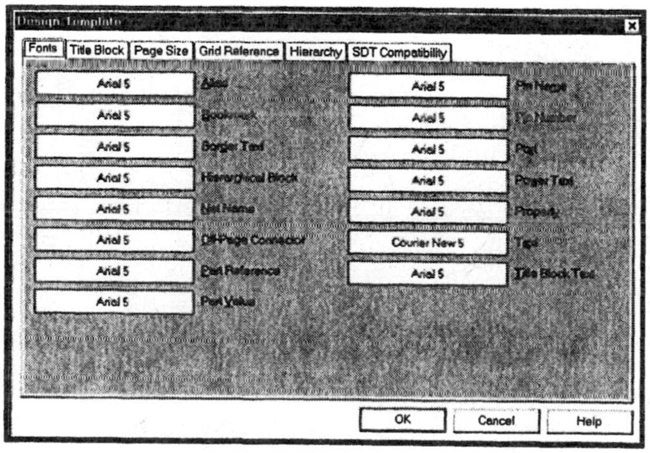

Design Template
Fonts | Title Block | Page Size | Grid Reference | Hierarchy | SDT Compatibility
Arial 5 Alias
Arial 5 Bookmark
Arial 5 Border Text
Arial 5 Hierarchical Block
Arial 5 Net Name
Arial 5 Off-Page Connector
Arial 5 Part Reference
Arial 5 Part Value
Arial 5 Pin Name
Arial 5 Pin Number
Arial 5 Port
Arial 5 Power Text
Arial 5 Property
Courier New 5 Text
Arial 5 Title Block Text
OK Cancel Help

Select the ***Title Block*** tab:

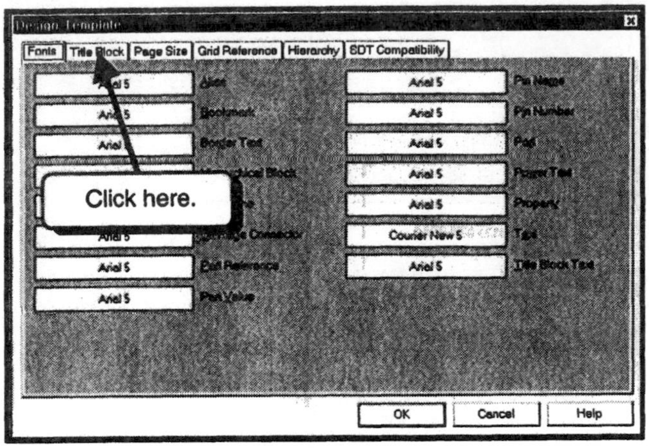

Design Template

Fonts | Title Block | Page Size | Grid Reference | Hierarchy | SDT Compatibility

Arial 5 Alias
Arial 5 Bookmark
Arial 5 Border Text
Hierarchical Block
Off-Page Connector
Arial 5 Part Reference
Arial 5 Part Value
Arial 5 Pin Name
Arial 5 Pin Number
Arial 5 Port
Arial 5 Power Text
Arial 5 Property
Courier New 5 Text
Arial 5 Title Block Text

Click here.

OK Cancel Help

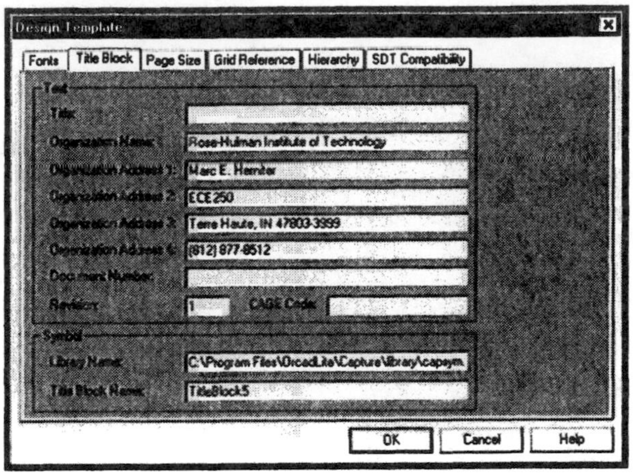

Design Template

Fonts | Title Block | Page Size | Grid Reference | Hierarchy | SDT Compatibility

Text
Title:
Organization Name: Rose-Hulman Institute of Technology
Organization Address 1: Marc E. Herniter
Organization Address 2: ECE 250
Organization Address 3: Terre Haute, IN 47803-3999
Organization Address 4: (812) 877-8512
Document Number:
Revision: 1 CAGE Code:
Symbol
Library Name: C:\Program Files\OrcadLite\Capture\library\capsym
Title Block Name: TitleBlock5

OK Cancel Help

The dialog box is filled in with information from our previous example. The last two fields in the dialog box specify the ***Library Name*** and the ***Title Block Name***. We need to change these to the appropriate values for the title block we just created. The information for the title block I created is:

- c:\Program Files\OrCADLite\Capture\library\PSpice\user.olb
- My_Title_Block

Your directory may be slightly different if you did not choose the default installation directory when you installed the software. I have entered this information in the screen capture shown below:

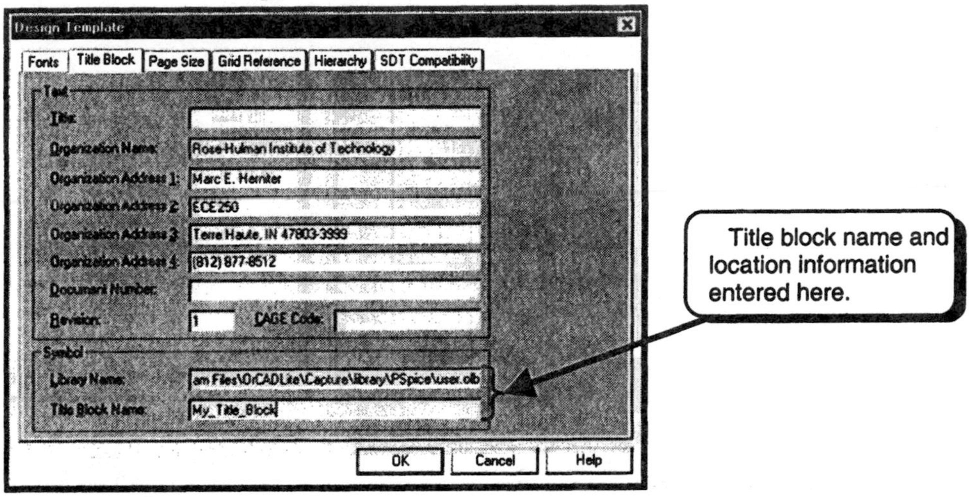

Design Template

Fonts | Title Block | Page Size | Grid Reference | Hierarchy | SDT Compatibility

Text
Title:
Organization Name: Rose-Hulman Institute of Technology
Organization Address 1: Marc E. Herniter
Organization Address 2: ECE 250
Organization Address 3: Terre Haute, IN 47803-3999
Organization Address 4: (812) 877-8512
Document Number:
Revision: 1 CAGE Code:
Symbol
Library Name: am Files\OrCADLite\Capture\library\PSpice\user.olb
Title Block Name: My_Title_Block

OK Cancel Help

Title block name and location information entered here.

If you enter the wrong path name or the wrong title block name in the dialog box above, new schematic pages will be created without a title block.

The title block we created has two fields that take information from this title block. The fields are ***<OrgAddr1>*** and ***<OrgAddr2>***. I will leave the information for these fields, and remove the data from the unused fields:

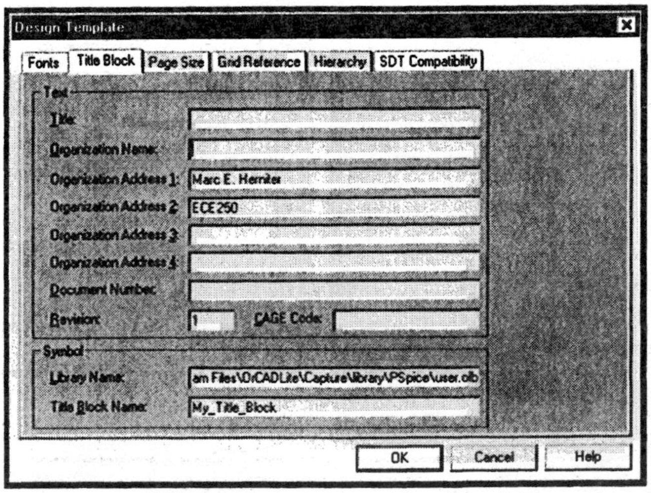

Design Template

Fonts | Title Block | Page Size | Grid Reference | Hierarchy | SDT Compatibility

Text
Title:
Organization Name:
Organization Address 1: Marc E. Herniter
Organization Address 2: ECE 250
Organization Address 3:
Organization Address 4:
Document Number:
Revision: 1 CAGE Code:
Symbol
Library Name: am Files\OrCADLite\Capture\library\PSpice\user.olb
Title Block Name: My_Title_Block

OK Cancel Help

Click the ***OK*** button when you are finished.

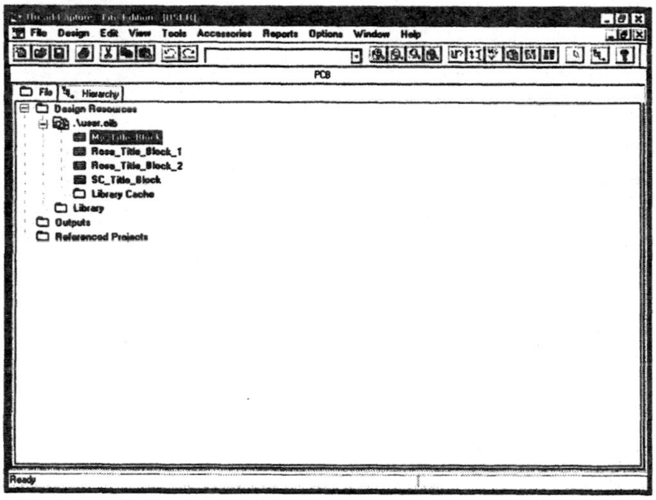

File Design Edit View Tools Accessories Reports Options Window Help
PCB
File Hierarchy
Design Resources
.\user.olb
Rose_Title_Block_1
Rose_Title_Block_2
SC_Title_Block
Library Cache
Library
Outputs
Referenced Projects
Ready

Select **File** and then **Close Project** from the menus:

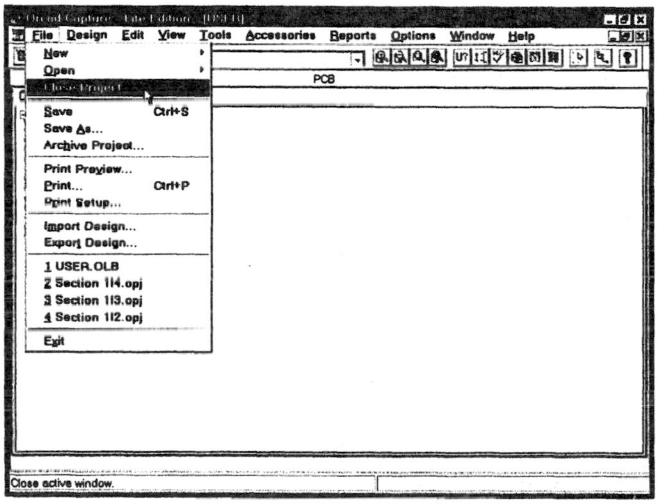

File Design Edit View Tools Accessories Reports Options Window Help
New
Open
Close Project
Save Ctrl+S
Save As...
Archive Project...
Print Preview...
Print... Ctrl+P
Print Setup...
Import Design...
Export Design...
1 USER.OLB
2 Section 114.opj
3 Section 113.opj
4 Section 112.opj
Exit
PCB
Close active window.

Save Files In Project
Save changes to .\user.olb?
(Yes All and No All affect all remaining open files in USER.)
Yes Yes All No No All Cancel Help

Select ***Yes All*** to save the changes to the libraries:

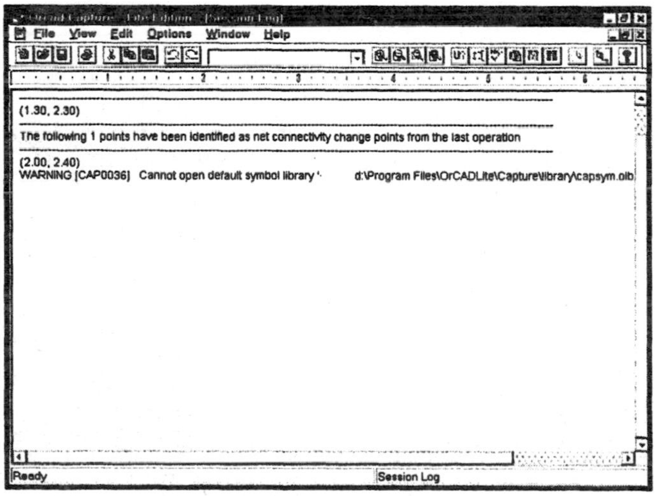

File View Edit Options Window Help
(1.30, 2.30)
The following 1 points have been identified as net connectivity change points from the last operation
(2.00, 2.40)
WARNING [CAP0036] Cannot open default symbol library ' d:\Program Files\OrCADLite\Capture\library\capsym.olb
Ready Session Log

When you create a new project, it should use the new title block we just created:

If you see no title block at all, you may have entered the wrong path name or the wrong title block name in the dialog box on page 62.

# 1.K. Adding Schematic Pages to a Project

You can have multi-page projects. We will demonstrate by adding a page to the first schematic we created. Close all open projects:

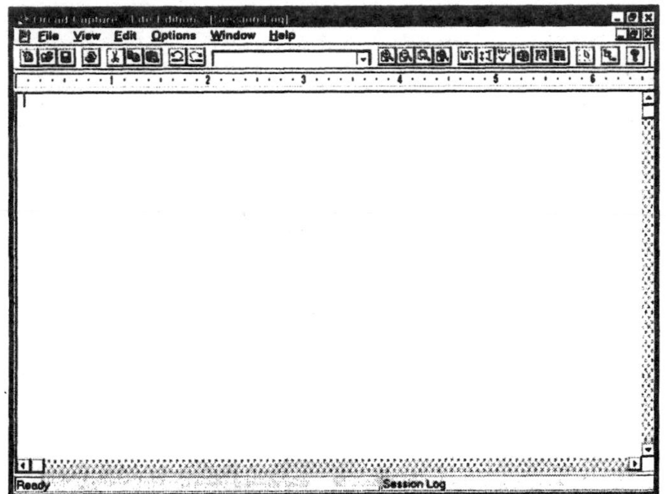


We will now open our first project. Select **File**, **Open**, and then **Project** from the menus:

Select the project named ***Part 1 Schematic*** and click the ***Open*** button:

Orcad Capture - Lite Edition - [Part 1 Schematic.opj]
File Design Edit View Tools PSpice Accessories Reports Options Window Help
Analog or Mixed A/D
File Hierarchy
Design Resources
.\part 1 schematic.dsn
Library
Outputs
PSpice Resources
Reading Design Variants Information...
Session Log

We must now expand the tree to list the schematic pages. Click the ***LEFT*** mouse button on the ⊞ next to the text ***.\part 1 schematic.dsn***:

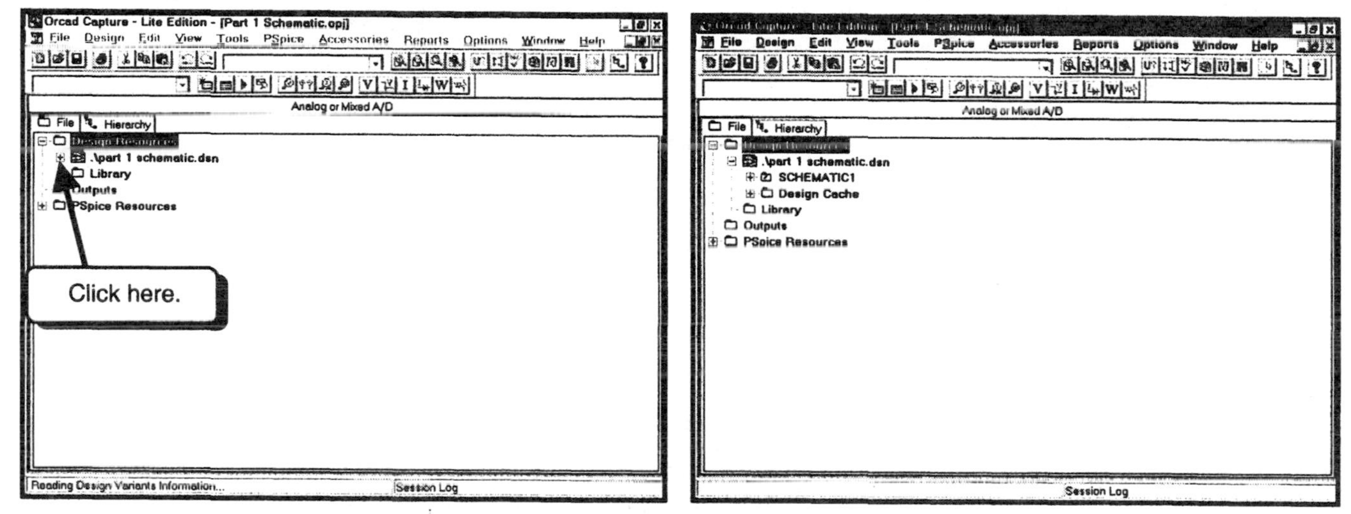


Click the ***LEFT*** mouse button on the ⊞ next to the text ***SCHEMATIC1***

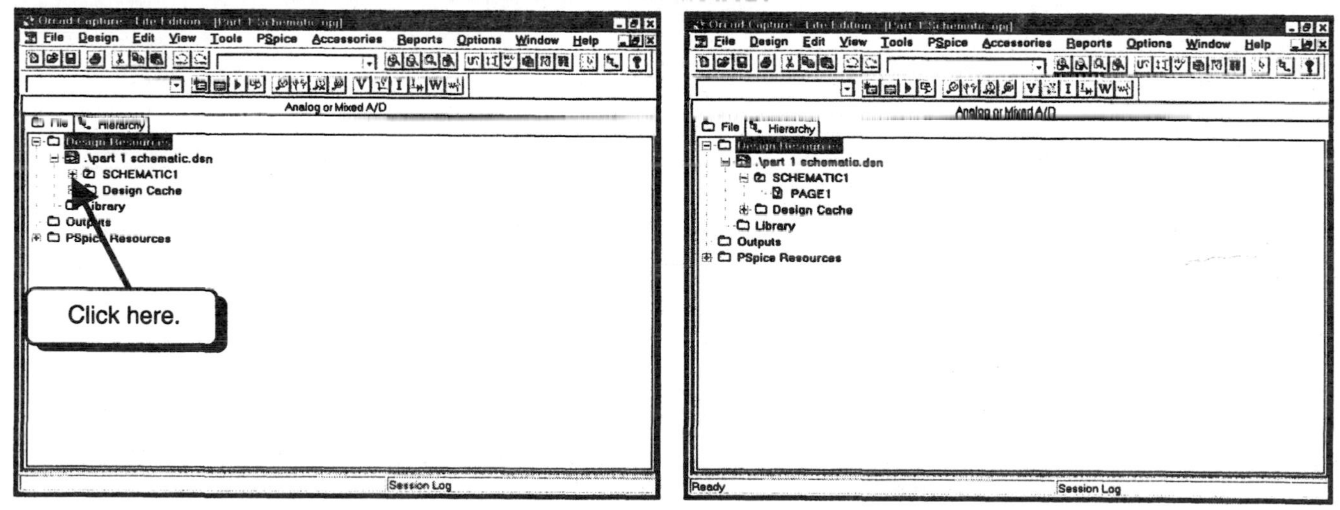


This project has one schematic drawing named ***PAGE1***.

To create a new page, click the ***RIGHT*** mouse button on the text ***SCHEMATIC1***:

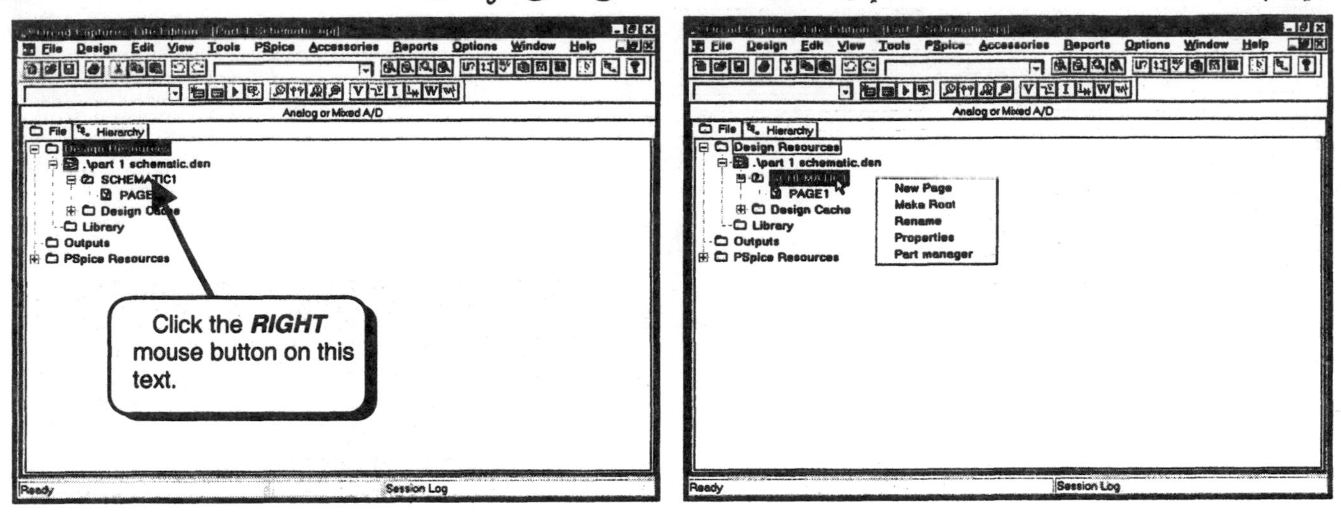


Select **New Page**:

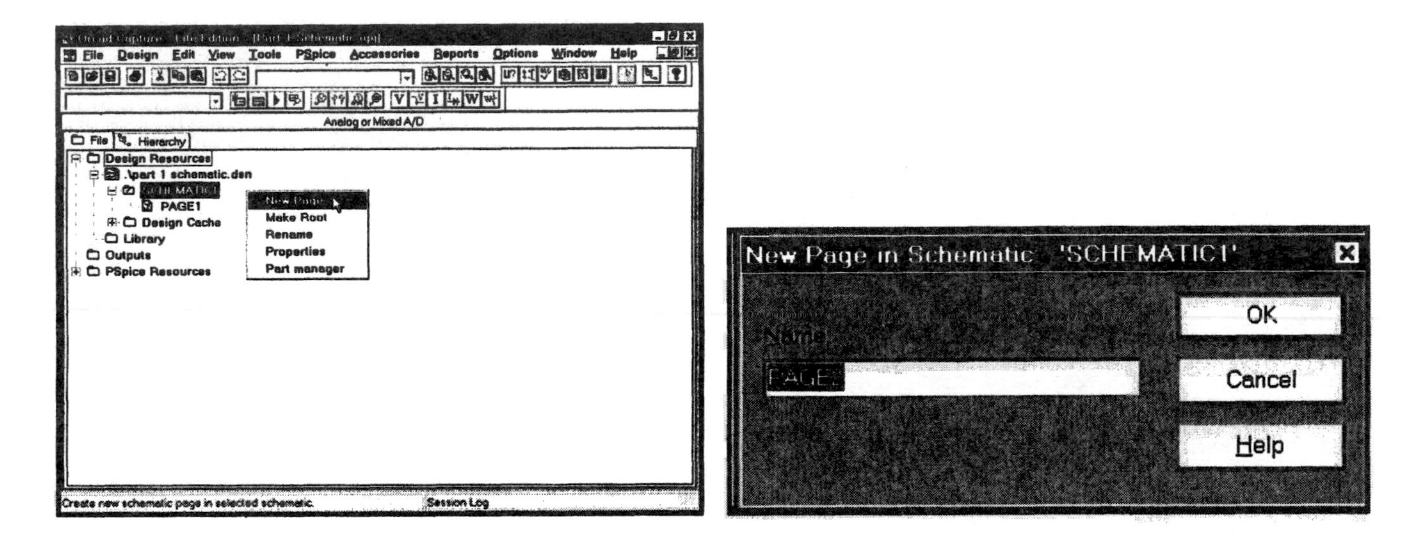


Enter the name for the page and click the ***OK*** button:

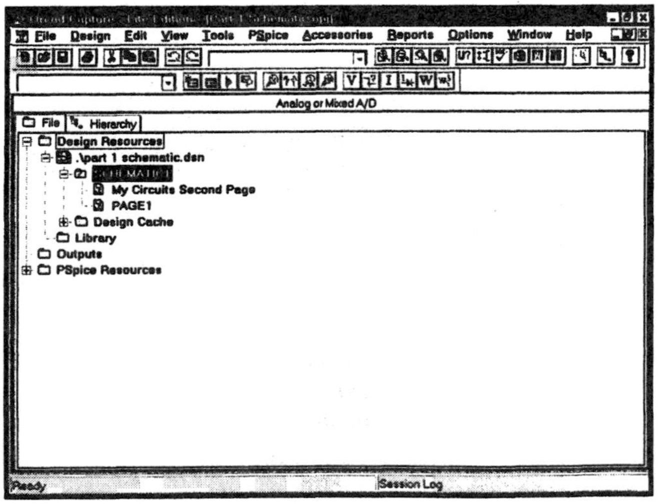

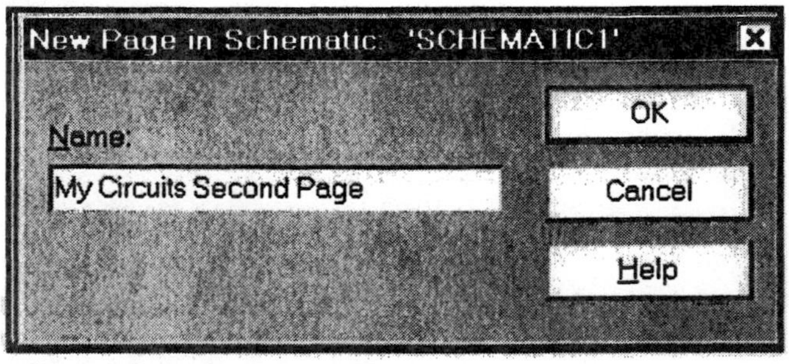

If you want to rename a page, click the ***RIGHT*** mouse button on the name of the page and then select **Rename**. For example, we will rename ***PAGE1***:

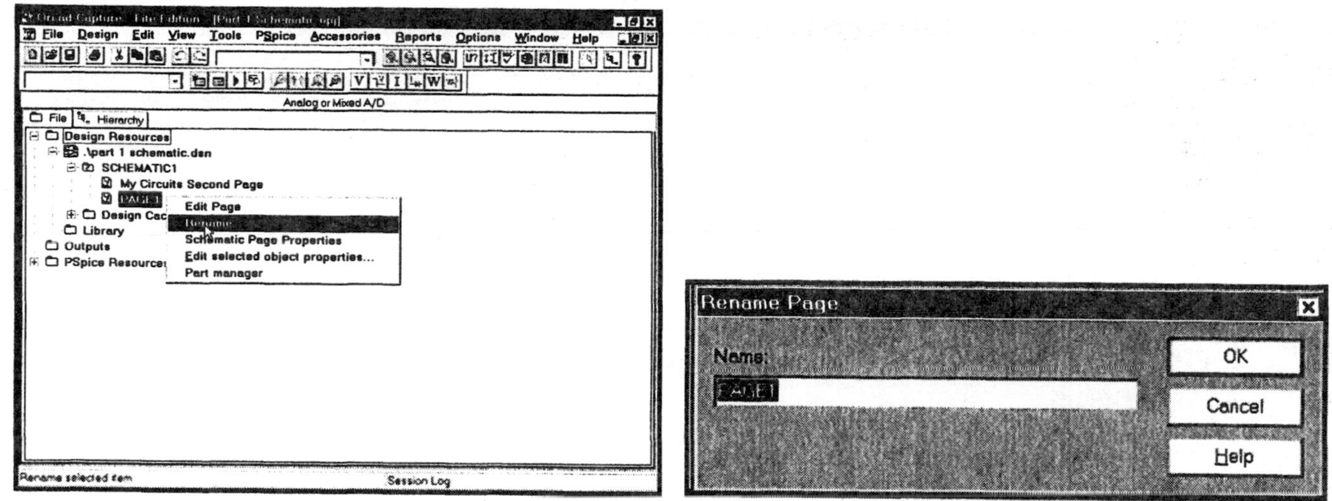


Enter a new name and click the ***OK*** button:

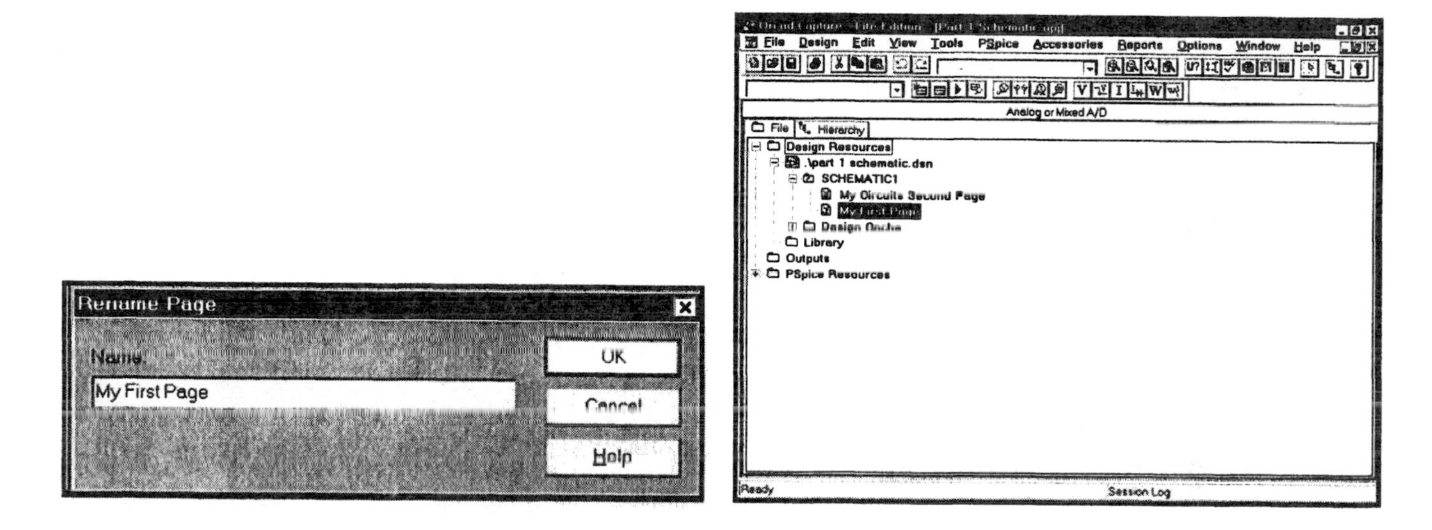


To edit a page you can either click the ***RIGHT*** mouse button on the page name and then select **Edit Page**, or you can double-click the ***LEFT*** mouse button on the page name.

Next, we will open a page and show how to connect nodes of a circuit on one page to nodes on another page. Double-click on the text ***My First Page*** to open the schematic:

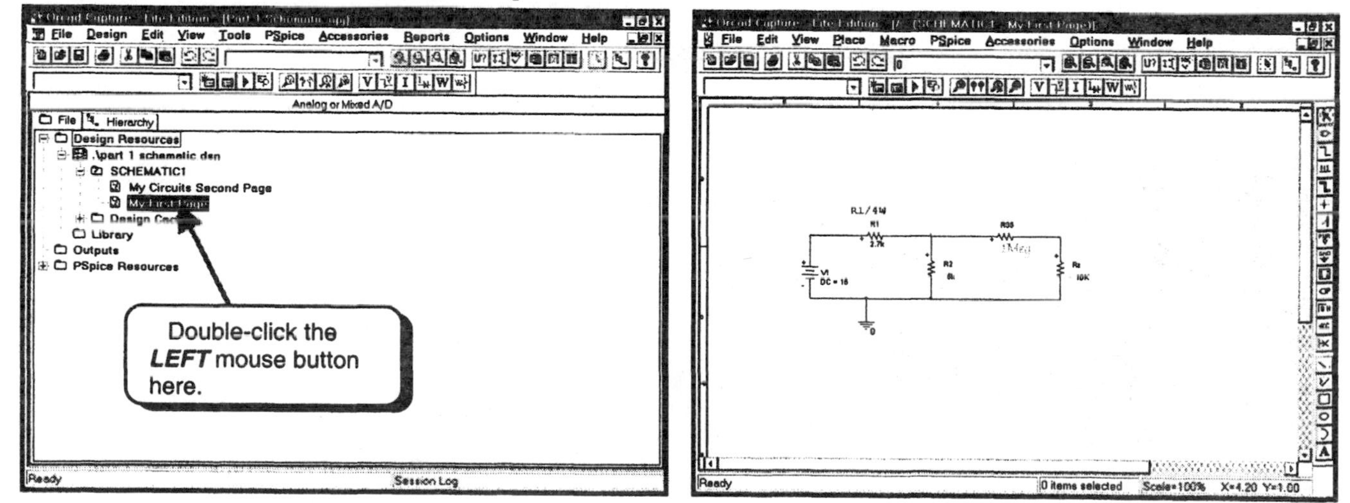


Place the mouse pointer over the circuit and press the **I** key to zoom in:

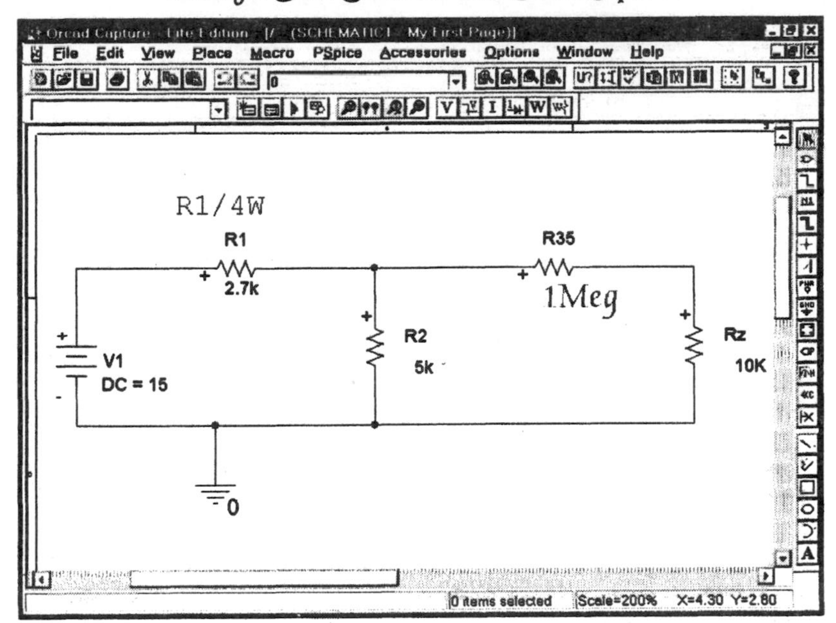


To connect nodes on different pages we use the offpage connectors. Select **Place** and then **Off-Page Connector** from the menus:

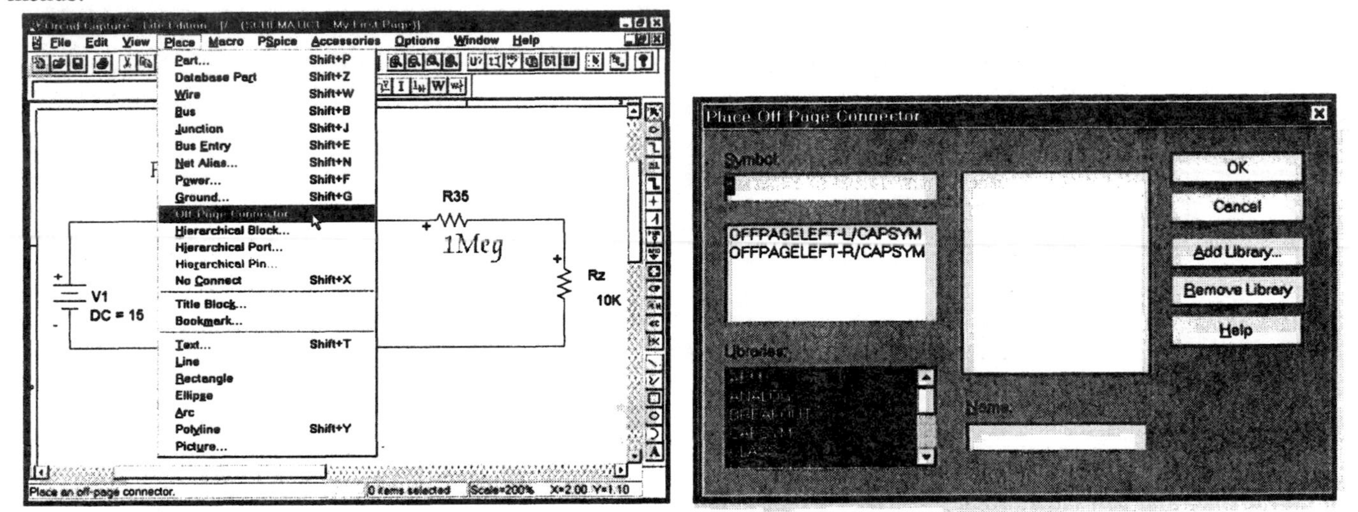


You can use either of these connectors. Select a connector and click the ***OK*** button. The part will appear attached to the mouse pointer.

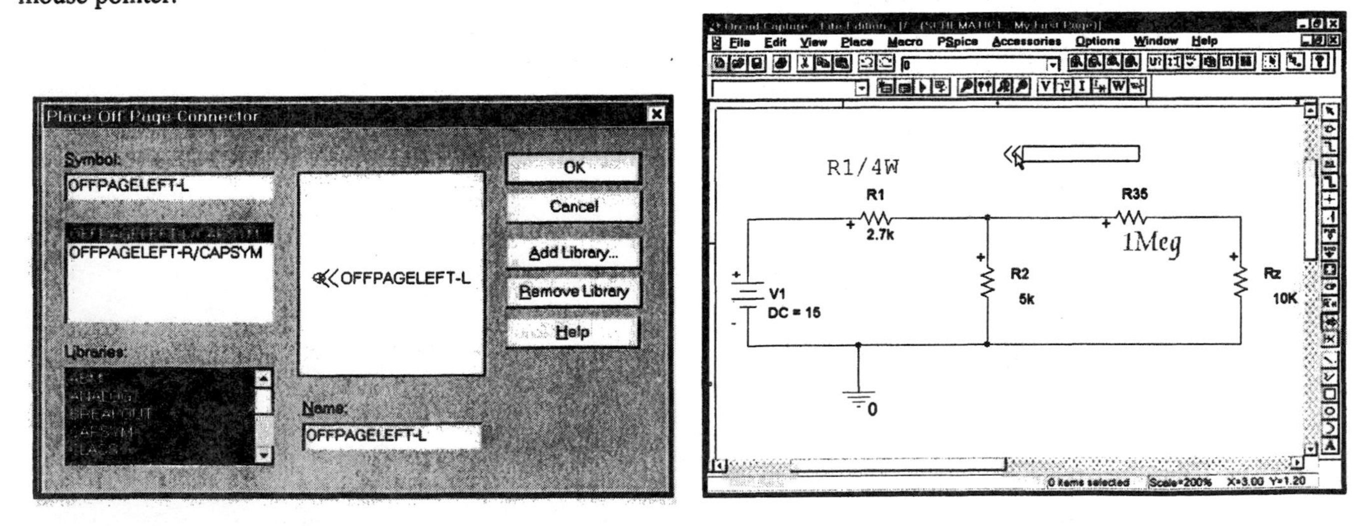


Type **R** to rotate the part and then move it to the location shown below:

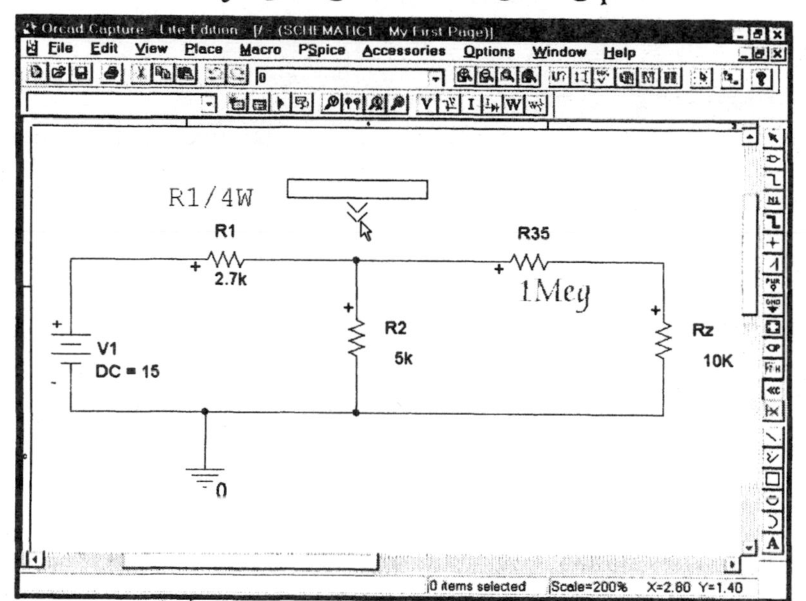


Click the ***LEFT*** mouse button to place the part and then move the mouse:

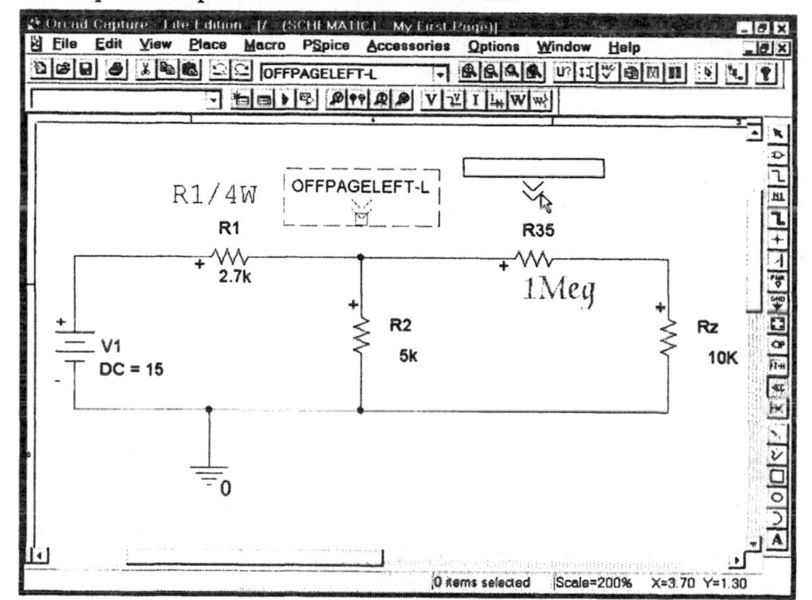


Place a second ***OFFPAGELEFT-L*** part as shown:

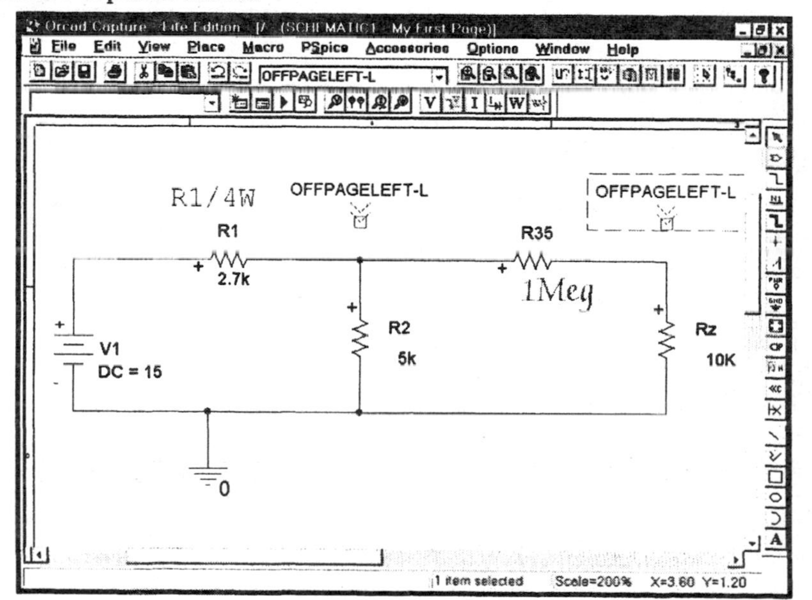

To stop placing parts, press the **ESC** key. We must now rename the parts. Double-click the ***LEFT*** mouse button on the text to edit it:

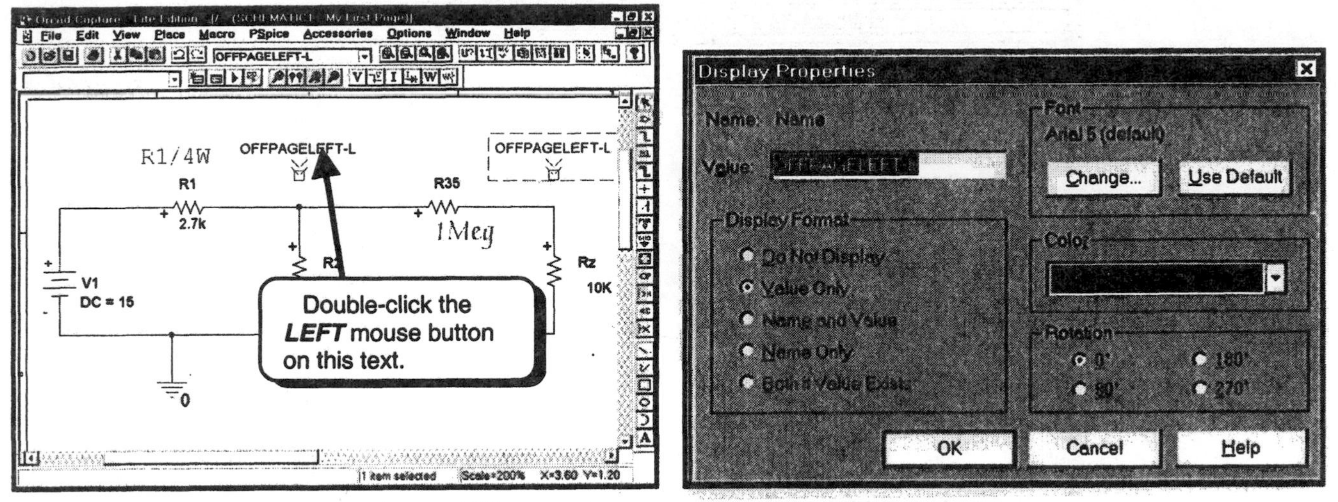


Change the name to **NODE1** and click the ***OK*** button. There must be no spaces in the name:

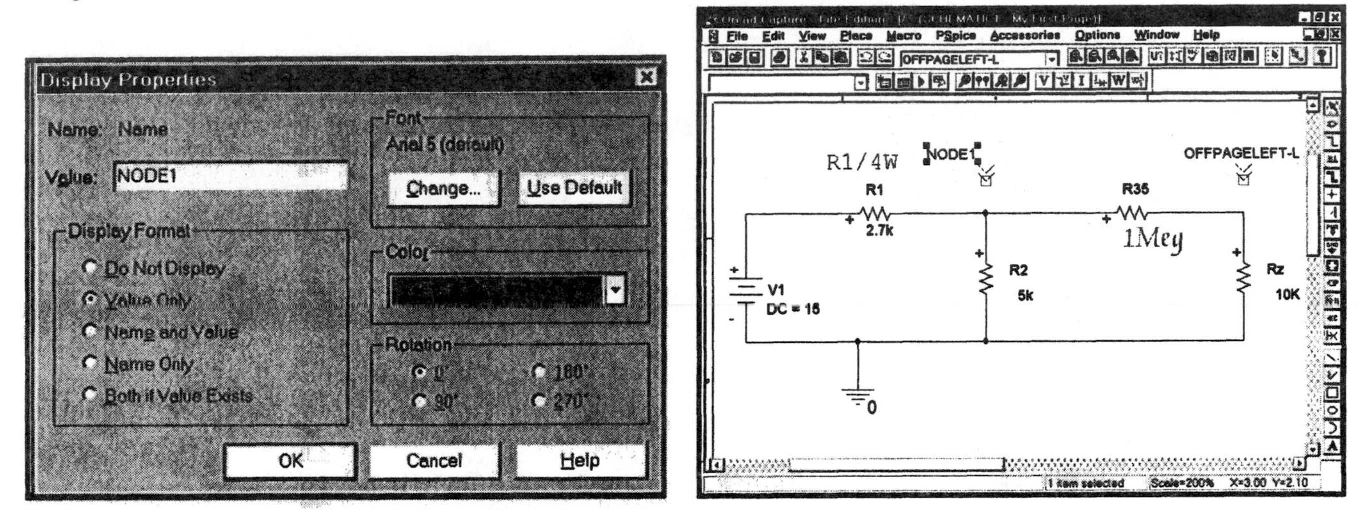


Use the mouse to drag the text to a better location. Use the same procedure to rename the second connector to **Vo**:

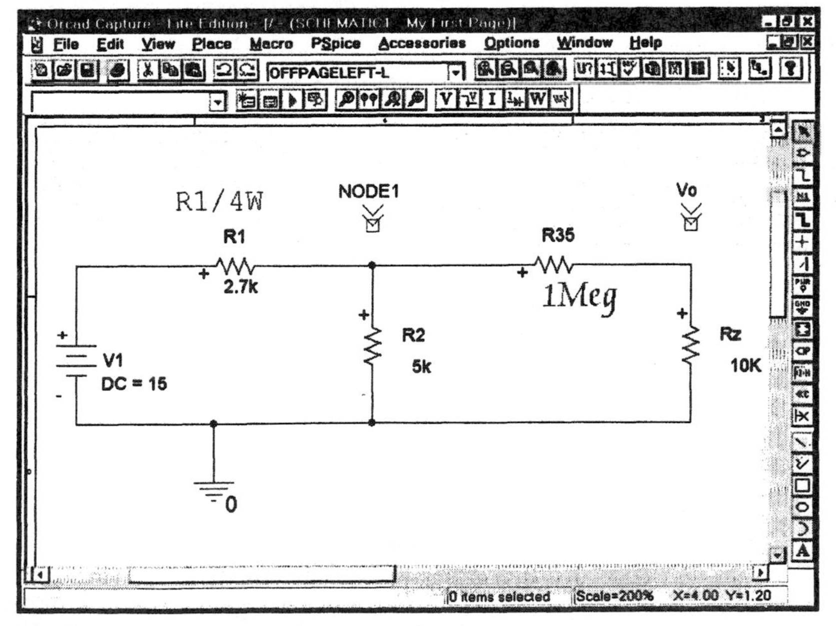


Type **W** to start wire mode and connect the connectors to the circuit:

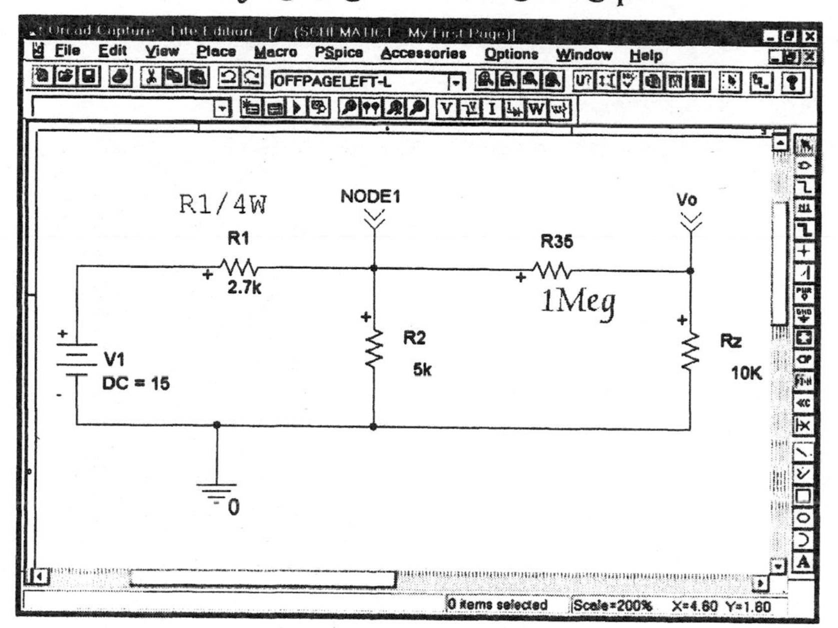


Press the **ESC** key to terminate wire mode.

Select **File** and then **Save** to save the changes. Save all changes if asked. Next, select **File** and then **Close** to close the page:

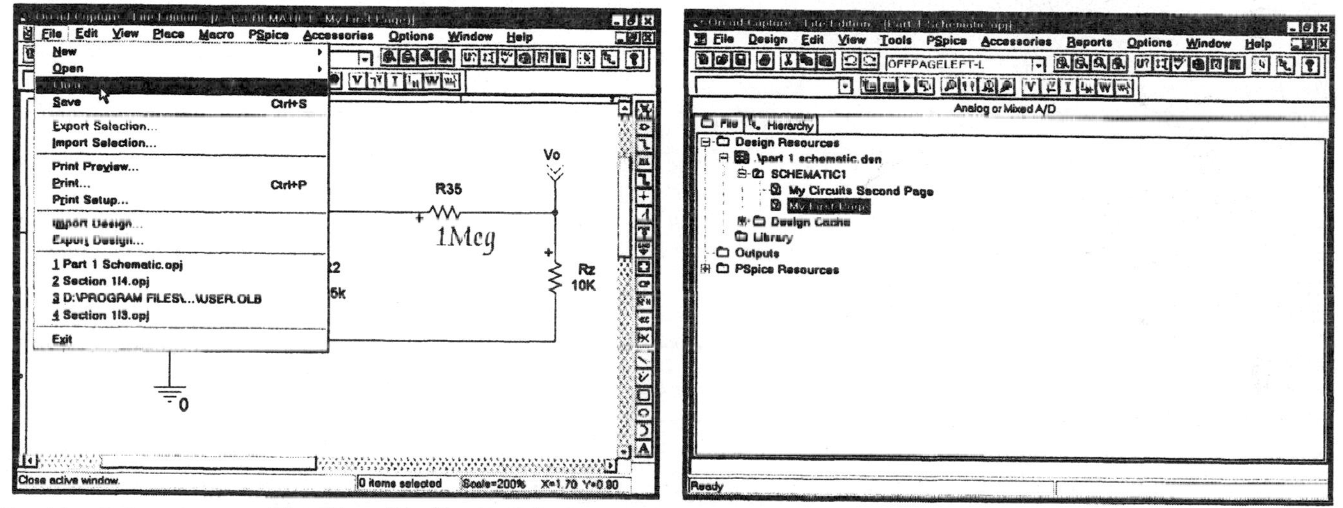


Double-click on the text ***My Circuits Second Page*** to open the page:

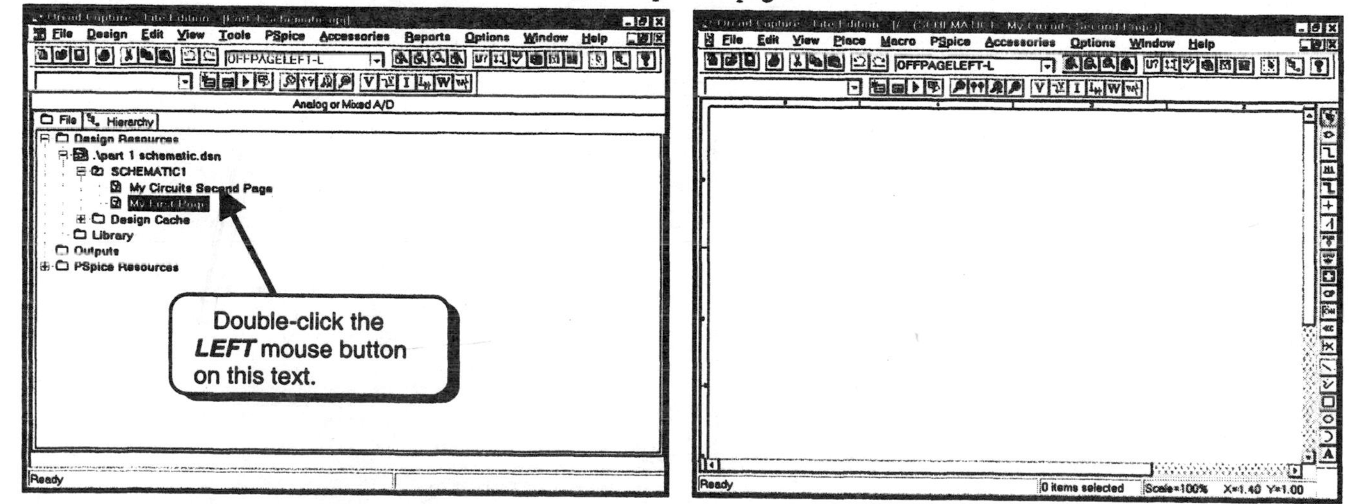


Place a resistor and add the offpage connectors as shown. You can use either of the offpage connectors.

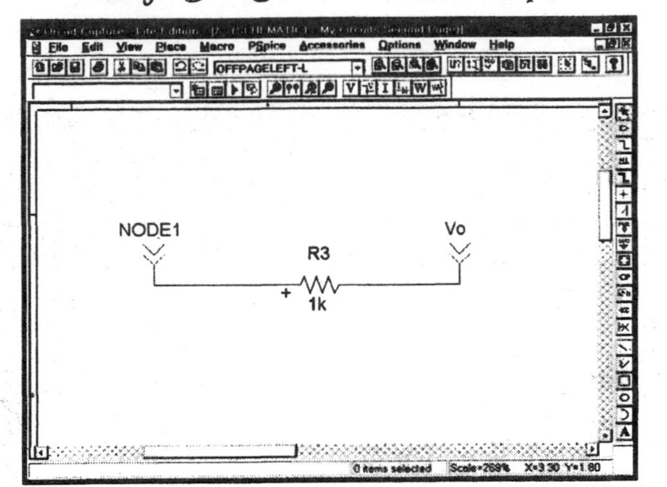


***R3*** is now connected to nodes ***NODE1*** and ***Vo*** of the circuit on the previous page.

# 1.L. Changing the Size of a Schematic

You may have noticed when drawing circuits and printing circuits that you cannot fit too many parts on an 8.5″ x 11″ page. The parts are sized to be very readable, but for complex schematics, you cannot fit many parts on a page. Reducing the part size can be accomplished by choosing a different page size. When you print your document, you can still print a schematic page on a single page of 8.5″ x 11″. We will double the width and length of the standard page size to increase the page area by a factor of four. When we print this larger page on an 8.5″ x 11″ piece of paper, the symbols will be reduced to 25% of their original size. The changes we make will not apply to the schematics we have already created. They will apply only to new schematics.

Close all open schematics or restart Orcad Capture:

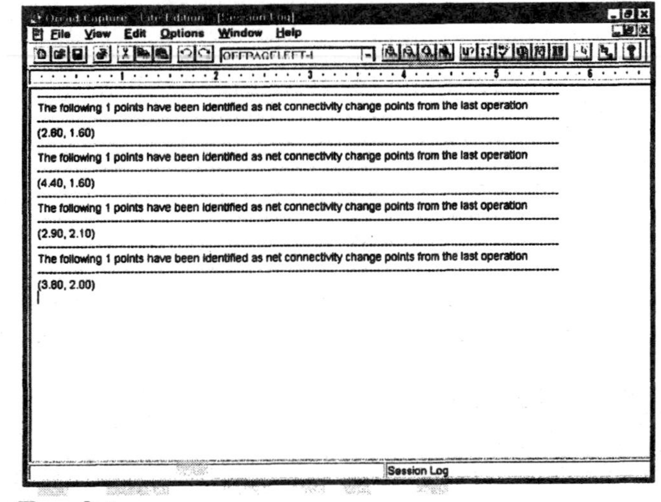


Select **Options** and then **Design Template**:

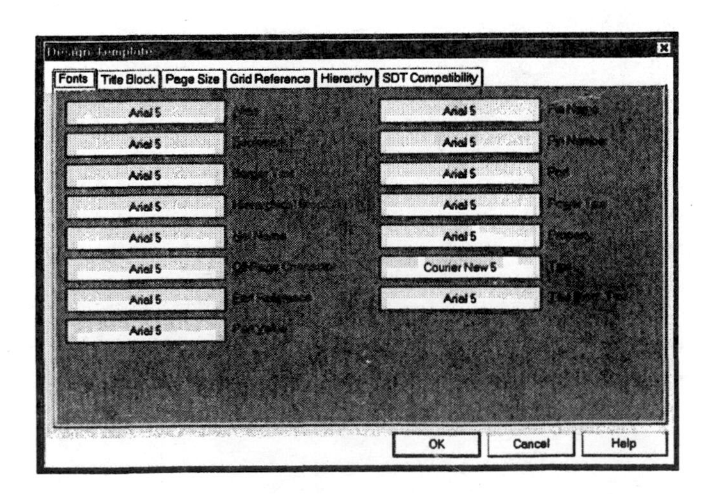

Select the ***Page Size*** tab:

Click here.

We are assuming that you are printing on A size paper (8.5" x 11") so we want to create a custom page size that is a scaled version of ***A*** size paper. Choose the ***Custom*** option and enter a ***Width*** and a ***Height*** that are multiples of the ***A*** size ***Width*** and ***Height***. I will double the size from 9.7 by 7.2 to 19.4 by 14.4:

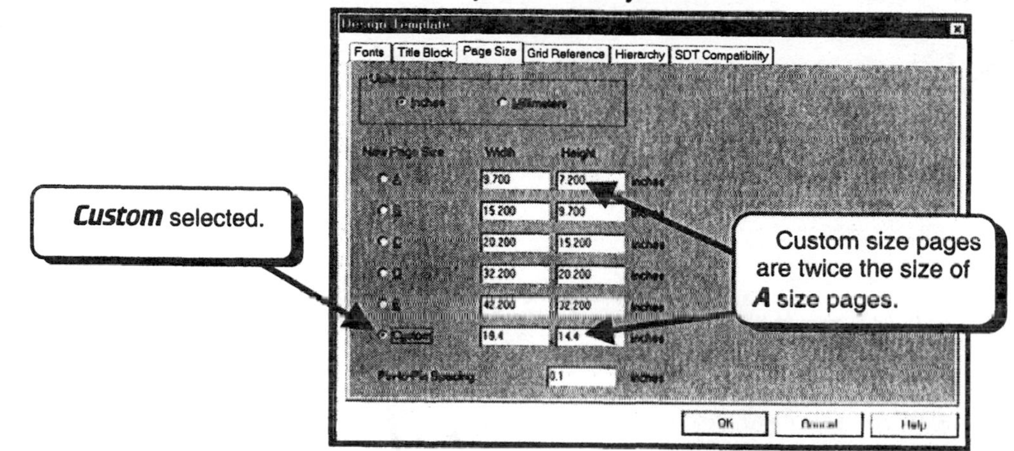


Click the ***OK*** button to accept the changes.

When you create new schematics, they will use this custom size. I have recreated the schematic we used in the previous section using the custom page size. Full page schematics of the original schematic and the new one with the custom page size are shown below:

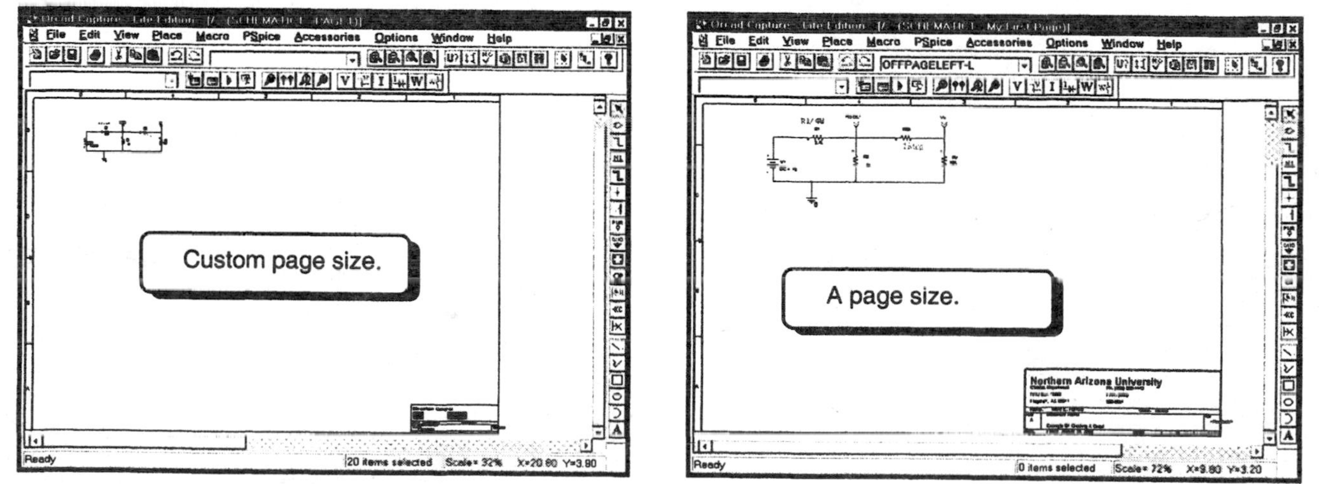


An example printed page from an A size schematic is shown on page 74. An example printed page from our custom sized page is shown on page 75.

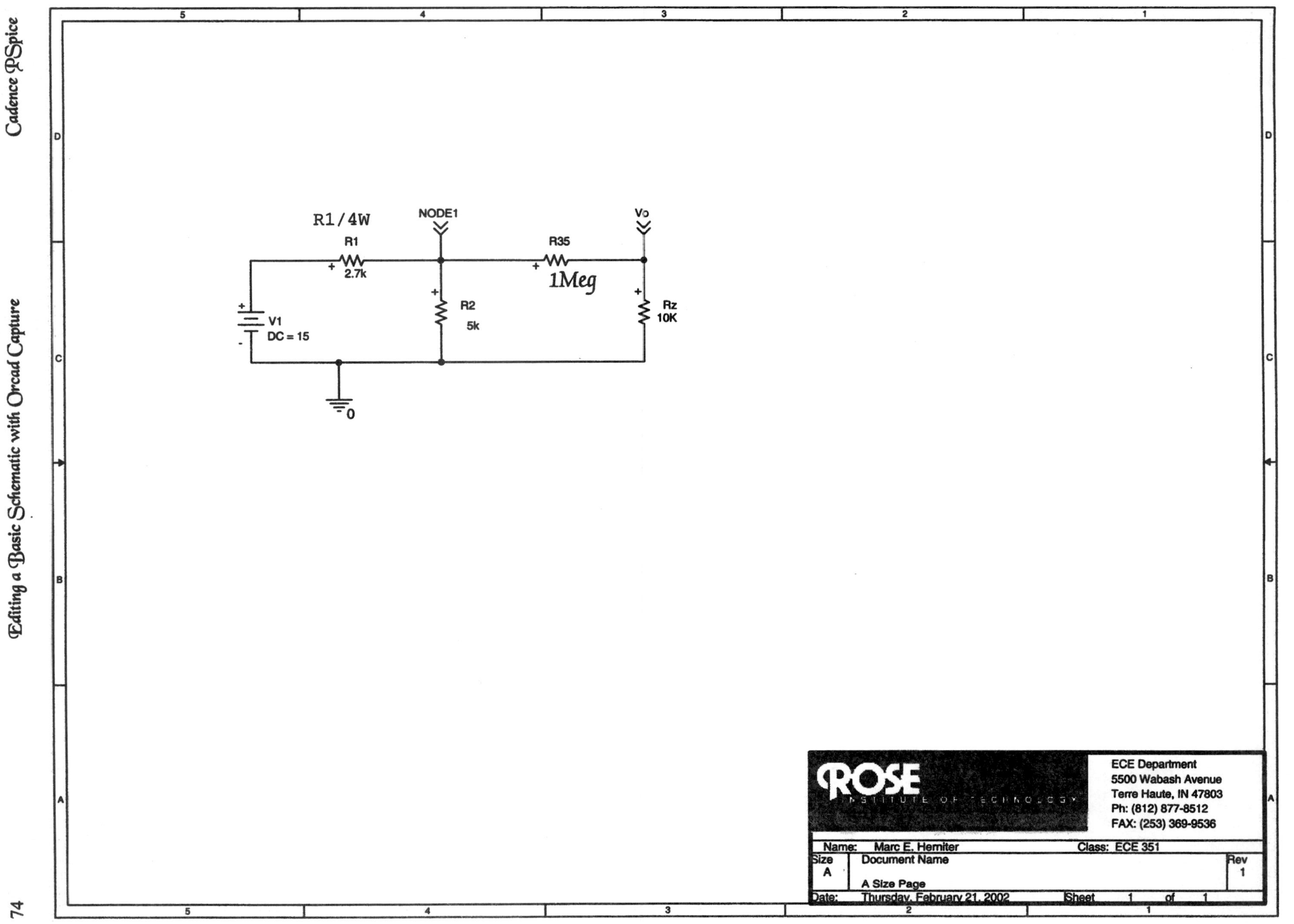
R1 / 4W
R1
2.7k
V1
DC = 15
0
NODE1
R2
5k
R35
1Meg
Vo
Rz
10K
ECE Department
5500 Wabash Avenue
Terre Haute, IN 47803
Ph: (812) 877-8512
FAX: (253) 369-9536
ROSE
INSTITUTE OF TECHNOLOGY
Name: Marc E. Herniter
Class: ECE 351
Size
A
Document Name
A Size Page
Rev
1
Date: Thursday, February 21, 2002
Sheet 1 of 1

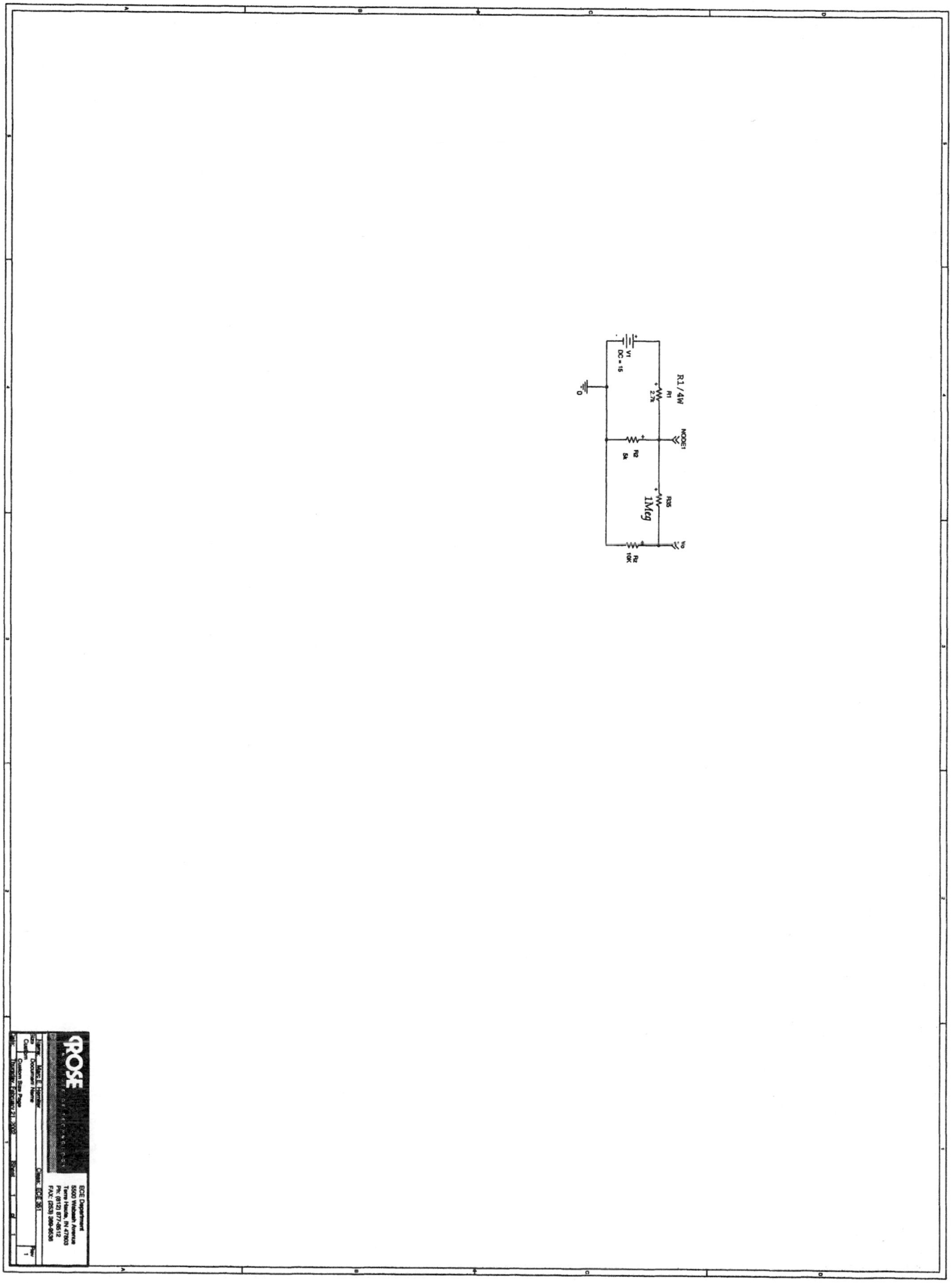
V1
DC = 15
R1
2.7k
R1/4W
NODE1
R2
5k
1Meg
10K
0

# 1.M. Creating a Hierarchical Design

Orcad Capture has many facilities for managing large projects. One of these facilities is creating a hierarchical design where schematics can contain hierarchical blocks. These blocks contain detailed circuits and can also contain hierarchical blocks as well. When you edit a block, a new schematic page opens and you can use the entire page for creating the schematic. This page can also contain a hierarchical block. Thus, you can have blocks within blocks within blocks. A hierarchical design allows you to break your schematic into smaller pieces that perform specific functions, and allows you to easily see the connections between the blocks.

For our example, we will create an amplifier which we will break up into the power supply, the pre-amplifier block, and the power amplifier block. The power amplifier will also contain a hierarchical block for the load. The circuits we show will be fairly trivial and could easily be placed on a single schematic page. The point of this exercise is to show how to use the hierarchical tools available in Orcad Capture.

This is an advanced topic. We assume that the reader has already mastered creating schematics with Capture, running PSpice simulations, and plotting with Probe. All we will cover here is how to use the hierarchical tools. We will use many of the drawing tools covered earlier in this chapter without much explanation.

We will start with a new empty project. Select **File**, **New**, and the **Project** from the menus:

Specify a name and directory for the file and click the ***OK*** button:

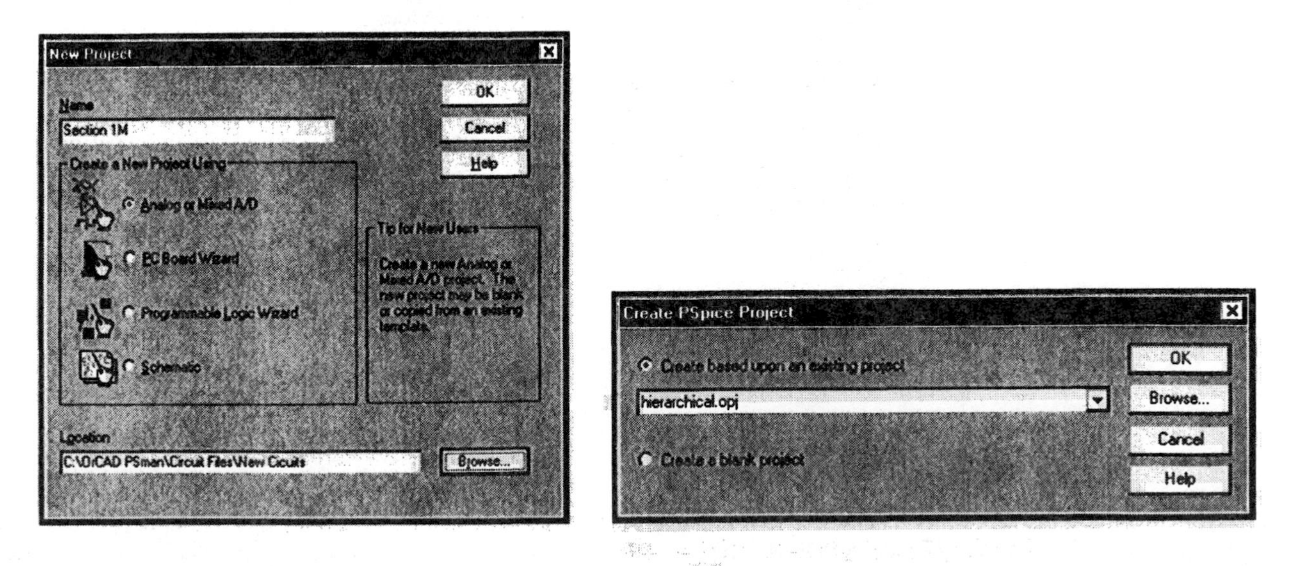


We will set up the hierarchical blocks ourselves, so select the option to ***Create a blank project*** and click the ***OK*** button:

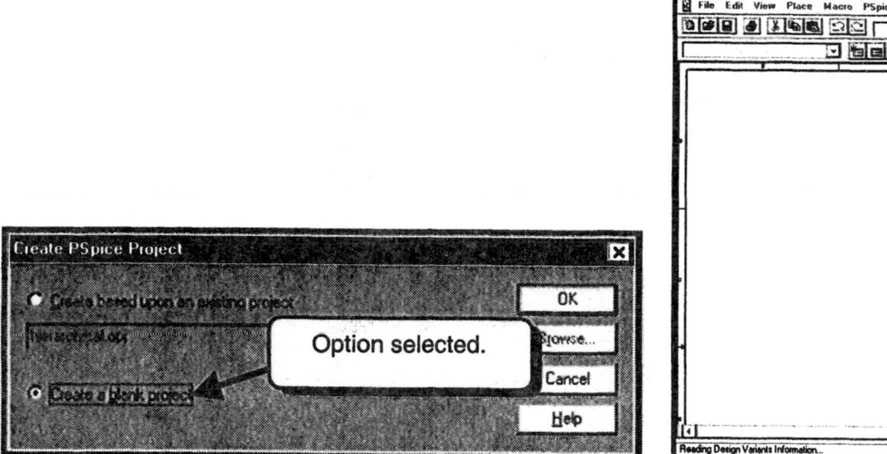


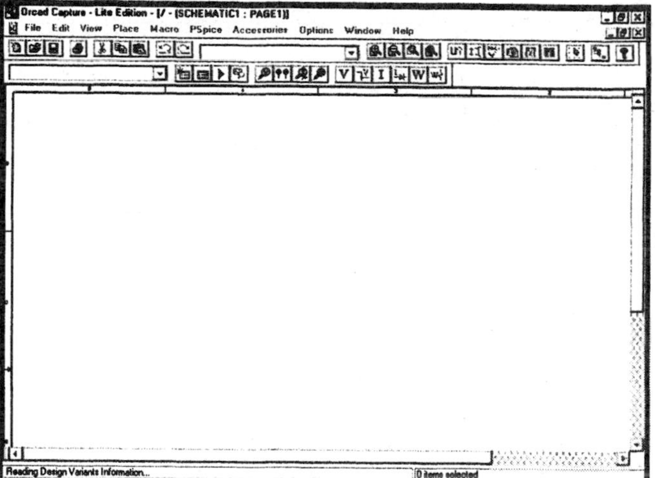


We can now start placing parts. We will have a top level drawing that only contains hierarchical blocks. We will first create these blocks. We will then descend into each block and create the specific circuit for that block. Select **Place** and then **Hierarchical Block** from the menus:

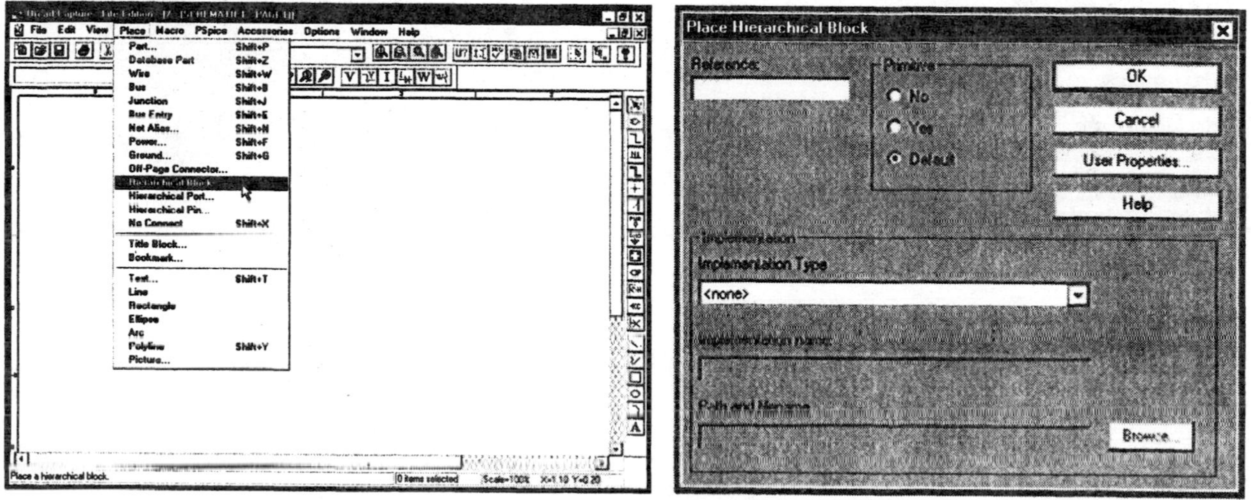


Fill in the dialog box as shown below:

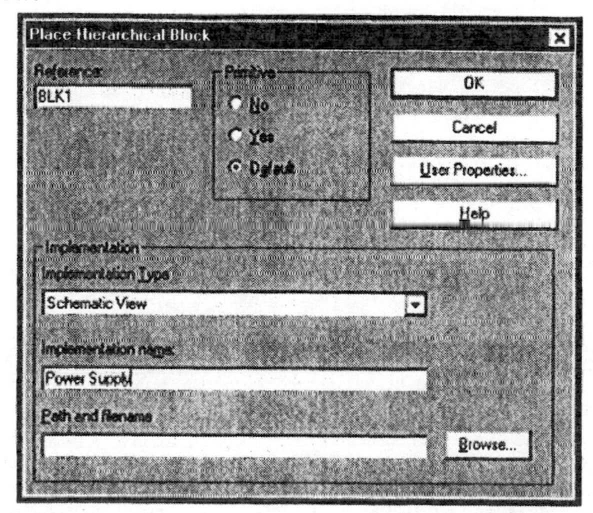


The ***Reference*** is set to ***BLK1***. This is similar to a part name such as R1 or C3 except that the reference will affect how nodes and parts within the block are named. For example, a resistor within BLK1 named R1 on the schematic would be referred to as R_BLK1_R1 in the PSpice netlist. A node labeled as vo in block 1 will be renamed as BLK1_vo if that node does not connect to components inside another block. An ***Implementation Type*** of ***Schematic View*** was chosen because we will create the contents of the block using OrCAD Capture. You can also specify the function of a block using other methods such as writing a VHDL description of the block (not available in Orcad Lite). The ***Implementation Name*** will become the name of the folder in the project tree where the schematic is located. When we look at the tree view of the project, the function of the name specified here will become apparent. Click the ***OK*** button. The mouse pointer will be replaced by crosshairs:

Draw a box using the mouse. When you release the mouse button, a block will be drawn and labeled:

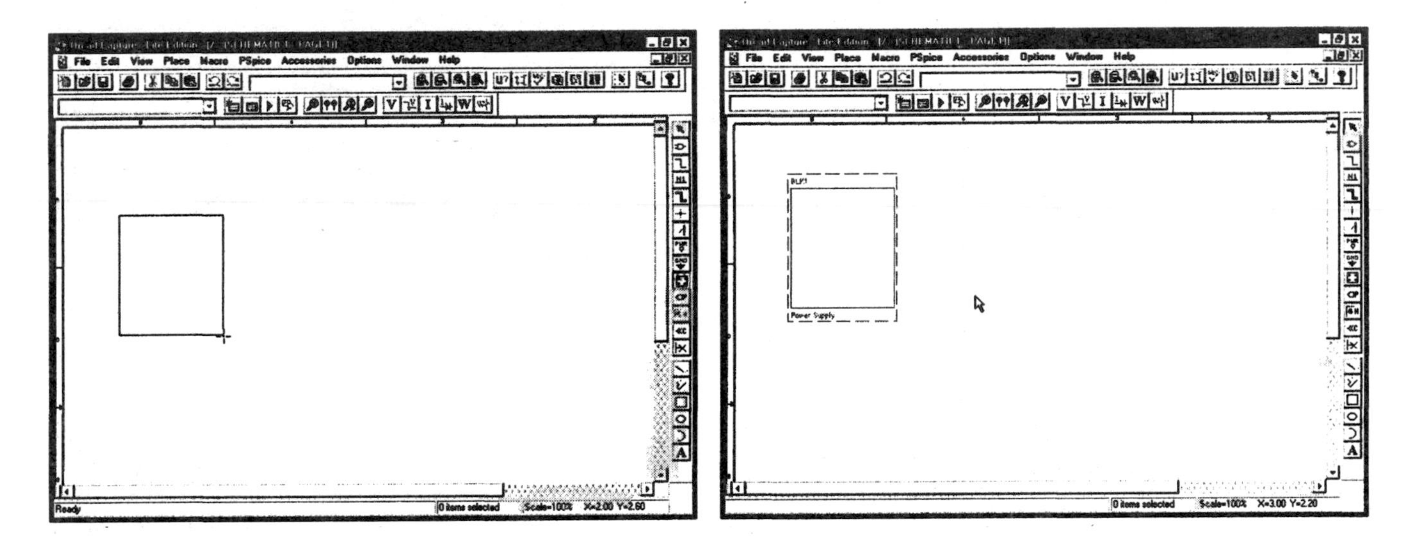

Use the same procedure to create two more hierarchical blocks for the Pre-Amplifier and the Power Amplifier. Set the ***Reference*** for these blocks to be ***BLK2*** and ***BLK3***:

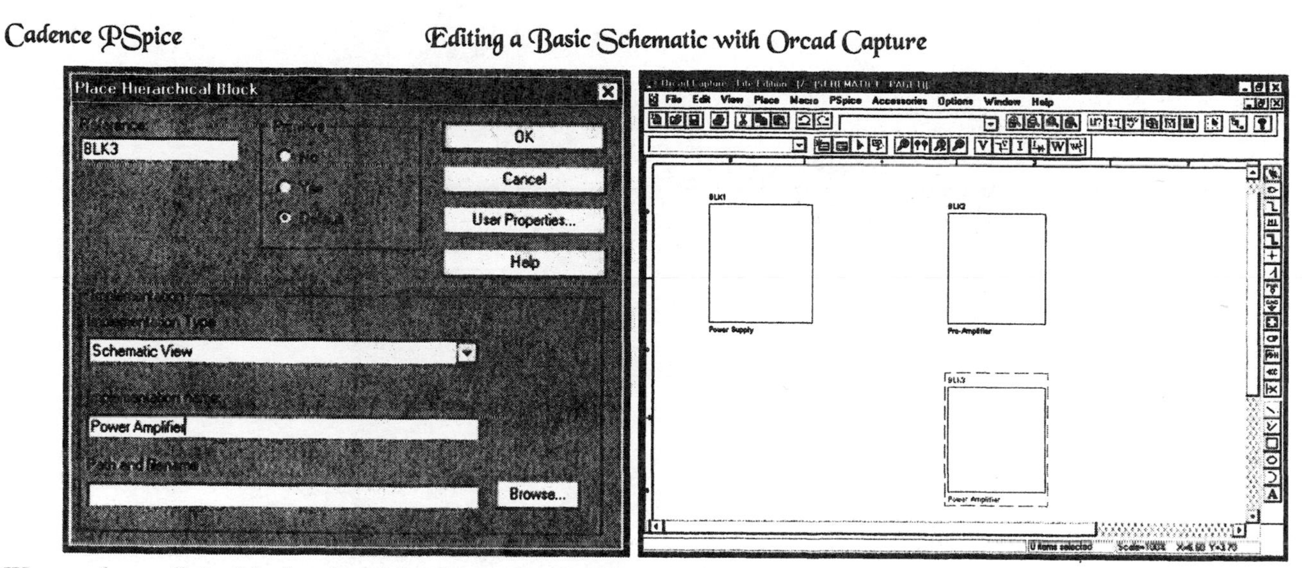

We now have all the blocks placed for the top level schematic. We are arranging this hierarchy so that it resembles a block diagram. Each block will contain a complicated schematic, but there will be very few signals passed between those blocks. The arrangement is shown so that we can easily group components together that make up a specific function and we can also easily identify the signals that need to be exchanged between the blocks.

Next, we need to specify pins for each hierarchical block. These pins will allow connections in and out of the blocks. First, click the ***LEFT*** mouse button on block BLK1 to select it:

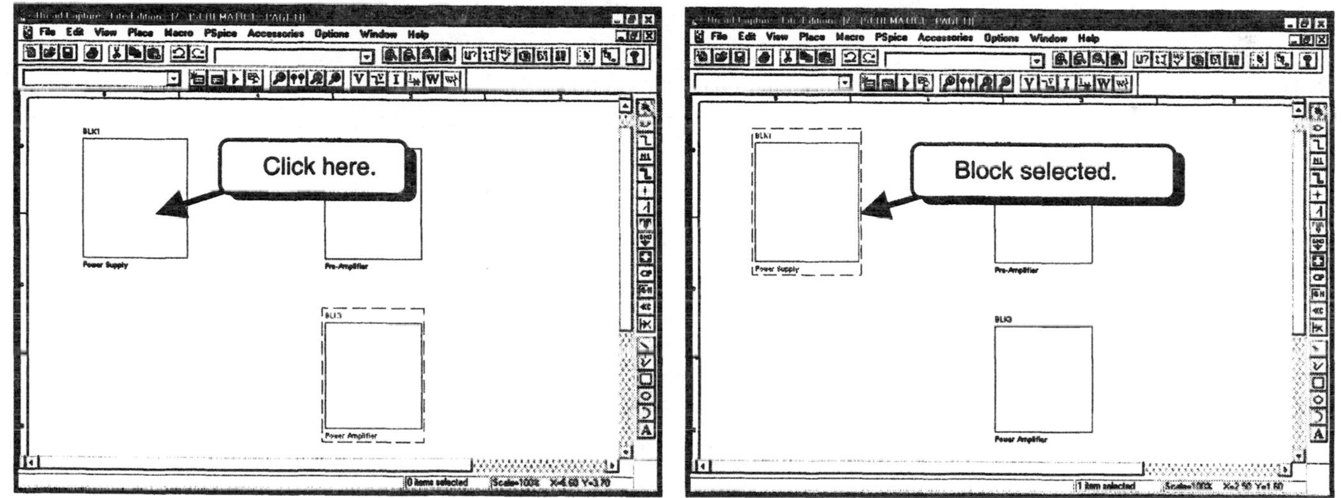


When the block is selected, it will become highlighted in pink. Select **Place** and then **Hierarchical Pin** from the Capture menus:

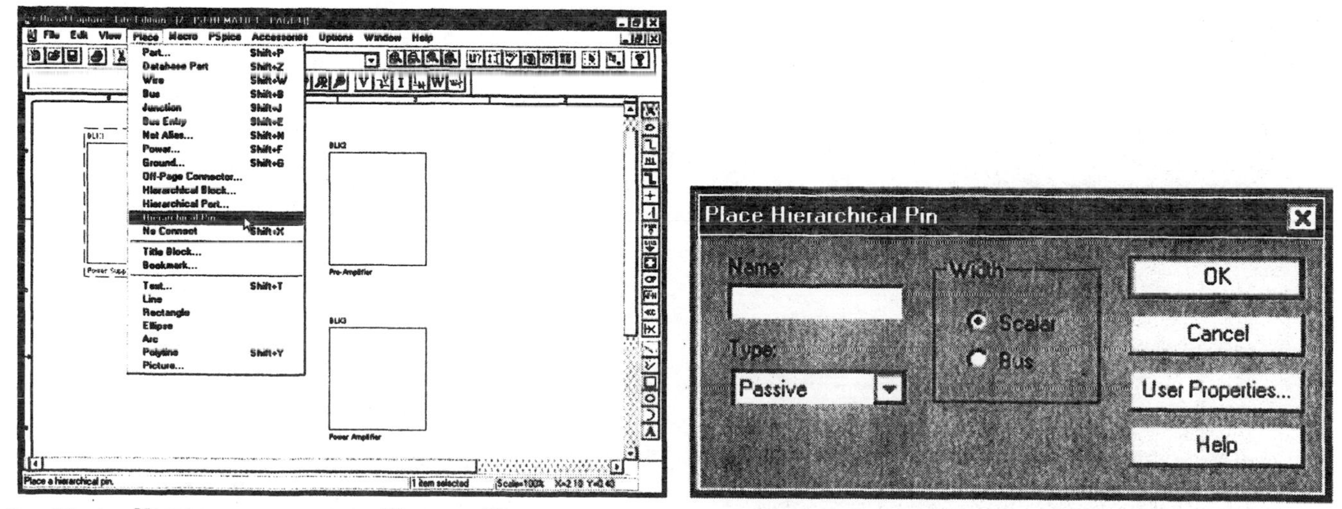

Specify the ***Name*** as vcc and the ***Type*** as ***Power***:

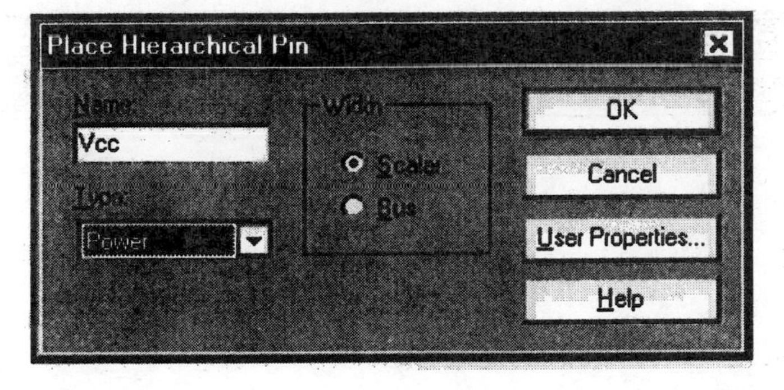


The ***Type*** you select determines the shape of the pin and also specifies connectivity requirements. Capture requires that some pin types have a connection, while other types can be left floating. Click the ***OK*** button. A pin will be placed inside the hierarchical block and move with the mouse:

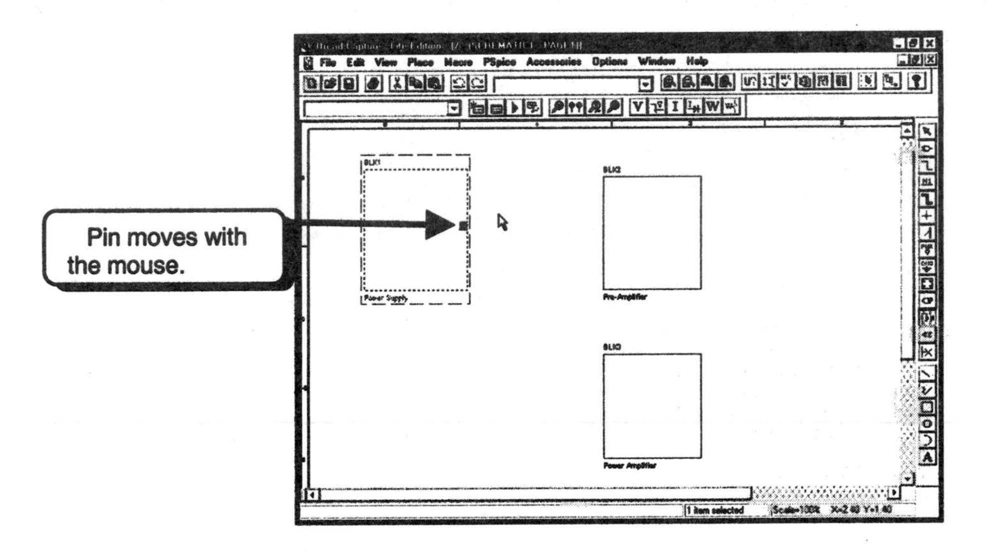


Move the pin to the location at which you want to place it and click the ***LEFT*** mouse button. The pin will be placed. When you move the mouse, you should notice that a second pin is attached to the mouse.

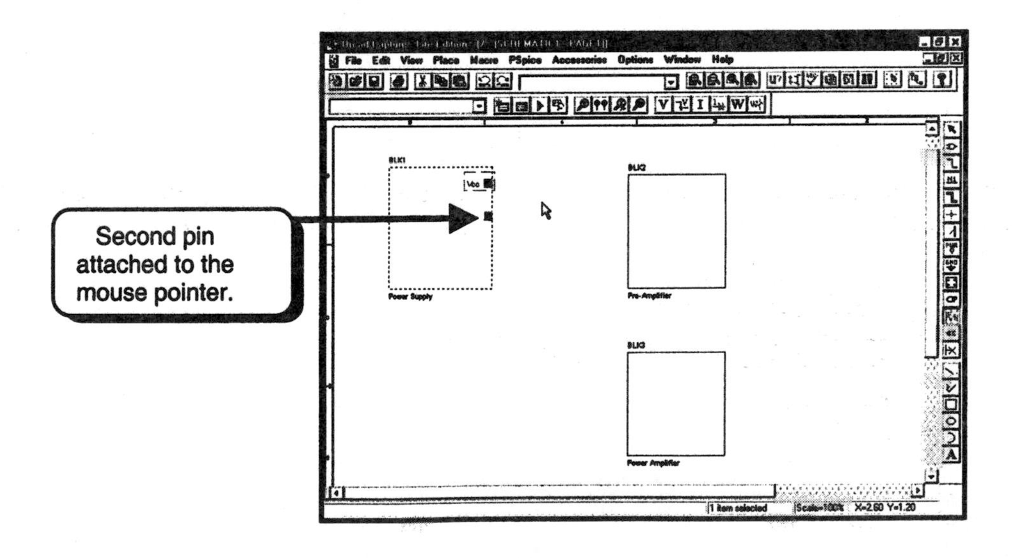


Press the **ESC** key to stop placing pins. Next, we want to place a pin with the same name and properties in BLK2. Click the ***LEFT*** mouse button at the center of hierarchical block BLK2 to select it and then select **Place** and then **Hierarchical Pin** from the Capture menus:

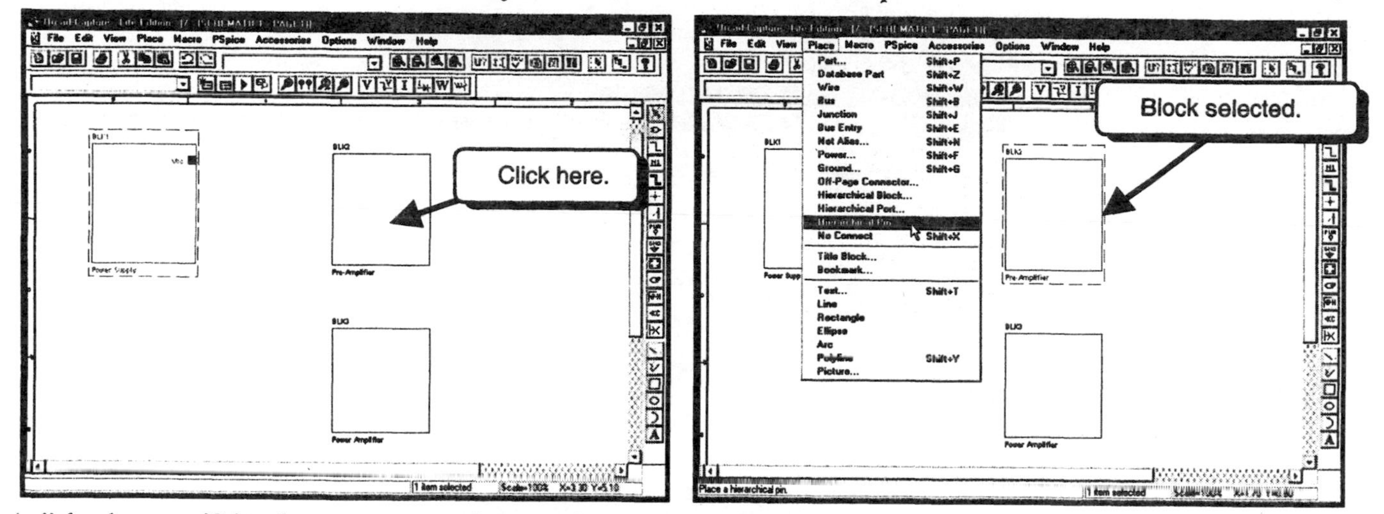


A dialog box specifying the same name and type as the previously placed pin will appear:

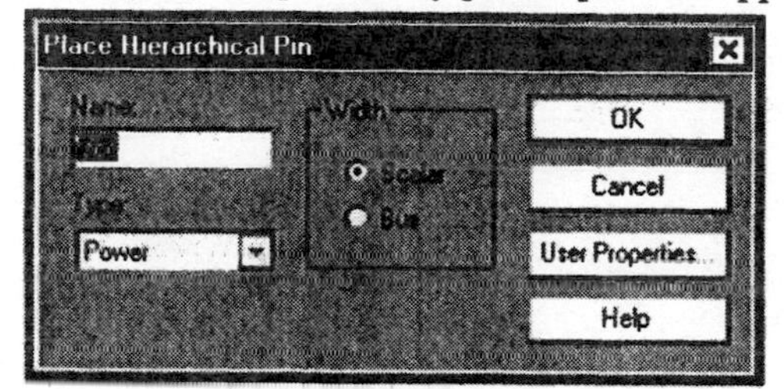


Click the ***OK*** button. Place the pin at a convenient location and then press the **ESC** key to stop placing pins:

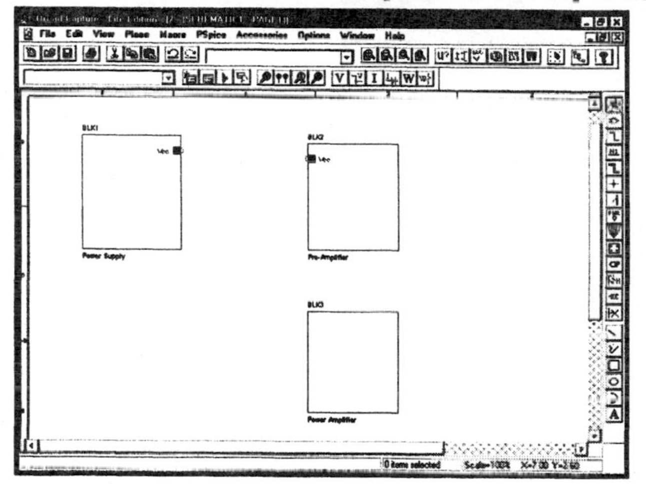

This method shows an easy way to place identical pins in different hierarchical blocks. Use the same method to place a Vcc power pin in BLK3. Remember, click the ***LEFT*** mouse button on the block to select the block and then select **Place** and then **Hierarchical Pin** to place a pin:

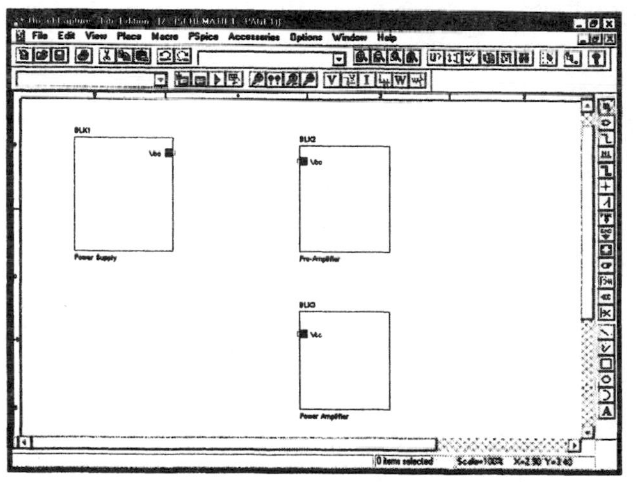

Each block will have two more power pins with the same name. Place power pins labeled "Ground" and "Vee" in all three blocks using the method just presented:

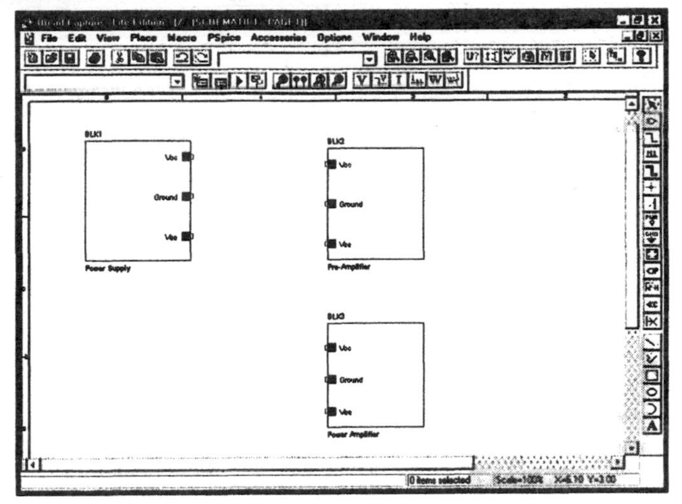

Next, hierarchical block BLK2 will have two output pins labeled as "Left" and "Right." Use the same method as before, except specify the ***Type*** to be ***Output*** rather than ***Power***:

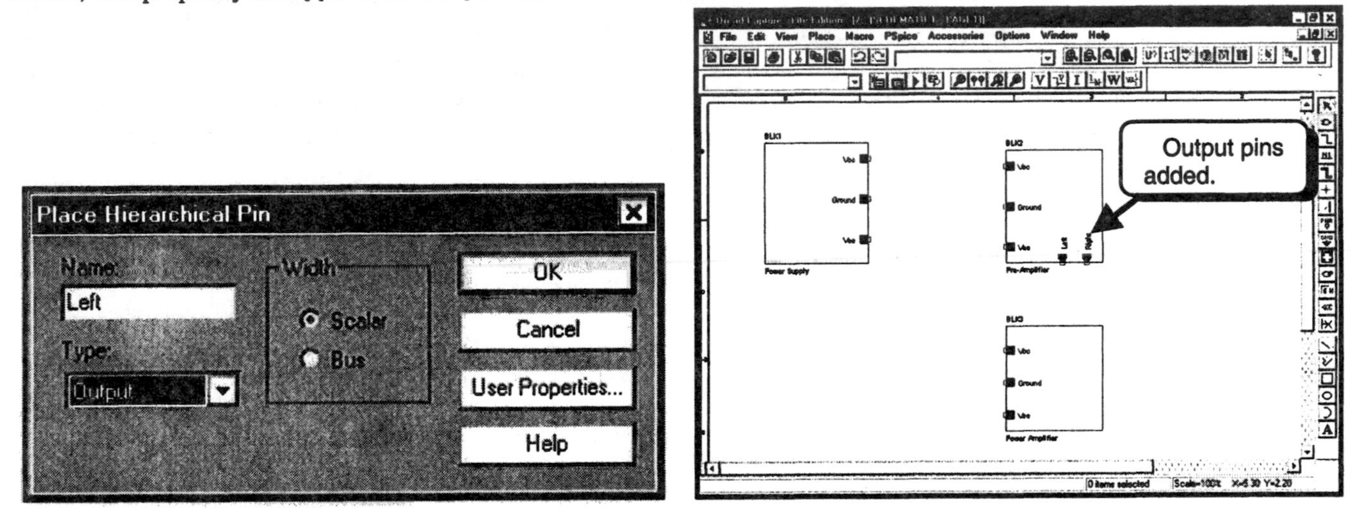


Notice that the output pins are shaped differently than the power pins. Next, we want to add two input pins to hierarchical block ***BLK3*** named "Left" and "Right." Use the same procedure as before, but specify the ***Type*** as ***Input***:

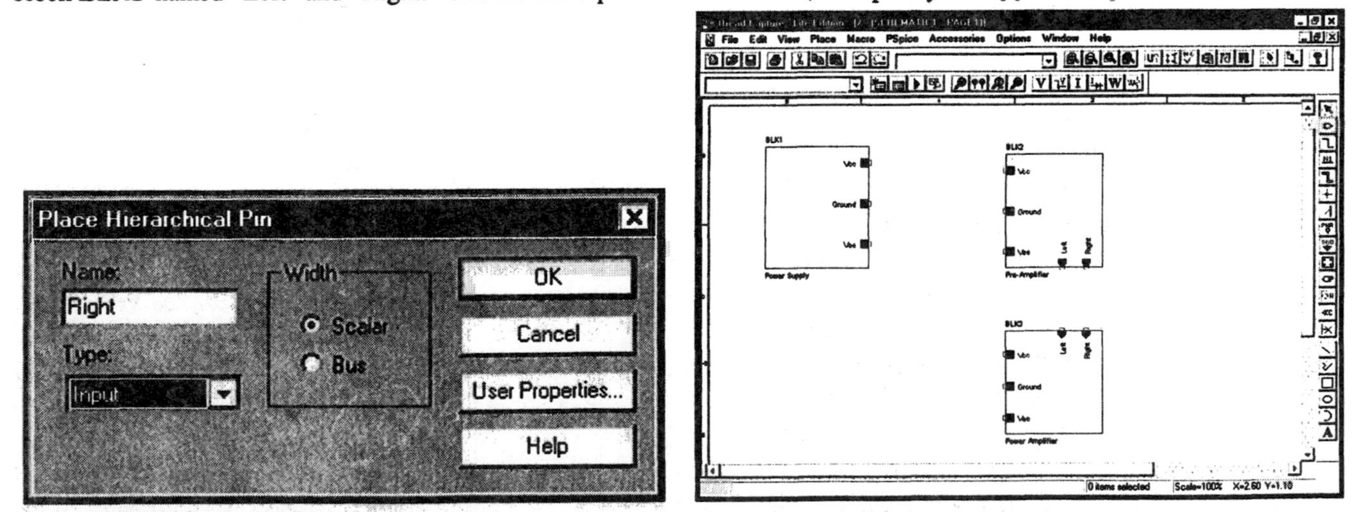


We have now labeled all of the pins. We will clean up the schematic a bit by moving the text labels. You can move any of the labels by dragging them. Move the text labels inside of the boxes so that we can easily wire the blocks together:

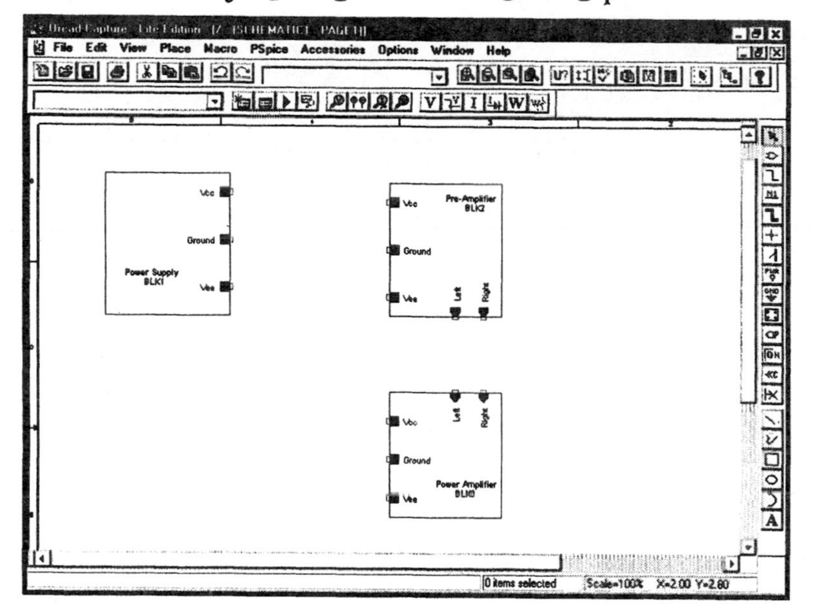


It is important to note that pin names do not specify net names or connectivity. In other words, commonly named pins are not automatically connected, and the net to which they are connected is not named the same as the pin names. Thus, as shown above, none of the blocks are connected. To connect the blocks, we need to wire them together. The schematic shown below shows the blocks wired together. Note that some wires are labeled with the text Vcc and Vee. This was done using net aliases as described in section 1.H. Note also the wire fragments used for Vee. These wires are all given the same net name (Vee) and thus are connected. This is a shorthand method for connecting components without drawing wires all over your schematic:

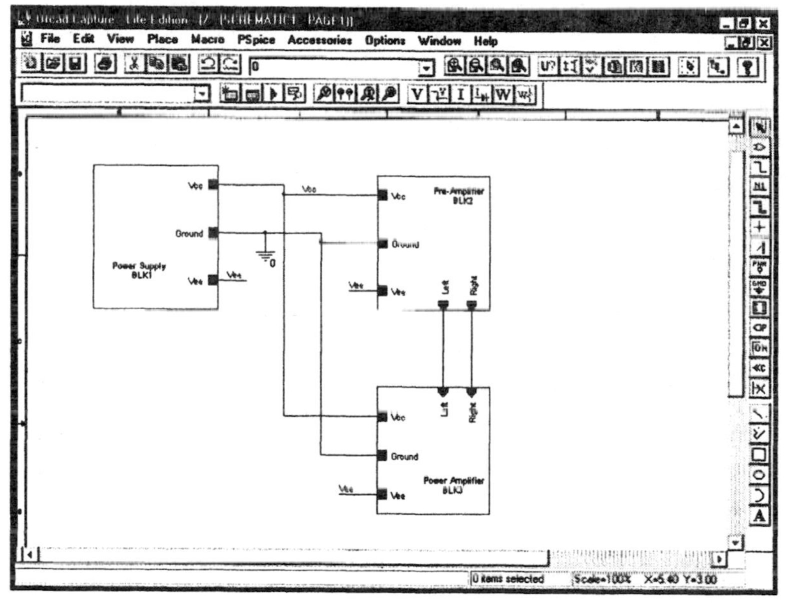


We have also placed a ground (part name 0/source) and connected it to the ground pins. This gives the ground net a label of **0** (node zero), which is required for PSpice simulations. If you are not going to simulate the circuit, you could use a different ground symbol, or you could label the net something other than **0**. As shown above, the net connecting the **Vcc** pins is named **Vcc**, the net connecting the **Vee** pins is named **Vee**, and the net connecting the **Ground** pins is named **0**. The **Left** pins are wired together and the **Right** pins are wired together, but we have not specified a name for the net. Capture will give these nodes a cryptic name such as N00127. If you want to be able to easily identify these nets, you should label them with a net alias the same way we named **Vcc** and **Vee**.

We are now finished with our top level block diagram and can now edit each of the blocks to create the detailed circuit for the block. We will start with BLK1. Click the ***LEFT*** mouse button on the block to select it, and then click the ***RIGHT*** mouse button on the block to display a menu:

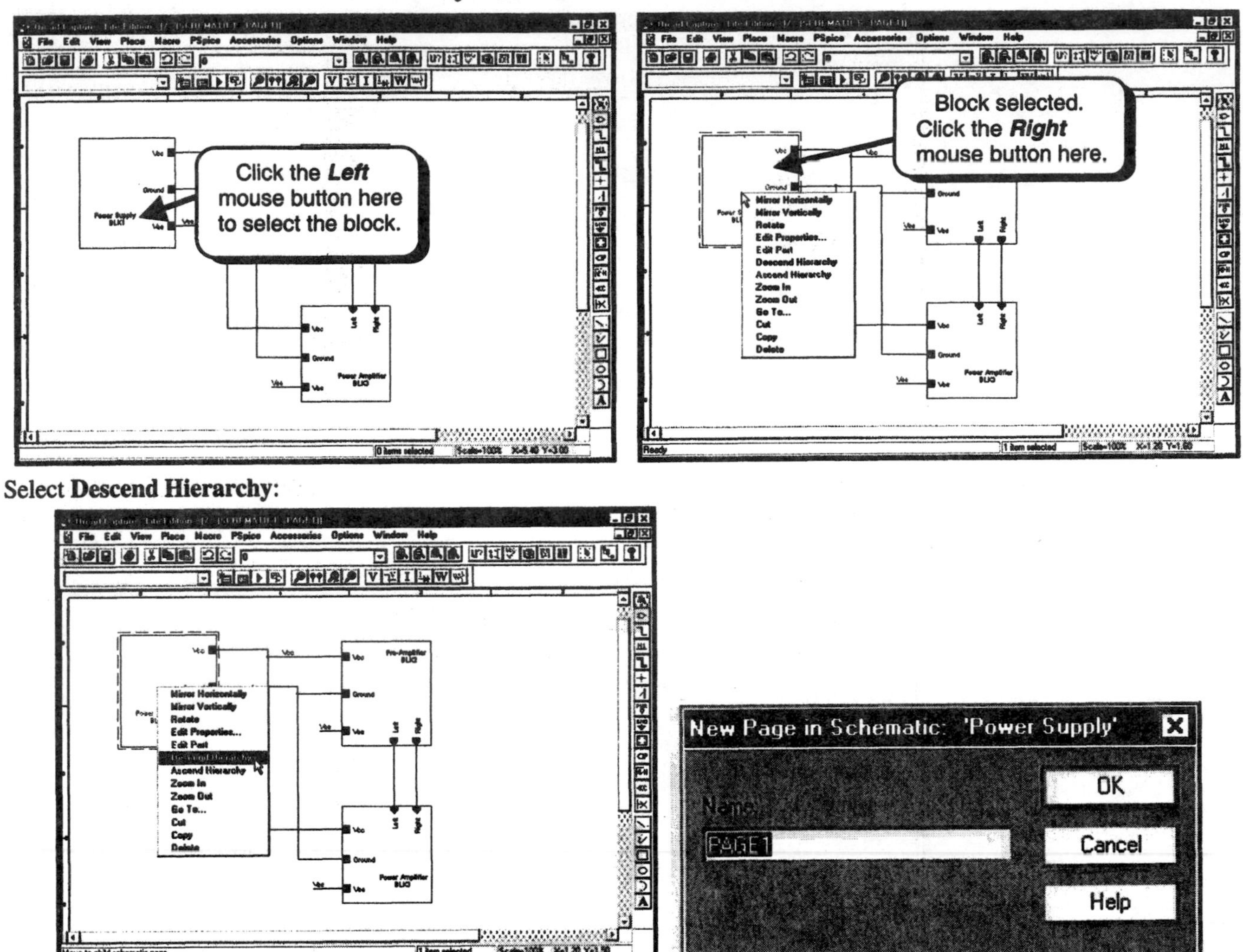


Select **Descend Hierarchy**:

The first time we descend the hierarchy we create a new schematic page for the block. The dialog box above is asking us what we want to name that page. You can change the name if you wish. Click the ***OK*** button to create the new schematic page:

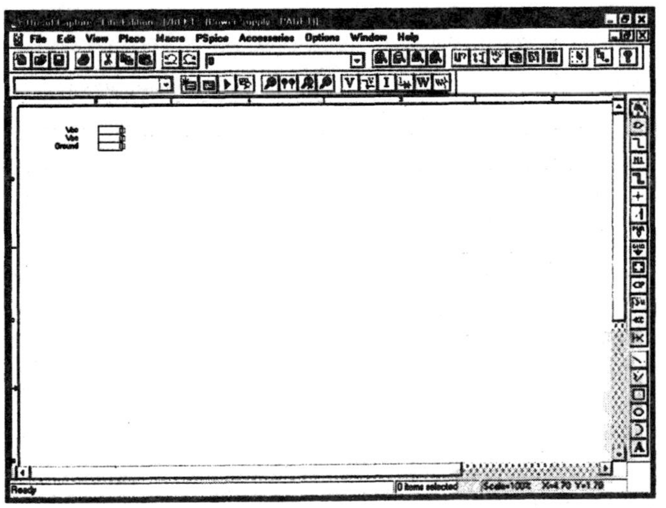

We see that a new schematic page is created and that the page is empty except for the three ports that correspond to the pins we placed in the hierarchical block. If we zoom* in around the ports, we see that they are named the same as the hierarchical pins we placed in the hierarchical block:

* To zoom, place the mouse pointer in the area in which you want to zoom, and type the letter **I**.

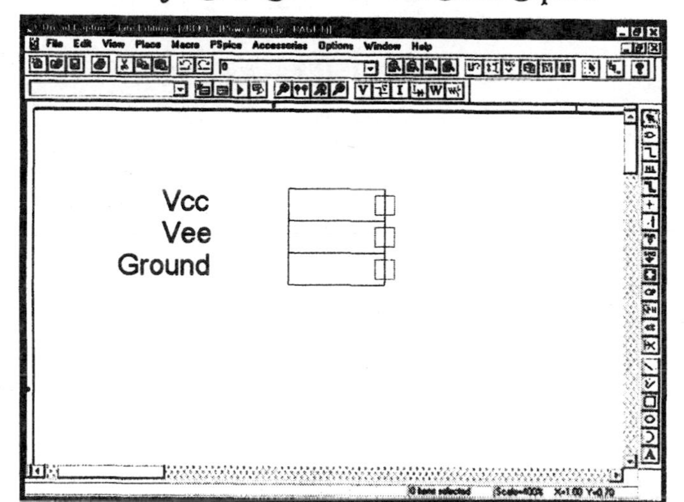


We will now create a schematic for this block. The point of this exercise is to create a hierarchical schematic, not examine a circuit for a DC supply. Thus, we will create a simple circuit to accomplish this task. You can create a more complicated DC power supply circuit if you wish:

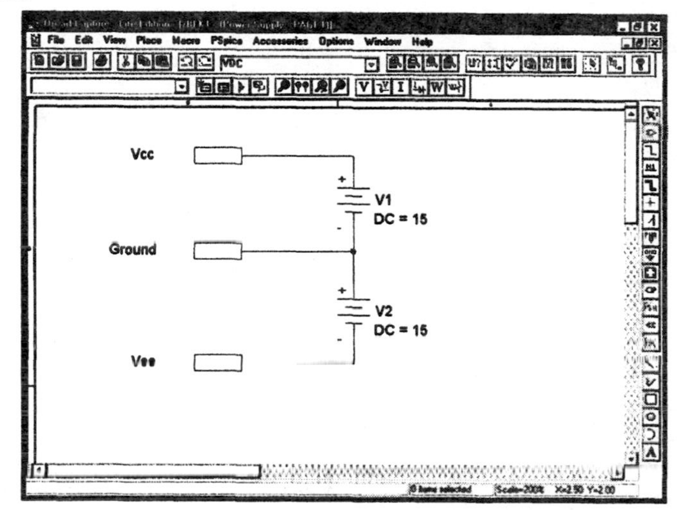


When you are finished with your circuit, select **File** and then **Save** from the Capture menus. This will save the changes to this page only:

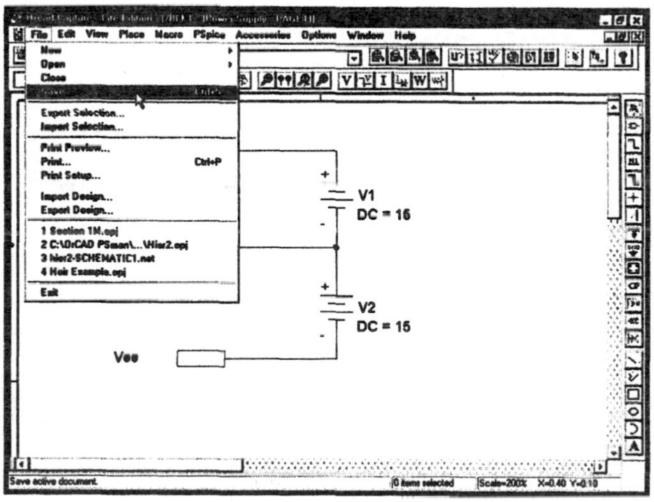

Next, we need to ascend the hierarchy and return to the top level block diagram. Click the ***RIGHT*** mouse button on the screen somewhere and then select **Ascend Hierarchy**:

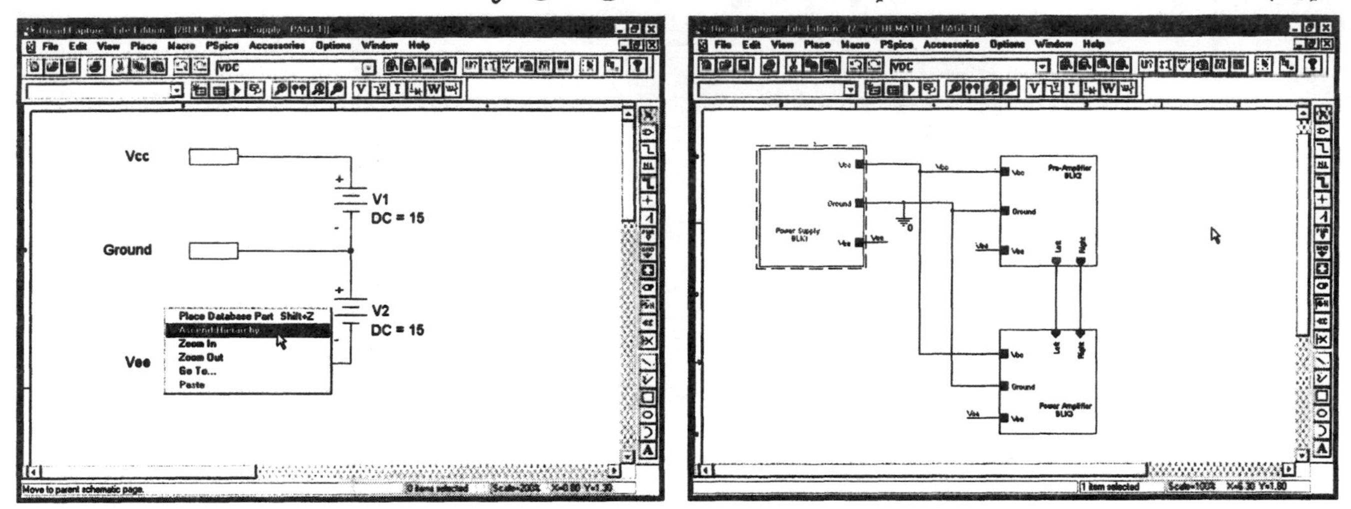


Next, we want to create the schematic for the ***Pre-Amplifier*** block. Click the ***LEFT*** mouse button on the block to select it:

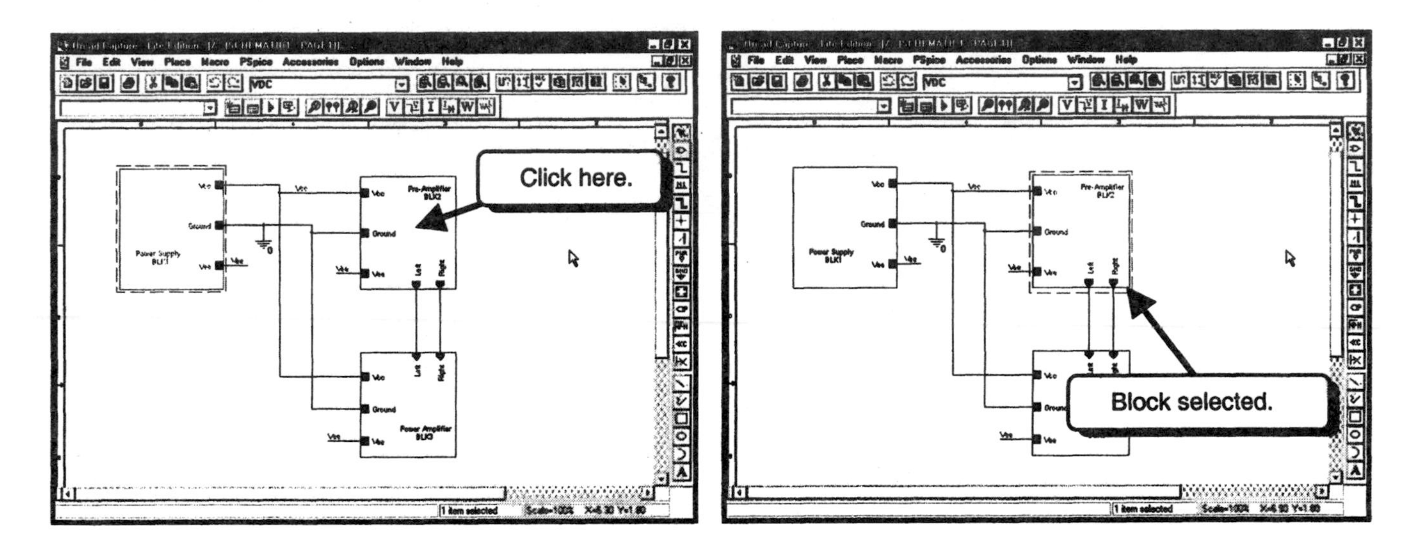


Click the ***RIGHT*** mouse button on ***BLK2*** and then select **Descend Hierarchy** from the menu to create a new schematic page for ***BLK2***:

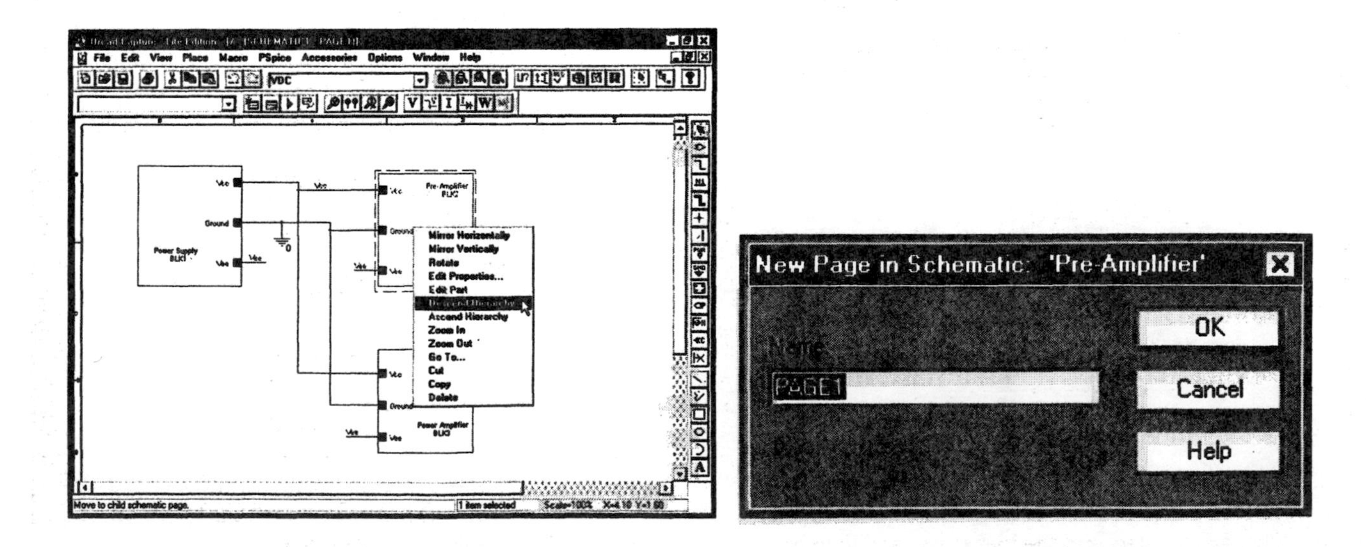


Click the ***OK*** button to create a new schematic page:

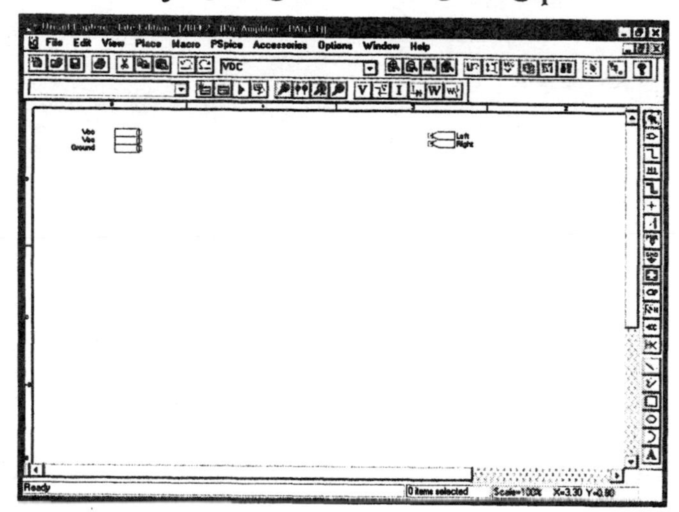

Notice that the page is empty except for the five hierarchical ports that correspond to the five hierarchical pins we placed in the Pre-Amplifier block. We can now create a circuit for the pre-amplifier. Once again, we will create a simple circuit since we are illustrating the hierarchical tools available in Capture, not the design of an amplifier. My completed circuit is shown below:

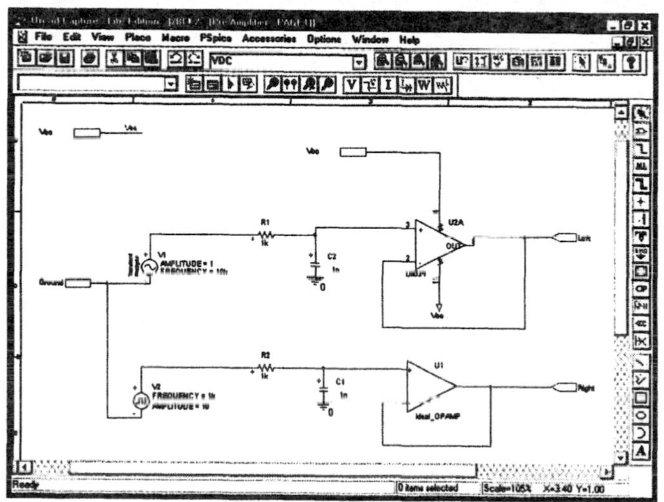

When you have finished drawing your circuit, select **File** and then **Save** to save your circuit. Next, click the ***RIGHT*** mouse button on the schematic and select **Ascend Hierarchy** to return to the top level schematic:

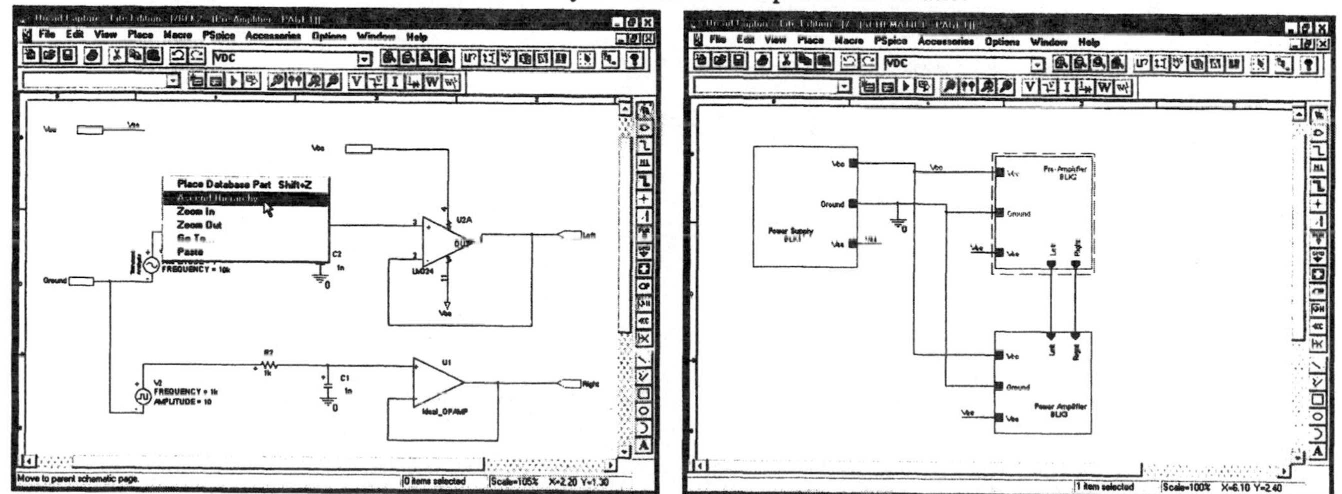

Next, select BLK3, descend the hierarchy, and then create the schematic below for the amplifier stage:

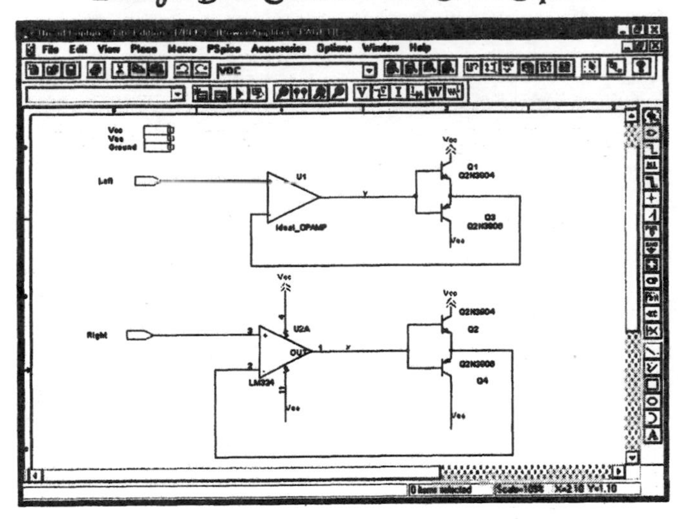


As an example of showing a hierarchical block within a hierarchical block, we will add a hierarchical block to this schematic. To place a block, select **Place** and then **Hierarchical Block**. To place a pin, select **Place** and then **Hierarchical Pin**. This block has three pins called ***Out-L***, ***Out-R***, and ***Ground***. The reference is ***BLK4***, and the name of the block is ***Load***. The block is shown below. Use the techniques covered earlier to create this block:

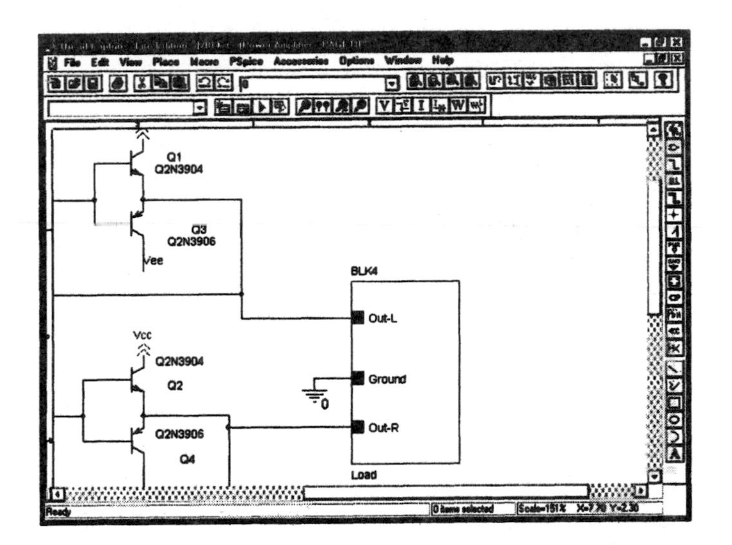


Next, we will create the schematic inside this block. Select the block and then descend the hierarchy. You will create the new schematic page below:

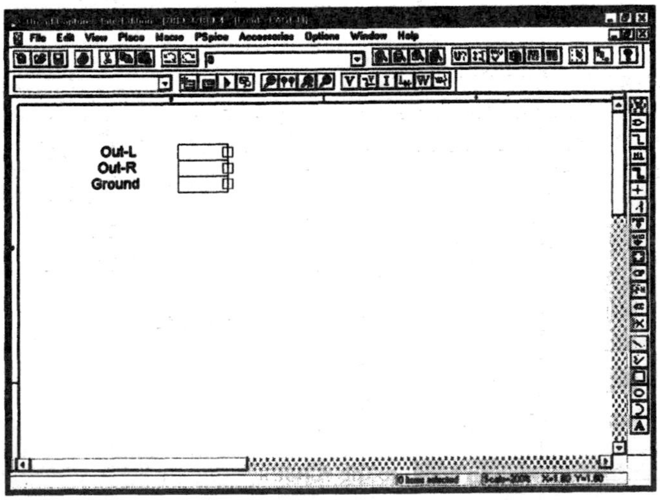

We will now create the schematic for the load. I will use the simple circuit shown below:

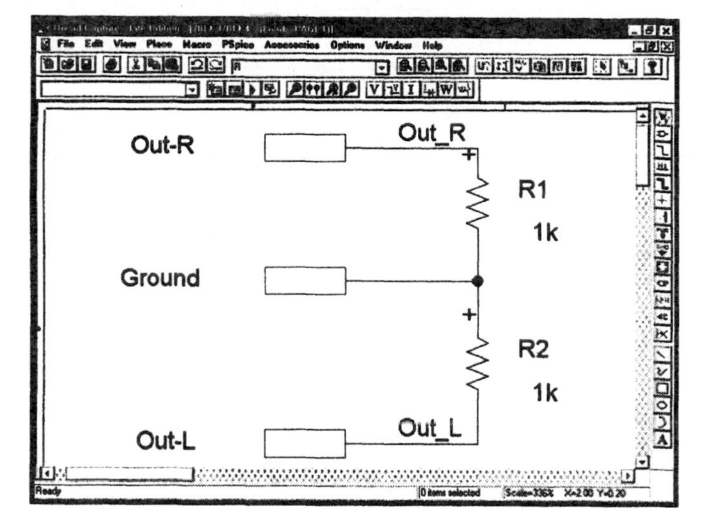


We are now done with the schematic. We will set up the transient analysis shown below. Once again, this is an advanced topic, and we assume that you are familiar with Capture, plotting traces with Probe, and running simulations in PSpice.

Simulation Settings - test

General | Analysis | Include Files | Libraries | Stimulus | Options | Data Collection | Probe Window

Analysis type:
Time Domain (Transient)

Options:
Monte Carlo/Worst Case
Parametric Sweep
Temperature (Sweep)
Save Bias Point
Load Bias Point

Run to time: 5m seconds (TSTOP)

Start saving data after: 0 seconds

Transient options
Maximum step size: 10u seconds
Skip the initial transient bias point calculation (SKIPBP)

Output File Options...

OK | Cancel | Apply | Help

Click the ***OK*** button when you have finished setting up the simulation settings.

Next, we will look at the netlist created by this hierarchical design. Select **PSpice** and then **Create Netlist** from the Capture menus to create the netlist, and then select **PSpice** and then **View Netlist** to display the netlist:

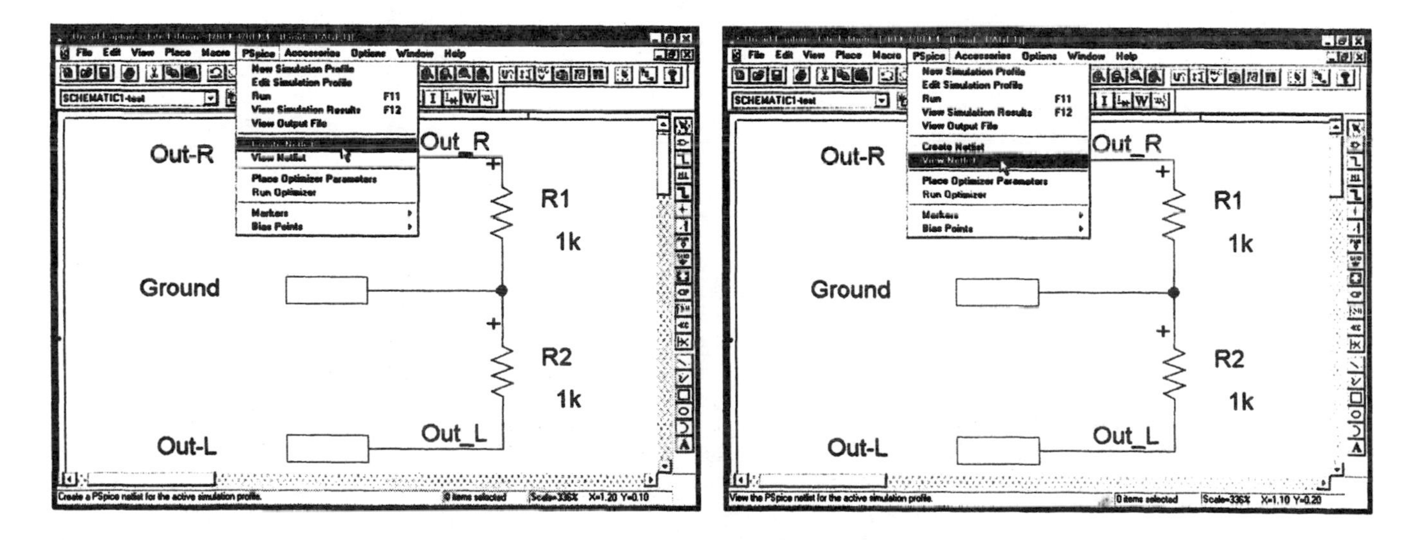

```
Orcad Capture - Lite Edition - [section 1m-SCHEMATIC1.net]
File  Edit  Options  Window  Help
SCHEMATIC1-test

 1: * source SECTION 1M
 2: V_BLK1_V1          VCC 0  DC 15
 3: V_BLK1_V2          0 VEE  DC 15
 4: V_BLK2_V1          BLK2_N00319 0
 5: +SIN 0 1 10k 0 0 0
 6: V_BLK2_V2          BLK2_N00285 0 PWL
 7: + TIME_SCALE_FACTOR=(1/1k)
 8: + VALUE_SCALE_FACTOR=10
 9: +repeat forever
10: +(0,0) (0.001,1) (0.499,1) (0.501,-1)(0.999,-1)(1,0)
11: +endrepeat
12: X_BLK2_U1          BLK2_N00235 N00258  N00258 Ideal_OPAMP
13: C_BLK2_C1          BLK2_N00235 0 1n
14: C_BLK2_C2          BLK2_N00179 0 1n
15: X_BLK2_U2A          BLK2_N00179 N00247 VCC VEE N00247 LM324
16: R_BLK2_R1          BLK2_N00319 BLK2_N00179 1k
17: R_BLK2_R2          BLK2_N00285 BLK2_N00235 1k
18: X_BLK3_U1          N00247 BLK3_N00479  BLK3_Y Ideal_OPAMP
19: Q_BLK3_Q1          VCC BLK3_Y BLK3_N00479 Q2N3904
20: X_BLK3_U2A          N00258 BLK3_N00347 VCC VEE BLK3_X LM324
21: R_BLK3_BLK4_R1          BLK3_N00347 0 1k
22: R_BLK3_BLK4_R2          0 BLK3_N00479 1k
23: Q_BLK3_Q2          VCC BLK3_X BLK3_N00347 Q2N3904
24: Q_BLK3_Q3          VEE BLK3_Y BLK3_N00479 Q2N3906
25: Q_BLK3_Q4          VEE BLK3_X BLK3_N00347 Q2N3906
26:

Ready                                   Line 15, Col 88   INS
```

From this netlist we can see the naming convention. Nodes that were named in our top level block diagram are used throughout the circuit. We see the nodes Vcc, Vee, and 0 throughout the netlist. Nodes inside a block are given a name with the block prefix, such as BLK2_N00319, BLK3_X, or BLK3_N00347. Circuit elements inside a block use the block as part of their name. Examples are V_BLK1_V1, C_BLK2_C1, and Q_BLK3_Q2. Our last block, BLK4, is a block within BLK3. Components inside this block use both blocks to name components, such as R_BLK3_BLK4_R1. We did not name any nodes inside BLK4, so this listing does not give us an example of a node name within BLK4. However, we would assume that it uses a similar naming convention as used by the circuit elements.

Select **File** and then **Close** to close the netlist:

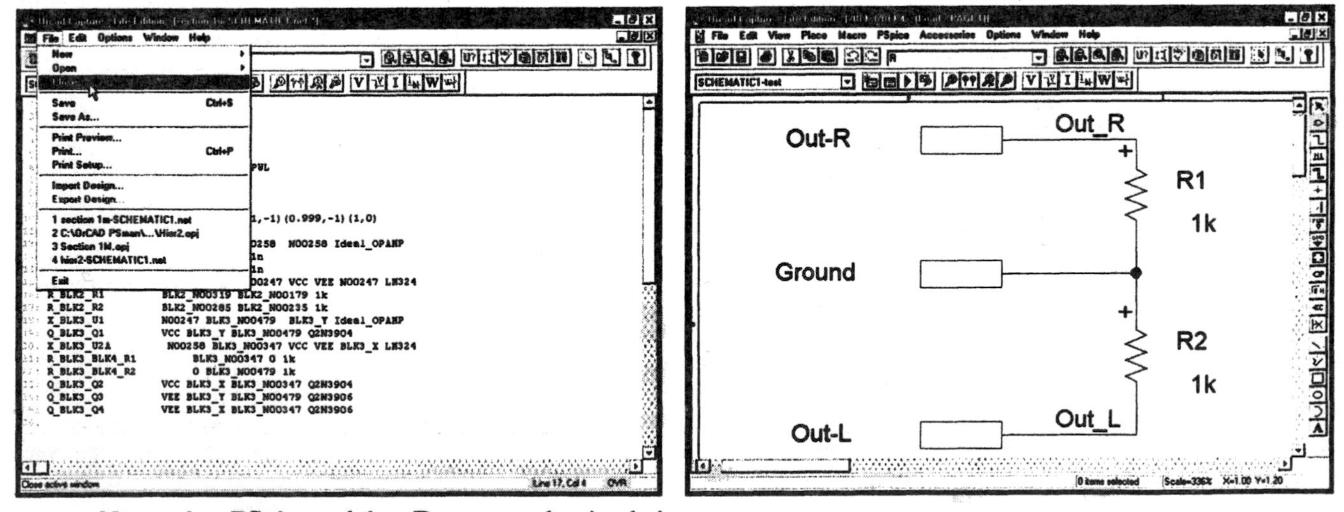


Next, select **PSpice** and then **Run** to run the simulation:

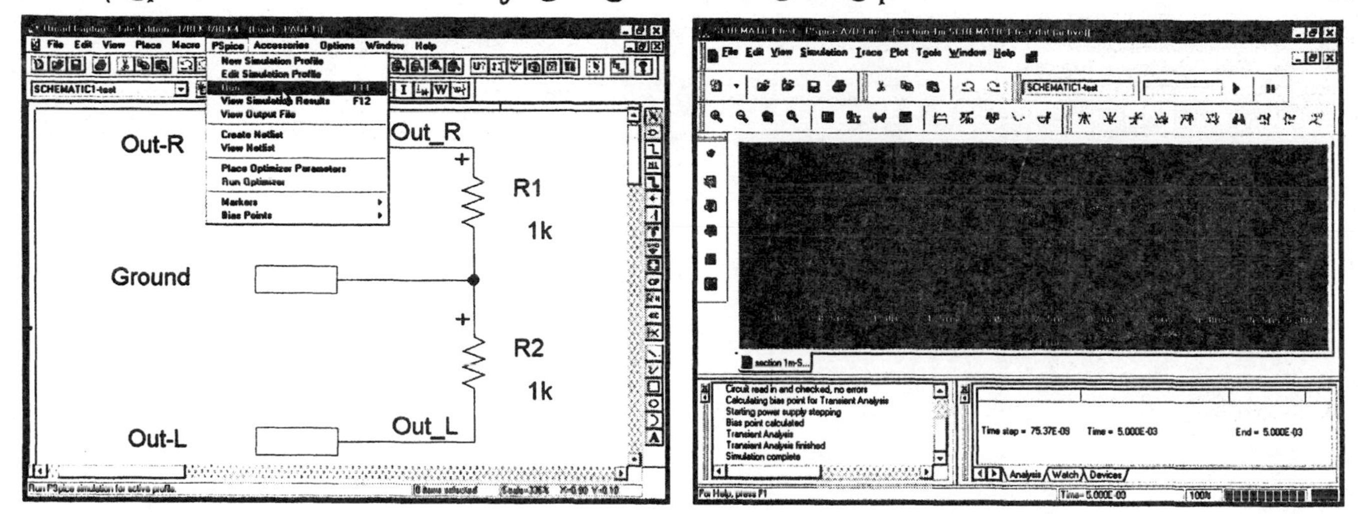


From the netlist we saw previously, the node names in this hierarchical circuit are rather cryptic. The easiest way to add a trace of a node that is inside a hierarchical block is to use the markers. See section 2.C for instructions on using markers. Switch back to the Capture window and add markers shown:

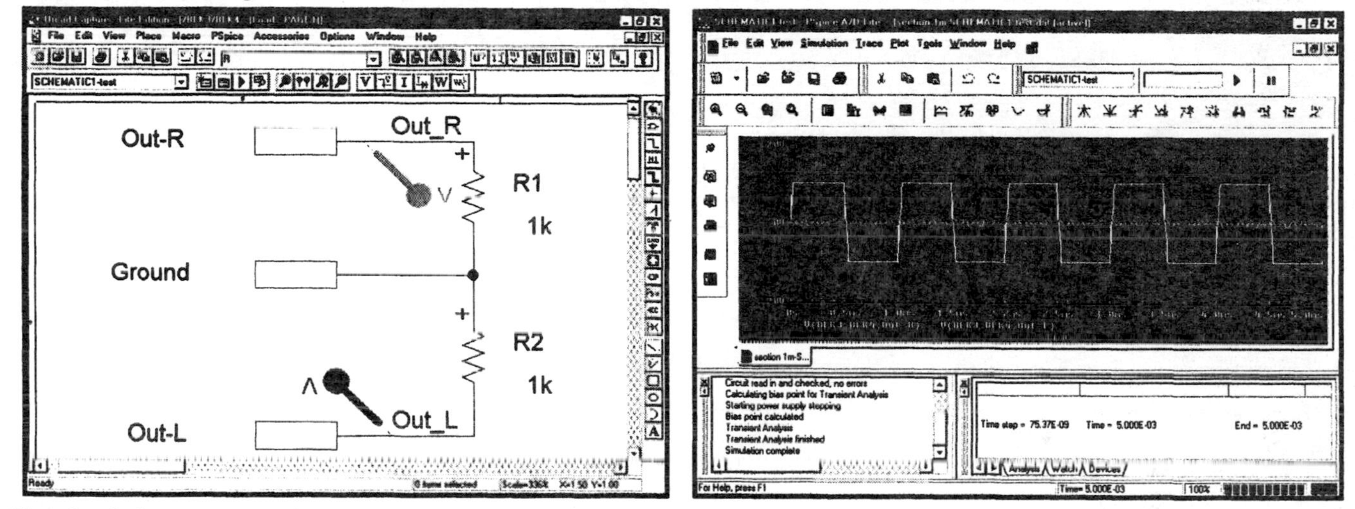


The simulation appears to have run successfully.

The last thing we will do is look at the project tree for this hierarchical design. Switch back to Capture and select the ***Project manager*** button .

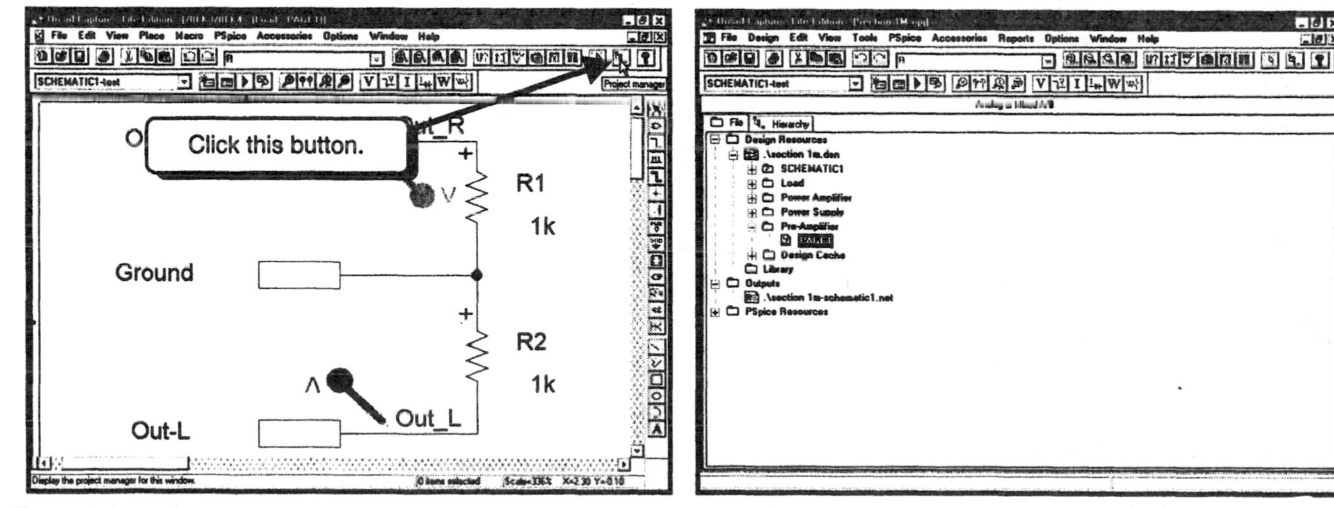


Expand the entire tree:

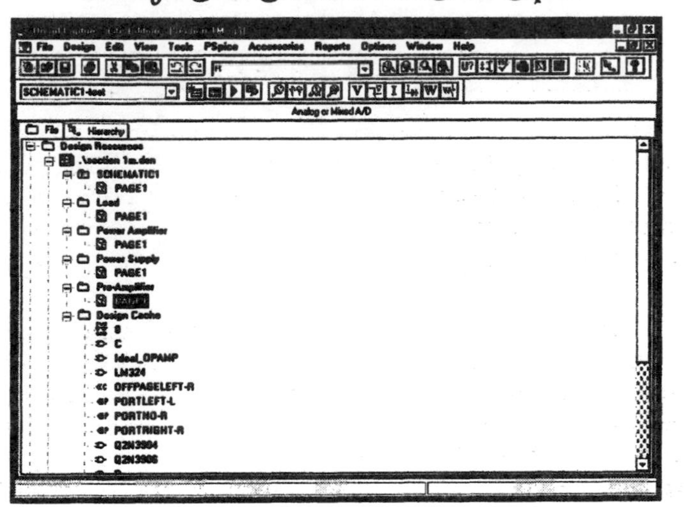

This view shows us all of the schematic pages in our circuit, but it does not show the hierarchical structure of the design. To see this structure, select the ***Hierarchy*** tab:

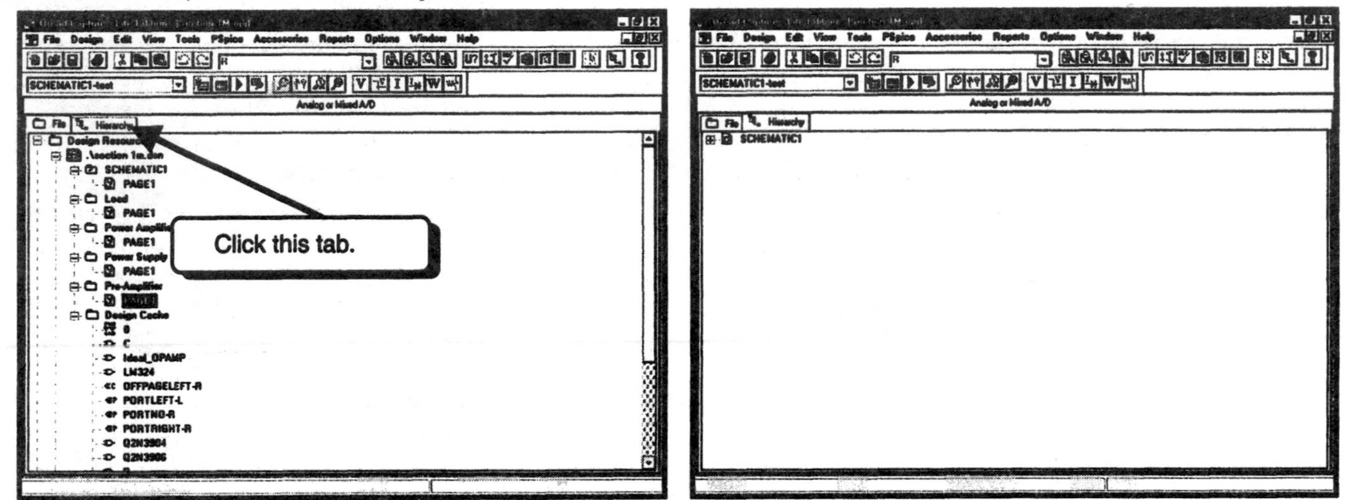


Expand the entire tree to see the structure:

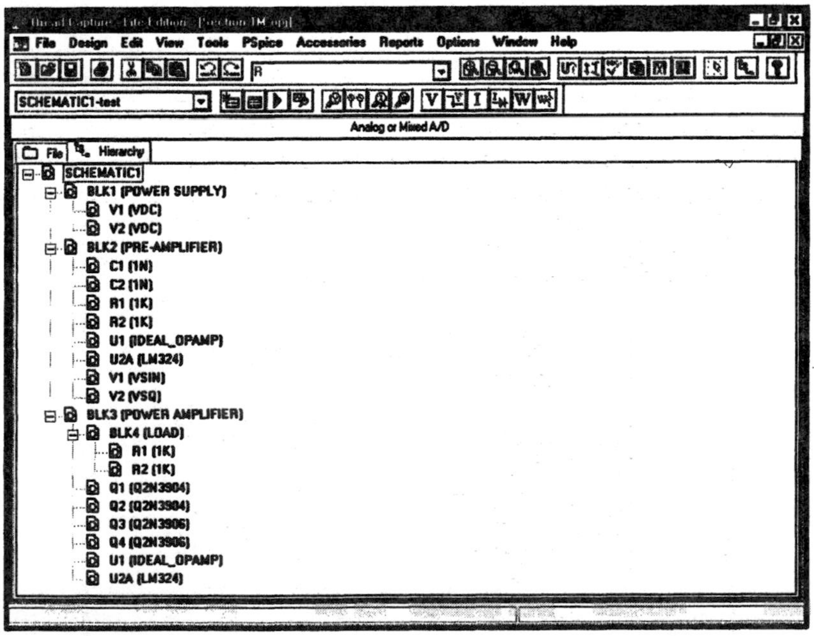

This view shows us the hierarchical structure as well as the components inside each hierarchical block. We can see that blocks ***BLK1***, ***BLK2***, and ***BLK3*** are inside ***SCHEMATIC1***, and that ***BLK4*** is inside ***BLK3***.

# PART 2
# Introduction to Probe

Probe is a program that will display the results obtained from PSpice graphically. We will be using Probe extensively throughout this manual to display the results of simulations. Various aspects of Probe are discussed in sections throughout this manual. However, if you pick specific examples you may miss those showing how to use some of the tools provided by Probe. The result may be that a section in the manual you are currently using refers to a tool in Probe that was discussed in a section that you did not cover. To avoid this problem, we will review how to use the most frequently used tools.

To demonstrate Probe, we will simulate a power supply circuit with a Transient Analysis. Although this may be too complicated a circuit for beginning students, the methods discussed in Probe can be used with any analysis. We are using this simulation because it provides many interesting waveforms. Open the project named Section 2.opj. This file is located on the CD-ROM that accompanies this text and may also be loaded on your hard drive. During installation of the library files, the circuit files are loaded into a subdirectory named "Book Circuits" in the Orcad Lite installation directory. If you chose the default installation options, this directory will be named C:\Program Files\Orcadlite\Book Circuits. You can use the files in this directory, or the files located on the CD-ROM. **Note: If you attempt to run this example directly from the CD-ROM, or open the file from the CD-ROM and then use the Save As command to save the file to the hard drive, you will get a write error.** To use the circuit from the CD-ROM, use the Windows Explorer to copy the file from the CD-ROM to your hard drive. Then use the Windows Explorer to change the properties of the file to not read-only. To change the properties with the Windows Explorer, run the Explorer and select the file. Next, select **File** and then **Properties** from the Explorer menus. You can then remove the checkmark from the box next to the text Read-Only. Note that the read-only property may or may not be selected when you first look at the properties and may not need to be changed. For more information on copying a single file from the CD-ROM, and then opening the file, see Section A.3 on page 604.

For this example, we will assume that you are opening the files that were copied to your hard drive during the installation. Select **File, Open**, and then **Project** from the Capture menu bar:

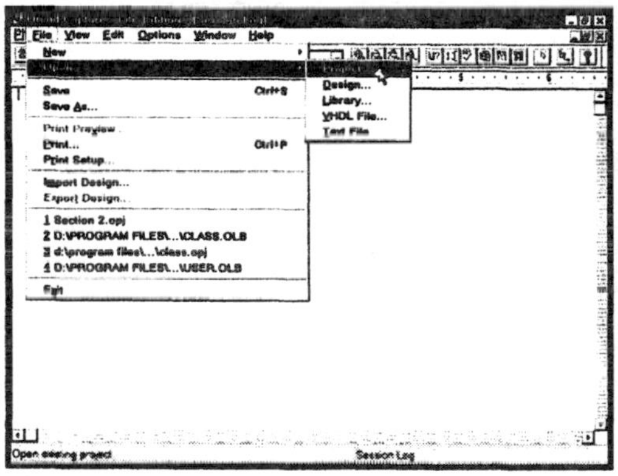

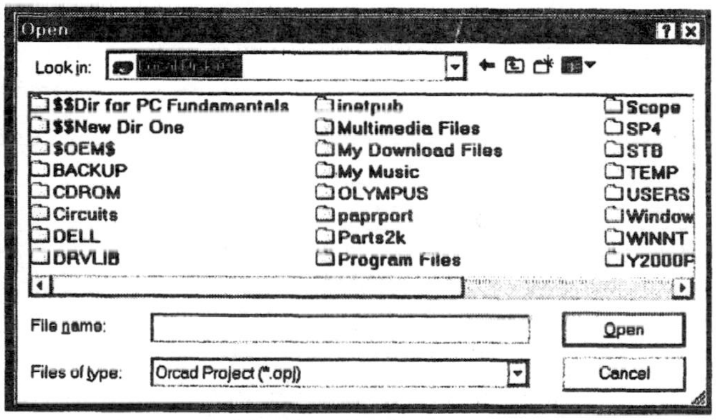

Select **Open**. We will assume that you installed Capture on drive C: in directory \Program Files\OrcadLite. (This was the default installation location.) For this installation directory the example circuits will be located in directory C:\Program Files\OrcadLite\Book Circuits. First, select the directory named C:\Program Files\OrcadLite\Book Circuits:

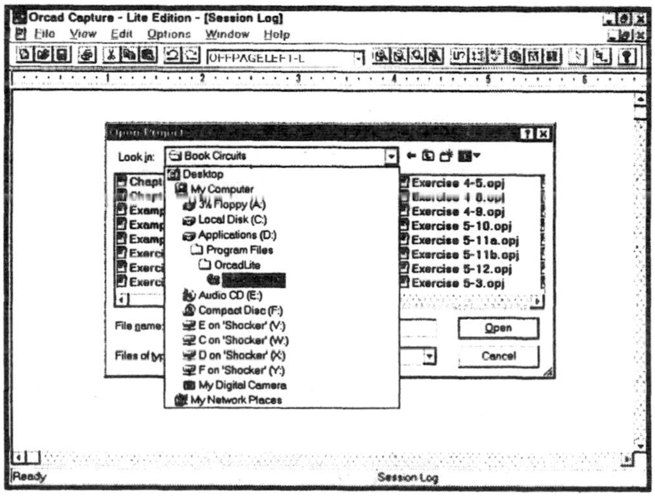

After you select the directory ***Book Circuits***, a dialog window will display the files in that directory:

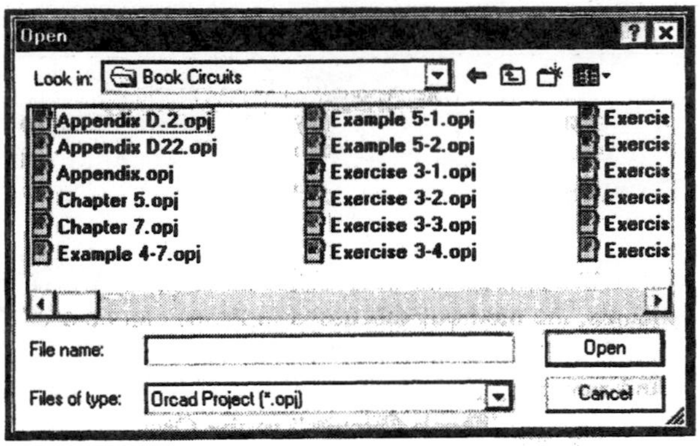

We would like to open the file named Section 2.opj so click on the text ***Section 2.opj*** as shown:

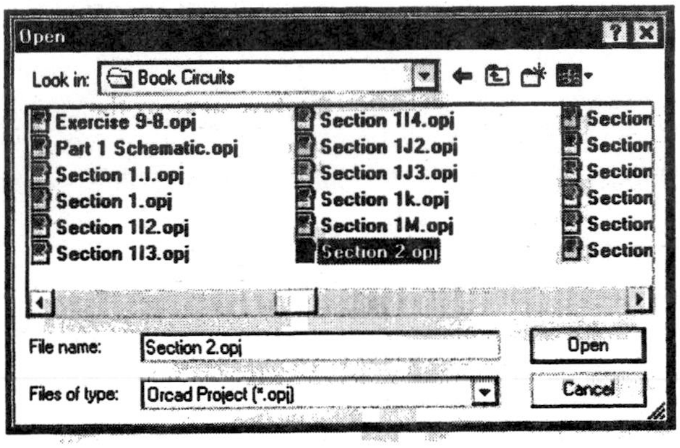

Click the ***Open*** button to open the file:

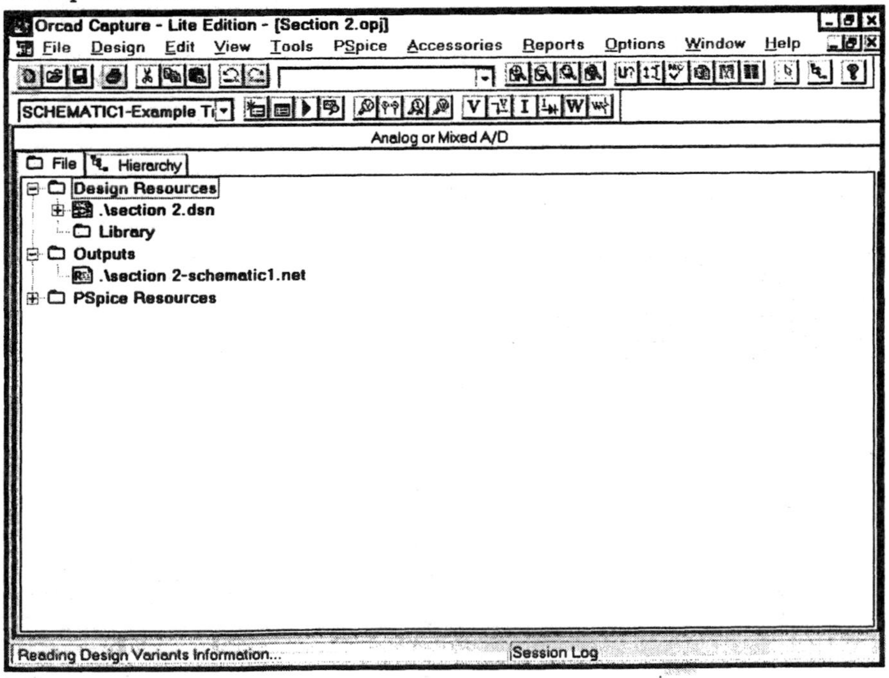

Next, we need to open the schematic page. Click the *LEFT* mouse button on the ⊞ as shown below to expand the tree:

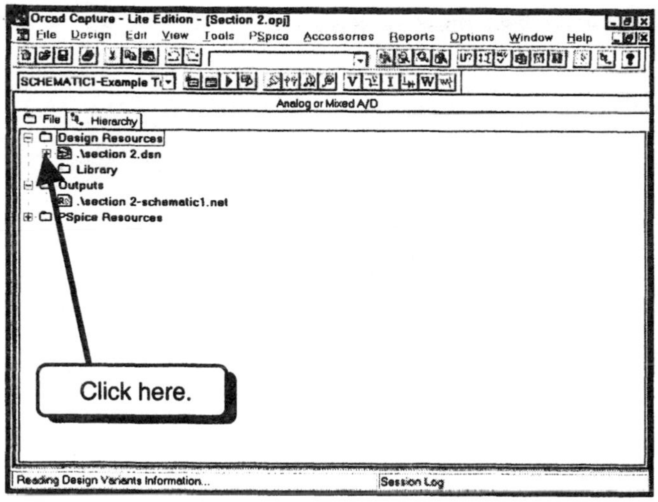

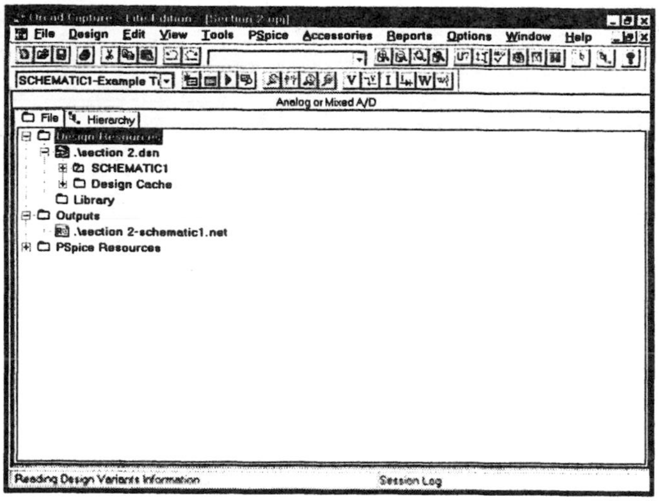

Orcad Capture - Lite Edition - [Section 2.opj]
File Design Edit View Tools PSpice Accessories Reports Options Window Help
SCHEMATIC1-Example T
Analog or Mixed A/D
File Hierarchy
Design Resources
.\section 2.dsn
Library
Outputs
.\section 2-schematic1.net
PSpice Resources
Click here.
Reading Design Variants Information... Session Log

File Design Edit View Tools PSpice Accessories Reports Options Window Help
SCHEMATIC1-Example T
Analog or Mixed A/D
File Hierarchy
Design Resources
.\section 2.dsn
SCHEMATIC1
Design Cache
Library
Outputs
.\section 2-schematic1.net
PSpice Resources
Reading Design Variants Information Session Log

Next, we need to expand the tree to list the schematic pages of the project. Click the ***LEFT*** mouse button on the ⊞ as shown below:

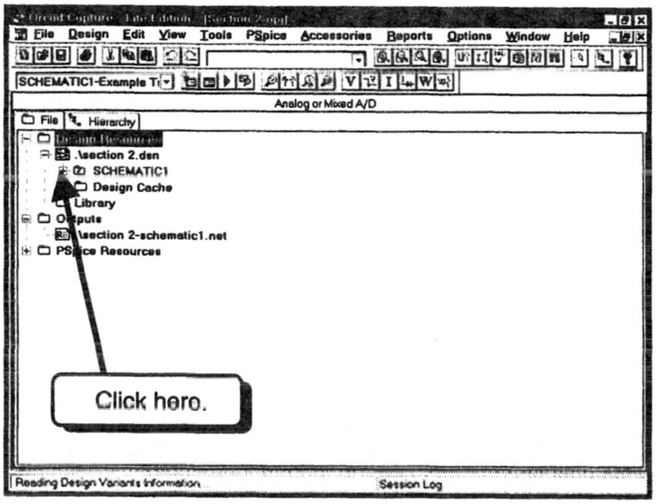

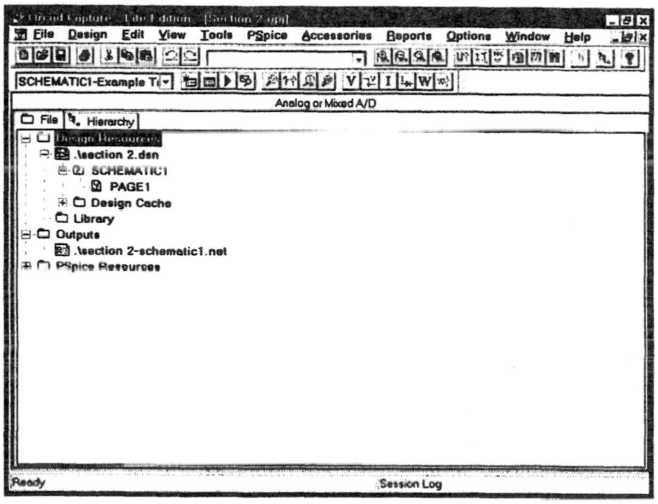

File Design Edit View Tools PSpice Accessories Reports Options Window Help
SCHEMATIC1-Example T
Analog or Mixed A/D
File Hierarchy
Design Resources
.\section 2.dsn
SCHEMATIC1
Design Cache
Library
Outputs
.\section 2-schematic1.net
PSpice Resources
Click here.
Reading Design Variants Information Session Log

File Design Edit View Tools PSpice Accessories Reports Options Window Help
SCHEMATIC1-Example T
Analog or Mixed A/D
File Hierarchy
Design Resources
.\section 2.dsn
SCHEMATIC1
PAGE1
Design Cache
Library
Outputs
.\section 2-schematic1.net
PSpice Resources
Ready Session Log

This project has one schematic page. To open the page, double-click the ***LEFT*** mouse button on the text ***PAGE1***.

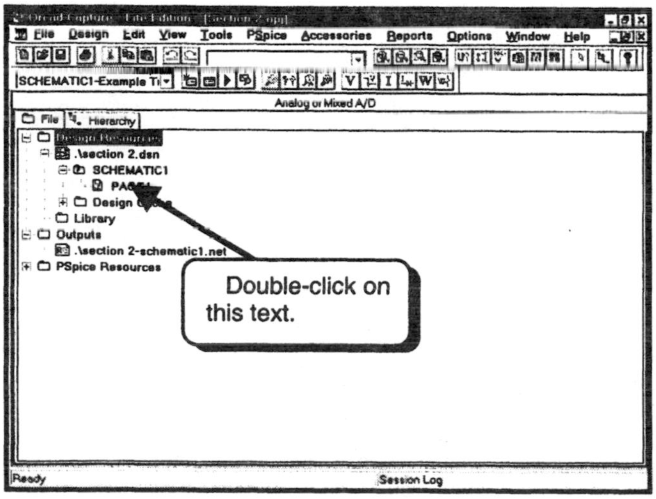

File Design Edit View Tools PSpice Accessories Reports Options Window Help
SCHEMATIC1-Example T
Analog or Mixed A/D
File Hierarchy
Design Resources
.\section 2.dsn
SCHEMATIC1
PAGE1
Design Cache
Library
Outputs
.\section 2-schematic1.net
PSpice Resources
Double-click on this text.
Ready Session Log

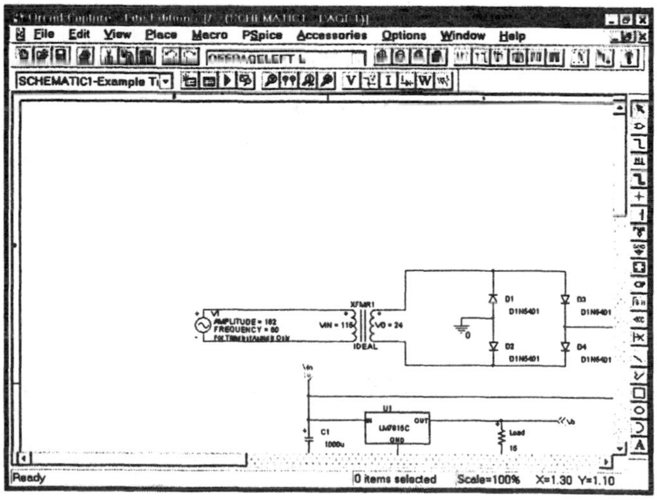

File Edit View Place Macro PSpice Accessories Options Window Help
SCHEMATIC1-Example T
XFMR1
D1 D1N5401
D3 D1N5401
D2 D1N5401
D4 D1N5401
IDEAL
U1
C1 1000u
Load 16
Ready 0 items selected Scale=100% X=1.30 Y=1.10

Use the scroll bars and zoom facilities to fill the screen with the circuit:

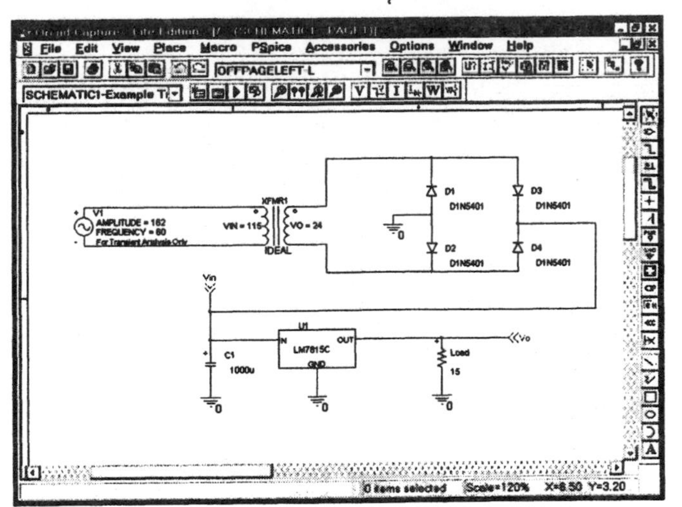


Notice that the file has two nodes labeled Vo and Vin.

To simulate the circuit, select **PSpice** and then **Run** from the Capture menus:

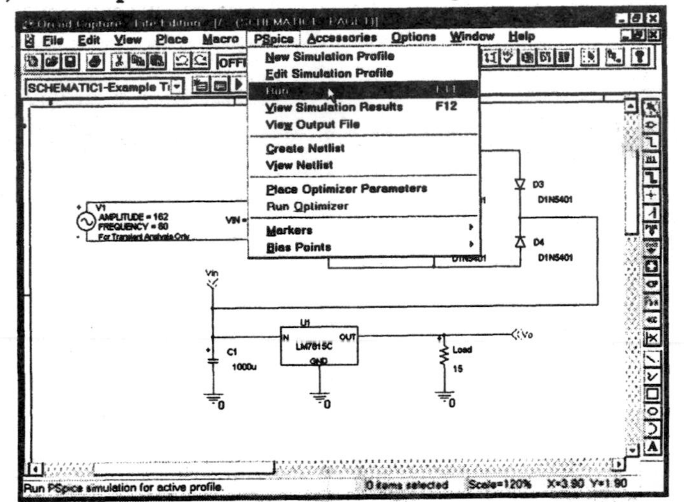


The PSpice A/D Lite window will open and the simulation will begin. You will see one of the two screens below:

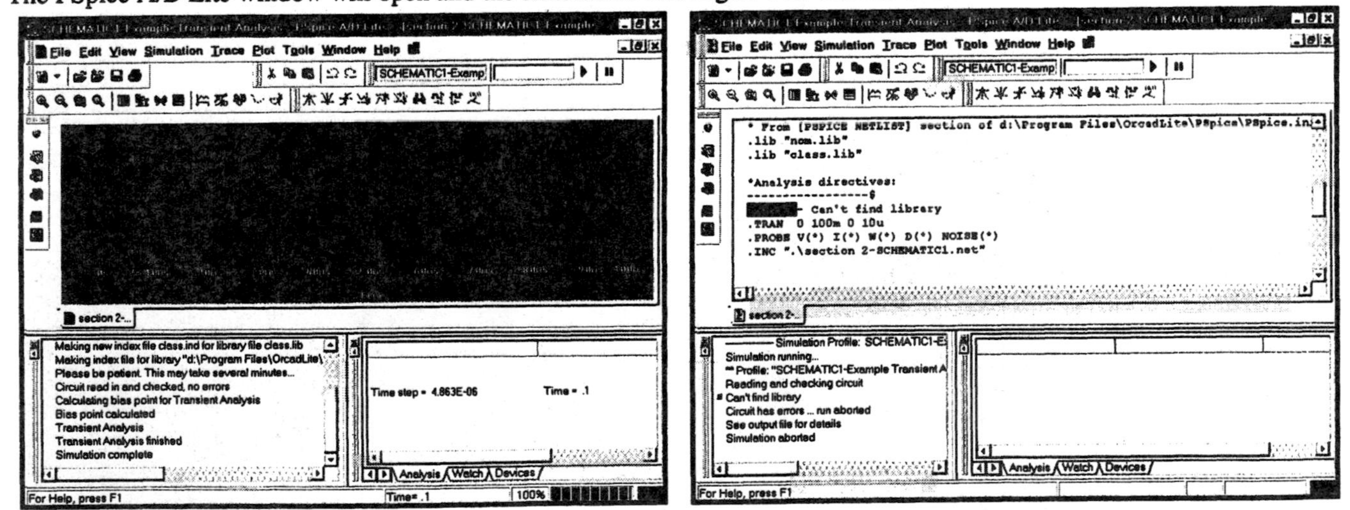


The above left screen shows that PSpice is working correctly. If your screen matches the above left screen, you can continue with the remainder of this example. The above right screen indicates that PSpice generated an error message. The error is that PSpice cannot find the library class.lib. If you see the above right screen on your computer, you may have not properly installed the libraries for this text. See Section A.2 on page 602 for the procedure for installing the libraries. After you have installed the libraries, attempt to run this example again.

We will discuss the PSpice A/D Lite window. It has three sections: a message window, the Probe display window, and the PSpice simulation status window.

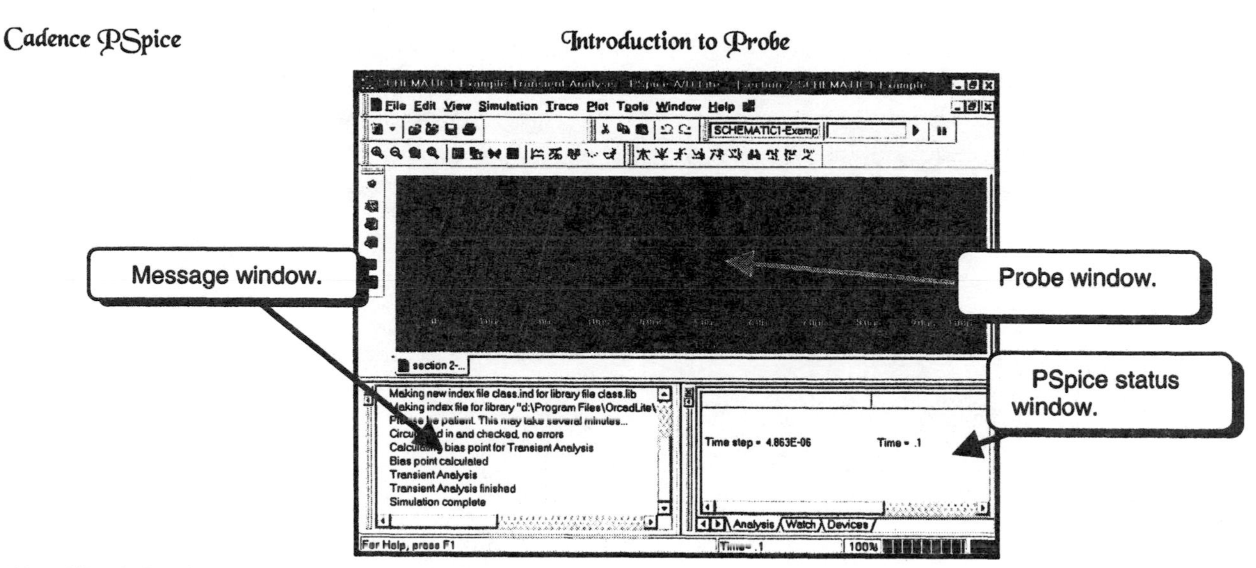


We will briefly show how to manipulate these windows. You can close either of the two bottom windows by clicking on the ☒ in the upper left corner of the window. For example, we will close the message window:

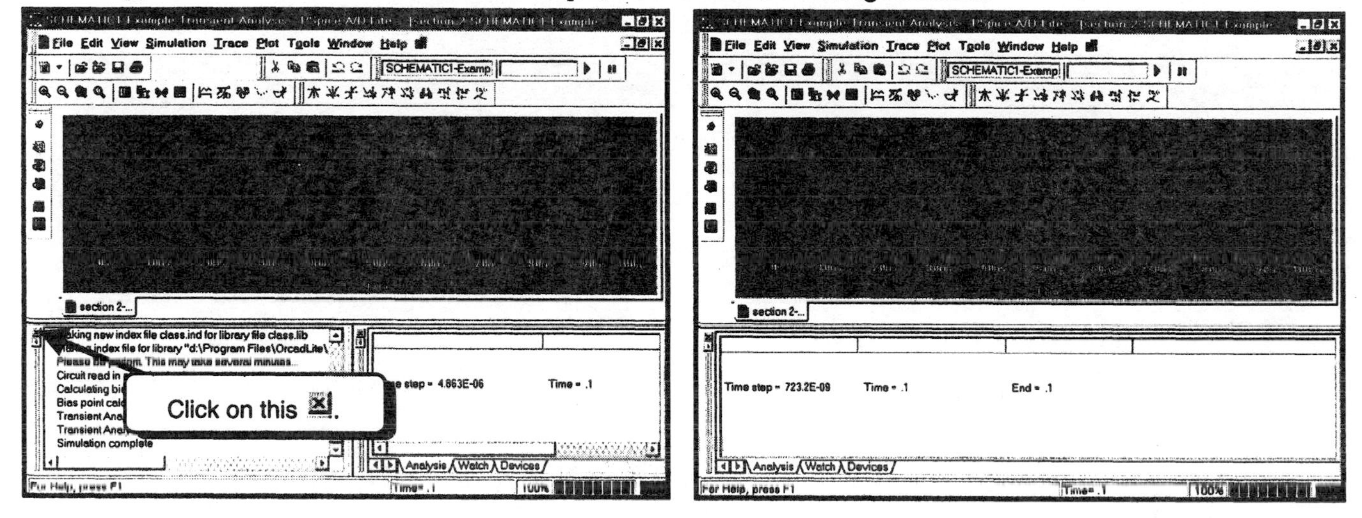


The message window is closed, and the PSpice simulation status window expands to fill the empty space. Next, we will close the simulation window:

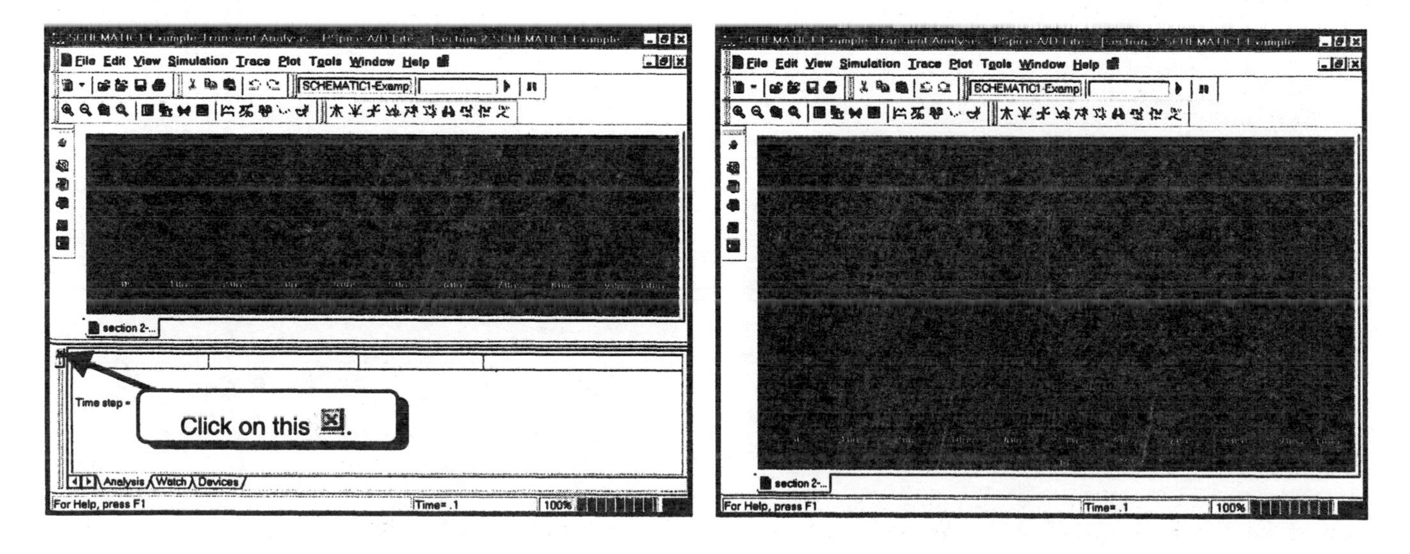


The Probe window now occupies the entire window.

To get the message and simulation status windows back, use the menus. To display the message window, select **View** and then **Output Window** from the menus:

To display the simulation status window, select **View** and then **Simulation Status Window** from the menus:

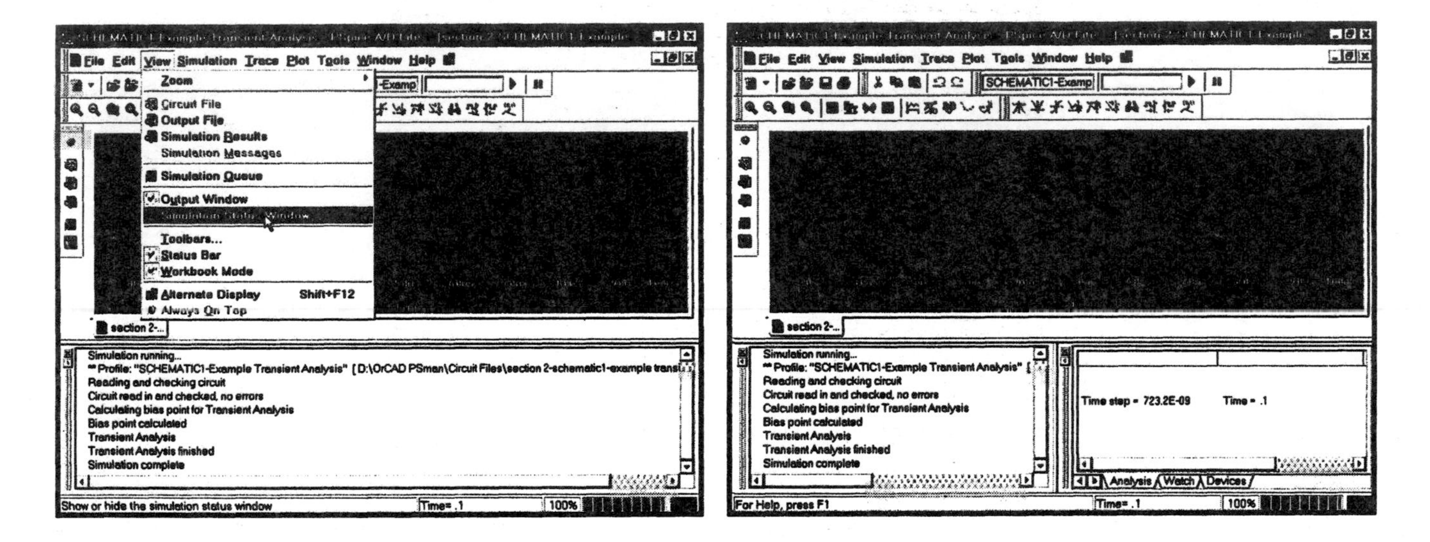


When we are running a simulation, we want the simulation status and the message windows to be visible because they give us information on problems with the simulations and indicate the progress of the simulation. When we are plotting traces, we would like Probe to use the full screen so that the traces are as large as possible. To toggle between these two conditions, select **View** and then **Alternate Display** from the menus:

To toggle back to the original display, select **View** and then **Alternate Display** again from the menus:

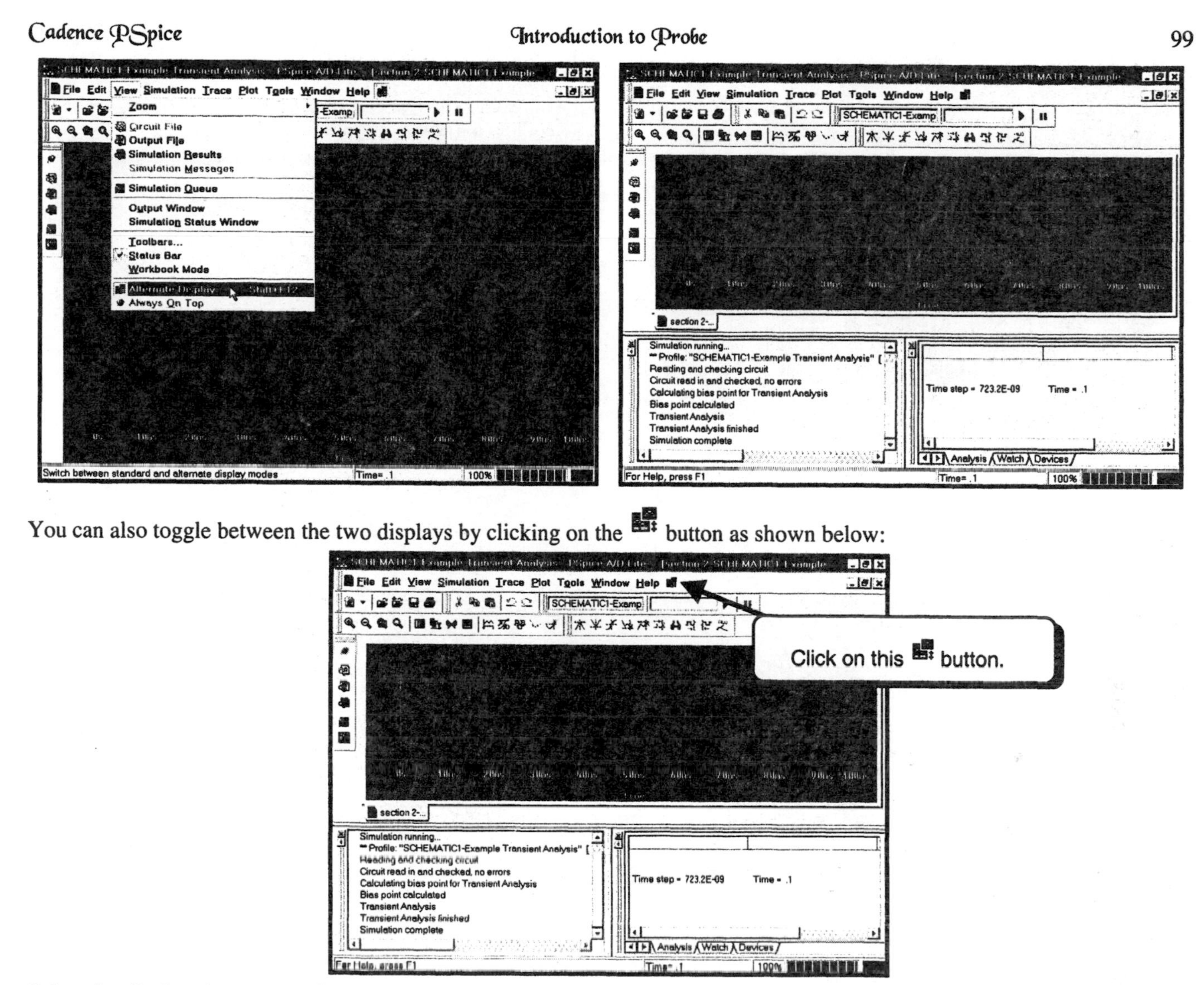
Click on this button.

You can also toggle between the two displays by clicking on the button as shown below:

Select the display that you prefer to use. For most traces shown in this text, we will show the Probe window in full screen. Note also that your probe screen will have a black background. To make the screen captures more readable, the background color of the probe screen has been changed to white for the remainder of this book. This was accomplished by modifying the [PROBE DISPLAY COLORS] section of the pspipce.ini file located in the OrcadLite/PSpice directory.

# 2.A. Plotting Traces

We will now add some traces. In the schematic we have labeled two nodes, Vin and Vout. Since we know the names of the nodes, we can easily plot the voltages. We will first plot Vin. To add a trace, select **Trace** and then **Add Trace** from the menu bar:

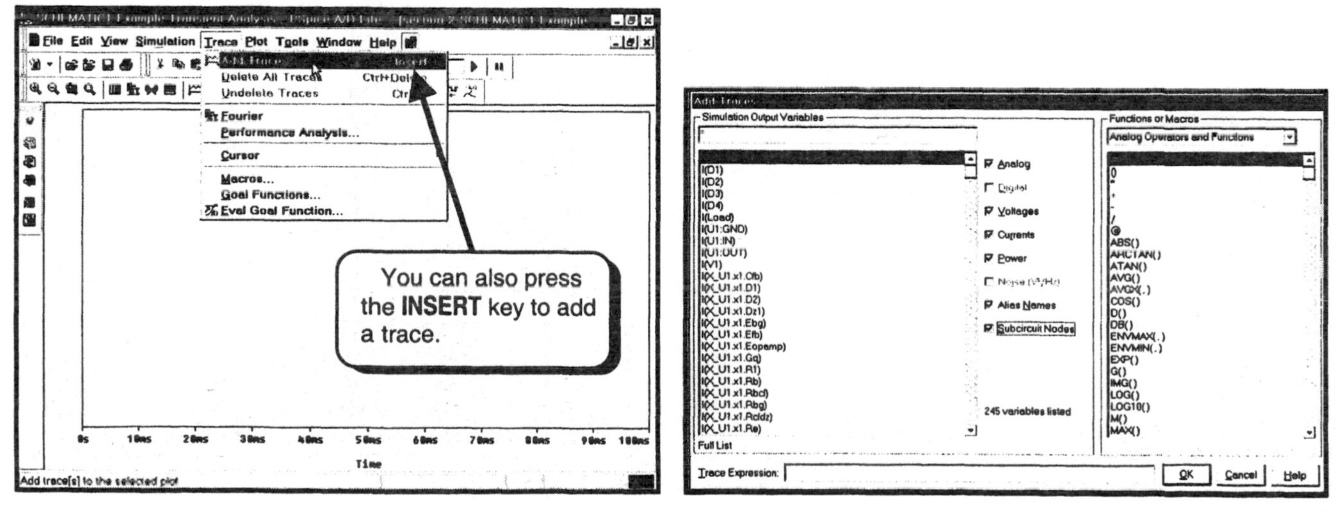
You can also press the INSERT key to add a trace.

The left pane lists all of the traces we can plot. The right pane displays mathematical operations we can perform on the traces. It lists digital waveforms (we do not have any in our circuit), voltages, currents, and power. It also lists what are referred to as alias names. Aliases are different names that refer to the same thing. For example, in our circuit the cathodes of D3 and D4 are connected to the input of the voltage regulator and the capacitor. The node is also labeled as Vin. We can refer to the voltage of this node in several ways. This node could be addressed as pin 1 of D4, pin 1 of D3, pin 1 of C1, pin IN of the voltage regulator, or as Vin. Thus, to display the voltage at this node, we can use any of the aliases mentioned. Each node will have many aliases, and thus the list shown in the left pane is very large. Note that almost all trace types are displayed: ***Analog***, ***Voltages***, ***Currents***, ***Power***, ***Alias Names***, and ***Subcircuit Nodes***:

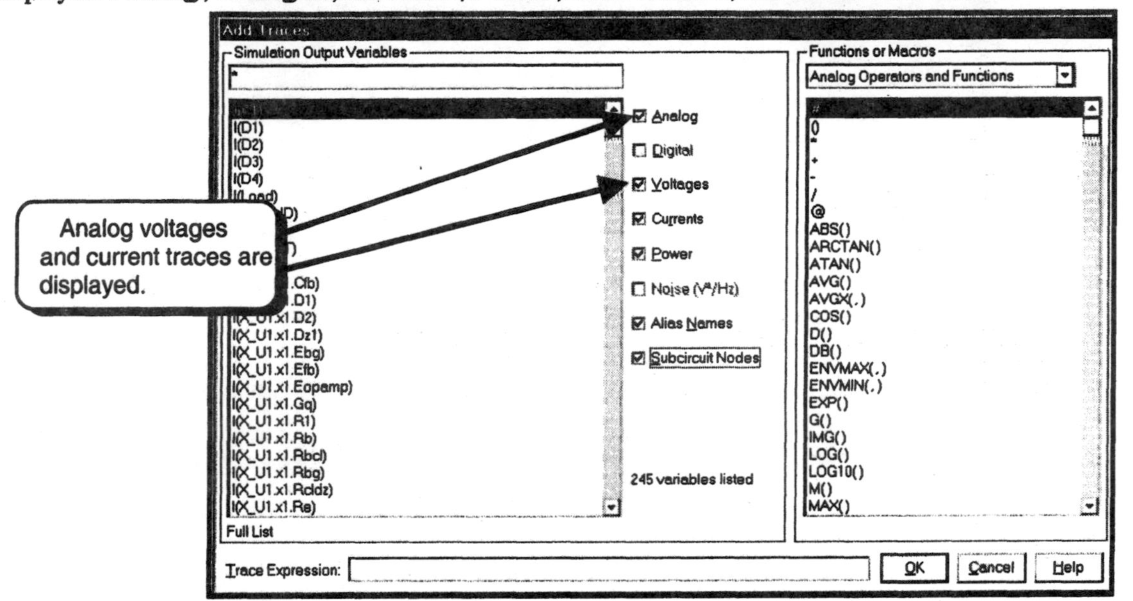


Because all of the options are selected, many names are shown in the left pane. Note that the voltage regulator is a subcircuit. A subcircuit is shown as a single block on the schematic, but it may be composed of several circuit elements within the subcircuit. The left pane is currently displaying all subcircuit nodes and the aliases for the subcircuit nodes. If you scroll through the list, you will see too many traces.

We would first like to display Vin and Vout. The list of traces is too long and the traces are not easily spotted. Vin and Vout are analog voltages so we will select only the ***Analog***, ***Voltages***, and ***Alias Names*** boxes:

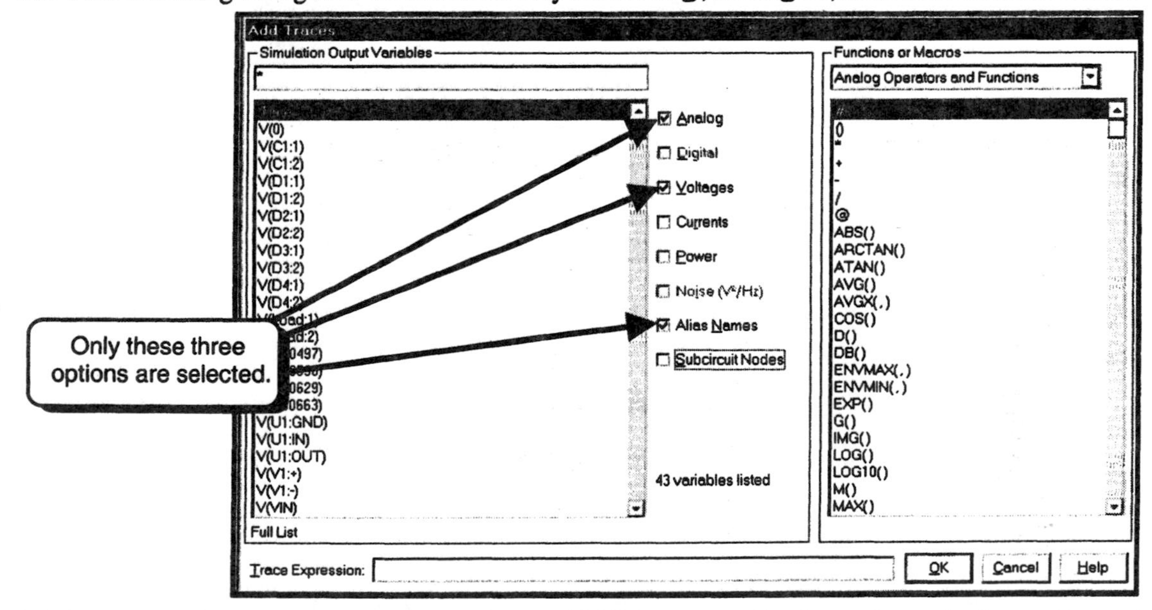


This list is now shorter and we can locate the line ***V(VIN)*** more easily. Scroll down the list until you find the trace labeled ***V(VIN)*** and then click on the text to select it. It should become highlighted:

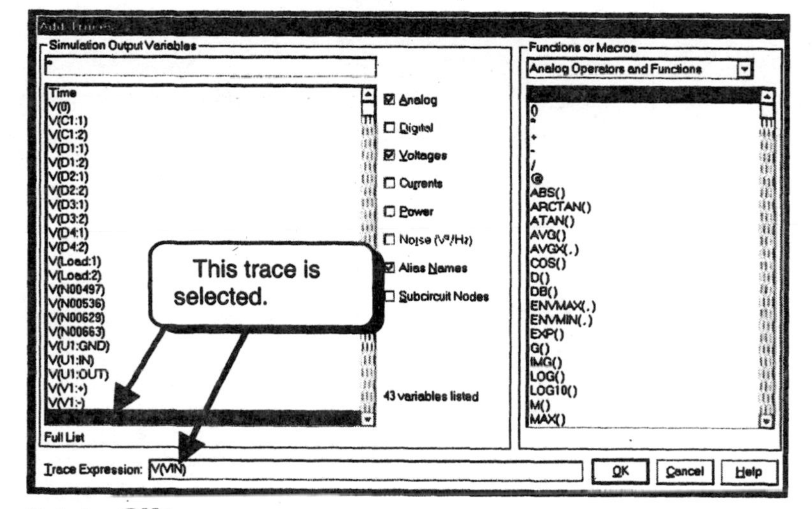


To plot the selected trace, click the ***OK*** button:

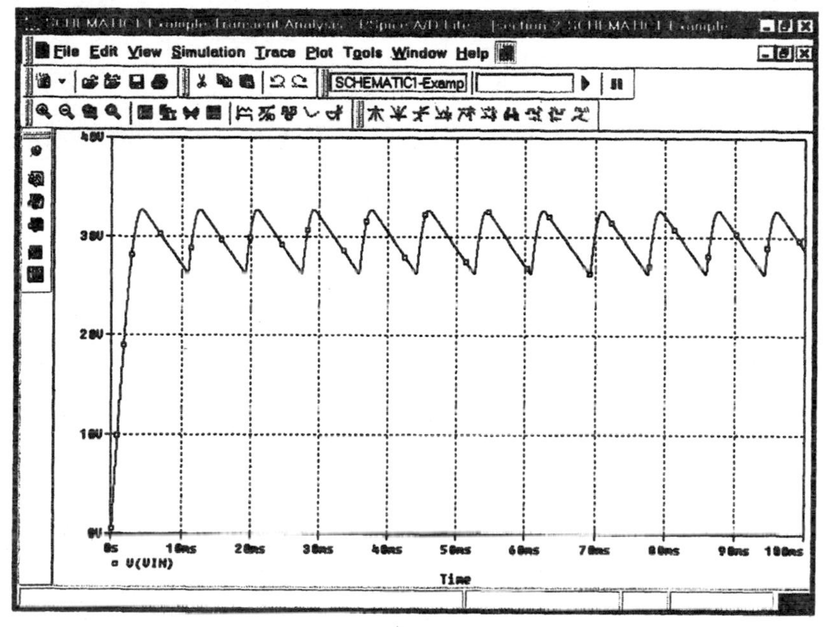


Next we will plot Vo. Press the **INSERT** key. This is a shortcut that will open the ***Add Traces*** dialog box:

Add Traces
Simulation Output Variables
*
Functions or Macros
Analog Operators and Functions
V(0)
V(C1:1)
V(C1:2)
V(D1:1)
V(D1:2)
V(D2:1)
V(D2:2)
V(D3:1)
V(D3:2)
V(D4:1)
V(D4:2)
V(Load:1)
V(Load:2)
V(N00497)
V(N00536)
V(N00629)
V(N00663)
V(U1:GND)
V(U1:IN)
V(U1:OUT)
V(V1:+)
V(V1:-)
V(VIN)
Full List
Analog
Digital
Voltages
Currents
Power
Noise (V²/Hz)
Alias Names
Subcircuit Nodes
43 variables listed
()
*
+
-
/
@
ABS()
ARCTAN()
ATAN()
AVG()
AVGX(,)
COS()
D()
DB()
ENVMAX(,)
ENVMIN(,)
EXP()
G()
IMG()
LOG()
LOG10()
M()
MAX()
Trace Expression:
OK
Cancel
Help

Click on the text ***V(VO)*** to select the trace:

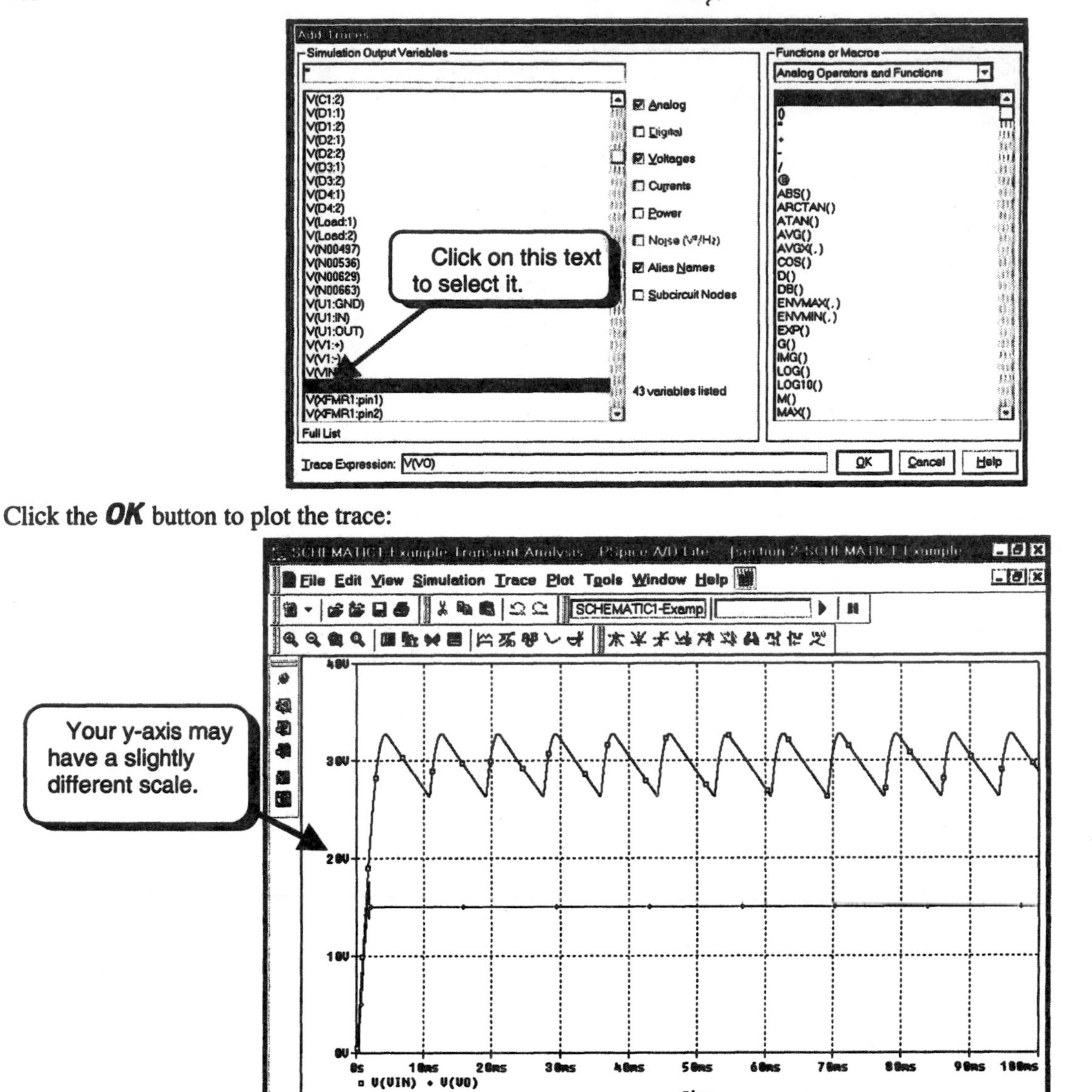


Click the ***OK*** button to plot the trace:

We can add many traces to the plot. Next we will display the current through D1. Press the **INSERT** key to obtain the ***Add Traces*** dialog box:

Add Traces
Simulation Output Variables
V(0)
V(C1:1)
V(C1:2)
V(D1:1)
V(D1:2)
V(D2:1)
V(D2:2)
V(D3:1)
V(D3:2)
V(D4:1)
V(D4:2)
V(Load:1)
V(Load:2)
V(N00497)
V(N00536)
V(N00644)
V(N00663)
V(U1:GND)
V(U1:IN)
V(U1:OUT)
V(V1:+)
V(V1:-)
V(VIN)
Full List
Analog
Digital
Voltages
Currents
Power
Noise (V²/Hz)
Alias Names
Subcircuit Nodes
43 variables listed
Functions or Macros
Analog Operators and Functions
ABS()
ARCTAN()
ATAN()
AVG()
AVGX(,)
COS()
D()
DB()
ENVMAX(,)
ENVMIN(,)
EXP()
G()
IMG()
LOG()
LOG10()
M()
MAX()
Trace Expression:
OK
Cancel
Help

Presently the dialog box shows only analog voltages. We wish to plot a current, so specify the options as shown:

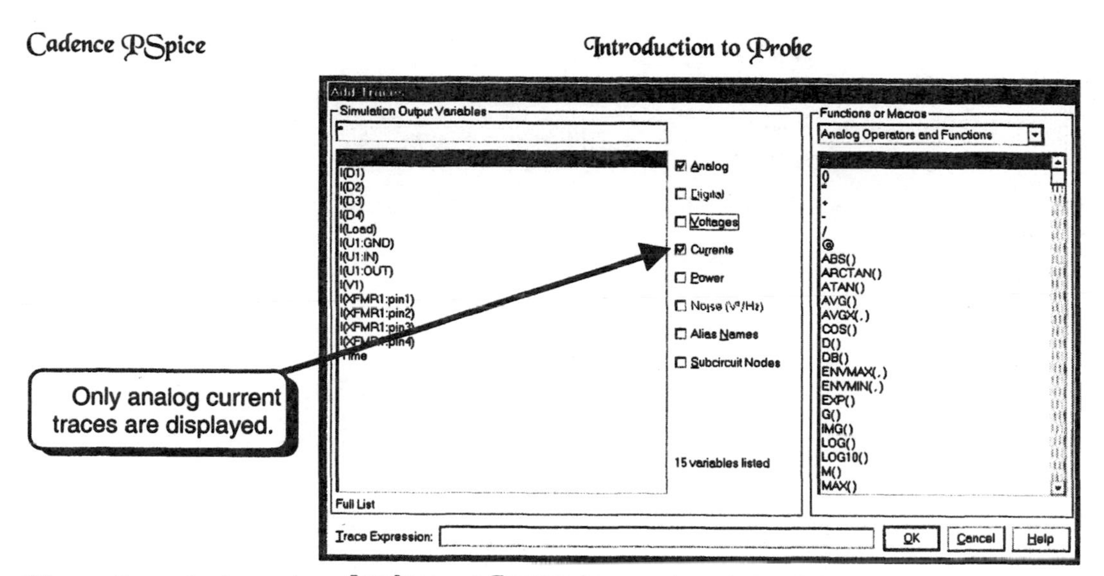
Only analog current traces are displayed.

We see that only the options ***Analog*** and ***Currents*** are selected. The left pane displays only the currents. Click on the text ***I(D1)*** to select the trace:

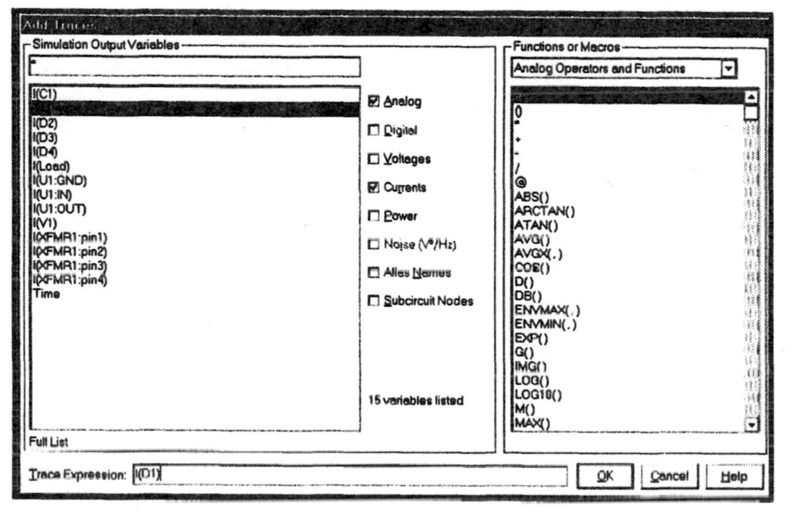

Click the ***OK*** button to plot the trace:

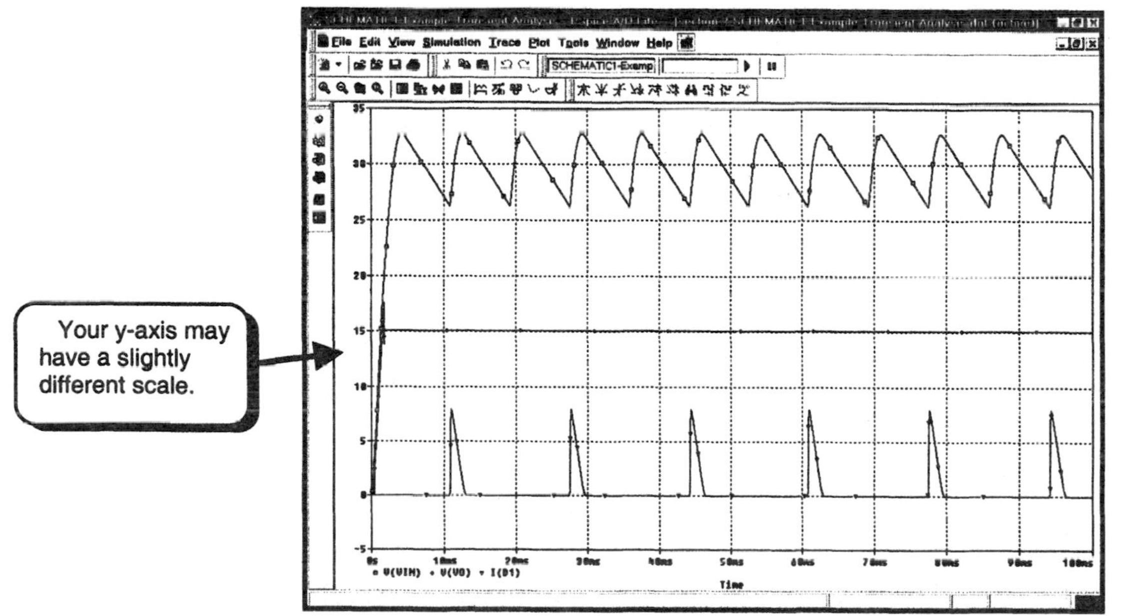
Your y-axis may have a slightly different scale.

As a last example, we will show how to use one of the mathematical operations in the right pane of the ***Add Traces*** dialog box. Press the **INSERT** key to obtain the dialog box:

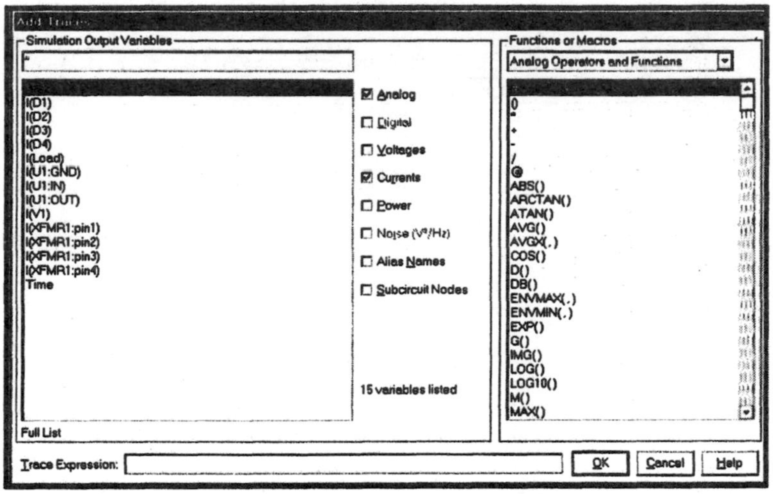

Add Traces
Simulation Output Variables
*
I(D1)
I(D2)
I(D3)
I(D4)
I(Load)
I(U1:GND)
I(U1:IN)
I(U1:OUT)
I(V1)
I(XFMR1:pin1)
I(XFMR1:pin2)
I(XFMR1:pin3)
I(XFMR1:pin4)
Time
Full List
Analog
Digital
Voltages
Currents
Power
Noise (V²/Hz)
Alias Names
Subcircuit Nodes
15 variables listed
Functions or Macros
Analog Operators and Functions
()
*
+
-
/
@
ABS()
ARCTAN()
ATAN()
AVG()
AVGX(,)
COS()
D()
DB()
ENVMAX(,)
ENVMIN(,)
EXP()
G()
IMG()
LOG()
LOG10()
M()
MAX()
Trace Expression:
OK
Cancel
Help

We will plot the time average current through D1. The ***AVG*** function will perform this function. Click the ***LEFT*** mouse button on the text ***AVG()*** to select the function:

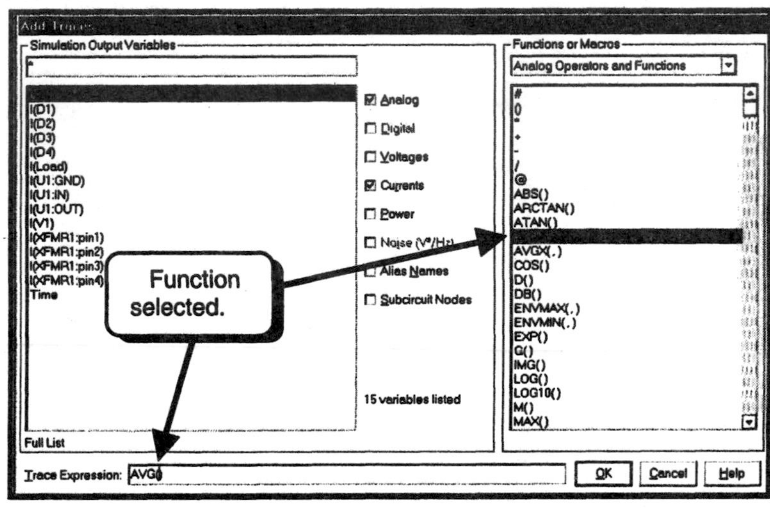


Notice that the text ***AVG()*** appears in the ***Trace Expression*** text field and that the cursor is positioned between the parentheses waiting for a trace. Next, click the ***LEFT*** mouse button on the text ***I(D1)***. This will select the trace and place it within the parentheses of the ***AVG*** function:

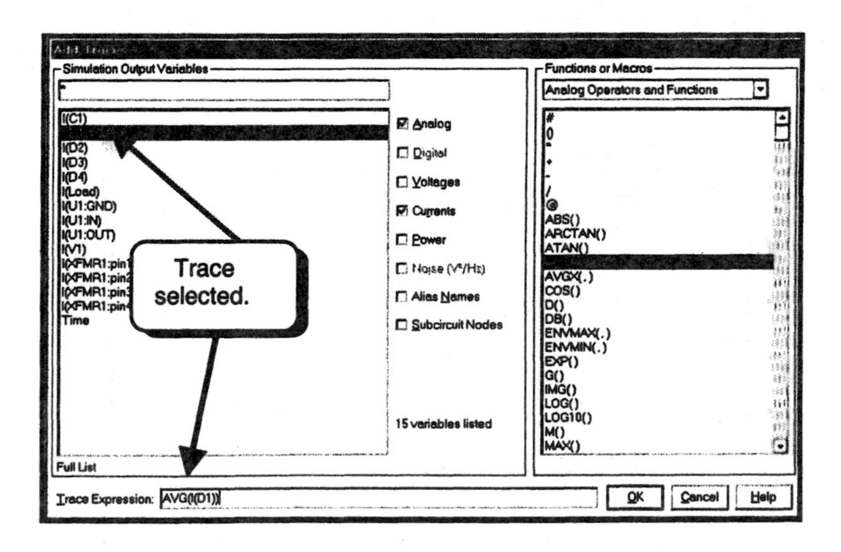


This trace is plotting the time average of the current through D1. Click the ***OK*** button to plot the trace:

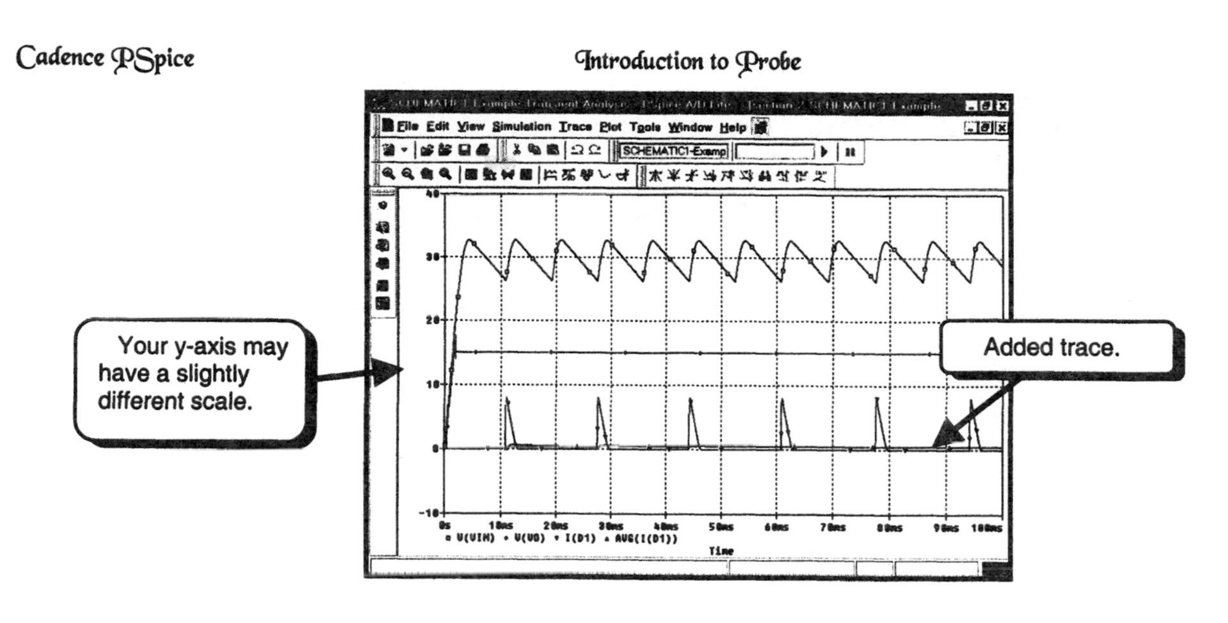


The trace starts out small but we can see the fourth trace.

## 2.B. Deleting Traces

We now have a number of traces on the plot. We can remove individual traces easily. We will remove trace ***V(VO)***. Click the ***LEFT*** mouse button on the text ***V(VO)*** as shown below:

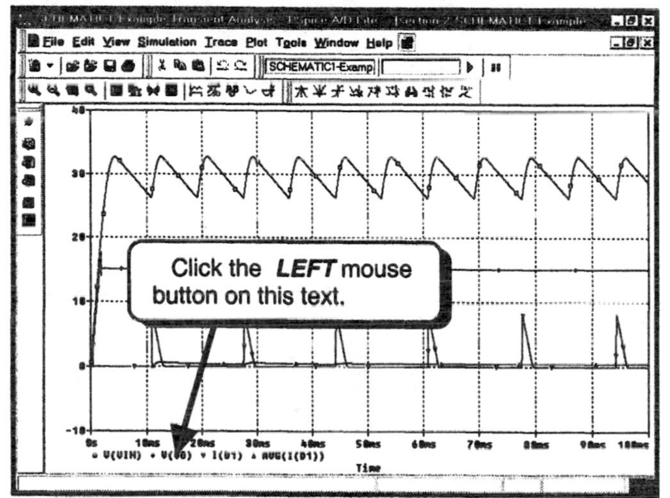


The text will become highlighted in red, indicating that it is selected. When the text ***V(VO)*** is highlighted in red, press the **DELETE** key. The trace will be removed:

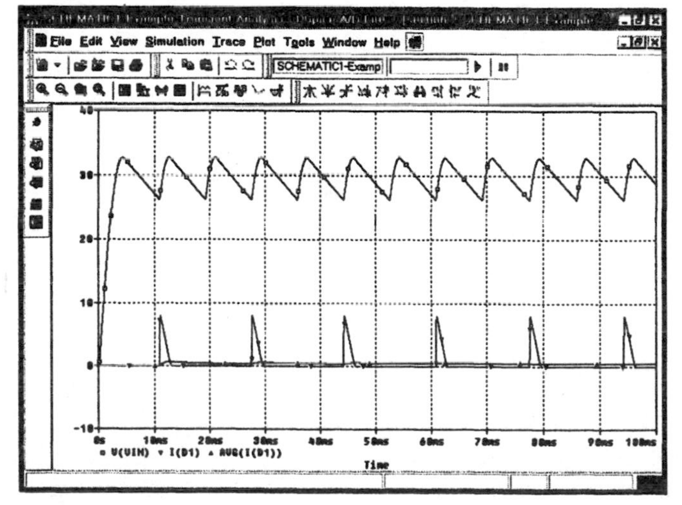

Delete all traces but V(Vin) using this method:

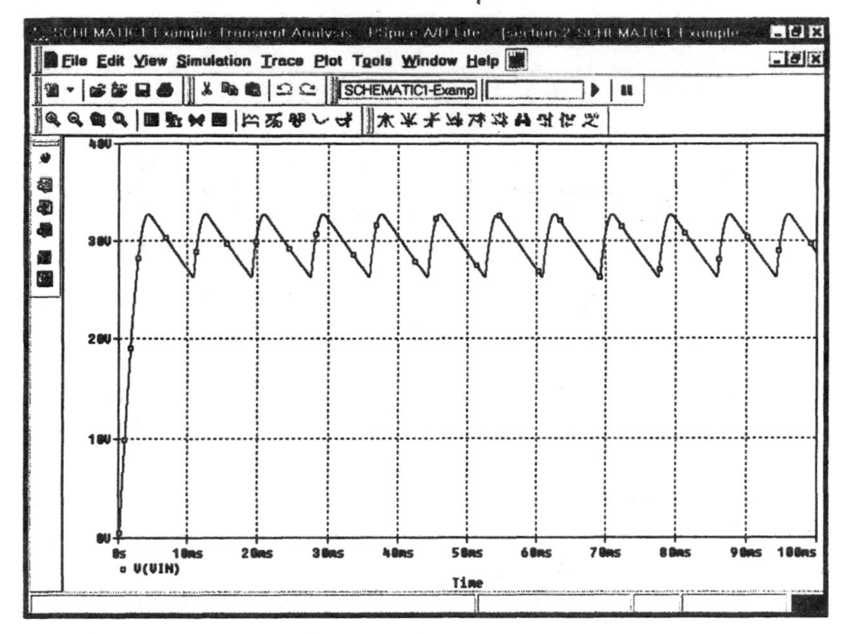


## 2.C. Using the Markers to Add Traces

In this example we are currently describing, both Probe and Capture are running. We would like to switch back to Capture. We can switch windows using two methods:

1. Hold down the **ALT** key and press the **TAB** key. Continue to hold down the **ALT** key and release the **TAB** key. When you do this, a window will pop up and display the icons of the programs currently running. One of the icons should be the Capture icon. Press the **TAB** key until the black box encloses the Capture icon. When the box encloses the Capture icon, release the **ALT** key. The Capture window will pop to the top.
2. You can use the Start menu to switch to the Capture window. If the Start menu is not displayed on the screen, bring the mouse to the bottom of the screen to display the Start menu, and then select the Capture button.

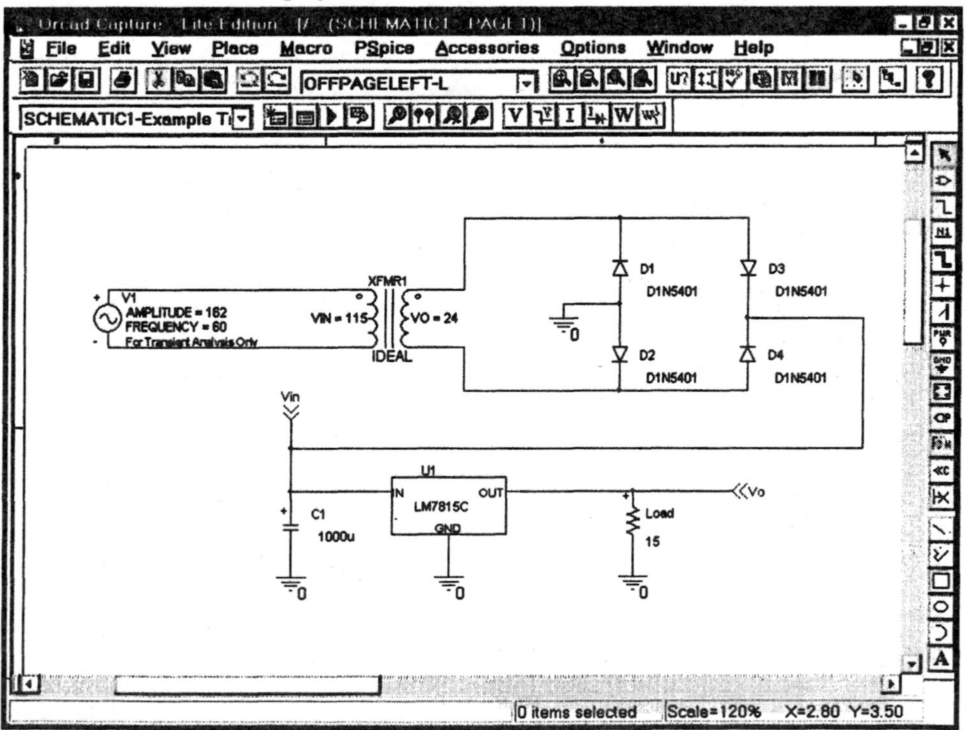


We can now use the markers to display the currents or voltages in the circuit. To obtain a voltage marker, select **PSpice** and then **Markers** from the Capture menu bar:

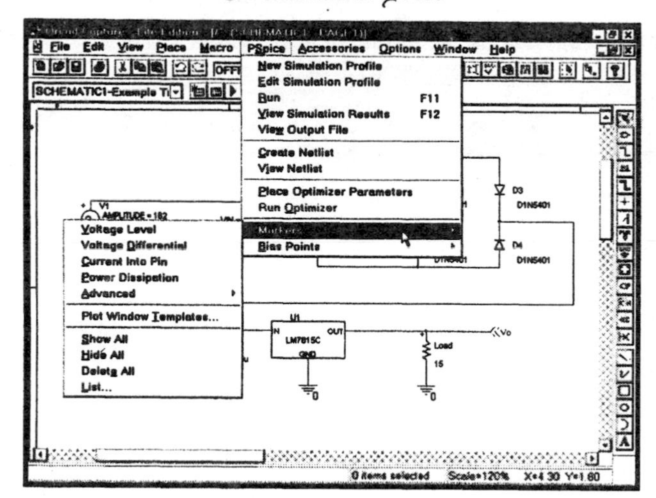

**Voltage Level** will display the voltage at a node relative to ground. **Voltage Differential** will display the voltage between any two points. With **Voltage Differential**, you will be required to place two markers. The first marker will designate the positive reference for the voltage difference, and the second marker will designate the negative reference for the voltage difference. **Current Into Pin** will display the current into a device. You will need to place the marker on one of the pins to the device.

We will look at the voltage at the cathode of D1 relative to ground. Select **Voltage Level** from the menu. A marker will become attached to the mouse pointer:

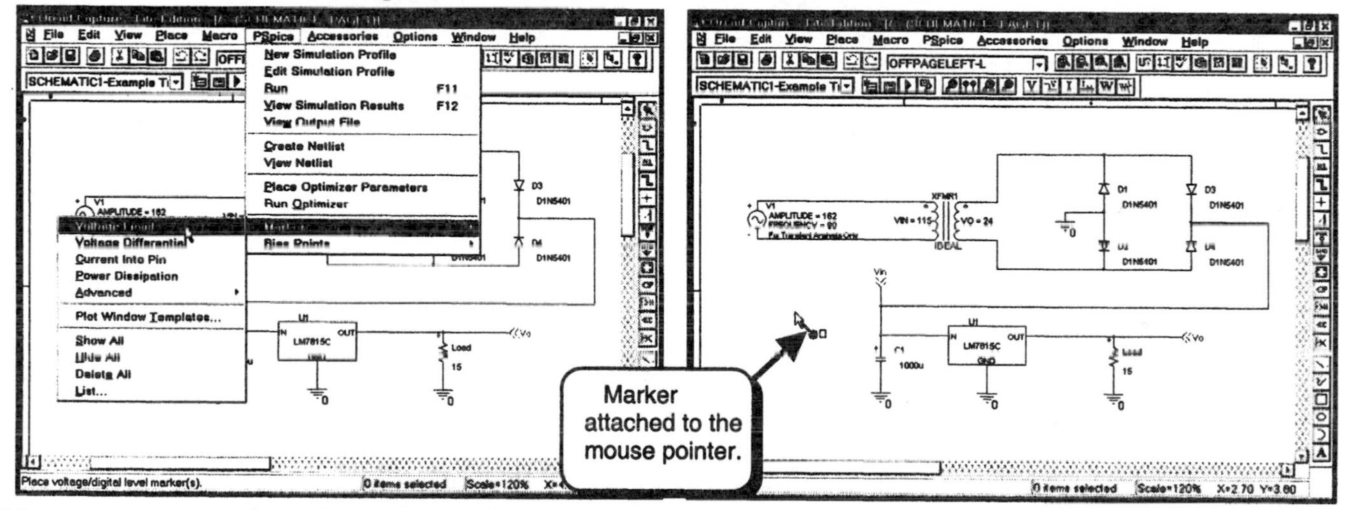


Move the mouse to position the pointer as shown:

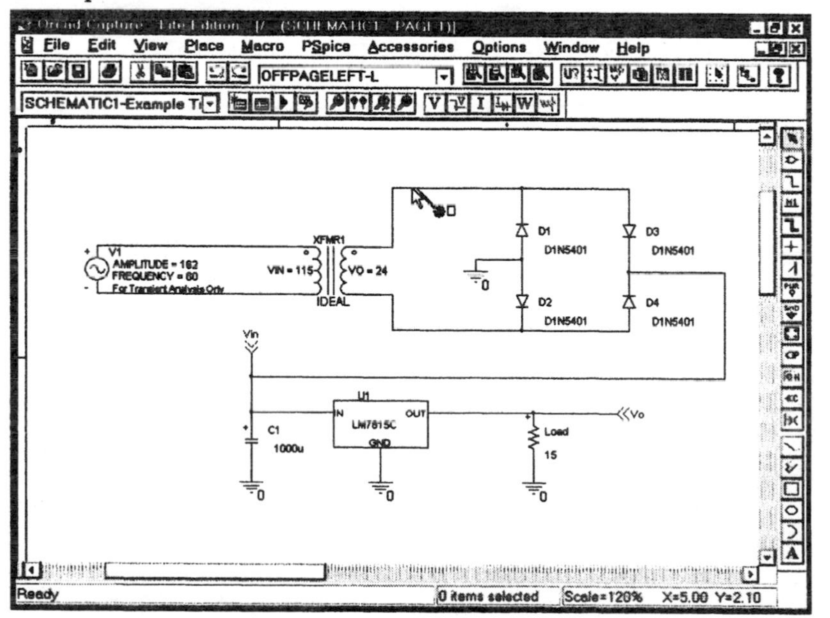

Click the ***LEFT*** mouse button to place the marker. Move the mouse away:

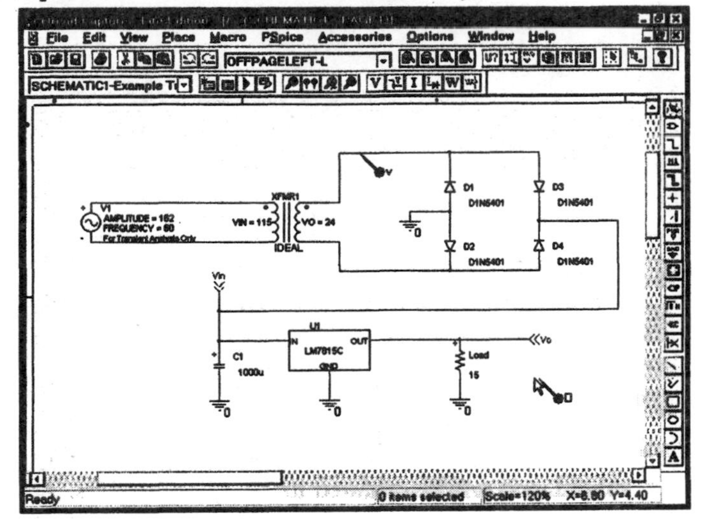


A marker is placed on the wire and a new marker is attached to the mouse pointer. We can place more markers if we want. Press the **ESC** key to terminate placing markers. The marker will change to the color the trace is displayed in Probe. On my computer, all Probe traces are displayed in black, so the marker is displayed in black. If your marker turns green, then the trace will be displayed in green on the Probe screen.

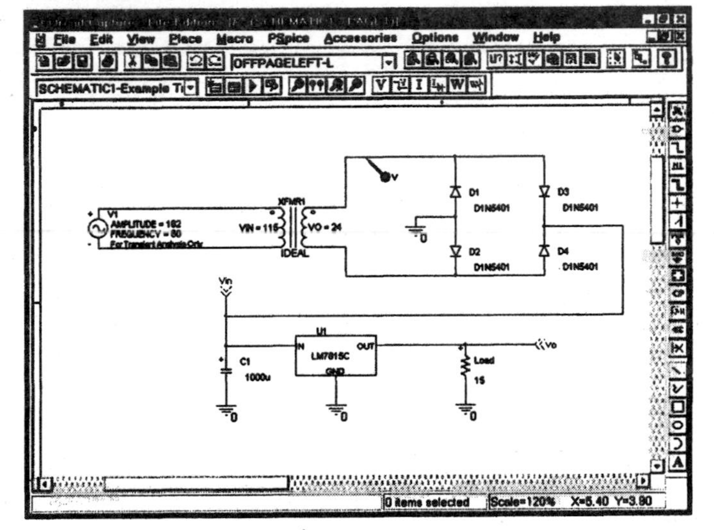


Use the **ALT - TAB** key sequence to switch back to Probe.* A new trace will be displayed:

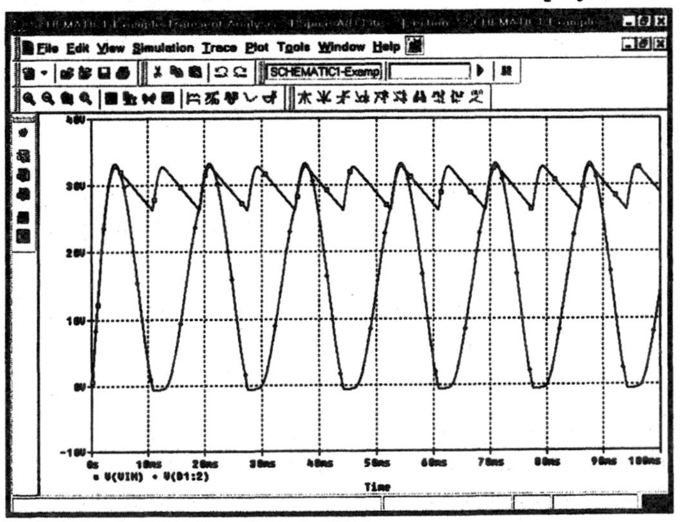


* See Page 106 for a more detailed description of how to switch between windows using this method.

We see that the voltage specified by the marker is displayed. Markers are convenient because we do not need to know the node names to plot a trace. Note that the name of this trace is ***V(D1:2)***, hardly an obvious name. Use the **ALT - TAB** key sequence to switch back to Capture:

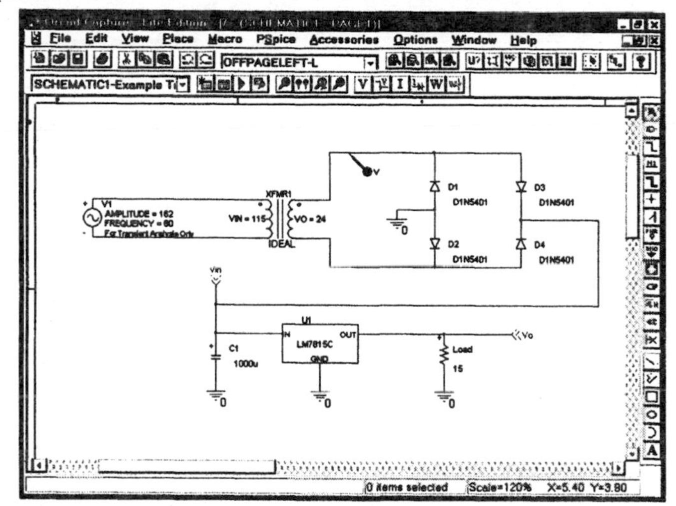


Next we will plot the voltage of source ***V1*** using the voltage differential markers. Select **PSpice, Markers**, and then **Voltage Differential**. A marker will appear:

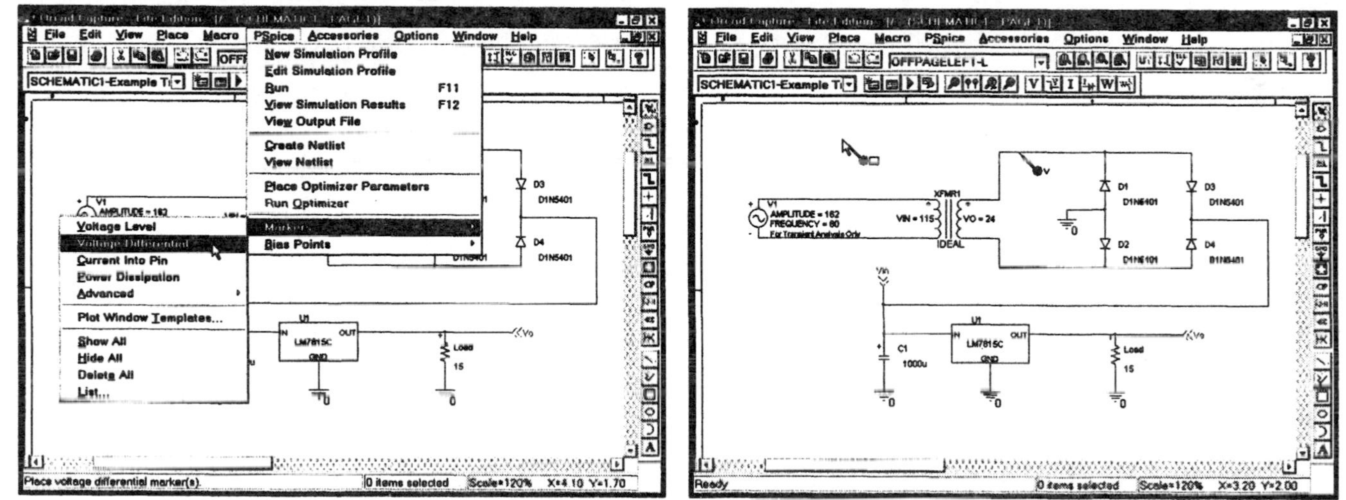


Move the mouse to position the marker as shown:

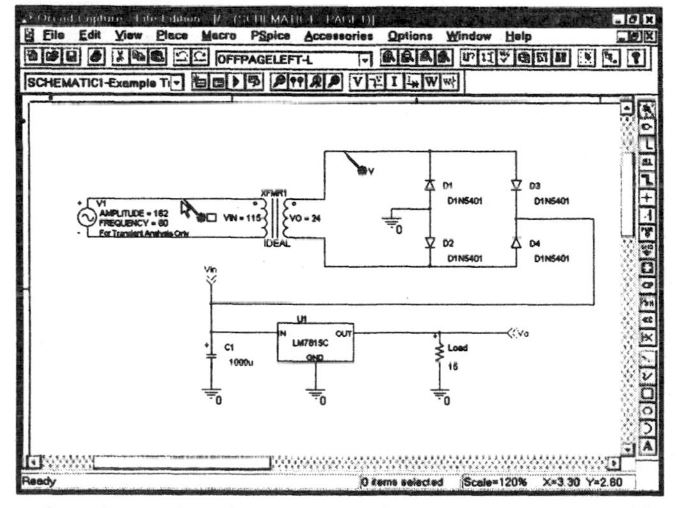


Click the ***LEFT*** mouse button to place the marker. As you move the mouse away, you will notice that a marker is placed on the wire and that a second marker is attached to the mouse pointer:

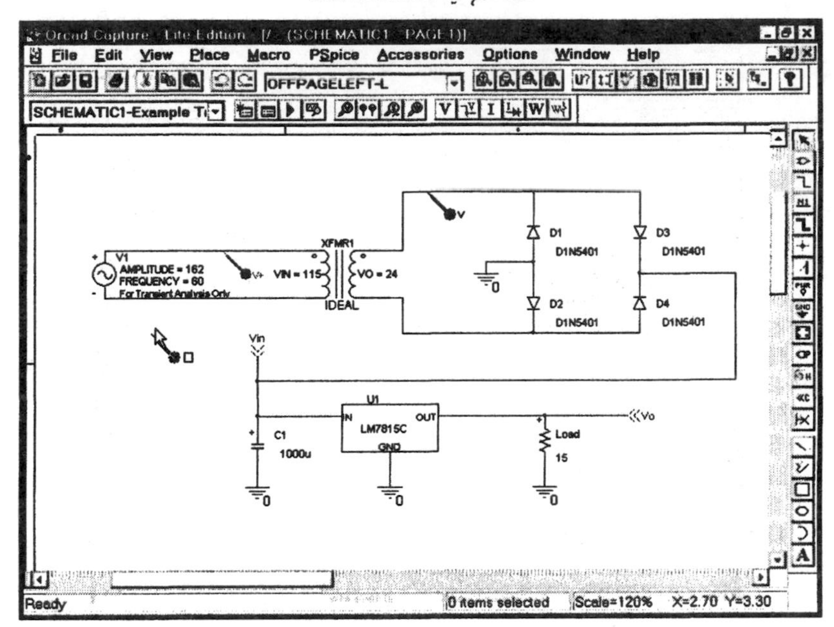


Move the mouse to position the marker as shown:

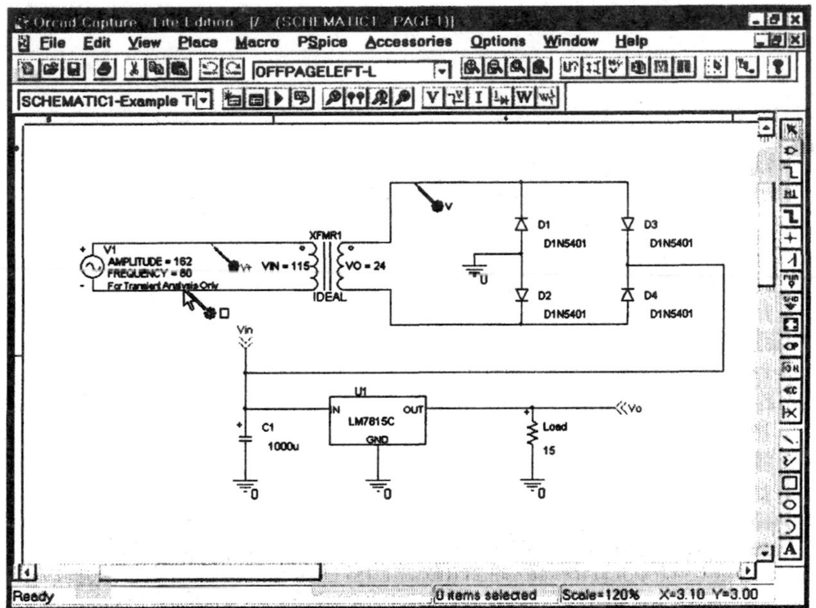


Click the ***LEFT*** mouse button to place the marker. Move the mouse pointer away:

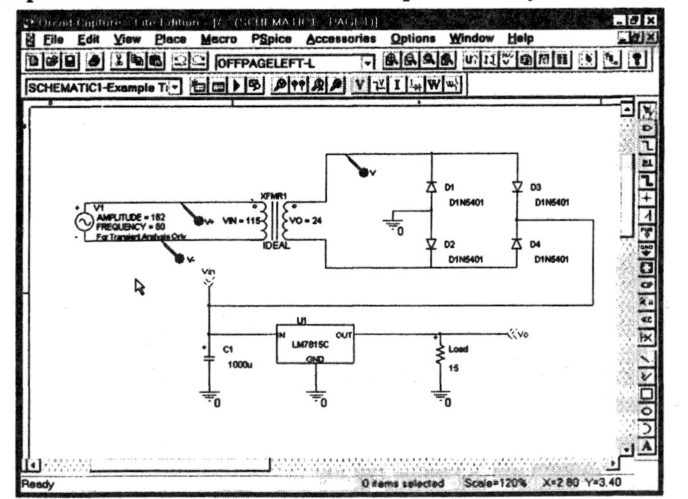


Press the **ESC** key to terminate placing markers. I will zoom in on the markers to show that one marker has a plus sign and the other has a minus sign:

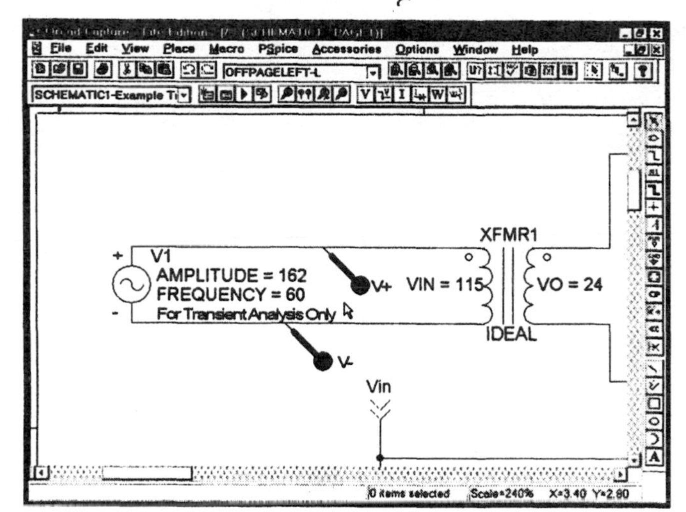


These markers display the voltage difference between the two markers. To view the trace, switch to the Probe window using the **ALT - TAB** key sequence.* The new trace will be displayed:

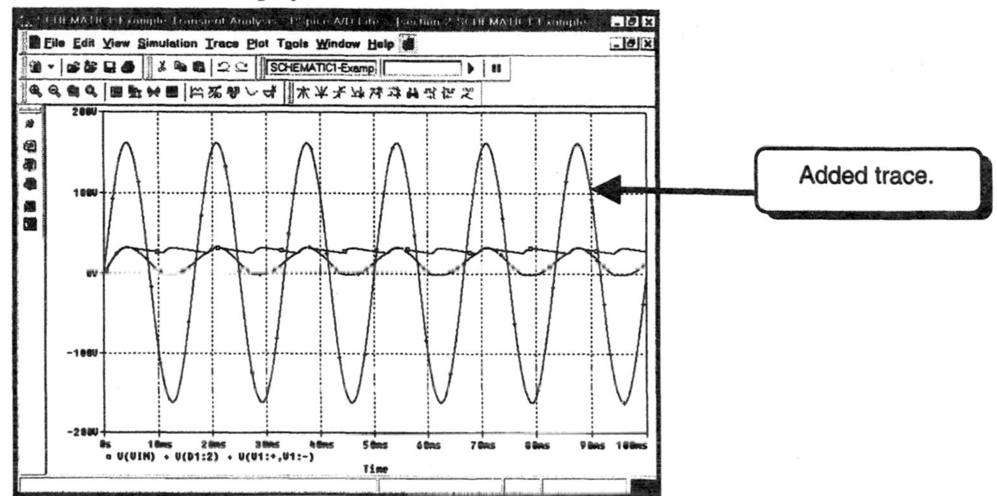


Last, we will use a current marker to display the current through an element. Use the **ALT - TAB** key sequence to switch back to Capture:

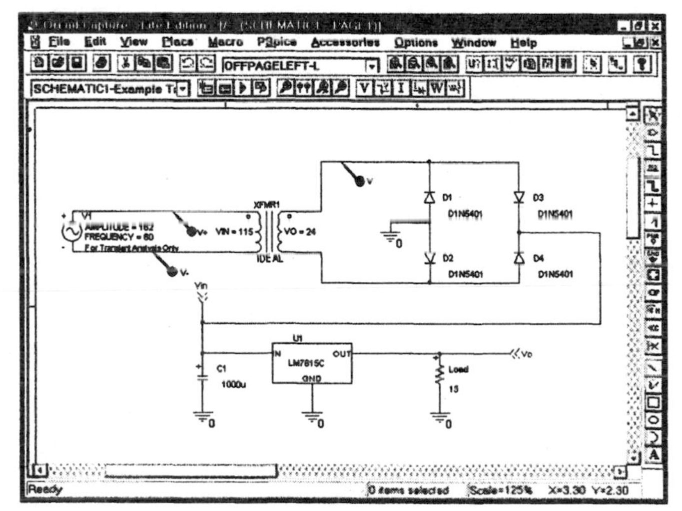


We would like to plot the capacitor current. Select **PSpice**, **Markers**, and then **Current Into Pin**. A marker with an I will become attached to the mouse pointer:

---

* See page 106 for a more detailed description of how to switch between windows using this method.

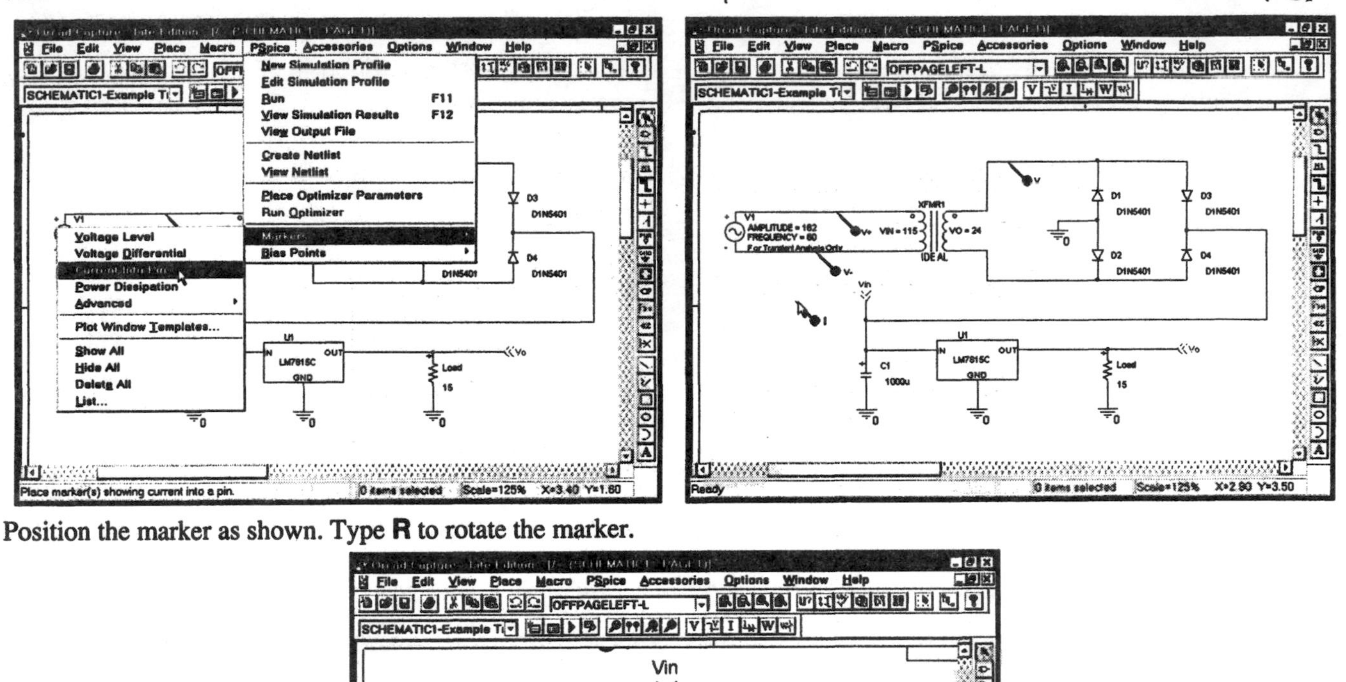


Position the marker as shown. Type **R** to rotate the marker.

Position the current marker here.

Click the ***LEFT*** mouse button to place the marker. If you get the message:

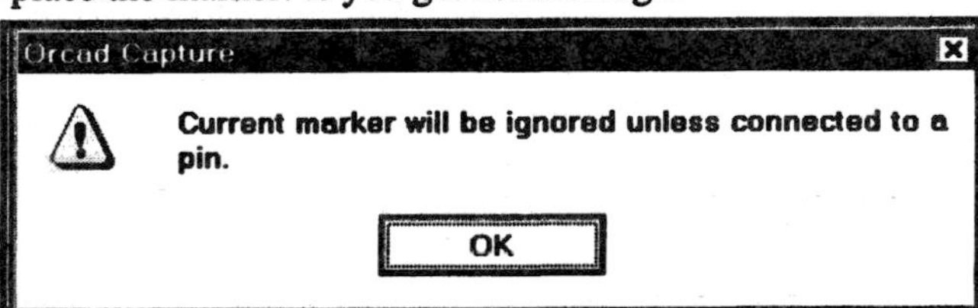


you missed the pin. Move the marker up or down one grid increment and try again. The marker must be placed at the end of a pin connected to a device. After you click the ***LEFT*** mouse button to place the marker and do not receive an error message, move the mouse pointer away:

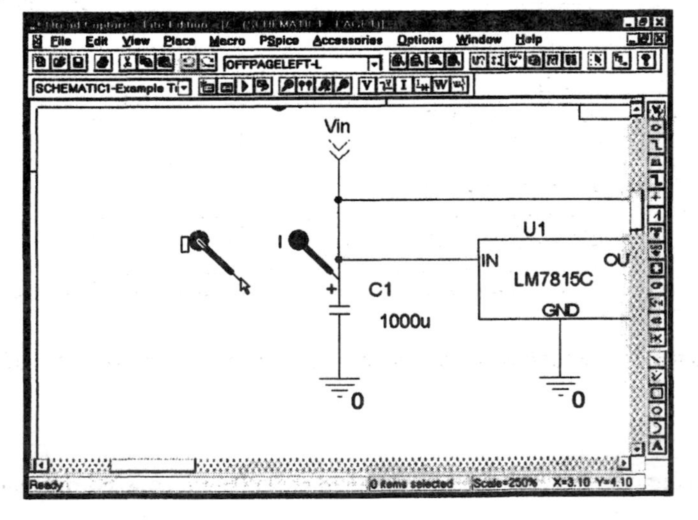

The marker is now attached to the end of the top pin of the capacitor. Press the **ESC** key to terminate placing markers. Use the **ALT - TAB** key sequence to switch back to Probe.* The current trace will be displayed:

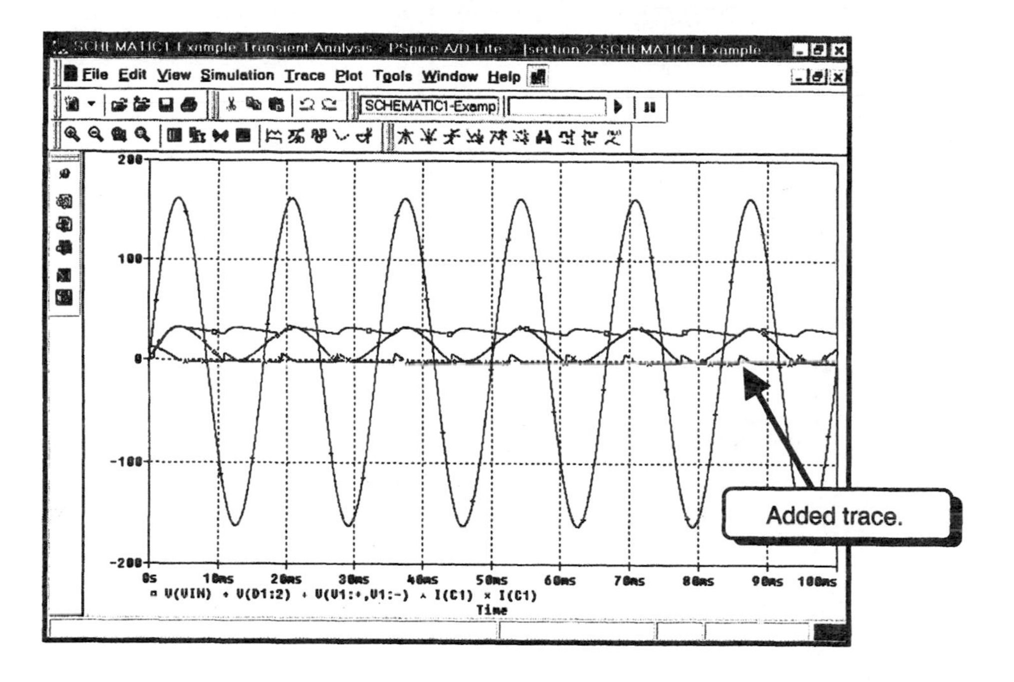


The current trace is small, but its peaked shape is easy to spot.

# 2.D. Zooming In and Out

We now have a number of traces displayed. However, the current trace is small. Suppose that we would like to look a little closer at a peak in the current waveform. We can do this by using some of the zoom features provided by Probe. Select **View** and then **Zoom** from the menu bar:

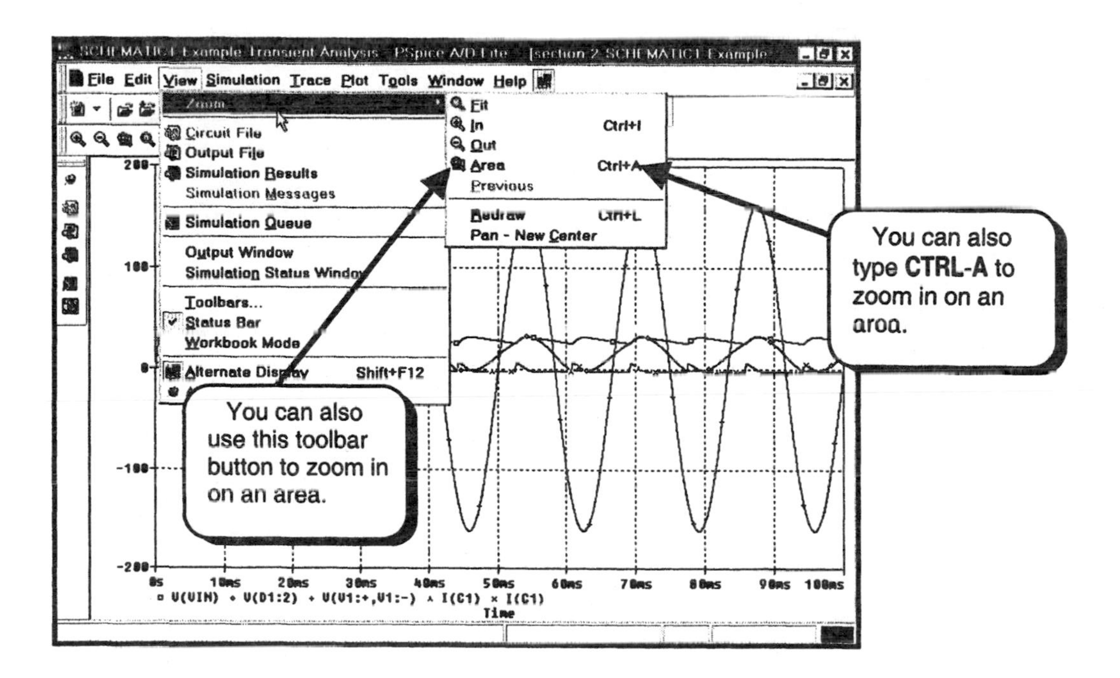


The menu lists 5 ways to zoom in and out. Select **Area.** The cursor will be replaced by crosshairs:

---

* See page 106 for a more detailed description of how to switch between windows using this method.

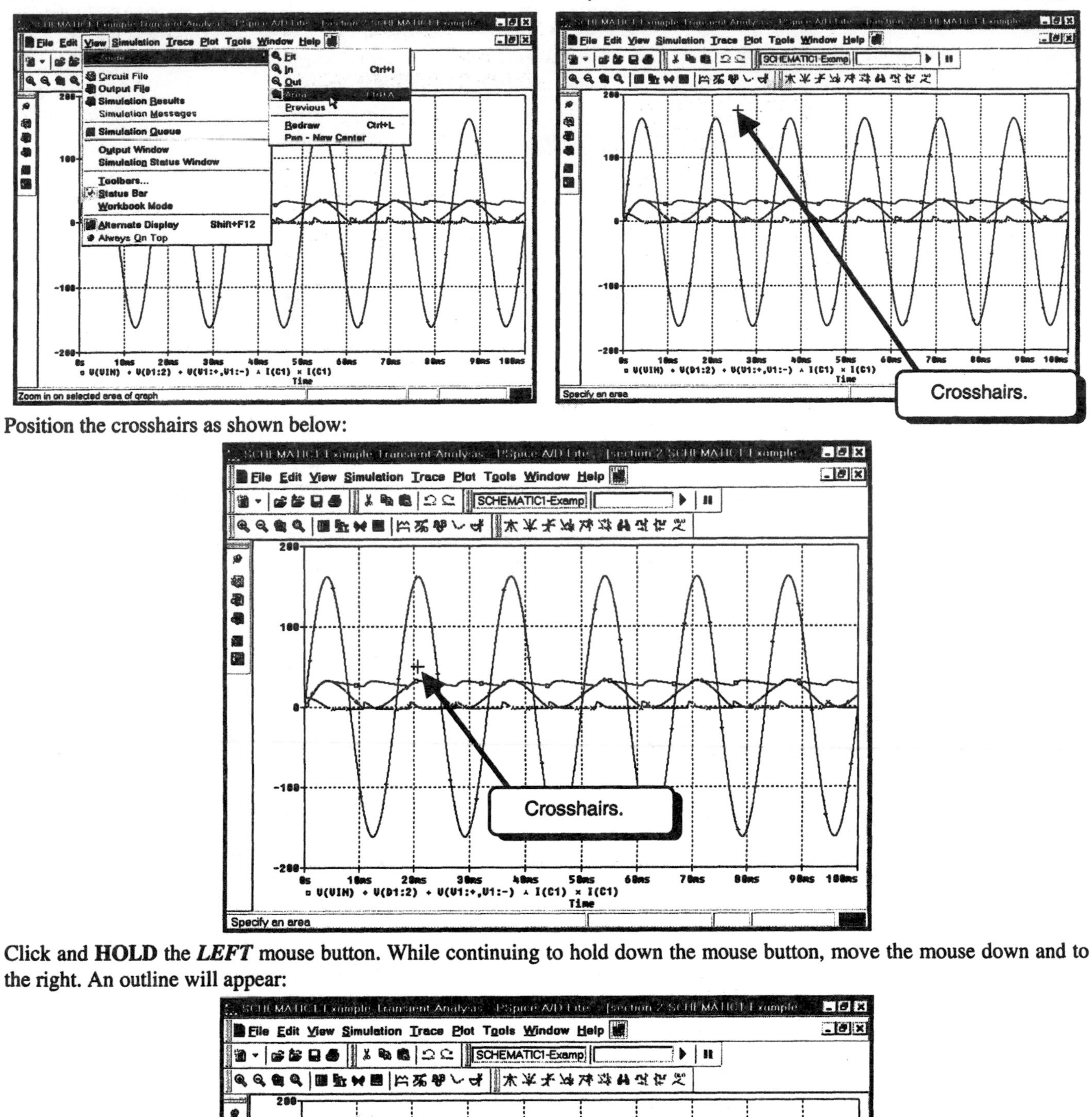


Position the crosshairs as shown below:

Click and **HOLD** the ***LEFT*** mouse button. While continuing to hold down the mouse button, move the mouse down and to the right. An outline will appear:

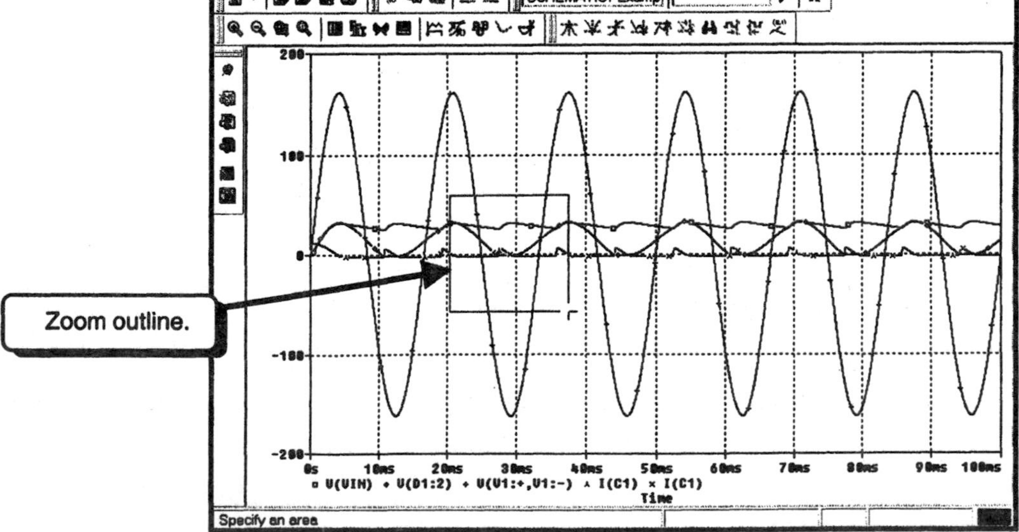

The portion of the plot inside the outline will be enlarged to fit the screen. Move the mouse to make an outline as shown above and release the mouse button. The display will zoom in on the area:

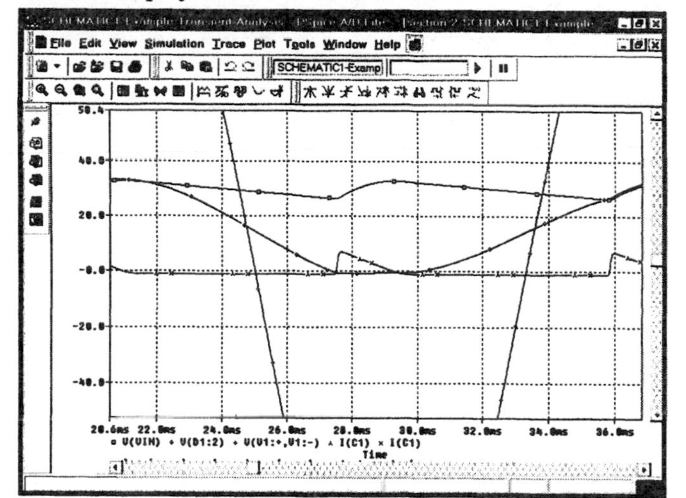


We can use the same technique to zoom in further. This time, type **CTRL-A** to zoom in again with the same method. Draw an outline as shown:

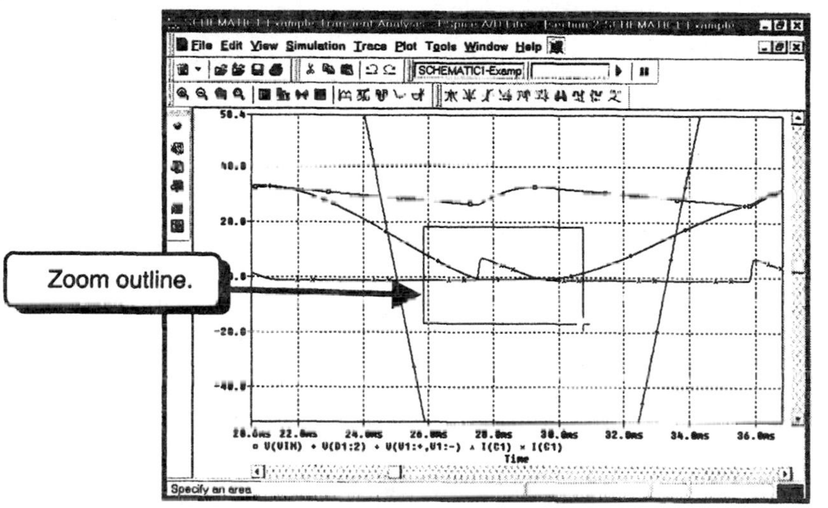


When you release the mouse button, Probe will zoom into the specified area:

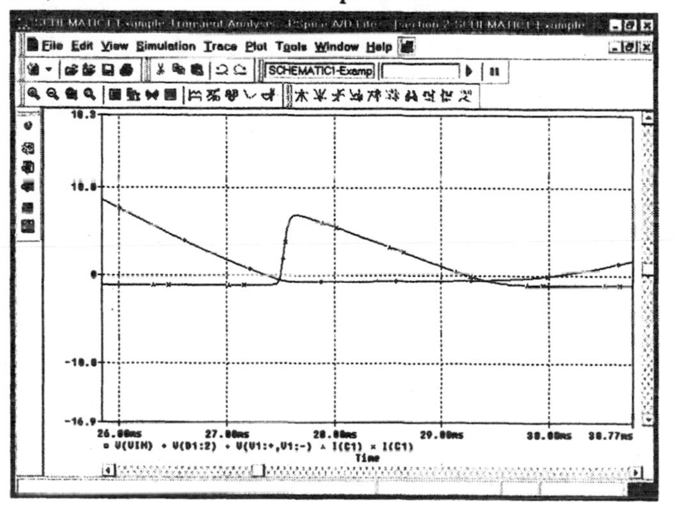


Suppose that we do not like the present view. To return to the previous view, select **View**, **Zoom**, and then **Previous**:

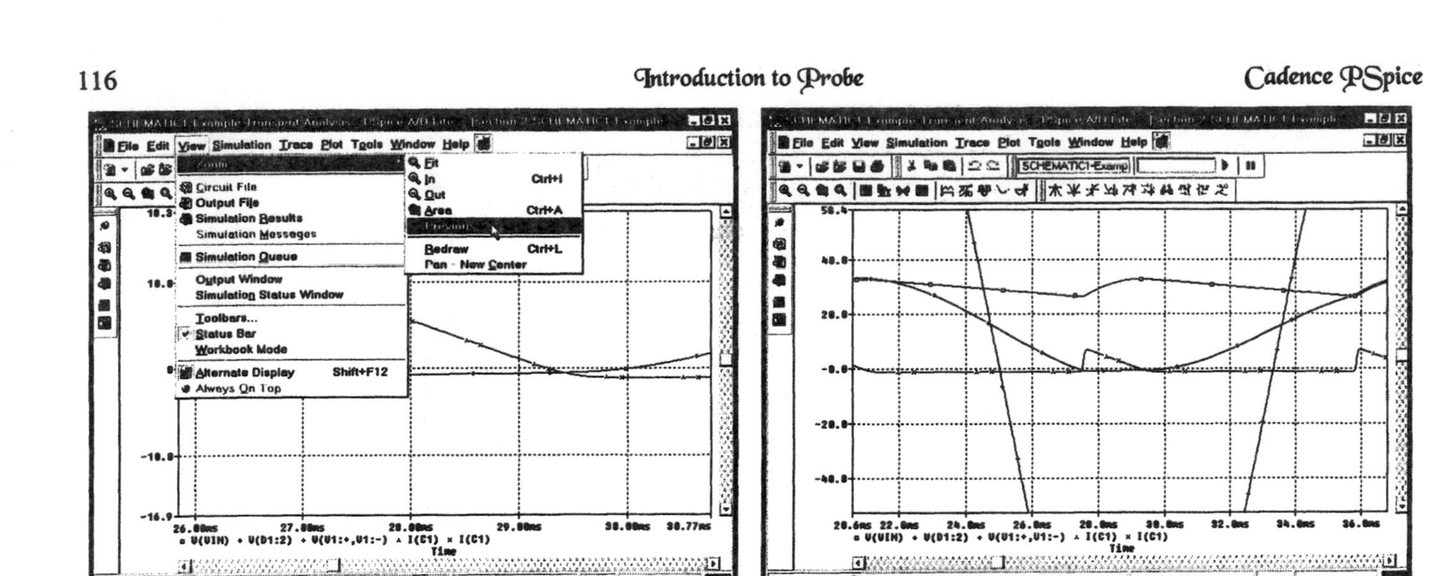


A second method for zooming in is selecting **View**, **Zoom**, and then **In**. This will zoom in around the cursor by a fixed percent. Select **View**, **Zoom**, and then **In**. The cursor will be replaced by crosshairs:

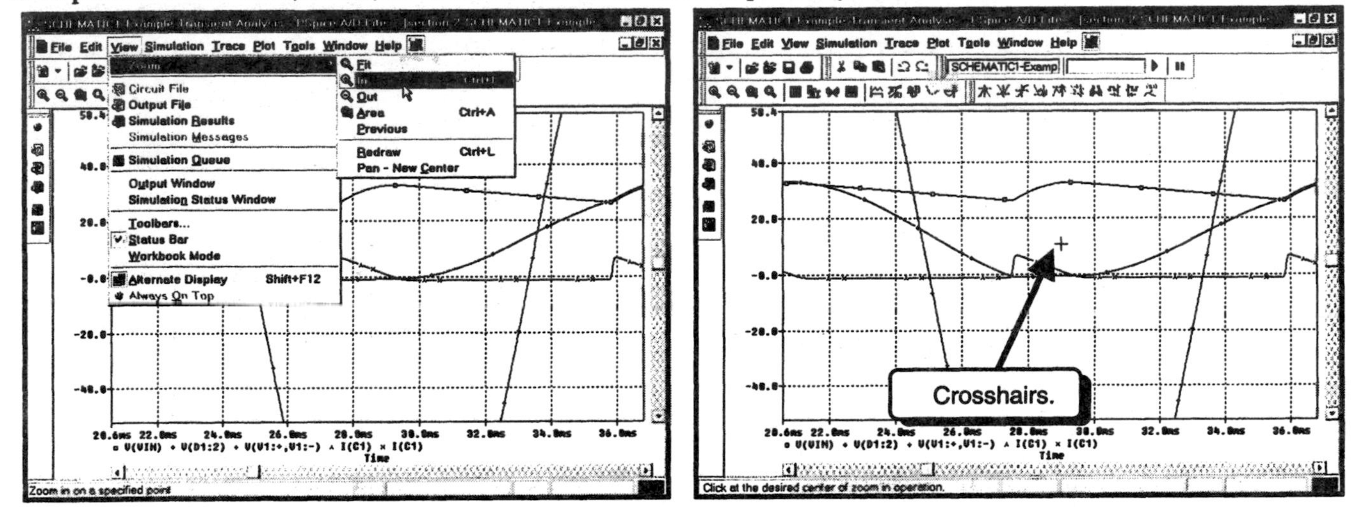


Place the crosshairs as shown above and click the ***LEFT*** mouse button. Probe will zoom in around the location of the crosshairs:

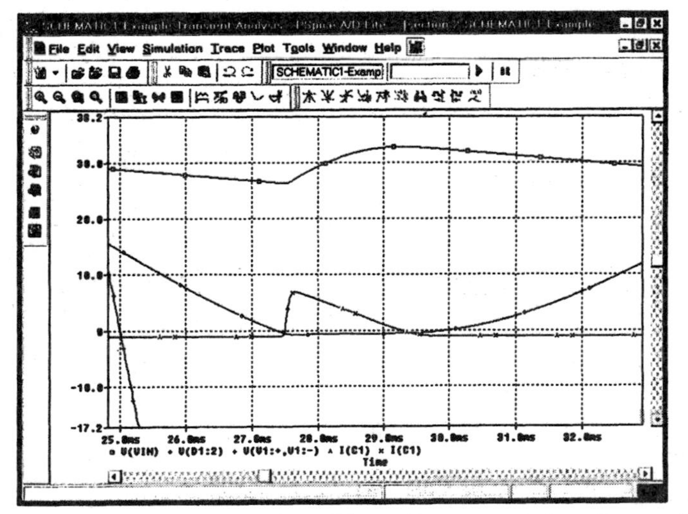

We can repeat the procedure and zoom in further. Select **View**, **Zoom**, and then **In** to obtain the crosshairs. Place the crosshairs where you would like to enlarge the plot and click the ***LEFT*** mouse button:

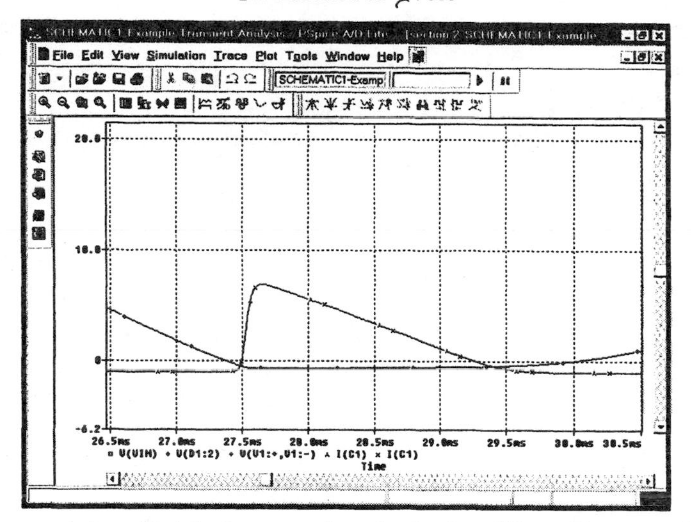

Selecting **View**, **Zoom**, and then **Out** is the opposite of selecting **View**, **Zoom**, and then **In**. Select **View**, **Zoom**, and then **Out**. Crosshairs will appear. Place the crosshairs where you would like to zoom out. When you click the ***LEFT*** mouse button, Probe will zoom out around the crosshairs:

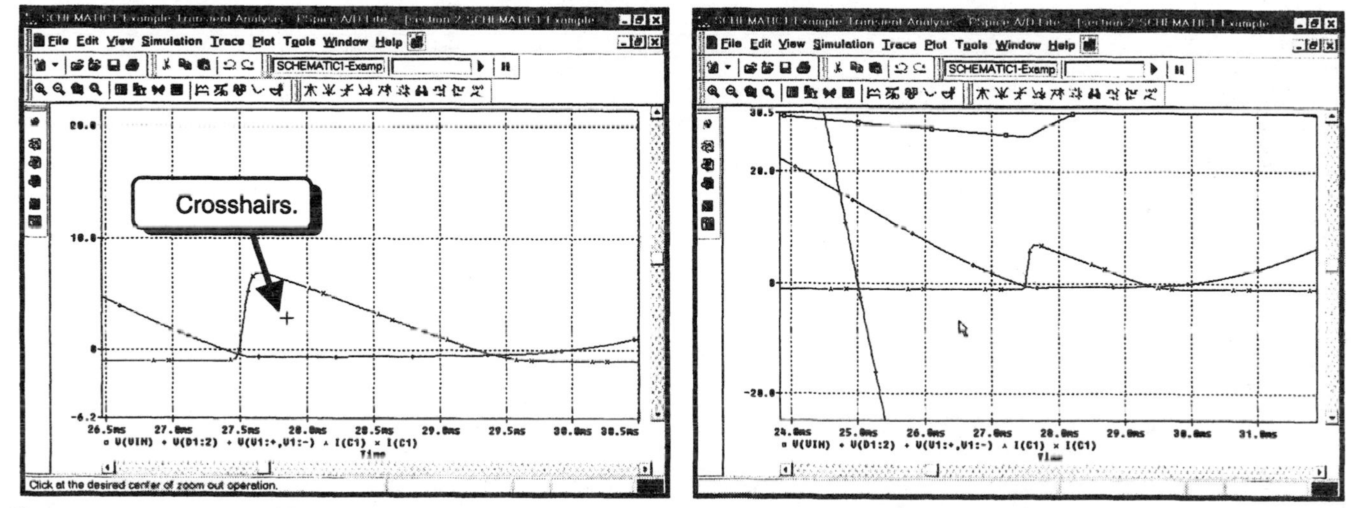


You can zoom out more if you wish.

To return the plot to its original view, select **View**, **Zoom**, and then **Fit**. This will fit the entire plot to the screen:

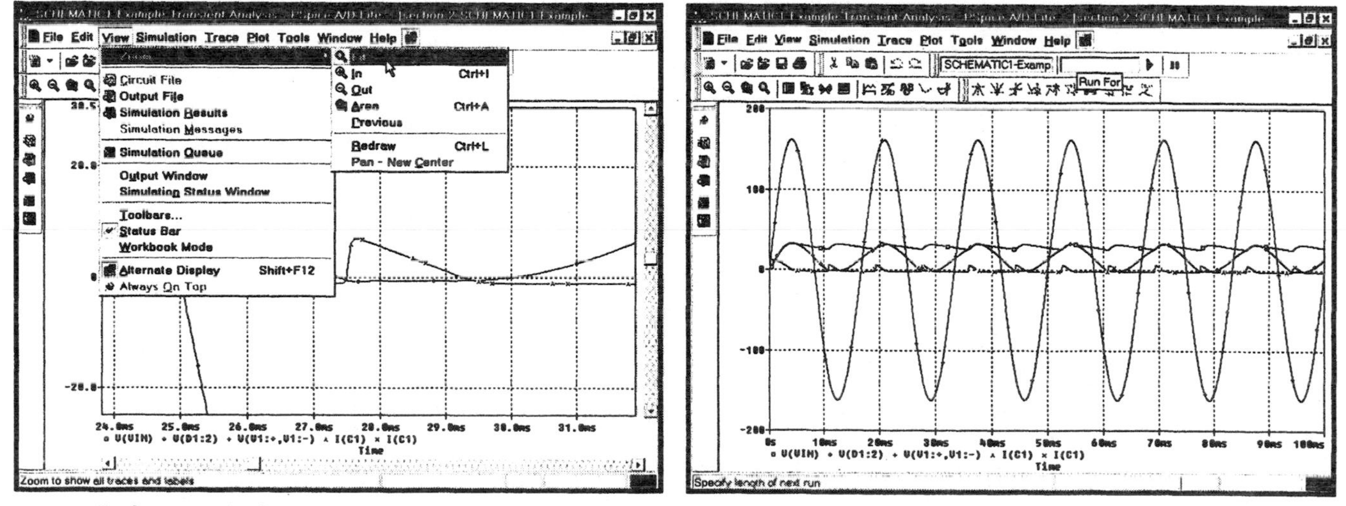

Before continuing we must close the current data file in Probe, and close the current schematic in Capture. In Probe, select **File** and then **Close** to close the data file.

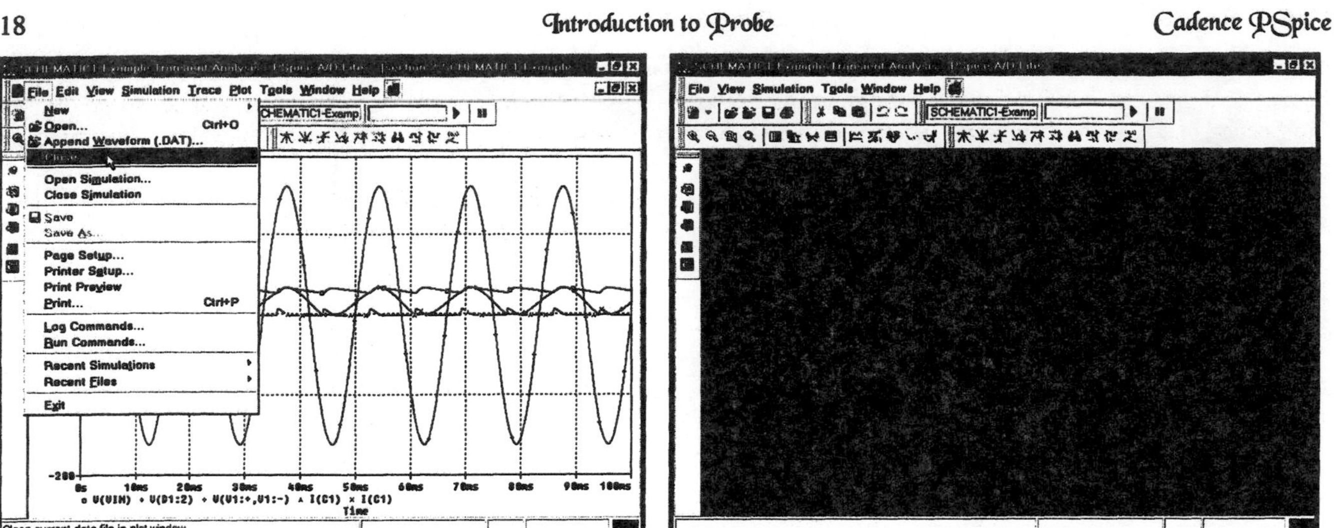


Next, select **File** and then **Close** in Capture to close the schematic. Save the changes if asked.

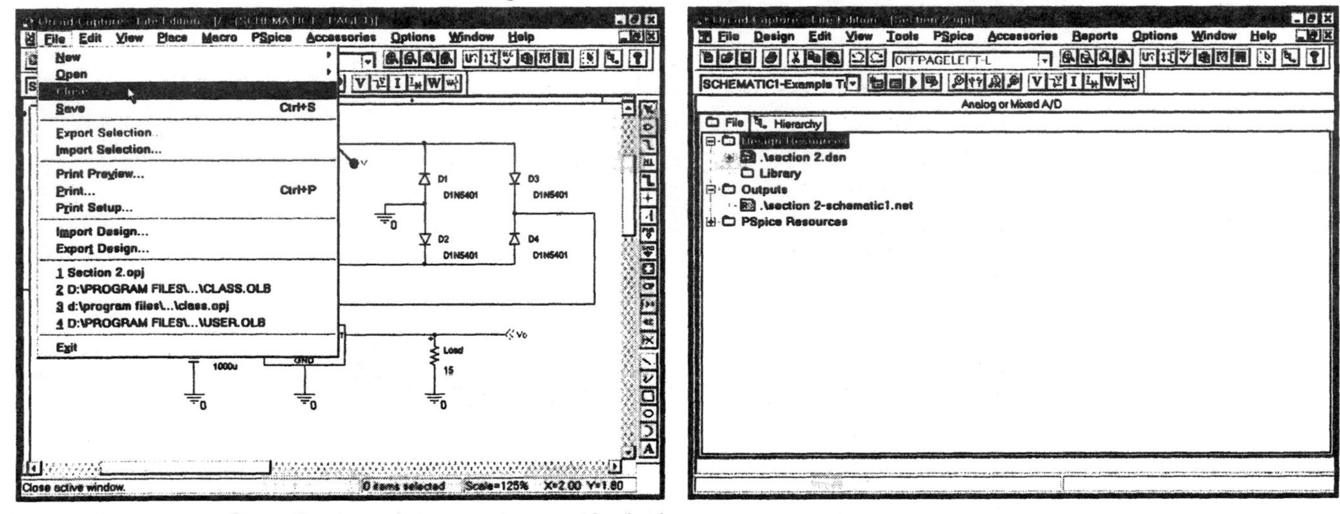


Select **File** and then **Close Project**. Save the changes if asked.

# 2.E. Adding a Second Y-Axis

In the previous example, we had several traces on a single plot. Some of the traces became hard to see when the magnitude of the numerical values of the traces differed by large amounts. A typical example would be plotting a voltage trace and a current trace on the same plot. Typically voltage traces may be in volts and currents may be in milliamperes or microamperes. When Probe plots traces with different units on the same plot, it plots the magnitude of the numerical values on the plot. If we plot a voltage that ranges from 0 to 5 volts and currents that range from 0 to 5 mA, both traces will be plotted with a y-axis that can accommodate numerical values from 0 to 5. With this scale, the current trace will be displayed close to zero and it will be hard to see any detail. We will use the circuit below to illustrate. The name of the circuit is Section 2e.opj. Open the file from the Program Files\OrcadLite\Book Circuits directory in the same manner as we did with file Section 2.opj at the beginning of this chapter. See pages 93–94 for a detailed procedure of opening a file from this directory. The schematic is shown below:

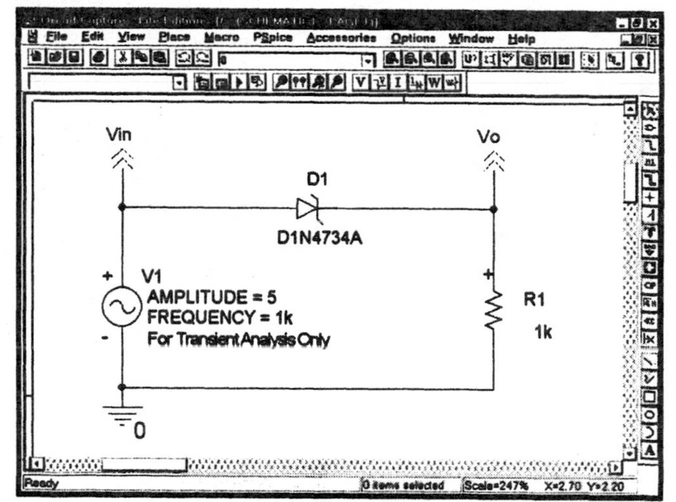


Press the **F11** key to simulate the circuit and run Probe:

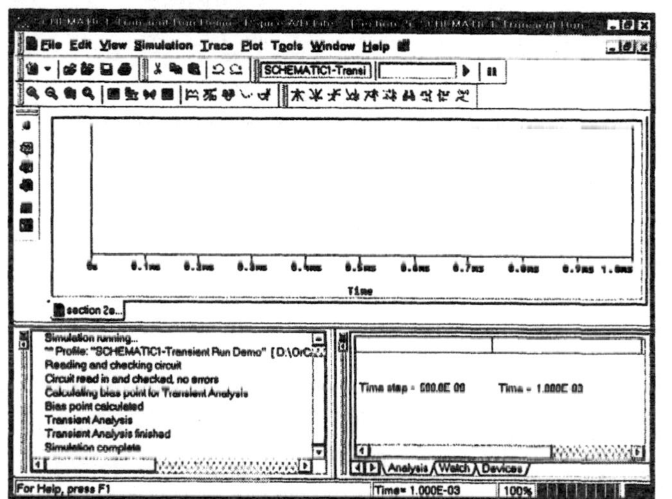

Add traces V(Vin) and I(R1). You should see the following plot:

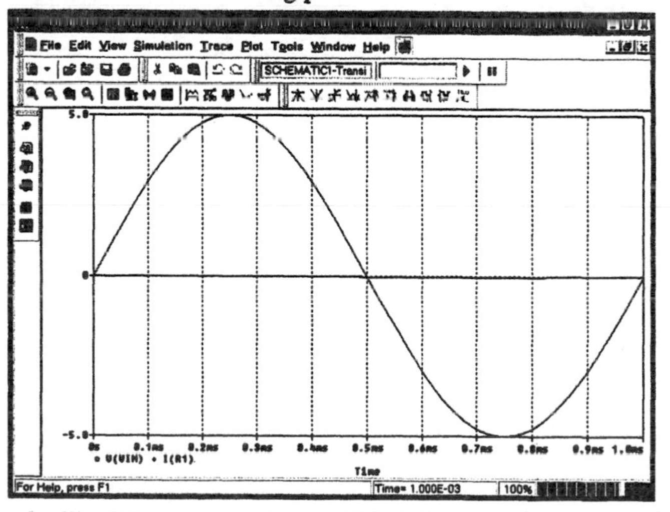

We see that the trace of I(R1) looks like it is constant at zero. This is because the numerical values of the current are 1,000 times smaller than those of the voltage. Delete the trace I(R1):*

---

*See page 105 for deleting traces.

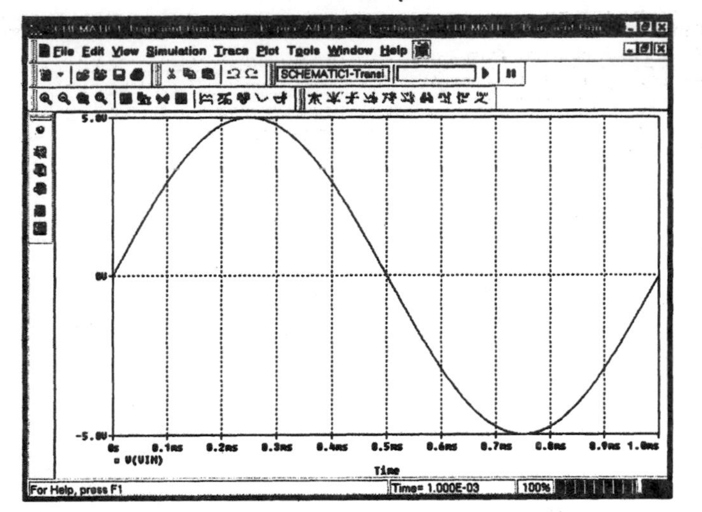

To fix this display problem, we will add a second y-axis. One y-axis will be used for the voltage trace and the second y-axis will be used for the current trace. The advantage of this arrangement is that the two y-axes can have different scales. To add another y-axis, select **Plot** and then **Add Y Axis**:

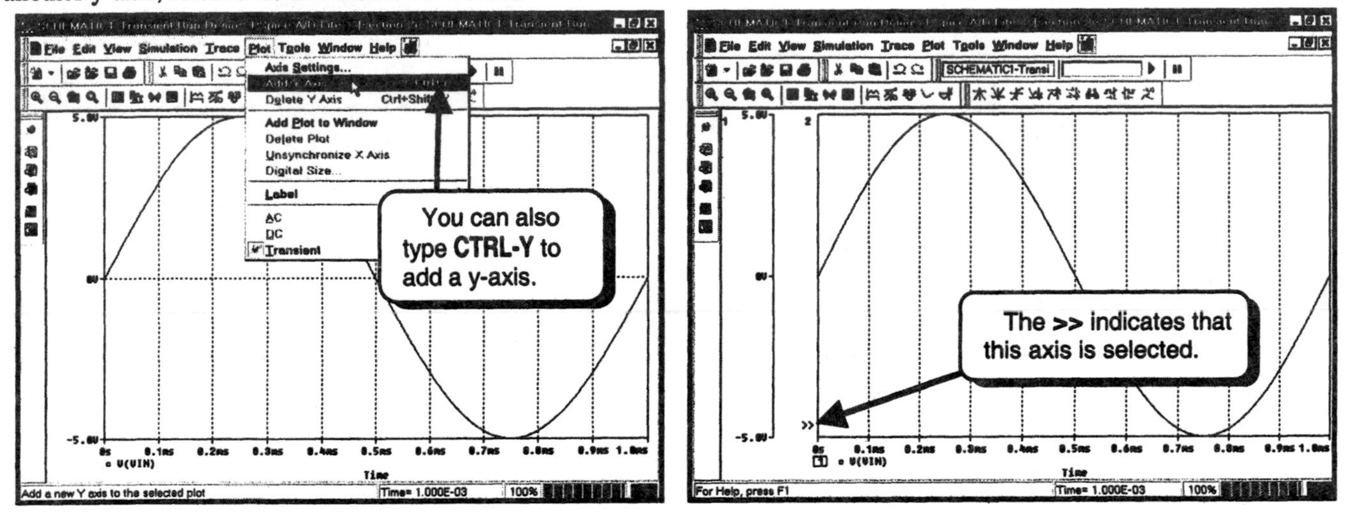


Note in the screen capture above that the axis we just added is selected. The next trace we add will be displayed using the selected axis, in this case, the new axis. Add the trace I(R1):

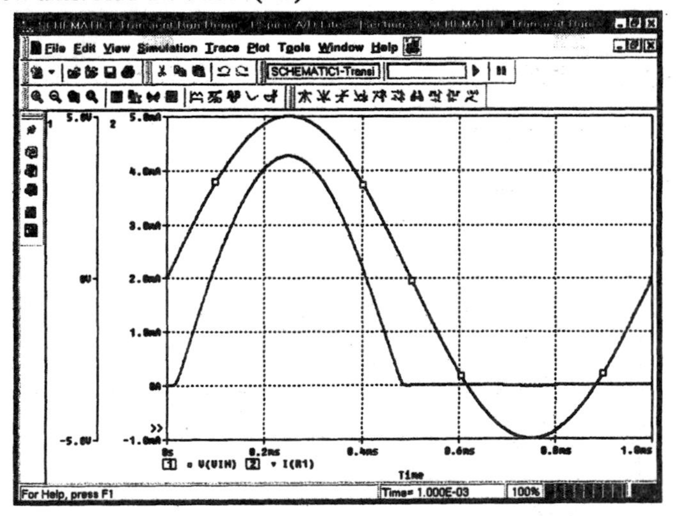

Traces that are added are placed on the selected y-axis. Below, the **>>** symbol indicates that the second y-axis is selected:

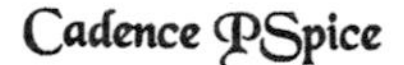


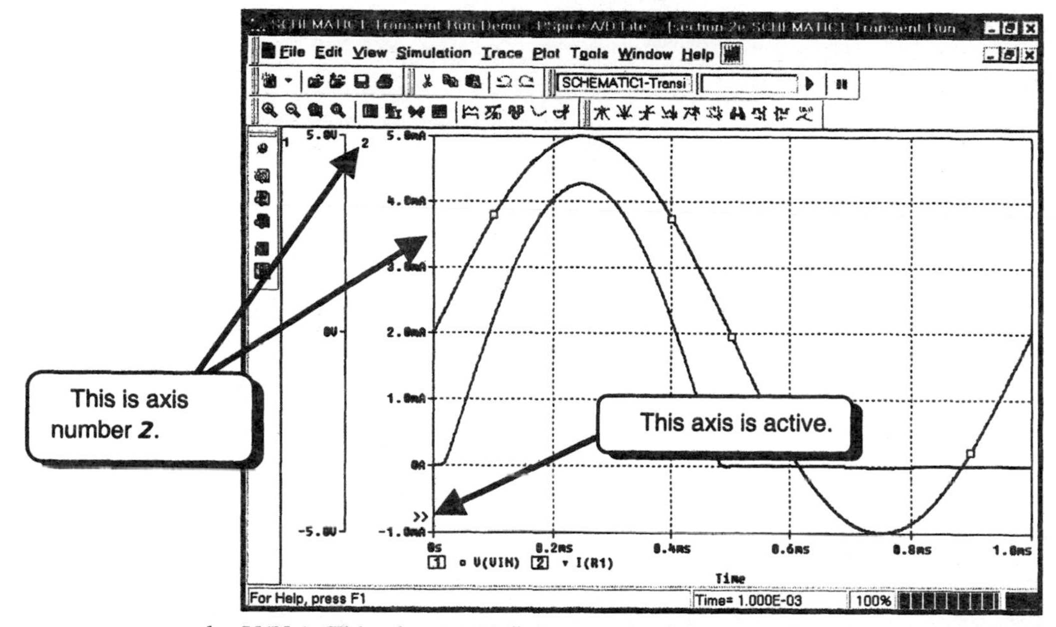


Suppose we want to plot V(Vo). This plot would fit best on the first y-axis. To select a y-axis, click the ***LEFT*** mouse button on the axis:

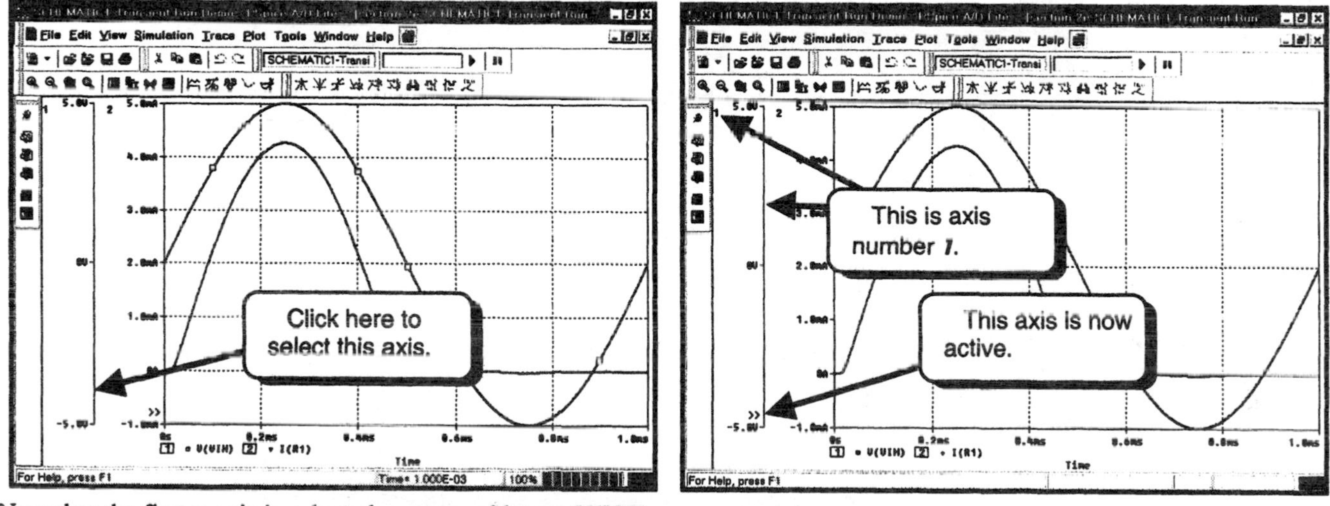

Now that the first y-axis is selected, we can add trace V(VO):

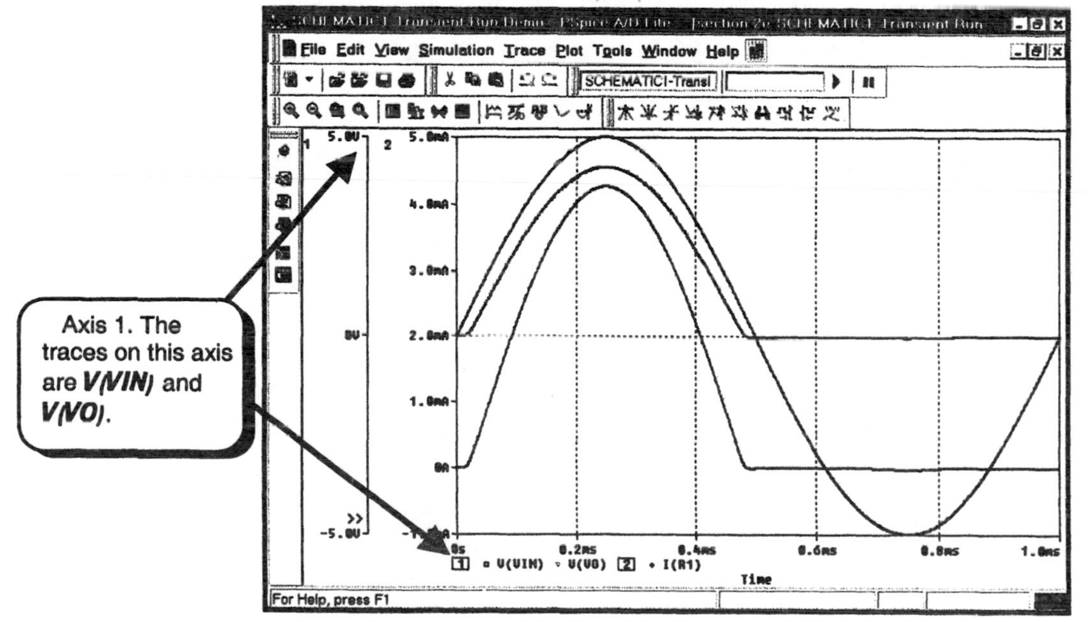

Note that trace ***V(VO)*** is placed in the list of traces for y-axis 1.

# 2.F. Adding Plots

We will use the circuit of the previous example to illustrate how to display multiple plots on the same window. Switch back to Capture using the **ALT - TAB** key sequence.* We will start with the schematic:

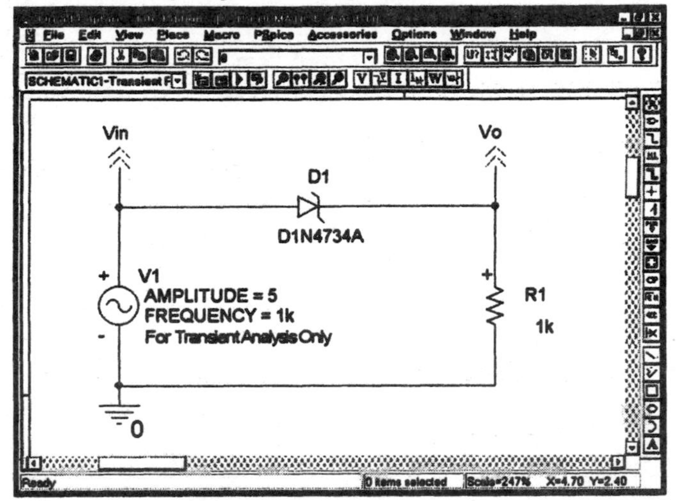


Press the **F11** key to simulate the circuit and run **Probe**:

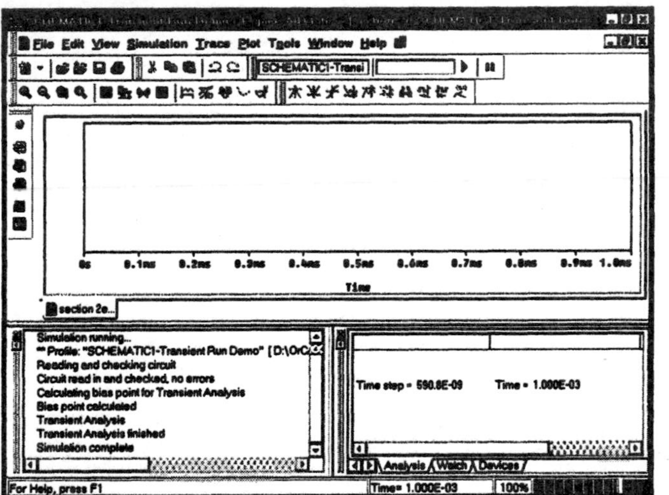


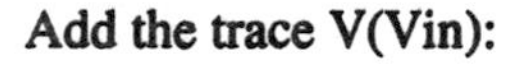
Add the trace V(Vin):

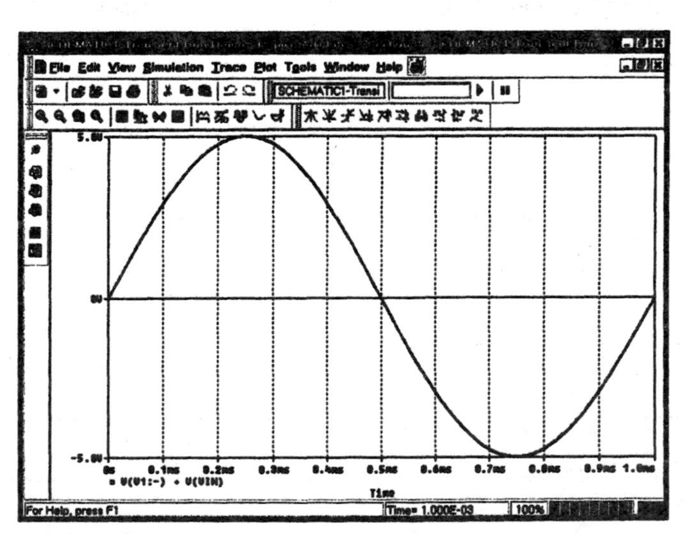

We will now create a second plot. Select **Plot** and then **Add Plot to Window**:

---

* See page 106 for a more detailed description of how to switch between windows using this method.

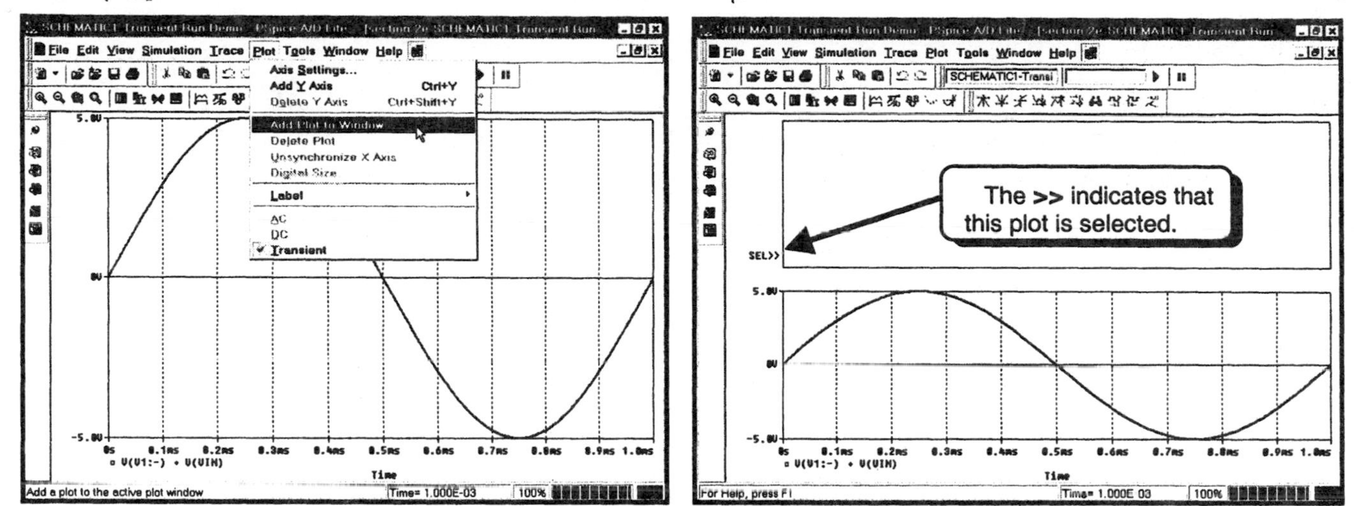


Notice that the top plot is selected. All new traces are added to the selected plot. Add the trace I(R1):

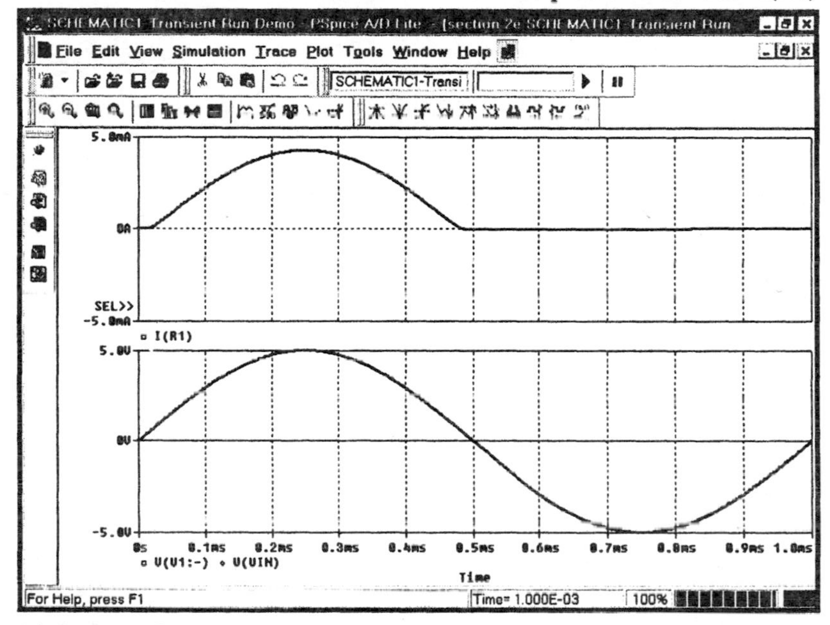


We would now like to add the trace V(Vo) to the lower plot. To select the lower plot and make it active, click the ***LEFT*** mouse button on the lower plot:

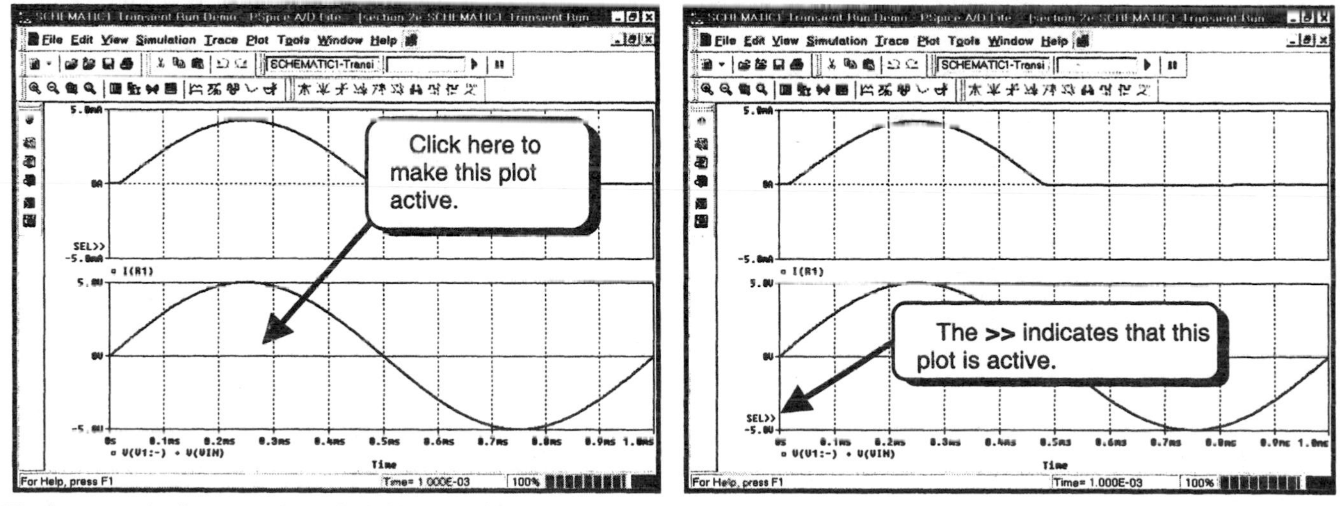


The bottom plot is now selected. Add trace V(Vo):

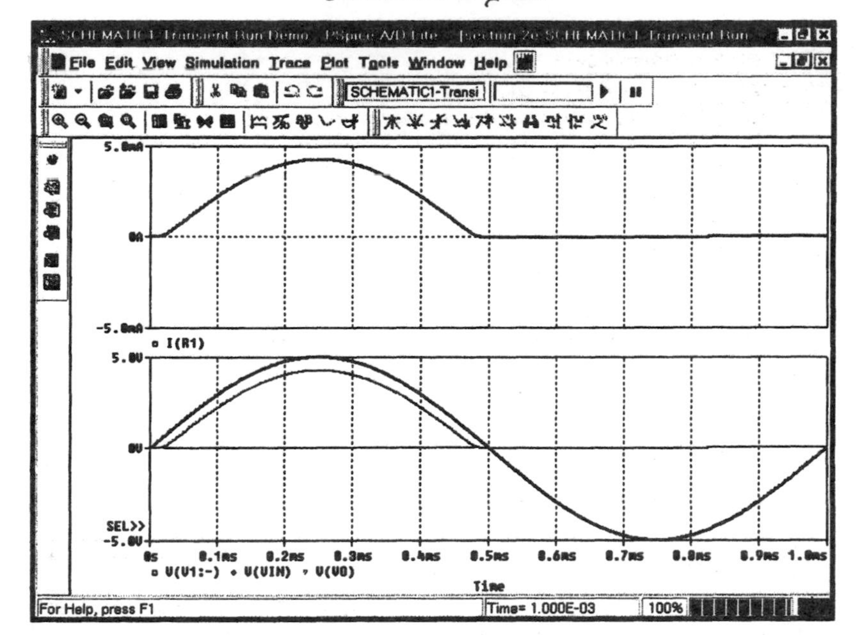

We see that the newly added trace is placed on the selected plot. We can add more plots to this page if we wish. Select **Plot** and then **Add Plot to Window**. A third plot will be added to the window:

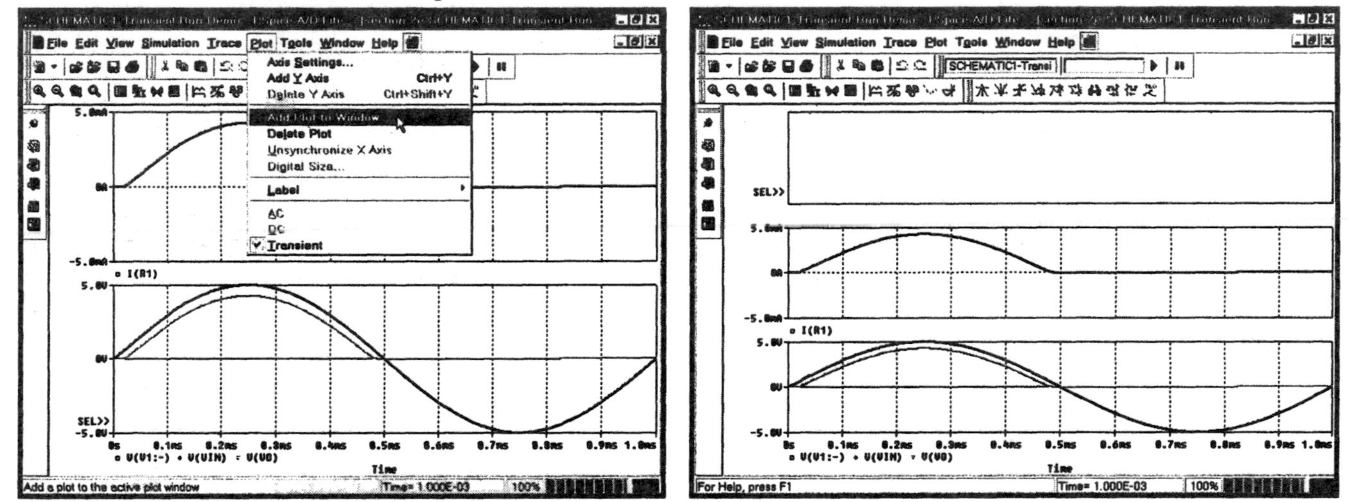

# 2.G. Adding a Window

Probe has the ability to display multiple windows. Each window can display different traces. We will continue now, starting at the end of the previous example. Select **Window** and then **New Window** to create a new window:

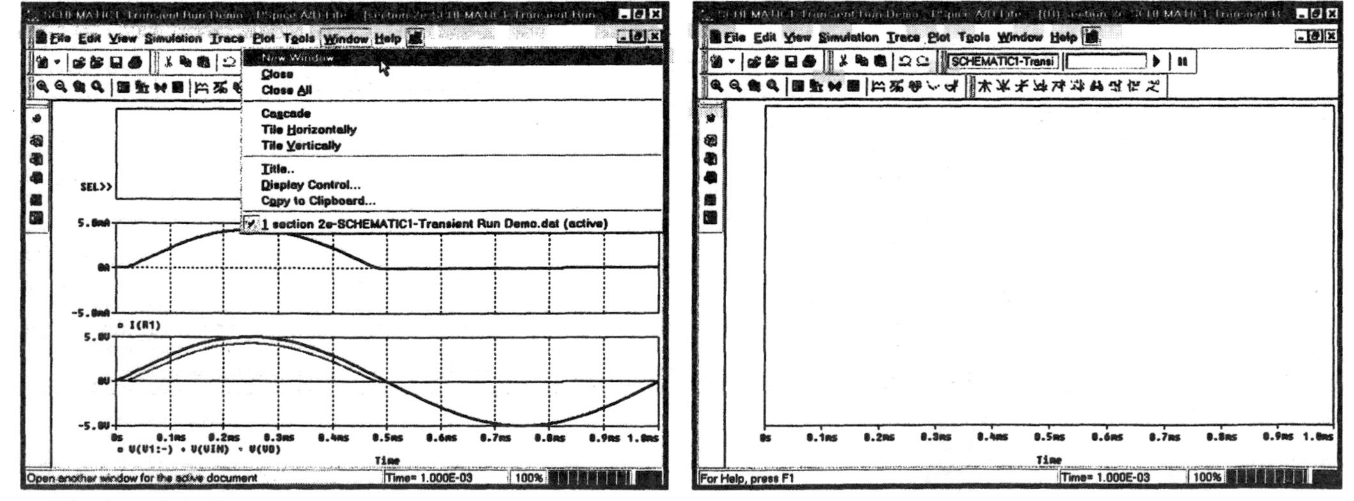

Add the trace V(Vin):

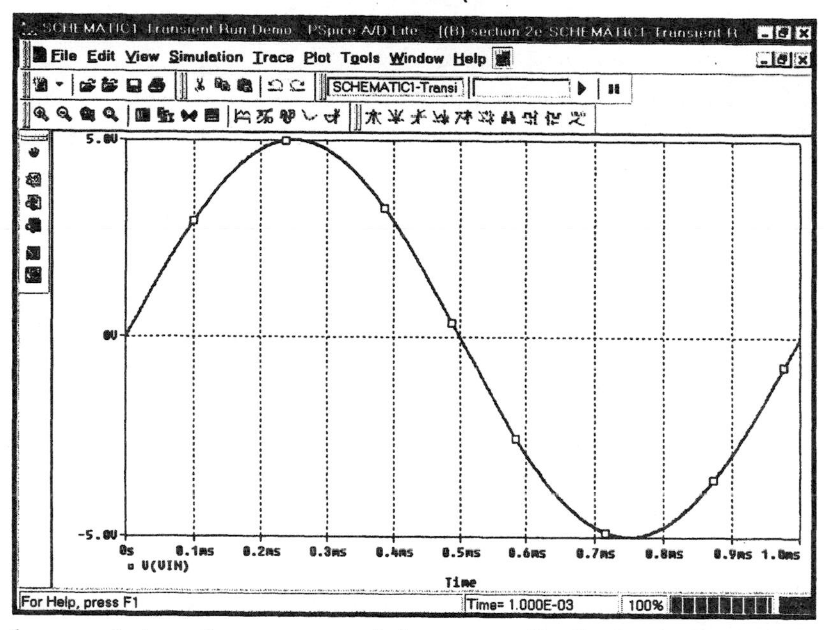


Although we can see only one window, there are two windows open. To display both windows at the same time, select **Window** and then **Tile Vertically**:

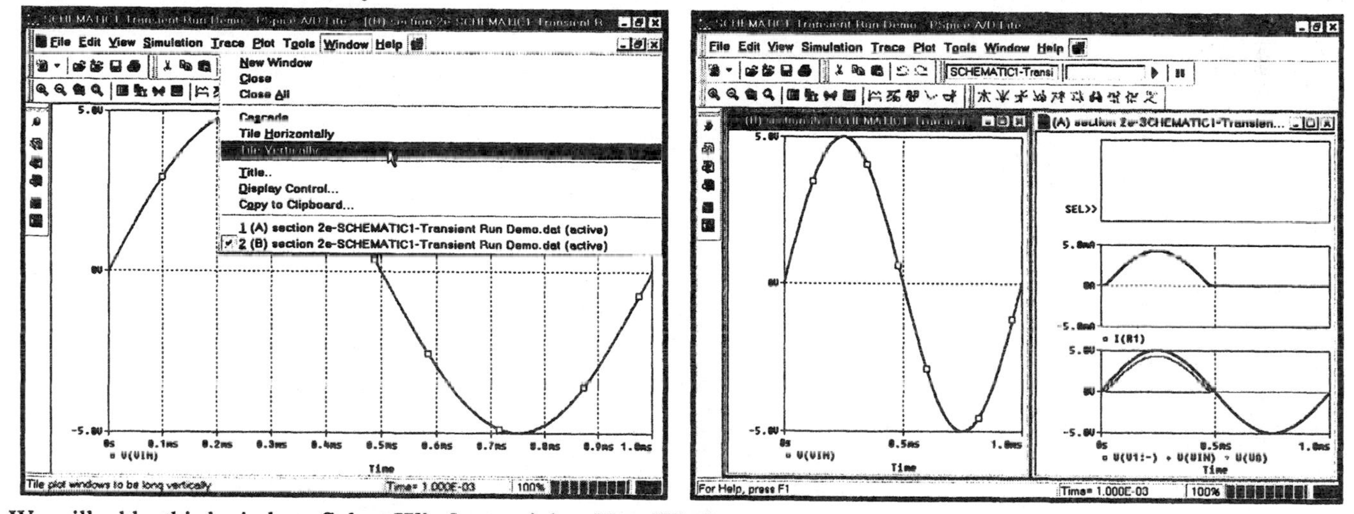


We will add a third window. Select **Window** and then **New Window**:

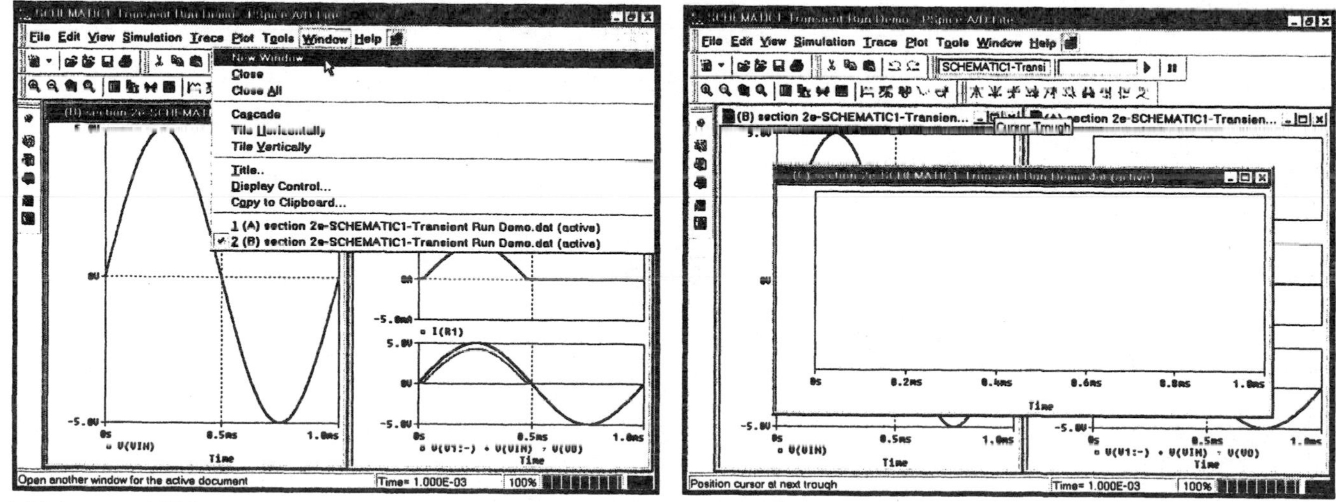


Add the trace I(R1):

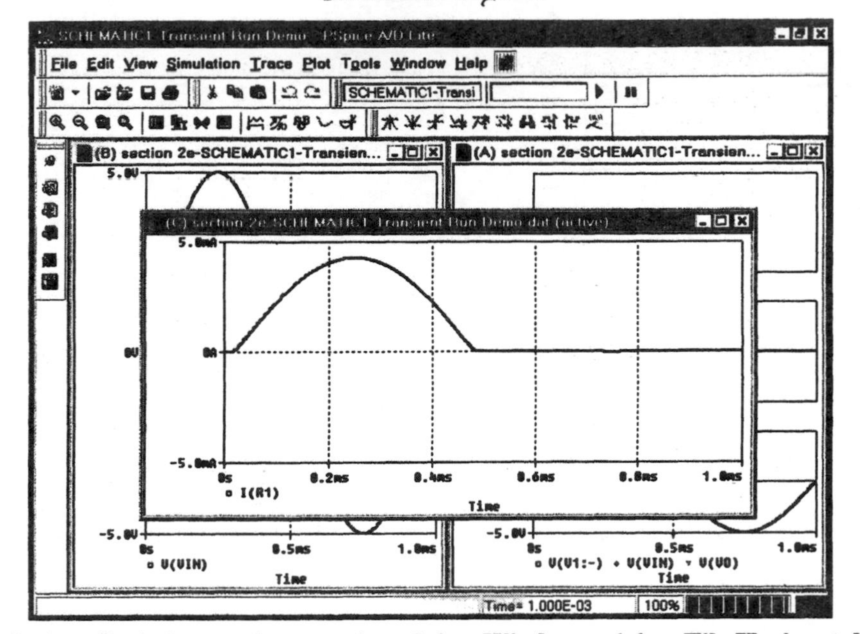

We would now like to display all windows at the same time. Select **Window** and then **Tile Horizontally**:

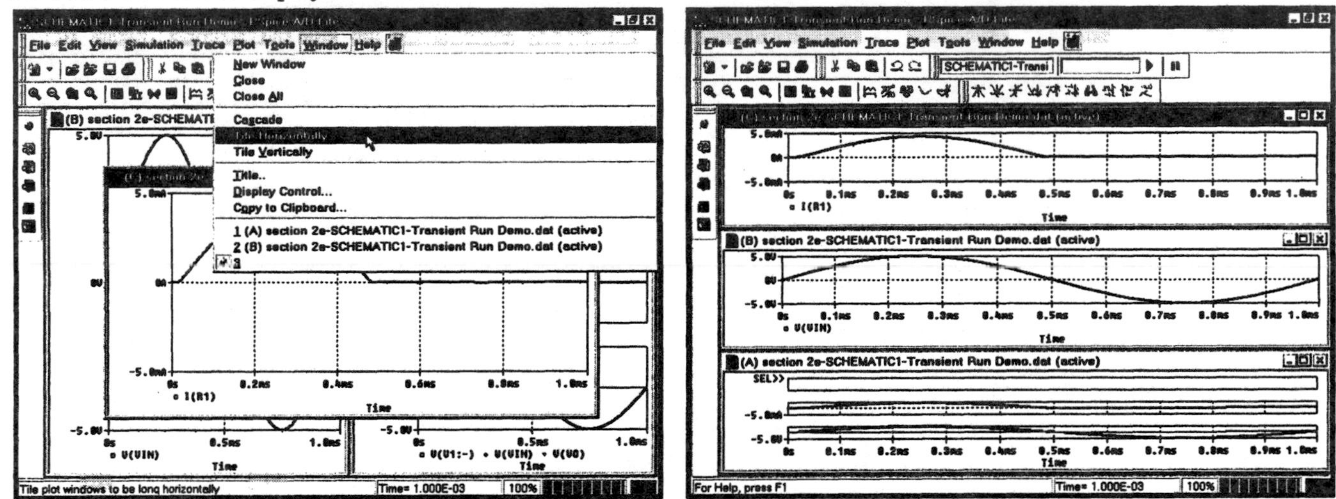

These plots do not look good when displayed in this manner. Select **Window** and then **Tile Vertically**:

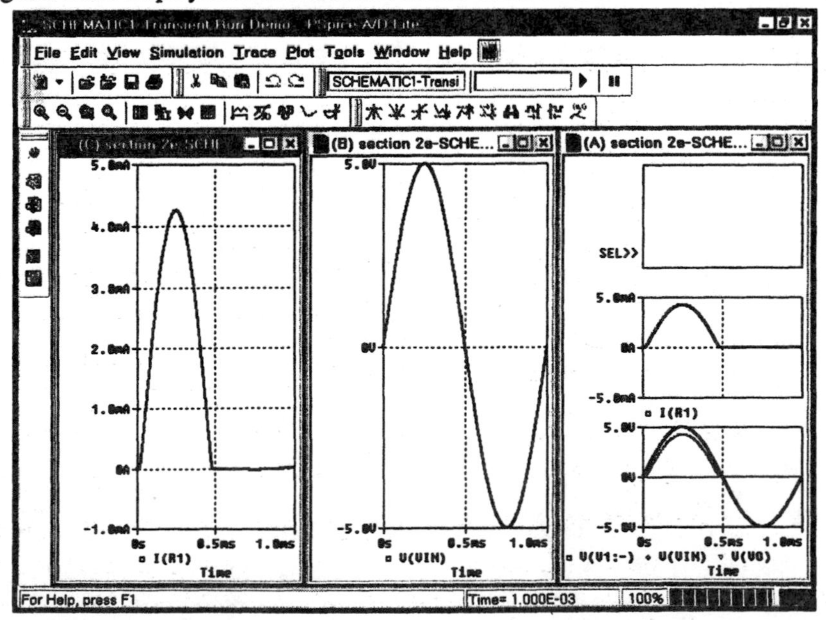

New traces are added to the selected window. To select a window, click the ***LEFT*** mouse button on the window you wish to use. For example, click the ***LEFT*** mouse button on the rightmost window. It will become selected:

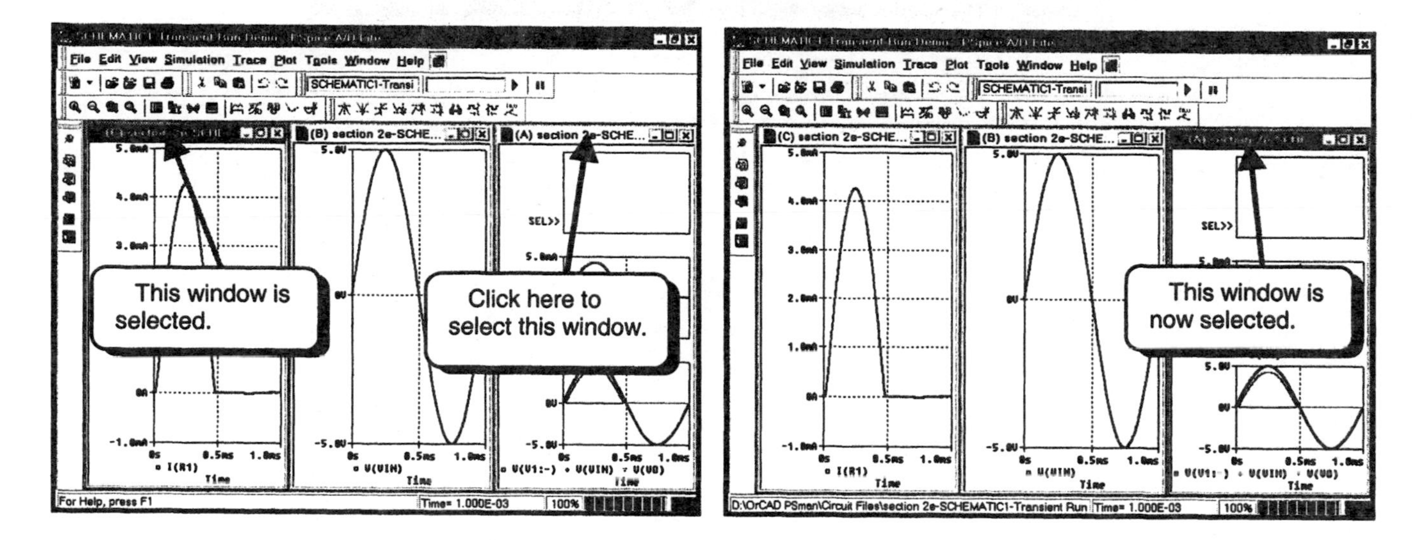


Add the trace V(Vo) to the selected window:

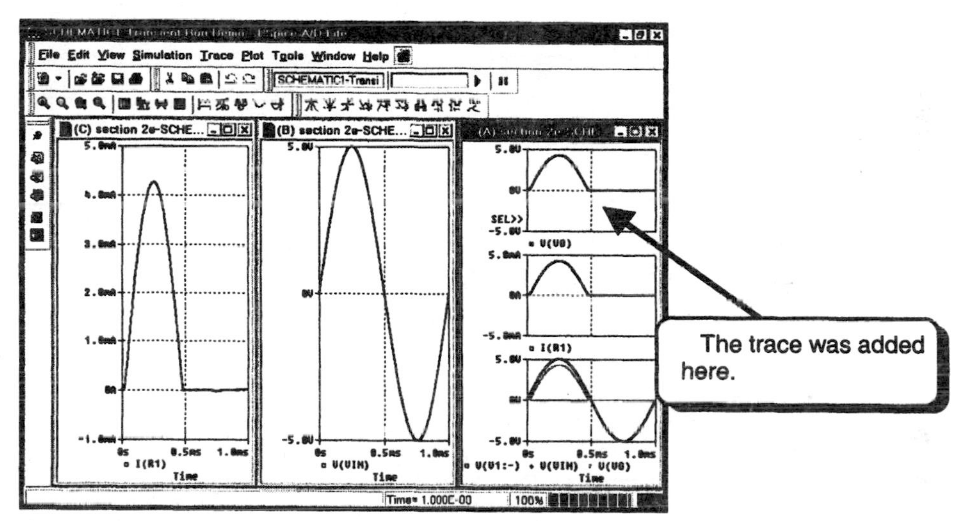


Notice that the trace is added to the active plot in the active window. To enlarge a window, click the ***LEFT*** mouse button on the maximize icon as shown below:

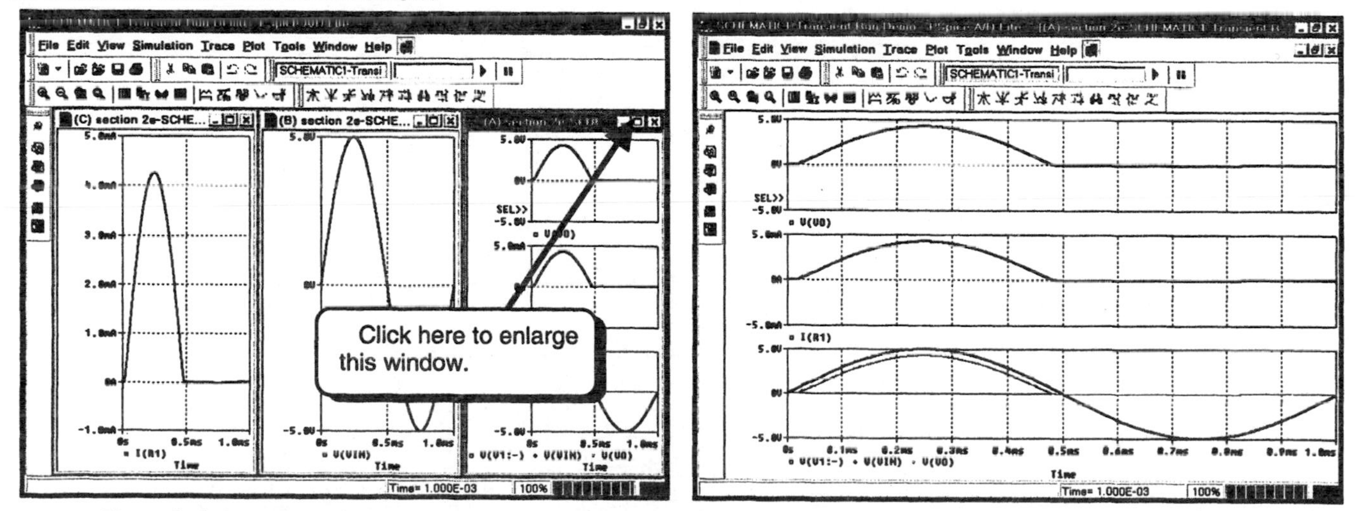


To switch between windows, we can use two methods. All of my screen captures shown recently show the Probe window occupying the entire window. In this mode we use the menus to switch between windows. Select **Window** from the Probe menus. The pull-down menu displays which window is currently selected and allows us to select a different window:

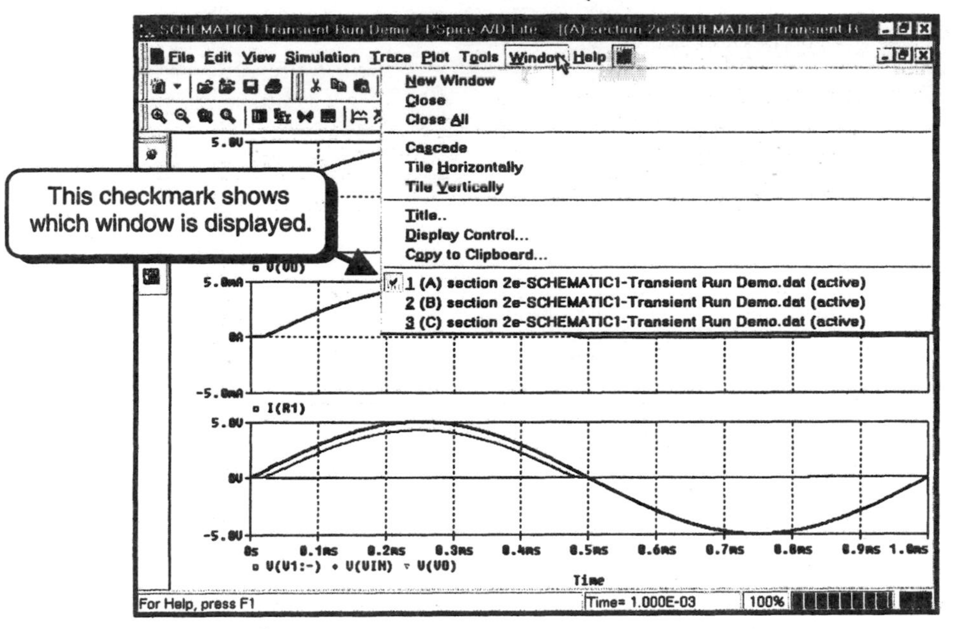


The checkmark shows that window ***A*** is selected. You can use the windows to switch to a different window. Below, we will display window ***B***:

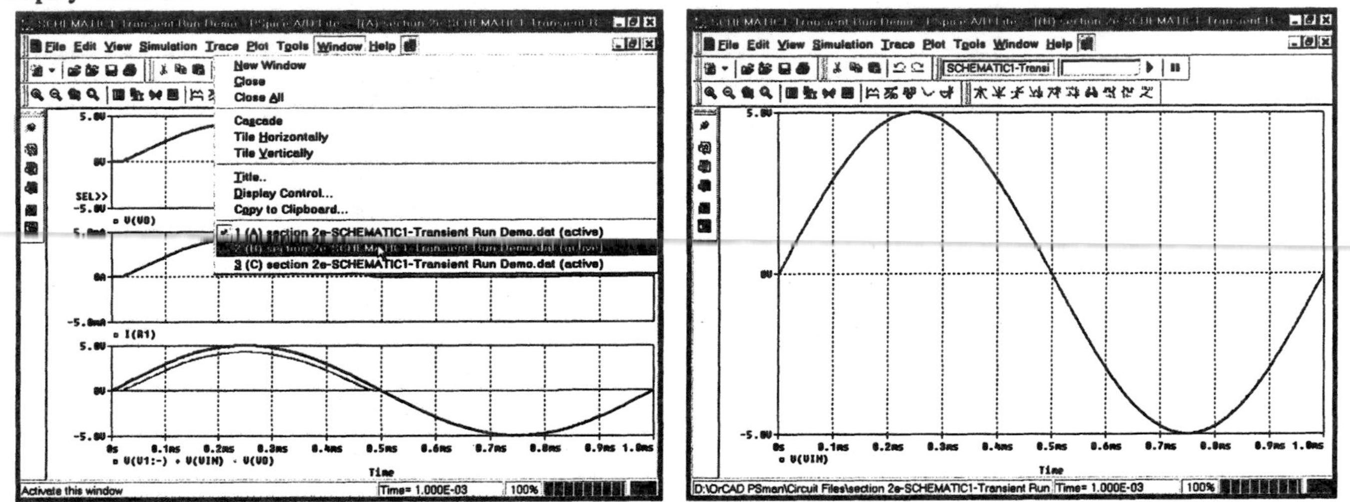

The second method of switching between windows requires us to use the alternate display. To toggle between displays, select **View** and then **Alternate Display** from the menus:

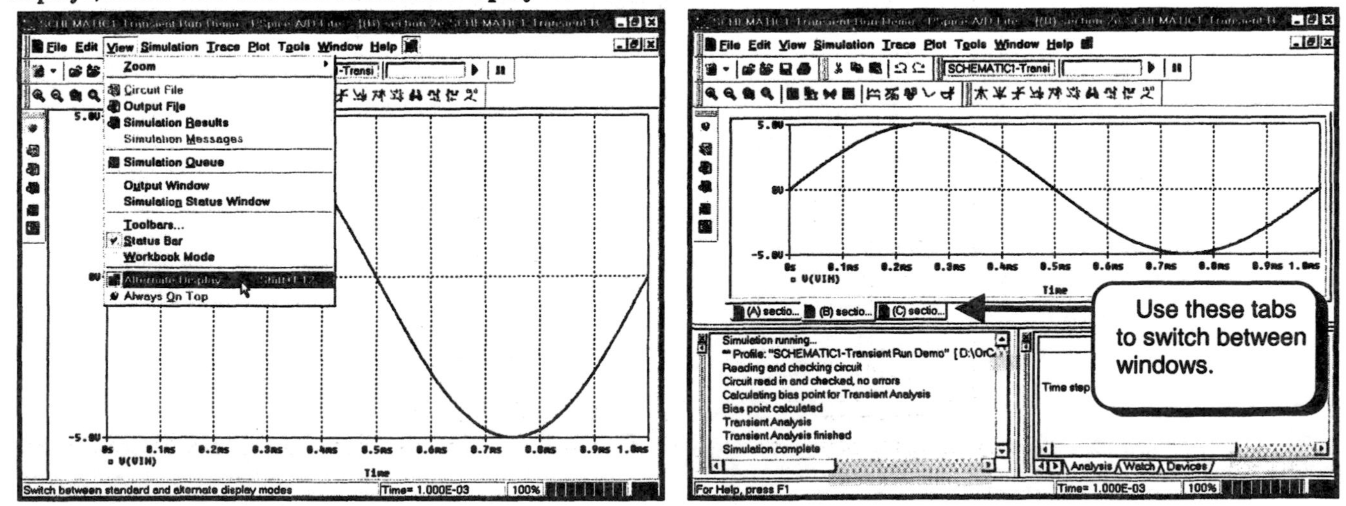


You can now use the tabs to switch between the windows. We will switch back to the previous display to show screen captures in full screen for the remainder of this text.

# 2.H. Placing Text on Probe's Screen

We will use the circuit named Section 2.opj. Open this file:*

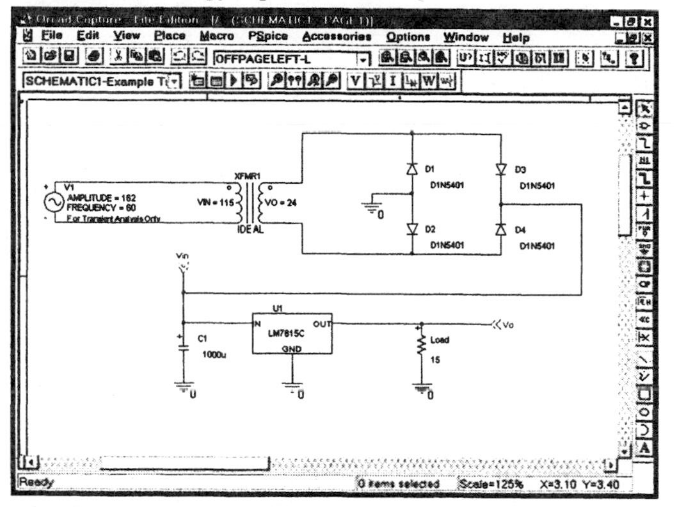


Press the **F11** key to simulate the circuit and then run Probe:

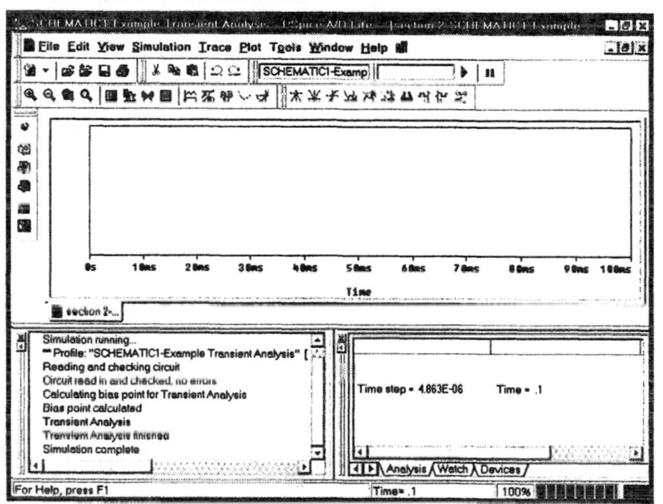


Add the trace V(Vin):

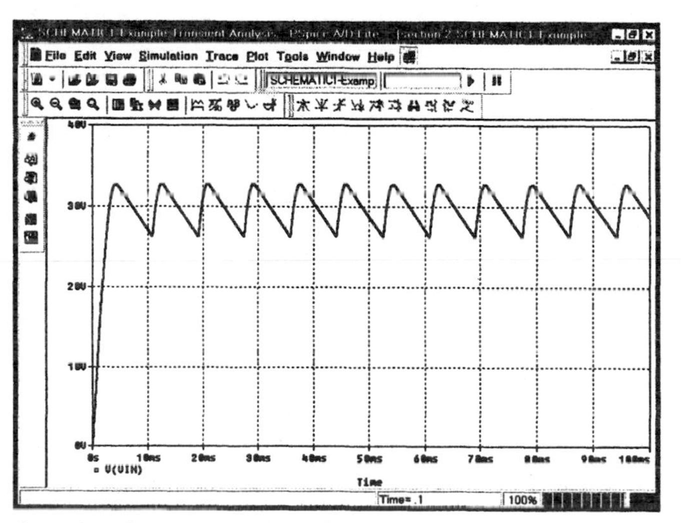


We can place text on the screen by using the menus or by using the button bar. We will first use the menus. Select **Plot** and then select **Label**:

---

*We used this circuit at the beginning of the chapter. Follow the instructions on pages 93–94 to open the file. If you saved changes to file Section 2.opj earlier, your Probe screen may display some waveforms when Probe starts.

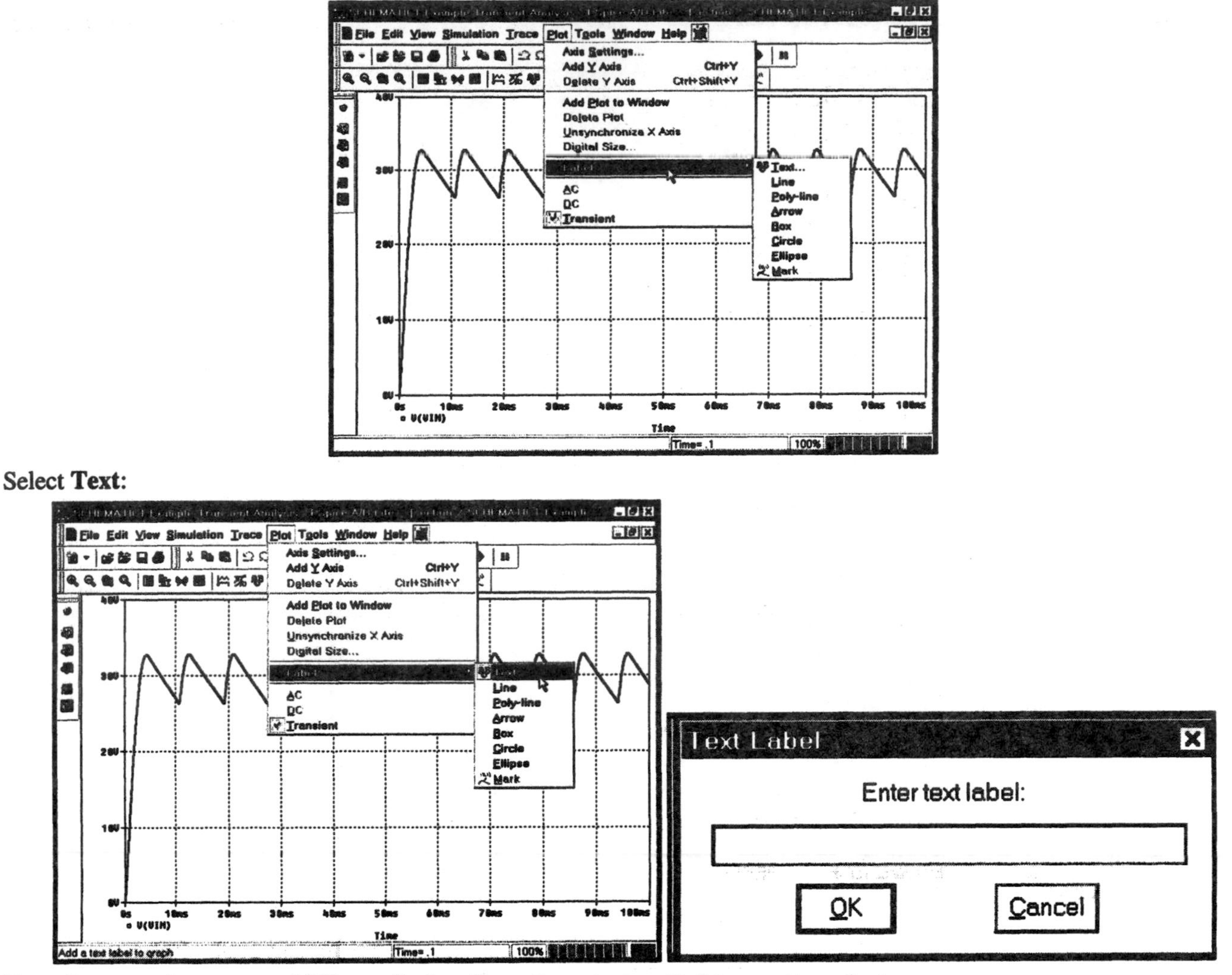


Select **Text**:

Type the text string you would like to display. Type **`Input to Voltage Regulator`**:

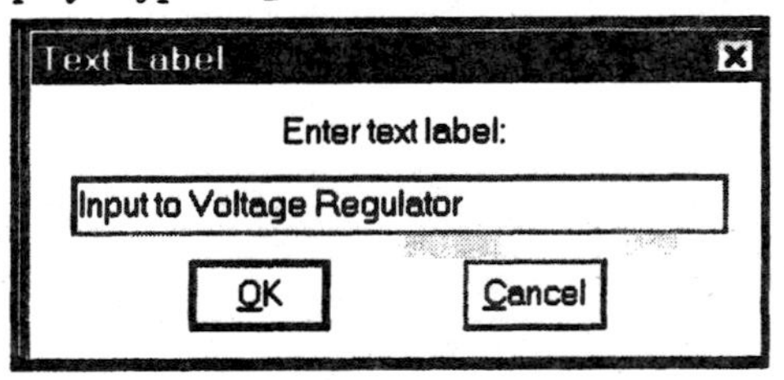


Click the ***OK*** button. The text string will replace the mouse pointer and move with the mouse:

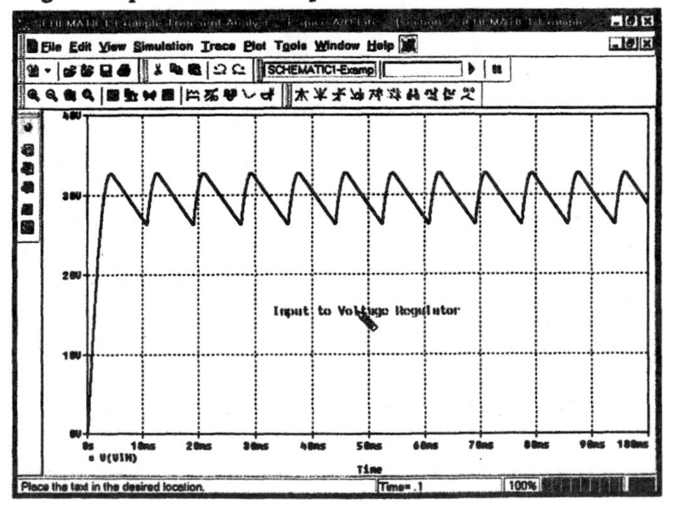

Position the text as shown below and click the ***LEFT*** mouse button. The text will be placed and the normal mouse pointer will return:

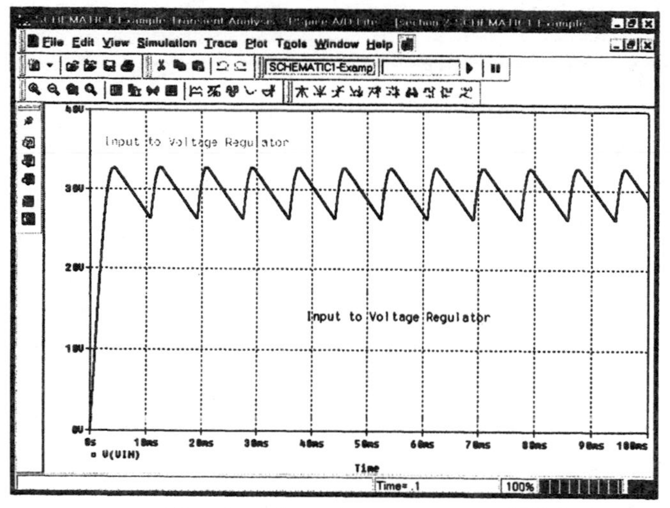


There may be fragments of text on your screen. To clear these fragments, type **CTRL-L**:

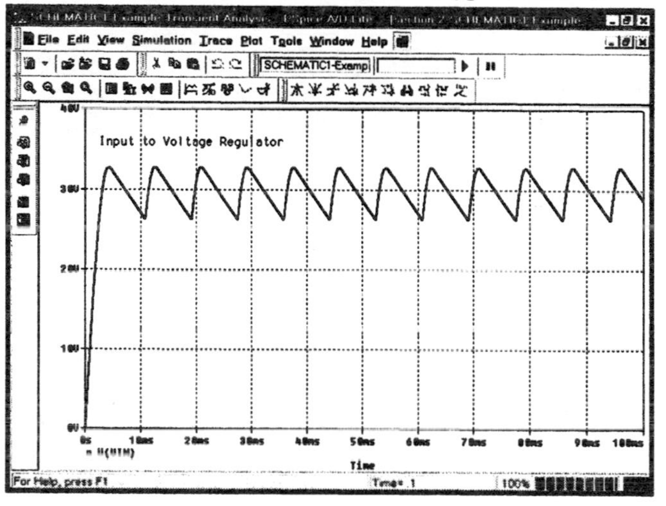


Any text now displayed on your screen is actually on your screen.

Next, add the trace for the output of the regulator, V(Vo), to the plot:

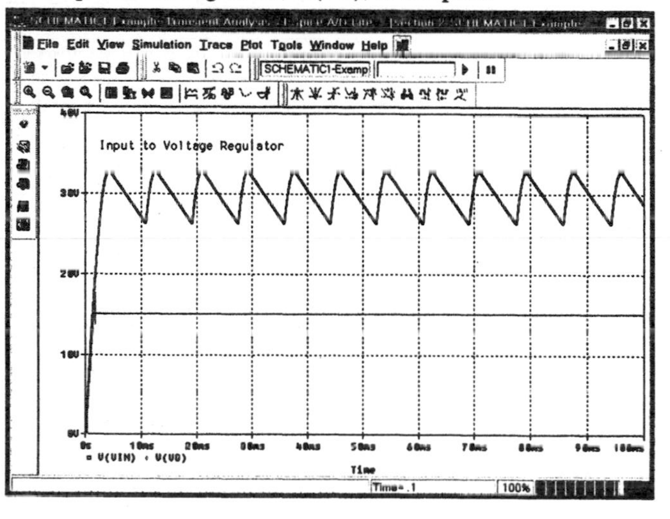


The second method of adding text is to click the ***ABC*** button as shown below:

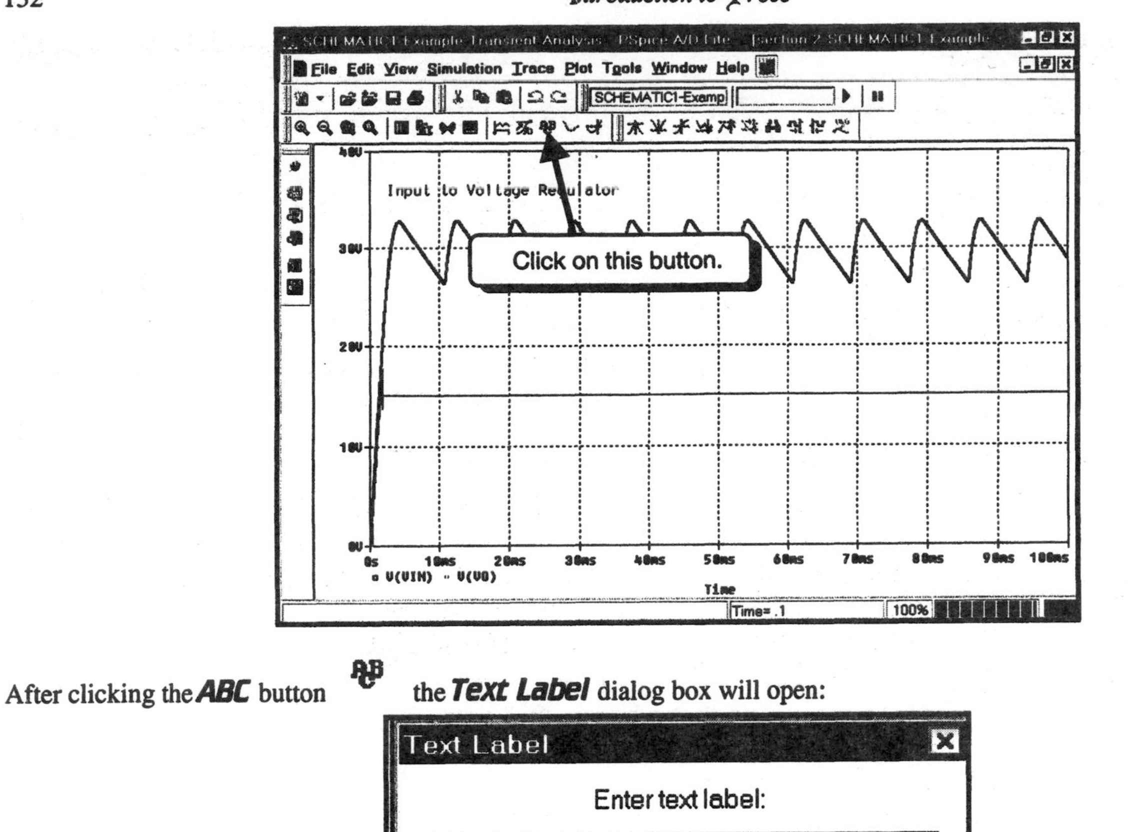


After clicking the ***ABC*** button the ***Text Label*** dialog box will open:

Text Label

Enter text label:

OK Cancel

Type the text **`Regulator Output Voltage`** and press the **ENTER** key. Position the text as shown and click the ***LEFT*** mouse button:

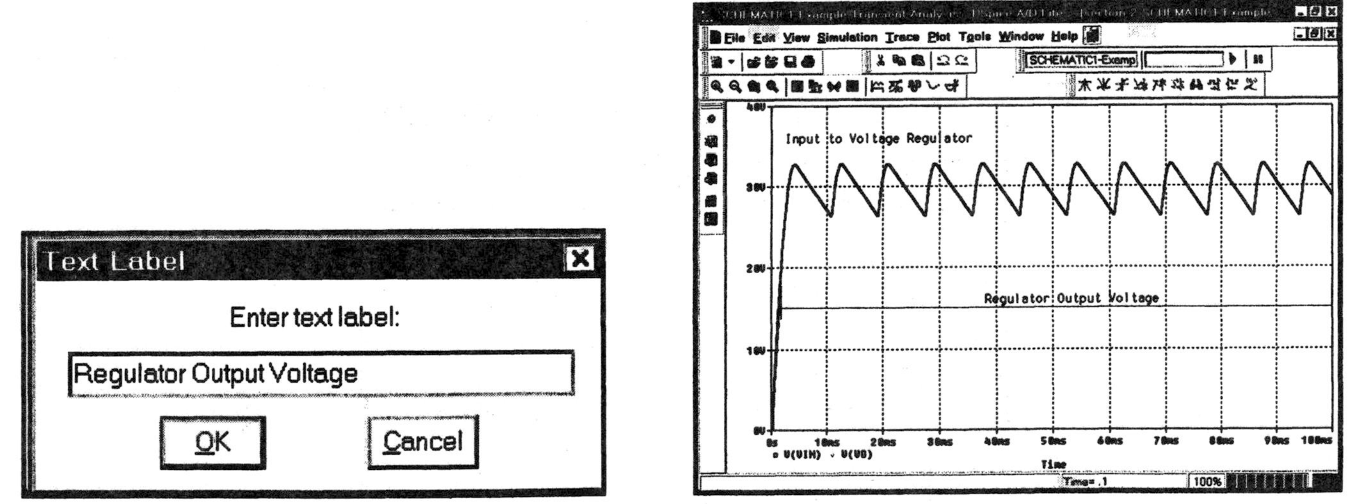


You may need to type **CTRL-L** to redraw the screen to remove text fragments.

# 2.I. Placing Arrows on the Screen

We will now place arrows on the previous plot to point from the text to the appropriate traces. To add an arrow select **Plot, Label,** and then **Arrow.** The mouse pointer will be replaced by a pencil :

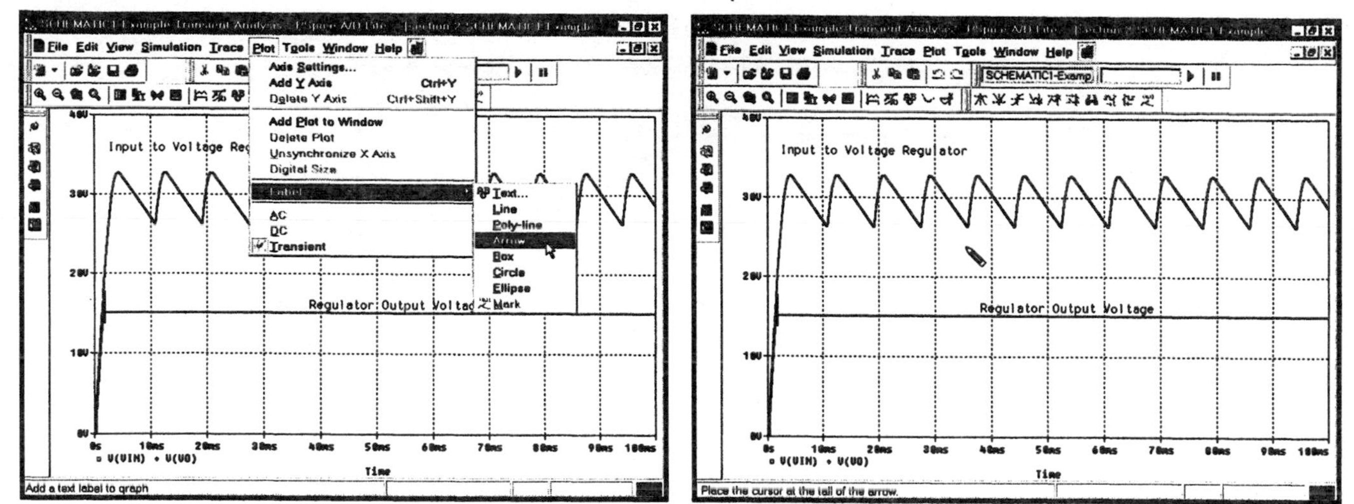

Position the point of the pencil at the place where you would like the tail of the arrow to start. In our example, we place the pencil to the right of the text ***Input to Voltage Regulator***. Click the ***LEFT*** mouse button (do not click and hold the button). An arrow will appear on the screen and change its length as you move the mouse:

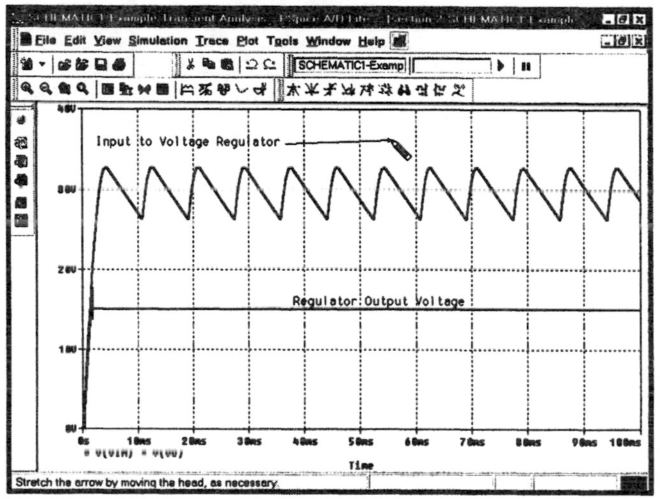

Position the head of the arrow as shown below and click the ***LEFT*** mouse button to place the arrow. When you click the ***LEFT*** mouse button the arrow will be drawn on the screen and the pencil will be replaced by the normal mouse pointer.

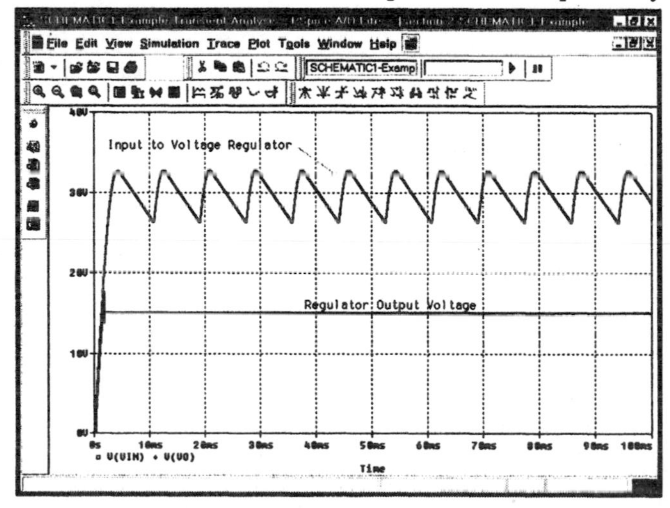

# 2.J. Moving Items on the Screen

Items that we place on the Probe screen can be moved with the same techniques we use to move items in Capture. We will first move the text ***Regulator Output Voltage***. Click the ***LEFT*** mouse button on the text ***Regulator Output Voltage***. The text should turn red, indicating that it is selected. Next, click and drag the selected text to a new location:

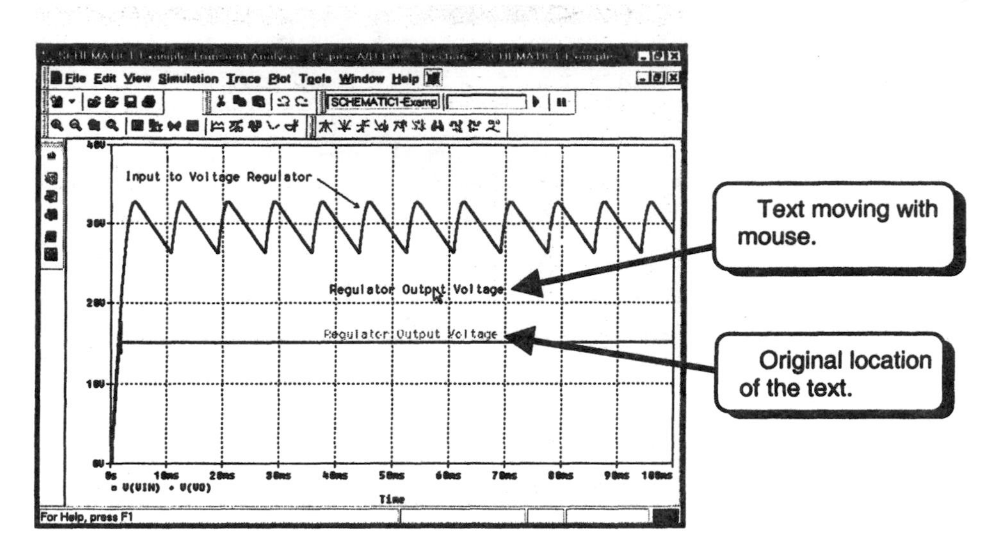


Position the text in a convenient location and release the mouse button to place the text:

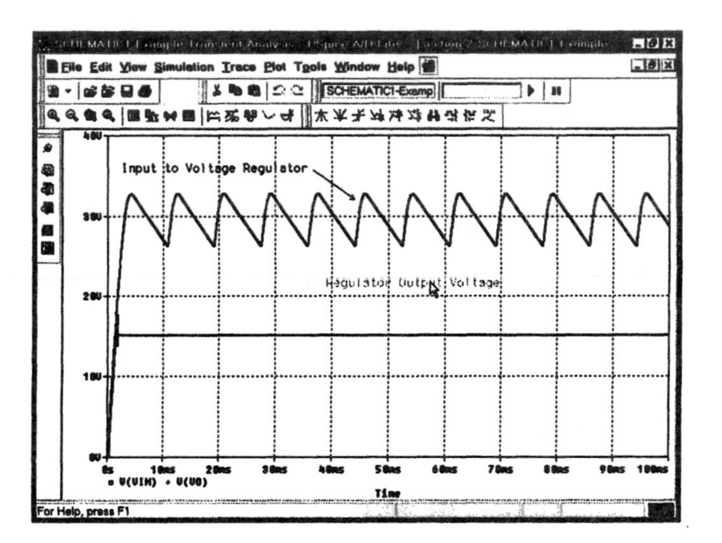


We can use the same technique to move the arrow. Click the ***LEFT*** mouse button on the arrow to select the arrow. It should turn red, indicating that it is selected. Next, place the mouse pointer at the center of the selected arrow and click and drag the arrow to a new location:

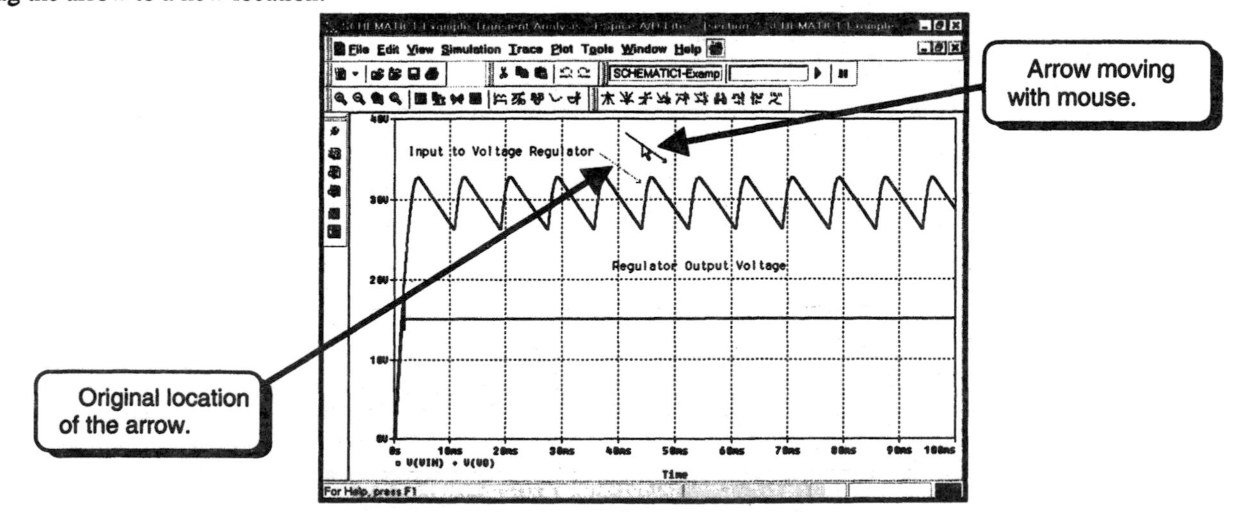


Position the arrow as shown below and release the mouse button to place the arrow:

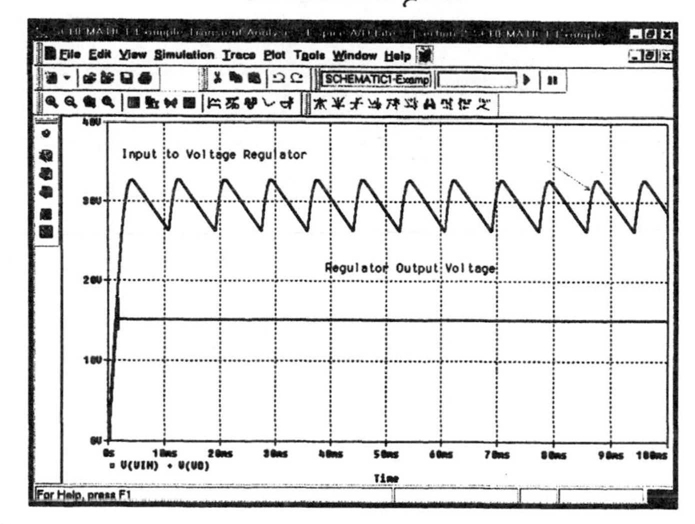


## 2.K. Using the Cursors

The cursors can be used to obtain numerical values from traces. To display the cursors, click the cursor button as shown below:

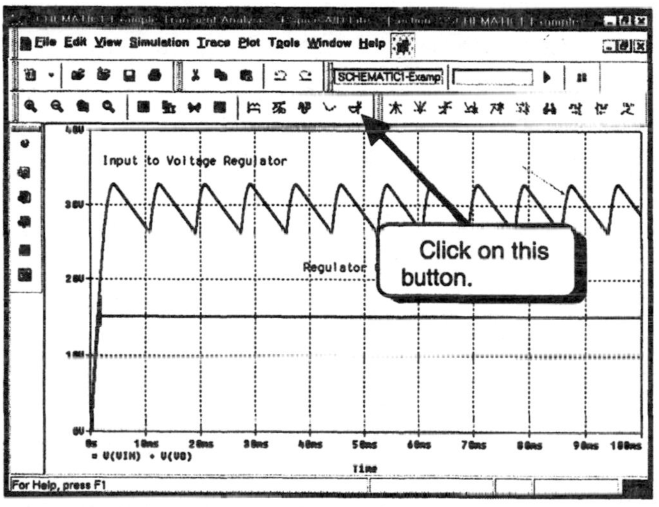


After clicking the button, the cursors will be displayed:

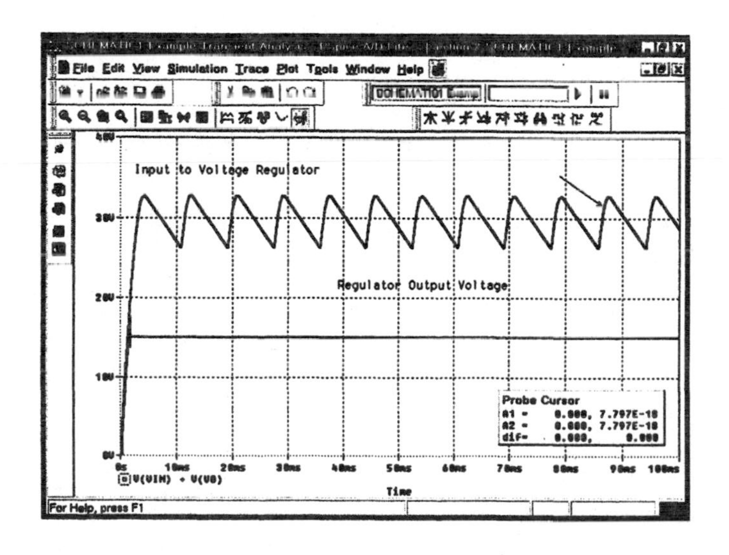

The cursors are positioned at the leftmost data point of the first trace so they cannot be easily seen. A new dialog box is displayed on the Probe screen. This dialog box displays the coordinates of each cursor and the difference between the two cursors. (There are two cursors.)

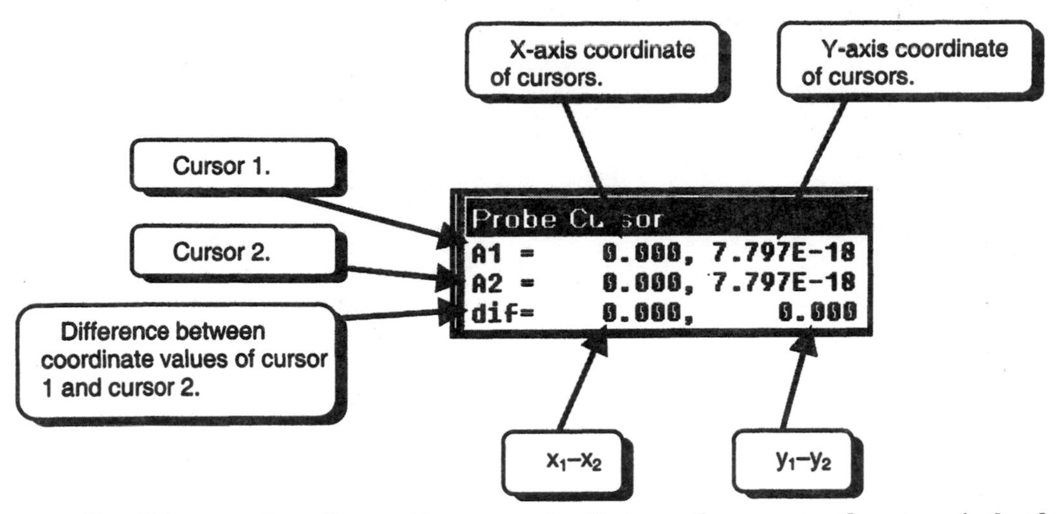


**For this example and any other examples that use the cursors, the numerical values of the coordinates of your cursors may be slightly different than those shown in the examples since the points you choose may be different than those chosen in the examples.**

The cursors can be controlled using the mouse buttons or the keyboard. The left mouse button moves cursor 1 and the right mouse button moves cursor 2. Also, the left and right arrow keys (←→) move cursor 1, and the **SHIFT** key plus the left and right arrow keys (←→) move cursor 2. Place the mouse pointer as shown below:

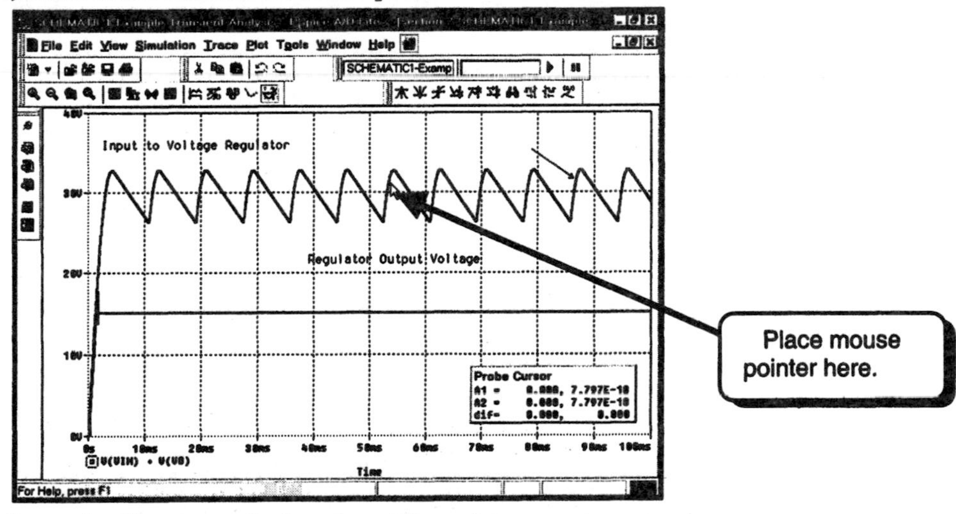


Click the ***LEFT*** mouse button. Cursor 1 will move to the location of the pointer:

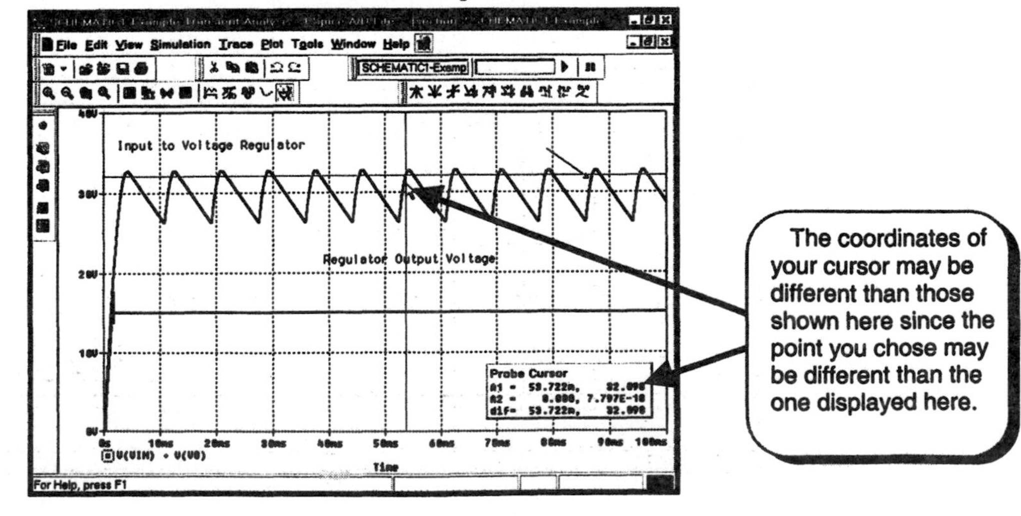

Next, press and **HOLD** the right arrow key (→). The cursor should move to the right:

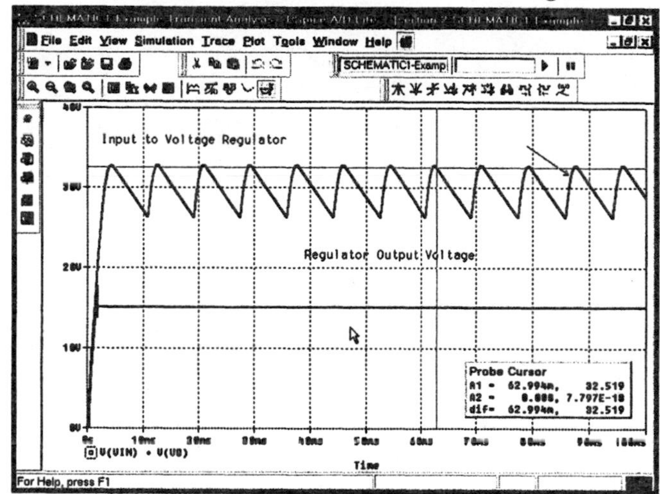

If you press the left arrow key (←), the cursor will move to the left. Note that as you move the cursor, the values in the ***Probe Cursor*** dialog box change.

Next, we will move cursor 2. Place the mouse pointer as shown:

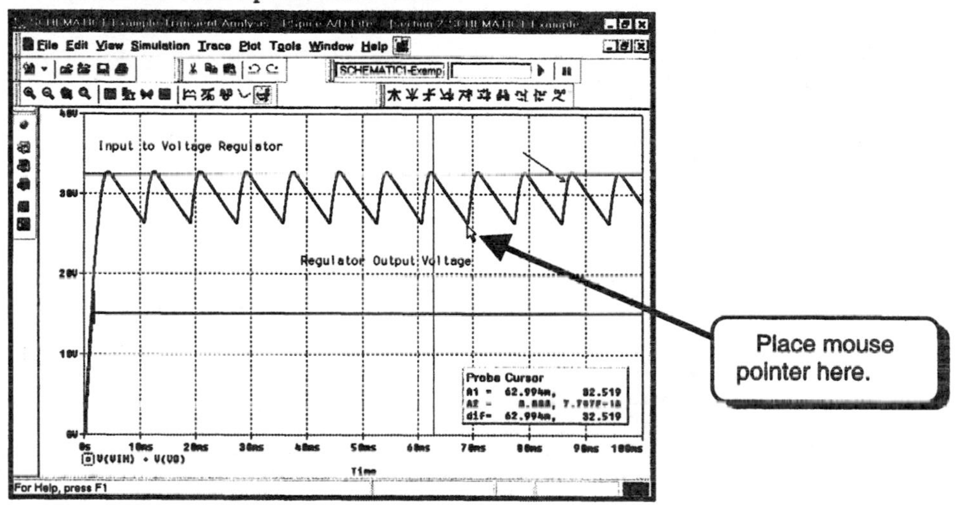


Click the ***RIGHT*** mouse button. Cursor 2 will move to the location of the pointer:

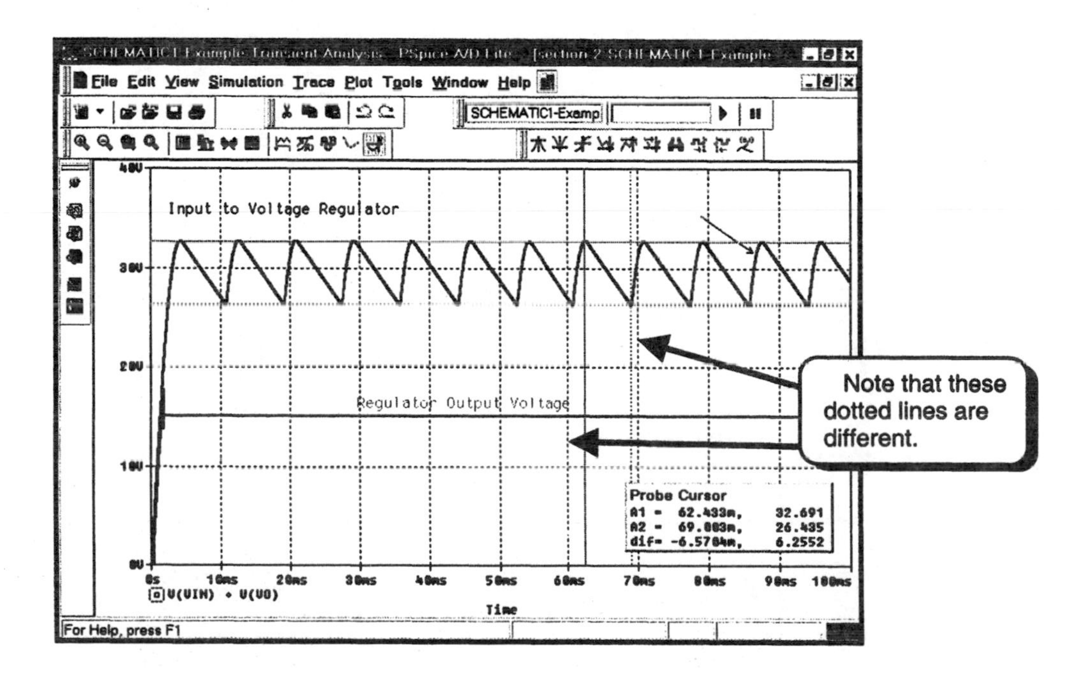

Notice that the dotted lines of the cursor are slightly different for the two cursors. Next, press and **HOLD** the **SHIFT** key and press and **HOLD** the right arrow key (**SHIFT-→**). Cursor 2 should move to the right:

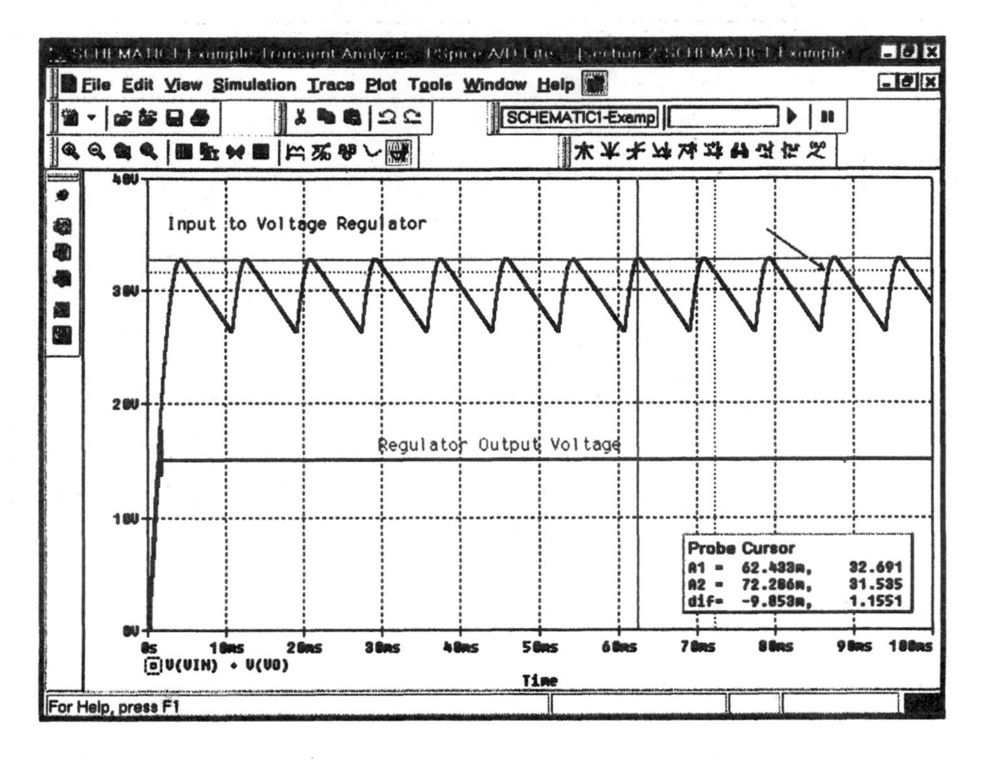


If you hold down the **SHIFT** key and press the left arrow key (**SHIFT-←**), cursor 2 will move to the left.

Presently, both cursors are displayed on the trace V(Vin). An indication of this is given by the dashed box around the symbol for V(Vin), as shown below:

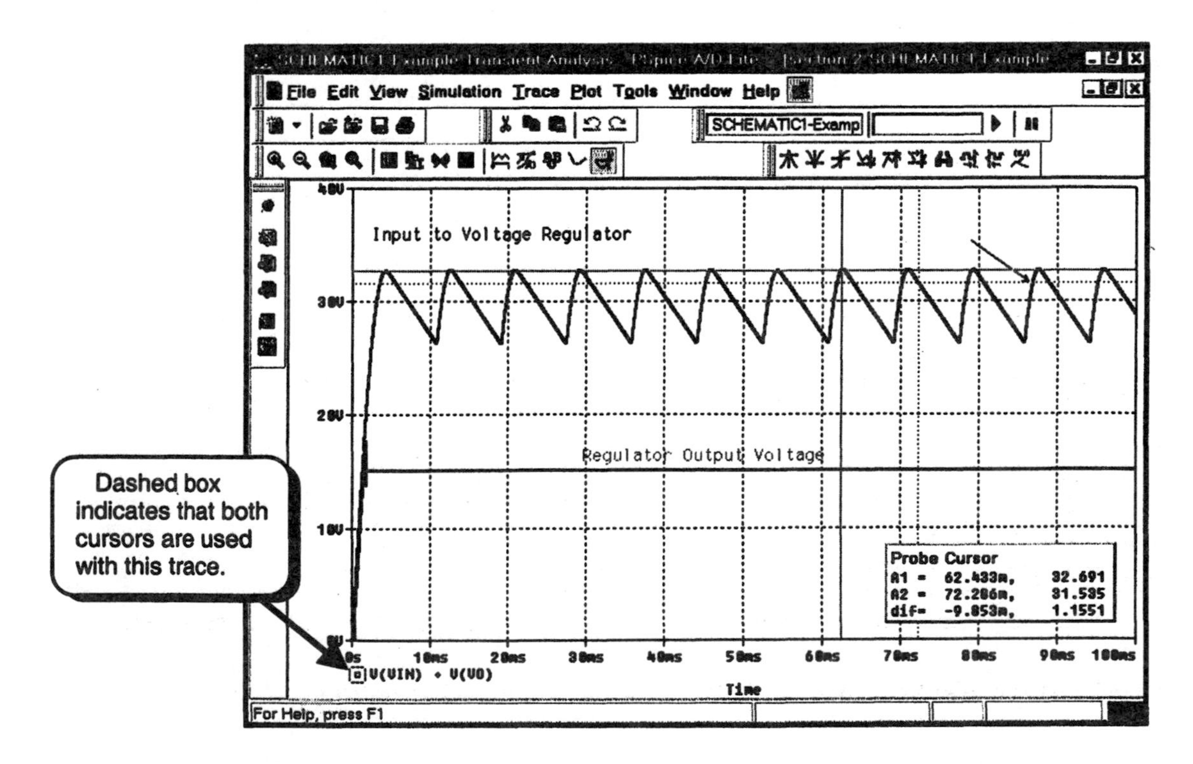


To place cursor 1 on trace V(Vo), click the ***LEFT*** mouse button on the marker for trace V(Vo):

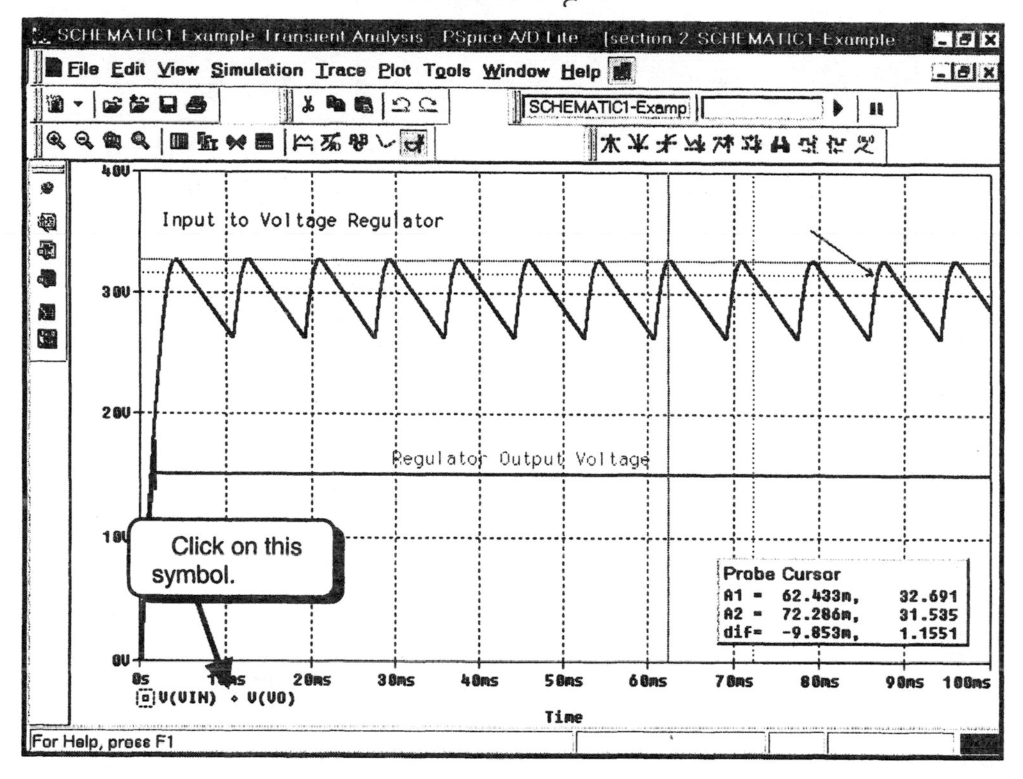


Cursor 1 will jump to trace V(Vo). Notice that a new dotted box encloses the symbol for trace V(Vin) and a different dotted box encloses the symbol for V(Vo).

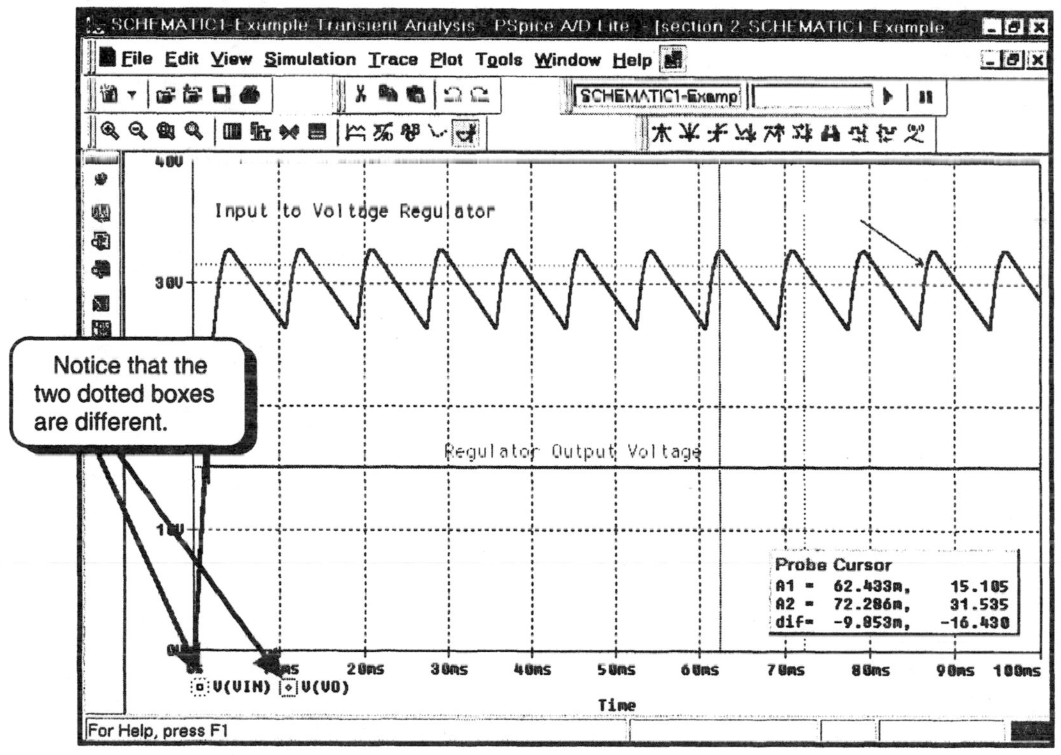


The dot patterns of the boxes match the dot patterns of the lines of two cursors. These dotted boxes indicate which cursor is attached to which trace.

To place cursor 2 on trace V(Vo), click the ***RIGHT*** mouse button on the marker for trace V(Vo):

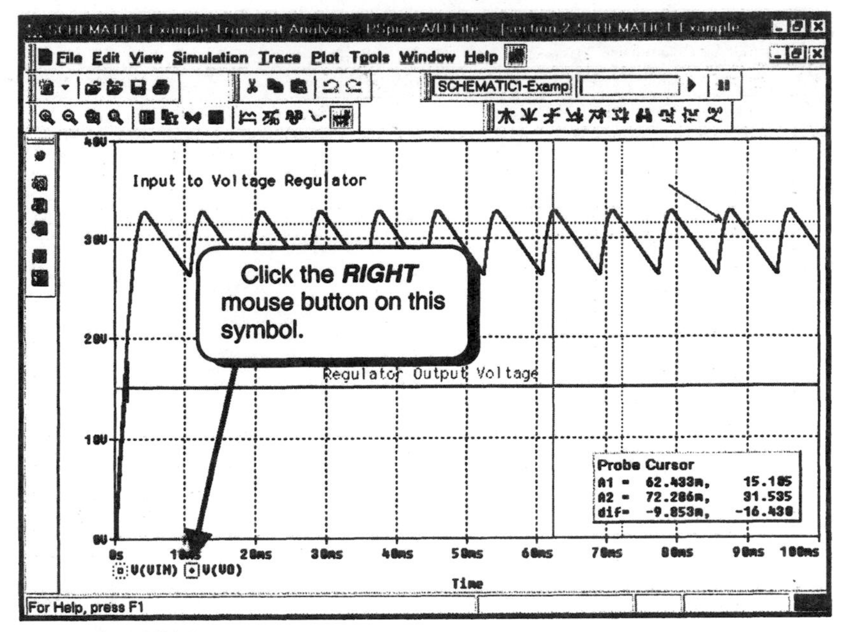


The second cursor will jump to the V(Vo) trace.

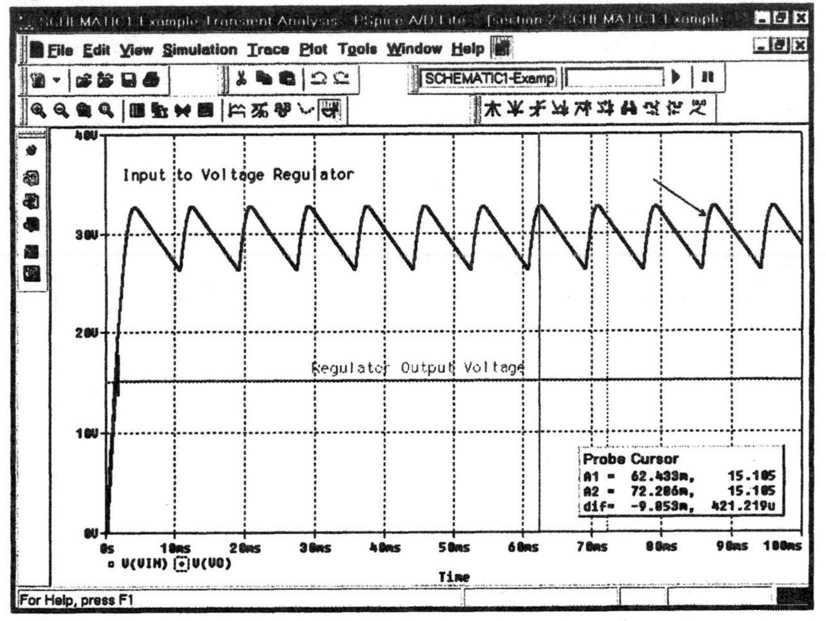


We can use the same method to move the cursors back to trace V(Vin).

The cursor that you used most recently (cursor 1 or cursor 2) is the active cursor. There are several tools in the button bar for positioning the active cursor. Some of the buttons are described below:

- Positions the cursor to the absolute maximum of the trace.

- Positions the cursor to the absolute minimum of the trace.

 - Positions the cursor to the next local maximum.

- Positions the cursor to the next local minimum.

# 2.L. Labeling Points

The cursors are used to find numerical values of a trace. Once important points are found, the coordinates of those points can be placed on the plot. The cursor that you used most recently is the active cursor. Place cursor 1 on trace V(Vin) at the point shown:

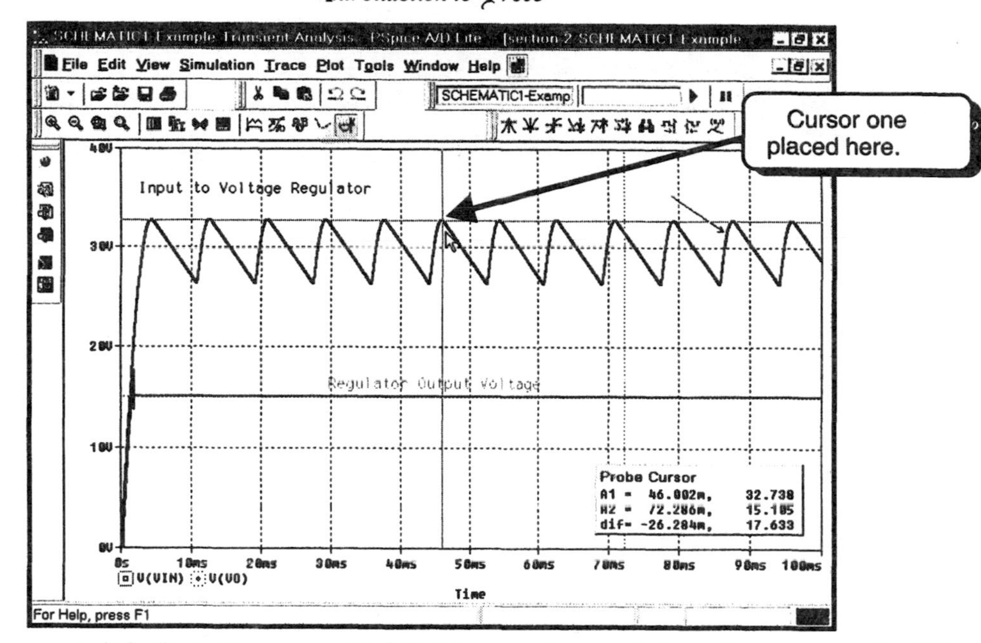


Since we just moved cursor 1, it is the active cursor. To label the coordinates of the active cursor, select **Plot**, **Label**, and then **Mark**. The coordinates will be displayed on the screen:

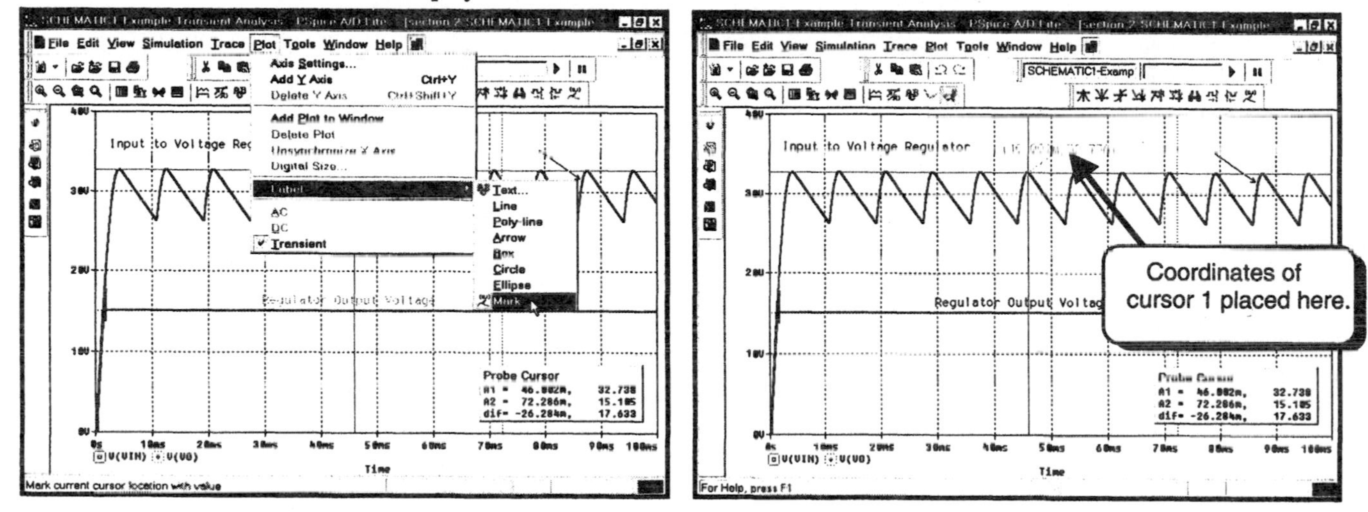


To hide the cursors, click the *LEFT* mouse button on the cursor button :

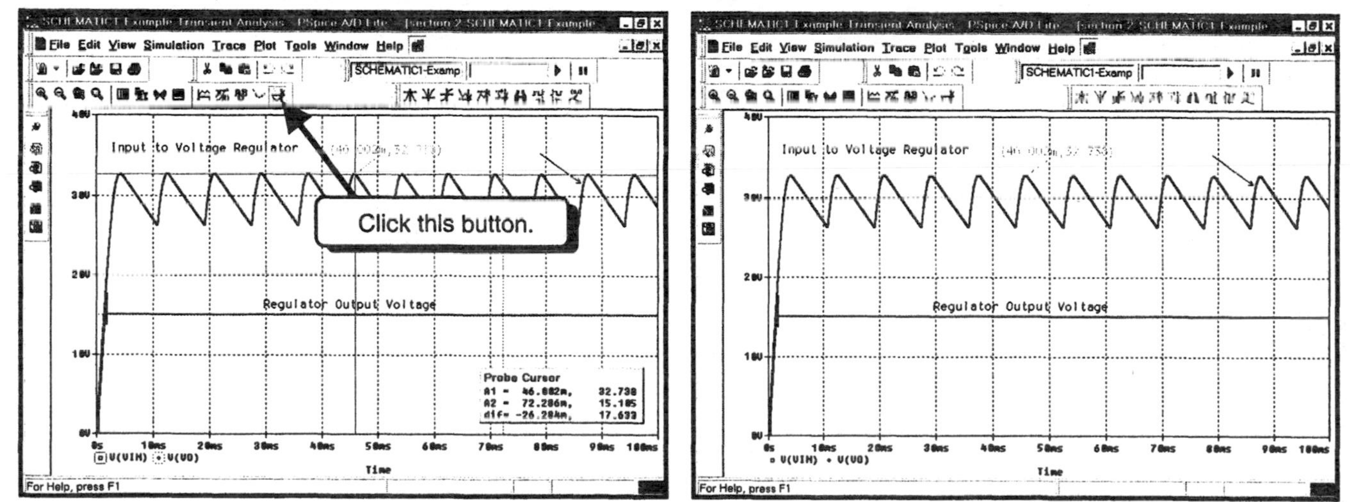


We can now move any of the items on the screen. The steps for moving any item are the same as shown in Section 2.J.

# 2.M. Modifying Probe Trace Colors and Line Widths

You may have noticed that the traces shown on most of the screen captures are thicker than on your screen and the colors are different colors. This is because I have changed the thickness and color of the traces to make the screen captures more readable. To change the properties of a trace, click the ***RIGHT*** mouse button on a trace. A menu will appear:

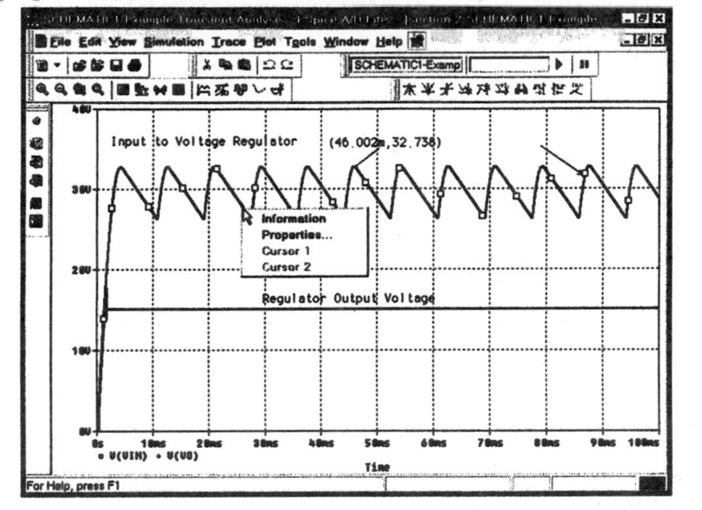


Select **Properties**:

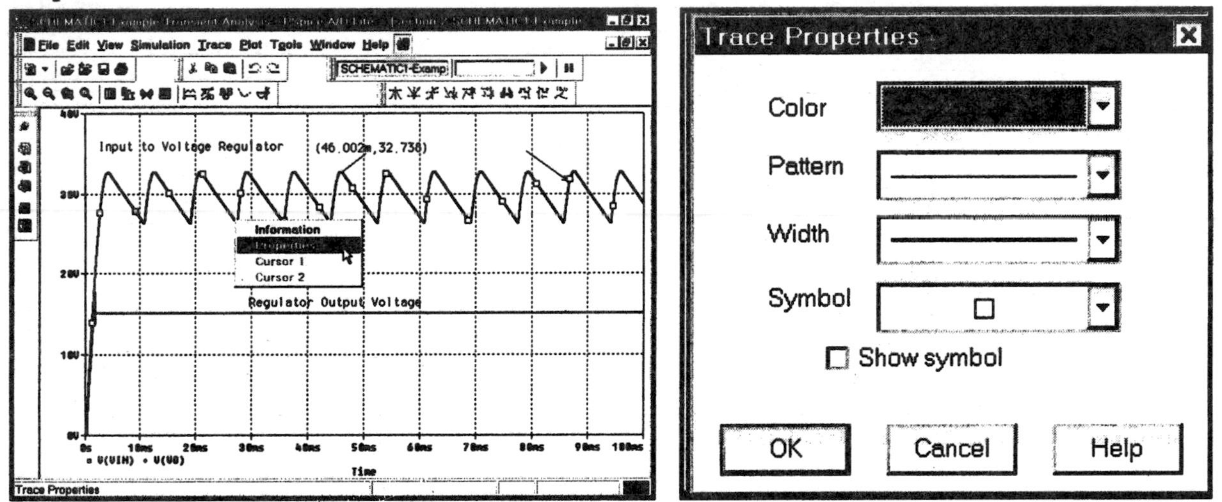


This dialog box allows us to choose the color, pattern, width, and symbol for the trace. I will choose a very thick dashed black line for this trace (you should choose something else):

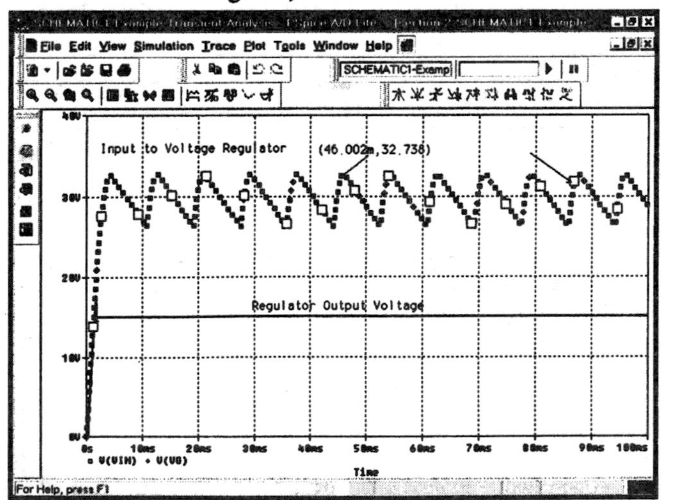


The trace is still selected and thus the square symbols are displayed on the trace. To deselect the trace, click the ***LEFT*** mouse button on the plot somewhere outside the trace:

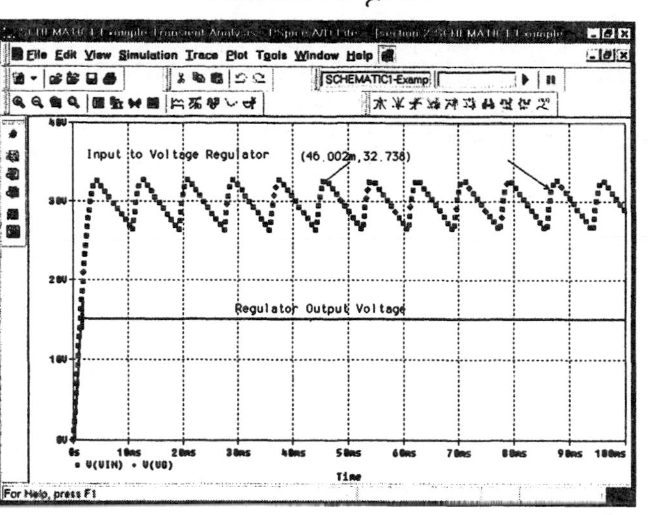


This is not the greatest trace pattern, but it shows what you can do.

Once you set up the trace properties to your liking, you can save the settings for all future uses of Probe. From the menus, select **Plot** and then **Axis Settings**:

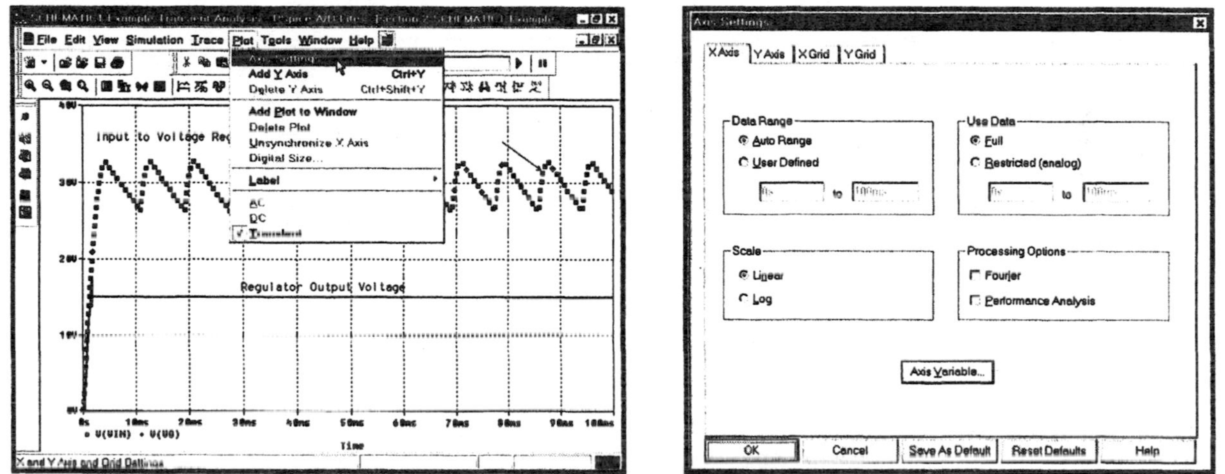


Click the ***Save As Default*** button. All traces will use the trace settings you chose in future uses of Probe. If you wish to return to the original settings, click the ***Reset Defaults*** button in the above dialog box.

# 2.N. Including Probe and Capture Graphics in MS Word

*Both Probe screens and Orcad Capture circuits can be copied and pasted into other documents. Here we will show how to paste them into a Microsoft Word document. We will start with the circuit of Part 2 on page 96:

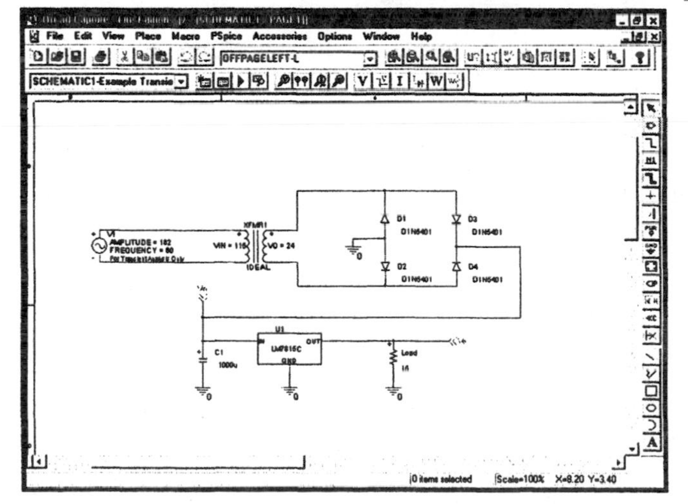


* Thanks to John Brews of the University of Arizona for suggesting this example.

Select the portion of the circuit that you would like to copy. Note that **CTRL-A** selects everything, including the title block:

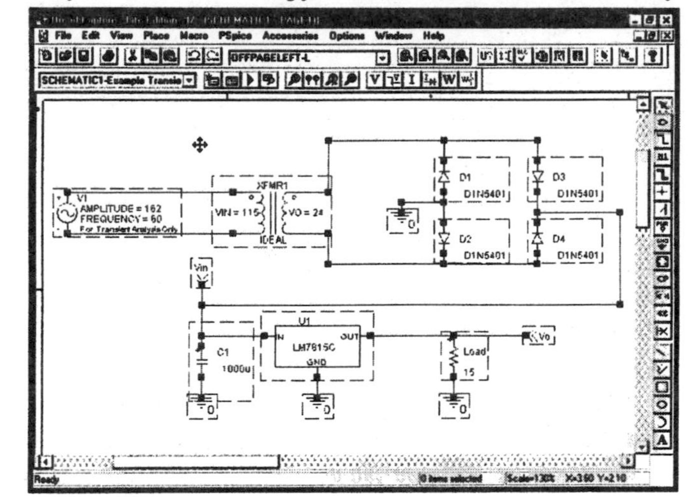


Select **Edit** and then **Copy** from the menus or type **CTRL-C** to copy the selected items to the clipboard. Next, switch to Microsoft Word and place the cursor at the location where you would like to place the circuit:

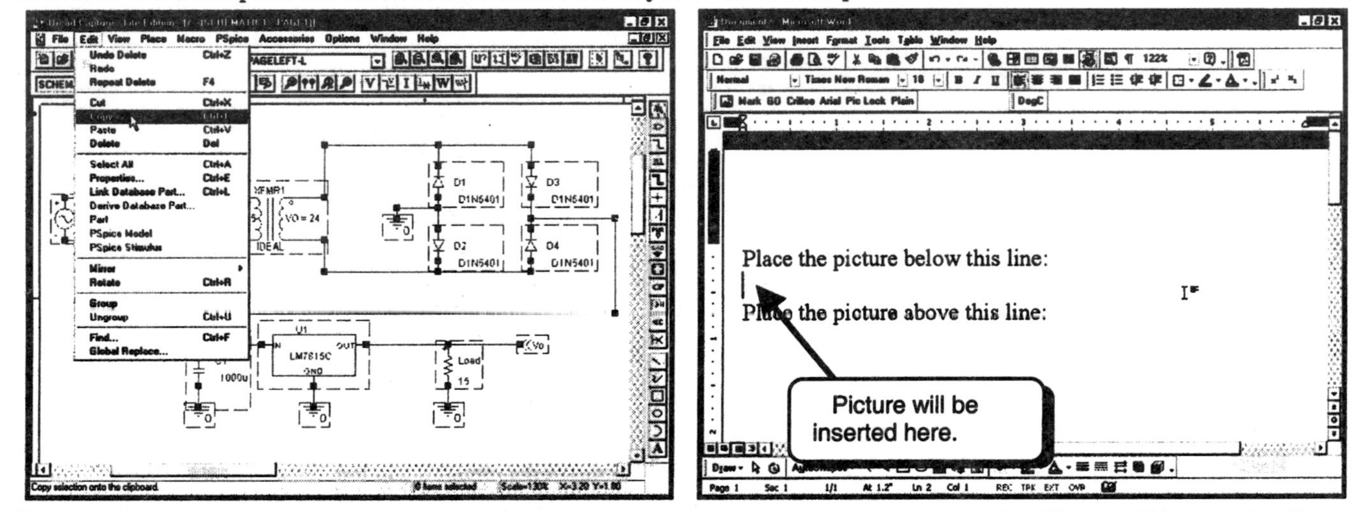


Within MS Word, you can use any of the techniques to paste a graphic into your document. I will select **Edit** and then **Paste**, but if you have experience with Word you can use the **Paste Special** menu selection or type **CTRL-V**.

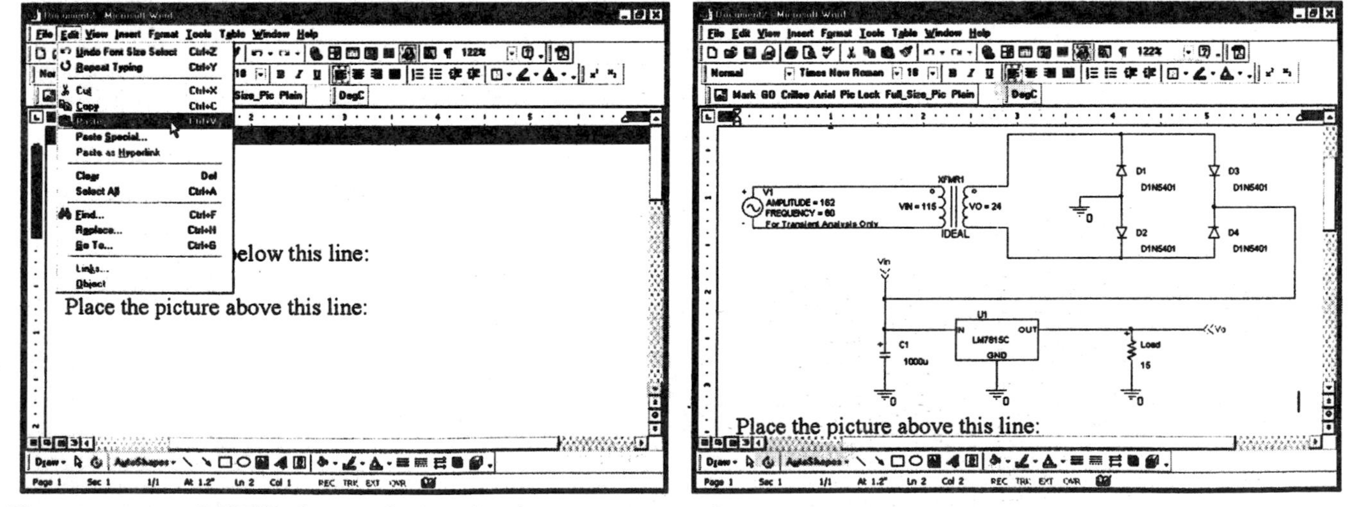


You can now use MS Word to manipulate the picture to your needs:

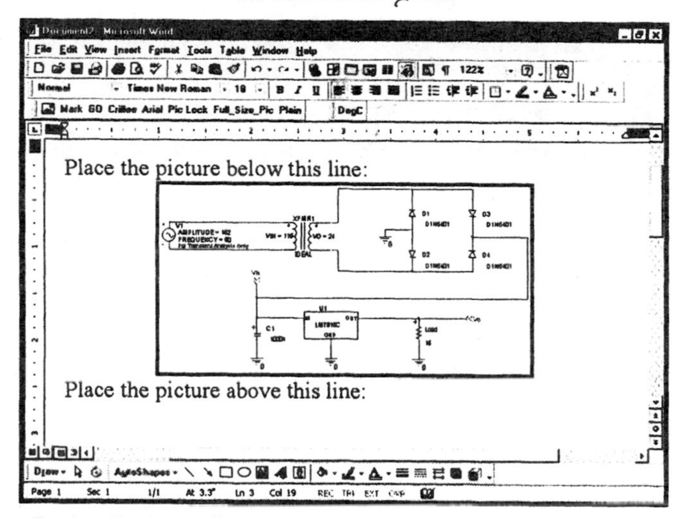


Copying screens from a Probe display is just as easy as it was to copy circuit drawings in Orcad Capture. However, there are a few options in Probe for copying the screen that we can explore. **Note in this section that your pasted screens will be slightly different than those shown here. This is because I have changed Probe to display plots with a white background to make the screen captures for this text easier to read. Your Probe screen displays plots with a black background.**

Run Probe, display some waveforms, and then manipulate the display so that the Probe displays what you want:

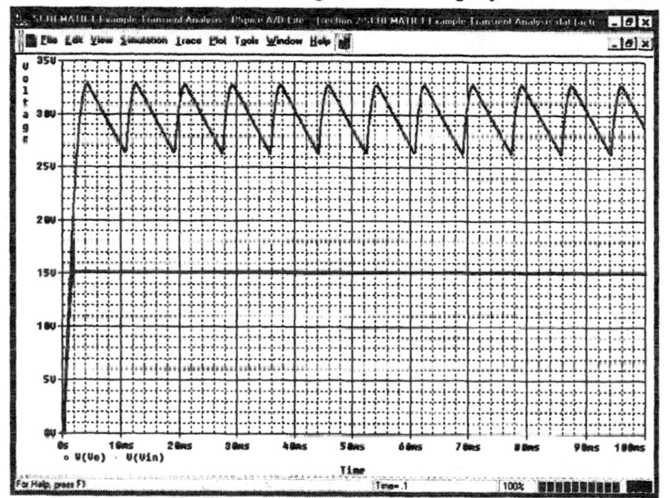


Select **Window** and then **Copy to Clipboard** from the Probe menus:

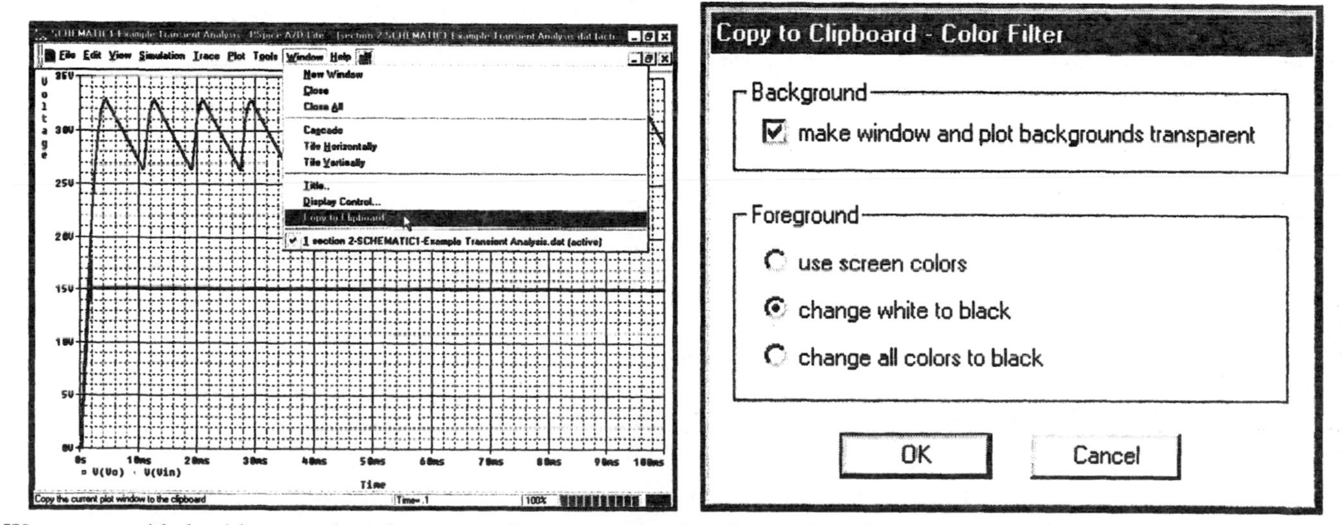


We are provided with several options as to how we copy the plot to the clipboard. We will first experiment with the background. Uncheck the option ***make window and plot backgrounds transparent*** and then click the ***OK*** button:

Copy to Clipboard - Color Filter

Background

Option not selected.

☐ make window and plot backgrounds transparent

Foreground

○ use screen colors

◉ change white to black

○ change all colors to black

OK Cancel

When you click the ***OK*** button, the image is copied to the clipboard. Switch to MS Word, place the cursor at the location at which you would like to insert the picture, and then select **Edit** and then **Paste** from the MS Word menus:

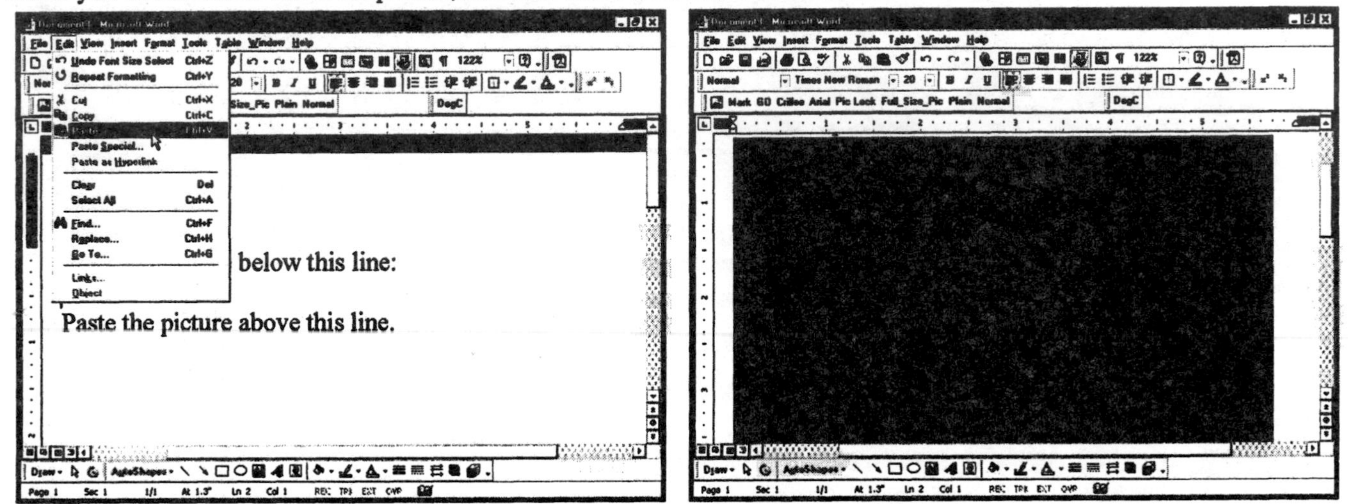


We notice that the plot background is black. This is the effect of not selecting the background to be transparent. Delete the picture and then switch back to Probe. Select **Window** and then **Copy to Clipboard**, and select the options as shown:

Copy to Clipboard - Color Filter

Background

Options selected.

☑ make window and plot backgrounds transparent

Foreground

◉ use screen colors

○ change white to black

○ change all colors to black

OK Cancel

Here we have selected options to make the background transparent and use the screen colors shown in Probe. Click the ***OK*** button to copy the Probe screen to the clipboard. Switch to MS Word and paste the image into your document:

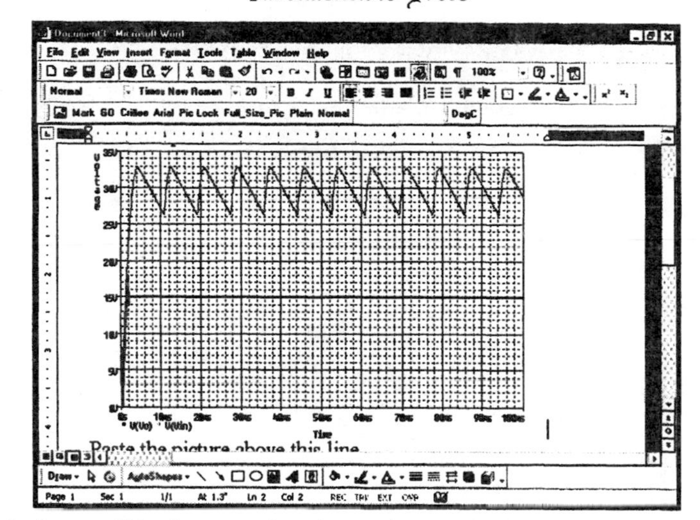

A transparent background looks better. We also note that the traces are displayed in color. Once you have the picture pasted into your document, you can Use MS Word to manipulate the picture for your needs.

The screens below show the effects of selecting the other two options for copying Probe screens. The difference between the screens shown here may not be obvious because the screen captures are shown in gray scale.

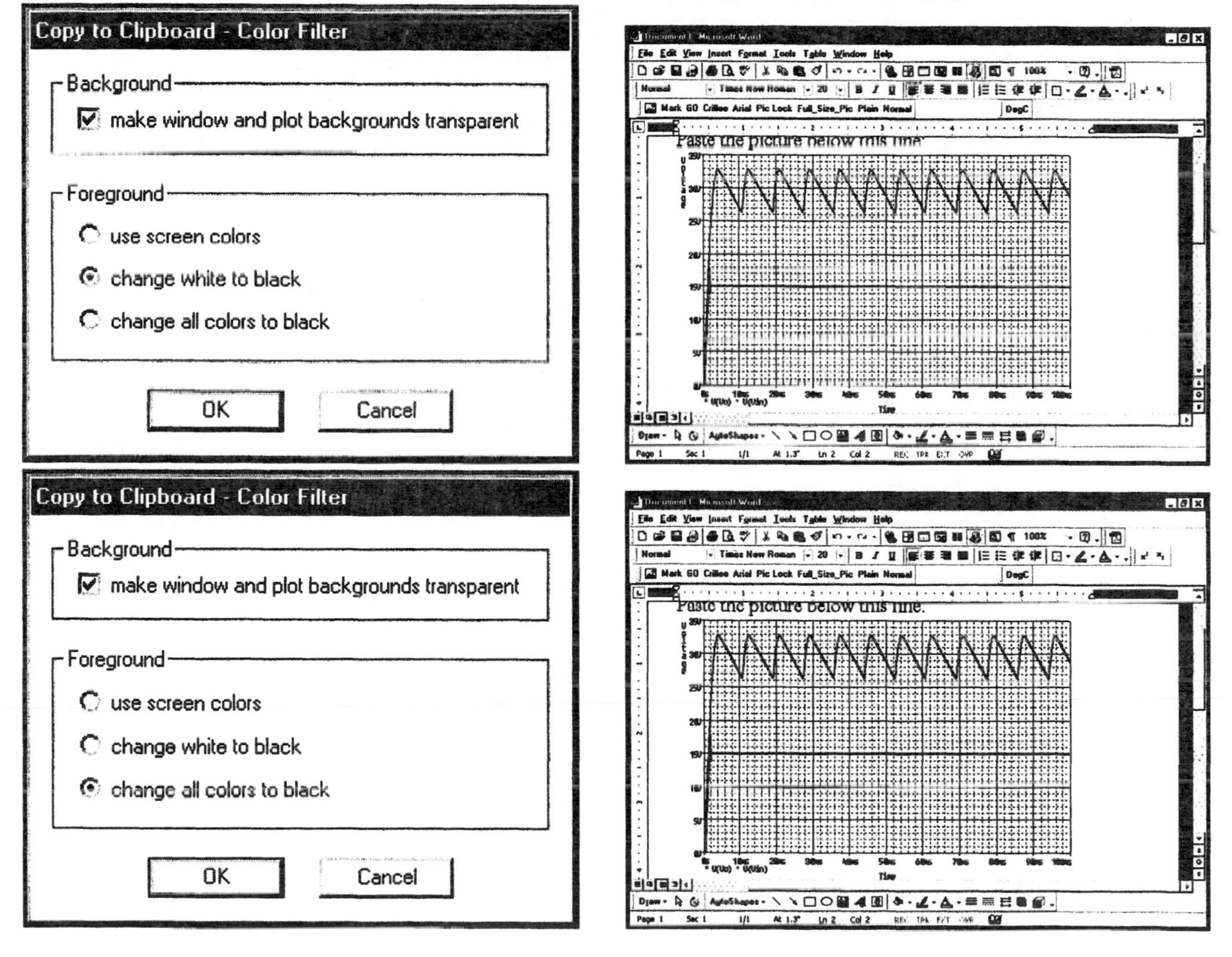

# 2.O. Exporting Data From PSpice and Probe to MS Excel

*Data from PSpice can easily be exported to other programs by copying the traces displayed by Probe. To copy data from PSpice, you must run the simulation, display the results using Probe, and then copy and paste the data from Probe into MS Excel or another spreadsheet program. Once the data is in your spreadsheet, you can manipulate it or export it to other programs such as MATLAB by using comma separated values (.CSV) formatted data files.

Run Probe and display some waveforms. I will use the simulation of section 2 and display Vin and Vo.

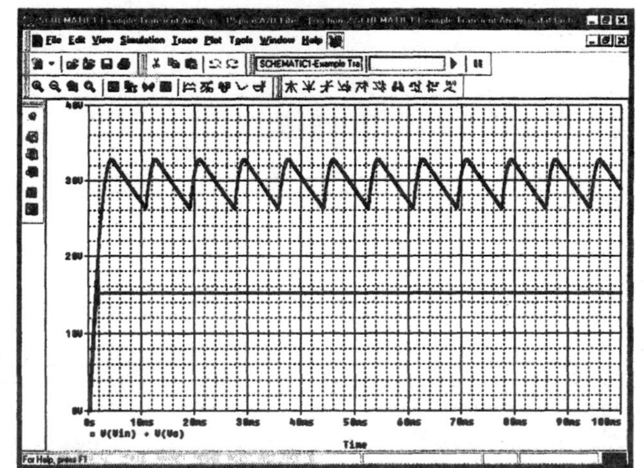

Click the ***LEFT*** mouse button on the text label of the trace you wish to copy, as shown below. When the text and trace are selected, the text should be highlighted in red:

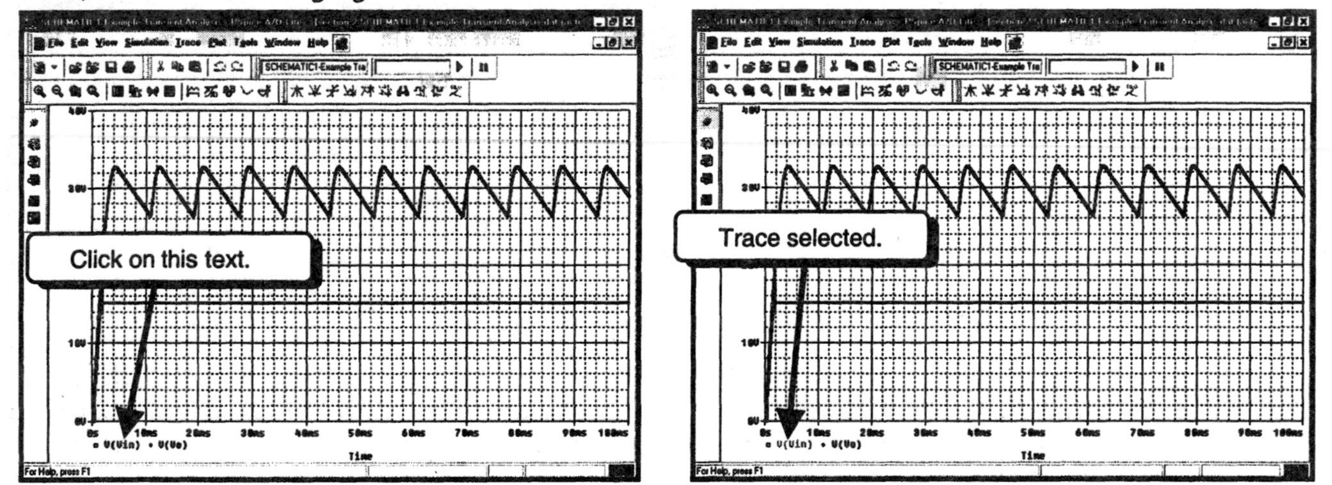


Once the text is highlighted in red, you can copy the data. Select **Edit** and then **Copy** from the Probe menus to copy the data of the selected trace:

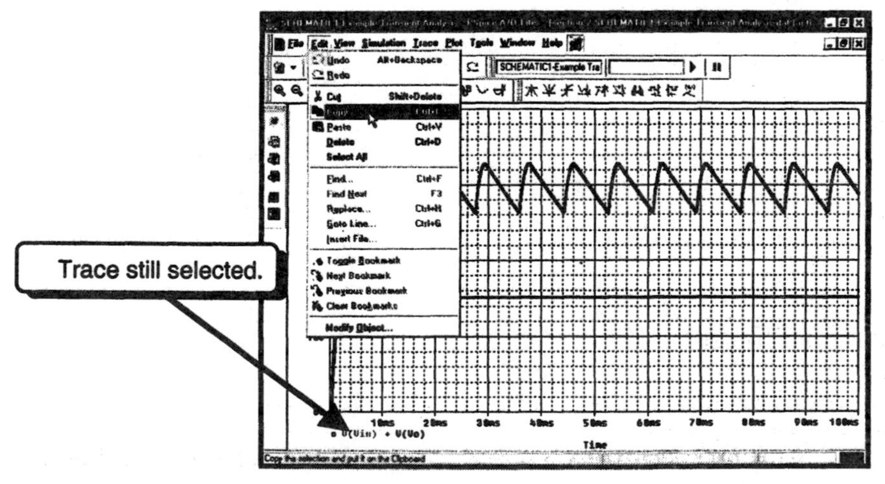


---

* Thanks to John Brews of the University of Arizona for suggesting this example.

The data for the trace has now been copied to the clipboard. Switch to your spreadsheet application and select the cell that you would like to contain the text label for the time data. The data will be pasted into two columns. The first column will contain the time coefficient of the data. The first row will contain the text labels of the data (Time and V(Vin) in this example). I will paste the data starting at cell A1:

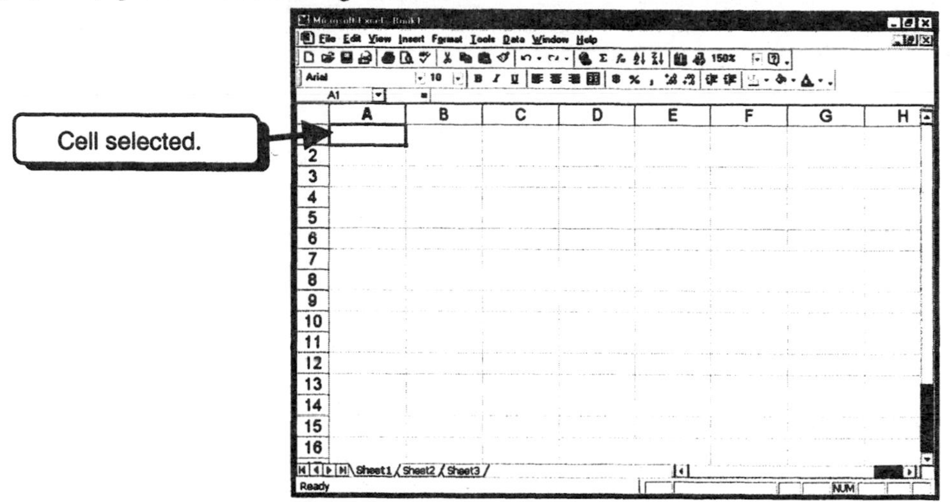


Select **Edit** and then **Paste** from the menus to paste the data into your spreadsheet:

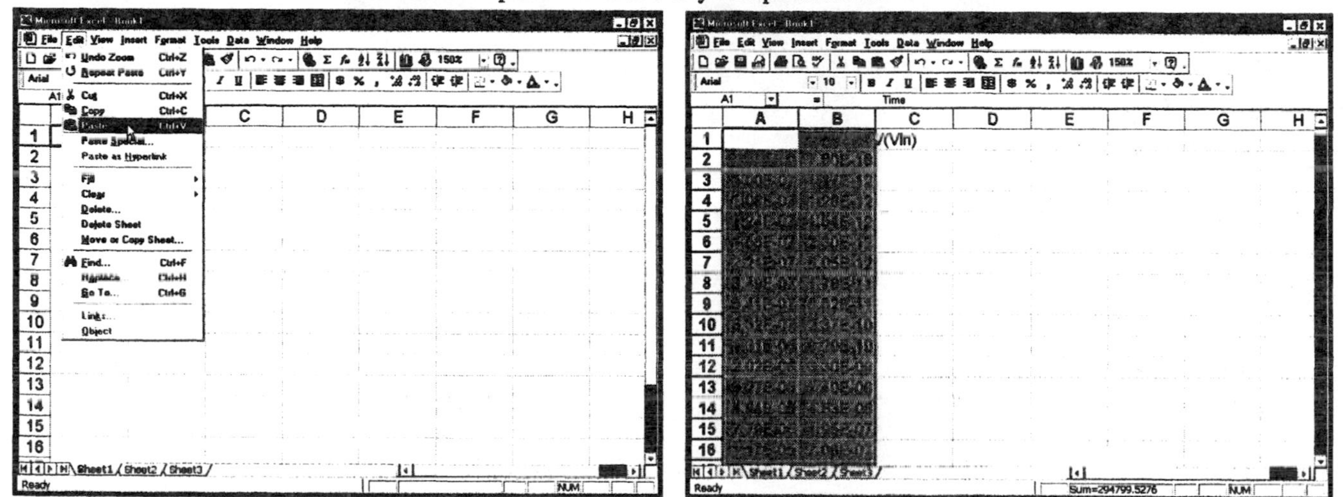

Cells A1 and B1 contain the names of the two columns of data. However, the text contains leading spaces and does not fit in the columns. If you select cell A1 you will notice that it contains a number of spaces before the text. Remove extra spaces before the text in cells A1 and B1:

| | A | B |
|---|---|---|
| 1 | Time | V(Vin) |
| 2 | 0 | 7.80E-18 |
| 3 | 1.00E-07 | 1.17E-12 |
| 4 | 1.08E-07 | 1.28E-12 |
| 5 | 1.24E-07 | 1.54E-12 |
| 6 | 1.56E-07 | 2.30E-12 |
| 7 | 2.21E-07 | 5.05E-12 |
| 8 | 3.49E-07 | 1.78E-11 |
| 9 | 5.41E-07 | 6.52E-11 |
| 10 | 8.32E-07 | 2.37E-10 |
| 11 | 1.31E-06 | 9.20E-10 |
| 12 | 2.02E-06 | 3.30E-09 |
| 13 | 3.27E-06 | 1.40E-08 |
| 14 | 4.94E-06 | 4.83E-08 |
| 15 | 7.79E-06 | 1.95E-07 |
| 16 | 1.17E-05 | 7.06E-07 |

The spreadsheet is now a little more readable. You can now manipulate and plot the data with your spreadsheet:

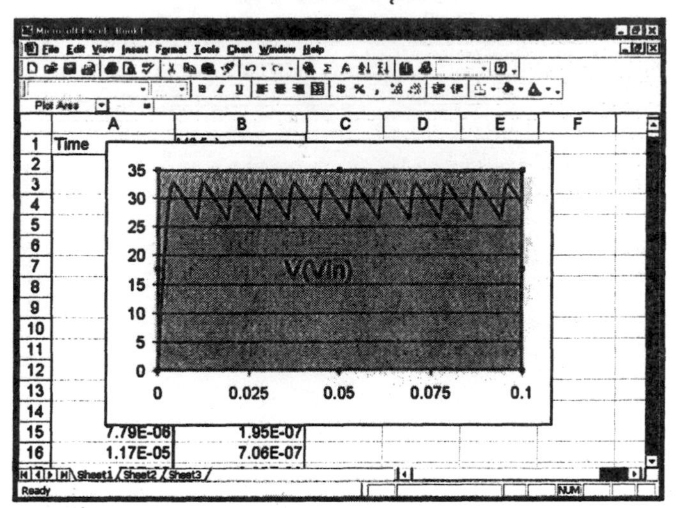


We can copy the data from several traces at the same time and then paste that data into a spreadsheet. In the Probe screen below, we have three traces displayed:

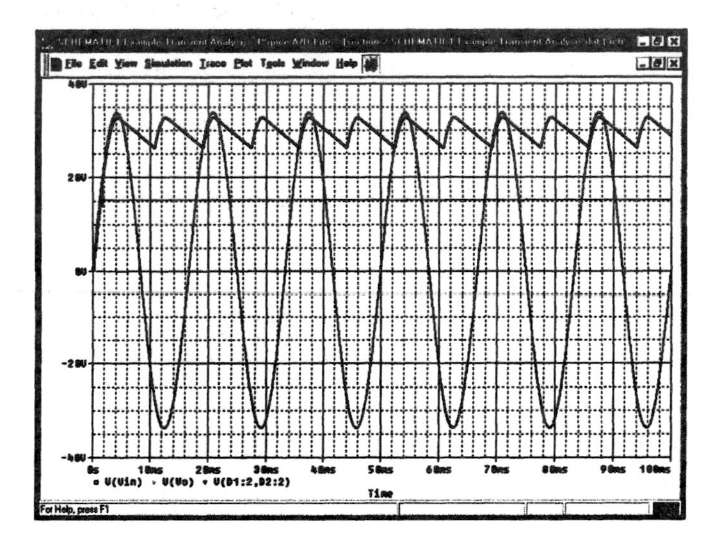

To select multiple traces, hold down the **CTRL** key and then click the ***LEFT*** mouse button on the text label of the traces you want to select:

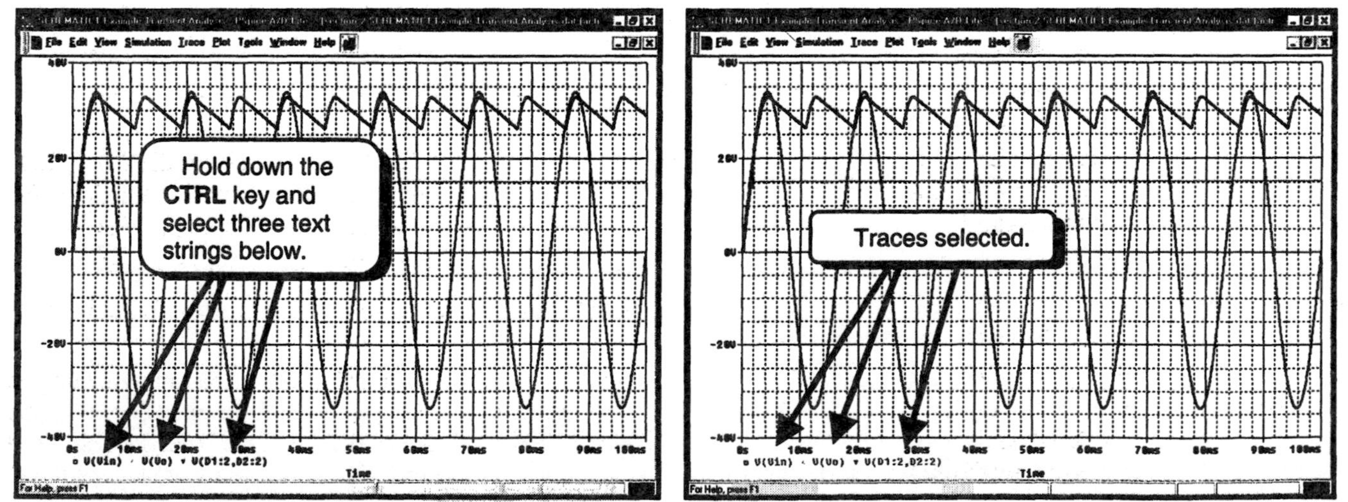


When a trace is selected, the text label should be highlighted in red. When you have selected all of the traces, select **Edit** and then **Copy** from the Probe menus. Switch to your spreadsheet and then paste the data:

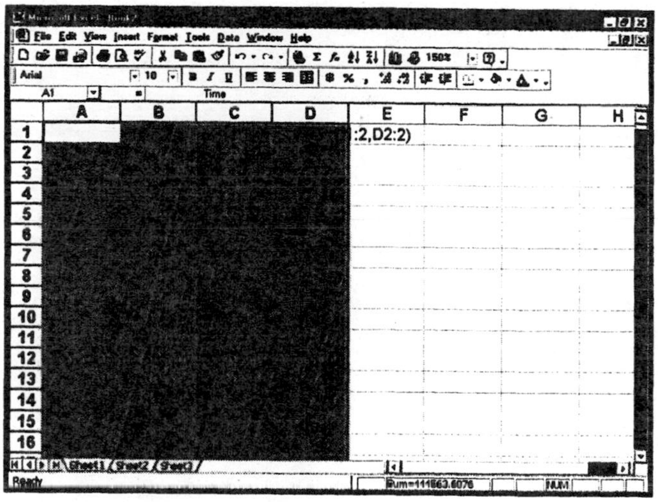

There are four columns of data, one column for time and the other three for the voltage data for the selected traces. Once again we need to modify the first row so that the text labels fit within the columns. When we remove the leading spaces from the text labels and we change the column widths in the spreadsheet, the labels for the columns are easily seen:

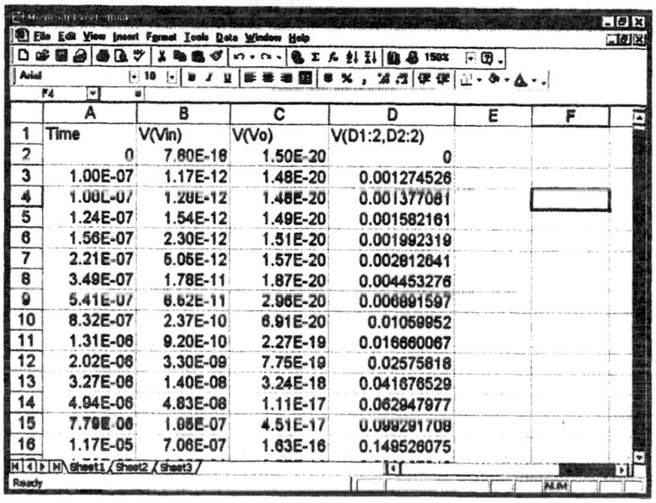

| | A | B | C | D |
|---|---|---|---|---|
| 1 | Time | V(Vin) | V(Vo) | V(D1:2,D2:2) |
| 2 | 0 | 7.80E-18 | 1.50E-20 | 0 |
| 3 | 1.00E-07 | 1.17E-12 | 1.48E-20 | 0.001274526 |
| 4 | 1.00E-07 | 1.20E-12 | 1.48E-20 | 0.001377081 |
| 5 | 1.24E-07 | 1.54E-12 | 1.49E-20 | 0.001582161 |
| 6 | 1.56E-07 | 2.30E-12 | 1.51E-20 | 0.001992319 |
| 7 | 2.21E-07 | 5.05E-12 | 1.57E-20 | 0.002812641 |
| 8 | 3.49E-07 | 1.78E-11 | 1.87E-20 | 0.004453276 |
| 9 | 5.41E-07 | 6.62E-11 | 2.96E-20 | 0.006891597 |
| 10 | 8.32E-07 | 2.37E-10 | 6.91E-20 | 0.01059952 |
| 11 | 1.31E-06 | 9.20E-10 | 2.27E-19 | 0.016660067 |
| 12 | 2.02E-06 | 3.30E-09 | 7.75E-19 | 0.02575818 |
| 13 | 3.27E-06 | 1.40E-08 | 3.24E-18 | 0.041676529 |
| 14 | 4.94E-06 | 4.83E-08 | 1.11E-17 | 0.062947977 |
| 15 | 7.79E-06 | 1.06E-07 | 4.51E-17 | 0.099291708 |
| 16 | 1.17E-05 | 7.06E-07 | 1.63E-16 | 0.149526075 |

We can now do just about anything with the data we have exported from PSpice and Probe.

# 2.P. Sending an Orcad Project Through Email

*One of the ways you may share your simulations and results with others is through email. If you have errors in a project and cannot find the answer, you may wish to contact me for help. If it is a question about a circuit, I may ask you to email me the project because discussing the problem through email can be nonproductive. Emailing projects is fairly easy to do, but you must follow a few guidelines:

- Keep file names short. Different versions of Windows allow different file name lengths. If you give a file a very long name that is valid in Windows 2000, and then send the project to someone with a computer running Windows 98, Windows 98 may truncate the file name. Orcad will then have trouble opening the file or some of its support files.
- Keep each project in a separate directory. This will make it easy for you to identify the files that belong to a project.
- Use an archiving program such as Winzip or WinRAR to archive all the files for a project into a single file, and then e-mail the archive. Do not email the individual files.
- Create a self-extracting archive so that the email recipient is not required to have the archive extraction program that you used to create the archive. A self-extracting archive has an ".exe" extension, meaning that the file can be executed (the user can run the file). When the recipient runs the file it will extract the files in the archive. The recipient does not need to know the specifics of how the archive was created.

* Thanks to John Brews of the University of Arizona for suggesting this example.

- Send all files for a project. Although some files are not required, such as the output data file (".dat" extension), if you forget a required file, the email recipient may not be able to open your project.

I will show a short example of the procedure. From the Capture menus, select **File, New,** and then **Project** to create a new project:

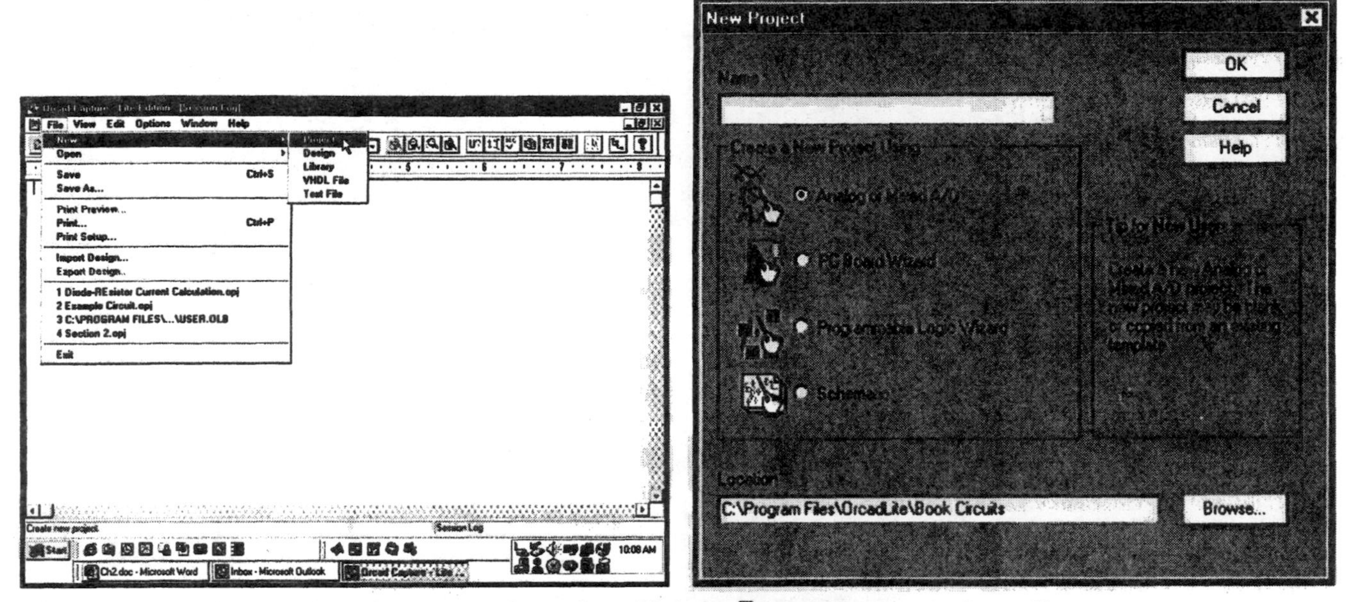


I will create a new directory in which to place the project. Click the ***Browse*** button to select a directory:

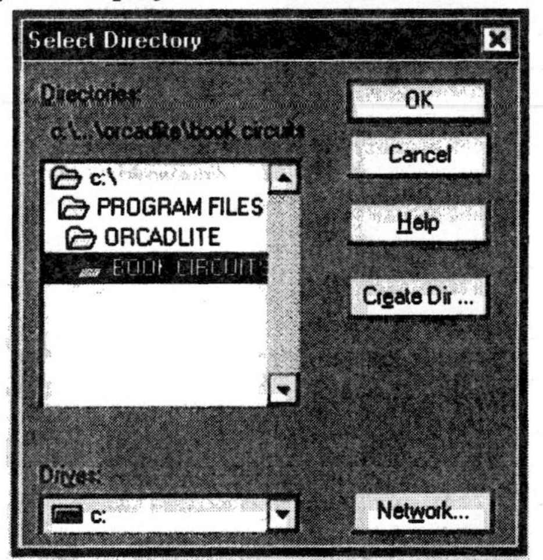


You can use this dialog box to select an exiting directory or to create a new directory. I will create a new directory within the ***BOOK CIRCUITS*** directory. You can create your directory somewhere else if you want:

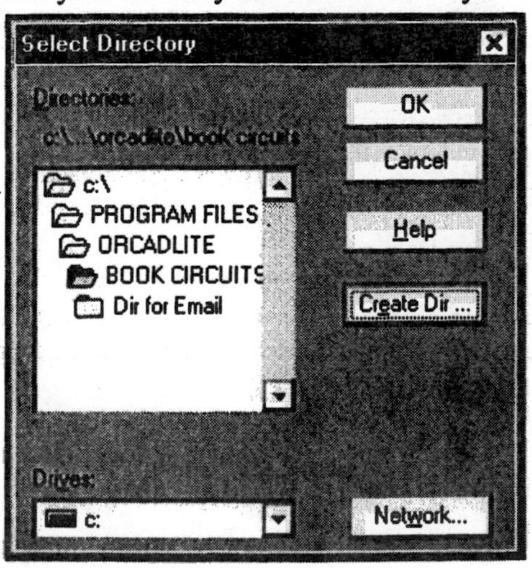

Before proceeding, you must select the directory you just created:

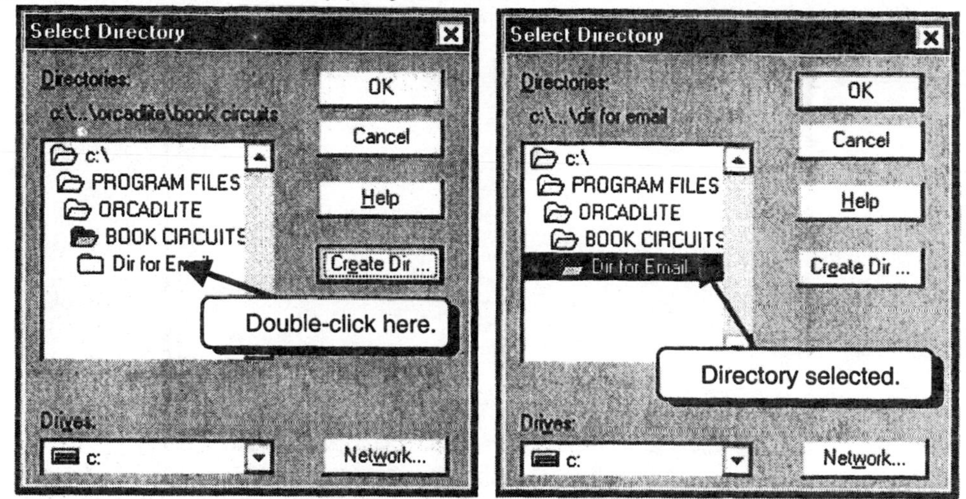


Click the ***OK*** button to accept the selected directory:

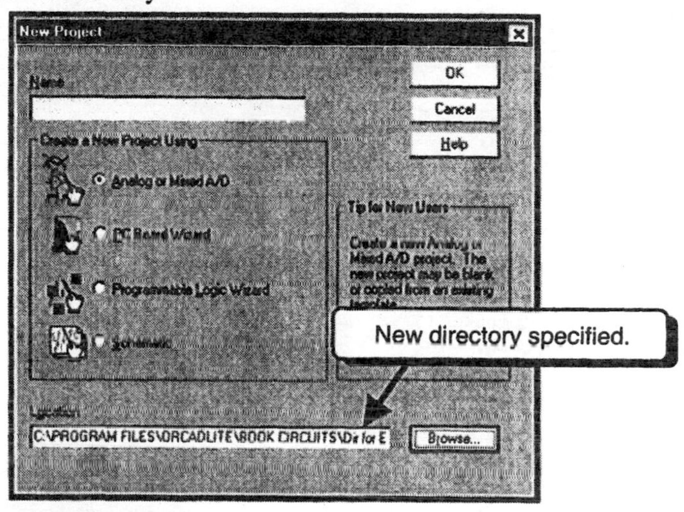


Give the project a name and then click the ***OK*** button:

Select the option to create a blank project and click the ***OK*** button. We can now draw our circuit. I will create a simple circuit and then set up and run a simulation. I am simulating the circuit so that you can see all of the files created for a project. My circuit is shown below:

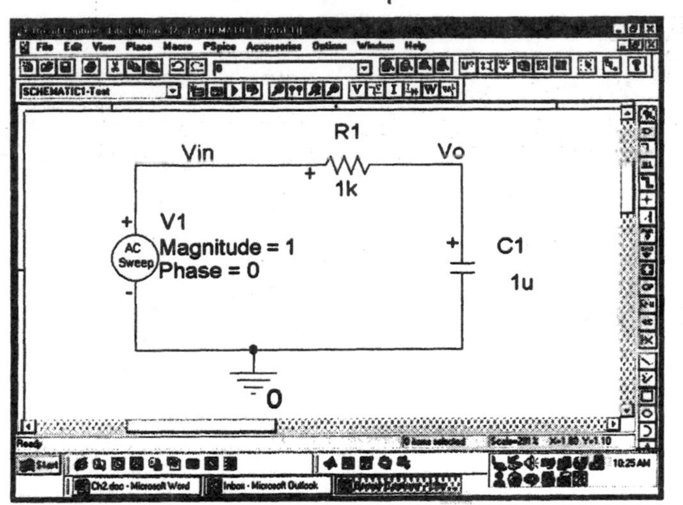


I now wish to email this project to someone. I will close all of the Orcad programs (Capture and PSpice).

Before we archive the files, we will use the Windows Explorer to look at the files for the project:

We see that a large number of files are created, and we do not know which ones are not required so that someone on another computer could open our project. The only files that you can safely exclude are the ".dat" and the ".out" files. Fortunately, these files are usually the largest files. We will delete the ".dat" and the ".out" files before we create the archive:

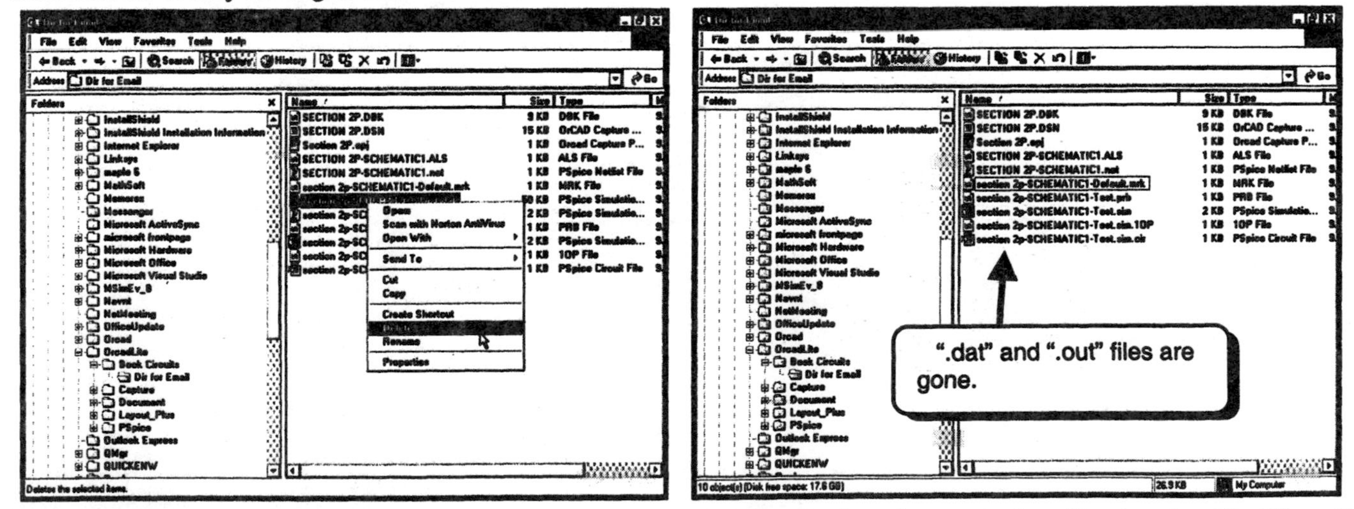


Next we will use an archiving program to compress the files and bundle them into a single self-extracting file. We will demonstrate the procedure using a program called WinRAR; however, you can use other programs such as Winzip. These shareware programs can be obtained at www.winsite.com and www.tucows.com.

Run the program and display the directory in which your project is stored:

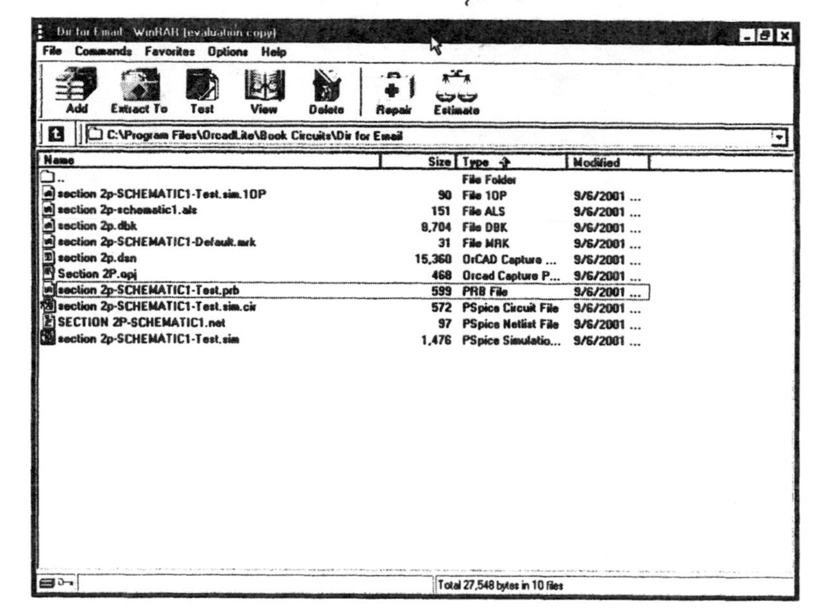


Select all of the files in the directory:

Select **Commands** and then **Add files to archive**:

You can experiment with the options if you wish. You must select the option to create an SFX file. This creates a self-extracting archive with an ".exe" extension.

Archive name and parameters

General | Advanced | Files | Backup | Comment

Archive — Browse...

Dir for Email.exe

Profiles... — Update mode: Add and replace files

Archive format: ◉ RAR ○ ZIP

Archiving options:
- ☐ Delete files after archiving
- ☑ Create SFX archive
- ☐ Create solid archive
- ☐ Put authenticity verification
- ☐ Use multimedia compression
- ☐ Put recovery record
- ☐ Lock archive

Compression method: Normal

Volume size, bytes

OK | Cancel | Help

Option selected.

I will change the name from ***Dir for Email.exe*** to `Section 2p.exe`:

Archive name and parameters

General | Advanced | Files | Backup | Comment

Archive — Browse...

Section 2p.exe

Profiles... — Update mode: Add and replace files

Archive format: ◉ RAR ○ ZIP

Archiving options:
- ☐ Delete files after archiving
- ☑ Create SFX archive
- ☐ Create solid archive
- ☐ Put authenticity verification
- ☐ Use multimedia compression
- ☐ Put recovery record
- ☐ Lock archive

Compression method: Normal

Volume size, bytes

OK | Cancel | Help

Click the ***OK*** button to create the archive. The ".exe" file is now displayed by WinRAR in the directory:

File Commands Favorites Options Help

Add | Extract To | Test | View | Delete | Repair | Estimate

C:\Program Files\OrcadLite\Book Circuits\Dir for Email

| Name | Size | Type | Modified |
|---|---|---|---|
| .. | | File Folder | |
| Section 2p.exe | [illegible] | Application | 9/6/2001 |
| section 2p-SCHEMATIC1-Test.sim.10P | 90 | File 10P | 9/6/2001 ... |
| section 2p-schematic1.als | 151 | File ALS | 9/6/2001 ... |
| section 2p.dbk | 8,704 | File DBK | 9/6/2001 ... |
| section 2p-SCHEMATIC1-Default.mrk | 31 | File MRK | 9/6/2001 ... |
| section 2p.dsn | 15,360 | OrCAD Capture ... | 9/6/2001 ... |
| Section 2P.opj | 468 | Orcad Capture P... | 9/6/2001 ... |
| section 2p-SCHEMATIC1-Test.prb | 596 | PRB File | 9/6/2001 ... |
| section 2p-SCHEMATIC1-Test.sim.cir | 572 | PSpice Circuit File | 9/6/2001 ... |
| SECTION 2P-SCHEMATIC1.net | 97 | PSpice Netlist File | 9/6/2001 ... |
| section 2p-SCHEMATIC1-Test.sim | 1,476 | PSpice Simulatio... | 9/6/2001 ... |

Selected 30,356 bytes in 1 file — Total 57,904 bytes in 11 files

Newly created archive file.

We can now close WinRAR. The archive file, ***Section 2p.exe***, can now be emailed as an attachment. The recipient can then place the file in its own directory and run the file. It will self extract and create all of the files that were placed in the archive. The recipient should then be able to open your project with Orcad Capture and run it.

# PART 3
# DC Nodal Analysis

The node voltage analysis performed by PSpice is for DC node voltages only. This analysis solves for the DC voltage at each node of the circuit. If any AC or transient sources are present in the circuit, those sources are set to zero. Only sources with an attribute of the form ***DC=value*** are used in the analysis. If you wish to find AC node voltages, you will need to run the AC Sweep described in Part 5. The node voltage analysis assumes that all capacitors are open circuits and that all inductors are short circuits.

## 3.A. Resistive Circuit Nodal Analysis

We will perform the nodal analysis on the circuit wired in Part 1 and shown on page 36. The circuit is repeated below:

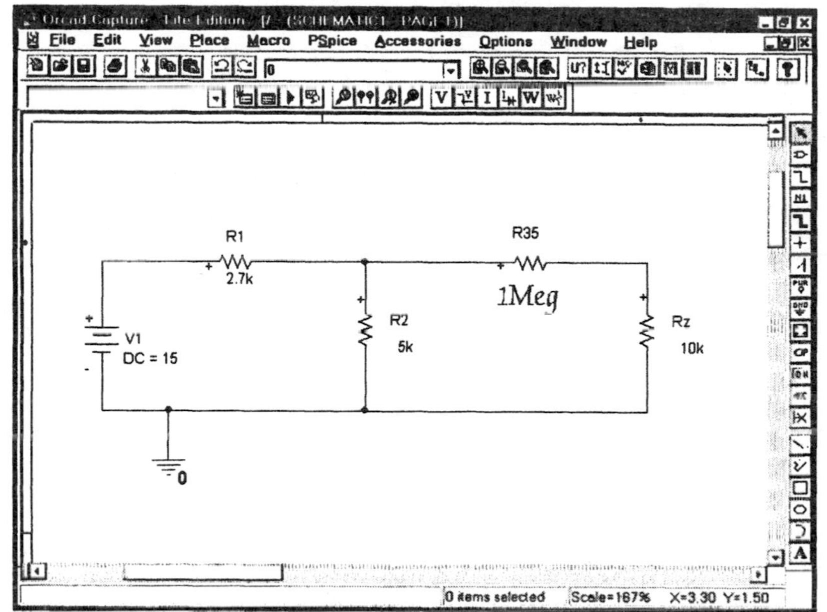


We will show several different ways of finding the node voltages and branch currents. First we will use the bias display markers provided by Capture.

We must first set up the analysis. Select **PSpice** and then **New Simulation Profile** from the menus:

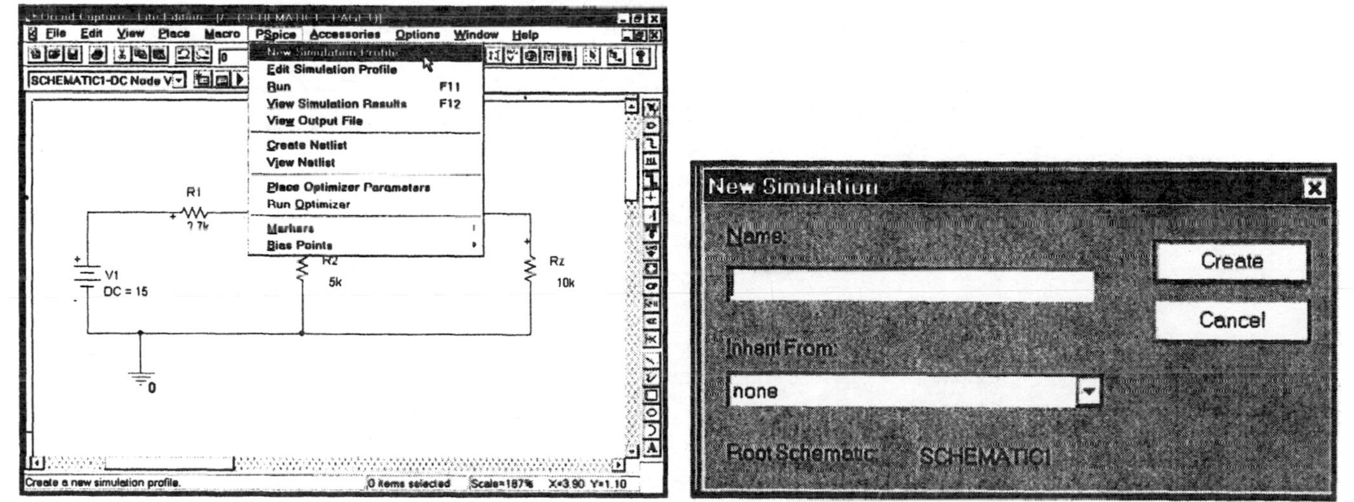


Enter a name for the simulation. I will type **DC Node Voltages** and click the ***Create*** button:

New Simulation
Name:
DC Node Voltages
Create
Cancel
Inherit From:
none
Root Schematic SCHEMATIC1

Simulation Settings - DC Node Voltages
General | Analysis | Include Files | Libraries | Stimulus | Options | Data Collection | Probe Window
Analysis type:
Run to time: 1000ns seconds (TSTOP)
Start saving data after: 0 seconds
Options:
Transient options
Monte Carlo/Worst Case
Maximum step size: seconds
Parametric Sweep
Skip the initial transient bias point calculation (SKIPBP)
Temperature (Sweep)
Save Bias Point
Load Bias Point
Output File Options...
OK | Cancel | Apply | Help

The bias point analysis is automatically run each time PSpice runs so we do not have to specify any settings. However, we will specify the simulation settings to only run a bias point calculation. Click the ***LEFT*** mouse button on the down triangle ▾ as shown to see the list of available simulations:

Simulation Settings - DC Node Voltages
General | Analysis | Include Files | Libraries | Stimulus | Options | Data Collection | Probe Window
Analysis type:
Run to time: 1000ns seconds (TSTOP)
Start saving data after: 0 seconds
Options:
Transient options
Monte Carlo/Worst Case
Maximum step size: seconds
Parametric Sweep
calculation (SKIPBP)
Temperature (Sweep)
Save Bias Point
Load Bias Point
Click here.
Output File Options...
OK | Cancel | Apply | Help

Orcad Capture - Lite Edition - [/ - (SCHEMATIC1 : PAGE1)]
File Edit View Place Macro PSpice Accessories Options Window Help
Simulation Settings - DC Node Voltages
General | Analysis | Include Files | Libraries | Stimulus | Options | Data Collection | Probe Window
Analysis type:
Time Domain (Transient)
Run to time: 1000ns seconds (TSTOP)
DC Sweep
AC Sweep/Noise
Bias Point
Start saving data after: 0 seconds
Transient options
Monte Carlo/Worst Case
Maximum step size: seconds
Parametric Sweep
Skip the initial transient bias point calculation (SKIPBP)
Temperature (Sweep)
Save Bias Point
Load Bias Point
Output File Options...
OK | Cancel | Apply | Help

Select the **Bias Point** selection:

Orcad Capture - Lite Edition - [/ - (SCHEMATIC1 : PAGE1)]
File Edit View Place Macro PSpice Accessories Options Window Help
Simulation Settings - DC Node Voltages
General | Analysis | Include Files | Libraries | Stimulus | Options | Data Collection | Probe Window
Analysis type:
Time Domain (Transient)
Run to time: 1000ns seconds (TSTOP)
Time Domain (Transient)
DC Sweep
AC Sweep/Noise
Start saving data after: 0 seconds
Transient options
Monte Carlo/Worst Case
Maximum step size: seconds
Parametric Sweep
Skip the initial transient bias point calculation (SKIPBP)
Temperature (Sweep)
Save Bias Point
Load Bias Point
Output File Options...
OK | Cancel | Apply | Help

Simulation Settings - DC Node Voltages
General | Analysis | Include Files | Libraries | Stimulus | Options | Data Collection | Probe Window
Analysis type:
Output File Options
Include detailed bias point information for nonlinear controlled sources and semiconductors (.OP)
Options:
Perform Sensitivity analysis (.SENS)
Temperature (Sweep)
Output variable(s)
Save Bias Point
Load Bias Point
Calculate small-signal DC gain (.TF)
From Input source name:
To Output variable:
OK | Cancel | Apply | Help

This selection specifies that only the Bias Point simulation will run. We do not need to specify any output file options because we will be displaying the results on the schematic. Click the ***OK*** button to save the simulation profile and return to the circuit.

We will run the analysis to calculate all node voltages and all branch currents. Select **PSpice** and then **Run** from the Capture menus:

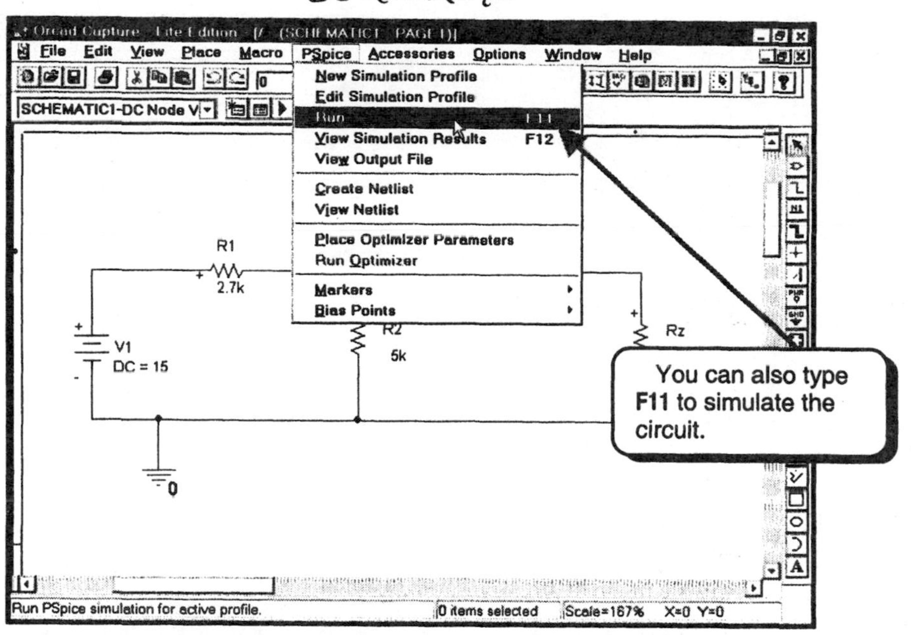


PSpice will run and the PSpice A/D Lite window will be displayed:

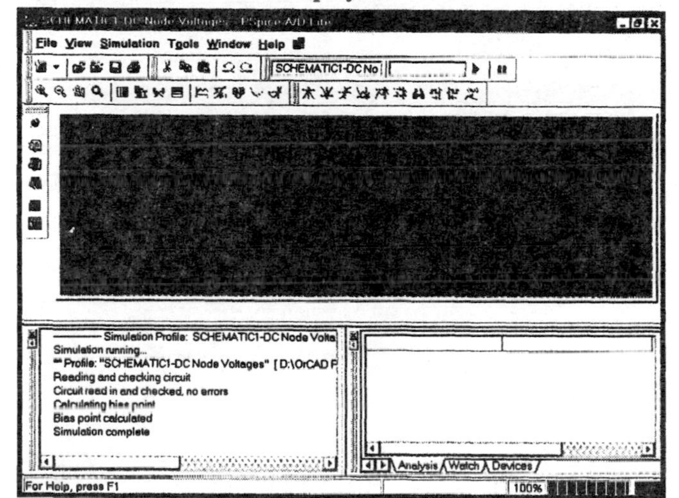

When the message window says ***Bias point calculated***, the simulation is complete. We can now display the voltages and currents. Switch back to Capture to view the circuit. On my circuit, no voltages or currents are displayed:

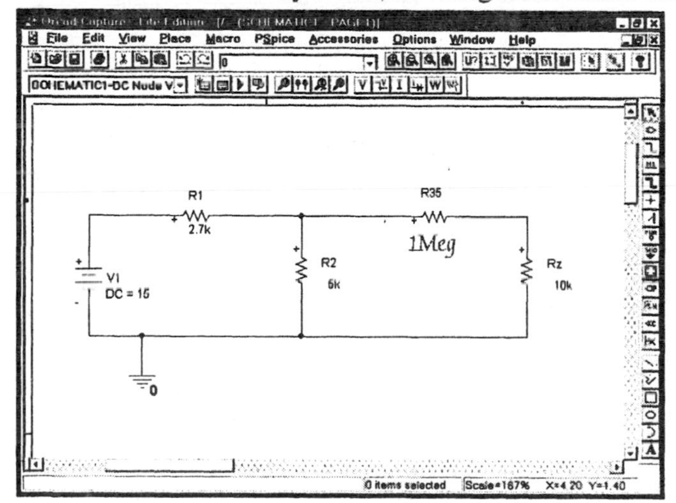

Before continuing, we will make sure that we enable the display of voltages, currents, and power on the schematic. Select **PSpice** and then **Bias Points**:

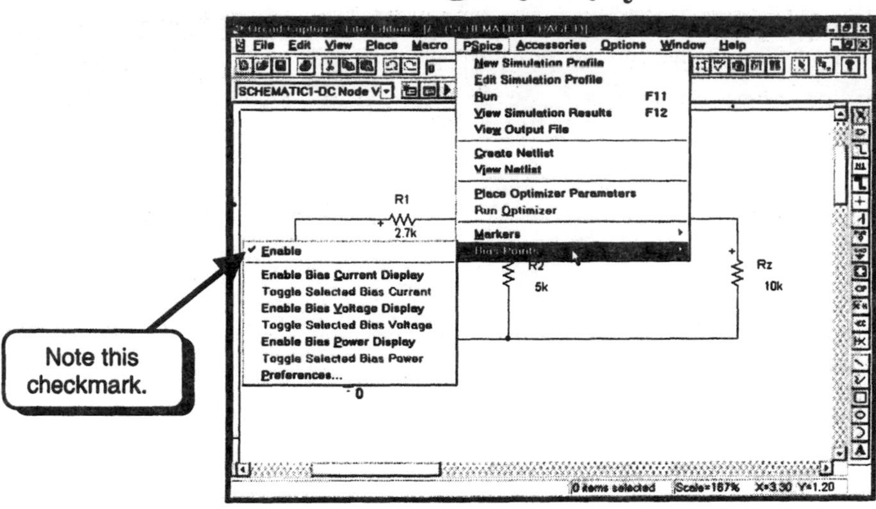


Note that there is a checkmark next to the text **Enable.** If your menu selection does not have a checkmark, as shown above, you will not be able to display voltages and currents on your schematic. **If you do not have a checkmark displayed next to menu selection Enable,** select **Enable** from the menu to toggle the checkmark on. If a checkmark is displayed, click the ***LEFT*** mouse button on the schematic somewhere to hide the menus.

We will first display the node voltages. To toggle the voltages on or off, click the ***LEFT*** mouse button on the V button **V** in the Capture button bar:

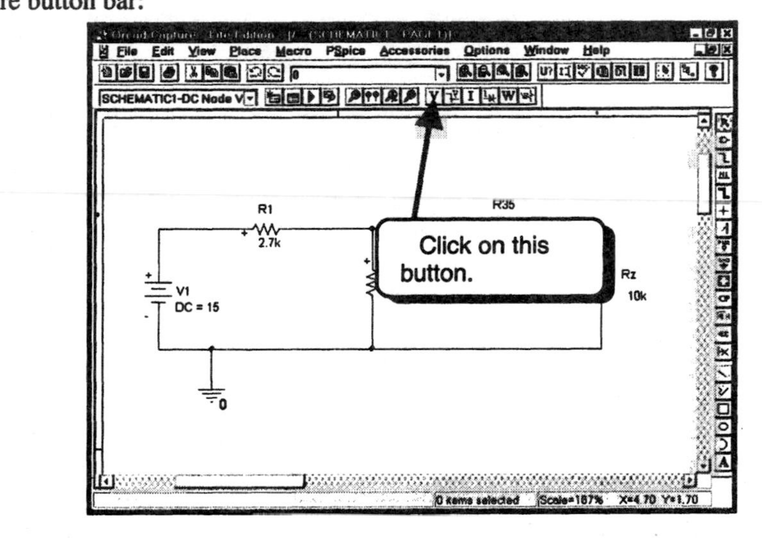


Clicking on the **V** button will toggle the display of node voltages on or off. Click on the **V** button until the node voltages are displayed:

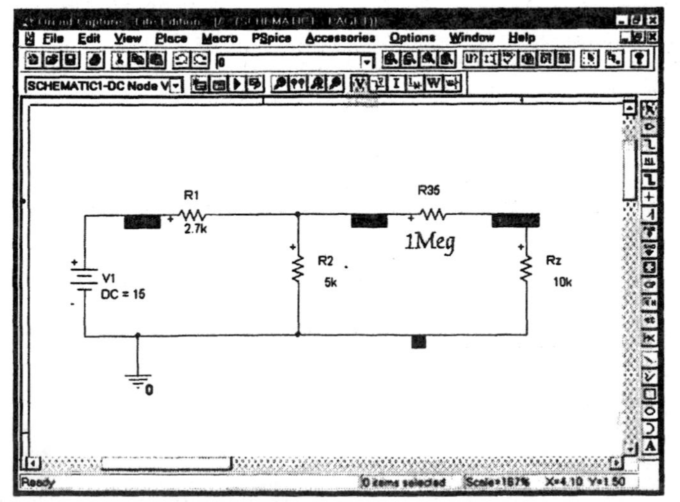


The screen capture above displays the voltages for every node. If your screen looks like one of the following:

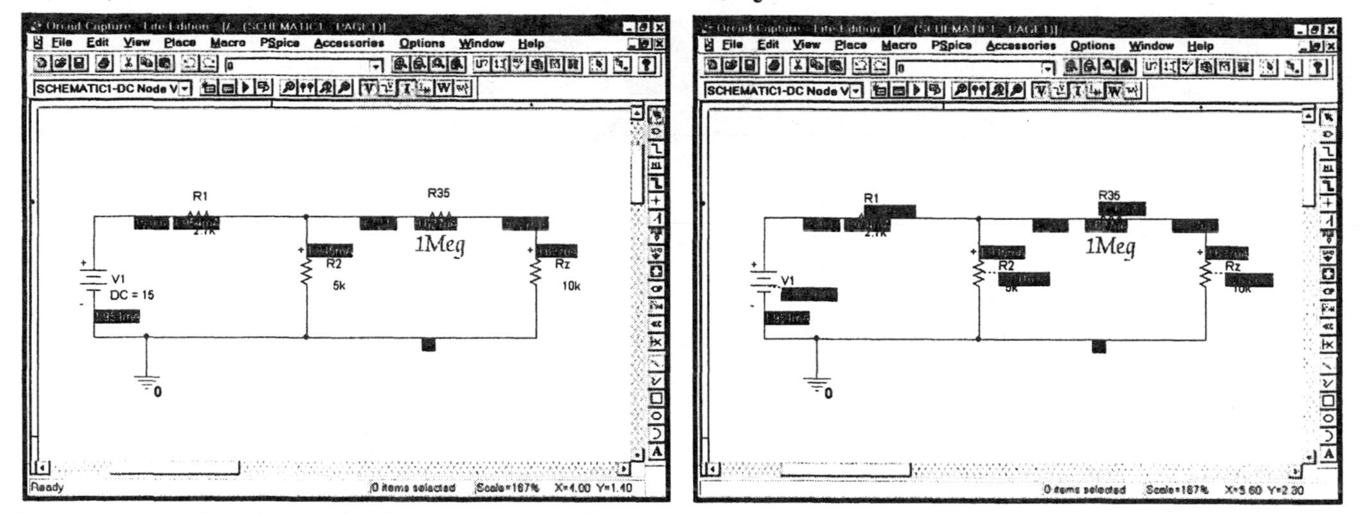


then you have node voltages, plus branch currents and/or power data displayed. To hide the branch currents, click the ***LEFT*** mouse button on the I button **I** as shown below:

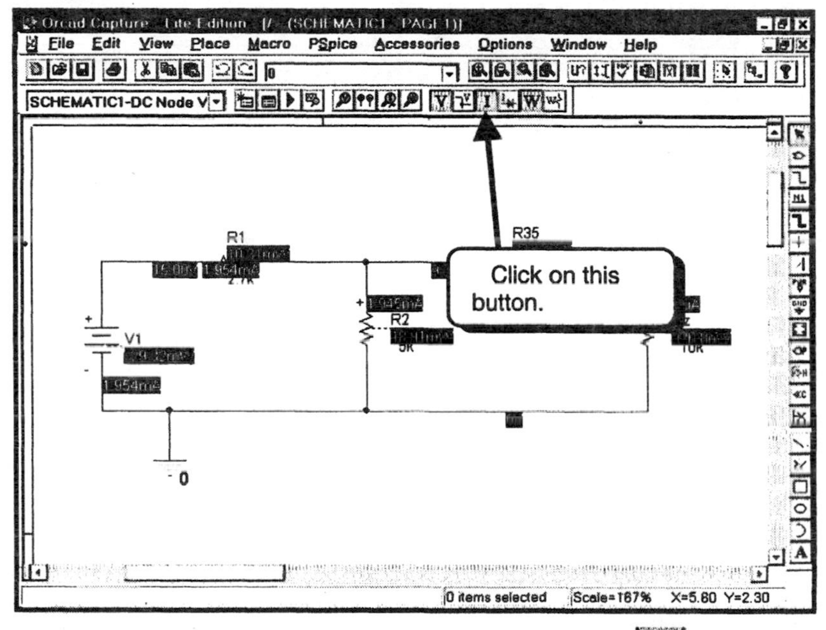


Clicking this button will toggle the display of branch currents on and off. Click the **I** button to hide the branch currents.

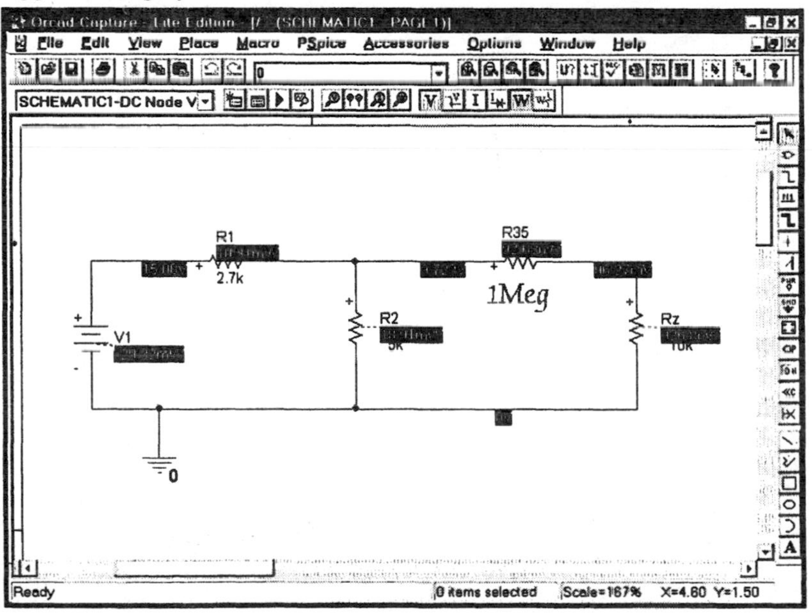

To hide the power data, click the ***LEFT*** mouse button on the W button [W] as shown below:

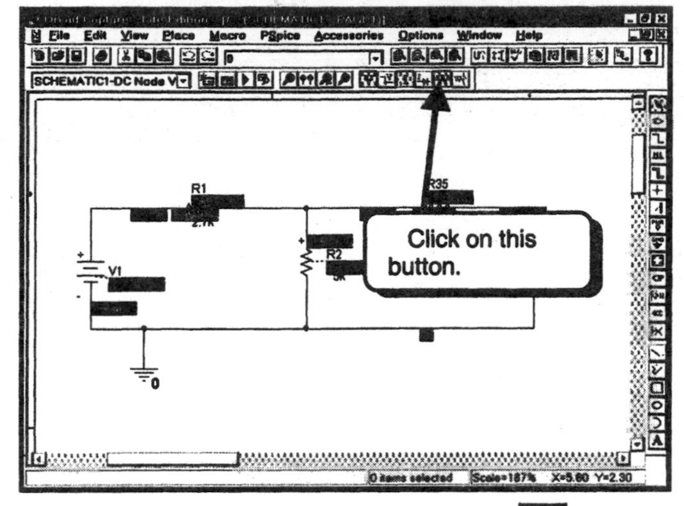


Clicking this button will toggle the display of power data on and off. Click the [W] button to hide the power data.

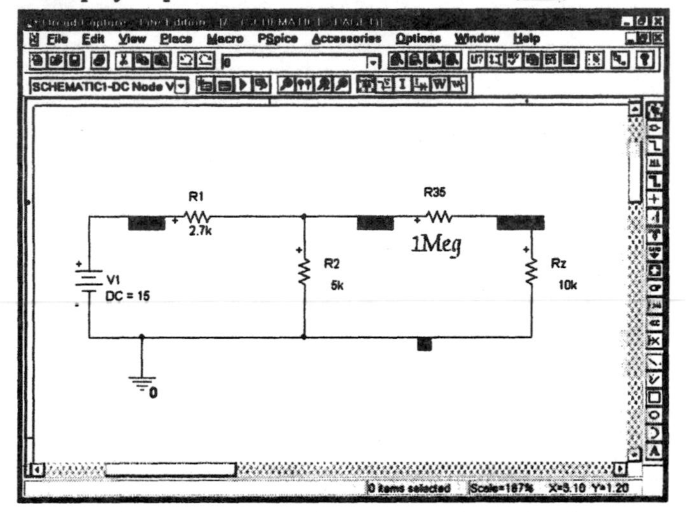


We now only have the node voltages displayed, so we can clean up the schematic a bit.

The node voltages are a little bit too close to some of the components. The numbers can be moved like any graphic on the screen. Move the numbers to clean up the circuit a little bit:

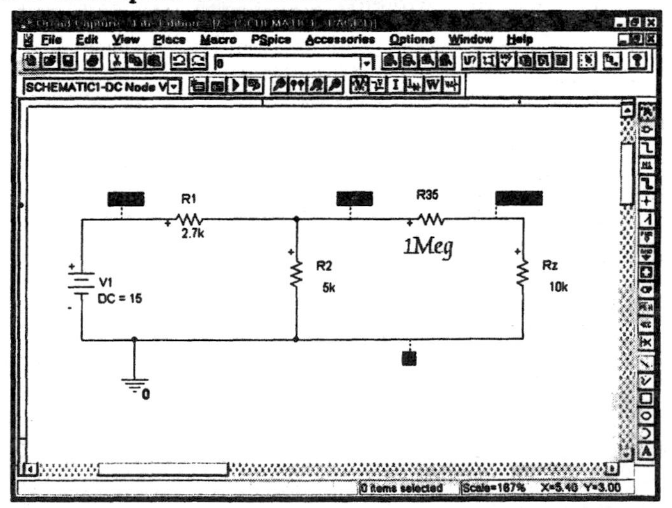


We see that dashed lines are drawn from the data to the node to which they refer. This makes it easy in complicated circuits to figure out which number belongs to which node.

Some of the node voltages displayed are unnecessary and we would like to hide them in the schematic. For example, there is a node voltage number telling us that ground is at zero volts. This is unnecessary and we will hide it. First, click the ***LEFT*** mouse button on the text ***0V*** as shown. It will become selected and displayed in pink:

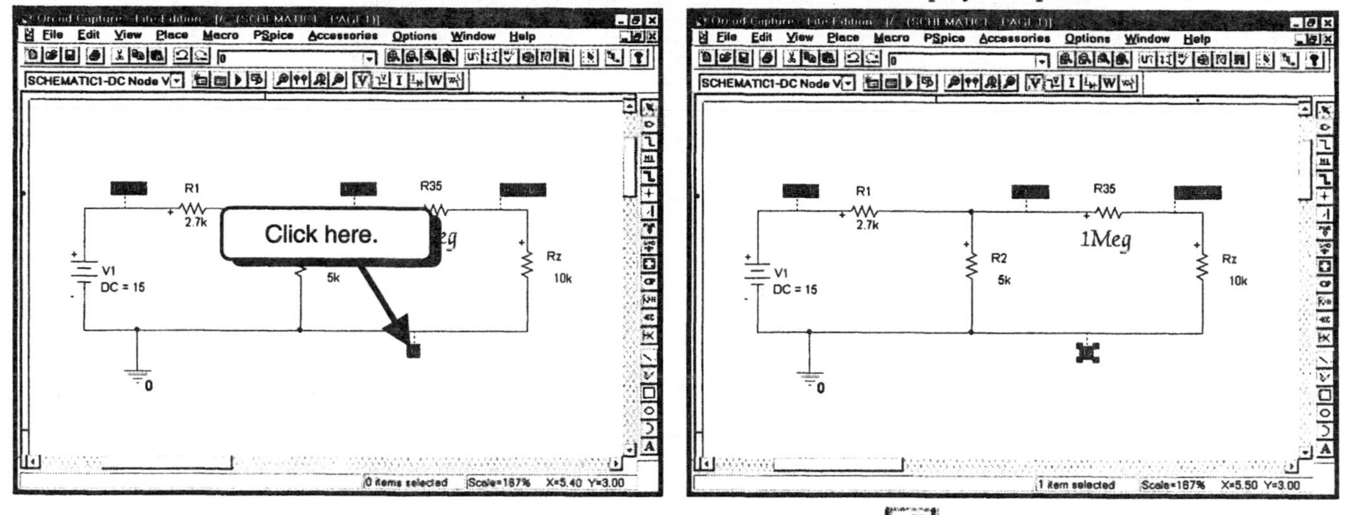


To hide the selected marker, click on the Toggle Voltages On Selected Net button:

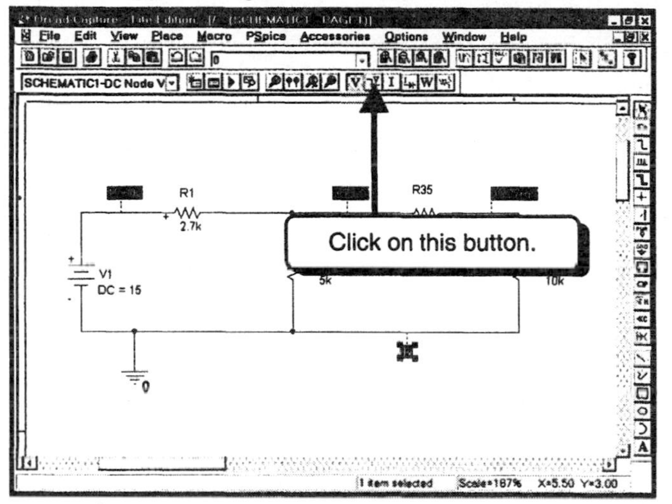


The selected node voltage will be removed:

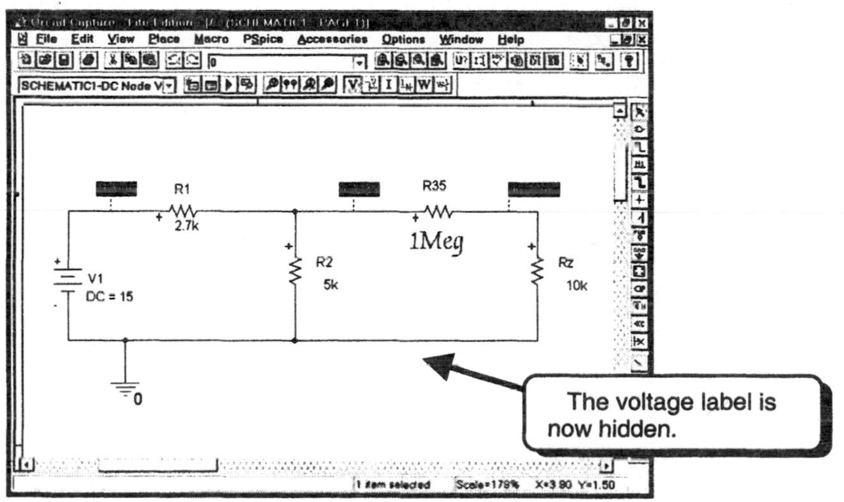


Use this technique to hide the ***15.00V*** number:

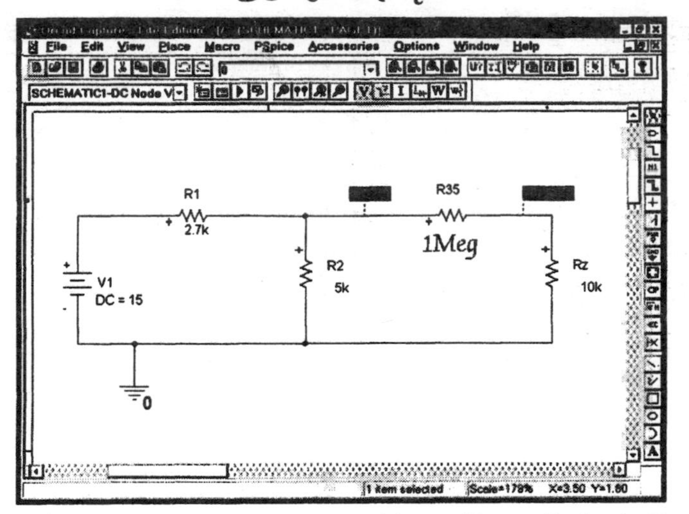


Suppose that we have hidden a node voltage and we would now like to display it. To do this, click the ***LEFT*** mouse button on the wire of which you would like to display the node voltage. The wire will turn red, indicating that it has been selected. I would like to display the voltage of the node between V1 and R1 (the one I removed previously) so I will select the wire shown below:

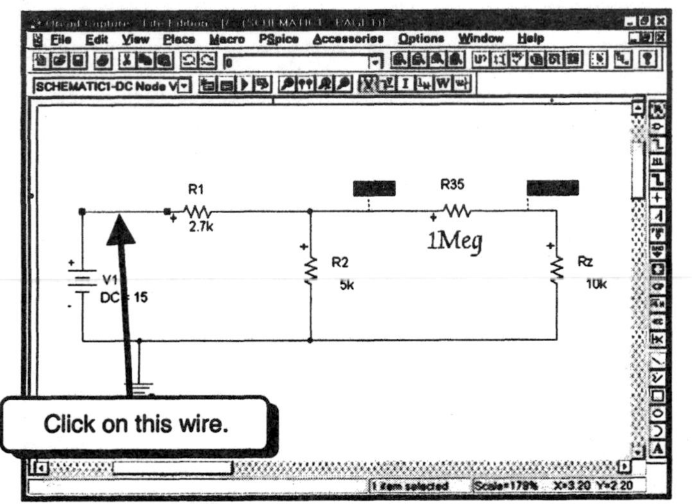


When the wire is highlighted in pink, click the ***LEFT*** mouse button on the Toggle Voltages On Selected Net button . The voltage for the node will be displayed:

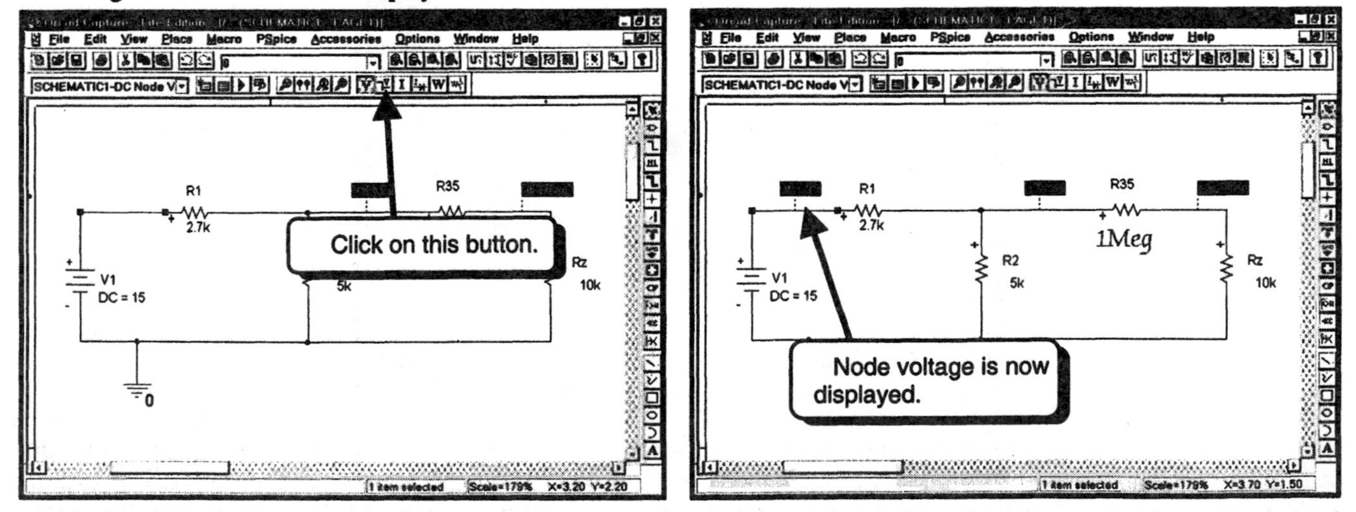


Now that we know all of the node voltages, we would like to display the branch currents. First, hide the node voltages by clicking the ***LEFT*** mouse button on the V button **V**. The voltages should no longer be displayed:

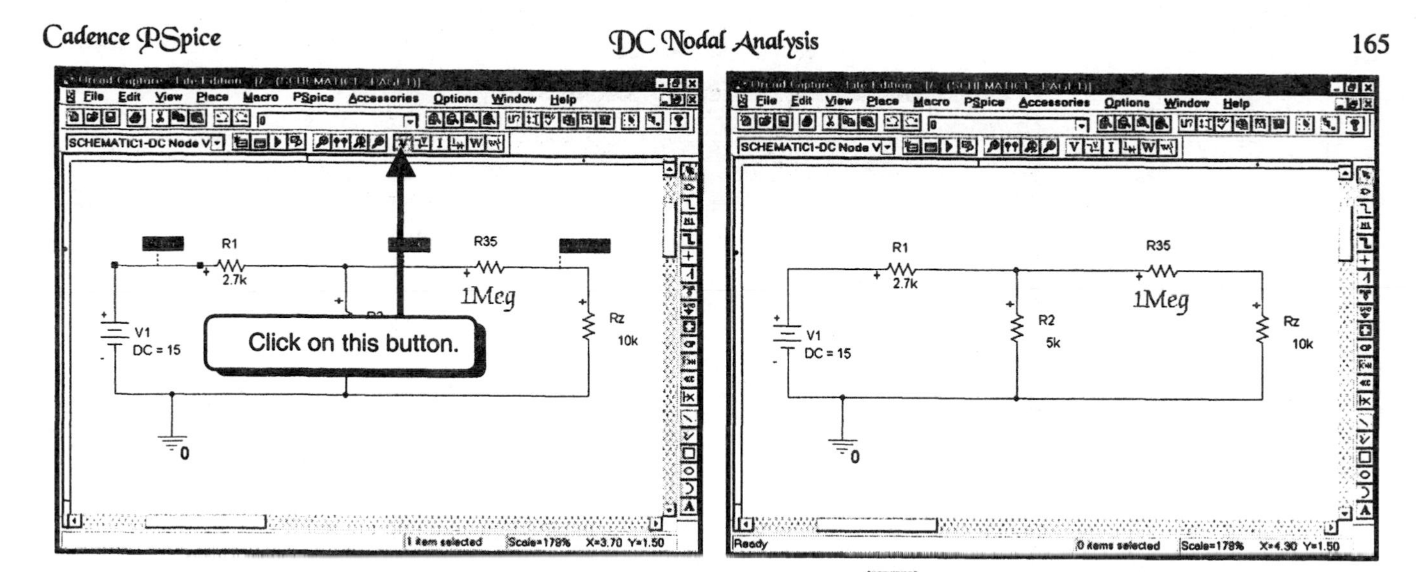


To display the branch currents, click the ***LEFT*** mouse button on the I button **I**. The branch currents will be displayed:

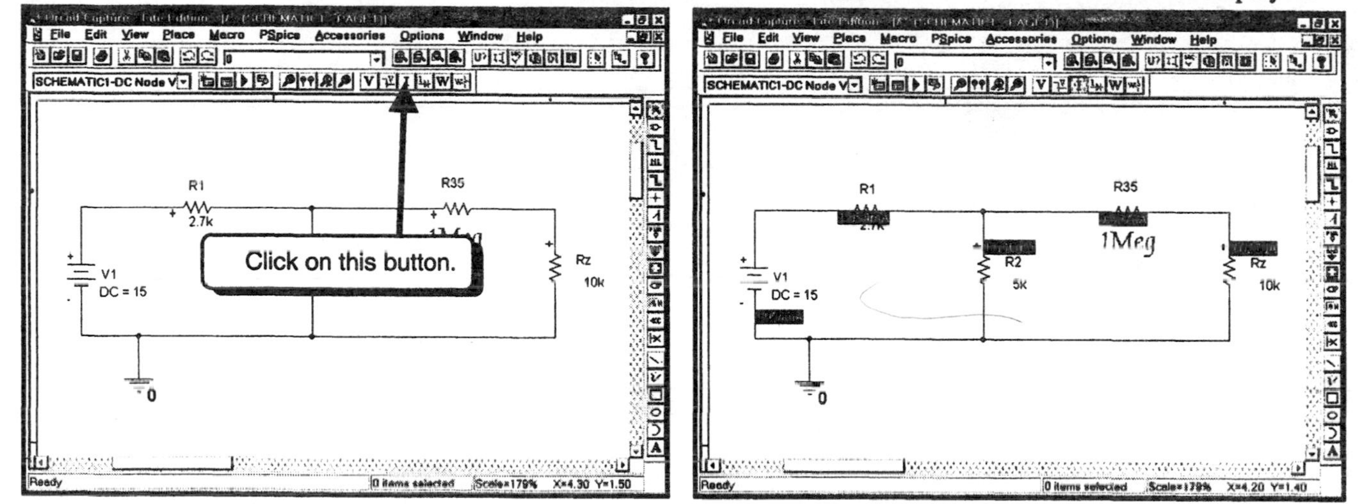


We can move the displayed numbers in the same manner as we did with the node voltages. Move the numbers so that the schematic is a little clearer:

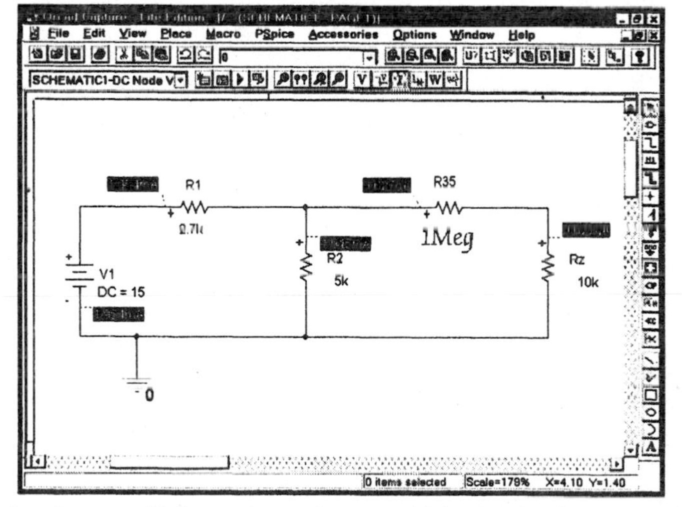

With currents we need to know which number refers to which circuit element and which direction the current is flowing. Dashed lines are shown connecting the numbers to a terminal of an element. The numbers shown are for positive currents entering the indicated terminals. For example, there is 1.954 mA entering the negative terminal of V1. This indicates that this voltage source is supplying power to the circuit. As another example, there is 1.954 mA entering the positive terminal of R1, and 9.627 μA entering the positive terminal of R35. In the drawing below, I have manually added arrows to the drawing to show the direction of the currents.

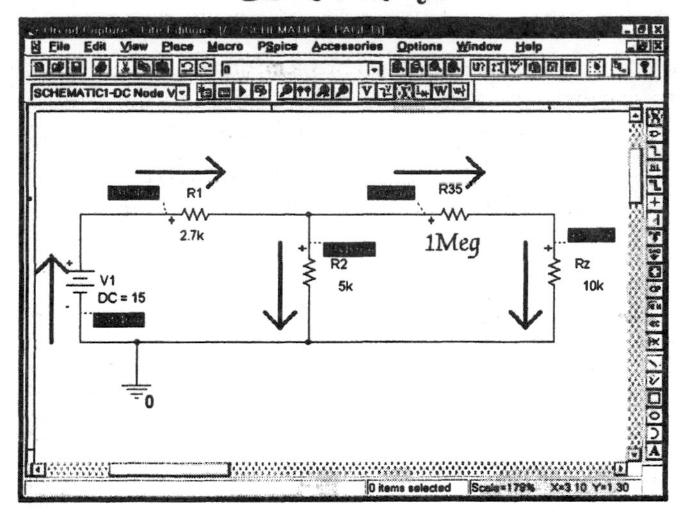


Branch currents can be hidden using a procedure similar to what we used to hide node voltages. Click the ***LEFT*** mouse button on the text ***1.954mA*** at the bottom of the screen. It will become highlighted in pink:

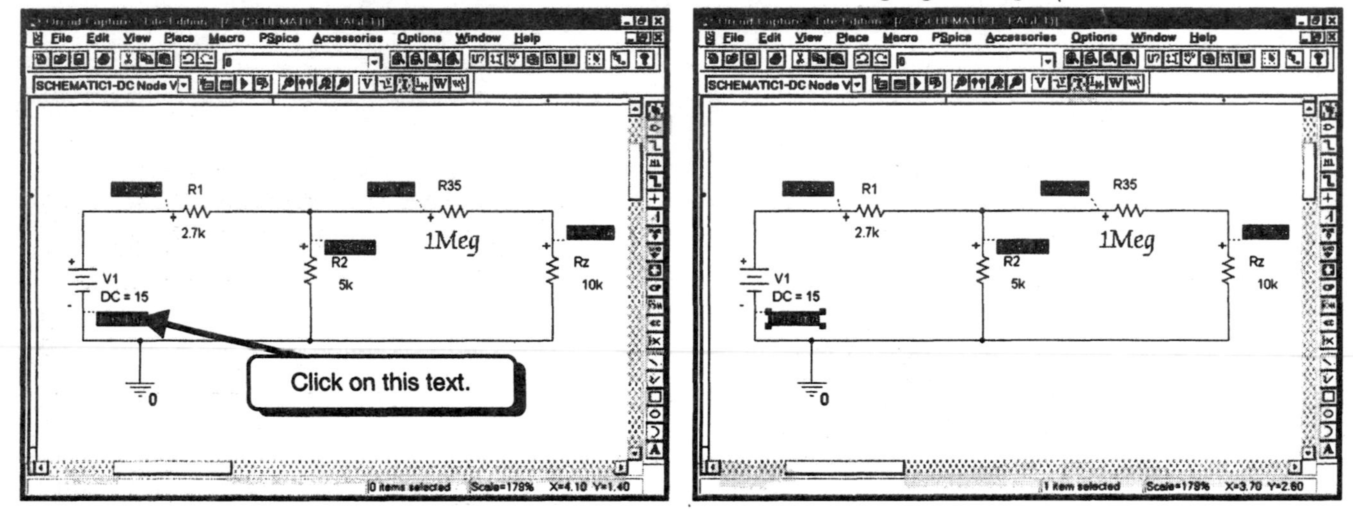


Click the ***LEFT*** mouse button on the Toggle Current On Selected Parts/Pins button . The selected current will disappear:

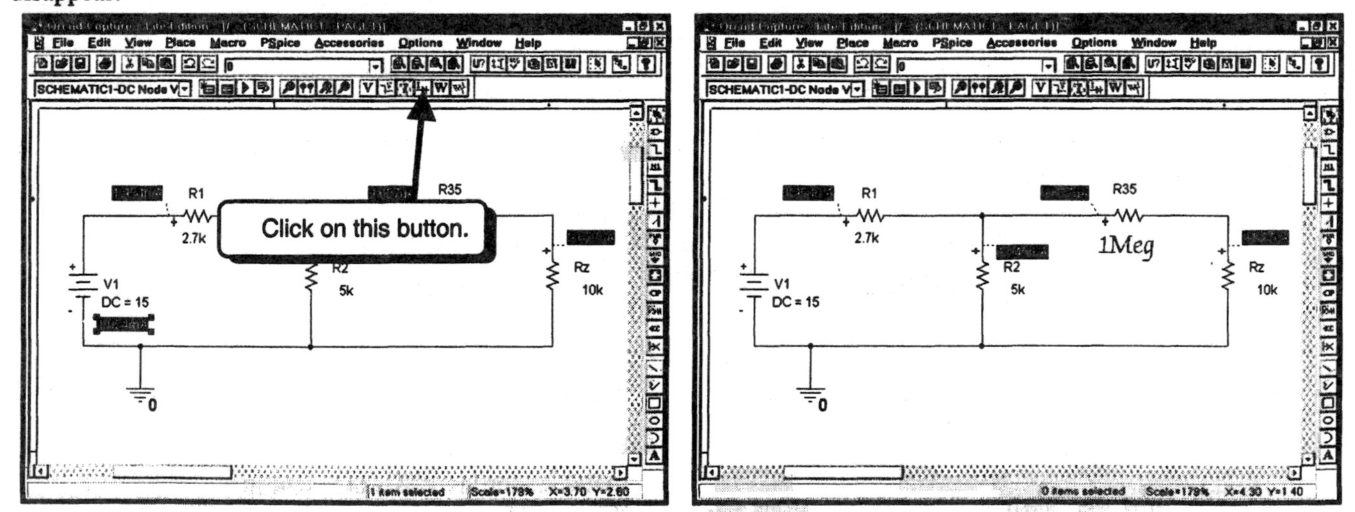


We can hide as many currents as we wish.

Suppose that we have hidden a number of the branch currents, and we now wish to display one of the hidden currents. Displaying hidden branch currents is a little different than displaying hidden node voltages. Node voltages are associated with wires, so, to display a node voltage, we selected a wire and then clicked the  button. Branch currents are associated with circuit elements. To display a current into an element, select the element and then click the [I] button. As an exercise, remove branch currents so that your circuit resembles the screen capture below:

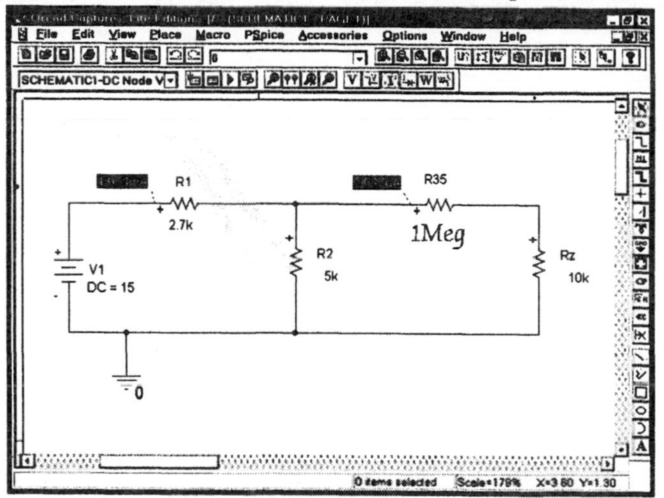


The currents through R2 and through Rz are not shown. We will display the current through R2. Click the ***LEFT*** mouse button on the graphic for R2 . It will become highlighted in pink, indicating that it has been selected. When it is selected, click the ***LEFT*** mouse button on the [I] button. The current for the selected element will be displayed:

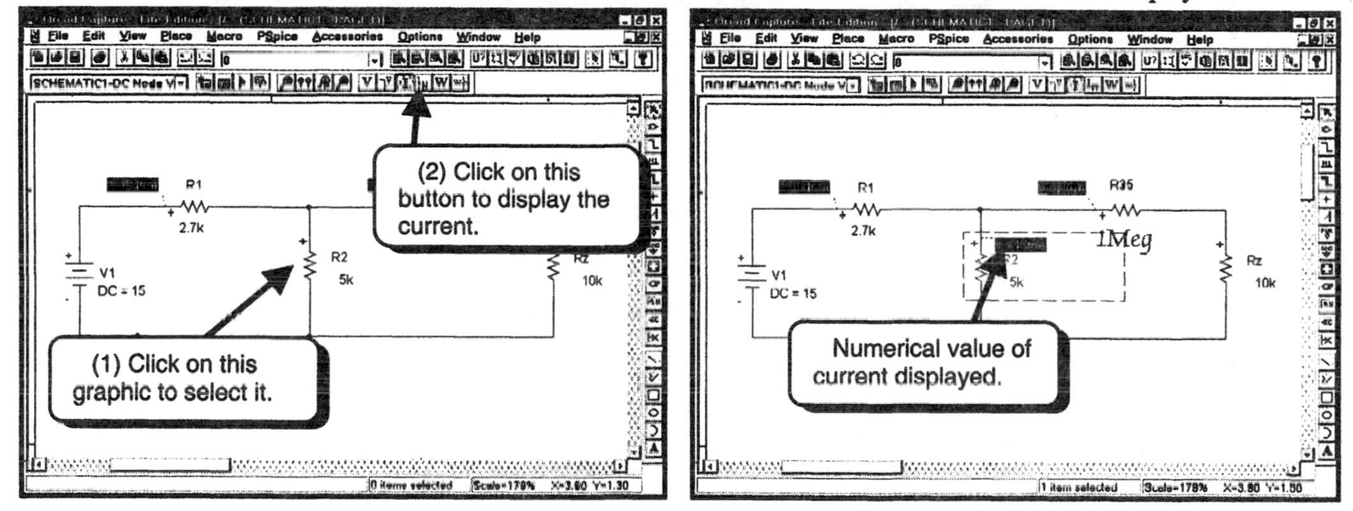


For the next example, we would like to hide all of the voltages and currents. Use the [V] and [I] buttons to hide all node voltages and branch currents:

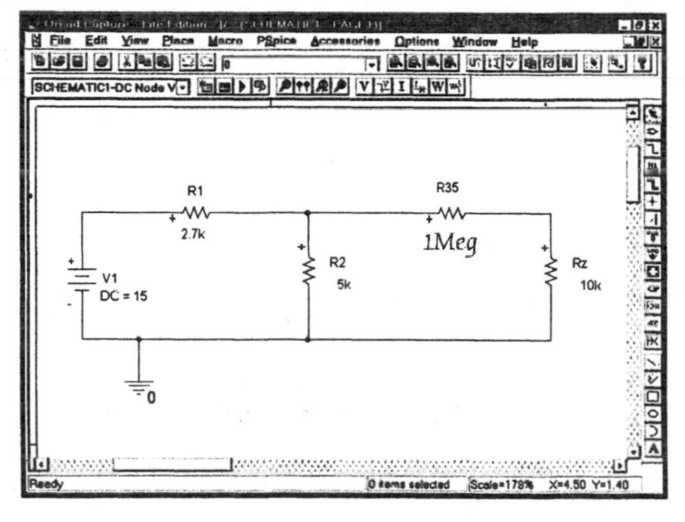

We would now like to display the power absorbed by all elements in the circuit. To display the power, click the ***LEFT*** mouse button on the **W** button:

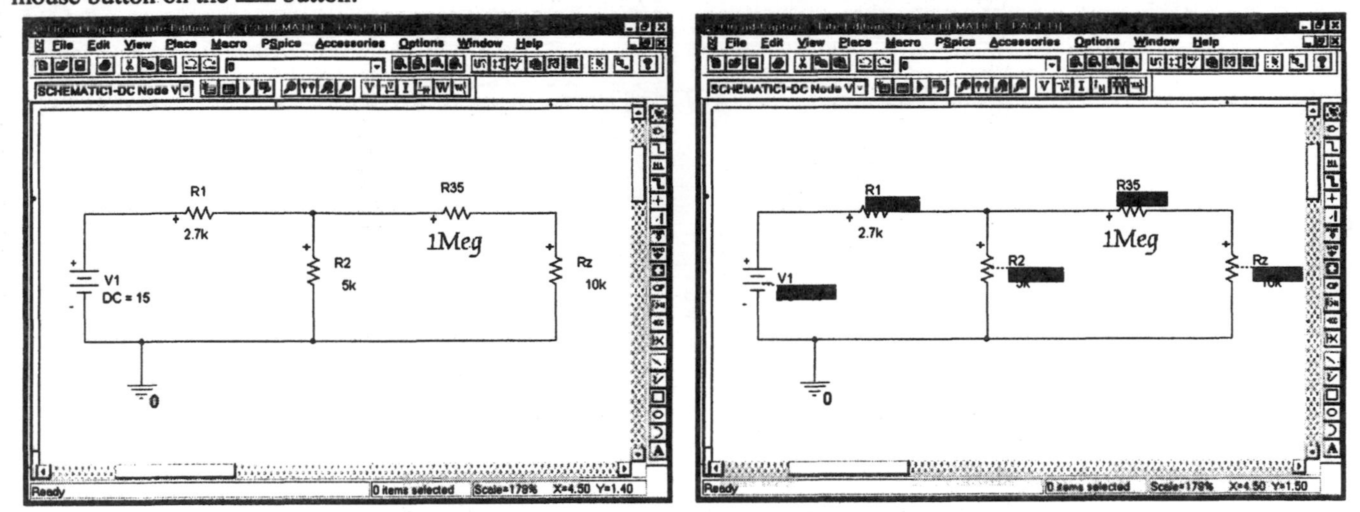


The numbers display the power absorbed or sourced by every element in the circuit. Move the numbers around to make the circuit more readable:

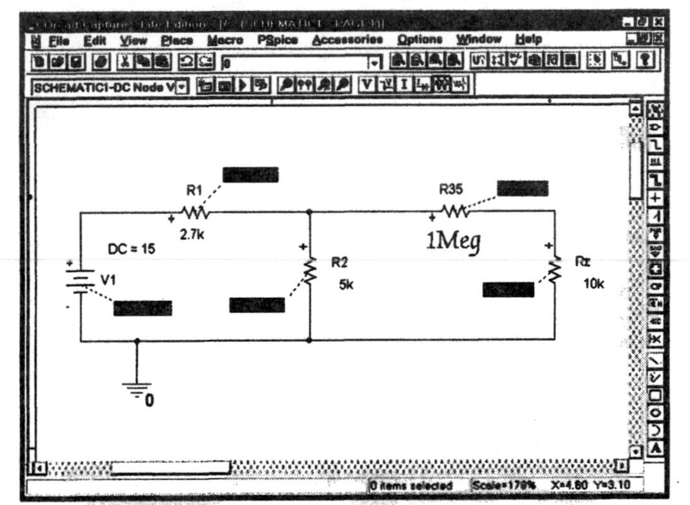


Notice that dashed lines associate a number with a circuit element. Note that the power numbers displayed for resistors are all positive. This means that those elements are absorbing that amount of power. The number displayed for the voltage source is negative, indicating that the element is sourcing power to the circuit.

**EXERCISE 3-1**: Find the DC node voltages for the circuit below:

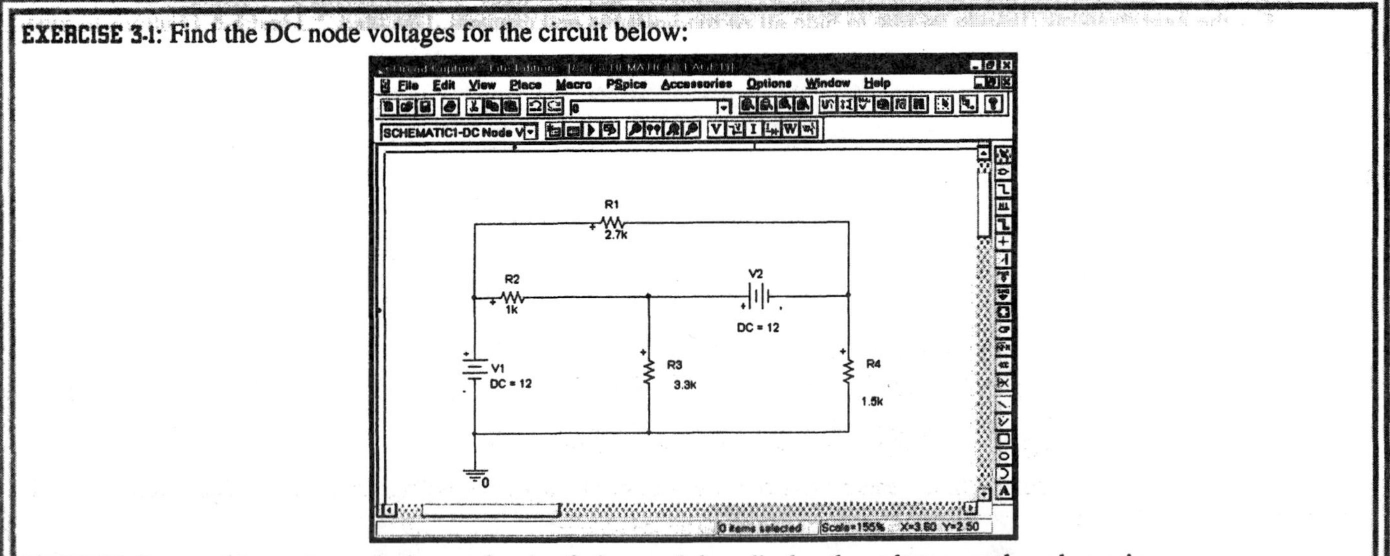


**SOLUTION**: Set up a bias point analysis, run the simulation, and then display the voltages on the schematic.

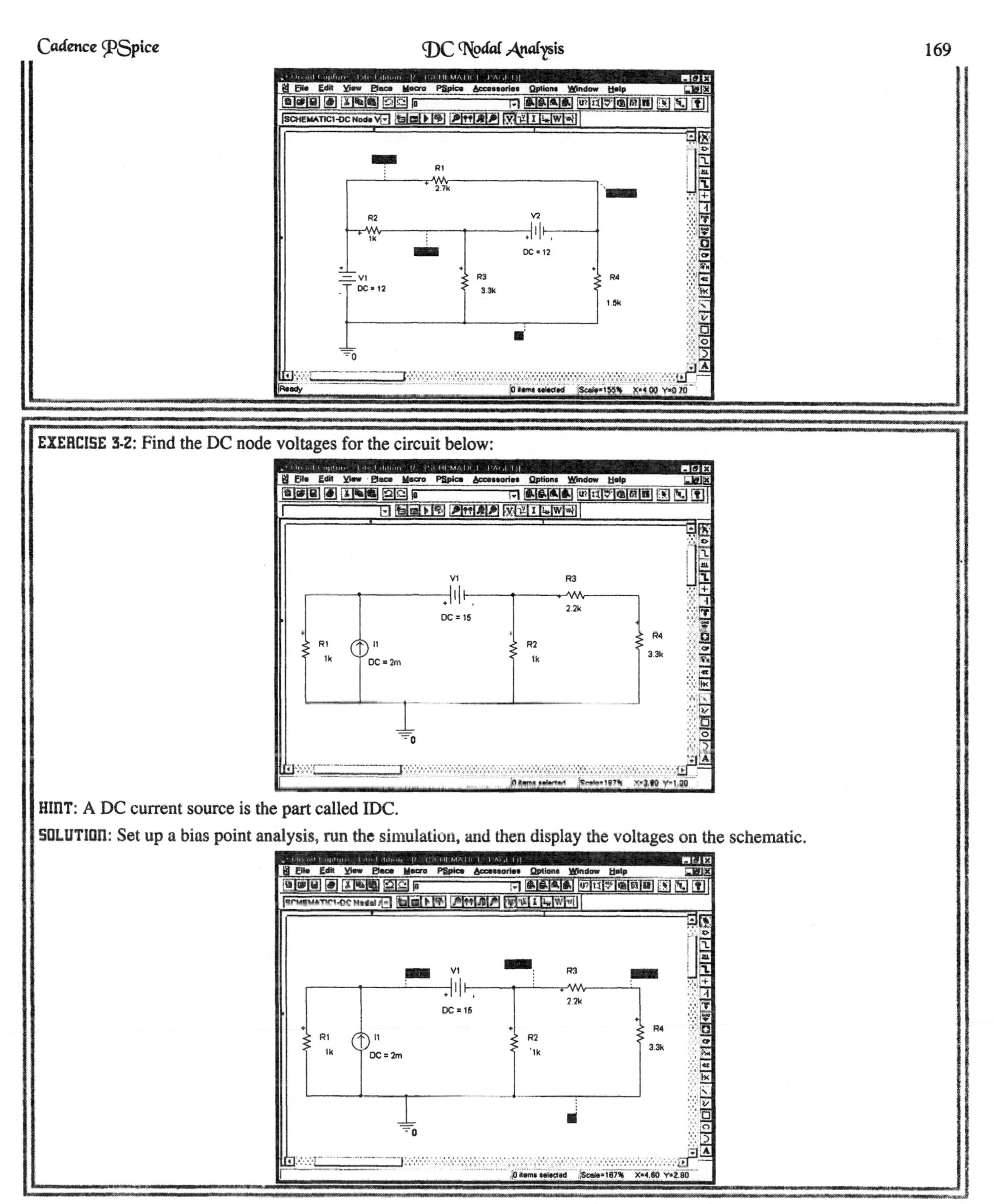


**EXERCISE 3-2**: Find the DC node voltages for the circuit below:

**HINT**: A DC current source is the part called IDC.

**SOLUTION**: Set up a bias point analysis, run the simulation, and then display the voltages on the schematic.

# 3.B. Nodal Analysis with Dependent Sources

To illustrate an example with dependent sources, we will perform a node voltage analysis on the circuit below. If you are unfamiliar with wiring a circuit, review Part 1. Create the circuit below.

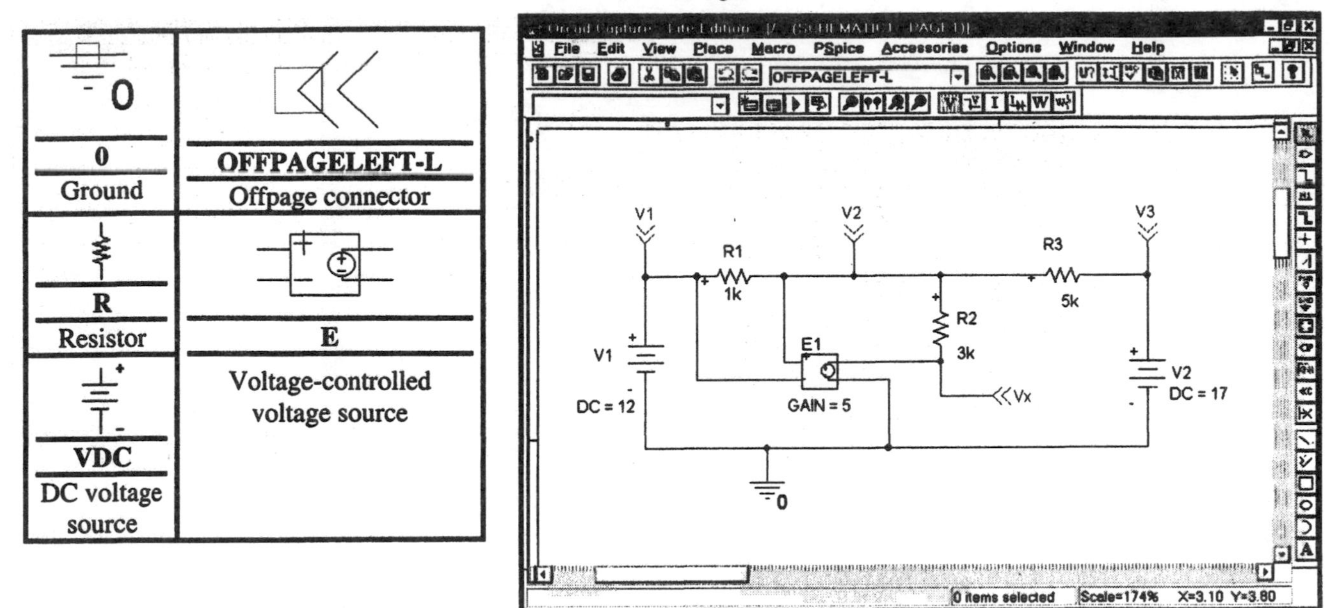

| | |
|---|---|
| 0<br>Ground | OFFPAGELEFT-L<br>Offpage connector |
| R<br>Resistor | E<br>Voltage-controlled voltage source |
| VDC<br>DC voltage source | |



The new part in this circuit is the voltage-controlled voltage source. The way this element is wired, the voltage at node ***VX*** is 5 times the voltage across ***R1***: $V_X = 5(V_2 - V_1)$. If you zoom in* on the voltage-controlled voltage source, you will see the graphic shown in Figure 3-1. Notice that the plus (+) and minus (–) terminals are open circuited. These connections draw no current and only sense the voltage of the nodes to which they are connected. The right half of the graphic contains the dependent source. The voltage of this source is the gain times the voltage at the sensing nodes.

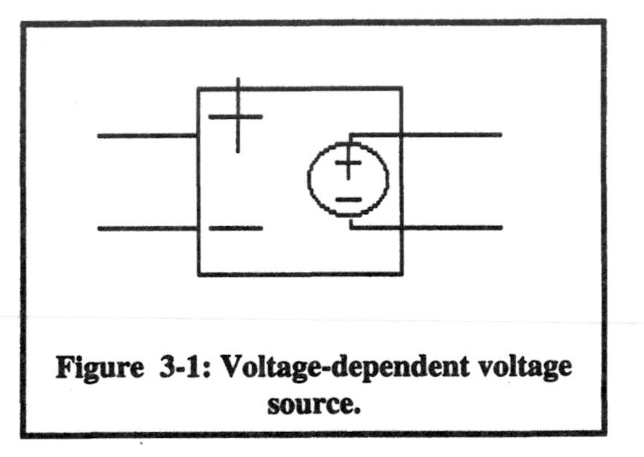

**Figure 3-1: Voltage-dependent voltage source.**

We must now set up a Bias Point simulation. Click on the New Simulation Profile button to create a new profile (in the last section, we selected **PSpice** and then **New Simulation Profile** from the menus):

New Simulation
Create
Cancel
none

Specify a name for the simulation profile and then click the ***Create*** button:

*To zoom in on a particular spot on the screen, select **View** and then **In** from the Capture main menu. The cursor will be replaced by crosshairs (+). Move the crosshairs to the spot on the screen where you want to zoom in. Click the ***LEFT*** mouse button. Repeat the steps to make the drawing larger if necessary.

New Simulation
Name:
Voltages at Nodes
Create
Cancel
Inherit From:
none
Root Schematic: SCHEMATIC1

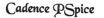

Specify a **Bias Point** ***Analysis type***:

Bias Point selected.

We will display the results on the schematic, so we do not need to specify any ***Output File Options***. Click the ***OK*** button to save the settings.

Select **PSpice** and then **Run** to simulate the circuit:

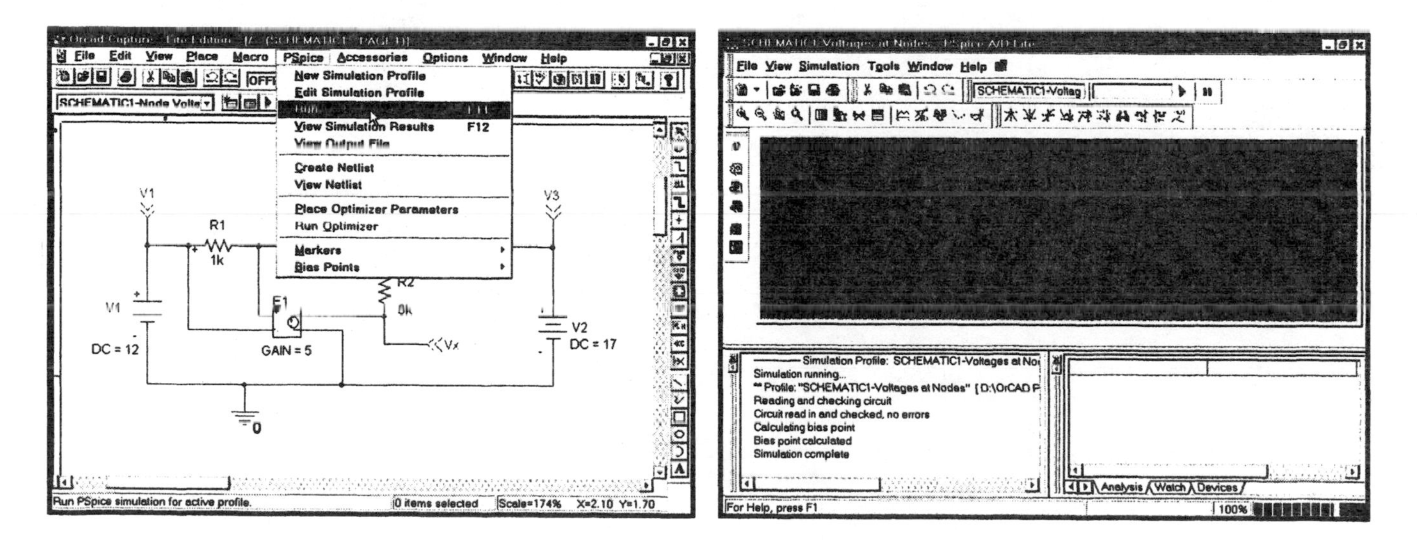


When the simulation is complete, switch to Capture to display the circuit:

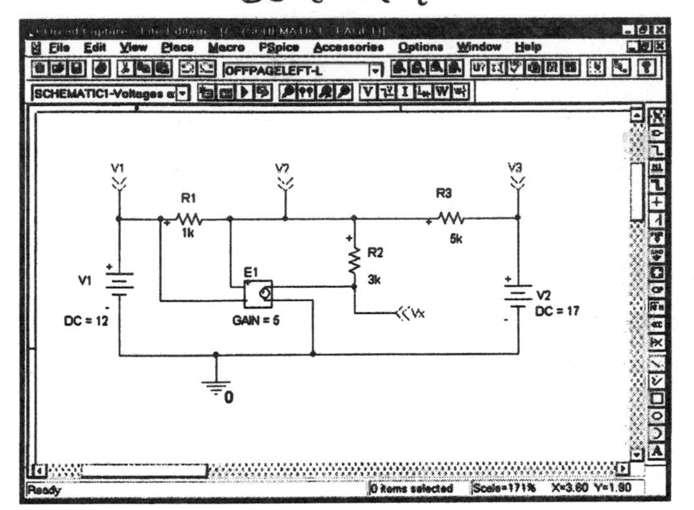


If the node voltages are not displayed, use the steps detailed in the previous section to display node voltages on your schematic:

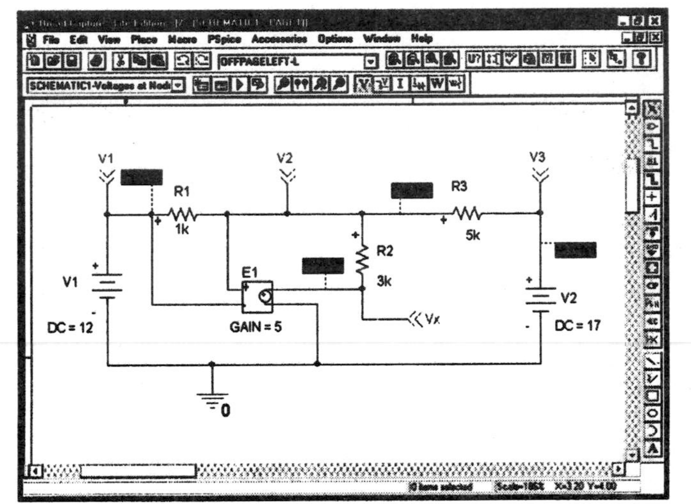


The results are also contained in the output file. To examine the output file, select **PSpice** and then **View Output File** from the Capture menus. The output file will be displayed by Capture:

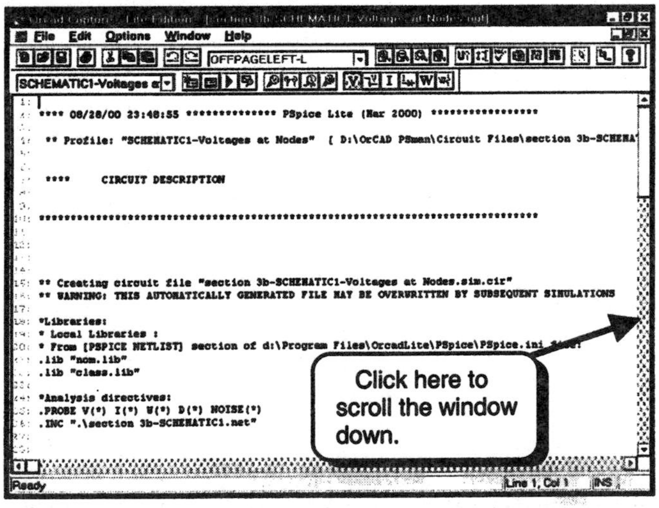

```
**** 08/28/00 23:48:55 ************** PSpice Lite (Mar 2000) ****************

 ** Profile: "SCHEMATIC1-Voltages at Nodes"  [ D:\OrCAD PSmmn\Circuit Files\section 3b-SCHEMA

****     CIRCUIT DESCRIPTION

******************************************************************************

** Creating circuit file "section 3b-SCHEMATIC1-Voltages at Nodes.sim.cir"
** WARNING: THIS AUTOMATICALLY GENERATED FILE MAY BE OVERWRITTEN BY SUBSEQUENT SIMULATIONS

*Libraries:
* Local Libraries :
* From [PSPICE NETLIST] section of d:\Program Files\OrcadLite\PSpice\PSpice.ini
.lib "nom.lib"
.lib "class.lib"

*Analysis directives:
.PROBE V(*) I(*) W(*) D(*) NOISE(*)
.INC ".\section 3b-SCHEMATIC1.net"
```

Click here to scroll the window down.

The node voltage results are contained near the bottom of the file. Click the ***LEFT*** mouse button on the vertical scroll bar until you see this text:

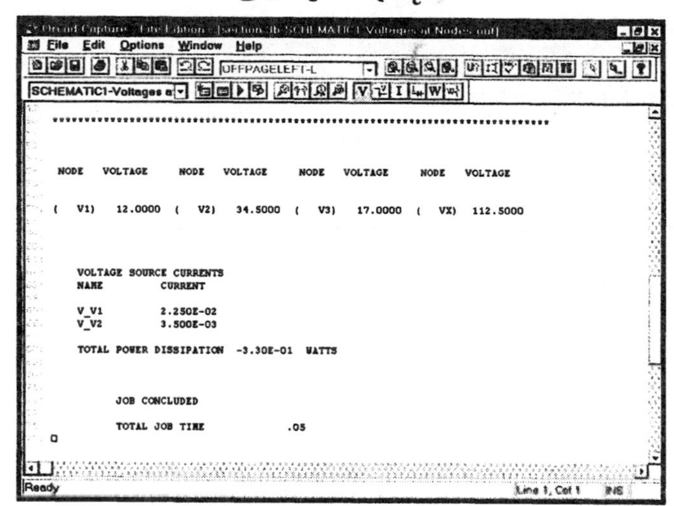


The results show the voltages at the specified nodes relative to ground, and are the same numerical results as displayed on the schematic. To close the text editor program, select **File** and then **Close** from the Capture menus. You will return to the schematic:

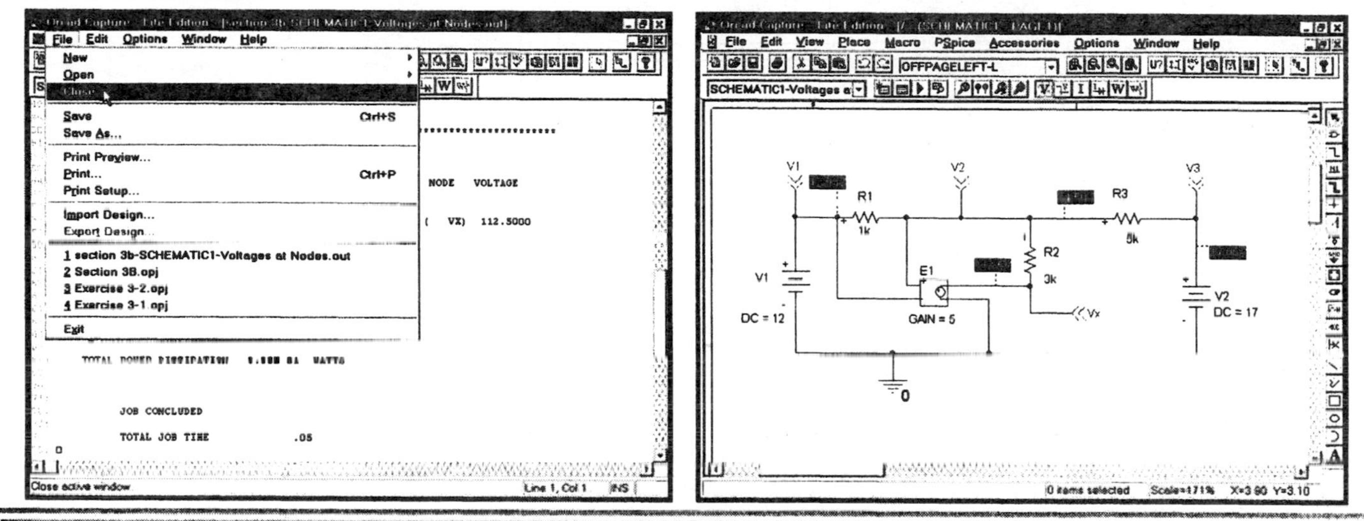


**EXERCISE 3-3**: Find the DC node voltages for the circuit below:

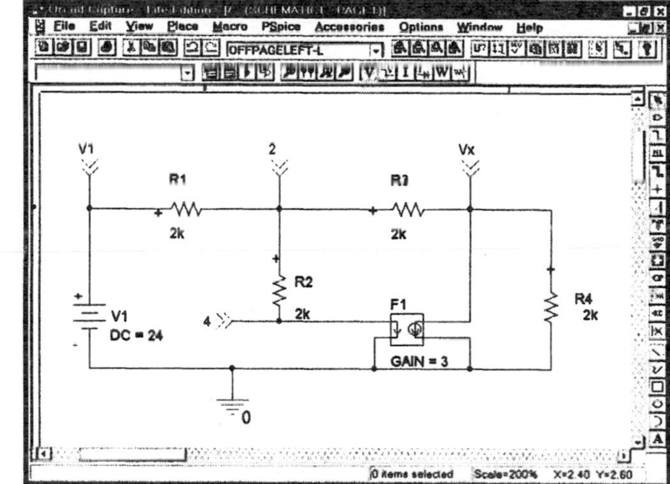


**HINT**: F1 is a current-controlled current source. Note that node 4 is connected to ground with a wire. Thus, the voltage of node 4 should be zero volts. Node 4 is necessary because it joins the lower terminal of R2 to the current-sensing terminal of F1.

**SOLUTION**: Set up a Bias Point analysis and then simulate the circuit. The results can be viewed on the schematic or in the output file:

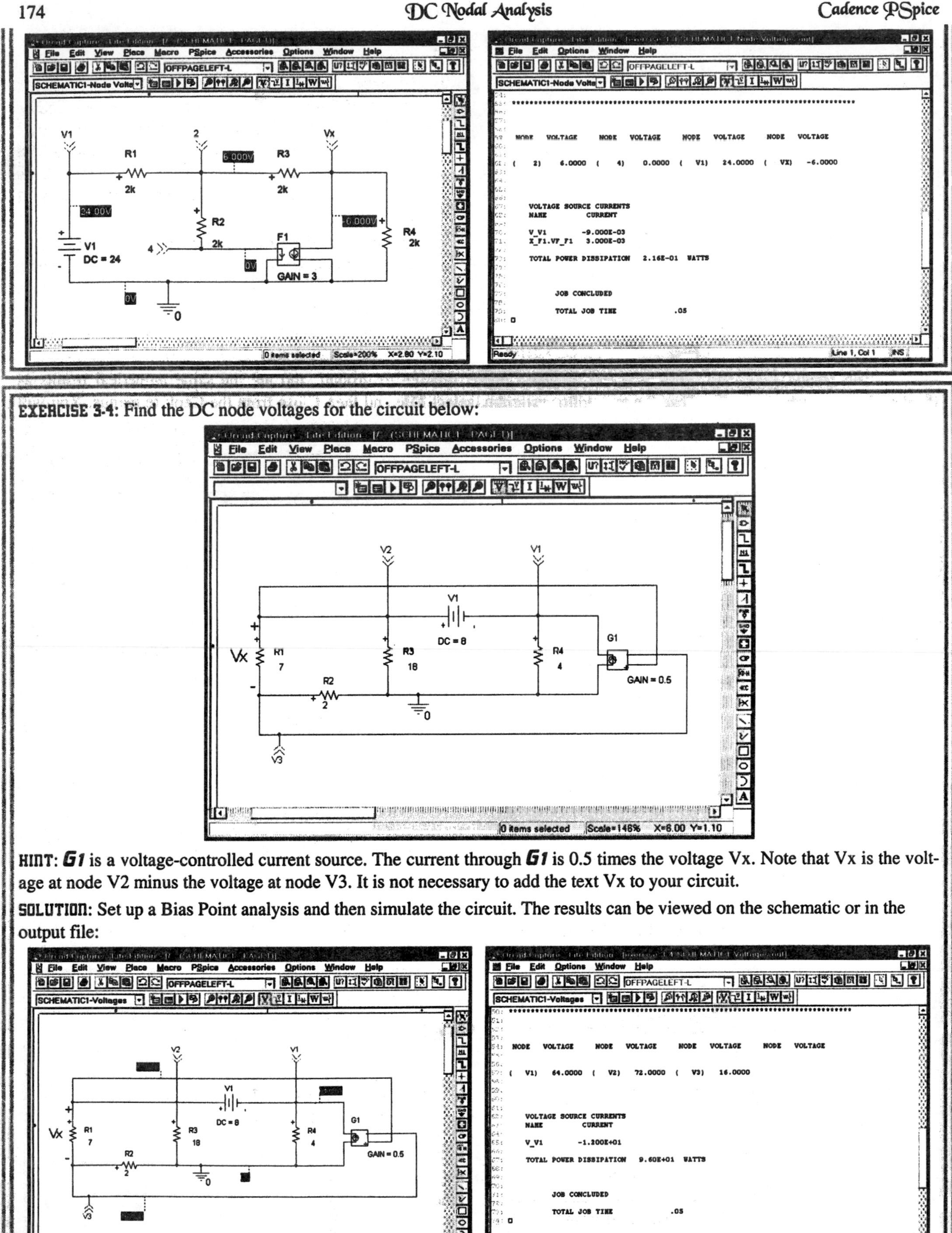


**EXERCISE 3-4**: Find the DC node voltages for the circuit below:

**HINT**: ***G1*** is a voltage-controlled current source. The current through ***G1*** is 0.5 times the voltage Vx. Note that Vx is the voltage at node V2 minus the voltage at node V3. It is not necessary to add the text Vx to your circuit.

**SOLUTION**: Set up a Bias Point analysis and then simulate the circuit. The results can be viewed on the schematic or in the output file:

# 3.C. Diode DC Current and Voltage

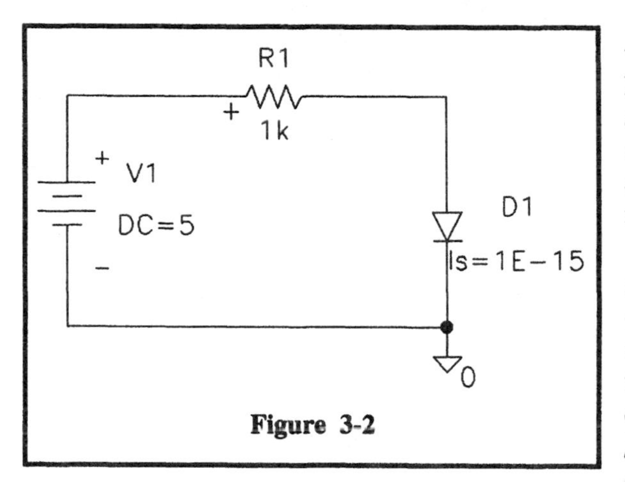


Figure 3-2

We will now use PSpice to find the diode current and voltage in the circuit of Figure 3-2. The diode current is given as $I_D = I_s[\exp(V_D/\eta V_T) - 1]$. $I_S$ is the diode saturation current and is $10^{-15}$ amps for this example. $V_T$ is the thermal voltage and is equal to 25.8 mV at room temperature. $\eta$ is the emission coefficient for the diode and its default value is 1. PSpice automatically runs all simulations at room temperature by default.

When you use a diode in a circuit, you will have to specify a model for the diode. In our case, the model will tell PSpice the value of $I_S$ for our diode. The class libraries have a number of predefined models that are usually used in a classroom environment. However, the model for this diode is not in our libraries so we will have to define a new model for it. The part for the diode you should use is ***Dbreak***. This diode is used to define your own model. Draw the circuit below:

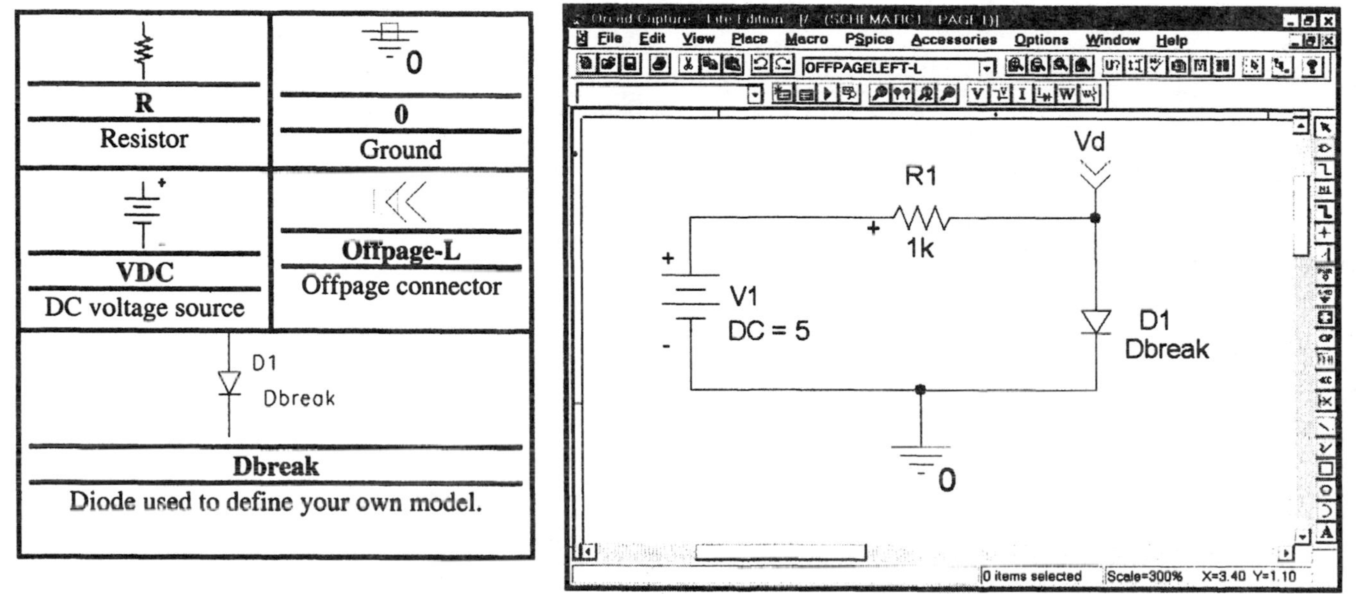


**Note for this example: If you use file Section 3C.opj from the CD-ROM or from the files copied from the CD-ROM to your hard drive during installation, you may have problems with the libraries during the simulation. For this example, you should draw the circuit from scratch and not use the provided example files.**

We must now define the model for the diode. Click the ***LEFT*** mouse button on the diode graphic, —▷|—. The graphic should turn pink, indicating that it has been selected. Next, select **Edit** and then **PSpice Model** from the Capture menus:

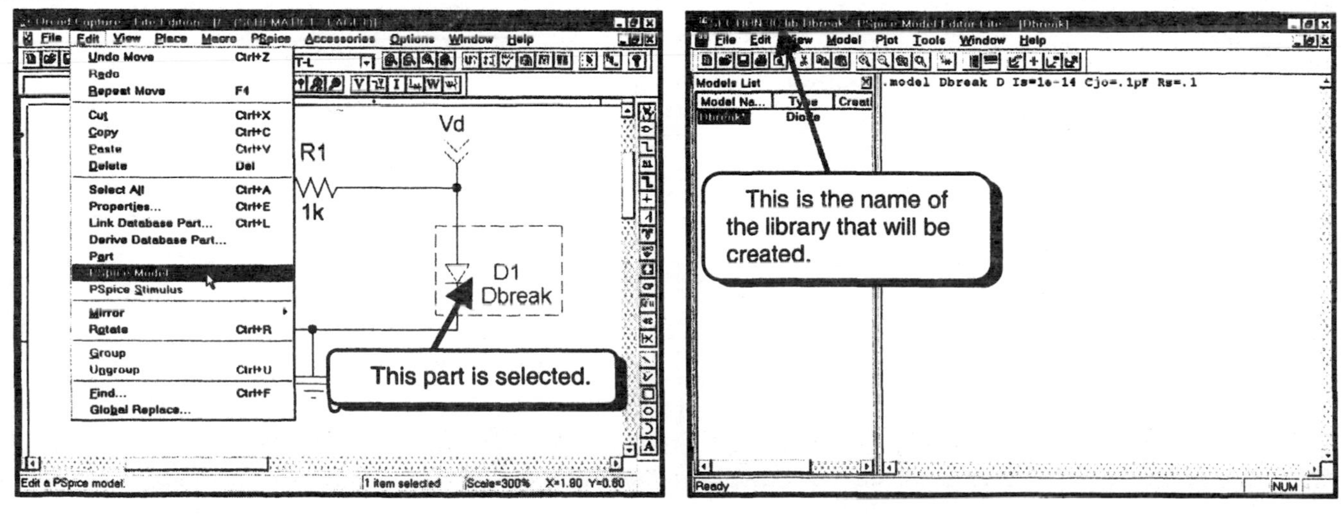

The right screen is the PSpice model editor and it tells us that the model we create will be saved in a file named ***SECTION 3C.lib***. The model for the part we selected is displayed by the model editor; in this case the model name is Dbreak. In this model, ***Rs*** is the series resistance of the diode and ***Cjo*** is the junction capacitance. The only parameter that we will change for this example is the saturation current ***Is***. Note in the screen capture above that the default value of $I_S$ is $1 \times 10^{-14}$ amps. We can use the model editor to create a new model. Change the name of the model to ***DX*** and change $I_S$ to 1e-15:

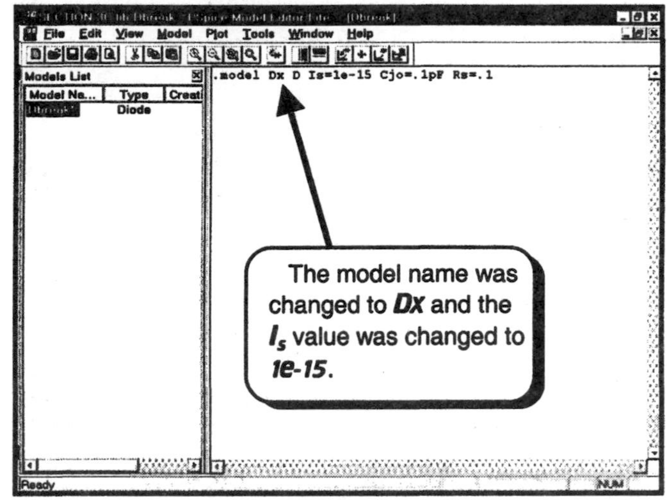


Select **File** and then **Save** to create and save the model:

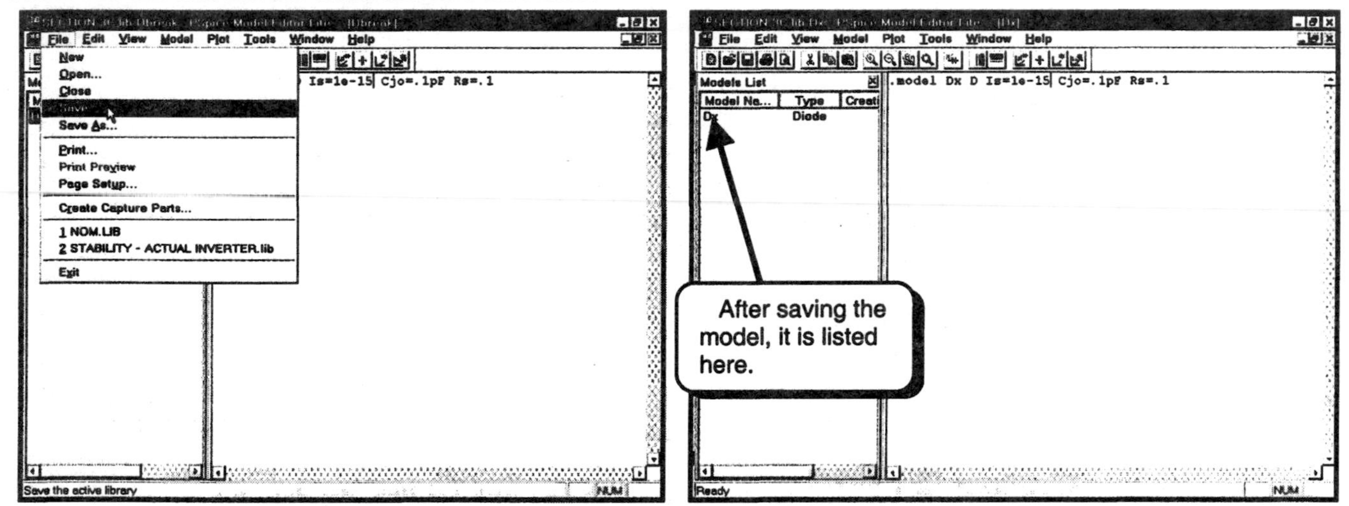


After we save the model, it is listed in the left window pane. This window pane lists all of the models contained in the library we are editing, in this case Section 3C.lib. Select **File** and then **Exit** to return to the schematic. Notice that the diode model name has changed from Dbreak to ***DX***:

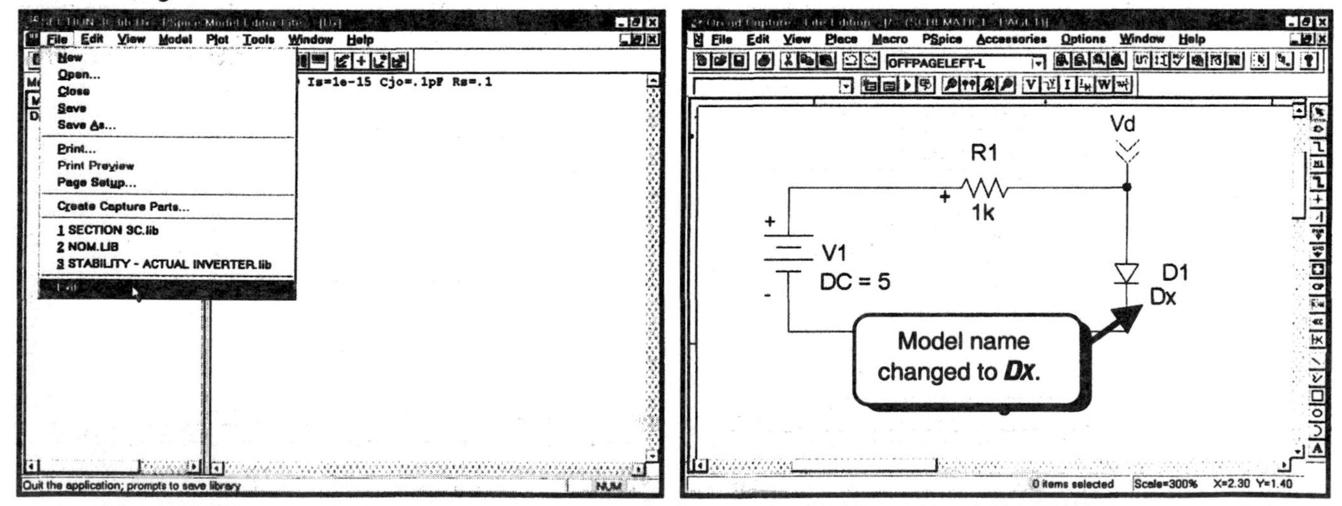


Next, we must set up the bias point simulation. Select **PSpice** and then **New Simulation Profile**:

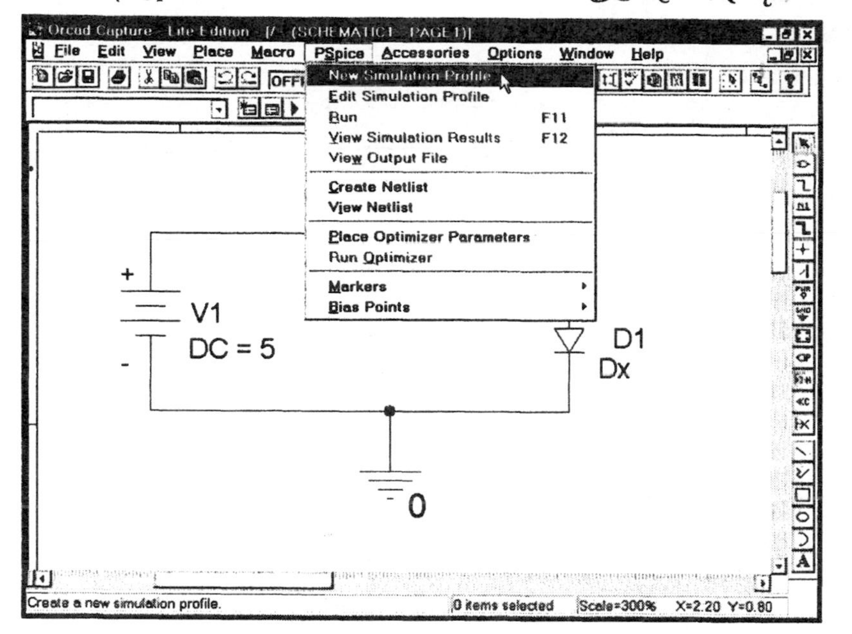


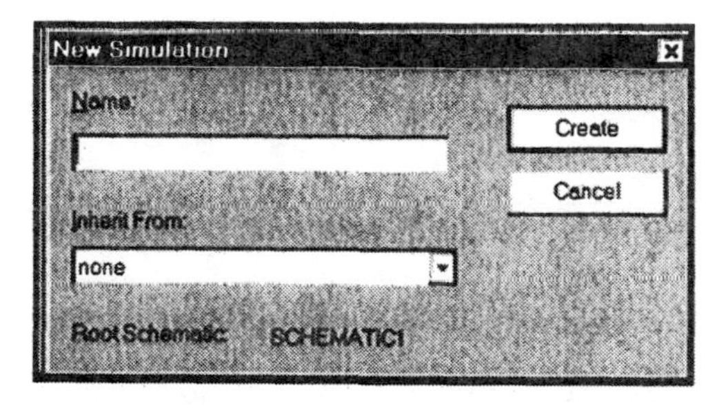


Enter a name for the profile and then click the ***Create*** button:

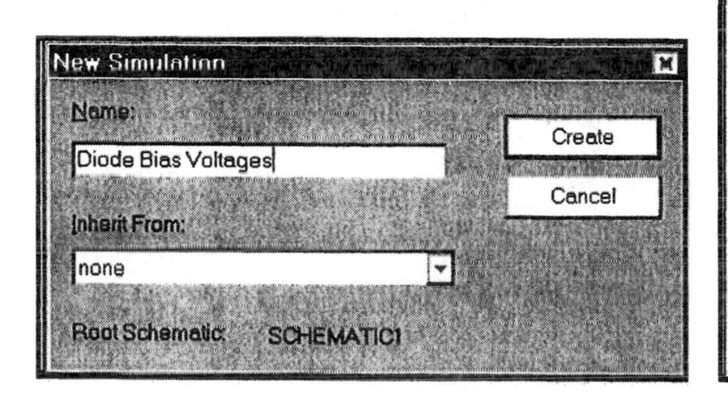


Select the ***Bias Point Analysis type*** and click the ***OK*** button:

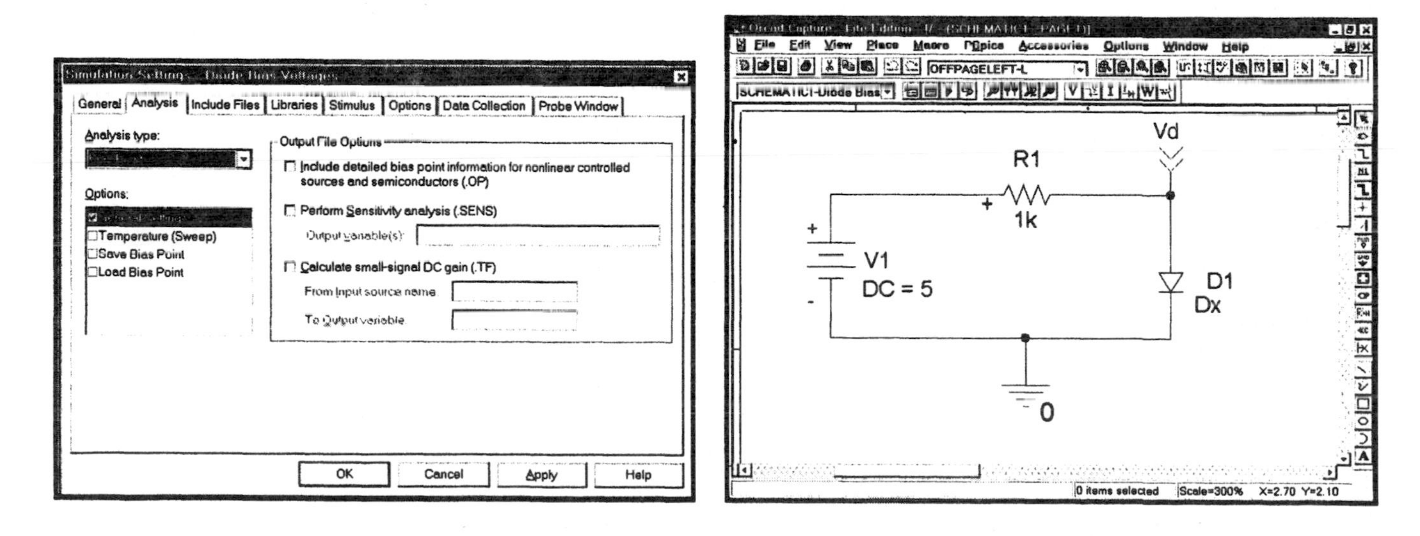


Select **PSpice** and then **Run** to simulate the circuit:

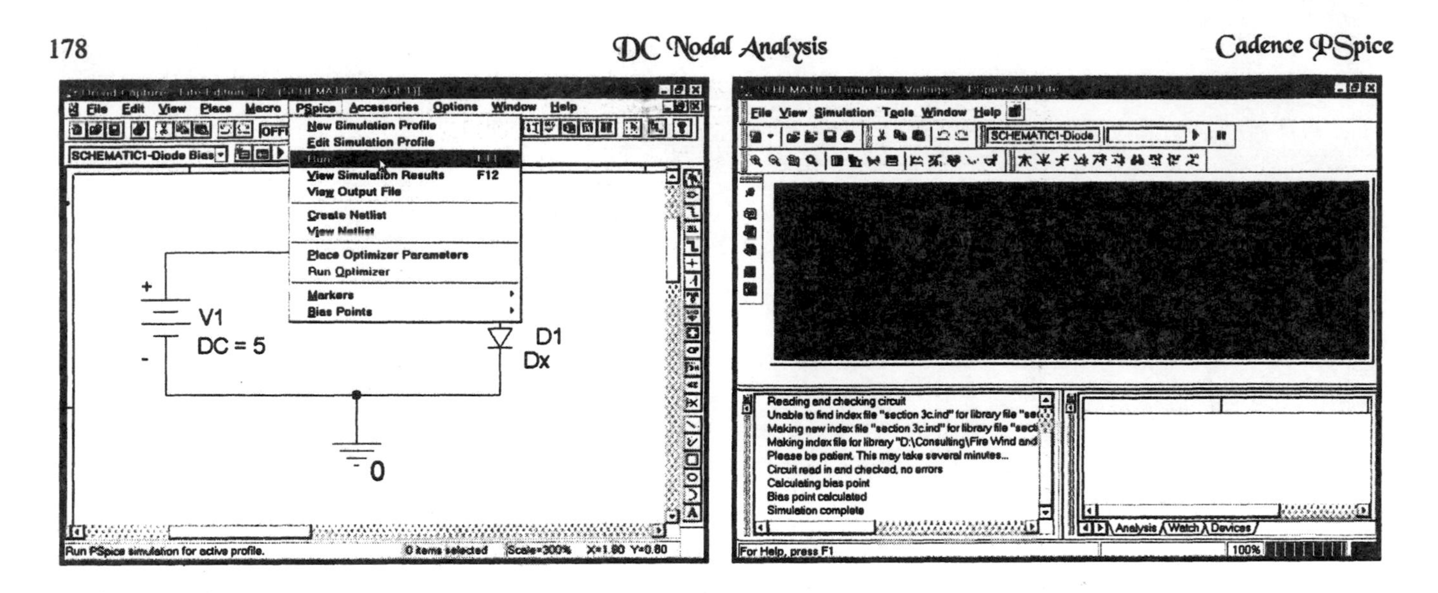

The messages indicate that an index file does not exist for the library file Section 3C.lib, and that PSpice will create the file. Library files (files with the .lib extension) are text files that contain model information. Some of the library files in the professional version of PSpice contain thousands of models. To reduce the simulation time, the libraries are compiled once into index files (files with extension .ind). During a simulation, PSpice reads the index file to get the model information it needs, not the library file. PSpice can read index files much more quickly than library files. The library file is provided because it is a text file and easy for humans to read. If you create a new library file or edit a model in an existing library file, PSpice will compile the .lib file into an .ind file. PSpice will leave the index file alone if no changes have been made to the library file. Compiling a library file into an index file can take a long time so it is done only when necessary. The warning messages are telling us that the index file for Section 3C.lib does not exist and it is creating the index file. If you run the simulation again, these warning messages will not be generated because the index file already exists. If you make changes to the diode model, you are changing the model in library Section 3C.lib. Warning messages will again be generated stating that file Section 3C.lib has changed and it must recreate the index file for that file.

In this case, the messages do not indicate a serious problem and we can view the results. Bring the Capture window to the top:

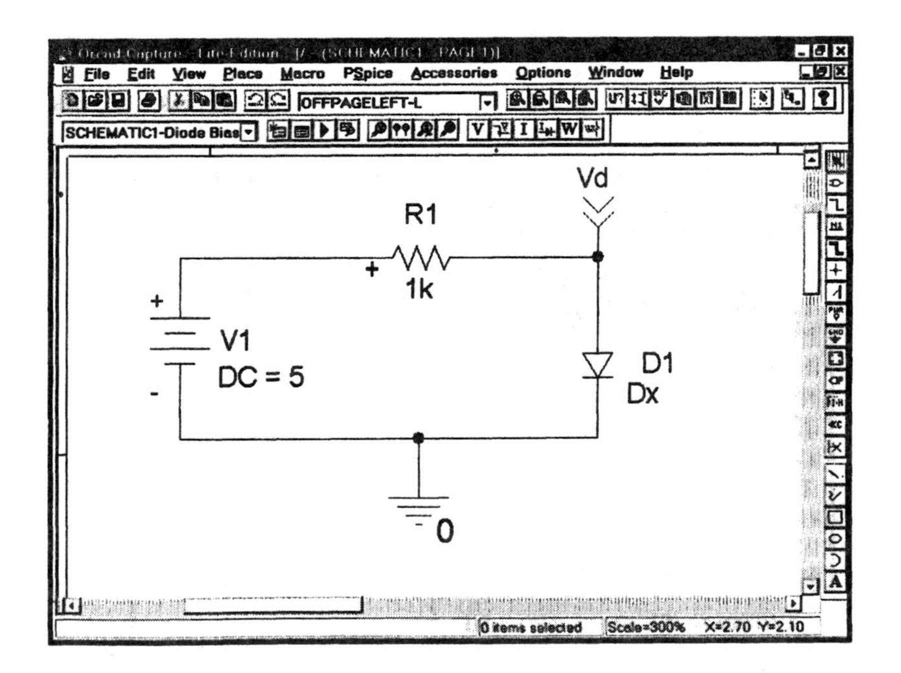


Use the techniques covered on pages 159 to 167 to display the node voltage at Vd and the DC current flowing into the diode:

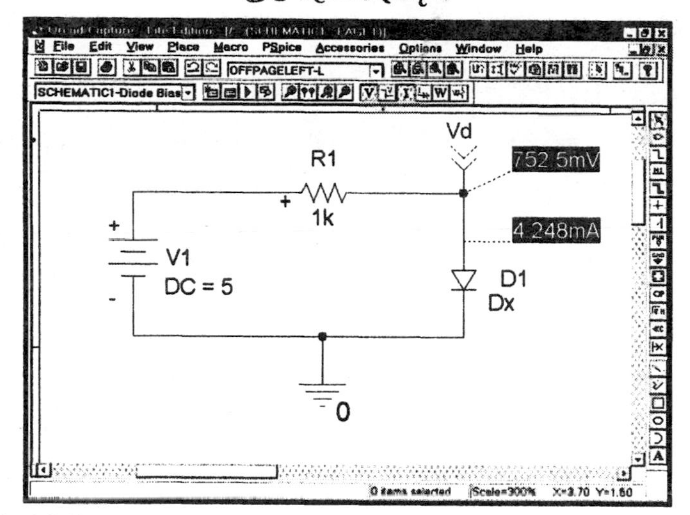


We see that the diode current is 4.248 mA, and the diode voltage is 0.7525 volts. Note that the node voltage displayed gives the voltage of the node relative to ground. Since the other side of the diode is grounded, the voltage displayed is also the diode voltage.

**EXERCISE 3-5**: Find the diode current and voltage in the circuit below:

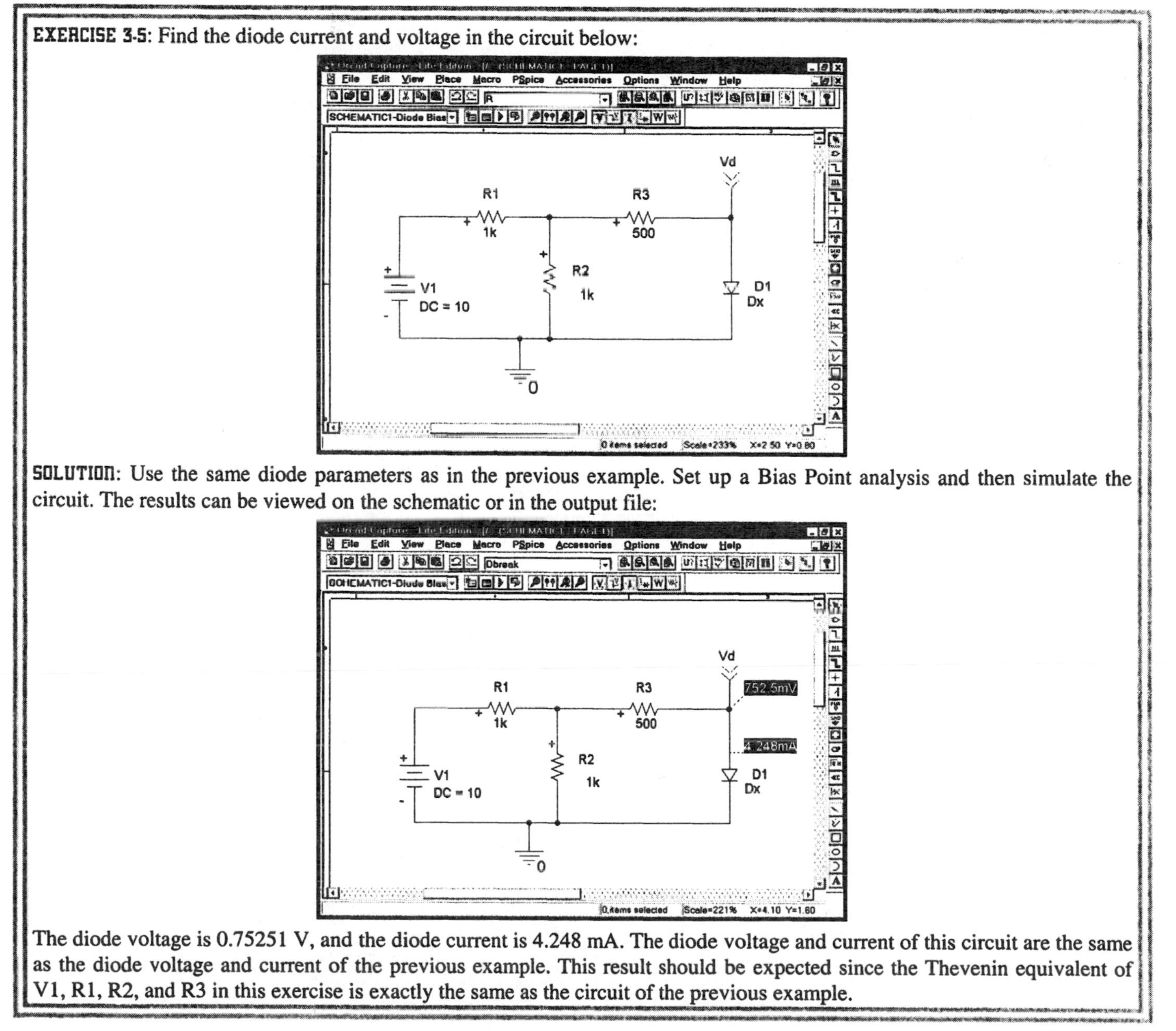


**SOLUTION**: Use the same diode parameters as in the previous example. Set up a Bias Point analysis and then simulate the circuit. The results can be viewed on the schematic or in the output file:

The diode voltage is 0.75251 V, and the diode current is 4.248 mA. The diode voltage and current of this circuit are the same as the diode voltage and current of the previous example. This result should be expected since the Thevenin equivalent of V1, R1, R2, and R3 in this exercise is exactly the same as the circuit of the previous example.

## 3.C.1. Changing the Temperature of the Simulation

In the last simulation we found the diode voltage and current at the default temperature of 25°C. Suppose we want to simulate the circuit at a different temperature? This can easily be done by selecting the temperature option in the simulation profile. We will continue with the circuit of the previous simulation:

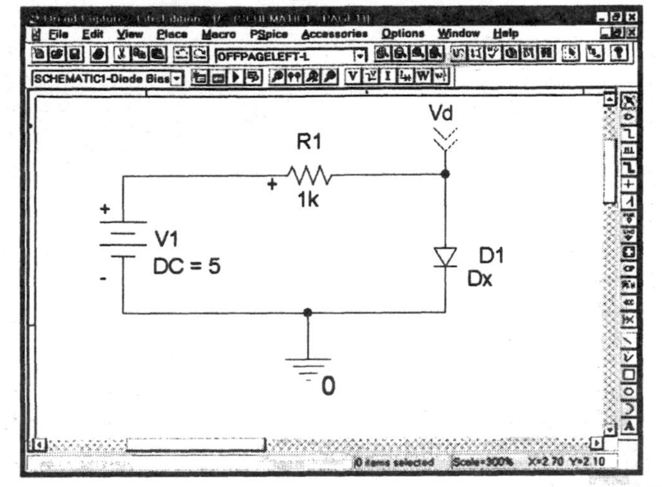


We now wish to edit the simulation profile. Since we are using a previously created profile, we need to open the profile rather than create a new profile. Select **PSpice** and then **Edit Simulation Profile** from the Capture menus to edit the existing profile:

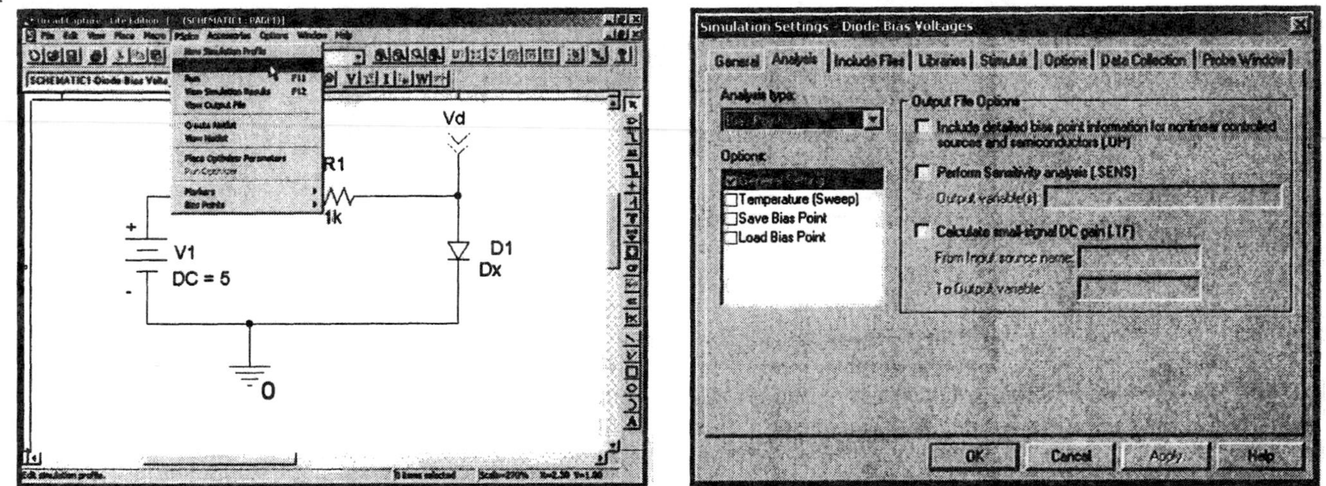


The ***Bias Point*** simulation is already selected from the previous example. Here we need to specify the temperature of the simulation. By default, all simulations are run at 25°C. To specify a temperature other than the default, select the ***Temperature (Sweep)*** option:

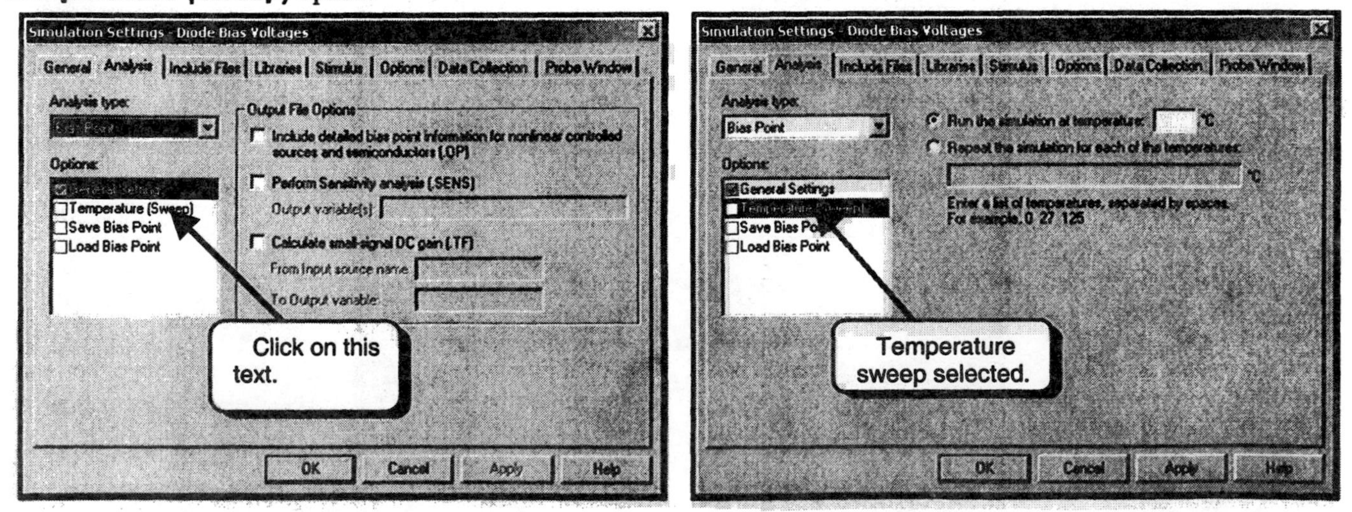

The ***Temperature (Sweep)*** allows us to specify a single temperature or a list of temperatures. If you specify a list of temperatures, the simulation will be run several times, once for each temperature you place in the list. We only need to run the simulation once, so fill in the dialog box as shown:

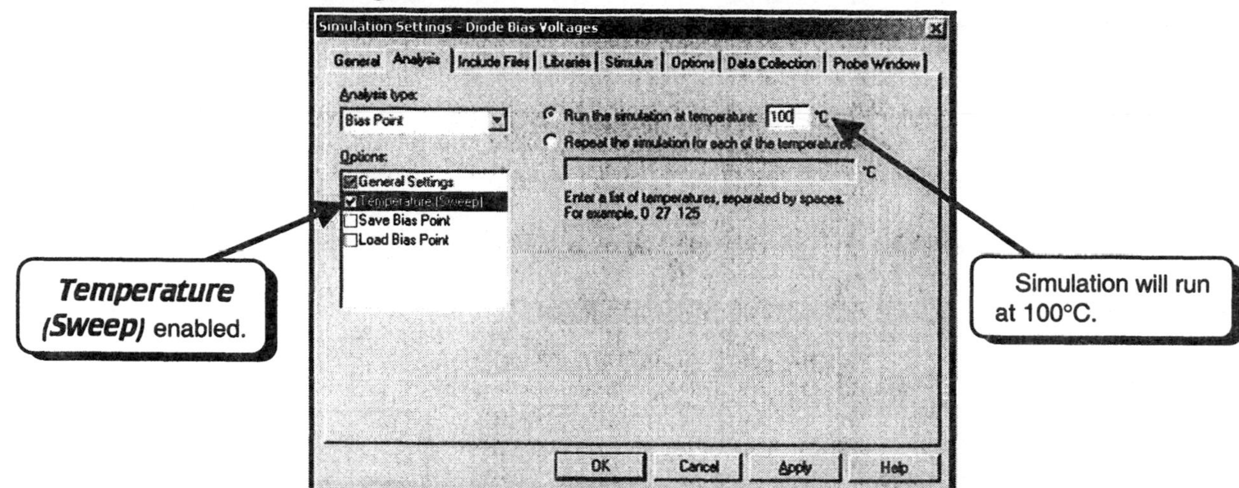


We have specified the simulation to run at 100°C. The simulation of the previous simulation did not specify a temperature and the simulation ran at the default temperature of 25°C. Notice in the screen capture above that there is a checkmark ☑ in the box next to the ***Temperature (Sweep)*** option. The temperature sweep will not run if this box is not checked. The option above specifies that the ***Bias Point*** simulation will run at 100°C. Click the ***OK*** button to return to OrCAD Capture.

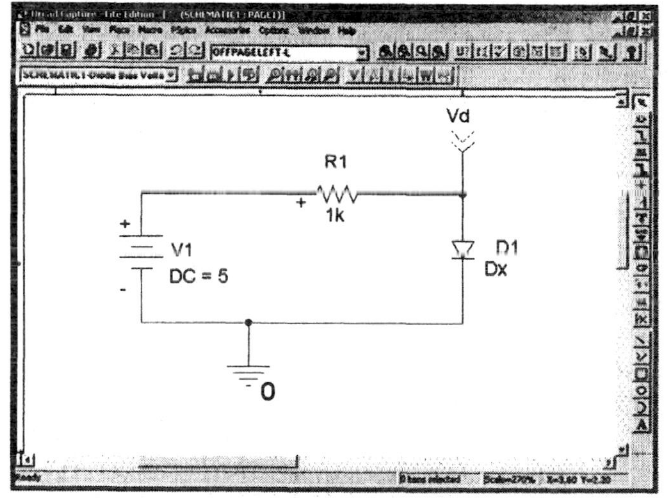


Select **PSpice** and then **Run** to run the simulation and then display the diode current and voltage on the schematic:

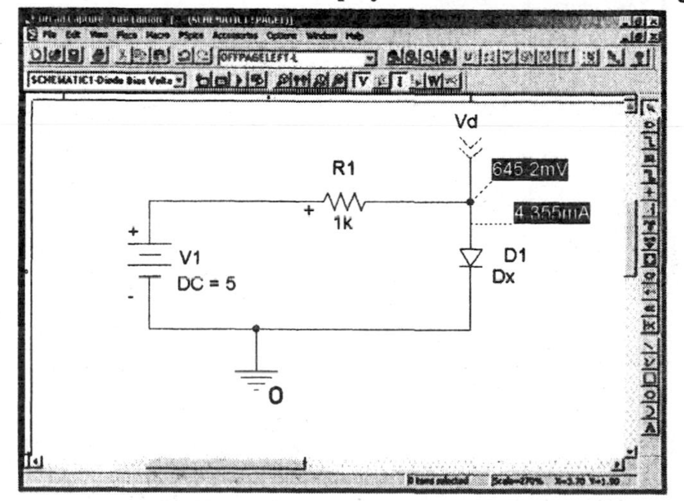

# 3.D. Finding the Thevenin and Norton Equivalents of a Circuit

Capture and PSpice can be used to easily calculate the Norton and Thevenin equivalents of a circuit. The method we will use is the same as if we were going to find the equivalent circuits in the lab. We will make two measurements, the open circuit voltage and the short circuit current. The Thevenin resistance is then the open circuit voltage divided by the short circuit current. This will require us to create two circuits, one to find the open circuit voltage, and the second to find the short circuit current. In this example, we will find the Norton and Thevenin equivalent circuits for a DC circuit. This same procedure can be used to find the equivalent circuits of an AC circuit (a circuit with capacitors or inductors). However, instead of finding the open circuit voltage and short circuit current using the DC Nodal Analysis, we would need to use the AC analysis.

For this example, we will find the Thevenin and Norton equivalent circuits for the circuit attached to the diode in **EXERCISE 3-5**. The circuit is repeated below:

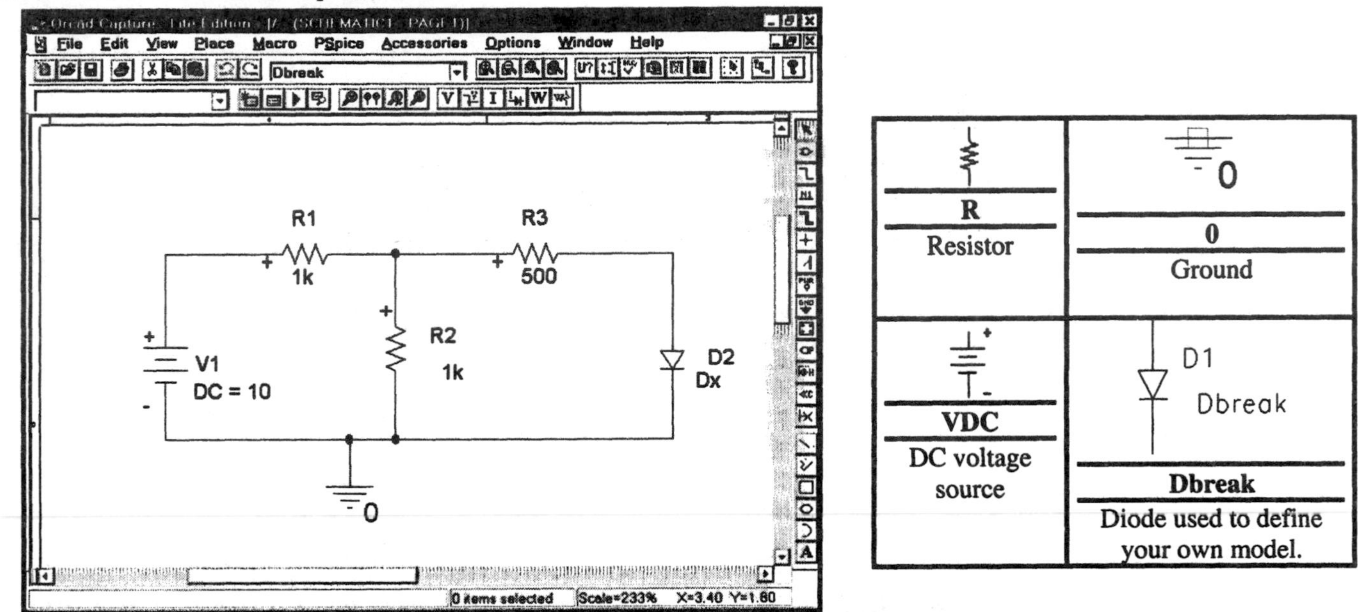


This circuit is difficult because it contains a nonlinear element (the diode) and a complex linear circuit. If we could replace V1, R1, R2, and R3 by a simpler circuit, the analysis of the nonlinear element would be much easier. To simplify the analysis of the diode, we will find the Thevenin and Norton equivalent circuits of the circuit connected to the diode; that is, we will find the Thevenin and Norton equivalents of the circuit below:

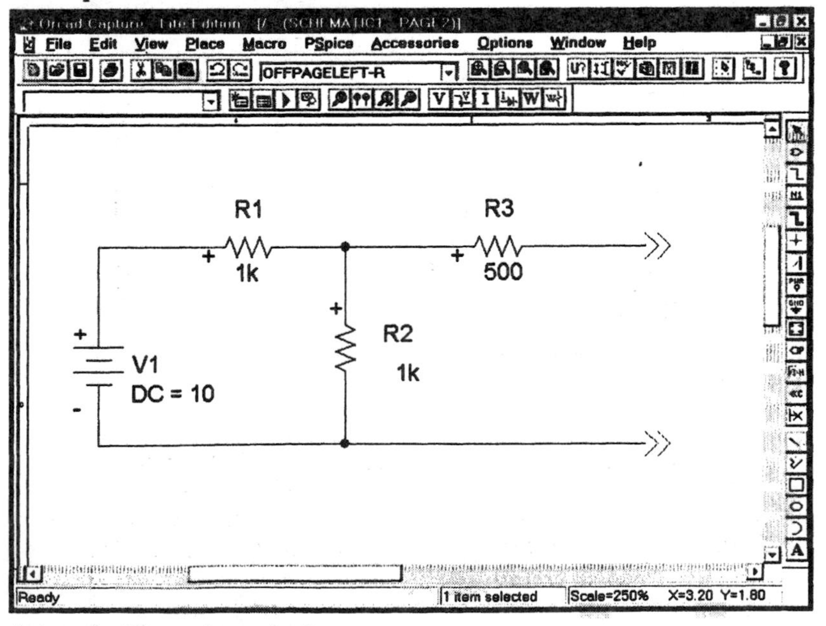


We will convert this circuit into the Thevenin equivalent :

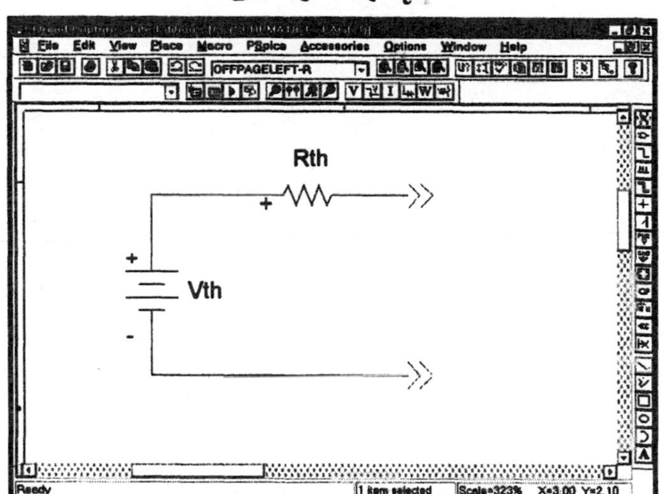


Once we find numerical values for Vth and Rth the entire circuit of **EXERCISE 3-5** reduces to:

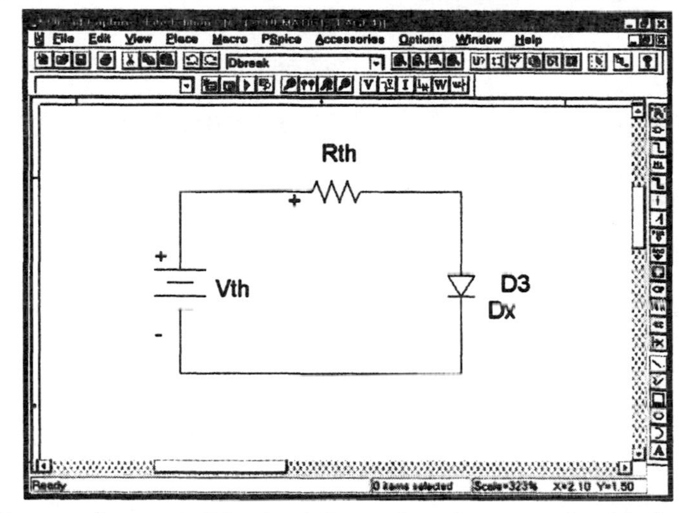


For determining the diode voltage and current, this circuit is much easier to work with than the original. This cxample is concerned with finding the numerical values of the equivalent circuit. The analysis of the circuit above was covered in Section 3.C. We will now find the Thevenin and Norton equivalent circuits of the circuit shown below:

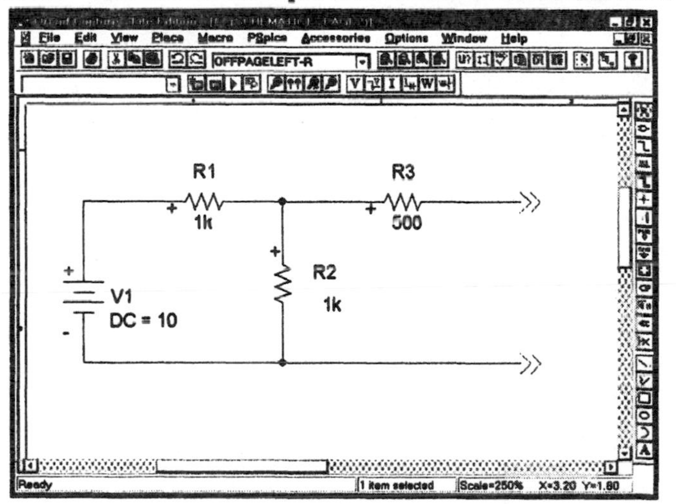


We will first find the open circuit voltage. This is just the voltage across the two terminals in the circuit shown above. First we must add a ground to the circuit. This is necessary because PSpice requires all circuits to have a ground reference:

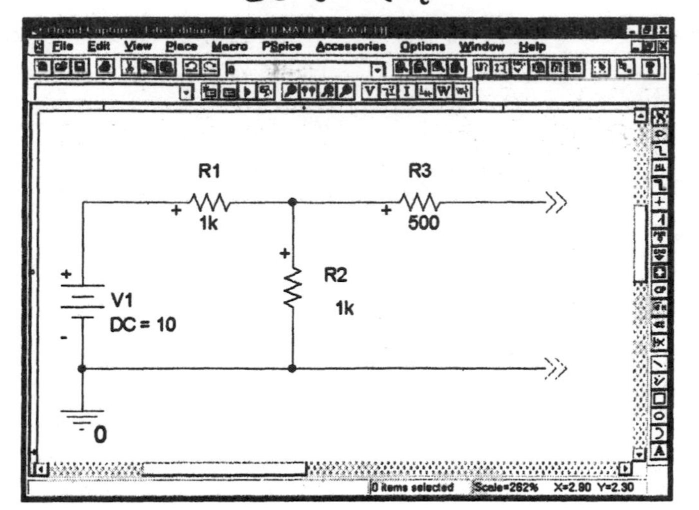


The lower terminal is now at ground potential, zero volts.

There are several errors in the circuit above. One is that there are two offpage connectors in the circuit and neither of them have labels. All bubbles must be labeled or errors will be generated and the simulation will not run. There are actually labels on the schematic: I just moved them far away to clean up the screen capture:

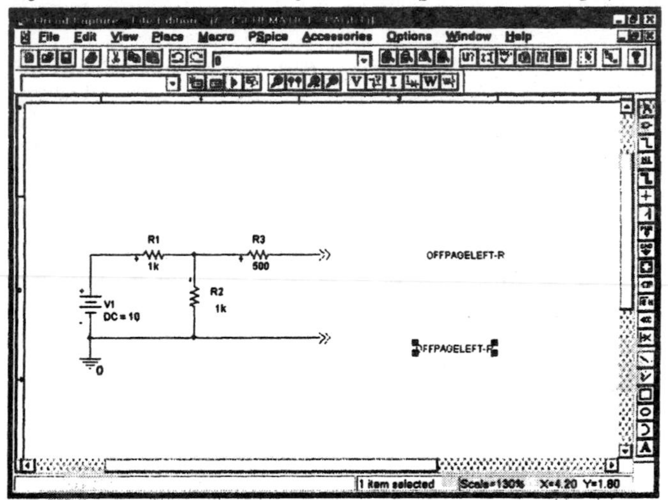


I will delete the lower offpage connector, and rename and move the upper offpage label:

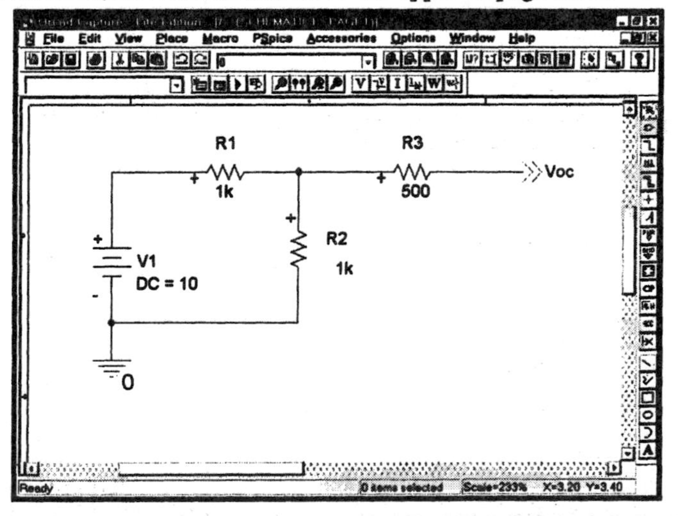


A second error is that there is only one element connected to the upper right node (now labeled ***Voc***). That is, nothing is connected to the right terminal of the 500 Ω resistor. PSpice requires that all nodes have at least two elements connected to them. To fix this problem, we must add another element to the circuit that does not affect the operation of the circuit. To simulate an open circuit, I will add a resistor of value 100T. The suffix T in PSpice is a multiplier with a value of

$10^{12}$. Thus, a resistor with the value 100T will have a resistance of $100\times10^{12}$ Ω. This value is significantly larger than all other resistors in the circuit and is an open circuit for all practical purposes. Thus, we will simulate the circuit below:

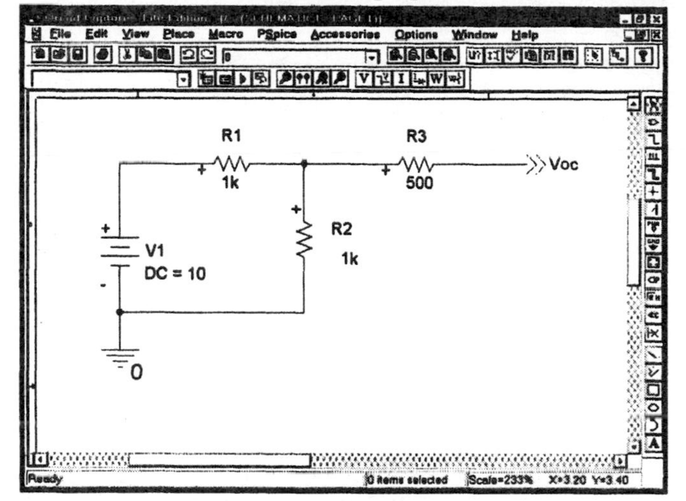


Set up the DC Bias simulation (select **PSpice** and then **New Simulation Profile**) and then run PSpice (**PSpice** and then **Run**). When the simulation is complete, display the node voltage at Voc on the schematic:

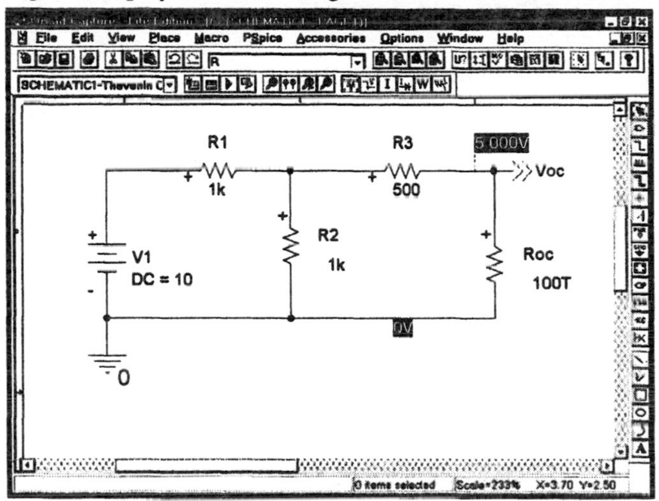


The open circuit voltage is ***5.000*** volts.

Next we must find the short circuit current. We will start with the original circuit as shown below:

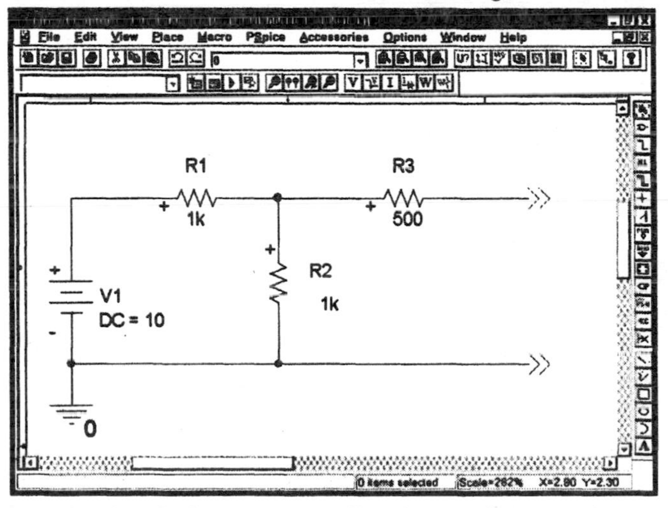


We can short the two terminals together by placing a very small resistance between the two terminals. The current through this resistance will be equal to the short circuit current. We will use a resistance of 1 fΩ or $1\times10^{-15}$ ohms. For all practical purposes this is zero resistance and a short. Modify the circuit as shown:

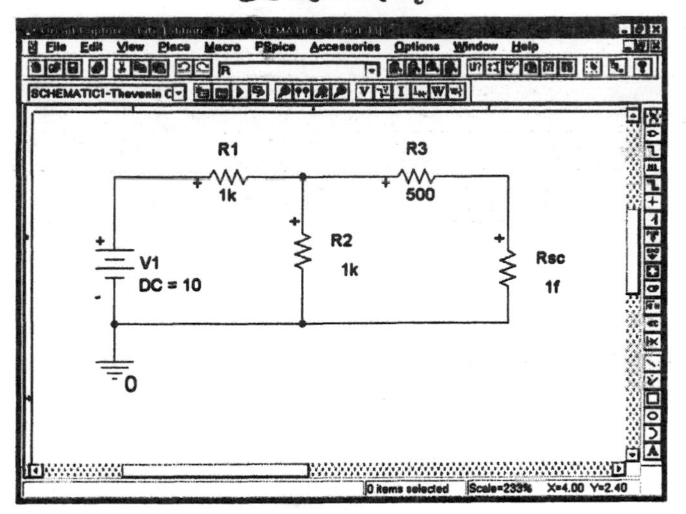


Simulate the circuit and display the current through ***Rsc*** on the screen::

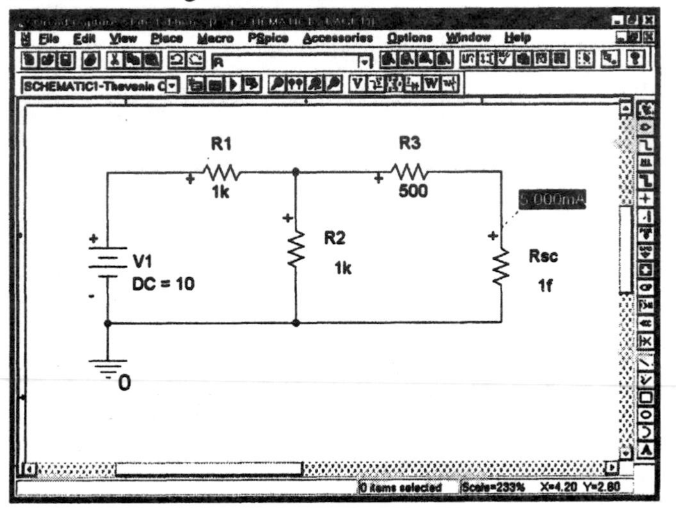


We see that the short circuit current is ***5.000*** mA.

We can now find the Thevenin resistance by dividing the open circuit voltage by the short circuit current:

$$R_{th} = \frac{Voc}{Isc} = \frac{5.000\ \text{V}}{5.00\ \text{mA}} = 1000\ \Omega$$

Our Thevenin and Norton equivalent circuits are shown below:

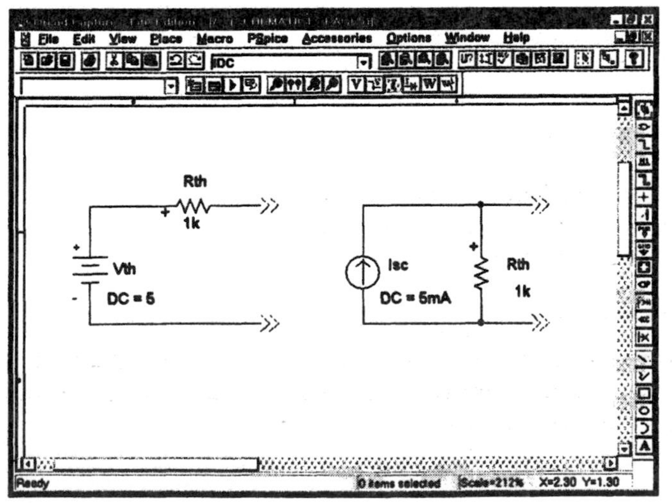

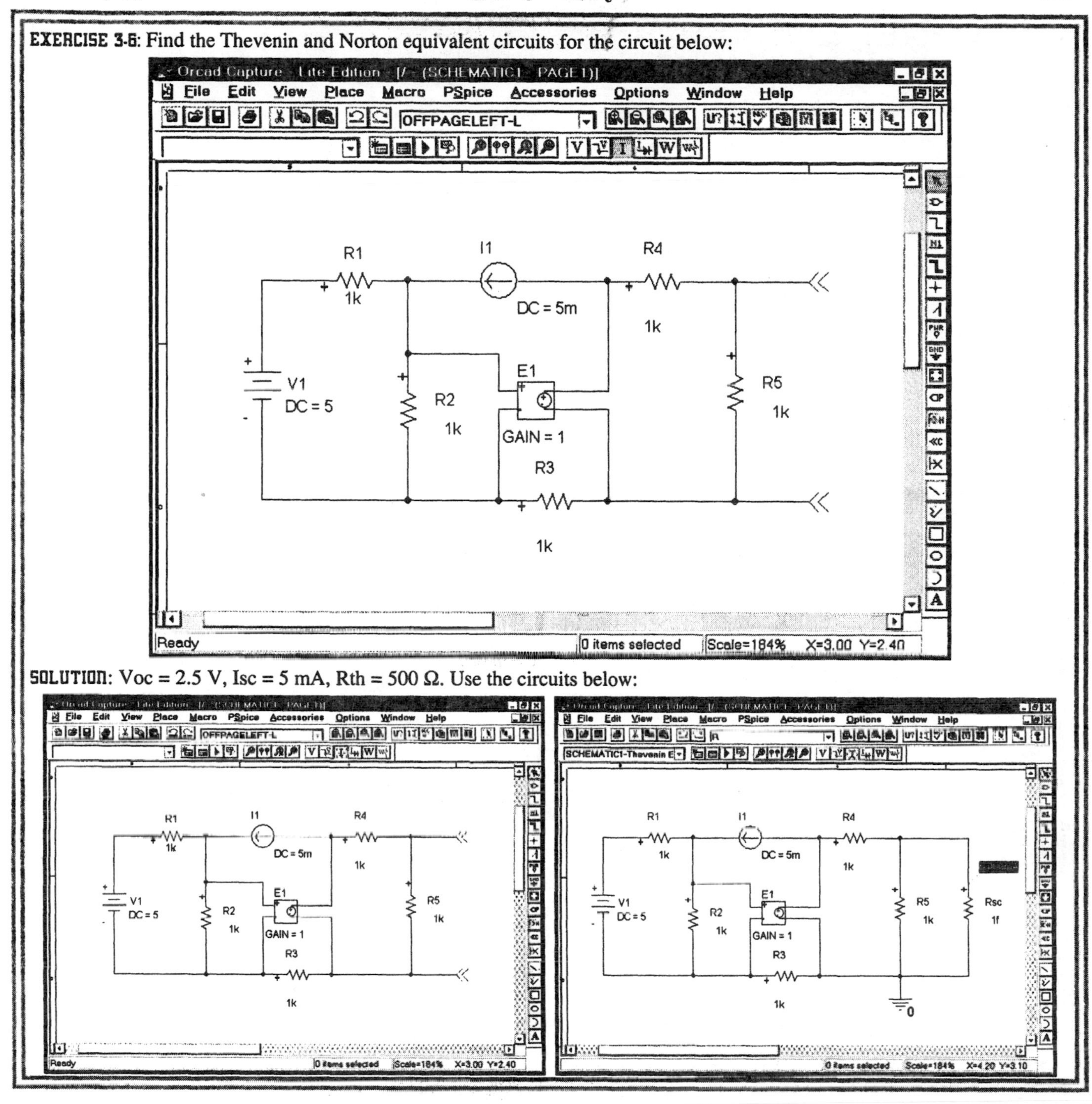

**EXERCISE 3-6:** Find the Thevenin and Norton equivalent circuits for the circuit below:

**SOLUTION:** Voc = 2.5 V, Isc = 5 mA, Rth = 500 Ω. Use the circuits below:



# 3.E. Transistor Bias Point Detail

One of the first things you should do when you are simulating an amplifier circuit is to check the transistor operating point. If the transistor bias is incorrect, none of the other analyses will be valid. If another analysis does not make sense, check the operating point. When PSpice finds the bias point, it assumes that all capacitors are open circuits and that all inductors are short circuits.

For a BJT, the Bias Point Detail gives the collector current, the collector-emitter voltage, and some small-signal parameters for the BJT at the bias point. For a jFET, the Bias Point Detail gives the drain current, the drain-source voltage, and some small-signal model parameters at the bias point. The results of the Bias Point Detail are contained in the output file. We will illustrate the Bias Point Detail analysis with the circuit below:

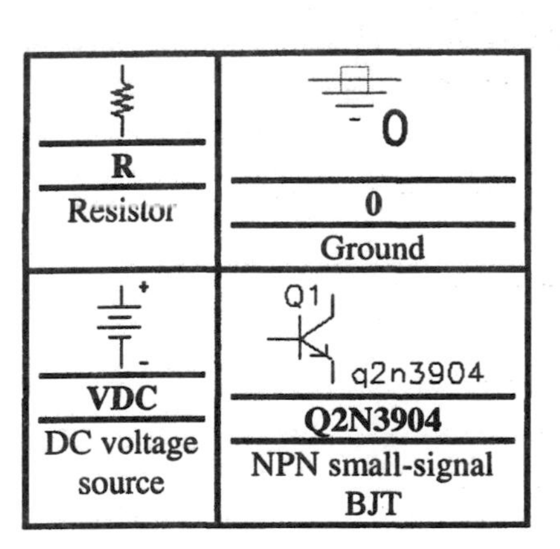


We must first set up the Bias Point simulation. Select **PSpice** and then **New Simulation Profile** from the menus:

Specify a name for the new profile and click the ***Create*** button:

Specify the ***Bias Point Analysis type*** and select the option to include detailed bias point information for nonlinear controlled sources such as BJTs.

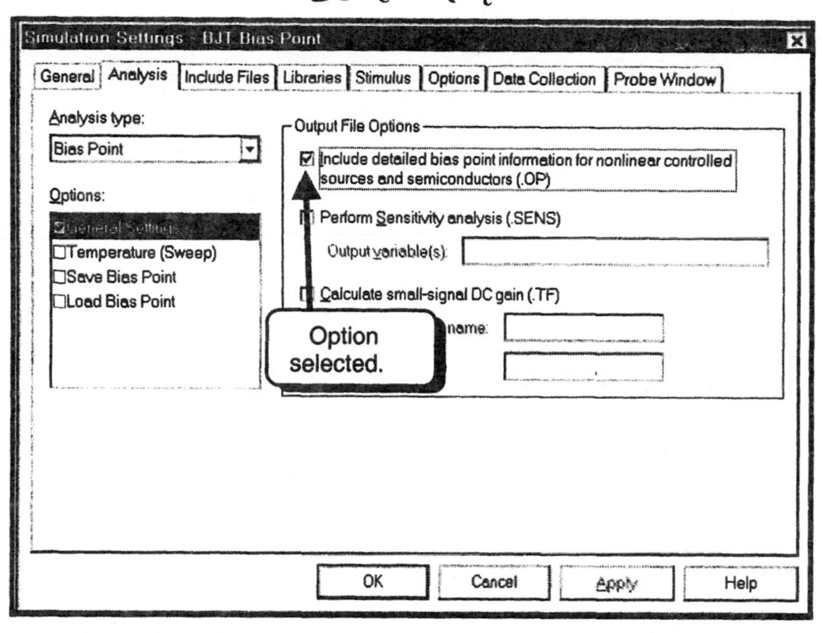


Click the ***OK*** button to accept the settings. We can now run the simulation. Select **PSpice** and then **Run** from the Capture menus or press the **F11** key. PSpice will run:

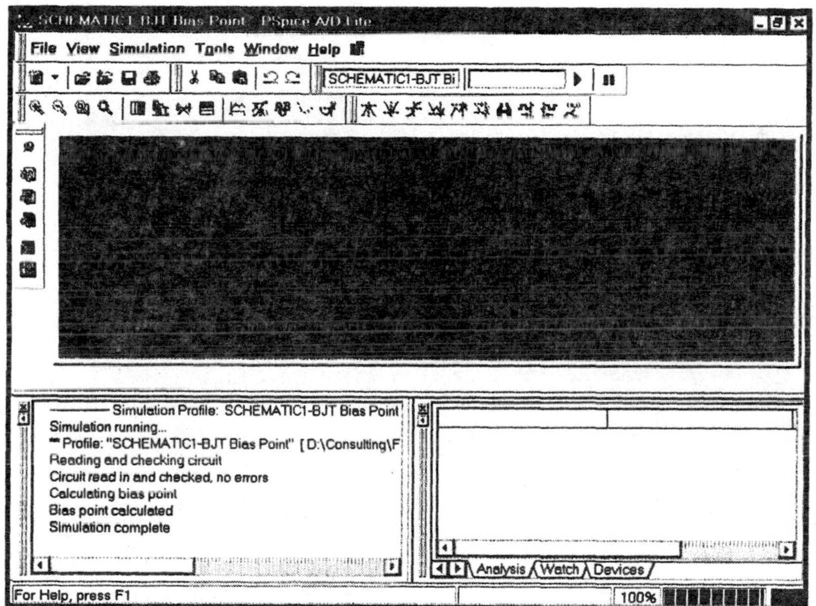


The results of the analysis are contained in the output file. Select **View** and then **Output File**:

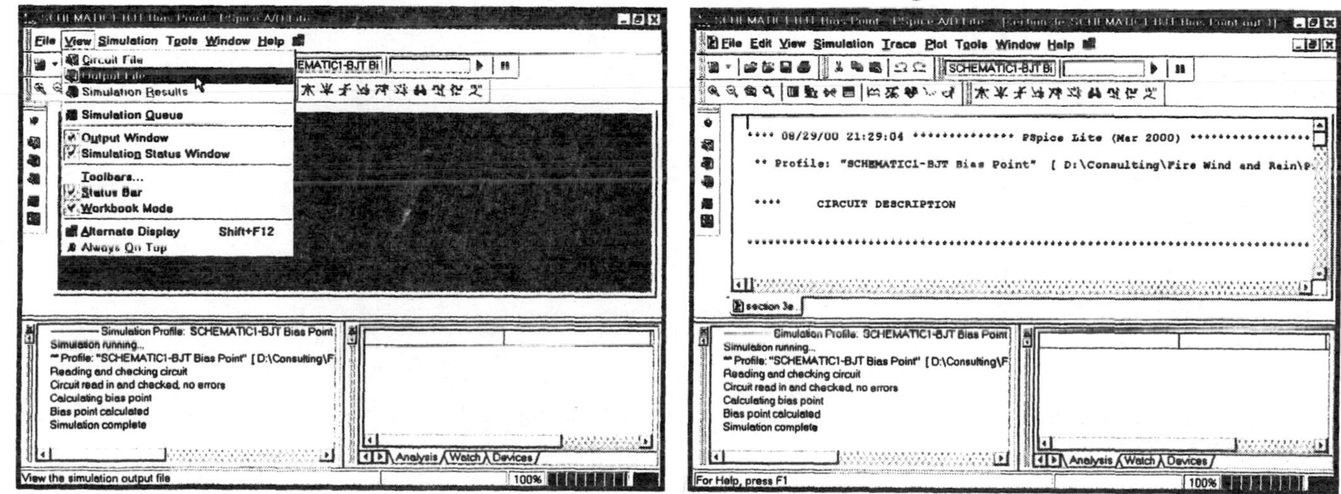


To view the text file in a full window, select **View** and then **Alternate Display**:

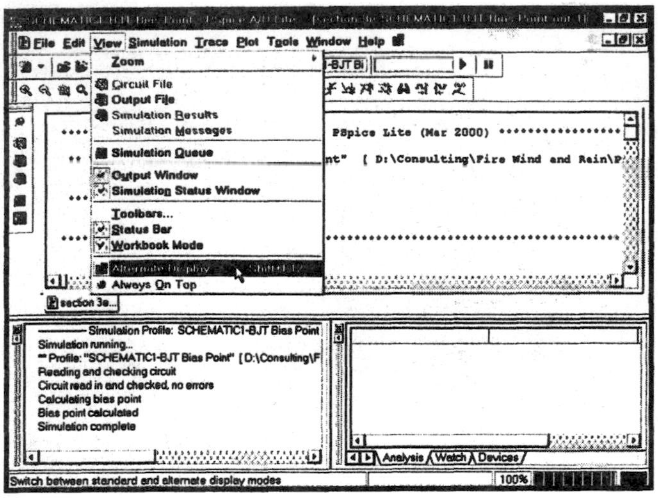

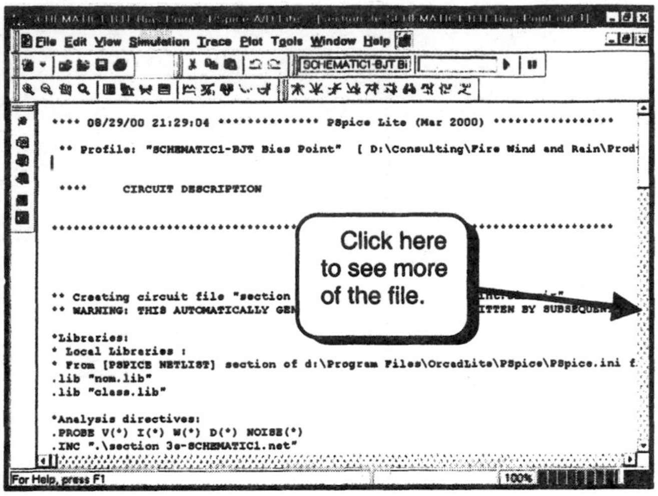

Click here to see more of the file.

To see the results, click the ***LEFT*** mouse button on the vertical scroll bar to scroll the window through the remainder of the file. You will first see the node voltages:

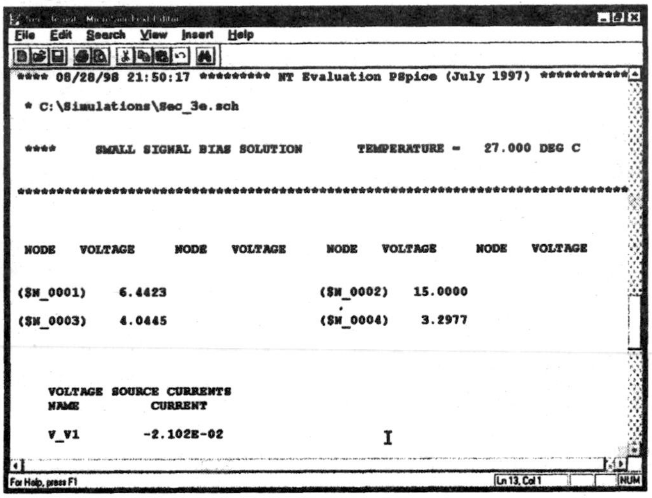

```
**** 08/28/98 21:50:17 ********* NT Evaluation PSpice (July 1997) ***********

 * C:\Simulations\Sec_3e.sch

 ****     SMALL SIGNAL BIAS SOLUTION        TEMPERATURE =   27.000 DEG C

******************************************************************************

 NODE   VOLTAGE     NODE   VOLTAGE     NODE   VOLTAGE     NODE   VOLTAGE

($N_0001)    6.4423                  ($N_0002)   15.0000

($N_0003)    4.0445                  ($N_0004)    3.2977

    VOLTAGE SOURCE CURRENTS
    NAME         CURRENT

    V_V1         -2.102E-02
```

Since we did not place any BUBBLEs or name any wires, the node names are a bit cryptic to us. If we were interested in the node voltages, we could have named some of the wires or placed a few BUBBLEs at the nodes in question.

Further down in the file we see the ***OPERATING POINT INFORMATION***:

```
 ****     OPERATING POINT INFORMATION      TEMPERATURE =   27.000 DEG C

 ******************************************************************************

 **** BIPOLAR JUNCTION TRANSISTORS

 NAME        Q_Q1
 MODEL       Q2N3904
 IB          1.13E-04
 IC          1.82E-02
 VBE         7.47E-01
 VBC        -2.40E+00
 VCE         3.14E+00
 BETADC      1.61E+02
 GM          5.82E-01
 RPI         2.57E+02
 RX          1.00E+01
 RO          4.20E+03
```

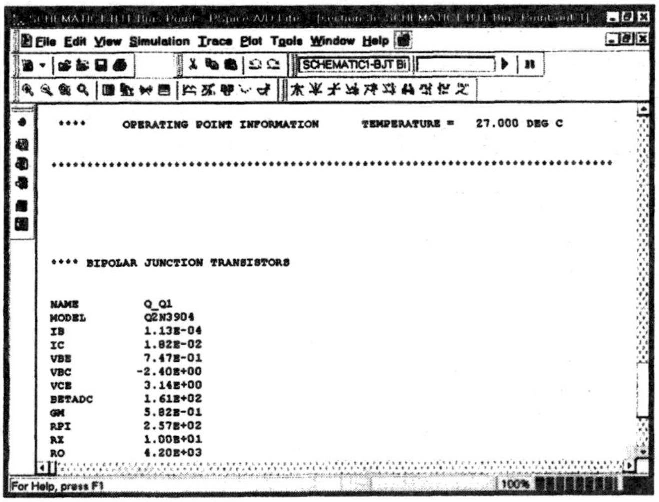

These results show several parameters for the BJT. In particular $I_C$ is 18.2 mA and $V_{CE}$ is 3.14 volts. Had there been more than one BJT in the circuit, the operating point information would be displayed for all BJTs. Similar information is shown for MOSFET, jFETs, and other three-terminal devices. To close the Text Editor program select **File** and then **Close** from the menus:

To switch back to the three-window display, select **View** and then **Alternate Display** from the menus:

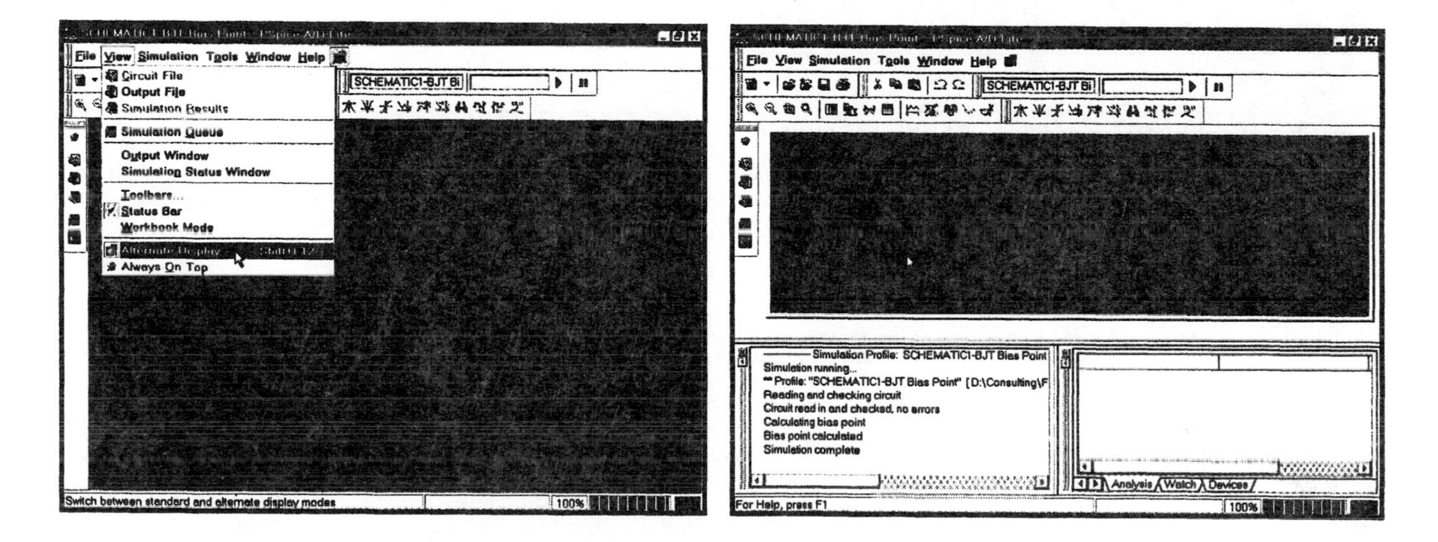


**EXERCISE 3-7**: Find the bias point for the transistor circuit:

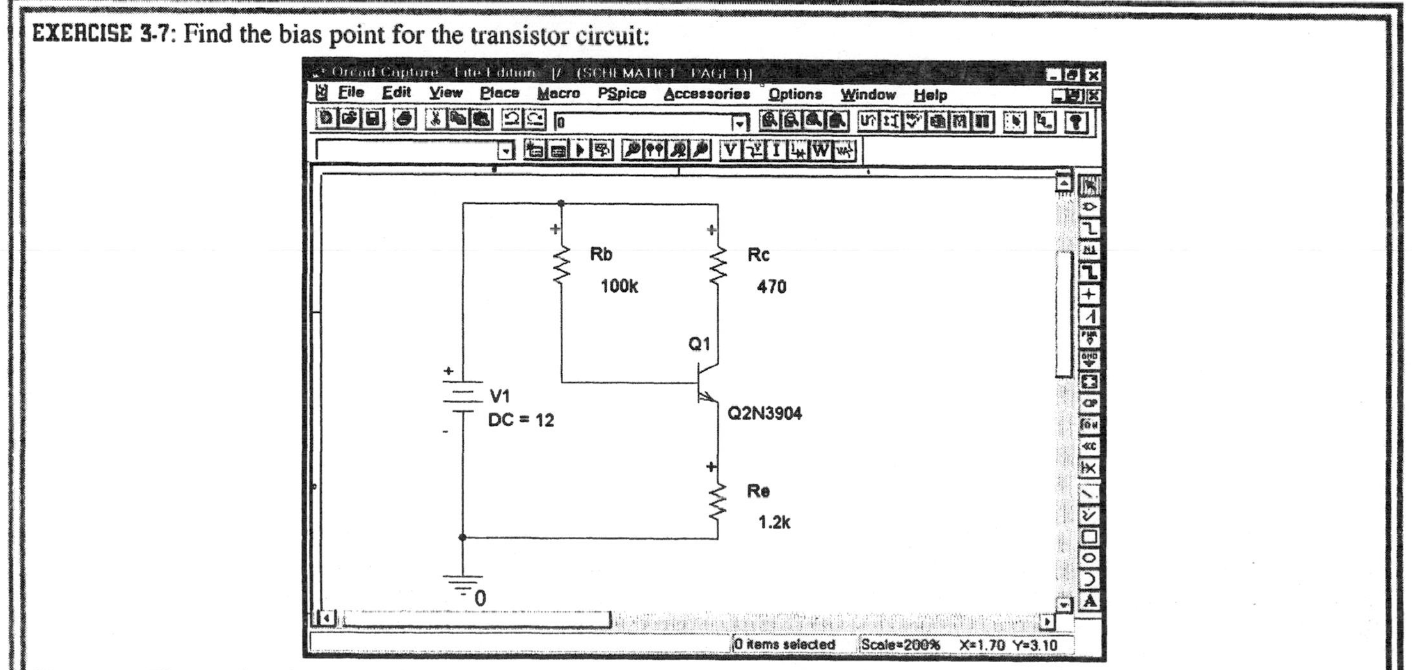


**SOLUTION**: The results of the Bias Point analysis are contained in the output file:

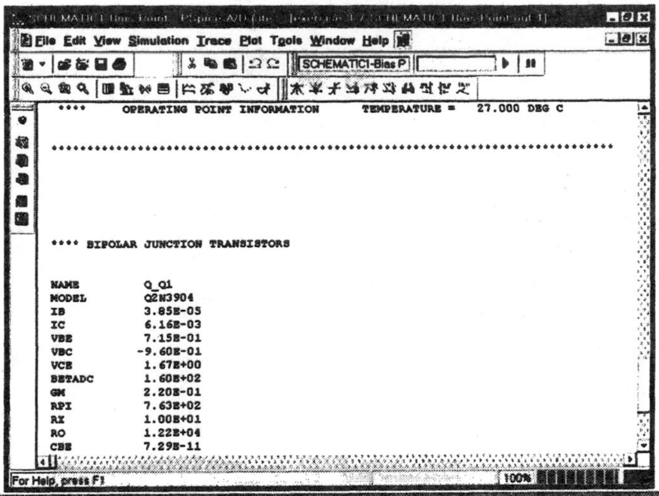

```
****     OPERATING POINT INFORMATION      TEMPERATURE =   27.000 DEG C

******************************************************************************

**** BIPOLAR JUNCTION TRANSISTORS

NAME          Q_Q1
MODEL         Q2N3904
IB            3.85E-05
IC            6.16E-03
VBE           7.15E-01
VBC          -9.60E-01
VCE           1.67E+00
BETADC        1.60E+02
GM            2.20E-01
RPI           7.63E+02
RX            1.00E+01
RO            1.22E+04
CBE           7.29E-11
```

# 3.F. Summary

- The DC Nodal Analysis finds the DC voltage at every node in the circuit. The voltages are relative to ground.
- The results are given in the output file. Use **Examine Output** from the **PSpice** menu to view the results.
- All capacitors are replaced by open circuits.
- All inductors are replaced by short circuits.
- All AC and time-varying sources are set to zero (IAC, VAC, Vsin, Isin, Vpulse, etc.).
- Use the part called "BUBBLE" to label the nodes of interest. If you do not label nodes with BUBBLEs, Capture will label the nodes for you and you will not know which node is which.
- The IPROBE and VIEWPOINT parts can be used to view the results of the Nodal Analysis on the schematic.
- Node voltages and element currents can be displayed on the schematic using the V and I buttons V I.

# PART 4
# DC Sweep

The DC Sweep can be used to find all DC voltages and currents of a circuit. The DC Sweep is similar to the node voltage analysis, but adds more flexibility. The added flexibility is the ability to allow DC sources to change voltages or currents. For example, the circuit on page 157 will give us results only for the single value of Vx = 15 V if the node voltage analysis is used. For each different value of Vx we are interested in, we must run the simulation again. If we use the DC Sweep, we can simulate the circuit for several different values of Vx in the same simulation. How node voltages vary for changing source voltages or how a BJT's bias collector current changes for different DC supply voltages would be example applications of the DC Sweep. As in the node voltage analysis, all capacitors are assumed to be open circuits, and all inductors are assumed to be short circuits.

## 4.A. Basic DC Analysis

We will first start with a modification of the circuit discussed in Part 1. Using Capture, create the schematic:

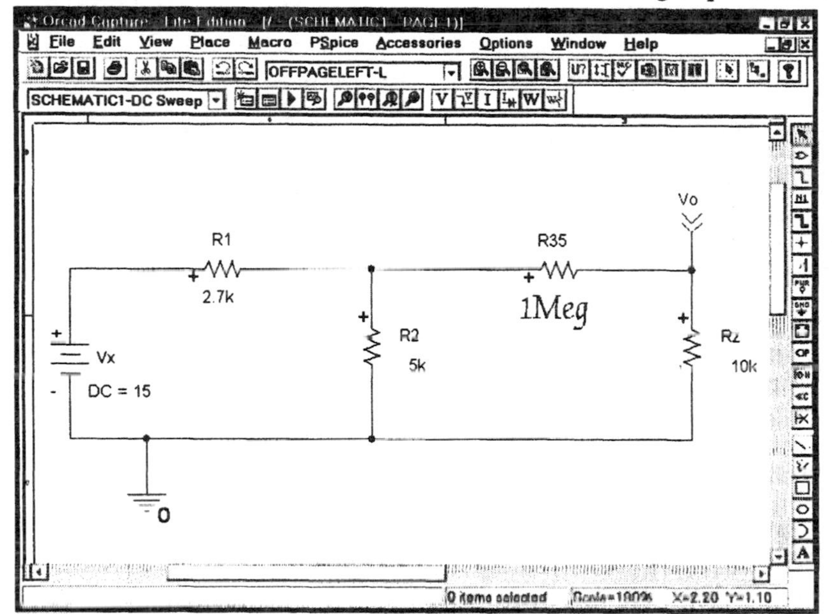


The question we will ask is: How does the voltage at ***Vo*** vary as ***Vx*** is raised from 0 to 25 volts? We will also view some of the currents through the components. Since this is a DC Sweep, all capacitors are assumed to be open circuits, and all inductors are assumed to be short circuits. We will now set up the DC Sweep. From the menu bar select **PSpice** and then **New Simulation Profile**:

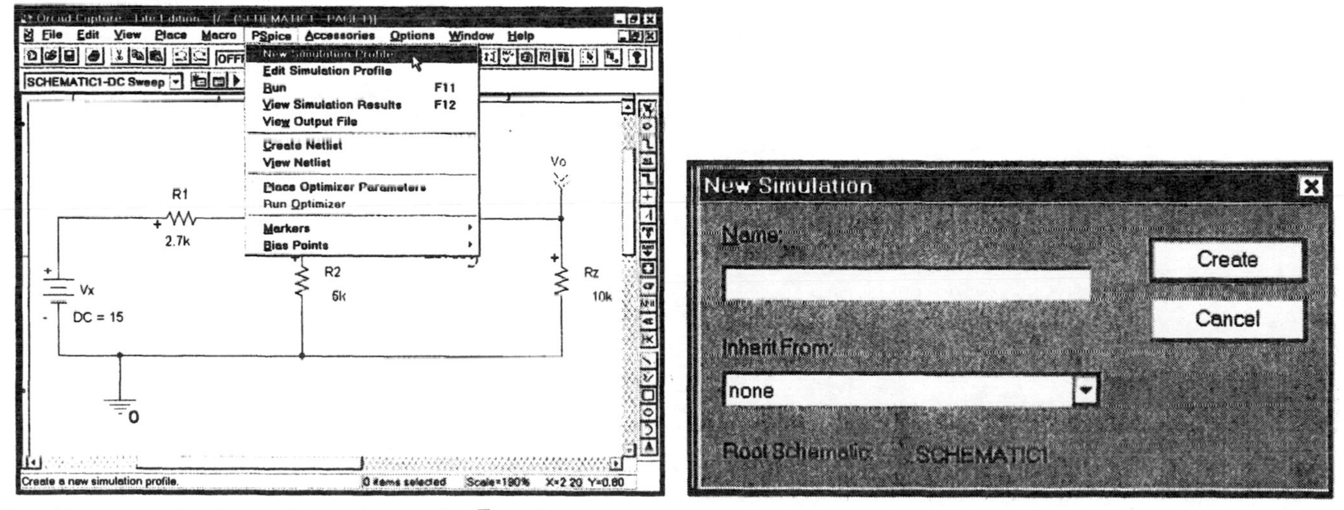


Specify a name for the profile and click the ***Create*** button:

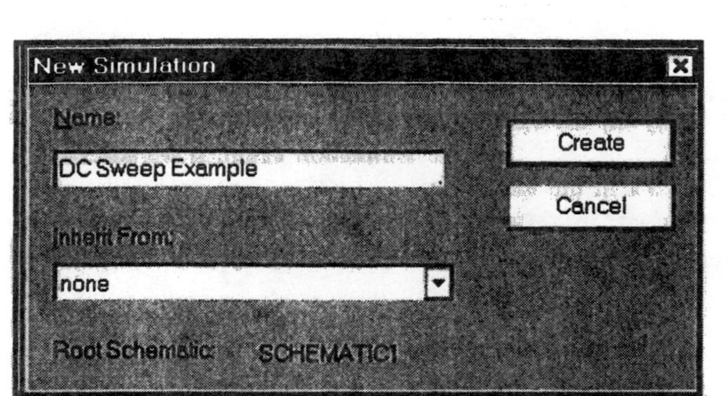

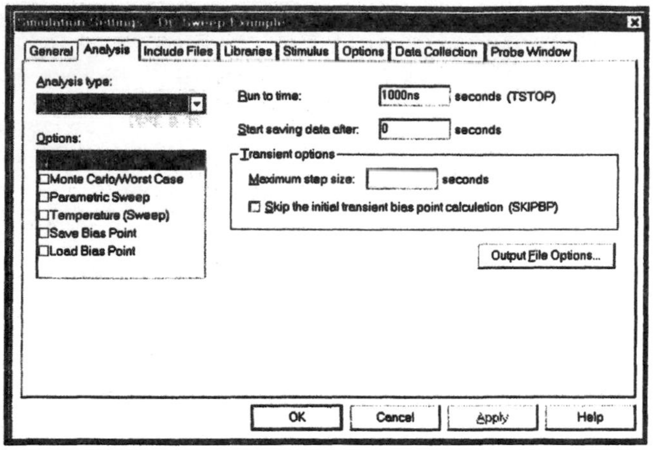

Click on the down triangle ▼ to display the list of available analyses:

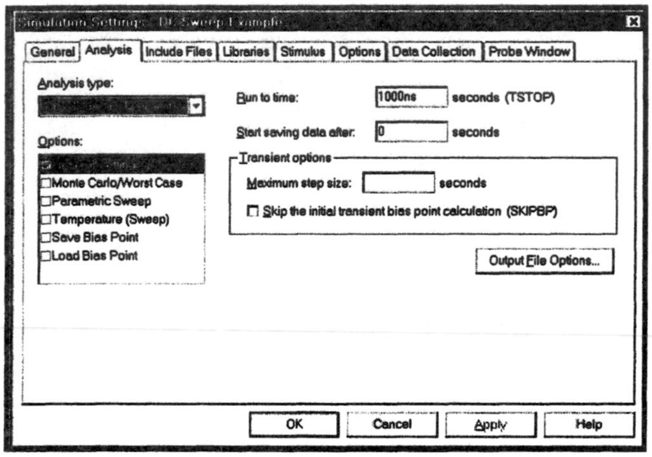

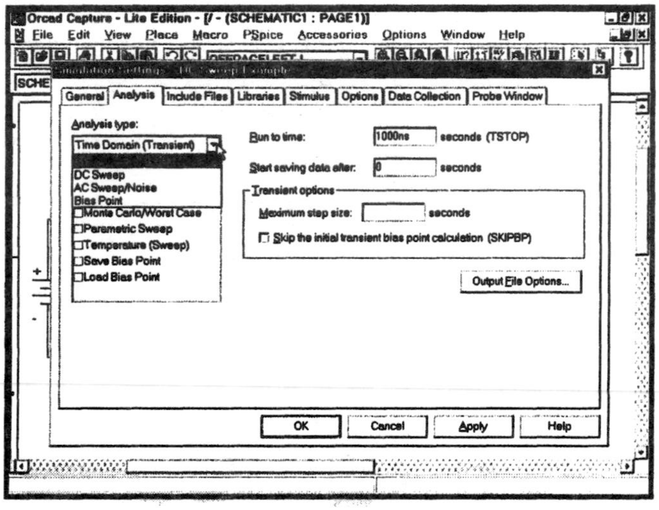

Select the **DC Sweep** selection:

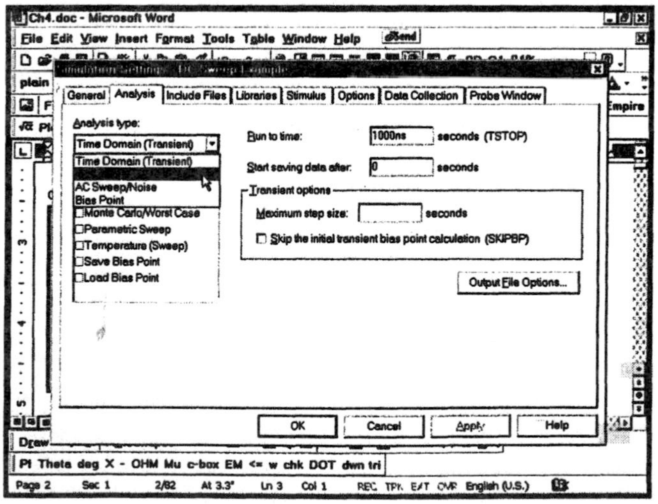

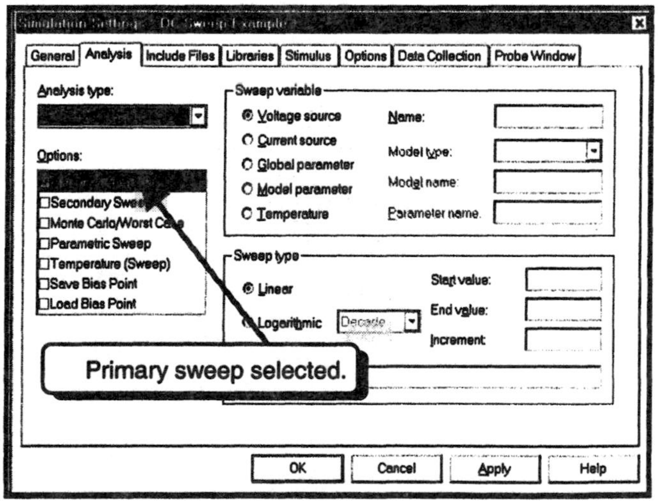

We are setting the parameters for the primary sweep. A primary sweep means that we are changing only one circuit parameter for this simulation. A secondary sweep can be performed inside a primary sweep. For a secondary sweep, we would be changing two circuit parameters in the same simulation. For this example we will demonstrate a linear sweep of one parameter. The voltage source ***VX*** will be swept from 0 volts to 25 volts in 1-volt increments. A linear sweep means that points between the beginning and ending values are equally spaced. Fill out the dialog box as shown below:

Simulation Settings - DC Sweep Example
General | Analysis | Include Files | Libraries | Stimulus | Options | Data Collection | Probe Window
Analysis type: DC Sweep
Options: Primary Sweep, Secondary Sweep, Monte Carlo/Worst Case, Parametric Sweep, Temperature (Sweep), Save Bias Point, Load Bias Point
Sweep variable: Voltage source, Current source, Global parameter, Model parameter, Temperature
Name: Vx
Model type:
Model name:
Parameter name:
Sweep type: Linear, Logarithmic (Decade), Value list
Start value: 0
End value: 25
Increment: 1
OK | Cancel | Apply | Help

When you have set all the parameters as shown above, click the ***LEFT*** mouse button on the ***OK*** button.

We are now ready to simulate the circuit. Select **PSpice** from the menu bar and then select **Run**. Capture will first create an updated netlist and then run PSpice. When the simulation is complete, the Probe window will display an empty plot:

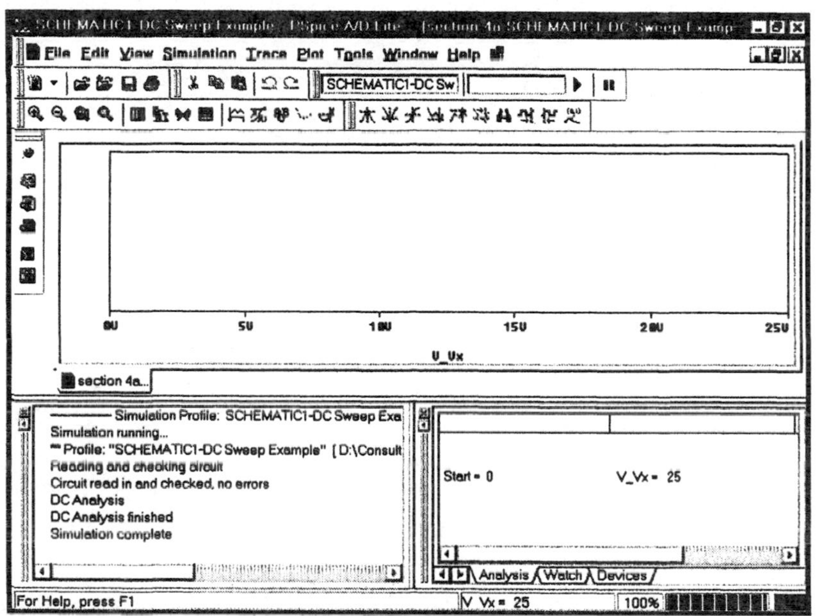


I will display Probe in full screen:

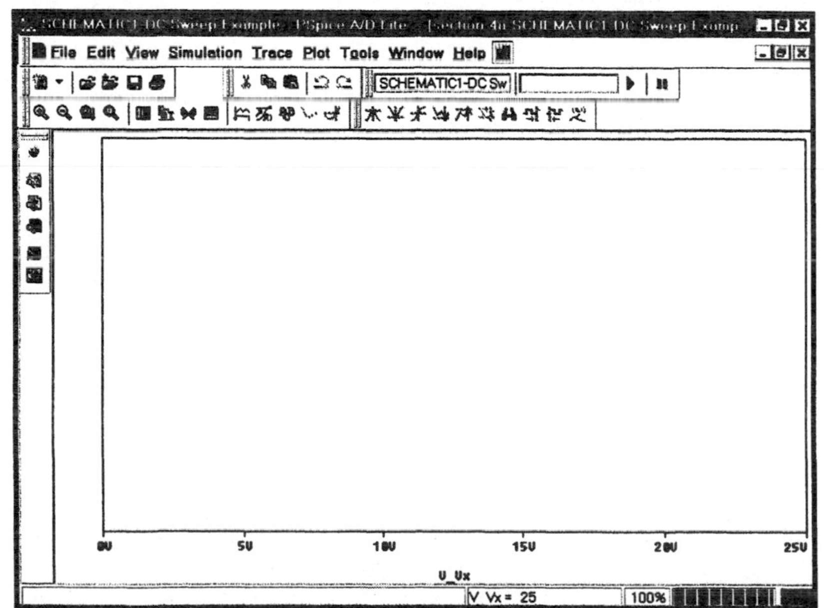

Press the **INSERT** key to add a trace. Add the trace V(Vo).

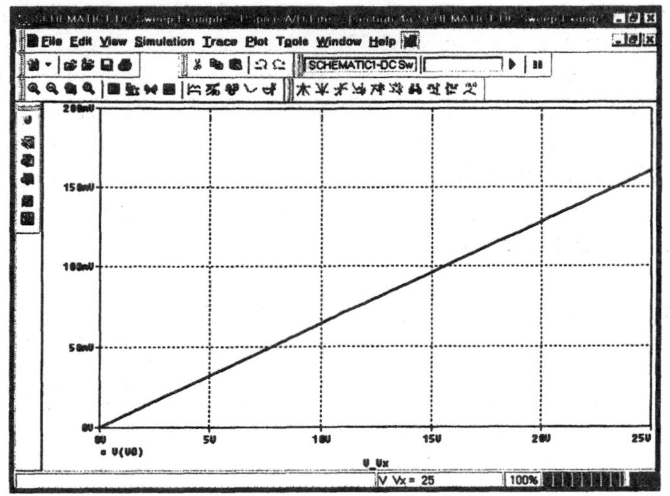

You can add as many traces to this graph as you wish. We will now add a second trace to show the capabilities of Probe. Select **Trace** and then **Add Trace** (or press the **INSERT** key) to get the ***Add Traces*** dialog box. Fill in the dialog box as shown below:

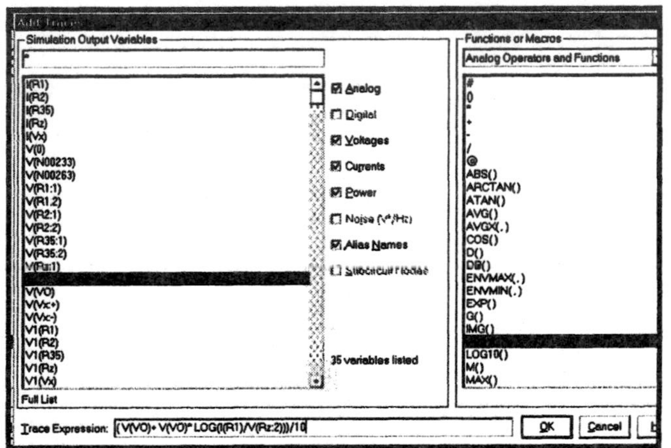

The ***Trace Expression*** was entered by a combination of typing operators and parentheses in the text field and clicking the ***LEFT*** mouse button on the desired voltage.* Although the displayed waveform may not have much meaning to us, it does show what can be displayed with Probe. Click the ***OK*** button to display the trace:

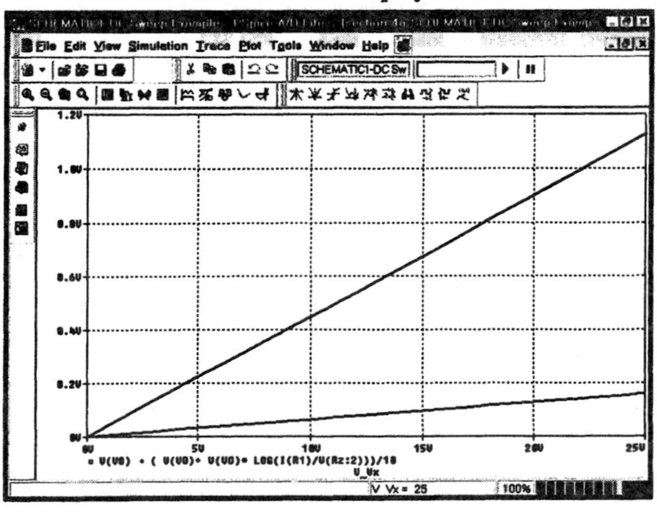

*The box next to ***Trace Expression*** is a text field. Select the text field by clicking the ***LEFT*** mouse button at the desired point in the text field. A vertical cursor will appear.

Currents through any device can also be displayed. Be careful, since the magnitude of currents is usually much smaller than that of voltages. Typically, currents are in milliamperes and voltages are in volts or tens of volts. If voltage and current traces are shown on the same screen, you will probably not see the current trace due to its small magnitude relative to that of the voltage*. It is usually better to display currents on a different plot, or to delete some of the traces. To select a trace for deletion, click the ***LEFT*** mouse button on the name or expression of the trace you wish to delete. When a trace is selected, its expression or name should turn red, indicating that it has been selected. Delete the selected trace by selecting **Edit** and then **Delete** from the Probe menu bar, or by pressing the **DELETE** key. The trace should disappear from the screen.

Instead of deleting a trace, we will display a current trace in a new window. Select **Window** and then **New Window** from the Probe menus. A new empty window will open:

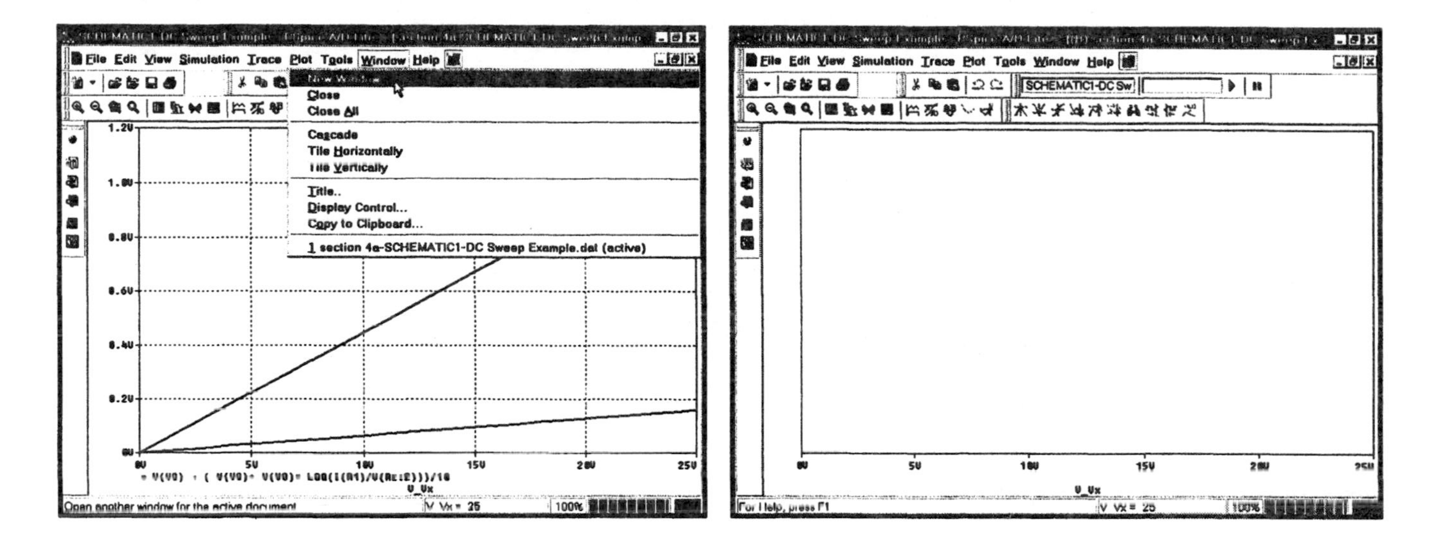


We can now use this window to display more traces. To add a trace select **Trace** and then **Add Trace** (or press the **INSERT** key), and add trace I(R1):

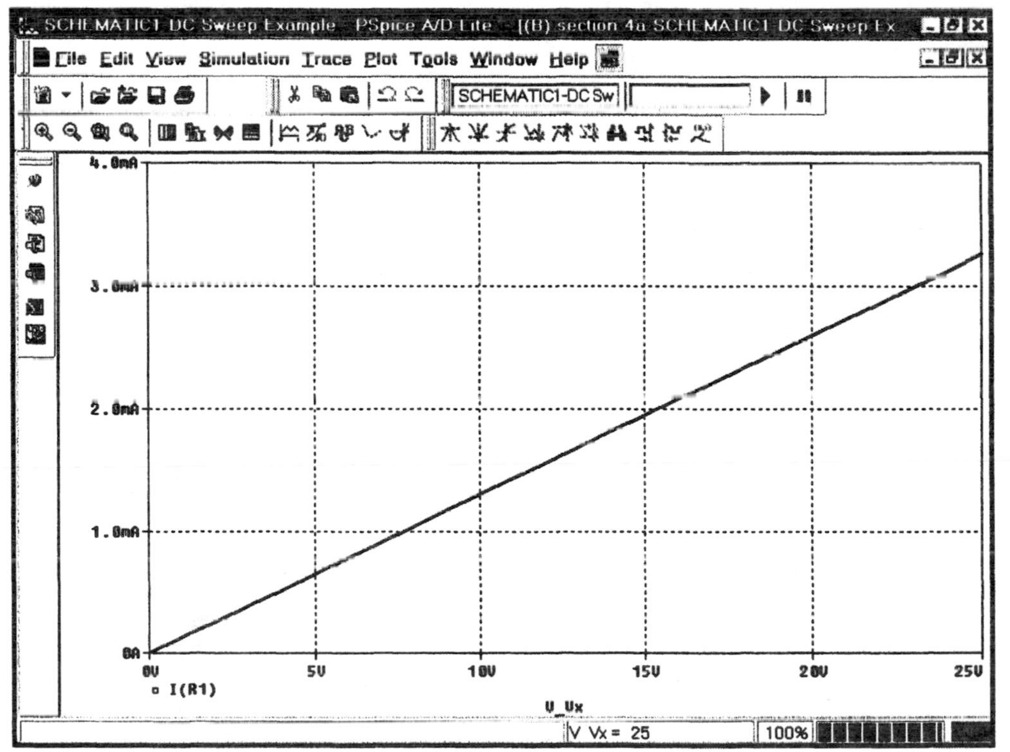


You can open several windows and add as many traces as you want.

---

* See section 2.E to add a second y-axis for viewing currents and voltages on the same plot.

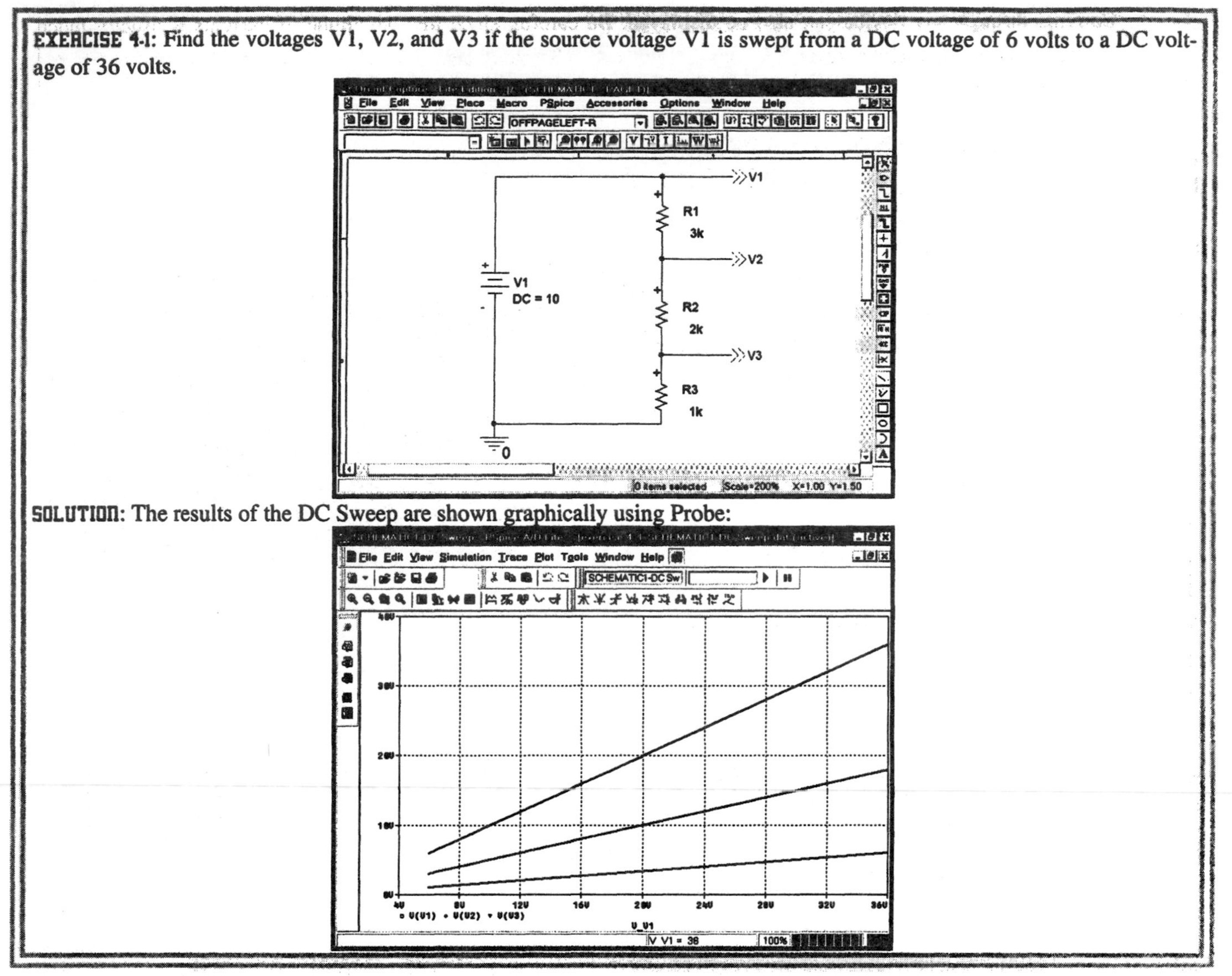

**EXERCISE 4-1**: Find the voltages V1, V2, and V3 if the source voltage V1 is swept from a DC voltage of 6 volts to a DC voltage of 36 volts.



**SOLUTION**: The results of the DC Sweep are shown graphically using Probe:



# 4.B. Diode I-V Characteristic

We would now like to use PSpice to obtain the I-V characteristic of a semiconductor diode. Wire the circuit shown below:

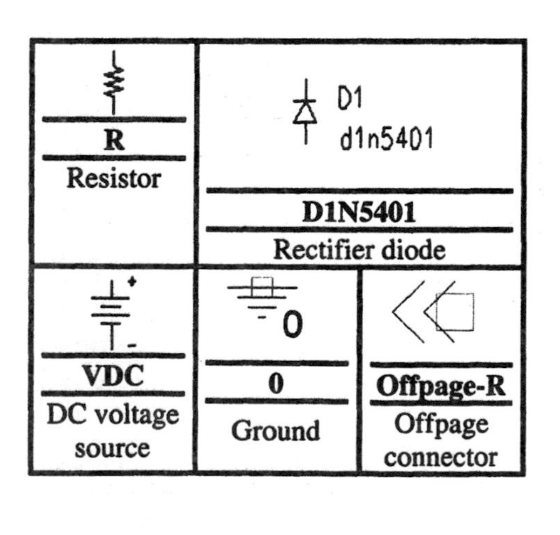


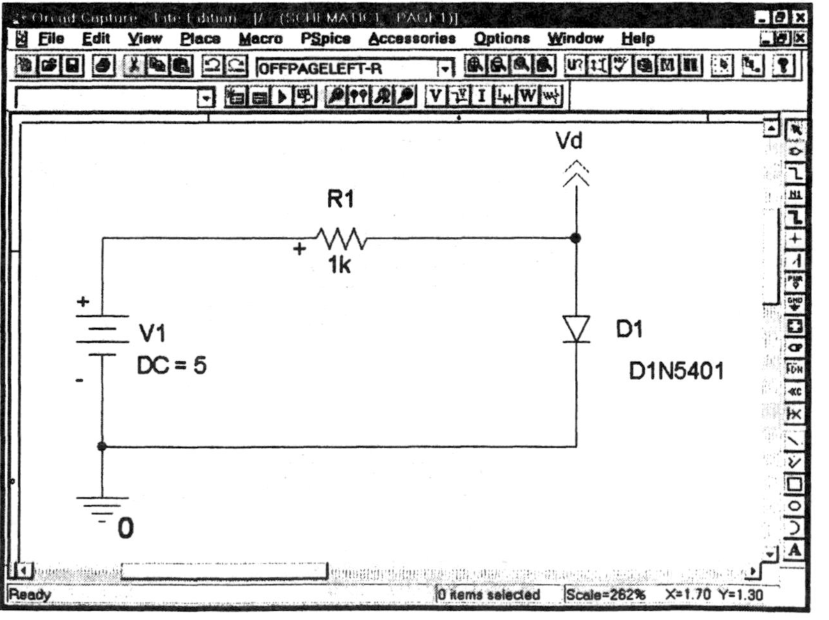

A DC Sweep will be run to sweep ***V1*** from −15 volts to +15 volts. Select **PSpice** and then **New Simulation Profile** from the Capture menus:

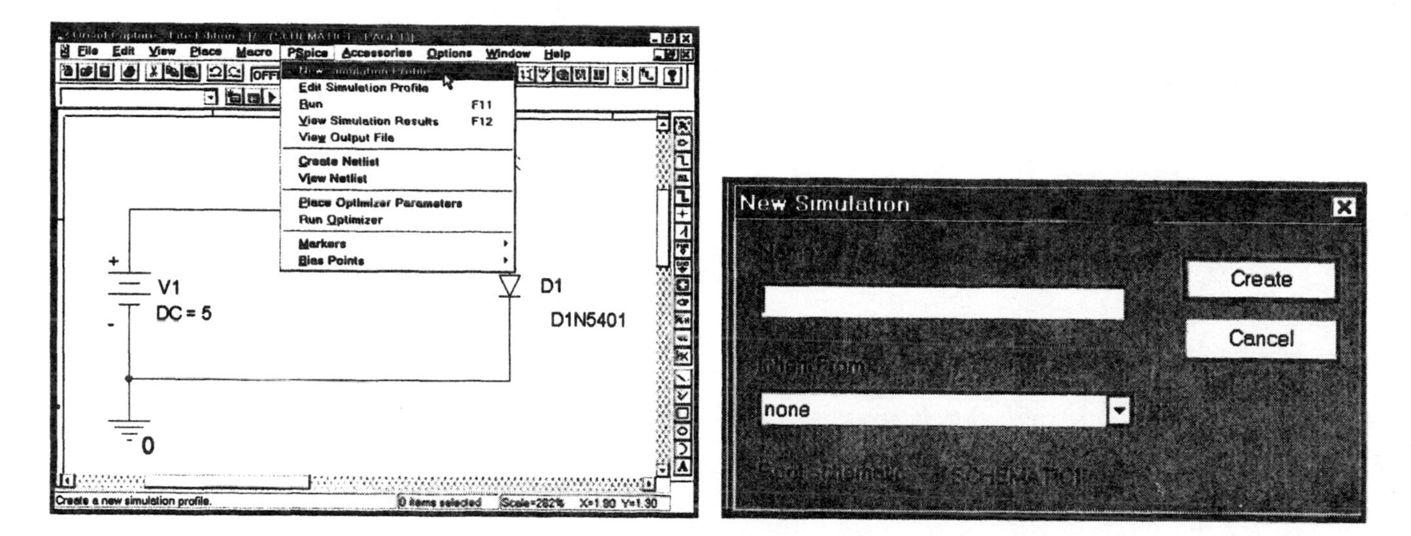


Enter a name for the profile and click the ***Create*** button:

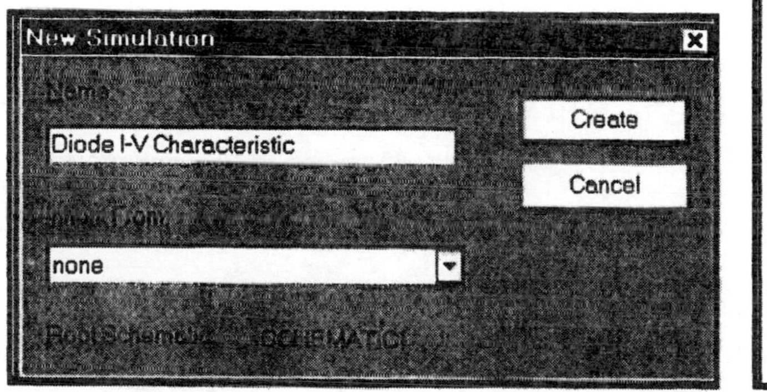


Click the ***LEFT*** mouse button on the down triangle ▾ under ***Analysis type*** to obtain the list of available analyses:

Select **DC Sweep** and fill in the settings as shown:

Simulation Settings - Diode I-V Characteristic

General | Analysis | Include Files | Libraries | Stimulus | Options | Data Collection | Probe Window

Analysis type: DC Sweep

Options: Primary Sweep, Secondary Sweep, Monte Carlo/Worst Case, Parametric Sweep, Temperature (Sweep), Save Bias Point, Load Bias Point

Sweep variable: Voltage source, Current source, Global parameter, Model parameter, Temperature — Name: V1; Model type:; Model name:; Parameter name:

Sweep type: Linear, Logarithmic (Decade), Value list — Start value: -15; End value: 15; Increment: 0.01

OK | Cancel | Apply | Help

This dialog box is set up to sweep the voltage source named ***V1*** from –15 V to +15 V in 0.01 V increments. The sweep type is linear, which means that the voltage points are equally spaced. Click the ***OK*** button to accept the settings:

Run the simulation (select **PSpice** and then **Run** or press the **F11** key):

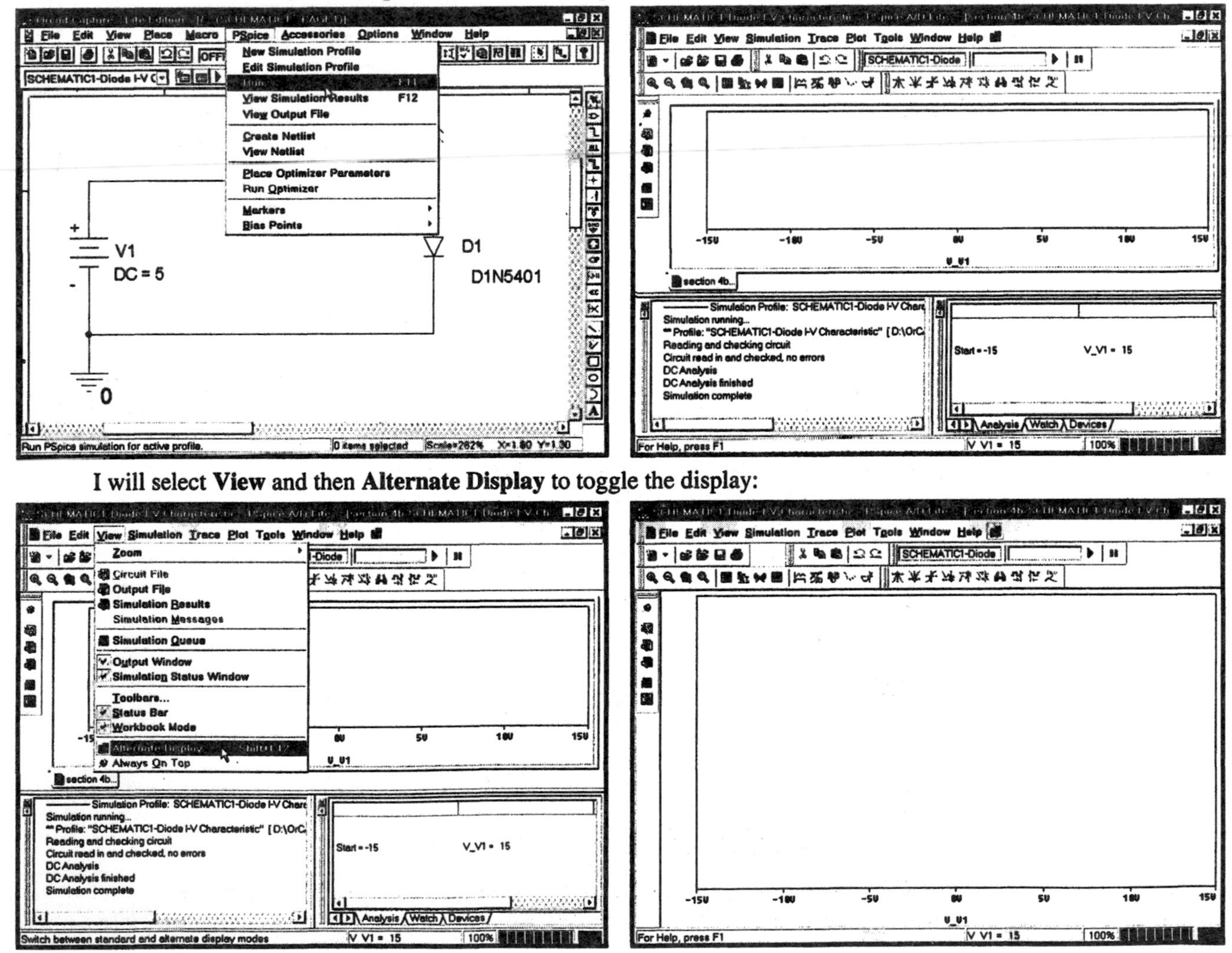


I will select **View** and then **Alternate Display** to toggle the display:

We now need to add a trace displaying the diode current. Select **Trace** and then **Add Trace** from the Probe menus (or press the **INSERT** key) and add the trace I(D1):

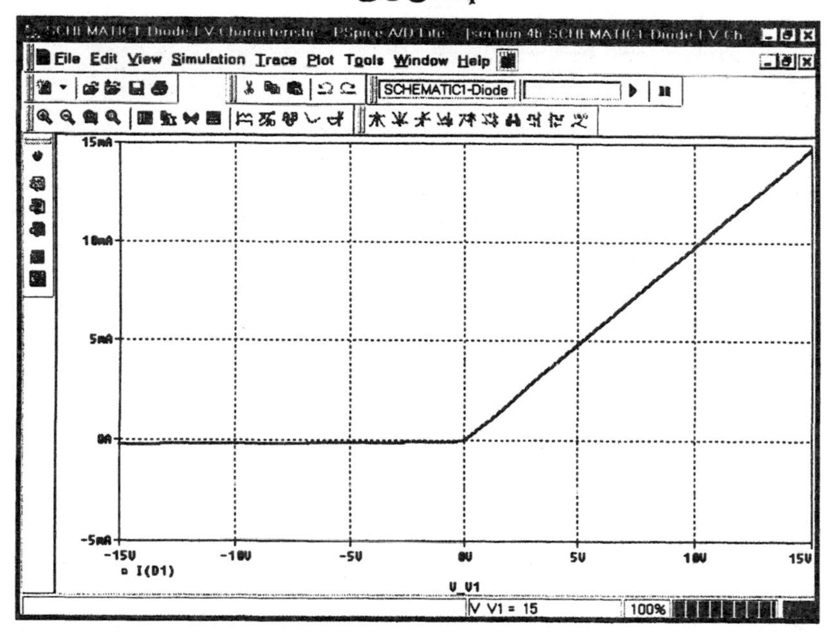


This is a trace of the diode current versus the voltage ***V1***. We now need to change the x-axis from the voltage ***V1*** to the diode voltage. Select **Plot** and then **Axis Settings** from the menus. The ***Axis Settings*** dialog box below will appear:

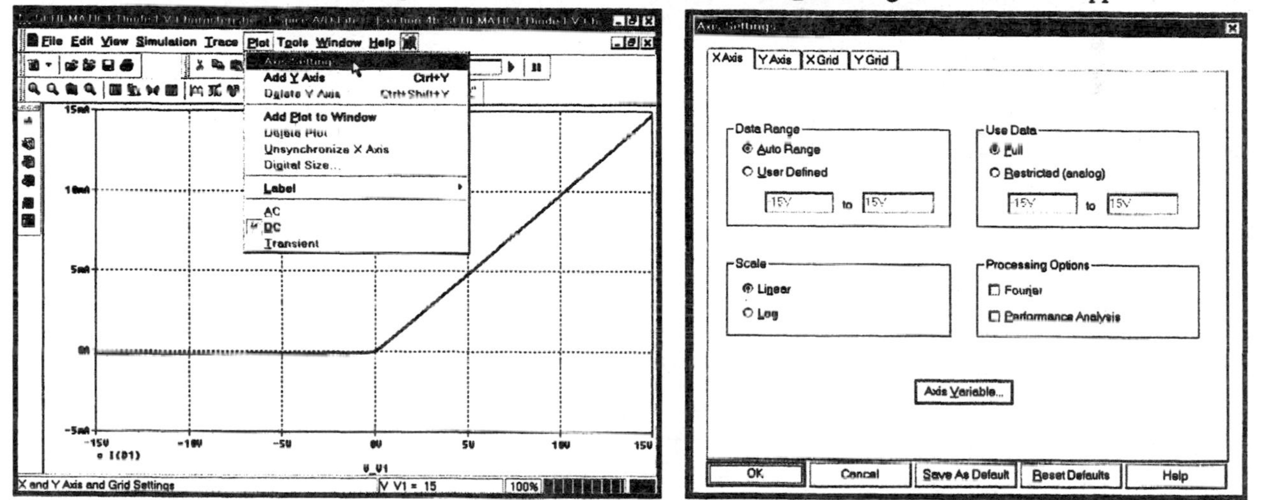


This dialog box allows us to change the settings for both the x- and y-axes. Presently, the tab for the x-axis is selected, which is the tab we need to use. Click the ***LEFT*** mouse button on the ***Axis Variable*** button:

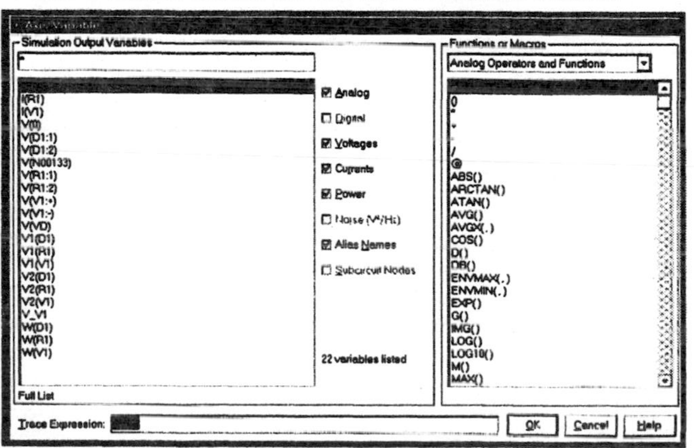


We see that the present ***Trace Expression*** is the voltage of our source ***V_V1***. We need to change the axis to the diode voltage. Click the ***LEFT*** mouse button on the text ***V(VD)***. The text ***V(VD)*** will be highlighted and will appear in the text field next to the text ***Trace Expression***:

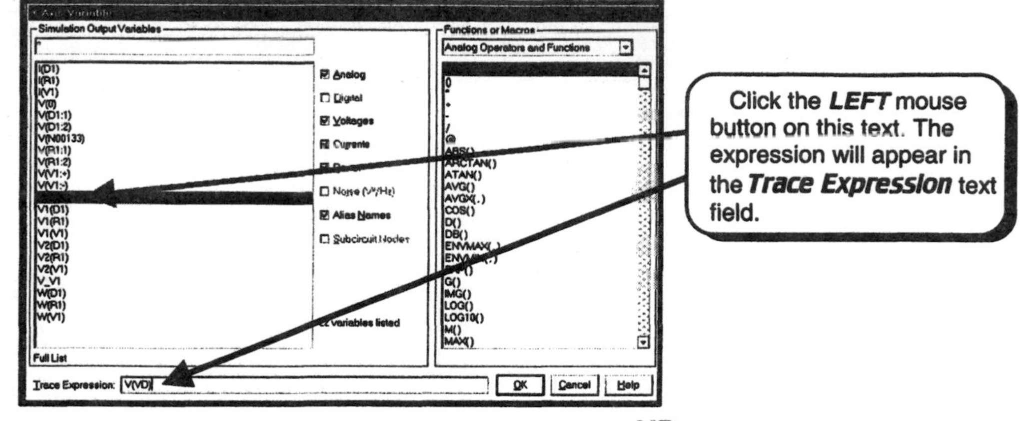


This is the voltage at node ***VD*** relative to ground. Since one side of the diode is grounded, ***VD*** is also the diode voltage. Click the ***OK*** button to accept the changes.

Click the ***OK*** button to display the I-V characteristic:

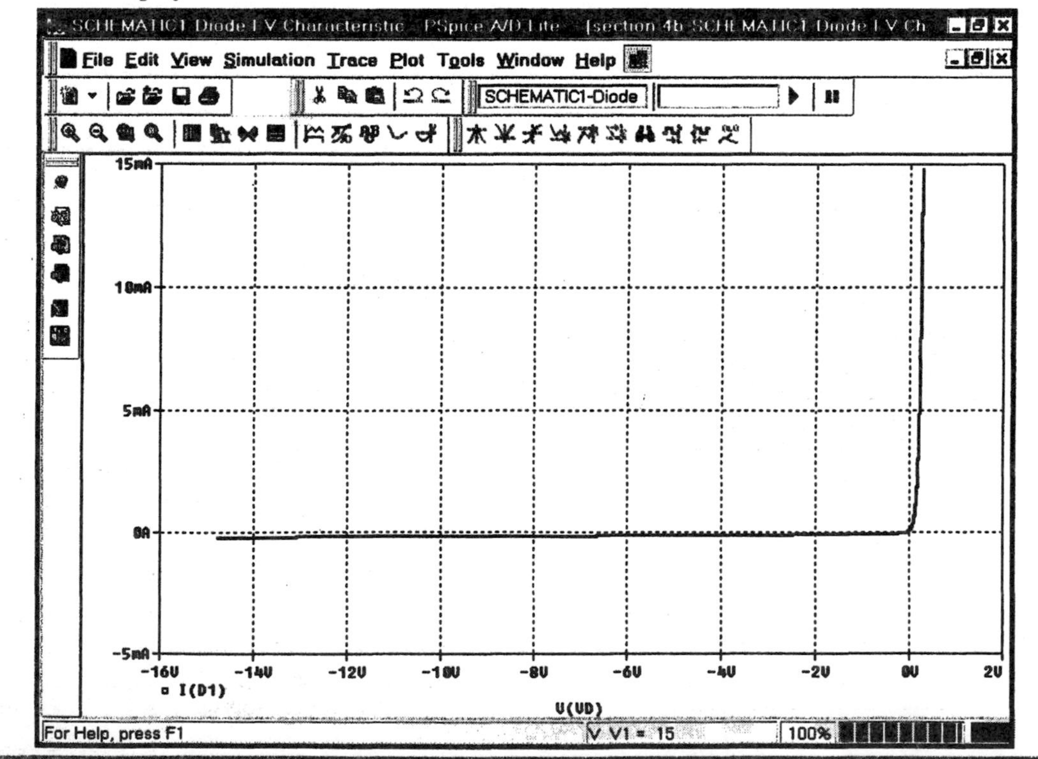


**EXERCISE 4-2:** Display the I-V characteristic of the Zener diode in the circuit below:

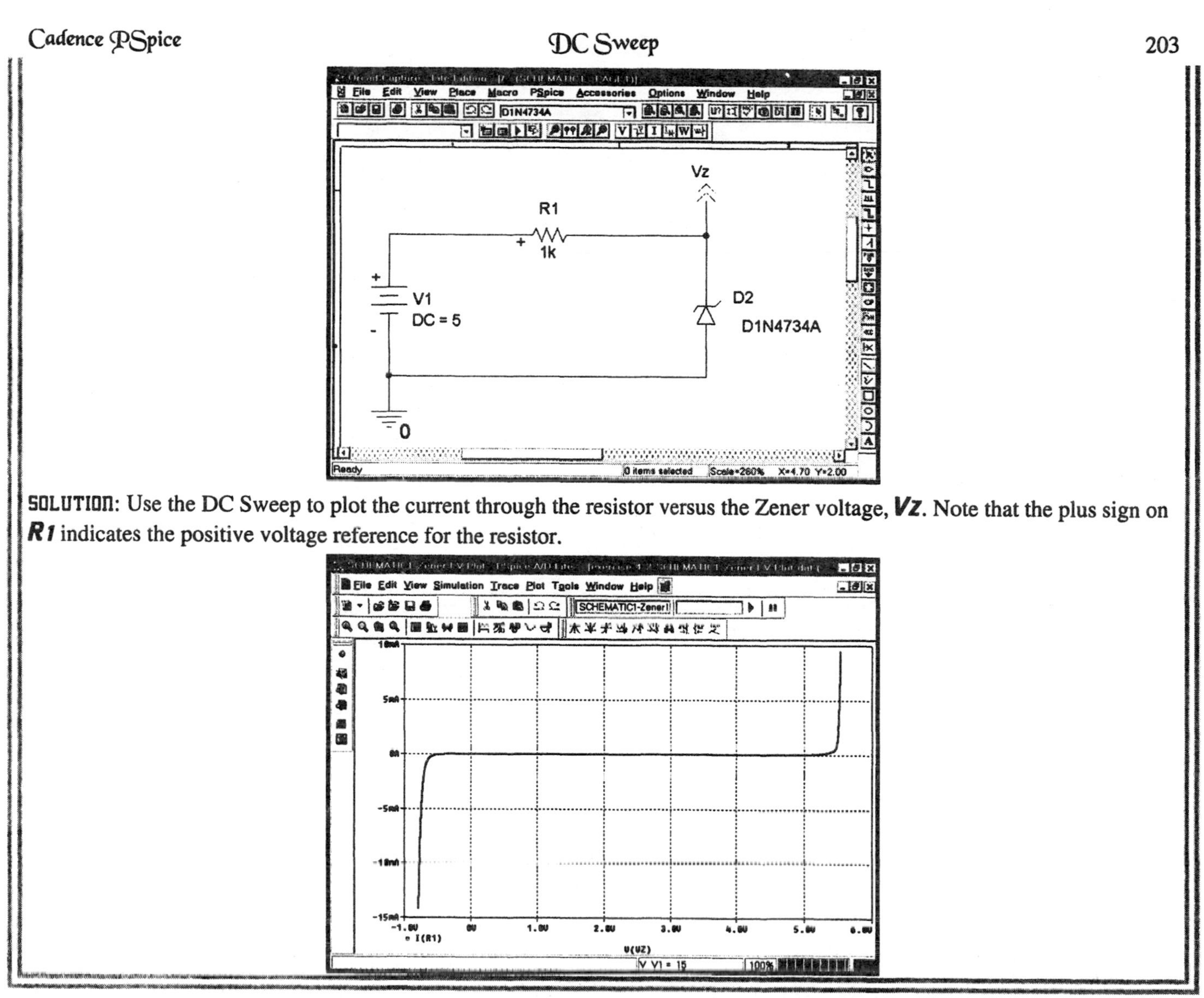


**SOLUTION**: Use the DC Sweep to plot the current through the resistor versus the Zener voltage, ***Vz***. Note that the plus sign on ***R1*** indicates the positive voltage reference for the resistor.

# 4.B.1. Temperature Sweep — Diode I-V Characteristic

Most of the devices used by PSpice can include temperature effects in the model. Most of the semiconductor models provided by Orcad include temperature dependence. By default, the passive devices such as resistors, capacitors, and inductors do not include temperature dependence. To make these items include temperature effects, you will need to create models that include temperature effects. The temperature dependence of resistors is discussed in Section 4.G.1. In this section, we will show only how the I-V characteristic of a 1N5401 diode is affected by temperature. The D1N5401 diode model already includes temperature effects so we will not need to modify the model. We will use the standard resistor, which does not include temperature effects. We will continue with the circuit of Section 4.B:

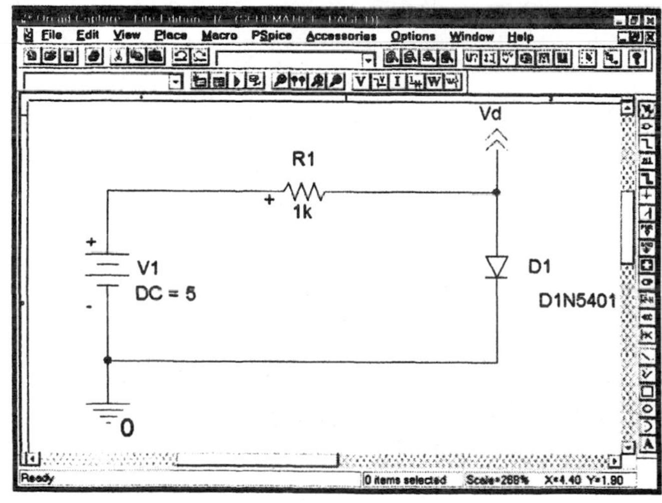


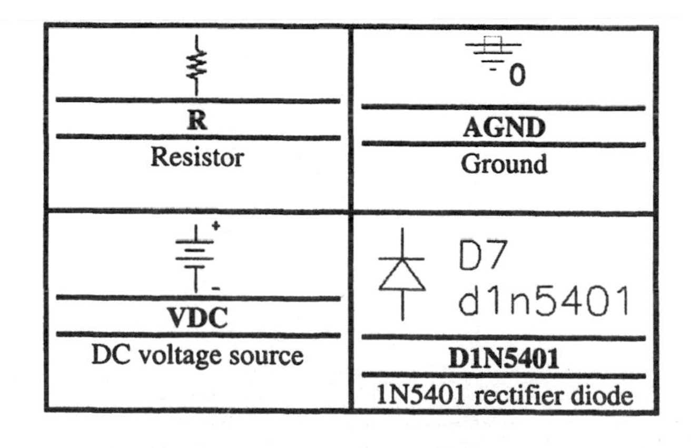

| | |
|---|---|
| **R**<br>Resistor | **AGND**<br>Ground |
| **VDC**<br>DC voltage source | **D1N5401**<br>1N5401 rectifier diode |

The DC Sweep is set up the same as in Section 4.B:

In general, we can run the temperature sweep as either a Parametric Sweep or a Secondary Sweep. However, to generate I-V characteristic curves for different temperatures, we must use the Secondary Sweep. Click on the square ☐ next to the ***Secondary Sweep*** selection:

We wish to generate three I-V curves, one at −25°C, one at 0°C, and one at 50°C. Select ***Temperature*** as the ***Sweep variable***. We will use the ***Value list*** to specify the three temperature values. Click on the circle ○ next to the text ***Value list*** to select the option:

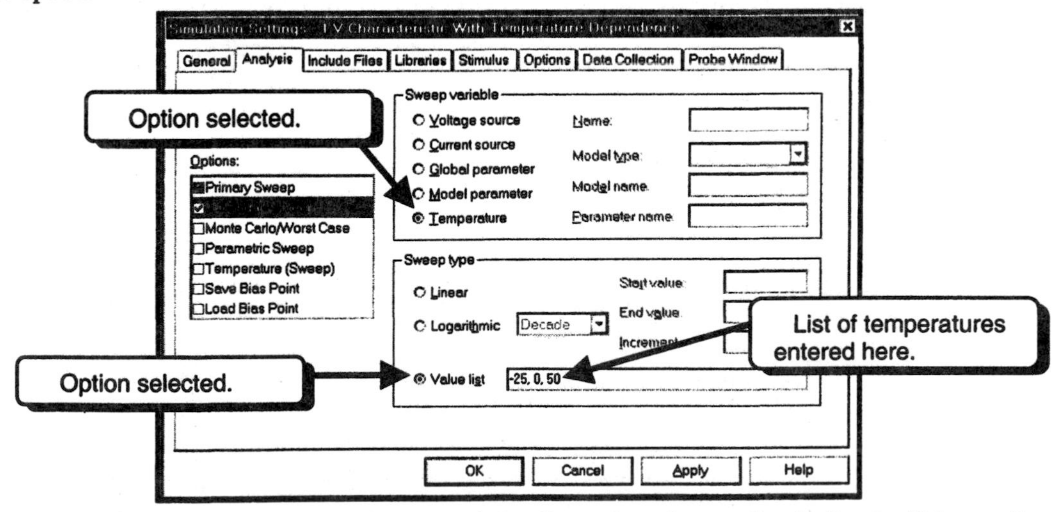


We will be running both the Primary Sweep and the Secondary Sweep. Logically the Primary Sweep executes inside the Secondary Sweep. That is, for our settings, the temperature is set to −25 °C and V1 is swept from −15 to 15. Then the temperature is set to 0 °C and V1 is swept from −15 to 15. Last, the temperature is set to 50 °C and V1 is swept from −15 to 15.

Click the ***OK*** button to return to the schematic:

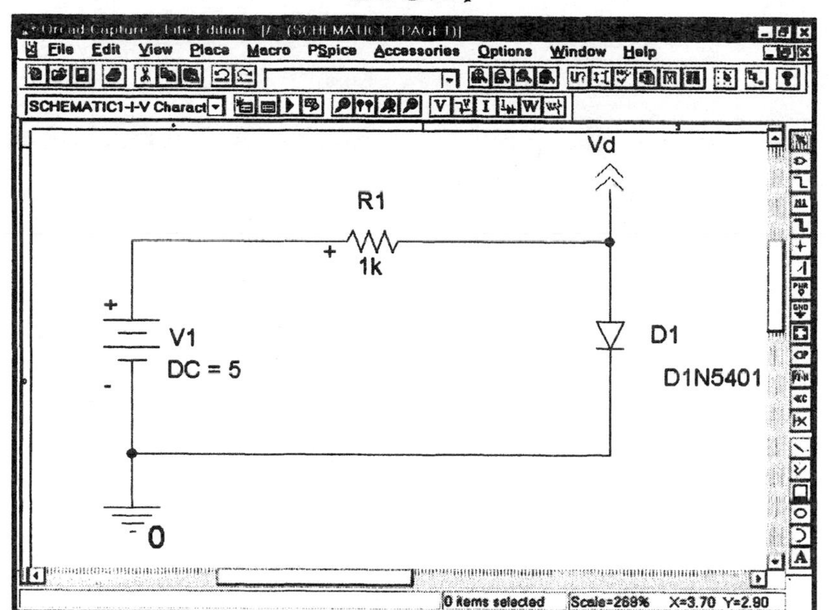


Simulate the circuit (select **PSpice** and then **Run** or press the **F11** key):

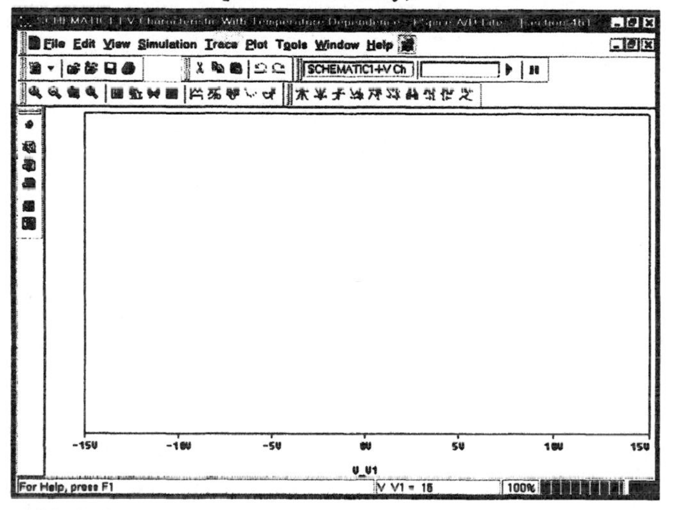


Follow the instructions on pages 200–202 to display the I-V characteristic of the diode, I(D1) versus V(Vd). Three plots will be generated instead of one:

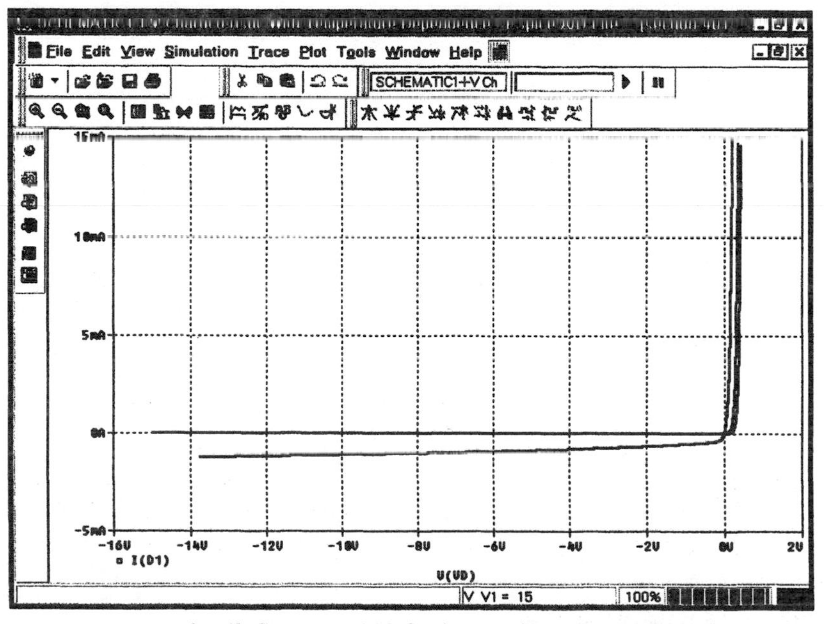


We can zoom in on the trace to see more detail. See page 113 for instructions on zooming in on a trace.

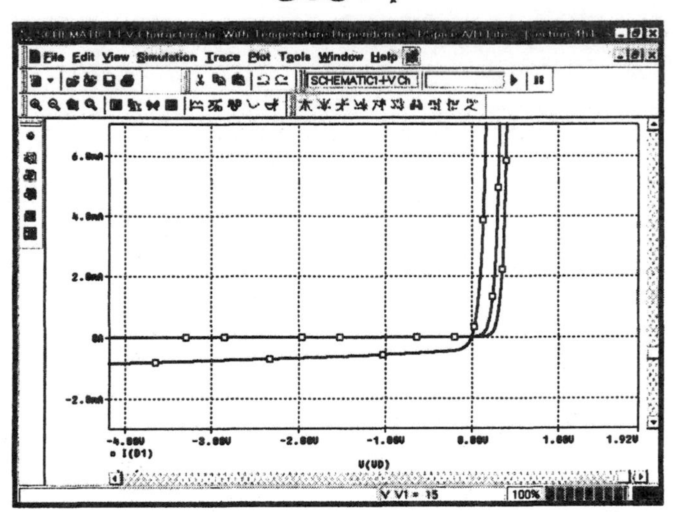


With the Secondary Sweep, we cannot easily identify which trace occurs with what temperature. To associate a trace with a specific temperature, we would need to run a single I-V characteristic at the specified temperature. If we ran a DC Sweep (sweep variable Vin) together with a Parametric Sweep (sweep variable temperature) we could then identify which trace was at what temperature. However, we cannot generate the I-V characteristic using the Parametric Sweep together with the DC Sweep.

**EXERCISE 4-3**: Display the I-V characteristic of the Zener diode in the circuit below for temperatures −25°C, 0°C, and 50°C.

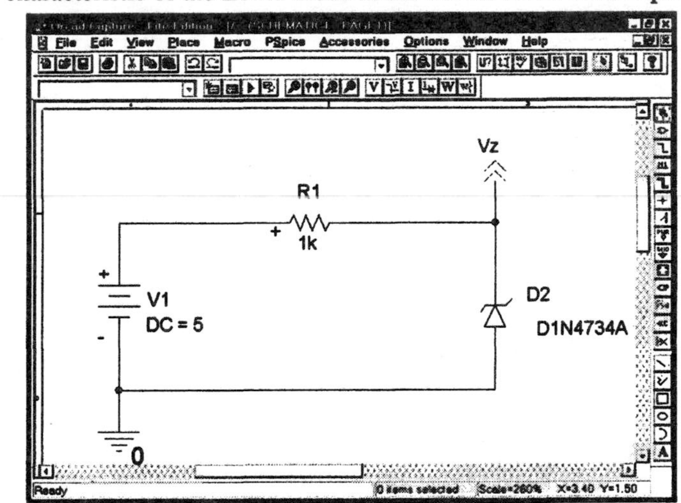


**SOLUTION**: Use the DC Sweep and Secondary Sweep to plot the current through the resistor versus the Zener voltage, ***Vz***. Note that the plus sign on ***R1*** indicates the positive voltage reference for the resistor.

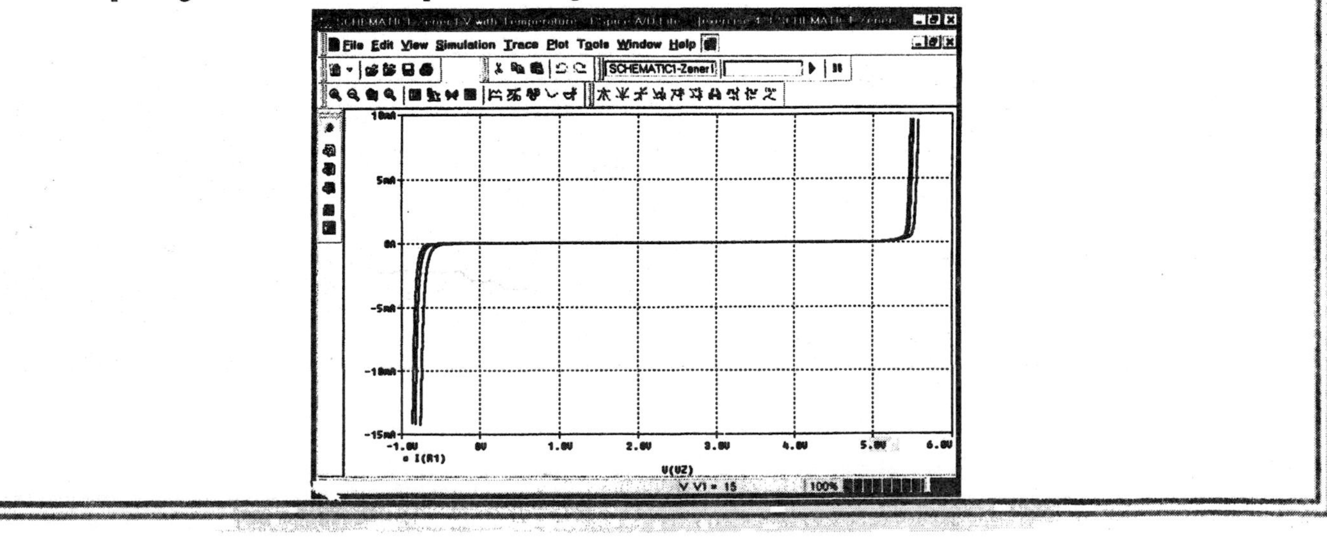

# 4.C. Parametric Sweep — Maximum Power Transfer

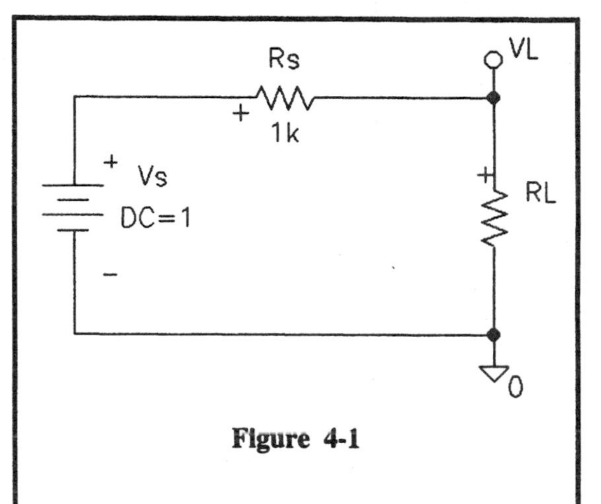

Figure 4-1

In circuit design, we are sometimes concerned with how a circuit parameter affects performance. There are two ways to vary parameters in PSpice. The first is the DC Sweep, where we vary a parameter rather than a DC voltage. This method generates a single curve. The second is a Parametric Sweep that is run in conjunction with another analysis such as an AC Sweep, DC Sweep, or a Transient Analysis. The second method generates a family of curves. In this section we will demonstrate only the DC Parametric Sweep. Throughout this manual there will be examples using the Parametric Sweep in conjunction with the other analyses.

A frequently demonstrated problem in beginning circuit analysis courses is, what value of RL in the circuit of Figure 4-1 will deliver maximum power to RL?* With a little bit of circuit analysis and some calculus, it can be shown that for fixed Rs, maximum power will be delivered to RL when RL is equal to Rs. We will demonstrate this result using PSpice. Wire the following circuit:

| R<br>Resistor | PARAMETERS:<br>**PARAM**<br>Part to define parameters | |
|---|---|---|
| **VDC**<br>DC voltage source | 0<br>**0**<br>Ground | **Offpage-R**<br>Offpage connector |

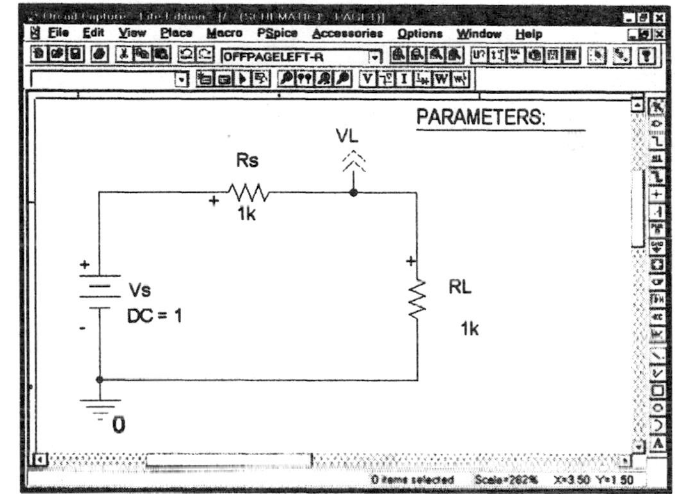


We would like to vary the value of ***RL***. To do this we need to define the value of ***RL*** as a parameter. Double-click on the text ***1k*** for resistor ***RL***:

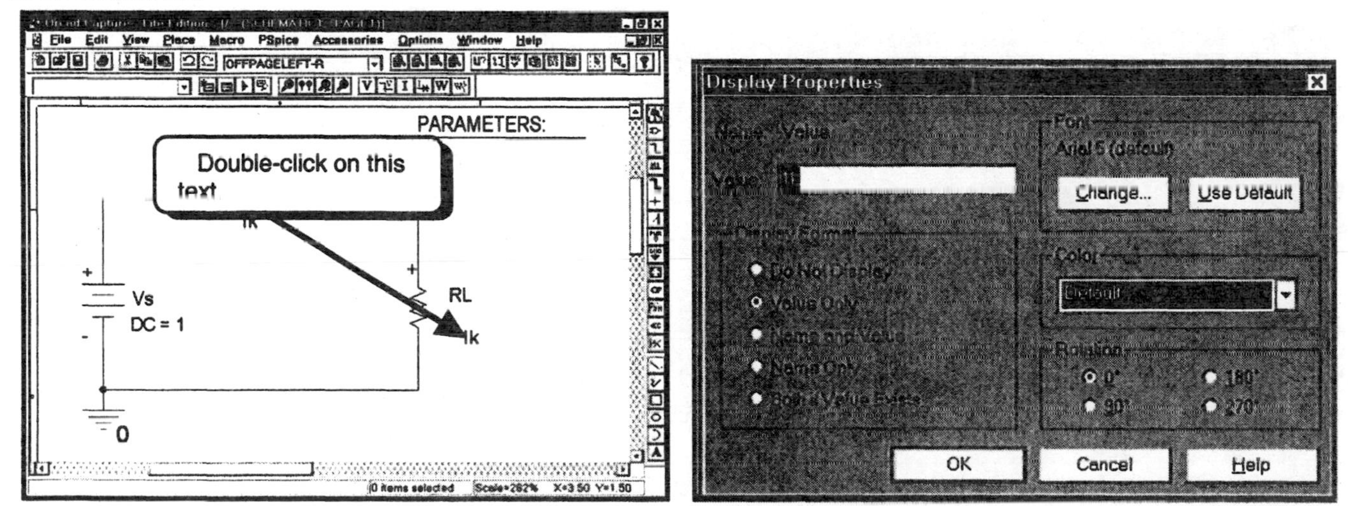


Type the text **{RL_val}**:

*Thanks to Dr. David M. Szmyd of Philips Semiconductors for this example.

Display Properties
Name: Value
Value: {RL_val}
Font
Arial 5 (default)
Change... Use Default
Display Format
Do Not Display
Value Only
Name and Value
Name Only
Both if Value Exists
Color
Default
Rotation
0° 180°
90° 270°
OK Cancel Help

Click the ***OK*** button to accept the value:

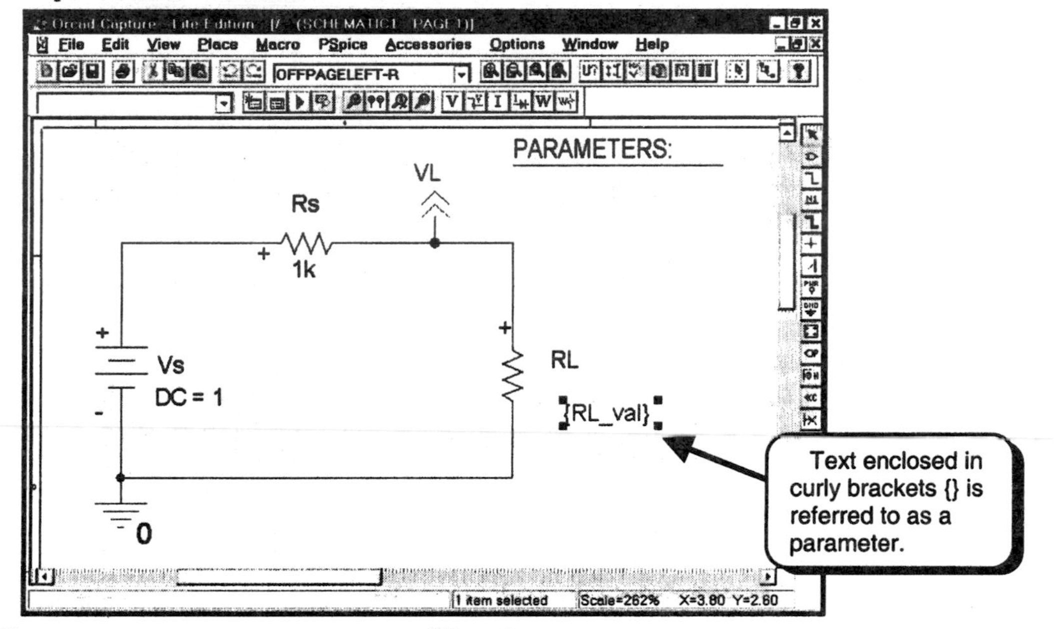


The value of resistor ***RL*** is now the value of the parameter ***{RL_val}***. We must now define the parameter and assign it a default value. Double-click the ***LEFT*** mouse button on the text ***PARAMETERS***:

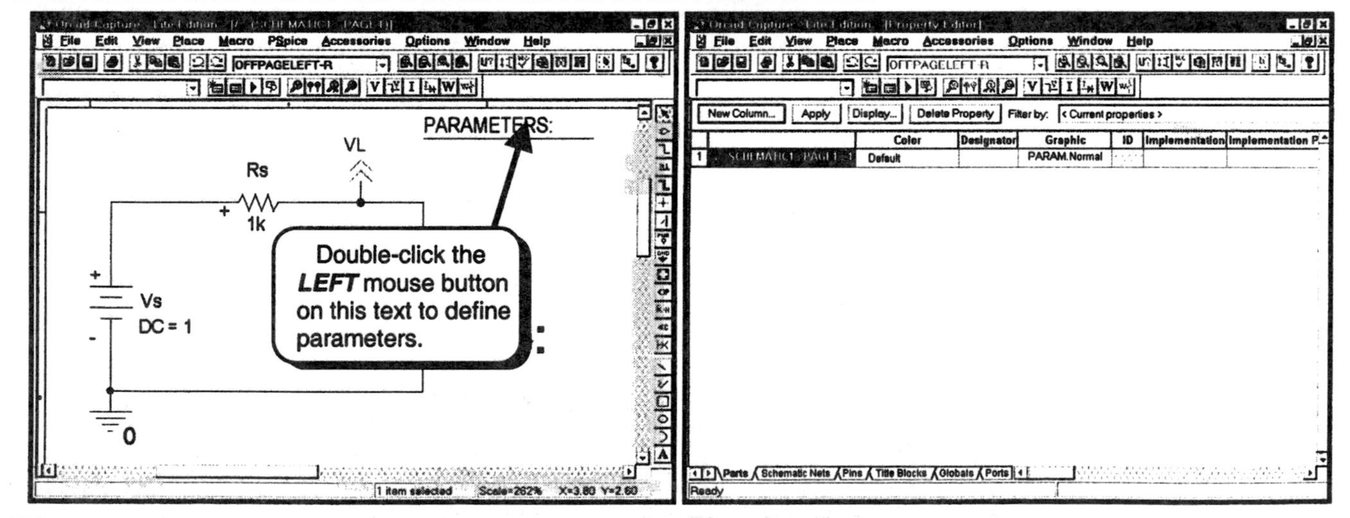


This attribute box is used to define parameters. We will define ***RL_val*** as a parameter and give it a default value of 1k. Click the ***New Column*** button to create a new parameter:

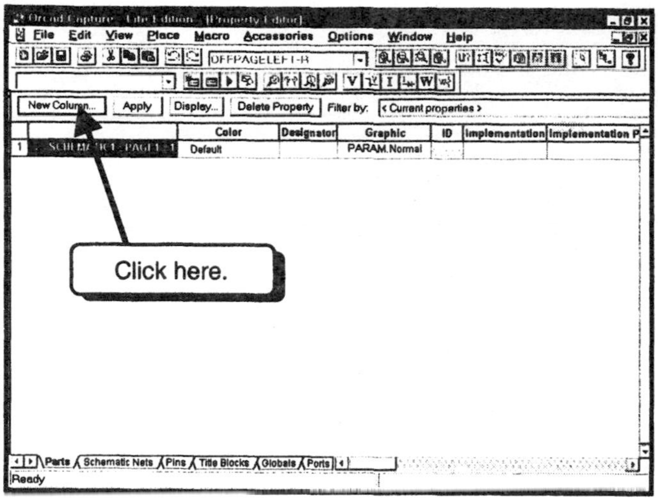

Orcad Capture - Lite Edition - [Property Editor]

File Edit View Place Macro Accessories Options Window Help

OFFPAGELEFT-R

New Column... | Apply | Display... | Delete Property | Filter by: < Current properties >

| | | Color | Designator | Graphic | ID | Implementation | Implementation P |
|---|---|---|---|---|---|---|---|
| 1 | SCHEMATIC1 : PAGE1 : 1 | Default | | PARAM.Normal | | | |

Parts / Schematic Nets / Pins / Title Blocks / Globals / Ports

Ready

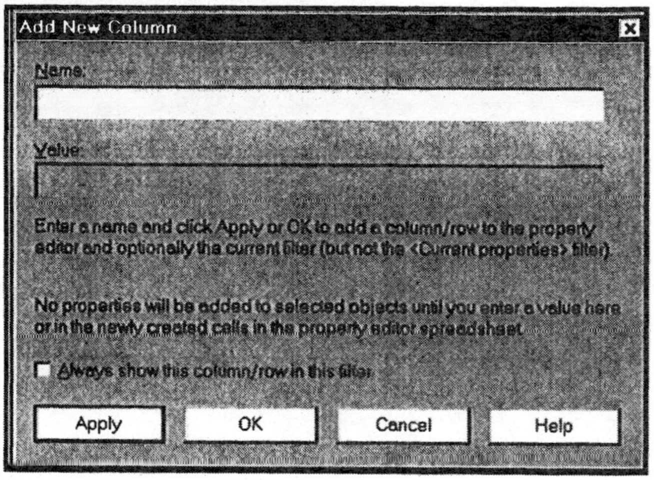

Click here.

Add New Column

Name:

Value:

Enter a name and click Apply or OK to add a column/row to the property editor and optionally the current filter (but not the <Current properties> filter).

No properties will be added to selected objects until you enter a value here or in the newly created cells in the property editor spreadsheet.

Always show this column/row in this filter

Apply | OK | Cancel | Help

Enter the text **RL_val**. Note that there are no curly brackets around the name:

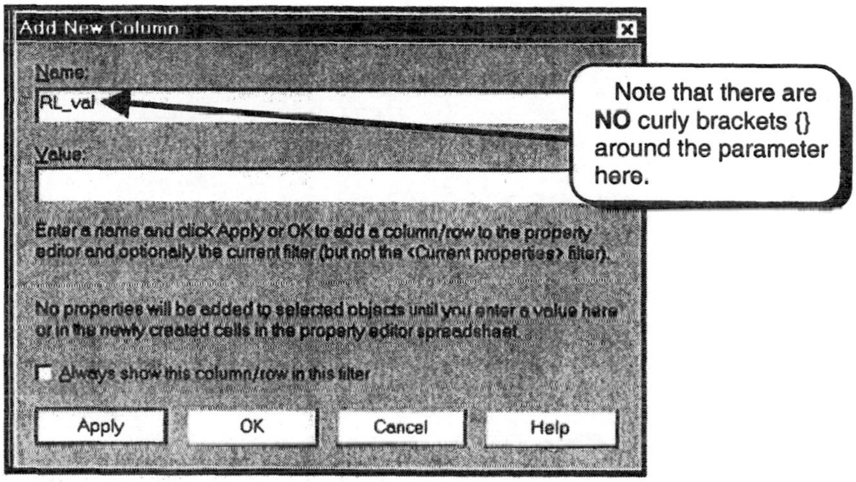

Add New Column

Name: RL_val

Value:

Enter a name and click Apply or OK to add a column/row to the property editor and optionally the current filter (but not the <Current properties> filter).

No properties will be added to selected objects until you enter a value here or in the newly created cells in the property editor spreadsheet.

Always show this column/row in this filter

Apply | OK | Cancel | Help

Note that there are **NO** curly brackets {} around the parameter here.

Once you enter a name for the parameter, the ***Value*** field becomes enabled and you can enter a value. Enter **1k** in the ***Value*** field:

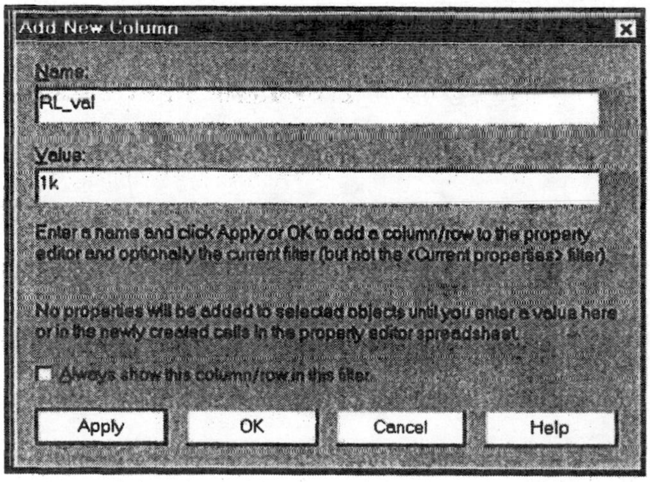

Add New Column

Name: RL_val

Value: 1k

Enter a name and click Apply or OK to add a column/row to the property editor and optionally the current filter (but not the <Current properties> filter).

No properties will be added to selected objects until you enter a value here or in the newly created cells in the property editor spreadsheet.

Always show this column/row in this filter

Apply | OK | Cancel | Help

Click the ***OK*** button to add the parameter to the spreadsheet. A new column will be created for the parameter:

Next, we need to change the display of the parameter so that it is displayed on the schematic. Click the ***LEFT*** mouse button on the text ***RL_val*** to select the column:

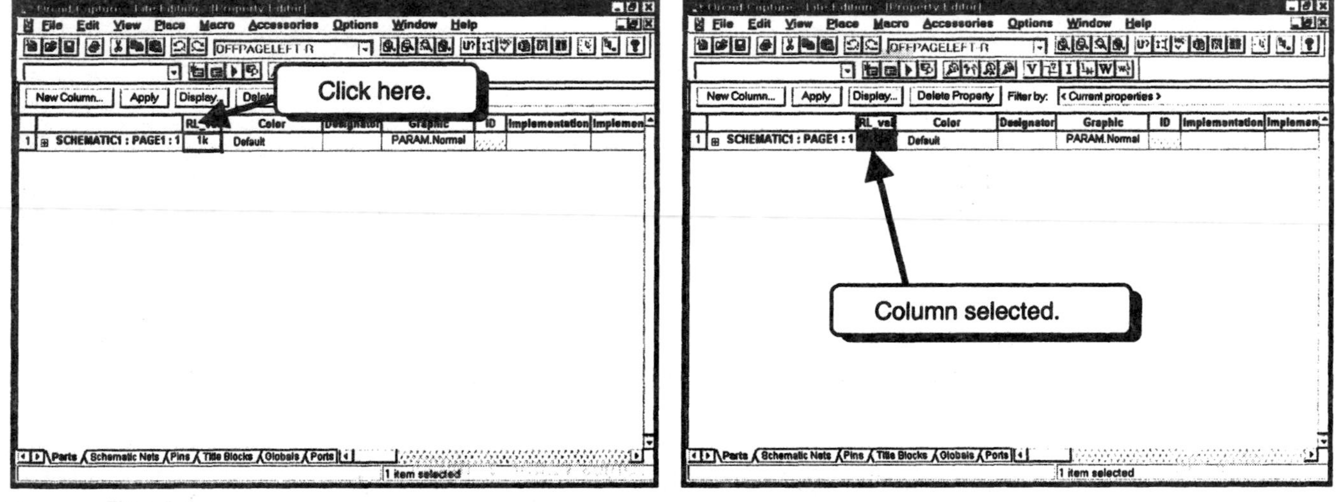


Click the ***Display*** button:

Presently, no information will be displayed on the schematic about this parameter. Select the ***Name and Value*** option and then click the ***OK*** button to return to the schematic:

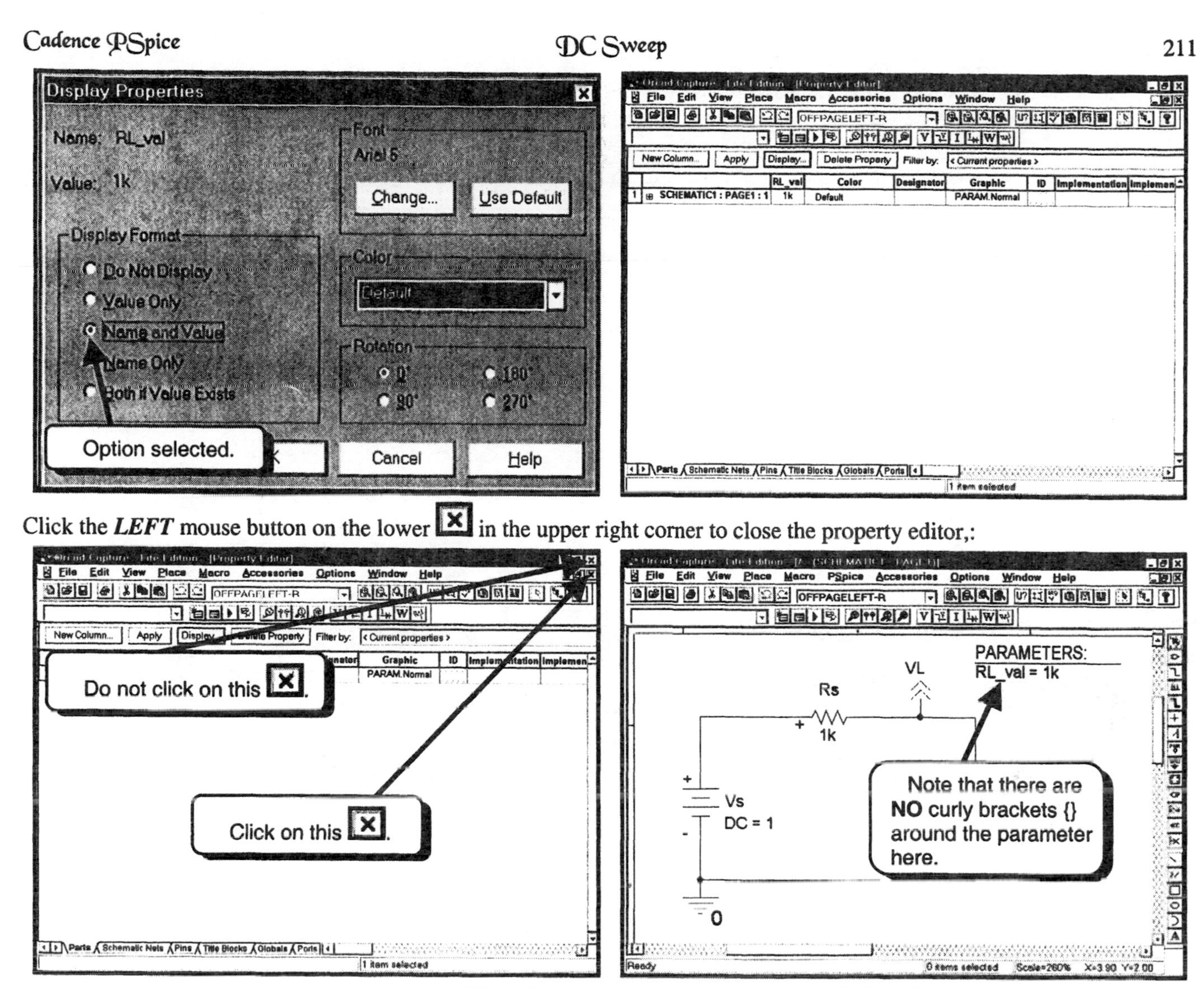


Click the ***LEFT*** mouse button on the lower [X] in the upper right corner to close the property editor,:



In the schematic we notice that the parameter ***RL_val*** is listed under the text ***PARAMETERS:*** and its default value is set to ***1k***.

We will now set up the DC Sweep to sweep the parameter value. Select **PSpice** and then **New Simulation Profile**. Choose a name for the profile and then click the ***Create*** button. Select the ***DC Sweep Analysis type*** and fill in the dialog box as shown:

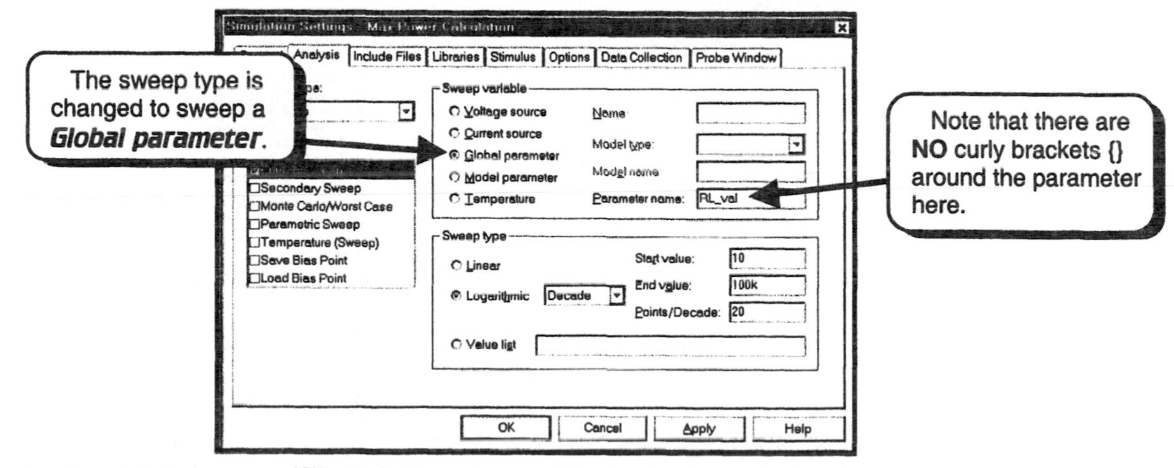


We are sweeping the global parameter ***RL_val***. The sweep will be in decades. The starting value is 10 Ω and the ending value is 100 kΩ. A decade is a factor of 10 and would be 1 to 10, 10 to 100, 100 to 1,000, and so on. The sweep is set up for 20 points per decade so there will be 20 values of ***RL_val*** from 10 to 100 ohms, 20 values from 100 to 1,000 ohms, 20 values from 1,000 to 10,000 ohms, and 20 values from 10,000 to 100,000 ohms. This should give us a fairly detailed plot. Click the ***OK*** button to accept the sweep and return to the schematic.

Run the analysis by selecting **PSpice** and then **Run**.

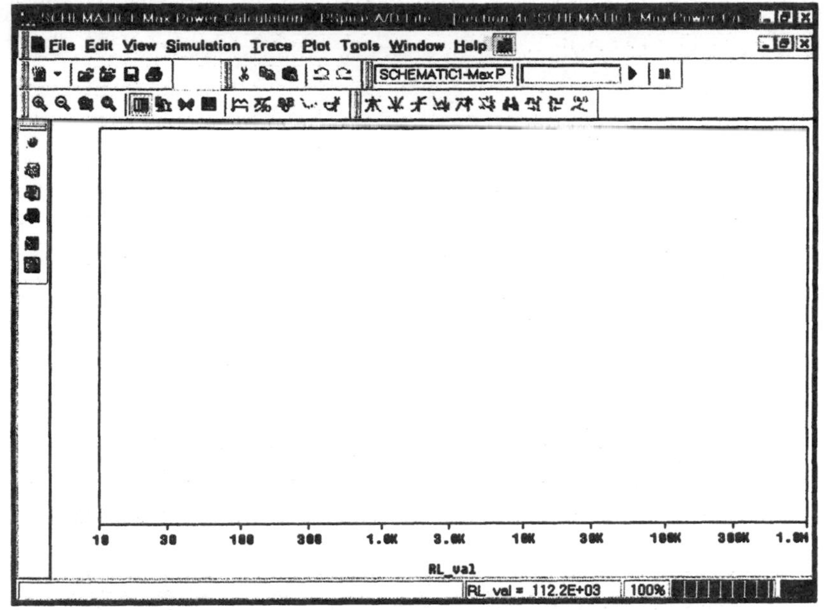


Notice that the x-axis is automatically set to ***RL_val***. We would like to plot the power absorbed by ***RL*** versus the value of ***RL***, ***RL_val***. Select **Trace** and then **Add Trace** to add a trace:

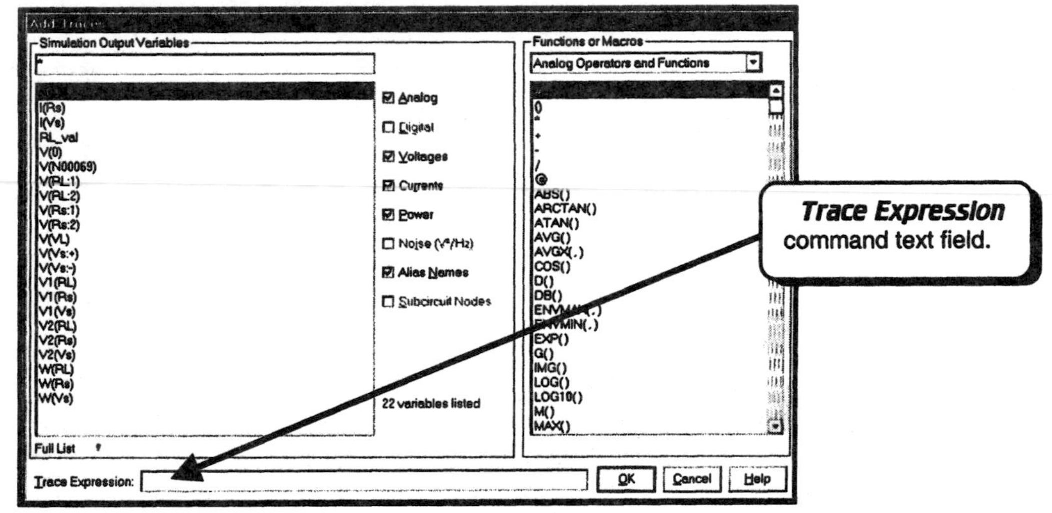


The power absorbed by ***RL*** can be expressed many ways, but I will use $V^2/R$. Type the text **`V(VL)*V(VL)/RL_val`** in the ***Trace Expression*** text field:

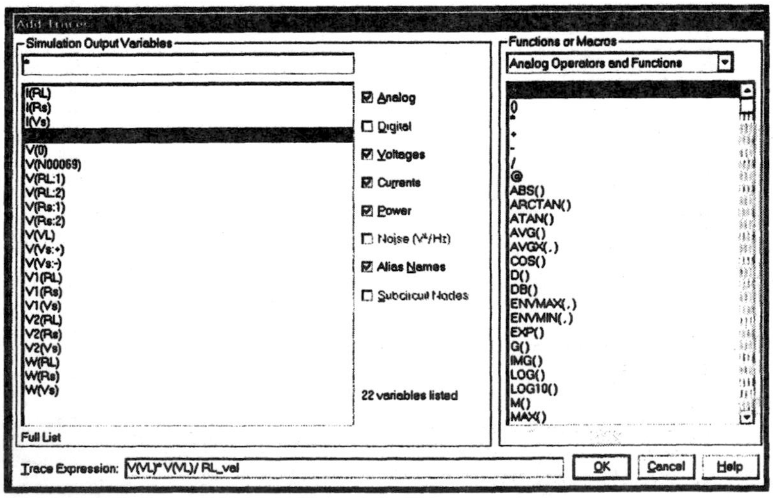

Add Traces

Simulation Output Variables

*

I(RL)
I(Rs)
I(Vs)
RL_val
V(0)
V(N00069)
V(RL:1)
V(RL:2)
V(Rs:1)
V(Rs:2)
V(VL)
V(Vs:+)
V(Vs:-)
V1(RL)
V1(Rs)
V1(Vs)
V2(RL)
V2(Rs)
V2(Vs)
W(RL)
W(Rs)
W(Vs)

Analog
Digital
Voltages
Currents
Power
Noise (V²/Hz)
Alias Names
Subcircuit Nodes

22 variables listed

Full List

Functions or Macros

Analog Operators and Functions

()
*
+
-
/
@
ABS()
ARCTAN()
ATAN()
AVG()
AVGX(,)
COS()
D()
DB()
ENVMAX(,)
ENVMIN(,)
EXP()
G()
IMG()
LOG()
LOG10()
M()
MAX()

Trace Expression: V(VL)*V(VL)/ RL_val

OK Cancel Help

Click the ***OK*** button to plot the trace:

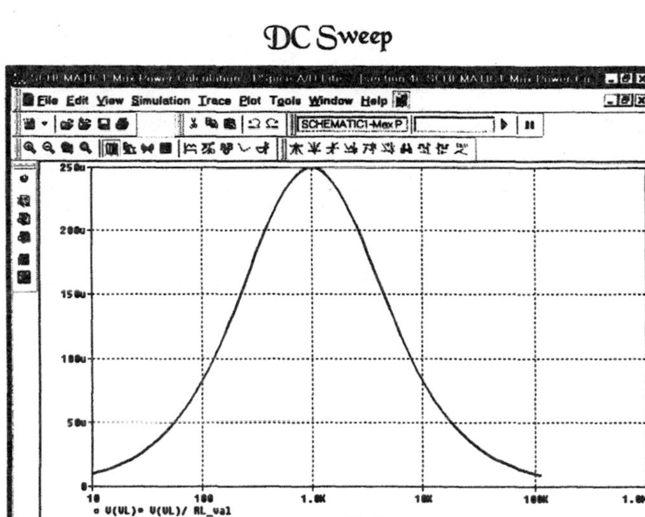


Notice that the x-axis is automatically displayed as a log scale because our sweep of RL_val was in decades.

Using the cursors to locate the maximum, we see that the maximum power to the load occurs when the load is 1000 Ω, equal to the source resistance:

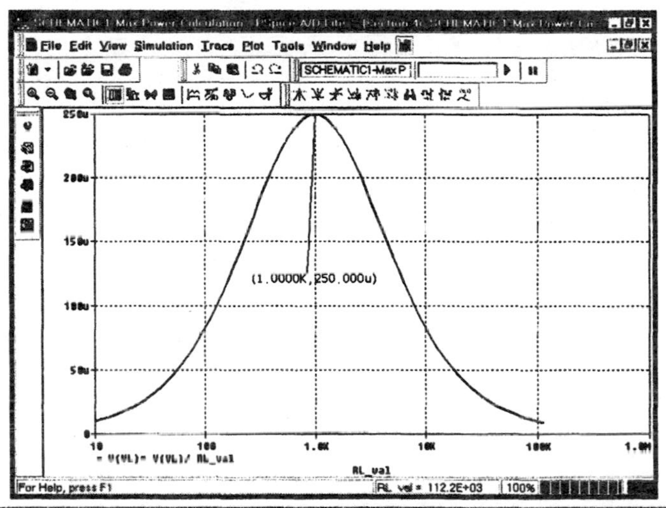


**EXERCISE 4-4**: Repeat the above example, but this time assume that the load is fixed and the source resistance is free to vary. With RL = 1 kΩ, find the value of Rs that delivers maximum power to RL.

**SOLUTION**: Set up the circuit the same as in the previous example, except let the value of Rs be the value of the parameter RS_val:

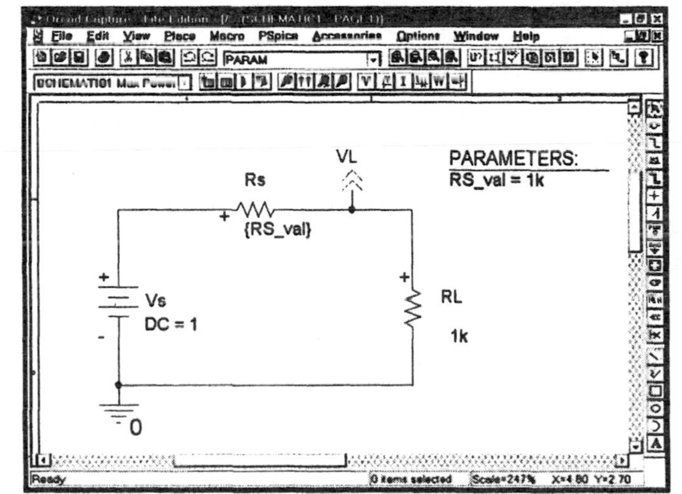


Set up the DC Sweep to sweep the global parameter RS_val from 0.001 Ω to 10 kΩ:

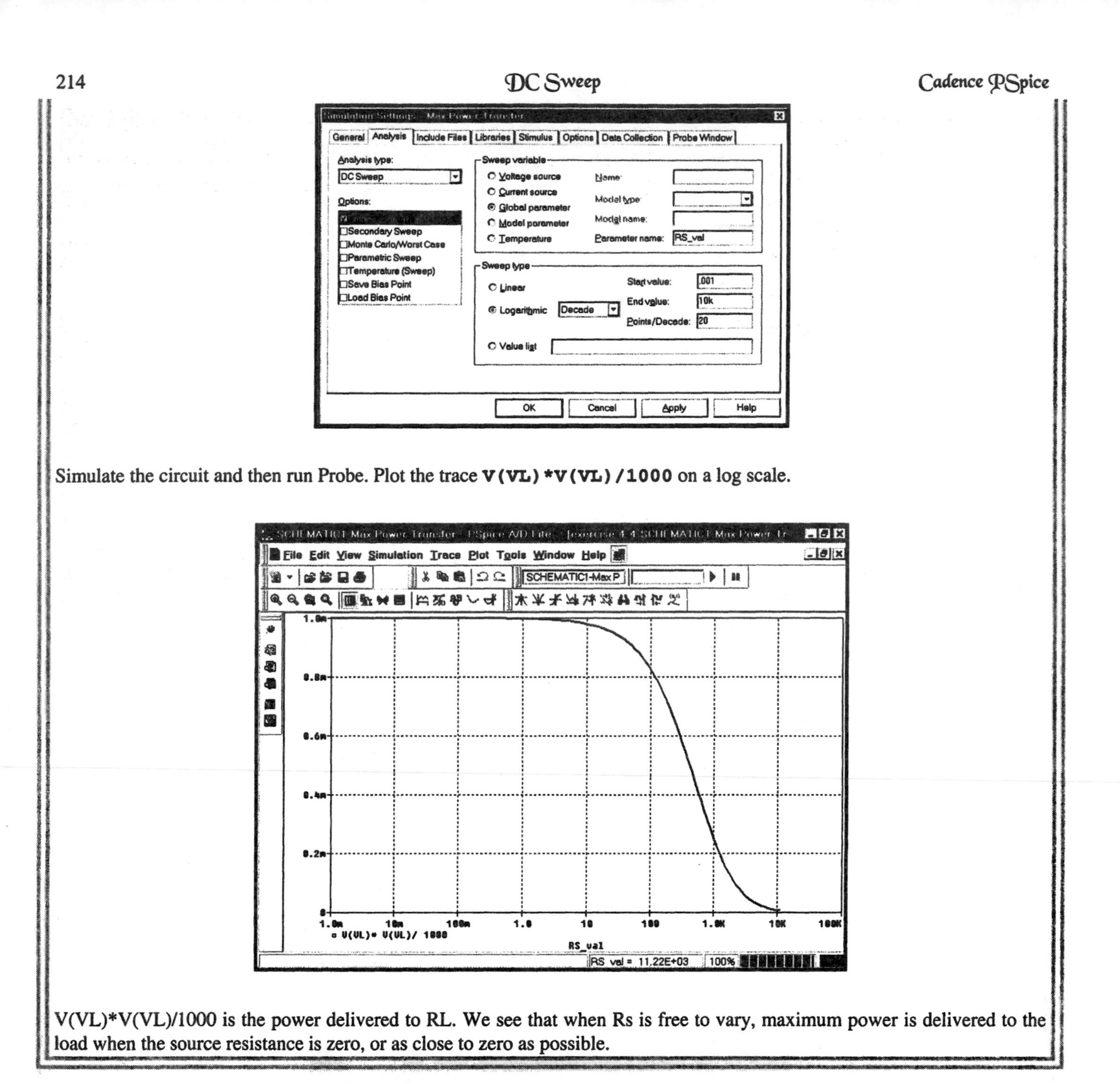


Simulate the circuit and then run Probe. Plot the trace **V(VL)*V(VL)/1000** on a log scale.

V(VL)*V(VL)/1000 is the power delivered to RL. We see that when Rs is free to vary, maximum power is delivered to the load when the source resistance is zero, or as close to zero as possible.

# 4.D. DC Transfer Curves

One of the more useful functions of the DC Sweep is to plot transfer curves. A transfer curve usually plots an input versus an output. A DC transfer curve plots an input versus an output, assuming all capacitors are open circuits and all inductors are short circuits. In a DC Sweep, all capacitors are replaced by open circuits and all inductors are replaced by short circuits. Thus the DC Sweep is ideal for DC transfer curves. The Transient Analysis can also be used for DC transfer curves, but you must run the analysis with low-frequency waveforms to eliminate the effects of capacitance and inductance. Usually a DC Sweep works better for a transfer curve. The one place where a transient analysis works better is plotting a hysteresis curve for a Schmitt Trigger. For a Schmitt Trigger, the input must go from positive to negative, and then from negative to positive to trace out the entire hysteresis loop. This is not possible with a DC Sweep.

## 4.D.1. Zener Clipping Circuit

The circuit below should clip positive voltages at the Zener breakdown voltage and negative voltages at the diode cut-in voltage, approximately 0.6 volts:

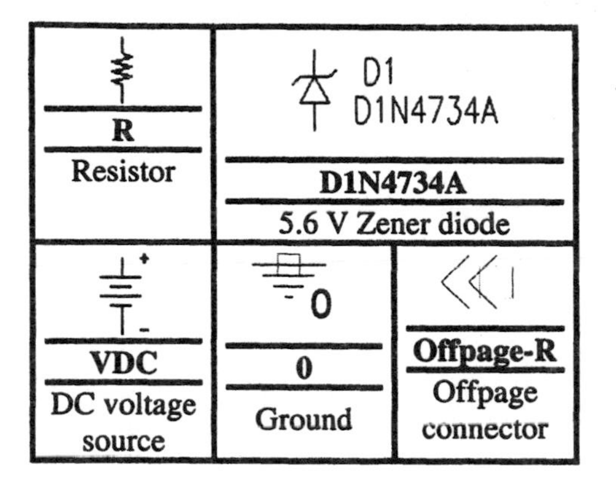


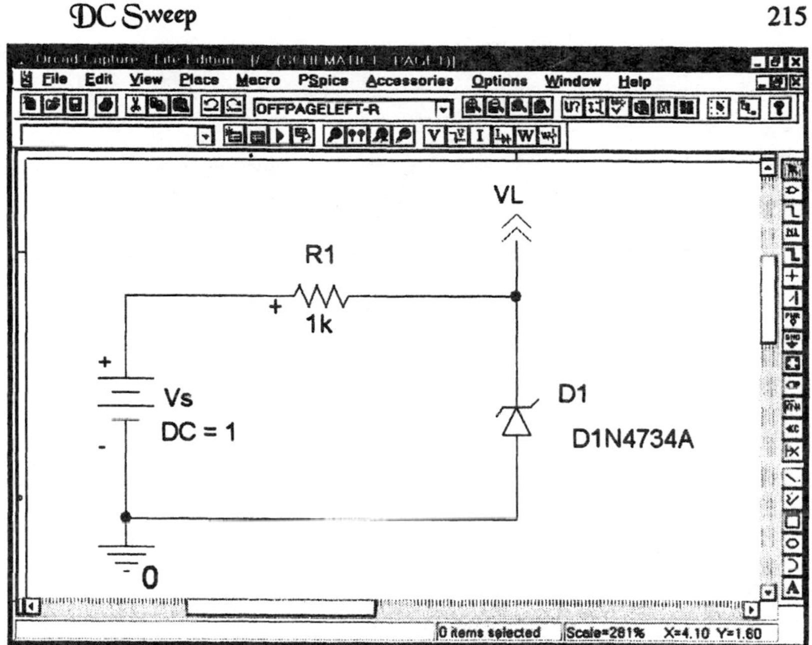


This circuit is also simulated in Section 6.F.1. We would like to sweep ***Vs*** from −15 volts to +15 volts and plot the output. Select **PSpice** and then **New Simulation Profile** from the Capture menus, select a name for the profile, and then click the ***Create*** button. Select the ***DC Sweep Analysis type*** and fill in the dialog box as shown:

General | Analysis | Include Files | Libraries | Stimulus | Options | Data Collection | Probe Window

Analysis type: DC Sweep

Options: Secondary Sweep, Monte Carlo/Worst Case, Parametric Sweep, Temperature (Sweep), Save Bias Point, Load Bias Point

Sweep variable: Voltage source, Current source, Global parameter, Model parameter, Temperature — Name: Vs; Model type; Model name; Parameter name

Sweep type: Linear, Logarithmic (Decade), Value list — Start value: -15; End value: 15; Increment: 0.01

OK | Cancel | Apply | Help

The dialog box is set to sweep ***Vs*** from ***-15*** volts to ***+15*** volts in ***0.01*** volt steps. Click the ***OK*** button to return to the schematic. Run the analysis by selecting **PSpice** and then **Run**:

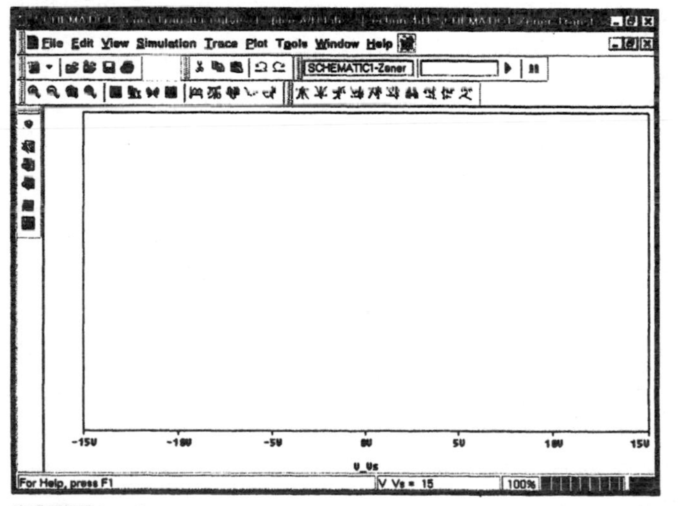


Add the trace `V(VL)` (press the **INSERT** key):

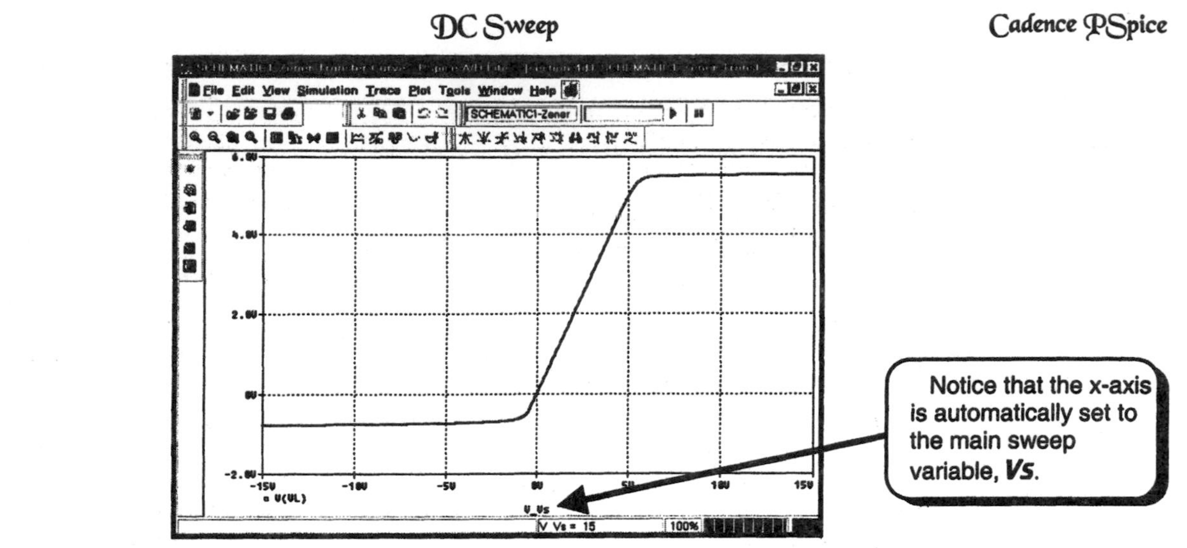

Notice that the x-axis is automatically set to the main sweep variable, *Vs*.
V(VL)
V_Vs


The screen capture above is the transfer curve for this circuit.

**EXERCISE 4-5**: Find the transfer curve of the circuit below:

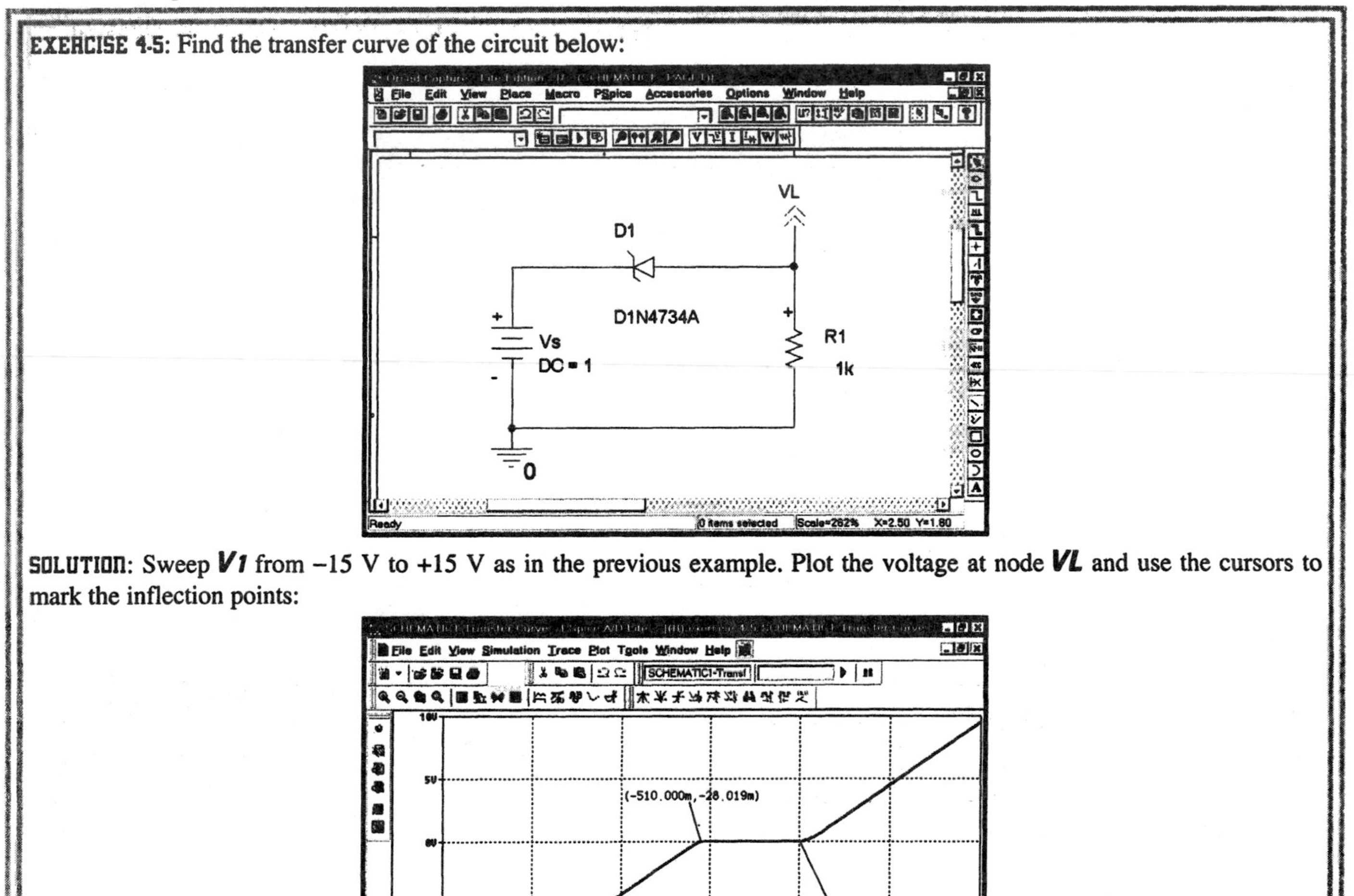


**SOLUTION**: Sweep ***V1*** from –15 V to +15 V as in the previous example. Plot the voltage at node ***VL*** and use the cursors to mark the inflection points:

Notice that the output is zero for input voltages from approximately –0.5 V to 5.5 V.

## 4.D.2. Secondary Sweep — Family of Transfer Curves

In the previous section, the question may arise as to how the Zener breakdown voltage affects the transfer curve. We will assume that you have followed the procedure of the previous section and have already set up the DC Sweep.

We will now set up a Secondary Sweep to sweep the breakdown voltage of the Zener diode. First we must look at the model for the Zener diode. Click the ***LEFT*** mouse button on the graphic symbol for the Zener, ⊣▷⊢. It should turn pink, indicating that it has been selected. Next, select **Edit** and then **PSpice Model** from the Capture menus:*

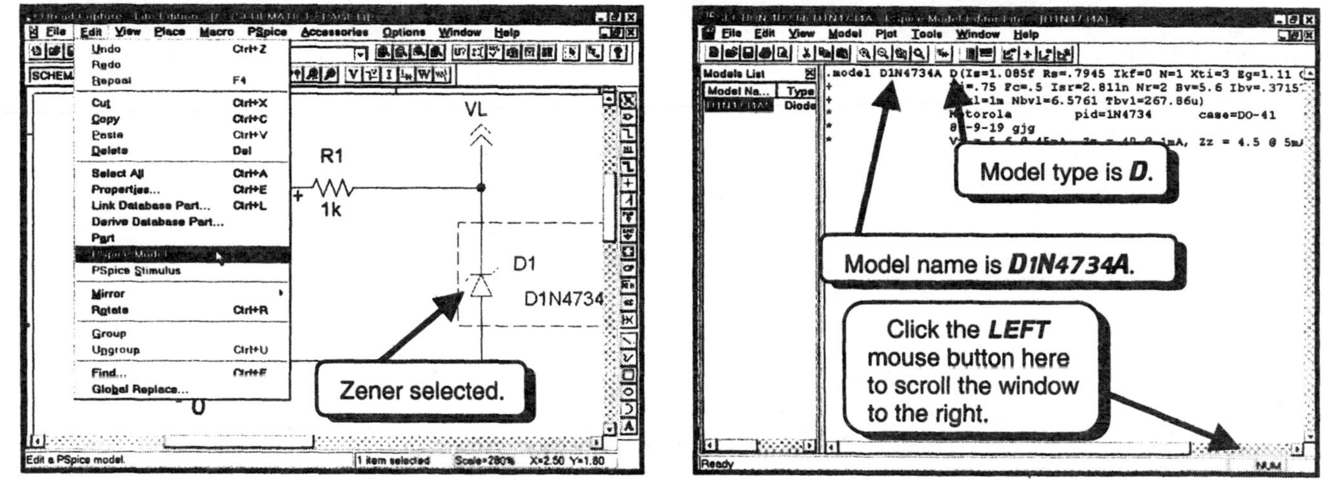


This screen shows that the model name is ***D1N4734A***. The parameter for breakdown voltage is Bv and cannot be seen on the screen. Scroll the text window to the right to see more of the model:

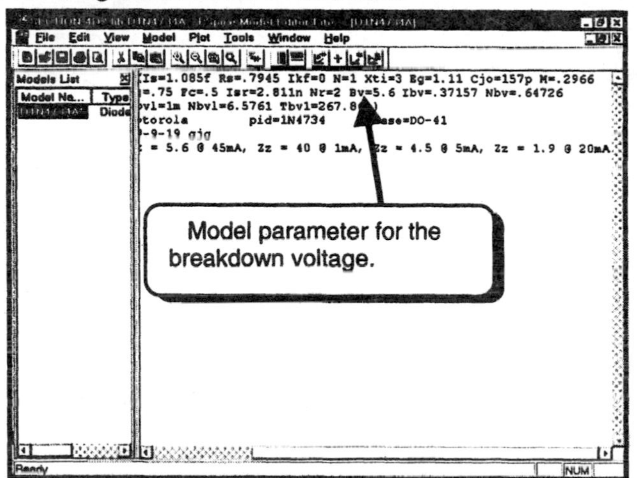


We see that the breakdown voltage is ***5.6*** volts. Select **File** and then **Exit** to close the model editor and return to Capture. Do not save the changes if asked.

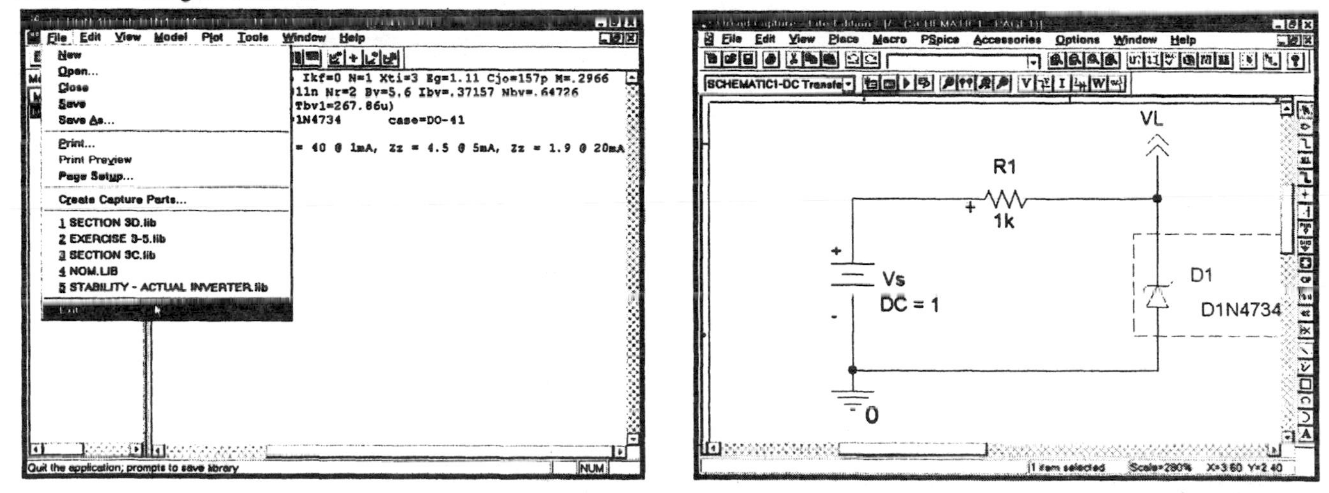

*If the menu selection **PSpice Model** appears grayed out in the menu (**PSpice Model**), you will not be able to choose the **PSpice Model** menu selection. Attempt to select the Zener graphic again until its graphic, ⊣▷⊢, is highlighted in pink.

Since we are working with the previous example, a simulation profile has already been created and we just need to modify it. To edit the existing profile, select **PSpice** and then **Edit Simulation Profile**:

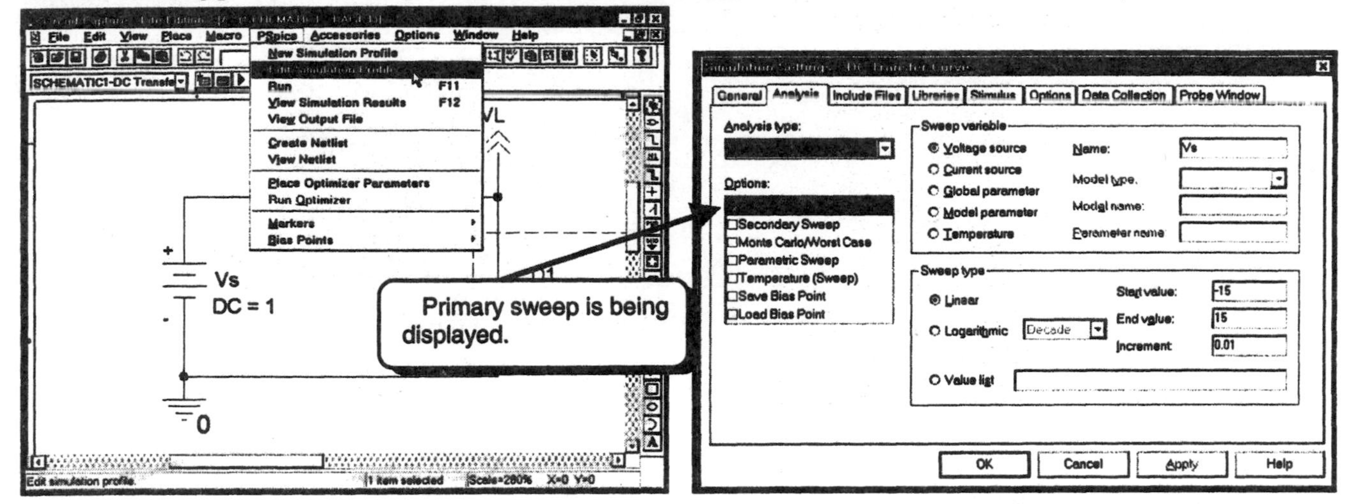


The dialog box is displaying the Primary Sweep, which we set up from the last section. We would like to set up a Secondary Sweep for this simulation. Click the ***LEFT*** mouse button on the □ next to the text ***Secondary Sweep*** to enable the sweep:

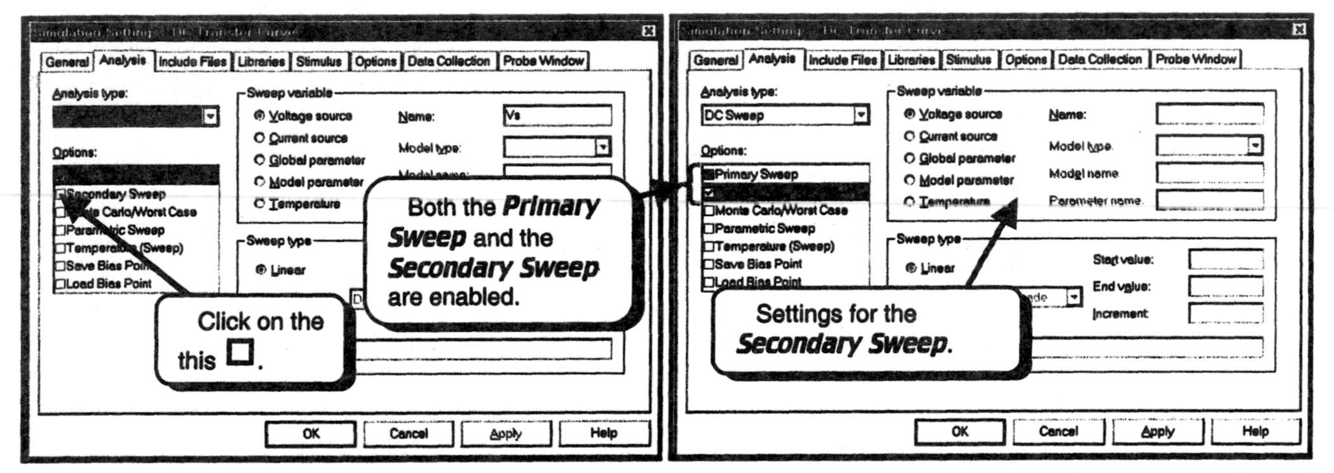


We would like to sweep the breakdown voltage of D1. As in the model shown on page 217, the model type was a diode (D), the model name was D1N4734A, and the parameter we wish to vary is BV. Fill in the dialog box as shown:

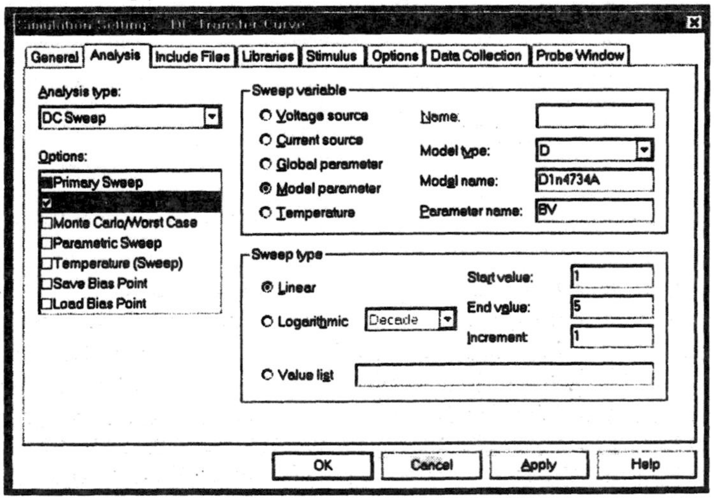

The box is set to sweep the model parameter ***BV*** from ***1*** volt to ***5*** volts in ***1*** volt steps. Click the ***OK*** button to the schematic. Run PSpice (select **PSpice** and then **Run** or press the **F11** key):

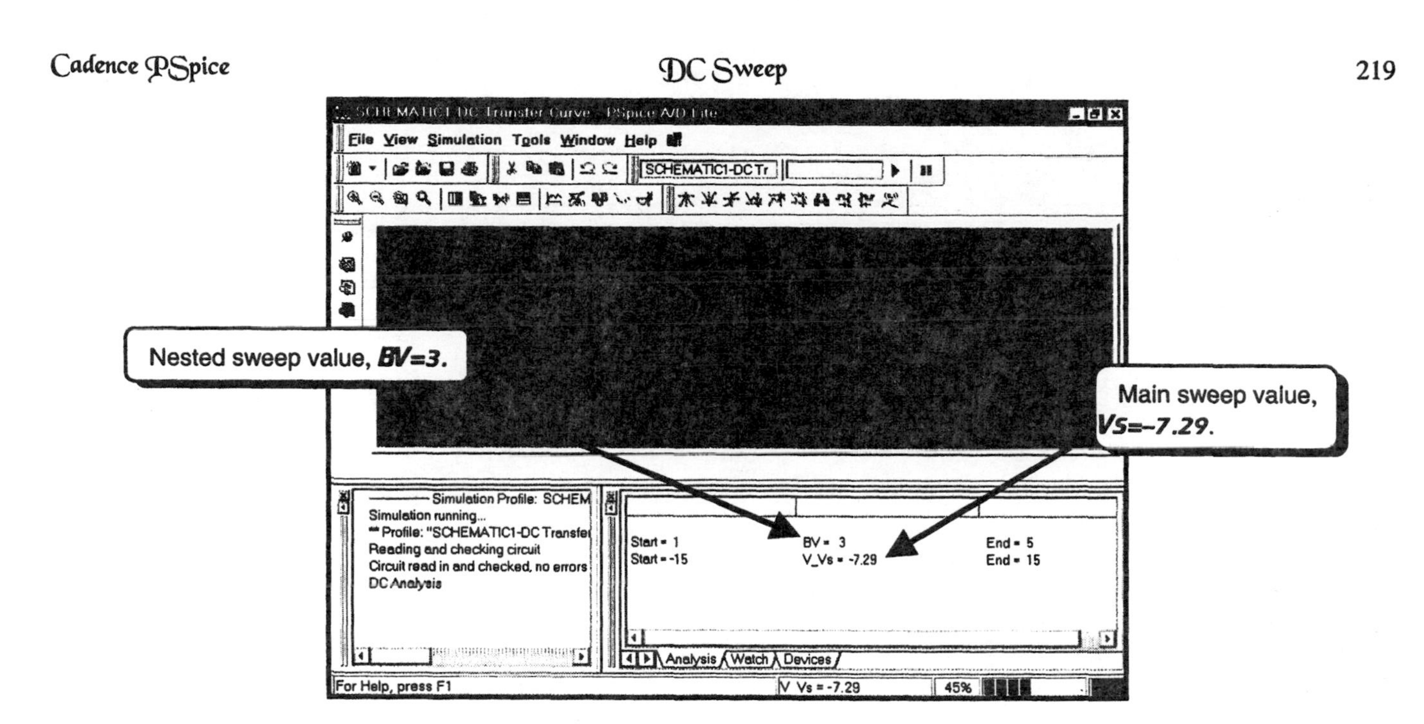


Notice that two values are displayed in the Simulation Status window. The bottom line changes quickly and is the main sweep variable, Vs. The line second to the bottom is the nested sweep variable, BV. Notice that for each value of BV, Vs is swept from −15 to +15. Logically, the primary sweep loop executes inside the secondary sweep loop. That is, BV is set to 1 and Vs is swept from −15 to +15. Then BV is set to 2, and Vs is swept from −15 to +15. Then BV is set to 3, and Vs is swept from −15 to +15, and so on. When the analysis is finished, Probe will run. Add the trace **V(VL)** (press the **INSERT** key):

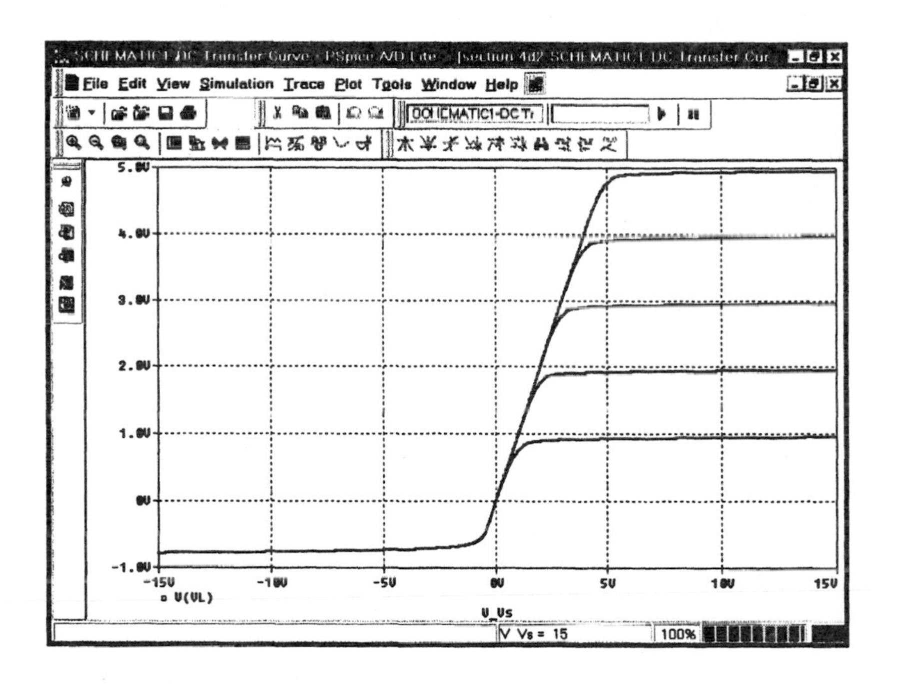


Using a Secondary Sweep, we cannot have Probe identify which trace occurs at which value of the breakdown voltage. We could have generated this same family of curves using a DC Sweep together with a Parametric Sweep. An example is shown for an NMOS inverter in Section 4.D.5. Follow the procedure of Section 4.D.5 and set the DC Sweep variable to Vs and the Parametric Sweep to the Model parameter BV. You can then follow the procedure at the end of Section 4.D.5 to identify which trace is associated with which value of the breakdown voltage (BV).

**EXERCISE 4-6**: For the circuit below, find the nested family of transfer curves if the breakdown voltage of the Zener is swept from 1 V to 6 V.

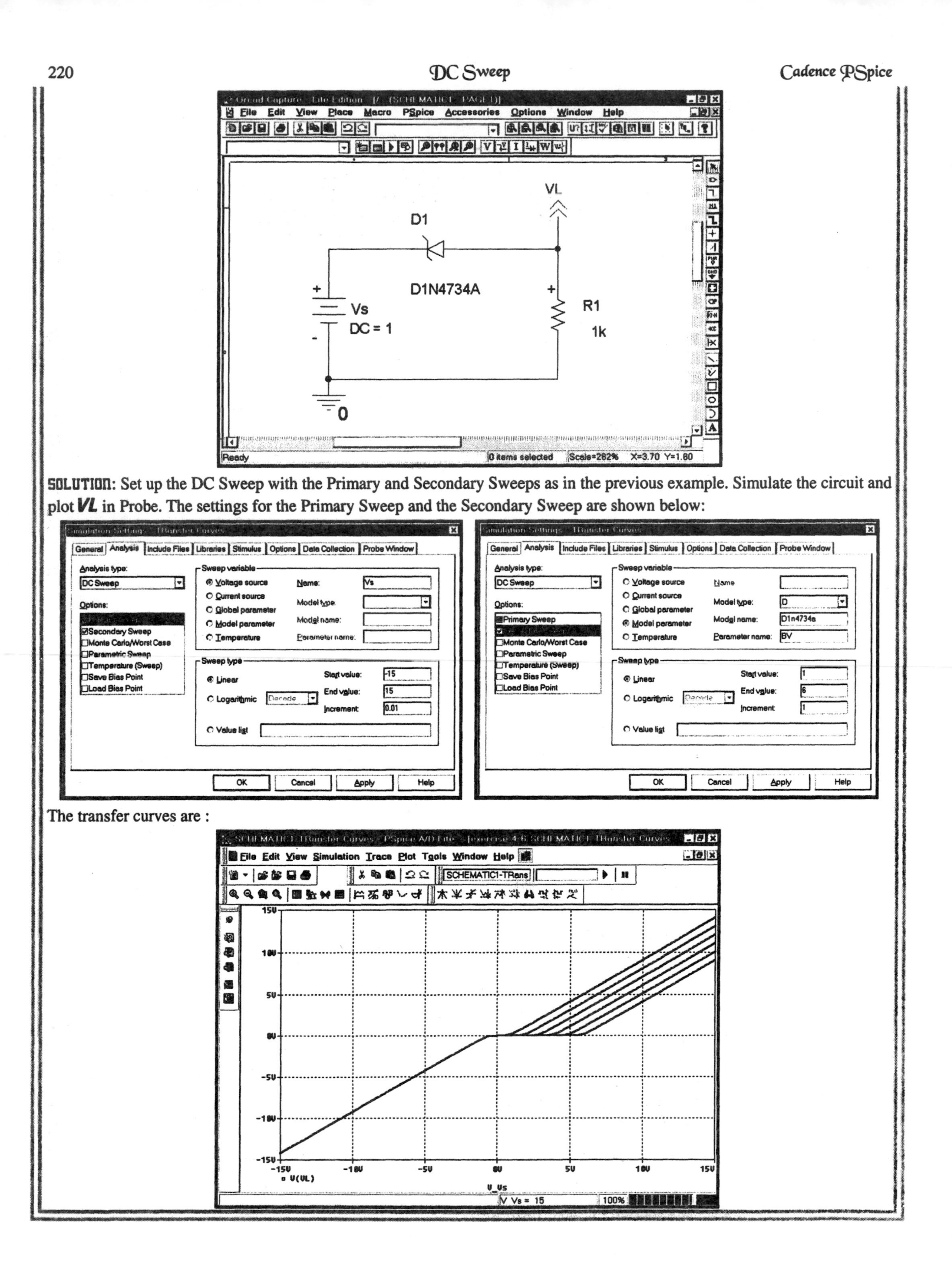


**SOLUTION:** Set up the DC Sweep with the Primary and Secondary Sweeps as in the previous example. Simulate the circuit and plot ***VL*** in Probe. The settings for the Primary Sweep and the Secondary Sweep are shown below:

The transfer curves are :

# 4.D.3. NMOS Inverter Transfer Curve

We would like to plot the transfer curve ***Vo*** versus ***Vin*** for the NMOS inverter below:

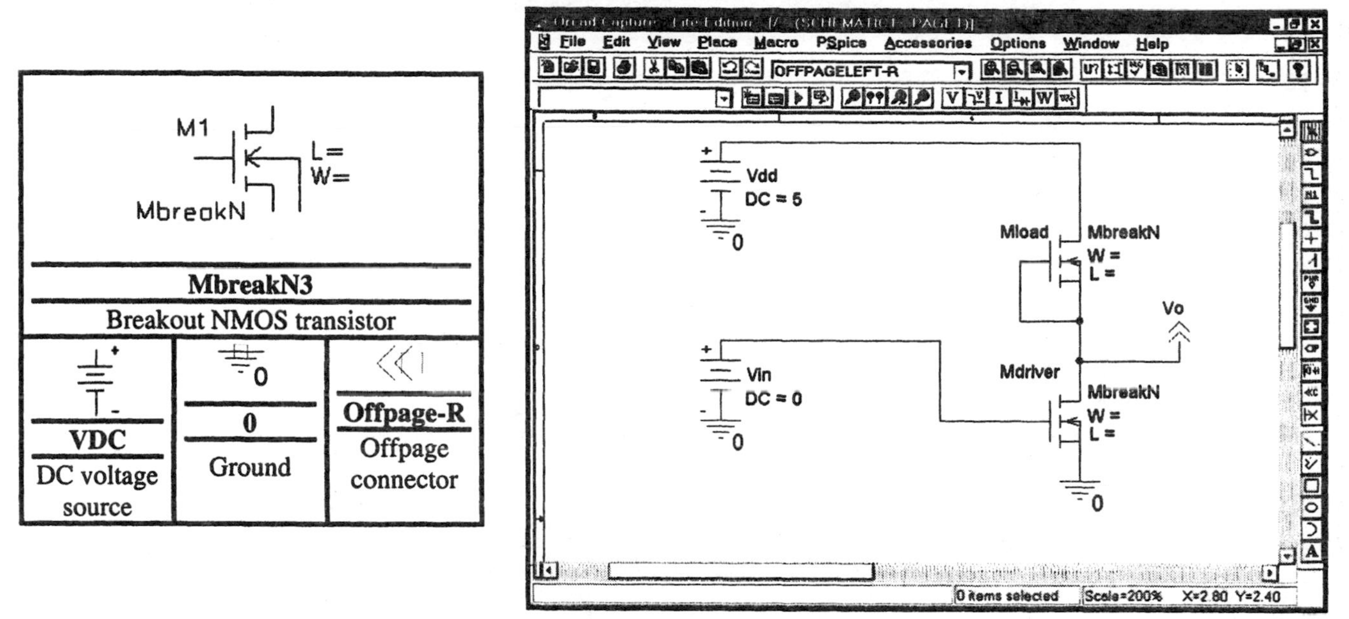


This circuit is an NMOS inverter with a depletion load. Many textbooks describe MOSFET operation by the following equations:

$$I_D = \left(\frac{W}{L}\right) K \left(V_{GS} - V_T\right)^2 \left(1 + \lambda V_{DS}\right) \qquad \text{saturation region}$$

$$I_D = \left(\frac{W}{L}\right) K \left[2\left(V_{GS} - V_T\right)V_{DS} - V_{DS}^2\right]\left(1 + \lambda V_{DS}\right) \qquad \text{linear region}$$

For our example, the load MOSFET is a depletion-mode NMOS transistor. Its parameters are $K = 20\ \mu A/V^2$, $V_T = -1.5$ V, and $\lambda = 0.05\ V^{-1}$. The driver is an enhancement-mode NMOS transistor. Its parameters are $K = 20\ \mu A/V^2$, $V_T = +1.5$ V, and $\lambda = 0.05\ V^{-1}$.

First, we need to create models to describe these transistors. When you look up the MOSFET models in the Cadence PSpice reference manual on the CD-ROM that accompanies this text, you will find the following equations to describe MOSFET operation:

$$I_D = \left(\frac{W}{L}\right) \frac{Kp}{2} \left(V_{GS} - Vto\right)^2 \left(1 + \lambda V_{DS}\right) \qquad \text{saturation region}$$

$$I_D = \left(\frac{W}{L}\right) \frac{Kp}{2} \left[2\left(V_{GS} - Vto\right)V_{DS} - V_{DS}^2\right]\left(1 + \lambda V_{DS}\right) \qquad \text{linear region}$$

The conversion between the two sets of equations is shown in Table 4-1. The PSpice model parameter names are close to the standard names used to represent MOSFET operation in many textbooks. One difference is that the PSpice model parameter Kp is twice the value of K. Thus, in our model we should set $K_P = 40\ \mu A/V^2$. All other model parameters will be the same.

| Table 4-1 | |
|---|---|
| **PSpice Model Parameter** | **Equation Variables** |
| Kp | 2K |
| Vto | $V_T$ |
| lambda | $\lambda$ |

We must now define the models for these two MOSFETs. Two separate models are required. Click the ***LEFT*** mouse button on the graphic symbol of the driver MOSFET, [MOSFET symbol]. It should turn pink, indicating that it has been selected.

Select **Edit** and then **PSpice Model:***

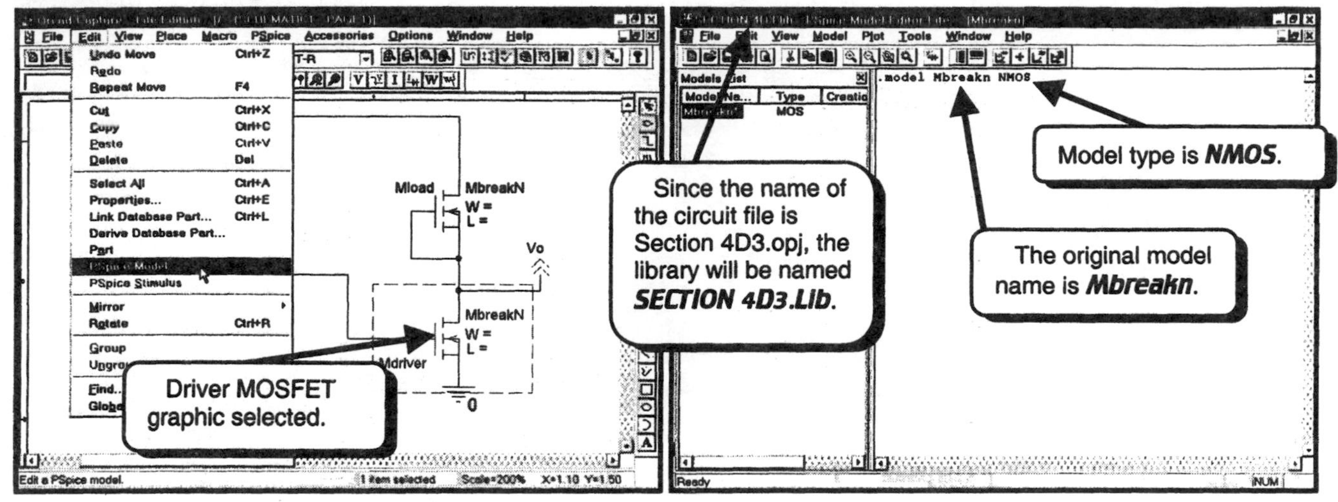


This is the default model for all n-channel MOSFETs (no model parameters specified). This window is a text editor. Change the model name as shown:

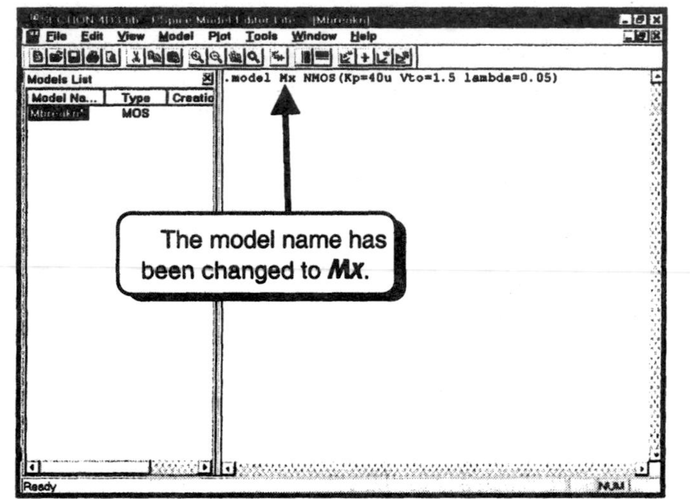


We are calling the new model ***Mx***. Select **File** and then **Save** to save the model in library ***SECTION 4D3.lib***:

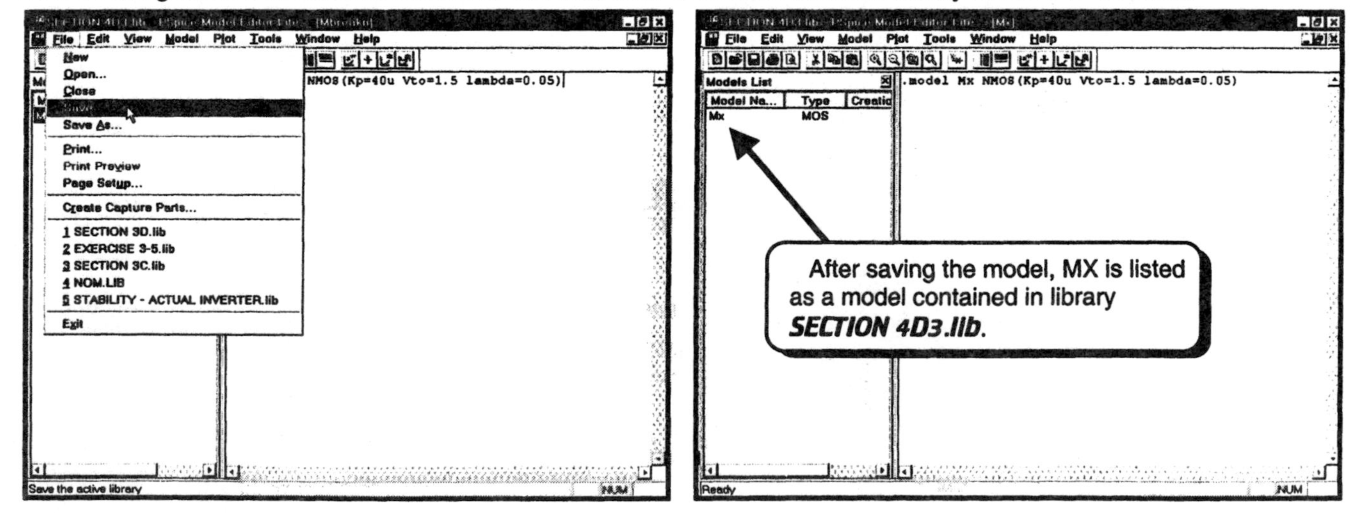


*If the menu selection **PSpice Model** appears grayed out in the menu (**PSpice Model**), you will not be able to choose the **PSpice Model** menu selection. Attempt to select the MOSFET graphic again until its graphic, is highlighted in pink.

After we save the library, model ***Mx*** is added to the list of models contained in library ***SECTION 4D3.lib***. Select **File** and then **Exit** to the schematic. Notice that the model for the driver MOSFET has changed from ***MbreakN*** to ***Mx***:

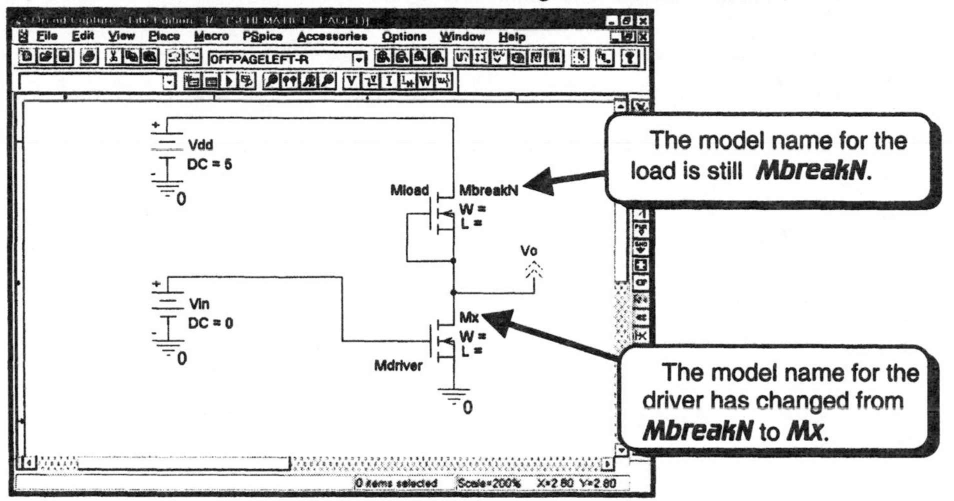


Next, we must change the model for the load MOSFET. Click the ***LEFT*** mouse button on the graphic symbol for the load MOSFET, . It should turn pink, indicating that it has been selected. Select **Edit** and then **PSpice Model**:

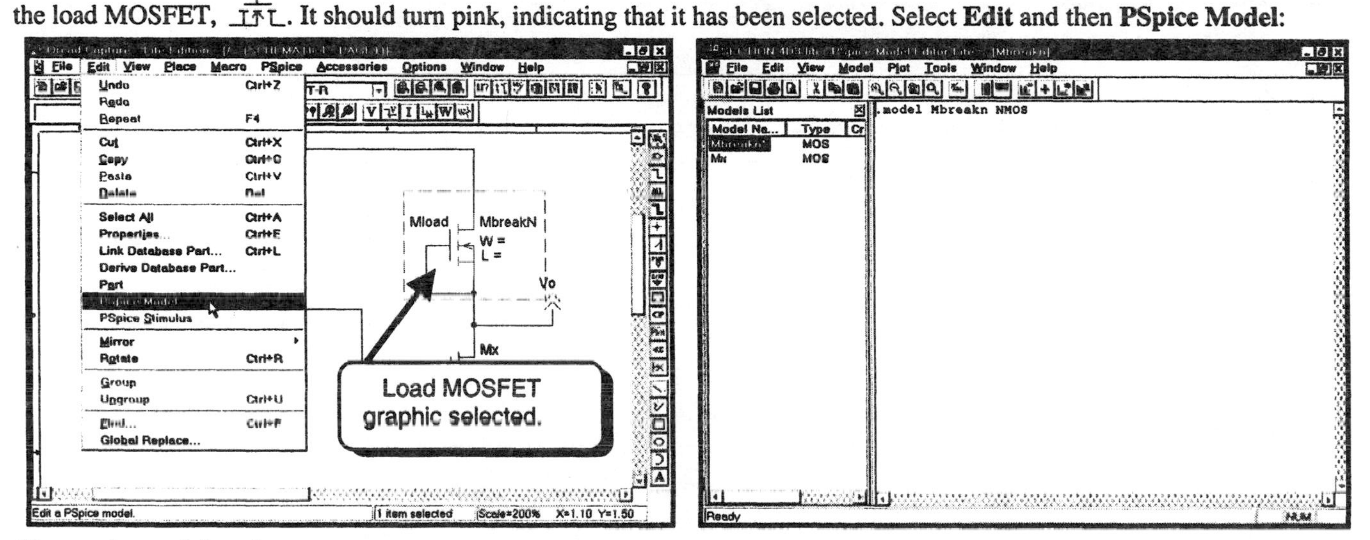


Change the model as shown:

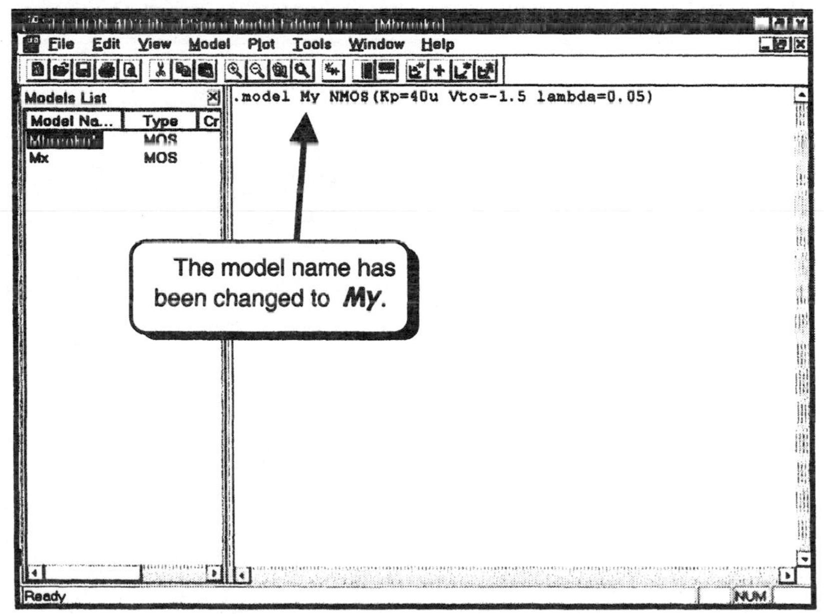

We are calling the new model ***My***. Note that the symbols for enhancement- and depletion-mode MOSFETs are the same. PSpice knows that a MOSFET is either an enhancement- or a depletion-mode MOSFET by the value of the threshold voltage parameter, ***Vto***.

Select File and then Save to save the model:

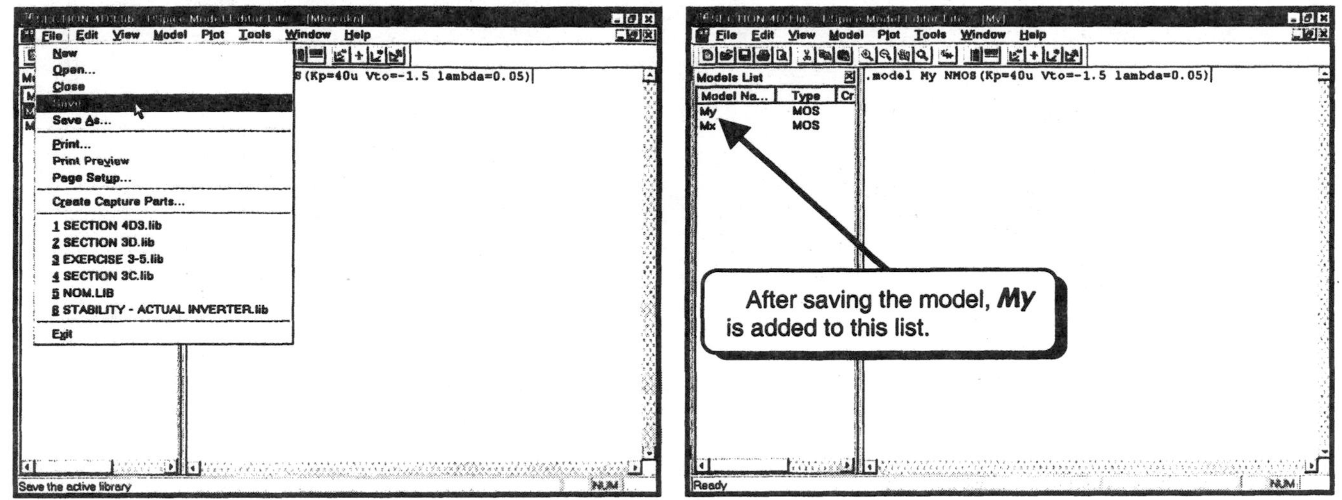


After saving the model, we see that the library now contains two models, ***My*** and ***Mx***. Select File and then Exit to close the model editor and return to the schematic. Notice that the model for the driver MOSFET has changed from MbreakN to ***My***:

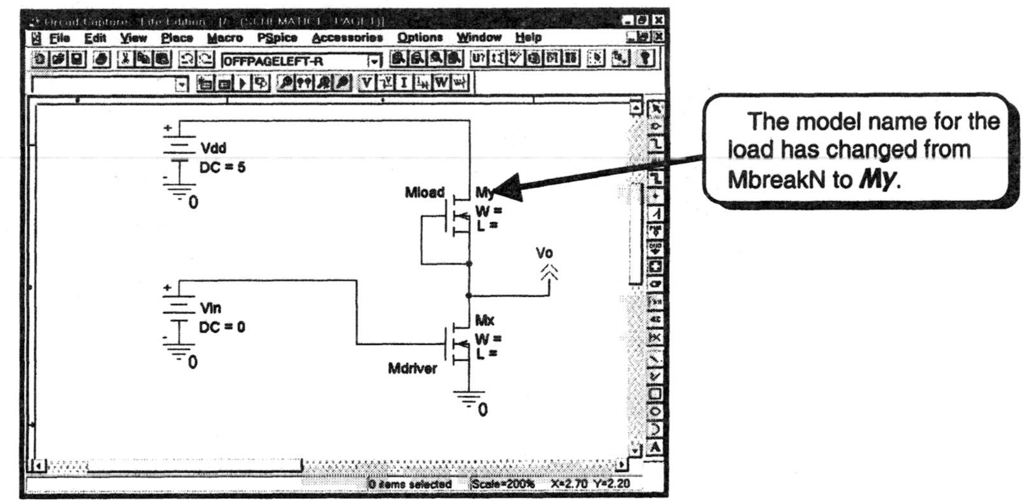


The last thing we need to do is to specify the widths and lengths of the transistors. You can do this by double-clicking on the text ***W=*** or ***L=*** and then changing their values. The units for the width and length are in meters so you usually need to specify lengths with a u suffix. Change all the widths and lengths to 1u (1u specifies 1 micron):

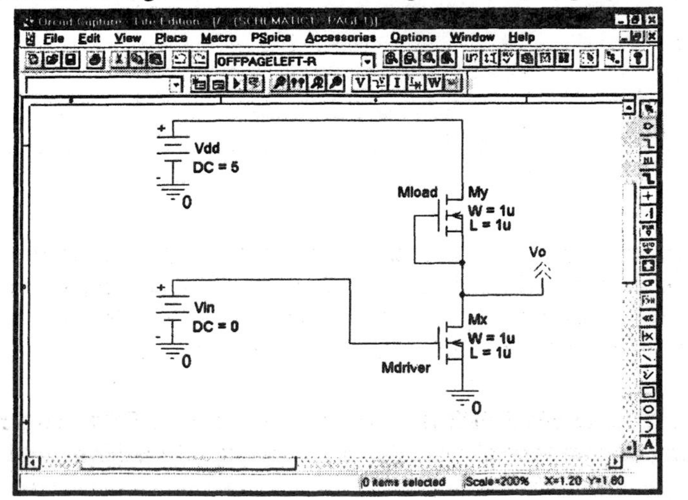

We are now ready to set up the DC Sweep. Select **PSpice** and then **New Simulation Profile**, give the profile a name, and then click the ***Create*** button. Select an analysis type of ***DC Sweep***. We would like to sweep Vin from 0 volts to 5 volts. Fill in the dialog box as shown:

Simulation Settings - MOSFET Inverter Transfer Curve

General | Analysis | Include Files | Libraries | Stimulus | Options | Data Collection | Probe Window

Analysis type: DC Sweep

Options:
- ☒ Primary Sweep
- ☐ Secondary Sweep
- ☐ Monte Carlo/Worst Case
- ☐ Parametric Sweep
- ☐ Temperature (Sweep)
- ☐ Save Bias Point
- ☐ Load Bias Point

Sweep variable
- ⦿ Voltage source — Name: Vin
- ○ Current source — Model type:
- ○ Global parameter — Model name:
- ○ Model parameter — Parameter name:
- ○ Temperature

Sweep type
- ⦿ Linear — Start value: 0
- ○ Logarithmic (Decade) — End value: 5
- Increment: 1m
- ○ Value list

OK | Cancel | Apply | Help

The dialog box is set to sweep ***Vin*** from 0 volts to 5 volts in 0.001 volt steps (1m = 0.001). A linear sweep means that points will be equally spaced. Click the ***OK*** button to accept the settings. Run PSpice (select **PSpice** and then **Run** or press the **F11** key). When the simulation is complete add the trace `V(Vo)` (select **Trace** and then **Add Trace**). The transfer curve is:

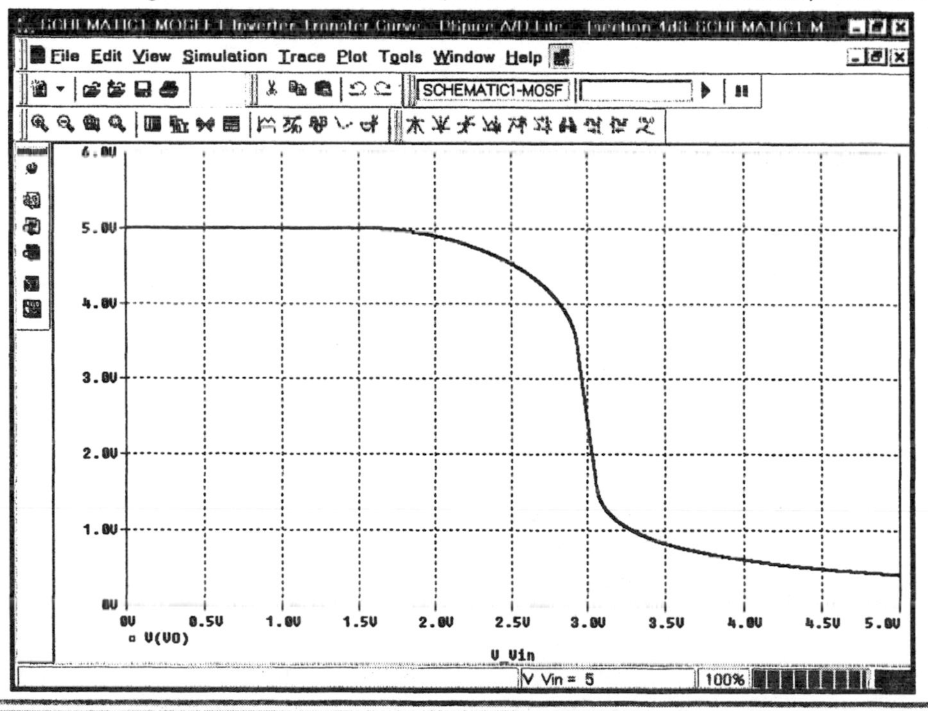


**EXERCISE 4-7**: Find the transfer curve for a CMOS inverter. For the NMOS transistor, let Kp = 24 $\mu A/V^2$, Vto = 1.5 V, and lambda = 0.01 $V^{-1}$. For the PMOS transistor, let Kp = 8 $\mu A/V^2$, Vto = −1.5 V, and lambda = 0.01 $V^{-1}$. To compensate for the differences in the transistors' transconductance (Kp), let the W/L ratio of the PMOS be three times the W/L ratio of the NMOS.

**SOLUTION**: The part name for a three-terminal PMOS transistor is Mbreakp3. Draw the circuit below. The model of the PMOS transistor has been changed from Mbreakp to Mp and the model of the NMOS transistor has been changed from Mbreakn to Mn:

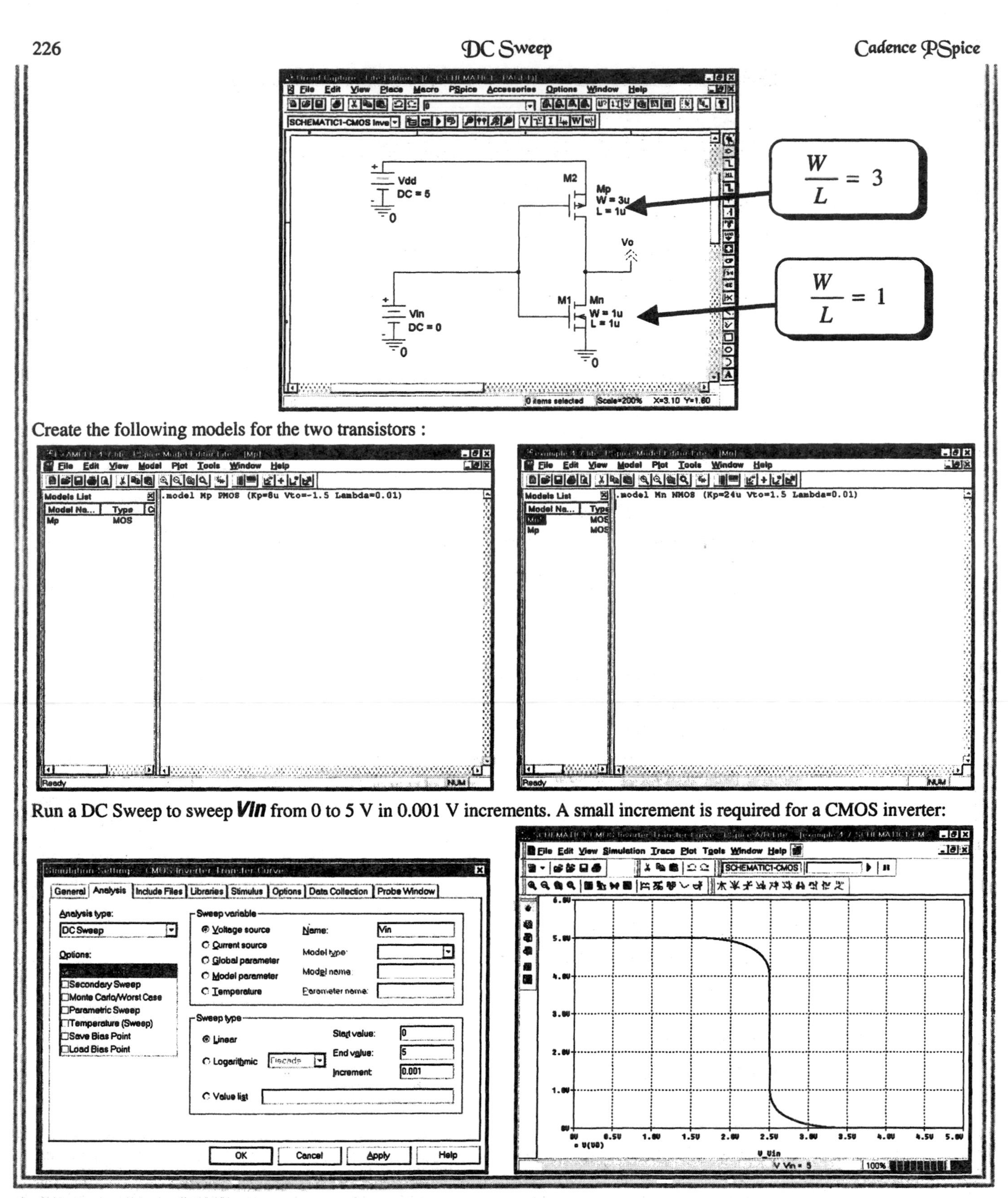


Create the following models for the two transistors :

Run a DC Sweep to sweep ***Vin*** from 0 to 5 V in 0.001 V increments. A small increment is required for a CMOS inverter:

# 4.D.4. Goal Functions — Inverter Analysis

We will use the circuit of the previous section and find the noise margins ($NM_H$ and $NM_L$) and input and output transition voltages ($V_{IL}$, $V_{IH}$, $V_{OL}$, $V_{OH}$). The transfer function of the NMOS inverter of the previous section was found on page 225, and is repeated below with approximate values of $V_{IL}$, $V_{IH}$, $V_{OL}$, $V_{OH}$ shown on the plot:

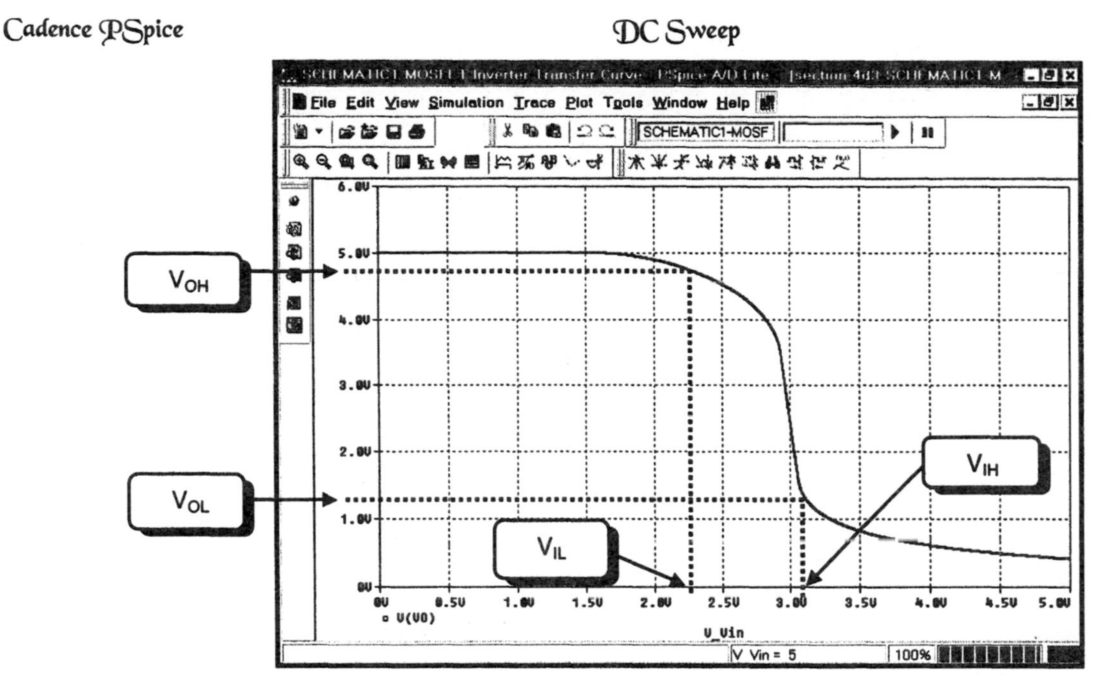


The definitions of $V_{IL}$, $V_{IH}$, $V_{OL}$, $V_{OH}$, $NM_H$, and $NM_L$ are:

- $V_{IL}$ - The maximum value of the input voltage that is considered a logic zero (low input). For $0 \le V_{IN} \le V_{IL}$, the input is considered a logic zero.
- $V_{IH}$ - The minimum value of the input voltage that is considered a logic one (high input). For $V_{IH} \le V_{IN} \le V_{max}$, the input is considered a logic one.
- $V_{OL}$ - The maximum output voltage of the gate that is considered to be a logic zero (low output).
- $V_{OH}$ - The minimum output voltage of the gate that is considered to be a logic one (high output).
- $NM_H$ - High input noise margin. $NM_H = V_{OH} - V_{IH}$.
- $NM_L$ - Low input noise margin. $NM_L = V_{IL} - V_{OL}$.

How to locate $V_{IL}$, $V_{IH}$, $V_{OL}$, and $V_{OH}$ for a particular gate is not always consistent and may vary for different types of logic. One definition of the location of the points $V_{IL}$, $V_{IH}$, $V_{OL}$, and $V_{OH}$ is where the slope of the transfer curve is one, $dV_O/dV_{IN} = 1$. We will use this definition in this example. The question is, how can we find these points on the graph? We can do this by using the goal functions available with Probe. The available goal functions are located in file C:\Program Files\OrcadLite\PSpice\COMMON\PSpice.prb. If you edit this file using the Windows Notepad, you will see the text below near the bottom of the file:

```
Voh(1,2)=y2
* This function finds Voh for a digital logic gate.
* The transition points are defined as where dVo/dVin = 1.
* Usage: Voh(d(V(Vo), V(Vo))
{
 1| search forward level (-1) !1;
 2| search forward xvalue(x1) !2;
}
```

The name of the function is **Voh**. It has 2 input arguments **(1,2)**. The first input will be the derivative of the trace, and the second input will be the trace itself. To use this function we would type Voh(d(V(Vo)), V(Vo)), where d(V(Vo)) is the derivative of the output trace and V(Vo) is the output trace. **1|search forward level** means search the first input forward and find a level. The level we are looking for is **-1**. When the point is found, the text **!1** designates its coordinates as x1 and y1. Since the first input is the derivative of the trace, **1| search forward level (-1) !1** finds where the slope of the trace is –1. **2| search forward xvalue(x1) !2** searches the second input forward and finds the point when the x-coordinate is equal to **x1**. This point is where the slope of the transfer curve is –1. **!2** marks the coordinates of this point as x2 and y2. The function returns the y-coordinate of this point **Voh(1,2)=y2**, which is equal to $V_{OH}$.

A second function is:

```
Vil(1)=x1
* This function finds Vil for a digital logic inverter.
* The transition points are defined as where dVo/dVin = 1.
* Usage: Vil(d(V(Vo)))
{
1| search forward level (-1) !1;
}
```

This function is similar to Voh except that it returns the x-coordinate of the point where the slope is –1. The x-coordinate of the point where the slope is –1 is our definition of $V_{IL}$.

A third function is:

```
Vol(1,2)=y2
* This function finds Vol for a digital logic gate.
* The transition points are defined as where dVo/dVin = 1.
* Usage: Vol(d(V(Vo),V(Vo)))
{
1| search backward /End/ level (-1) !1;
2| search backward /End/ xvalue(x1) !2;
}
```

This function is used to find $V_{OL}$ and is similar to the $V_{OH}$ function. Note that $V_{OL}$ is also defined where the slope of the transfer curve is –1. To distinguish $V_{OH}$ and $V_{OL}$, this function finds the point starting from the end of the trace and searching backwards. There are two points on the trace where the slope is –1. This function finds the point closest to the end of the trace.

The goal functions are used in Probe. Follow the procedure of Section 4.D.3. When you obtain the Probe plot of the transfer curve V(Vo) versus V_Vin below, continue with this section:



Next, we would like to find the values of $V_{IL}$, $V_{IH}$, $V_{OL}$, $V_{OH}$, $NM_H$, and $NM_L$. Before we continue, we will instruct Probe to display the points used in the evaluation of goal functions. Select **Tools** and then **Options** from the Probe menu bar:

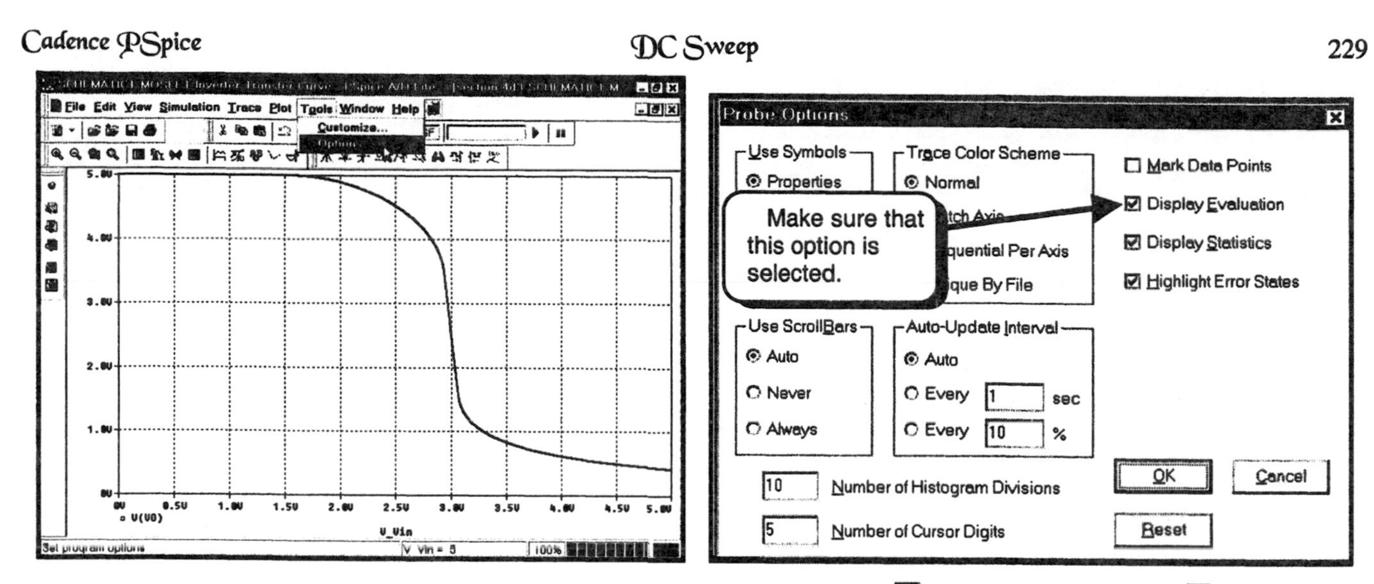


Make sure that the box next to the text ***Display Evaluation*** has a check in it ☑, as shown above. The ☑ indicates that the option is selected. If this option is not selected, you will get a different display than shown below. Click the ***OK*** button to accept the settings.

We will now continue with evaluating the goal function. Select **Window** and then **New Window** to create an empty window:

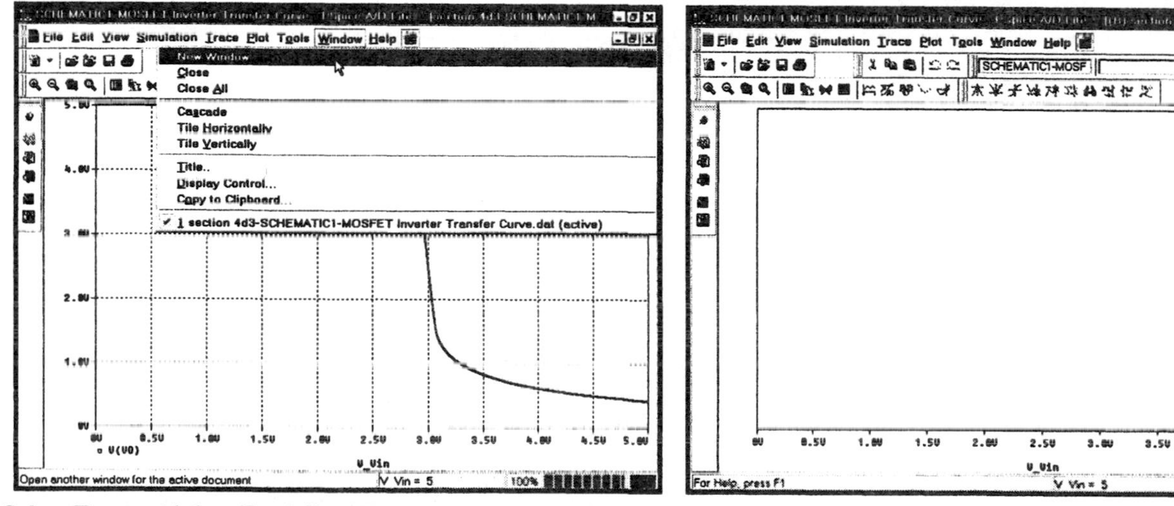


Select Trace and then **Eval Goal Function** from the menus:

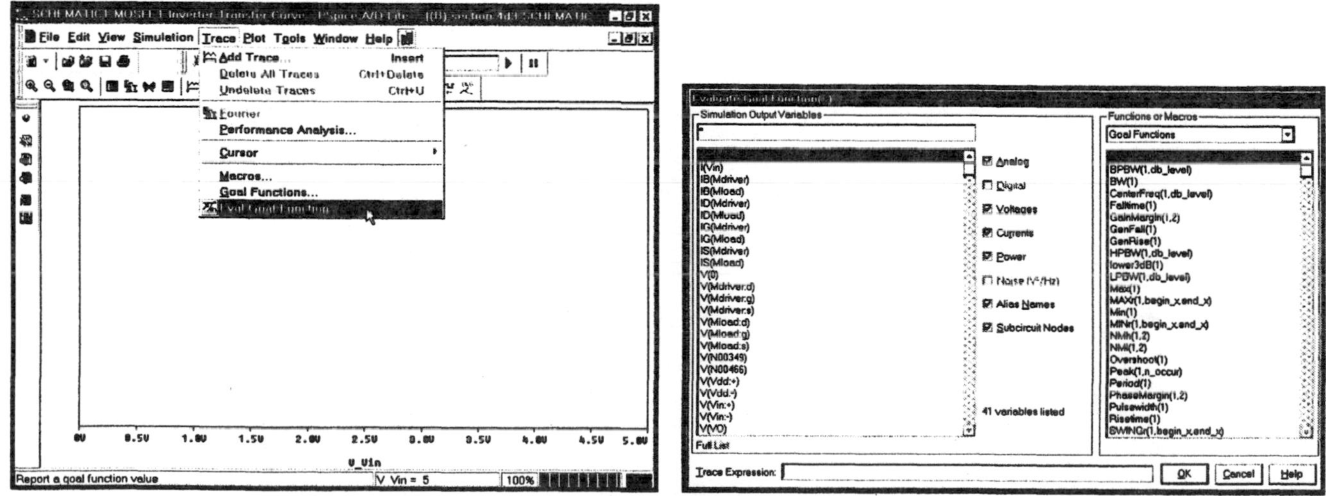


The left pane lists the voltage and current traces that we have seen previously in the Add Traces dialog box. The right pane lists the available goal functions. Click the ***LEFT*** mouse button on the vertical scroll bar in the right pane to view more goal functions:

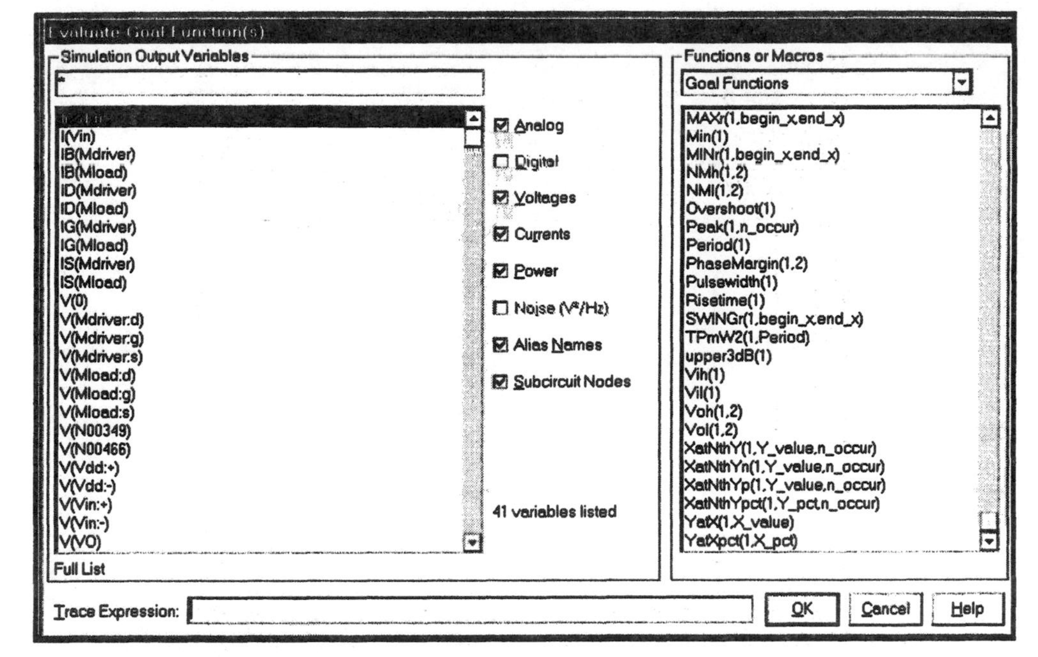


We see the goal functions discussed earlier on the right half of the dialog box. To obtain a value for $V_{OH}$, enter the ***Trace Expression*** `voh(d(V(Vo)),V(Vo))`

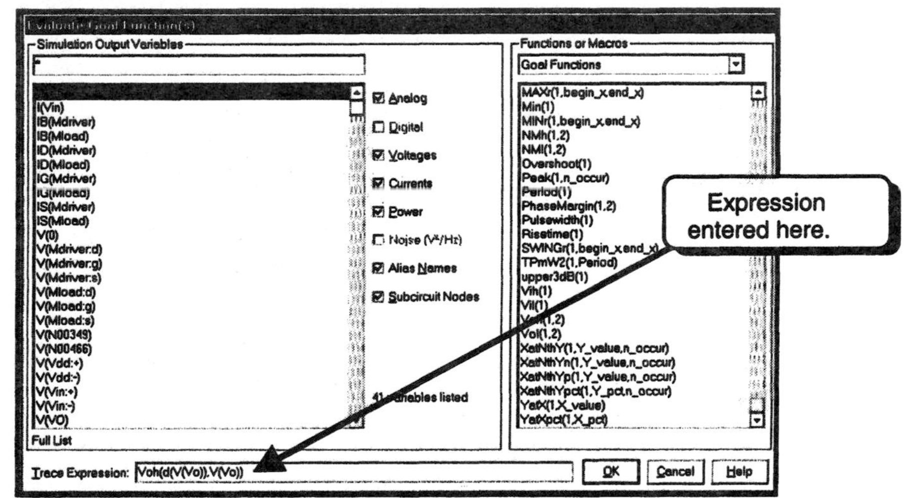


Click the ***OK*** button to view the value:

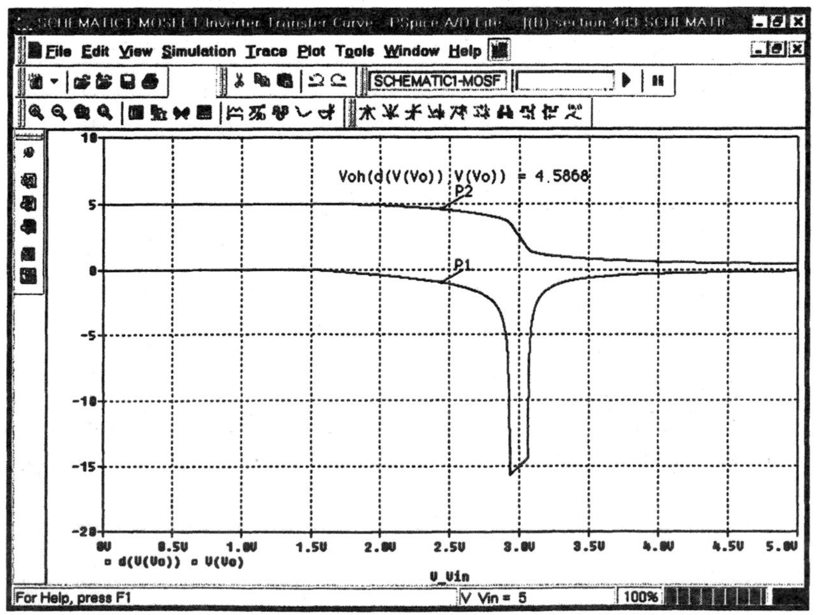

Probe draws a plot of the traces used in the goal function (***d(V(Vo))*** and ***V(Vo)*** in this example), labels the locations of the points used in the goal function (***P1*** and ***P2*** in this example), and then displays the value of the goal function. The value of $V_{OH}$ is ***4.5868*** volts. Select **Window** and then **New** to obtain a new Probe window:

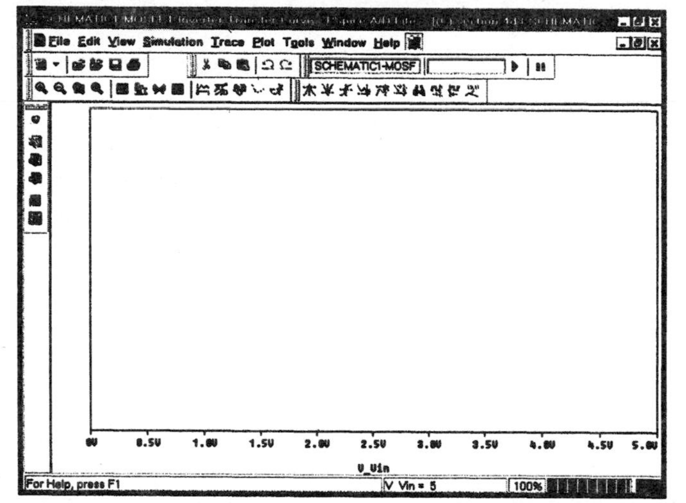


To find $V_{IL}$, select **Trace** and then **Eval Goal Function**, and enter the trace **Vil(d(V(Vo)))**:

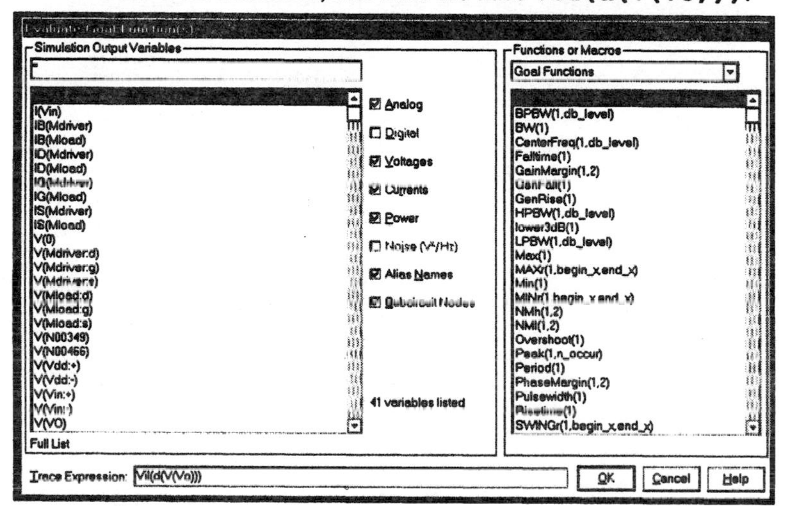


Click the ***OK*** button to view the value of $V_{IL}$:

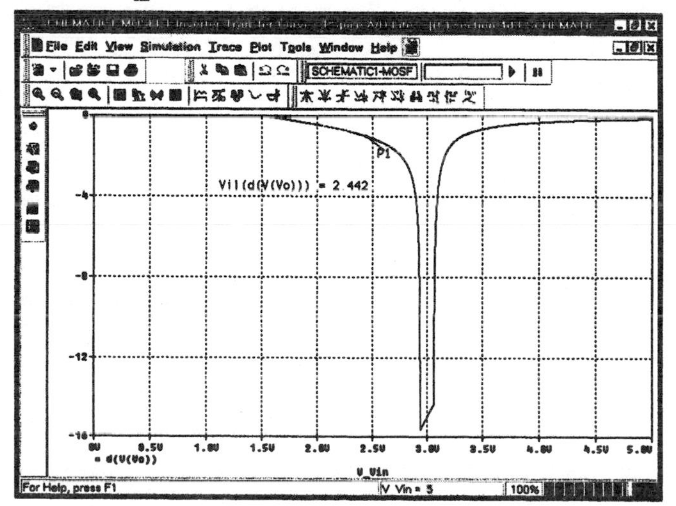


The value of $V_{IL}$ is ***2.442*** volts. Repeat the procedure above to find values for $V_{IH}$, $V_{OL}$, $NM_H$, and $NM_L$. The values are summarized in Table 4-2.

Suppose we wanted to show the values of $V_{IL}$, $V_{IH}$, $V_{OL}$, and $V_{OH}$ on the Probe screen. This can be done using the cursors. First, select **Window** from the menus and then select the first plot we generated to display a trace of V(Vo) by itself on the Probe window:

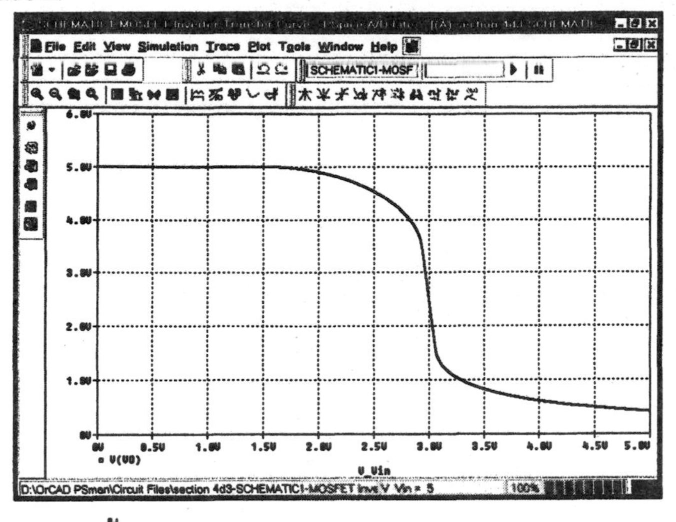

| Table 4-2 | |
|---|---|
| | Value |
| $V_{IH}$ | 3.32 V |
| $V_{OL}$ | 0.966 V |
| $NM_H$ | 1.27 V |
| $NM_L$ | 1.48 V |

Click the cursor button in the Probe toolbar to display the cursors:

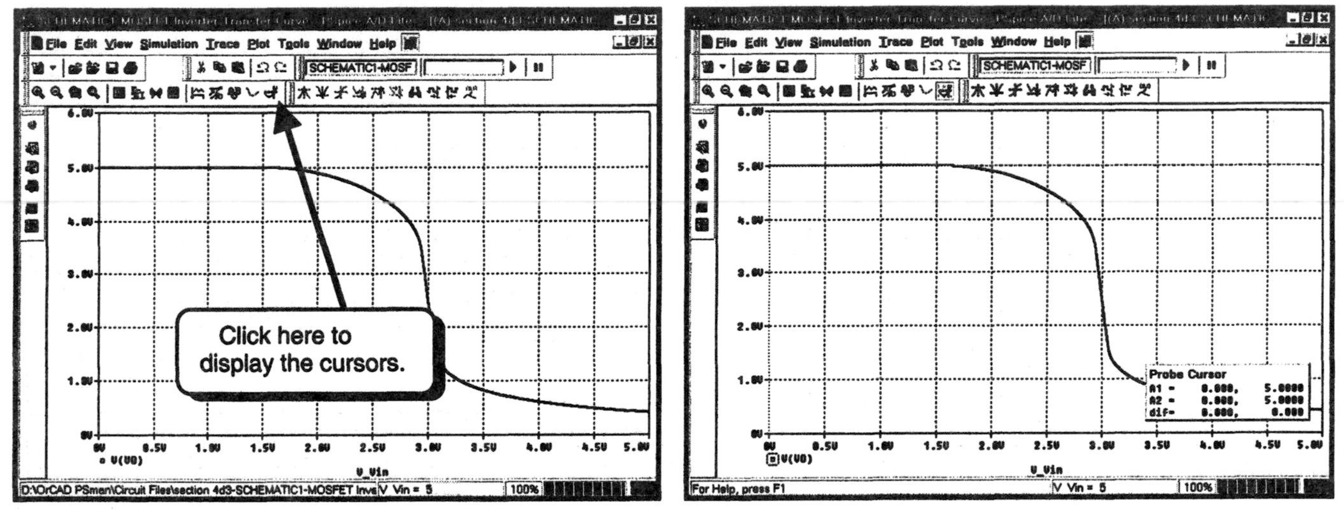


Crosshairs appear on the Probe screen, and a dialog box displays the coordinates of the cursors. We would like to mark the point $V_{OH}$ and $V_{IL}$. We will search for the x-coordinate $x = V_{IL} = 2.44$. Select **Trace, Cursor,** and then **Search Commands** from the Probe menus:

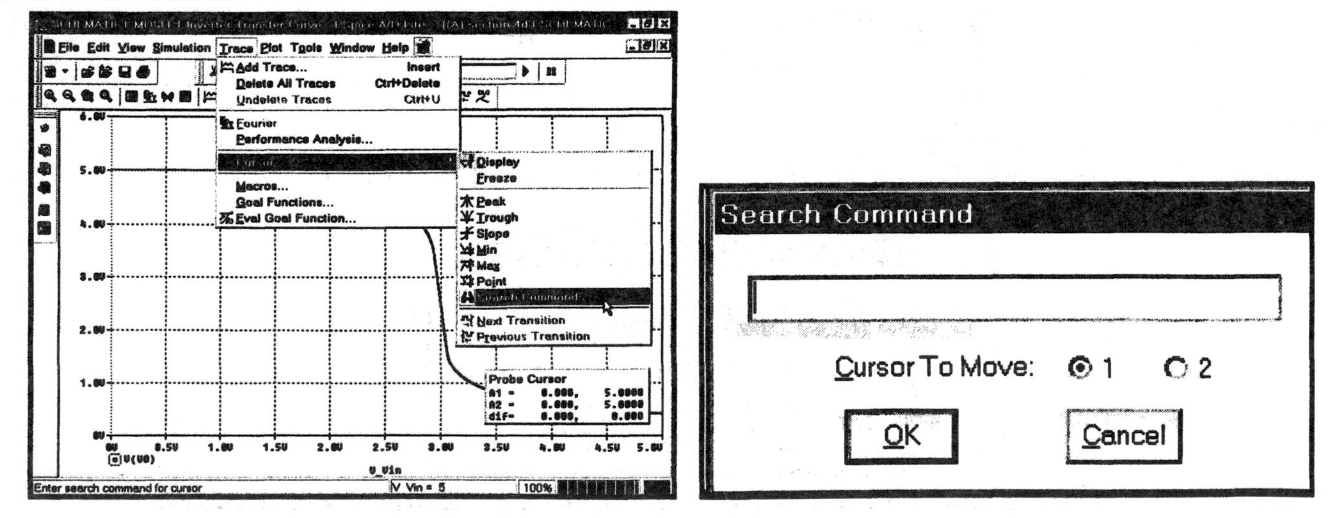

We will use a search command similar to the goal functions discussed earlier. Enter the text **`search forward xvalue(2.44201)`**:

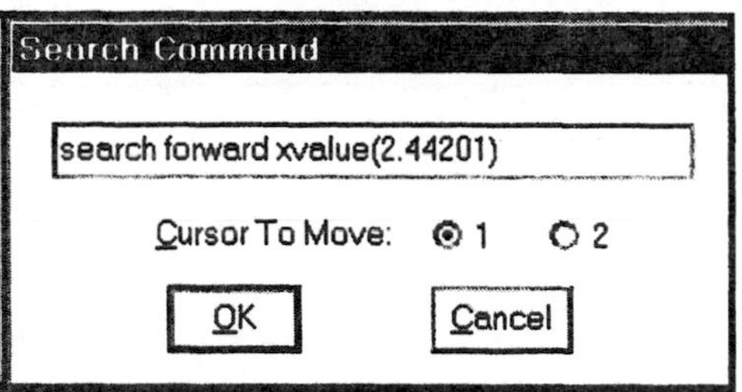


Click the ***OK*** button. The cursor will move to the coordinate x = 2.44201:

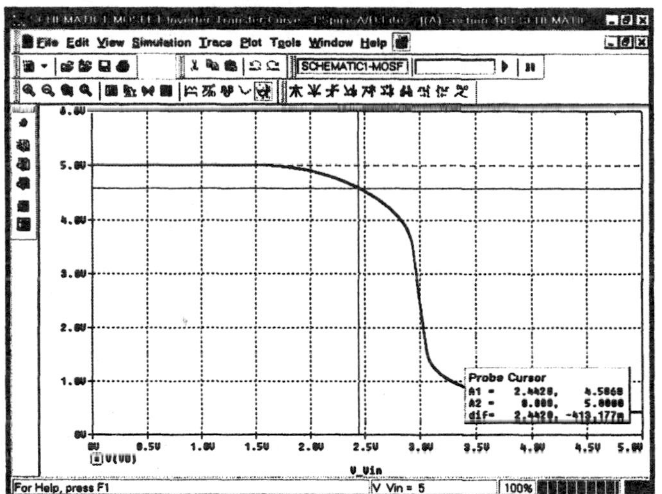


We would like to mark this point on the screen, so select **Plot**, **Label**, and then **Mark** from the Probe menus:

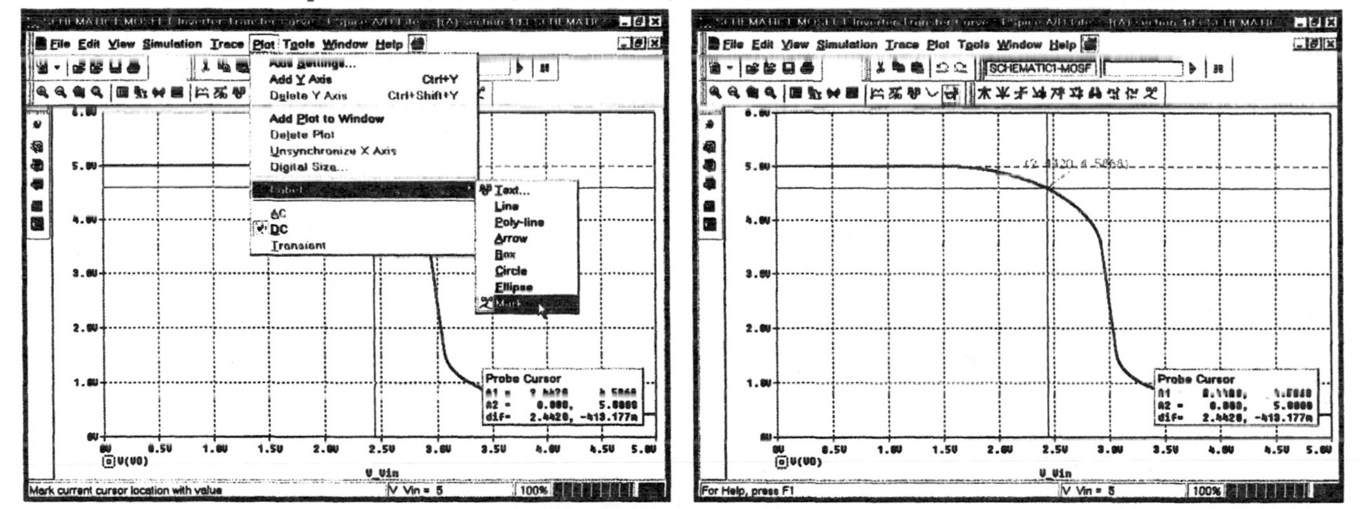


Next, we will mark the point $V_{OL}$ and $V_{IH}$. We will search for the x-coordinate x = $V_{IH}$ = 3.32. Select **Trace**, **Cursor**, and then **Search Commands** from the Probe menus:

Search Command

search forward xvalue(2.44201)

Cursor To Move: 1 2

OK Cancel

The dialog box contains our last search command. Modify the text to **`search forward xvalue(3.32)`**:

Search Command

search forward xvalue(3.32)

Cursor To Move: ◉ 1 ○ 2

OK Cancel

Click the ***OK*** button. The cursor will move to the coordinates of x = 3.32:

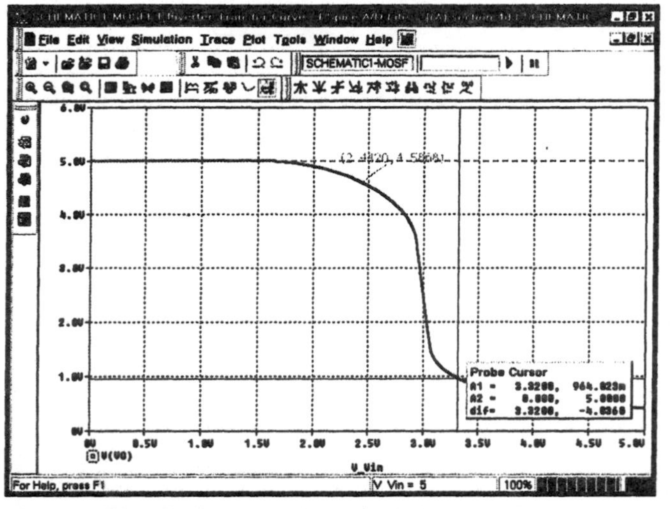


Mark the point on the screen by selecting **Plot**, **Label**, and then **Mark** from the Probe menus:

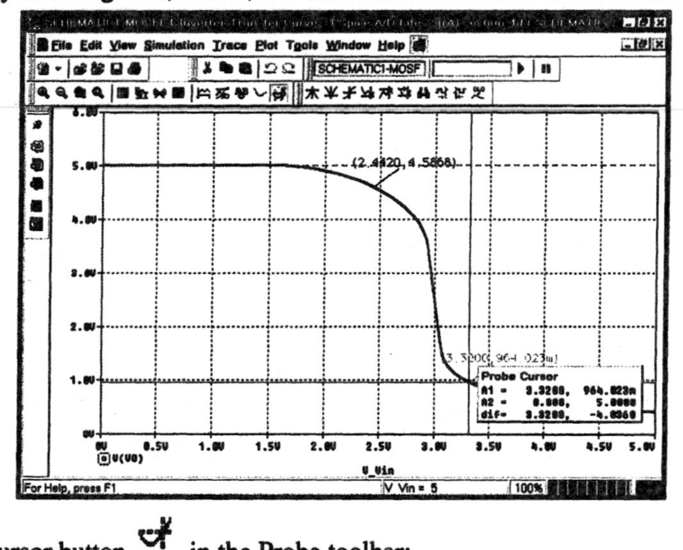


To hide the cursors, click the cursor button in the Probe toolbar:

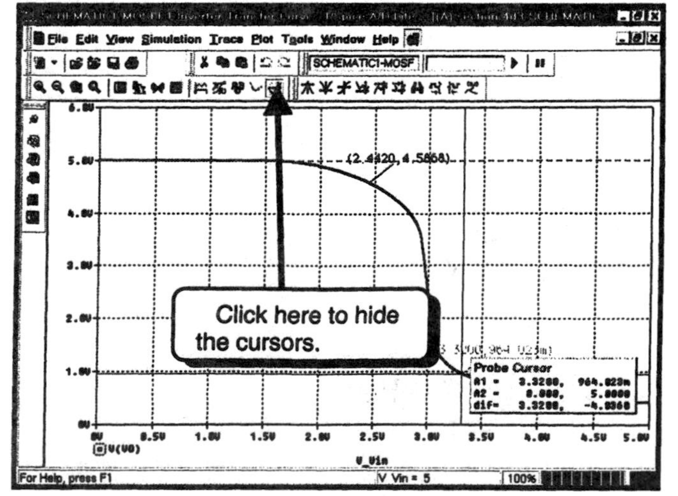


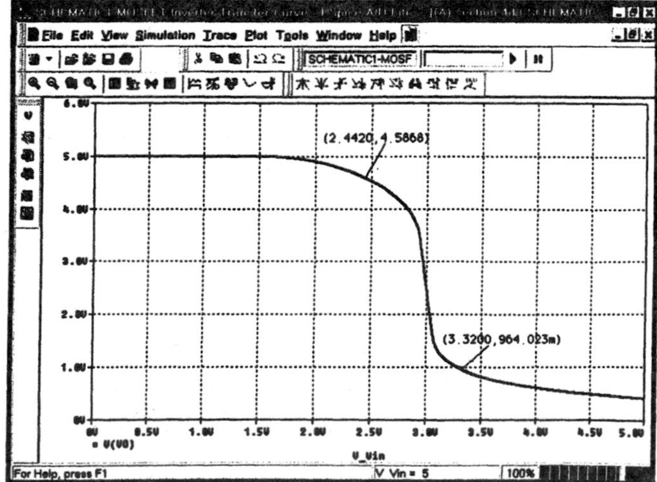

**EXERCISE 4-8**: Find values of $V_{IL}$, $V_{IH}$, $V_{OL}$, $V_{OH}$, $NM_H$, and $NM_L$ for the CMOS inverter of **EXERCISE 4-7**.

**SOLUTION**: $V_{IL}$ = 2.237 V, $V_{IH}$ = 2.763 V, $V_{OL}$ = 0.249 V, $V_{OH}$ = 4.751 V, $NM_H$ = 1.988 V, and $NM_L$ = 1.988 V.

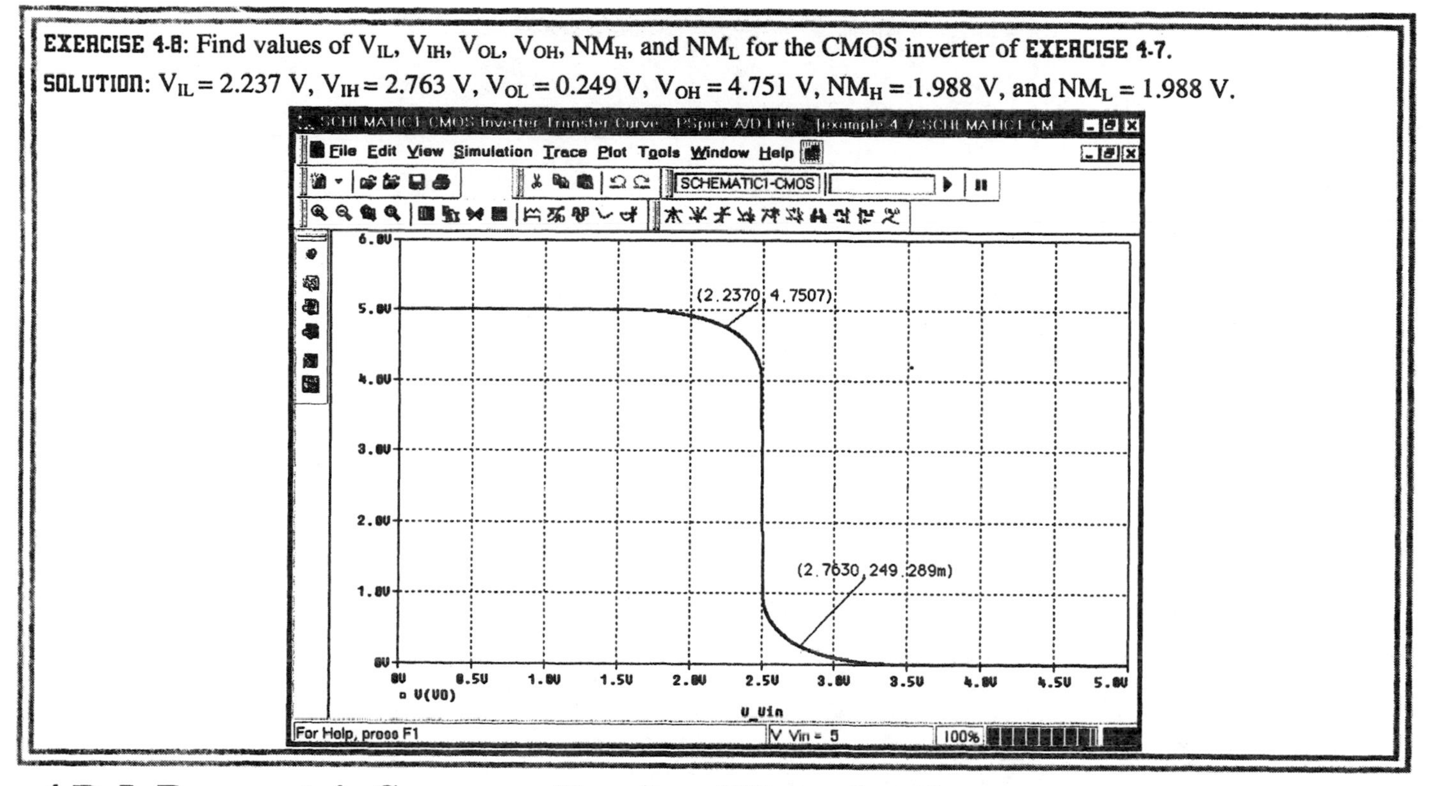


## 4.D.5. Parametric Sweep — Family of Transfer Curves

A question we would like to ask is, how does the transfer curve of the circuit from the previous section change as we change the driver MOSFET width-to-length ratio? We would like a family of curves that show the effect of changing the driver width. Families of transfer curves can be generated using the Parametric Sweep in conjunction with the DC Sweep or by using the Secondary Sweep. The Secondary Sweep was demonstrated in Section 4.D.2, so we will demonstrate the Parametric Sweep here.

A Parametric Sweep allows us to sweep a parameter. We would like to sweep the width of the driver MOSFET in the circuit of the previous section. We are assuming that you have followed that procedure and have set up the MOSFET models and DC Sweep. We must first set up a parameter for the width of the driver. We will start with the completed circuit from the previous section. It is repeated here for convenience.

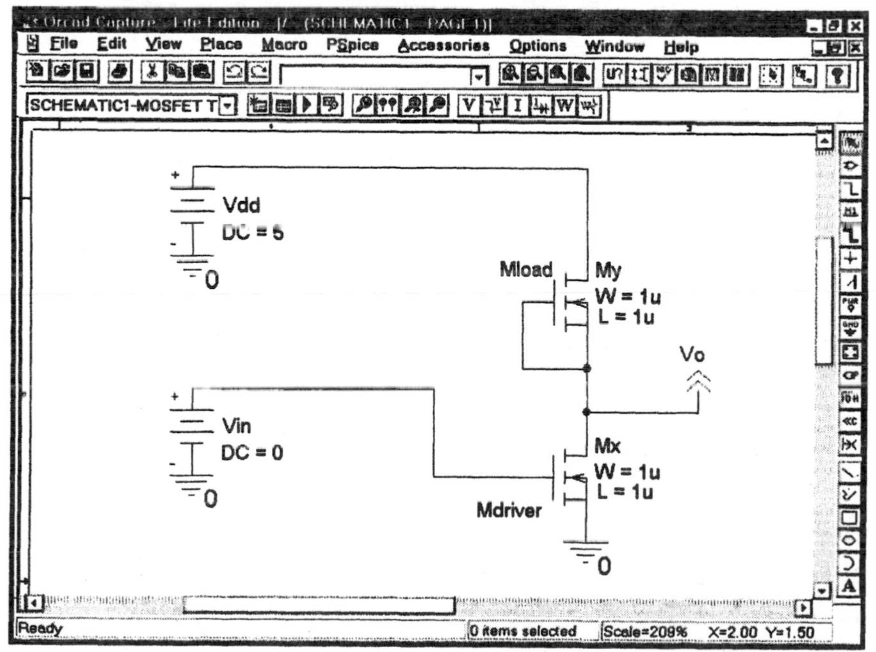


Double-click the ***LEFT*** mouse button on the text ***W = 1U*** next to the driver MOSFET:

Display Properties

Name: W

Value: 1u

Display Format

Do Not Display

Value Only

Name and Value

Name Only

Both if Value Exists

Font

Arial 5 (default)

Change... Use Default

Color

Default

Rotation

0° 180°

90° 270°

OK Cancel Help

We wish to enter a parameter for the value instead of ***1U***. Type the text **{W_val}**:

Display Properties

Name: W

Value: {W_val}

Display Format

Do Not Display

Value Only

Name and Value

Name Only

Both if Value Exists

Font

Arial 5 (default)

Change... Use Default

Color

Default

Rotation

0° 180°

90° 270°

OK Cancel Help

Click the ***OK*** button to accept the change:

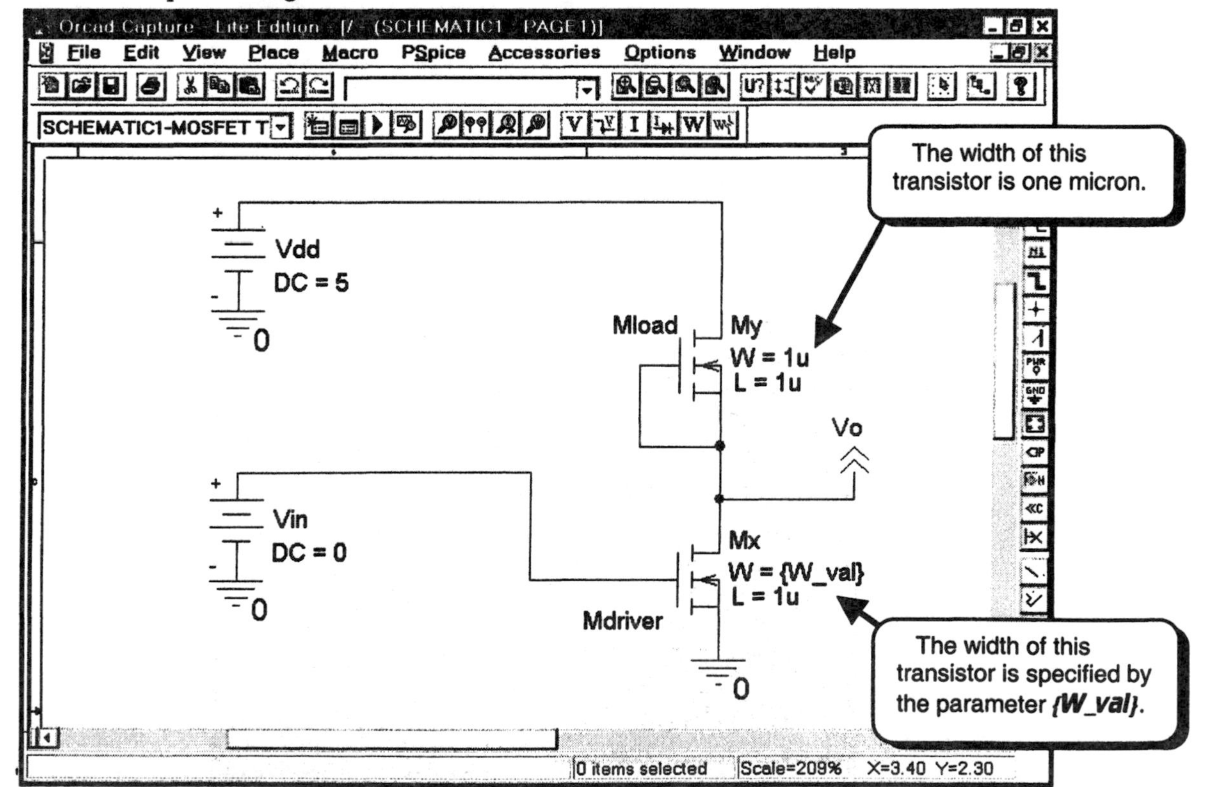


The driver width is now the value ***{W_val}***. In PSpice, any text enclosed in curly brackets { } is referred to as a parameter. A parameter can be assigned different values and can be changed during a simulation.

Next, we must declare ***{W_val}*** as a parameter. Get a part called PARAM and place it on your schematic.

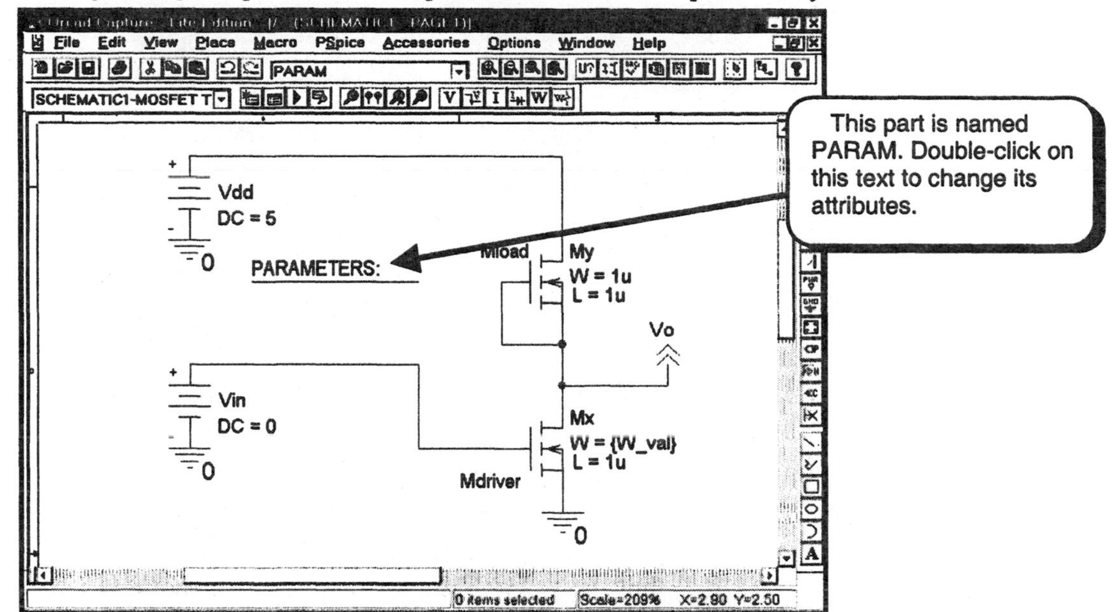


When you have placed the part, double-click the ***LEFT*** mouse button on the text ***PARAMETERS***: to obtain the spreadsheet for the part:

We need to create a new parameter, so click the ***New Column*** button:

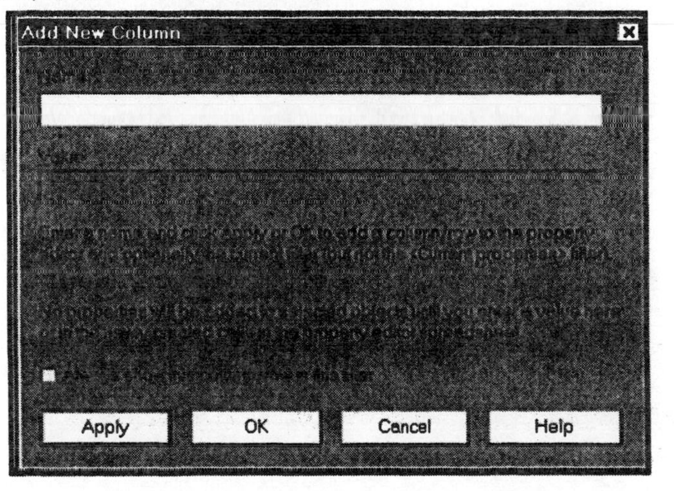

Enter the text **W_val** in the Name field:

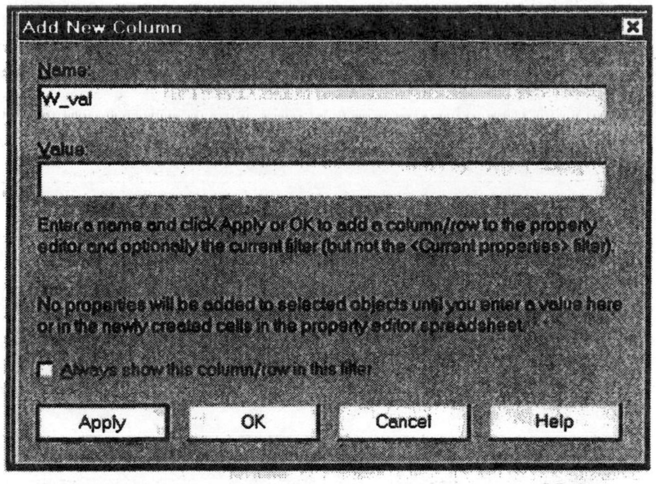

Add New Column

Name:

W_val

Value:

Enter a name and click Apply or OK to add a column/row to the property editor and optionally the current filter (but not the <Current properties> filter).

No properties will be added to selected objects until you enter a value here or in the newly created cells in the property editor spreadsheet.

Always show this column/row in this filter

Apply OK Cancel Help

When we enter a name, the ***Value*** field becomes enabled. Enter the text **1u** in the ***Value*** field:

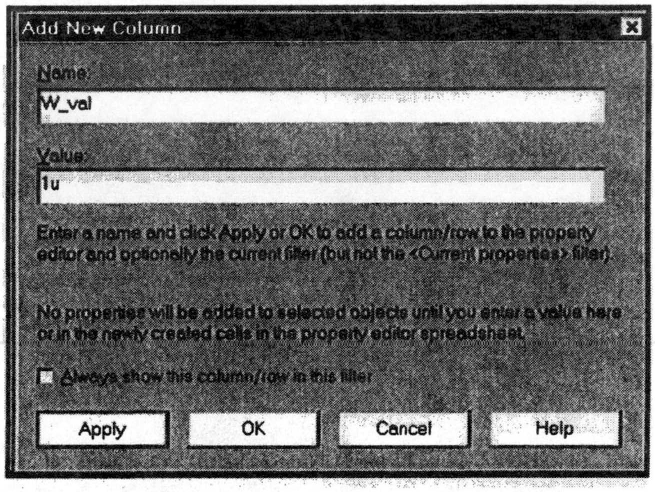

Add New Column

Name:

W_val

Value:

1u

Enter a name and click Apply or OK to add a column/row to the property editor and optionally the current filter (but not the <Current properties> filter).

No properties will be added to selected objects until you enter a value here or in the newly created cells in the property editor spreadsheet.

Always show this column/row in this filter

Apply OK Cancel Help

Click the ***OK*** button to create a new parameter and add a new column to the spreadsheet:

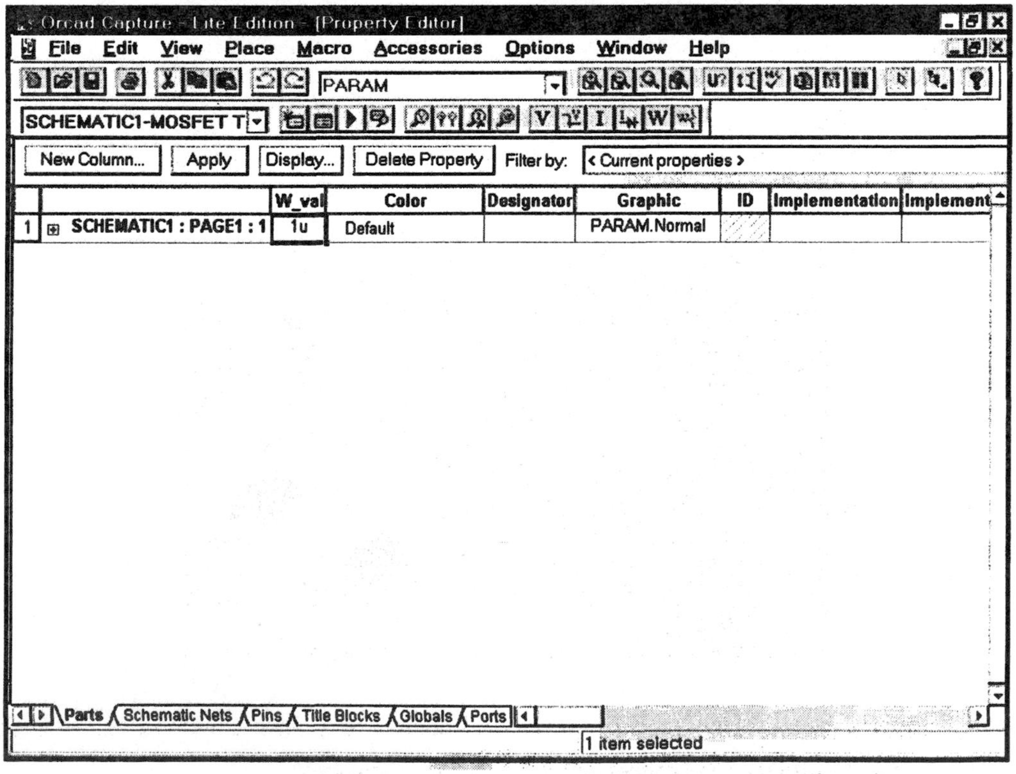

Orcad Capture - Lite Edition - [Property Editor]

File Edit View Place Macro Accessories Options Window Help

PARAM

SCHEMATIC1-MOSFET T

New Column... Apply Display... Delete Property Filter by: < Current properties >

| | | W_val | Color | Designator | Graphic | ID | Implementation | Implement |
|---|---|---|---|---|---|---|---|---|
| 1 | SCHEMATIC1 : PAGE1 : 1 | 1u | Default | | PARAM.Normal | | | |

Parts Schematic Nets Pins Title Blocks Globals Ports

1 item selected

Next, we would like to display this parameter on the schematic. Click the ***LEFT*** mouse button on the text ***W_val*** to select the column:

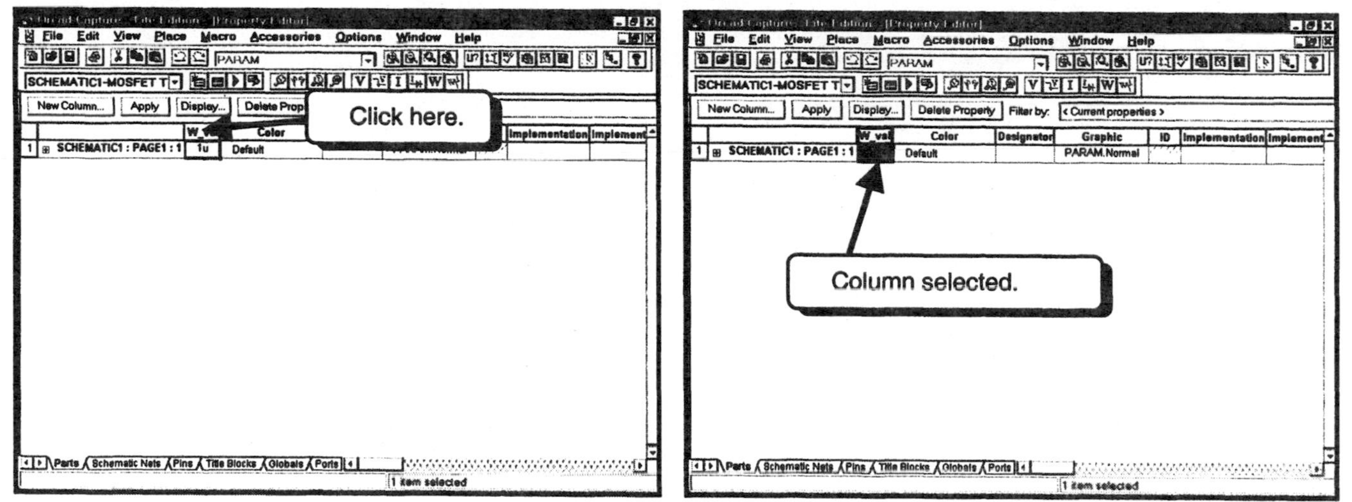


Once the column is selected, click the ***Display*** button to change the display properties for the selected column:

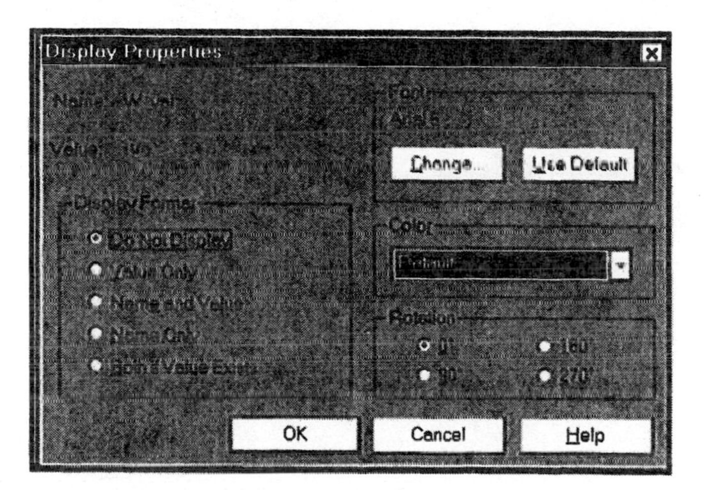


The present ***Display Format*** is for nothing in the column to be displayed. Select option ***Name and Value*** to display both the name of the parameter (W_val) and the value of the parameter (1u) on the schematic:

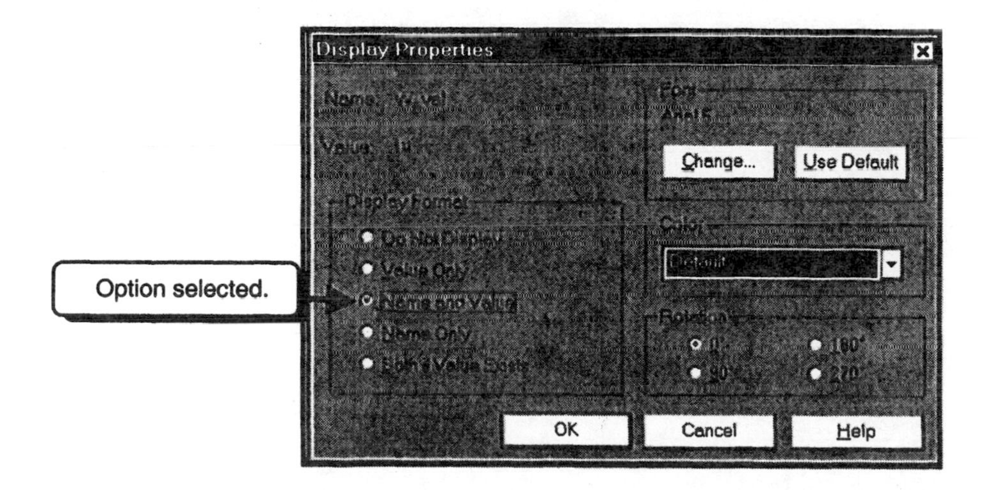


Click the ***OK*** button to accept the changes. You will return to the spreadsheet. Click the ***LEFT*** mouse button on the lower ☒ in the upper right corner to close the spreadsheet and return to the schematic:

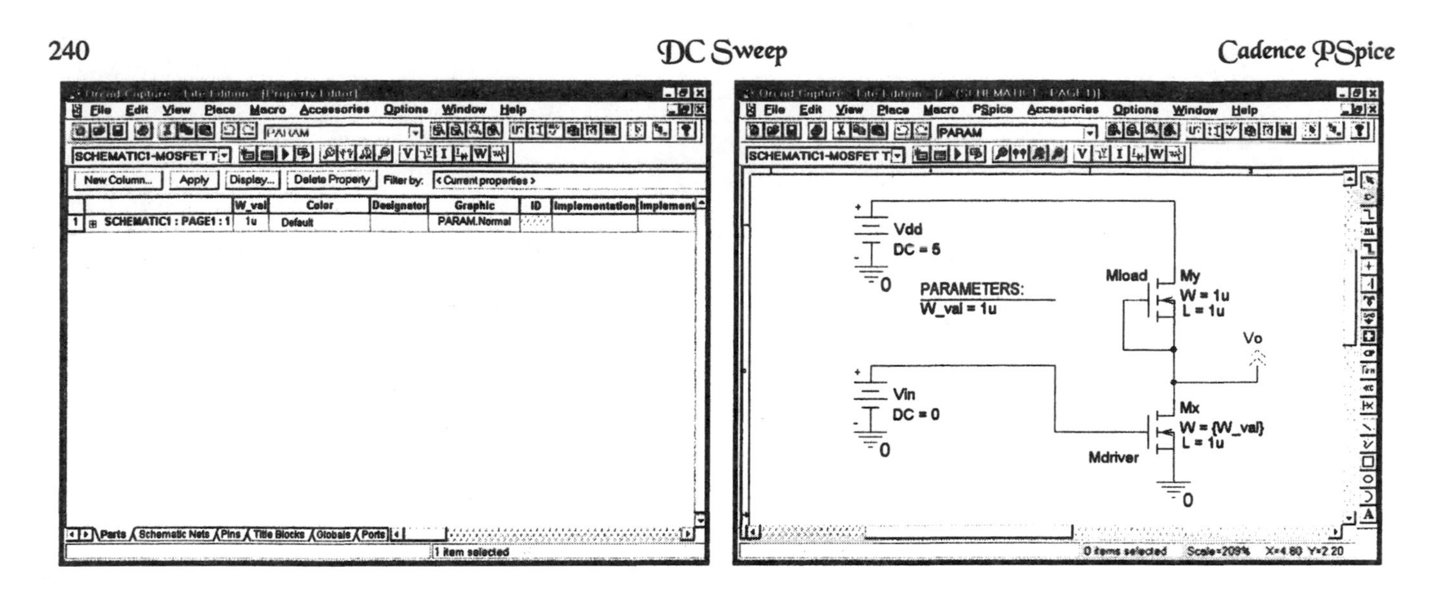


We must now set up the Parametric Sweep. If you are continuing from the previous example, you will already have created a simulation profile for the DC Sweep. To open the profile, select **PSpice** and then **Edit Simulation Profile** from the menus. If you started this example as a new circuit, select **PSpice** and then **New Simulation Profile**, specify a name for the new profile, and then click the ***Create*** button. Set up a DC Sweep as shown below. Using either procedure, you should have the screen below:

The ***Primary Sweep*** should be set up from the previous example.

The parameters for the Primary Sweep are set up and the sweep is enabled. We must now set up the ***Parametric Sweep***. Click the ***LEFT*** mouse button on the □ next to the text ***Parametric Sweep*** to toggle to the Parametric Sweep settings:

We would like to sweep the parameter W_val from 1 micron to 32 microns in factors of 2. We will use an ***Octave Sweep type***. An octave is a factor of 2. An octave would be from 1 to 2, 2 to 4, 4 to 8, 8 to 16, and so on. Fill in the dialog box as shown:

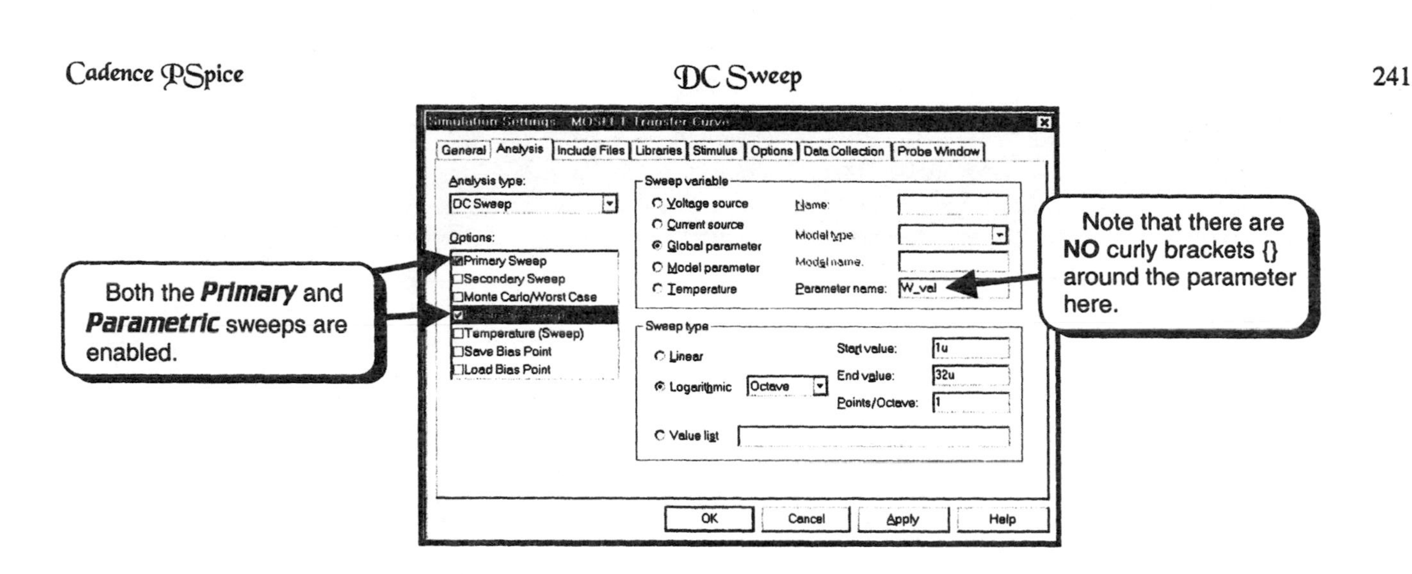


The dialog box is set to sweep the global parameter ***W_val***. The sweep is from ***1u*** ($1\times10^{-6}$ meters = 1 micron) to ***32u*** ($32\times10^{-6}$ meters = 32 microns). The sweep is in octaves, 1u to 2u, 2u to 4u, 4u to 8u, 8u to 16u, and 16u to 32u. ***1*** point per octave is specified so the values of ***W_val*** will be 1u, 2u, 4u, 8u, 16u, and 32u. Notice in the above dialog box that both the ***Primary Sweep*** and the ***Parametric Sweep*** are enabled. Click the ***OK*** button to accept the setup and return to the schematic:

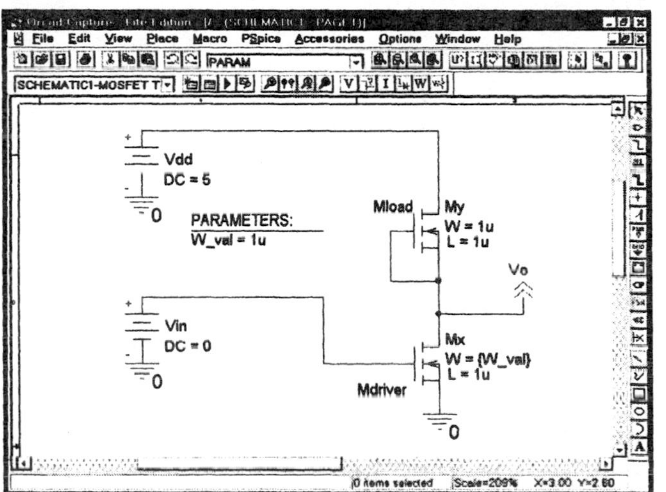


Run PSpice (select **PSpice** and then **Run** or press the **F11** key):

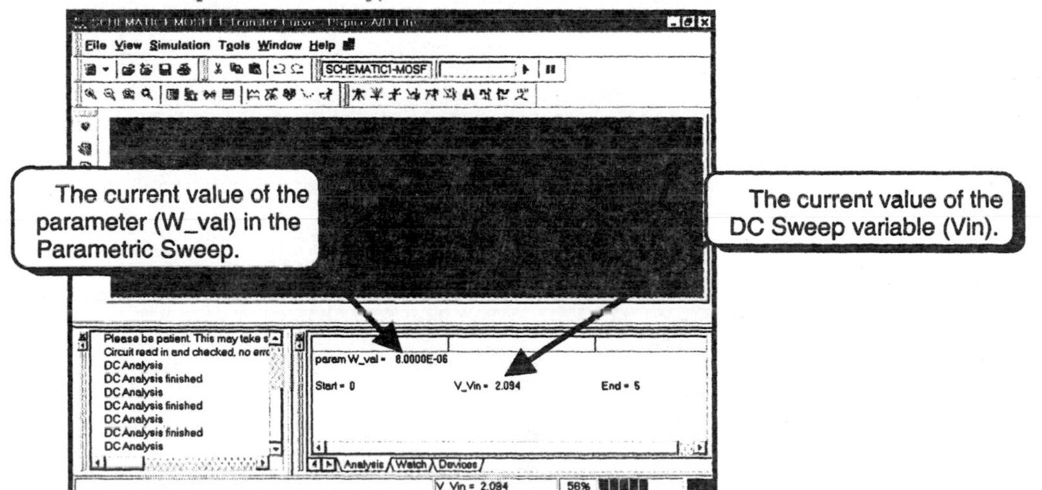


The PSpice window indicates the current value of the parameter. Notice that for each value of the parameter, the input voltage is swept from 0 to 5 volts. Logically, the DC Sweep loop is executed inside the Parametric Sweep loop. When the simulation is complete, you will be asked which values of ***W_val*** you would like to view. By default, all runs are selected:

Available Sections

All None OK Cancel

We would like to look at all of the curves, so click the ***OK*** button:

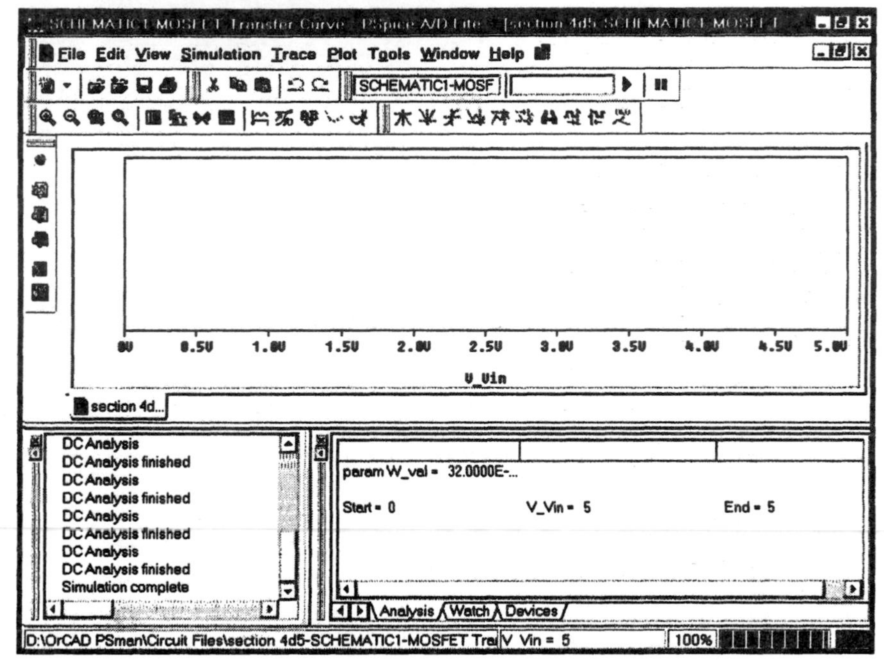


I will select **View** and then **Alternate Display** to switch the Probe window to full screen. We would like to plot the output voltage. Select **Trace** and then **Add Trace** from the Probe menus (or press the **INSERT** key), and then add the trace **V(Vo)**:

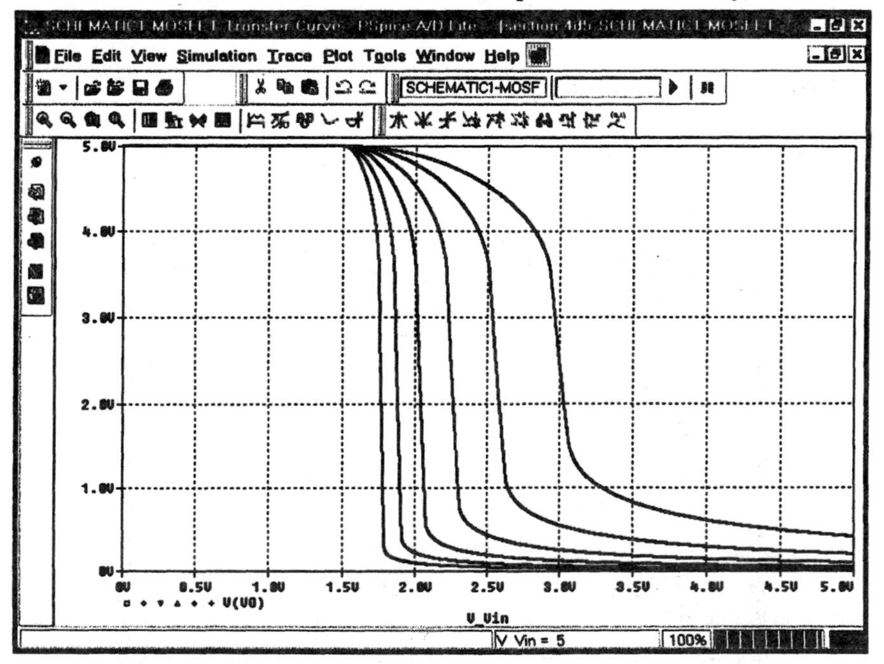


The question may arise as to which trace is for which value of the parameter, W_val. We can obtain this information easily. Click the ***RIGHT*** mouse button on one of the traces. A menu will appear:

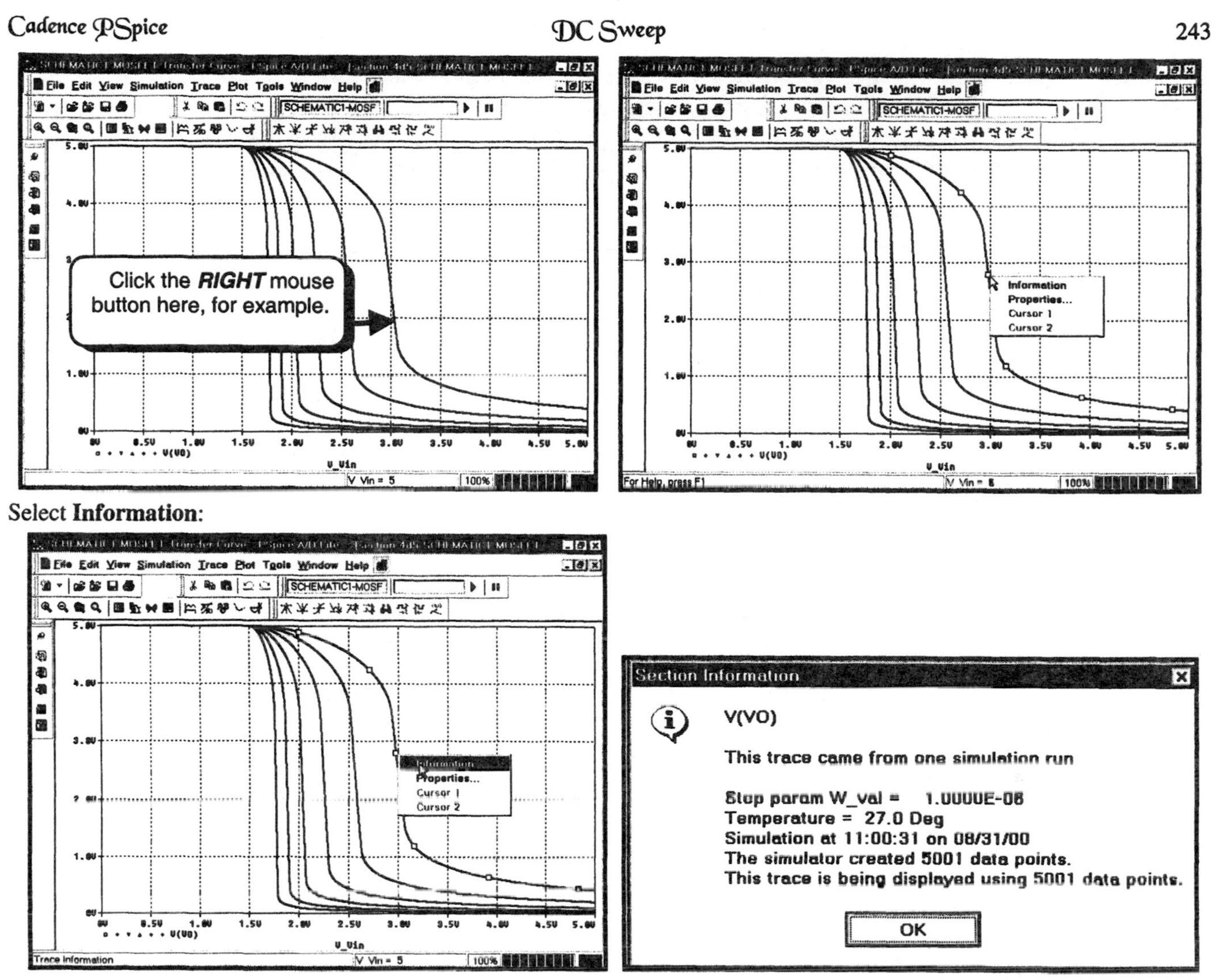


Select **Information**:

We see that the rightmost trace is the transfer curve for ***W_val*** = 1 micron.

---

**EXERCISE 4-9**: In **EXERCISE 4-7** we looked at the operation of a CMOS inverter. We will now investigate how the (W/L) ratio of the PMOS transistor affects the transfer curve of the inverter. Let the (W/L) ratio of the PMOS transistor have values 1, 3, 6, 9, 12, and 15.

**SOLUTION**: Starting with the circuit of **EXERCISE 4-7**, let the W value of the PMOS transistor be set by the parameter Wp_val. Also, add the "PARAM" part:

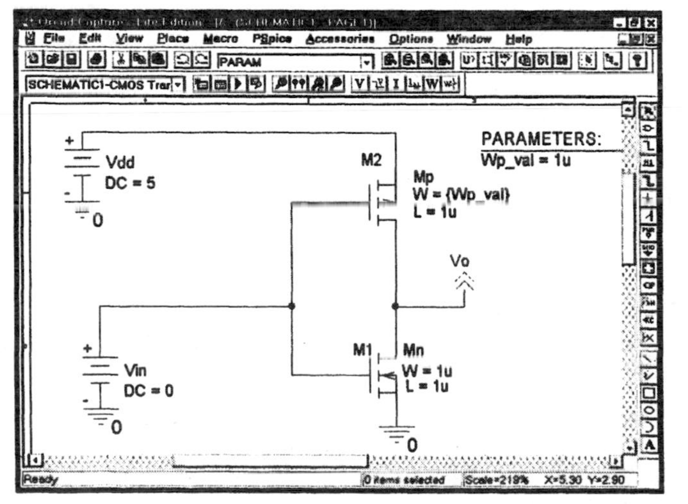

Set up the ***Primary Sweep*** to sweep ***Vin*** from 0 V to 5 V in 0.001 V steps as shown in the left dialog box below. Set up the ***Parametric Sweep*** to use the value list as shown in the right dialog box:

Run the analysis and then display V(Vo) in Probe:

# 4.E. DC Secondary Sweep — BJT Characteristic Curves

The nested DC Sweep can be used to generate characteristic curves for transistors. We will illustrate generating these curves using a BJT. Wire the circuit below:

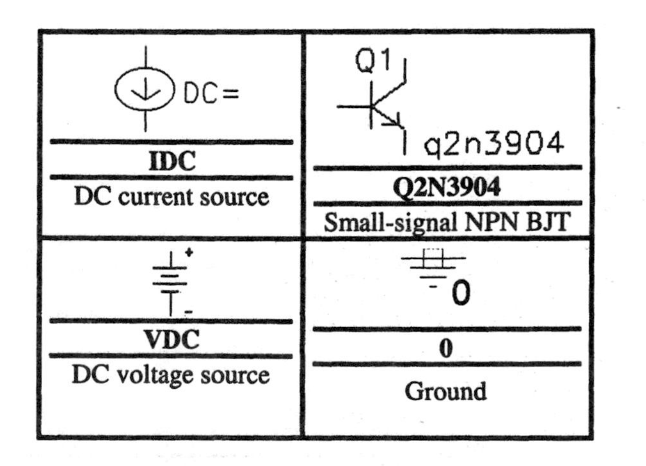

| | |
|---|---|
| IDC<br>DC current source | Q2N3904<br>Small-signal NPN BJT |
| VDC<br>DC voltage source | 0<br>Ground |

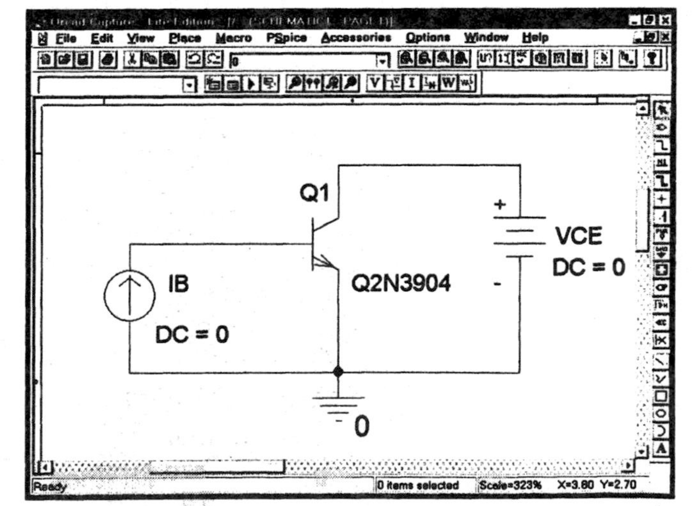


We need to set up the Primary DC Sweep and the Secondary Sweep. Select **PSpice** and then **New Simulation Profile** from the Capture menus, enter a name for the profile, and click the ***Create*** button. Select the ***DC Sweep Analysis type*** and fill in the ***Primary Sweep*** as shown:

The dialog box lists the settings for the Primary DC Sweep. The ***Primary Sweep*** sweeps ***VCE*** from ***0*** volts to ***15*** volts in ***0.01*** volt increments. The ***Sweep type*** is linear, which means that the voltage points are equally spaced.

Next, we need to set up the ***Secondary Sweep***. Click the *LEFT* mouse button on the box □ next to ***Secondary Sweep*** to select the option. The dialog box will change to display the settings for the ***Secondary Sweep***. We would like to sweep the base current from 0 mA to 1 mA in 100 μA increments. Fill in the dialog box as shown:

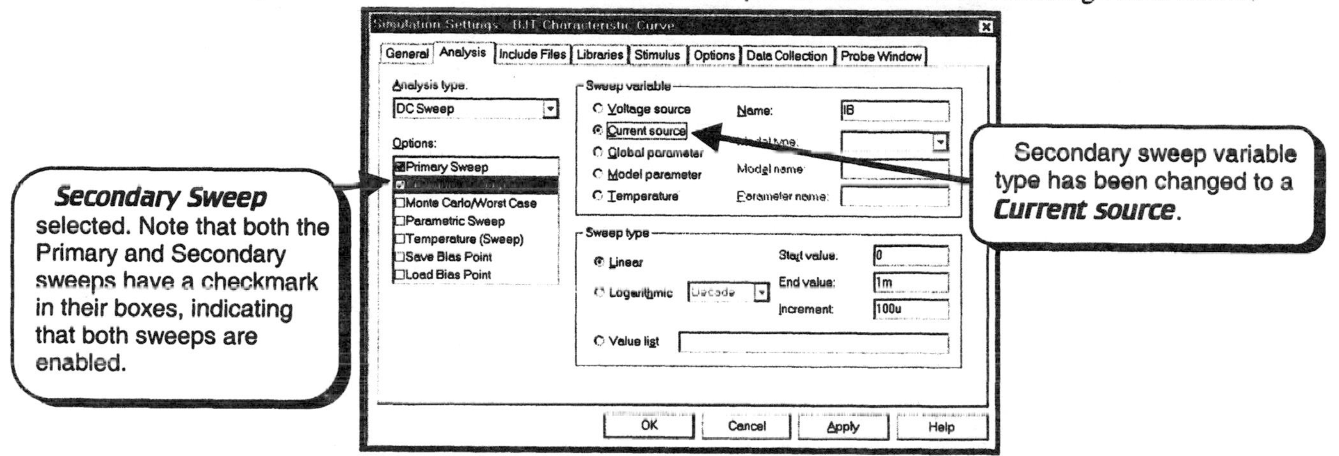


Notice that a ***Current source*** is set as the ***Sweep variable***, and the ***Sweep type*** is set to ***Linear***.

Logically, the ***Primary Sweep*** loop is executed inside the ***Secondary Sweep*** loop; that is, for each value of the base current, VCE is swept from 0 to 15 volts. Return to the schematic by clicking the ***OK*** button. Run PSpice (select **PSpice** and then **Run**):

The simulation status window indicates the progress of each loop. Notice that for each value of the nested variable, the main variable is swept from 0 to 15 volts. When the simulation is finished, the Probe window will display a blank window. I will switch Probe to occupy the entire window. Press the **INSERT** key and add the trace `IC(Q1)`:

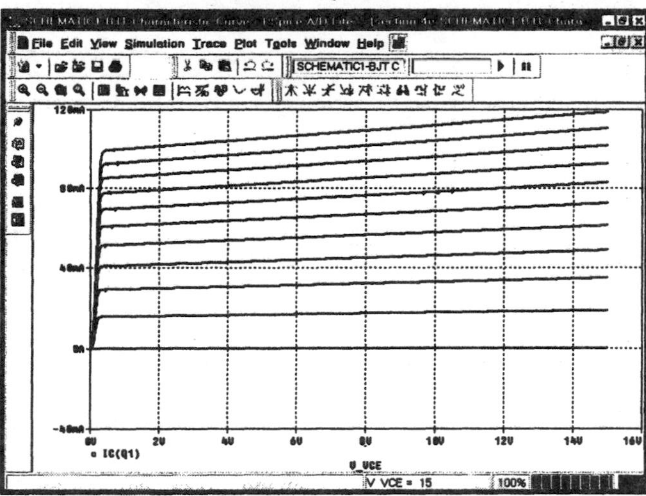


This is the characteristic family of curves for a 2N3904 NPN bipolar junction transistor.

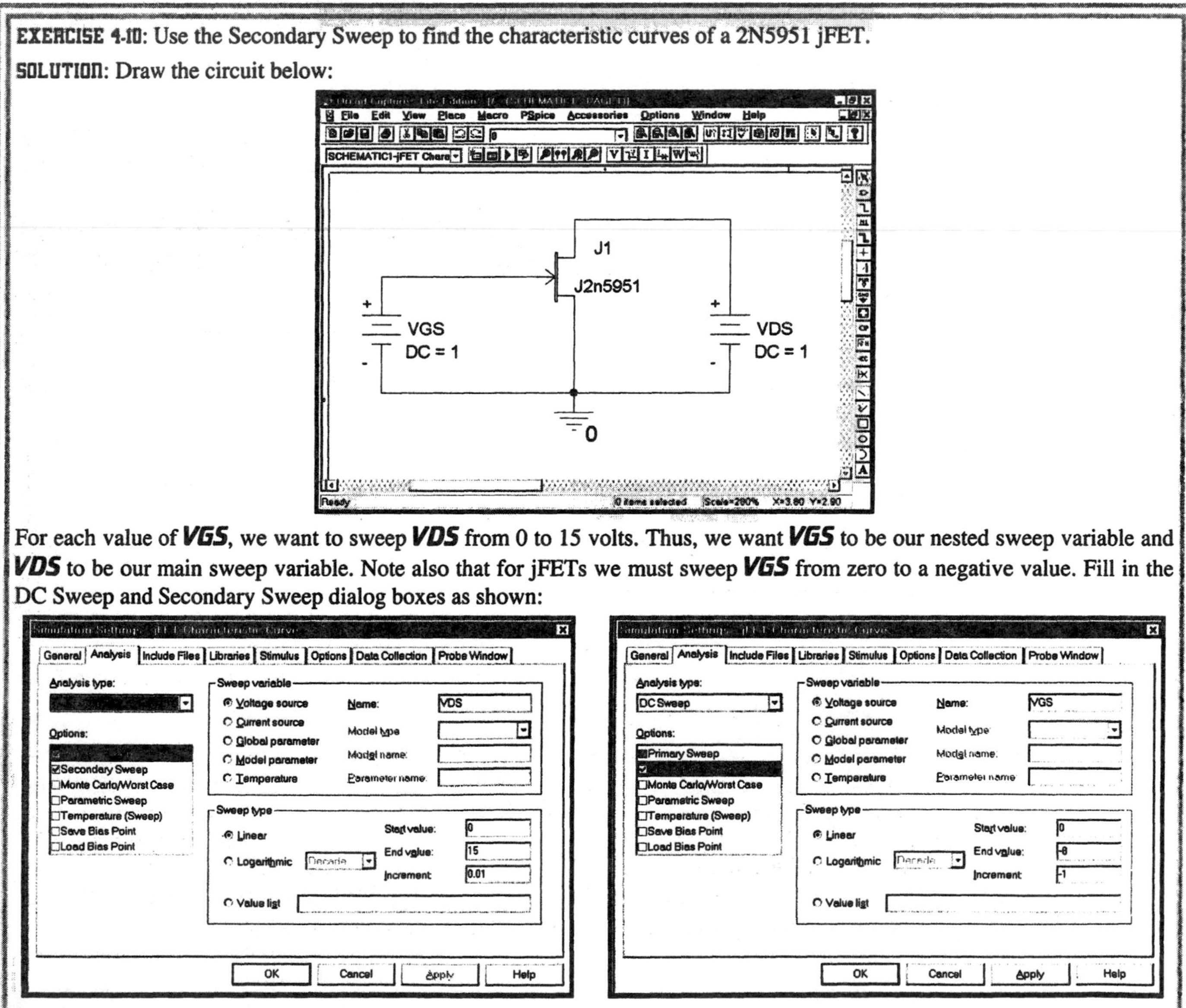


**EXERCISE 4-10**: Use the Secondary Sweep to find the characteristic curves of a 2N5951 jFET.

**SOLUTION**: Draw the circuit below:

For each value of ***VGS***, we want to sweep ***VDS*** from 0 to 15 volts. Thus, we want ***VGS*** to be our nested sweep variable and ***VDS*** to be our main sweep variable. Note also that for jFETs we must sweep ***VGS*** from zero to a negative value. Fill in the DC Sweep and Secondary Sweep dialog boxes as shown:

Run PSpice and plot the results with Probe. The x-axis will automatically be VDS. To plot the characteristic curves, plot the drain current ID(J1):

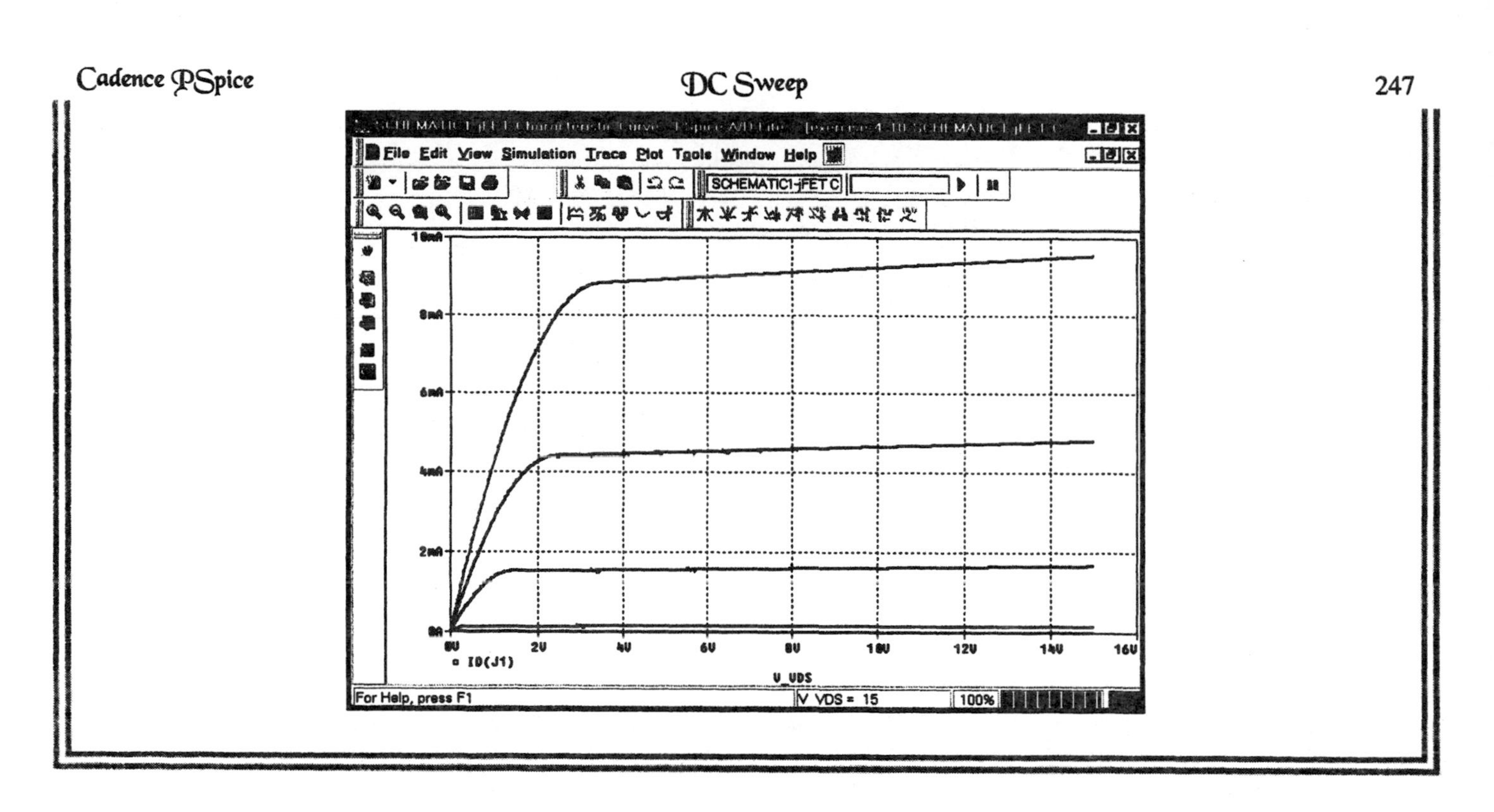

# 4.F. DC Current Gain of a BJT

In this section we will investigate how the DC current gain ($H_{FE}$) of a bipolar junction transistor varies with DC bias collector current $I_{CQ}$, DC bias collector-emitter voltage $V_{CEQ}$, and temperature. We will use the basic circuit shown below for all simulations:

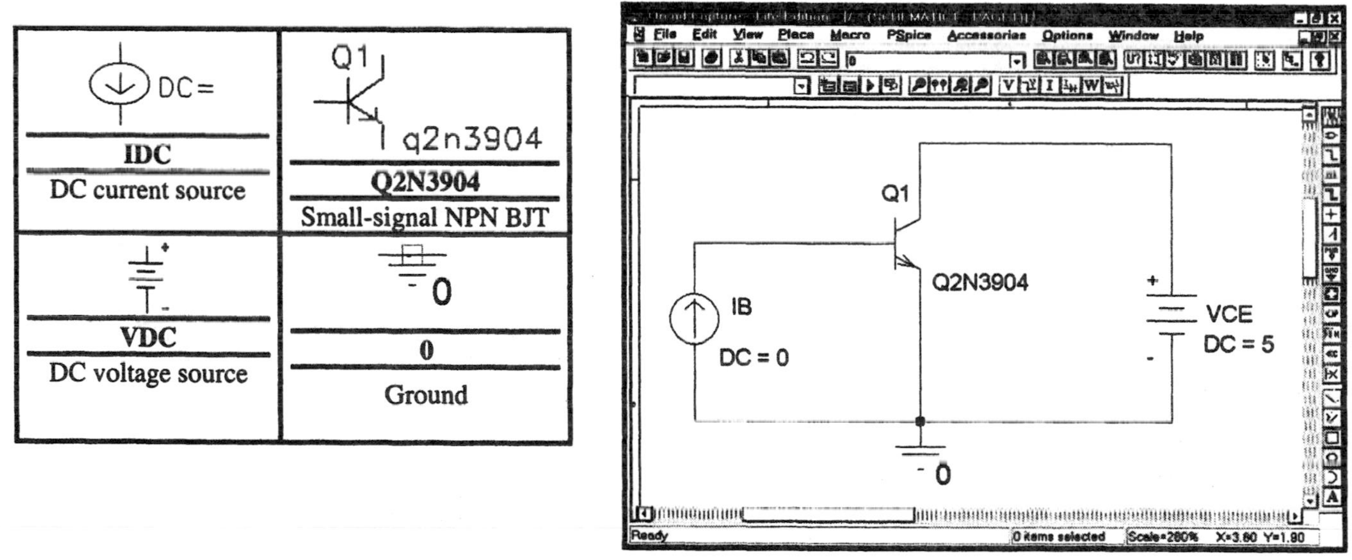


## 4.F.1. $H_{FE}$ Versus Collector Current

Our first analysis will display how $H_{FE}$ of a transistor varies with DC collector current.* In many applications we want to use a transistor where it has its maximum current gain. This plot will tell us how to choose the collector current for maximum current gain. We can then bias the transistor at this collector current. This plot is easily generated with a DC Sweep. We will generate this curve with $V_{CE}$ constant at 5 V. Note in the circuit above that $V_{CE}$ is a DC voltage source of 5 volts, and we will not change it during the simulation. Select **PSpice** and then **New Simulation Profile** from the Capture menus, specify a name for the profile, and then click the ***Create*** button. Select the ***DC Sweep Analysis type***. We will sweep IB from 1 µA to 1 mA in ***Decade***s with 20 points per decade. Fill in the dialog box as shown:

*Thanks to John D. Welkes of Arizona State University for this example.

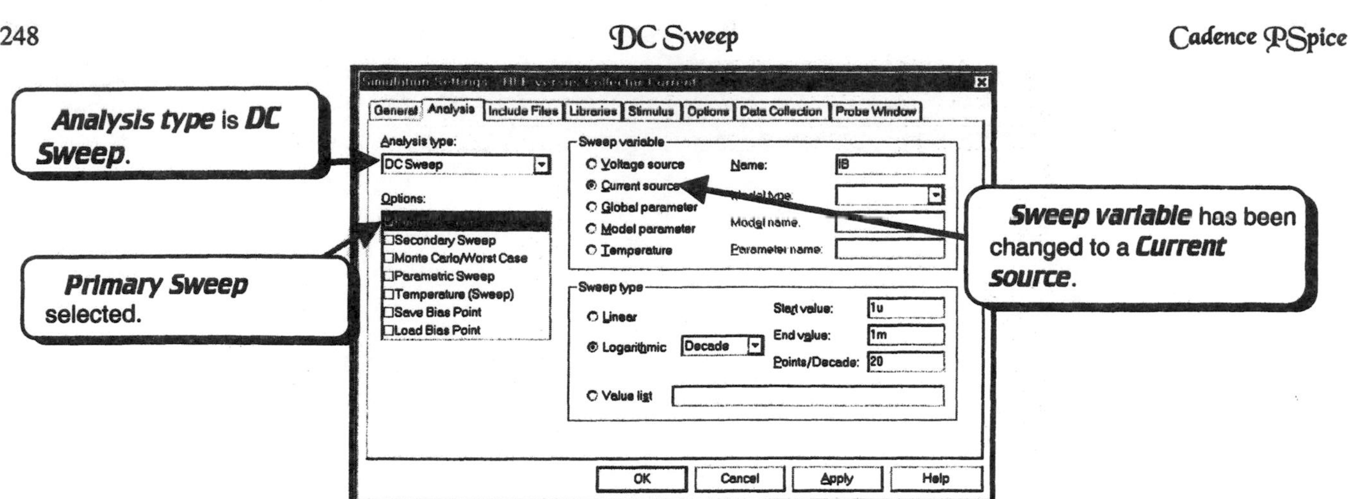


Click the **OK** button to return to the schematic. Run the simulation. We would like to plot the current gain, $H_{FE}$. When we go to add a trace, we do not see $H_{FE}$ listed. However, $H_{FE}$ can be calculated as the ratio of IC/IB for the transistor. Thus, we will plot the trace **`IC(Q1)/IB(Q1)`**:

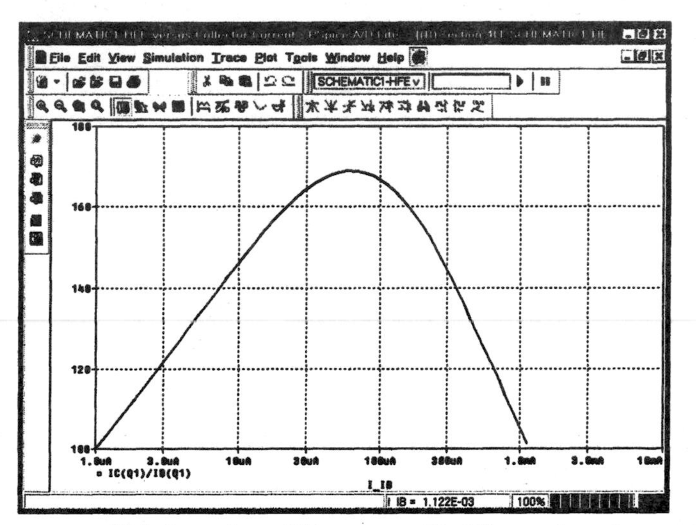

Note that the x-axis is the base current. This is a plot of $H_{FE}$ versus IB. We need to change the x-axis to collector current. From the Probe menus, select **Plot** and then **Axis Settings**:

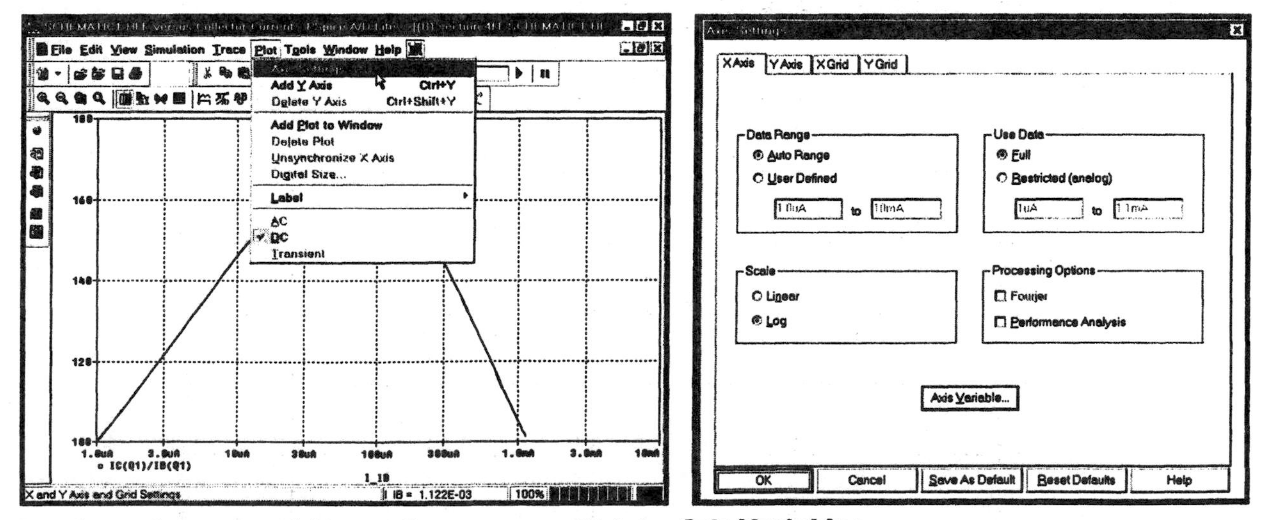

We must change the x-axis variable to collector current. Click the **Axis Variable** button:

Axis Variable

Simulation Output Variables

*

I(VCE)
I_IB
IB(Q1)
IC(Q1)
IE(Q1)
IS(Q1)
V(0)
V(IB:+)
V(IB:-)
V(N00121)
V(N00153)
V(Q1:b)
V(Q1:c)
V(Q1:e)
V(VCE:+)
V(VCE:-)
V1(IB)
V1(VCE)
V2(IB)
V2(VCE)
VB(Q1)
VC(Q1)
VE(Q1)

Full List

☑ Analog
☐ Digital
☑ Voltages
☑ Currents
☑ Power
☐ Noise (V²/Hz)
☑ Alias Names
☐ Subcircuit Nodes

28 variables listed

Functions or Macros

Analog Operators and Functions

()
*
+
-
/
@
ABS()
ARCTAN()
ATAN()
AVG()
AVGX(,)
COS()
D()
DB()
ENVMAX(,)
ENVMIN(,)
EXP()
G()
IMG()
LOG()
LOG10()
M()
MAX()

Trace Expression: [OK] [Cancel] [Help]

Select trace IC(Q1):

Axis Variable

Simulation Output Variables

*

I(IB)
I(VCE)
I_IB
IB(Q1)
IE(Q1)
IS(Q1)

Full List

☑ Analog
☐ Digital
☐ Voltages
☑ Currents
☐ Power
☐ Noise (V²/Hz)
☑ Alias Names
☐ Subcircuit Nodes

7 variables listed

Functions or Macros

Analog Operators and Functions

()
*
+
-
/
@
ABS()
ARCTAN()
ATAN()
AVG()
AVGX(,)
COS()
D()
DB()
ENVMAX(,)
ENVMIN(,)
EXP()
G()
IMG()
LOG()
LOG10()
M()
MAX()

Trace Expression: IC(Q1) [OK] [Cancel] [Help]

Click the ***OK*** button to accept the trace and then click the ***OK*** button again to plot the trace with the new settings:

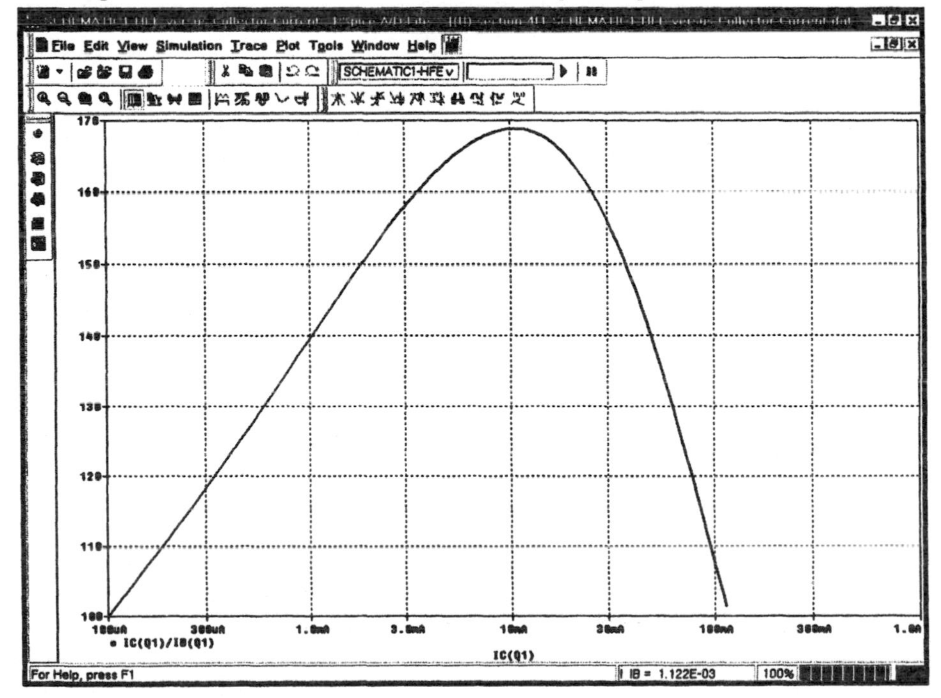


We see that the 2N3904 BJT has a maximum DC current gain at a collector current of about 10 mA. Use the cursors to find and label the maximum value of current gain:

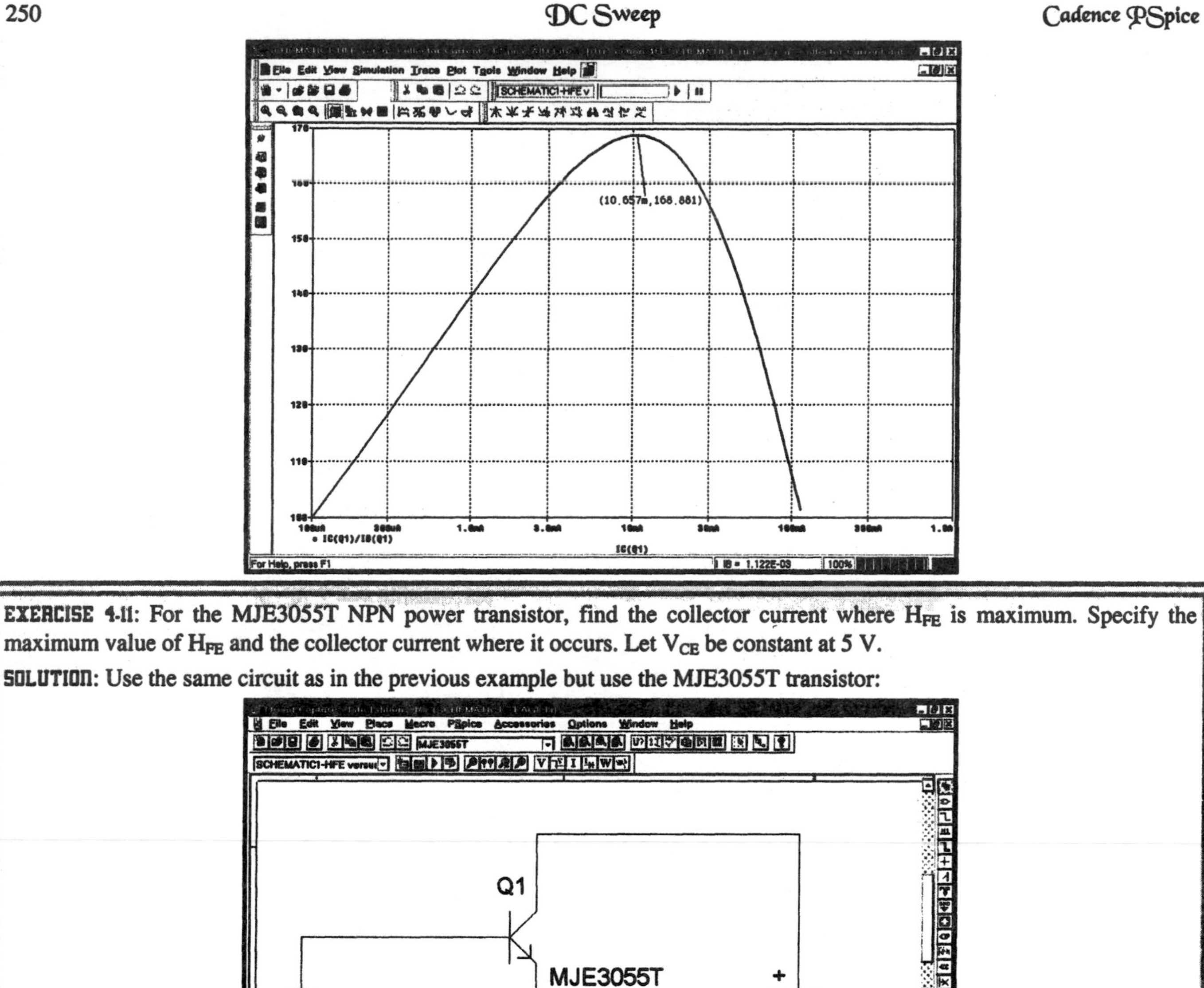


**EXERCISE 4-11**: For the MJE3055T NPN power transistor, find the collector current where $H_{FE}$ is maximum. Specify the maximum value of $H_{FE}$ and the collector current where it occurs. Let $V_{CE}$ be constant at 5 V.

**SOLUTION**: Use the same circuit as in the previous example but use the MJE3055T transistor:

Since this is a power transistor, it can handle much higher collector currents. Sweep ***IB*** from 100 μA to 100 mA:

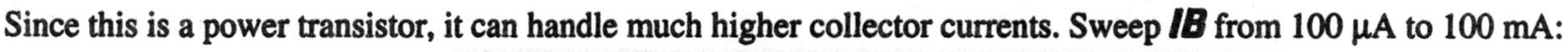

Use Probe to plot $H_{FE}$ versus $I_C$ and use the cursors to display the maximum value:

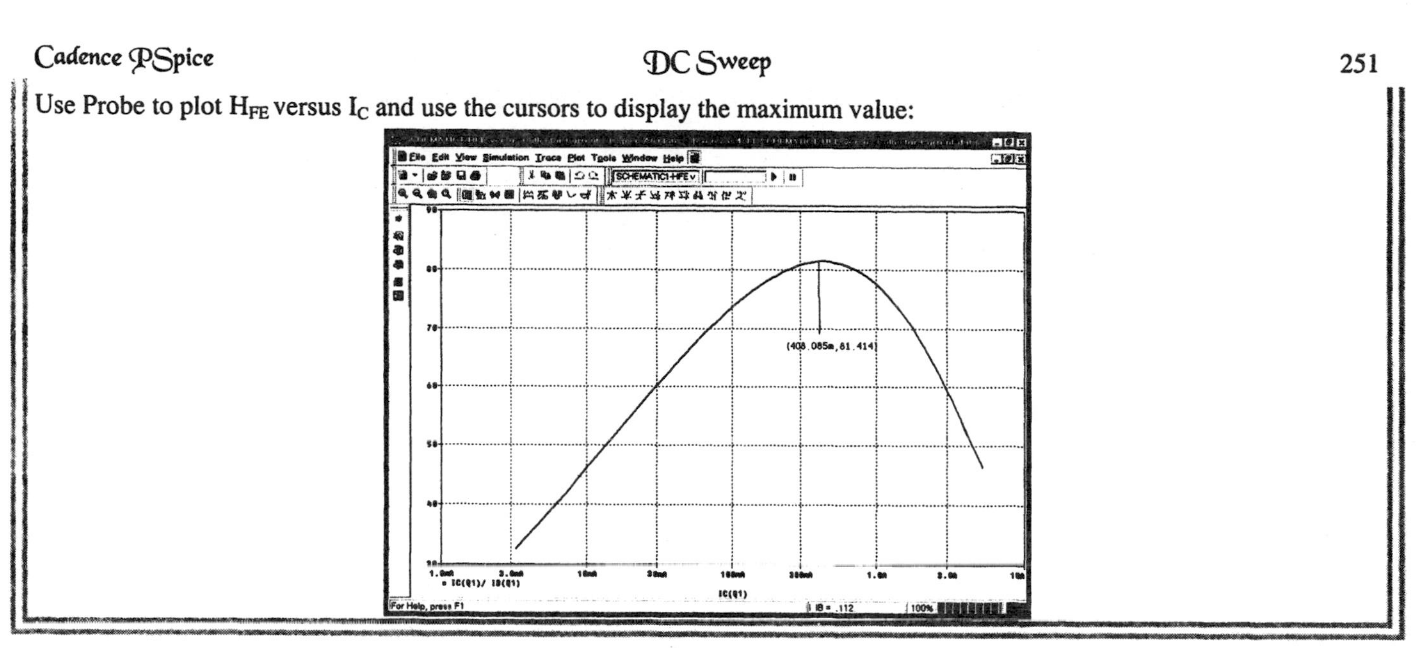


## 4.F.2. $H_{FE}$ Versus $I_C$ for Different Values of $V_{CE}$

The next thing we would like to do is to see how the $H_{FE}$ versus $I_C$ curve is affected for different values of DC collector-emitter voltage. The curve in the previous example was generated at $V_{CE}$ = 5 V. We would now like to generate four curves at different values of $V_{CE}$ and plot them all on the same graph. We will generate curves at $V_{CE}$ = 2 V, 5 V, 10 V, and 15 V. We will use the same circuit and simulation profile as in the previous section:

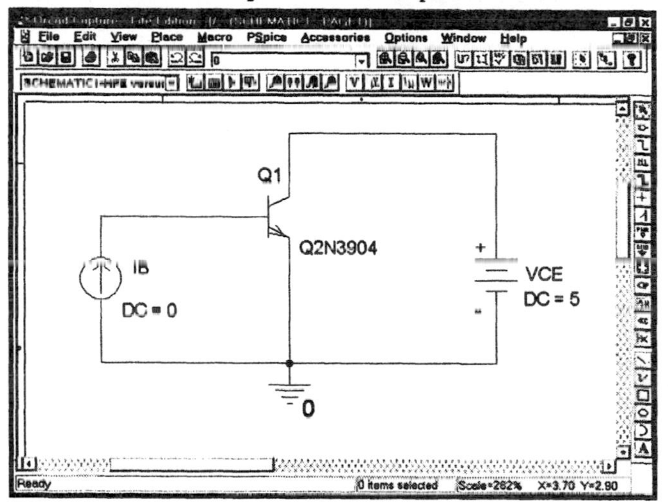


Since we have already created a simulation profile from the previous example, we do not need to create a new profile. To open the previous profile, select **PSpice** and then ***Edit Simulation Profile*** from the Capture menus:

The ***Primary Sweep*** is set up from the previous example. It is set to sweep IB from 1 µA to 1 mA at 20 points per decade. To change $V_{CE}$, we will use a ***Secondary Sweep***. Click the box ☐ next to ***Secondary Sweep*** and fill in the dialog box as shown:

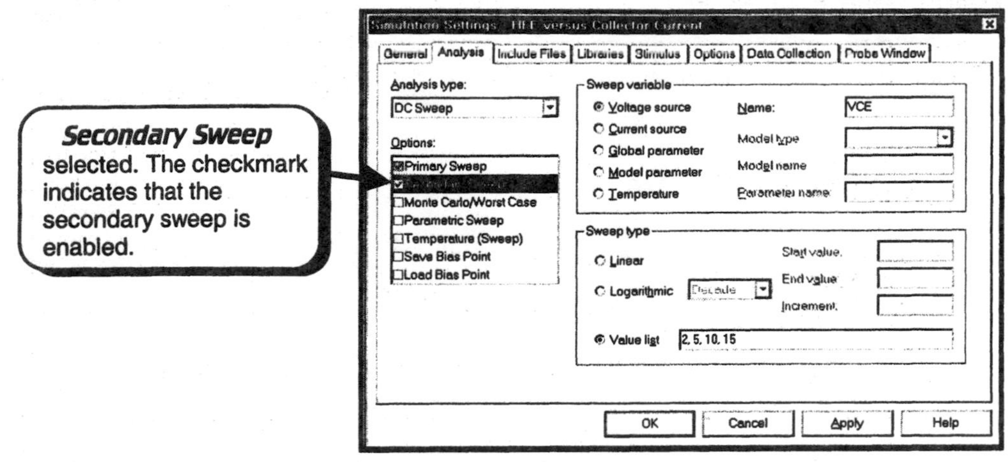


Note that the ***Sweep variable*** is a ***Voltage source***, and that there is a checkmark in the squares ☑ to enable both the Secondary and Primary sweeps. The ***Secondary Sweep*** uses the ***Value list***, which sets VCE to 2, 5, 10, and 15 volts. Click the ***OK*** button to return to the schematic. Simulate the circuit (select **PSpice** and then **Run** or press the **F11** key) and then view the results with Probe. Plot $H_{FE}$ versus $I_C$. Use the same procedure to generate the plot that we used on pages 248–249. Four curves will be displayed:

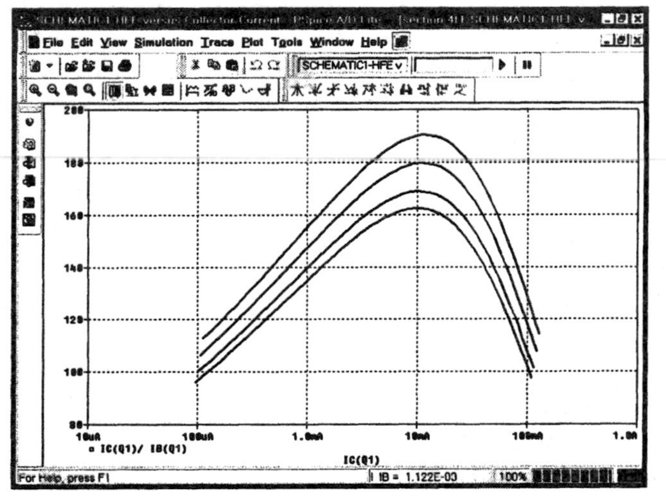


A question may arise as to how to determine which curve represents $V_{CE}$ when it equals 2 V, 5 V, 10 V, and 15 V. In Section 4.F.1. we generated a single plot at $V_{CE}$ = 5 V. This plot had a maximum of $H_{FE}$ = 168.881. One of the curves on this plot is at $V_{CE}$ equals 5 V. If we can identify the plot with this maximum, we can determine all of the other curves. Display the cursors (select **Trace**, **Cursor**, and then **Display**):

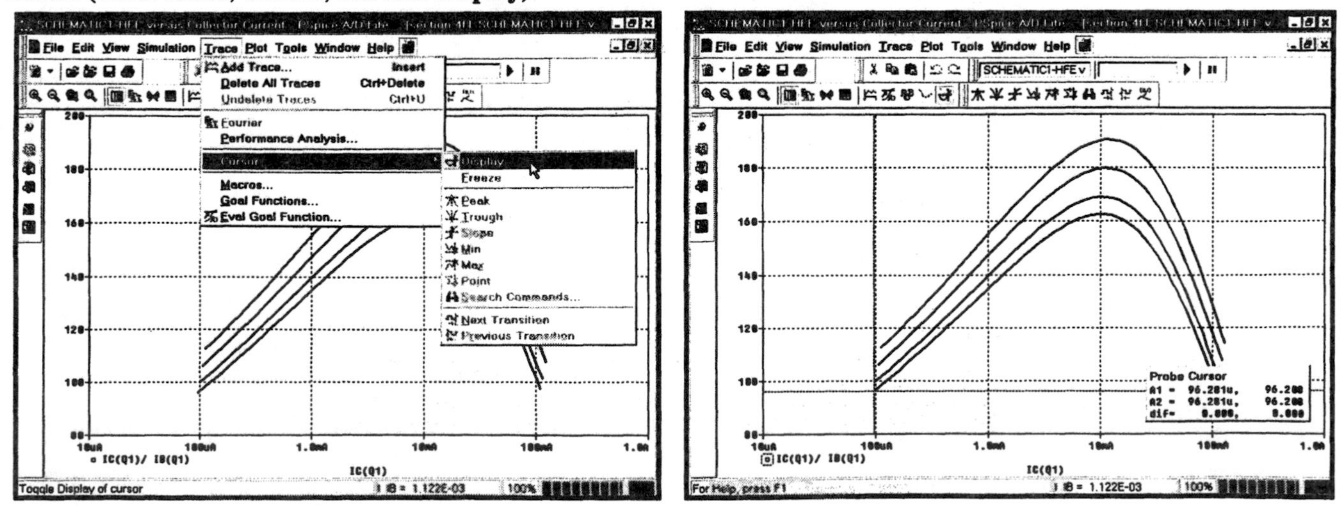

The cursor is now on the bottom curve. To find the maximum value of this curve, select **Trace, Cursor**, and then **Peak**. Do not select **Max**. Peak will move the cursor to the next local maximum while Max will find the absolute maximum for the entire plot. After selecting **Trace, Cursor**, and then **Peak**, the cursor moves to the peak of the lowest curve:

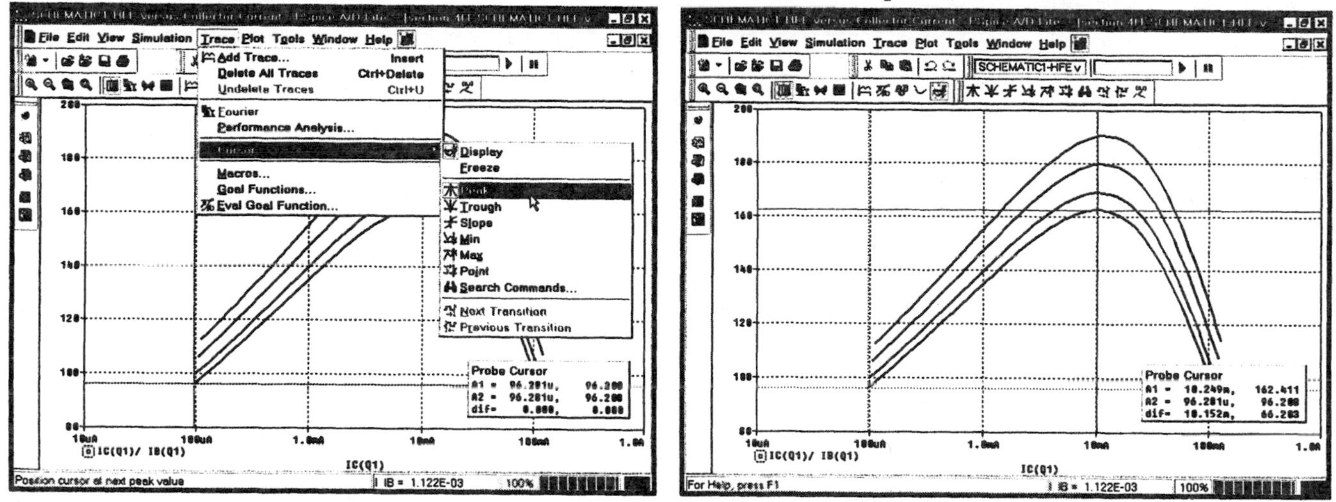


The cursor coordinates show that this curve has a maximum $H_{FE}$ of 162.411. This is not the curve at $V_{CE}$ = 5 V. Select **Trace, Cursor**, and then **Peak**. This will find the next local maximum, or the peak of the second curve:

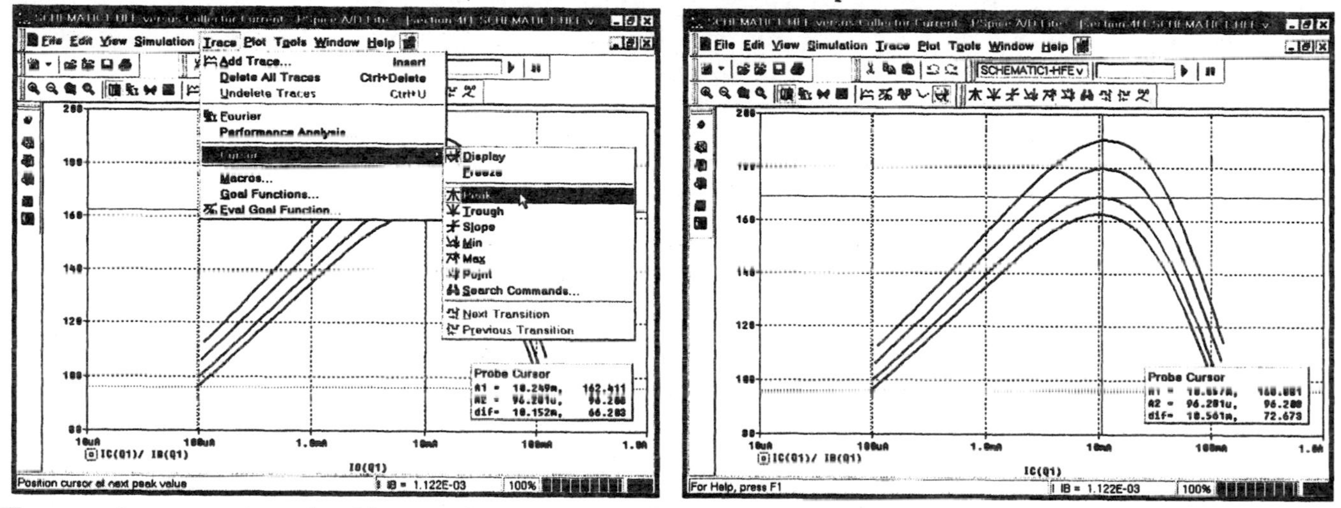


The example cursors show that this curve has a maximum $H_{FE}$ of 168.881 and thus this is the curve at $V_{CE}$ = 5 V. We can conclude that the bottom curve is at $V_{CE}$ = 2 V, the second curve is at $V_{CE}$ = 5 V, and the top curve is at $V_{CE}$ = 15 V. To be absolutely certain, you should run each case separately and verify the identity of each curve. Place text on the plot to identify each curve:

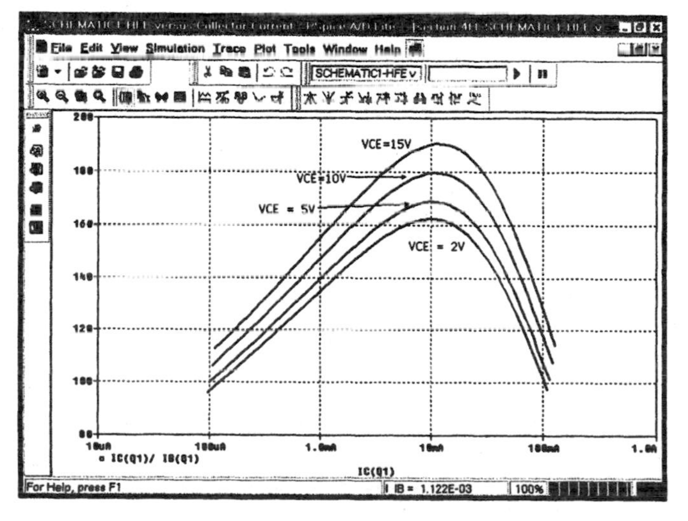

## 4.F.3. $H_{FE}$ Versus Temperature

We will now generate $H_{FE}$ versus $I_C$ curves at different temperatures. This is a typical curve found in most data sheets for BJTs. We will use almost the same Secondary Sweep that we used in Section 4.F.2, except that for the Secondary Sweep we will vary temperature rather than $V_{CE}$. We will use the same circuit and setup as in Section 4.F.2, but we will modify the Nested Sweep. We will start with the circuit from the previous example:

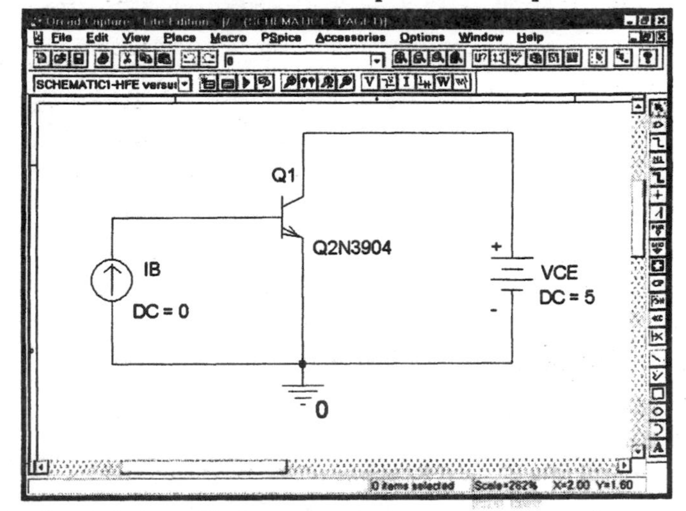


Since we have already created a simulation profile from the previous example, we do not need to create a new profile. To open the previous profile, select **PSpice** and then ***Edit Simulation Profile*** from the Capture menus:

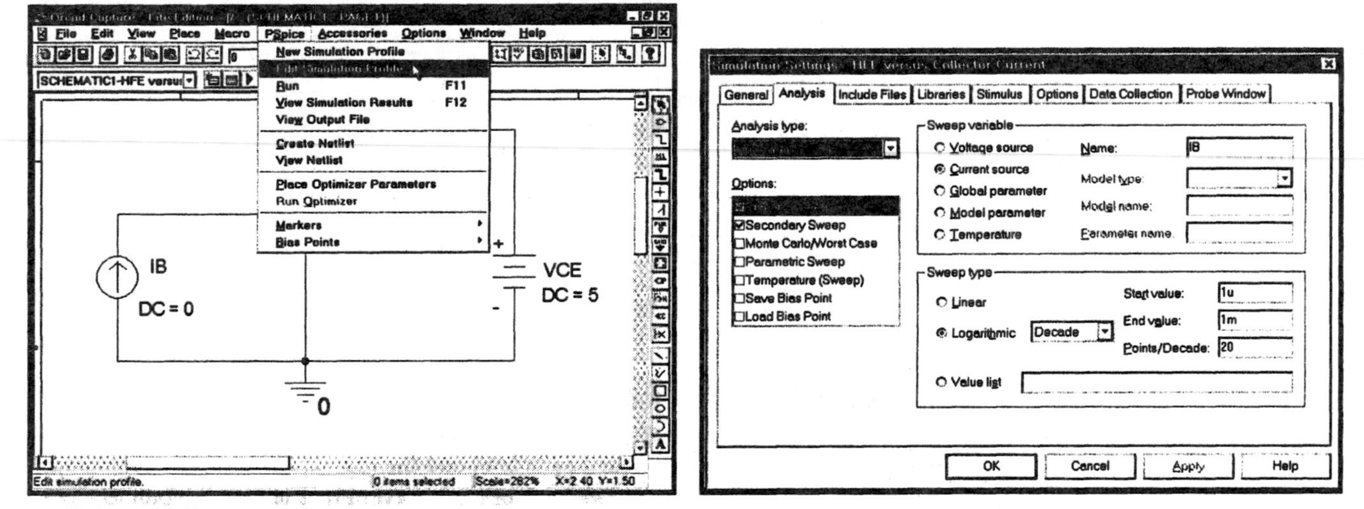


The ***Primary Sweep*** is set up from the previous example. It is set to sweep IB from 1 μA to 1 mA at 20 points per decade. Click the ***LEFT*** mouse button on the text ***Secondary Sweep*** to display its setup:

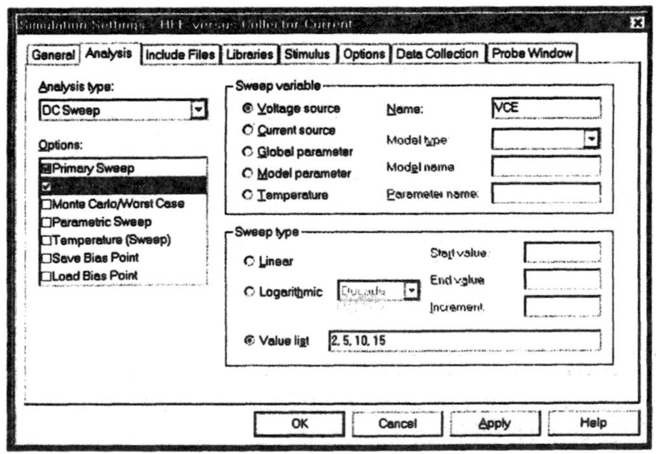


From the previous example, the Secondary Sweep is set to sweep VCE. We would simulate the circuit at –25°C, +25°C, and +125°C, so we need to modify the settings as shown:

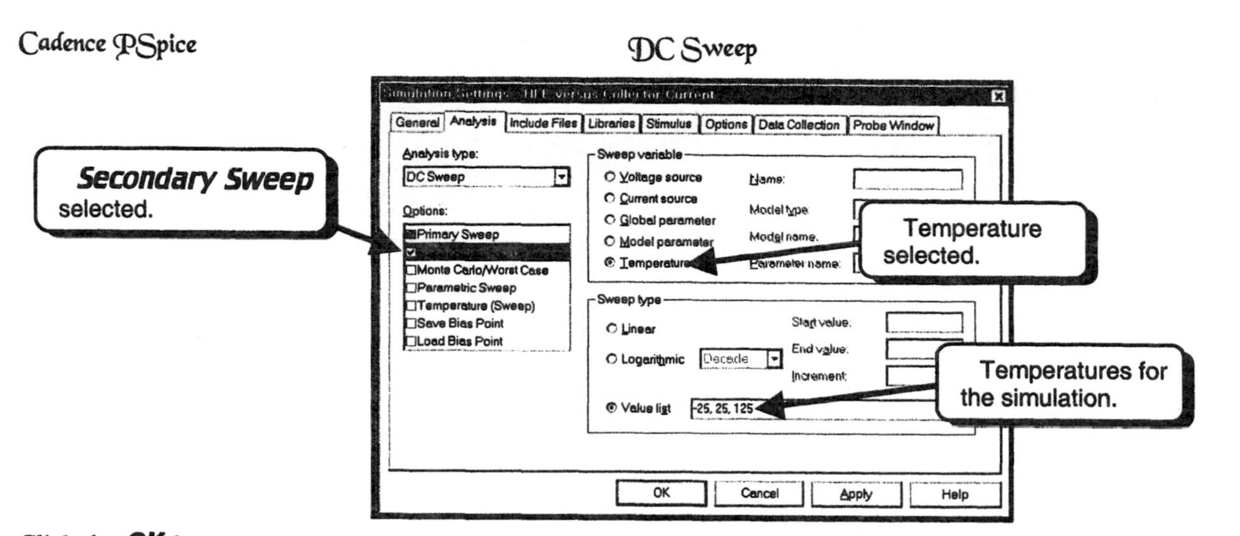


Click the ***OK*** button to accept the settings and return to the schematic. Run the simulation and then display the results with Probe. Generate a plot of $H_{FE}$ versus $I_C$. Three curves will be shown:

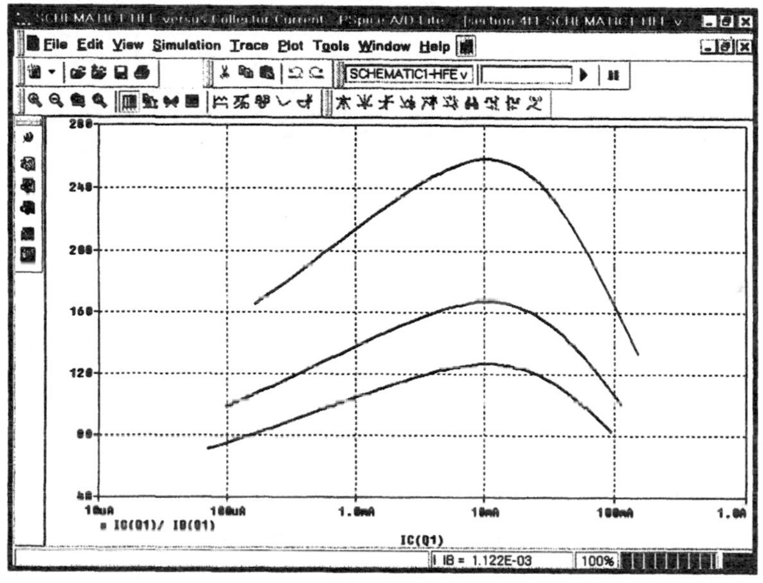

$H_{FE}$ increases with temperature so the plot should be labeled as:

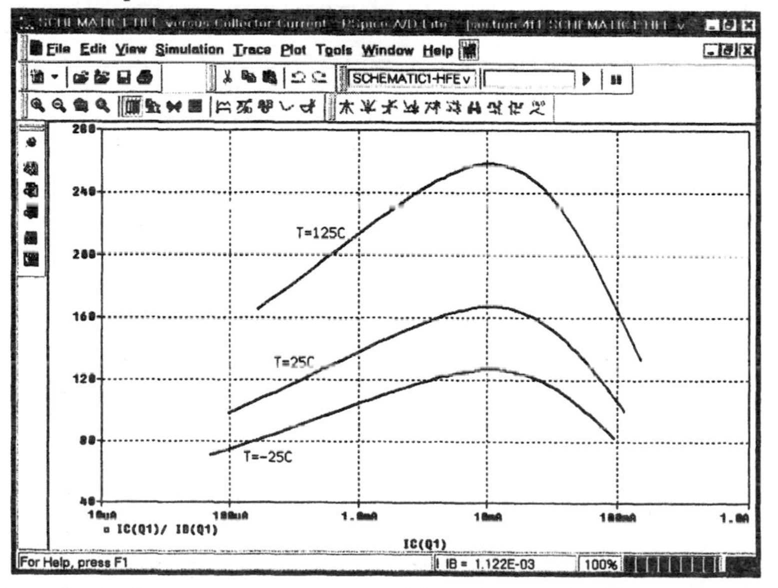


If you do not know which trace is at what temperature, you would need to run the $H_{FE}$ versus $I_C$ simulation three times, once at each temperature, to identify which curve is which.

The above plot shows the interesting result that at any particular temperature, the maximum value of $H_{FE}$ always occurs at approximately the same value of $I_C$ (10 mA in the screen capture below).

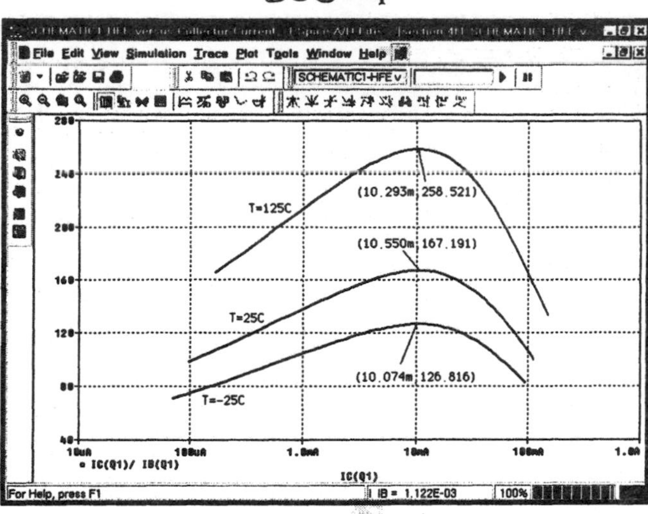

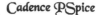

An interesting question is, how does $H_{FE}$ vary with temperature when $V_{CE}$ is constant at 5 V and $I_C$ is constant at 10 mA? We will need a new circuit to perform this simulation:

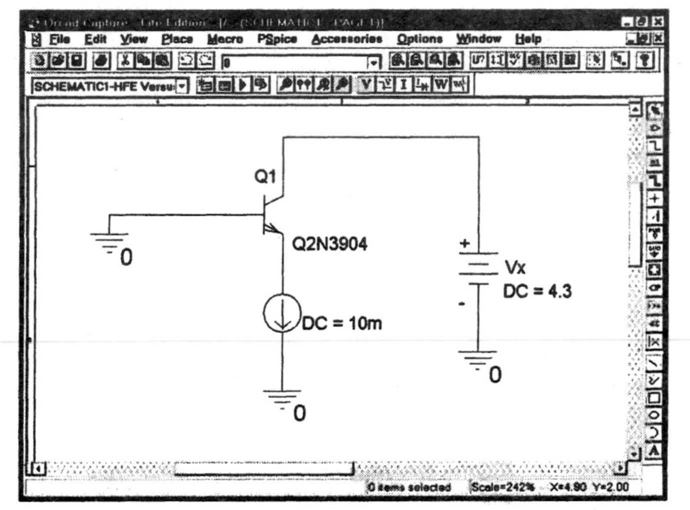


The constant current source keeps the emitter current constant at 10 mA. Since the collector current is approximately equal to the emitter current, the constant current source keeps the collector current approximately constant at 10 mA. Since the base is grounded, the emitter voltage is a diode drop below ground:

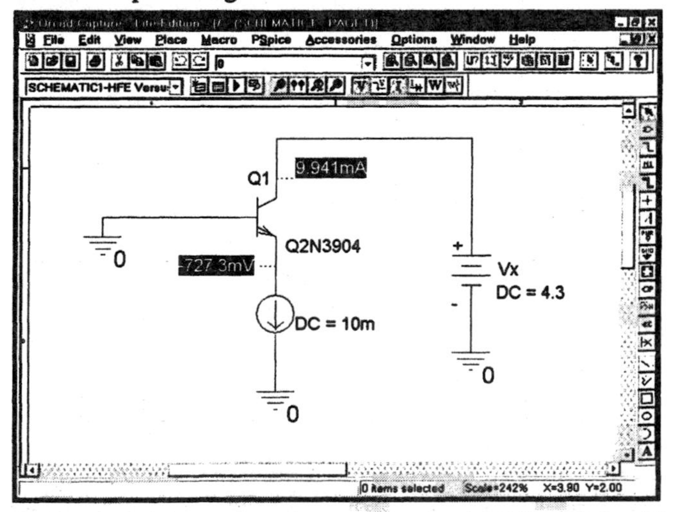


Since the emitter is approximately 0.7 volts below ground, we set Vx to 4.3 volts so that $V_{CE}$ is approximately 5 V. As the temperature changes, $V_{BE}$ will change, and $V_{CE}$ will drift away from 5 V. However, $V_{CE}$ is approximately 5 V for the simulation.

We now need to set up a DC Sweep to vary the temperature from –25 to 125°C. We need only the Primary Sweep, not both the Primary and Secondary sweeps. Set up the DC Sweep dialog box as shown:

Run the simulation and the display the results with Probe. To show that the circuit keeps $I_C$ and $V_{CE}$ relatively constant versus temperature, we will plot $I_C$ and $V_{CE}$ versus temperature:

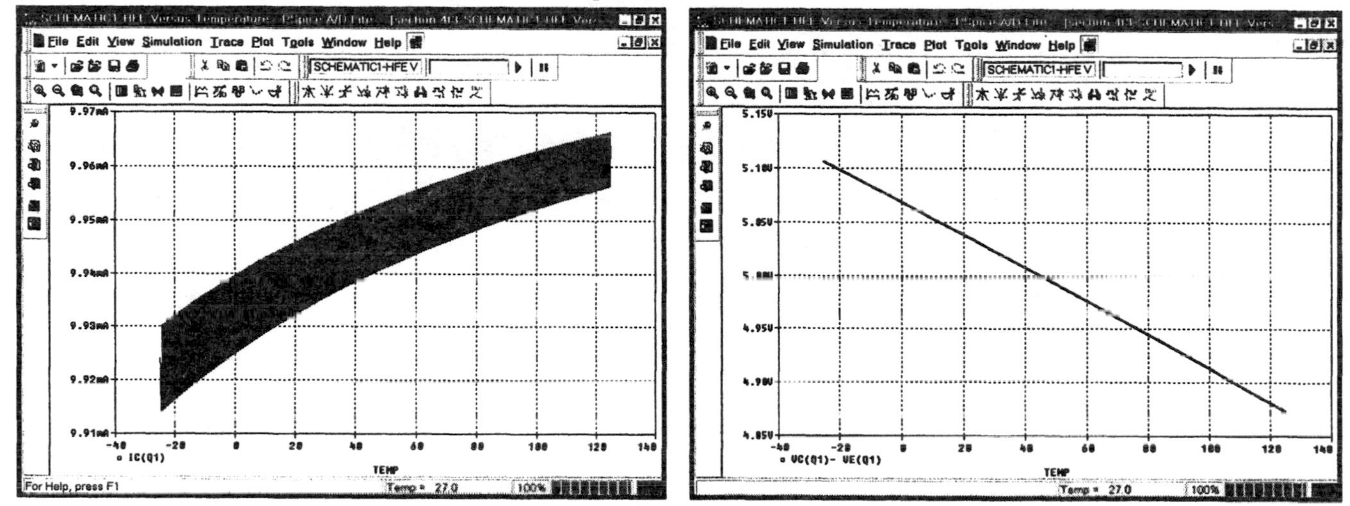


We see that $I_C$ varies by about 40 μA and $V_{CE}$ varies by about 200 mV over the range of the simulation. For the purposes of this simulation, these quantities will be assumed to be constant. Notice that the x-axis is automatically set to temperature. To plot $H_{FE}$ versus temperature all we need to do is add the trace IC(Q1)/IB(Q1). The x-axis will automatically be set to temperature:

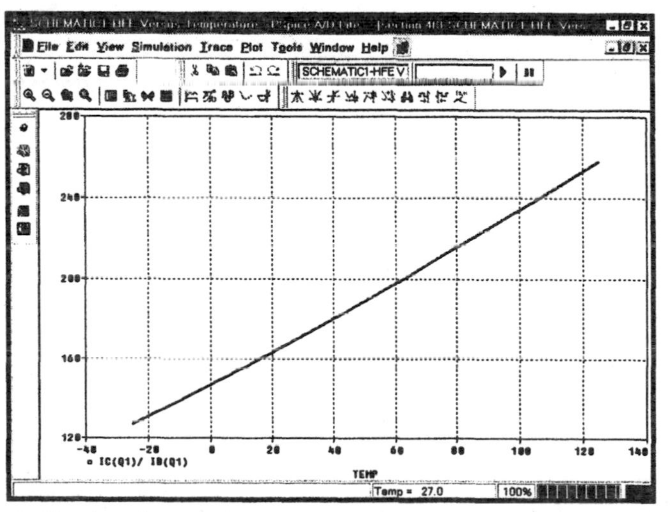


The plot is almost a straight line, indicating that $H_{FE}$ is approximately proportional to temperature.

**EXERCISE 4-12**: For a 2N3904 transistor, if $I_C$ is held constant at 10 mA, how does the base-emitter voltage vary with temperature?

**SOLUTION**: Use the simulation of the previous example and plot the base-emitter voltage:

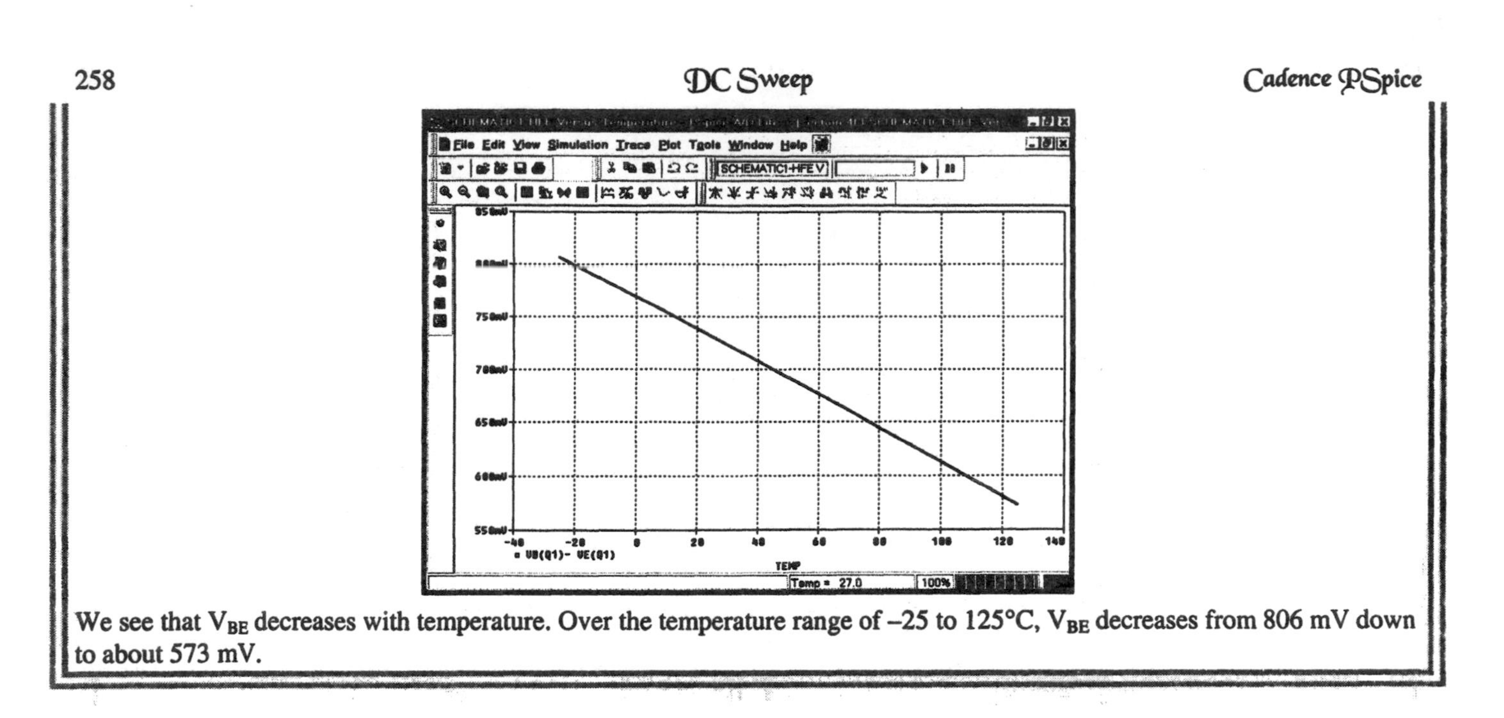


We see that $V_{BE}$ decreases with temperature. Over the temperature range of –25 to 125°C, $V_{BE}$ decreases from 806 mV down to about 573 mV.

# 4.G. Temperature Analysis — Constant Current Sources

To demonstrate temperature effects, we will look at two circuits that can be used as constant current sources. One circuit will be greatly affected by temperature and the other is designed to be relatively independent of temperature. The circuits use transistors and resistors. The temperature dependence of transistors has already been discussed in detail in Section 4.F. Before we look at the circuits, we will look at how PSpice handles temperature characteristics of resistors.

## 4.G.1. Temperature Characteristics of a Resistor

When you place a resistor part called R ( + ) in your circuit, you are using an ideal resistor; that is, the resistor has no tolerance and no temperature dependence. The resistor is exactly the value you specify and the resistance does not change as the temperature changes. In practice, all resistors have tolerance and temperature variations. If you use a 1 kΩ resistor with ±5% tolerance, the resistor will not be exactly equal to 1000 Ω. Its value could be anywhere between 950 Ω and 1050 Ω. Temperature variations are anywhere from 50 parts per million (ppm) to 400 ppm. In PSpice we can specify both temperature variations and resistor tolerance. Tolerance will be covered in Part 9.

PSpice has three ways to specify temperature dependence. Three model parameters named TC1, TC2, and TCE are available. TC1 and TC2 may be used together. If you use TCE, then TC1 and TC2 are ignored. Suppose we have a resistor with a resistance equal to Rval. If TCE is specified, then the resistance is:

$$\text{resistance} = \text{Rval}\left[1.01^{\text{TCE}(\text{T}-\text{Tnom})}\right]$$

TCE is called the exponential temperature coefficient. If TCE is specified, then the values of TC1 and TC2 are ignored. If TCE is not specified, then the resistance is

$$\text{resistance} = \text{Rval}\left[1+\text{TC1}(\text{T}-\text{Tnom})+\text{TC2}(\text{T}-\text{Tnom})^2\right]$$

TC1 is referred to as the linear temperature coefficient and TC2 is referred to as the quadratic temperature coefficient. Tnom is the nominal temperature. Its default value is 27°C.

The default values of TCE, TC1, and TC2 are zero. Thus, if you do not specify any of the coefficients, there will be no temperature dependence. In all of our previous simulations the resistors had no temperature dependence. Thus, if we varied temperature in any of the simulations, the resistors would not be affected. To add temperature dependence we must specify the temperature coefficients.

There are two ways to specify the temperature coefficients. One way is to create a resistor model and specify TC1, TC2, or TCE. The second way is to specify numerical values for the TC1 and TC2 attributes of a resistor in the circuit. We will look first at the resistor attributes. Place a resistor part (R) in a schematic:

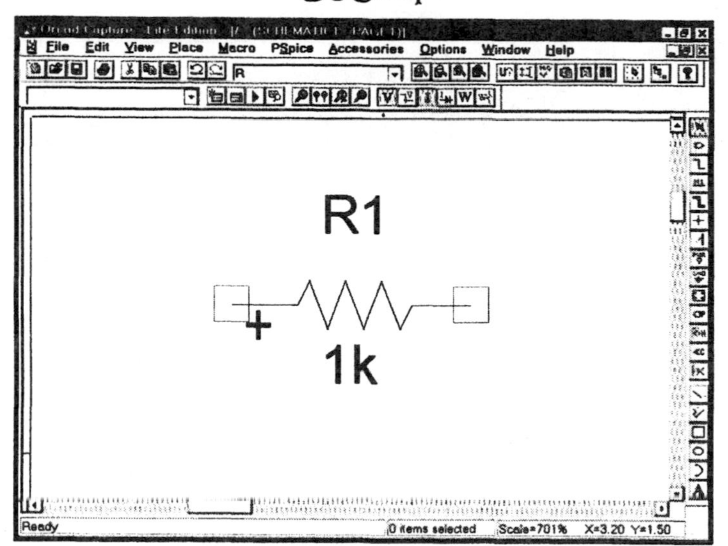


Double-click on the resistor graphic to edit its spreadsheet:

New Column... Apply Display... Delete Property Filter by: < Current properties >

| | | Color | COMPONENT | Designator | Graphic | ID | Implementation | Impleme |
|---|---|---|---|---|---|---|---|---|
| 1 | SCHEMATIC1 : PAGE1 : R1 | Default | RES | | R.Normal | | | |

Parts / Schematic Nets / Pins / Title Blocks / Globals / Ports

Ready

Scroll the sheet to the right until you see the ***TC*** column:

New Column... Apply Display... Delete Property Filter by: < Current properties >

| | | Source Package | Source Part | TC | Value |
|---|---|---|---|---|---|
| 1 | SCHEMATIC1 : PAGE1 : R1 | R | R.Normal | | 1k |

Parts / Schematic Nets / Pins / Title Blocks / Globals / Ports

Ready

The attribute ***TC*** is used to specify TC1 and TC2. TC2 is optional. To specify TC1 = .0005 and TC2 = 0 we would set TC = **0.0005** as shown below:

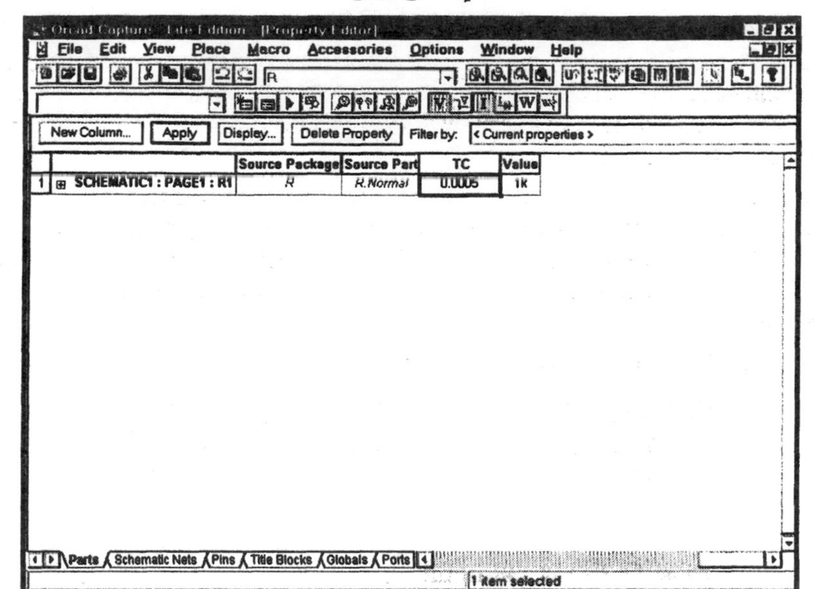


To specify TC1 = 0 and TC2 = 0.00003 we would specify TC = **0,0.00003**

To specify TC1 = 0.0005 and TC2 = 0.00003 we would use the line TC = **0.0005,0.00003**

The advantage to using this method is that each resistor in the circuit can have different values for TC1 and TC2. Thus some resistors can have no temperature dependence, some resistors can have the same temperature dependence, and some can have a different temperature dependence. The disadvantage to using this method is that if you have a circuit with many resistors, all from the same manufacturer and all with the same temperature characteristics, you must specify the coefficients for each resistor individually even though they are the same.

The second method is a resistor model that specifies temperature dependence. Every resistor that uses the model will have the same temperature dependence. Thus, if you have a circuit with many resistors, all from the same manufacturer and all with the same temperature characteristics, the same model can be used to specify the temperature characteristics of all of the resistors. Models will be covered in more detail in Part 7. Here we will just discuss making a model for resistors that includes temperature dependence.

To create a resistor model, place a part called Rbreak in your circuit:

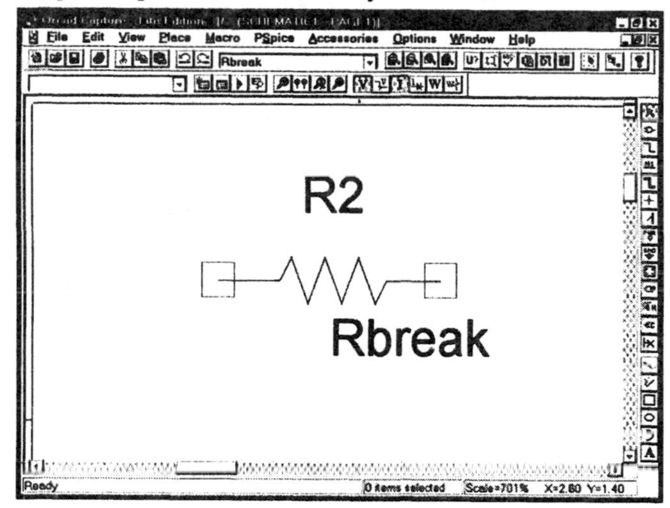


Click the ***LEFT*** mouse button on the resistor graphic to select it. The graphic should turn pink, indicating that it has been selected. Select **Edit** and then **PSpice Model** from the Capture menus*:

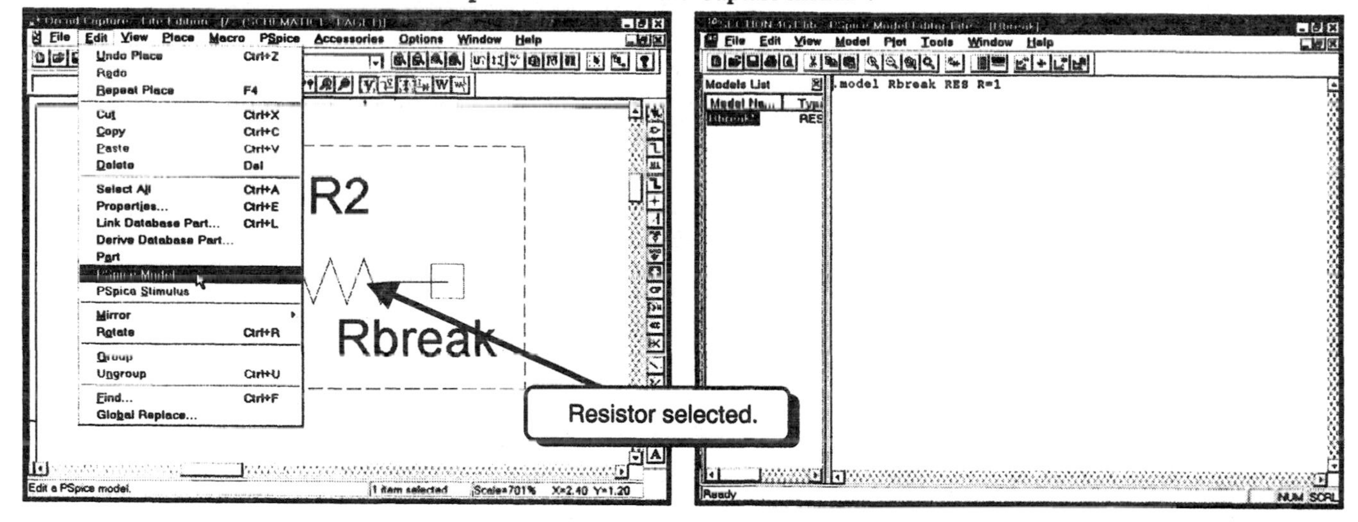


This window is a text editor. The name of the model is Rbreak. The only parameter specified in the model is ***R=1***. R is a model parameter used to specify the value of the resistance. If the value of the resistor specified in the schematic is x, then the actual value of the resistor used for calculations is x•R. Since R is set to 1, the value of resistance specified in the schematic is also the value used in calculations. We can delete the model parameter ***R=1*** or we can leave it in the model. It will have no effect on the simulations. We will now add the temperature coefficients.

---

*If the menu selection **PSpice Model** appears grayed out in the menu (**PSpice Model**), you will not be able to choose the **PSpice Model** menu selection. Attempt to select the diode graphic again until its graphic, , is highlighted in pink.

To specify a resistor with an exponential temperature coefficient of 0.0007 we would modify the model as follows:

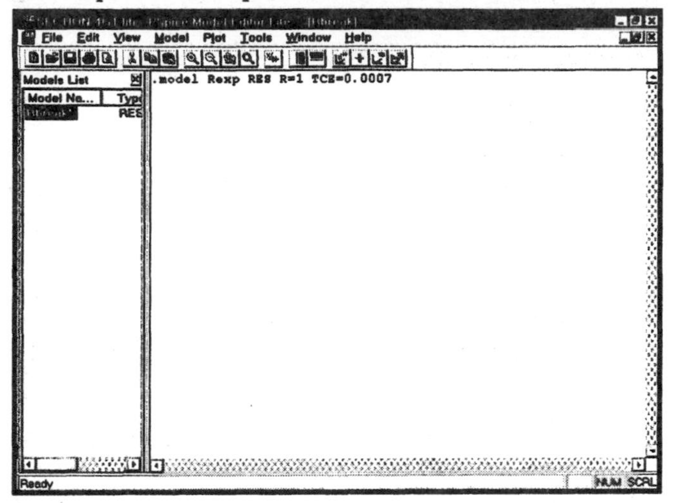
.model Rexp RES R=1 TCE=0.0007

We have renamed the model to ***Rexp*** and added the model parameter TCE. Select File and then Save from the menus to save the new model and add it to the list:

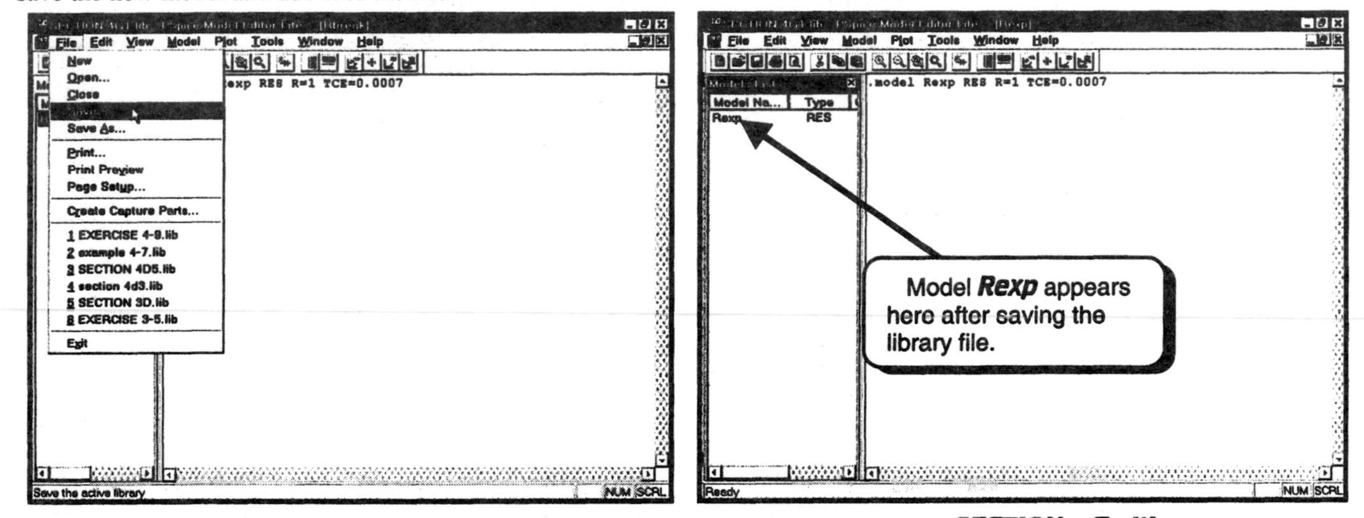
Model Rexp appears here after saving the library file.

The name of this project is Section 4G1.opj so the model is saved in a file named ***SECTION 4G1.lib***. After saving the model, the name of the model appears in the list of models saved in library ***SECTION 4G1.lib***. Presently, only one model is saved in this library, ***Rexp***. Select **File** and then **Exit** to accept the model and return to the schematic:

Notice that the model name ***Rexp*** is now displayed next to the resistor. Rbreak is referred to as a breakout part. Since breakout parts are specifically used to create models, the name of the model used by the part is displayed on the schematic.

Place three more Rbreak parts in your schematic:

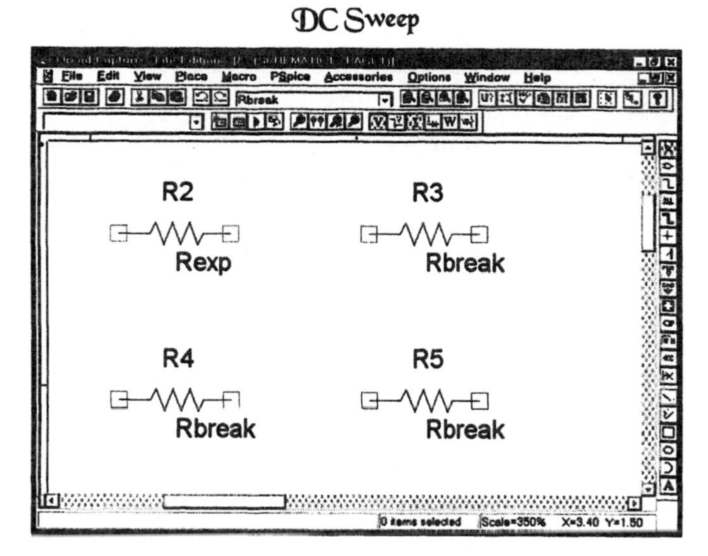


We will now create two more models. Click the ***LEFT*** mouse button on the resistor graphic of ***R3*** to select it. The graphic should turn pink, indicating that it has been selected. Select **Edit** and then **PSpice Model** from the Capture menus. This will allow us to edit the model for part Rbreak and create another new model:

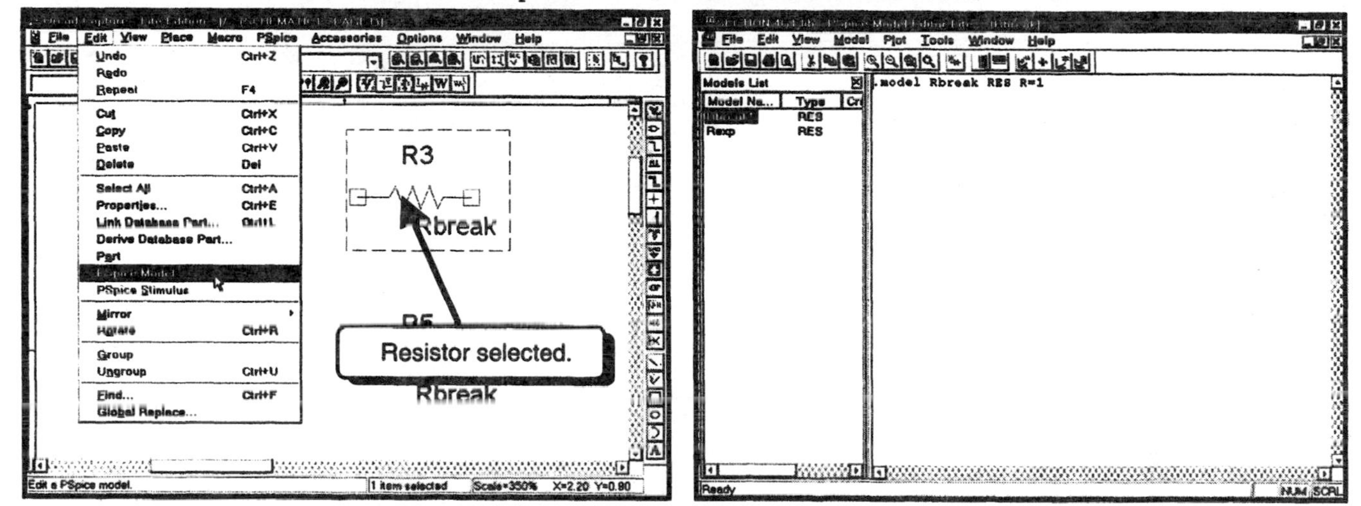


We will now create a new model with TC1 = 0.0001. TC2 and TCE will be zero by default. We will call this model Rlinear:

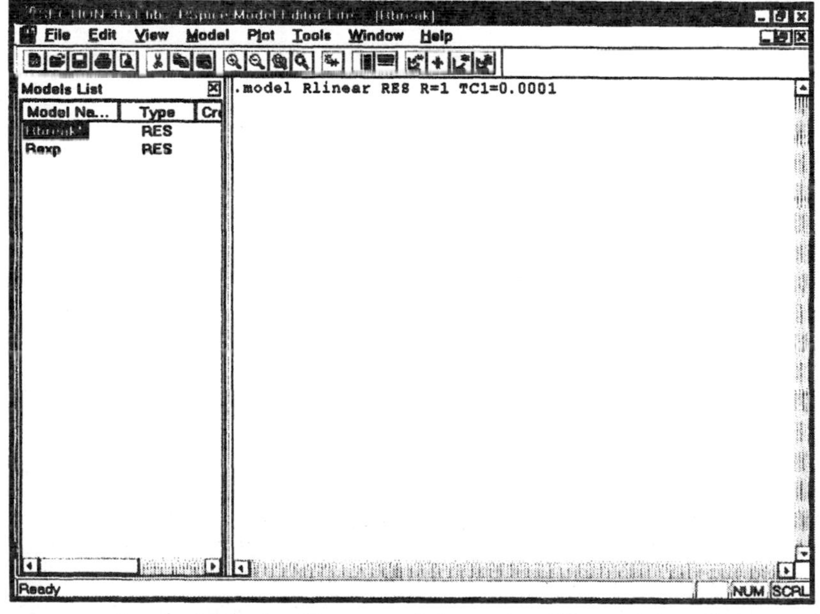


Select **File** and then **Save** to save the model in the library, and then select **File** and then **Exit** to return to the schematic. R3 should be displayed with the new model name next to it:

Next we will create a model for R4 that has quadratic temperature dependence. Click the ***LEFT*** mouse button on the resistor graphic for R4 to select it. The graphic should turn red, indicating that it has been selected. Select **Edit** and then **PSpice Model** from the Capture menus. This will allow us to edit the model for part Rbreak and create another new model:

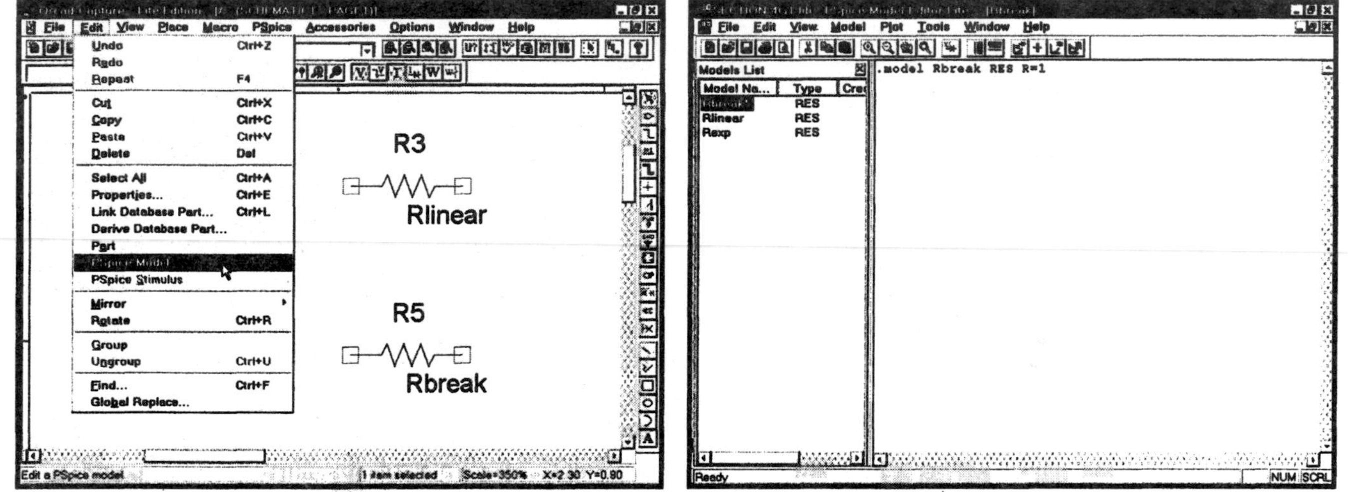


We will now create a new model with TC1 = 0.0001 and TC2 = 0.00005. We will call this model Rquad:

Select **File** and then **Save** to save the model in the library, and then select **File** and then **Exit** to return to the schematic. R4 should be displayed with the new model name next to it:

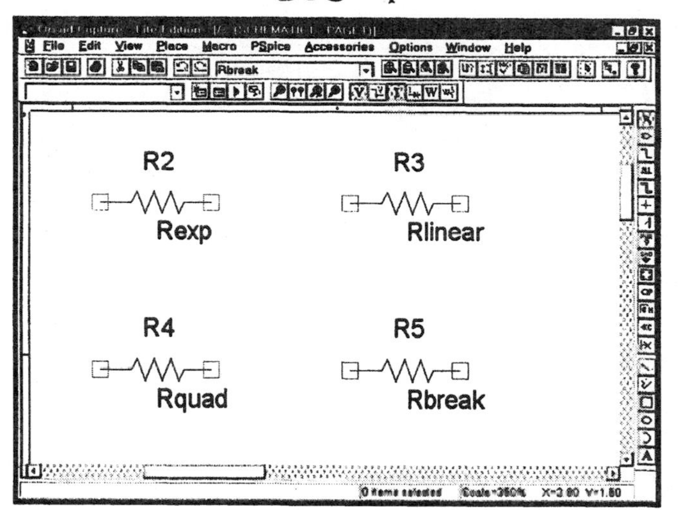


Last, we will show how to use one of the new models for ***R5***. Double-click on the text Rbreak:

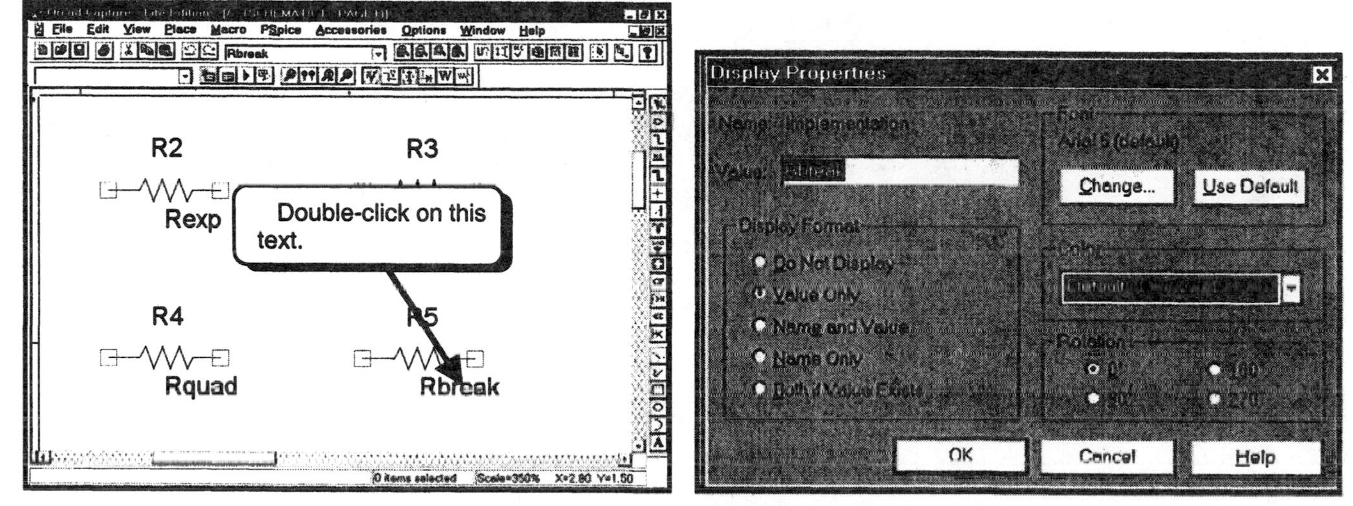


Enter a name for the model you would like to use and then click the ***OK*** button:

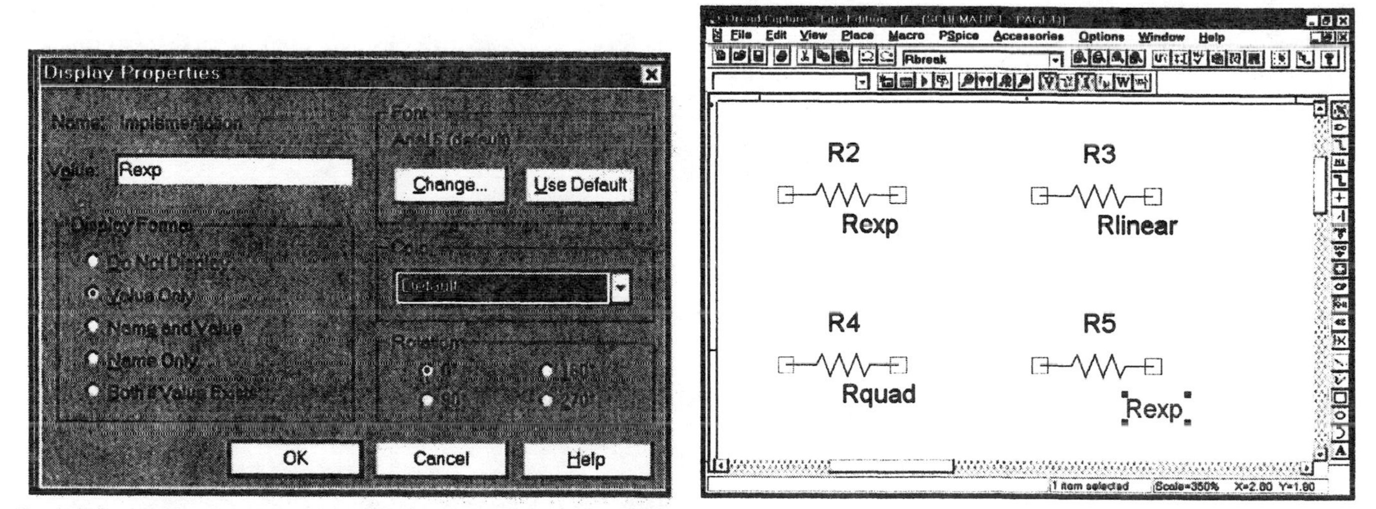


Both R2 and R5 use the same model. If we had 100 resistors, we would use this last method to make them all use the same model. Thus we would have to create only a single model, and then change the model reference of each resistor.

To get an idea of how these resistors change with temperature, we will plot their resistance from –25°C to 125°C. Create the circuit below. Resistors R2, R3, and R4 use the models we just created. R5 has been deleted. Rx is a regular resistor (get a part named R) that has no temperature dependence.

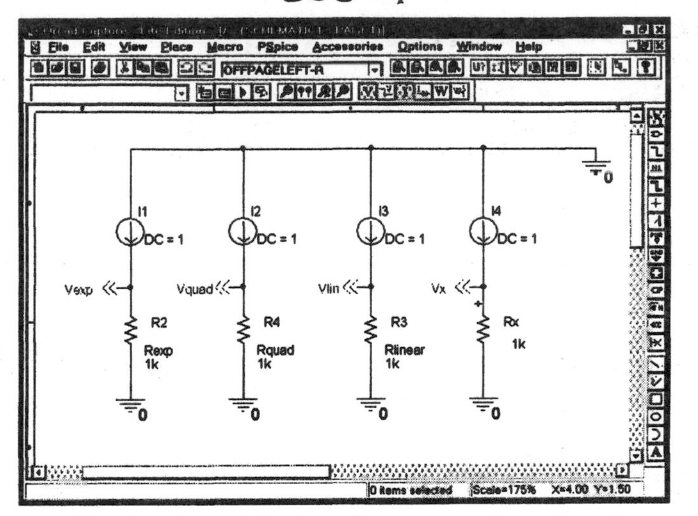


Since each resistor has 1 A of current flowing through it, the voltage across the resistor is numerically equal to the resistance. We will set up a DC Sweep to vary the temperature. Select **PSpice** and then **New Simulation Profile,** give the profile a name, and then click the ***Create*** button. Select an ***Analysis type*** of ***DC Sweep***. Fill in the dialog box as shown:

General | Analysis | Include Files | Libraries | Stimulus | Options | Data Collection | Probe Window

Analysis type: DC Sweep

Options: Secondary Sweep, Monte Carlo/Worst Case, Parametric Sweep, Temperature (Sweep), Save Bias Point, Load Bias Point

Sweep variable: Voltage source, Current source, Global parameter, Model parameter, Temperature (selected)

Name: / Model type: / Model name: / Parameter name:

Sweep type: Linear (selected), Logarithmic (Decade), Value list

Start value: -25 / End value: 125 / Increment: 1

OK | Cancel | Apply | Help

The ***Sweep variable*** is set to ***Temperature***. We have chosen a ***Linear*** sweep with temperature going from –25°C to 125°C in 1°C increments. Click the ***OK*** button to return to the schematic. Select **PSpice** and then **Run** to simulate the circuit. When the simulation is complete, display the traces V(Vexp), V(Vquad), V(Vlin), and V(Vx) to display the value of each resistor (to add traces press the **INSERT** key):

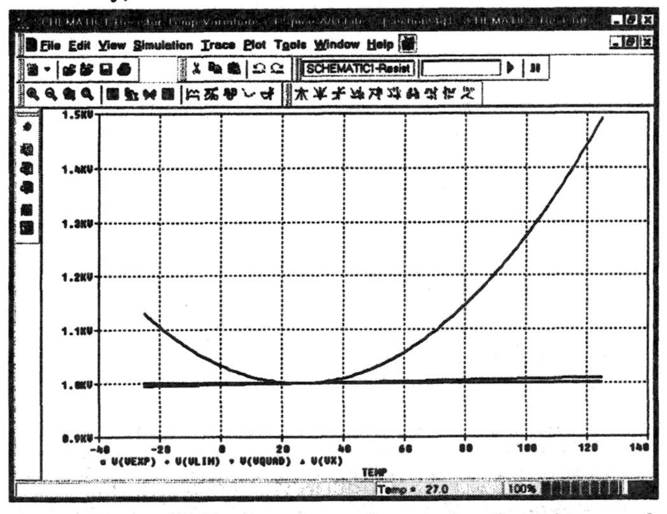


The resistor with the quadratic temperature coefficient appears to have a drastic resistance change. This is unrealistic because the coefficient we chose was too large. Remove trace V(Vquad) so that we can see the other traces more clearly:

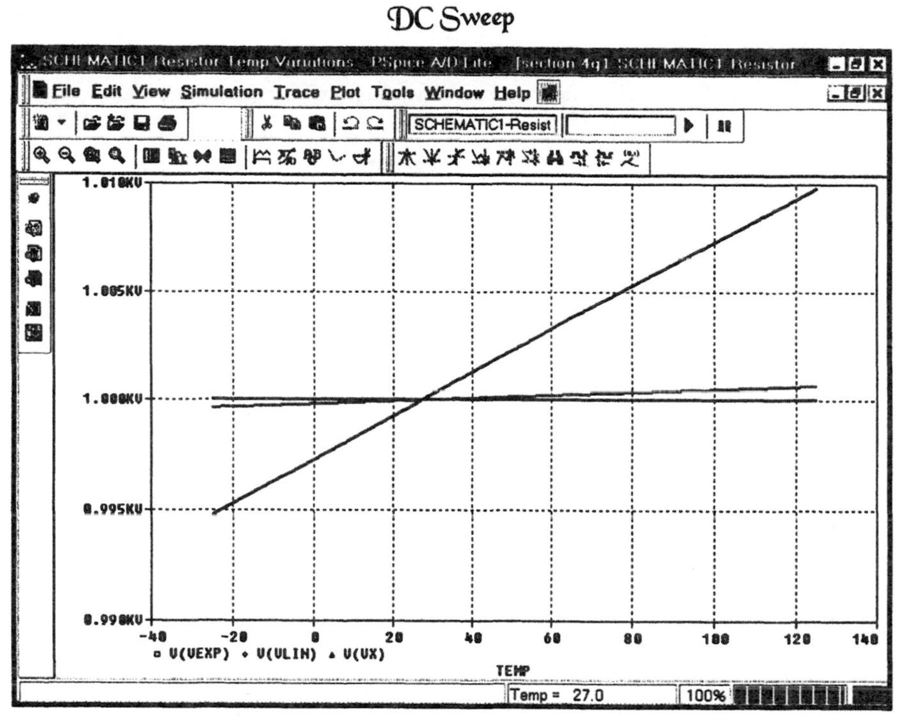


We can now see the temperature variations more clearly. Note that V(Vx) was constant. V(Vx) was the voltage across the ideal resistor and has no temperature variation.

The curves generated here are arbitrary because we just randomly picked the temperature coefficients. To accurately model your resistors, you would need to get a data sheet on the resistors you are using and find out if the temperature dependence is linear, quadratic, or exponential, and also find the correct coefficients. The coefficients used here were just for illustration.

## 4.G.2. BJT Constant Current Source

For an example of a temperature dependence, we will look at the circuit below:

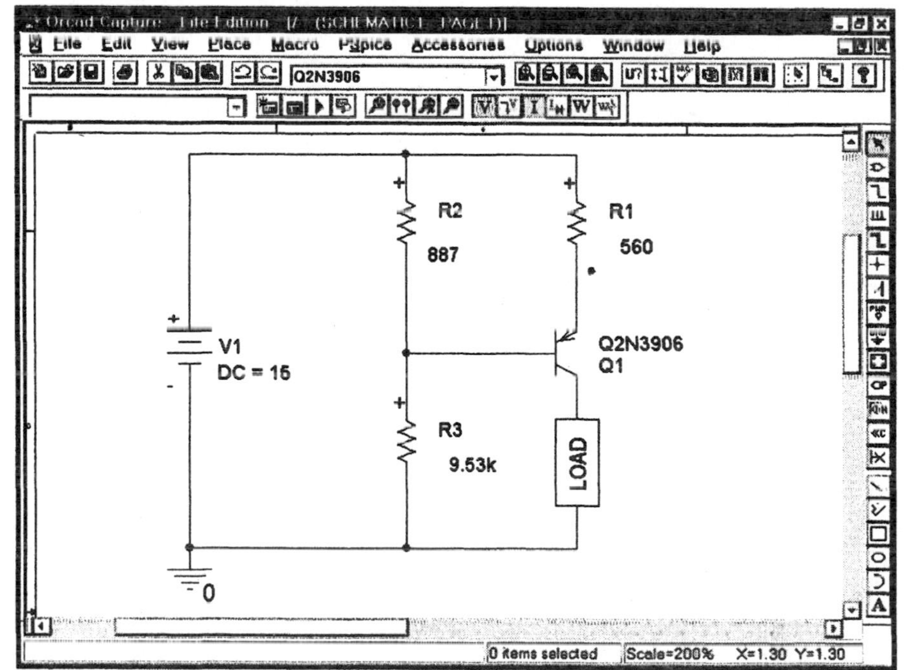


There are two ways to look at this circuit. One is that this is a self-biasing circuit for a BJT amplifier. If this were an amplifier, the load would most likely be a resistor. (The term "self-bias" means that the goal of the circuit is to make the collector current independent of device parameters such as $H_{FE}$ and $V_{BE}$.). The second way to look at the circuit is that as far as the load is concerned, Q1, R1, R2, and R3 form a current source; that is, the current through the load is determined by Q1, R1, R2, and R3. If this circuit were designed as a current source for the load, we could view the circuit as:

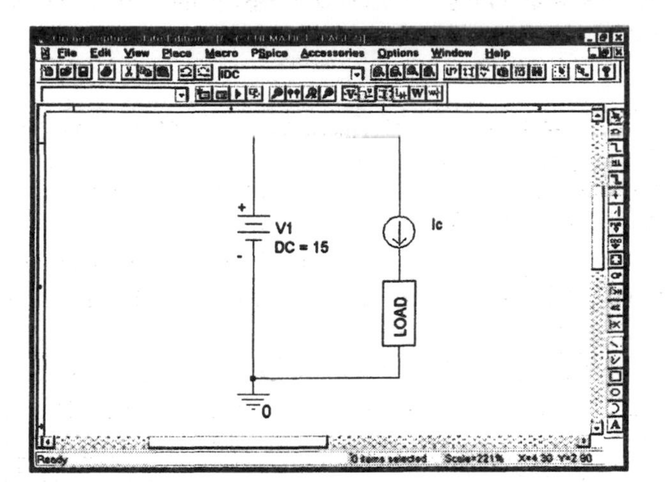


No matter what the application, the circuit is designed to make the collector current independent of $V_{BE}$ and $H_{FE}$. We will assume that the resistors used have no tolerance and have their exact values specified in the circuit. To see how the tolerance of resistors affects the collector current, see Section 9.C. From previous sections we know that the resistance of a resistor, and $H_{FE}$ and $V_{BE}$ of a BJT, are affected by temperature. The question is, if the temperature changes, how much does the collector current change?

We will perform two analyses on the circuit. The first will be a plot of collector current versus temperature with ideal resistors. The resistors will have no temperature dependence. The only temperature-dependent device will be the BJT. The second simulation will use temperature dependence for resistors and the BJT. For simulation, we will use the circuit below:

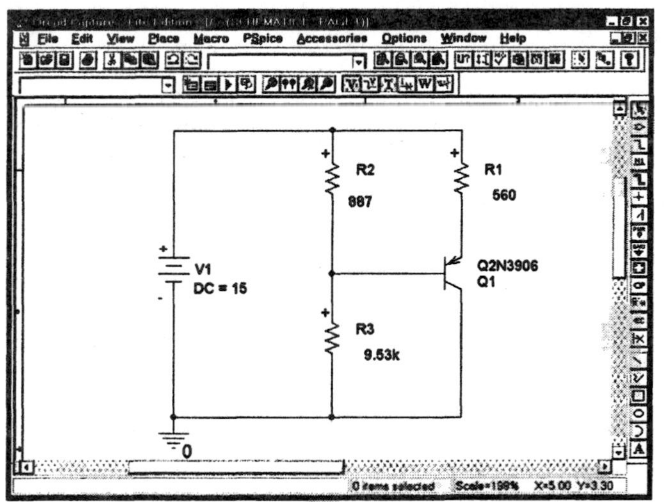


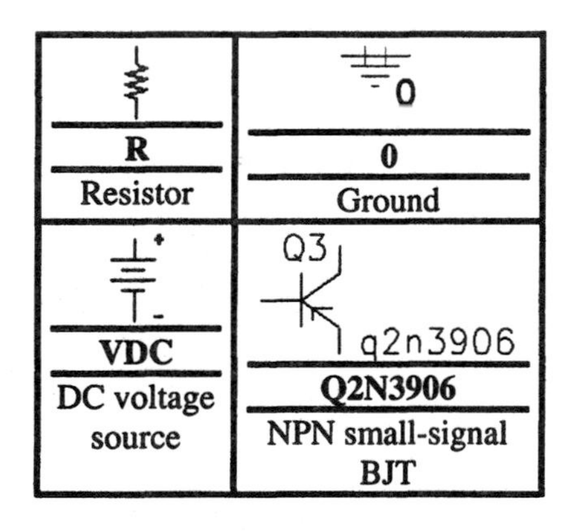

| R<br>Resistor | 0<br>Ground |
|---|---|
| VDC<br>DC voltage source | Q2N3906<br>NPN small-signal BJT |

First, we will use PSpice to find the collector current without temperature dependence using a bias point analysis and display the collector current on the schematic. See Section 3.A for the procedure to display currents on the schematic:

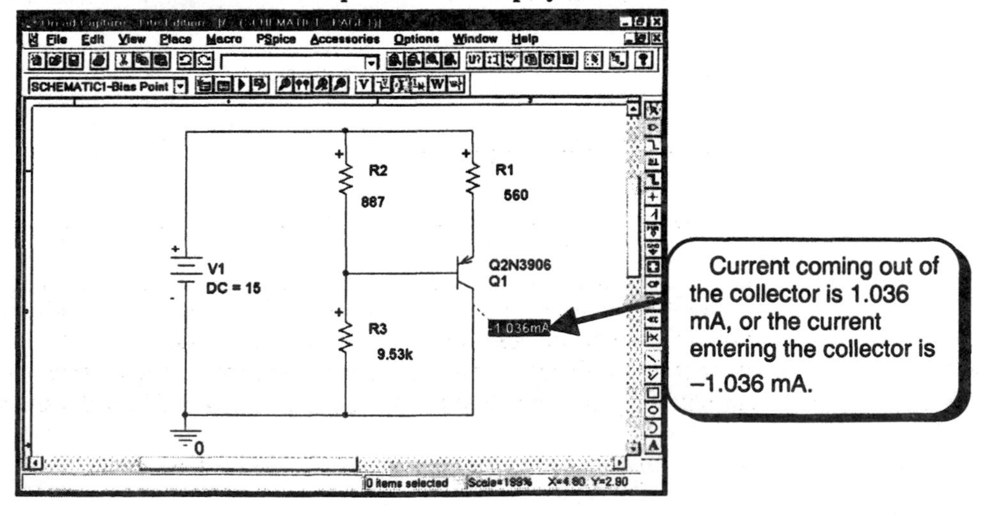

The resistors in this example were tweaked slightly to achieve a current close to 1 mA. We will now see how this circuit performs with temperature changes.

## 4.G.2.a. Current Versus Temperature — No Resistor Temperature Dependence

Our first simulation will not specify temperature coefficients for any of the resistors. This will make the resistors independent of temperature. The model for the 2N3906 includes temperature dependence and will be the only device in the circuit that varies with temperature. Both $H_{FE}$ and $V_{BE}$ are functions of temperature, and these parameters will affect $I_C$.

We will set up a temperature sweep from –25°C to +125°C. Select **PSpice** and then **New Simulation Profile** from the Capture menus, specify a name for the profile, and then click the ***Create*** button. Select the ***DC Sweep Analysis type*** and fill in the dialog box as shown:

We have specified a linear sweep from –25°C to 125°C with 1-degree increments. Click the ***OK*** button to the schematic. Simulate the circuit and display the results with Probe. Add the trace `IC(Q1)` to plot the collector current. Use the cursors to label the end points of the range.

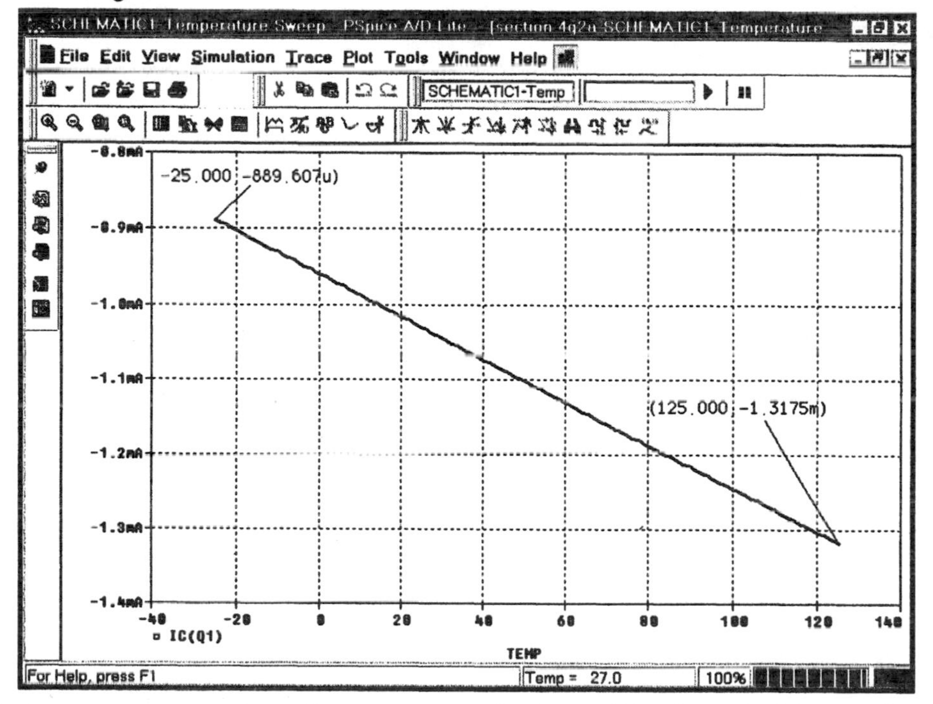


We see that, over the temperature range, the collector current varies from –889 μA to –1.317 mA. It is not very independent of temperature. Remember that this variation is without resistor temperature dependence. We could make the circuit more temperature independent by choosing smaller values for R2 and R3. However, this will consume more power.

## 4.G.2.b. Current Versus Temperature — With Resistor Temperature Dependence

We will now add temperature dependence to the previous circuit. A typical 1% resistor has a linear temperature coefficient of 100 parts per million, or $100\times10^{-6}$. We will create a resistor model with this temperature coefficient. First, we will replace all resistors in the above circuit with the Rbreak part:

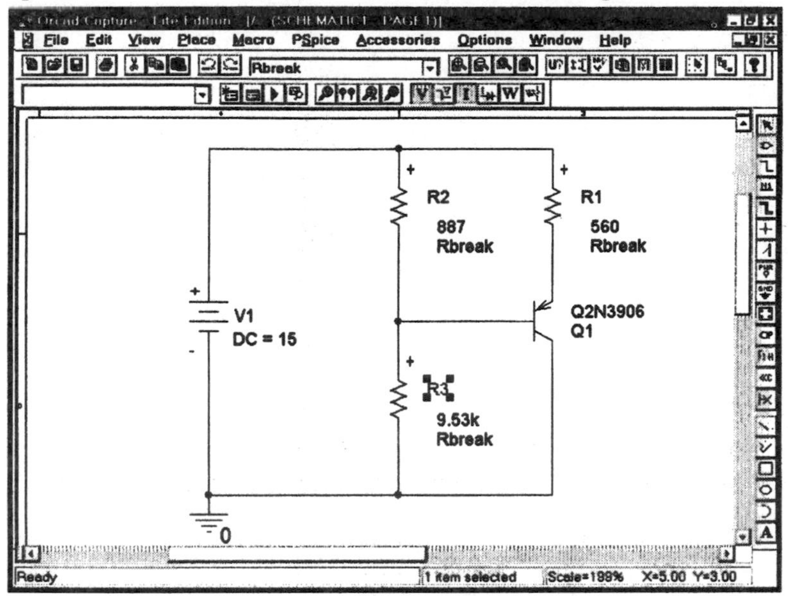


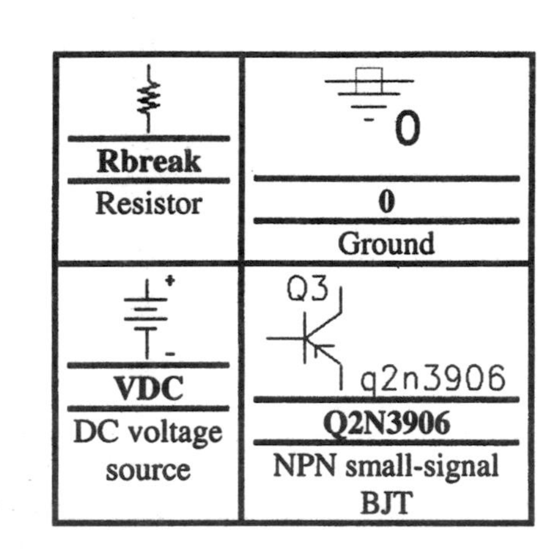


We will now edit the Rbreak model and create a model with the specified temperature coefficient. Click on the graphic for one of the resistors to select it. It should turn pink, indicating that it has been selected. After a resistor is selected, select **Edit** and then **PSpice Model** from the menus:

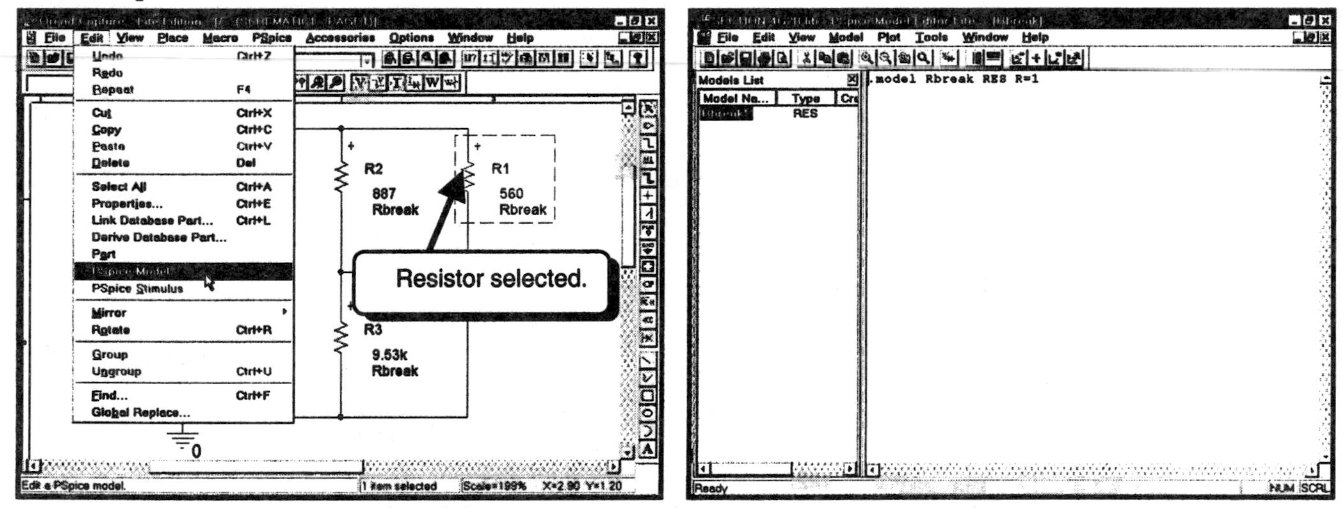


Modify the model as shown:

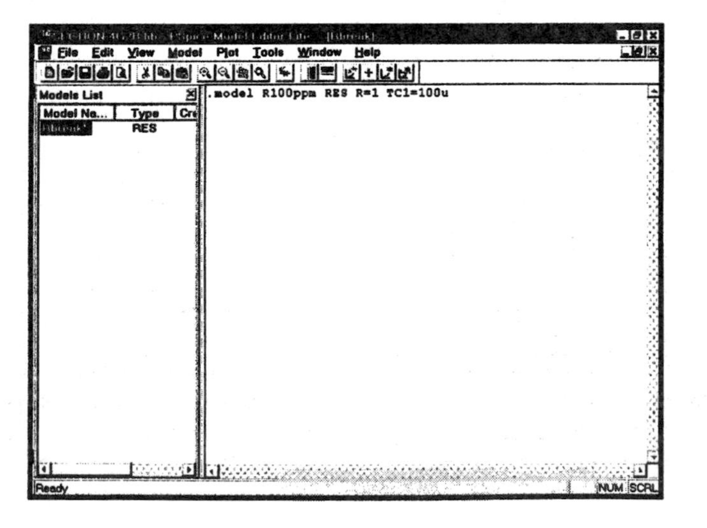

We have named the model ***R100ppm*** and specified the value of TC1 to be 100u, or $100 \times 10^{-6}$. We will use this model for all of the resistors. Select **File** and then **Save** to save the model, and then select **File** and then **Exit** to close the model editor. When you return to the schematic, the resistor model name should be changed to ***R100ppm***:

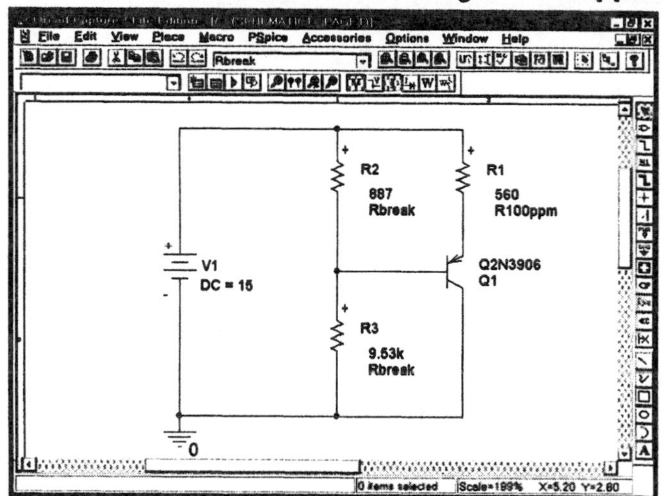


Change all of the other model names from Rbreak to R100ppm. To do this, double-click the ***LEFT*** mouse button on the text Rbreak and then enter the new model name in the dialog box.

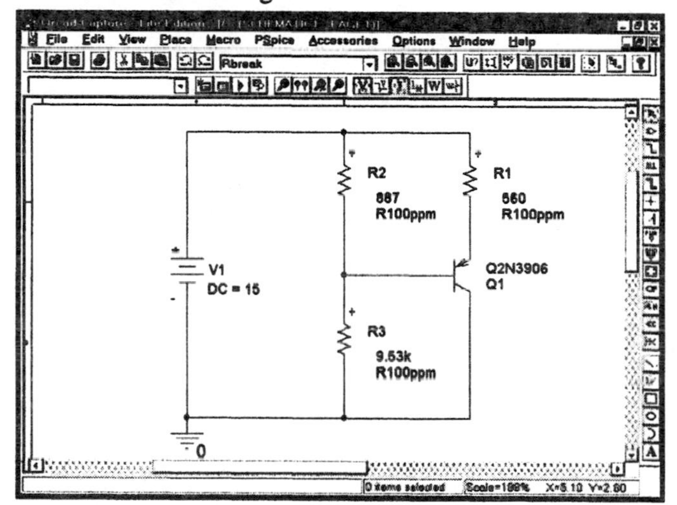


We will now run the same temperature sweep that we did in the previous section. Plot IC(Q1) and use the cursors to label the values at the end points of the trace.

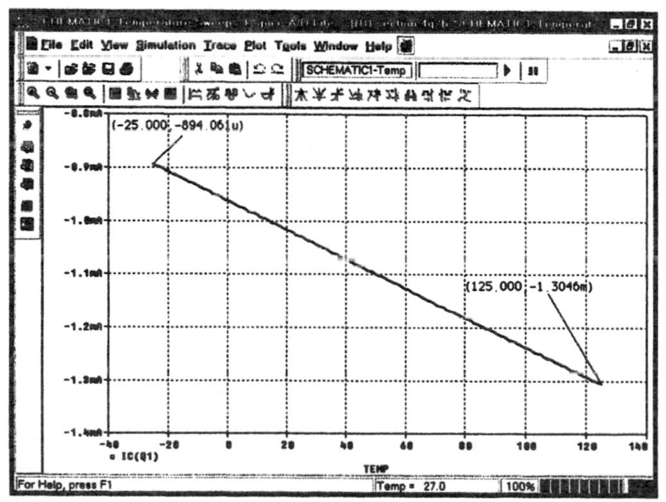


We see that with both resistor and BJT temperature dependence, the collector varies from –894 μA to –1.304 mA. In our simulation with only BJT temperature dependence, the collector current ranged from –889 μA to –1.317 mA. Our conclusion is that most of the temperature dependence in this circuit is due to the BJT.

# 4.G.3. Op-Amp Constant Current Source

The op-amp constant current source below is designed to eliminate the effects of temperature on the BJT used in the current source. This current source is a very accurate and temperature-independent current source:

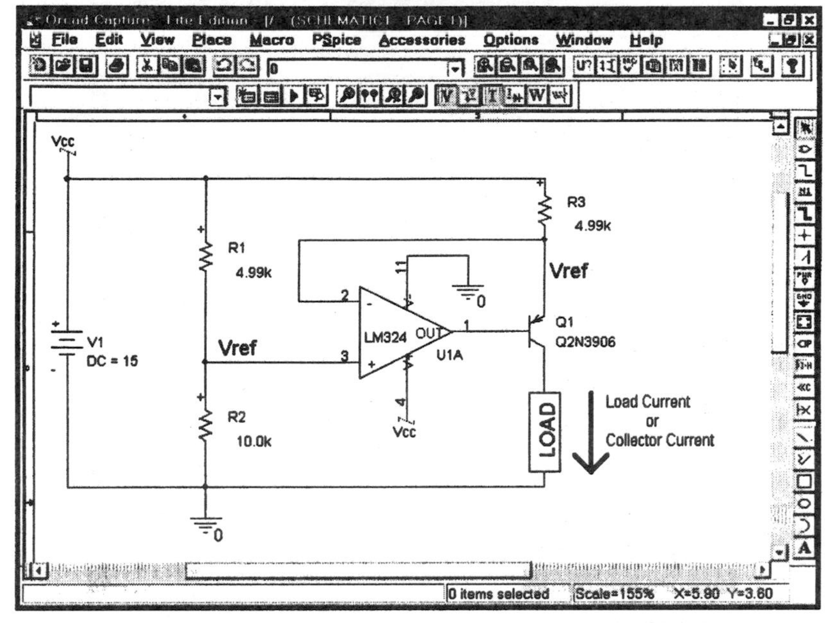


Resistors R1 and R2 form a voltage divider. The voltage at the positive terminal of R2 is Vref = 10 V. Due to the negative feedback of the circuit, the voltage at the negative terminal of R3 is also Vref. The positive terminal of R3 is hooked to the supply so the voltage across R3 is held constant at 5 V. If the voltage across R3 is constant at 5 V, then the current through R3 is constant at 1 mA. This circuit is designed to keep the voltage across R3 constant. By keeping the voltage constant, the current through R3 is held constant.

The current through R3 is also the emitter current of Q1. Since for a BJT, the collector current is approximately equal to the emitter current, the collector current and load current are held constant at approximately 1 mA. We will run a Bias Point analysis on the slightly modified circuit below to see the voltage Vref and the collector current:

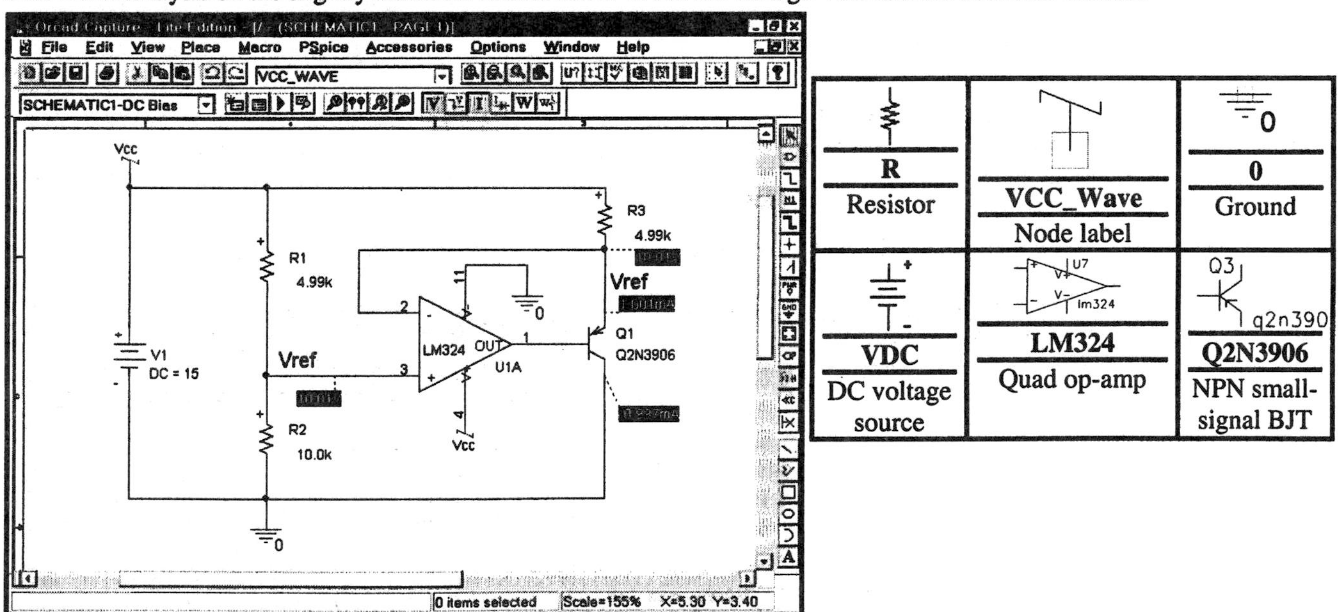


We see that both nodes labeled Vref are at a voltage of 10.01 volts. The negative feedback attempts to keep these two voltages the same. The emitter current is 1.001 mA and the collector current is 997 μA. Since $I_C = [\beta/(\beta+1)]I_E$, we expect the collector current to be slightly less than 1 mA, in this case 3 μA.

We will now run a linear temperature sweep from –25 to 125°C to see how this circuit is affected by temperature. For the first simulation only the BJT will have temperature dependence. We will use normal resistors (part name R), which are independent of temperature. Note that the Q2n3906 BJT model includes temperature dependence. Set up the DC Sweep as shown:

Run the simulation and plot the collector current :

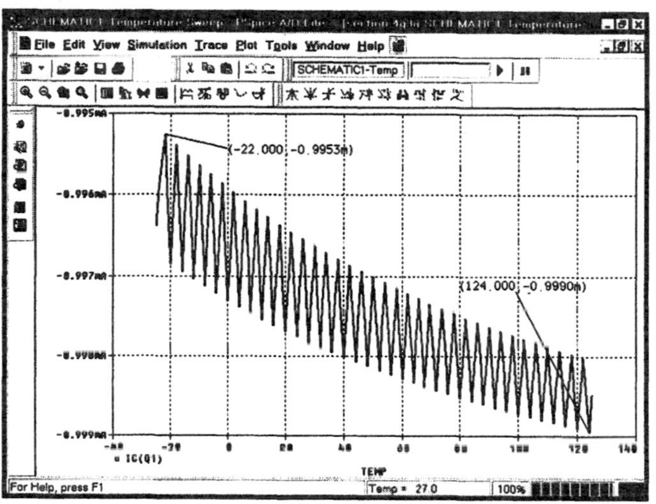


Over the entire temperature range the current varies by only 3.7 μA. We have eliminated the temperature effects of the BJT.

The circuit is not designed to eliminate temperature effects of resistors. The voltage Vref will be very independent of temperature if resistors R1 and R2 are all maintained at the same temperature and have the same temperature coefficients. This will be the case in this simulation. Thus, the voltage Vref will remain relatively constant. However, the resistance of R3 will change with temperature. We are maintaining constant voltage across R3, but R3's resistance will increase with temperature, so the current through it will decrease. Thus, we should expect the current of this source to decrease with increasing temperature.

We will create a resistor model with a 100 parts per million temperature coefficient. This model will be used by all resistors in the circuit:

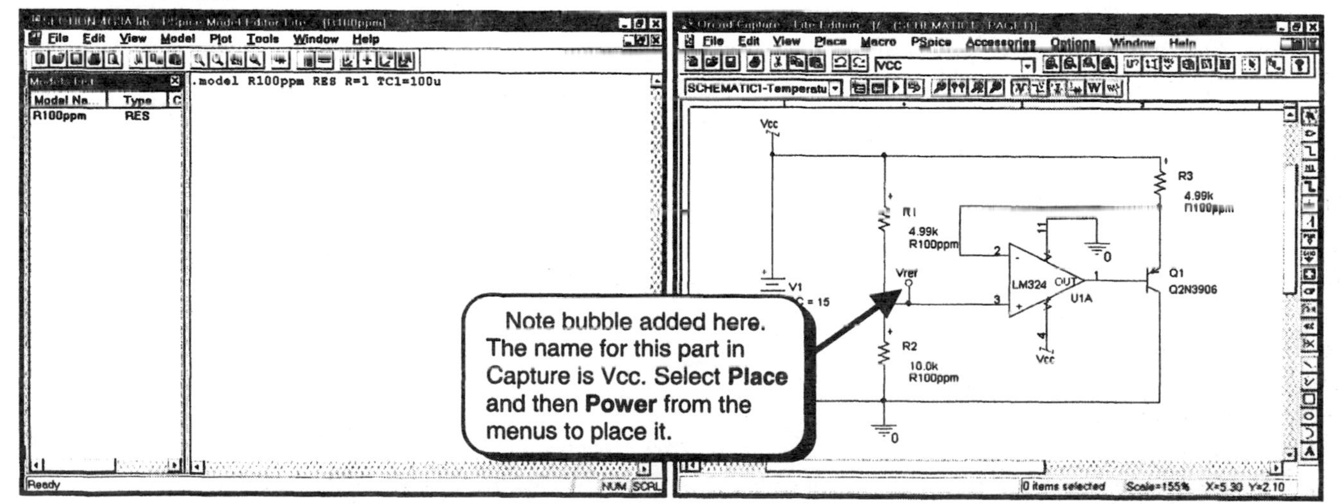


Notice that we have added a bubble at the voltage reference node so that we can easily plot the voltage. The name for this part in Capture is Vcc. Select **Place** and then **Power** from the menus to place it. We will run the same temperature analysis as we did before. Sweep the temperature from –25 to +125°C. We will first plot the reference voltage to see if it is independent of temperature:

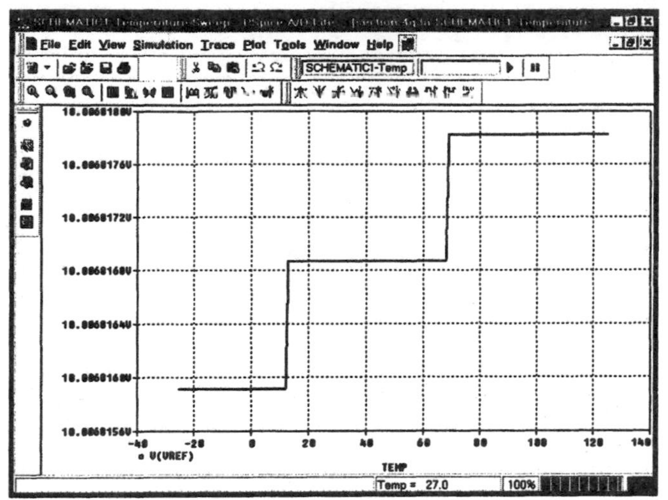


We see that the voltage is, for all practical purposes, constant at 10 V. Next we will plot the collector current of the BJT:

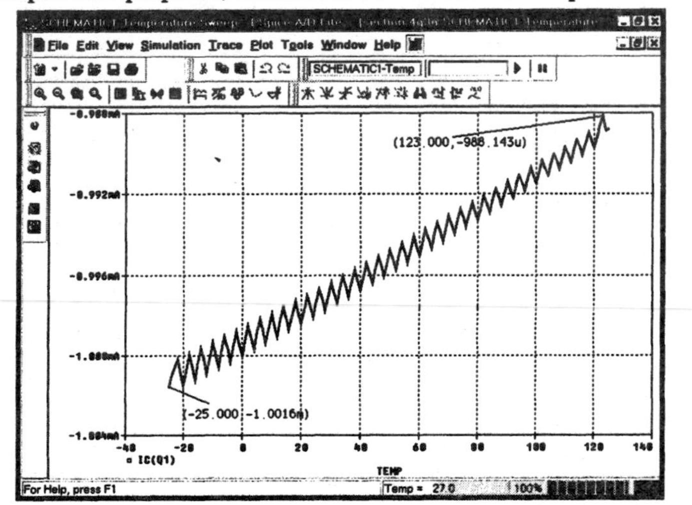


Even with resistor temperature dependence, the collector current changes by only about 13 μA over the entire temperature range. This circuit is a very temperature-independent constant current source, as long as the temperatures of R1 and R2 are the same and they have similar temperature coefficients.

**EXERCISE 4-13**: The circuit below is a current mirror and is a constant current source of about 50 μA. Display how the collector current of Q2 varies with temperature. Let resistors have a linear temperature coefficient of 200 ppm. The part name of both transistors is LM3046-Q.

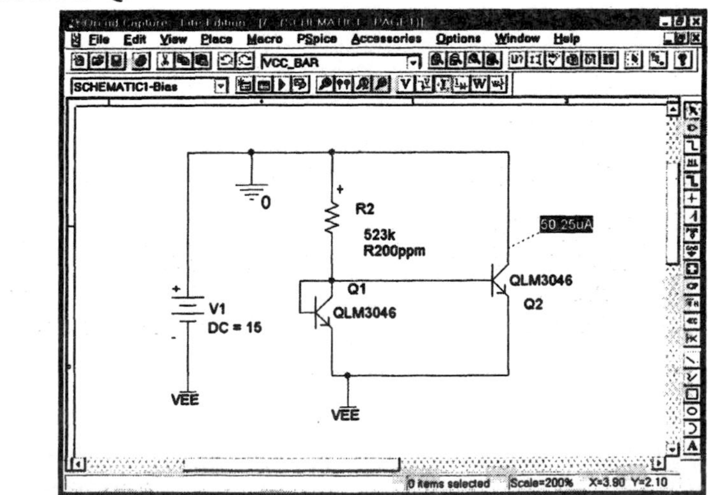


**SOLUTION**: Edit the resistor model to create a model with temperature dependence:

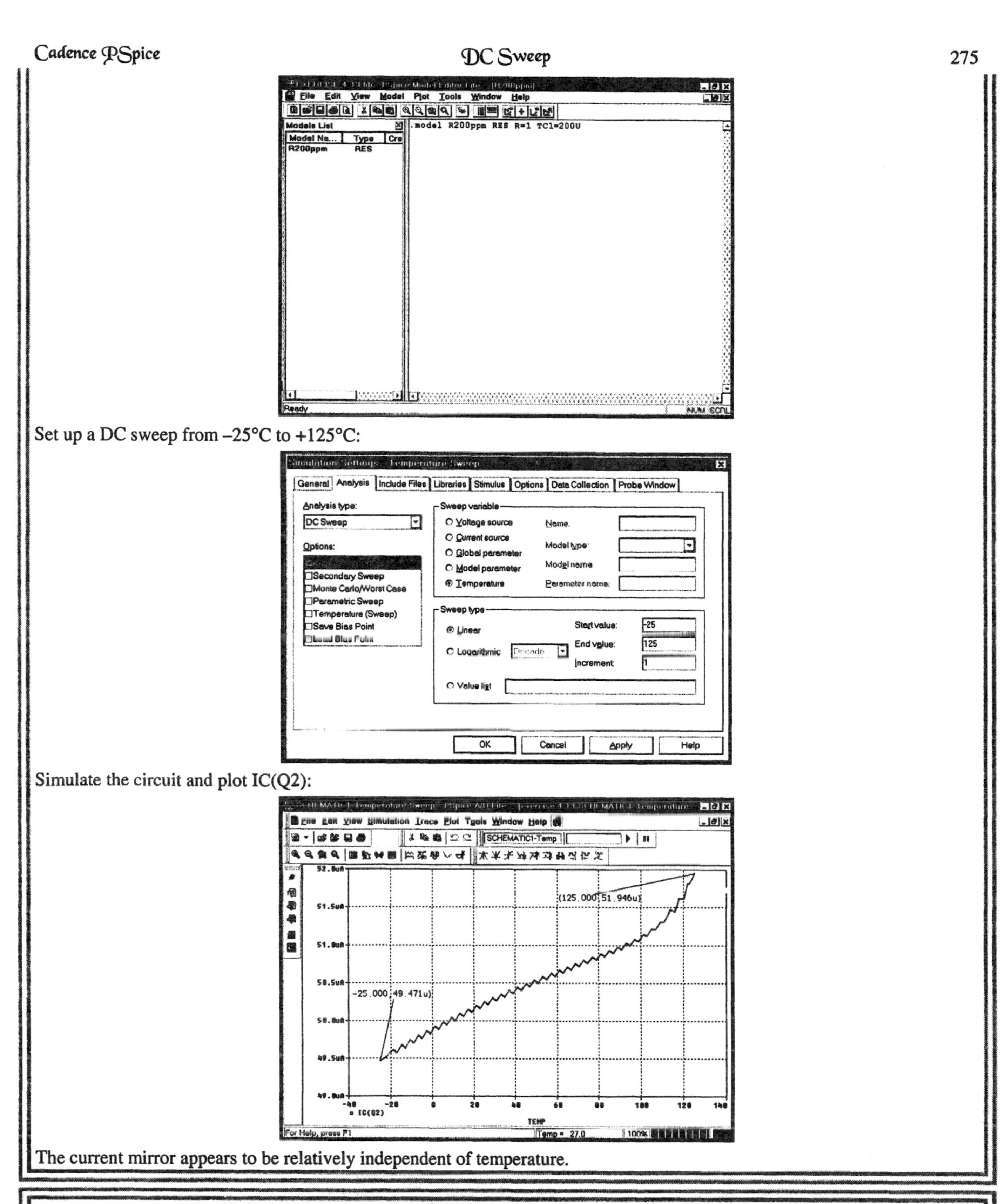


Set up a DC sweep from –25°C to +125°C:

Simulate the circuit and plot IC(Q2):

The current mirror appears to be relatively independent of temperature.

**EXERCISE 4-14**: The circuit below is a Widlar current source of about 50 μA. Display how the collector current of Q2 varies with temperature. Let resistors have a linear temperature coefficient of 200 ppm. The part name of both transistors is LM3046-Q.

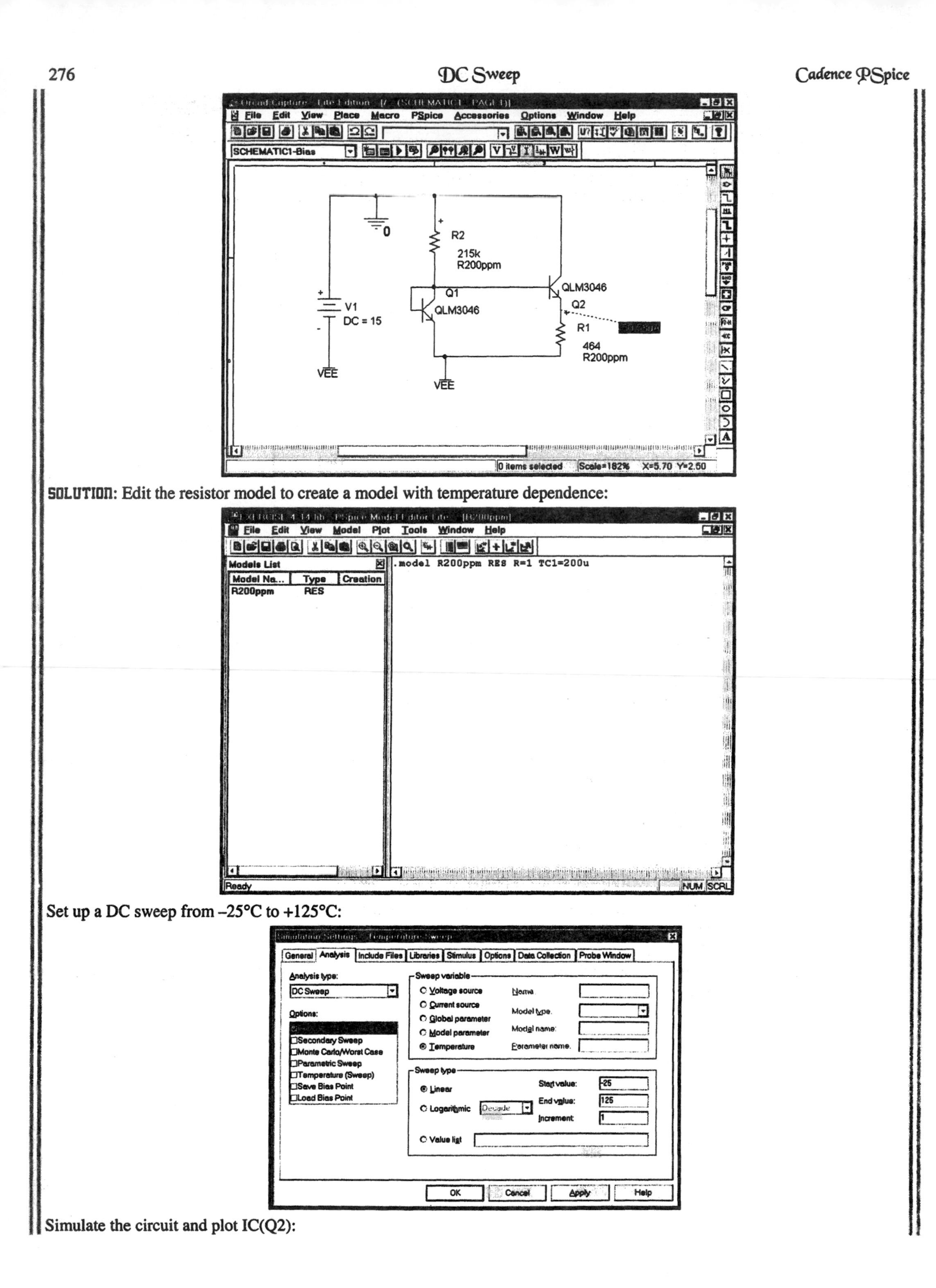


**SOLUTION**: Edit the resistor model to create a model with temperature dependence:

Set up a DC sweep from –25°C to +125°C:

Simulate the circuit and plot IC(Q2):

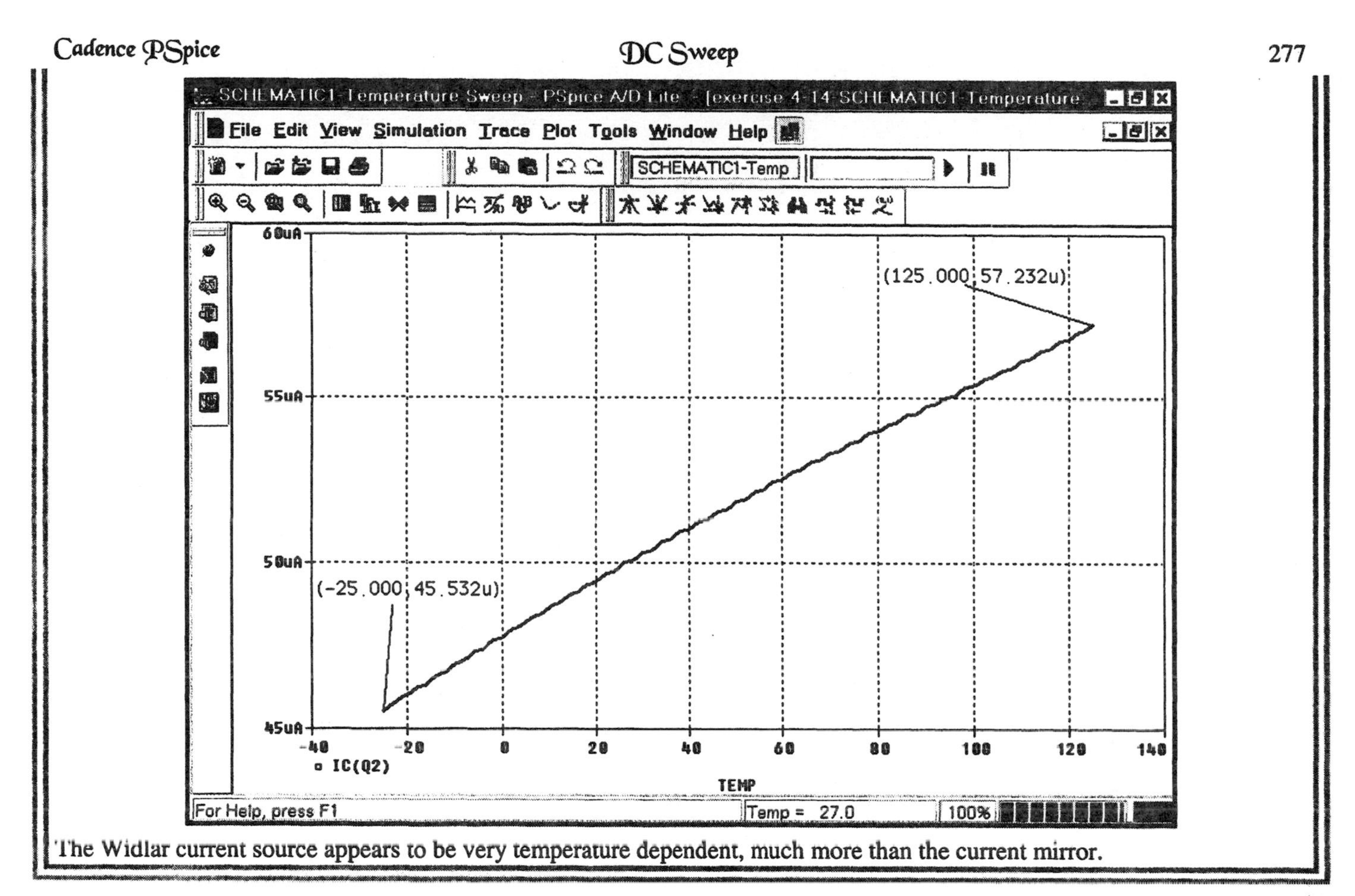


The Widlar current source appears to be very temperature dependent, much more than the current mirror.

# 4.H. Summary

- The DC Sweep is used to find DC voltages and currents in a circuit.
- The output can be viewed graphically using Probe.
- The DC Sweep can be used to find DC voltages and currents for multiple values of DC sources. It can answer the following question in one simulation: "What is the voltage at node 1 for V1 = 10 VDC, V1 = 11 VDC, and V1 = 12 VDC?" This question could be answered using the DC Nodal Analysis if the DC Nodal Analysis were run three times.
- All capacitors are replaced by open circuits.
- All inductors are replaced by short circuits.
- All AC and time-varying sources are set to zero (IAC, VAC, Vsin, Isin, Vpulse, etc.).
- The DC Sweep may be used to view I-V characteristics of a device.
- The Secondary DC Sweep and the Parametric Sweep can be used to see how values of devices affect the performance of a circuit.
- The Secondary DC Sweep and the Parametric Sweep can be used to generate families of curves.
- Goal functions can be used to obtain numerical data from Probe graphs. The result of evaluating a goal function is a single numerical value.

# PART 5
# AC Sweep

The AC Sweep is used for Bode plots, gain and phase plots, and phasor analysis. The circuit can be analyzed at a single frequency or at multiple frequencies. In this part we will illustrate its use at a single frequency for magnitude and phase results (phasors), and at multiple frequencies for Bode plots.

It is important to realize the difference between the AC Sweep and the Transient Analysis discussed in Part 6. The AC Sweep is used to find the magnitude and phase of voltages and currents. The Transient Analysis is used to look at waveforms versus time. An example of a waveform versus time is:

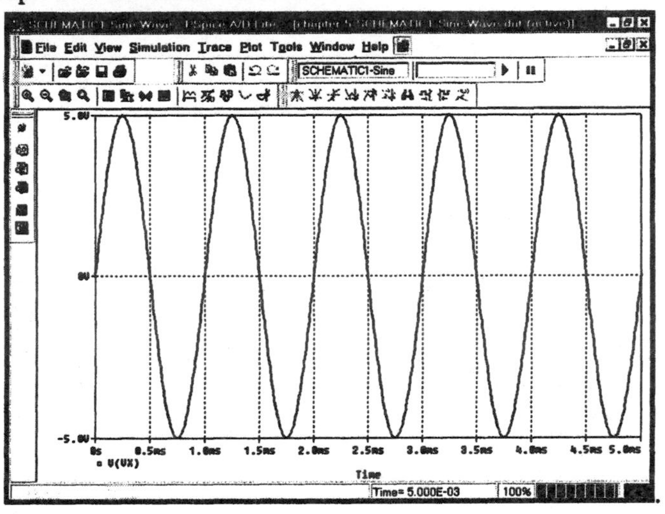


This graph shows us a voltage versus time. If you want to look at the magnitude and phase of voltages and currents, use the AC Sweep. The magnitude of the above waveform is 5 V and the phase is zero degrees—in phasor notation, 5∠0°. The AC Sweep will give a result such as 5∠0°.

The AC Sweep uses the sources VAC and IAC. These sources are functions of magnitude and frequency. **Do not use the source VSIN for the AC Sweep.**

## 5.A. Magnitude and Phase (Phasors) Text Output

Wire the circuit shown below:

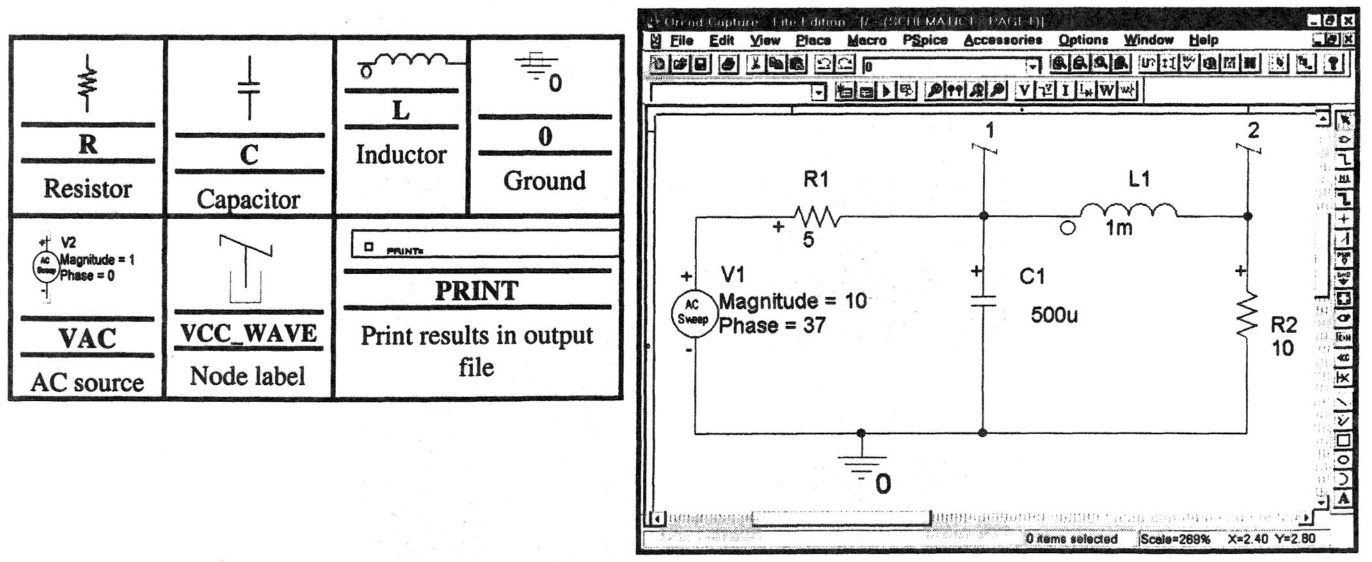


The source is an AC source with a magnitude of 10 volts and a phase of 37 degrees, 10∠37° in phasor notation. This magnitude can be interpreted as either peak or RMS. If you specify 10∠37° as the magnitude of the source, and the number

10 is an RMS value, then the magnitudes of all of your results will be RMS values. If you specify 10∠37° as the magnitude of the source, and the number 10 is a peak value, then the magnitudes of all of your results will be peak values.

We must now add a part to the circuit that prints the results we would like. Get the part called ***PRINT***. When you get the ***PRINT*** part, the cursor will be attached to a rectangle, [rectangle]. Move the rectangle to where you want to put the ***PRINT*** statement and click the ***LEFT*** mouse button. When you click the button, a second rectangle appears. Move the rectangle away from the first ***PRINT*** part and click the ***LEFT*** mouse button. Since we need only two ***PRINT*** parts, press the **ESC** key to stop placing parts. You should have a schematic similar to the one below:

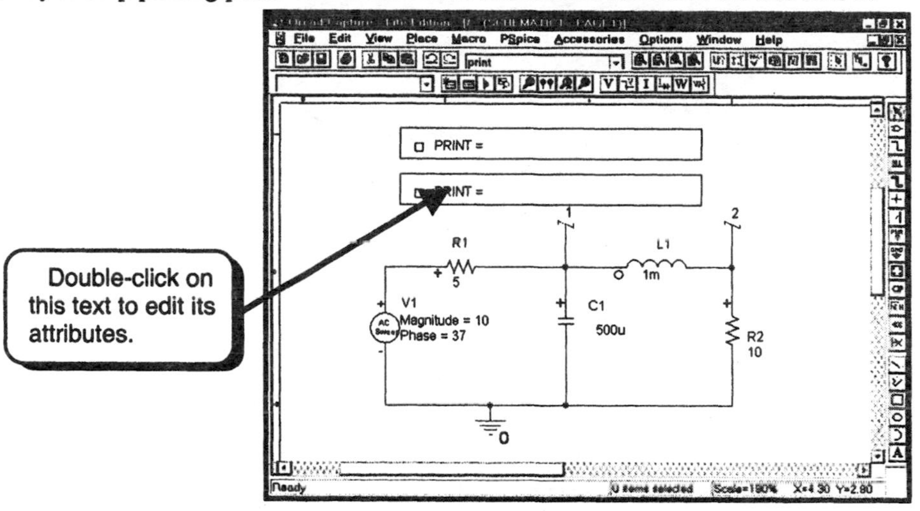


To print voltages and currents we must edit the attributes of the ***PRINT*** part. Double-click the ***LEFT*** mouse button on one of the ***PRINT=*** texts. The dialog box below will appear:

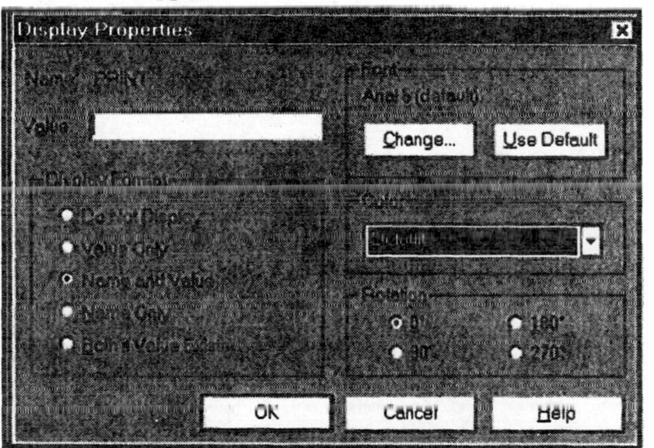


Type in the text **`AC V(1,0) I(R_R1) I(C_C1)`**:

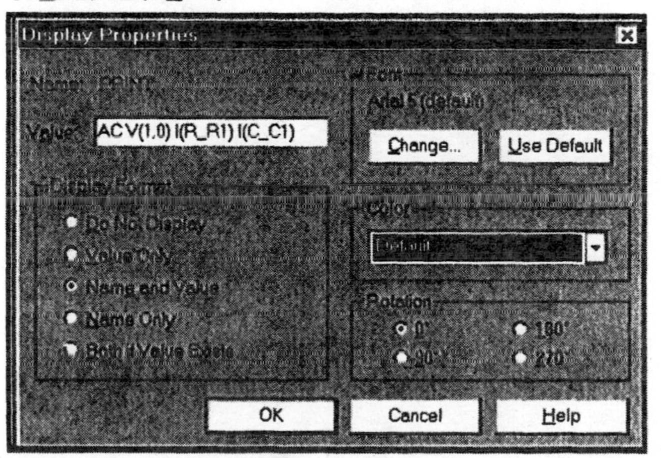


Click the ***OK*** button to accept the attribute. This ***PRINT*** statement prints the magnitudes of the specified voltages and currents. For example, ***V(1,0)*** prints the magnitude of the voltage at node ***1*** relative to node ***0*** (ground), and ***I(C_C1)*** prints the magnitude of the current through capacitor ***C1***. The ***AC*** in the attribute tells the ***PRINT*** statement to print the results from the AC Sweep. If we run an analysis other than the AC Sweep, this print part will generate no output because it specifies output for the AC Sweep.

Double-click the ***LEFT*** mouse button on the other ***PRINT=*** text. Enter the text **`AC VP(2,0) IP(C_C1) IP(L_L1)`**:

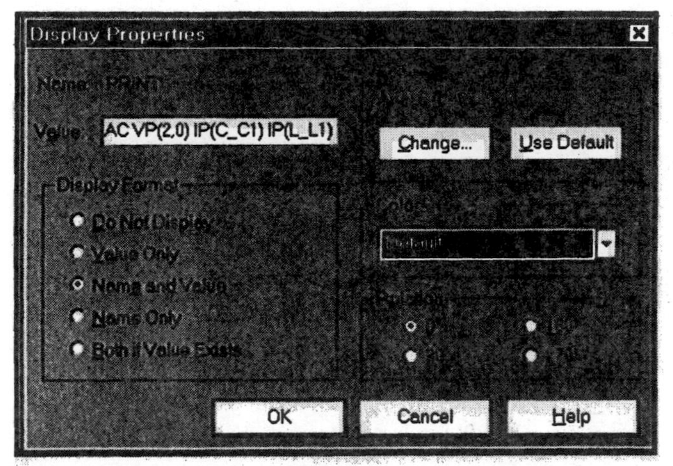


This ***PRINT*** statement prints the phases of the specified voltages and currents. For example, ***VP(2,0)*** prints the phase of the voltage at node ***2*** relative to node ***0*** (ground), and ***IP(L_L1)*** prints the phase of the current through inductor ***L1***. The ***AC*** in the attribute tells the ***PRINT*** statement to print the results of the AC Sweep. When you change the attributes you should have the following schematic:

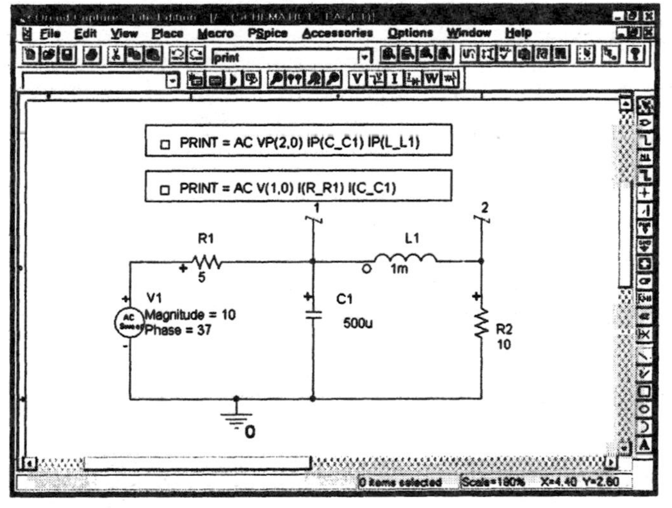


We are now ready to set up the simulation. From the Capture menu bar select **PSpice** and then **New Simulation Profile**:

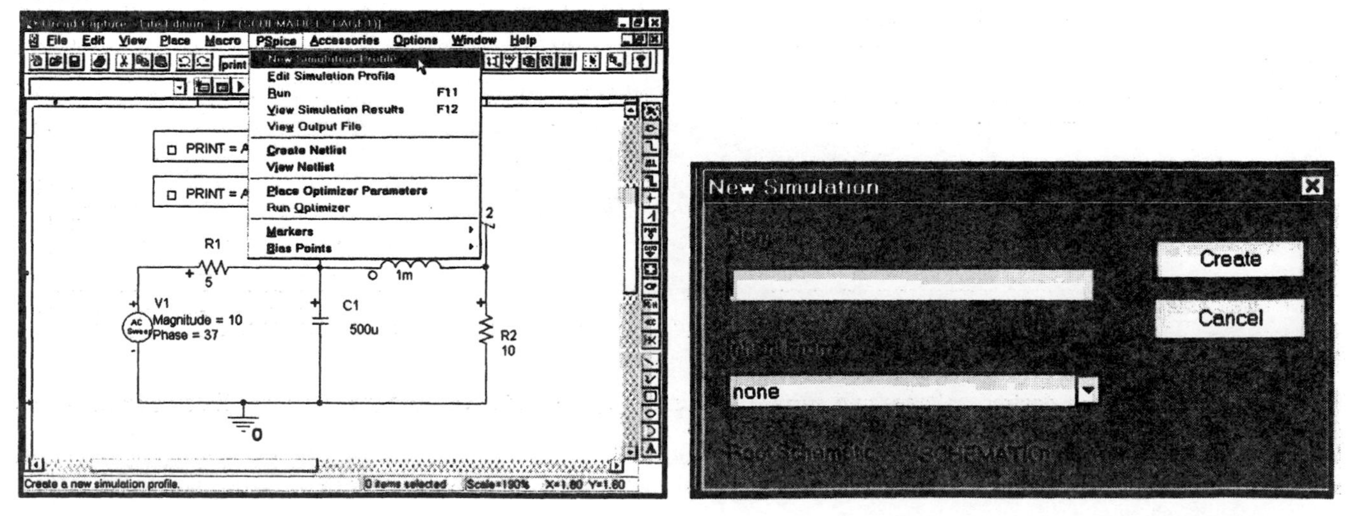

Enter a name for the profile and then click the ***Create*** button:

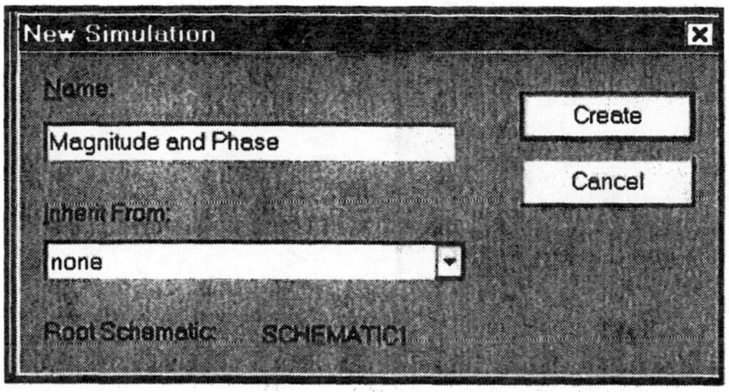

New Simulation
Name:
Magnitude and Phase
Create
Cancel
Inherit From:
none
Root Schematic: SCHEMATIC1

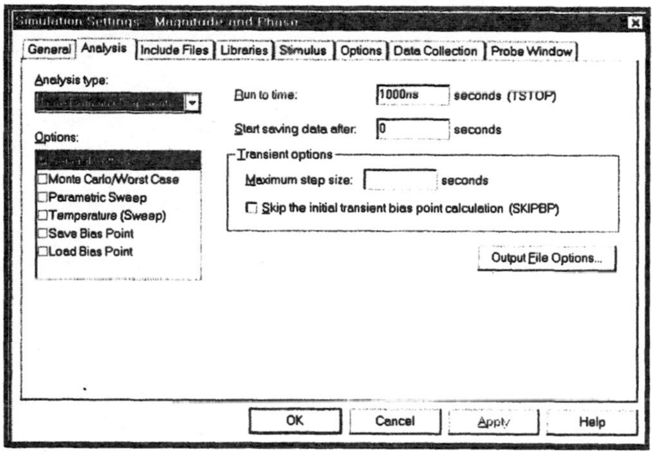

Simulation Settings - Magnitude and Phase
General | Analysis | Include Files | Libraries | Stimulus | Options | Data Collection | Probe Window
Analysis type:
Options:
Monte Carlo/Worst Case
Parametric Sweep
Temperature (Sweep)
Save Bias Point
Load Bias Point
Run to time: 1000ns seconds (TSTOP)
Start saving data after: 0 seconds
Transient options
Maximum step size: seconds
Skip the initial transient bias point calculation (SKIPBP)
Output File Options...
OK | Cancel | Apply | Help

Click the ***LEFT*** mouse button on the down triangle to view the list of available ***Analysis type***s:

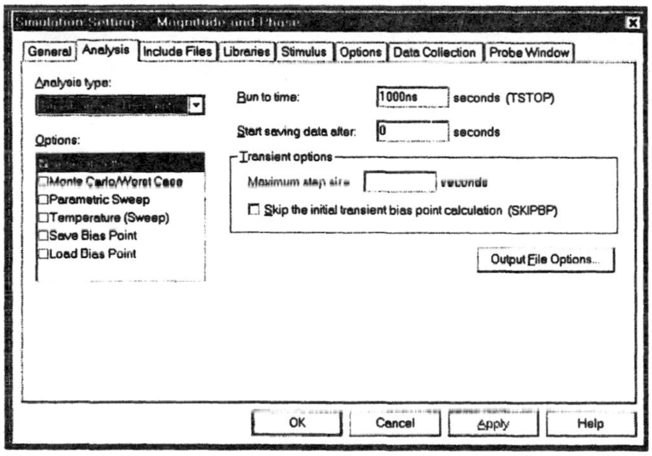

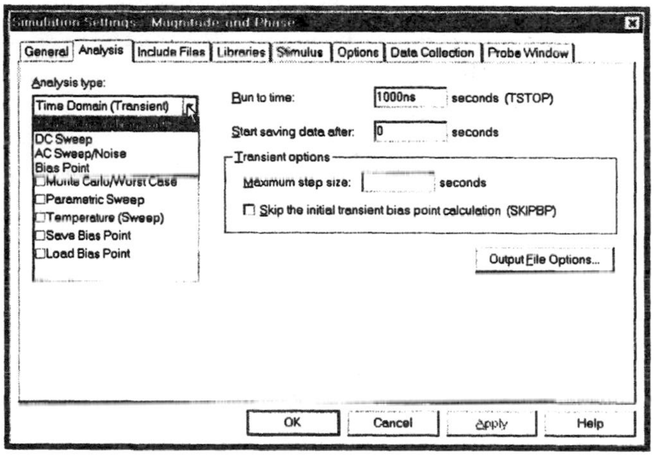

Simulation Settings - Magnitude and Phase
General | Analysis | Include Files | Libraries | Stimulus | Options | Data Collection | Probe Window
Analysis type:
Options:
Monte Carlo/Worst Case
Parametric Sweep
Temperature (Sweep)
Save Bias Point
Load Bias Point
Run to time: 1000ns seconds (TSTOP)
Start saving data after: 0 seconds
Transient options
Maximum step size: seconds
Skip the initial transient bias point calculation (SKIPBP)
Output File Options...
OK | Cancel | Apply | Help

Simulation Settings - Magnitude and Phase
General | Analysis | Include Files | Libraries | Stimulus | Options | Data Collection | Probe Window
Analysis type:
Time Domain (Transient)
DC Sweep
AC Sweep/Noise
Bias Point
Monte Carlo/Worst Case
Parametric Sweep
Temperature (Sweep)
Save Bias Point
Load Bias Point
Run to time: 1000ns seconds (TSTOP)
Start saving data after: 0 seconds
Transient options
Maximum step size: seconds
Skip the initial transient bias point calculation (SKIPBP)
Output File Options...
OK | Cancel | Apply | Help

Select the ***AC Sweep/Noise*** menu selection:

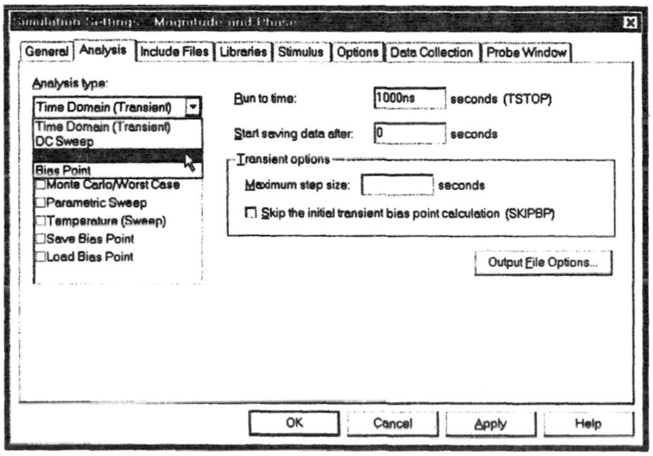

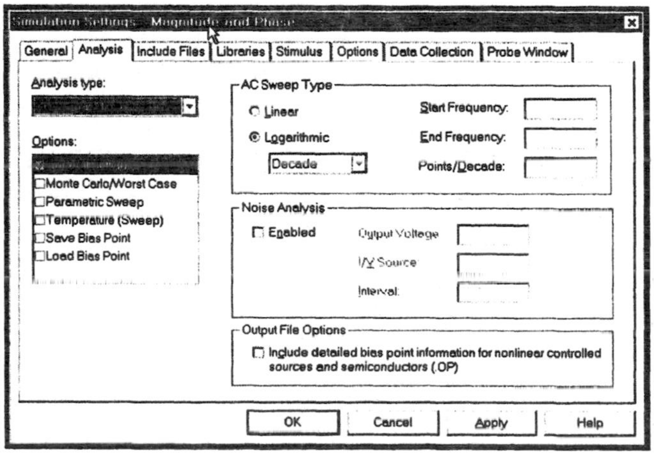

Simulation Settings - Magnitude and Phase
General | Analysis | Include Files | Libraries | Stimulus | Options | Data Collection | Probe Window
Analysis type:
Time Domain (Transient)
Time Domain (Transient)
DC Sweep
Bias Point
Monte Carlo/Worst Case
Parametric Sweep
Temperature (Sweep)
Save Bias Point
Load Bias Point
Run to time: 1000ns seconds (TSTOP)
Start saving data after: 0 seconds
Transient options
Maximum step size: seconds
Skip the initial transient bias point calculation (SKIPBP)
Output File Options...
OK | Cancel | Apply | Help

Simulation Settings - Magnitude and Phase
General | Analysis | Include Files | Libraries | Stimulus | Options | Data Collection | Probe Window
Analysis type:
Options:
Monte Carlo/Worst Case
Parametric Sweep
Temperature (Sweep)
Save Bias Point
Load Bias Point
AC Sweep Type
Linear
Logarithmic
Decade
Start Frequency:
End Frequency:
Points/Decade:
Noise Analysis
Enabled
Output Voltage
I/V Source
Interval
Output File Options
Include detailed bias point information for nonlinear controlled sources and semiconductors (.OP)
OK | Cancel | Apply | Help

We would like to find the desired voltages and currents at frequencies of 100, 200, and 300 Hz. The dialog box below has been set up for these frequencies:

A linear sweep means that points are equally spaced between the starting and ending frequencies. If you were only interested in a single frequency, you would set the total points to 1 and set the starting and ending frequencies to the same value.

Since we are using a print statement, the results will be saved in the output file and there is no need to run Probe to view the results. We will disable Probe. Select the ***Probe Window*** tab:

In the dialog box above, Probe is set to run after the simulation is complete. We do not want Probe to run, so disable this option:

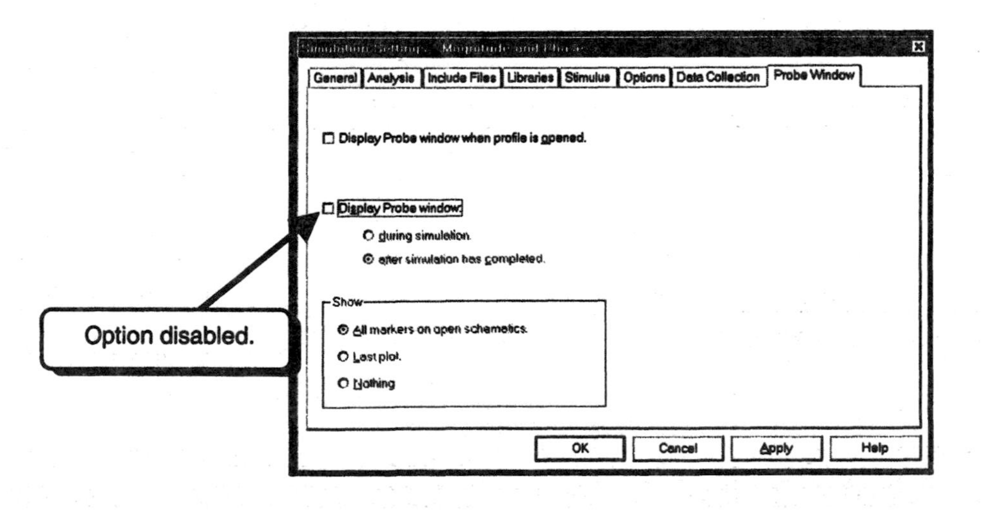


Click the ***OK*** button to return to the schematic. Select **PSpice** and then **Run** to simulate the circuit:

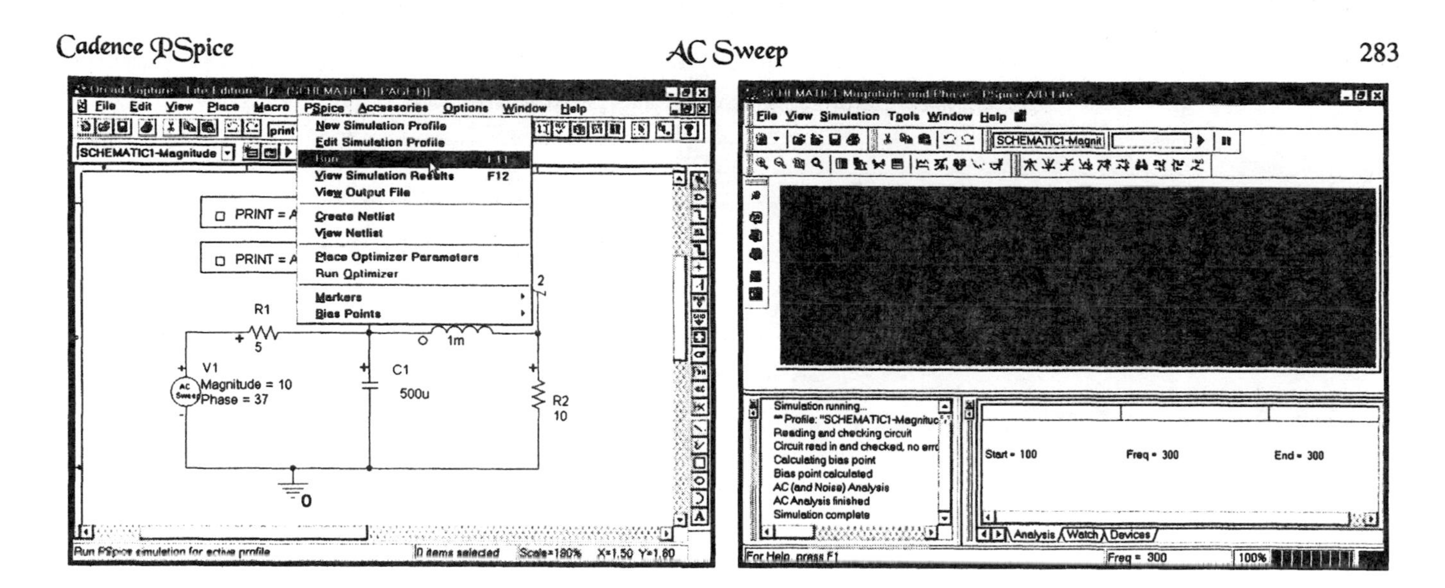


When the simulation is finished, select **View** and then **Output File** from the menus:

We would like to see the output file occupy a bit more space in the window. Click on the [X] in the simulation status window and the message window to close the windows:

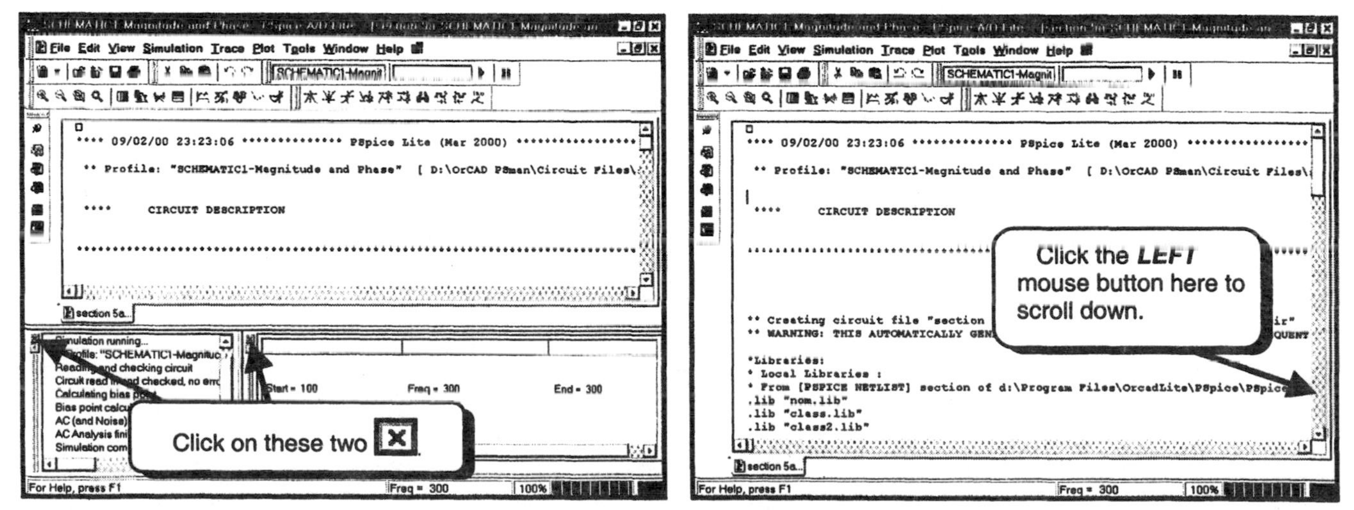


The results are printed near the bottom of the output file. Click the *LEFT* mouse button on the vertical scroll bar until you see the text shown below:

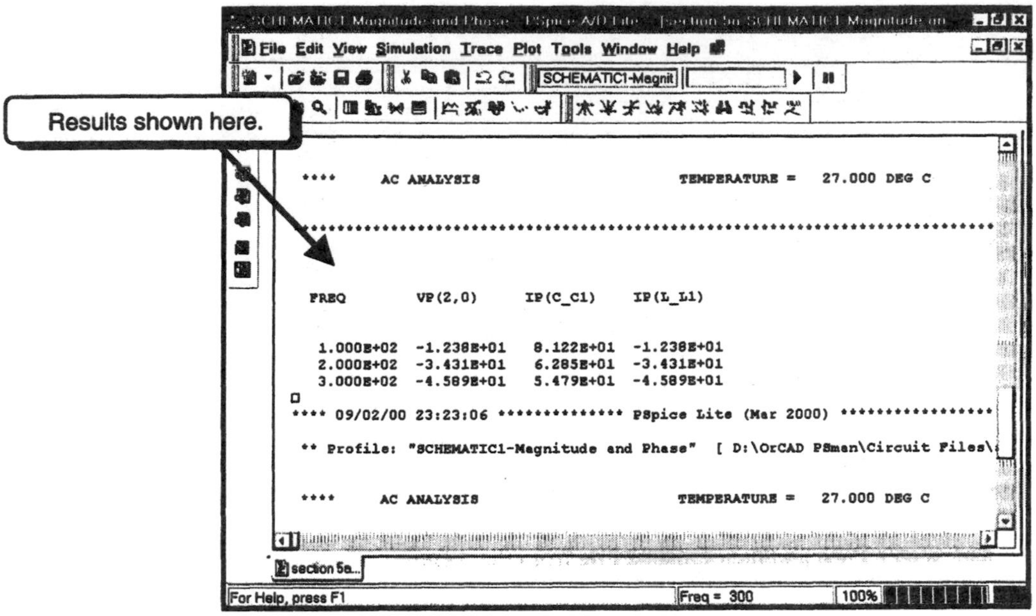

Results shown here.

```
****     AC ANALYSIS                         TEMPERATURE =   27.000 DEG C

******************************************************************************

  FREQ        VP(2,0)      IP(C_C1)     IP(L_L1)

  1.000E+02  -1.238E+01   8.122E+01   -1.238E+01
  2.000E+02  -3.431E+01   6.285E+01   -3.431E+01
  3.000E+02  -4.589E+01   5.479E+01   -4.589E+01

**** 09/02/00 23:23:06 ************** PSpice Lite (Mar 2000) ****************

 ** Profile: "SCHEMATIC1-Magnitude and Phase"  [ D:\OrCAD PSmen\Circuit Files\

****     AC ANALYSIS                         TEMPERATURE =   27.000 DEG C
```

The text shows the results of one of the print statements: At a frequency of 100 Hz the phase of the voltage at node 2 is –12.38 degrees; at 200 Hz the phase of the current through C1 is 62.85 degrees; at 300 Hz the phase of the current through L1 is –45.89 degrees.

If you click the ***LEFT*** mouse button on the down arrow several more times, you will reach the text of the second print statement:

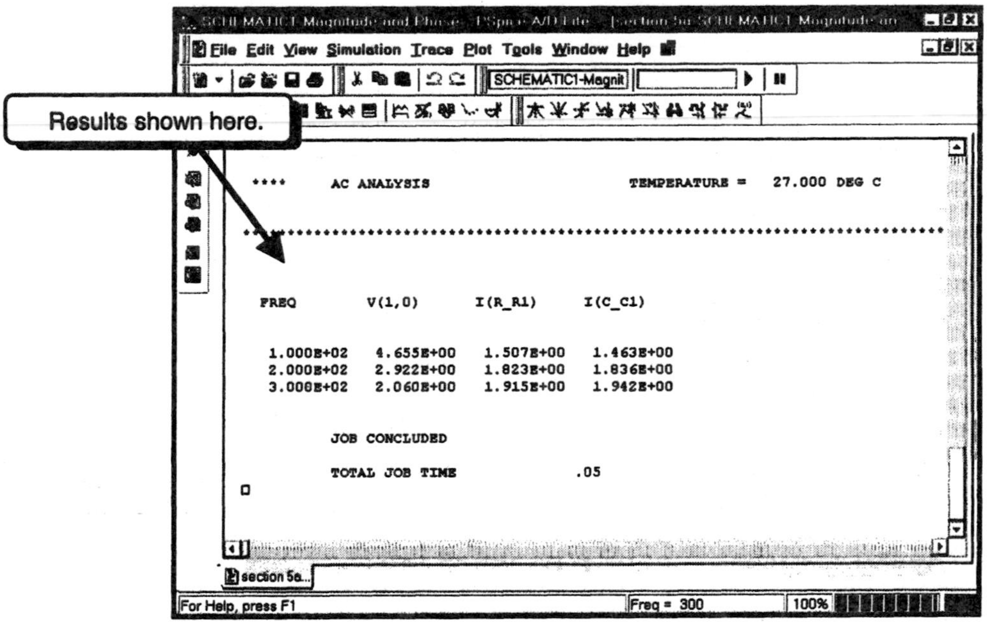

Results shown here.

```
****     AC ANALYSIS                         TEMPERATURE =   27.000 DEG C

******************************************************************************

  FREQ        V(1,0)       I(R_R1)      I(C_C1)

  1.000E+02   4.655E+00   1.507E+00   1.463E+00
  2.000E+02   2.922E+00   1.823E+00   1.836E+00
  3.000E+02   2.060E+00   1.915E+00   1.942E+00

          JOB CONCLUDED

          TOTAL JOB TIME            .05
```

The text shows the results of the other print statement: at 100 Hz the voltage at node 1 is ***4.655*** V. At 200 Hz the current through R1 is ***1.823*** A. At 300 Hz the current through C1 is ***1.942*** A.

The order in which the results of the print statement are written to the output file may be different for your simulation. The order of the displayed results depends on the order in which you placed the print statements in your schematic.

**EXERCISE 5-1**: Find the magnitude and phase of the voltages at nodes 1, 2, and 3 at a frequency of 1 kHz:

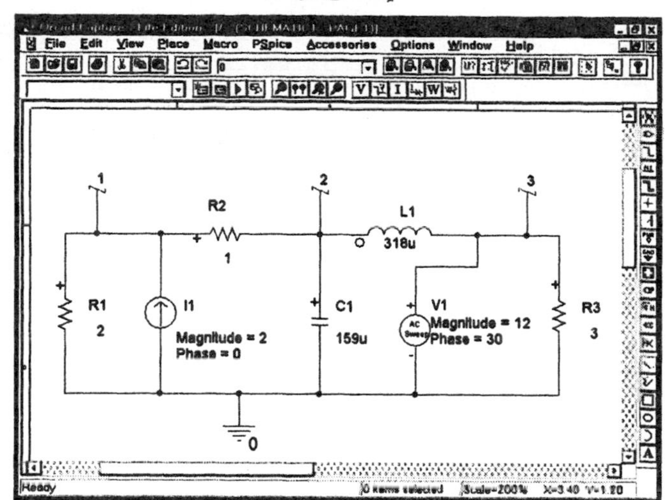


**SOLUTION**: Add two print parts and fill them in as shown:

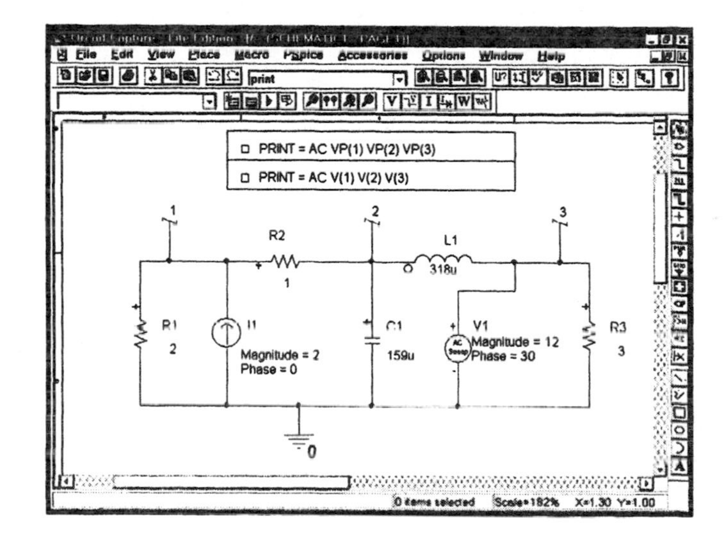


The AC Sweep setup is shown below. The results are stored in the output file:

```
FREQ        VP(1)        VP(2)        VP(3)

1.000E+03   -9.627E+01   -1.064E+02   3.000E+01

****  AC ANALYSIS                 TEMPERATURE =   27.000 DEG C

FREQ        V(1)         V(2)         V(3)

1.000E+03   7.264E+00    1.129E+01    1.200E+01

JOB CONCLUDED
```

**EXERCISE 5-2**: Find the magnitude and phase of the voltage at node 1 and the current through ***R1*** at a frequency of 1 Hz:

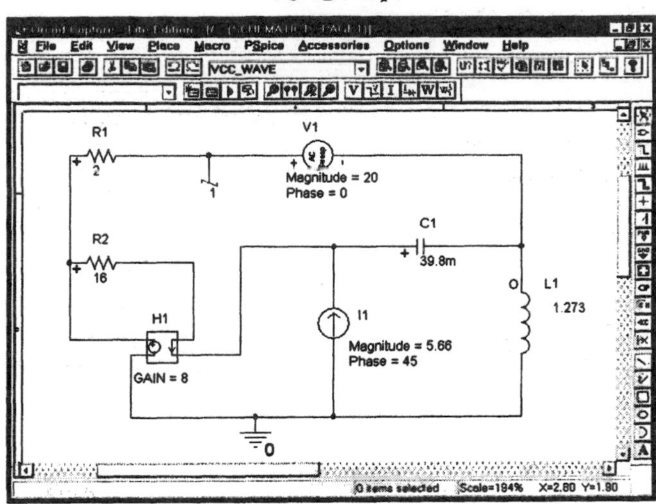


**HINT**: ***H1*** is a current-controlled voltage source. The voltage of ***H1*** is ***8*** times the current through ***R2***.

**SOLUTION**: Add a print part and fill it in as shown:

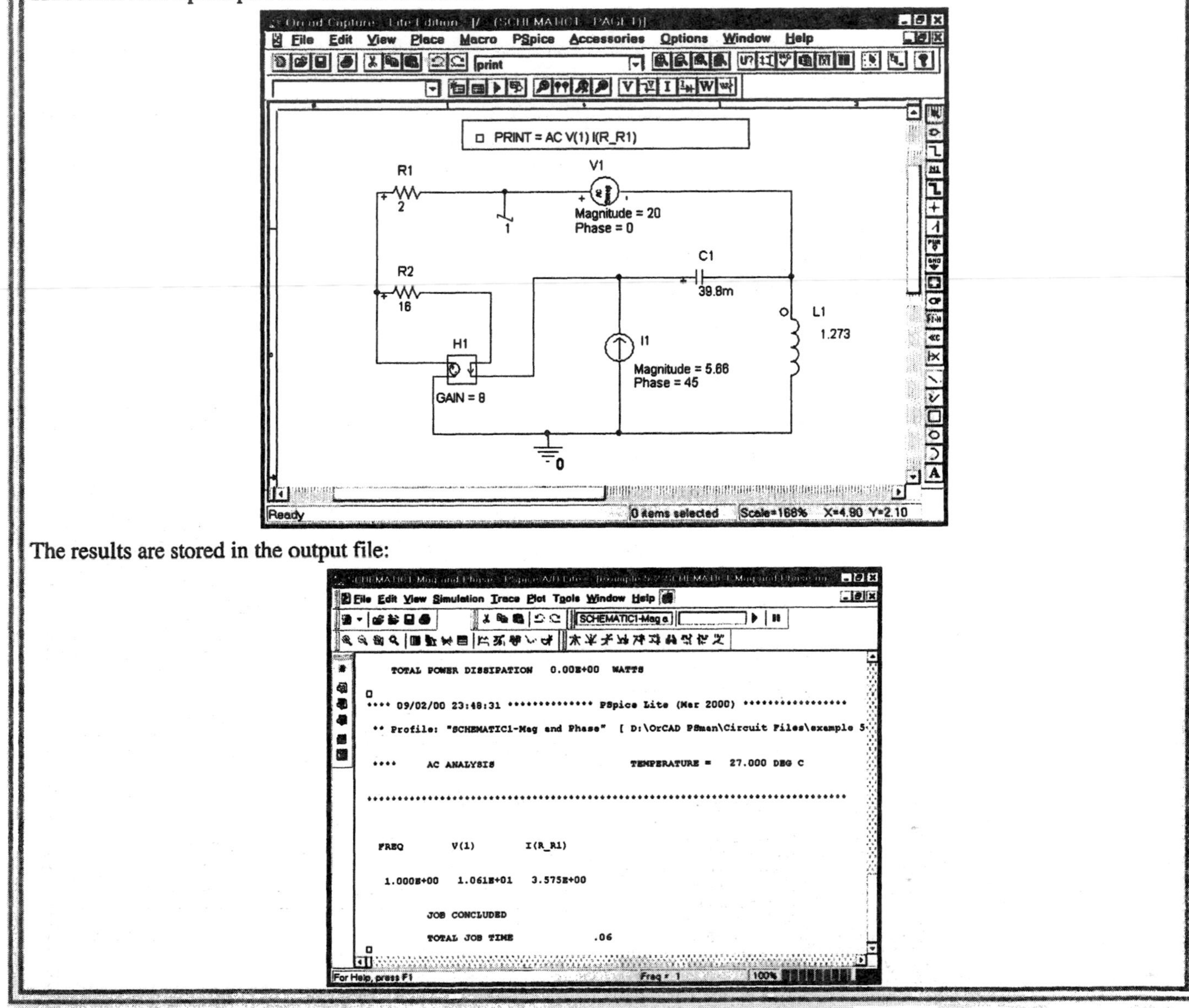


The results are stored in the output file:

## 5.B. Magnitude and Phase (Phasors) Graphical Output

Wire the circuit shown below:

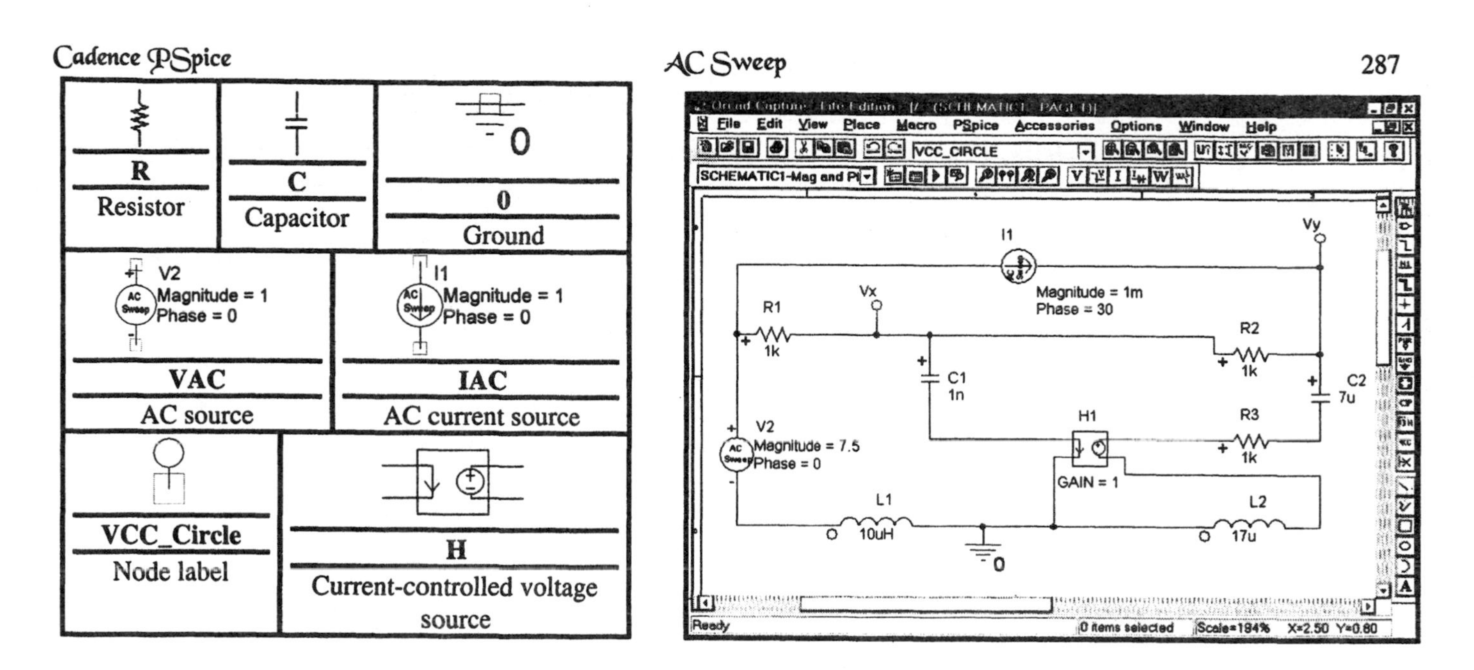


In PSpice, the suffixes ***U*** and ***U*** stand for micro, and the suffix ***n*** stands for nano. Thus, ***L1*** is $10^{-5}$ H and ***C1*** is $10^{-9}$ F. ***I1*** is an independent current source. Its magnitude is 1 mA and its phase is 30 degrees.

We will now set up the AC Sweep for this circuit. Remember that the AC Sweep gives us the phase and magnitude of sinusoidal waveforms at specified frequencies. For example, if ***Vy*** = 5sin(1000t + 30), the result of the AC Sweep will be 5 for the magnitude and 30 for the phase. If you want to see ***Vy*** displayed as a function of time, you must run a Transient Analysis.

To set up the analysis select **PSpice** and then **New Simulation Profile**:

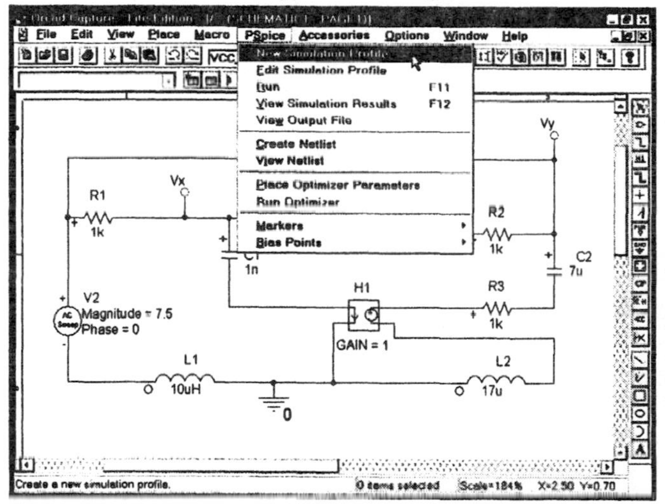


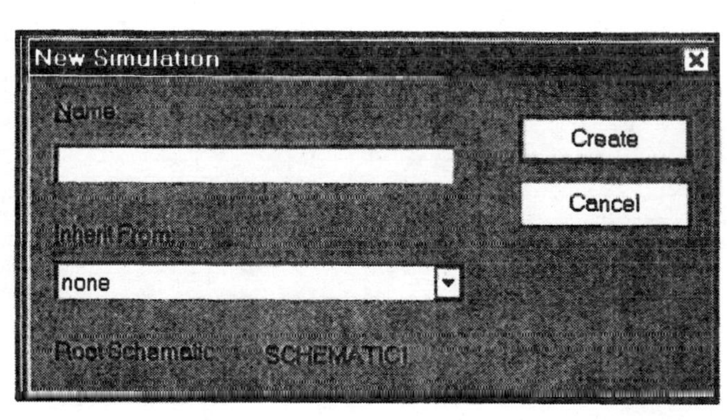


Enter a name for the new profile and click the ***Create*** button:

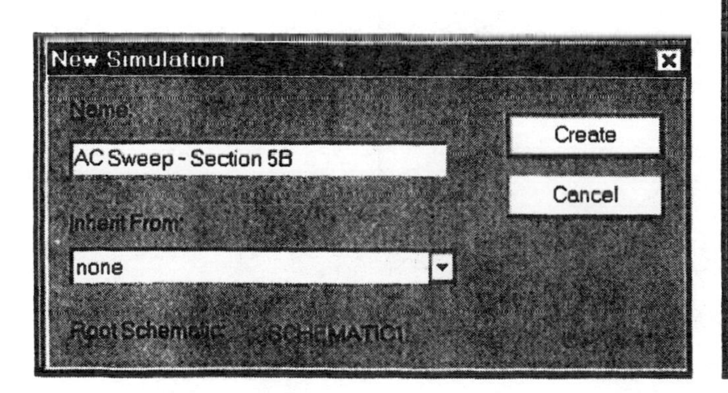


General | Analysis | Include Files | Libraries | Stimulus | Options | Data Collection | Probe Window

Analysis type:

Run to time: 1000ns seconds (TSTOP)

Start saving data after: 0 seconds

Options:

Transient options

Maximum step size: seconds

Monte Carlo/Worst Case

Parametric Sweep

Temperature (Sweep)

Save Bias Point

Load Bias Point

Skip the initial transient bias point calculation (SKIPBP)

Output File Options...

OK | Cancel | Apply | Help

Select the **AC Sweep/Noise** ***Analysis type*** and fill in the dialog box as shown below.

This AC Sweep is set up to simulate the circuit for frequencies of 1 Hz to $10^6$ Hz. The analysis is set to sweep the frequency in decades. A decade is a factor of 10 in frequency and is from 1 Hz to 10 Hz, from 10 Hz to 100 Hz, from 100 Hz to 1,000 Hz, and so on. The analysis is set for 100 frequency points per decade, so there will be 100 points between 1 and 10 Hz, 100 points between 10 and 100 Hz, and so on. Click the ***OK*** button when you have set up the parameters. You will return to the schematic.

We are now set to run the analysis. Select **PSpice** and then **Run** from the Capture menus:

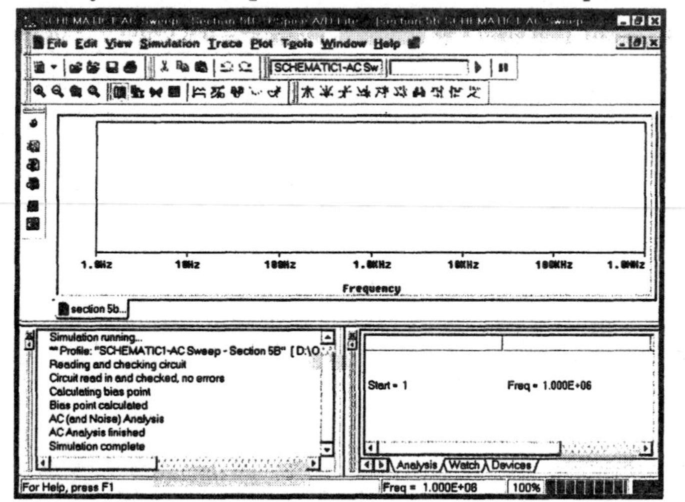


I will switch the display so that the Probe Window occupies the entire window. See page 98 for the procedure:

Add the trace **V(VX)**:

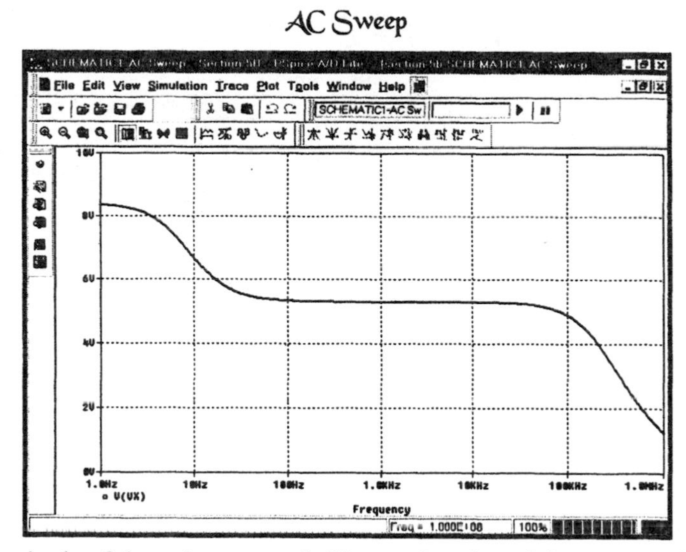


The trace above shows the magnitude of the voltage at node Vx as a function of frequency. If we ask for a voltage or current trace like V(Vx) or I(L2), the plot will be the magnitude of the voltage or current. To plot the phase of a voltage or current we would plot the traces Vp(Vx) or Ip(L2). Vp means "voltage phase" and Ip means "current phase." A second way of plotting phase is to use the phase operator *P(expression)* in Probe. This function displays the phase of the function you are evaluating. For example, *P(V(Vx))* will give the phase of the voltage at node Vx. *P(V(Vx)/I(C1))* will give the phase of the ratio V(Vx)/I(C1). We will now plot the phase of the voltage at node Vx. Since the phase of any voltage or current will be between 0 and 360, we will add a new plot because the y-axis scale needed to plot the magnitude of a voltage is very different from the scale needed to plot the phase.

To add a new plot to the current window, select **Plot** and then **Add Plot to Window** from the Probe menus:

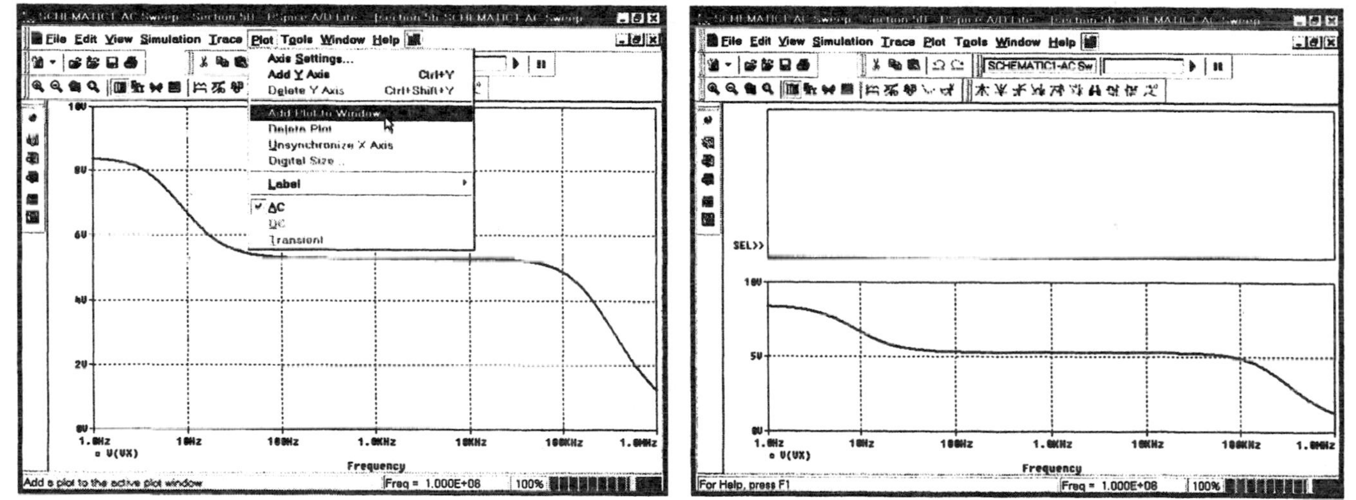


We will now add a trace that shows the phase of the voltage at node ***Vx***. Press the **INSERT** key from the Probe menu bar. The dialog box below will appear. The phase operator is located at the end of the list in the right window pane:

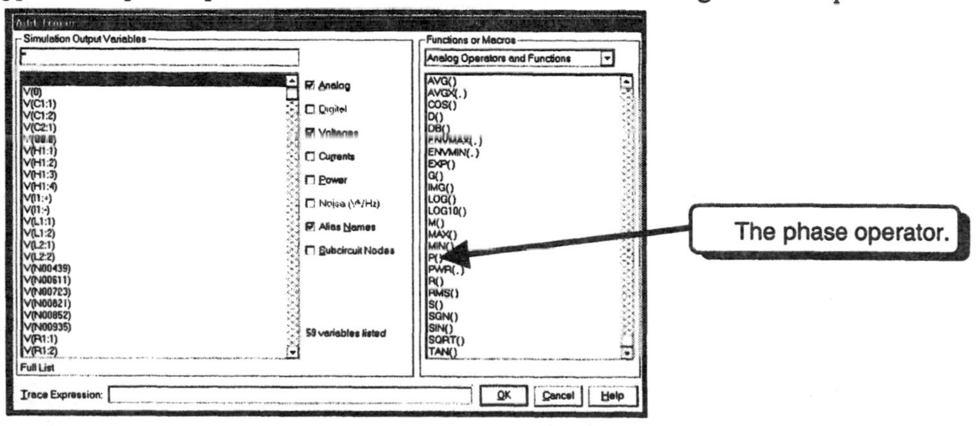


In the text field next to ***Trace Expression***: enter the text **Vp(Vx)** and press the **ENTER** key:

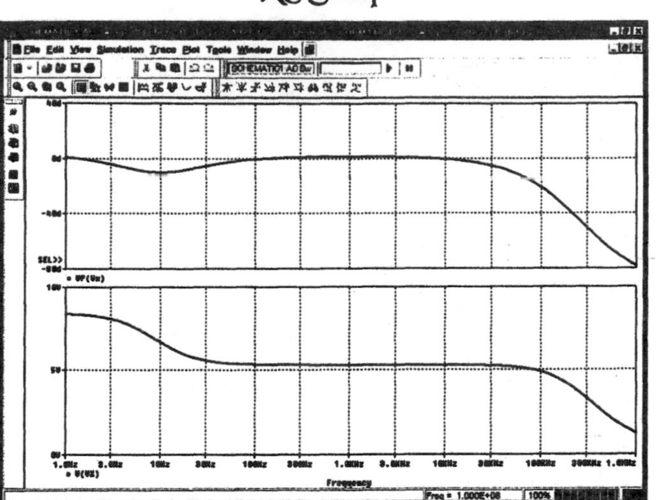

Next, press the **INSERT** key to obtain the ***Add Traces*** dialog box and add the trace **P(V(Vx))**:

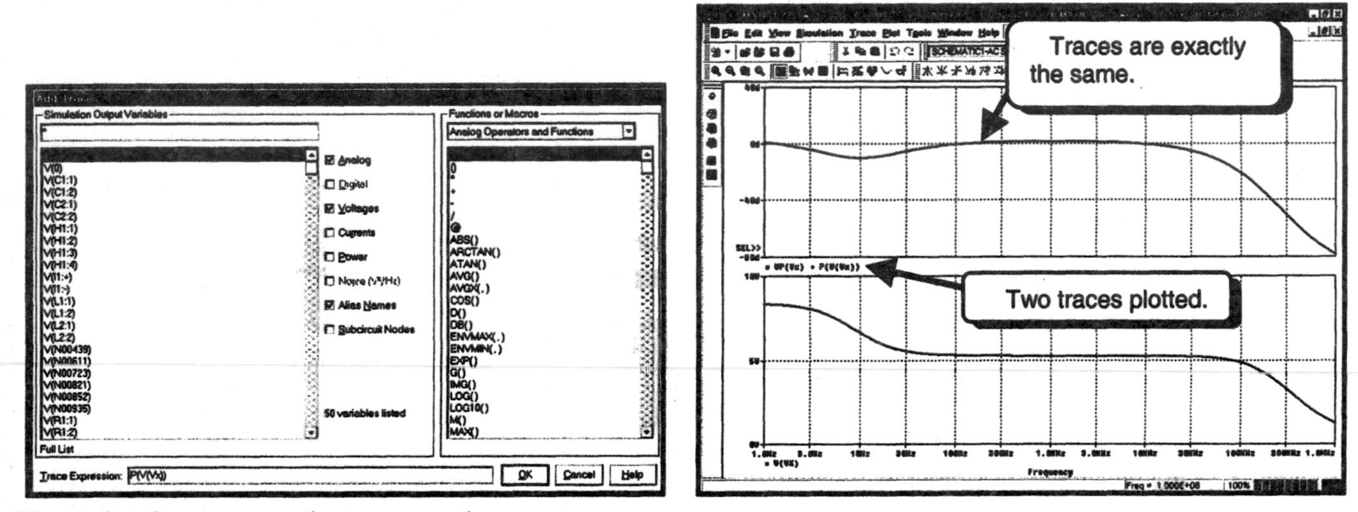


We see that the two expressions generate the same trace.

**EXERCISE 5-3**: Find the magnitude and phase of the voltage at node 1 for frequencies from 1 Hz to 100 Hz. Set your AC Sweep to decades with 100 points per decade.

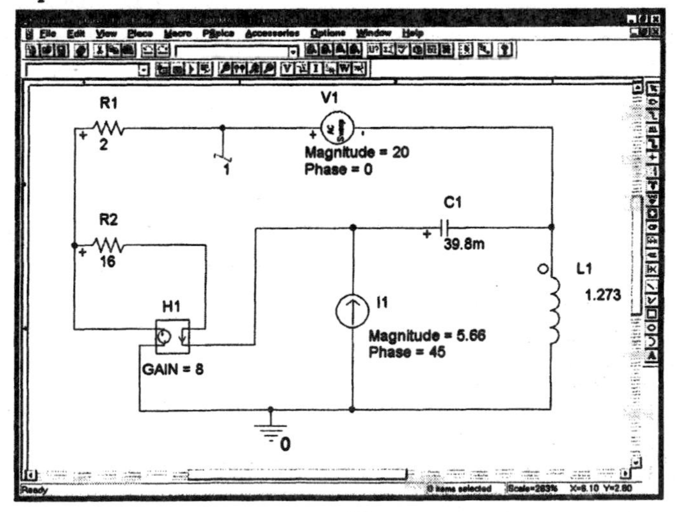


**HINT**: ***H1*** is a current-controlled voltage source. The voltage of ***H1*** is ***8*** times the current through ***R2***.

**SOLUTION**: The results are displayed using Probe. Obtain two separate plots, since the scale of the magnitude of V(1) is much smaller than the scale of the phase of V(1):

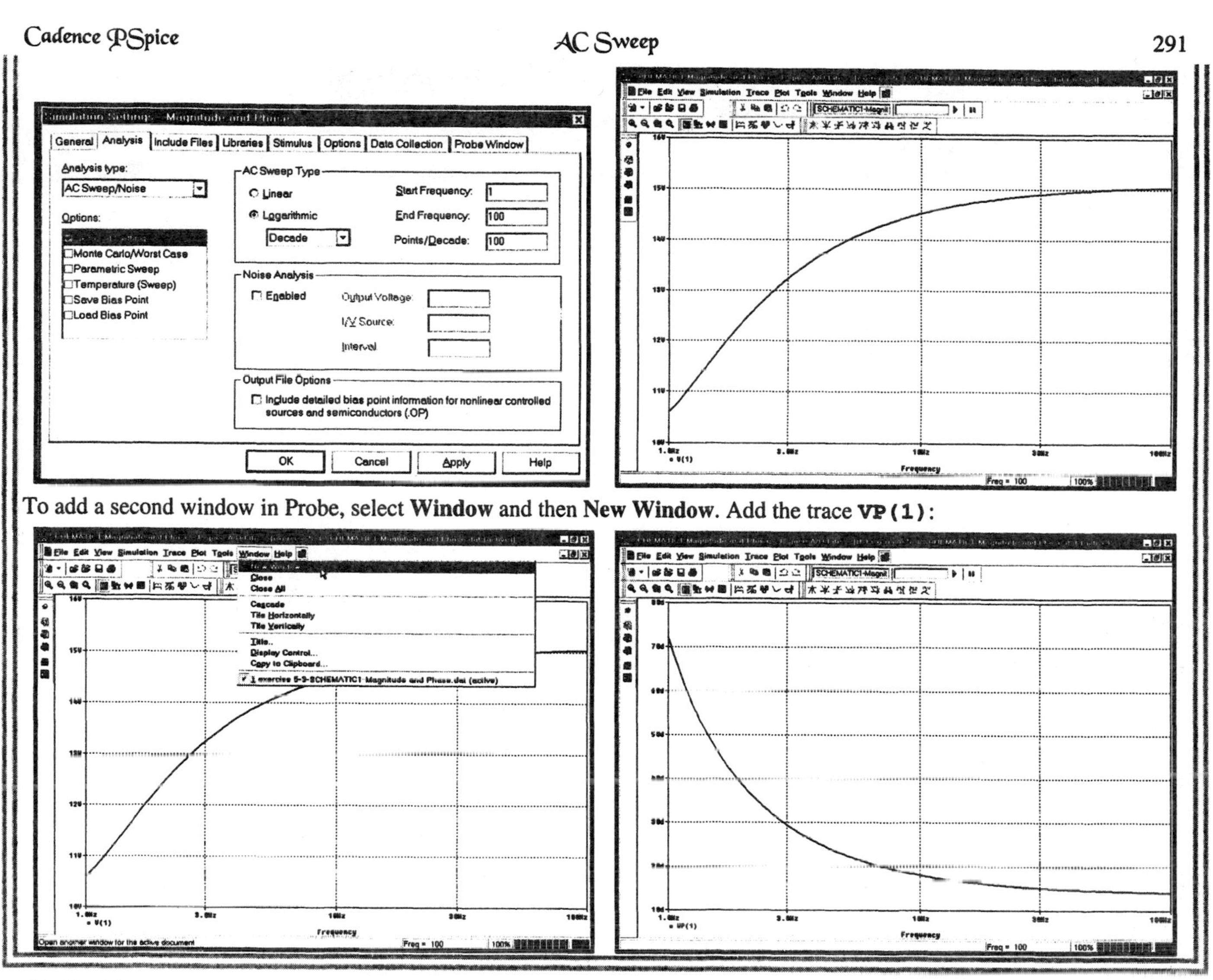


To add a second window in Probe, select **Window** and then **New Window**. Add the trace **`VP(1)`**:

# 5.C. Bode Plots

Bode plots are plots of magnitude and phase versus frequency. Since we are usually interested in the magnitude of the gain, an AC 1-volt source will be used. All AC analyses assume a linear network: If the output for a 1 V source is 3 V, the output for a 10 V source will be 30 V. Since gain is the ratio of an output to the source, and since the networks are linear, the magnitude of the input does not matter. For convenience we will use a 1-volt magnitude source. Bode phase plots will give us the phase of any voltage or current relative to the phase of the source. For simplicity we will set the phase of the source to zero.

Wire the low-pass filter shown:

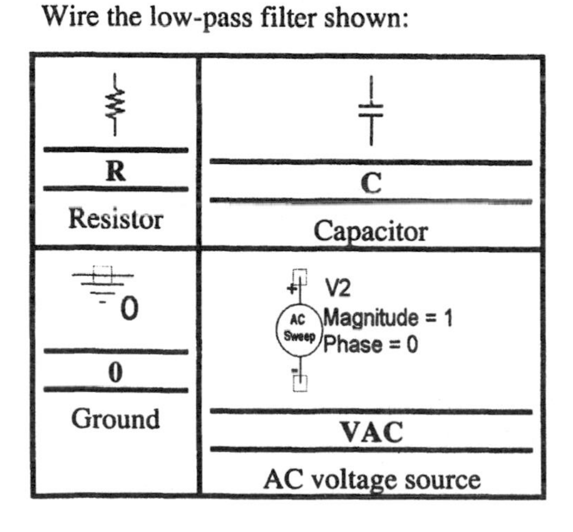


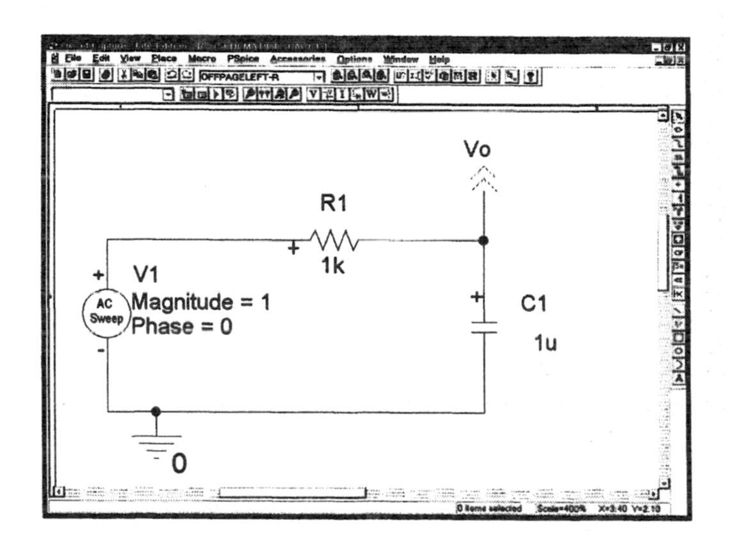

At low frequencies the capacitor is an open circuit and ***Vo*** should equal ***V1***. At high frequencies the capacitor becomes a short, and the gain goes toward zero. The 3 dB frequency of the circuit is $\omega = 1/RC = 1{,}000$ rad/s $\cong$ 159 Hz. We will set up an AC Sweep to sweep the frequency from 1 Hz to 10 kHz. Select **PSpice** and then **New Simulation Profile** from the Capture menus and then enter a name for the profile and click the ***Create*** button. Select the ***AC Sweep/Noise Analysis type*** and fill in the parameters as shown in the AC Sweep dialog box below:

Click the ***OK*** to accept the settings. Run the analysis by selecting **PSpice** and then **Run** from the Capture menus:

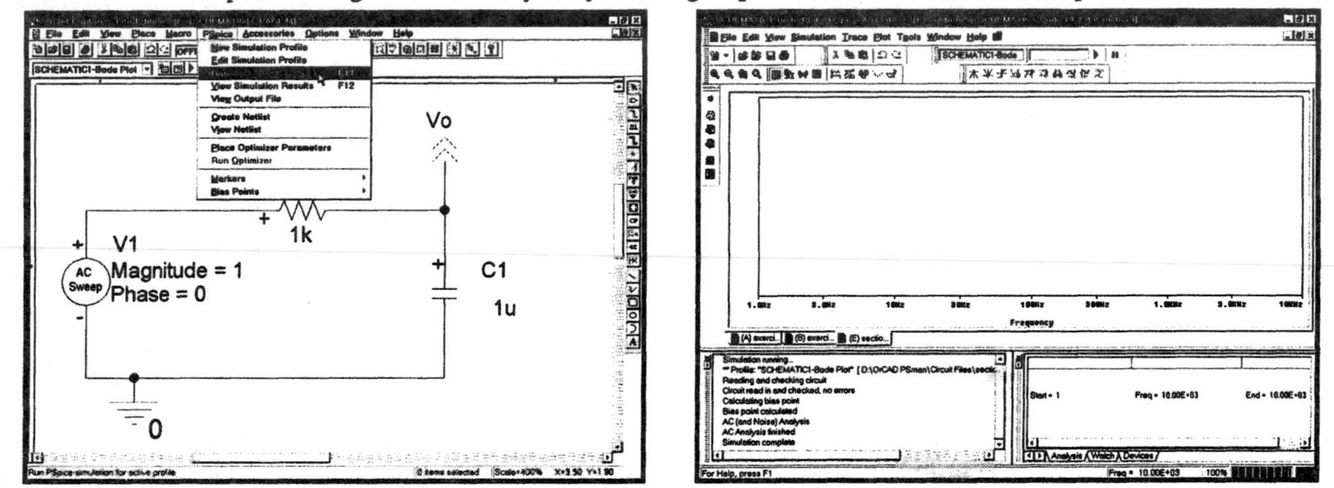


The phase and magnitude can be displayed as shown in previous sections. In this example we would like to show a Bode plot that displays the magnitude in decibels. To display Vo in decibels, we need to display the trace dB(V(VO)). Select **Trace** and then **Add Trace** from the Probe menu bar. In the text field next to ***Trace Expression*** enter the text **DB(V(VO))**:

The ***DB*** command takes $20\log_{10}$ of the specified voltage. Thus, ***DB(V(VO))*** is equivalent to 20log(V(Vo)). Since our source voltage (***V1***) had a magnitude of 1 V, ***DB(V(VO))*** is equivalent to DB(V(Vo)/V1),* which is the Bode plot of the gain of this circuit. Click the ***OK*** button to plot the trace:

*Note that "V1" is not a valid expression for the ***Trace*** command. It is used only for clarity. Instead of "V1" you would have to enter ***V(R1:1)***. Note that ***V(R1:1)*** is listed as one of the choices in the ***Add Traces*** dialog box.

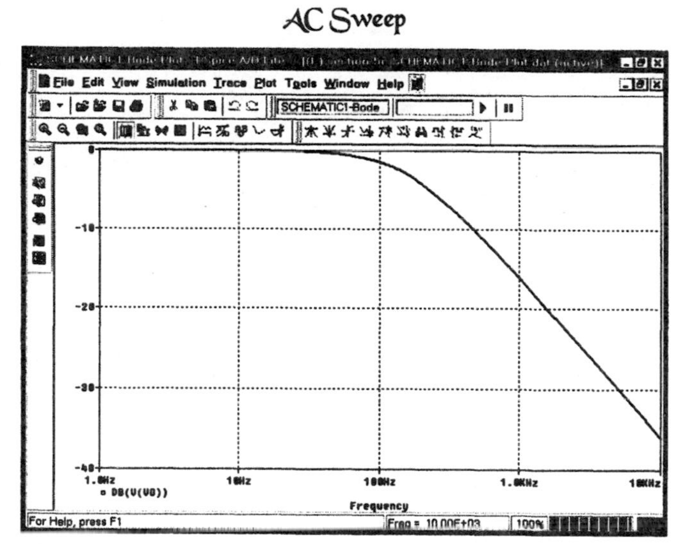


Note that the y-axis is in dB. A gain of 1 is equal to 0 dB.

We can now use the cursors to find the 3 dB frequency. (See Section 2.K for a full explanation of cursors.) To display the cursors, select **Trace, Cursor**, and then **Display** from the Probe menus. The cursors will appear:

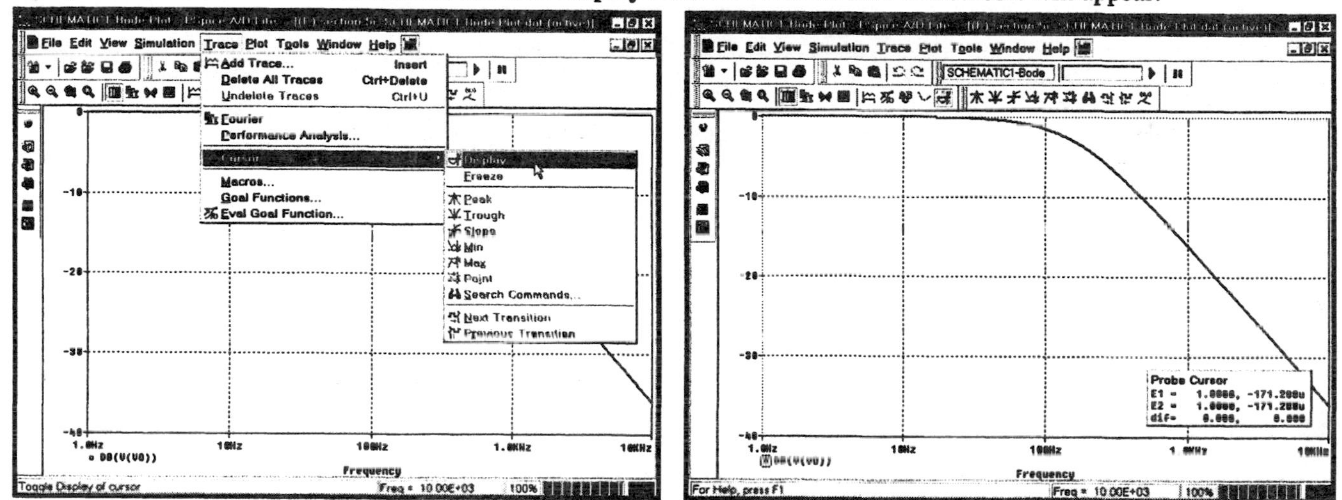


Notice that the values displayed by the cursors are in hertz and decibels. The first cursor is controlled by the ***LEFT*** mouse button or the arrow keys (←→). The second cursor is controlled by the ***RIGHT*** mouse button or the shift key plus the arrow keys. The difference, ***dif***, is the difference between the values of the first cursor and the values of the second cursor. To find the 3 dB point, we want to leave the second cursor in its original place and move the first cursor until the difference in magnitude is –3. To move the first cursor press and hold the **RIGHT ARROW** key. Move the first cursor until the difference is approximately –3:

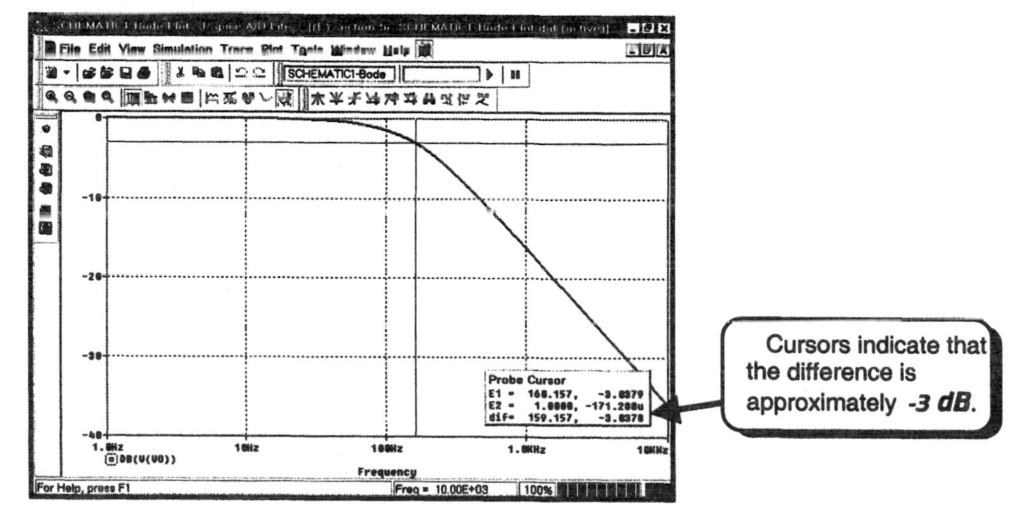

The cursors show that the 3 dB frequency is approximately at 159.157 Hz. Note that your cursor values may be slightly different than those shown here.

## 5.C.1. Using Goal Functions to Find the Upper 3 dB Frequency

A second method can be used to find the upper 3 dB frequency found in the previous section. We will continue from the end of the previous section. First, hide the cursors by selecting **Trace**, **Cursor**, and then **Display** from the Probe menus. Next, select **Window** and then **New Window** to open a new Probe window:

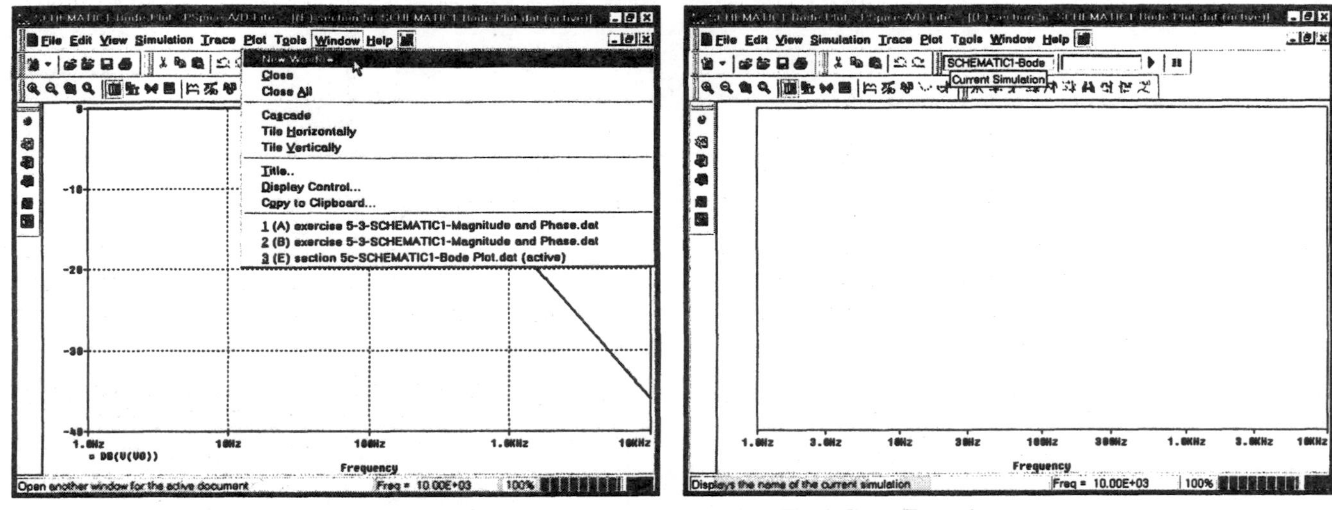


We will now evaluate the goal function. In Probe, select **Trace** and then **Eval Goal Function**:

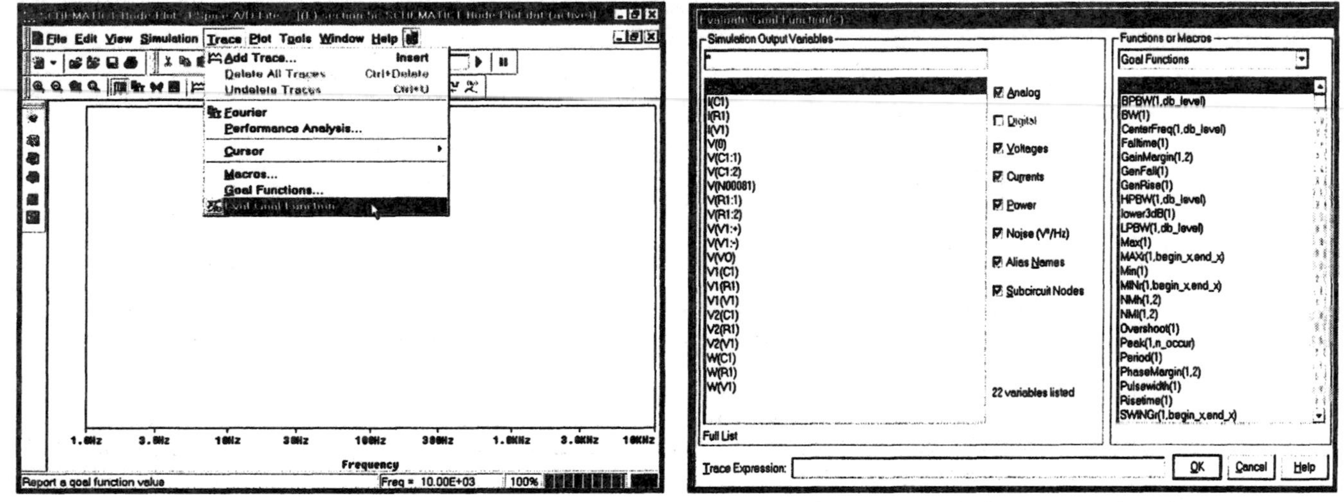


The right window pane lists the available goal functions. Use this list to locate the Upper3dB goal function. We would like to find the upper 3 dB frequency of the voltage trace V(VO), so enter the following ***Trace Expression***, **`upper3dB(V(VO))`**:

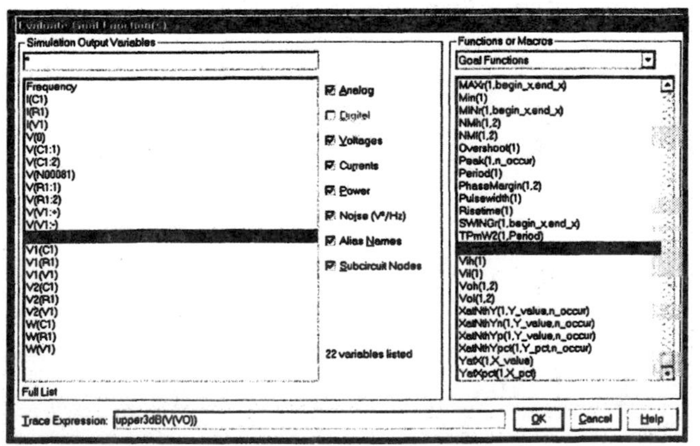


Click the ***OK*** button to evaluate the goal function:

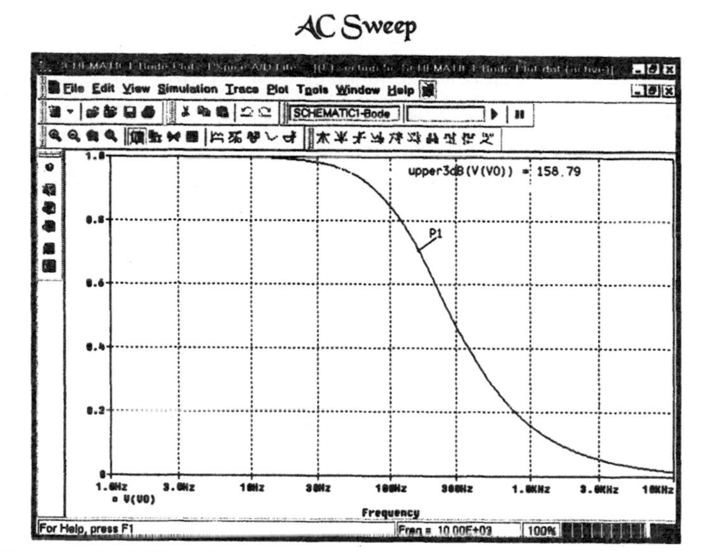

Probe plots the trace V(VO), locates the 3dB point, and then marks the point on the plot. Note that the plot of V(VO) is not in decibels. The plot is generated only if you have Probe set to display the plot used for the evaluation. If this option is not set, a small dialog box will appear and tell you the result of the goal function. To check the setting, select **Tools** and then **Options** from the Probe menus:

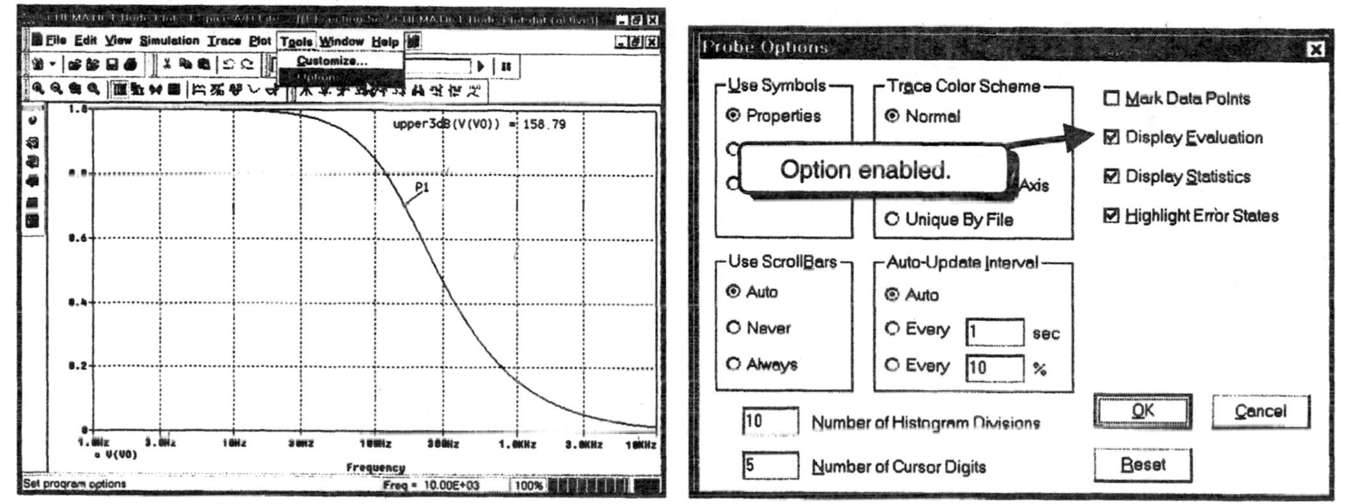


During installation, this option is enabled. However, it is possible that someone else that used Probe before you could have disabled this option.

**EXERCISE 5-4**: Plot the Bode phase and magnitude plots for frequencies from 1 Hz to 1 MHz. Use the cursors to find the frequencies of the poles and zeros.

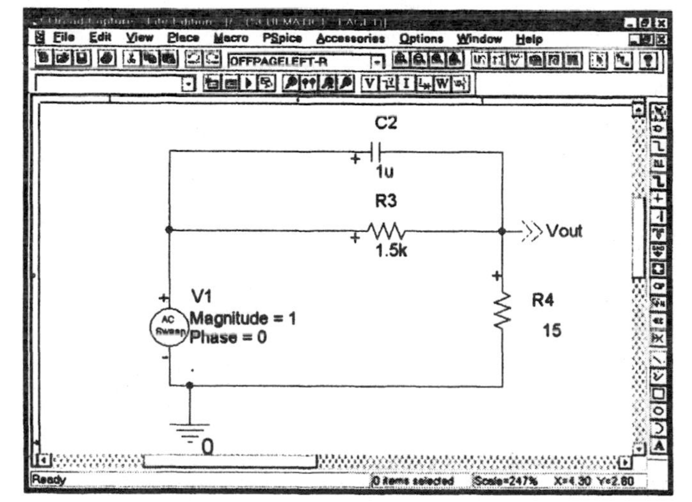

**ANALYTIC SOLUTION**: At low frequencies, the capacitor is an open circuit and the voltage at ***Vout*** can be obtained by the voltage divider of ***R3*** and ***R4***:

$$\frac{V_{out}}{V_{in}} = \frac{R_4}{R_3 + R_4} = 0.0099 = -40 \text{ dB}$$

At high frequencies, the capacitor is a short circuit and ***Vout*** = ***VIn*** so

$$\frac{V_{out}}{V_{in}} = 1 = 0 \text{ dB}$$

Thus, we should expect the magnitude plot to start at –40 dB and finish at 0 dB. The next question is, where are the poles and zeros? Let

$$Z = Z_c \| R_3 = \frac{(1/j\omega C_2)R_3}{(1/j\omega C_2) + R_3} = \frac{R_3}{1 + j\omega R_3 C_2}$$

Substituting Z for $Z_c \| R_3$ yields the equivalent circuit:

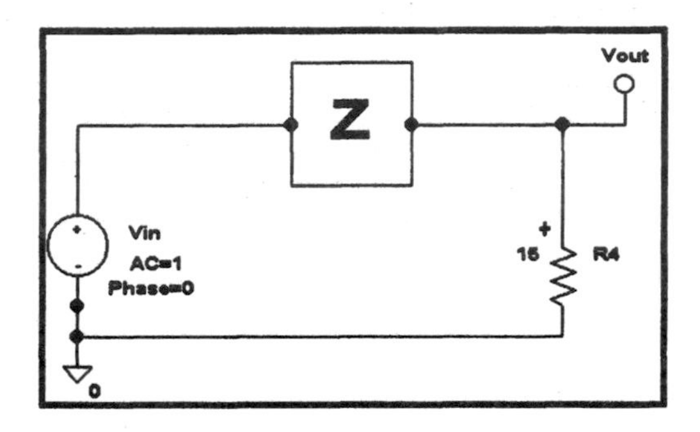


An expression for the gain can now be obtained from the voltage divider of ***R4*** and ***Z***:

$$\frac{V_{out}}{V_{in}} = \frac{R_4}{Z + R_4} = \frac{R_4}{\left(\frac{R_3}{1 + j\omega R_3 C_2}\right) + R_4}$$

Multiplying the numerator and denominator by $1 + j\omega R_3 C_2$ gives

$$\frac{V_{out}}{V_{in}} = \frac{R_4(1 + j\omega R_3 C_2)}{R_3 + R_4 + j\omega C_2 R_3 R_4}$$

We see that the zero is at $1 = \omega R_3 C_2$, or

$$\omega_z = \frac{1}{R_3 C_2} = \frac{1}{(1500\ \Omega)(1\ \mu F)} = 666 \text{ rad/s} = 106 \text{ Hz}$$

The pole is at $R_3 + R_4 = \omega C_2 R_3 R_4$, or

$$\omega_p = \frac{R_3 + R_4}{C_2 R_3 R_4} = \frac{1}{C_2\left(\frac{R_3 R_4}{R_3 + R_4}\right)} = \frac{1}{C_2 (R_3 \| R_4)}$$

$$= \frac{1}{(1 \mu F)(1500\ \Omega \| 15\ \Omega)} = 66.6 \text{ krad/sec} = 10.6 \text{ kHz}$$

**SOLUTION**: Set up an AC Sweep from 1 Hz to 1 MHz. Simulate the circuit and then run Probe:

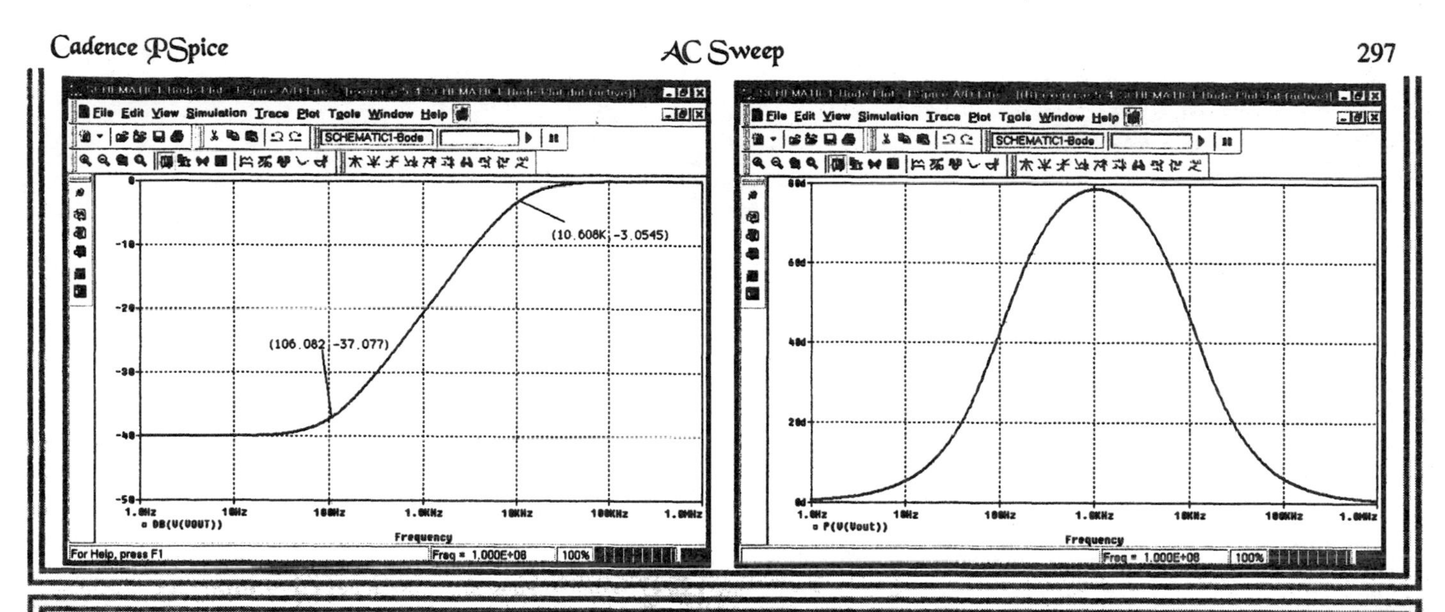


**EXERCISE 5-5**: Plot the Bode phase and magnitude plots for frequencies from 1 Hz to 1 MHz. Use the cursors to find the frequencies of the poles and zeros.

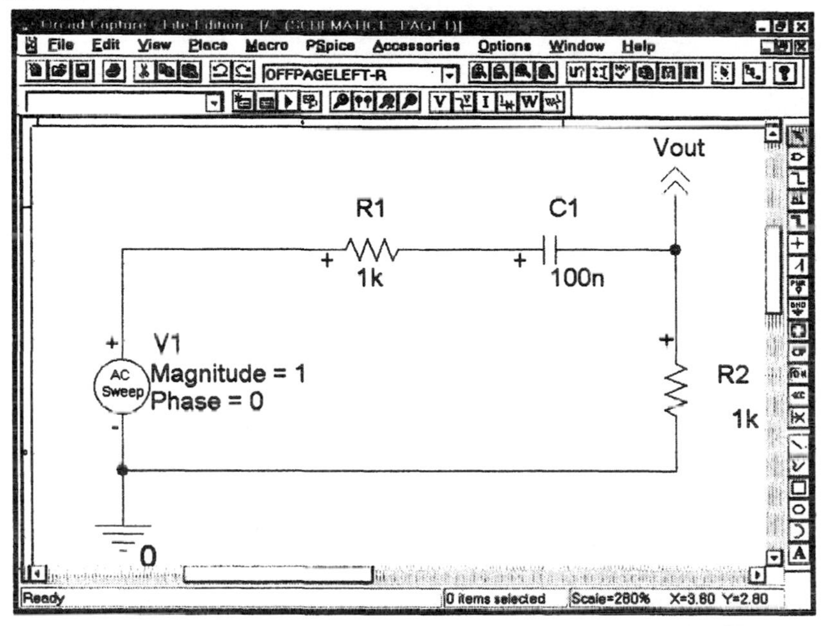


**ANALYTIC SOLUTION**: At low frequencies, the capacitor is an open circuit and the voltage at ***Vout*** is zero. As the frequency is increased, the impedance of the capacitor will decrease and the voltage at ***Vout*** will increase. Thus, at low frequencies we should expect the gain in decibels to be a large negative number and to increase at a rate of 20 dB/decade. At high frequencies, the capacitor is a short circuit and ***Vout*** can be obtained from the voltage divider of R1 and R2:

$$\frac{V_{out}}{V_{in}} = \frac{R_2}{R_2 + R_1} = 0.5 = -6 \text{ dB}$$

Thus, we should expect the magnitude plot to start at a negative value in decibels and finish at –6 dB. The next question is, where are the poles and zeros? The gain of the circuit can be obtained from the voltage divider of ***R1***, ***R2***, and ***C1***:

$$\frac{V_{out}}{V_{in}} = \frac{R_2}{R_1 + R_2 + Z_c} = \frac{R_2}{R_1 + R_2 + \left(\dfrac{1}{j\omega C_1}\right)} = \frac{j\omega C_1 R_2}{1 + j\omega C_1 (R_1 + R_2)}$$

We see that there is a zero at ω = 0 rad/sec and a pole at $1 = \omega C_1(R_1 + R_2)$, or

$$\omega_p = \frac{1}{C_1(R_1 + R_2)} = \frac{1}{(100 \text{ nF})(1000\ \Omega + 1000\ \Omega)} = 5000 \text{ rad/s} = 795.8 \text{ Hz}$$

**SOLUTION**: The results are displayed using Probe:

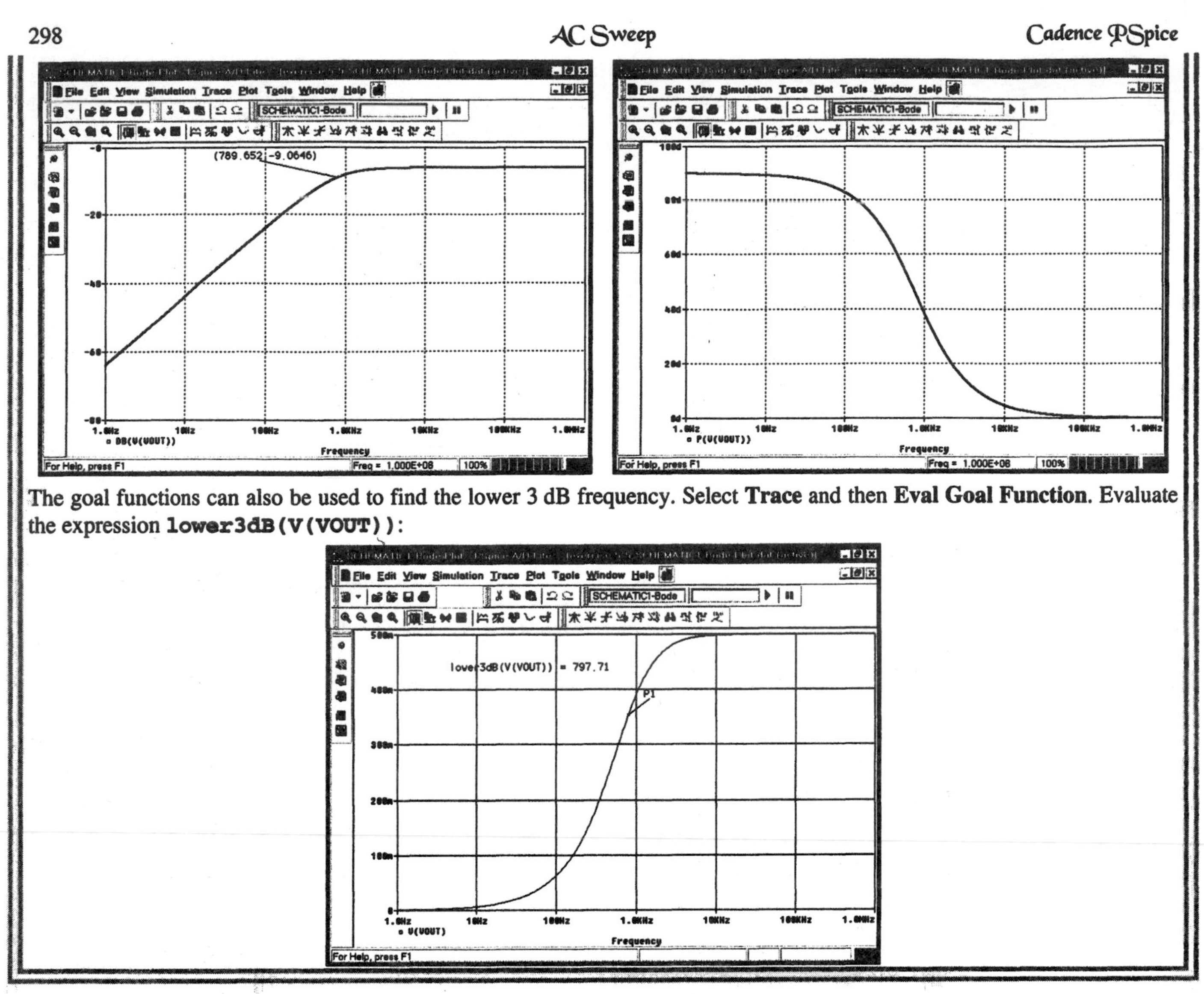


The goal functions can also be used to find the lower 3 dB frequency. Select **Trace** and then **Eval Goal Function**. Evaluate the expression `lower3dB(V(VOUT))`:

# 5.D. Amplifier Gain Analysis

One of the most important applications of the AC Sweep is to see the frequency response of an amplifier. If an AC Sweep is performed on a circuit with a transistor, the DC bias point is calculated and the transistor is replaced by a small-signal model around the bias point. The AC Sweep is then performed on the linearized model of the transistor. The AC Sweep can only be used to find the small-signal gain and frequency response. Voltage swing, clipping, and saturation information must be obtained from the transient simulation, or by using the operating point information.

Wire the amplifier circuit as shown:

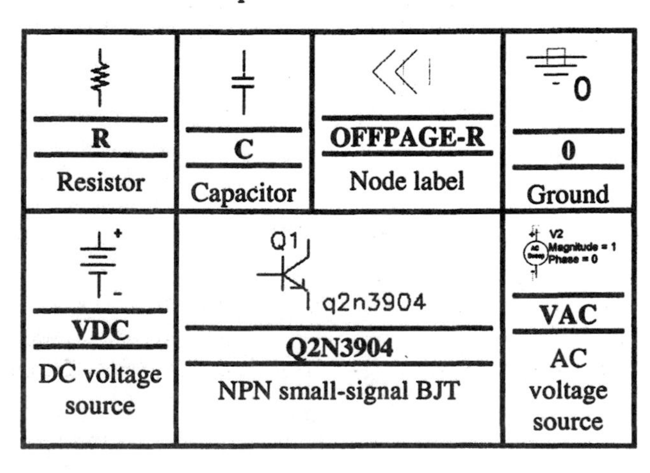

| R | C | OFFPAGE-R | 0 |
|---|---|---|---|
| Resistor | Capacitor | Node label | Ground |
| **VDC** | **Q2N3904** | | **VAC** |
| DC voltage source | NPN small-signal BJT | | AC voltage source |

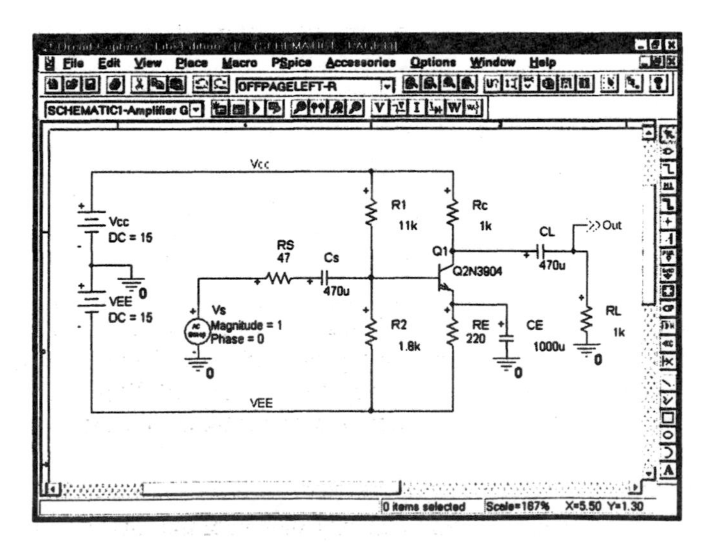

We would like to see the gain of this amplifier from 1 Hz to 100 MHz. Set up the AC Sweep as shown below. Select **PSpice** and then **New Simulation Profile** from the Capture menus and then enter a name for the profile and click the ***Create*** button. Select the ***AC Sweep/Noise Analysis type*** and fill in the parameters as shown in the AC Sweep dialog box below:

Whenever you simulate an amplifier with transistors you should always check the bias point. (See Section 3.E, "**Transistor Bias Point Detail**," on page 187.) If the bias is not correct, then the AC Sweep results are meaningless. Always check the bias point first.

Run the PSpice simulation: Select **PSpice** and then **Run** from the Capture menu bar. When the simulation is complete, the Probe window will open. Add the trace `DB(V(VO))`* to plot the gain in decibels. Use the cursors to label the mid-band gain, and upper and lower –3 dB frequencies. Your cursor values may be slightly different than those shown here.

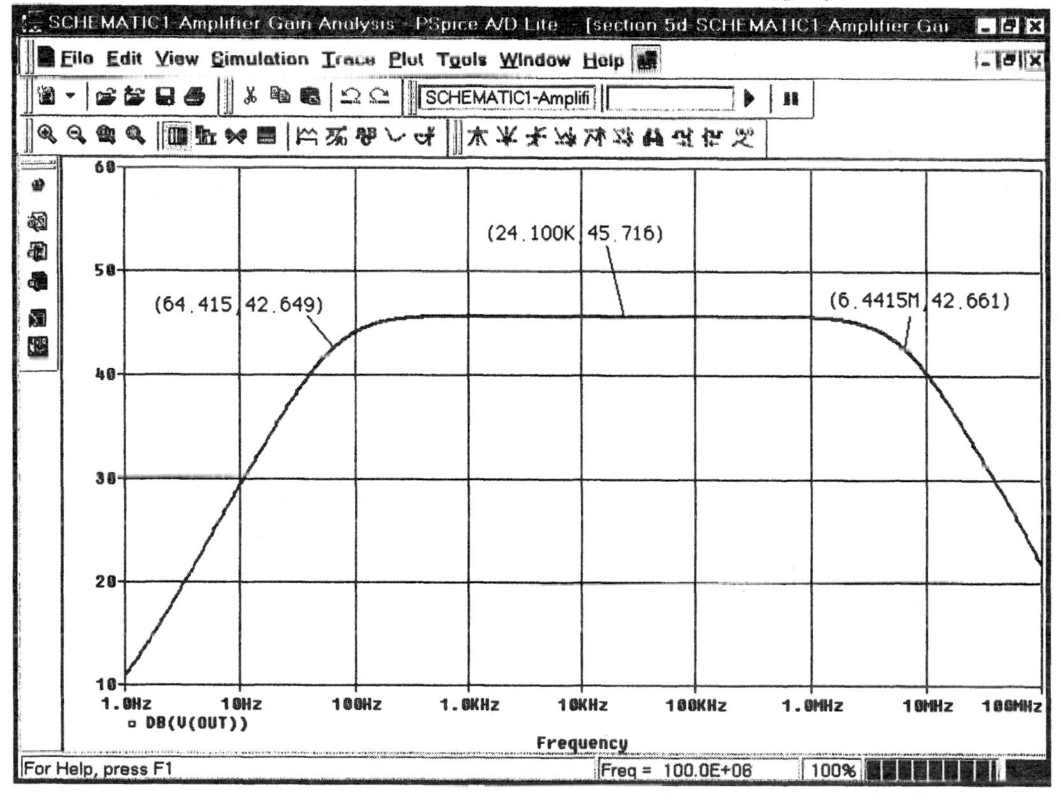


We can see that the mid-band gain is 45.7 dB, the upper 3 dB frequency is 6.4 MHz, and the lower 3 dB frequency is 64.4 Hz. You can also use the Upper3dB and Lower3dB goal functions to find upper and lower –3dB frequencies. Select **Trace** and then **Eval Goal Function** to evaluate these functions:

*See page 292 for an explanation of the "dB" command.

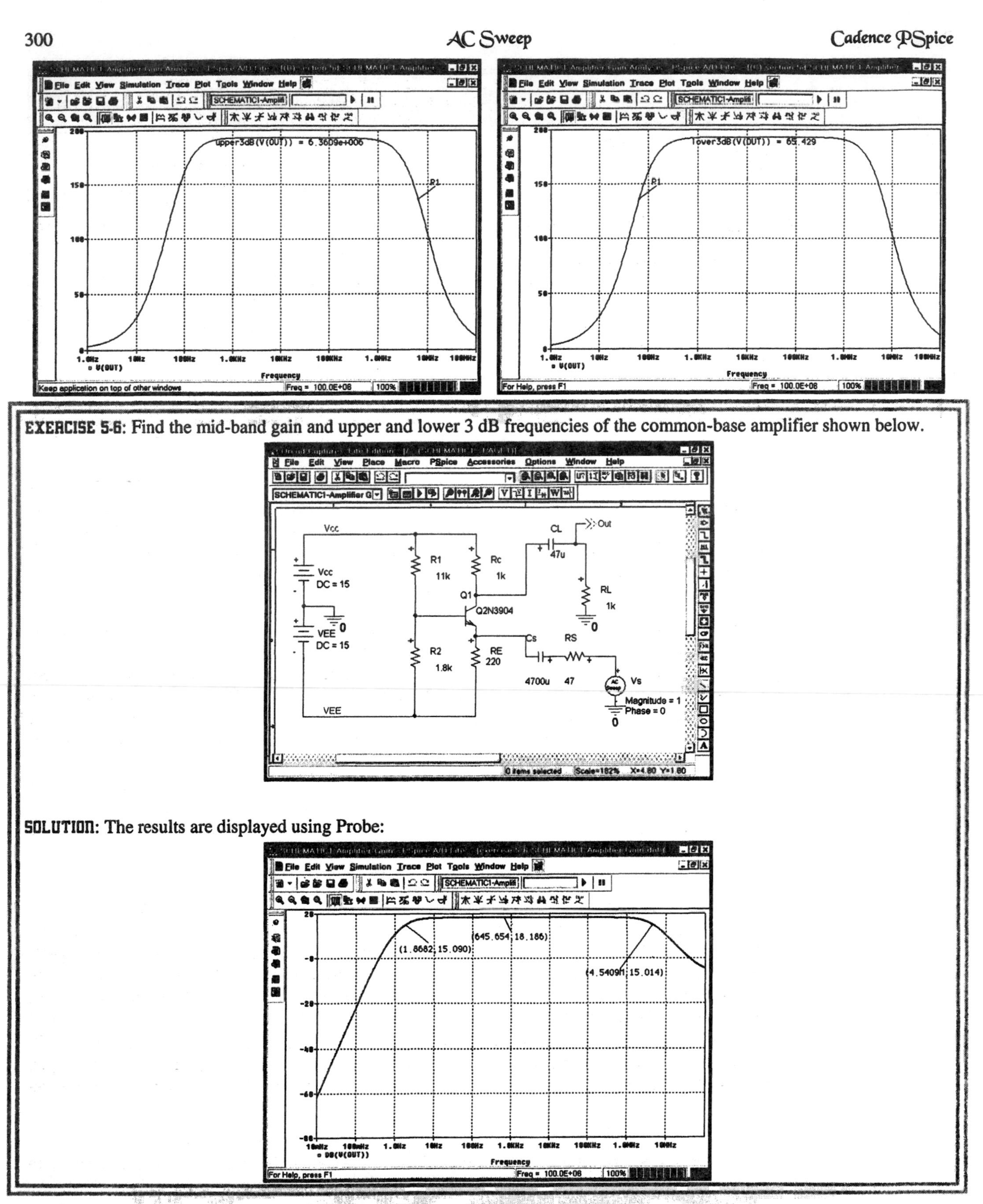


**EXERCISE 5-6**: Find the mid-band gain and upper and lower 3 dB frequencies of the common-base amplifier shown below.

**SOLUTION**: The results are displayed using Probe:

# 5.E. Operational Amplifier Gain

In this section we will use PSpice to determine the gain and bandwidth of an operational amplifier with negative feedback. Wire the circuit shown below:

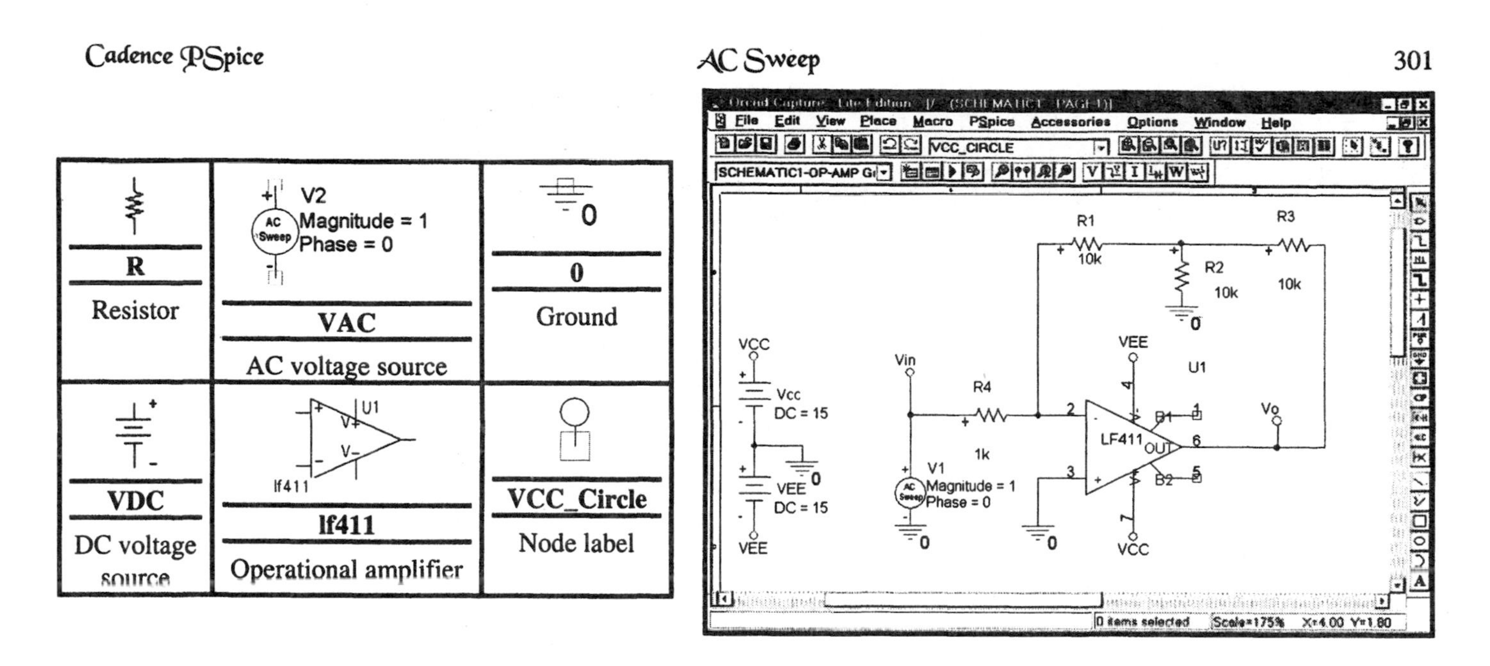


This example uses the non-ideal op-amp model of an LF411C. Since there is only one op-amp in this circuit, we will not reach the component limit of the Lite version. The method described here can be used for circuits with several op-amps. However, the component limit of the Lite version of PSpice limits us to only two non-ideal op-amp models. If more than two op-amps are needed, the Ideal_OPAMP model can be used. Since the Ideal_OPAMP model has no frequency limitations, it cannot be used to find the bandwidth, but it can be used to find the gain.

Notice that the ***lf411*** has power supply terminals ***V+*** and ***V-***. They must be connected to the appropriate DC supplies. In the circuit, node labels are used to make connections without drawing wires. These labels are referred to as power connectors in Capture, but they can be used to label any nodes and connect components together that use a node label with the same text. To place a node label, select **Place** and then **Power** from the Capture menus. You will be presented with a dialog box that lists the available power connectors. The graphics for all of the power connectors are different, but any of them can be used to label nodes. Two nodes in a circuit will be connected if they use any power connector as long as the text label of the power connector is the same. If two power connectors use a different graphic symbol, but the text label is the same, the two nodes will be connected. Conversely, if two nodes use the same graphic, but the text label for the power connector is different, the two nodes will not be connected. In the circuit above, the two power connectors labeled VCC are connected and the two power connectors labeled VEE are connected. Nodes Vin and Vo are not connected. Power connectors were used to label the nodes so that we can easily identify them.

We would like to find the gain of this amplifier, $V_o/V_{in}$. Since the magnitude of $V_{in}$ is one volt, the magnitude of $V_o$ is the gain. If the gain of the amplifier is 200, then the magnitude of the output will be 200 V. This may seem a little unreasonable since the DC supplies are ±15 V. When an AC Sweep is performed, the operating points of all non-linear parts are found, and the parts are replaced by their linear models. Since the models are linear, the voltages and currents in the circuit are not limited by the DC supplies. For a linear circuit, if an input of 1 V produces an output of 10 V, then an input of 1,000,000 V will produce an output of 10,000,000 V. Thus, for gain purposes the magnitude of the input does not matter and a 1 V input is chosen for convenience. For an AC simulation a magnitude of 200 V is not unreasonable. A magnitude of 200 V is not physically possible for our op-amp, but it is a valid number when using the AC Sweep to calculate the gain. If you wish to observe the maximum voltage output of a circuit, you must run a Transient Analysis. This is done in Section 6.G on page 361.

We must set up an AC Sweep. Select **PSpice** and then **New Simulation Profile** from the Capture menus and then enter a name for the profile and click the ***Create*** button. Select the ***AC Sweep/Noise Analysis type*** and fill in the parameters as shown in the AC Sweep dialog box below:

Simulation Settings - OP AMP Gain

General | Analysis | Include Files | Libraries | Stimulus | Options | Data Collection | Probe Window

Analysis type:

Options:
- Monte Carlo/Worst Case
- Parametric Sweep
- Temperature (Sweep)
- Save Bias Point
- Load Bias Point

AC Sweep Type
- Linear
- Logarithmic: Decade

Start Frequency: 1
End Frequency: 1meg
Points/Decade: 100

Noise Analysis
- Enabled
Output Voltage:
I/V Source:
Interval:

Output File Options
- Include detailed bias point information for nonlinear controlled sources and semiconductors (.OP)

OK | Cancel | Apply | Help

The sweep is set up to simulate the circuit for frequencies from 1 Hz to 1 MHz at 100 points per decade. Click the ***OK*** button to accept the settings, and then click the ***Close*** button to return to the schematic. Run PSpice by selecting **PSpice** and then **Run** from the Capture menu bar. When Probe runs, add the trace **V(VO)**. To add a trace, select **Trace** and then **Add Trace** from the Probe menu bar or press the **INSERT** key. You will see the amplifier gain as a function of frequency:

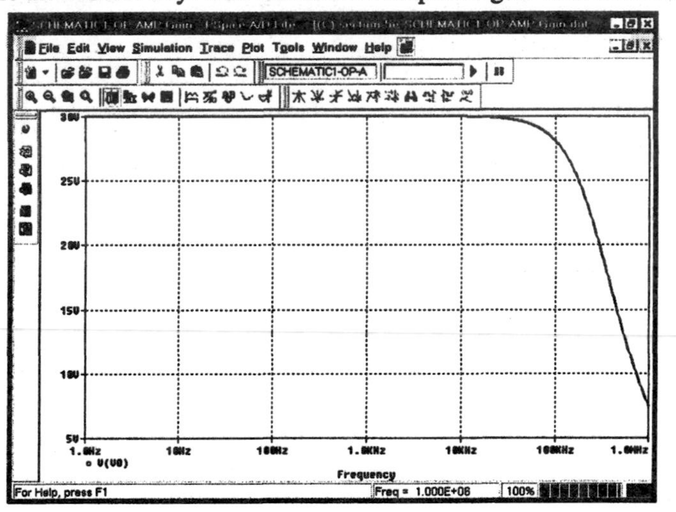


This plot shows us that the gain of the amplifier is 30. If we wish to find the 3 dB frequency for this amplifier, we must display the trace in dB.* We will add a second window to display the new trace. Select **Window** and then **New Window** to open a new window and then add the trace **DB(V(VO))**. Use the cursors to locate the –3 dB frequency:

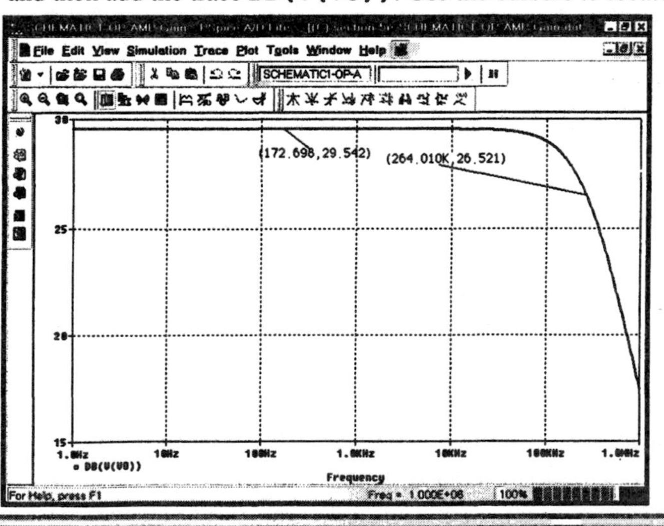


**EXERCISE 5-7**: Find the mid-band gain and upper 3 dB frequency for the amplifier shown below:

---

*See page 292 for an explanation of the "dB" command.

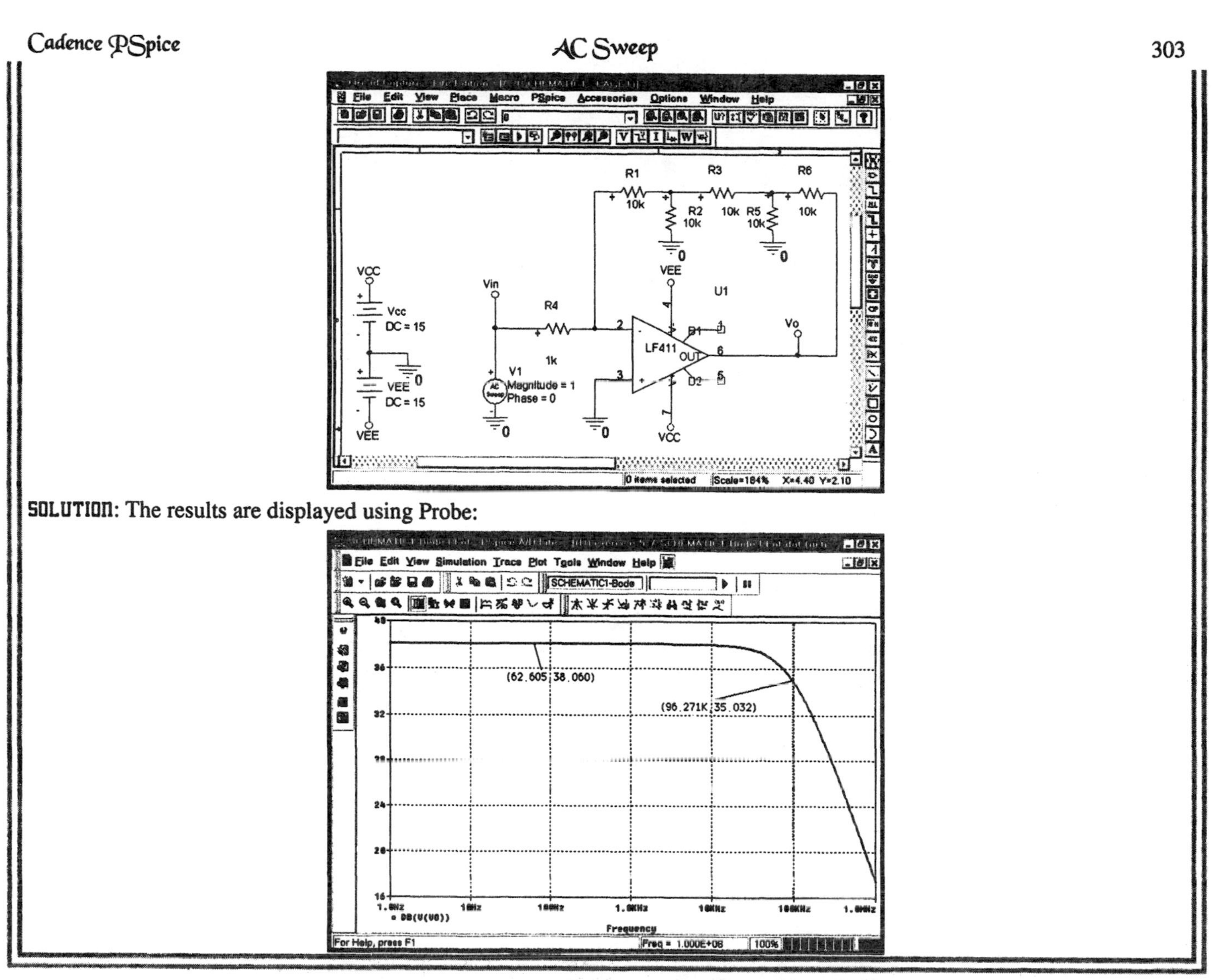


**SOLUTION**: The results are displayed using Probe:

# 5.F. Parametric Sweep — Op-Amp Gain Bandwidth

In this section we will use PSpice to determine the bandwidth of an op-amp circuit with varying amounts of negative feedback. For an op-amp circuit, the closed-loop gain times the bandwidth is approximately constant. To observe this property, we will run a simulation that creates a Bode plot for several different closed-loop gains. We will use the circuit below:

| R | PARAM | VCC_Circle | 0 |
|---|---|---|---|
| Resistor | PSpice parameter | Node label | Ground |
| **VDC** | **LF411** | | **VAC** |
| DC voltage source | Operational amplifier | | AC voltage source |

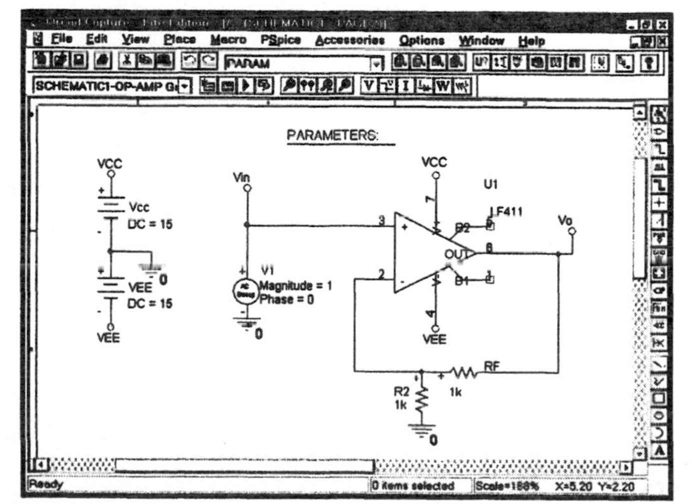


This circuit uses a ***LF411*** op-amp macro model. All op-amp models, except the ideal op-amp model, include bias currents, offset voltages, slew rate limitations, and frequency limitations. Also note that the op-amp model requires DC supplies. The model will not work without the supplies. In the Lite version, only two op-amp models can be used before reaching the component limitation. If you need more than two op-amps, use the ideal op-amp model.

In this simulation we would like to change the gain of the circuit by changing the value of RF. To do this, we will define a parameter called RF_val and use this as the value for RF. First, we must define the parameter RF_val using the PARAMETERS part. Double-click the *LEFT* mouse button on the text **PARAMETERS** to obtain the spreadsheet for this part:

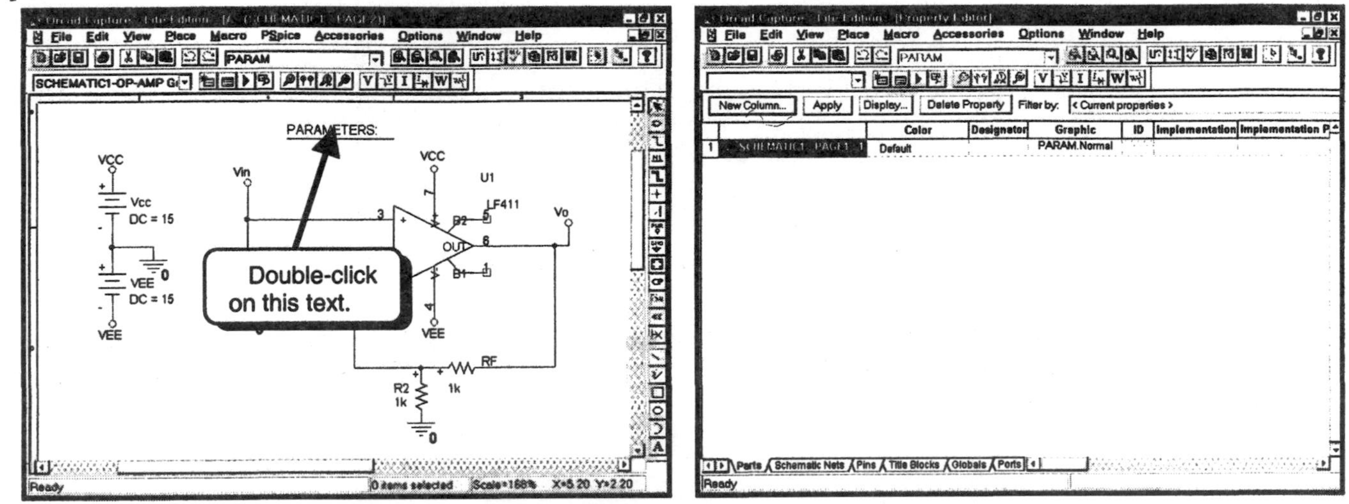


Click the **New Column** button to create a new parameter and add it as a new column to the spreadsheet:

Add New Column

Name:

Value:

Enter a name and click Apply or OK to add a column/row to the property editor and optionally the current filter (but not the <Current properties> filter).

No properties will be added to selected objects until you enter a value here or in the newly created cells in the property editor spreadsheet.

Always show this column/row in this filter

Apply OK Cancel Help

Type the name of the parameter, `RF_val`. As you enter the name, the **Value** field is enabled:

Add New Column

Name:
RF_val

Value:

Enter a name and click Apply or OK to add a column/row to the property editor and optionally the current filter (but not the <Current properties> filter).

No properties will be added to selected objects until you enter a value here or in the newly created cells in the property editor spreadsheet.

Always show this column/row in this filter

Apply OK Cancel Help

We can now enter a value for the parameter. Enter `1`:

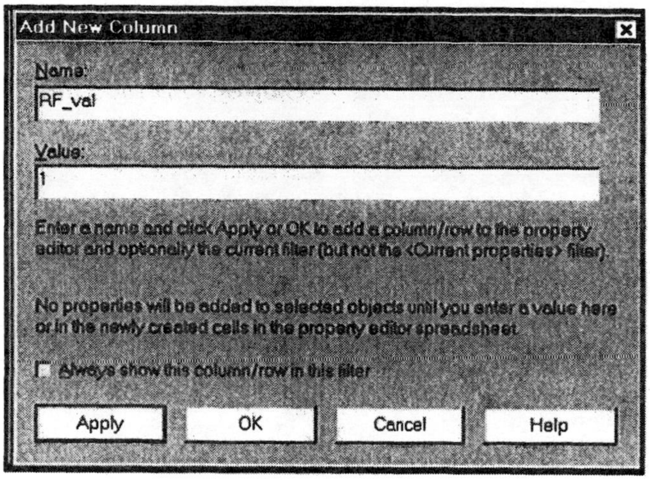

Add New Column

Name:
RF_val

Value:
1

Enter a name and click Apply or OK to add a column/row to the property editor and optionally the current filter (but not the <Current properties> filter).

No properties will be added to selected objects until you enter a value here or in the newly created cells in the property editor spreadsheet.

Always show this column/row in this filter

Apply OK Cancel Help

The value specified here is the default value for the parameter. If we do not modify the parameter in a DC Sweep or in a Parametric Sweep, the value of the parameter will be the one specified here. Click the ***OK*** button to accept the settings and add a new column to the spreadsheet:

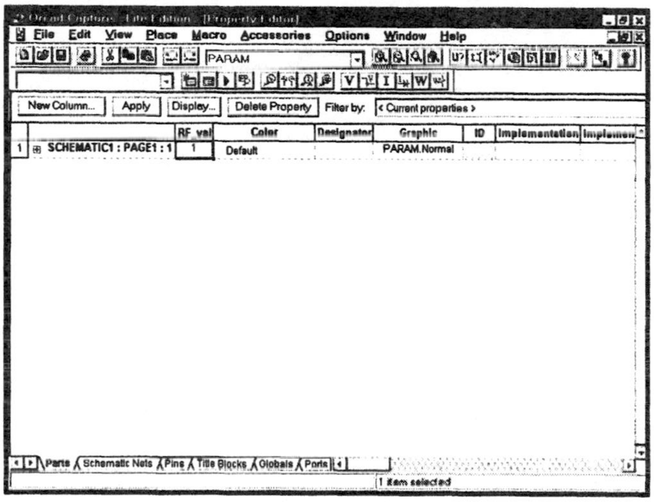

File Edit View Place Macro Accessories Options Window Help

PARAM

New Column... Apply Display... Delete Property Filter by: < Current properties >

| | RF_val | Color | Designator | Graphic | ID | Implementation | Implemen |
|---|---|---|---|---|---|---|---|
| 1 SCHEMATIC1 : PAGE1 : 1 | 1 | Default | | PARAM.Normal | | | |

Parts Schematic Nets Pins Title Blocks Globals Ports

1 item selected

Next, we need to modify the display properties of the parameter. Click the ***LEFT*** mouse button on the text RF_val as shown to select the column:

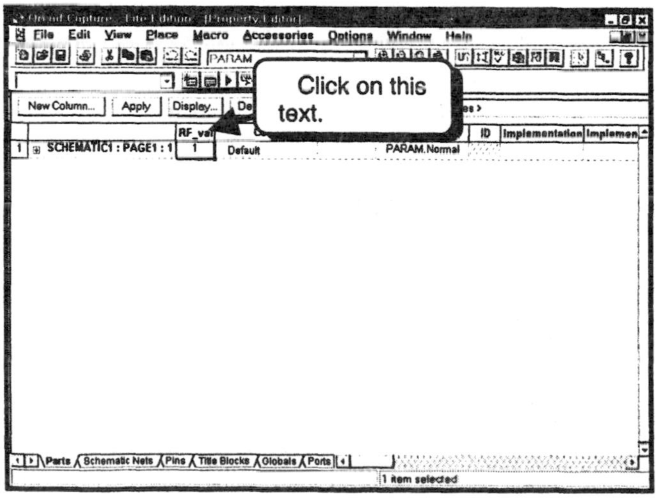

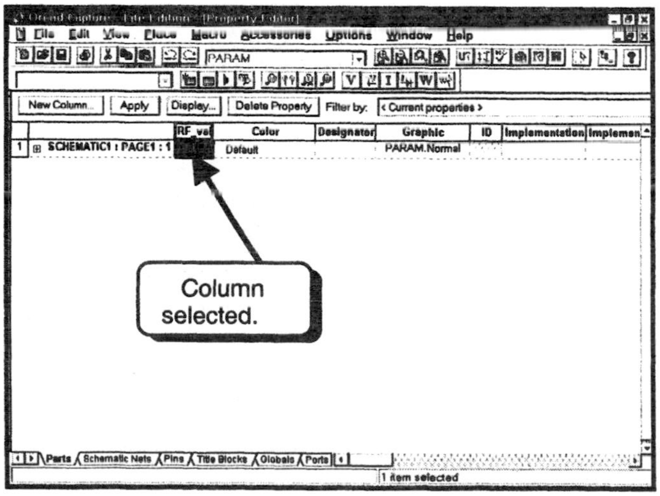

File Edit View Place Macro Accessories Options Window Help

PARAM

New Column... Apply Display...

| | RF_val | | ID | Implementation | Implemen |
|---|---|---|---|---|---|
| 1 SCHEMATIC1 : PAGE1 : 1 | 1 | Default | PARAM.Normal | | |

Click on this text.

Parts Schematic Nets Pins Title Blocks Globals Ports

1 item selected

File Edit View Place Macro Accessories Options Window Help

PARAM

New Column... Apply Display... Delete Property Filter by: < Current properties >

| | RF_val | Color | Designator | Graphic | ID | Implementation | Implemen |
|---|---|---|---|---|---|---|---|
| 1 SCHEMATIC1 : PAGE1 : 1 | | Default | | PARAM.Normal | | | |

Column selected.

Parts Schematic Nets Pins Title Blocks Globals Ports

1 item selected

Once the column is selected, click the ***Display*** button:

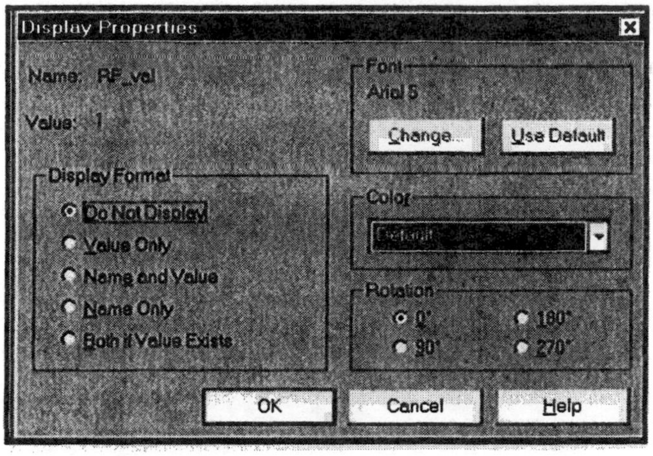

Presently, nothing about the parameter will be displayed on the schematic. Select option ***Name and Value***:

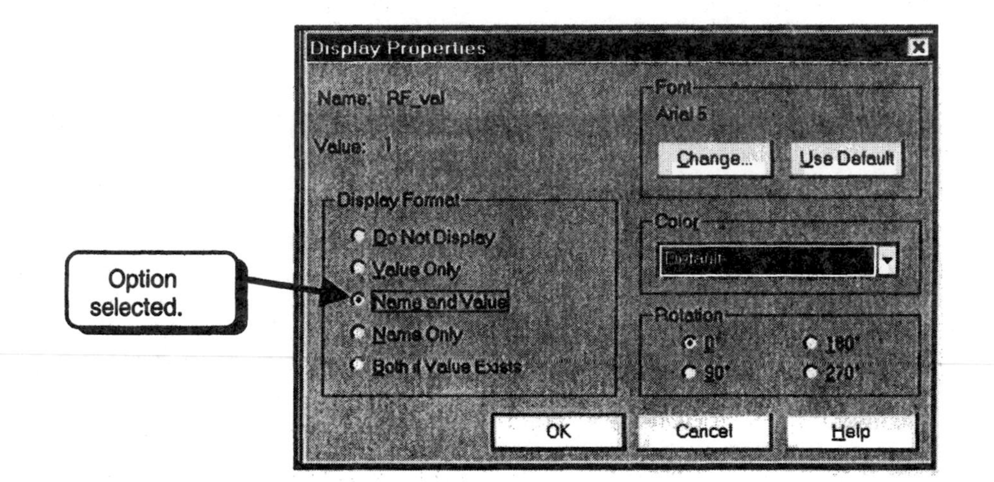


Click the ***OK*** button to return to the spreadsheet:

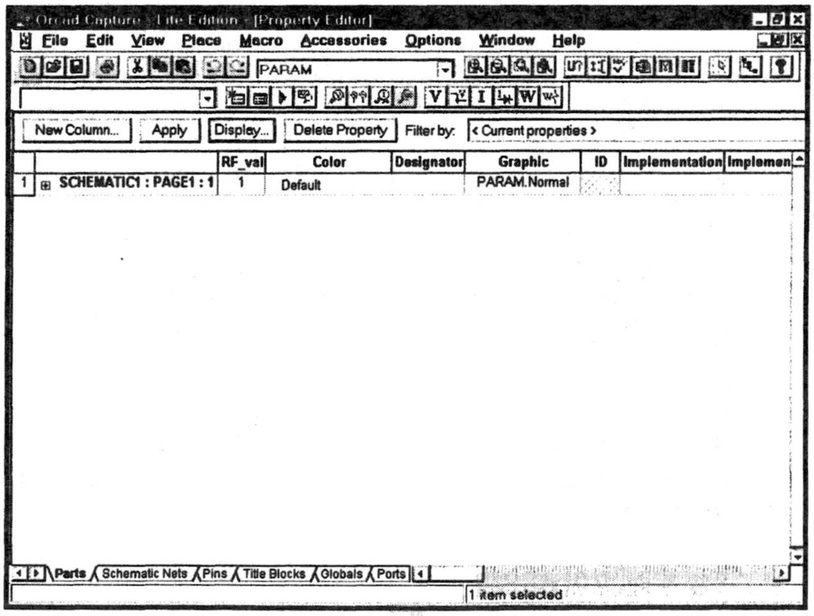

Click the ***LEFT*** mouse button on the lower ☒ as shown to close the spreadsheet (or type **CTRL F4**):

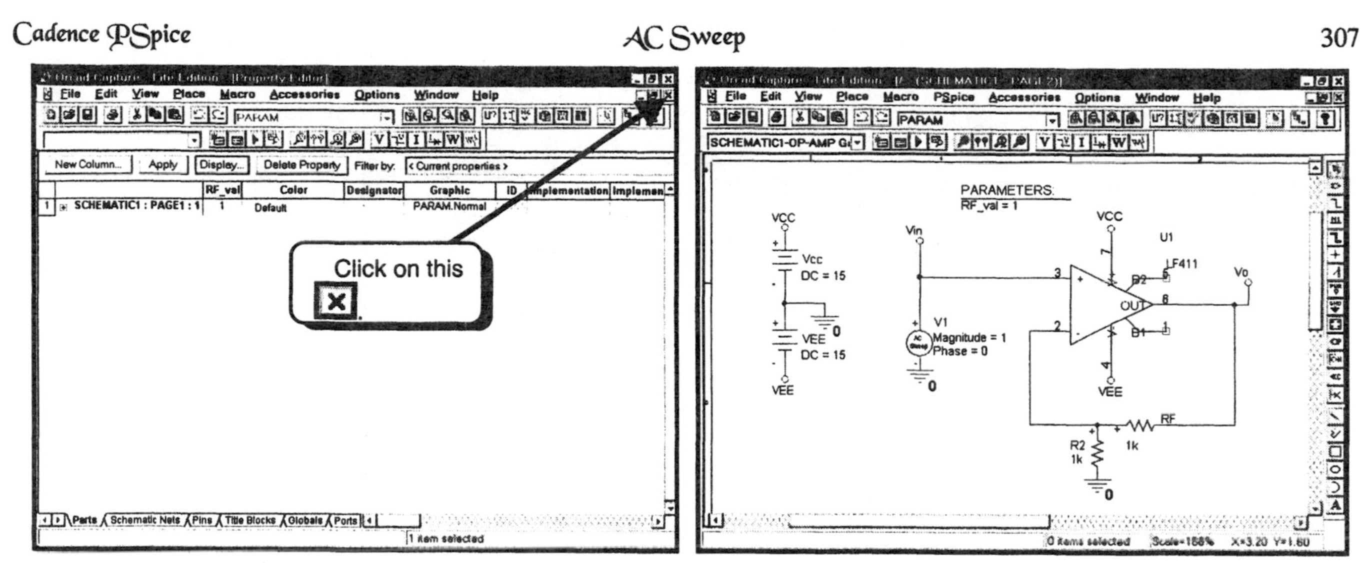


We see that the parameter is now displayed on the schematic.

The next thing we need to do is to use the parameter as the value of RF. Double-click the ***LEFT*** mouse button on the text ***1k*** for resistor ***RF***:

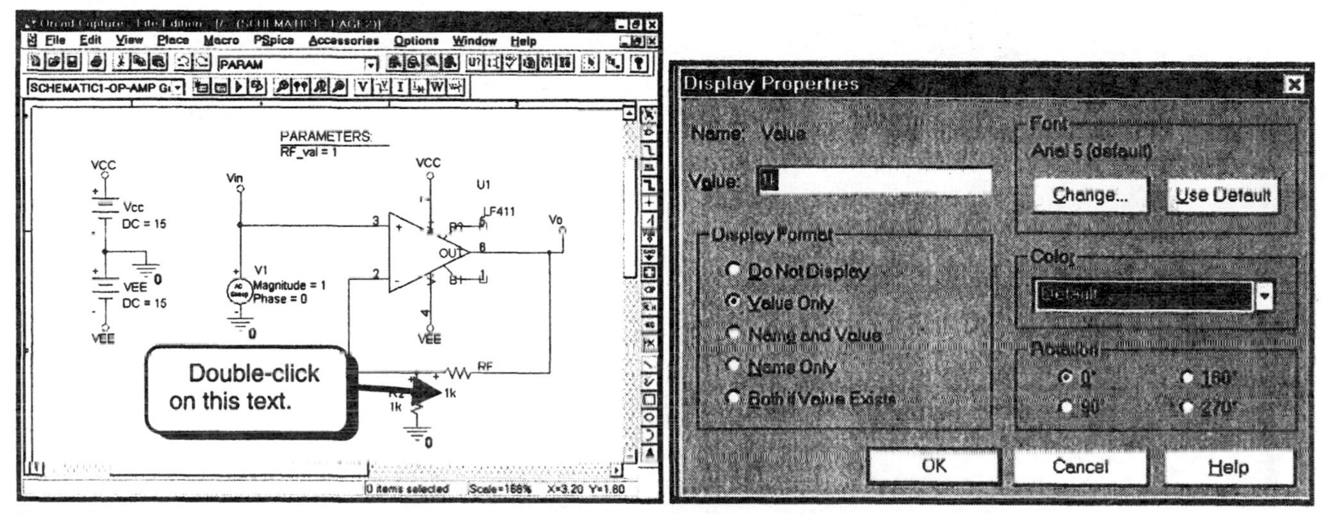


Type the text **{RF_val}**:

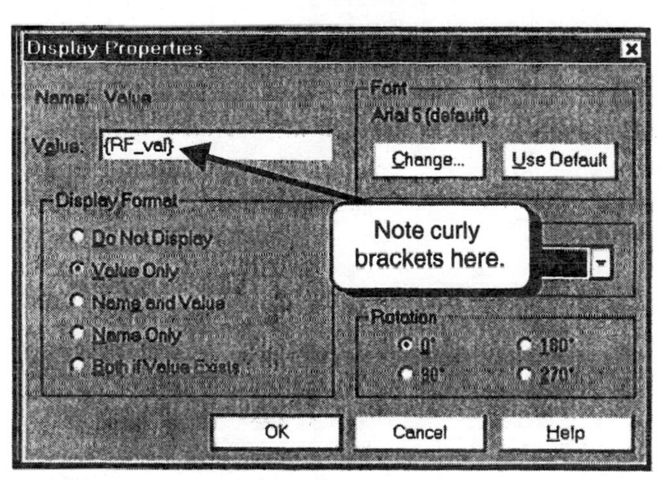


Note that there are curly brackets around the parameter. If you omit the curly brackets, the simulation will generate errors. Click the ***OK*** button to accept the changes:

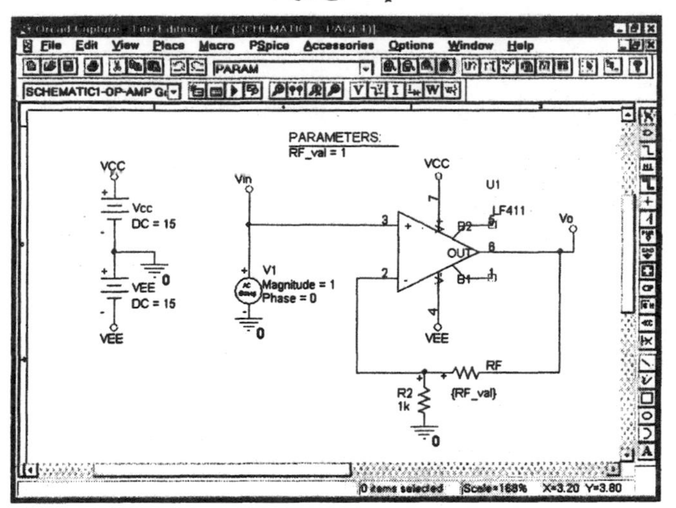


We have now defined the parameter with the PARAM part and used the parameter as the value for RF. Parameters are used to easily change values in a circuit, pass values to a subcircuit, or change the values of components during a simulation. We will use the parameter ***{RF_val}*** to change the value of the resistor ***RF*** during the simulation. This will allow us to change the gain of the op-amp circuit.

We would like to see the frequency response of the op-amp circuit for different values of the gain, so we must set up an AC Sweep to run in conjunction with a Parametric Sweep. First we will set up the AC Sweep. Select **PSpice** and then **New Simulation Profile** from the Capture menus and then enter a name for the profile and click the ***Create*** button. Select the ***AC Sweep/Noise Analysis type*** and fill in the parameters as shown in the AC Sweep dialog box below:

Simulation Settings - OP-AMP Gain Bandwidth
General | Analysis | Include Files | Libraries | Stimulus | Options | Data Collection | Probe Window
Analysis type: Time Domain (Transient) — Time Domain (Transient), DC Sweep, Bias Point
Run to time: 1000ns seconds (TSTOP)
Start saving data after: 0 seconds
Transient options — Maximum step size: seconds; Skip the initial transient bias point calculation (SKIPBP)
Monte Carlo/Worst Case, Parametric Sweep, Temperature (Sweep), Save Bias Point, Load Bias Point
Output File Options...
OK Cancel Apply Help

Simulation Settings - OP-AMP Gain Bandwidth
General | Analysis | Include Files | Libraries | Stimulus | Options | Data Collection | Probe Window
Analysis type: AC Sweep/Noise
Options: Monte Carlo/Worst Case, Parametric Sweep, Temperature (Sweep), Save Bias Point, Load Bias Point
AC Sweep Type — Linear / Logarithmic: Decade; Start Frequency: 1; End Frequency: 100Meg; Points/Decade: 100
Noise Analysis — Enabled; Output Voltage; I/V Source; Interval
Output File Options — Include detailed bias point information for nonlinear controlled sources and semiconductors (.OP)
OK Cancel Apply Help

The dialog box is set to sweep frequency from 1 Hz to 100 MHz at 100 points per decade.

We would now like to set up values for the parameter ***RF_val***. Parameters can be changed using the Parametric setup dialog box. Click the box next to the text ***Parametric Sweep***:

Simulation Settings - OP-AMP Gain Bandwidth
General | Analysis | Include Files | Libraries | Stimulus | Options | Data Collection | Probe Window
Analysis type: AC Sweep/Noise
Options: Monte Carlo/Worst Case, Parametric Sweep, Temperature (Sweep), Save Bias Point, Load Bias Point
AC Sweep Type — Linear / Logarithmic: Decade; Start Frequency: 1; End Frequency: 100Meg; Points/Decade: 100
Noise Analysis — Enabled; Output Voltage; I/V Source; Interval
Output File Options — Include detailed bias point information for nonlinear controlled sources and semiconductors (.OP)
Click on this square.
OK Cancel Apply Help

Simulation Settings - OP-AMP Gain Bandwidth
General | Analysis | Include Files | Libraries | Stimulus | Options | Data Collection | Probe Window
Analysis type: AC Sweep/Noise
Options: General Settings, Monte Carlo/Worst Case, Parametric Sweep, Temperature (Sweep), Save Bias Point, Load Bias Point
Sweep variable — Voltage source, Current source, Global parameter, Model parameter, Temperature; Name; Model type; Model name; Parameter name
Sweep type — Linear, Logarithmic: Decade; Start value; End value; Increment; Value list
OK Cancel Apply Help

The square fills with a checkmark ☑ indicating that the option is selected, and the dialog box for the parametric sweep is displayed. Fill in the dialog box as shown:

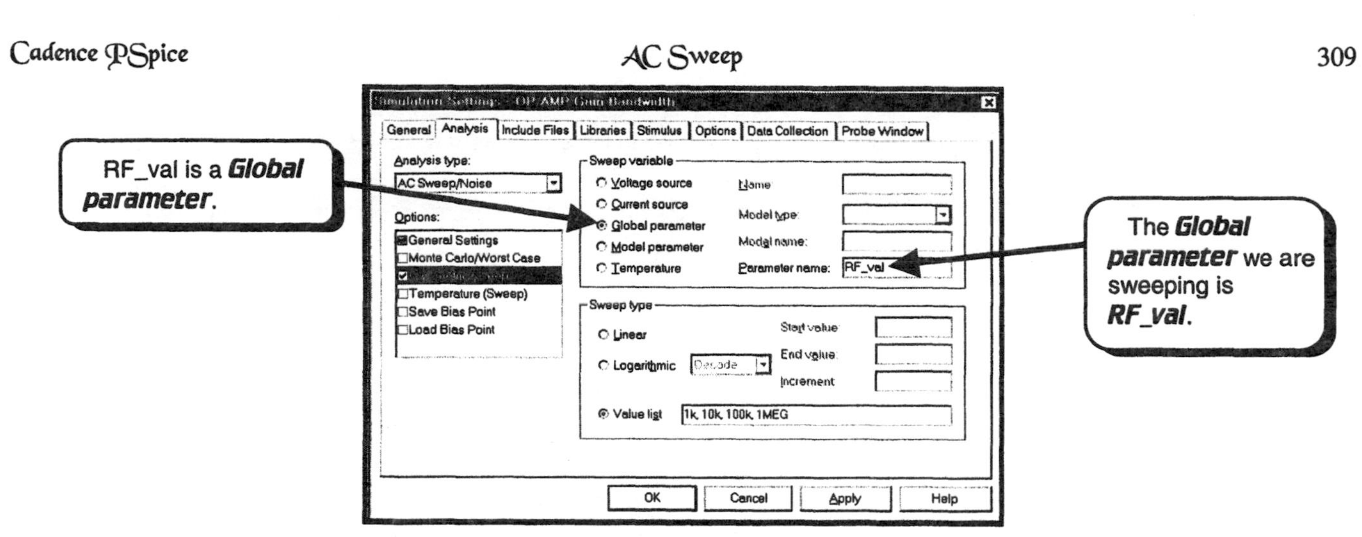


The parameter we have defined (***RF_val***) is a ***Global parameter***. The sweep settings are similar to the DC and AC Sweeps discussed previously. We would like specific values for the parameter, so we will use the value list. Logically, the parametric sweep is executed outside the AC Sweep. First, ***RF_val*** will be set to ***1k*** and then the AC Sweep will be performed. Next, ***RF_val*** will be set to ***10k*** and then the AC Sweep will be performed. Then, ***RF_val*** will be set to ***100k***, and so on. Click the ***OK*** button to accept the settings to return to the schematic.

We are now ready to run the simulation. Select **PSpice** and then **Run** from the Capture menu bar. The PSpice simulation window will appear:

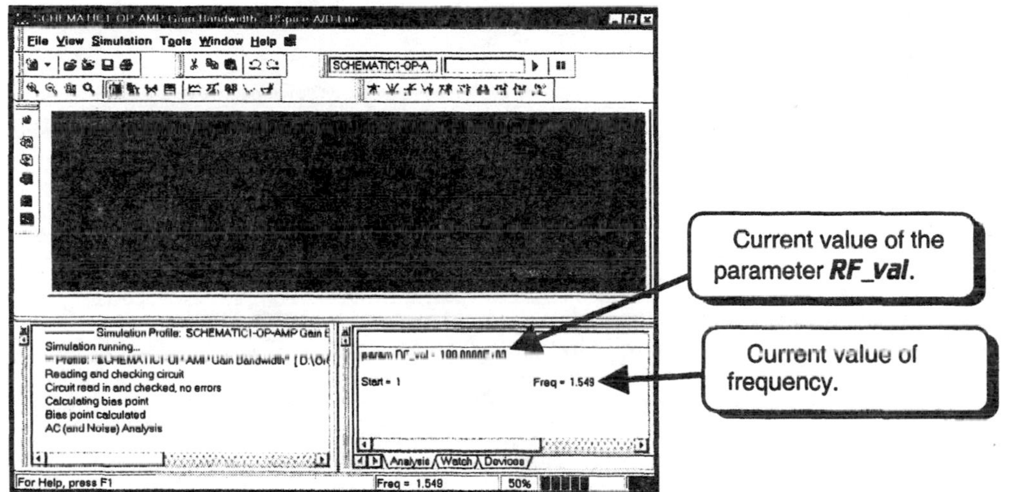


As you watch the simulation's progress you will see that the AC Sweep runs four times, once for each value of the parameter ***{RF_val}***. In the window above, the current AC Sweep is for ***{RF_val}*** = 100,000.

When the simulation is complete you will be presented with the following screen:

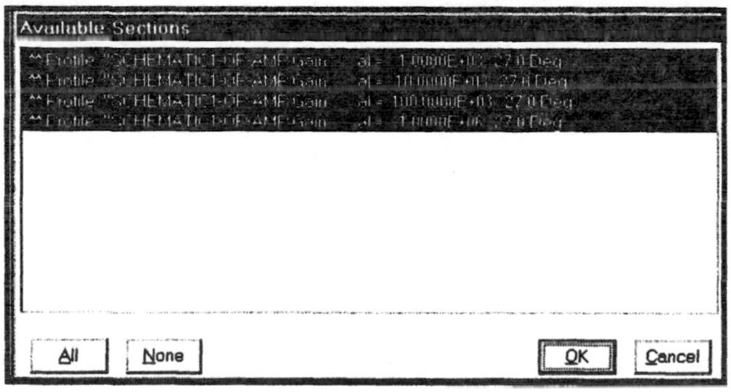

Available Sections

All None OK Cancel

Since there were four simulations, one for each value of parameter RF_val, Probe is asking which of the simulations we would like to view. You can look at any combination of the runs. By default, all of the simulations are selected. We would like to look at all of the simulations at the same time, so click the ***OK*** button to accept the runs. You will be presented with the standard blank Probe screen. Add the trace **DB(V(VO))**.* To add the trace, select **Trace** and then **Add Trace** from the Probe menu bar or press the **INSERT** key. This trace shows the Bode magnitude plot of the gain. You will see the screen below:

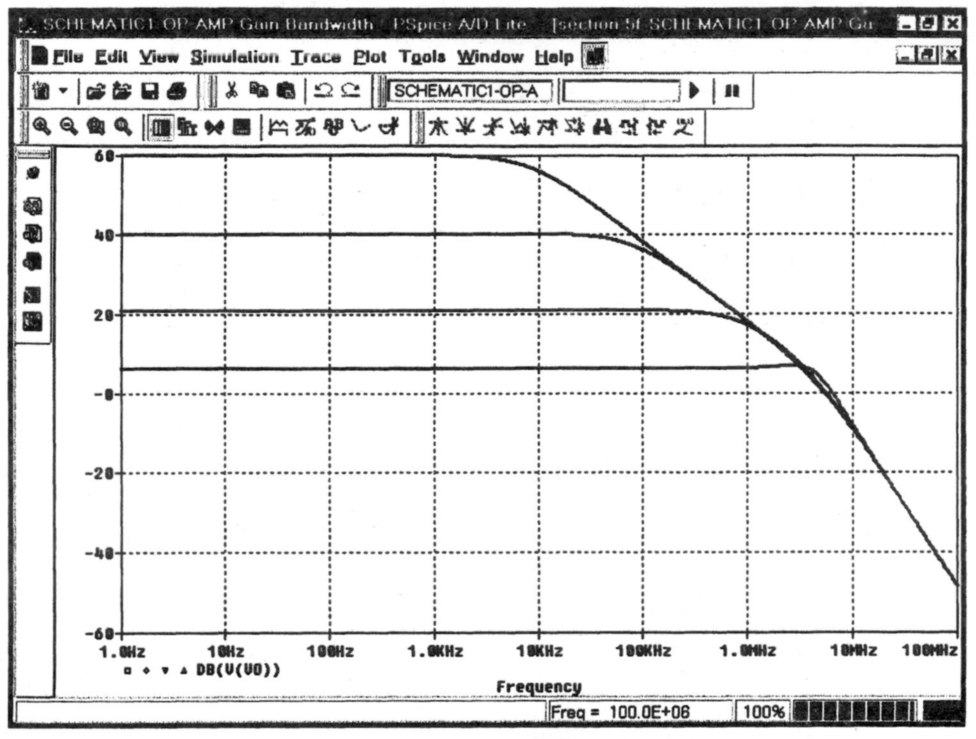


The plot shows four Bode plots, one for each value of ***{RF_val}***. We see that for larger gains the bandwidth is reduced. This example will be continued in the next section (Section 5.G), which starts with the screen capture above. You may wish to keep the Probe window above open so that you can easily continue with the next section.

**EXERCISE 5-8**: For the inverting op-amp below, find the upper 3 dB frequency for mid-band gains of –1, –10, –100, and –1000. Show the Bode magnitude plots for all of the gains on the same graph.

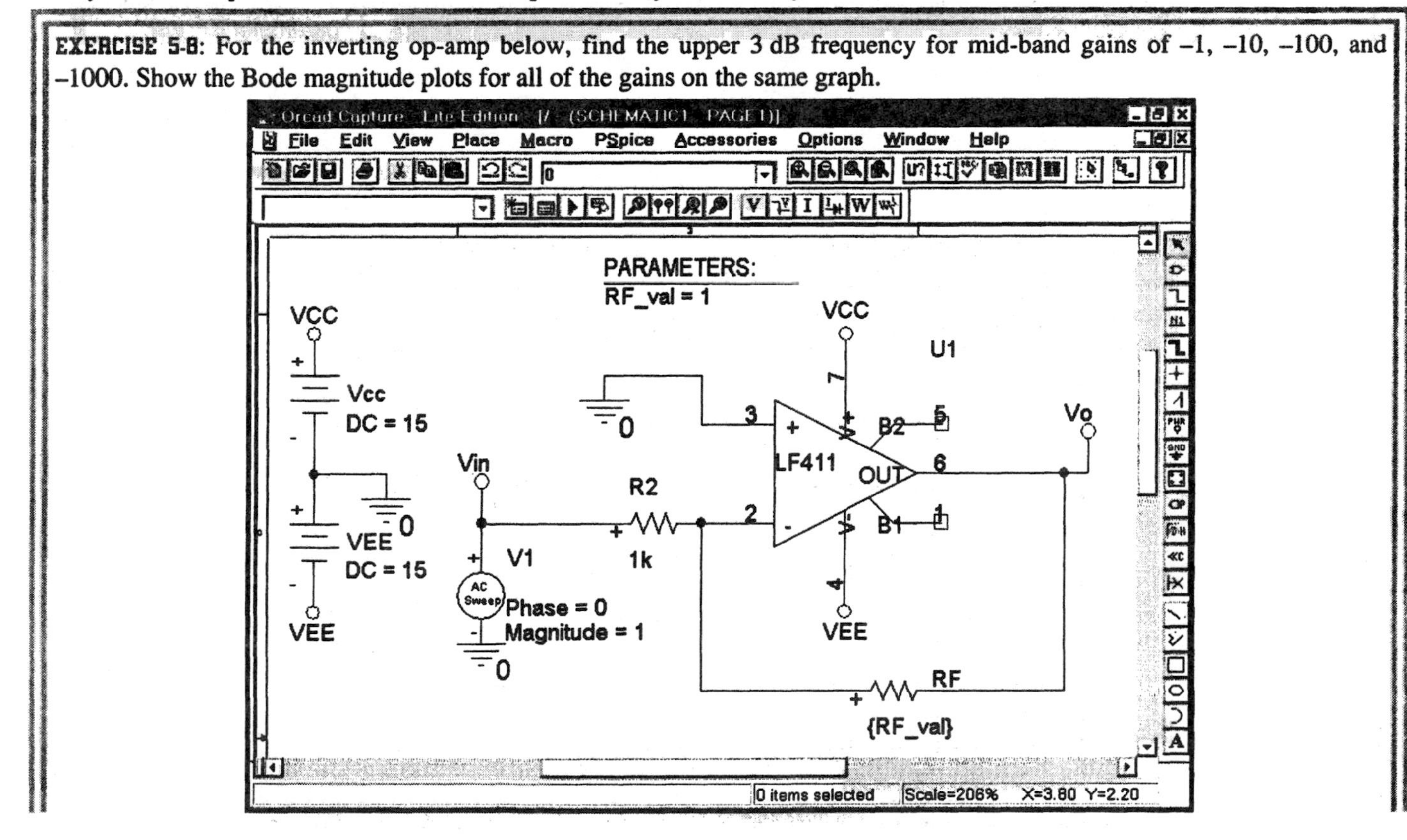


*See page 292 for an explanation of the "dB" command.

**SOLUTION**: The results are displayed using Probe:

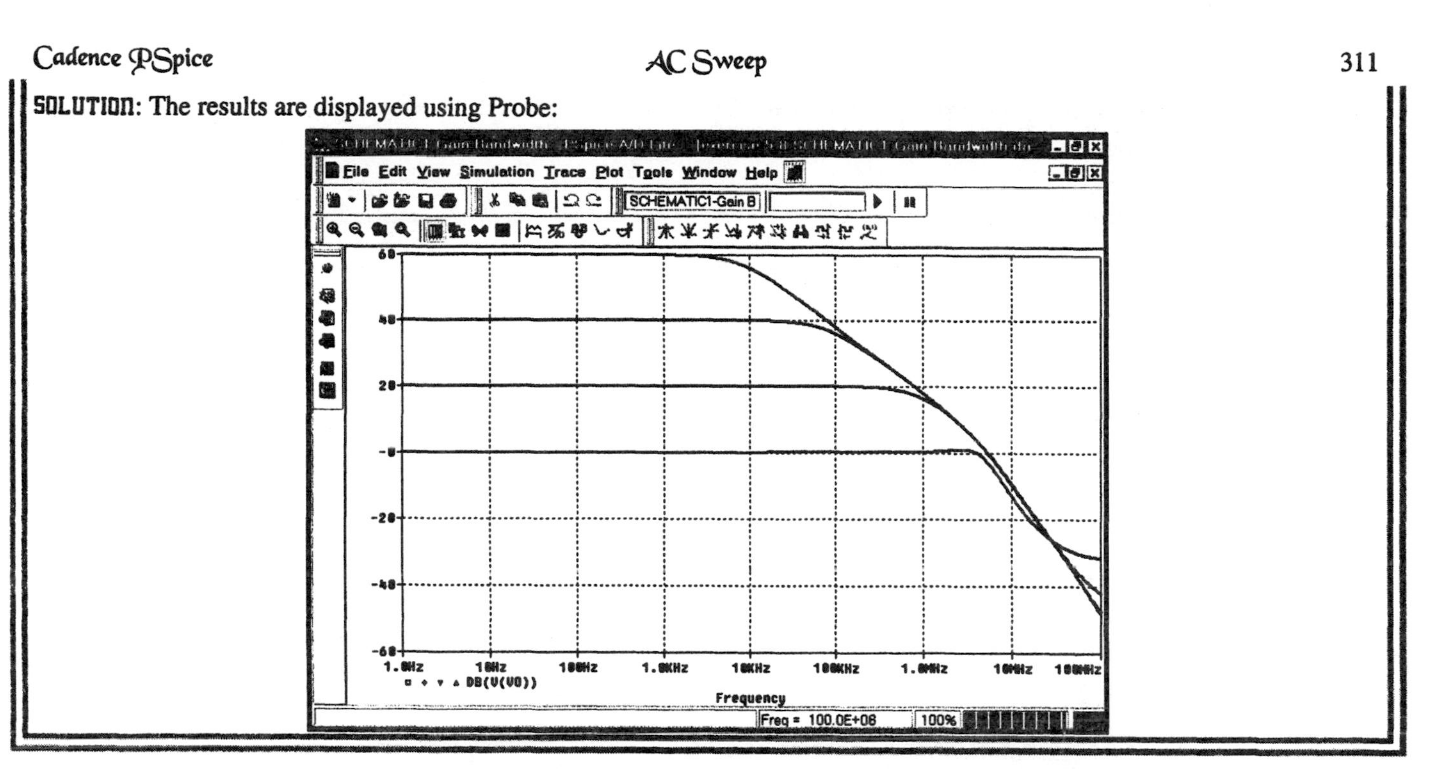


# 5.G. Performance Analysis — Op-Amp Gain Bandwidth

The Performance Analysis capabilities of Probe are used to view properties of waveforms that are not easily described. Examples are amplifier bandwidth, rise time, and overshoot. To calculate the bandwidth of a circuit, you must find the maximum gain, and then find the frequency where the gain is down by 3 dB. To calculate rise time, you must find the 10% and 90% points, and then find the time difference between the points. The Performance Analysis gives us the capability to plot these properties versus a parameter or device tolerances. The Performance Analysis is used in conjunction with the Parametric Sweep to see how the properties vary versus a parameter. The Performance Analysis is used in conjunction with the Monte Carlo analysis to see how the properties vary with device tolerances. In this section we will plot the bandwidth of an amplifier versus the value of the feedback resistor. See Sections 9.B.3 and 9.E to see how to use the Performance Analysis in conjunction with the Monte Carlo analysis.

Suppose that for the example of Section 5.F, we would like to see a plot of how the upper 3 dB frequency is affected by the value of the feedback resistor, ***Rf***. This plot can be accomplished using the Performance Analysis capabilities of Probe. Repeat the procedure of Section 5.F. When you obtain the plot on page 310, you may continue with this section. The plot on page 310 is repeated as follows:

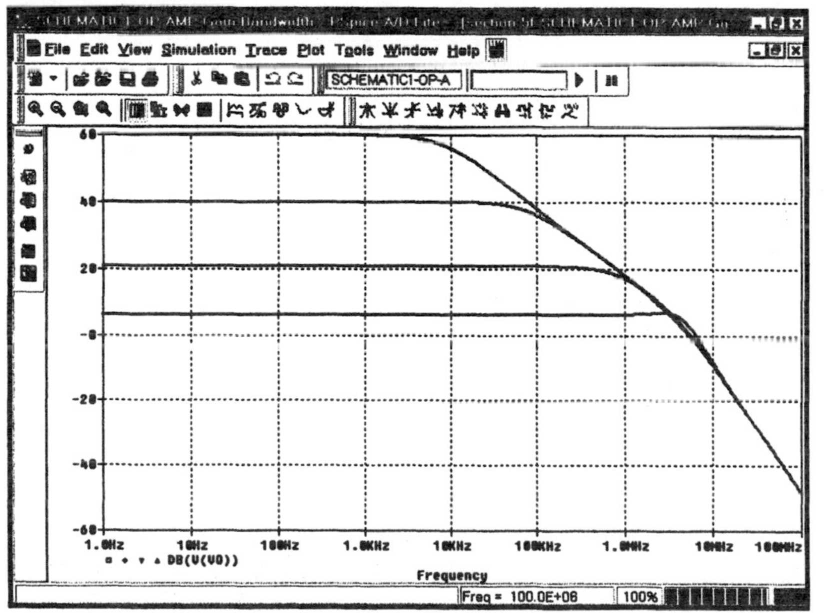


Before we look at the Performance Analysis we would like to create a new plot window. Select **Window** and then **New Window**. A second empty window will appear:

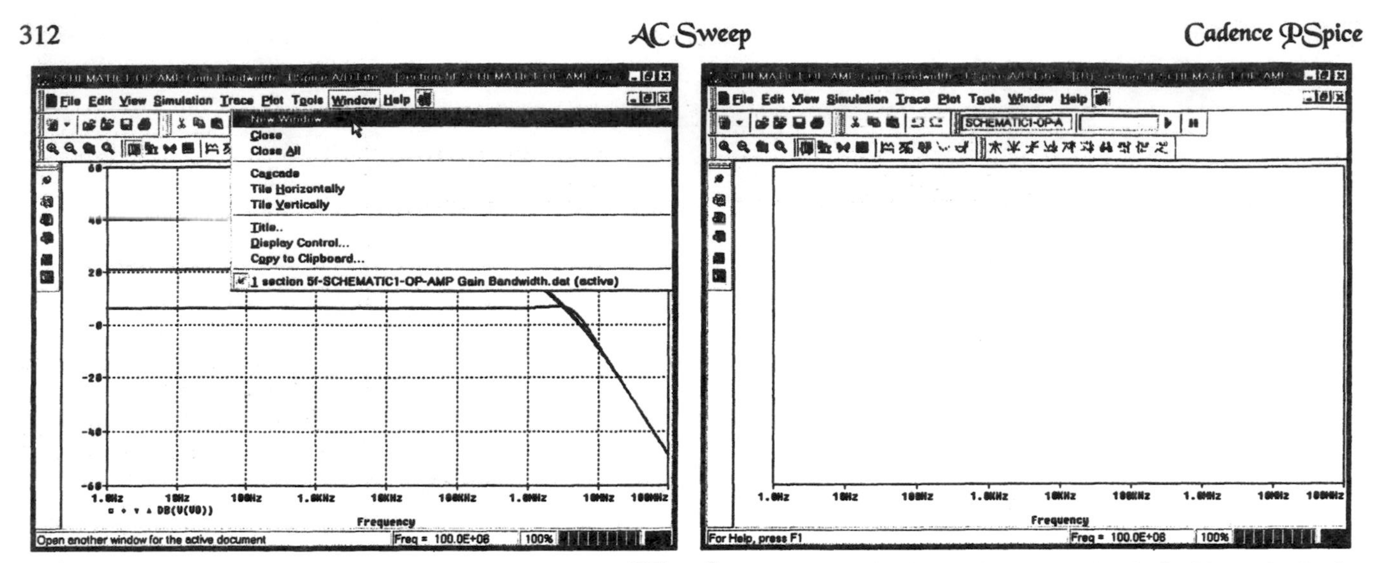


To plot the upper 3 dB bandwidth versus the parameter ***RF_val*** we must enable the Performance Analysis. From the Probe menus select **Plot** and then **Axis Settings**. The dialog box shows the settings for the x-axis:

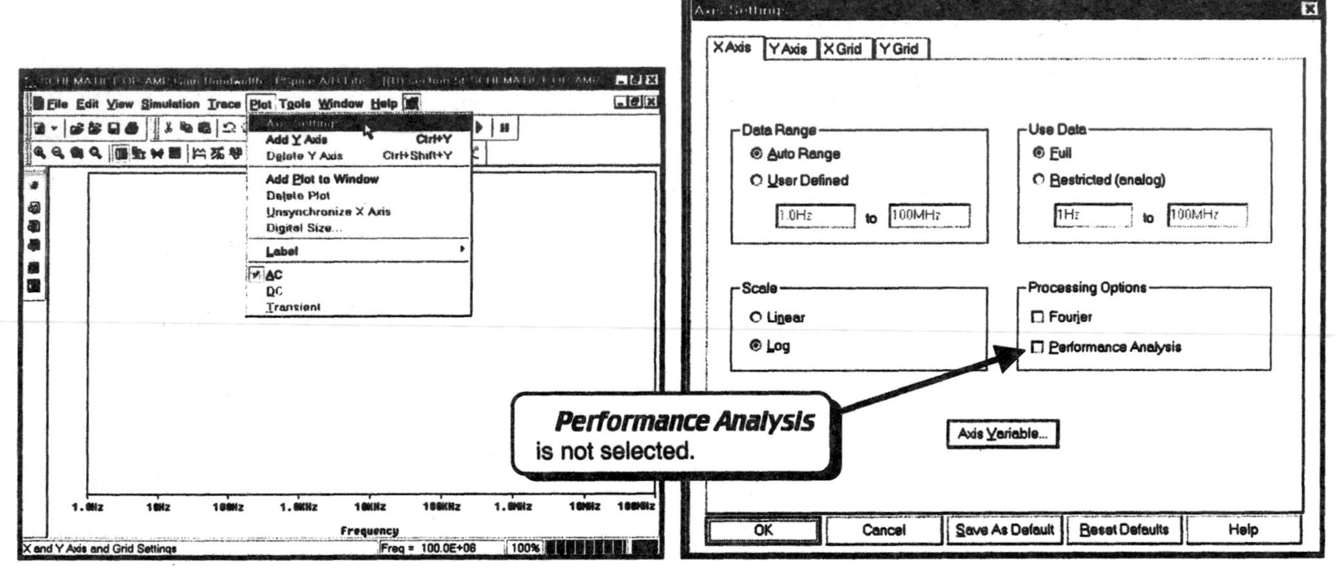


Under ***Processing Options*** we see that the square next to ***Performance Analysis*** does not have a checkmark in it, indicating that it is not enabled. To enable the ***Performance Analysis*** click the ***LEFT*** mouse button on the text ***Performance Analysis***. The square should fill with a checkmark ☑, indicating that the ***Performance Analysis*** is enabled:

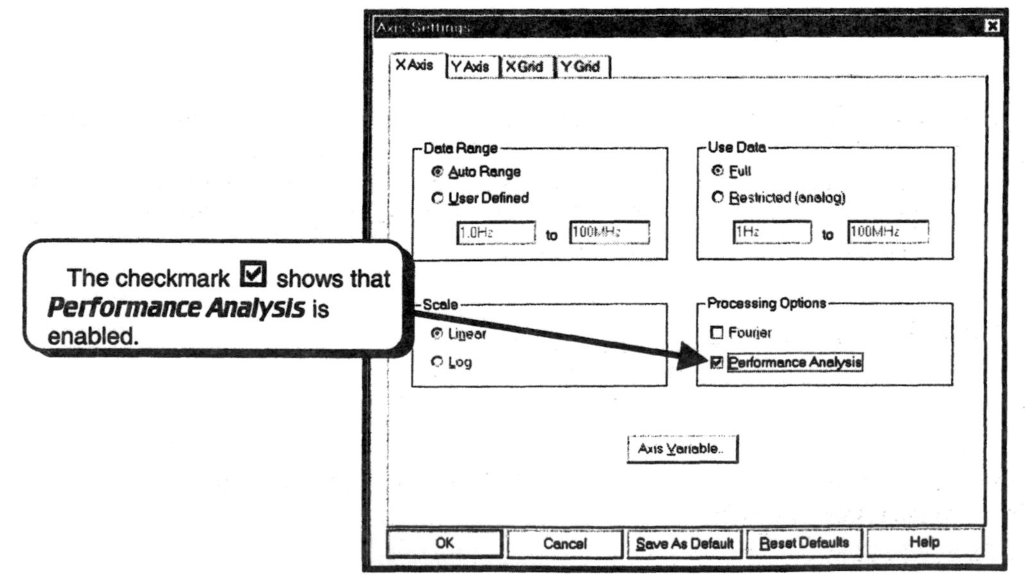

Click the ***OK*** button to return to the Probe screen:

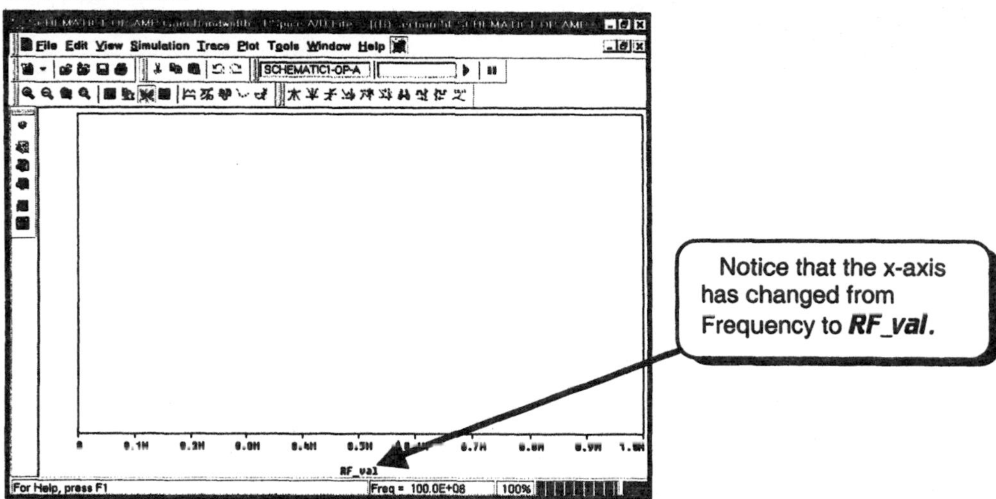


We now wish to plot the upper 3 dB frequency versus ***RF_val***. Select ***Trace*** and then ***Add Trace***:

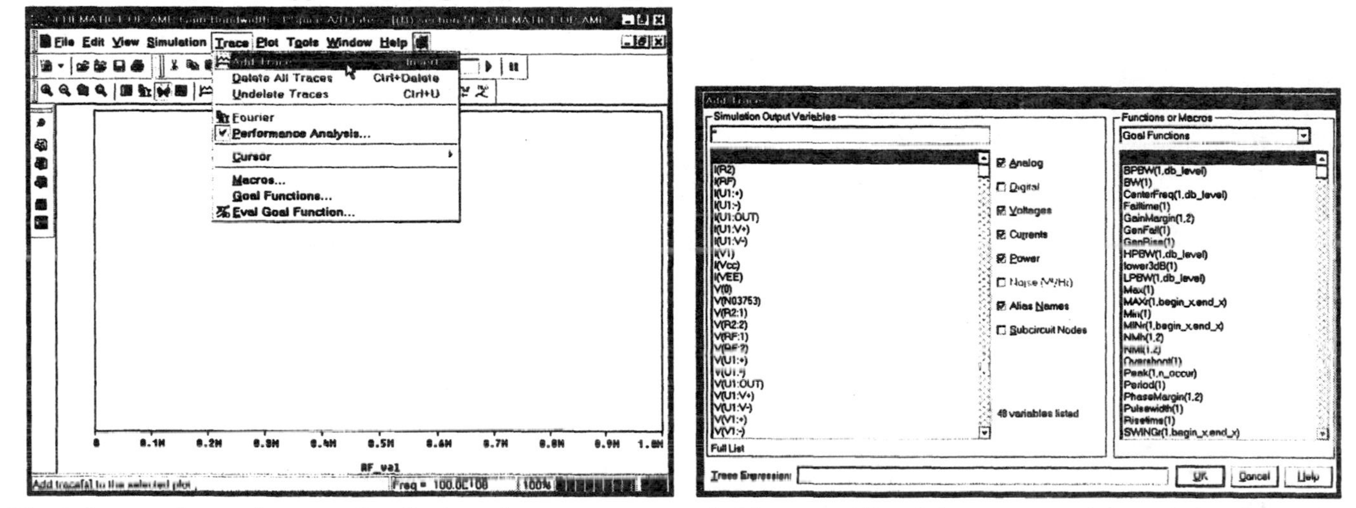


The left pane shows the normal voltage and current traces available to us. The right pane contains goal functions. Goal functions are defined in a file called c:\Program Files\OrcadLite\PSpice\Common\pspice.prb. If you view this file using Windows Notepad, you will see the following function in the file:

```
upper3dB(1) = x1
* Find the upper 3 dB frequency
{
1|sfle(max-3dB,n) !1;
}
```

The name of the function is **upper3dB**. It has **1** input argument. **1|sfle** means search the first input forward and find a level. The level we are looking for is 3 dB less than the maximum (**max-3**). The **n** means find the specified level when the trace has a negative-going slope. When the point is found, the text **!1** designates its coordinates as x1 and y1. The function returns the x-coordinate of the point (**upper3dB(1) = x1**). The x axis of a frequency trace is frequency, so this function returns the frequency of the upper 3 dB point. A second function is:

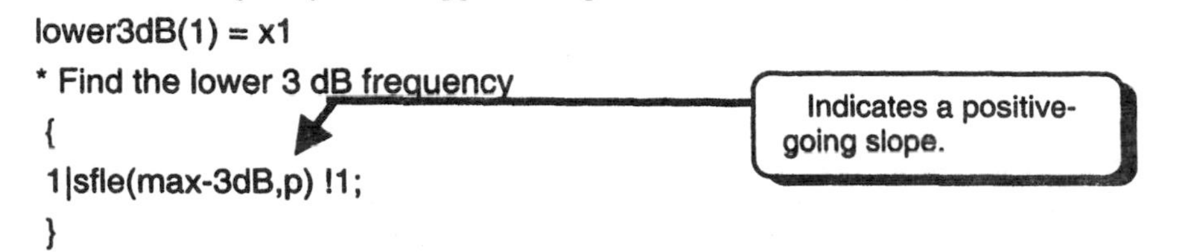


```
lower3dB(1) = x1
* Find the lower 3 dB frequency
{
1|sfle(max-3dB,p) !1;
}
```

This function is similar to the **upper3dB** function, except that it finds the 3 dB point when the trace has a positive-going slope. This will mark the coordinates of the lower 3 dB point.

Type in the trace `upper3dB(V(VO))`:

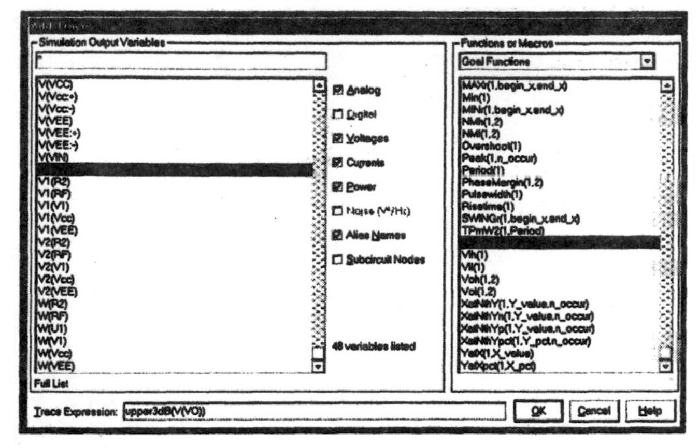


Click the ***OK*** button to accept the trace:

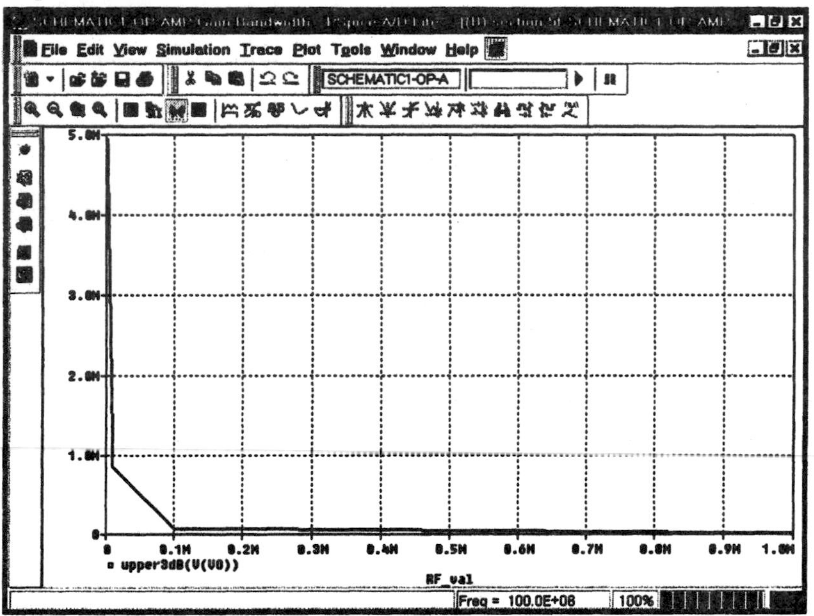


This plot is not too informative when plotted on a linear scale. A log-log plot is much more useful. To change the x-axis to a log scale select **Plot** and then **Axis Settings** from the Probe menu:

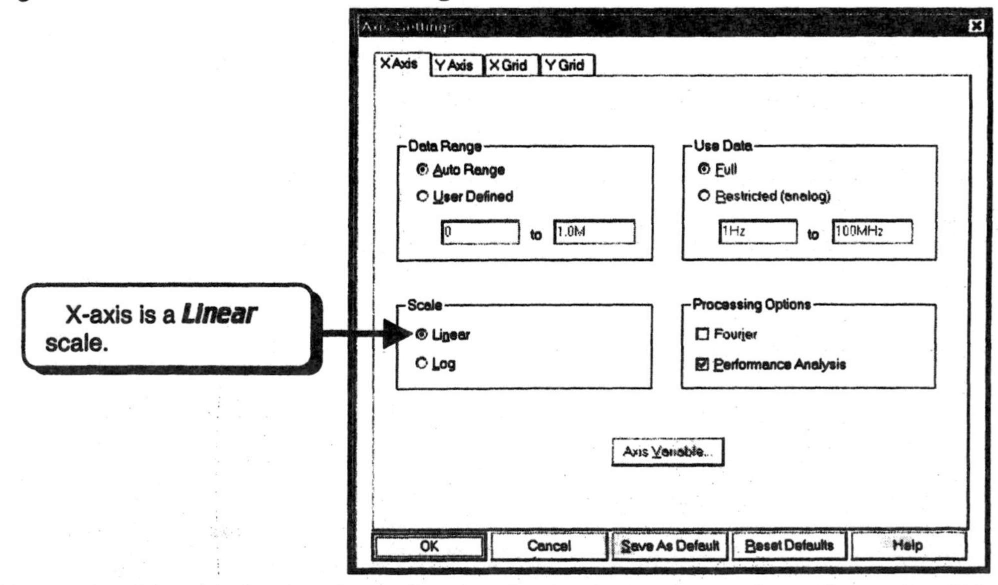


We see that the x-axis is plotted on a linear scale. To change the x-axis to a log scale, click the ***LEFT*** mouse button on the text ***Log***. The circle next to the text ***Log*** should fill with a dot ◉, indicating that a log scale is selected:

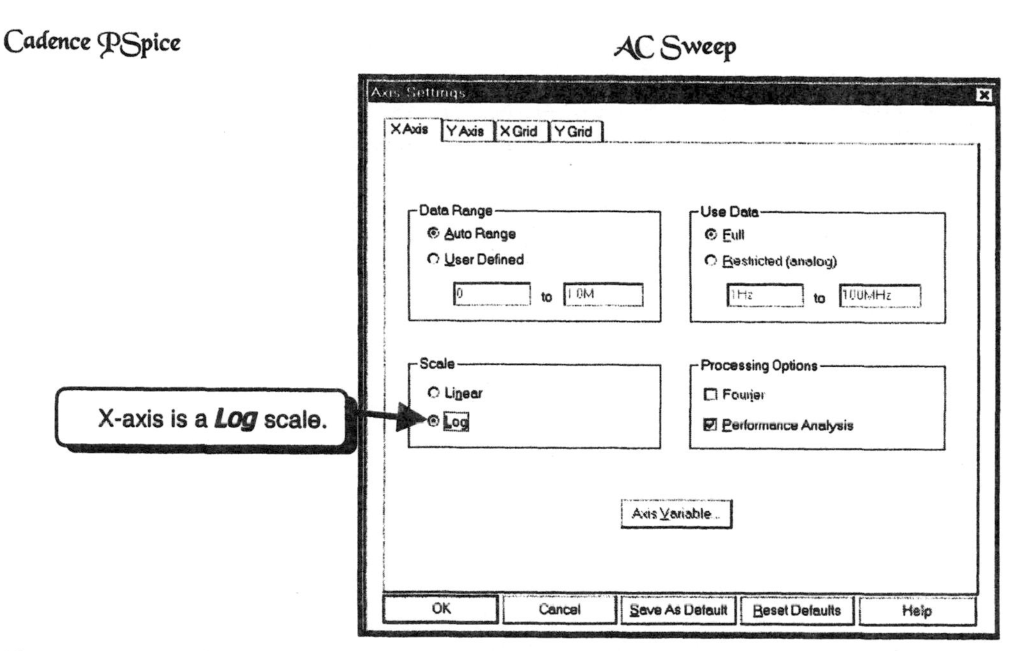


Next we need to change the y-axis to a log scale. Click the ***LEFT*** mouse button on the ***Y Axis*** tab and change the axis to a log scale:

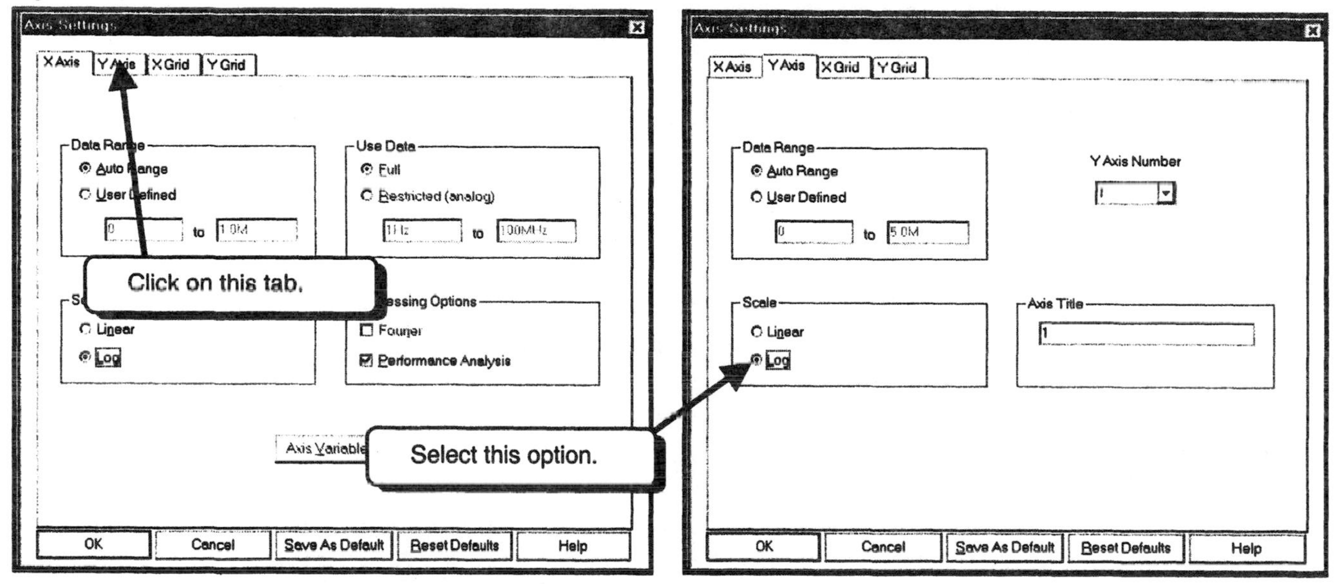


After you have selected a log scale for the y-axis, click the ***OK*** button to return to the plot:

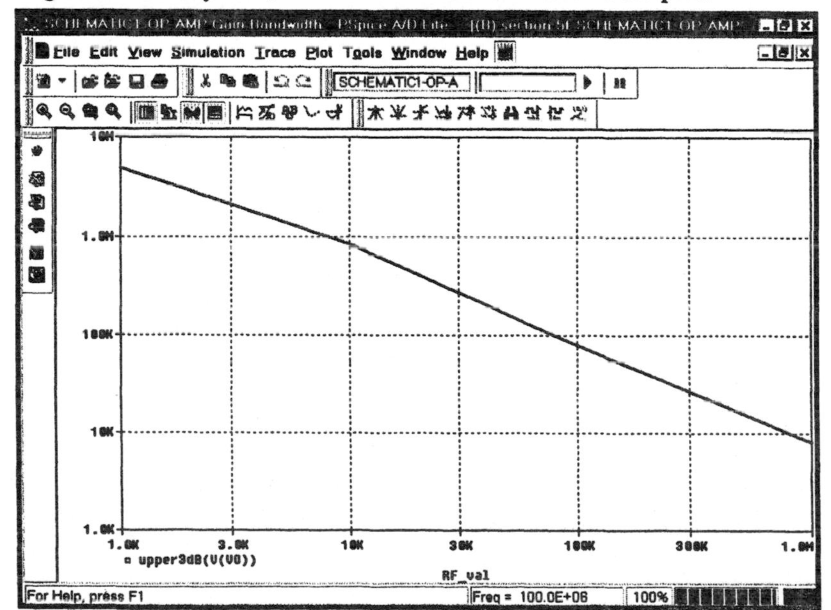

When plotted on a log-log scale, we can easily see the relationship between the value of ***Rf*** and the upper 3 dB frequency. We see that the upper –3dB frequency is proportional to RF_val. The gain for this amplifier is $1+{R_f}/{R_2}$, which is approximately equal to ${R_f}/{R_2}$. Thus the gain is approximately proportional to RF. As RF increases, the gain increases. From this plot, we see that as RF increases the upper –3 dB frequency goes down, or as the gain increases the upper –3 dB frequency goes down. Another way to state this is that the upper –3 dB frequency times the circuit gain is approximately constant. We shall show this in the next exercise.

**EXERCISE 5-9:** Plot the gain bandwidth of the amplifier in the previous example versus the feedback resistor $R_F$.

**SOLUTION:** Rerun the simulation with more detail in the Parametric Sweep:

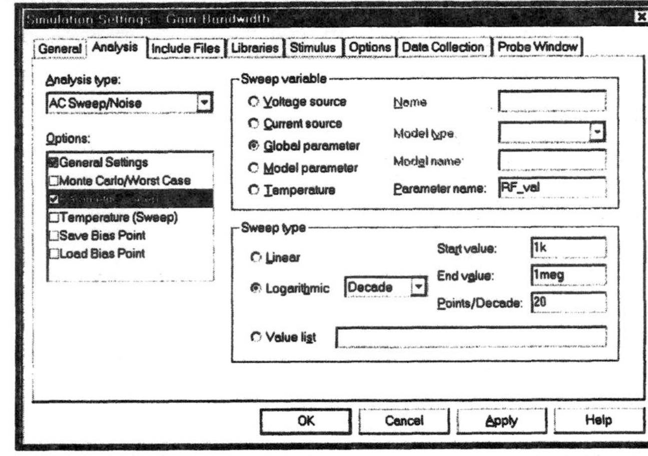


In Probe use the Performance Analysis and add the trace **`Max(V(VO))*upper3dB(V(VO))`**. Change the x-axis to a log scale:

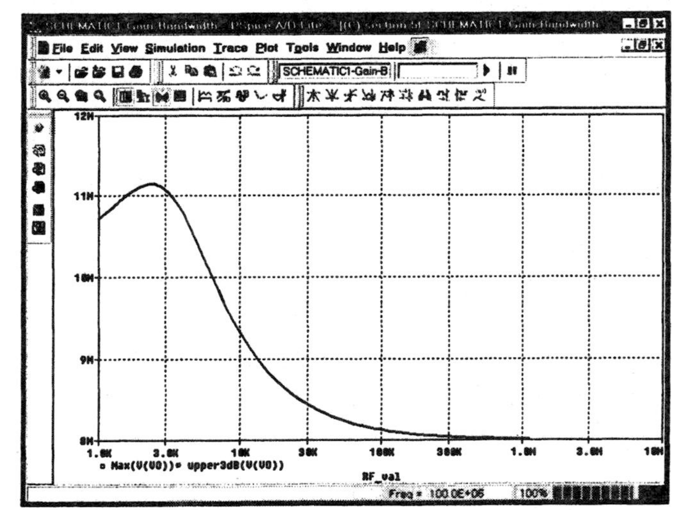


The function Max(V(Vo)) gives us the maximum value of Vo for each trace. Since the input was one volt, the Max(V(Vo)) is also the maximum gain for each trace. Thus Max(V(Vo))*upper3dB(V(Vo)) is the maximum gain for each trace times the upper 3 dB frequency for each trace, or the gain times bandwidth for each trace. For this circuit, the maximum gain occurs at midband frequencies.

We see from the plot that as ***RF_val*** changes by a factor of 1000 (the gain changes by a factor of 1000), the gain-bandwidth product is relatively constant, ranging between 8 MHz and 11 MHz. This value for the gain-bandwidth seems to be a bit high for an LF411 opamp. The datasheets specify the gain-bandwidth product for this part to be between 3 and 4 MHz. This result brings into question the accuracy of the LF411 opamp we are using.

# 5.H. Mutual Inductance

Mutual inductance requires two parts: the inductors (L) and the coupling between the inductors (K). We will illustrate the use of the coupling part K with two circuits. The first circuit will have three inductors with unequal coupling. The second circuit will have four inductors with equal coupling. Wire the circuit shown below. The dots on the inductors are critical since they indicate the polarity of the mutual coupling. Make sure the dots on your schematic agree with the ones on the schematic shown.

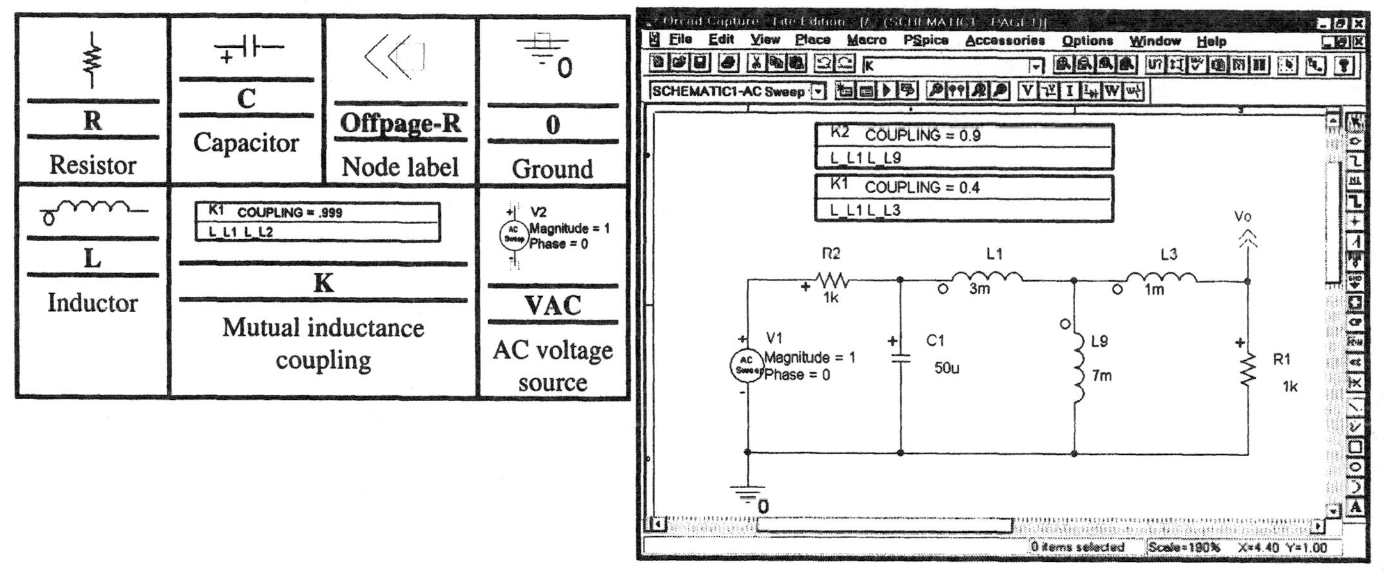


Note that there is coupling only between ***L1*** and ***L3*** and between ***L1*** and ***L9***. If you want coupling between ***L3*** and ***L9***, you will have to add another mutual inductance coupling part (K). To change the value of the coupling, double-click the ***LEFT*** mouse button on the text ***Coupling=***. To change the inductor names in the coupling part, double-click the ***LEFT*** mouse button on the inductor names inside the coupling part box. Make sure that you precede the inductor names by the text L_ or the part will not work correctly. This circuit required two coupling parts (K) because the coupling between the pairs of inductors is different.

We would like to run an AC Sweep to see how the magnitude and phase of ***Vo*** change with frequency. Set up an AC Sweep for frequencies from 1 Hz to 1 MHz at 100 points per decade. To set up the AC Sweep, select **PSpice** and then **New Simulation Profile** from the Capture menus and then enter a name for the profile and click the ***Create*** button. Select the ***AC Sweep/Noise Analysis type*** and fill in the parameters as shown in the AC Sweep dialog box below:

Simulation Settings - AC Sweep with Mutual Inductance

General | Analysis | Include Files | Libraries | Stimulus | Options | Data Collection | Probe Window

Analysis type: AC Sweep/Noise

Options: General Settings; Monte Carlo/Worst Case; Parametric Sweep; Temperature (Sweep); Save Bias Point; Load Bias Point

AC Sweep Type: Linear; Logarithmic (selected) — Decade; Start Frequency: 1; End Frequency: 1meg; Points/Decade: 100

Noise Analysis: Enabled; Output Voltage; I/V Source; Interval

Output File Options: Include detailed bias point information for nonlinear controlled sources and semiconductors (.OP)

OK | Cancel | Apply | Help

Click the ***OK*** button to accept the settings and then run PSpice: Select **PSpice** and then **Run** from the Capture menu bar. When the simulation is complete, display the results with Probe. Add the trace `V(VO)` (press the **INSERT** key). The trace shown below will appear:

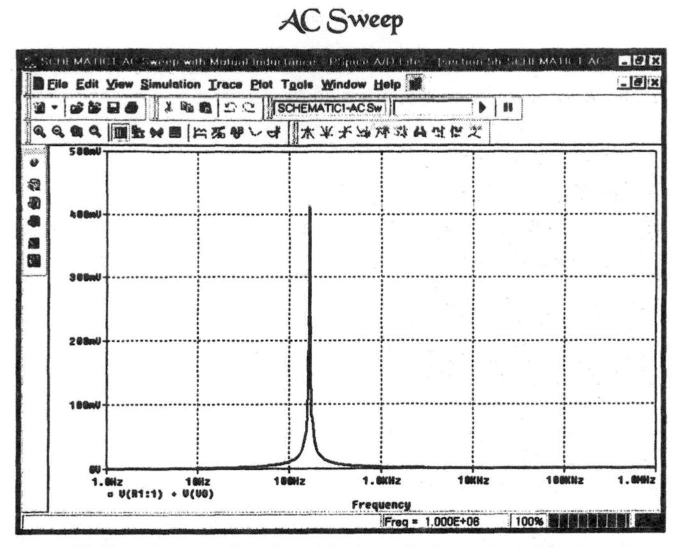


As a second example, we will illustrate a circuit with four inductors with equal coupling. In the previous example, two coupling parts were required because the coupling between the various inductors was different. In this circuit, since the coupling between all inductors is to be the same, only one coupling part will be needed. We will not simulate the circuit. We will only illustrate how to use the coupling part. Wire the circuit below:

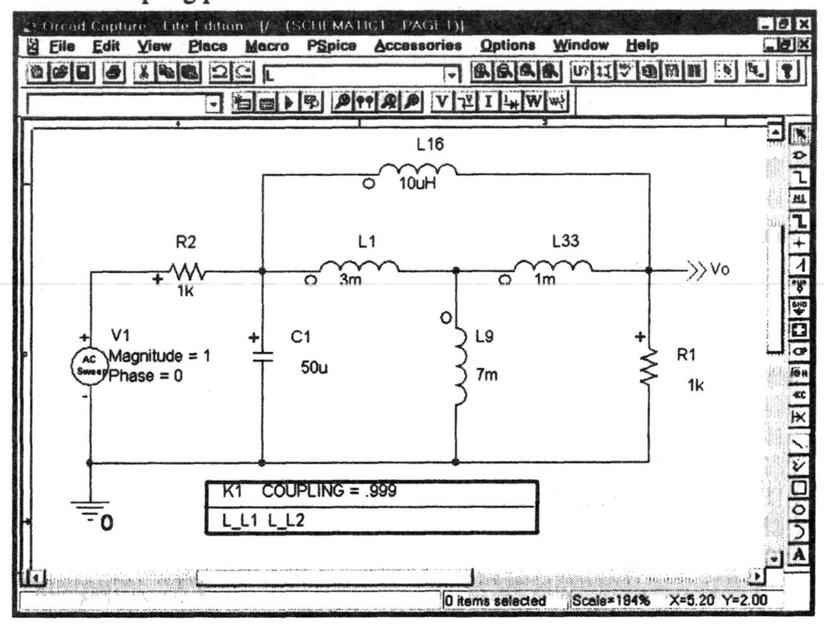


When you get the coupling part it will show only two inductors, L1 and L2, as being coupled. Double-click the ***LEFT*** mouse button on the text L_L1 L_L2 to change its value:

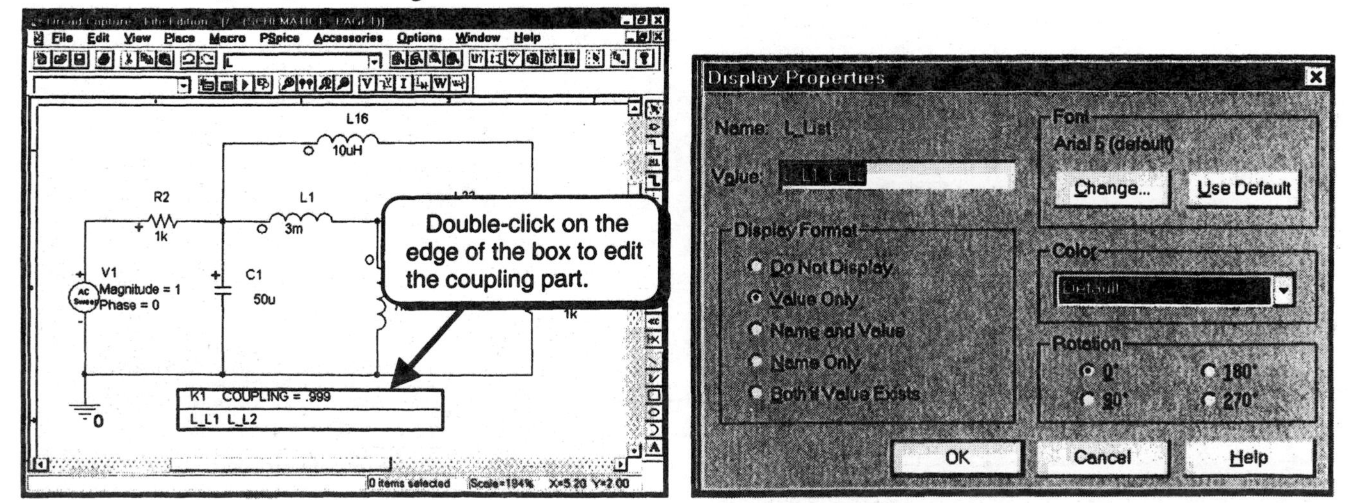

If you scroll down through the dialog box, you will notice that you can name up to nine inductors. This part can be used to couple several inductors, all with the same coefficient of coupling. We would like to specify coupling between all inductors so we must enter the names of all of the inductors, each preceded by the text L_. Type the text **`L_L1 L_L9 L_L16 L_L33`**. Note that we separate each value by a space. Enter the text exactly as shown:

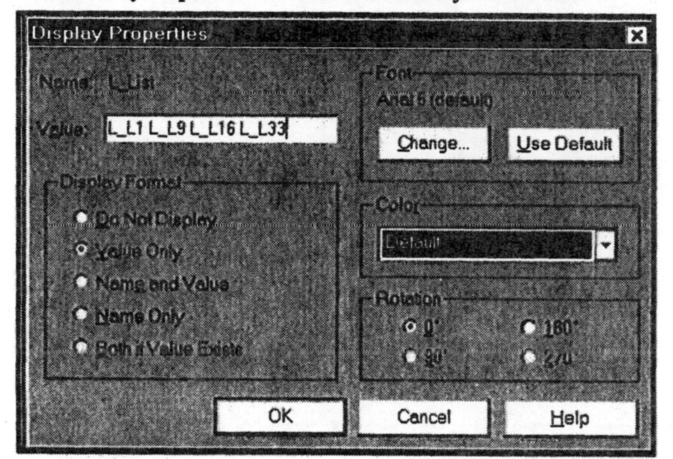


Click the ***OK*** button to return to the schematic:

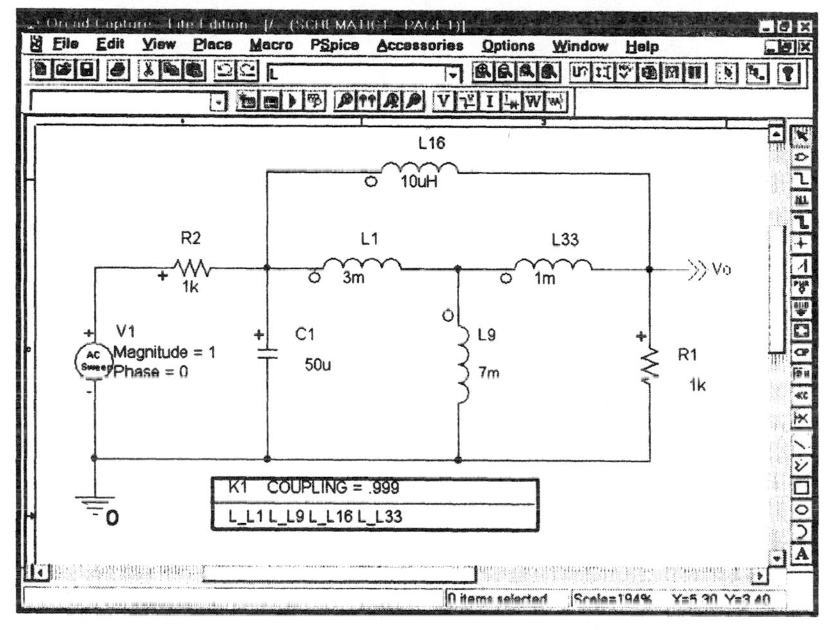


Next, we would like to specify the coefficient of coupling. Double-click the ***LEFT*** mouse button on the text ***Coupling =***:

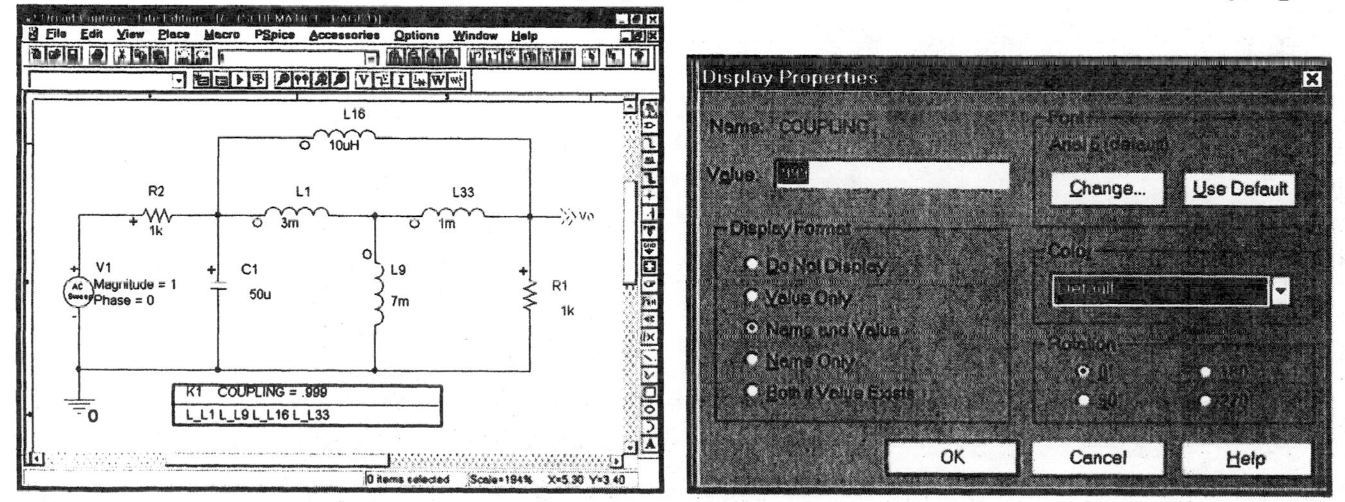


Enter the value you would like to use and click the ***OK*** button:

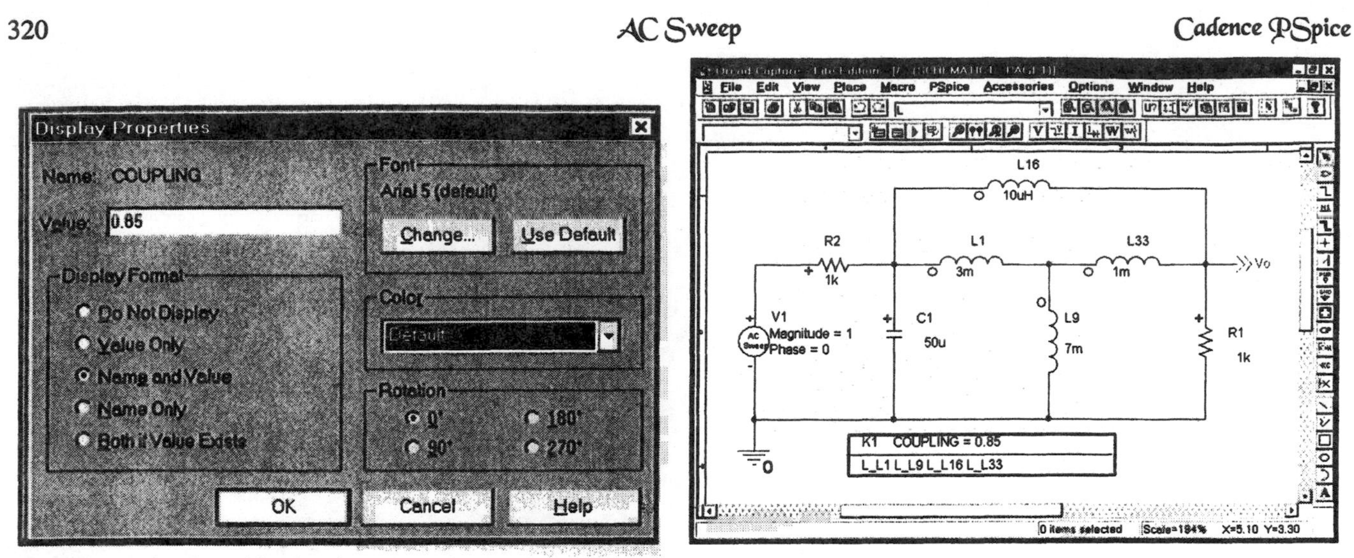


We will not simulate this circuit.

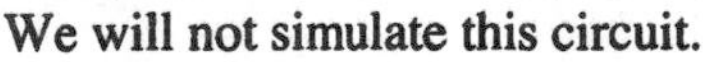

**EXERCISE 5-10**: Find the magnitude and phase of the voltage Vo for frequencies from 10 Hz to 1 kHz. Note that L1 is coupled to L2, but L3 is not coupled to either L1 or L2. The 100 GΩ resistor is used to isolate the two loops. In an actual circuit there would not be a connection between the two loops. However, PSpice requires all portions of a circuit to be referenced to ground. If the 100 GΩ resistor were replaced by an open circuit, the circuit would still function in the same way if tested in the lab, but the right loop would not have a ground reference. Without the 100 GΩ resistor, PSpice will generate an error message and will not simulate the circuit.

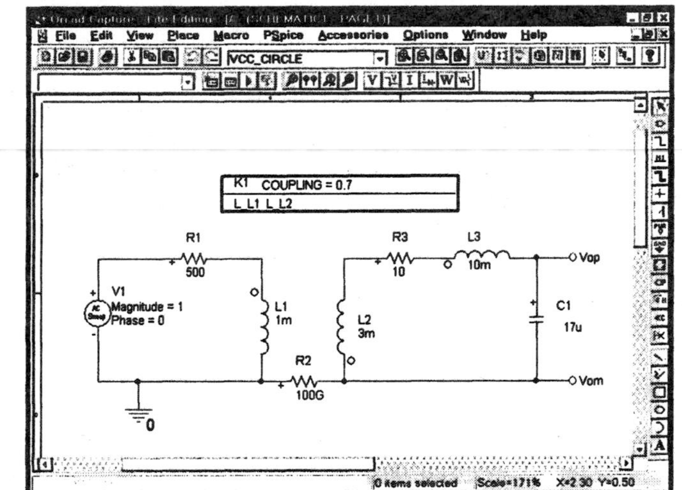


**SOLUTION**: Use Probe to display magnitude and phase plots of the voltage across *C1*.

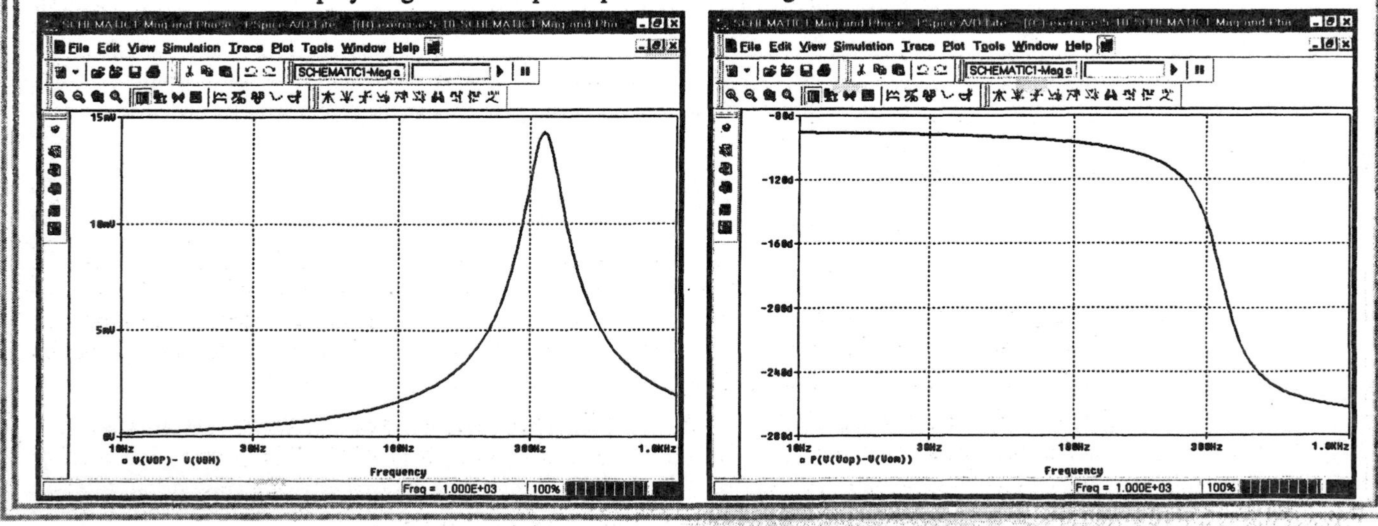

# 5.I. Measuring Impedance

The technique presented here can be used to measure the impedance or resistance between any two nodes. We will find the AC impedance between two nodes. We will illustrate using two examples. The first will be a passive circuit with resistors only. The second will be a jFET source follower.

## 5.I.1. Impedance Measurement of a Passive Circuit

We will find the impedance between nodes ***1*** and ***0*** in the circuit below:

Add a test current source as shown. The test current source is added between nodes 1 and 0 because we are interested in the impedance between those two nodes. Either an AC current or an AC voltage source can be used. Wire the circuit shown. Note that ground now serves as node 0.

| R | IAC |
|---|---|
| Resistor | AC current source |
| VCC_Circle | 0 |
| Node label | Ground |

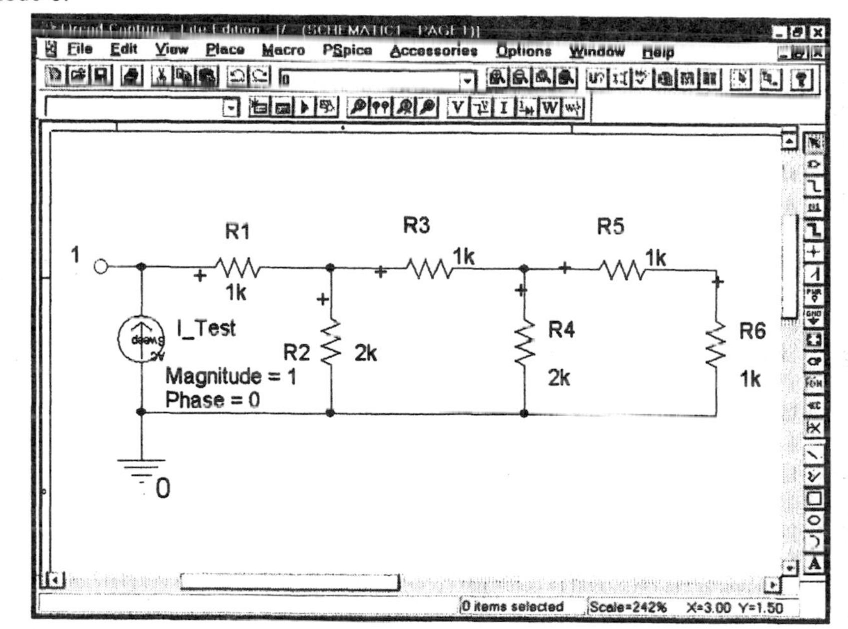


Although this circuit has no dependent sources, inductors, or capacitors, the circuit could contain any circuit element and we could use this method to find the impedance. This circuit was chosen because the impedance is easily calculated and can be compared to the PSpice result. For this circuit, the calculated impedance between nodes ***1*** and ***0*** is 2 kΩ.

Set up an AC Sweep (**PSpice, New Simulation Profile, AC Sweep/Noise**) to sweep frequencies from 1 Hz to 1 MHz at 100 points per decade. Run PSpice (**PSpice, Run**). When Probe runs, add the trace `V(1)/I(I_test)`. Remember that voltage divided by current is impedance. We are dividing the voltage between nodes ***1*** and ***0*** by the current flowing into and out of those nodes. This is the impedance between those nodes. You will see this trace:

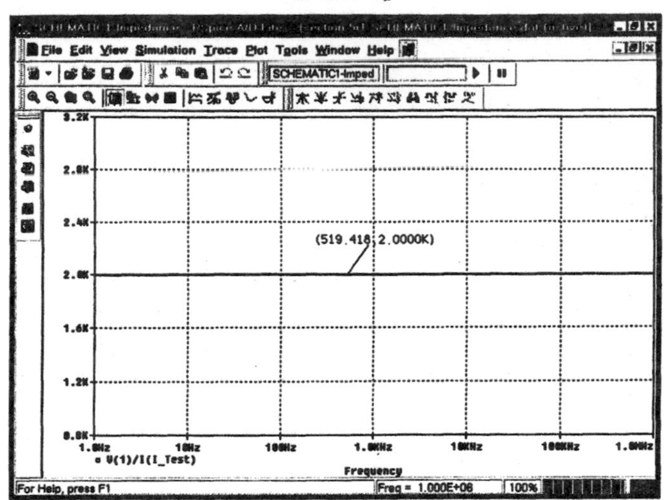


This trace shows that the impedance is constant at the expected value of 2 kΩ. If we had inductors or capacitors in the circuit, the impedance would have changed with frequency.

**EXERCISE 5-11:** Find the equivalent impedance of the following resistor network:

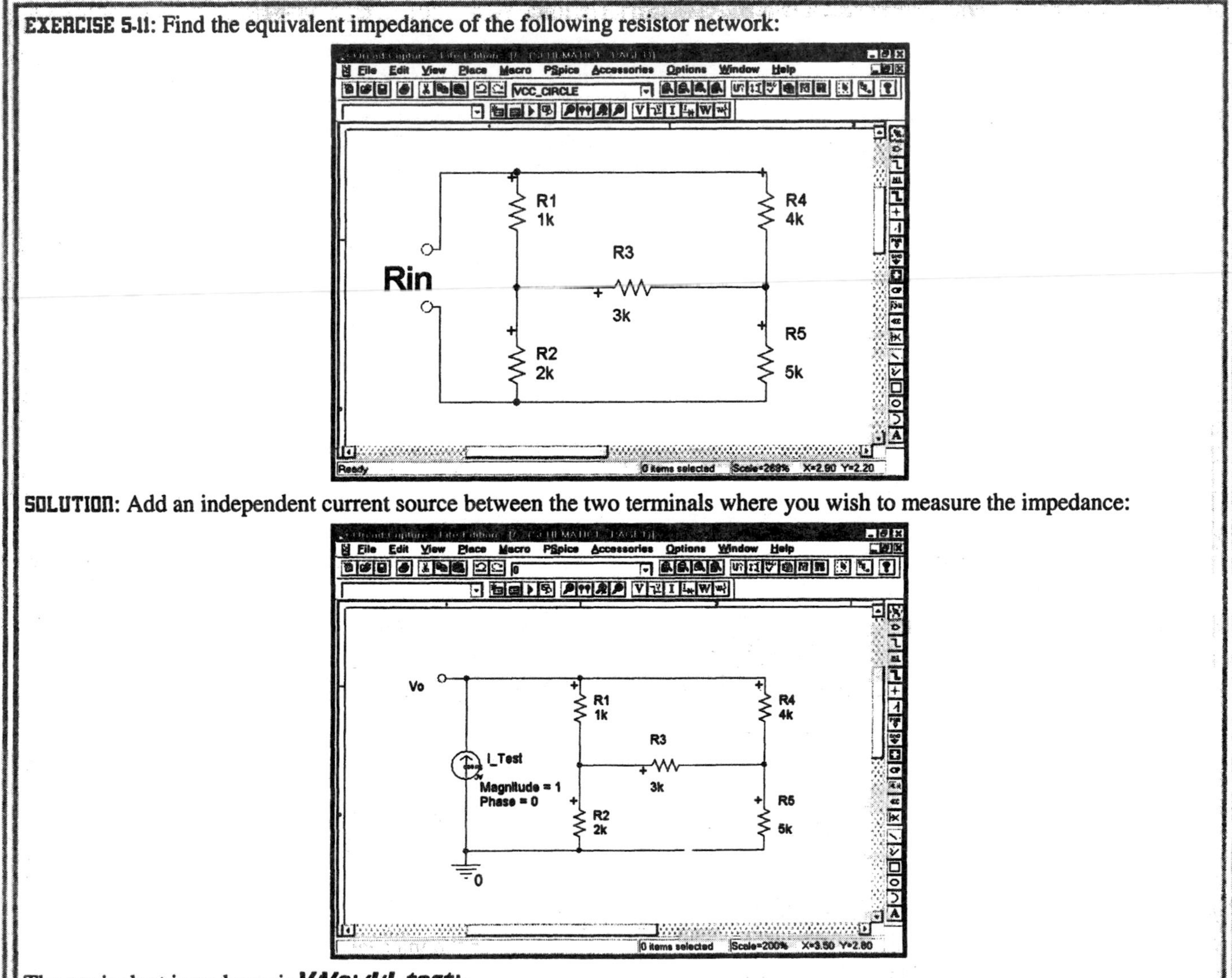


**SOLUTION:** Add an independent current source between the two terminals where you wish to measure the impedance:

The equivalent impedance is ***V(Vo)/I(I_test)***:

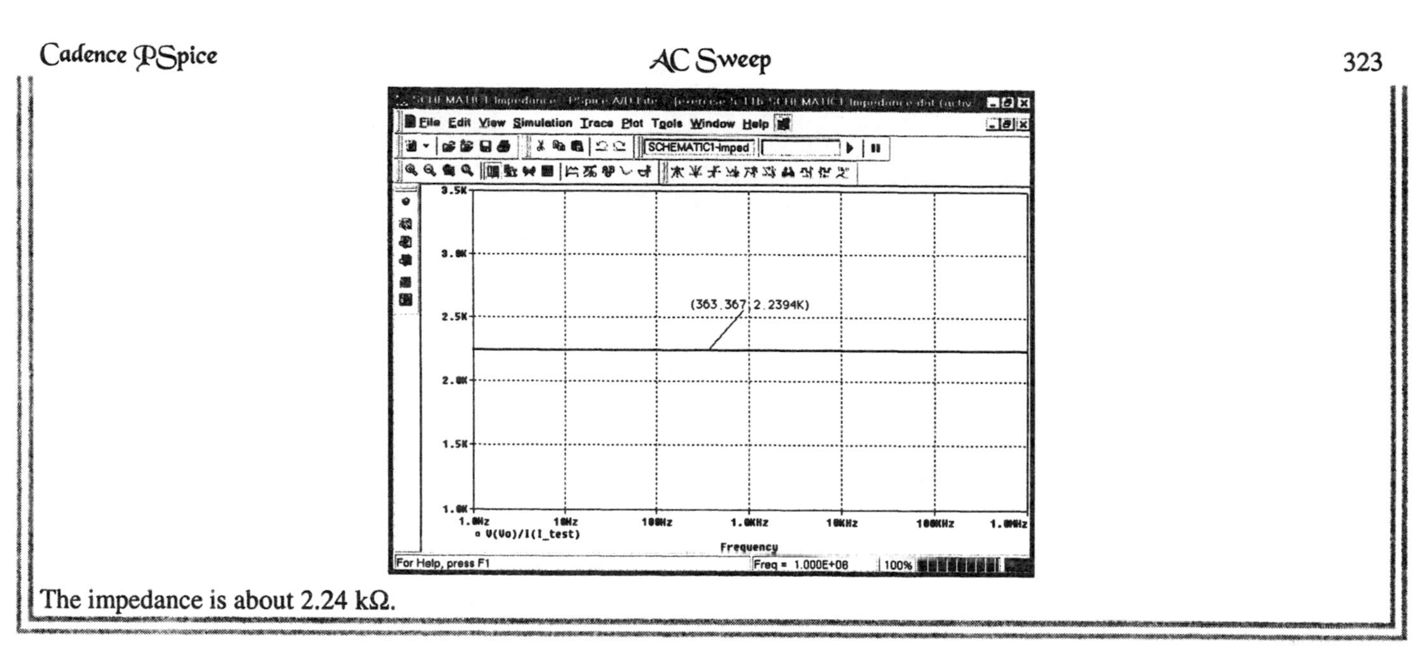


The impedance is about 2.24 kΩ.

## 5.I.2. Impedance Measurement of an Active Circuit

We would now like to find the output impedance of the jFET source follower circuit shown below. The output is taken at the source terminal of the jFET, so the impedance we are interested in is between nodes ***Vo*** and ***0*** (ground).

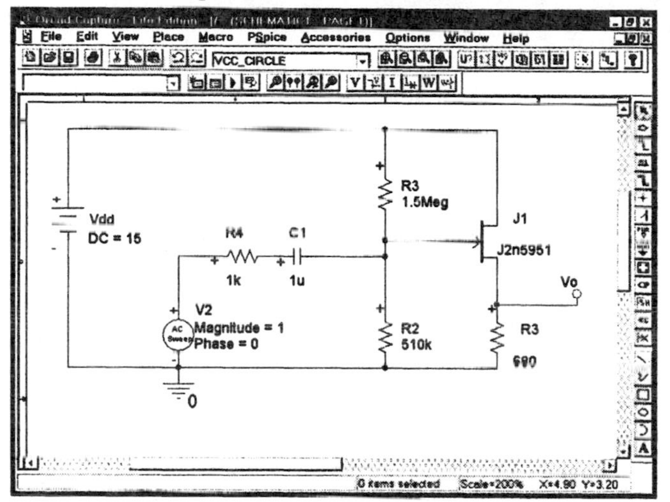


When you are finding output impedance you must set the input AC source to zero. Since the input is a voltage source, we replace ***Vs*** with a short:

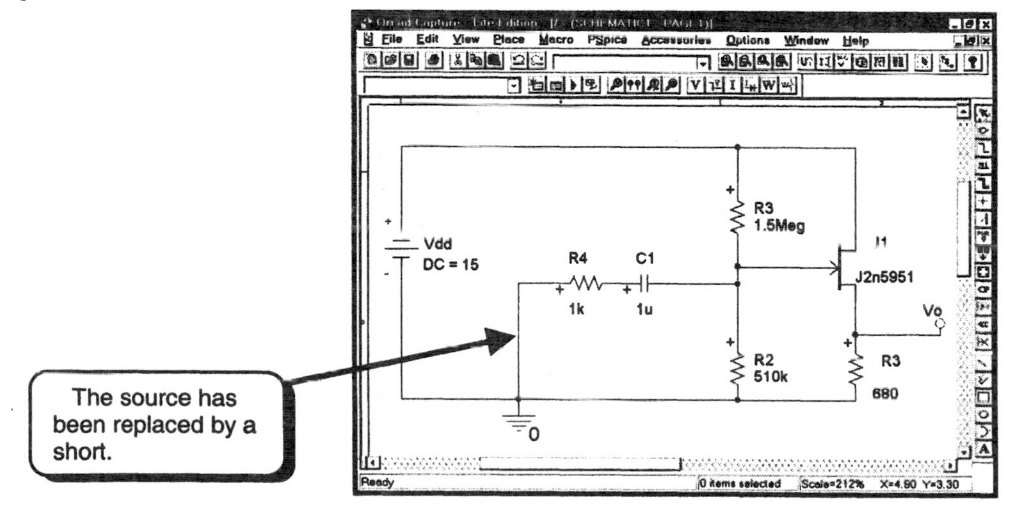


Next we need to add an AC source at the output to measure the impedance. We could use either an AC voltage source or an AC current source. If you use a voltage source, wire the circuit shown below:

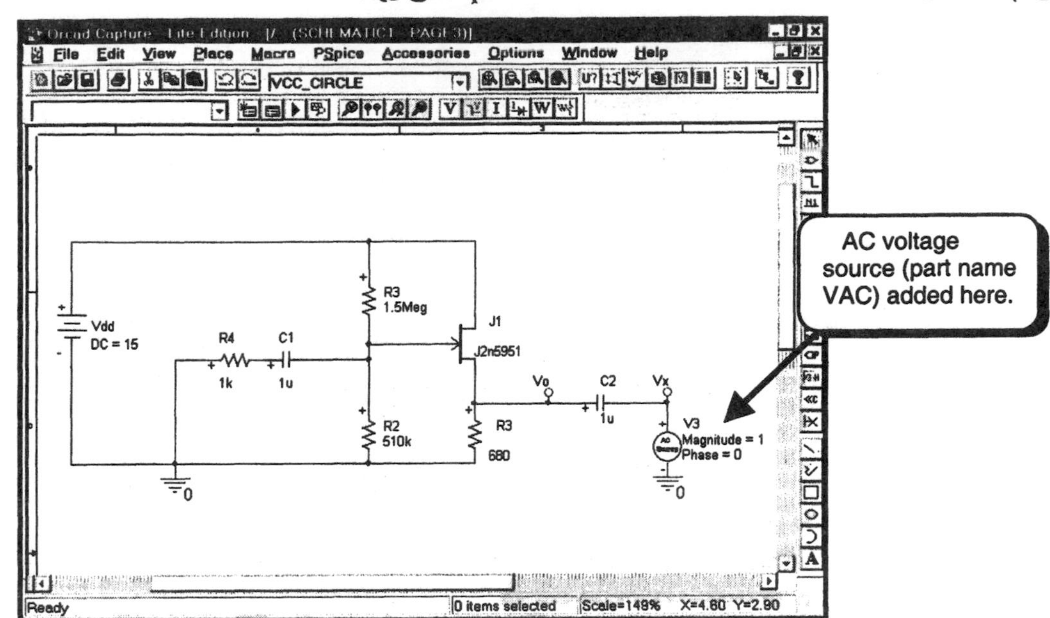


Capacitor ***C2*** is necessary to preserve the bias of the amplifier. Remember that ***V3*** is an AC source. For biasing, all AC sources are set to zero. Since ***V3*** is a voltage source, it would be replaced by a short. Without ***C2***, the source terminal of the jFET would be grounded when calculating the bias. This would destroy the bias and render the impedance measurement invalid.

The output impedance of the source follower can be observed by plotting the ratio **`V(Vo)/I(V3)`** in the circuit above. It is important to note that the ratio V(Vx)/I(V3) is not the output impedance of the source follower, but the output impedance plus the impedance of the capacitor.

The second method to measure the impedance of a circuit uses an AC current source:

| **R**<br>Resistor | **C**<br>Capacitor | **VCC_Circle**<br>Node label | **0**<br>Ground |
|---|---|---|---|
| **VDC**<br>DC voltage source | **j2N5951**<br>N-channel jFET | | **IAC**<br>AC current source |

AC current source (part name IAC) added here.

***I1*** is an AC current source. Note that there is no blocking capacitor between the current source and the amplifier. When the bias is calculated, all AC sources are set to zero. When a current source is zero, it is replaced by an open circuit. An open circuit at ***Vo*** is equivalent to the original circuit for bias calculations. The output impedance of the circuit can be calculated by the ratio **`V(Vo)/I(I1)`**.

Set up an AC Sweep (**PSpice, New Simulation Profile, AC Sweep/Noise**) to sweep frequencies from 1 Hz to 100 MHz at 100 points per decade. Run PSpice (**PSpice, Run**). When Probe runs, add the trace (**Trace, Add Trace**) **`V(Vo)/I(I1)`**. Remember that voltage divided by current is impedance. We are dividing the voltage between nodes ***Vo*** and ***0*** (ground) by the current flowing into and out of those nodes. This is the impedance between those nodes. You will see this trace:

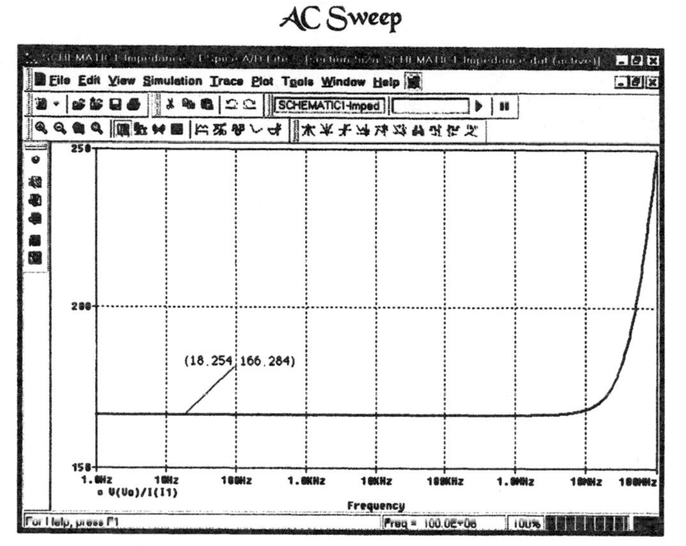


The trace shows that the impedance is 166 Ω for frequencies up to 1 MHz.

**EXERCISE 5-12**: Find the output impedance of the emitter-follower:

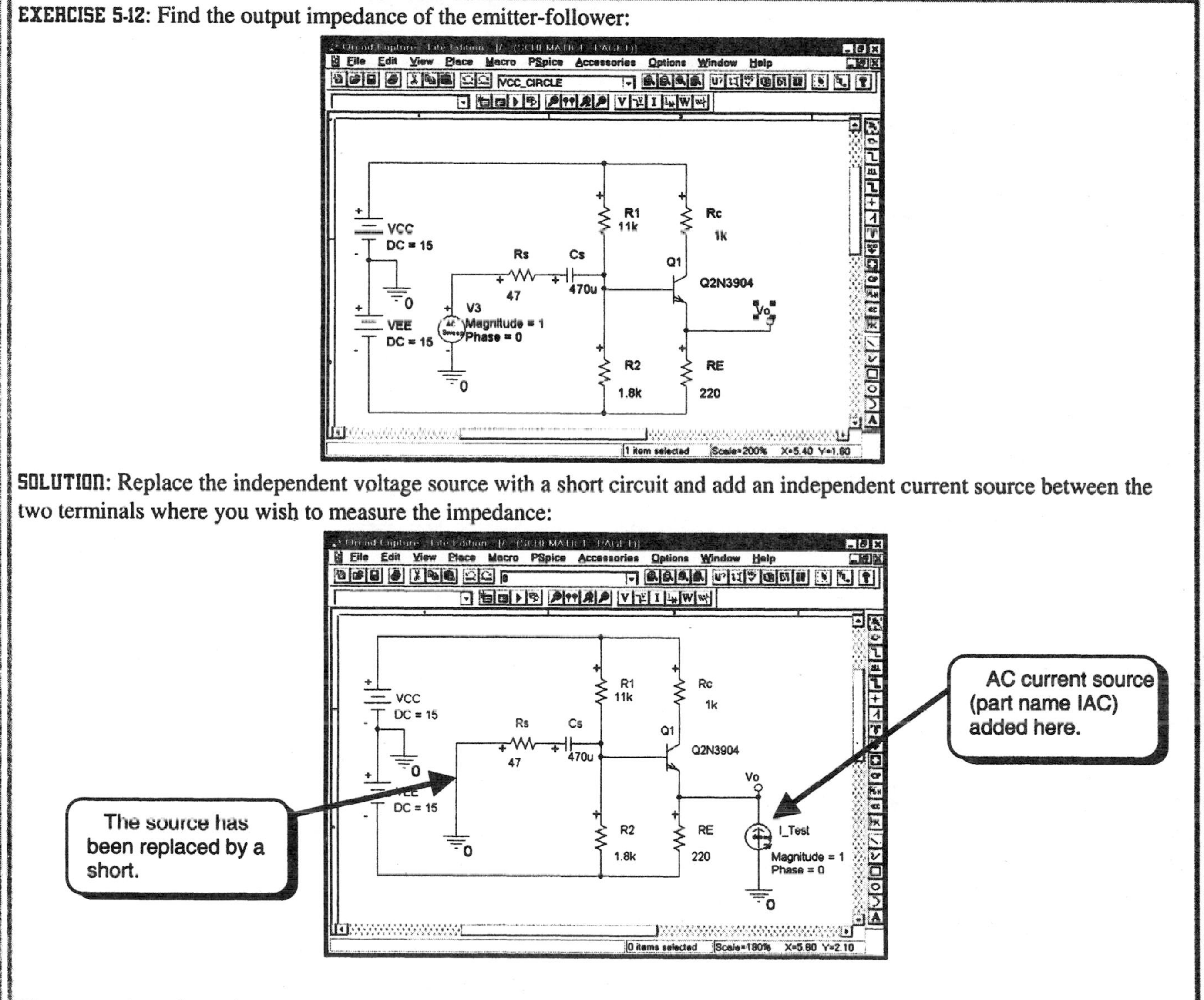


**SOLUTION**: Replace the independent voltage source with a short circuit and add an independent current source between the two terminals where you wish to measure the impedance:

The output impedance is `V(Vo)/I(I_test)`:

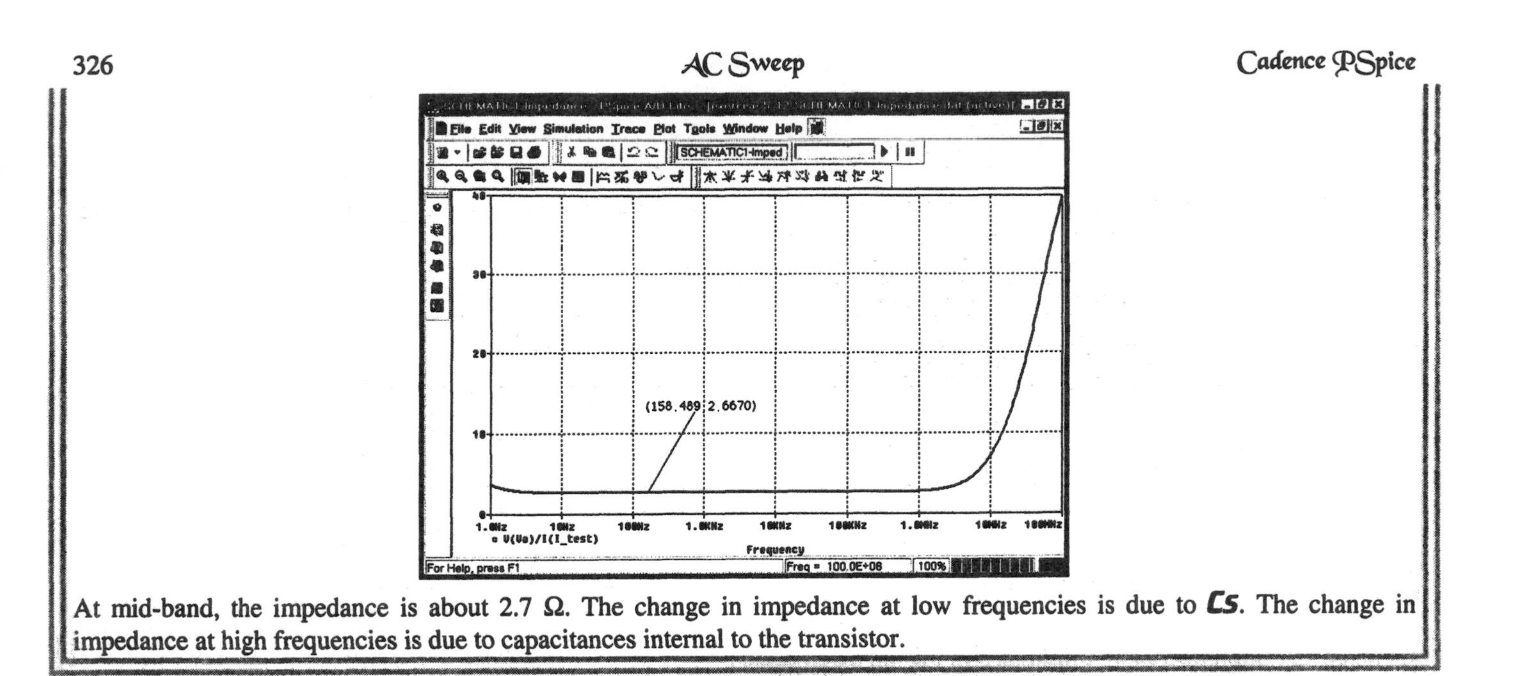


At mid-band, the impedance is about 2.7 Ω. The change in impedance at low frequencies is due to **Cs**. The change in impedance at high frequencies is due to capacitances internal to the transistor.

# 5.J. Summary

- The AC Sweep is used to find the magnitude and phase of voltages and currents. It is not used to look at waveforms versus time. It is used with the sources VAC and IAC only.
- All DC sources, such as VDC and IDC, and all time-varying sources, such as Vsin, Isin, Vpulse, and Ipulse, are set to zero.
- The results can be obtained for a single frequency or multiple frequencies.
- For multiple frequencies, the results can be viewed graphically using Probe. For a single frequency, the results can be viewed as text in the output file. Use the "Print" part to generate text output.
- The AC Sweep is used to generate plots of magnitude versus frequency, or plots of phase versus frequency.
- To plot the magnitude of a voltage, specify V(*expression*). For example, the magnitude of the voltage at node 1 would be specified as V(1).
- To plot the phase of a voltage, specify Vp(*expression*). For example, the phase of the voltage at node Vx would be specified as Vp(Vx).
- To plot the magnitude of a current, specify I(*expression*). For example, the magnitude of the current through R5 would be specified as I(R5).
- To plot the phase of a current, specify Ip(*expression*). For example, the phase of the current through C6 would be specified as Ip(C6).
- Use the dB command to plot a trace in decibels. The command dB(*expression*) is equivalent to $20\log_{10}$(*expression*).
- The AC Sweep and the dB command are used to generate Bode plots.
- The AC Sweep is used to find the gain and frequency response of an amplifier.
- Mutual inductance requires two parts: the inductors (L) and the coupling (K).
- The Parametric Sweep can be used to see how values of devices affect the performance of a circuit.
- Goal functions can be used to obtain numerical data from Probe graphs. The result of evaluating a goal function is a single numerical value.
- The Performance Analysis is the use of a goal function in conjunction with a Parametric Sweep. The result of the goal function is plotted versus the swept parameter.

# PART 6
# Transient Analysis

The Transient Analysis is used to look at waveforms versus time. Waveforms are displayed as you would see them on an oscilloscope screen. An example of a waveform versus time is:

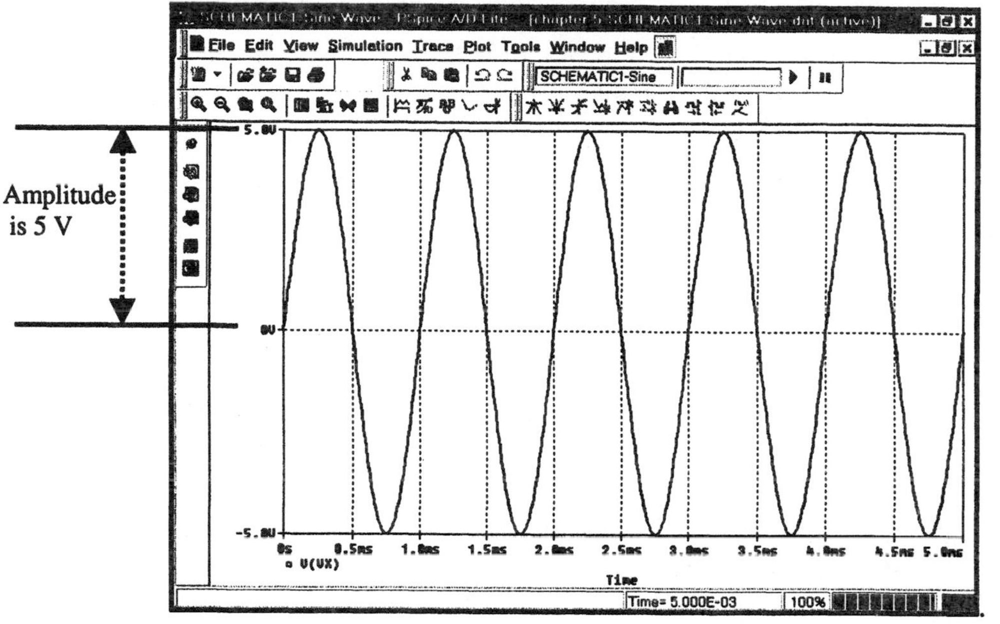


This graph shows us a voltage versus time. Use the Transient Analysis to obtain a waveform versus time. In the time domain the equation for this waveform is $v_x(t) = 5\sin(2\pi \bullet 1000t + 0^\circ)$. This waveform has an amplitude of 5 V and a frequency of 1000 Hz. If you want to look at the magnitude and phase of voltages and currents, use the AC Sweep. The magnitude of the above waveform is 5 V and the phase is zero degrees—in phasor notation, $5\angle 0^\circ$. The AC Sweep analysis will give us the result $5\angle 0^\circ$. The Transient Analysis will give us the graph above. The Transient Analysis uses voltage and current sources that are functions of time. AC sources such as VAC and IAC are functions of magnitude and frequency. These sources are used for the AC Sweep only. The AC sources are set to zero for the Transient Analysis.

## 6.A. Introduction

This section covers three topics that are important when using the Transient Analysis. The topics covered apply to all examples covered in this part. The topics may not be necessary, but they are good to keep in the back of your mind if problems arise.

## 6.A.1. Sources Used with the Transient Analysis

The sources below are meant to be used with the Transient Analysis:

- VSIN, ISIN - Sinusoidal voltage or current source. Typical voltage waveform: v(t) = 5 sin(2000t + 30°)
- VEXP, IEXP - Can be used to create an exponential waveform. Typical current waveform: i(t) = 5[1 – exp(t/τ)].
- Vpulse, IPULSE - Pulse waveform. Can be used to create a square wave by specifying the pulse width and period.
- VPWL, IPWL - Can create any arbitrary waveform that is made up of straight lines. PWL stands for piecewise-linear.
- VSFFM, ISFFM - Used to create a frequency-modulated sine wave.
- Vsq - A square wave voltage source. This source uses the pulsed voltage source to make a square wave. It is a special case of Vpulse.
- Vtri - A triangle wave voltage source. This source uses the pulsed voltage source to make a triangle wave. It is a special case of Vpulse.
- Vramp - A saw-tooth voltage source. This source uses the pulsed voltage source to make a saw-tooth wave. It is a special case of Vpulse.

- V_ttl - A 0 to 5 V square wave with adjustable frequency and duty cycle. This source uses the pulsed voltage source to make a square wave. It is a special case of Vpulse.

## 6.A.2. Maximum Step Size

One of the features of the Transient Analysis that causes much confusion is the Maximum step size argument in the Transient dialog box. Suppose we wish to run a circuit that has a sinusoidal source. We expect the voltages and currents to look sinusoidal, as follows:

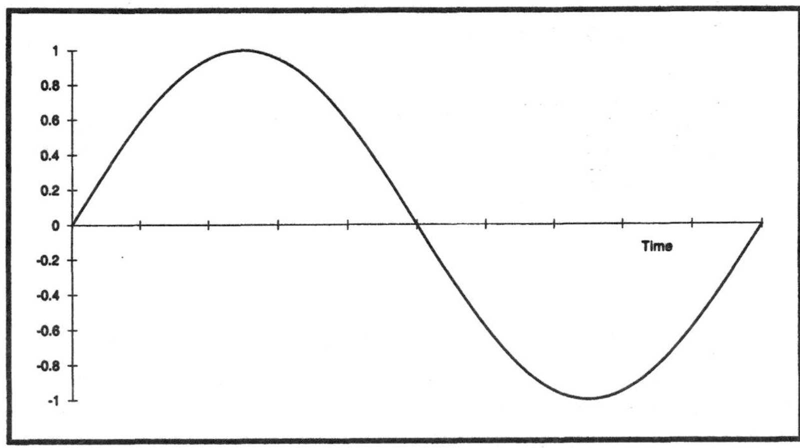


When PSpice runs a Transient Analysis, it solves differential equations to find voltages and currents versus time. The time between simulation points is chosen to be as large as possible while keeping the simulation error below a specified maximum. In some cases, where PSpice can take large time steps, you may get a graph that does not look sinusoidal:

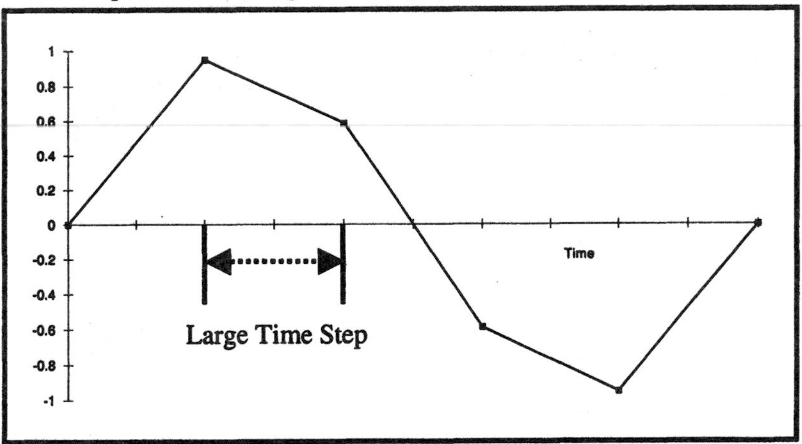


The graph does not look much like a sine wave because the time between points is so large. If we decrease the time between points, we see that the points do lie on a sinusoidal curve. The graph below is an overlay plot of the detailed sine wave of the top graph (the trace had a very small time step) and the center graph showing a sine wave trace with a large time step :

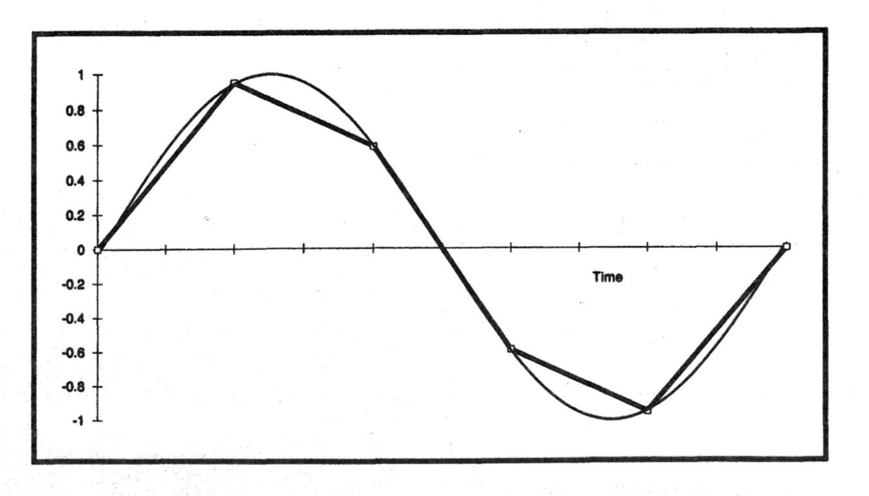

PSpice chooses as large a time step (time between simulation points) as possible to reduce simulation time, but the graphs constructed with these points may not look aesthetically pleasing. To increase the number of points, PSpice provides a Maximum step size argument for the Transient Analysis. **The Maximum step size is the maximum time between simulation points.** Reducing the Maximum step size increases the number of simulation points. The following Transient dialog box shows the location of the ***Maximum step size*** argument:

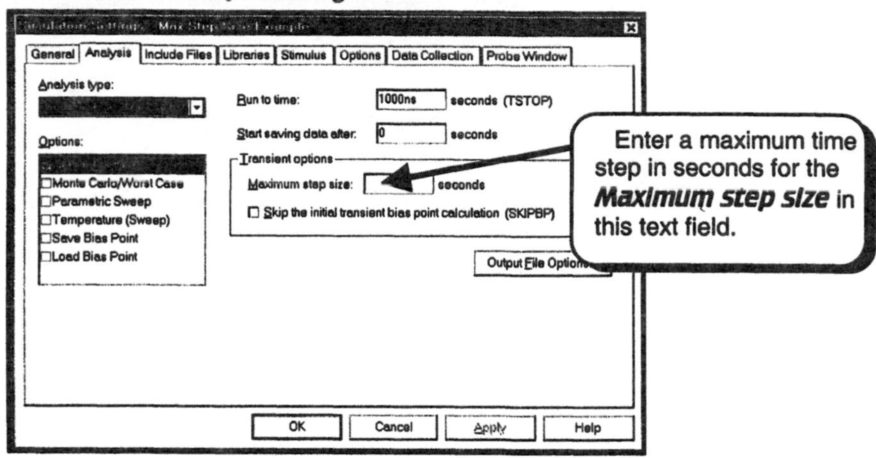


The ***Maximum step size*** is initially blank so that PSpice will choose as large a time step as possible. If you need more points in your simulation, you must enter a number for the ***Maximum step size***.

As an example, we will run the DC power supply below. A more detailed example of a DC power supply is shown in Section 6.E.

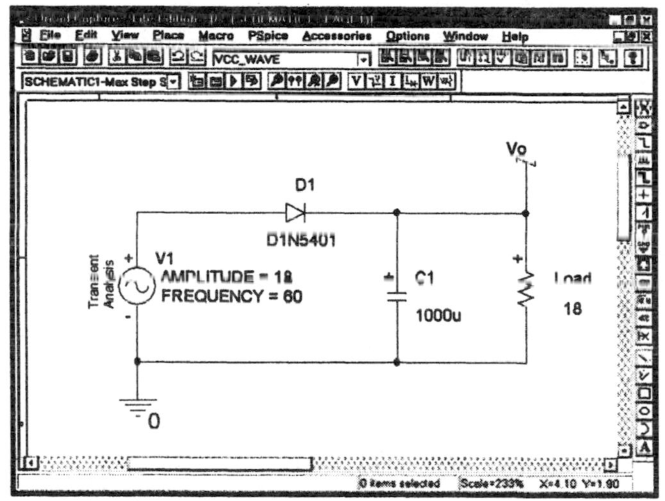


The source has a frequency of 60 Hz, corresponding to a period of 16.67 ms. We will run the circuit for three 60 Hz cycles, or 50 ms.

We must now create a new simulation profile. Select **PSpice** and then **New Simulation Profile** from the menus:

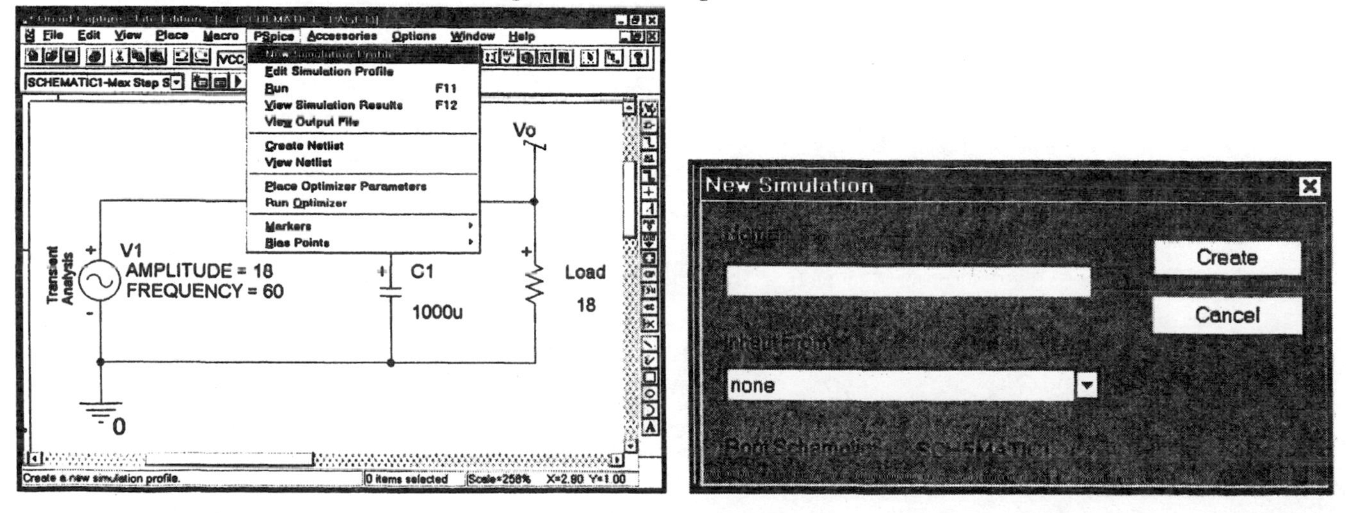

Enter a name for the profile and click the ***Create*** button:

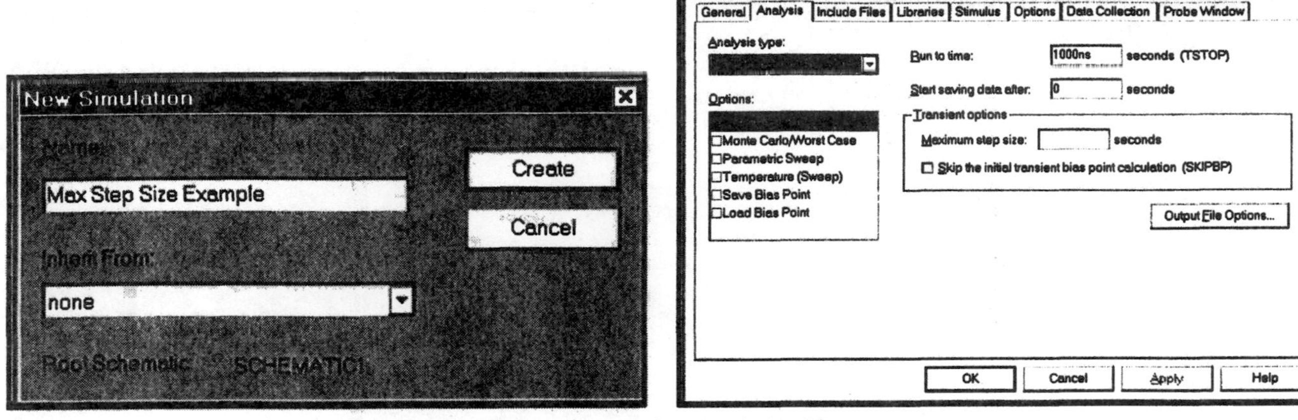


Fill in the dialog box below as shown below. It is set to run the circuit for 50 ms:

Note that the ***Maximum step size*** has not been specified, allowing PSpice to take as much time between simulation points as possible. If we run the circuit and plot the capacitor current, we obtain the plot:

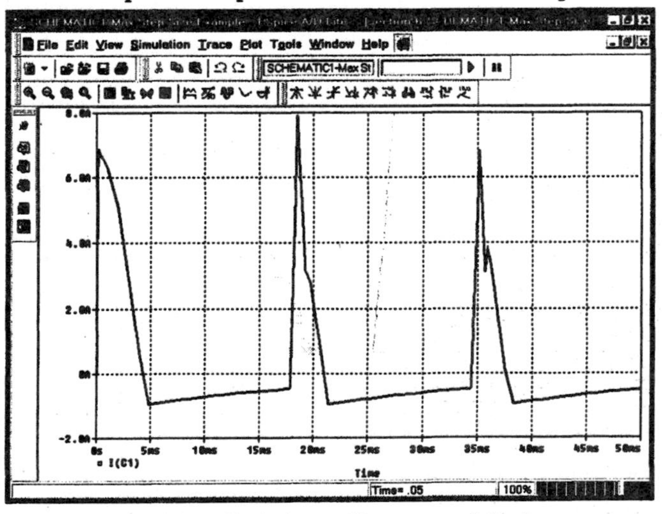


This plot is not what we expect for this circuit. The plot is jagged because of the large amount of time between points.

We will now set the ***Maximum step size*** to obtain a smoother plot. The amount of time for one 60 Hz cycle is 16.67 ms. We would like to see 1000 points in each 60 Hz cycle. For 1000 points per cycle, the time between points is:

$$\text{Time Between Points} = \frac{16.67\,\text{ms}}{1000} = 16.67\,\mu s$$

We need to set the ***Maximum step size*** to this value. The dialog box below specifies the value of ***Maximum step size***:

General | Analysis | Include Files | Libraries | Stimulus | Options | Data Collection | Probe Window

Analysis type: Time Domain (Transient)

Options: Monte Carlo/Worst Case, Parametric Sweep, Temperature (Sweep), Save Bias Point, Load Bias Point

Run to time: 50m seconds (TSTOP)

Start saving data after: 0 seconds

Transient options — Maximum step size: 16.67u seconds

Skip the initial transient bias point calculation (SKIPBP)

Output File Options...

OK | Cancel | Apply | Help

When we run the simulation and plot the capacitor current, we obtain the plot:

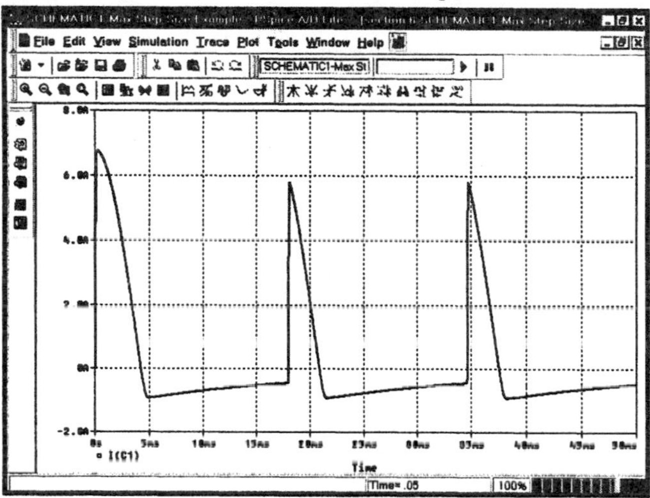


This plot agrees with what we expect the capacitor current to look like. However, the simulation does take longer to run.

## 6.A.3. Convergence

One of the most frustrating problems that you may encounter using the Transient Analysis is convergence problems. When PSpice is simulating a differential equation, it calculates a data point and estimates the error associated with the calculation. If the error is larger than a specified maximum, PSpice reduces the time step and recalculates the point and the error for the new point. Reducing the time step usually reduces the error. PSpice will continue reducing the time step until the error is within acceptable limits, or until PSpice reaches the limit on the number of times it is allowed to reduce the time step.

If PSpice reaches the limit on reducing the time step, PSpice will announce that the simulation failed to converge. The PSpice window will display the message:

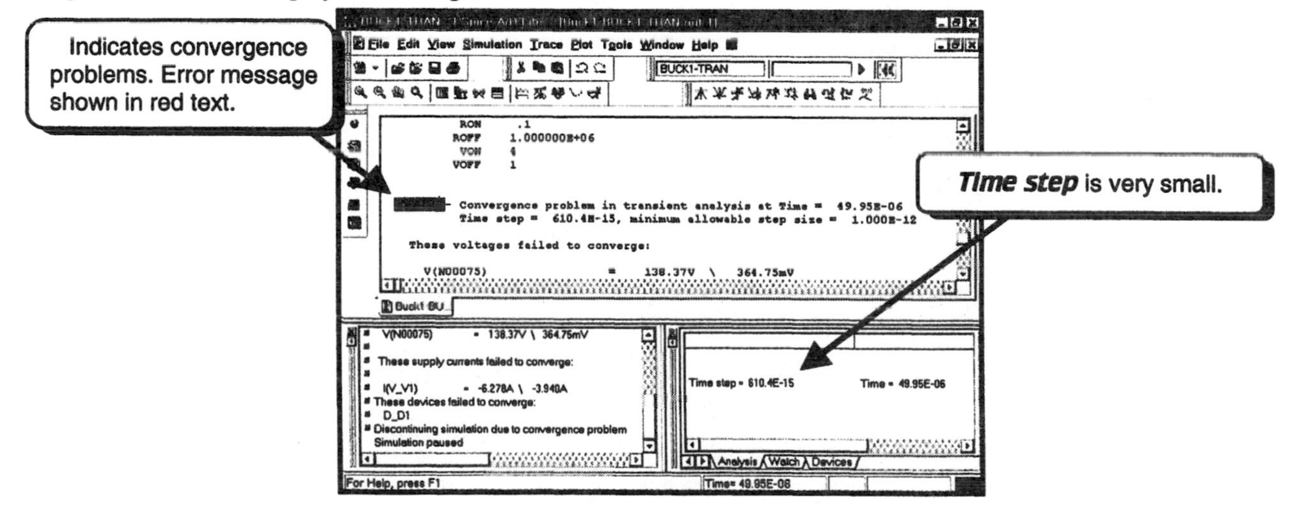

Aside from the message in the center of the screen, a second indication that the simulation did not converge is that the ***Time step*** is very small. It is small because PSpice went through the process of reducing the time step in order to reduce the error until it reached the limit on reducing the time step. Along with the information above, you will receive the dialog box:

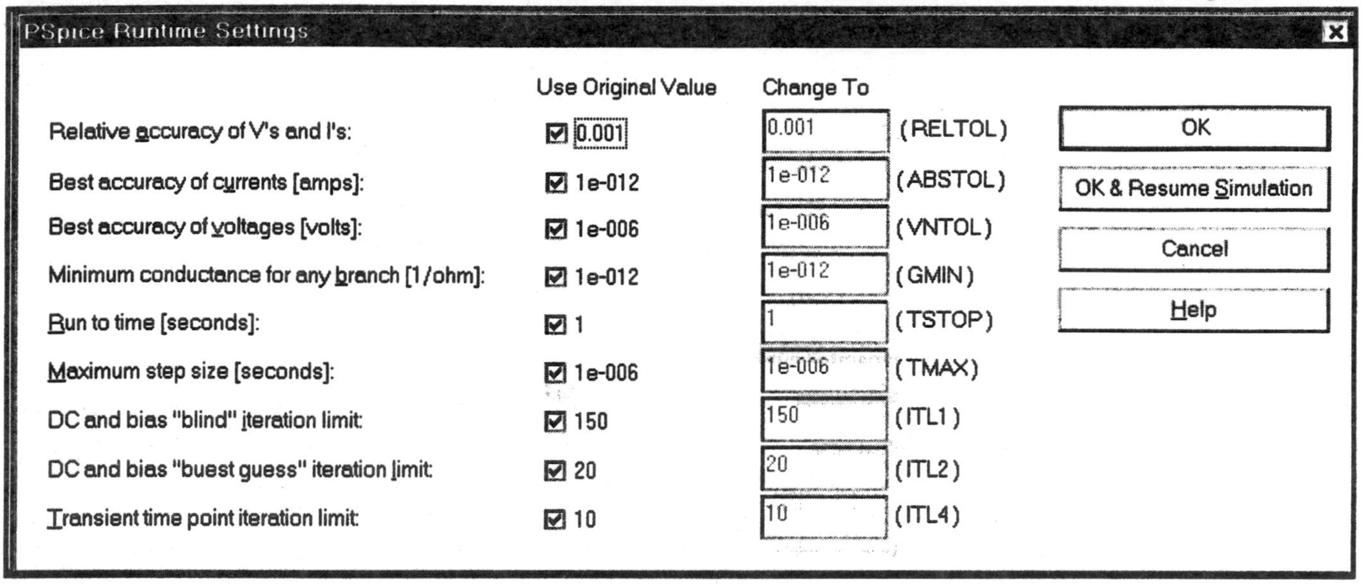

PSpice Runtime Settings

| | Use Original Value | Change To | |
|---|---|---|---|
| Relative accuracy of V's and I's: | ☑ 0.001 | 0.001 | (RELTOL) |
| Best accuracy of currents [amps]: | ☑ 1e-012 | 1e-012 | (ABSTOL) |
| Best accuracy of voltages [volts]: | ☑ 1e-006 | 1e-006 | (VNTOL) |
| Minimum conductance for any branch [1/ohm]: | ☑ 1e-012 | 1e-012 | (GMIN) |
| Run to time [seconds]: | ☑ 1 | 1 | (TSTOP) |
| Maximum step size [seconds]: | ☑ 1e-006 | 1e-006 | (TMAX) |
| DC and bias "blind" iteration limit: | ☑ 150 | 150 | (ITL1) |
| DC and bias "buest guess" iteration limit: | ☑ 20 | 20 | (ITL2) |
| Transient time point iteration limit: | ☑ 10 | 10 | (ITL4) |

OK | OK & Resume Simulation | Cancel | Help

This dialog box shows the settings that affect convergence problems. You can change some of these settings and then resume the simulation.

There are several ways to help prevent convergence problems in the Transient Analysis. We will discuss three here and, hopefully, one of the methods or a combination of the methods will solve the problem.

**Method 1: Reduce the Run to time.**

When you set up a Transient Analysis, you must specify the ***Run to time*** of the simulation:

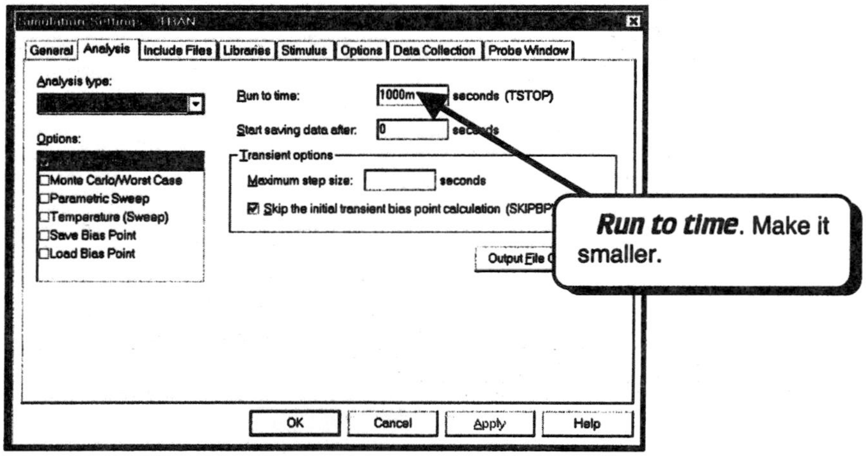

If the ***Maximum step size*** is not specified, as in the dialog box above, PSpice must make a guess for an appropriate value for the time step. The guess it makes is the ***Run to time*** divided by some number. It starts with this time step and then reduces the time step until the error is acceptable. If the ***Run to time*** is a large number, the initial time step is large and will yield a large amount of error. After reducing the time step a fixed number of times, the time step will still be large because it started out large, and PSpice will quit because the error is large and the limit on reducing the time step was reached. To avoid the problem, reduce the ***Run to time***.

**Method 2: Specify a value for the Maximum step size and make it small.**

If you do not wish to change the value of the ***Run to time***, you can specify a value for the ***Maximum step size***.

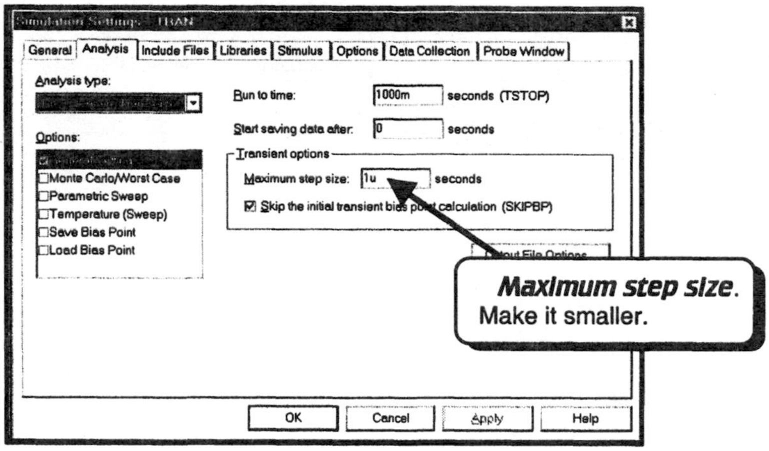

The ***Maximum step size*** is the largest value allowed for the time step. The initial value of the time step will be the value specified by ***Maximum step size***. If the error is too large, PSpice will reduce the time step. Since the step size can start with a small value (the value of ***Maximum step size***), after a number of reductions, the step size will become small enough to make the error acceptable. If, after specifying a value for the ***Maximum step size***, the simulation fails to converge, make the ***Maximum step size*** smaller. Note that making the ***Maximum step size*** smaller increases the simulation time.

**Method 3: Increasing the number of time step reductions.**

If the two methods above fail to solve the problem, you can increase the limit on the number of times the time step can be reduced. To increase the limit, select the **Options** tab in the ***Simulation Settings*** dialog box:

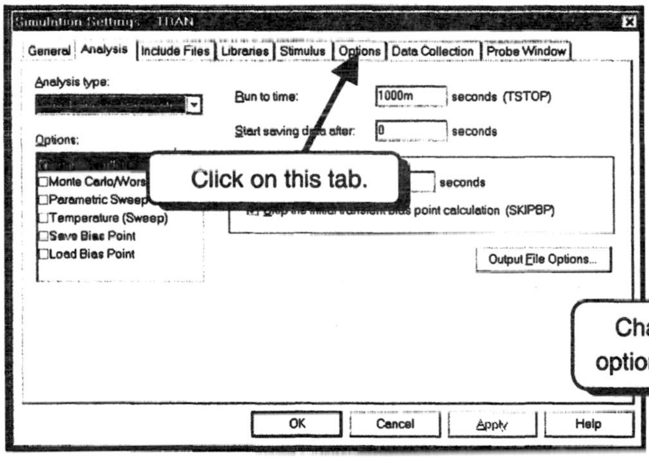

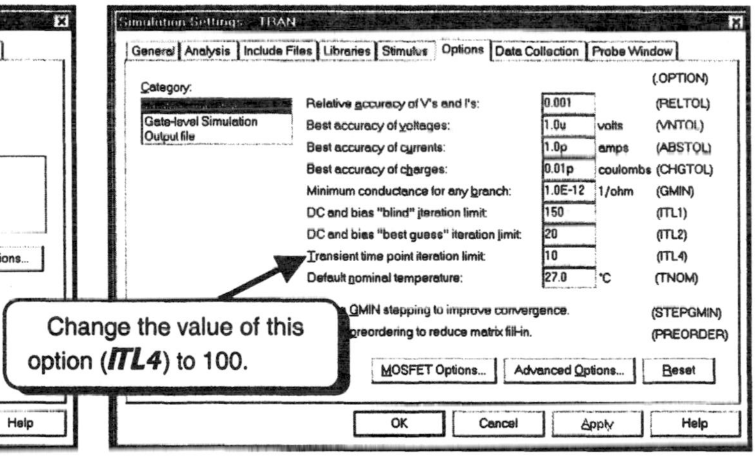

The ITL4 variable controls how many times the step size can be reduced during a simulation. The default value is 10, so after the step size has been reduced 10 times the simulation will stop. If the value of the option is not set to 100 then change the value to 100.

# 6.B. Capacitor Circuit with Initial Conditions

In this section we will observe the transient response of a capacitor circuit with an initial condition. We will use the circuit below:

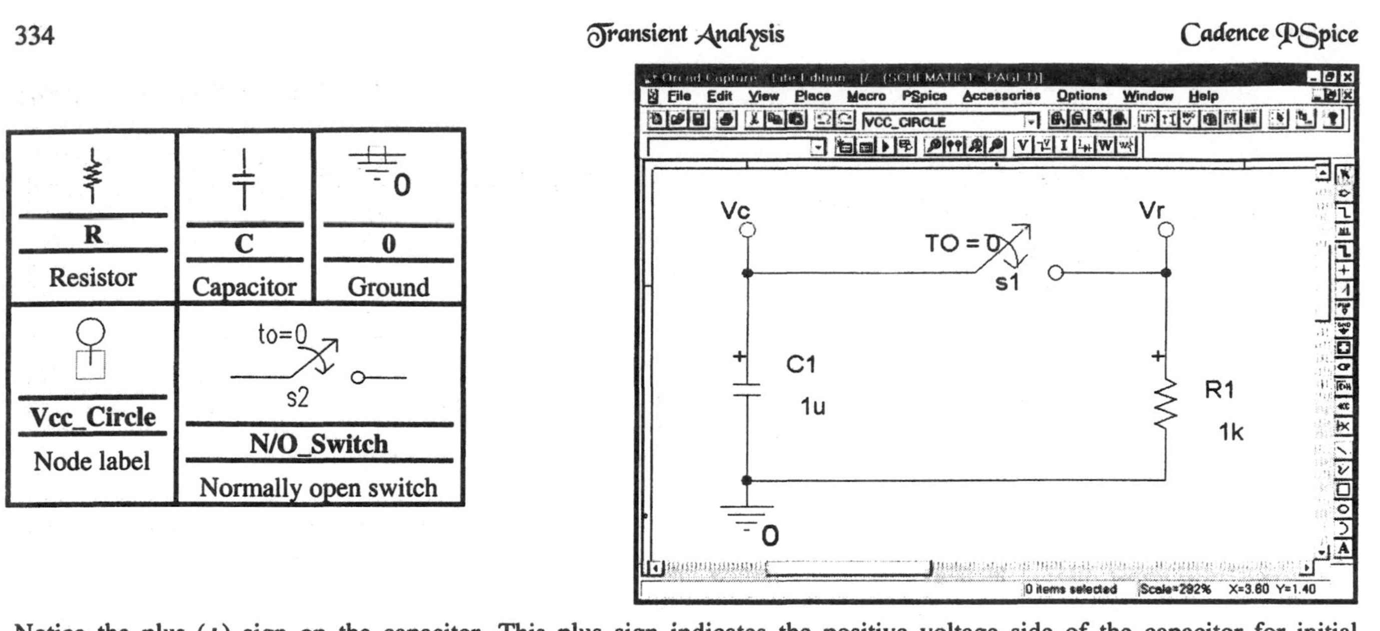


Notice the plus (+) sign on the capacitor. This plus sign indicates the positive voltage side of the capacitor for initial conditions. Make sure the plus sign is oriented as shown in the schematic. We would like the capacitor to have an initial condition of 5 V, and we would like the switch to close at t = 1 ms. The initial condition of the capacitor is not one of the displayed attributes, so we must edit all the attributes of the capacitor. Double-click the ***LEFT*** mouse button on the capacitor graphic, ⊣⊢, to obtain the spreadsheet for ***C1***:

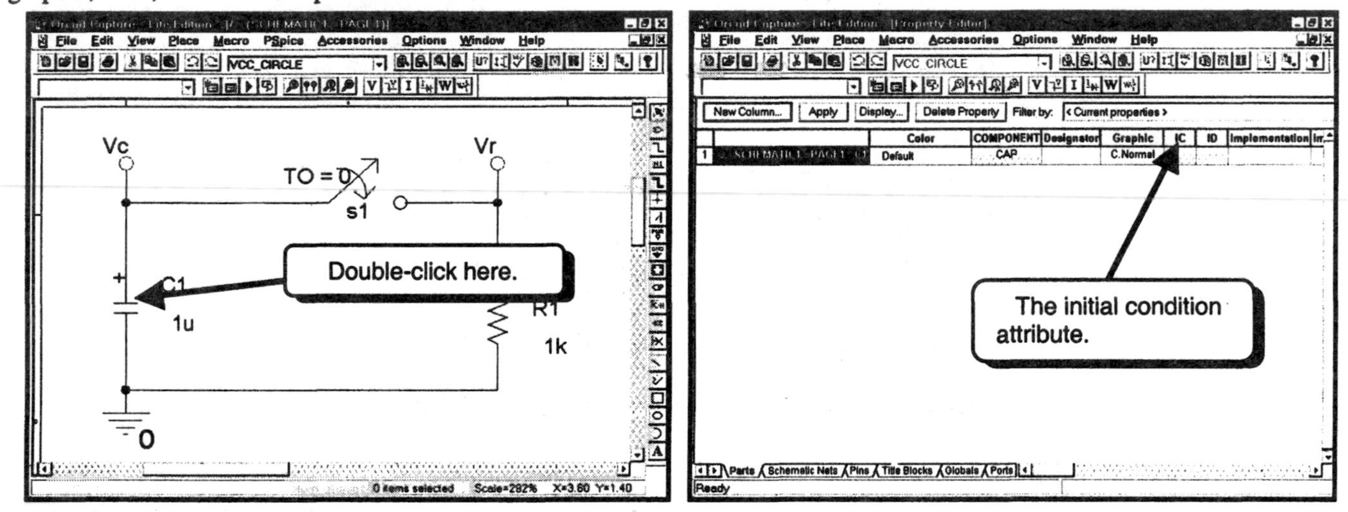


This box shows all the attributes for the capacitor. Of all the attributes, only the **Value** (1μ) and the **Part Reference** (C1) are displayed. All other attributes have their specified values, but are not displayed.

We would like to give the capacitor an initial voltage of 5 V. Click the ***LEFT*** mouse button on the cell below the text ***IC*** to select the cell:

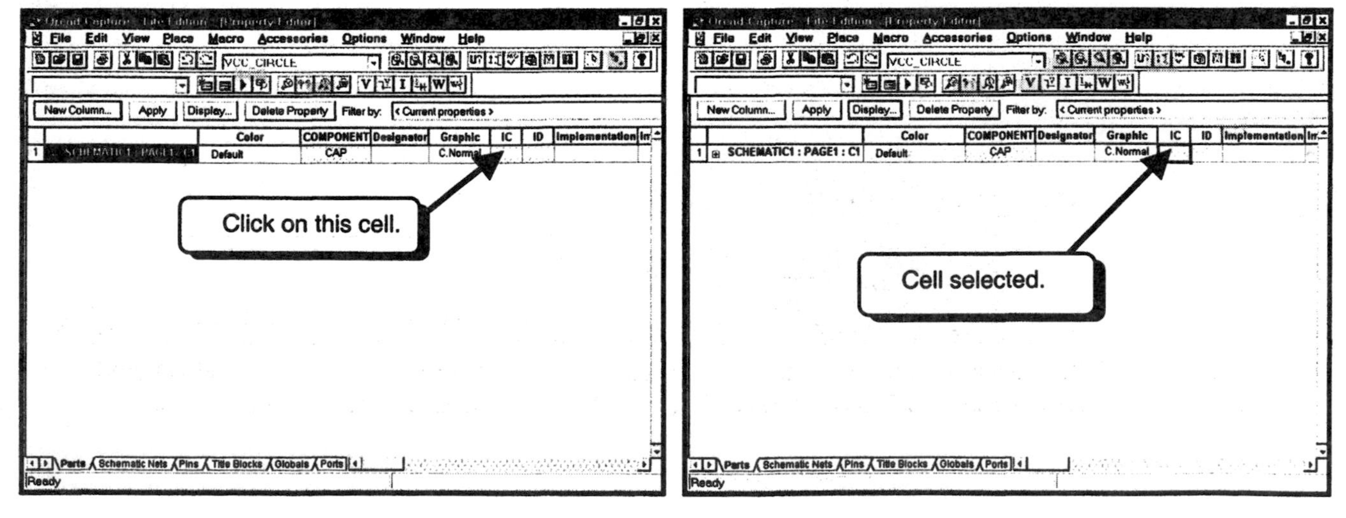

Type **5** and then click on the ☒ as shown to close the spreadsheet. When you return to the schematic, the initial condition attribute will be displayed.

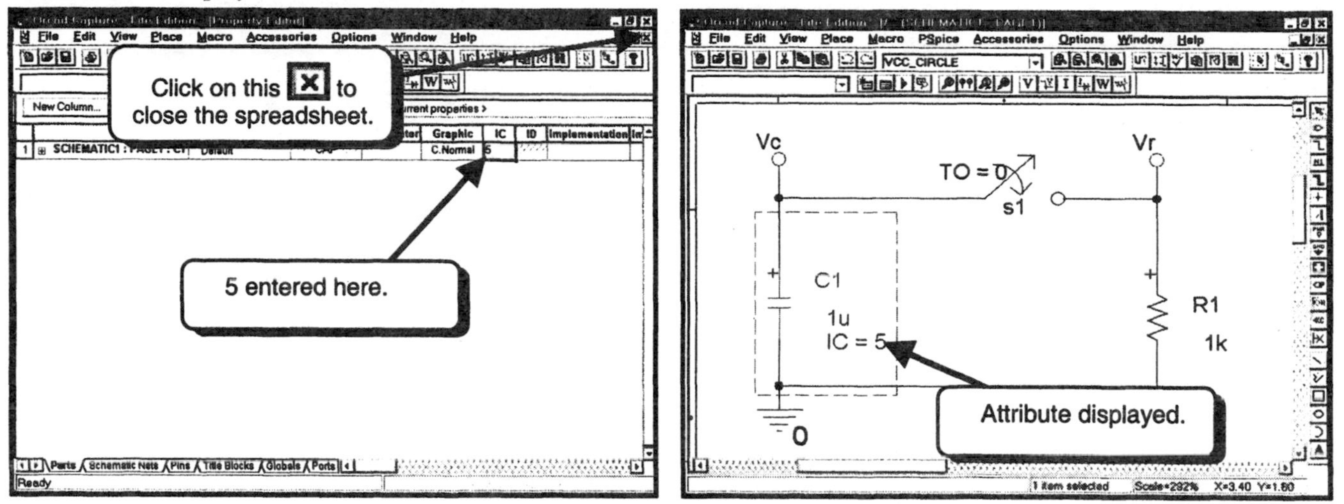


The next thing we need to do is change the attributes of the switch so that it closes at t = 1 ms. Double-click the ***LEFT*** mouse button on the text ***TO = 0***. The dialog box below will appear.

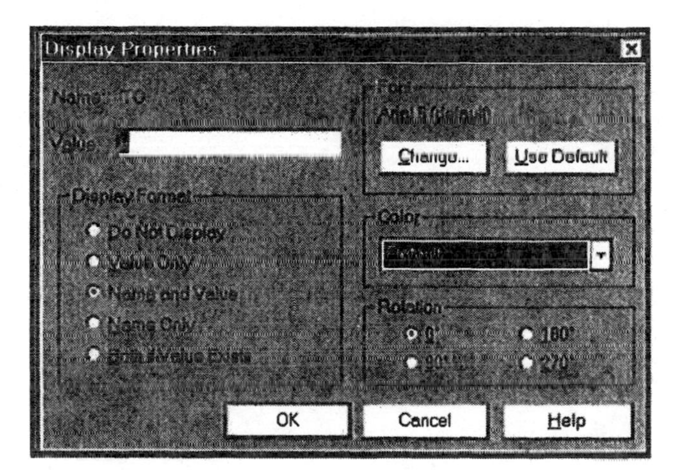


Enter the text **1m** and click the ***OK*** button. You should have the completed schematic as shown:

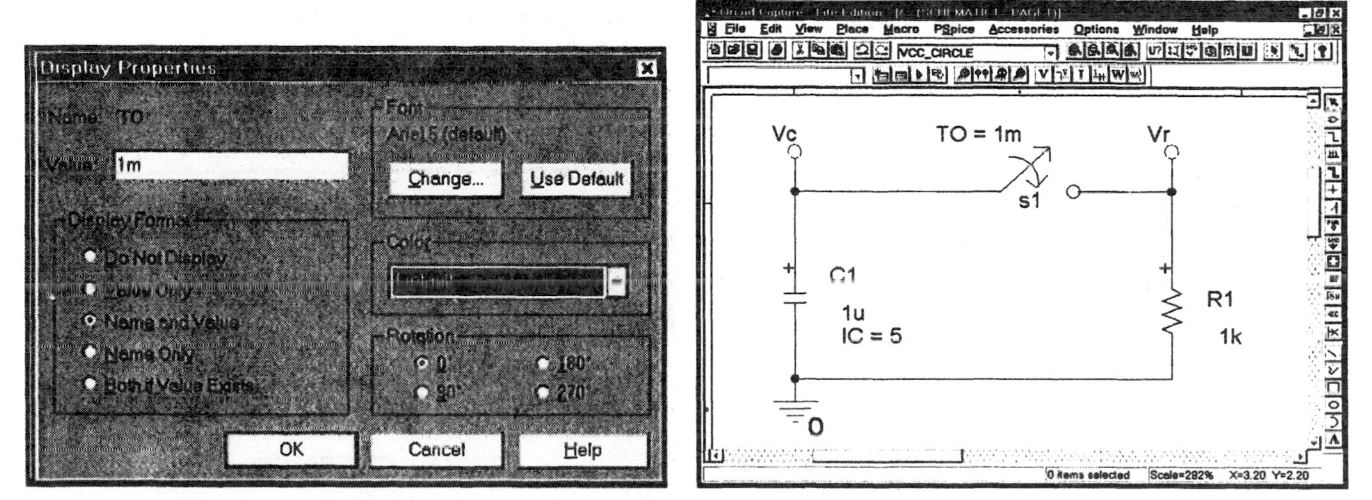


Note that the initial condition of the capacitor is ***5*** (5 volts) and the switch closes at time equals ***1m*** (1 millisecond). The next thing we need to do is set up the Transient Analysis. Select **PSpice** and then **New Simulation Profile**:

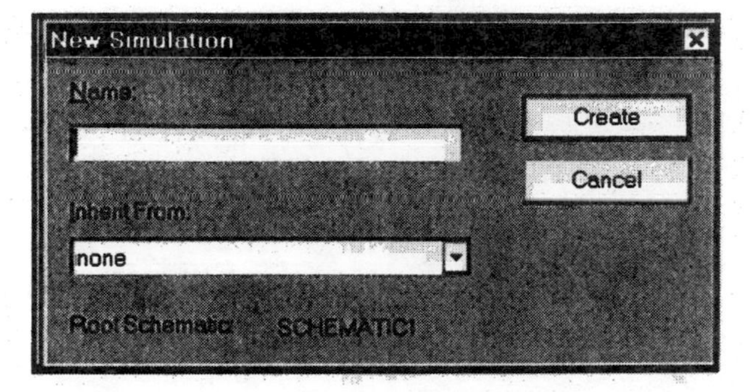


Enter a name for the profile and then click the ***Create*** button:

The ***Time Domain (Transient)*** dialog box is selected by default.

The ***Run to time*** option is the length of the simulation. The simulation runs from time equals zero to the ***Run to time***. The ***Start saving data after*** value is used for long simulations. If you ran a simulation for 1 second and you were interested only in the data from 990 ms to 1 s, then you would set the ***Start saving data after*** value to 990 ms. PSpice would not save any data from time zero to 990 ms. This parameter is used to reduce the size of the output data file for large circuits with long simulation times. We will ignore this argument since this is a short run.

The ***Maximum step size*** value is important. When PSpice runs a circuit, it chooses the time between simulation points to be as large as possible while keeping the error below a specified maximum. The larger the time step, the less time the simulation takes to complete. To keep simulation time to a minimum, leave the ***Maximum step size*** value blank. However, the step size PSpice takes may be too large for you. If this is the case, specify a value for ***Maximum step size***. ***Maximum step size*** is the longest time between simulation points PSpice will take. A smaller value of ***Maximum step size*** will give you more points in your simulation, but it will take more time to complete the simulation.

The time constant for our circuit is $\tau = R_1C_1 = 1\,\mu\text{F}\cdot 1\,\text{k}\Omega = 1\,\text{ms}$. After five time constants the circuit should reach steady state, so we will run the simulation for five time constants. Remember that the switch closes at 1 ms. To let the capacitor transient run for 5 ms we need a total simulation time of 6 ms. We would like to see at least 500 points during the capacitor transient, so set the ***Maximum step size*** to 5 ms/500 or 0.01 ms.*

You should have the ***Simulation Settings*** dialog box filled out as shown below. The simulation is set to run for 6 ms. The maximum time between simulation points is 0.01 ms, so the minimum number of points in the simulation is $\frac{6\text{ ms}}{0.01\text{ ms}} = 600$ points.

---

*Convergence problems have been observed in some transient simulations of circuits with switches, such as the circuit on page 334. You will notice this error because the transient simulation terminates prematurely. If you look at the output file, an error message near the end of the file will indicate a convergence problem. The problem can be fixed by choosing a smaller value of ***Maximum step size***, or by using one of the methods mentioned in Section 6.A.3.

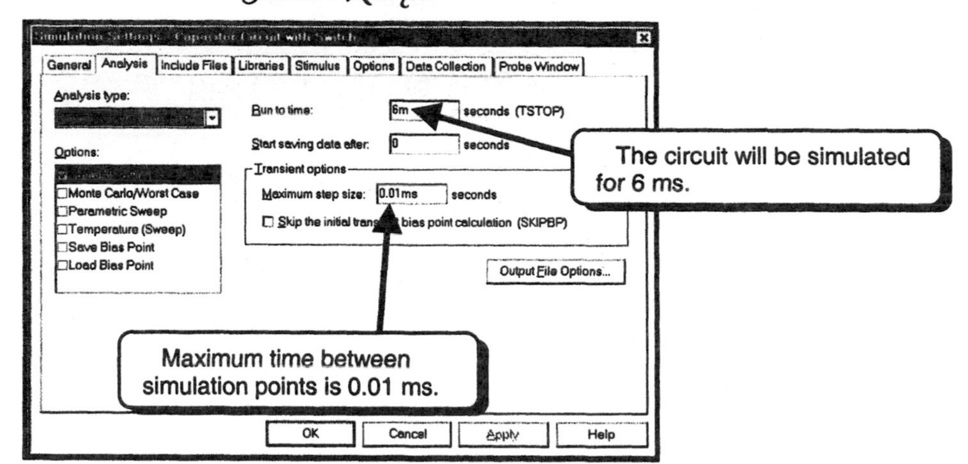
The circuit will be simulated for 6 ms.
Maximum time between simulation points is 0.01 ms.

Click the ***OK*** box to accept the settings and return to the schematic. To run the simulation select **PSpice** and then **Run** from the Capture menus:

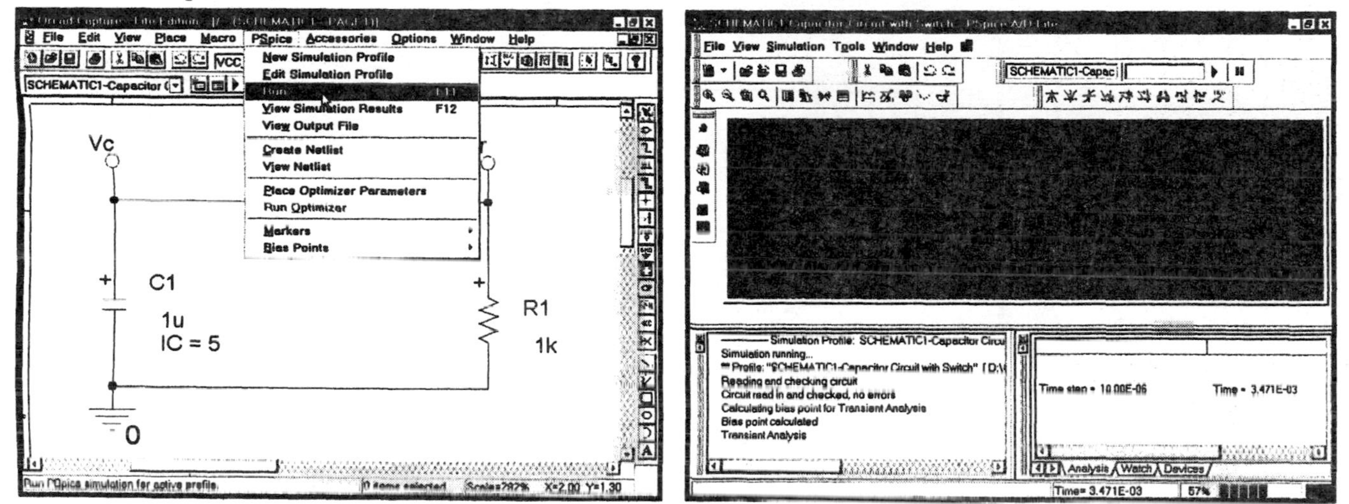

When the simulation is complete, Probe will automatically run. Add the trace **V(VR)** (select **Trace** and then **Add Trace** from the Probe menu bar). The trace of the resistor voltage will appear as shown below. See page 98 to change the display to have Probe use the entire window:

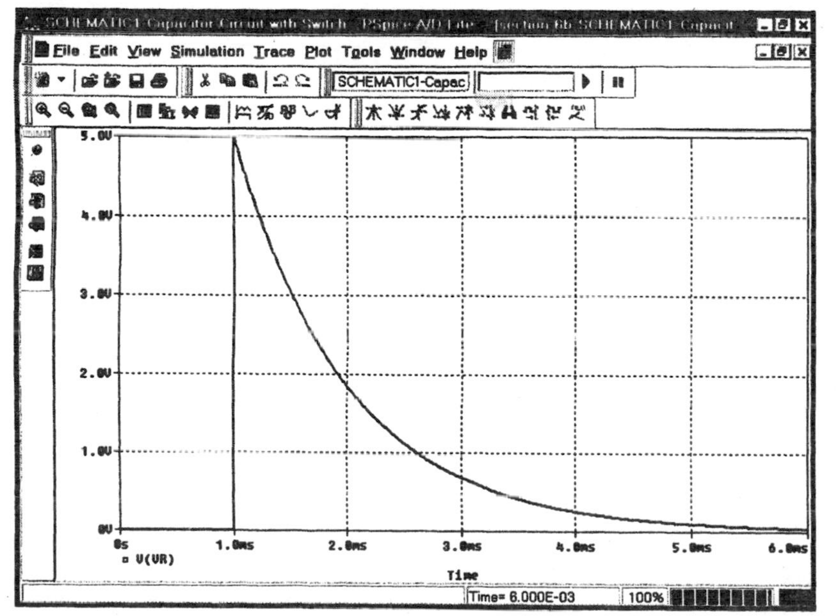

Next we would like to look at the capacitor voltage. Add a new window in Probe by selecting **Window** and then **New Window**:

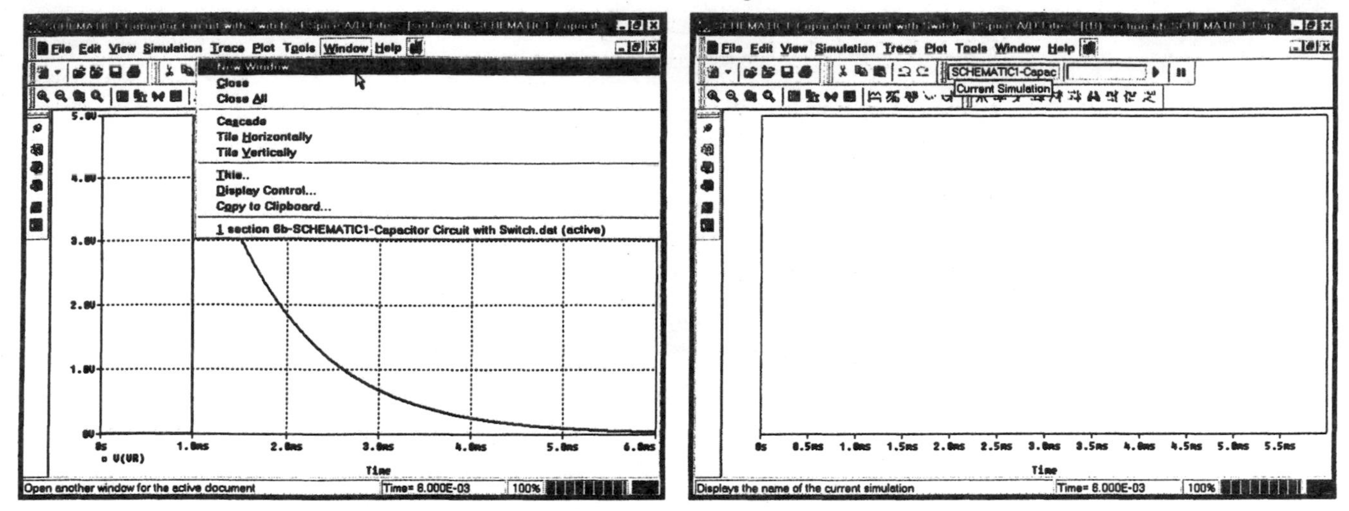

Add the capacitor voltage trace **V(VC)** by selecting **Trace** and then **Add Trace** from the Probe menu bar:

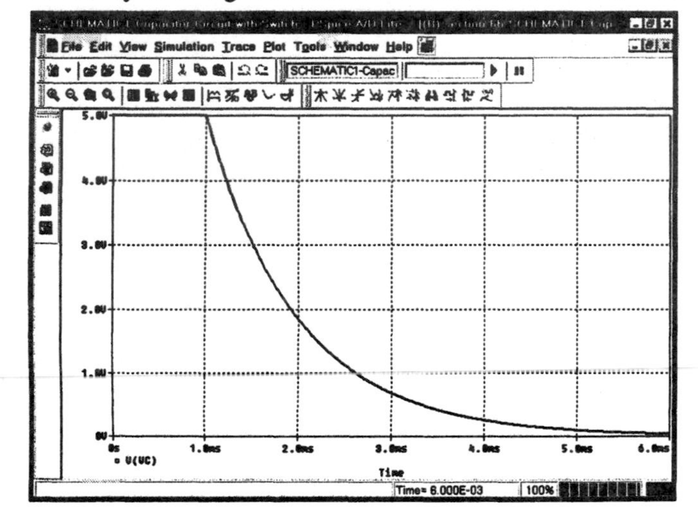

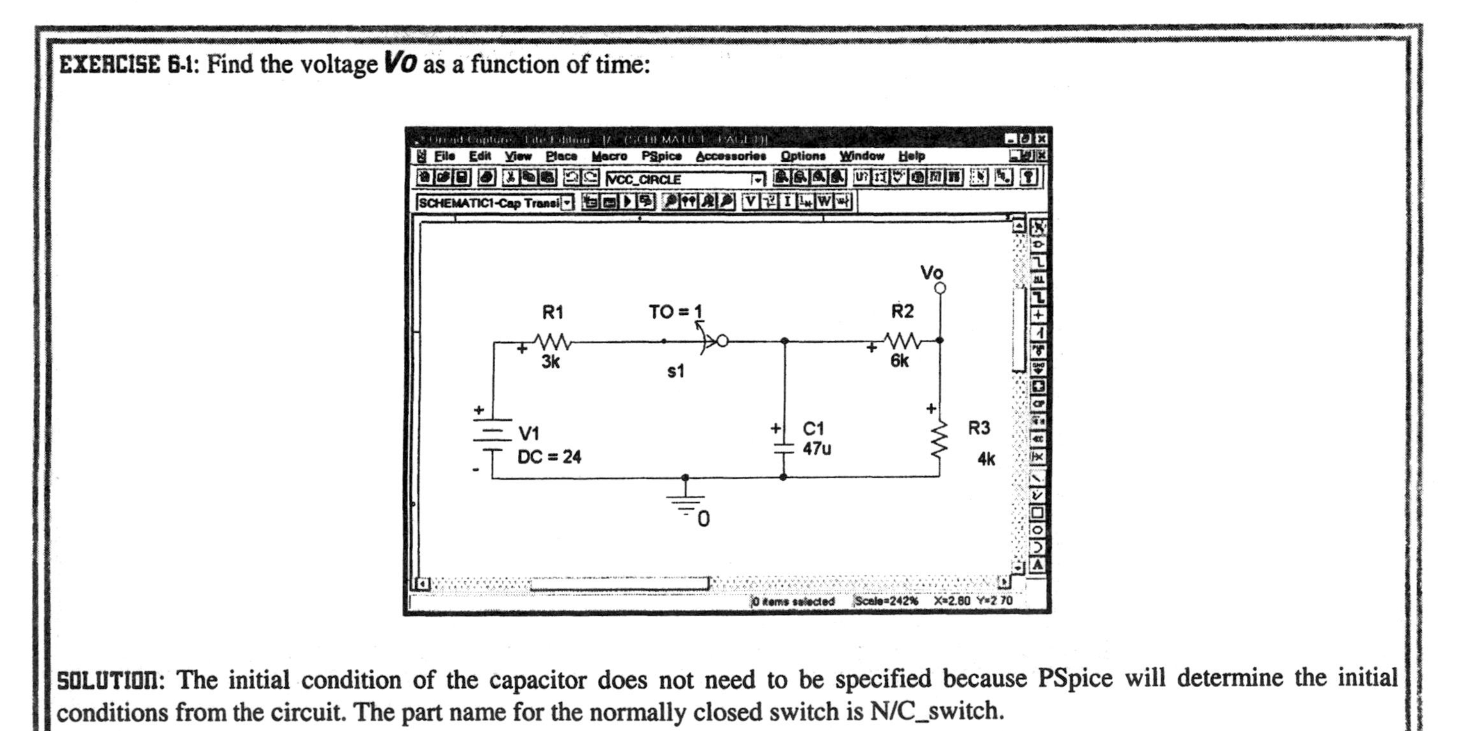


**EXERCISE 6-1**: Find the voltage ***Vo*** as a function of time:

**SOLUTION**: The initial condition of the capacitor does not need to be specified because PSpice will determine the initial conditions from the circuit. The part name for the normally closed switch is N/C_switch.

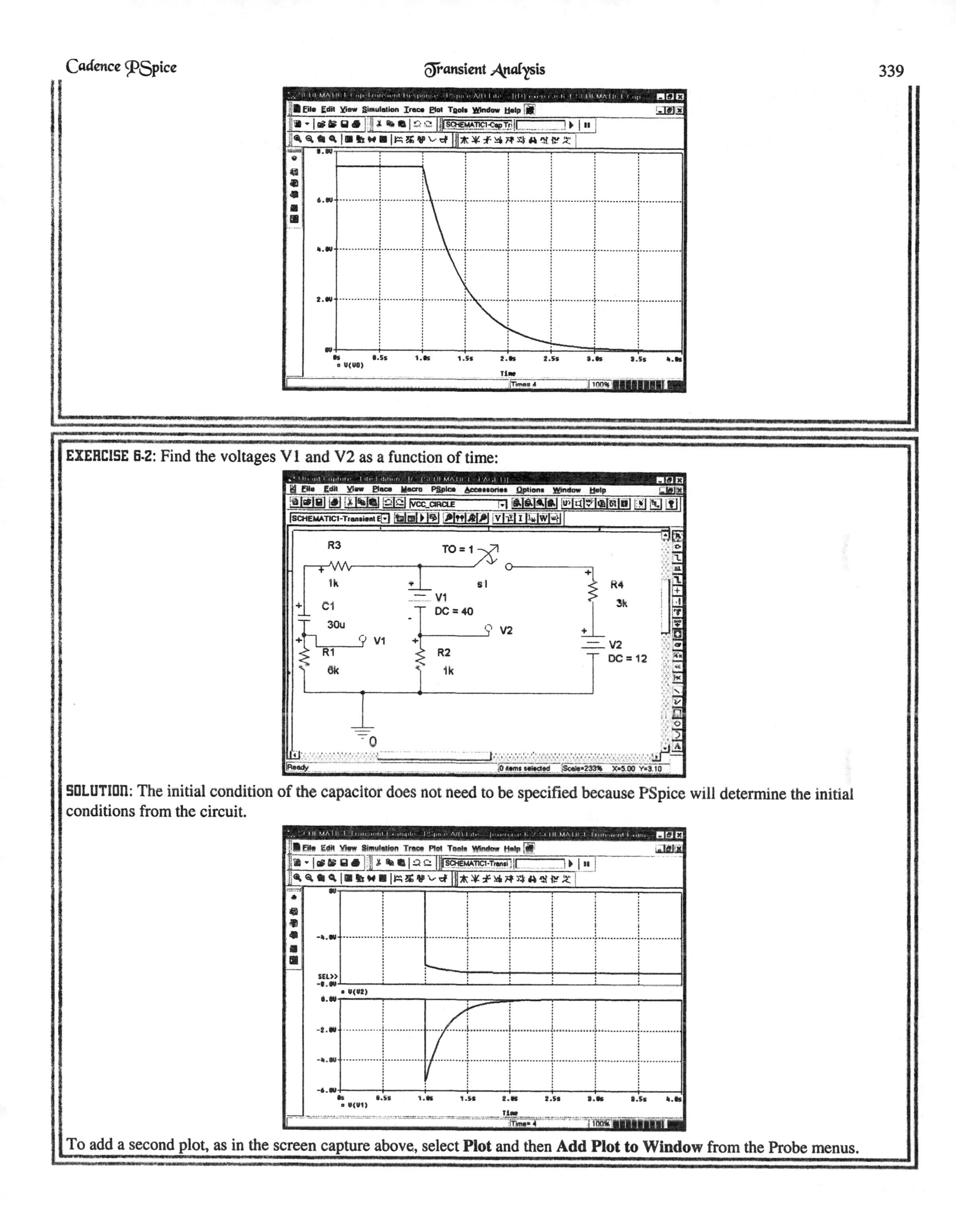


**EXERCISE 6-2**: Find the voltages V1 and V2 as a function of time:

**SOLUTION**: The initial condition of the capacitor does not need to be specified because PSpice will determine the initial conditions from the circuit.

To add a second plot, as in the screen capture above, select **Plot** and then **Add Plot to Window** from the Probe menus.

## 6.B.1. Note on Initial Conditions

Instead of specifying an initial condition on the capacitor, the initial voltages of nodes can be specified using the parts IC1 and IC2. For the circuit of Section 6.B, we will use the part IC1 to specify the initial condition of the capacitor. Get a part called IC1 and place it in your circuit:

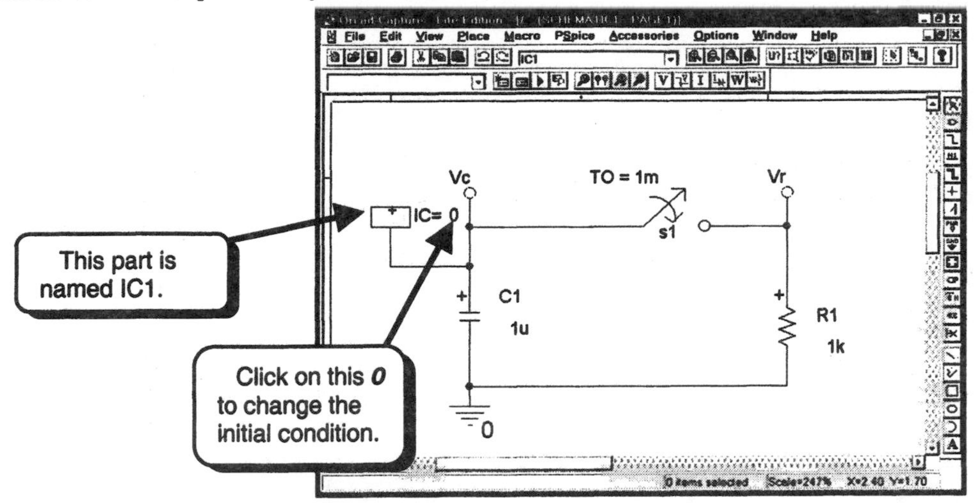


The default value of the initial condition is ***0*** V. To change the initial condition, double-click on the text ***0***. The dialog box below will open*:

Note that the dialog box is changing the attribute ***Value***. Type in the text for the initial condition, **5** in this case:

Click the ***OK*** button to accept the change. The initial condition will be changed in the circuit:

---

* Do not click on the text ***IC=***. Doing so will change the text from ***IC=*** to something else and will yield unpredictable results.

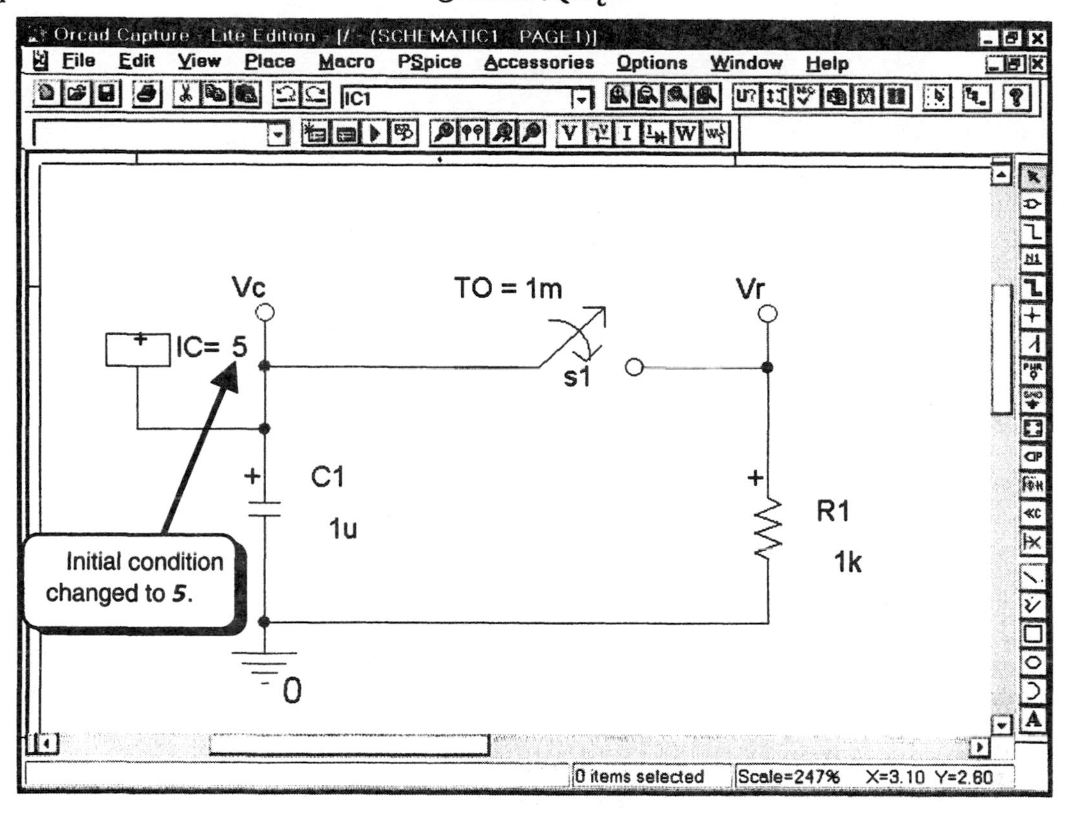


Note that the value specified is the voltage of the node relative to ground. The part IC2 allows you to specify the initial voltage between two nodes. You can use the IC2 part to specify the initial voltage of a capacitor when the capacitor does not have one of its leads grounded.

# 6.C. Capacitor Step Response

In this section we will look at the response of an RC circuit to a pulse input. We will work with the circuit below:

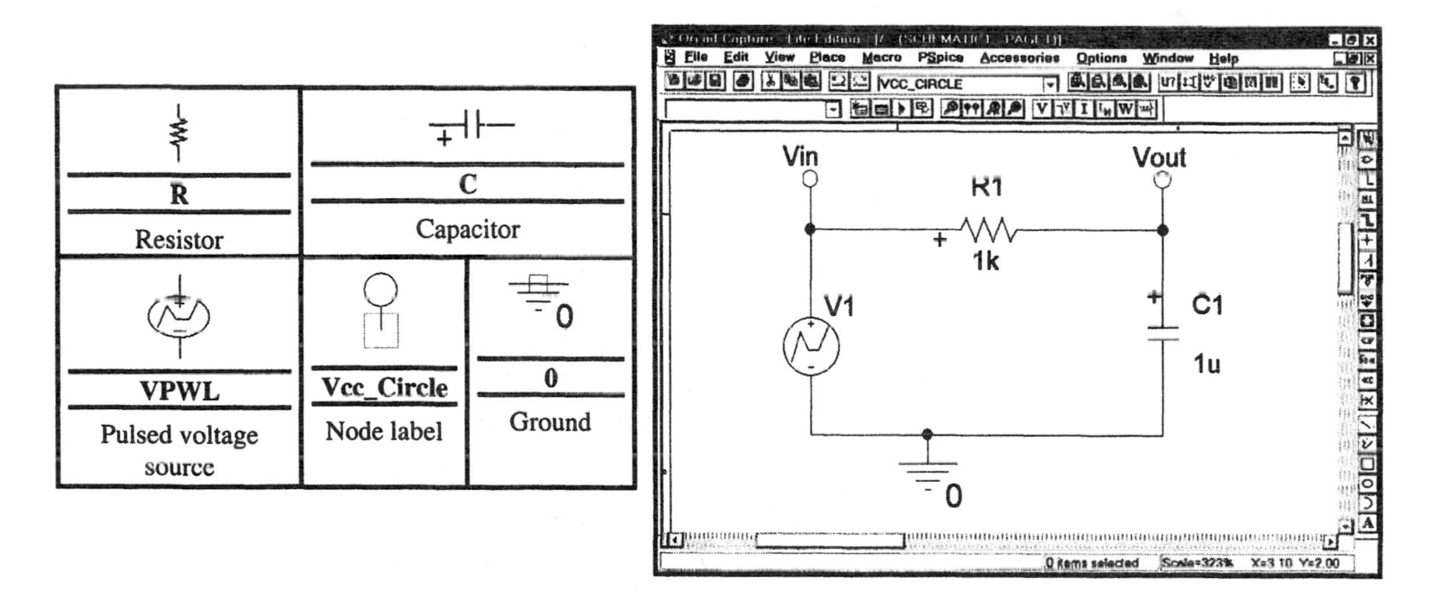


The time constant of the circuit is $\tau = RC = (1\,\text{k}\Omega)(1\,\mu\text{F}) = 1\,\text{ms}$. The response of an inductive or capacitive circuit will reach 99% of its final value in 3 time constants, and will reach steady state in about 5 time constants. We will run two simulations. The first will let the capacitor charge and discharge for 3 time constants (3 ms in this case) and the second will let the capacitor charge and discharge for 5 time constants (5 ms in this example). For the first simulation we will use the waveform below:

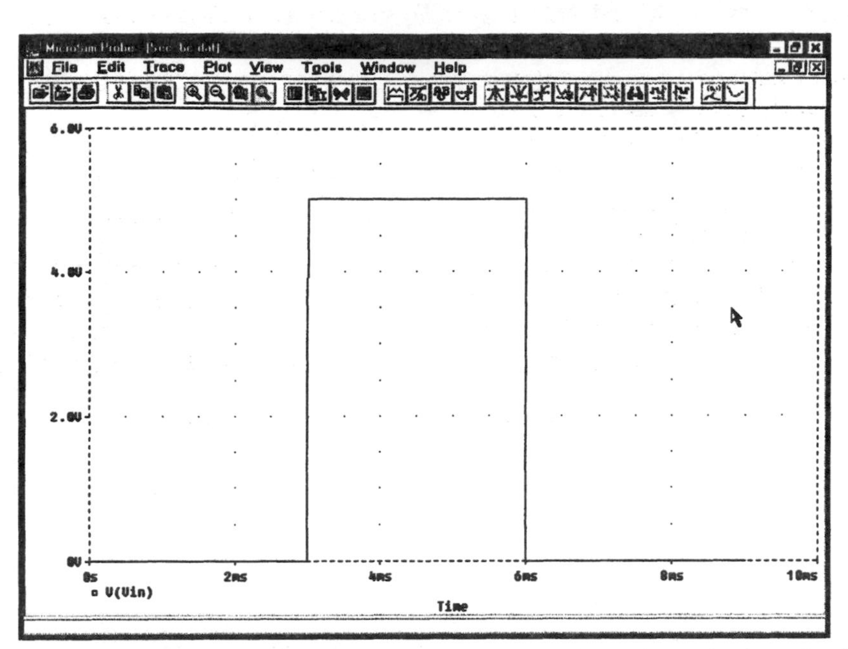


The length of the simulation is 9 ms. The pulse will start at zero volts and remain there for 3 ms, then go to 5 V for 3 ms, and then return to 0 for another 3 ms. To create this waveform we can use either a pulsed voltage source (VPULSE) or a piecewise linear voltage source (VPWL). For this example we will use the VPWL source. To edit the attributes of the PWL source, double-click the ***LEFT*** mouse button on the PWL graphic, . The dialog box below will appear:

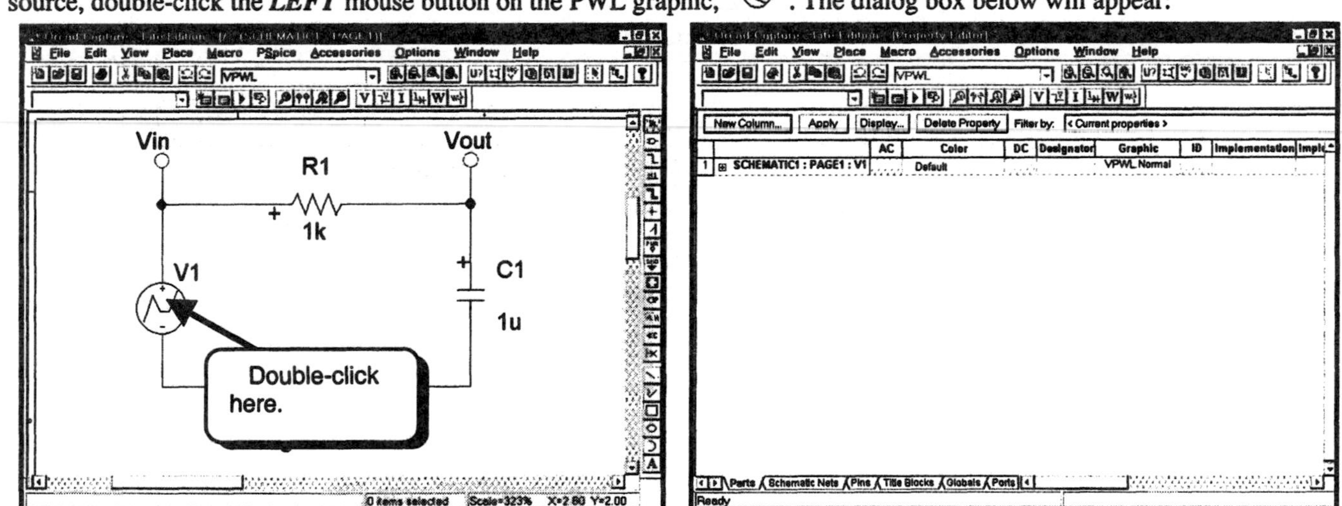


Scroll the window to the right until you see the time and voltage attributes:

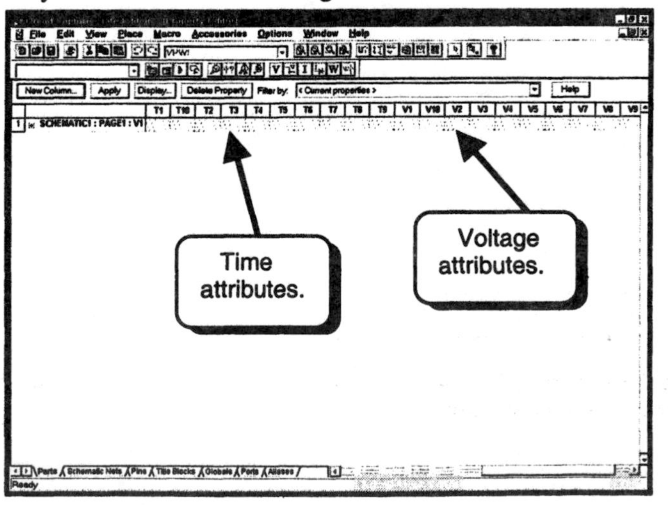

The DC and AC attributes are for the DC and AC analyses. We are running a Transient Analysis, so they will be ignored. The operation of the PWL source is the following: At t1 the voltage is v1; at t2 the voltage is v2; between t1 and t2 the voltage changes linearly from v1 to v2; at t3 the voltage is v3; between t2 and t3 the voltage changes linearly from v2 to v3; at t4 the voltage is v4; and so on. We would like to set up the pulsed waveform shown above. Set the PWL attributes as shown in Table 6-1. These settings specify that from 0 to 2.999 ms the voltage will be zero. From 2.999 ms to 3 ms the voltage changes from 0 to 5 V. From 3 ms to 5.999 ms the voltage is constant at 5 V. From 5.999 ms to 6 ms the voltage changes from 5 to 0 V. For time greater than 6 ms, the voltage will remain constant at 0 V. A plot showing the points listed in Table 6-1 and straight lines connecting those points is shown below:

**Table 6-1**

| t1 | t2 | t3 | t4 | t5 |
|---|---|---|---|---|
| 0 | 2.999m | 3m | 5.999m | 6m |
| **v1** | **v2** | **v3** | **v4** | **v5** |
| 0 | 0 | 5 | 5 | 0 |

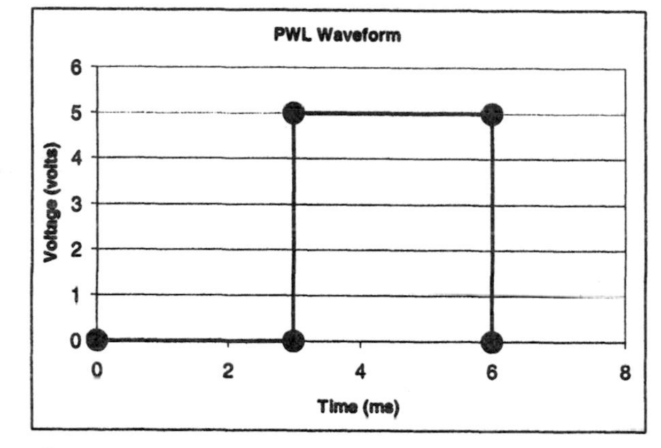


Fill in the spreadsheet with these settings:

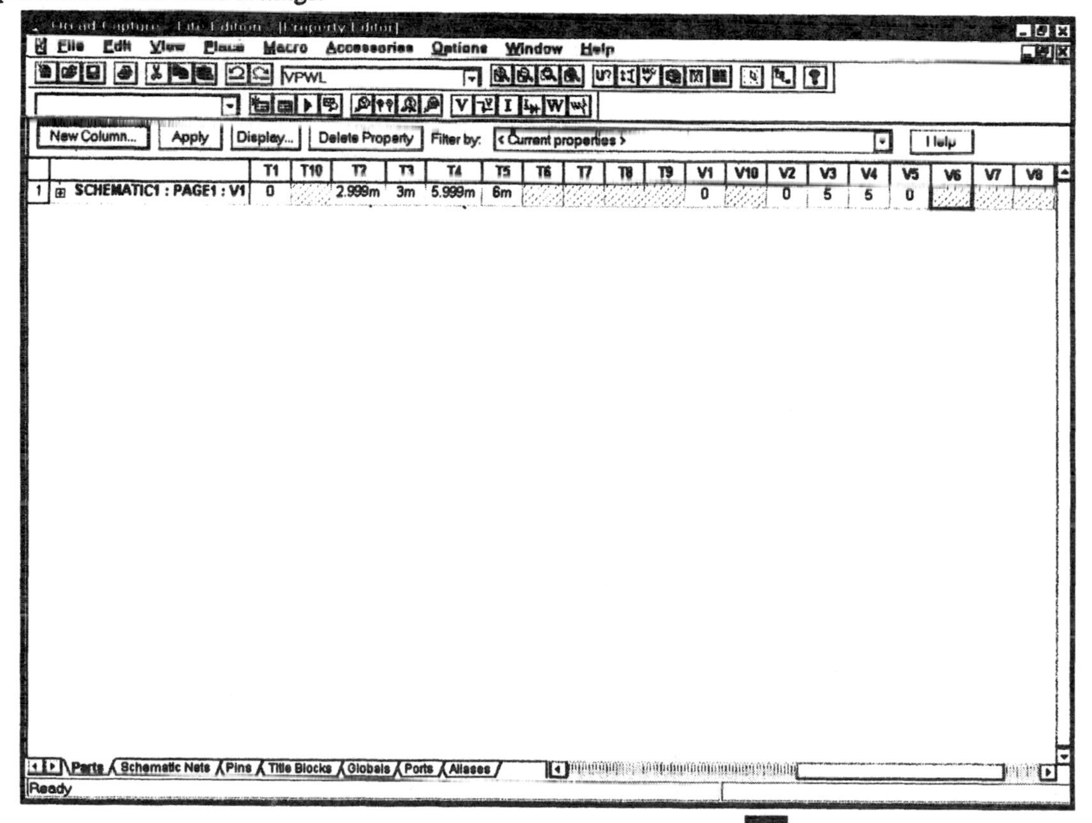

When you have made the changes to the attributes of the PWL source, click the [X] as shown below (or type **CTRL F4**) to close the spreadsheet and to return to the schematic:

Click on this [X].

Vin
Vout
R1
1k
V1
C1
1u
0

We will now set up the parameters for the transient analysis. Select **PSpice** and then **New Simulation Profile**:

New Simulation
Name:
Create
Cancel
Inherit From:
none
Root Schematic: SCHEMATIC1

Enter a name for the profile and then click the ***Create*** button:

New Simulation
Name:
Capacitor Step Response
Create
Cancel
Inherit From:
none
Root Schematic: SCHEMATIC1

General | Analysis | Include Files | Libraries | Stimulus | Options | Data Collection | Probe Window
Analysis type:
Options:
Monte Carlo/Worst Case
Parametric Sweep
Temperature (Sweep)
Save Bias Point
Load Bias Point
Run to time: 1000ns seconds (TSTOP)
Start saving data after: 0 seconds
Transient options
Maximum step size: seconds
Skip the initial transient bias point calculation (SKIPBP)
Output File Options...
OK Cancel Apply Help

The ***Time Domain (Transient)*** dialog box is selected by default. Fill in the dialog box as shown:

Simulation Settings - Capacitor Step Response
General | Analysis | Include Files | Libraries | Stimulus | Options | Data Collection | Probe Window
Analysis type:
Time Domain (Transient)
Options:
Monte Carlo/Worst Case
Parametric Sweep
Temperature (Sweep)
Save Bias Point
Load Bias Point
Run to time: 9m seconds (TSTOP)
Start saving data after: 0 seconds
Transient options
Maximum step size: seconds
Skip the initial transient bias point calculation (SKIPBP)
Output File Options...
OK Cancel Apply Help

The ***Run to time*** is set to 9 ms. This will run the simulation for 9 ms. Click the ***OK*** button to accept the settings.

Note that we did not specify initial conditions. When you do not specify initial conditions, PSpice will determine them itself. PSpice will determine the initial conditions from the state of the circuit at t = 0. In this case, the voltage of the PWL source is zero at t = 0. The initial conditions will be determined with this source set to zero. Since this is the only source in the circuit, all sources are zero at t = 0, so the initial condition of the capacitor must be zero at t = 0.

Run the simulation (select **PSpice** and then **Run** or press the **F11** key). When the simulation is complete, Probe will automatically run. Add the traces `V(Vin)` and `V(VOUT)` :

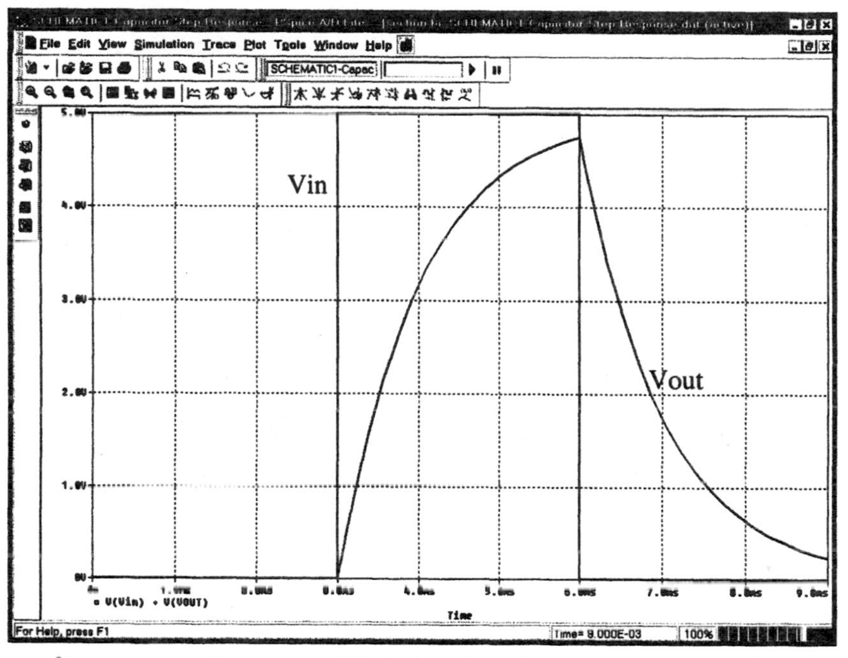


We see that the capacitor voltage never quite reaches 5 V during the pulse, and never quite discharges to zero after the pulse returns to zero. This is because we let the capacitor charge and discharge for only 3 time constants.

**EXERCISE 6-3**: For the circuit in the previous example, allow the capacitor to charge and discharge for 5 ms rather than 3 ms.

**SOLUTION**: Change the PWL source to lengthen the amount of time the pulse is at 5 V:

| | | T1 | T10 | T2 | T3 | T4 | T5 | T6 | T7 | T8 | T9 | V1 | V10 | V2 | V3 | V4 | V5 | V6 | V7 | V8 | V9 | Value |
|---|---|---|---|---|---|---|---|---|---|---|---|---|---|---|---|---|---|---|---|---|---|---|
| 1 | SCHEMATIC1 : PAGE1 : V1 | 0 | | 2.999m | 3m | 7.999m | 8m | | | | | 0 | | 0 | 5 | 5 | 0 | | | | | VPWL |

To allow the capacitor to discharge for 5 ms, we must allow the Transient Analysis to run for 5 ms after the pulse returns to zero. Fill in the ***Transient*** dialog box as shown below:

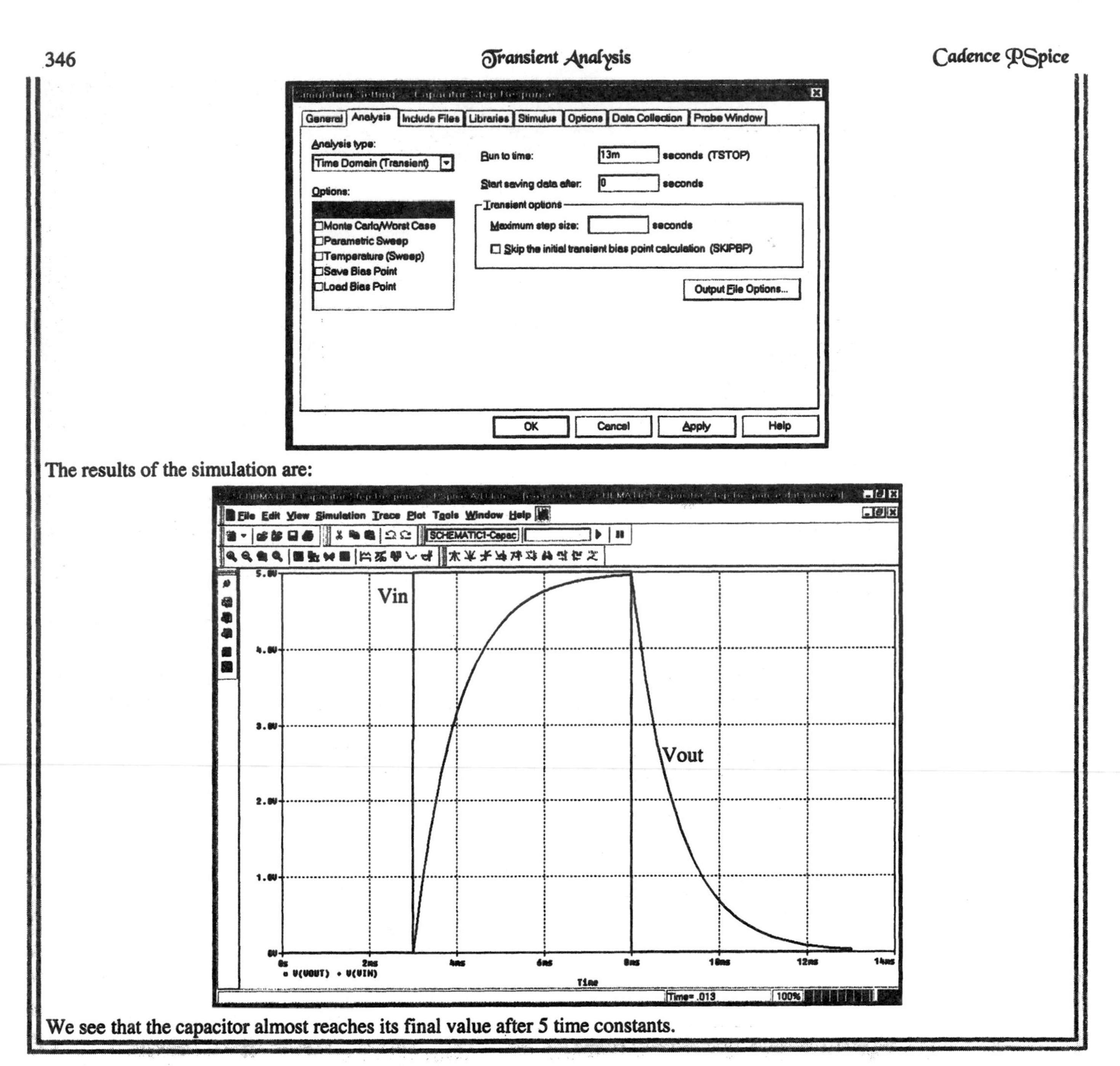


The results of the simulation are:

We see that the capacitor almost reaches its final value after 5 time constants.

# 6.D. Inductor Transient Response

In this section we will demonstrate the transient response of an inductor circuit with a switch that is normally closed. The initial condition of the inductor will not be specified by an "IC=" line in the circuit. Instead, the initial condition will be determined by PSpice from the initial state of the circuit before the switch changes position. If you wish to specify the initial condition of the inductor, it is specified in the same way as the initial condition of a capacitor. For an inductor, the direction for positive current is into the dotted terminal, as shown in Figure 6.1. The dot is always shown on the inductor graphic. The graphic should be rotated to obtain the desired direction of positive current flow.

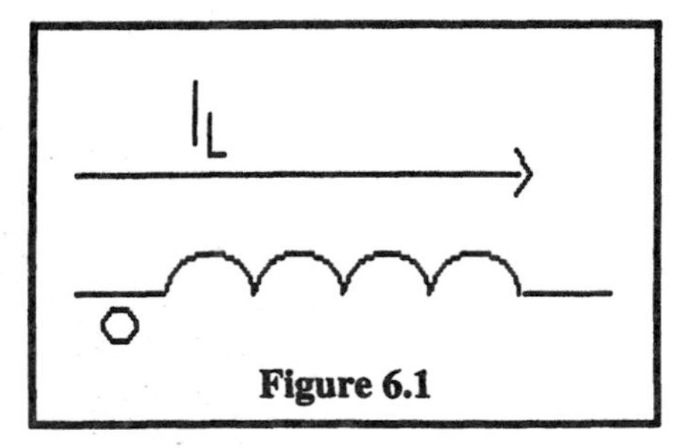

Figure 6.1

We will run the circuit below:

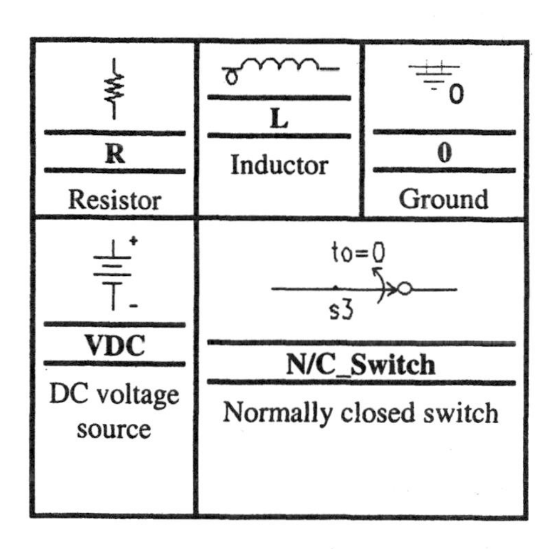

R
Resistor
L
Inductor
0
0
Ground
VDC
DC voltage source
to=0
s3
N/C_Switch
Normally closed switch


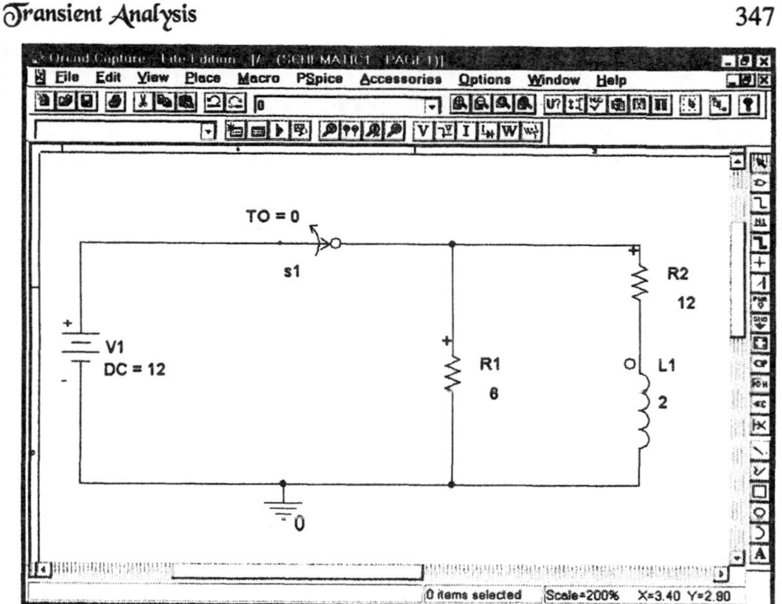


For this circuit, the initial current through the inductor is 1 A. The current will decay to zero with a time constant of $\tau = \dfrac{L}{R_1 + R_2} = \dfrac{2\,\mathrm{H}}{6\Omega + 12\Omega} = 111\ \mathrm{ms}$. We will run this circuit for approximately five time constants (550 ms). Note for this circuit that the switch is normally closed, and opens at t = 0 ms. It is important that the dot in your schematic match the dot in the schematic above. If the dots do not match, your results will be –1 times the results presented here.

We now need to set up a Transient Analysis. Select **PSpice** and then **New Simulation Profile** from the Capture menus and then enter a name for the profile and click the ***Create*** button. By default the ***Time Domain (Transient) Analysis type*** is selected:

General | Analysis | Include Files | Libraries | Stimulus | Options | Data Collection | Probe Window

Analysis type:

Run to time: 1000ns seconds (TSTOP)

Start saving data after: 0 seconds

Transient options

Maximum step size: seconds

Skip the initial transient bias point calculation (SKIPBP)

Monte Carlo/Worst Case
Parametric Sweep
Temperature (Sweep)
Save Bias Point
Load Bias Point

Output File Options...

OK Cancel Apply Help

We would like to run the transient simulation for 550 ms, so the ***Run to time*** should be set to **550m**. I would like to see at least 550 points in the simulation, so I will set the ***Maximum step size*** to **1m**. Fill in the dialog box as shown:

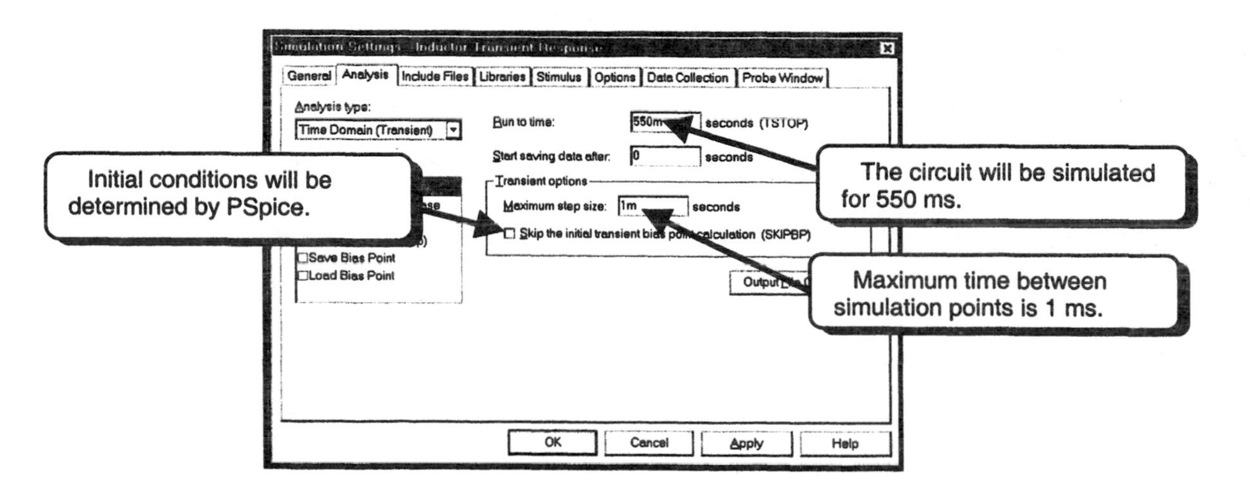

Click the ***OK*** button to accept the settings and return to the schematic.

To run the simulation, select **PSpice** and then **Run** from the Capture menu bar. When Probe runs, add the trace `I(L1)`. To add a trace select **Trace** and then **Add Trace** from the Probe menu bar or press the **INSERT** key. You will see the following Probe window:

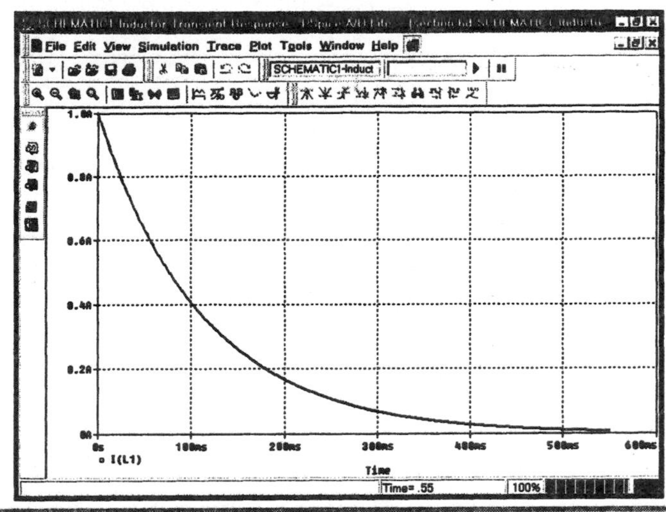


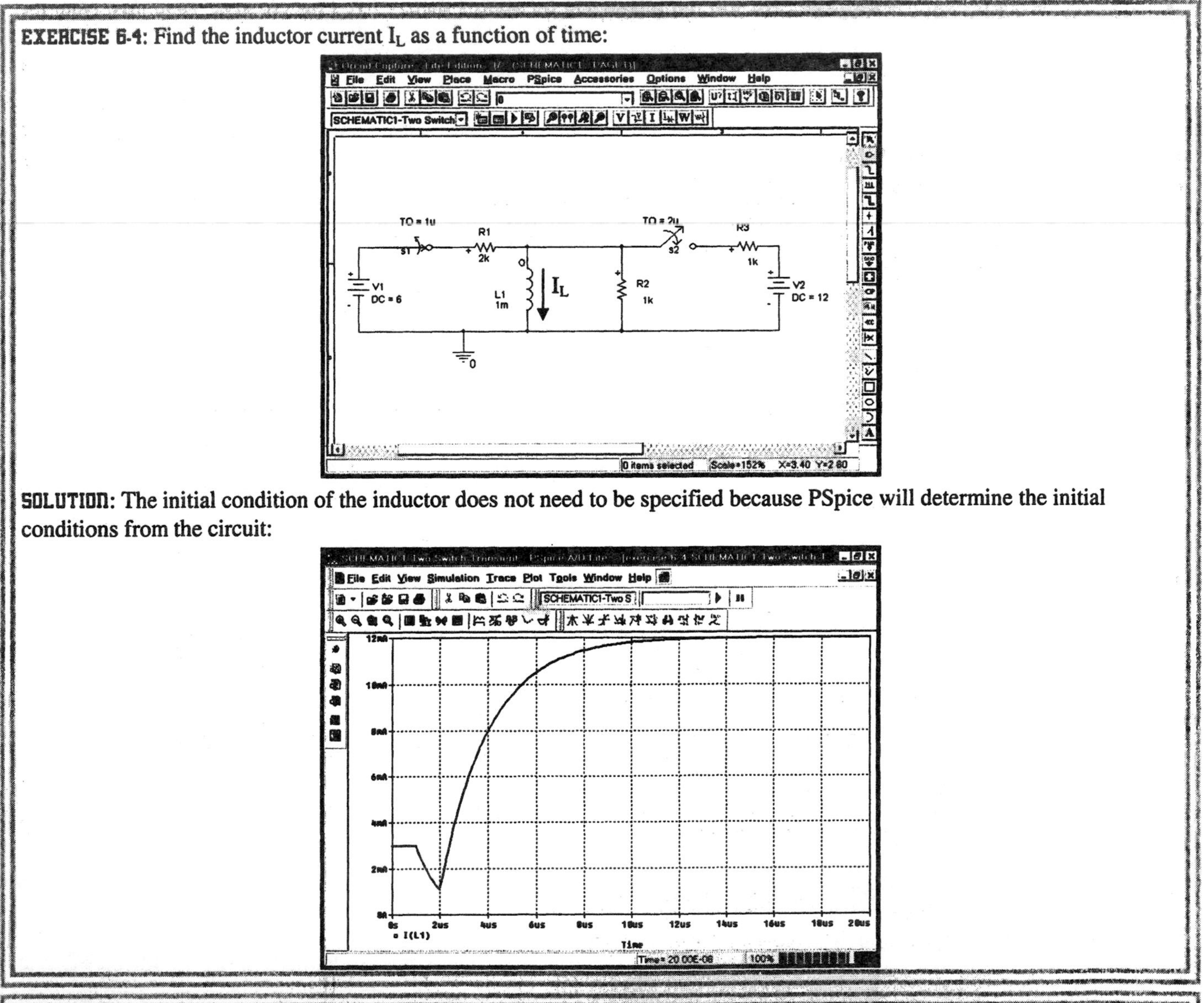


**EXERCISE 6-4**: Find the inductor current $I_L$ as a function of time:

**SOLUTION**: The initial condition of the inductor does not need to be specified because PSpice will determine the initial conditions from the circuit:

**EXERCISE 6-5**: Find the voltages V1 and V2 as a function of time:

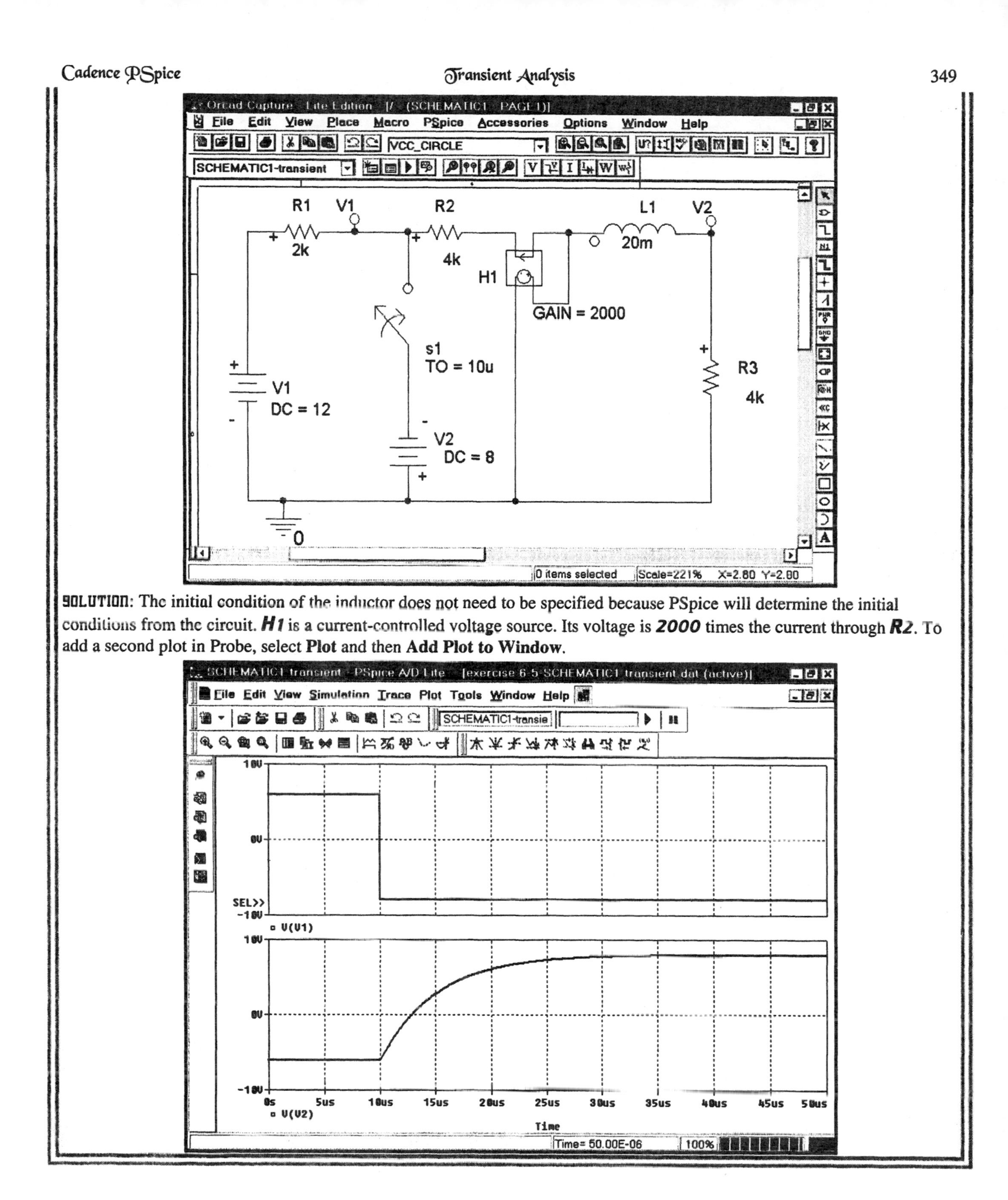


**SOLUTION**: The initial condition of the inductor does not need to be specified because PSpice will determine the initial conditions from the circuit. ***H1*** is a current-controlled voltage source. Its voltage is ***2000*** times the current through ***R2***. To add a second plot in Probe, select **Plot** and then **Add Plot to Window**.



# 6.E. Regulated DC Power Supply

Wire the following circuit. The parts are listed in **Table 6-2** on page 351.

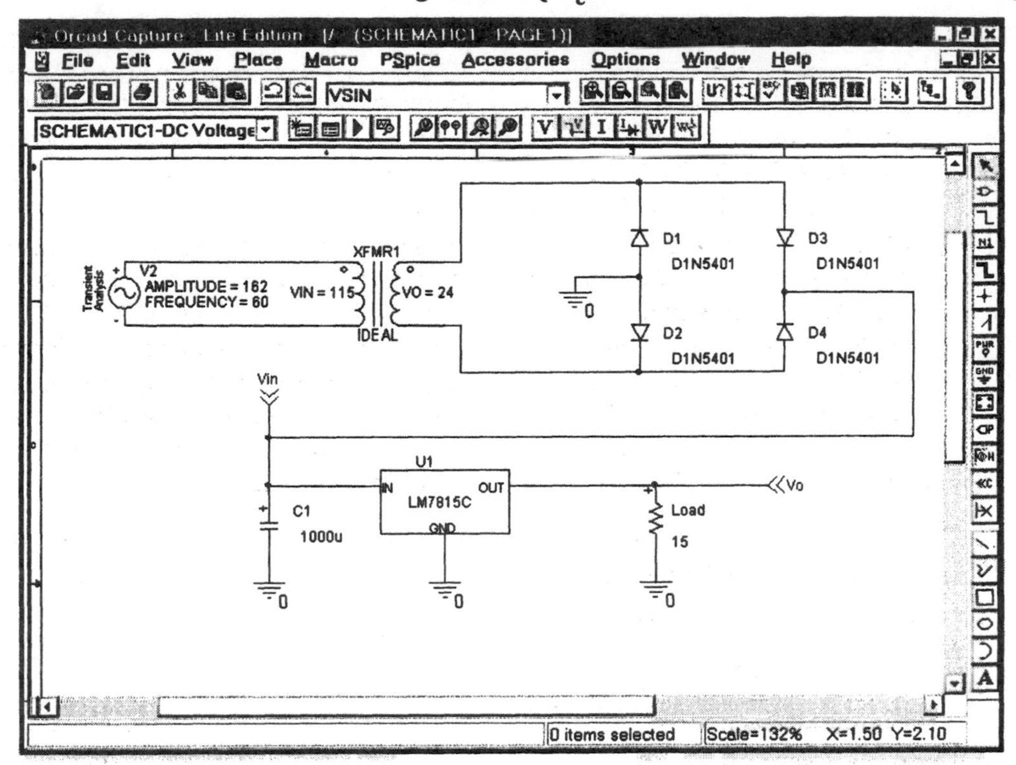


The Transient Analysis uses the sinusoidal voltage waveform when it performs the simulation. This circuit would not work if an AC Sweep were used. To set up a Transient Analysis, select **PSpice** and then **New Simulation Profile** from the Capture menus:

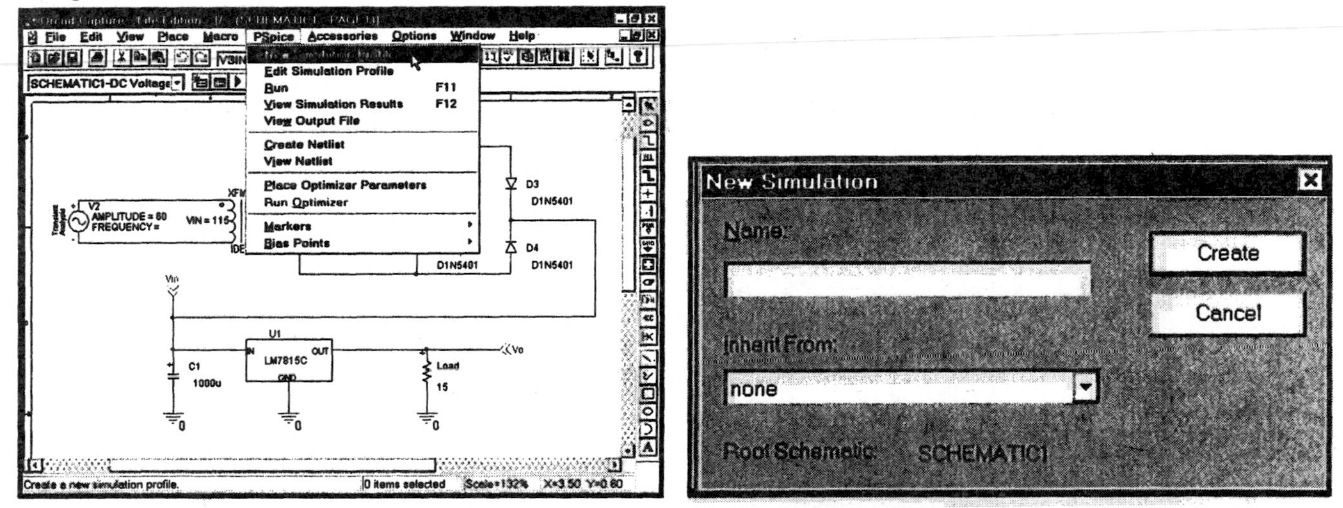


Enter a name for the profile and click the ***Create*** button. By default the ***Time Domain (Transient) Analysis type*** is selected.

The ***Run to time*** is the total time of the simulation, which runs from time equals zero to the ***Run to time***. The ***Start saving data after*** value is used for long simulations. If you ran a simulation for 1 second and you were only interested in the data from 990 ms to 1 s, then you would set the ***Start saving data after*** value to 990 ms. PSpice then would not save any data from time zero to 990 ms. This parameter is used to reduce the size of the output data file for large circuits with long simulation times. The ***Start saving data after*** is also used to allow circuits to reach steady state before data is collected. We will ignore this argument since this is a short run.

The ***Maximum step size*** value is important. When PSpice runs a circuit it chooses the time between simulation points to be as large as possible while keeping the error below a specified maximum. The larger the time step, the less time the simulation takes to complete. To keep simulation time to a minimum, leave the ***Maximum step size*** value blank. However, with the ***Maximum step size*** unspecified, the step size PSpice takes may be too large for you. If this is the case, specify a value for ***Maximum step size***. ***Maximum step size*** is the longest time between simulation points PSpice will take. A smaller value of step size will give you more points in your simulation, but will take more time. In our simulation, the voltage source is a 60 Hz sinusoid. Each cycle of the sinusoid takes 16 ms. I would like to see a minimum of 20 points per cycle, so I will choose the value for ***Maximum step size*** to be 0.8 ms.

**Table 6-2: Part list for circuit on page 350.**

| | | | |
|---|---|---|---|
| **Offpage-L**<br>Node label | XFMR3 Vin=115 Vo=12 IDEAL<br>**Ideal_XFMR_Vo/Vin**<br>Ideal transformer | **R**<br>Resistor | D1 d1n5401<br>**D1N5401**<br>Rectifier diode |
| 0<br>**0**<br>Ground | V3 AMPLITUDE = FREQUENCY = Transient Analysis<br>**VSIN**<br>Sinusoidal voltage source | **C**<br>Capacitor | IN OUT LM7815C U2 GND<br>**LM7815C**<br>+15 volt regulator |

Fill out the dialog box as shown below. Since each cycle takes 16 ms, I will run the simulation for 10 cycles or 160 ms.

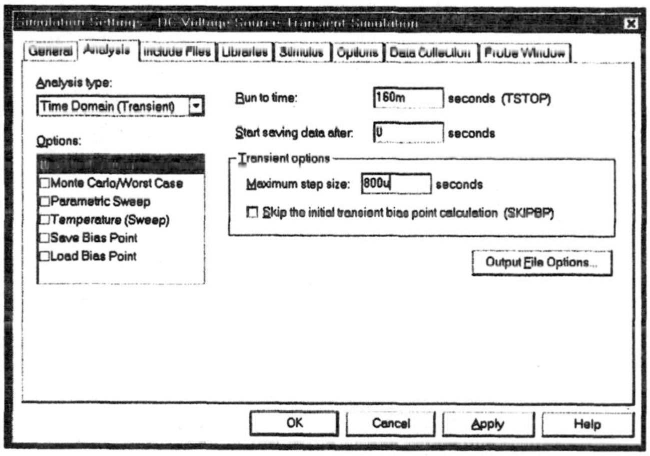

Simulation Settings - DC Voltage Source Transient Simulation

General | Analysis | Include Files | Libraries | Stimulus | Options | Data Collection | Probe Window

Analysis type: Time Domain (Transient)

Run to time: 160m seconds (TSTOP)

Start saving data after: 0 seconds

Options:
- Monte Carlo/Worst Case
- Parametric Sweep
- Temperature (Sweep)
- Save Bias Point
- Load Bias Point

Transient options
- Maximum step size: 800u seconds
- Skip the initial transient bias point calculation (SKIPBP)

Output File Options...

OK | Cancel | Apply | Help

Click the ***OK*** button to accept the settings to return to the schematic. Run PSpice (**PSpice, Run**).

When the simulation is finished, Probe will run. When the Probe window opens, notice that the x-axis is now the time axis. Add traces **V(VIN)** and **V(VO)** to observe the input and output of the regulator. To add a trace select **Trace** and then **Add Trace** from the Probe menus or press the **INSERT** key. You will see the following plot:

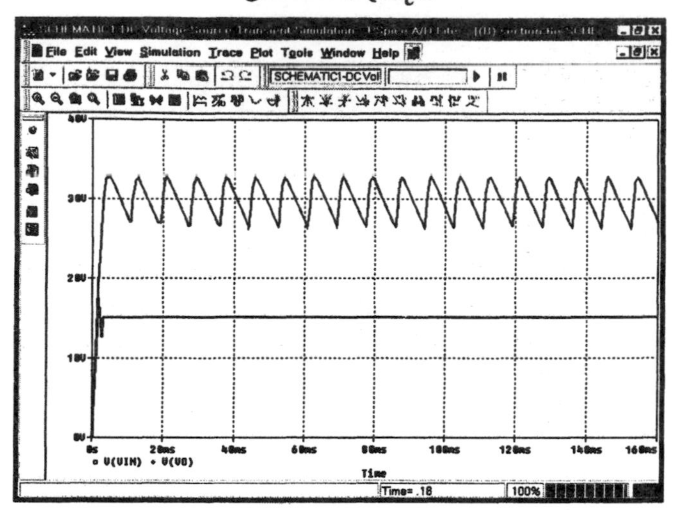

To see what happens at the beginning of the trace we can change the scale of the x-axis. To change the scale, select **Plot** and then **Axis Settings**. By default, the ***X Axis*** tab is selected:

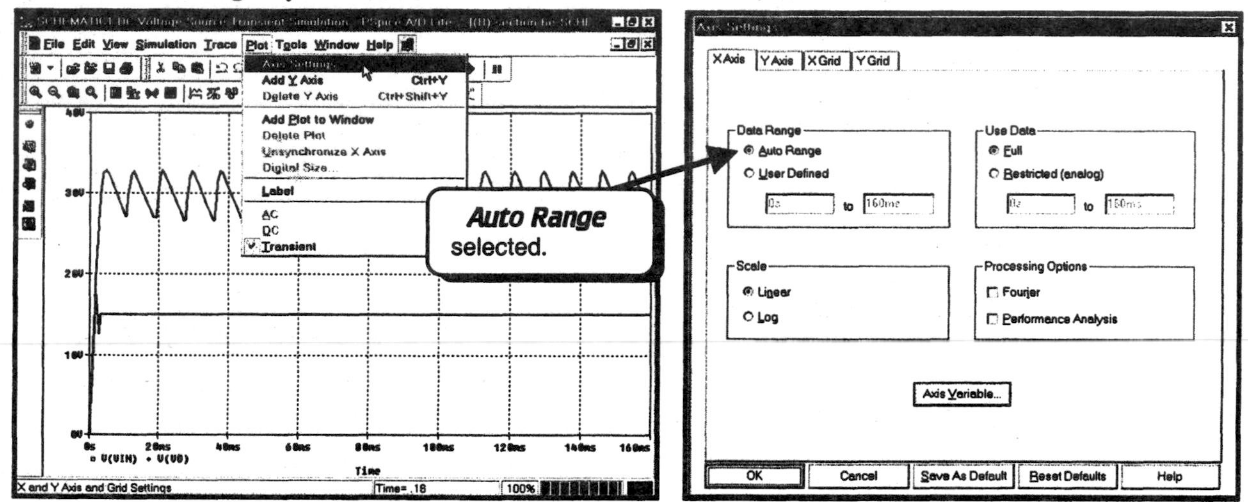


The ***Data Range*** portion of this dialog box allows you to change the x-axis range. Presently, the ***Auto Range*** option is selected, which means that Probe will determine the x-axis range. We would like to specify the x-axis range, so we must select the ***User Defined*** option and specify the range. We would like to set the x-axis to display the first 10 ms of the trace. Fill in the dialog box as shown:

Click the ***OK*** button to accept the range:

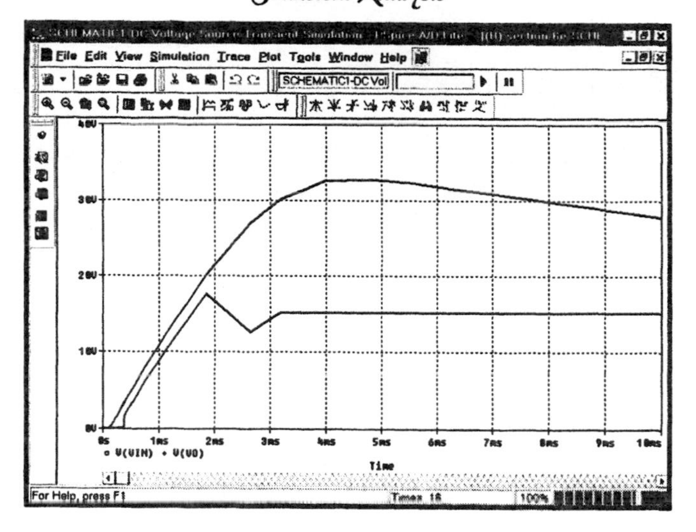


This plot shows us the effect of the ***Maximum step size*** in the Transient Analysis. Instead of seeing curved lines, we see straight lines connecting simulation points. If we wish to see a nicer curve for this time scale, we need to reduce the ***Maximum step size***. If we change the ***Maximum step size*** in the Transient Analysis dialog box shown on page 351 to 8 μs rather than 0.8 ms, the simulation will contain 100 times as many points. This will give us much more resolution in the plots. Since we have reduced the step size by a factor of 100, the simulation time will increase. We are interested in more detail for only the first 10 ms, so we will run the simulation for only 10 ms. To obtain the Transient setup dialog box, select **PSpice** and then **Edit Simulation Profile** from the Capture menus. The modifications to the Transient Analysis dialog box are shown below:

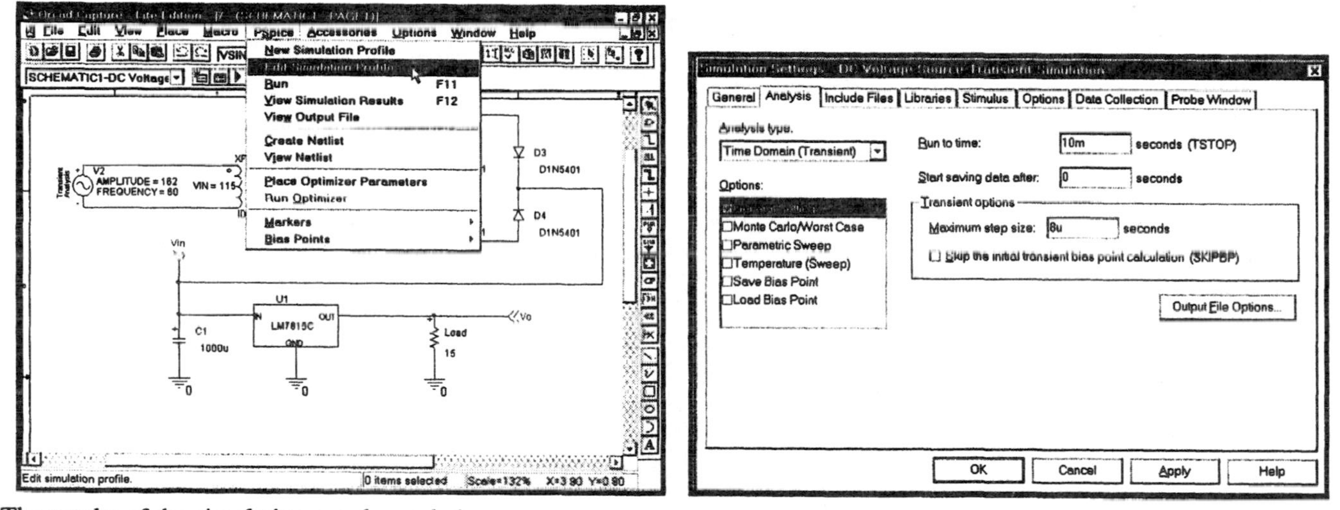


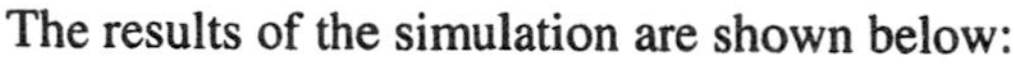
The results of the simulation are shown below:

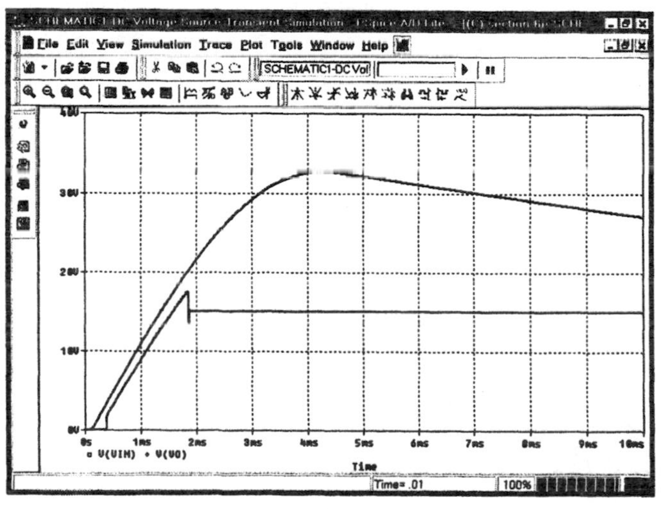

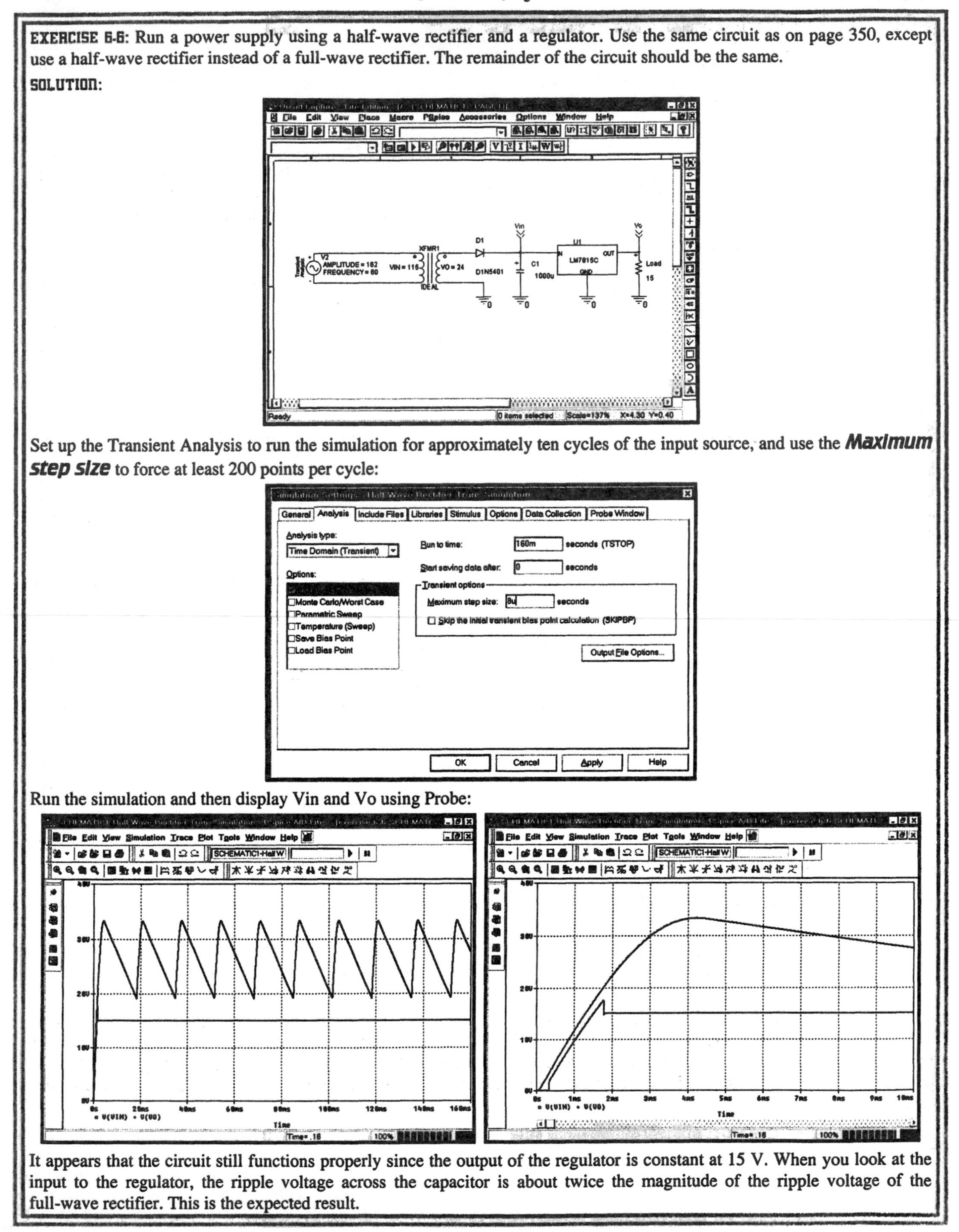


**EXERCISE 6-6**: Run a power supply using a half-wave rectifier and a regulator. Use the same circuit as on page 350, except use a half-wave rectifier instead of a full-wave rectifier. The remainder of the circuit should be the same.

**SOLUTION:**

Set up the Transient Analysis to run the simulation for approximately ten cycles of the input source, and use the ***Maximum step size*** to force at least 200 points per cycle:

Run the simulation and then display Vin and Vo using Probe:

It appears that the circuit still functions properly since the output of the regulator is constant at 15 V. When you look at the input to the regulator, the ripple voltage across the capacitor is about twice the magnitude of the ripple voltage of the full-wave rectifier. This is the expected result.

# 6.F. Zener Clipping Circuit

In this section we will demonstrate the use of a Zener diode, as well as the piece-wise linear (PWL) waveform voltage source. The PWL source can be used to create an arbitrary voltage waveform that is connected by straight lines between voltage points. Wire the circuit shown:

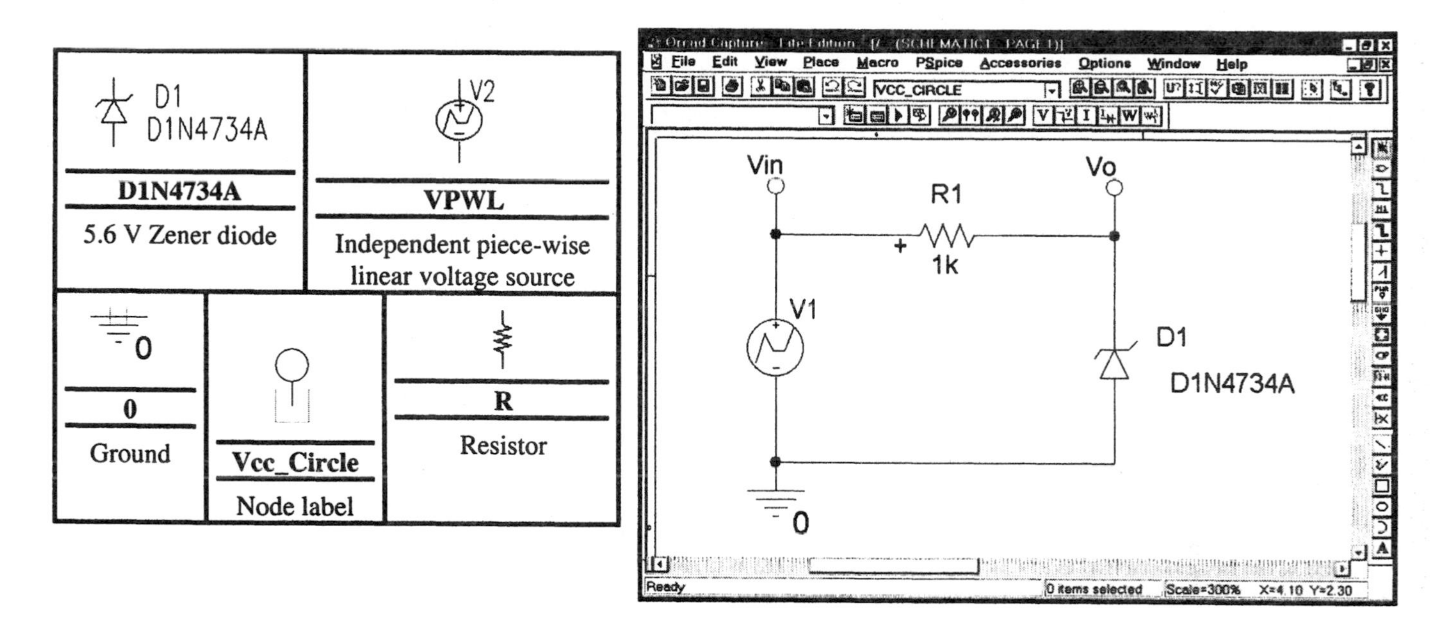


To edit the attributes of the PWL source, double-click the ***LEFT*** mouse button on the PWL graphic, to obtain the spreadsheet for the source:

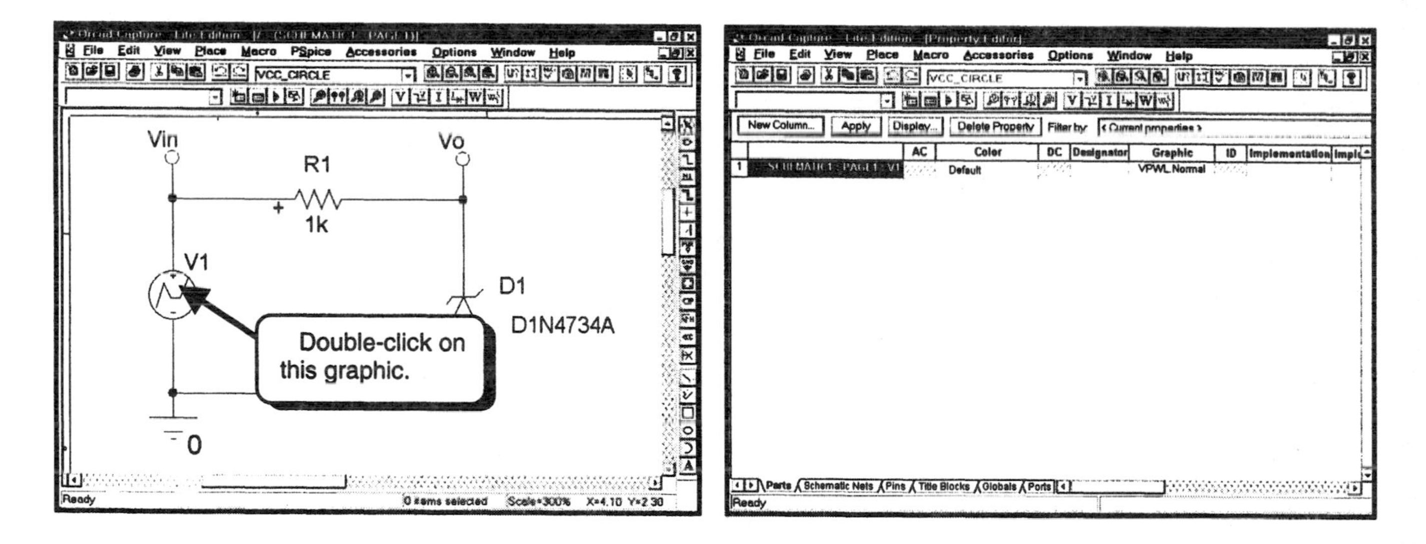


The DC and AC attributes are for the DC and AC analyses. We are running a Transient Analysis, so they will be ignored. The operation of the PWL source is the following: At t1 the voltage is v1; at t2 the voltage is v2; between t1 and t2 the voltage changes linearly from v1 to v2; at t3 the voltage is v3; between t2 and t3 the voltage changes linearly from v2 to v3; at t4 the voltage is v4; and so on. If you scroll the spreadsheet to the right, you will see the time and voltage attributes for the source:

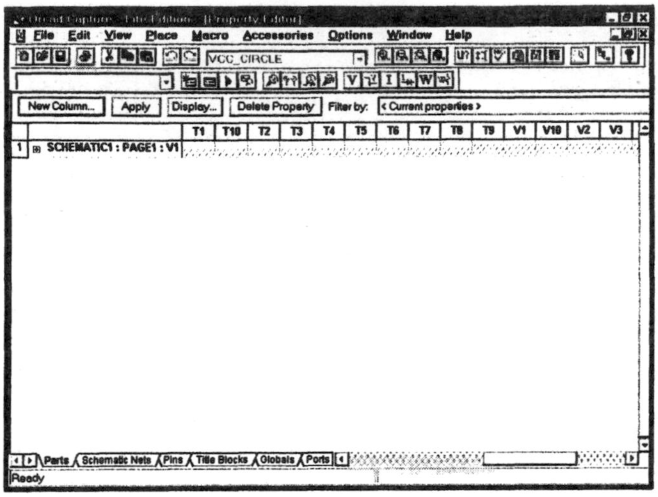

We would like to set up a ±15 V triangle wave. A plot showing the points listed in Table 6-3 and straight lines connecting those points is shown below:

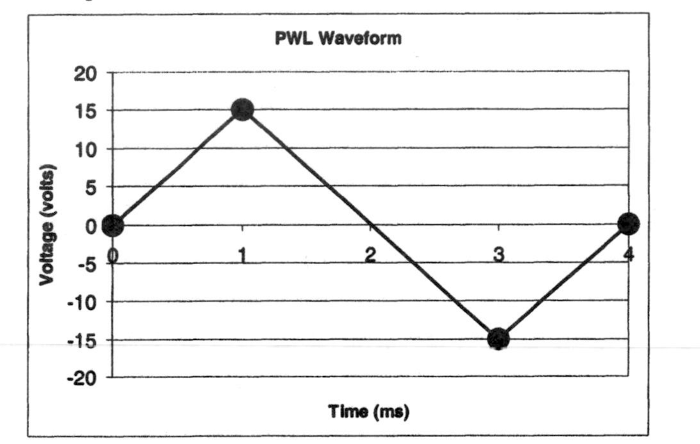


| Table 6-3 | | | |
|---|---|---|---|
| t1 | t2 | t3 | t4 |
| 0 | 1m | 3m | 4m |
| v1 | v2 | v3 | v4 |
| 0 | 15 | -15 | 0 |

This is the type of waveform that we can specify with the PWL voltage source. Set up the time and voltage attributes as specified in Table 6-3:

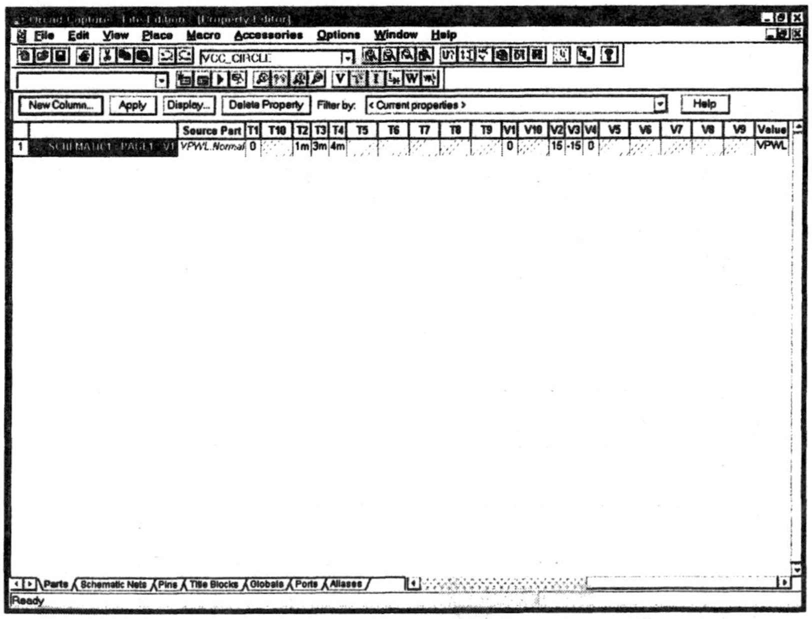

Click on the lower in the upper right corner of the window as shown below (or type **CTRL F4**) to close the spreadsheet and return to the schematic:

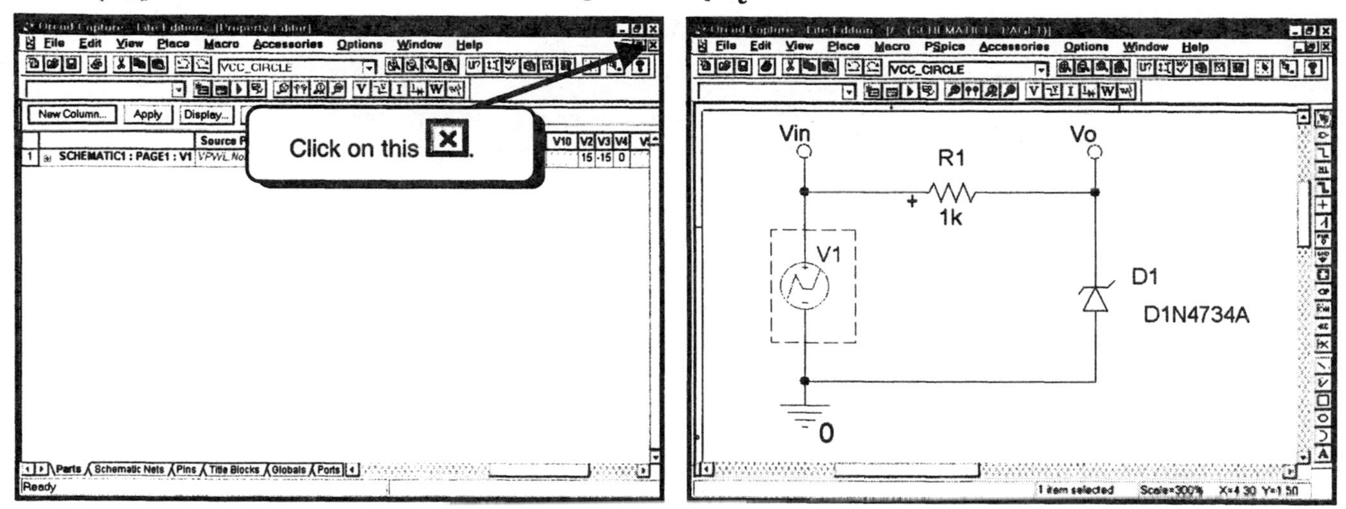


We would like to run a Transient Analysis for 4 ms. To set up the analysis select **PSpice** and then **New Simulation Profile** from the Capture menus, enter a name for the profile, and then click the ***Create*** button. By default the ***Time Domain (Transient) Analysis type*** is selected. Fill in the dialog box as shown below:

Simulation Settings - Zener Clipping Circuit

General | Analysis | Include Files | Libraries | Stimulus | Options | Data Collection | Probe Window

Analysis type: Time Domain (Transient)

Run to time: 4m seconds (TSTOP)

Start saving data after: 0 seconds

Options:
- Monte Carlo/Worst Case
- Parametric Sweep
- Temperature (Sweep)
- Save Bias Point
- Load Bias Point

Transient options
- Maximum step size: 10u seconds
- Skip the initial transient bias point calculation (SKIPBP)

Output File Options...

OK | Cancel | Apply | Help

The analysis is set up to run the circuit for 4 ms with a maximum time between simulation points of .01 ms. Thus, this simulation will have a minimum of 400 points (Run to time / Maximum step size = 4 ms/0.01 ms = 400). Click the ***OK*** button to return to the schematic.

Run the simulation: select **PSpice** and then **Run**. When the simulation is finished add the trace **V(VIN)** (select **Trace** and then **Add Trace**). You will see the triangle wave:

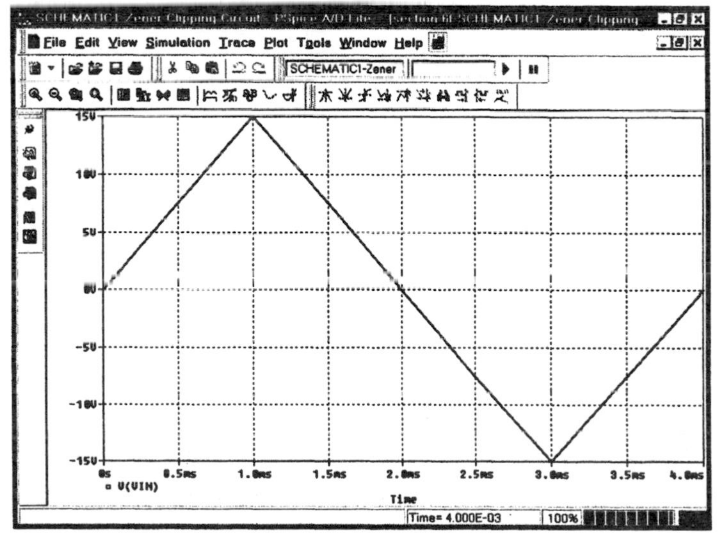


To view the output, add the trace **V(VO)**:

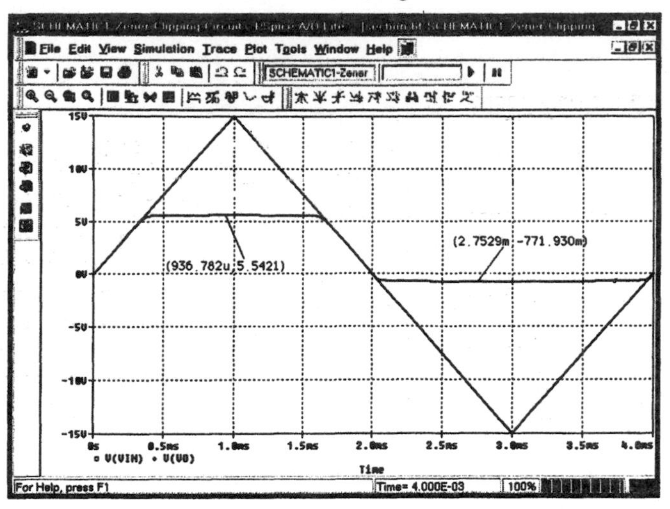


Note that the output of the circuit is limited to a maximum of about 5.5 V and a minimum of about –0.78 V.

## 6.F.1. Plotting Transfer Curves

When you have a clipping circuit, as in the last example, a transfer curve may be desired. This can easily be done by changing the x-axis. A transfer curve usually plots $V_o$ versus $V_{in}$, so we need to change the x-axis from ***Time*** to ***V(VIN)***. We must first delete the trace ***V(VIN)*** from the plot above. Click the ***LEFT*** mouse button on the text ***V(VIN)***. The text will appear in red, indicating that it has been selected. Once the trace is selected, press the **DELETE** key to delete the trace. You can delete the other text items and lines on the screen using the same method. Click the ***LEFT*** mouse button on an item to select it and then press the **DELETE** key to delete it. You should have a screen with a single trace:

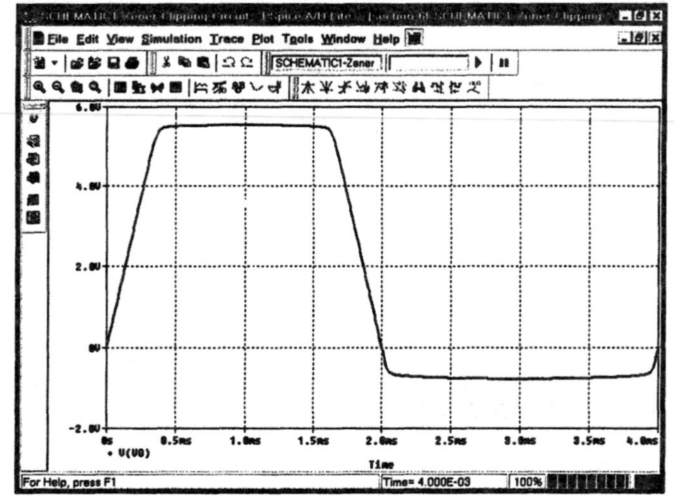


We will now change the x-axis. Select **Plot** and then **Axis Settings** from the Probe menus. By default, the dialog box displays the settings for the x-axis:

Axis Settings

X Axis | Y Axis | X Grid | Y Grid

Data Range: Auto Range; User Defined; 0s to 4.0ms

Use Data: Full; Restricted (analog); 0s to 4ms

Scale: Linear; Log

Processing Options: Fourier; Performance Analysis

Axis Variable...

OK | Cancel | Save As Default | Reset Defaults | Help

Click the ***Axis Variable*** button:

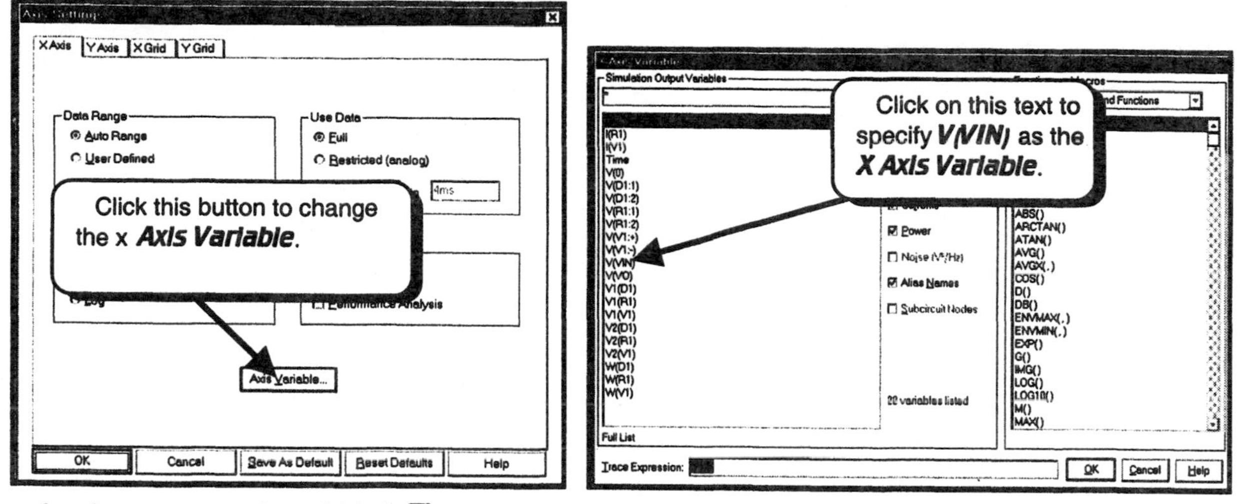


We see that the present x-axis variable is ***Time***. We would like to change the x-axis variable to V(Vin) so click the ***LEFT*** mouse button on the text ***V(VIN)***:

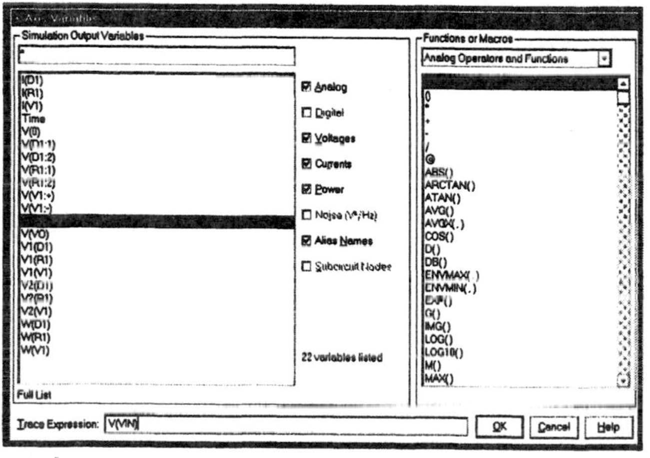

Click the ***OK*** button twice to plot the transfer curve:

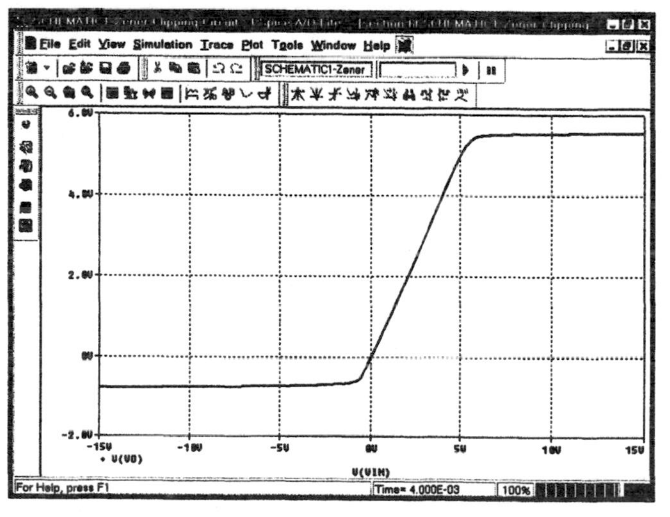

**EXERCISE 6-7:** Find the output voltage waveform and the transfer curve for the circuit below. Let the input be a ±15 volt triangle wave. Use the source Vtri to create a 1 Hz triangle wave.

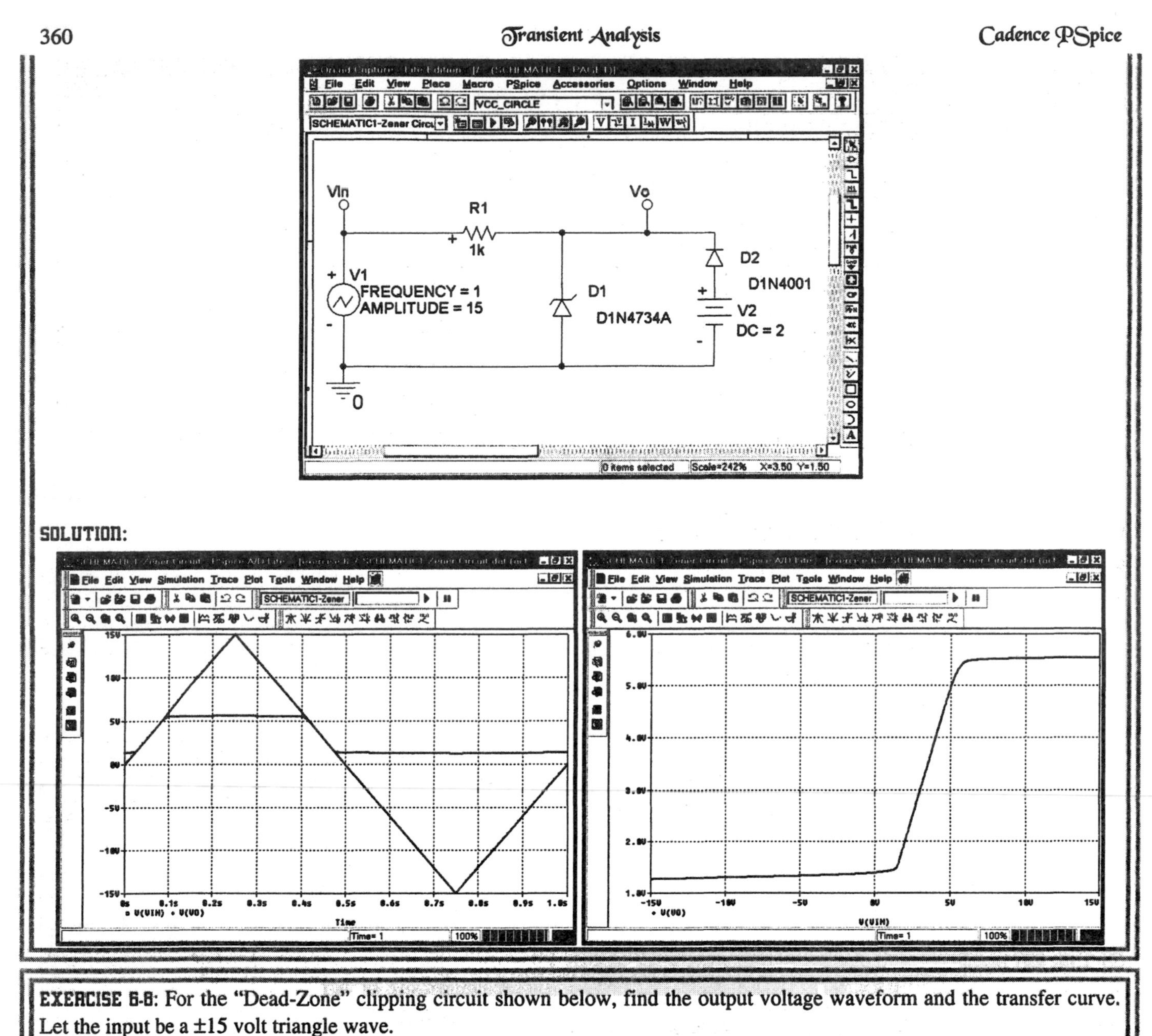


**SOLUTION:**

**EXERCISE 6-8**: For the "Dead-Zone" clipping circuit shown below, find the output voltage waveform and the transfer curve. Let the input be a ±15 volt triangle wave.

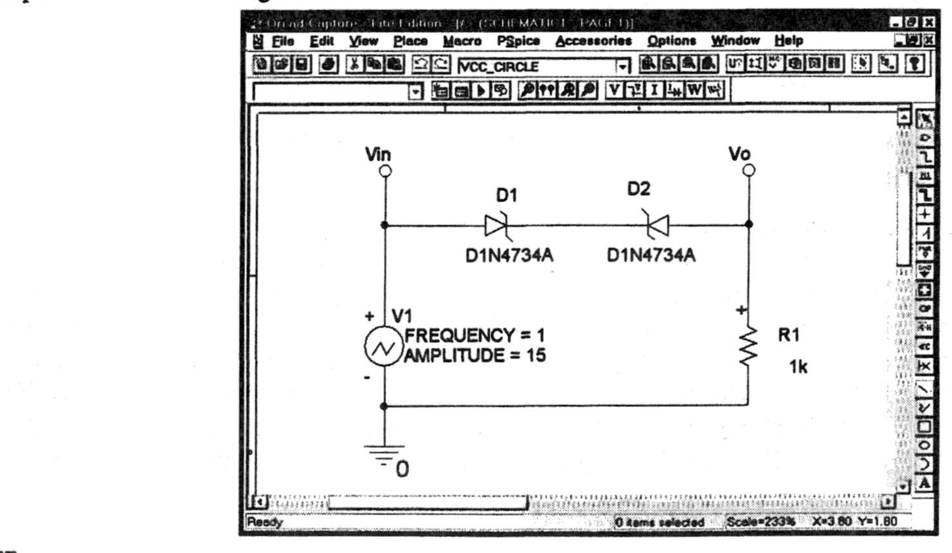


**SOLUTION:**

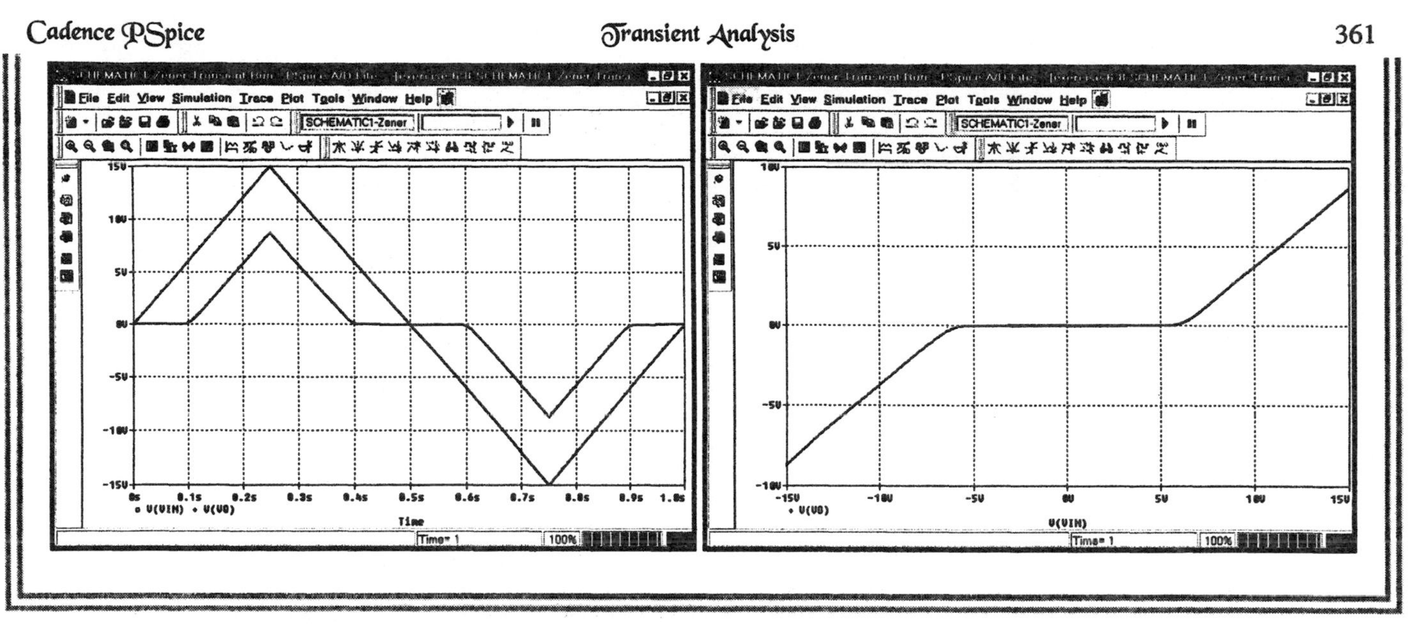


# 6.G. Amplifier Voltage Swing

In this section we will examine the voltage swing of the transistor amplifier used in Section 5.D. In Section 5.D we found that the gain of the amplifier was about 45 dB, which corresponds to a gain of 193 V/V. We also found that mid-band for this amplifier was from 42 Hz to 6.3 MHz. We would like to see what the maximum voltage swing of this amplifier looks like at mid-band. We will use the same circuit as in Section 5.D, except we will change the signal source from an AC source to a sinusoidal source at a frequency of 1 kHz.

Wire the following circuit:

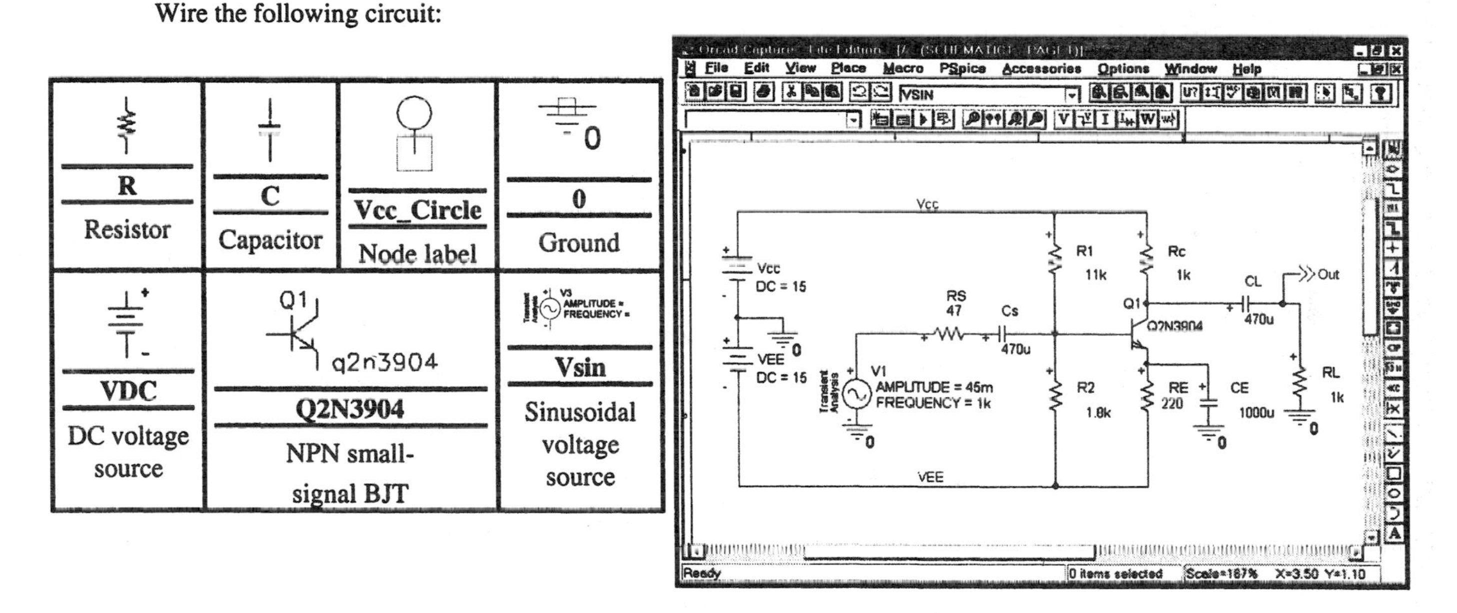


Note that the voltage source part is Vsin, not VAC. The source is now a sinusoidal voltage of amplitude 45 mV and frequency 1 kHz, $V_4 = 0.045\sin(2\pi*1000t)$. The amplitude was chosen to get the maximum swing out of this amplifier. You may have to have to run the simulation several times with different amplitude sine waves for Vin to see what amplitude you need for Vin to observe the maximum swing on the output. Too small an amplitude will cause a small output swing. Too large an output will cause the amplifier to go into saturation or cut-off. The amplitude should be chosen to cause the output of the amplifier to just saturate or cut-off. We must now set up the parameters for the Transient analysis. Select **PSpice** and then **New Simulation Profile** from the Capture menus, enter a name for the profile, and then click the ***Create*** button. By default the ***Time Domain (Transient) Analysis type*** is selected. Set up the Transient simulation dialog box as shown below:

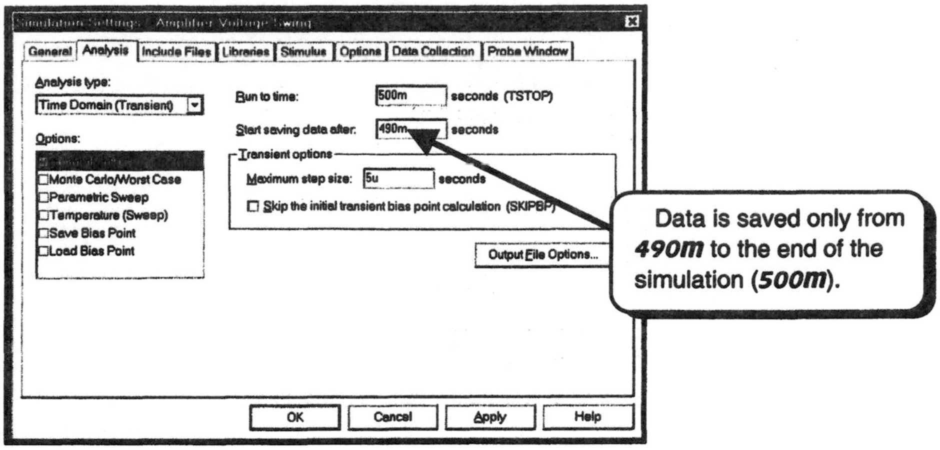

Data is saved only from *490m* to the end of the simulation (*500m*).

Leave this dialog box open as we will need it again in a few moments.

Note that we are running the simulation for 500 ms (***Run to time: 500m***). Since the source is at a frequency of 1 kHz, we are simulating the amplifier for 500 cycles. This is done to allow the amplifier to reach steady-state. If we look at the amplifier output at the beginning of the simulation, we will see the output voltage waveform slowly drift because the capacitors are charging in response to the DC average of the output voltage waveform. Since we are interested in the steady-state response of the amplifier, the simulation is run for 500 ms. Since we are not interested in the data at the start of the simulation, we will set the ***Start saving data after*** value to 490 ms. This tells PSpice not to save data for the first 490 ms of the simulation. PSpice will save data only from the last 10 ms of the simulation, 490 ms to 500 ms. If we left the ***Start saving data after*** value blank, PSpice would save data from time equals zero to 500 ms, resulting in a huge data file. The ***Maximum step size*** is set to 5 μs so that there will be 200 points in every 1 kHz cycle. This is unnecessary and increases the run time a fair amount if we are only looking at the waveform versus time. It will, however, give us a better looking output waveform and much more accurate results when we look at the Fourier components of the waveform in the following sections. To decrease the simulation time you may wish to increase the ***Maximum step size*** value or leave it blank.

Since this simulation will run for a long time and we have specified a small ***Maximum step size***, a lot of data will be collected. PSpice normally collects voltage data at every node and current data through every circuit component. This results in a large Probe data file that can take a long time to load and may cause memory problems. Since we are interested only in the input and output voltages, we will tell PSpice to collect data only at the input and output nodes, which will be marked with markers.

We will first change the setup to collect data only at markers and to display those waveforms when Probe starts. Click the *LEFT* mouse button on the ***Data Collection*** tab of the ***Simulation Settings*** dialog box:

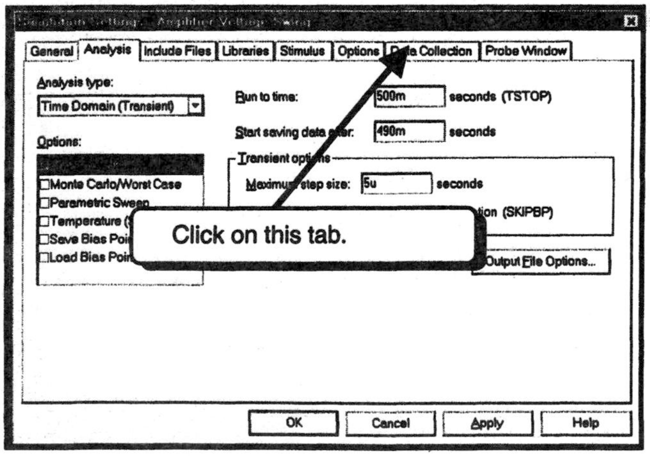

Click on this tab.

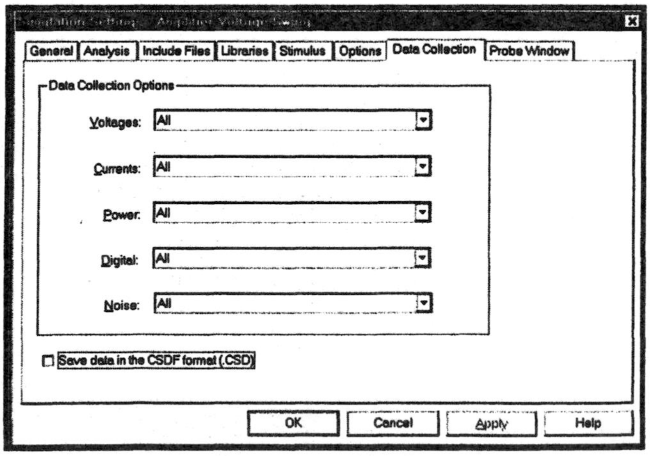

We only want to collect voltage data at nodes indicated by markers. Change the settings as shown below:

Next, we want to specify that when Probe starts, it should display all waveforms specified by markers. Click the ***LEFT*** mouse button on the ***Probe Window*** tab:

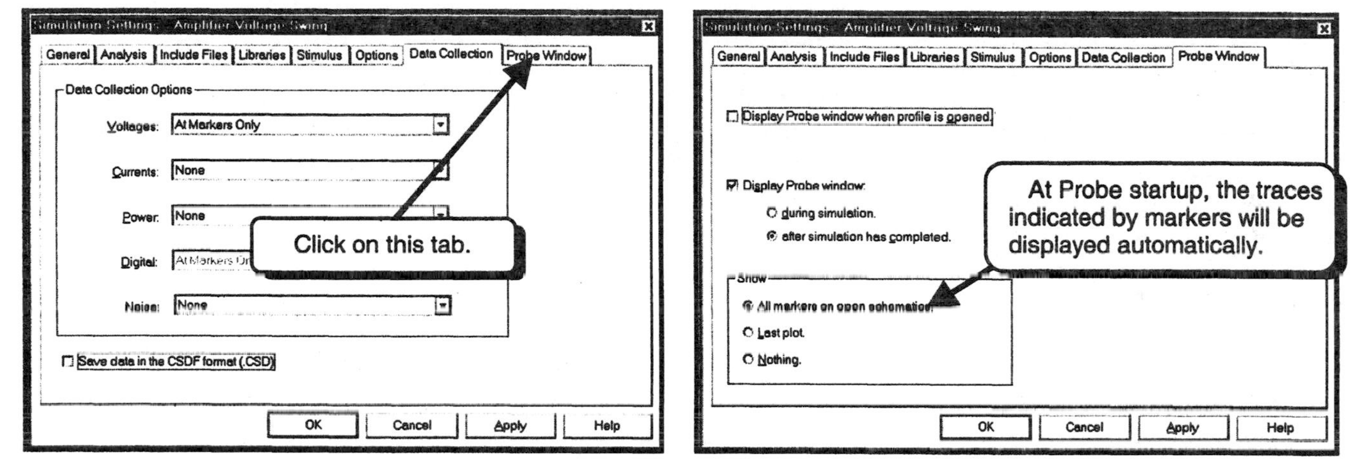


My dialog box is set up so that Probe will run after the simulation is complete and Probe will display all traces indicated by markers. Change your settings to match the dialog box above and click the ***OK*** button to return to the schematic.

Next, we must place markers on the schematic. In Capture select **PSpice**, **Markers**, and then **Voltage Level**. Place markers at ***Vin*** and ***Vo*** as shown:

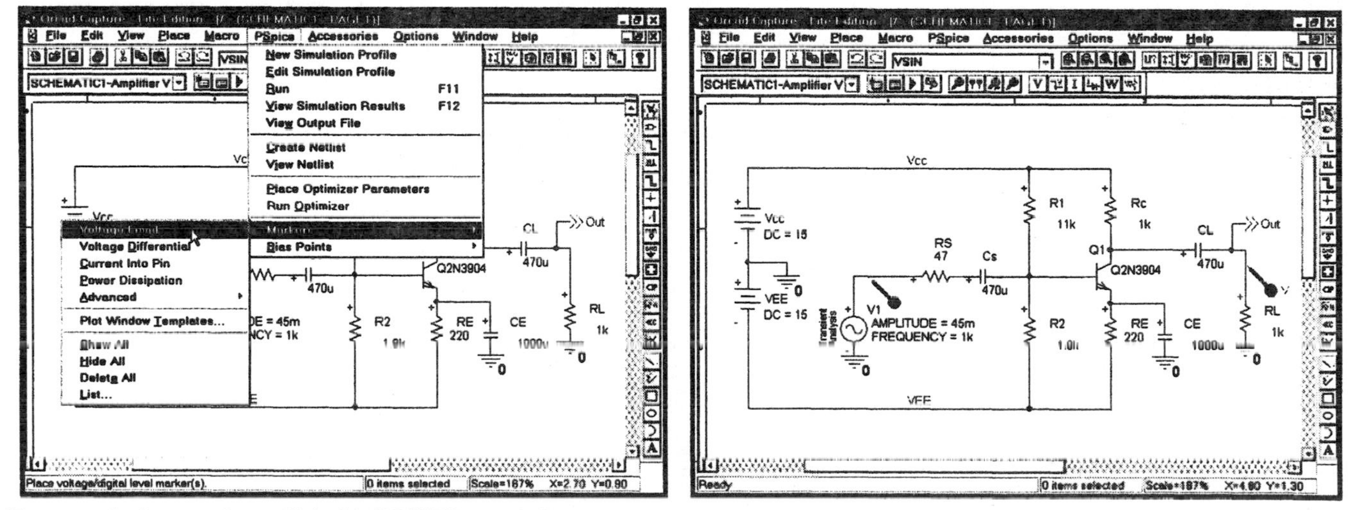


To stop placing markers click the ***RIGHT*** mouse button.

Run the circuit: Select **PSpice** and then **Run** from the Capture menu bar. When Probe runs, the traces Vin and Vo will be displayed. Note that the time axis only displays times from 490 ms to 500 ms. (This was specified in the Time Domain (Transient) setup dialog box when we set the Run to time to 500 ms and the Start saving data after to 490 ms.)

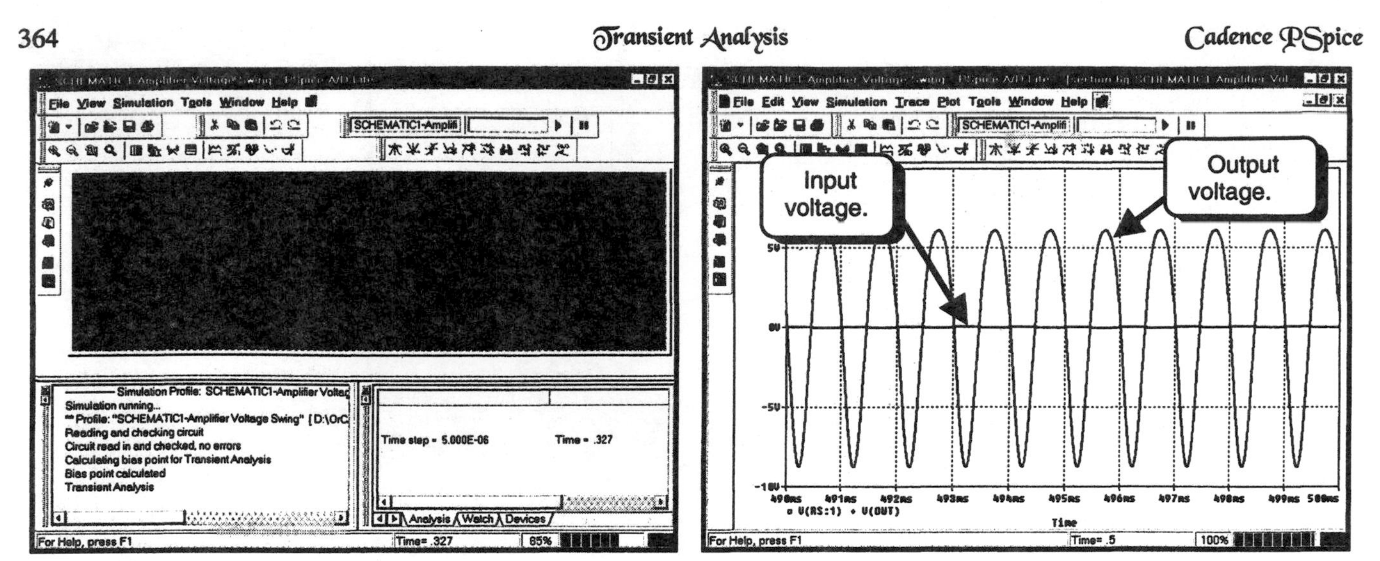


The output is so much larger than the input that you can hardly see any variation on the input. We also notice that the output does look sinusoidal, but seems to be rounded at the top and narrowed at the bottom. The output is not really a pure sine wave and contains some distortion. Distortion means that the output of the amplifier contains frequencies not present in the input. For a distortion-free amplifier, if we input a frequency of 1 kHz, the only frequency contained in the output will be 1 kHz. The output of our amplifier does not look like the input, thus it contains additional frequencies not present in the input. We have two ways to identify these frequencies using PSpice.

## 6.G.1. Fourier Analysis with Probe

Since we have already spent a good deal of simulation time and we are already in Probe, we can use Probe to view the frequency components of a signal. We will first view the components of the input. Delete the trace ***V(OUT)*** so that only the input voltage trace is displayed:

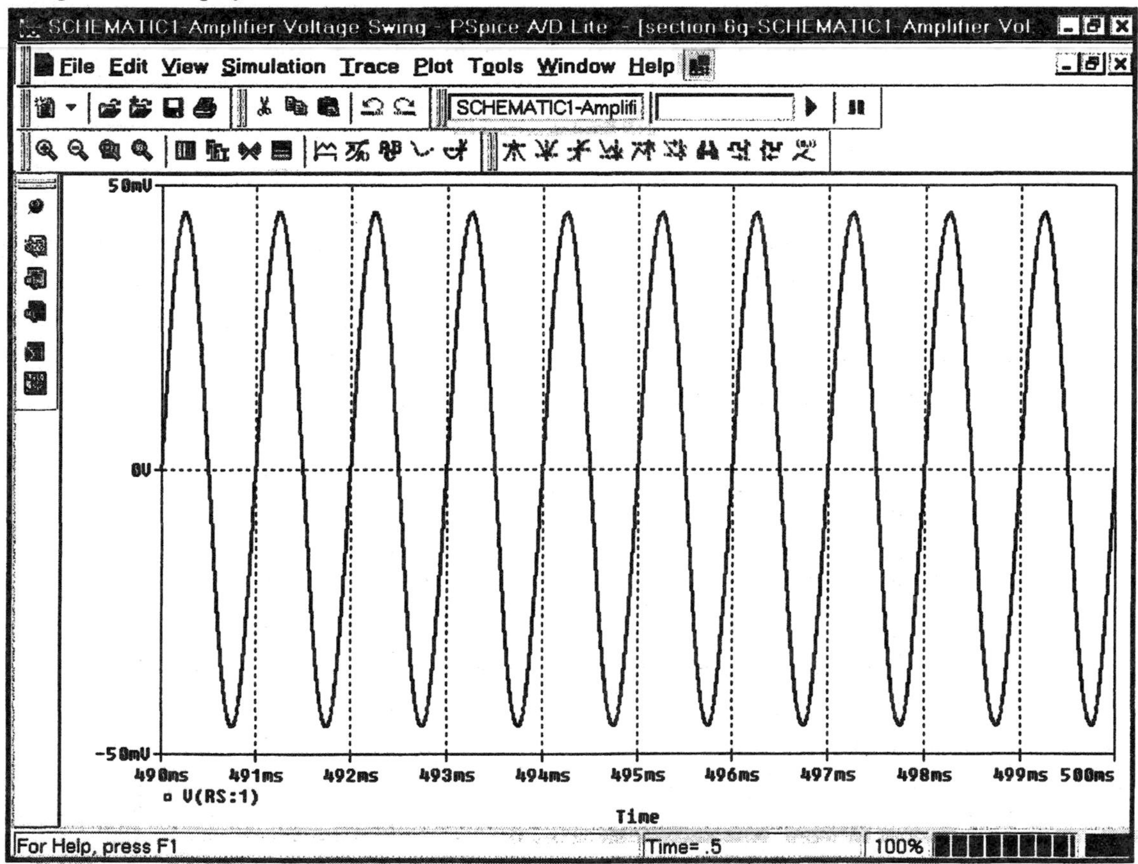


The input should be a sine wave with only one frequency, 1 kHz. To view the frequencies contained in this waveform we need to select the Fourier Processing option. Select **Plot** and then **Axis Settings** from the Probe menus. The ***X Axis*** tab is automatically selected:

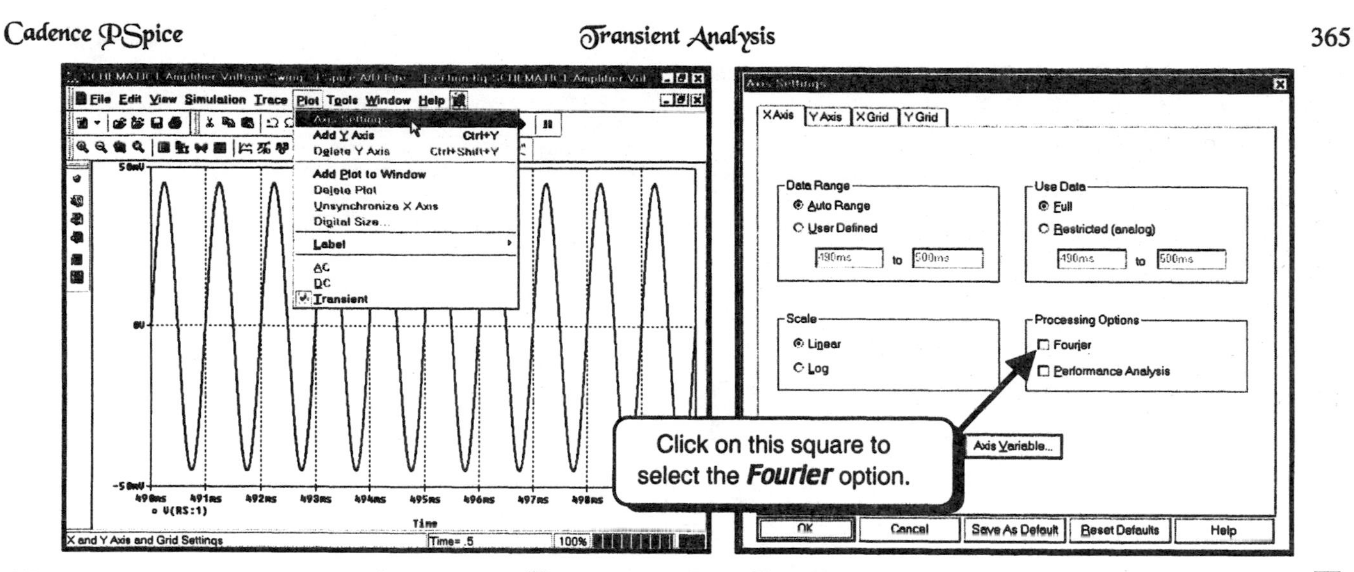


Click the ***LEFT*** mouse button on the square ☐ next to the text ***Fourier***. The square will fill with a checkmark, ☑, indicating that the option is selected:

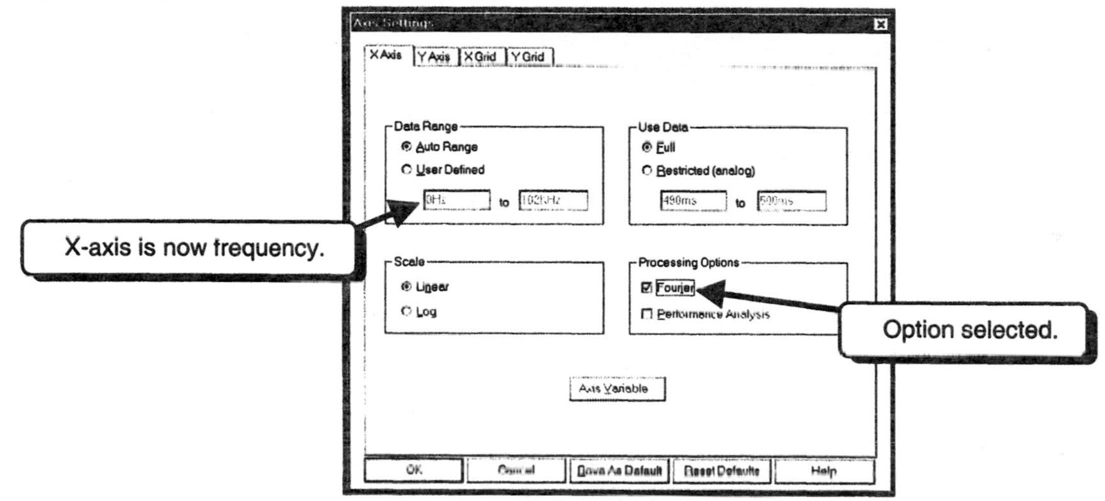


When we select the ***Fourier*** option, the x-axis scale changes to frequency. Click the ***OK*** button to accept the setting:

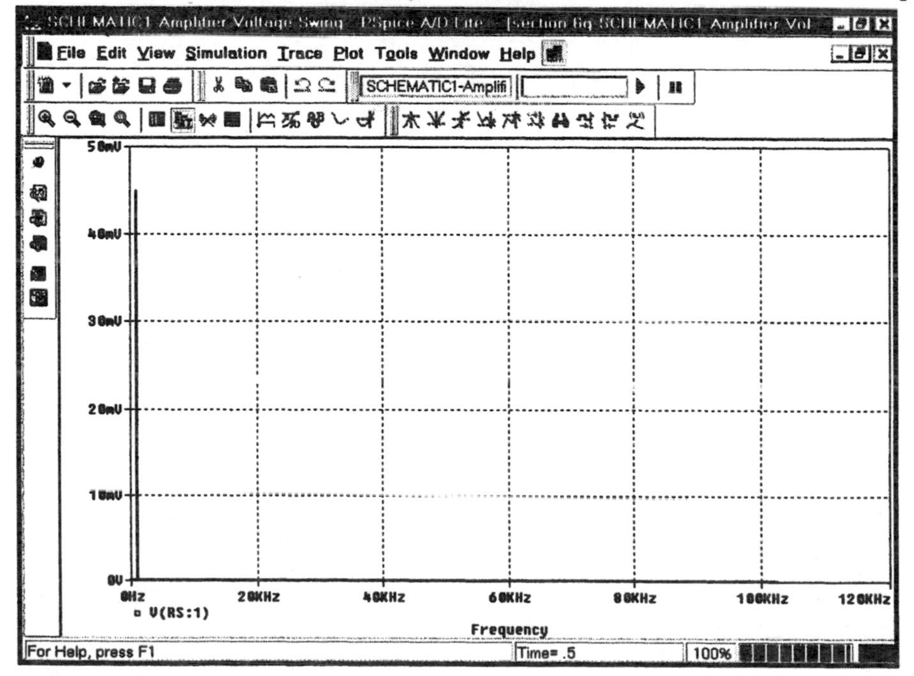


There is only a single spike near 0 Hz. The spike is hard to see because the x-axis ranges from 0 to ***120 KHz***. We need to change this scale. Select **Plot** and then **Axis Settings** from the Probe menus and specify a range of 0 to 10 kHz:

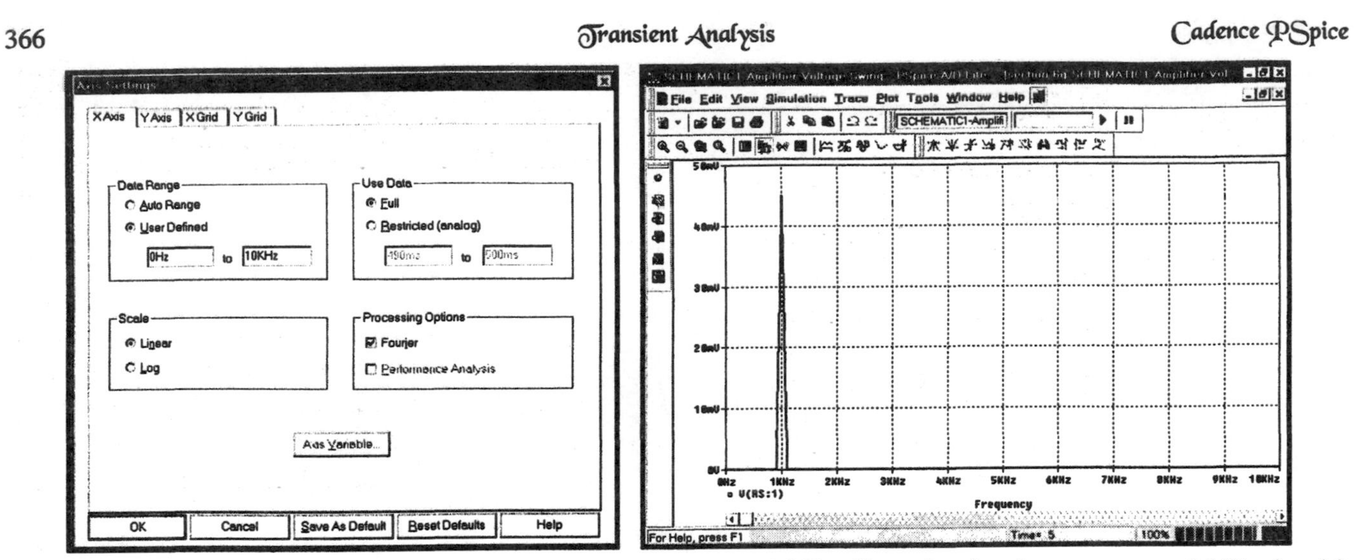


We see that the input voltage signal contains only one frequency at 1 kHz. Since the signal source was a 1 kHz signal it should contain only one frequency. If you look at the schematic, you will remember that the amplitude of Vin was a 45 mV amplitude sine wave. We can use the cursors to find the value of the frequency component at 1 kHz:

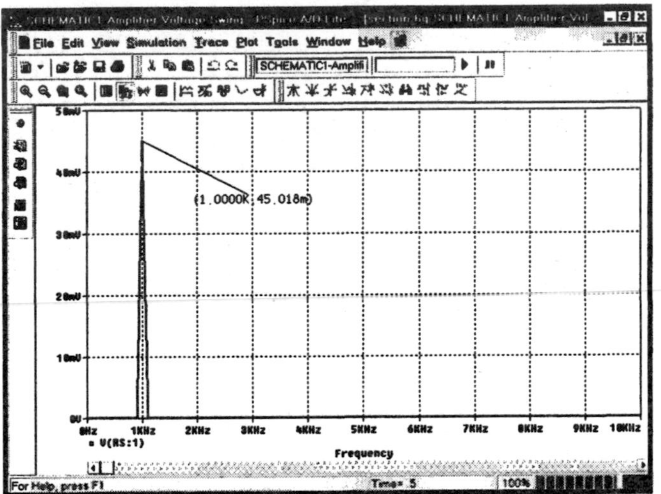


We see that the magnitude of the component at 1 kHz is 45.018 mV, fairly close to the amplitude of the input source. Switch back to the time domain by deselecting the Fourier option in the x ***Axis Settings*** dialog box.

We would now like to look at the frequencies at the output. Delete the input voltage trace and add the trace ***V(OUT)***:

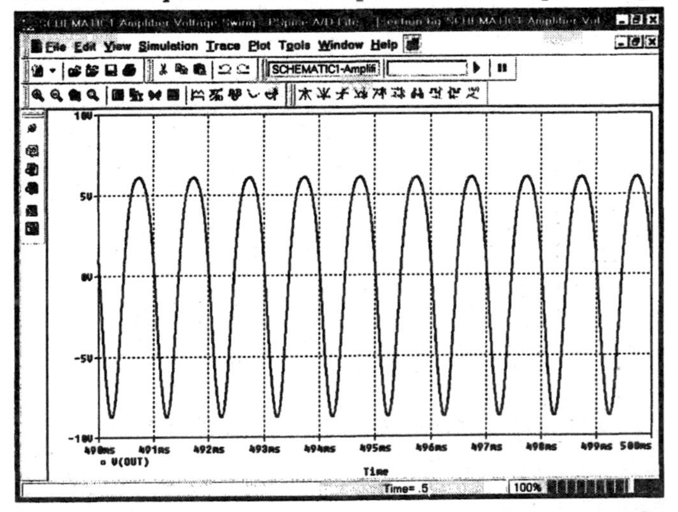


This is what the output looks like versus time. The output is not a perfect sine wave and contains some distortion. To see what frequencies the output waveform contains, we would like to create a second plot that displays the Fourier components of the waveform. Select **Plot** and then **Add Plot to Window** to add a second plot to the same window:

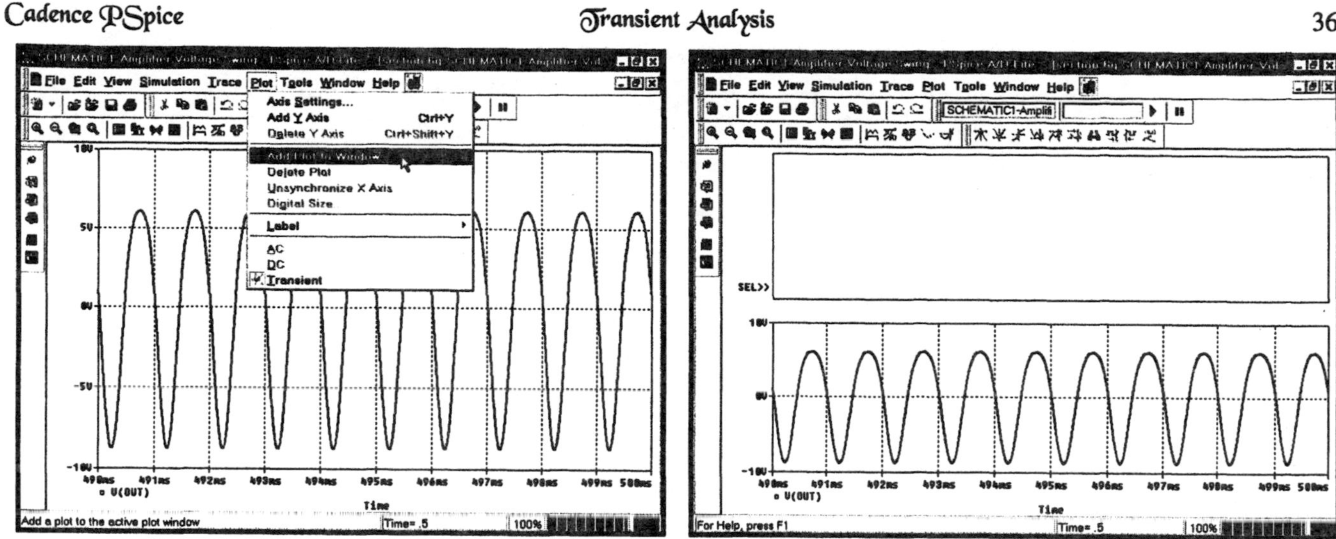


Add the trace **V(OUT)**:

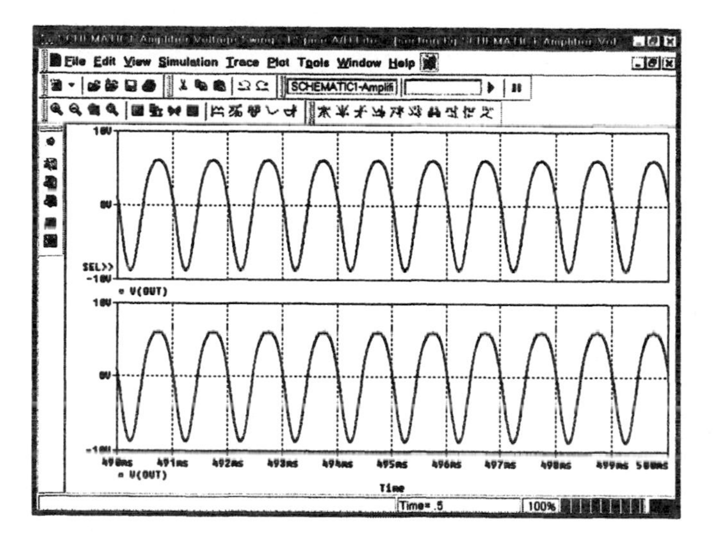


We would like the top plot to display the Fourier components and the bottom plot to display the waveform versus time. For a Fourier plot the x-axis is frequency. For a time plot the x-axis is time. Presently both plots use the same x-axis. To allow the plots to have different x-axes, select **Plot** and then **Unsynchronize X Axis**:

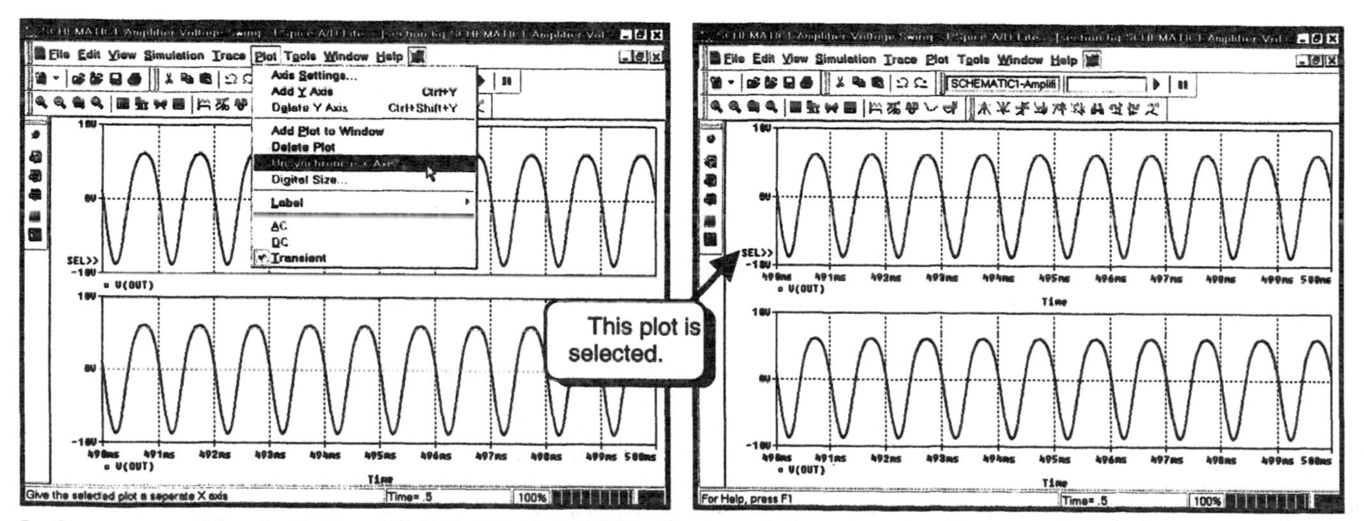


In the screen capture above the top plot is selected. Follow the procedure outlined previously to select the Fourier processing option for the selected plot (select **Plot**, **Axis Settings**, and then ***Fourier***):

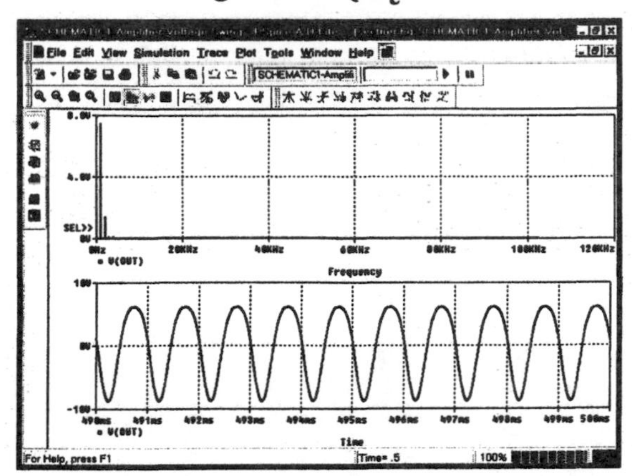


Once again there are some frequency spikes near 0 Hz and we need to change the x-axis range for the top plot to get a better view. Select **Plot** and then **Axis Settings** and change the x-axis range of the top plot to `0Hz` to `10kHz`:

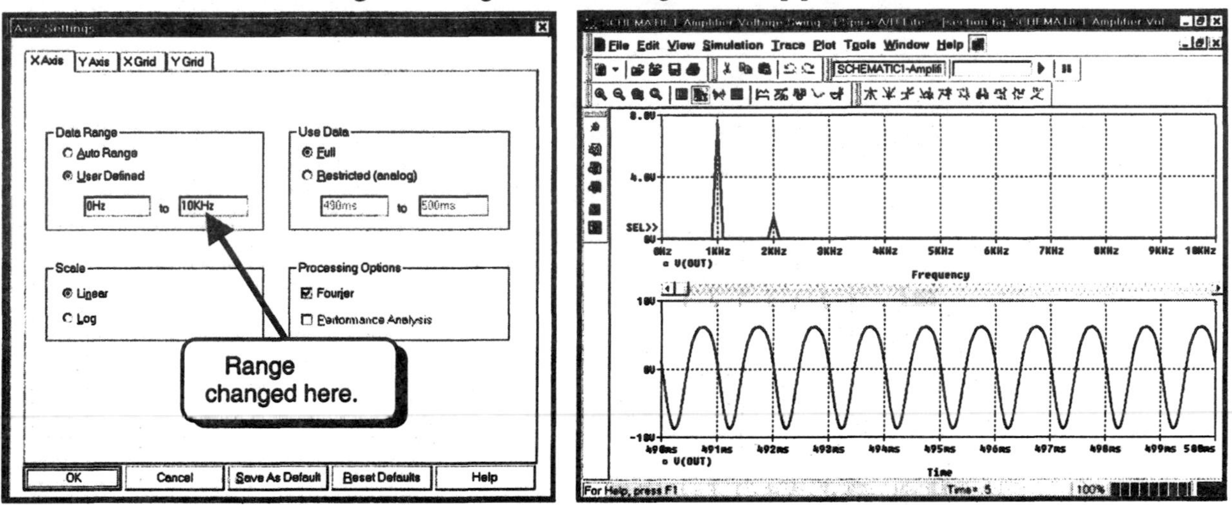


Notice that the largest frequency component contained in the output is at the fundamental (1 kHz), but there is also a large component at the second harmonic (2 kHz). There are other frequencies contained in the output, but they are too small to be seen on the graph. Using the cursors we find that the magnitude of the component at 1 kHz is 7.4665 V, and the magnitude of the component at 2 kHz is 1.3958 V.

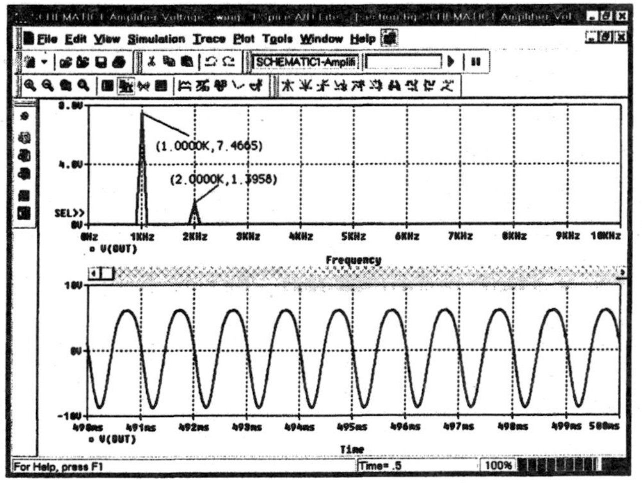


An equation for the output voltage would be:

$$Vo = 7.4665 \sin(2\pi \bullet 1000t) + 1.3958 \sin(2\pi \bullet 2000t)$$

This graph tells us that the output contains frequencies of 1 kHz and 2 kHz when the input contains only a single frequency of 1 kHz. This additional frequency is caused by the non-linearity of the amplifier.

**Important Note: To get more accurate results using the Fourier option, you should let the simulation run for many cycles and use a small step size (small value of Maximum step size in the Transient New Simulation Profile dialog box). This increases the simulation time but will improve the accuracy of your simulation results.**

**EXERCISE 6-9**: Find the magnitude of the first seven harmonics of a ±1 volt, 1 kHz square wave using Probe.

**SOLUTION**: Use the pulsed voltage source. Set the rise and fall times to 1 μs so that the rise and fall times are much shorter than the pulse width and period of the square wave. Wire the circuit:

The square wave voltage and Fourier results are:

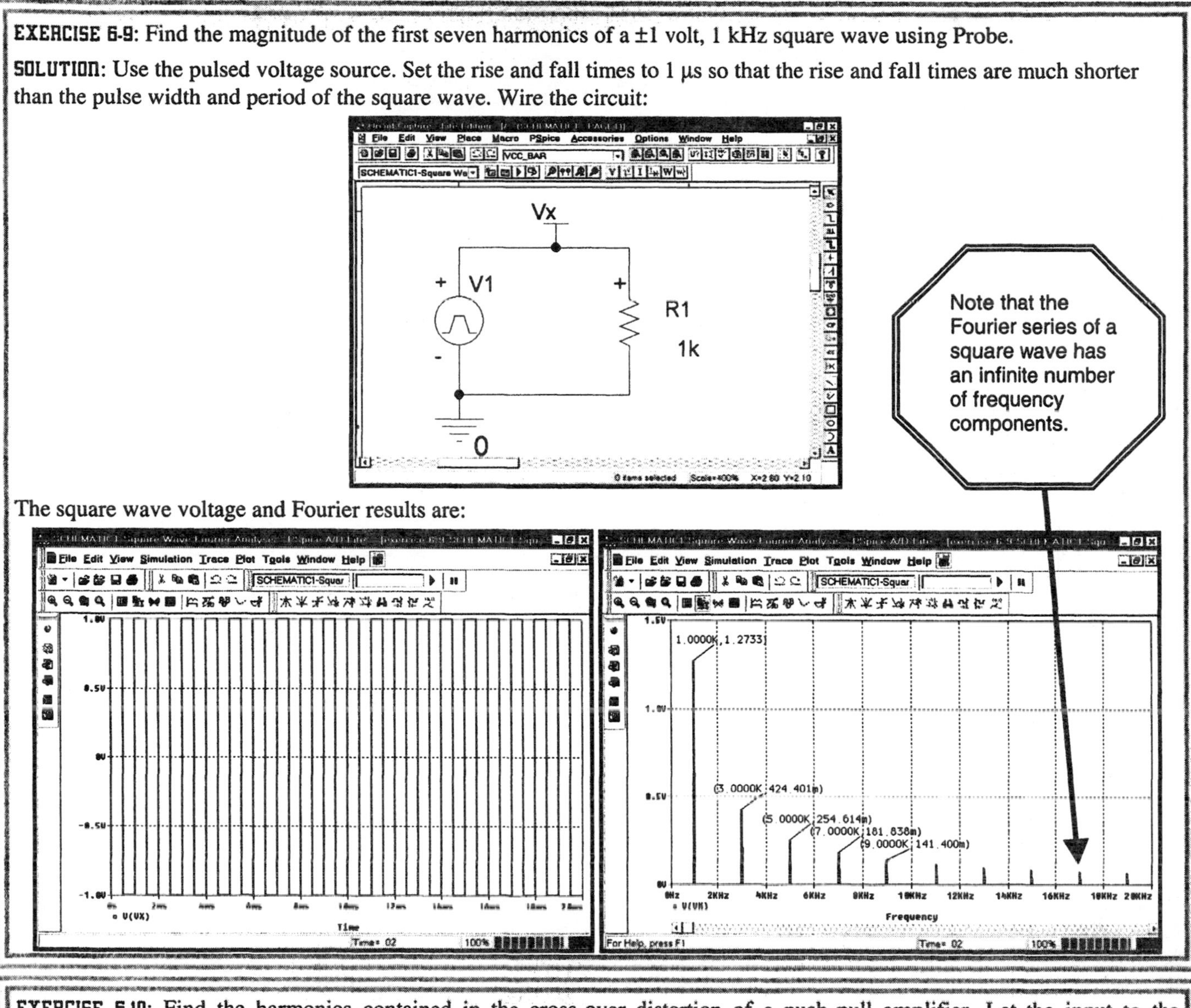


**EXERCISE 6-10**: Find the harmonics contained in the cross-over distortion of a push-pull amplifier. Let the input to the amplifier be a 2 V amplitude, 1 kHz sine wave. The push-pull amplifier drives an 8 Ω load.

**SOLUTION**: Wire the push-pull amplifier below. Run a Transient Analysis for several cycles with many points per cycle:

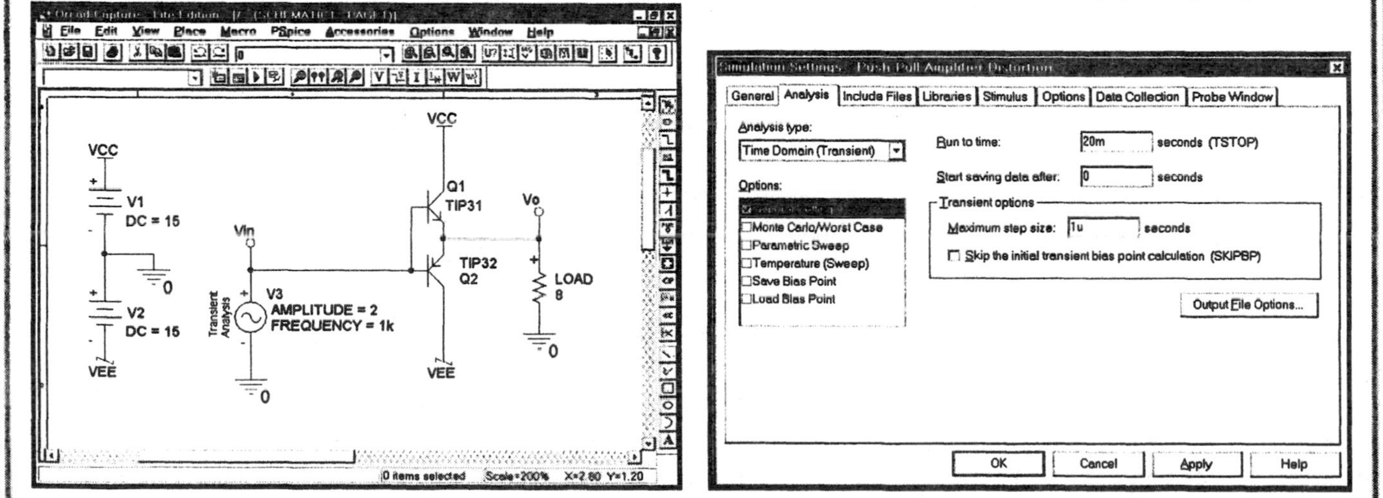


The input and output waveforms are shown on the following left screen capture. Twenty cycles were simulated, but only one is shown in the following left screen capture to make the distortion easily seen. The right screen capture shows the Fourier components of the output voltage waveform.

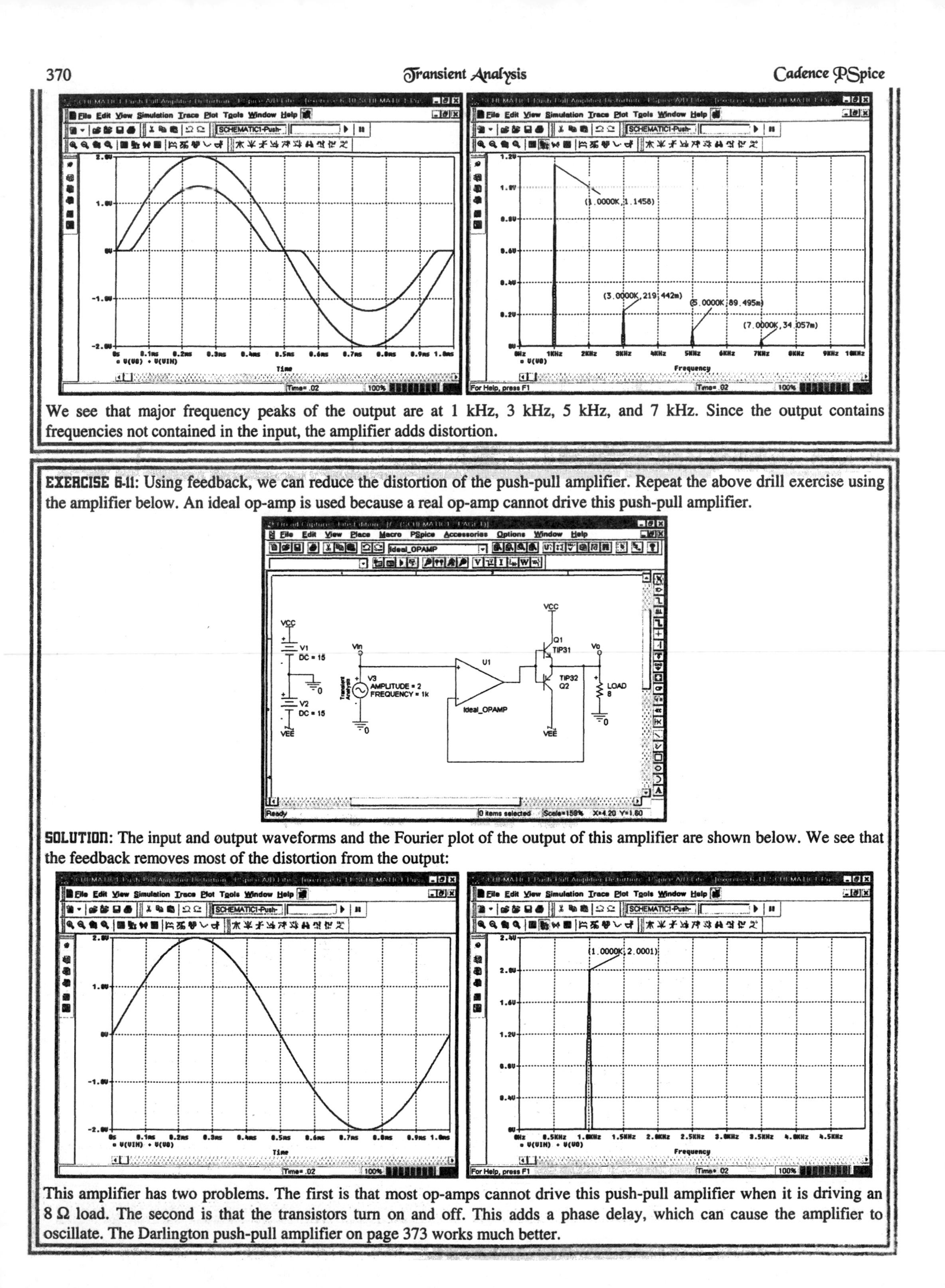


We see that major frequency peaks of the output are at 1 kHz, 3 kHz, 5 kHz, and 7 kHz. Since the output contains frequencies not contained in the input, the amplifier adds distortion.

**EXERCISE 6-11:** Using feedback, we can reduce the distortion of the push-pull amplifier. Repeat the above drill exercise using the amplifier below. An ideal op-amp is used because a real op-amp cannot drive this push-pull amplifier.

**SOLUTION:** The input and output waveforms and the Fourier plot of the output of this amplifier are shown below. We see that the feedback removes most of the distortion from the output:

This amplifier has two problems. The first is that most op-amps cannot drive this push-pull amplifier when it is driving an 8 Ω load. The second is that the transistors turn on and off. This adds a phase delay, which can cause the amplifier to oscillate. The Darlington push-pull amplifier on page 373 works much better.

## 6.G.2. Fourier Analysis with PSpice

The frequency components of a signal can be obtained directly from PSpice by enabling the Fourier option in the Time Domain (Transient) setup. We will use the common-emitter amplifier circuit shown on page 361. To modify the Time Domain (Transient) setup select **PSpice** and then **Edit Simulation Profile**.

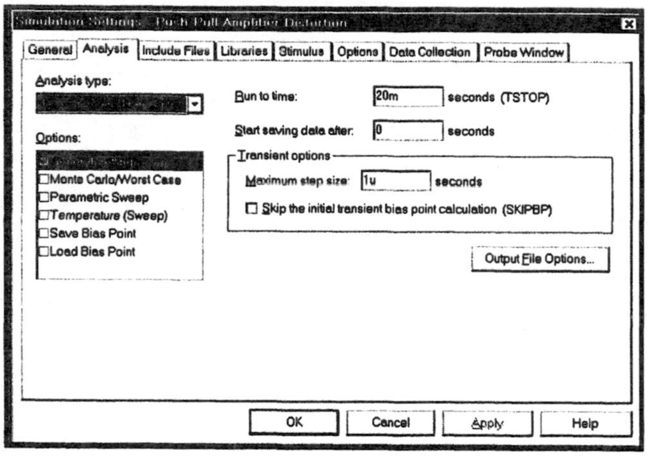

Select the ***Output File Options*** button and fill out the dialog box as shown:

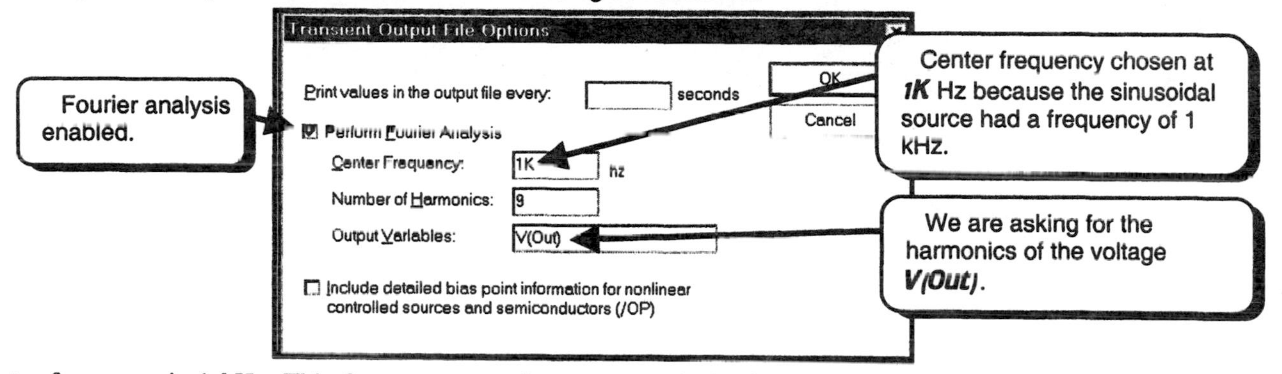


The center frequency is 1 kHz. This frequency was chosen to match the frequency of the sinusoidal input voltage. The harmonics that will be calculated are the first nine: 1 kHz, 2 kHz, 3 kHz, 4 kHz, 5 kHz, 6 kHz, 7 kHz, 8 kHz, and 9 kHz. There may be others, but we want numerical values for only the first nine. The output variable for the Fourier analysis is the voltage at node Out, ***V(Out)***. This is the output of the amplifier. We could look at the frequency components of any voltage or current, but for this example we are interested only in the output. Click the ***OK*** button twice to accept the settings and then run PSpice.

The results of the Fourier analysis will be saved in the output file. When PSpice has finished the simulation, Probe will run and display the waveforms at the markers (we set this up in the previous section). Select **View** and then **Output File** from the Probe menus:

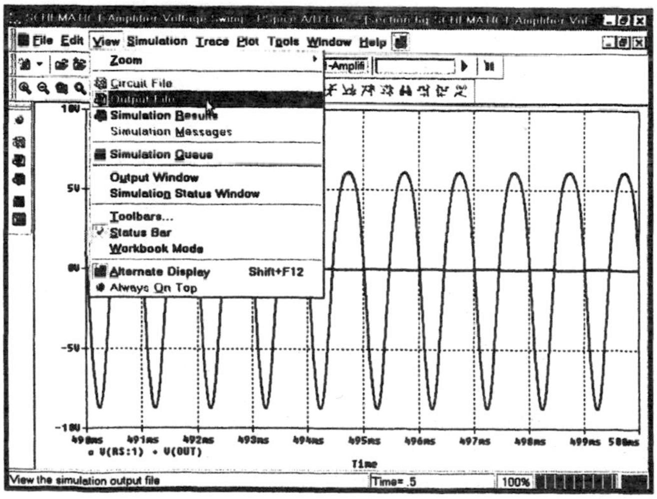

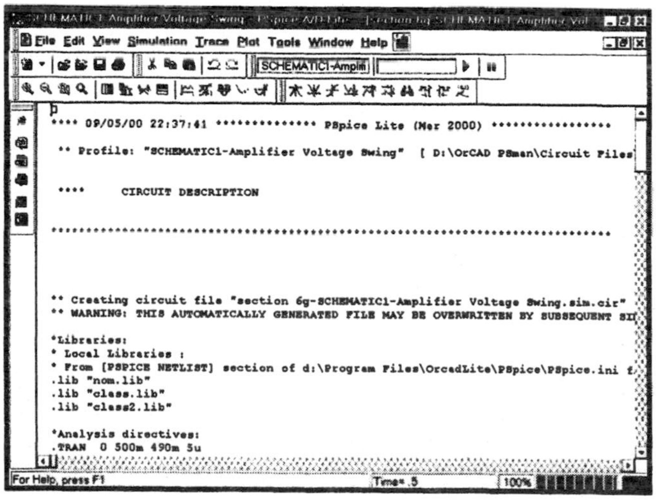

The results of the Fourier analysis are at the end of the output file. Click the ***LEFT*** mouse button on the down arrow until you reach the following text:

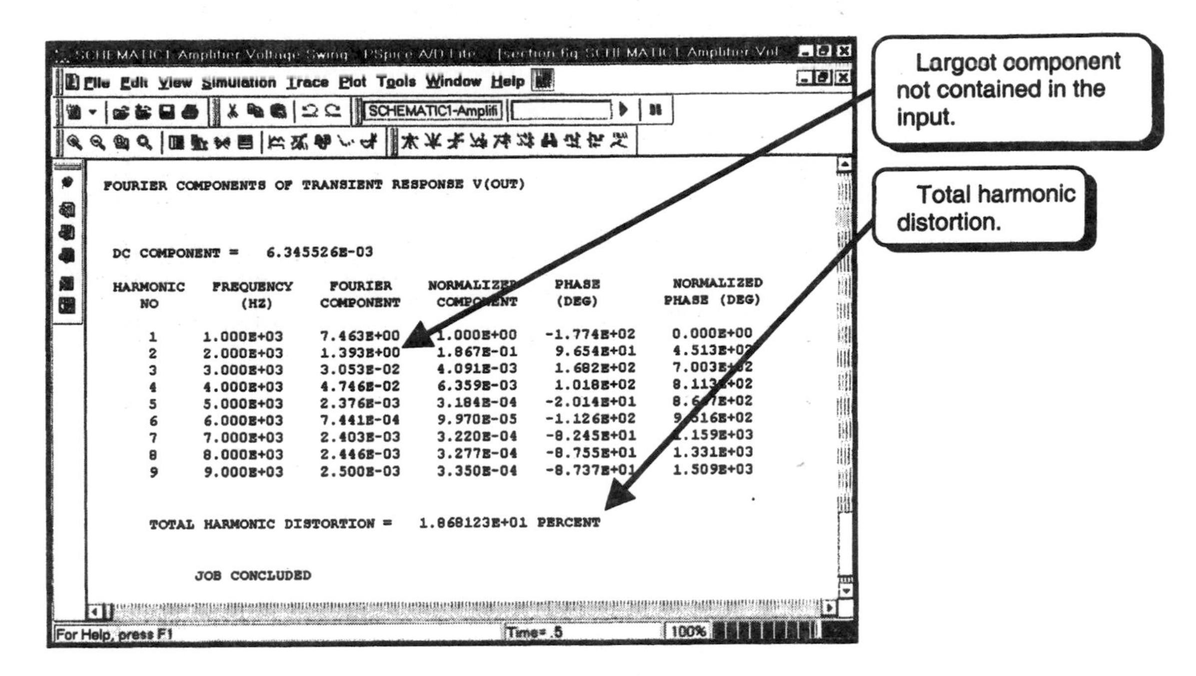


FOURIER COMPONENTS OF TRANSIENT RESPONSE V(OUT)

DC COMPONENT = 6.345526E-03

| HARMONIC NO | FREQUENCY (HZ) | FOURIER COMPONENT | NORMALIZED COMPONENT | PHASE (DEG) | NORMALIZED PHASE (DEG) |
|---|---|---|---|---|---|
| 1 | 1.000E+03 | 7.463E+00 | 1.000E+00 | -1.774E+02 | 0.000E+00 |
| 2 | 2.000E+03 | 1.393E+00 | 1.867E-01 | 9.654E+01 | 4.513E+02 |
| 3 | 3.000E+03 | 3.053E-02 | 4.091E-03 | 1.682E+02 | 7.003E+02 |
| 4 | 4.000E+03 | 4.746E-02 | 6.359E-03 | 1.018E+02 | 8.113E+02 |
| 5 | 5.000E+03 | 2.376E-03 | 3.184E-04 | -2.014E+01 | 8.67E+02 |
| 6 | 6.000E+03 | 7.441E-04 | 9.970E-05 | -1.126E+02 | 9.516E+02 |
| 7 | 7.000E+03 | 2.403E-03 | 3.220E-04 | -8.245E+01 | 1.159E+03 |
| 8 | 8.000E+03 | 2.446E-03 | 3.277E-04 | -8.755E+01 | 1.331E+03 |
| 9 | 9.000E+03 | 2.500E-03 | 3.350E-04 | -8.737E+01 | 1.509E+03 |

TOTAL HARMONIC DISTORTION = 1.868123E+01 PERCENT

JOB CONCLUDED

The results of the Fourier analysis show that the magnitude of the sine wave at 1 kHz is 7.463 V, and the magnitude of the sine wave at 2 kHz is 1.393. These results are similar to the results obtained using Probe. This method also gives us the magnitude of the frequency components too small to see on the Probe graph, the phase of each frequency component, and the total harmonic distortion. From the data above, an equation for the output voltage is:

$$V_o = 7.463 \sin(2\pi \cdot 1000t - 177.4°) + 1.393 \sin(2\pi \cdot 2000t + 96.54°) + 0.03053 \sin(2\pi \cdot 3000t + 168.2°) + \ldots$$

**EXERCISE 6-12:** Find the magnitude of the harmonics of a ±1 volt, 1 kHz square wave using PSpice.

**SOLUTION:** Use the pulsed voltage source. Set the rise and fall times to 1 μs so that the rise and fall times are much shorter than the pulse width and period of the square wave. Wire the circuit:

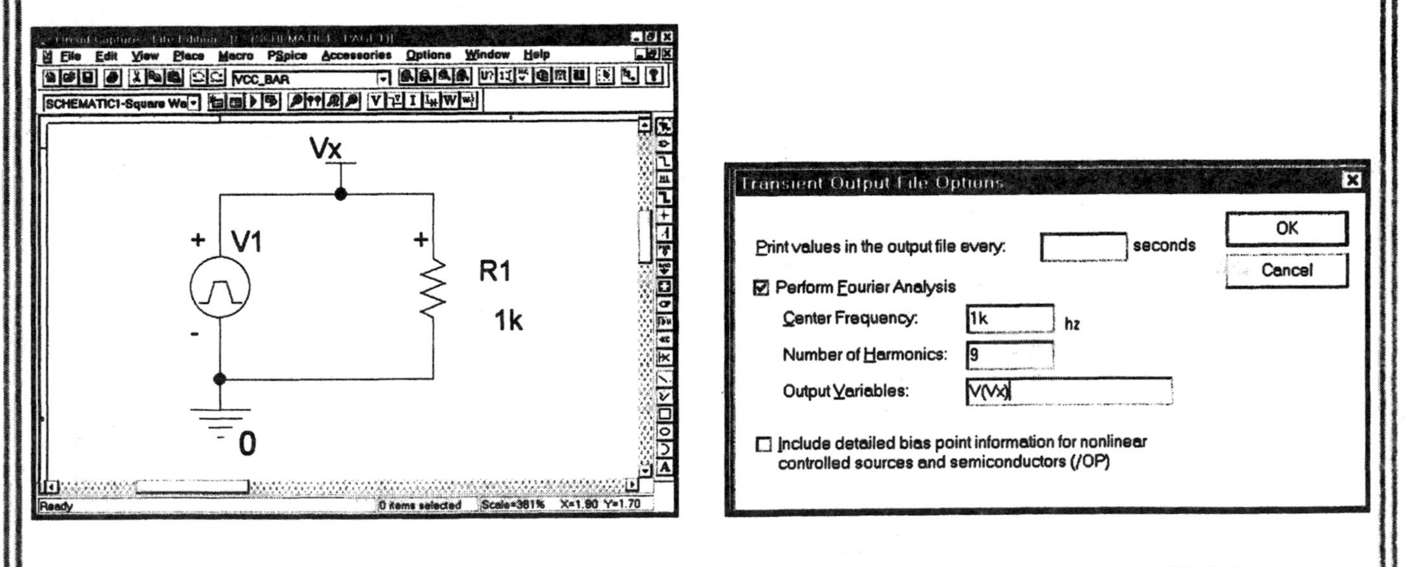


The results are contained in the output file. Your results should be similar to the ones shown in the output file below:

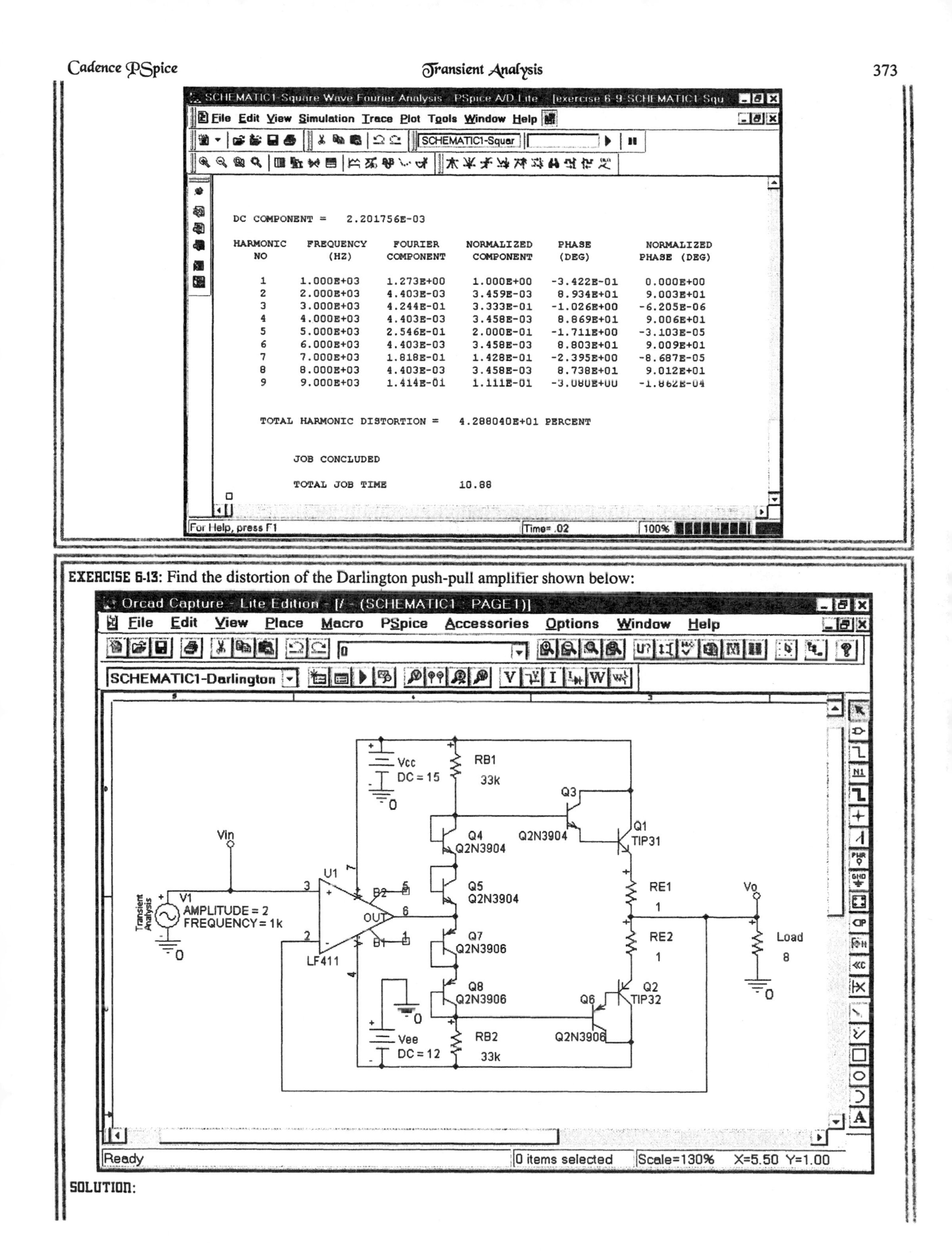

SCHEMATIC1-Square Wave Fourier Analysis - PSpice A/D Lite - [exercise 6-9-SCHEMATIC1-Squ

DC COMPONENT = 2.201756E-03

| HARMONIC NO | FREQUENCY (HZ) | FOURIER COMPONENT | NORMALIZED COMPONENT | PHASE (DEG) | NORMALIZED PHASE (DEG) |
|---|---|---|---|---|---|
| 1 | 1.000E+03 | 1.273E+00 | 1.000E+00 | -3.422E-01 | 0.000E+00 |
| 2 | 2.000E+03 | 4.403E-03 | 3.459E-03 | 8.934E+01 | 9.003E+01 |
| 3 | 3.000E+03 | 4.244E-01 | 3.333E-01 | -1.026E+00 | -6.205E-06 |
| 4 | 4.000E+03 | 4.403E-03 | 3.458E-03 | 8.869E+01 | 9.006E+01 |
| 5 | 5.000E+03 | 2.546E-01 | 2.000E-01 | -1.711E+00 | -3.103E-05 |
| 6 | 6.000E+03 | 4.403E-03 | 3.458E-03 | 8.803E+01 | 9.009E+01 |
| 7 | 7.000E+03 | 1.818E-01 | 1.428E-01 | -2.395E+00 | -8.687E-05 |
| 8 | 8.000E+03 | 4.403E-03 | 3.458E-03 | 8.738E+01 | 9.012E+01 |
| 9 | 9.000E+03 | 1.414E-01 | 1.111E-01 | -3.080E+00 | -1.862E-04 |

TOTAL HARMONIC DISTORTION = 4.288040E+01 PERCENT

JOB CONCLUDED

TOTAL JOB TIME 10.88

For Help, press F1 Time= .02 100%

**EXERCISE 6-13**: Find the distortion of the Darlington push-pull amplifier shown below:



**SOLUTION:**

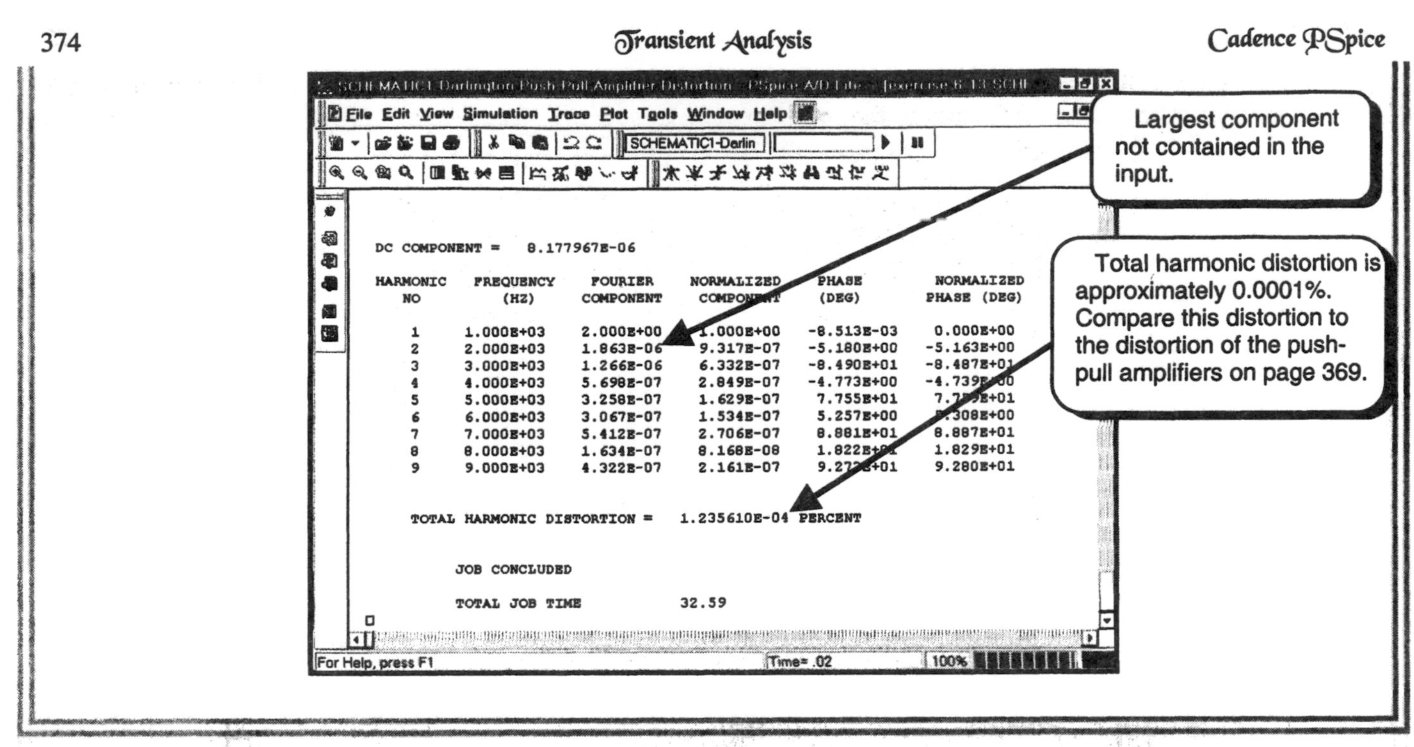


# 6.H. Ideal Operational Amplifier Integrator

In this section we will demonstrate the use of an ideal operational amplifier and the pulsed voltage waveform. Wire the circuit shown below.

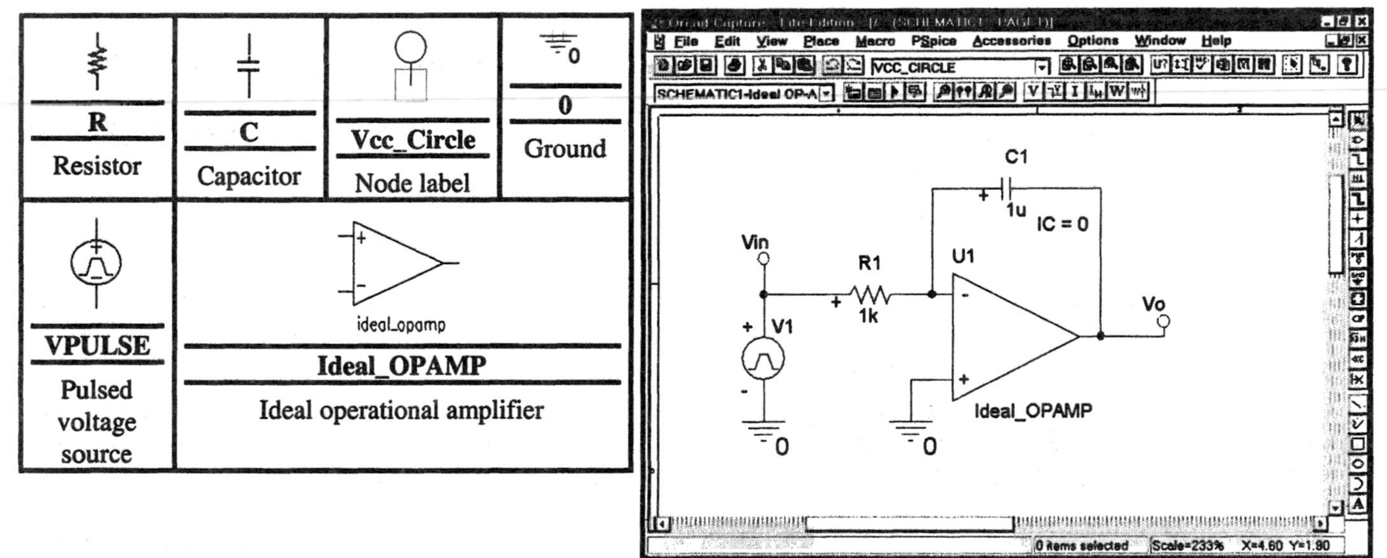


The ideal operational amplifier is very useful in the Lite version of PSpice. The ideal model has only three components in the subcircuit. This small number of components allows many ideal op-amps to be used before the component limit of the Lite version is reached. In the Lite version of PSpice, only two non-ideal op-amp models can be used before reaching the component limit. If you have a circuit with a large number of op-amps, you will be forced to use ideal op-amps in the Lite version.

The drawback of the ideal op-amp model is that none of the non-ideal properties are modeled. In this example, if a non-ideal op-amp model were used in the simulation, the integrator would not work because of bias currents. If this circuit were tested in the laboratory, it also would not work because of bias currents. Thus, the circuit simulation with a non-ideal op-amp matches the results in the lab, but the circuit simulation with an ideal op-amp does not match the lab results. For this example, the ideal model is not a good choice for simulation because it does not match the results in the lab. We will use it here for demonstration purposes only. See **EXERCISE 6-15** to learn how this integrator performs using non-ideal op-amps. **In general, you should always use the non-ideal op-amp models if possible**. The only reason you should use the ideal op-amp model is if the circuit is too large for the Lite version of Capture.

The pulsed voltage source can be used to create an arbitrary pulse-shaped waveform. We will use it to create a 1 kHz square wave. The rise and fall times of the square wave will be 1 μs. Double-click the ***LEFT*** mouse button on the pulsed voltage source graphic, -⊕-, to obtain its spreadsheet and edit its attributes:

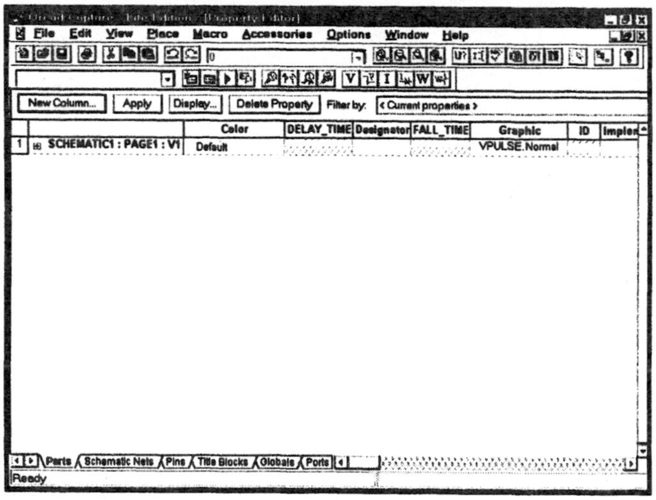

| | | Color | DELAY_TIME | Designator | FALL_TIME | Graphic | ID | Implen |
|---|---|---|---|---|---|---|---|---|
| 1 | ⊞ SCHEMATIC1 : PAGE1 : V1 | Default | | | | VPULSE.Normal | | |

The attributes are displayed alphabetically in the spreadsheet. The attributes we need to configure the pulsed source are:

- ***PERIOD*** - The period of the pulse waveform. The pulse shape will repeat itself every ***PERIOD*** seconds. The frequency of the pulse waveform is 1/***PERIOD***.
- ***RISE_TIME*** - The amount of time the voltage source takes to go from the initial voltage to the pulsed voltage, in seconds.
- ***FALL_TIME*** - The amount of time the voltage source takes to go from the pulsed voltage to the initial voltage, in seconds.
- ***PULSE_WIDTH*** - The amount of time the voltage spends at the pulsed value. It must be true that ***PULSE_WIDTH*** < ***PERIOD***. For a square wave, ***PULSE_WIDTH*** = ***PERIOD***/2.
- ***INITIAL_VOLTAGE*** - The value of the voltage at t = 0.
- ***PULSED_VOLTAGE*** - The value of the voltage source during the ***PULSE_WIDTH***.
- ***DELAY_TIME*** - At the start of the analysis, the voltage source stays at the initial voltage for an amount of time equal to the ***DELAY_TIME***. After the ***DELAY_TIME***, the voltage changes from the initial voltage to the pulsed voltage in the time specified by the ***RISE_TIME***.

We would like to create a ±5 V square wave at a frequency of 1 kHz. The rise and fall times will be 1 μs. We will specify the following values for the attributes: Period = 1m, rise_time = 1u, fall_time = 1u, pulse_width = 0.5m, initial_voltage = 5, pulsed_voltage = -5, delay_time = 0. A few of these settings are shown in the spreadsheets below:

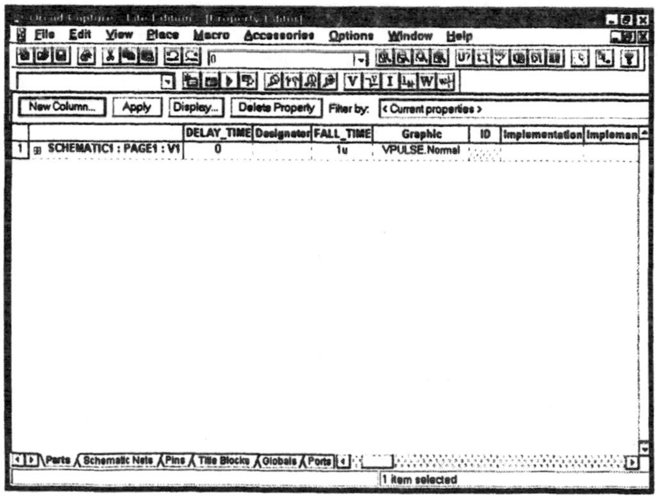

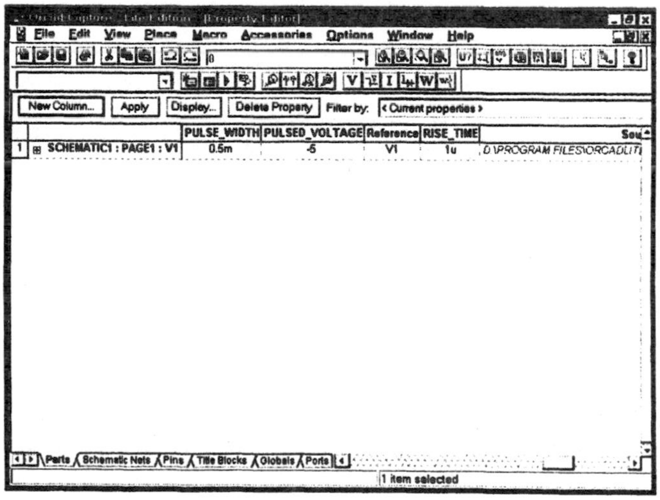

| | | DELAY_TIME | Designator | FALL_TIME | Graphic | ID | Implementation | Implemen |
|---|---|---|---|---|---|---|---|---|
| 1 | ⊞ SCHEMATIC1 : PAGE1 : V1 | 0 | | 1u | VPULSE.Normal | | | |

| | | PULSE_WIDTH | PULSED_VOLTAGE | Reference | RISE_TIME | Sou |
|---|---|---|---|---|---|---|
| 1 | ⊞ SCHEMATIC1 : PAGE1 : V1 | 0.5m | -5 | V1 | 1u | D:\PROGRAM FILES\ORCAD\LIT |

The voltage waveform described by this dialog box will produce the following waveform: At the start of the simulation, the voltage will be set to the initial voltage of 5 volts. After the delay time (zero seconds in this case) the voltage will flip to the pulsed voltage of –5 volts. From the delay time (t = 0+) to 0.0005 s, the voltage will be equal to the pulsed voltage of –5 volts. From 0.0005 s to t = 0.001 s, the voltage will be equal to the initial voltage of 5 V. At t = 0.001 s, the voltage will switch back to the pulsed voltage and repeat the cycle. Close the spreadsheet by clicking the ***LEFT*** mouse button on the lower [X] in the upper right corner of the window.

It is important to note that when PSpice calculates the bias to determine the initial condition of the capacitor, it sets the voltage of the pulsed source to its initial voltage. For our pulsed waveform, PSpice will calculate the initial capacitor voltage assuming that Vin = 5 V. This will cause the integrator to saturate initially. To prevent saturation, we must set the initial condition of the capacitor to 0 V. See Section 6.B, pages 333–337, for details on setting the capacitor initial condition.

We would like to run the circuit for 10 cycles of the square wave, or 10 ms. Select **PSpice** and then **New Simulation Profile** from the Capture menus, enter a name for the profile, and then click the ***Create*** button. By default the ***Time Domain (Transient) Analysis type*** is selected. Fill in the parameters as shown in the Time Domain dialog box below:

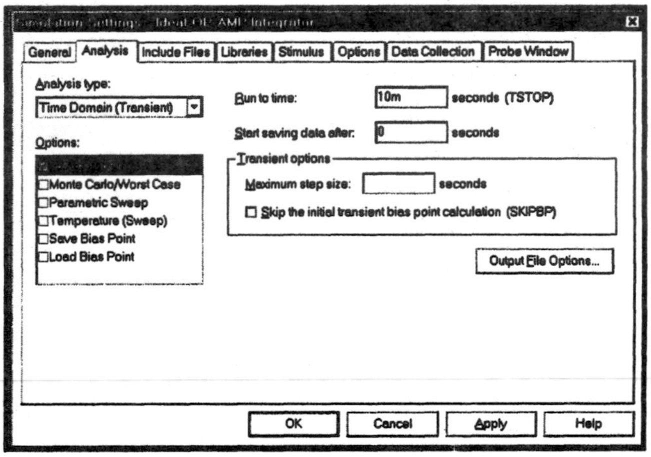

Click the ***OK*** button to return to the schematic. Run PSpice (**PSpice, Run**). In Probe, add the trace **V(VIN)** to observe the square wave voltage source we created:

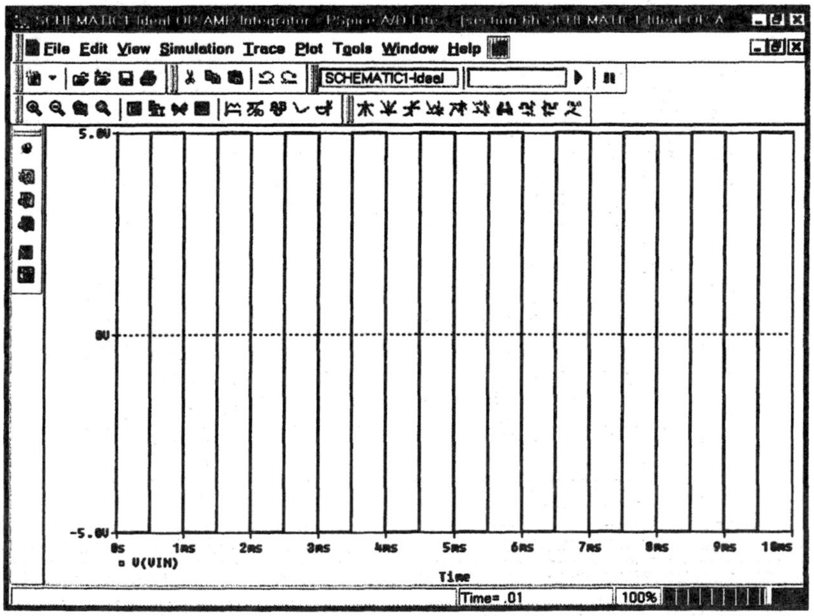

The output of the integrator should be a triangle wave. Add the trace **V(VO)**. You should see a triangle wave:

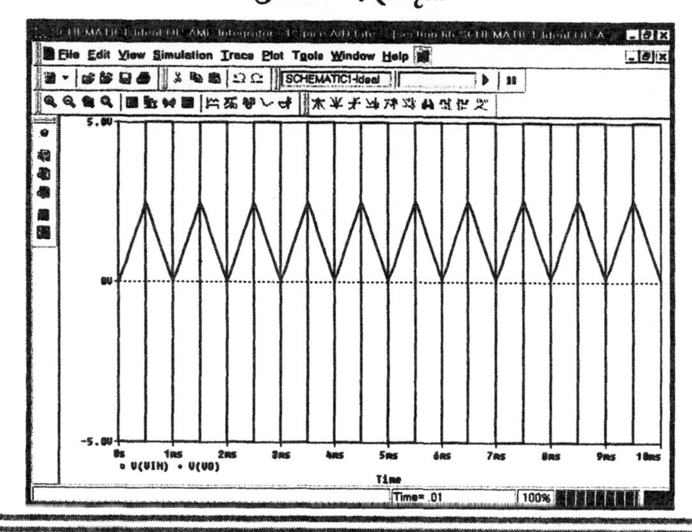

**EXERCISE 6-14**: Run the integrator shown below. Set the initial condition of the capacitor to zero volts. Show the output waveform for a 1 kHz square wave input voltage. Note that this circuit will not work if the resistors are not matched exactly. This is a good circuit for simulation, but never use it in practice. You may also wish to run this circuit using non-ideal op-amps to see the effects of bias currents and offset voltages.

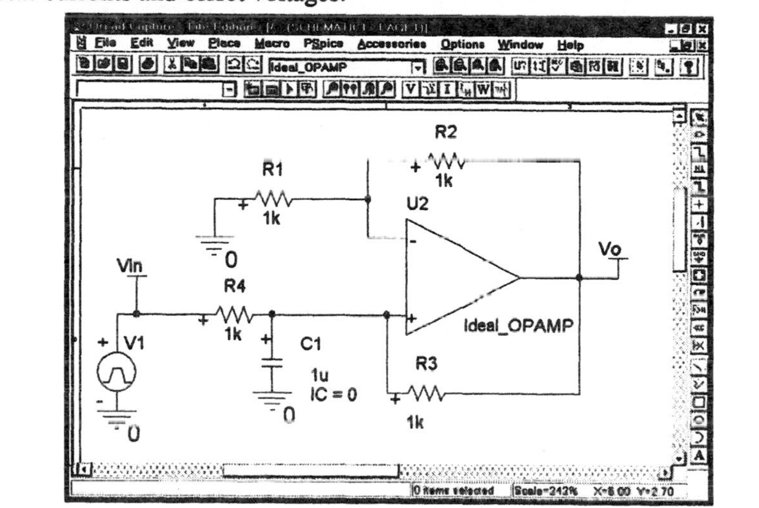


**SOLUTION**: Use the pulsed voltage source. Set the rise and fall times to 1 μs:

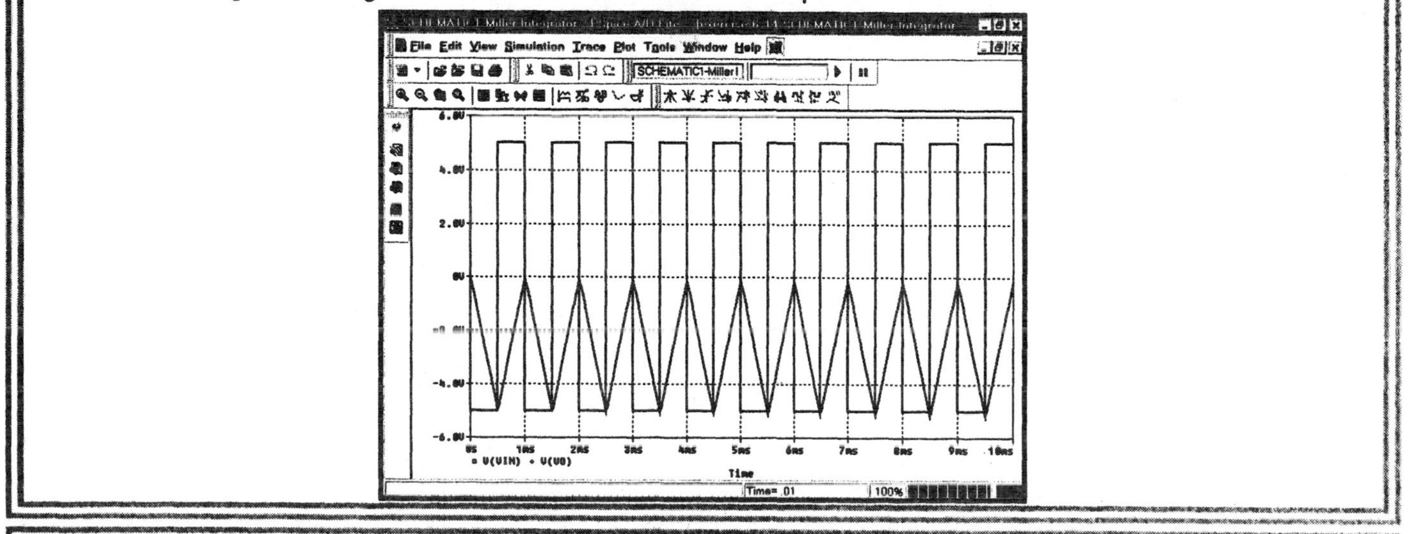

**EXERCISE 6-15**: Run the integrator below using a non-ideal op-amp model like the UA741. For the source, use the square wave voltage source "Vsq."

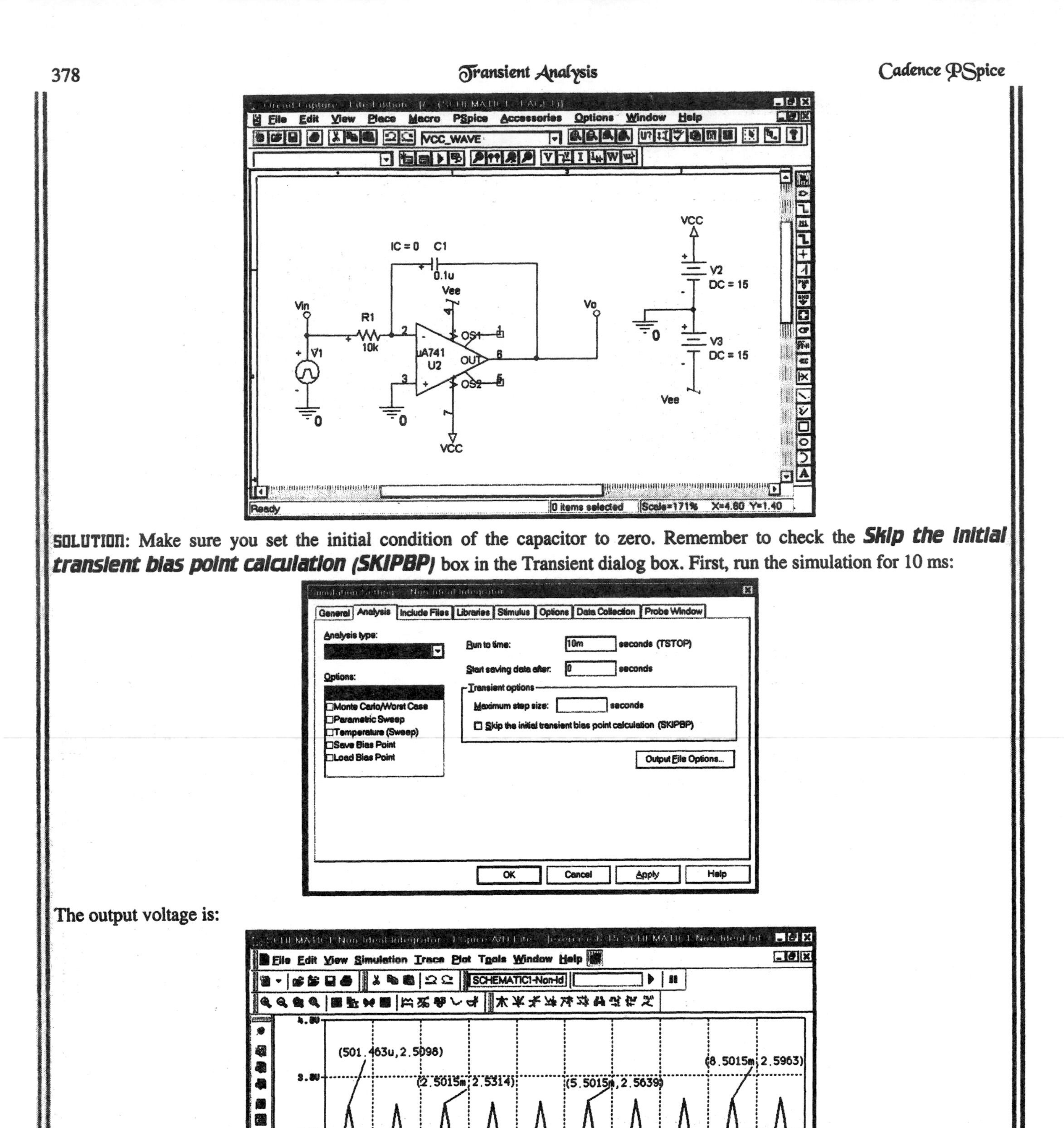


**SOLUTION**: Make sure you set the initial condition of the capacitor to zero. Remember to check the ***Skip the initial transient bias point calculation (SKIPBP)*** box in the Transient dialog box. First, run the simulation for 10 ms:



The output voltage is:



The positive peaks were marked so that we can see that the trace is slowly drifting positive. Set the maximum step size to 10 μs, run the simulation again for 100 ms, and plot the output:

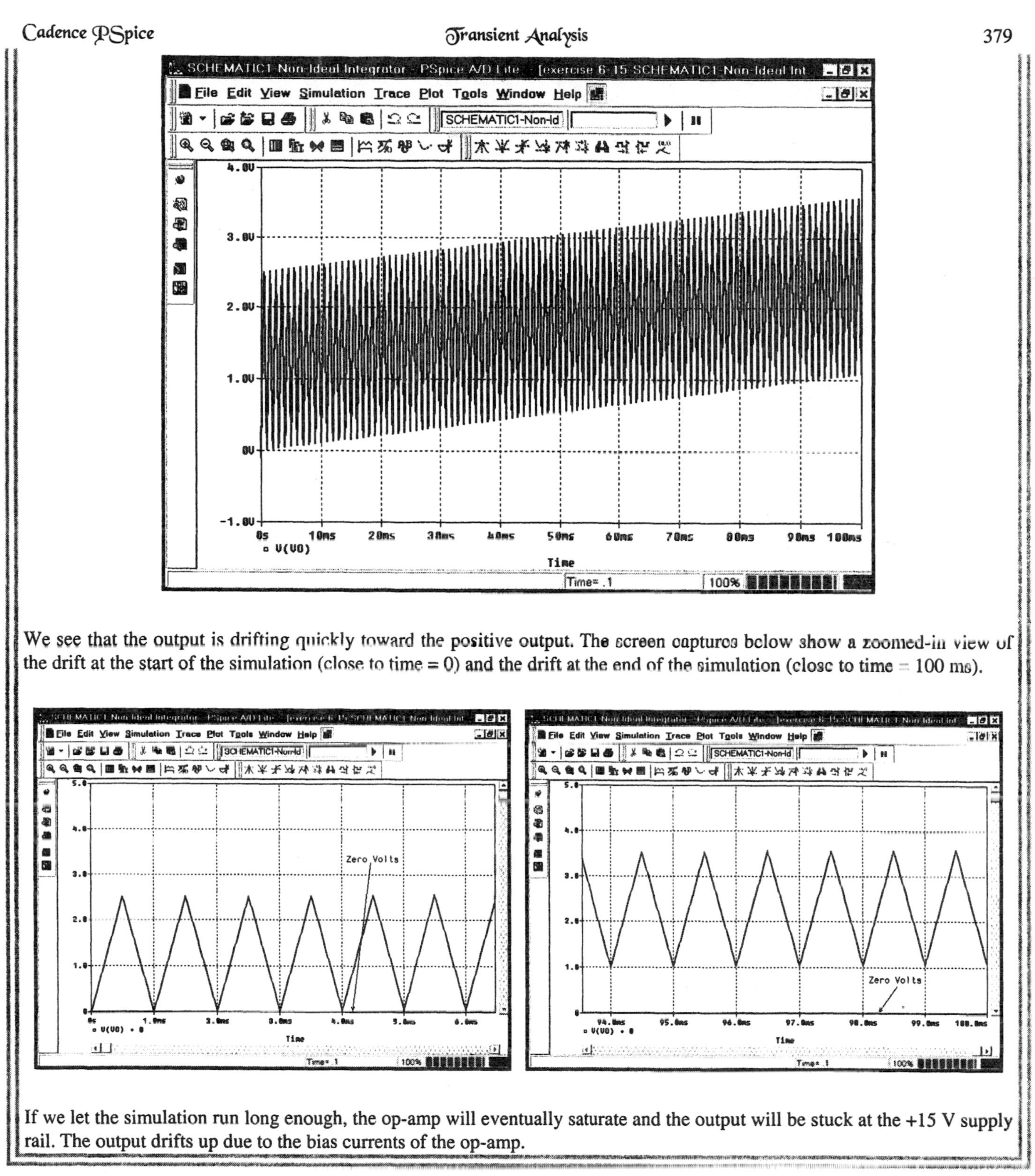


We see that the output is drifting quickly toward the positive output. The screen captures below show a zoomed-in view of the drift at the start of the simulation (close to time = 0) and the drift at the end of the simulation (close to time = 100 ms).

If we let the simulation run long enough, the op-amp will eventually saturate and the output will be stuck at the +15 V supply rail. The output drifts up due to the bias currents of the op-amp.

# 6.I. Multiple Operational Amplifier Circuit

We will now run a circuit with three ideal operational amplifiers. With the Lite version, the component limitation of PSpice limits us to two or three non-ideal operational amplifiers, depending on the complexity of the op-amp model. You may not be able to simulate the circuit of this section depending on the op-amp model you use. The ideal operational amplifier model was created so that a circuit with several operational amplifiers could be simulated using the Lite version. Simulation with ideal op-amps will give you a good idea about what the circuit is supposed to do, but it will not simulate any of the non-ideal properties that may cause your circuit to function improperly, or not meet certain specifications. Always use the non-ideal models when possible. For circuits with lots of op-amps, you will need the professional version of PSpice to accurately simulate the circuit if you want to include the non-ideal properties. Wire the circuit shown below.

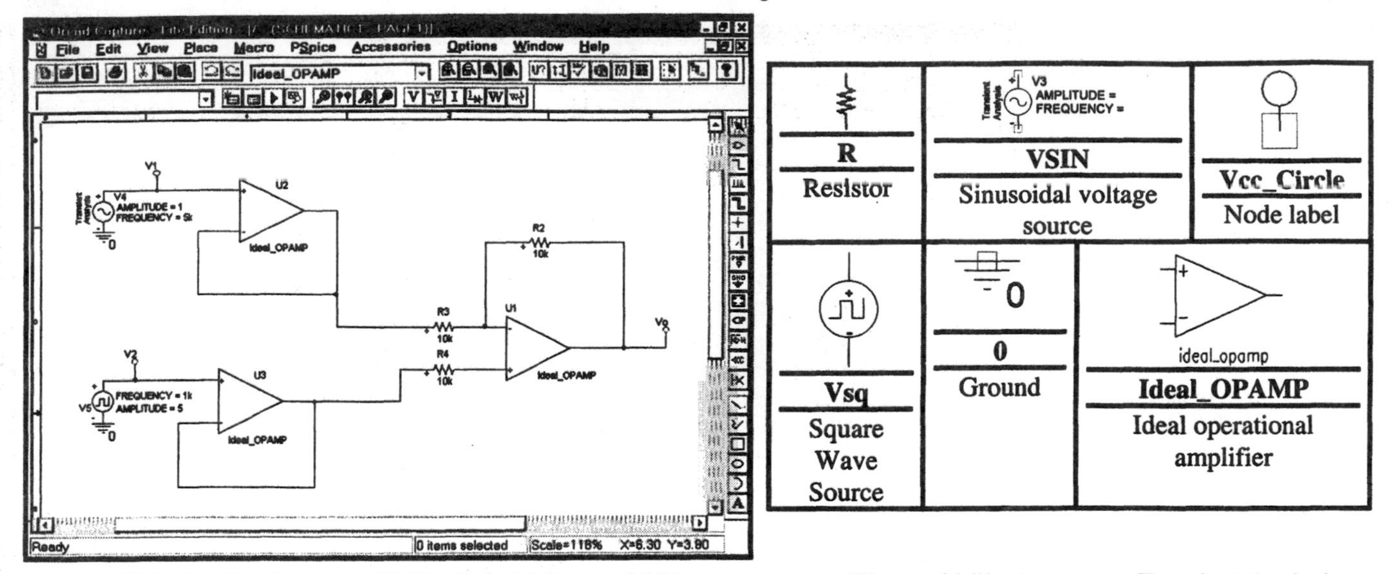


This circuit subtracts a 5 kHz sinusoid from a 1 kHz square wave. We would like to set up a Transient Analysis to run this circuit for 5 ms. Select **PSpice** and then **New Simulation Profile** from the Capture menus, enter a name for the profile, and then click the **Create** button. By default the ***Time Domain (Transient) Analysis type*** is selected. The dialog box below is set up to run the circuit for 5 ms (***Run to time: 5m***). The maximum step size (***Maximum step size***) is not specified so that the simulation will finish as soon as possible.

General | Analysis | Include Files | Libraries | Stimulus | Options | Data Collection | Probe Window

Analysis type: Time Domain (Transient)

Run to time: 5m seconds (TSTOP)

Start saving data after: 0 seconds

Transient options

Maximum step size: seconds

Skip the initial transient bias point calculation (SKIPBP)

Options: Monte Carlo/Worst Case, Parametric Sweep, Temperature (Sweep), Save Bias Point, Load Bias Point

Output File Options...

OK | Cancel | Apply | Help

Run the simulation by selecting **PSpice** and then **Run** from the Capture menu bar. When Probe runs, add the trace **V(VO)**. You will see the following Probe window:

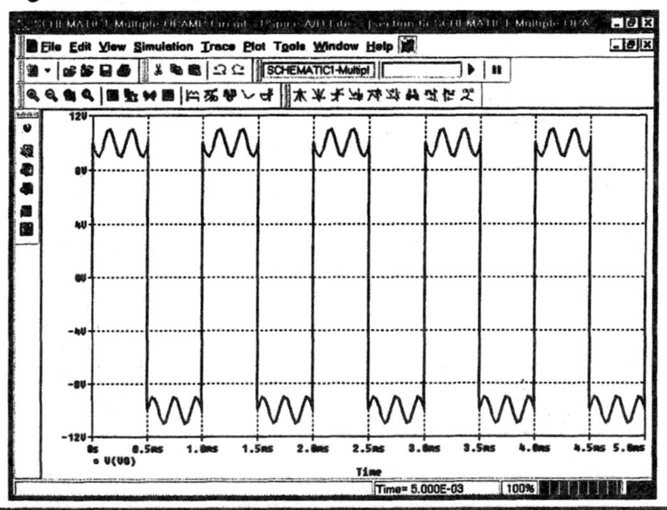


**EXERCISE 6-16**: Find the output of the circuit below if the input is a 5 V amplitude, 1 kHz square wave. The initial voltage of the capacitor is zero volts.

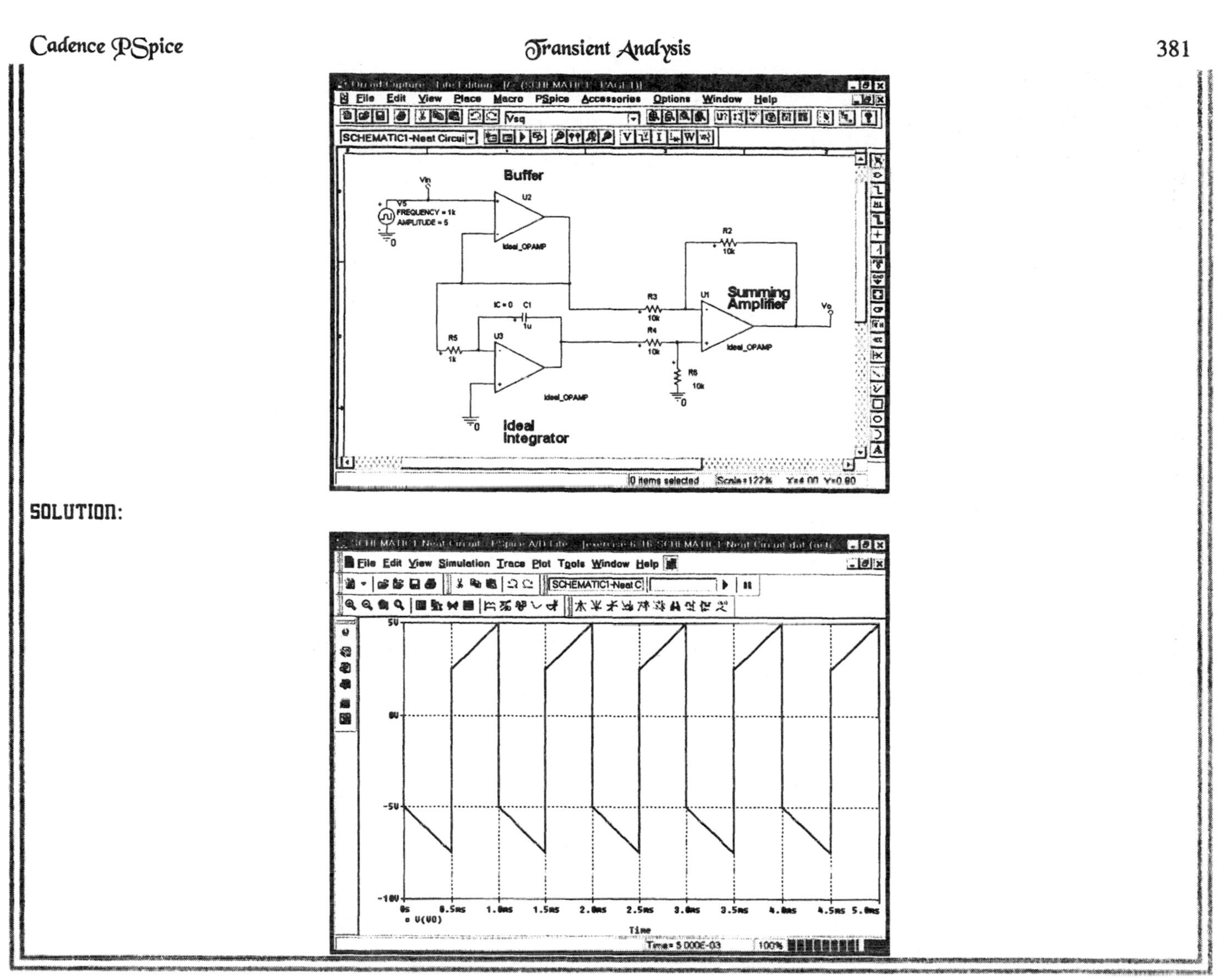


SOLUTION:

# 6.J. Operational Amplifier Schmitt Trigger

In this section we will use an operational amplifier to create a Schmitt Trigger. A non-ideal operational amplifier must be used because the ideal op-amp model has trouble converging when it is used as a Schmitt Trigger. Wire the circuit:

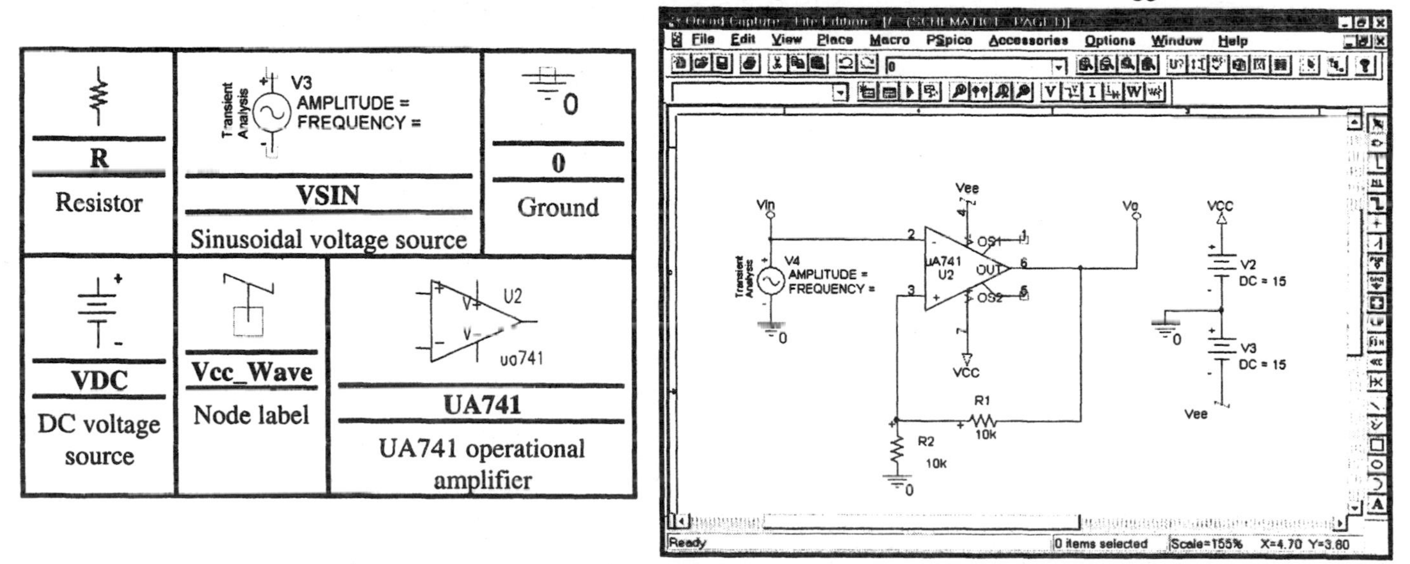

| R | VSIN | 0 |
|---|---|---|
| Resistor | Sinusoidal voltage source | Ground |

| VDC | Vcc_Wave | UA741 |
|---|---|---|
| DC voltage source | Node label | UA741 operational amplifier |



This op-amp circuit is a Schmitt Trigger with trigger points at approximately ±7.5 V. A sinusoidal voltage source will be used to swing the input from +14 V to –14 V and from –14 V to +14 V a few times. The frequency of the source is 1 Hz. This low frequency is chosen to eliminate the effects of the op-amp slew rate on the Schmitt Trigger performance. If you wish to observe slew rate effects, a higher frequency should be chosen. We would also like the sinusoidal source to start at +14 V instead of zero. This can be done by setting the phase of the source to 90°. The phase of the sinusoidal source is not automatically displayed, so it must be changed by editing the phase attribute of the sinusoidal source. Double-click the ***LEFT*** mouse button on the sinusoidal voltage source graphic, , to obtain its spreadsheet:

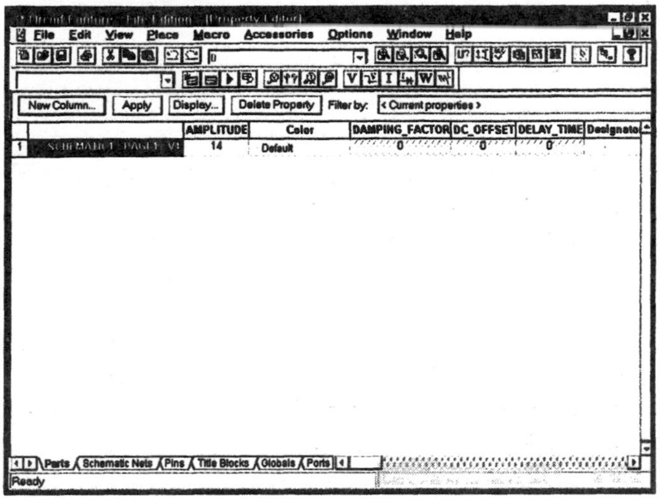

This dialog box shows us all of the possible attributes available for the sinusoidal source. The attributes are displayed alphabetically so not all of the attributes we can set are displayed on the screen capture above. We would like to change the phase of the source, so scroll the spreadsheet until you see the ***PHASE*** attribute and set its value to **90**:

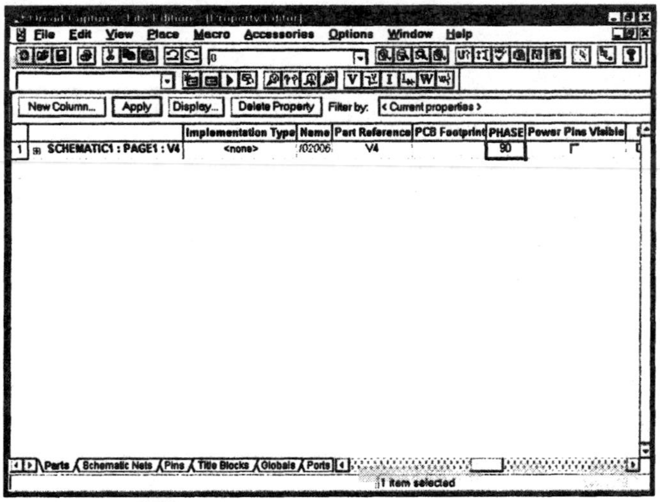

Click the ***LEFT*** mouse button on the lower in the upper right-hand corner to close the spreadsheet. The source we have created is: $V_1 = 14 \sin(2\pi t + 90°)$.

Since many of the attributes of the sinusoidal source have not been previously described, we will give a brief description here. The sinusoidal source is best described by an equation. We will use the following abbreviations:

| Attribute | Symbol | Units |
|---|---|---|
| Phase | $\phi$ | degrees |
| DC_Offset | B | volts |
| Amplitude | A | volts |
| Frequency | F | Hz |
| Damping_factor | $\beta$ | $sec^{-1}$ |
| Delay_time | $\tau_d$ | seconds |

The equation for the sinusoidal source is:

This term is a constant.

$$V_{\sin} = \begin{cases} B + A\sin\left(\left(\dfrac{2\pi}{360}\right)\varphi\right) & \text{for} \quad 0 \le t < \tau_d \\ B + A\sin\left[2\pi\ F(t-\tau_d) + \left(\dfrac{2\pi}{360}\right)\varphi\right]e^{-\beta(t-\tau_d)} & \text{for} \quad t \ge \tau_d \end{cases}$$

This may not be too clear. The full sinusoidal source can be exponentially damped, have a DC offset, and have a time delay as well as a phase delay. In the above equation, the phase ($\phi$) is specified in degrees and is converted into radians by the constant $2\pi/360$. We note that for $\tau_d \ge t \ge 0$, $V_{\sin}$ is constant. The sinusoid does not start until $t = \tau_d$. If there are no time or phase delays, the above equation reduces to the exponentially damped sine wave:

$$V_{\sin} = B + A\sin[2\pi\ Ft]e^{-\beta t}$$

The sources available for use in the Transient Analysis are very flexible to provide the user with many possible waveforms for simulation. Unfortunately, this flexibility can also lead to confusion when using a source for the first time. To make creating waveforms as easy as possible, Orcad has provided a program for creating and viewing waveforms before running a simulation. The program is called the Stimulus Editor and is discussed in Section 6.M. This program makes understanding the description of the sinusoidal source much easier.

We would now like to set up a Transient Analysis that allows the sinusoidal source to complete two cycles. Since the source has a frequency of 1 Hz, two cycles will take two seconds of simulation time. Select **PSpice** and then **New Simulation Profile** from the Capture menus, enter a name for the profile, and then click the ***Create*** button. By default the ***Time Domain (Transient) Analysis type*** is selected. Fill in the parameters as shown in the Time Domain dialog box below:

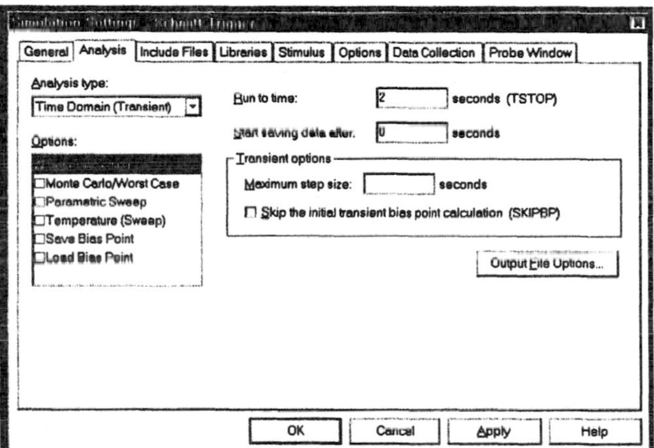


Run PSpice and then Probe. Add the traces **V(VIN)** and **V(VO)**. You should have the Probe screen shown:

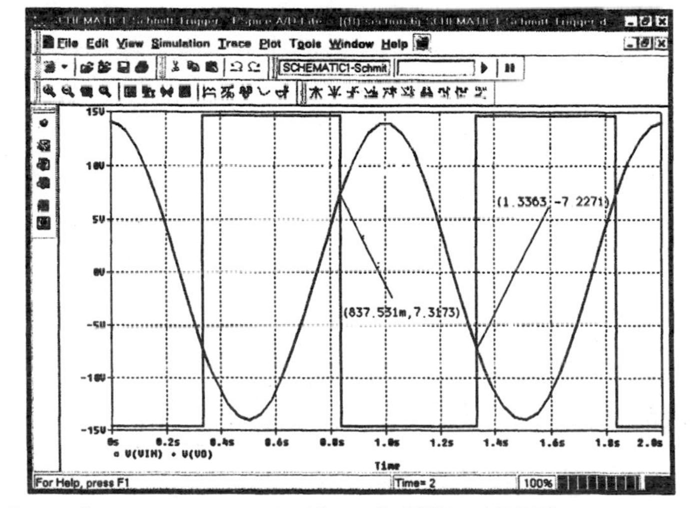


We can see that the Schmitt Trigger changes state at approximately VIN = ±7.2 V.

The next thing we would like to do is plot the hysteresis curve for this Schmitt Trigger. The hysteresis curve is a plot of $V_O$ versus $V_{IN}$. We must delete the trace ***V(VIN)***. Click the ***LEFT*** mouse button on the text ***V(VIN)***. It should turn red, indicating that it has been selected. To delete the trace, press the **DELETE** key. You should have the Probe screen shown below:

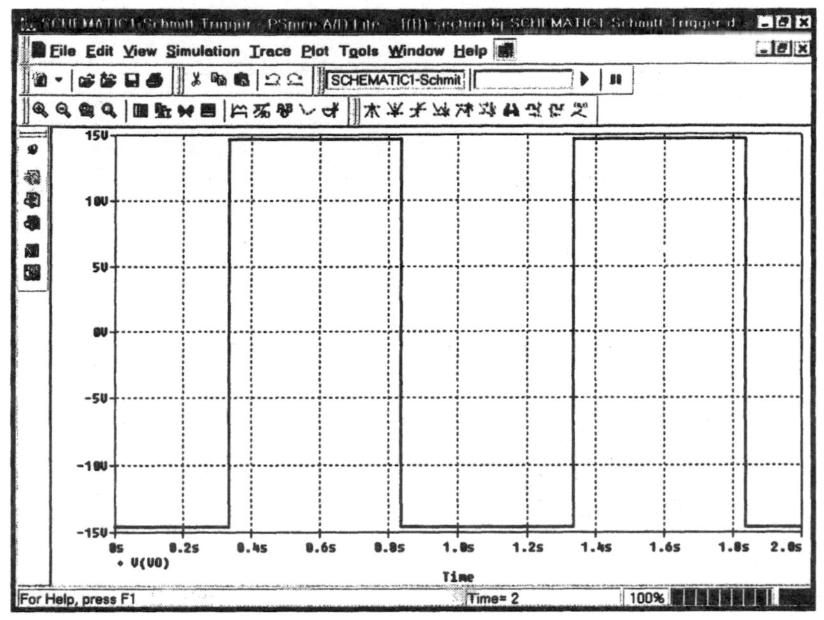

We must now change the x-axis from time to the input voltage, $V_{in}$. Select **Plot** and then **Axis Settings** from the Probe menu bar. The ***X Axis*** tab is displayed by default:

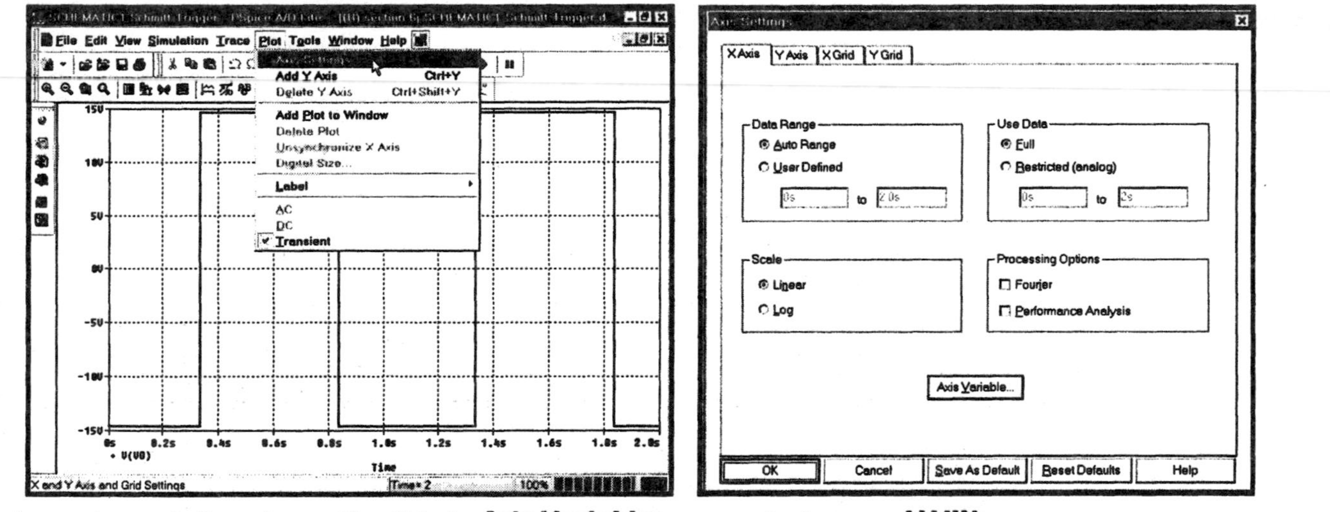

To change the x-axis from time to $V_{in}$, click the ***Axis Variable*** button and select trace ***V(VIN)***

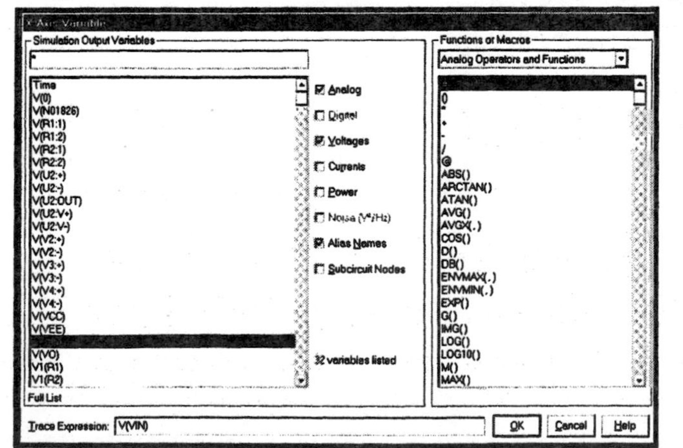

Click the ***OK*** button twice to accept the changes and view the trace. You should see the hysteresis curve:

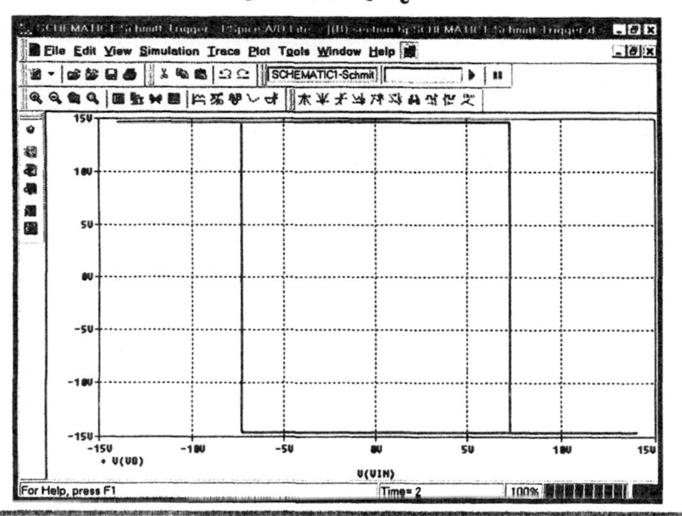

**EXERCISE 6-17:** Run a non-inverting Schmitt Trigger with a ±14 volt sine wave input.

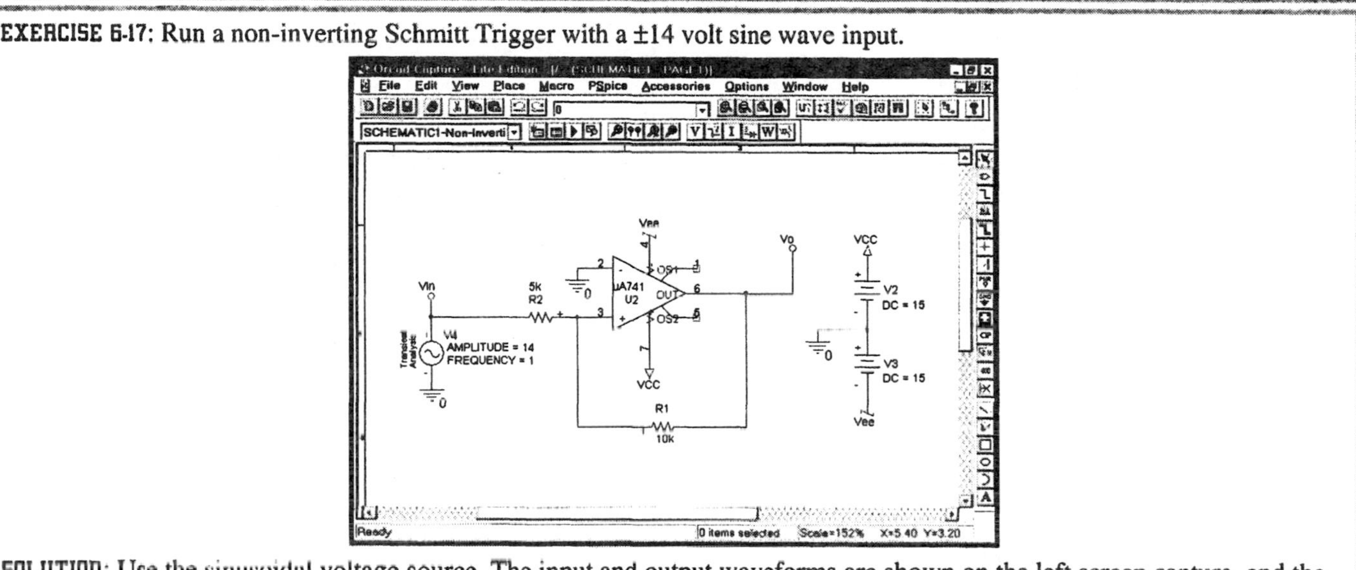


**SOLUTION:** Use the sinusoidal voltage source. The input and output waveforms are shown on the left screen capture, and the hysteresis curve is shown on the right screen capture.

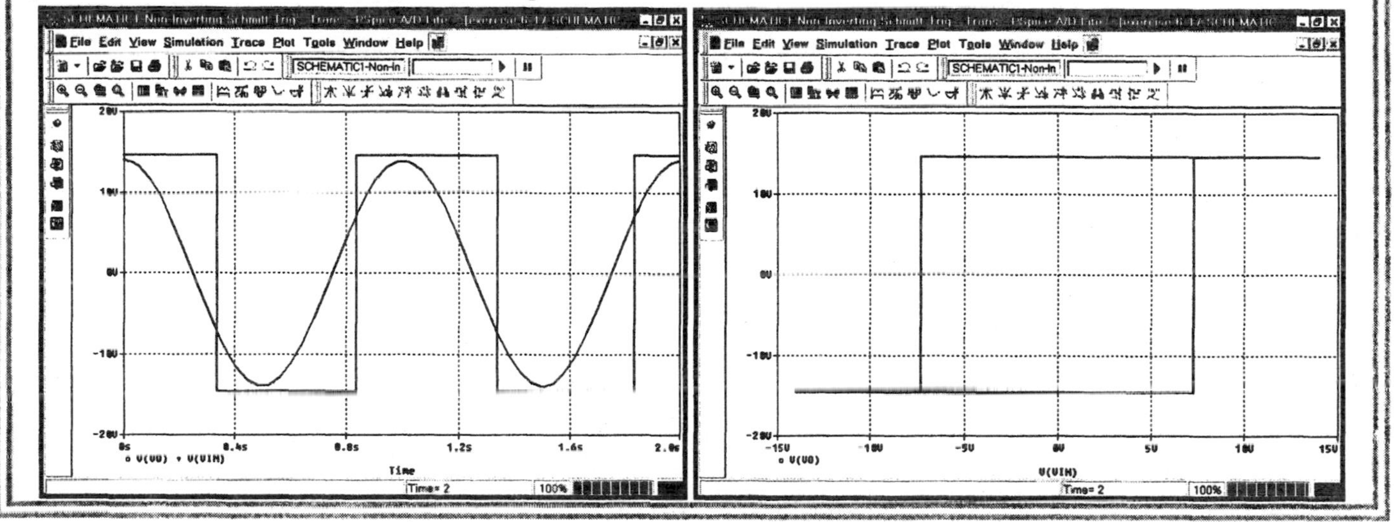

# 6.K. Parametric Sweep — Inverter Switching Speed

When designing digital circuits we are concerned with how different circuit elements affect the operation of the circuit. In this section we will look at switching speed. First, we will look at a basic BJT inverter and observe its operation. Wire the circuit below:

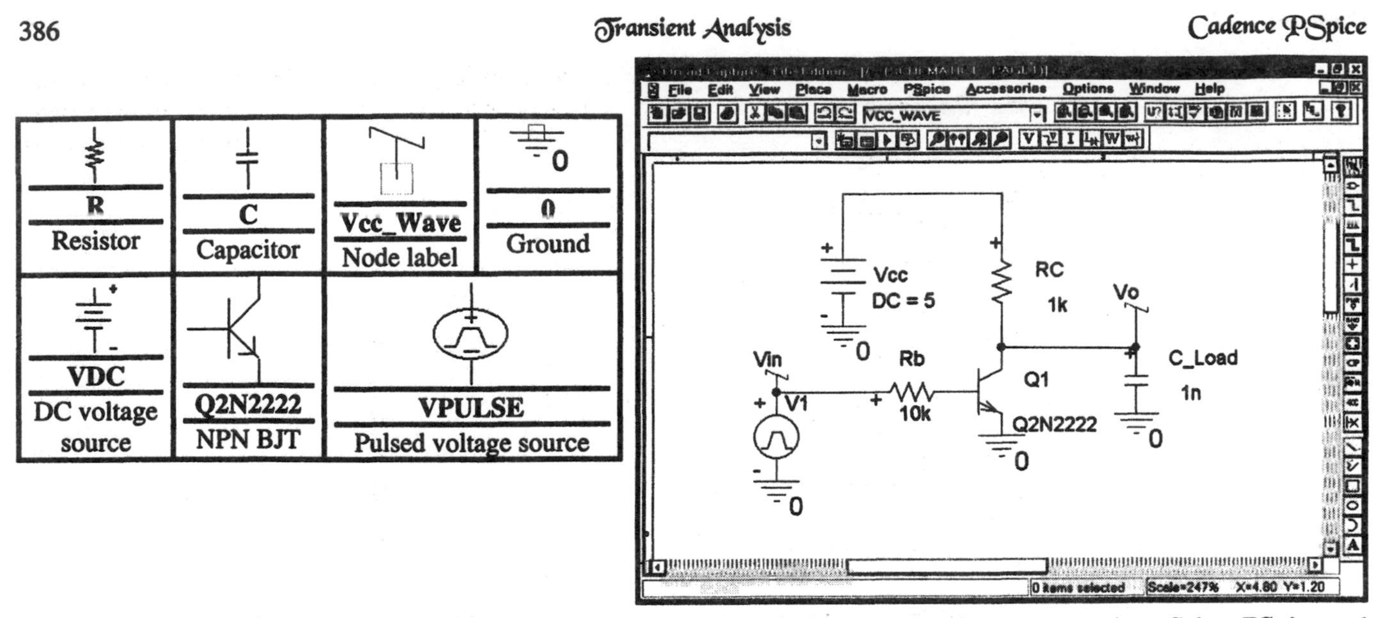


We would like to run a Transient Analysis because we are looking at waveforms versus time. Select **PSpice** and then **New Simulation Profile** from the Capture menus, enter a name for the profile, and then click the ***Create*** button. By default, the ***Time Domain (Transient) Analysis type*** is selected. Fill in the parameters as shown in the Time Domain dialog box below:

General | Analysis | Include Files | Libraries | Stimulus | Options | Data Collection | Probe Window

Analysis type: Time Domain (Transient)

Run to time: 50u seconds (TSTOP)

Start saving data after: 0 seconds

Transient options

Maximum step size: 10n seconds

Skip the initial transient bias point calculation (SKIPBP)

Options: Monte Carlo/Worst Case, Parametric Sweep, Temperature (Sweep), Save Bias Point, Load Bias Point

Output File Options...

OK | Cancel | Apply | Help

Click the ***OK*** button when you have made the changes to return to the schematic. The input to the inverter will be a short 1 μs pulse. The attributes of the pulsed voltage source are: Period = 50u, rise_time = 1n, fall_time = 1n, Pulse_width = 1u, initial_voltage = 0, Pulsed_voltage = 5, delay_time = 1u. Double-click the ***LEFT*** mouse button on the pulsed voltage source graphic to obtain the spreadsheet for the source:

New Column... | Apply | Display... | Delete Property | Filter by: < Current properties >

| | Color | DELAY_TIME | Designator | FALL_TIME | Graphic | ID | Implem |
|---|---|---|---|---|---|---|---|
| 1 | Default | | | | VPULSE.Normal | | |

Parts / Schematic Nets / Pins / Title Blocks / Globals / Ports

Ready

Fill in the specified parameters and then type **CTRL-F4** to close the spreadsheet and return to the schematic.

We see that the pulse width is 1 μs and the period is 50 μs long to let the capacitor come into steady state after the pulse. Note that we are simulating the circuit for only one period. We could have used the PWL source, but the pulsed voltage source is easier to set up. Run the simulation and then run Probe. We will first look at the input to see if the pulse is correct. Add the trace **`V(VIN)`**:

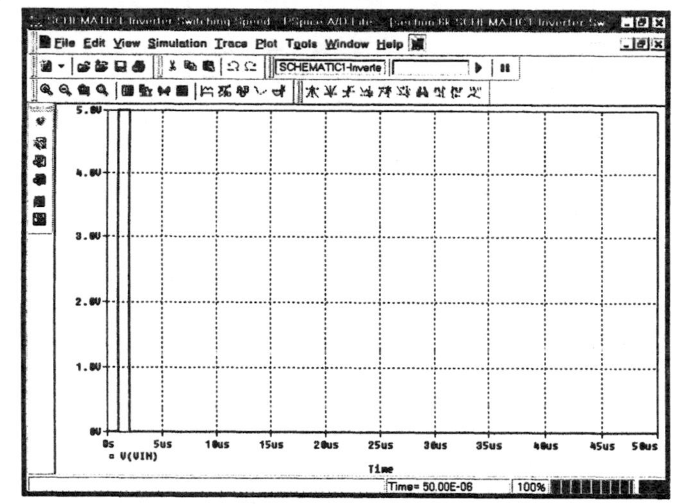


The pulse looks correct. Usually waveforms from a digital circuit have edges close together. If we plot both Vin and Vo on the same graph, the two traces may be hard to distinguish. To see both waveforms clearly, we will add a plot. Select **Plot** and then **Add Plot to Window** from the Probe menus. A second plot will appear on the same window. Add the trace **`V(VO)`**:

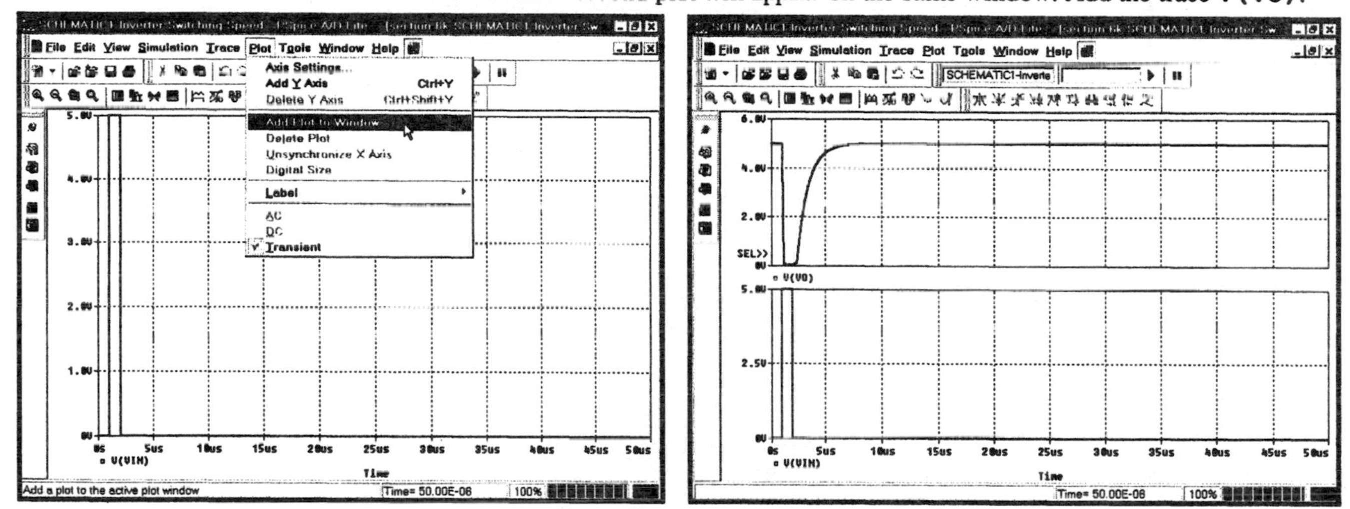


Now that we know what the input and output waveforms look like, we will see how changing the collector resistor ***RC*** affects the waveform. Return to the schematic. The Parametric Sweep can be used to change the value of any circuit parameter. First we must define the parameter we want to change. Get a part called "PARAM" and place it in your circuit:

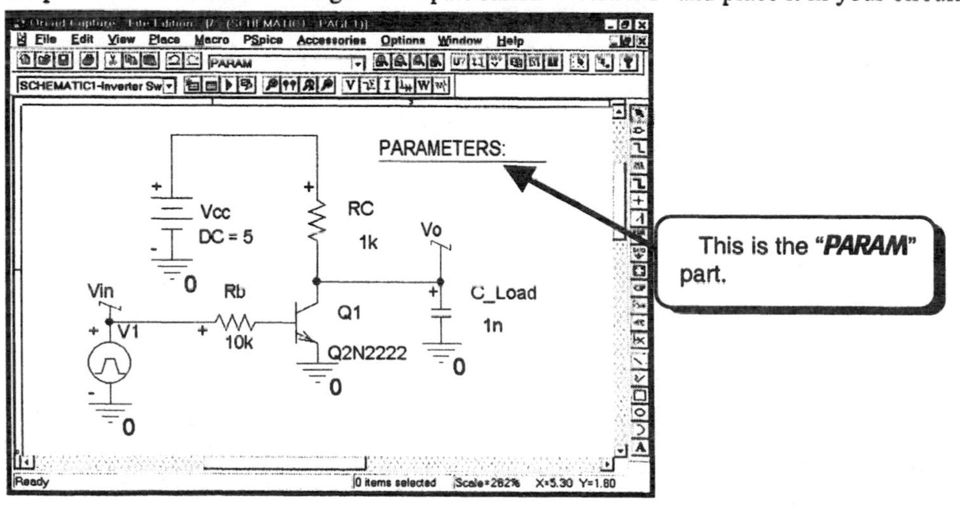

To edit the attributes of the ***PARAM*** part, double-click the *LEFT* mouse button on the text ***PARAMETERS:***. The spreadsheet for the ***PARAM*** part will open:



We need to create a new parameter, so click the ***New Column*** button:

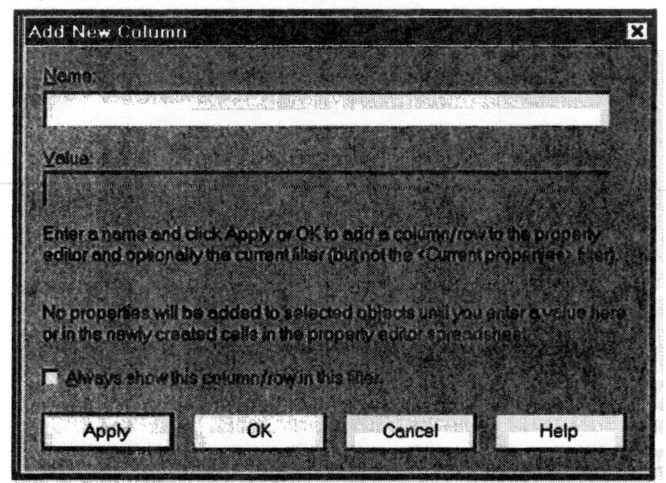


Enter **R_val** for the name and **1k** for the value of the parameter:



Click the ***OK*** button to add the parameter as a new column to the spreadsheet:

Orcad Capture - Lite Edition - [Property Editor]
File Edit View Place Macro Accessories Options Window Help
PARAM
SCHEMATIC1-Inverter Sw
New Column... Apply Display... Delete Property Filter by: < Current properties >

| | | R_val | Color | Designator | Graphic | ID | Implementation | Implement |
|---|---|---|---|---|---|---|---|---|
| 1 | ⊞ SCHEMATIC1 : PAGE1 : 1 | 1k | Default | | PARAM.Normal | | | |

Parts Schematic Nets Pins Title Blocks Globals Ports
1 item selected

Click the *LEFT* mouse button on the R_val column to select the column and then click the Display button. Change the option to ***Name and Value***:

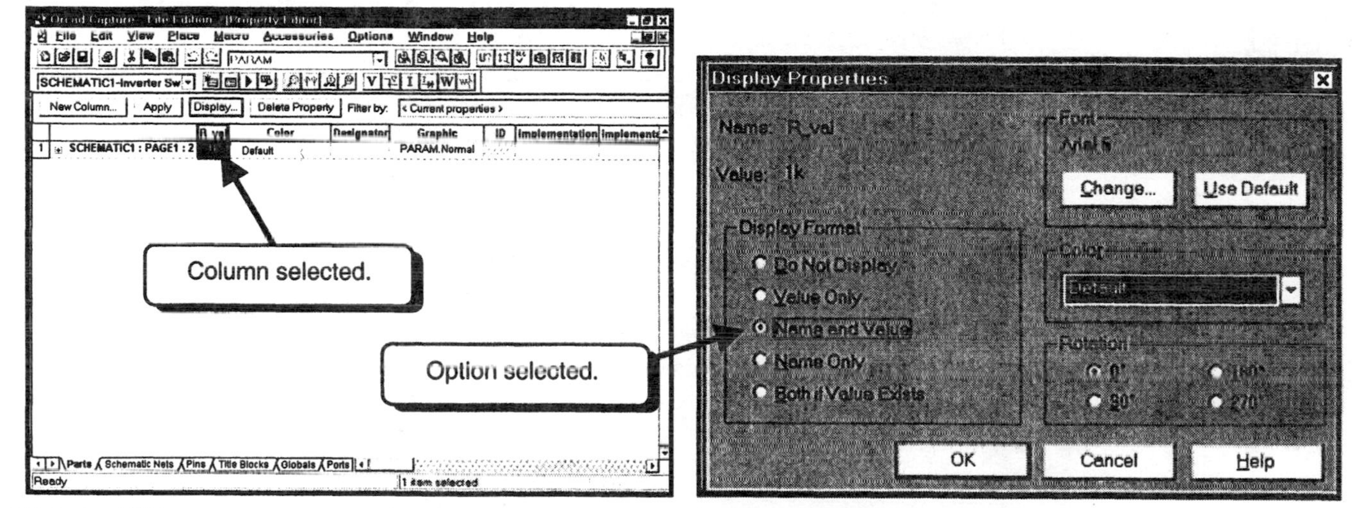


Click the ***OK*** button to return to the spreadsheet:

Orcad Capture - Lite Edition - [Property Editor]
File Edit View Place Macro Accessories Options Window Help
PARAM
SCHEMATIC1-Inverter Sw
New Column... Apply Display... Delete Property Filter by: < Current properties >

| | | R_val | Color | Designator | Graphic | ID | Implementation | Implement |
|---|---|---|---|---|---|---|---|---|
| 1 | ⊞ SCHEMATIC1 : PAGE1 : 2 | 1k | Default | | PARAM.Normal | | | |

Parts Schematic Nets Pins Title Blocks Globals Ports
1 item selected

Type **CTRL-F4** to close the spreadsheet and return to the schematic:

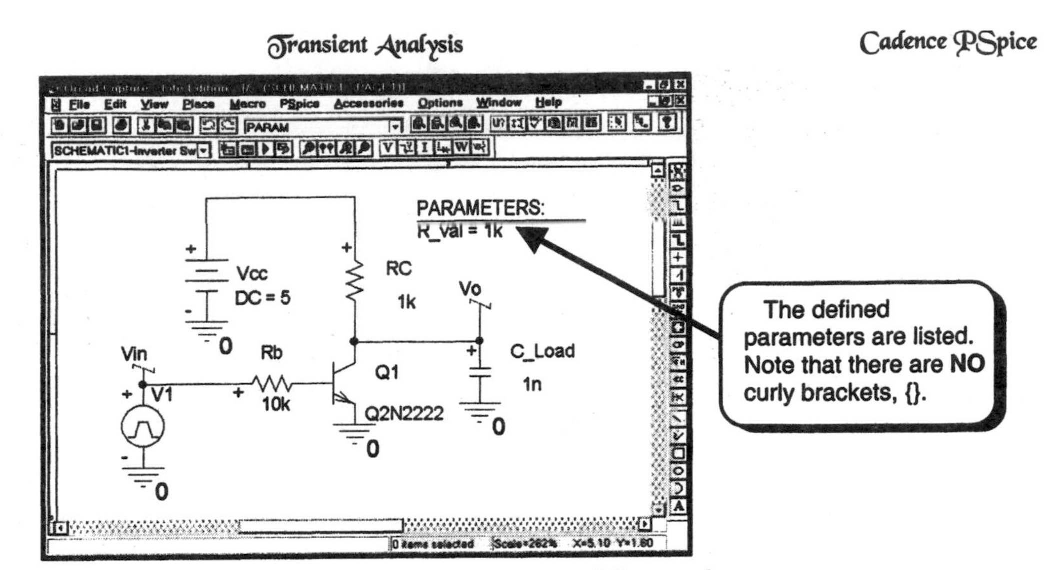


In a schematic a parameter is treated as a value. We must now change the value of ***RC*** from ***1k*** to the name of the parameter, ***R_val***. Double-click on the text ***1k*** next to resistor ***RC***:

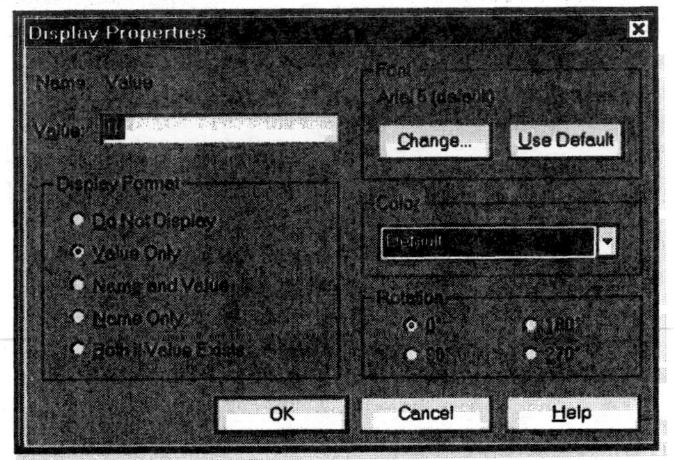


Type in the text **{R_val}** and click the ***OK*** button. **When parameters are used as component values they are enclosed in curly brackets.** The value of the resistor in the schematic should change to ***{R_val}***:

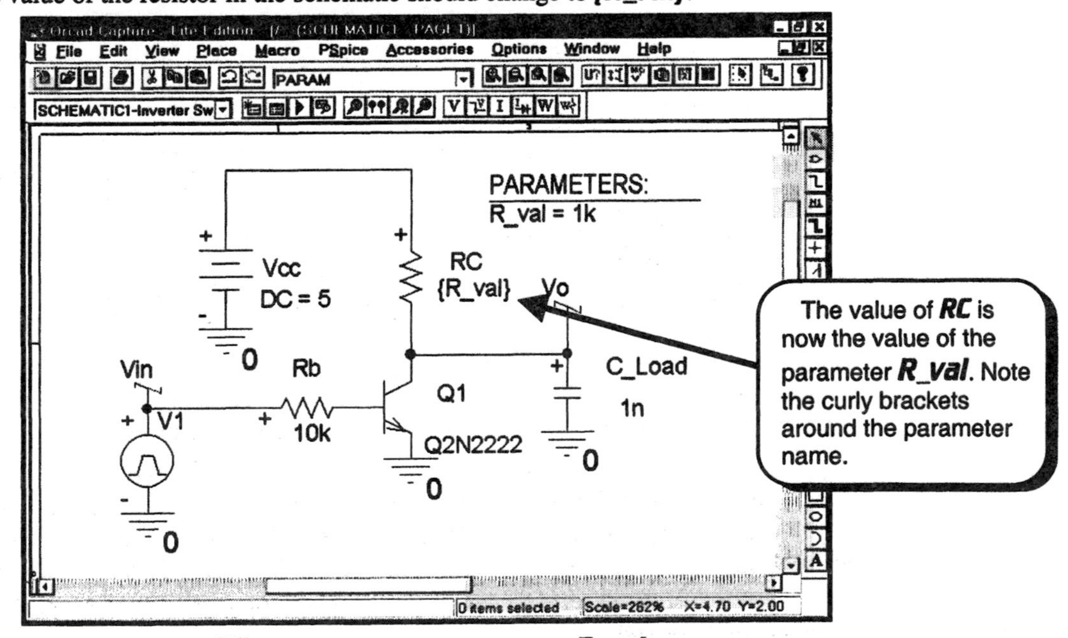


When the circuit is simulated, the value of ***RC*** is the value of the parameter ***{R_val}***.

We will now set up a Parametric Sweep to change the value of the parameter. Suppose we want to see the performance of the circuit for values of ***RC*** from 1 kΩ to 10 kΩ. We need to set up a Parametric Sweep. Select **PSpice** and then **Edit Simulation Profile** from the Capture menus:

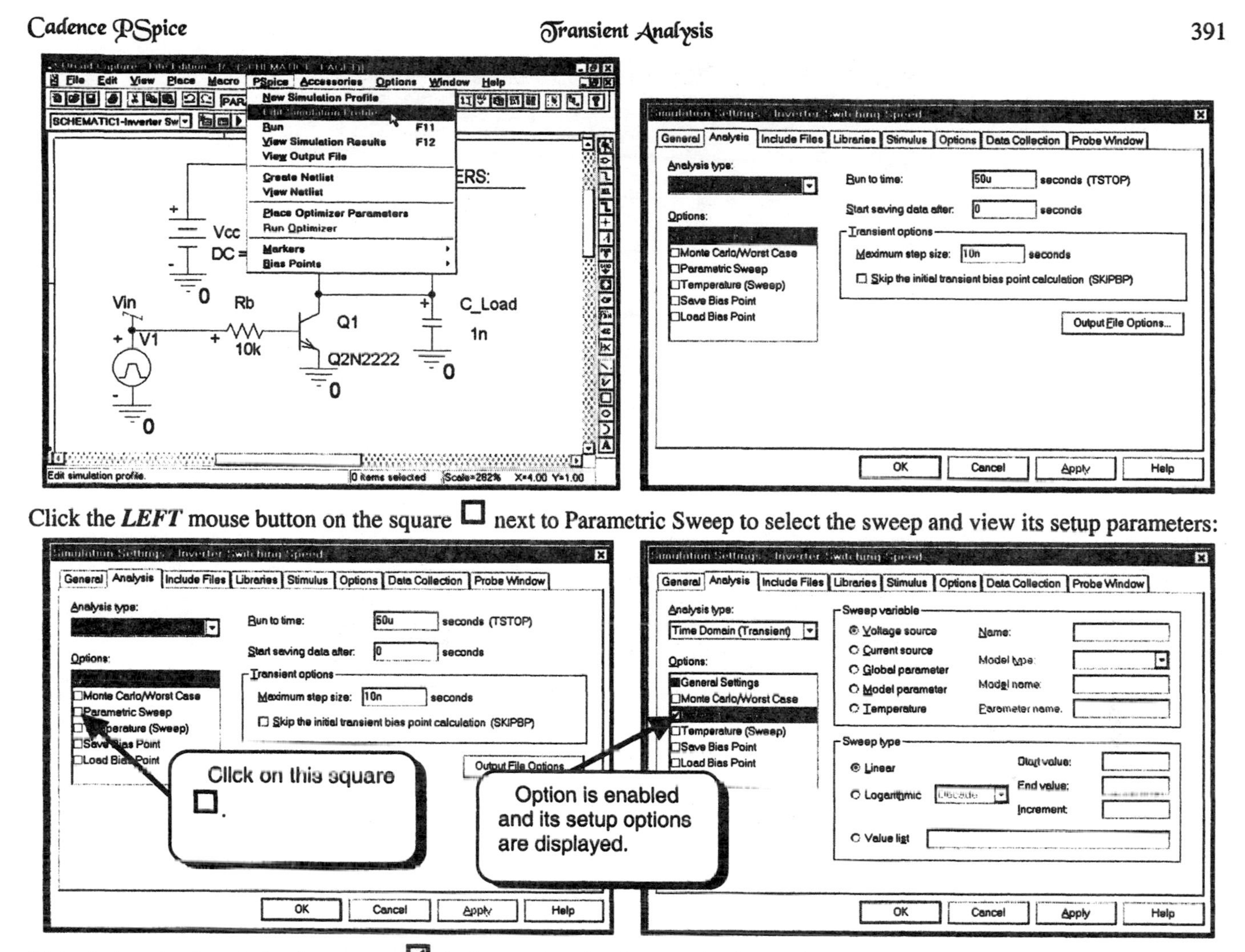


Click the ***LEFT*** mouse button on the square ☐ next to Parametric Sweep to select the sweep and view its setup parameters:

Note that a checkmark in the square ☑ indicates that the Parametric Sweep is enabled. A parameter is referred to as a ***Global parameter*** in this dialog box. Fill in the dialog box as shown:

Note that there are NO curly brackets, {}, in the parameter name.

Note that there are no curly brackets around the parameter. Click the ***OK*** button to accept the value and return to the schematic. Logically, the Transient Analysis executes inside the Parametric Sweep. That is, for each value of the parameter, the Transient Analysis is run. Thus, for this setup, ten Transient Analyses will be run. Since we chose a small Maximum time step in the Transient Analysis setup, this simulation will take a long time to run. To speed up the simulation, you may want to increase the value of the Maximum time step.

Run the simulation (**PSpice, Run**). When the simulation is complete you will see the dialog box:

Available Sections

All None OK Cancel

Since the Transient Analysis was run several times, Probe is asking which of the runs we want to view. We would like to see all results on the same graph. By default, all of the runs are selected, so click the ***OK*** button. Next, add the trace **`V(VIN)`**:

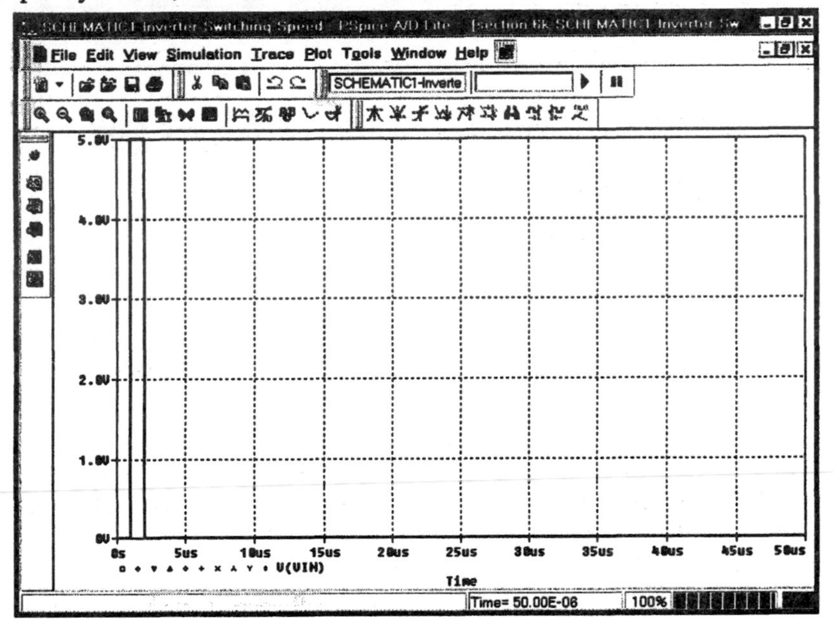


Ten traces are shown because the Transient Analysis was run ten times and ***V(VIN)*** was in each simulation. Next, we will add a plot to make seeing the traces easier. Select ***Plot*** followed by ***Add Plot to Window***, and then add the trace **`V(VO)`**:

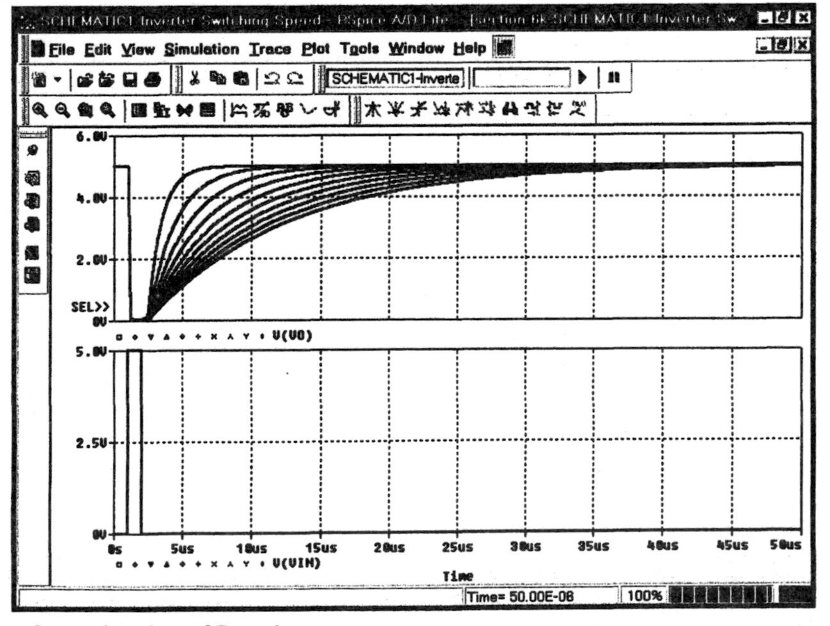


Ten traces are plotted, one for each value of R_val.

To identify which trace corresponds to which value of R_val, click the ***RIGHT*** mouse button on one of the traces and select **Information**:

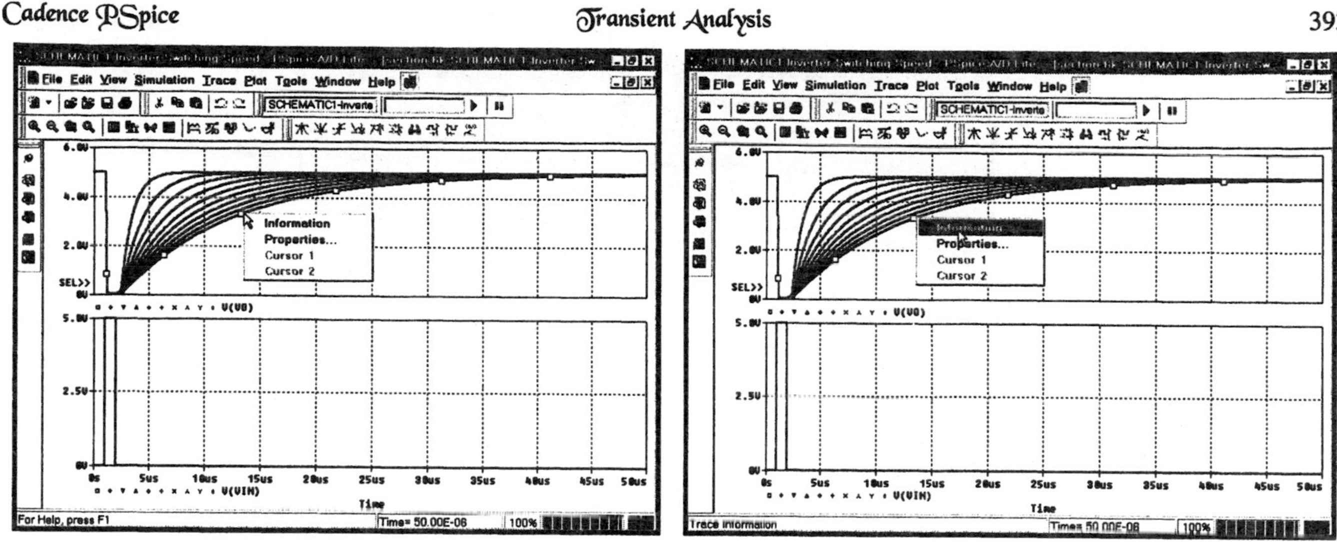


After double-clicking on the square marker, the information for that trace will appear:

**Section Information**

V(VO)

This trace came from one simulation run

Step param R_val = 10.0000E+03
Temperature = 27.0 Deg
Simulation at 14:18:44 on 09/07/00
The simulator created 5034 data points.
This trace is being displayed using 5036 data points.

OK

We see that the slowest rise time occurred for a value of ***R_val*** equal to 10 kΩ. This example will be continued in Section 6.L. You may wish to save this circuit before running the circuit of **EXERCISE 6-18**.

**EXERCISE 6-18**: For the BJT inverter, run the circuit to observe how ***Rb*** affects switching speed. Let ***RC*** remain constant at 1 kΩ.

**SOLUTION**: Add a second parameter called ***RB_val***. Change the value of ***Rb*** from 10k to ***{RB_val}***:

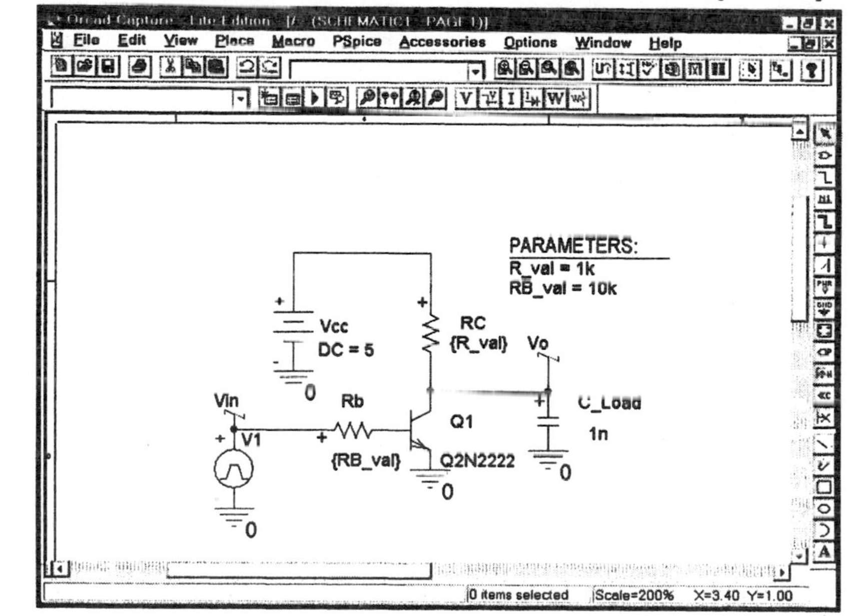


Modify the Parametric Sweep to sweep parameter ***RB_val*** from 5 kΩ to 30 kΩ in 5 kΩ steps. Note that the Transient Analysis will run six times.

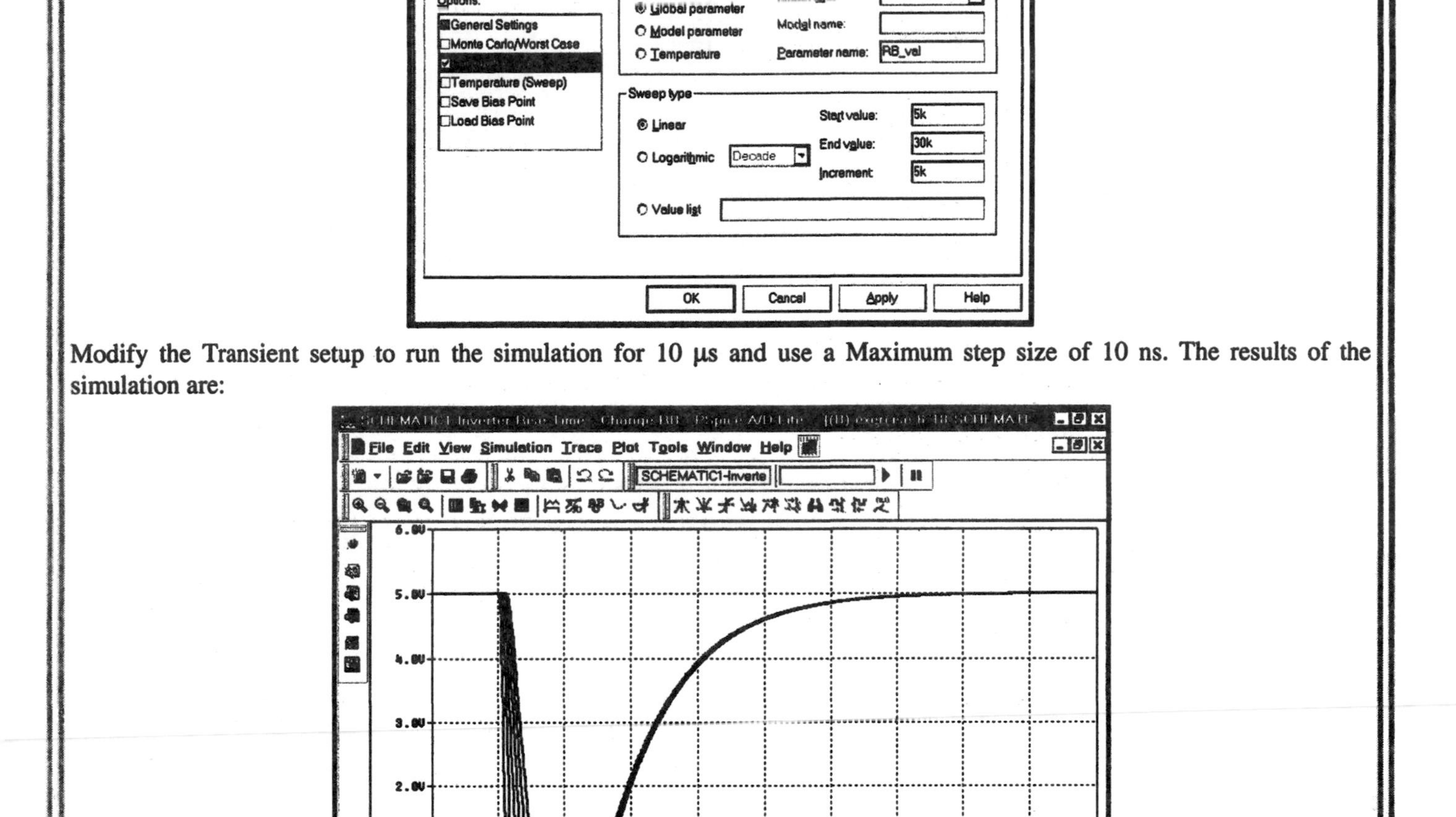


Modify the Transient setup to run the simulation for 10 µs and use a Maximum step size of 10 ns. The results of the simulation are:

The results show that the base resistance affects the fall time and the rise time. See **EXERCISE 6-19** to generate plots of fall time versus ***Rb***.

# 6.L. Performance Analysis — Inverter Rise Time

The Performance Analysis capabilities of Probe are used to view properties of waveforms that are not easily described. Amplifier bandwidth, rise time, and overshoot are examples. To calculate the bandwidth of a circuit, you must find the maximum gain, and then find the frequency where the gain is down by 3 dB. To calculate rise time, you must find the 10% and 90% points, and then find the time difference between the points. The Performance Analysis gives us the capability to plot these properties versus a parameter or device tolerances. The Performance Analysis is used in conjunction with the Parametric Sweep to see how the properties vary versus a parameter. The Performance Analysis is used in conjunction with the Monte Carlo analysis to see how the properties vary with device tolerances. In this section we will plot the rise time of a BJT inverter versus the value of the collector resistor. See Section 9.G to learn how to use the Performance Analysis in conjunction with the Monte Carlo analysis.

Suppose that for the example of Section 6.K we would like to see a plot of how the rise time is affected by the value of the collector resistor. This plot can be accomplished using the Performance Analysis capabilities of Probe. Repeat the procedure of Section 6.K. When Probe runs, select all of the runs and add the trace **V(VO)**:

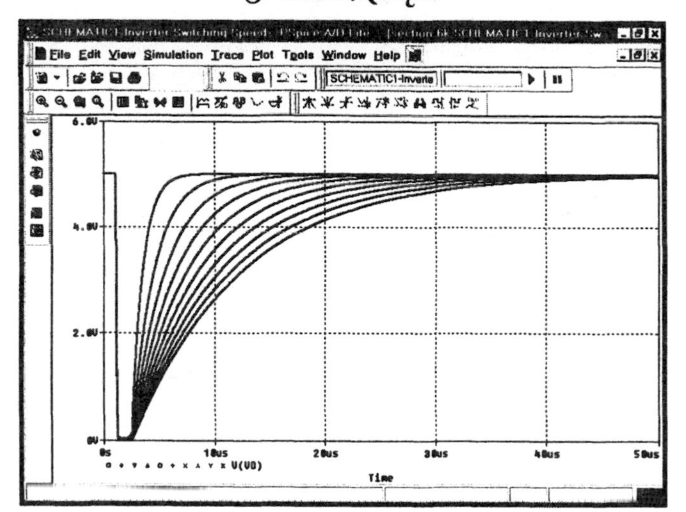


We see that all of the traces reach the final value of 5 V. To plot the rise time versus the parameter {R_val} we must select the Performance Analysis. From the Probe menu select **Plot** and then **Axis Settings**. By default, the ***X Axis*** tab is selected:

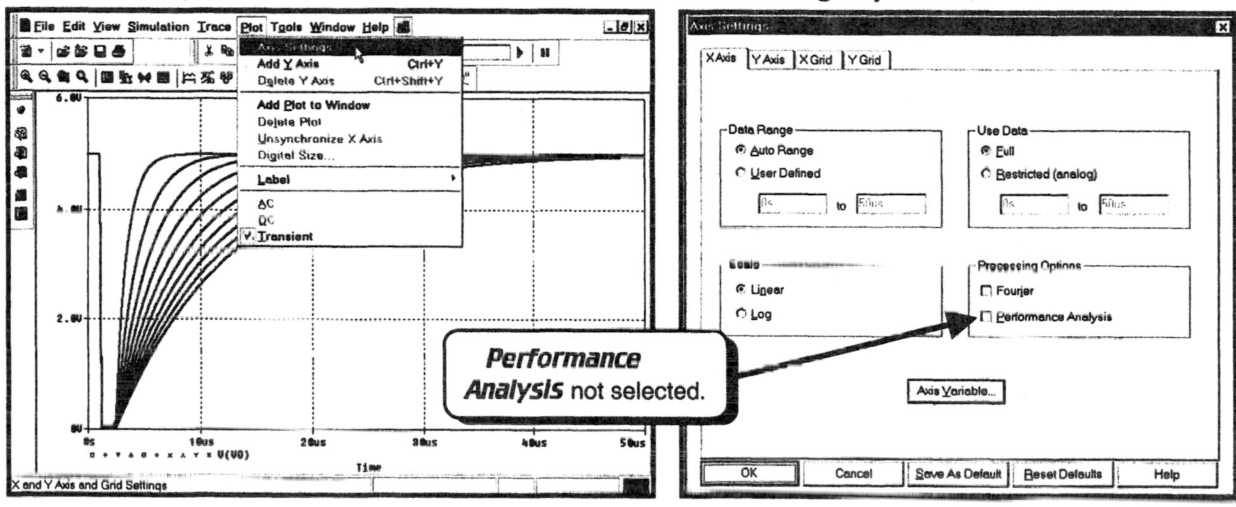


Under ***Processing Options*** we see that the square ☐ next to ***Performance Analysis*** does not have a checkmark in it, indicating that it is not enabled. To enable the ***Performance Analysis*** click the *LEFT* mouse button on the text ***Performance Analysis***. The square should fill with a checkmark ☑, indicating that the ***Performance Analysis*** is enabled:

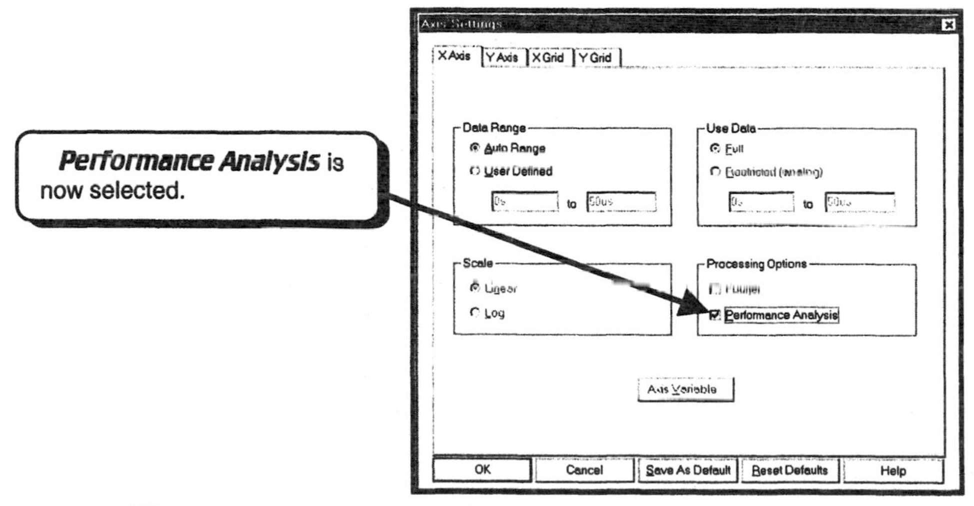


Click the ***OK*** button to return to Probe. You should have two plots on the screen:

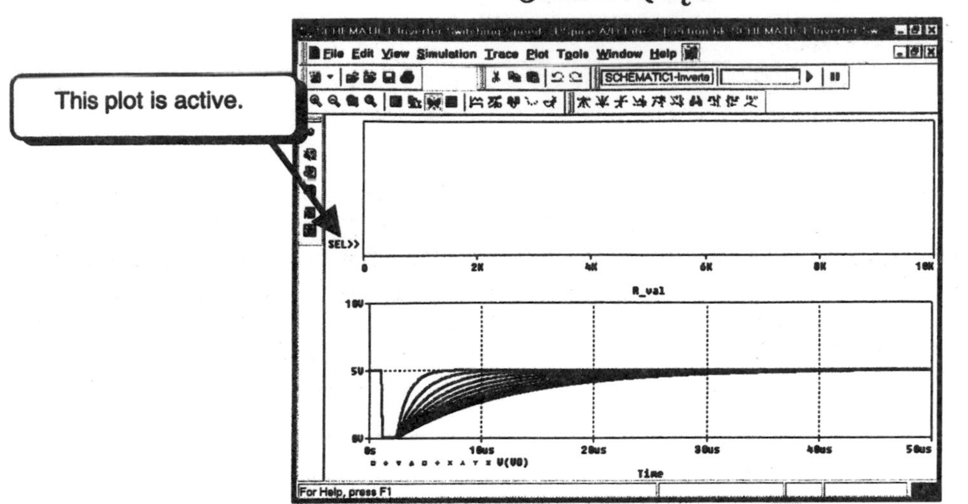


Notice that the top plot has the text ***SEL>>*** next to it. This means that the top plot is the active plot. Any traces that are added will appear on the active plot.

We now wish to add the rise time plot. Select **Trace** and then **Add Trace**:

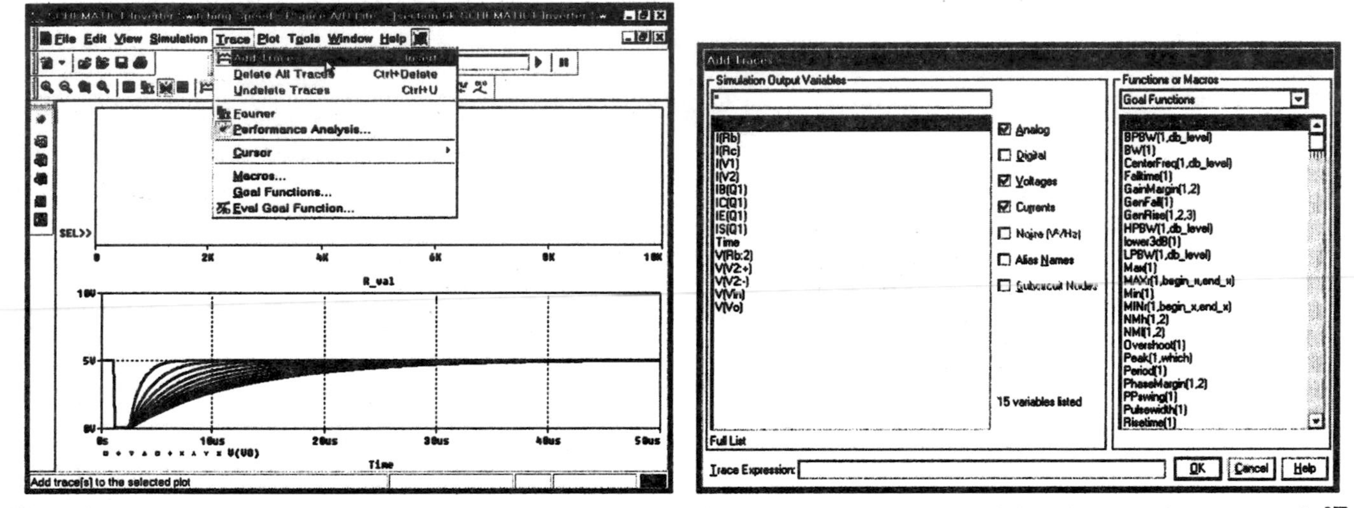

The left pane contains the normal voltage and current traces. The right pane contains goal functions, such as ***upper3dB***, ***lower3dB***, and ***Risetime***. These functions are defined using the Performance Analysis capabilities of Probe. To see how the functions are defined, use the Notepad program to edit the file c:\Program Files\OrcadLite\PSpice\common\pspice.prb. Near the bottom of the file you will see the text:

```
Risetime(1) = x2-x1
*
*#Desc#* Find the difference between the X values where the trace first
*#Desc#* crosses 10% and then 90% of its maximum value with a positive
*#Desc#* slope.
*#Desc#* (i.e. Find the risetime of a step response curve with no
*#Desc#* overshoot. If the signal has overshoot, use GenRise().)
*
*#Arg1#* Name of trace to search
*
* Usage:
*       Risetime(<trace name>)
*
 {
```

```
  1|Search forward level(10%, p) !1
   Search forward level(90%, p) !2;
}
```

The name of the function is **Risetime**. It has **1** input argument. **1|Search forward level** means search the first input forward and find a level. The level we are looking for is the 10% voltage level. 0% is defined as the minimum level of the trace; 100% is defined as the maximum. The **p** means find the specified level when the trace has a positive-going slope. When the point is found, the text **!1** designates its coordinates as x1 and y1. **Search forward level(90%, p) !2** means search the first input forward and find a point on the positive-going slope that is at the 90% level. When the point is found, the text **!2** designates its coordinates as x2 and y2. The function returns **x2-x1**, which is the time difference between the two points. Since the x-axis is time in a Transient Analysis, **x2-x1** is the time difference between the two points, or the rise time. A second function is :

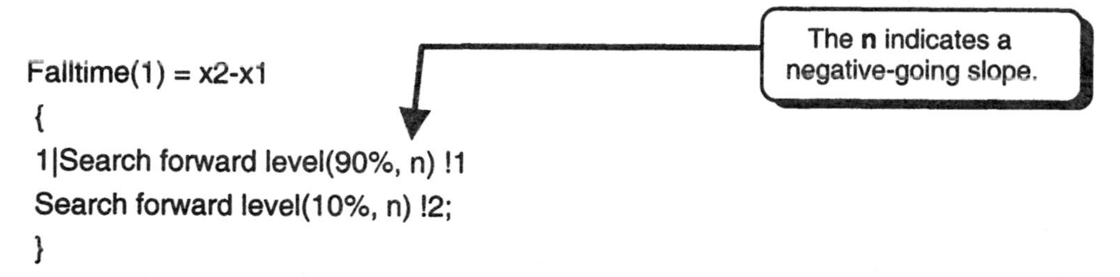


```
Falltime(1) = x2-x1
{
1|Search forward level(90%, n) !1
Search forward level(10%, n) !2;
}
```

This function is similar to the Risetime function, except that it finds the time between the 10% and 90% points when the points lie on the negative-going slope.

Enter the text `Risetime(V(VO))`. Instead of typing, a quick way to do this is to click on the ***Risetime(1)*** text in the right pane and then click on ***V(VO)*** in the left pane:

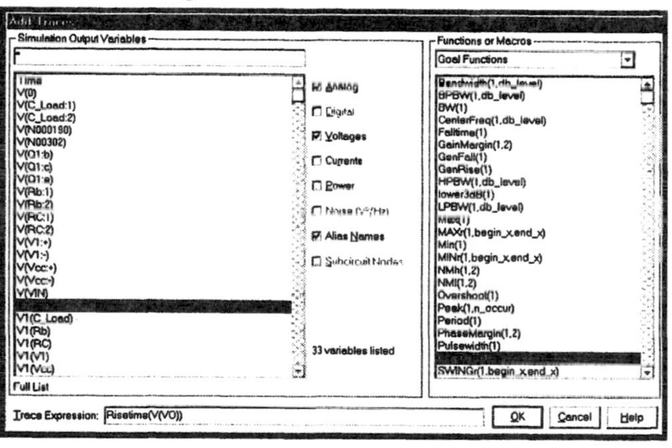


Click the ***OK*** button to view the plot:

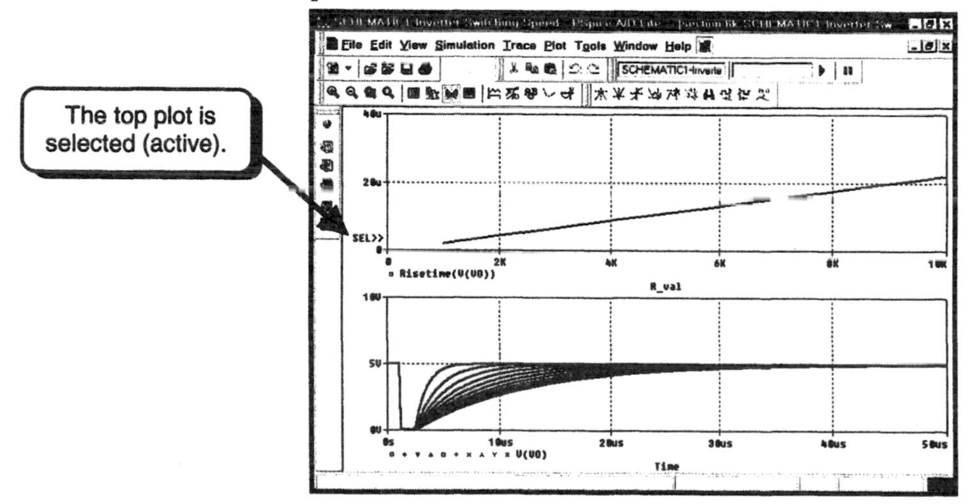

Depending on your setup the plots may or may not be easy to read. We will delete the lower plot so that the top plot will fill the screen. At the moment the top plot is active. Click the ***LEFT*** mouse button on the bottom plot to select it. The text ***SEL>>*** should toggle to the bottom plot:

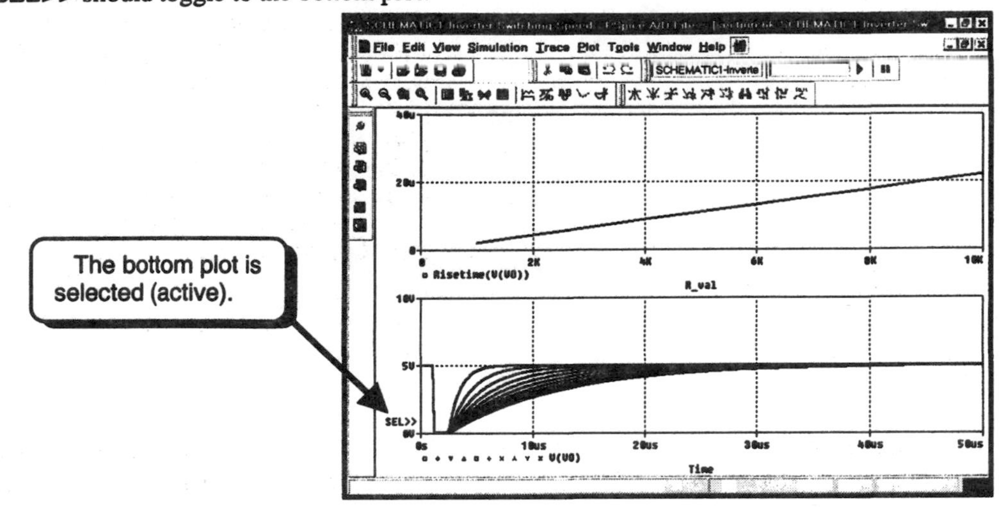


To delete the active plot, select **Plot** and then **Delete Plot** from the Probe menus:

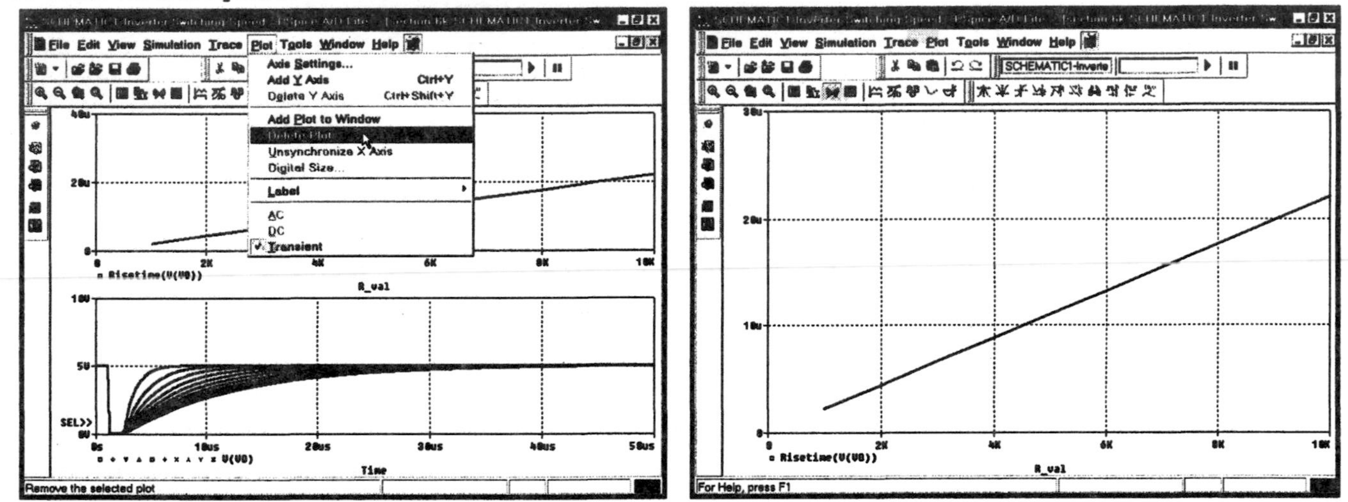


This plot is a little easier to read. For this type of plot it may be necessary to show the individual data points on the line. Thus, we would like to display the markers. Select **Tools** and then **Options** from the Probe menus:

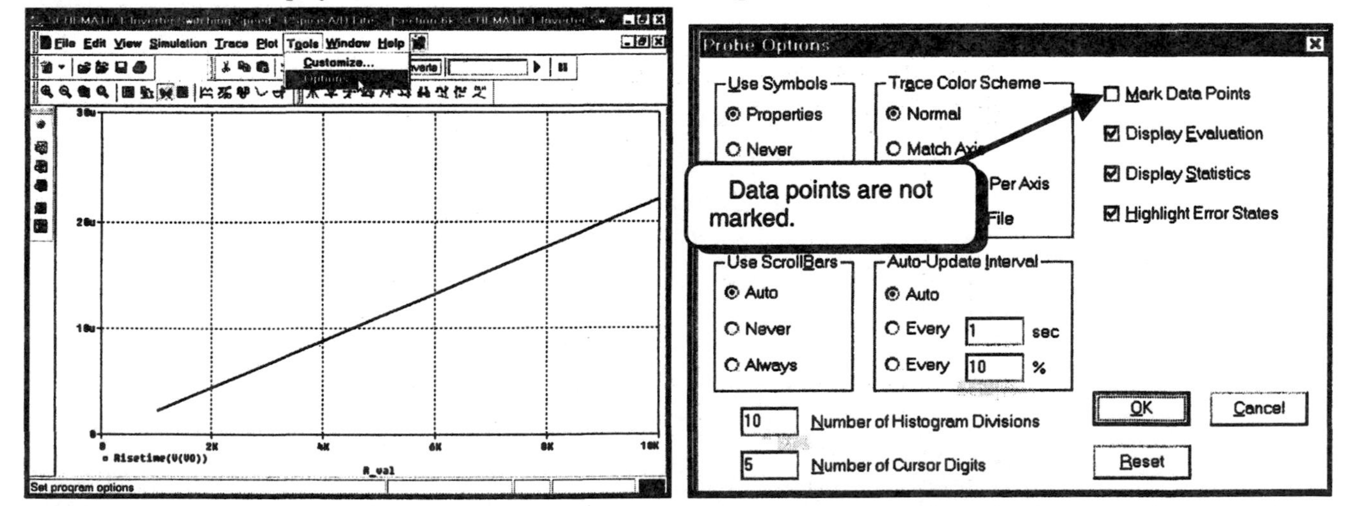


We notice that the individual data points are not displayed. To mark the points, click the ***LEFT*** mouse button on the text ***Mark Data Points***. The square next to the text ***Mark Data Points*** should fill with a checkmark ☑, indicating that the option is enabled:

Probe Options

Use Symbols: Properties / Never / Always
Trace Color Scheme: Normal
Mark Data Points
Display Evaluation
Display Statistics
Highlight Error States
Data points are marked.
Use ScrollBars: Auto / Never / Always
Auto-Update Interval: Auto / Every 1 sec / Every 10 %
10 Number of Histogram Divisions
5 Number of Cursor Digits
OK Cancel Reset

**Important note: By clicking the *OK* button, you will change the settings for all future plots. All plots in this session as well as plots in the future will mark the data points. If you do not want to make a setting permanent, you must change the setting back before you exit Probe.**

Click the ***OK*** button to change the setting. **Important note: By clicking the *OK* button, you will change the settings for all future plots. All plots in this session as well as plots in the future will mark the data points. If you do not want to make a setting permanent, you must change the setting back before you exit Probe.**

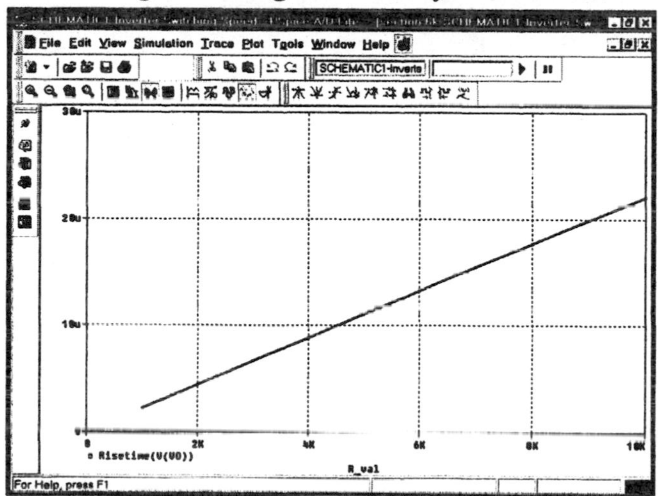


We cannot see the data points because I have made the trace width for my plots thicker to make the screen captures easier to read. We will show how to change the properties of a trace, and in doing so, make the trace thinner so that we can see the data points.

Click the ***RIGHT*** mouse button on the trace:

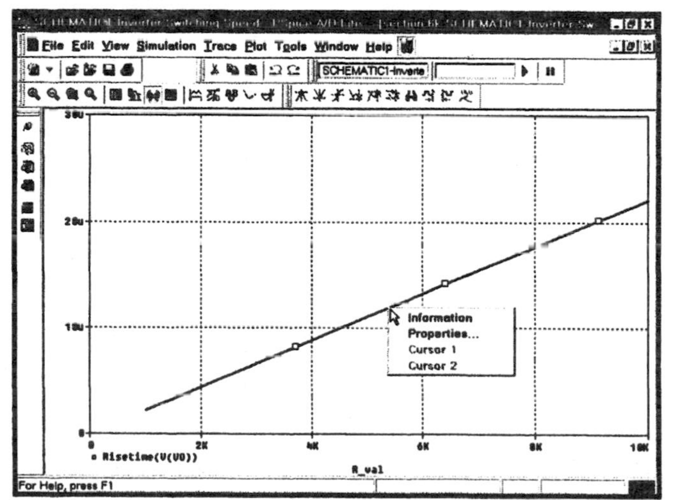


Select **Properties**:

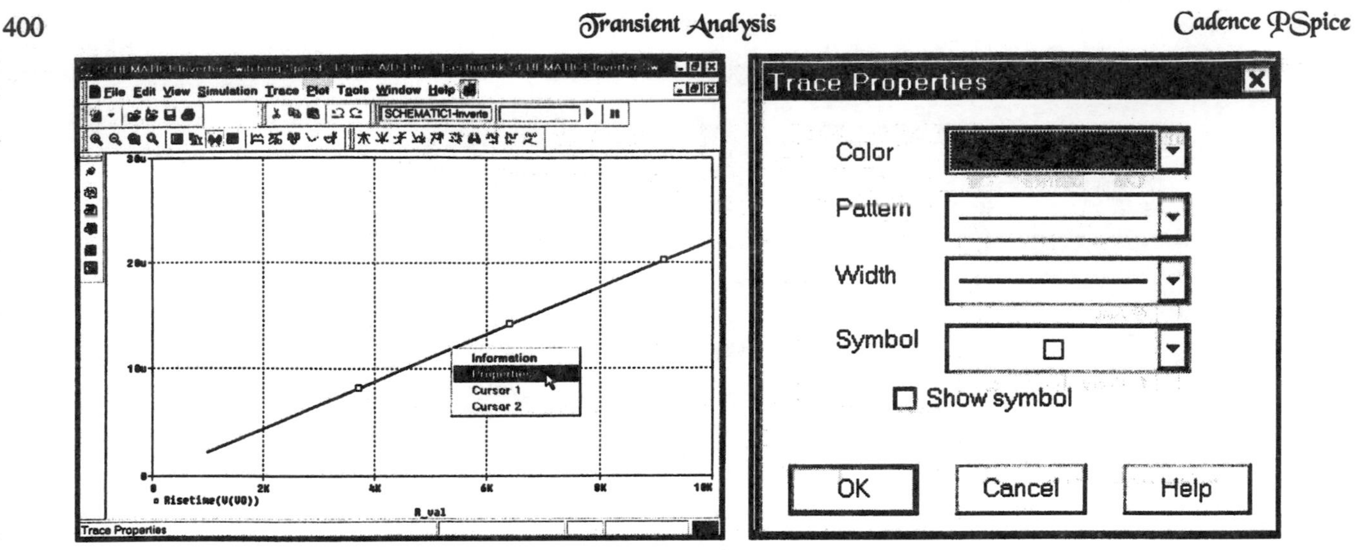


This dialog box allows us to change all of the properties of a trace. I will change the width to the minimum available and then click the ***OK*** button. You may want to experiment with the other settings.

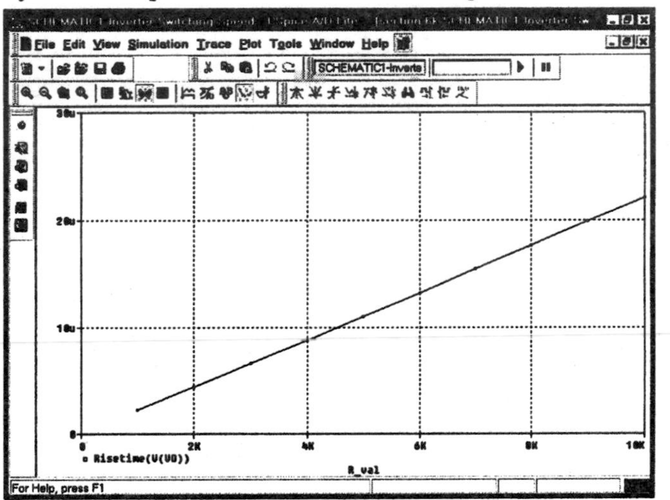


We can now see small marks on the trace that indicate the locations of the data points. This plot shows us that the rise time of the circuit is a linear function of the collector resistor value.

**Before continuing, you should disable the Mark Data Points option. In the future, if you display a trace with thousands of points, the trace will be cluttered with little data point markers.**

**EXERCISE 6-19**: Continuing with **EXERCISE 6-18**, we would like to see how the base resistor ***Rb*** affects the rise and fall times.

**SOLUTION**: Rerun **EXERCISE 6-18**. When Probe runs, enable the Performance Analysis. First add the trace **`Falltime(V(VO))`**:

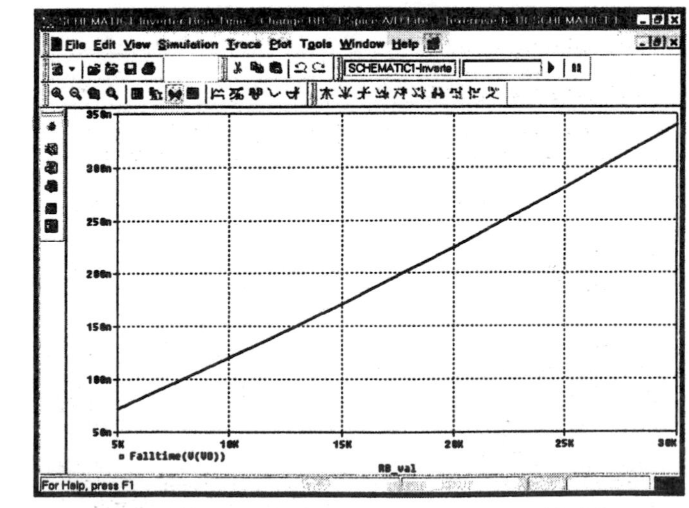

We see that the fall time is approximately a linear function of the base resistor value. Delete the trace and add the trace **Risetime(V(VO))**:

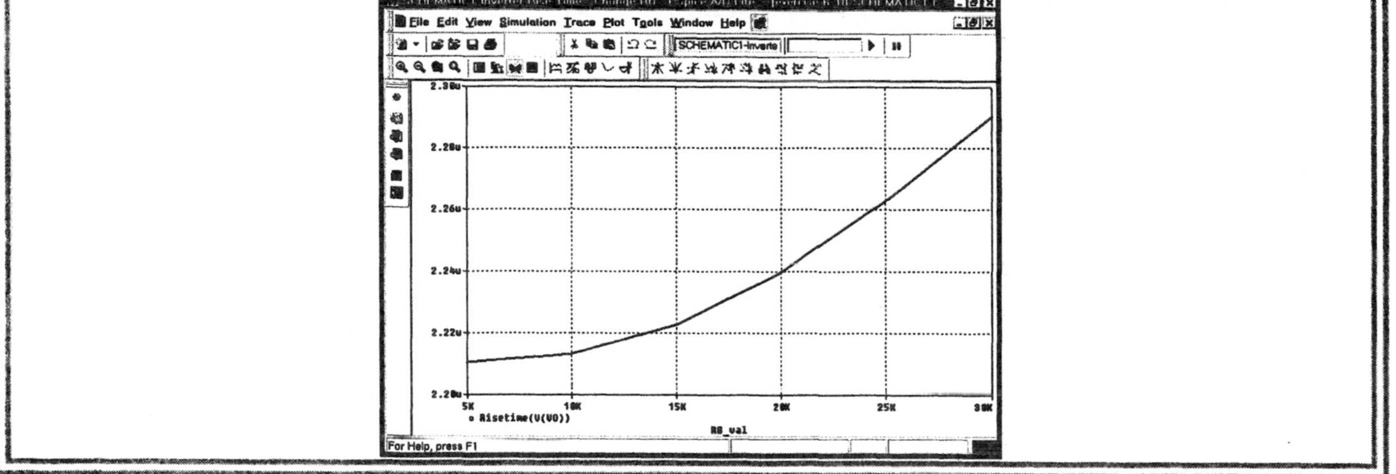

# 6.M. Stimulus Editor

The Stimulus Editor is a tool that allows us to create and view a signal source before we run a simulation. It can be used with analog voltage sources, analog current sources, and digital sources. Here, we will demonstrate its use with an analog voltage source. The Stimulus Editor is a very useful tool for creating waveforms if you are not familiar with the various sources available with PSpice. It can be used to create sources such as VPWL, VSIN, VPULSE, VSFFM, and VEXP. Unfortunately, the Lite version of the Stimulus Editor is limited to sinusoidal voltages. However, it is still a very useful tool.

The parts used for analog voltages and currents are called ISTIM and ***VSTIM*** and are located in the ***SOURCSTM***.olb library:

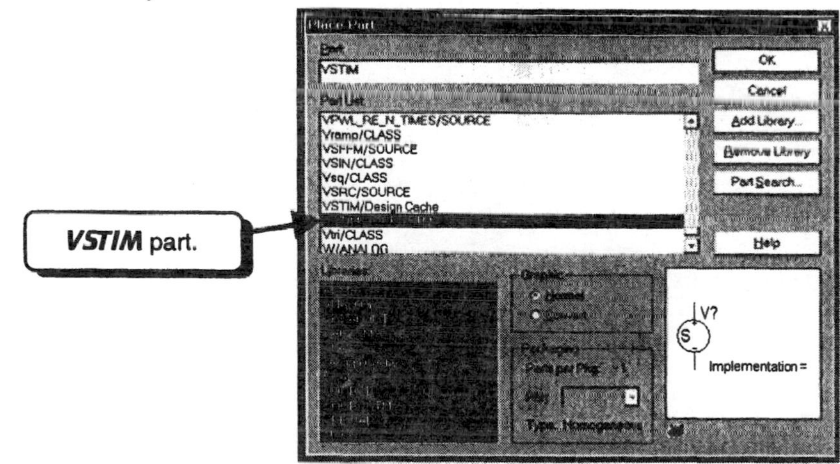


Place the ***VSTIM*** part in your schematic with a resistor load:

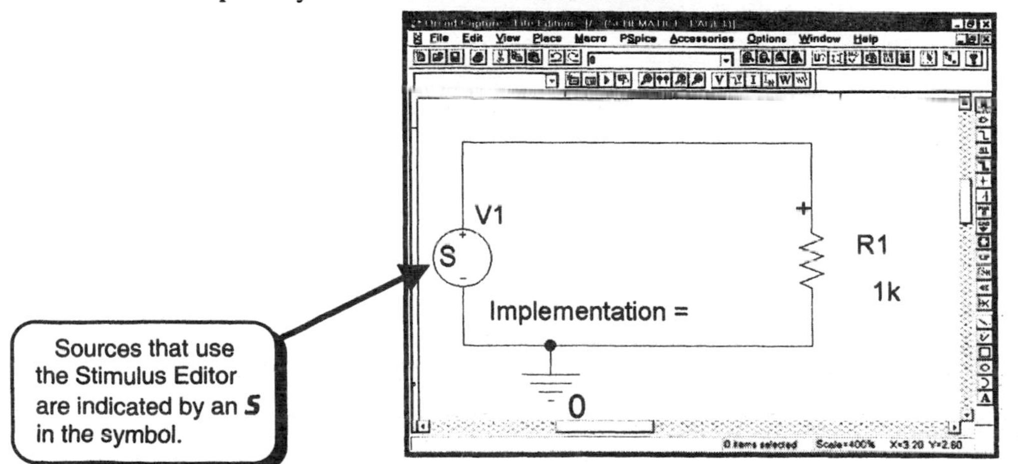

Notice that the graphic has the attribute ***Implementation=***. When you create a waveform with the Stimulus Editor, you will give the waveform a name. The name specified with the Stimulus Editor will be specified in the schematic using the ***Implementation=*** line.

To create the waveform for the part, click the ***LEFT*** mouse button on the VSTIM graphic, to select it:

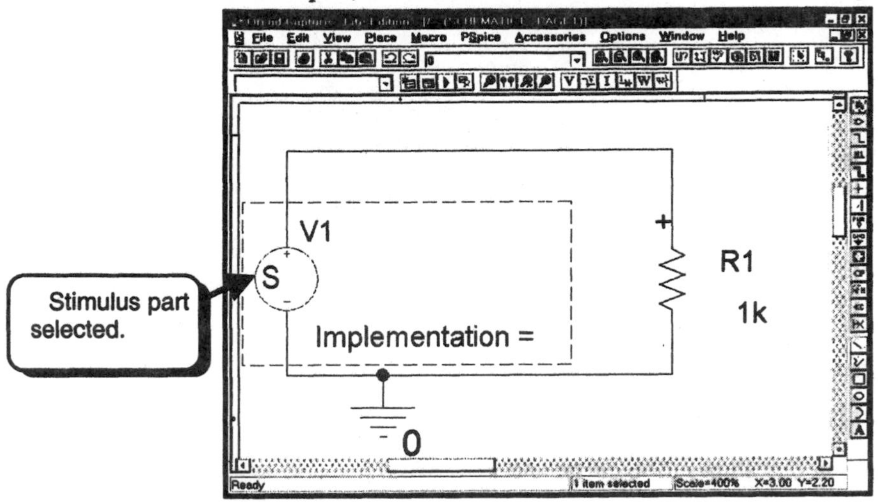


Select **Edit** and then **PSpice Stimulus**:

We must now specify a name for the stimulus and select the type of waveform. In the Lite edition, only the sinusoidal waveform is available. Enter a name for the stimulus waveform and click the ***OK*** button. This name will appear on the ***Implementation=*** line in the schematic:

New Stimulus
Name: Sinewave
Analog
EXP (exponential)
PULSE
PWL (piecewise linear)
SFFM (single-frequency FM)
SIN (sinusoidal)
Digital
Clock
Signal
Bus Width:
Initial Value:
OK Cancel

SIN Attributes
Name: Sine
Offset value
Amplitude
Frequency (Hz)
Time delay (sec) 0
Damping factor (1/sec) 0
Phase angle (degrees) 0
OK Cancel Apply

The Stimulus Editor window allows us to view the waveform. The ***SIN Attributes*** dialog box allows us to change the properties of the waveform.

We have a lot of flexibility since we can specify the ***Offset value*** (DC Offset), ***Amplitude***, ***Frequency***, ***Time delay***, ***Damping factor***, and ***Phase angle***. The dialog box below specifies a 1 kHz sine wave with a 1 V amplitude and 1 V offset:

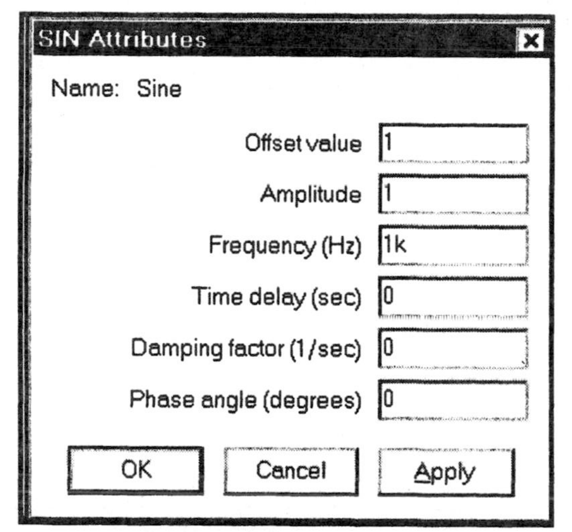


To view the waveform with the Stimulus Editor click the ***Apply*** button:

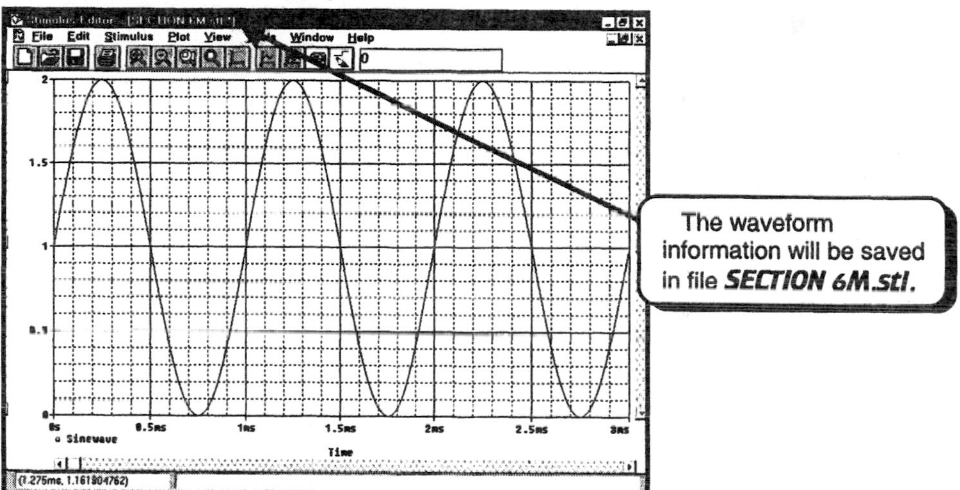


The waveform information will be saved in file ***SECTION 6M.stl***. If you copy this circuit to a floppy, make sure you copy the files ***SECTION 6M***.opj and ***SECTION 6M.stl***. If you do not copy the file ***SECTION 6M.stl***, you will lose the waveform information when you copy the files to a floppy disk.

Suppose we wish to see the effect of the other parameters. You may change any of the parameters in the ***SIN Attributes*** dialog box and then click the ***Apply*** button to view the waveform. Modify the paramcters as shown below:

SIN Attributes

Name: Sinewave

Offset value 1

Amplitude 1

Frequency (Hz) 1k

Time delay (sec) 1m

Damping factor (1/sec) 300

Phase angle (degrees) 90

OK Cancel Apply

The additional parameters add a time delay, exponential damping, and a phase angle to the waveform. Click the ***Apply*** button to view the waveform:

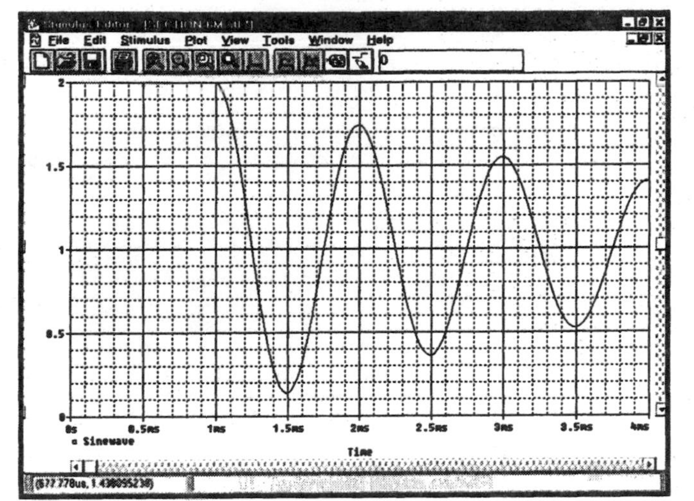


The Stimulus Editor allows us to see the effect of each of the parameters. As a last example, change the parameters as shown:

SIN Attributes

Name: Sinewave

Offset value 1

Amplitude 1

Frequency (Hz) 1k

Time delay (sec) 1m

Damping factor (1/sec) -500

Phase angle (degrees) 0

OK Cancel Apply

We will accept these parameters as the final waveform settings, so click the ***OK*** button. The waveform will be displayed, and the ***SIN Attributes*** dialog box will disappear:

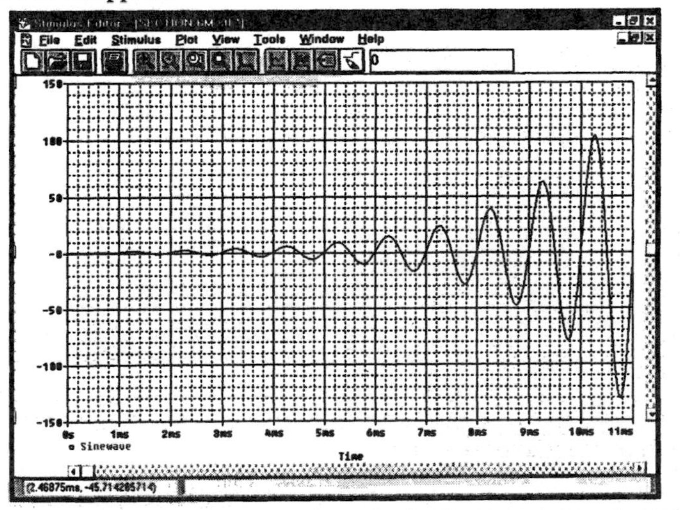


If you need to make further changes to the waveform, select **Edit** and then **Attributes** from the Stimulus Editor menus. The ***SIN Attributes*** dialog box will reappear:

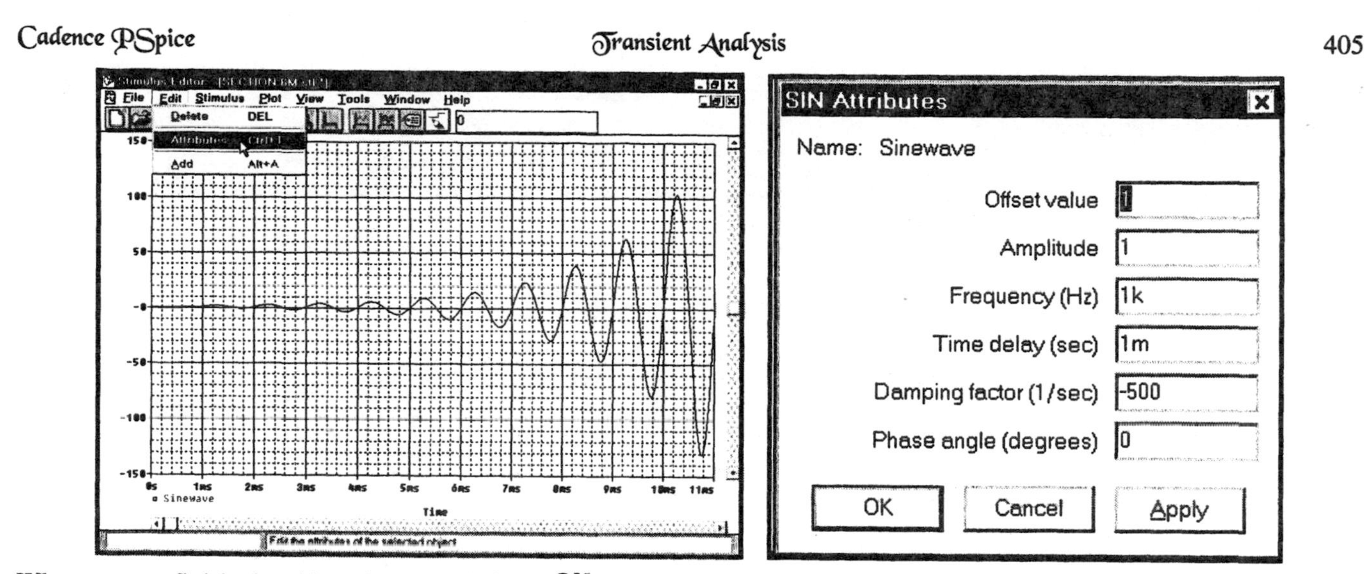


When you are finished making changes, click the ***OK*** button. The updated waveform will be displayed:

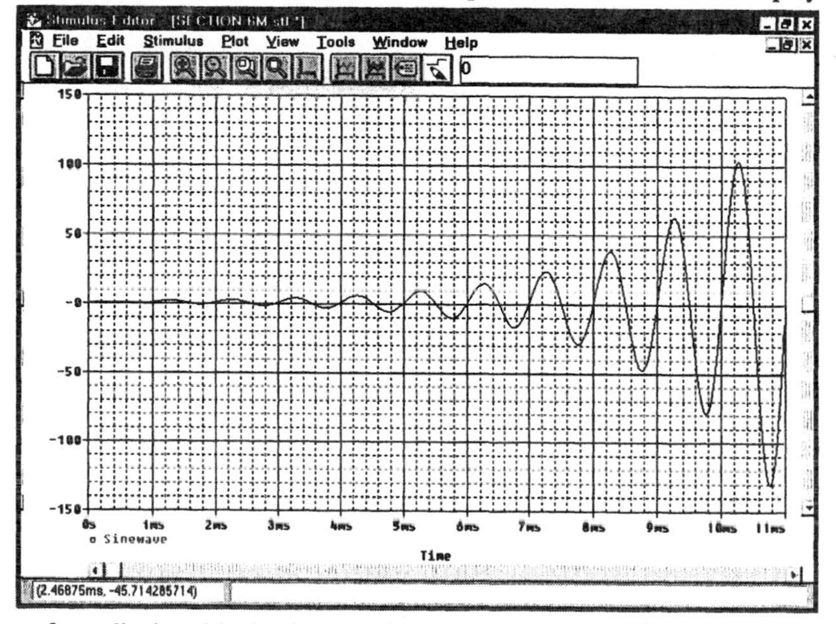


To apply the waveform displayed in the Stimulus Editor window to the V1 part in the schematic and return to the schematic, select **File** and then **Exit** from the Stimulus Editor menu bar:

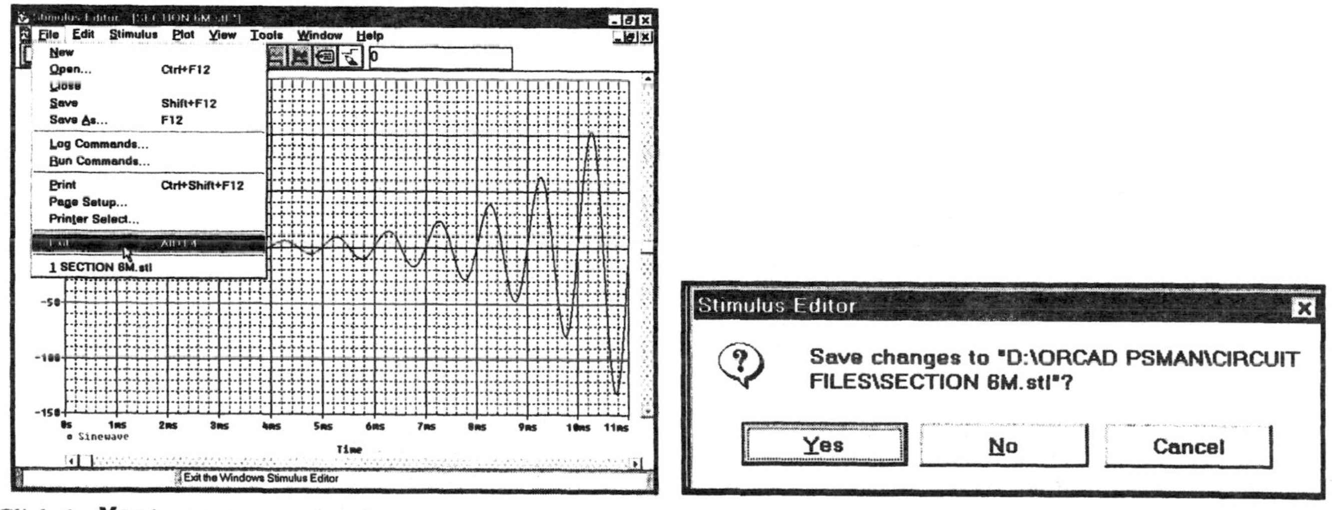


Click the ***Yes*** button to save the changes:

Stimulus Editor

Update schematic?

Yes No

Click the ***Yes*** button to use the stimulus we just created with the Vstim part in our schematic:

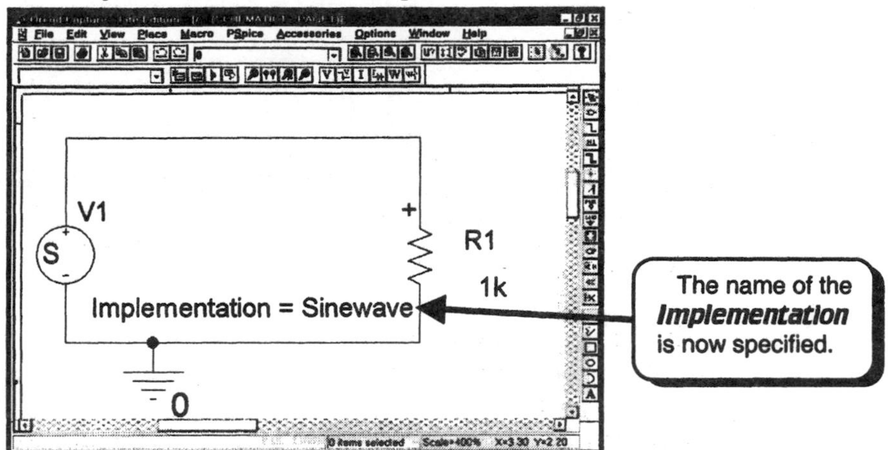


Notice that the name of the stimulus, ***Implementation = Sinewave***, is now specified in the schematic. If you run a Transient Analysis and plot the voltage across R1, the voltage will be the same as we saw in the Stimulus Editor.

# 6.N. Temperature Sweep — Linear Regulator

A temperature sweep can be used in conjunction with most of the analyses. Here we will show how temperature affects the performance of a linear voltage regulator. Create the circuit shown below:

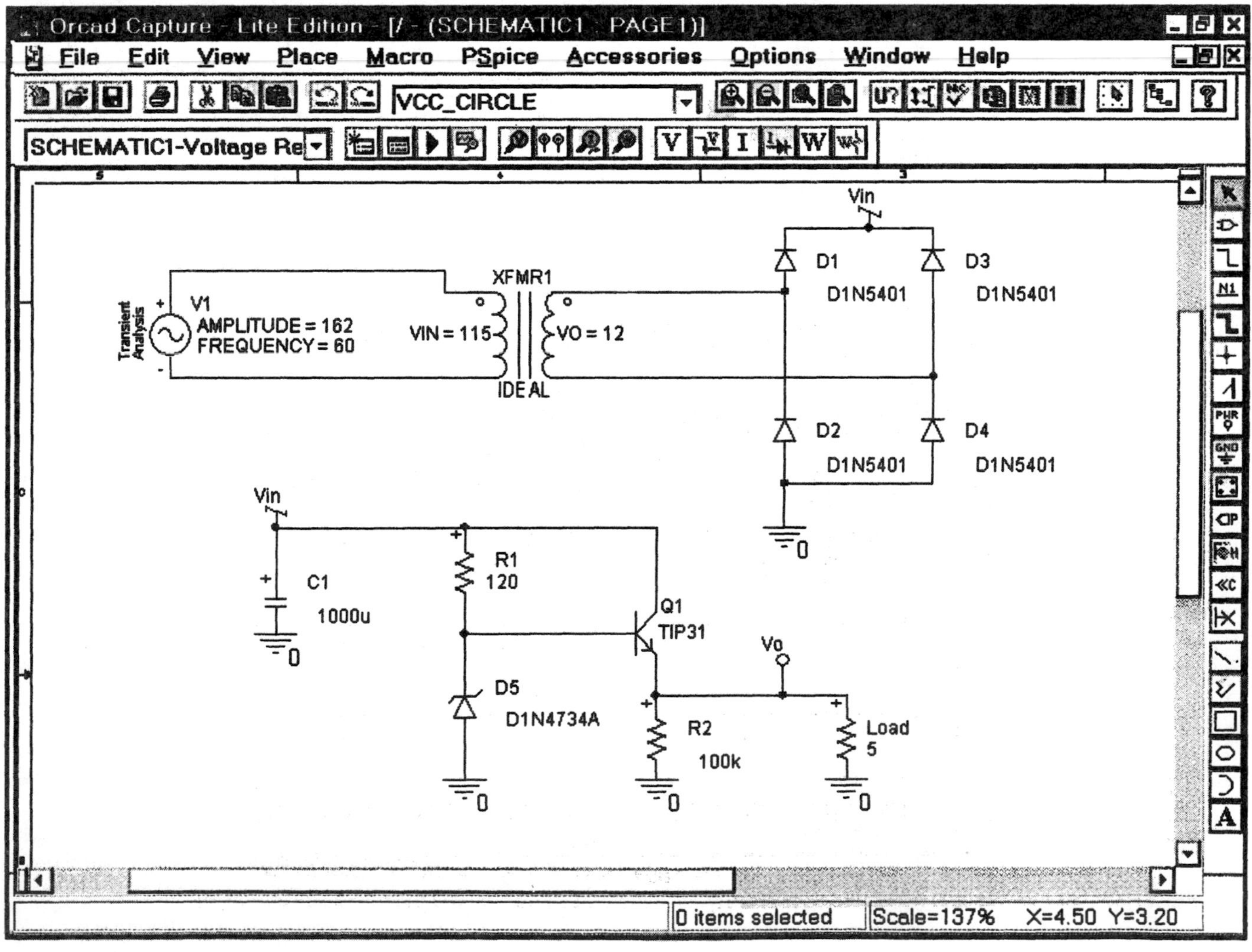

| Vcc_Circle | Vcc_Wave | R | XFMR3 Vin=115 Vo=12 IDEAL | D1 d1n5401 |
|---|---|---|---|---|
| **Vcc_Circle**<br>Node label | **Vcc_Wave**<br>Node label | **R**<br>Resistor | **Ideal_XFMR_Vo/Vin**<br>Ideal transformer | **D1N5401**<br>Rectifier diode |
| D1 d1n4734A<br>**D1N4734A**<br>Zener diode | 0<br>**0**<br>Ground | **C**<br>Capacitor | V3 AMPLITUDE = FREQUENCY = Transient Analysis<br>**VSIN**<br>Sinusoidal voltage source | Q5 tip31<br>**TIP31**<br>Power BJT |

This circuit is a bridge rectifier followed by a filter capacitor to produce a DC voltage with ripple at Vin. Connected to Vin is a linear regulator made from a Zener voltage reference and an NPN pass transistor. We will first run a Transient Analysis to see the operation of the circuit at room temperature (27°C). To set up a Transient Analysis, select **PSpice** and then **New Simulation Profile** from the Capture menus, enter a name for the profile and then click the ***Create*** button. By default the ***Time Domain (Transient) Analysis type*** is selected. Fill in the parameters as shown in the Time Domain dialog box below:

Simulation Settings - Voltage Regulator Transient Run

General | Analysis | Include Files | Libraries | Stimulus | Options | Data Collection | Probe Window

Analysis type:

Options:
- Monte Carlo/Worst Case
- Parametric Sweep
- Temperature (Sweep)
- Save Bias Point
- Load Bias Point

Run to time: 160m seconds (TSTOP)

Start saving data after: 0 seconds

Transient options

Maximum step size: 10u seconds

Skip the initial transient bias point calculation (SKIPBP)

Output File Options...

OK | Cancel | Apply | Help

Run the simulation and plot Vin and Vo:

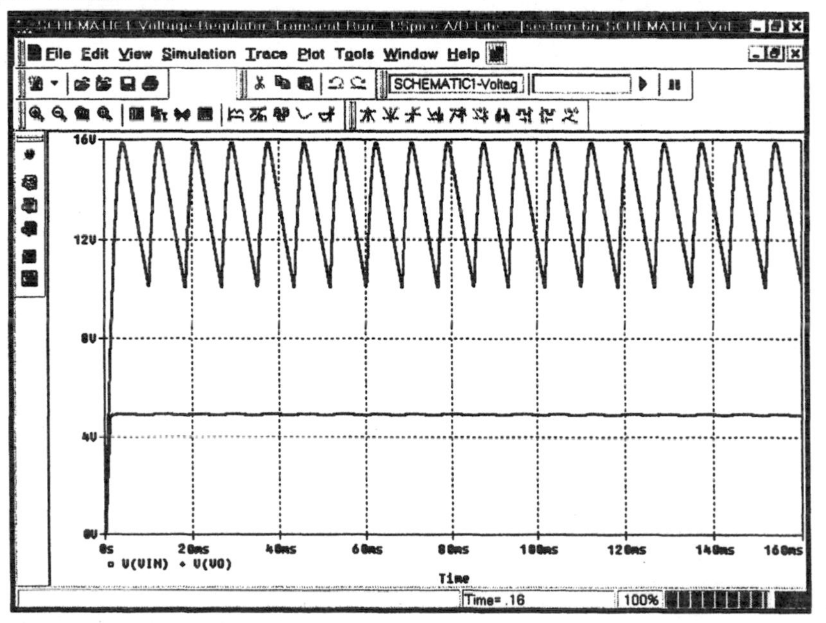


We will zoom in on the Vo trace to see it more closely. Use the cursors to find the magnitude of the ripple.

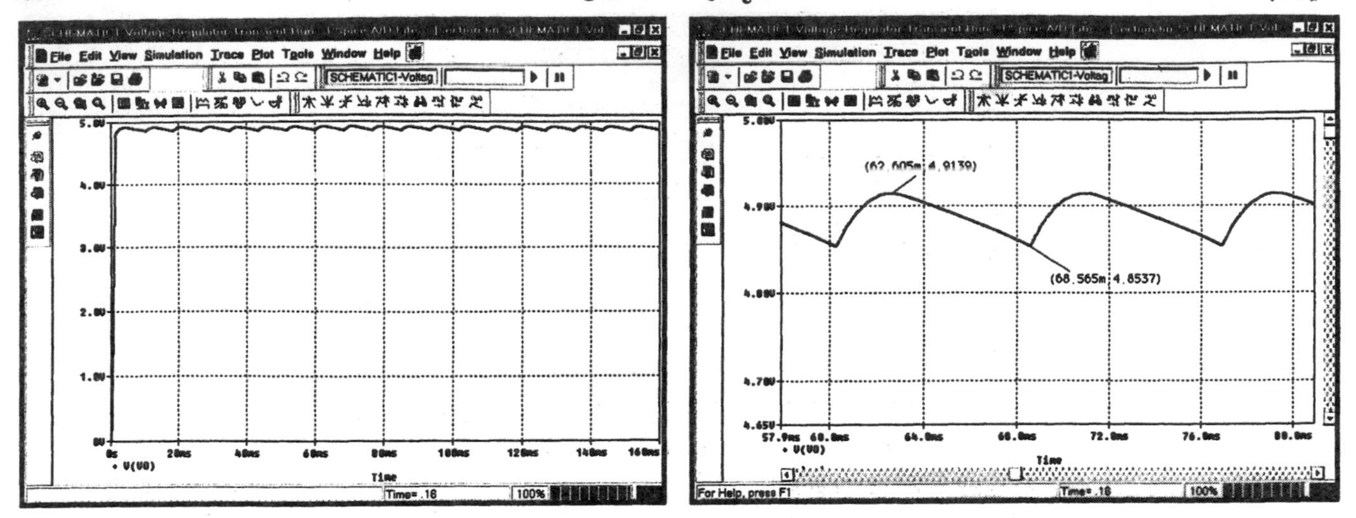

We see that the output has about 60 mV of ripple.

Next we will see how the output changes with temperature. We will run a Transient Analysis at –25°C, 25°C, and 125°C. Return to the schematic and select **PSpice**, and then select **Edit Simulation Profile** to obtain the Transient setup dialog box:

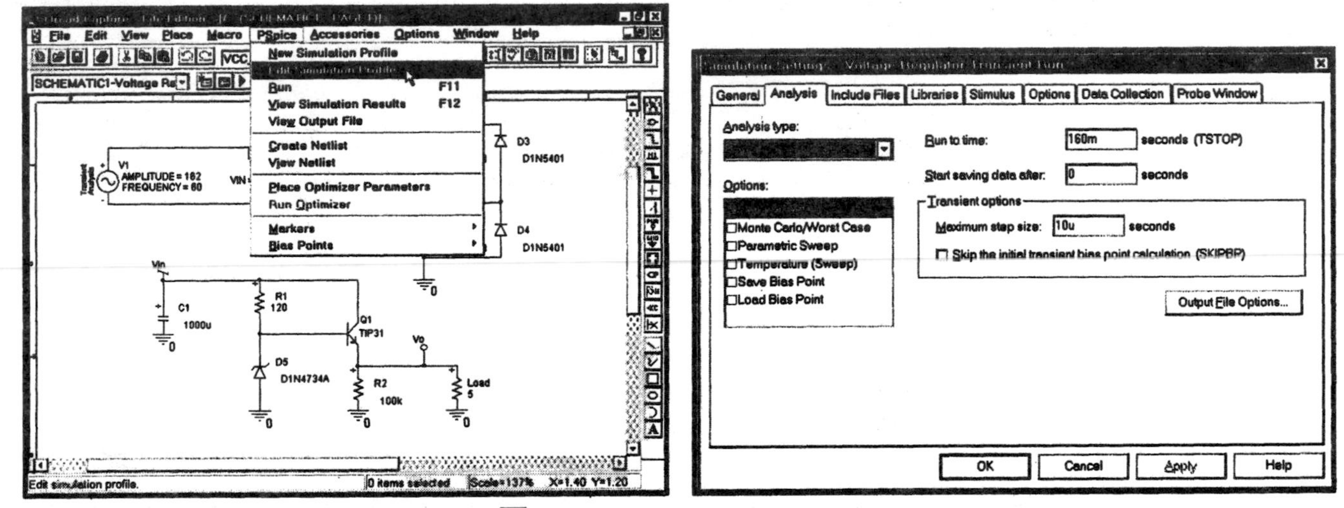

Click the ***LEFT*** mouse button on the square ☐ next to the text ***Temperature (Sweep)*** to enable the sweep and display the options for the sweep:

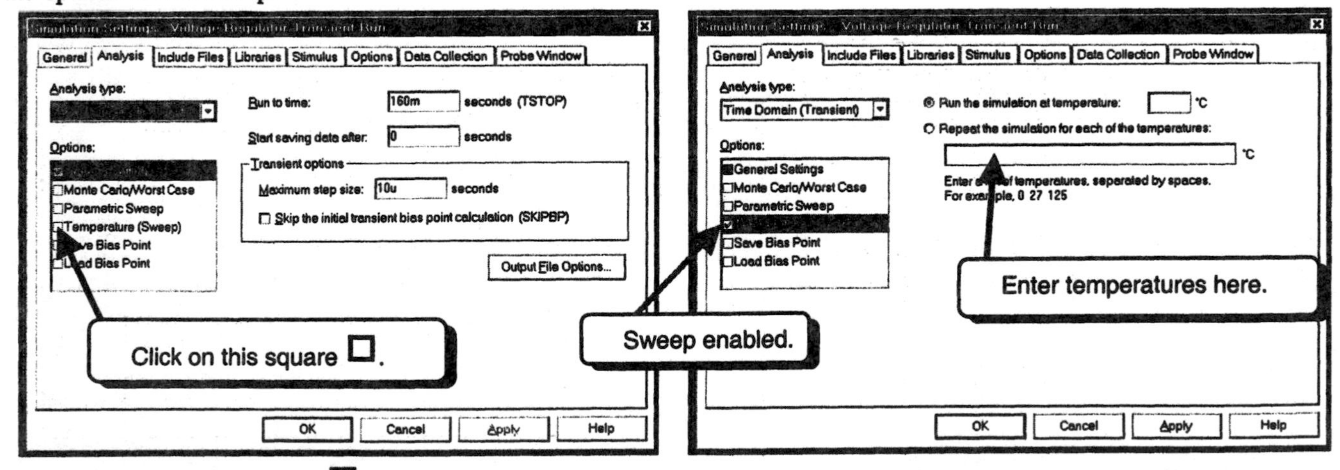


The checkmark in the square ☑ indicates that the sweep is enabled. Note that both the ***Temperature (Sweep)*** and the ***Time Domain*** analysis are enabled. Fill in the ***Temperature (Sweep)*** parameters as shown:

Click the ***OK*** button and then run the simulation. It will take three times as long as the previous simulation because the Transient Analysis will run three times. Logically, the Transient Analysis runs inside the Temperature Sweep. For this example, the temperature will be set to –25°C, and then the Transient Analysis will be run. Next, the temperature will be set to 25°C, and then the Transient Analysis will be run. Finally, the temperature will be set to 125°C, and then the Transient Analysis will be run.

When the three simulations are complete, Probe will run automatically. You will have a choice of which runs you would like to view. By default, all of the runs are selected:

Click the ***OK*** button to select all of the runs. First we will plot the input voltage waveform. Add the trace **V(VIN)**. Zoom in on the traces to see the ripple in more detail:

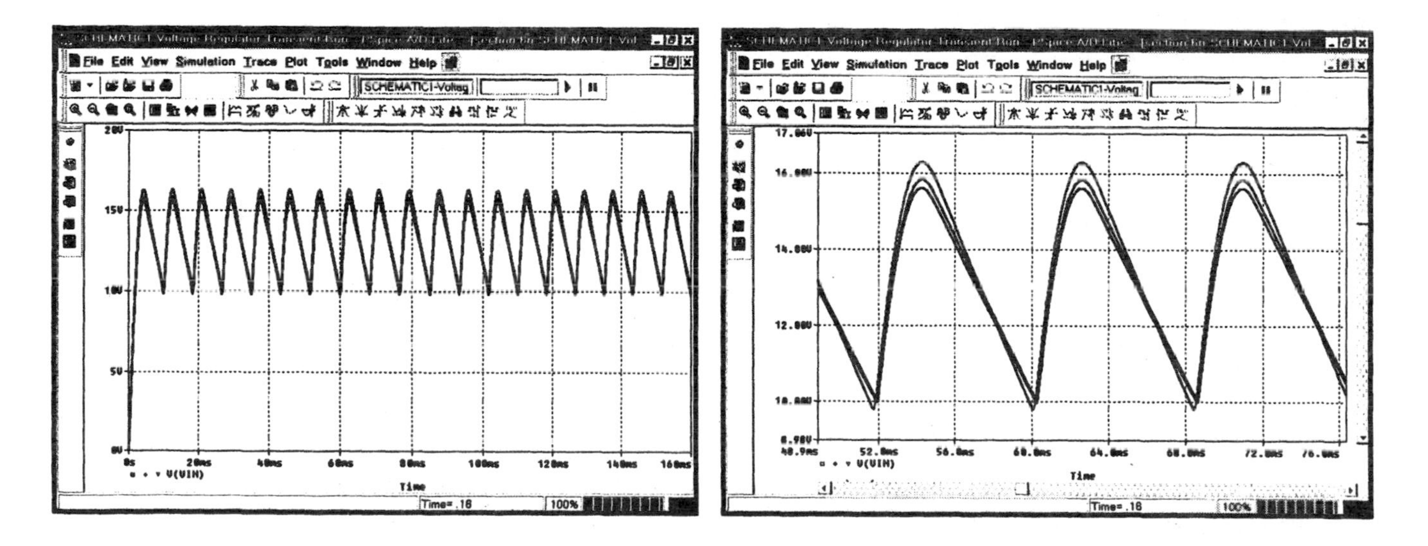

Three traces are shown, one for each temperature. The result shows that temperature does not have too much of an effect on the input voltage. Remove the trace and plot the output voltage. Zoom in on the traces and use the cursors to display the magnitude of the ripple.

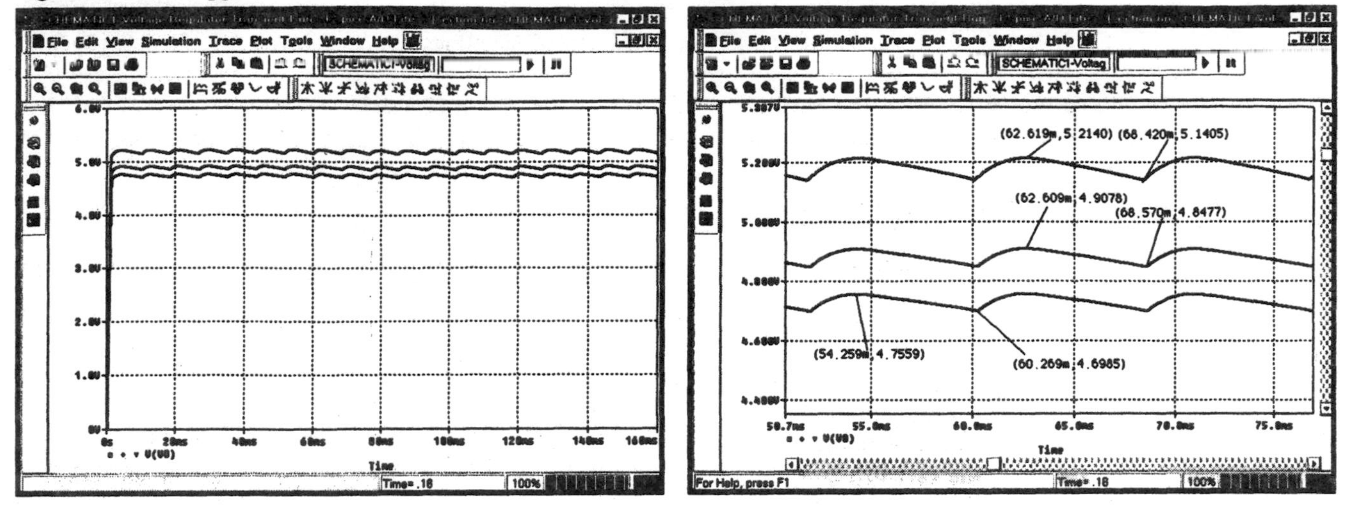

The top trace has 73.5 mV of ripple and the bottom trace has 57.4 mV of ripple.

The question may arise as to how to tell which trace is at what temperature. We can find out more information about each trace. Click the ***RIGHT*** mouse button on one of the traces:

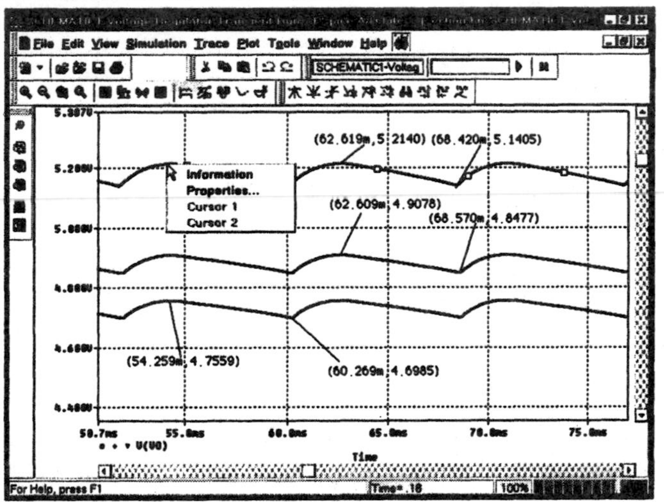

Select the **Information** menu selection:

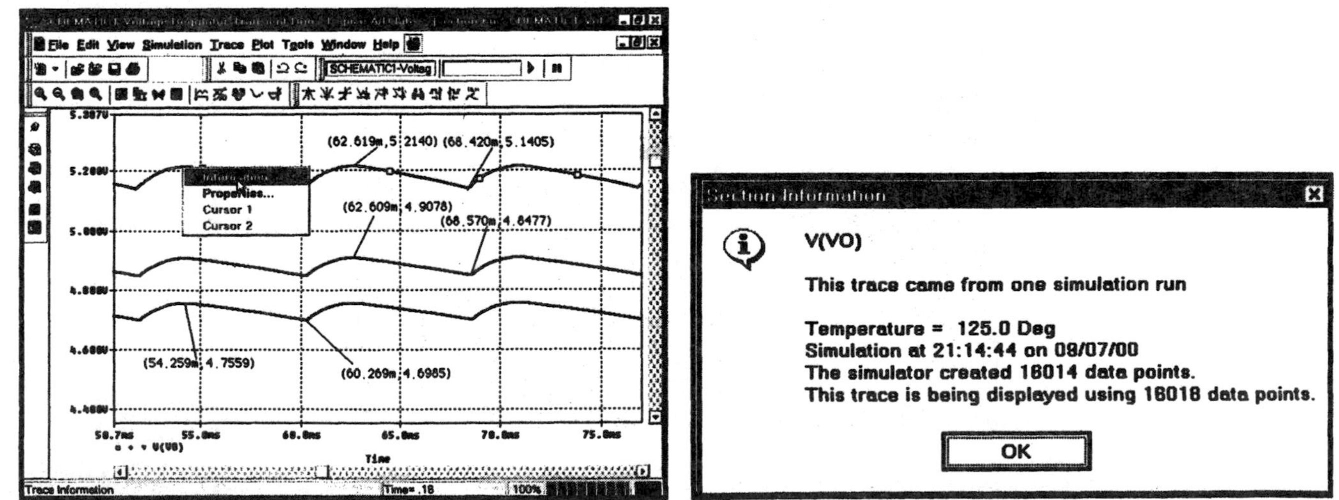


We see that the top trace is at a temperature of 125°C. We can use this procedure to determine the identity of each trace.

# 6.O. Analog Behavioral Modeling

The Analog Behavioral Modeling (ABM) parts give you an easy way to include function blocks in your circuit without having to create a circuit that implements the function. You can implement simple math functions like addition, subtraction, multiplication, and division of waveforms. More complicated functions include power, log, sin, and absolute value. Circuit functions such as gain with limits and Chebyshev filters are also provided. These blocks are usually used instead of creating a circuit to perform the function. In the Lite version, they can be used to replace a circuit block with a large number of components by a single part. This enables the simulation of more complicated circuits by reducing the component count.

## 6.O.1. Examples of ABM Parts

The ABM parts are located in library ABM.olb. We will look at some of the parts. Type **P** to place a part and then select the ***ABM*** library:

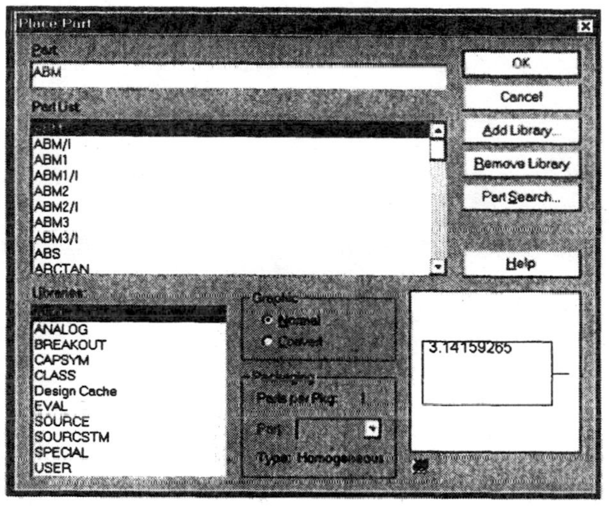


The first few parts are uncommitted ABM parts for general use. These blocks do not perform specific functions. To see what functions can be performed, refer to the ABM section of the reference manuals provided on the CD-ROM. Further down the list we see some specific functions such as ***ABS***, ***ARCTAN***, ***BANDPASS***, and so on. To see what function a block performs, click the ***LEFT*** mouse button on the name of the block to select it. For example, click on ***DIFF***:

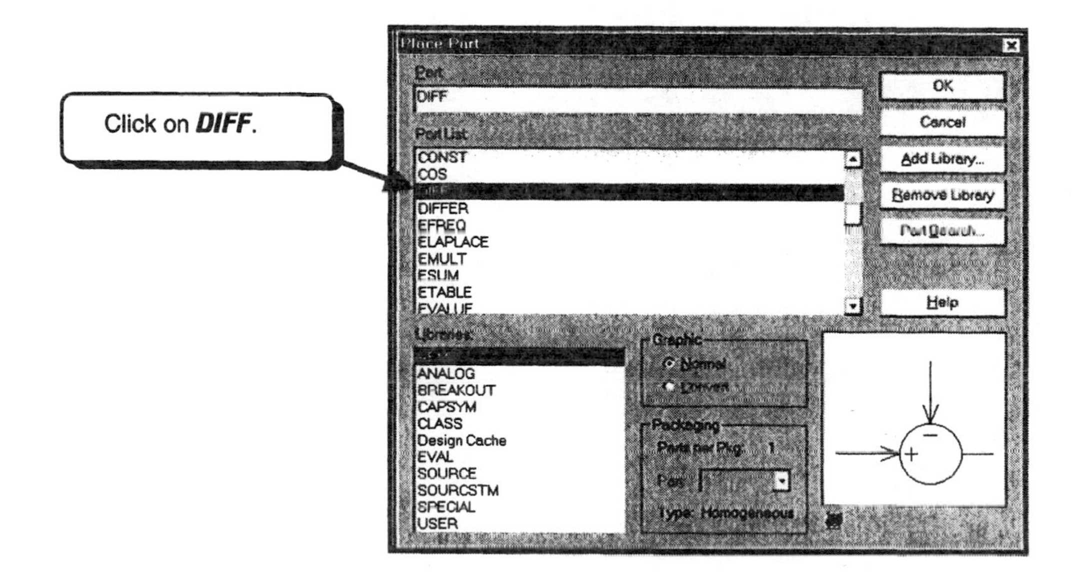


The graphic of this block indicates that it is a difference junction; that is, this block will subtract two waveforms. Next, select the part named ***DIFFER***:

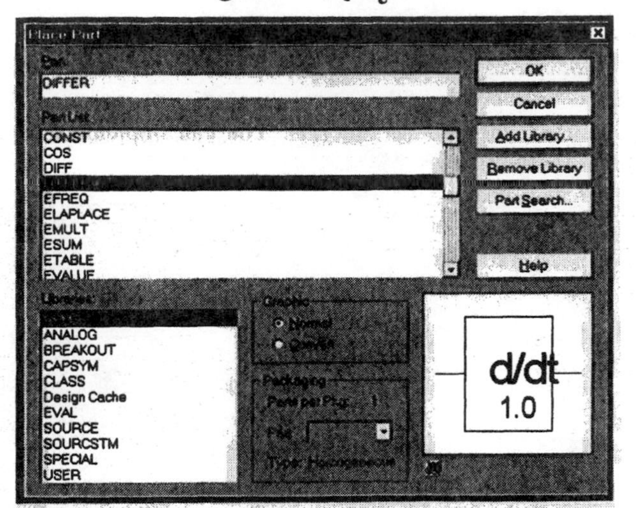

The block indicates that this is a differentiator. The output of this block is the time derivative of the input. Scroll down the ***Part List*** pane to see more parts:

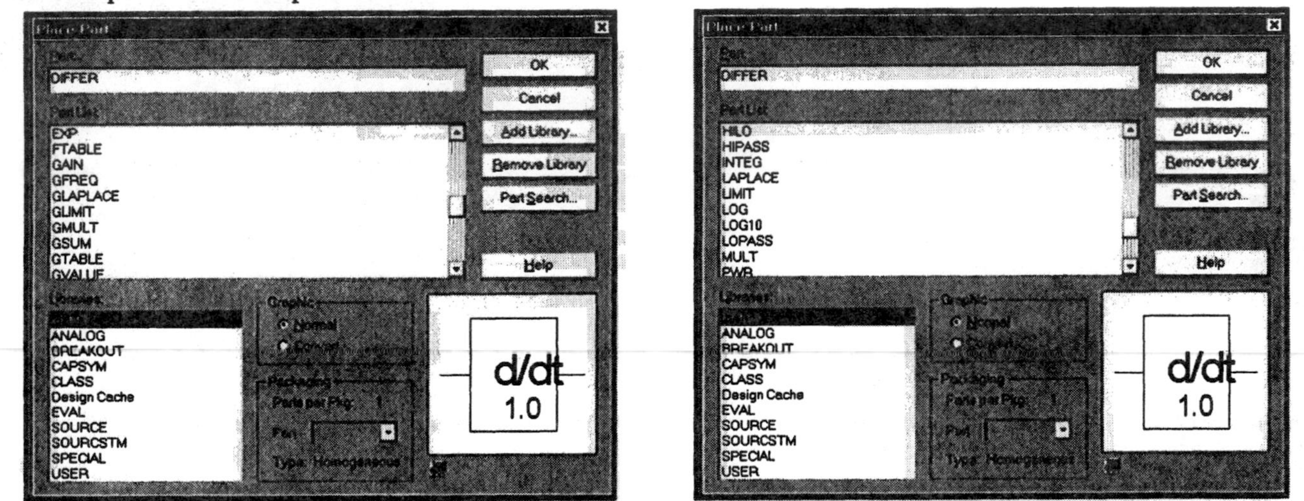

We see such parts as ***EXP***, ***GAIN***, ***HIPASS***, ***INTEG***, ***MULT***, ***SIN***, and ***SQRT***. The blocks perform the stated function on the input waveform. For example, with the ***SQRT*** function, the output voltage is the square root of the input voltage. With the ***INTEG*** function, the output waveform is the integral over time of the input waveform.

These blocks can be used together with any circuit components in a Transient Analysis. We will demonstrate the function of a few blocks. Create the circuit below. The parts are listed in **Table 6-4** on page 413:

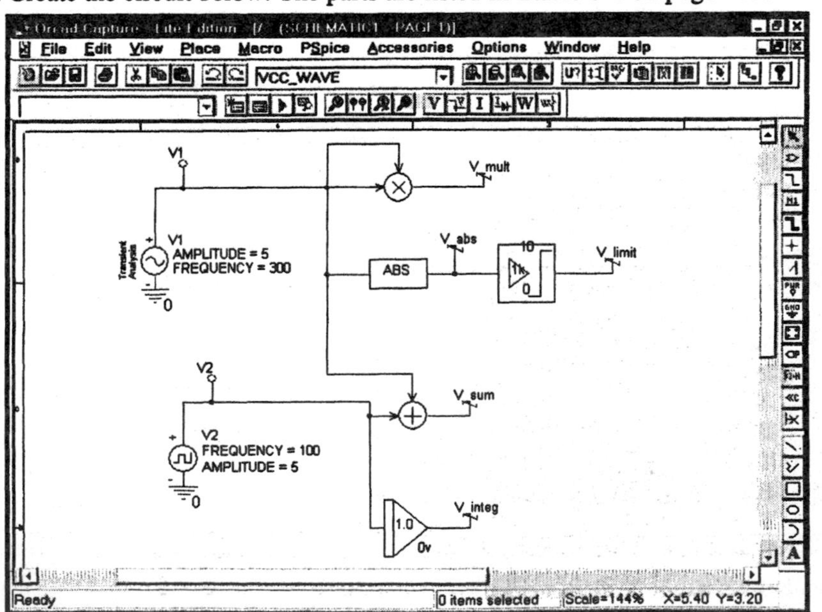

Part ***ABS*** is an absolute value block. The output voltage is the absolute value of the input voltage. Part ***GLIMIT*** is a gain block with output voltage limits. The part placed has a gain of ***1k*** with the output voltage constrained between ***0*** V and ***10*** V. This block can be viewed as a single-sided op-amp with one supply pin tied to ground and the other to 10 V. We can change the limits and the gain. To change the gain, double-click on the text ***1k***. A dialog box will open:

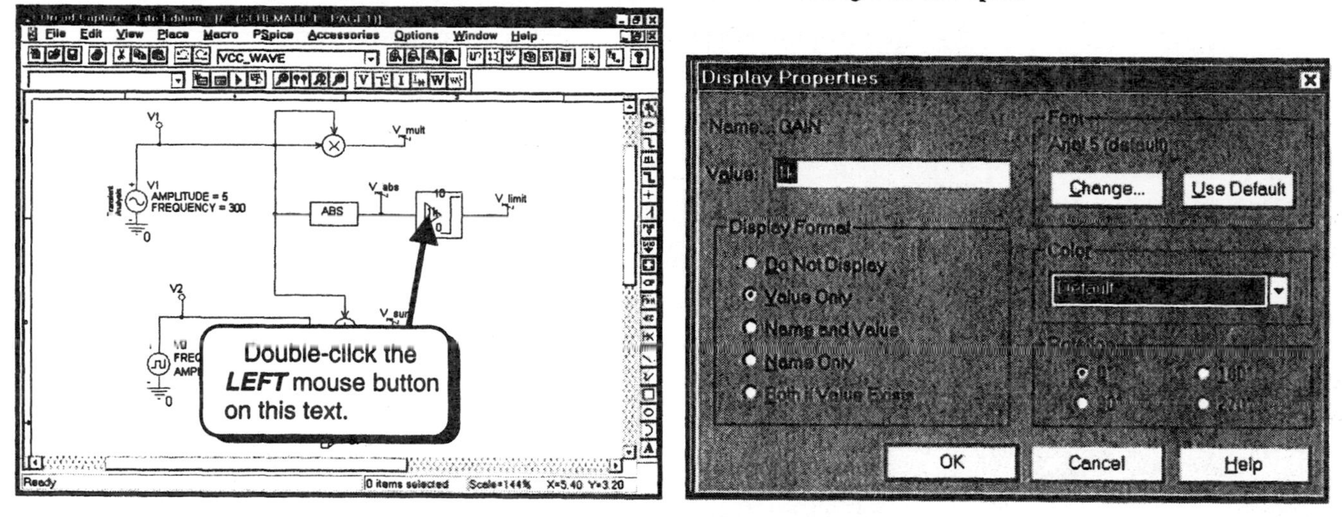


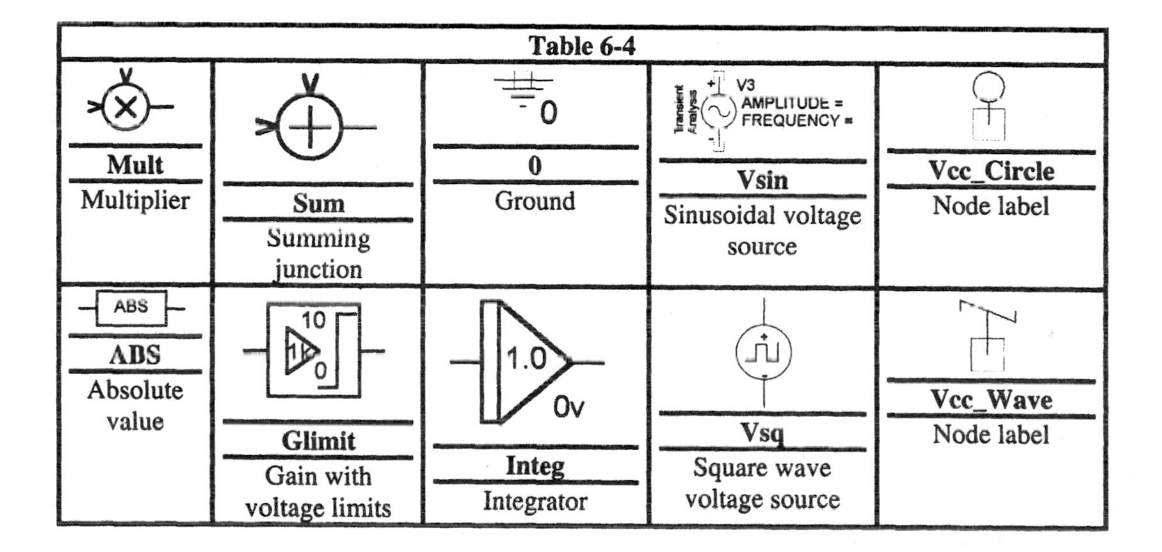

**Table 6-4**

| Mult | Sum | 0 | Vsin | Vcc_Circle |
|---|---|---|---|---|
| Multiplier | Summing junction | Ground | Sinusoidal voltage source | Node label |
| **ABS** | **Glimit** | **Integ** | **Vsq** | **Vcc_Wave** |
| Absolute value | Gain with voltage limits | Integrator | Square wave voltage source | Node label |

Type **5** and click the ***OK*** button to accept the change and return to the schematic. The gain will now be displayed as ***5***:

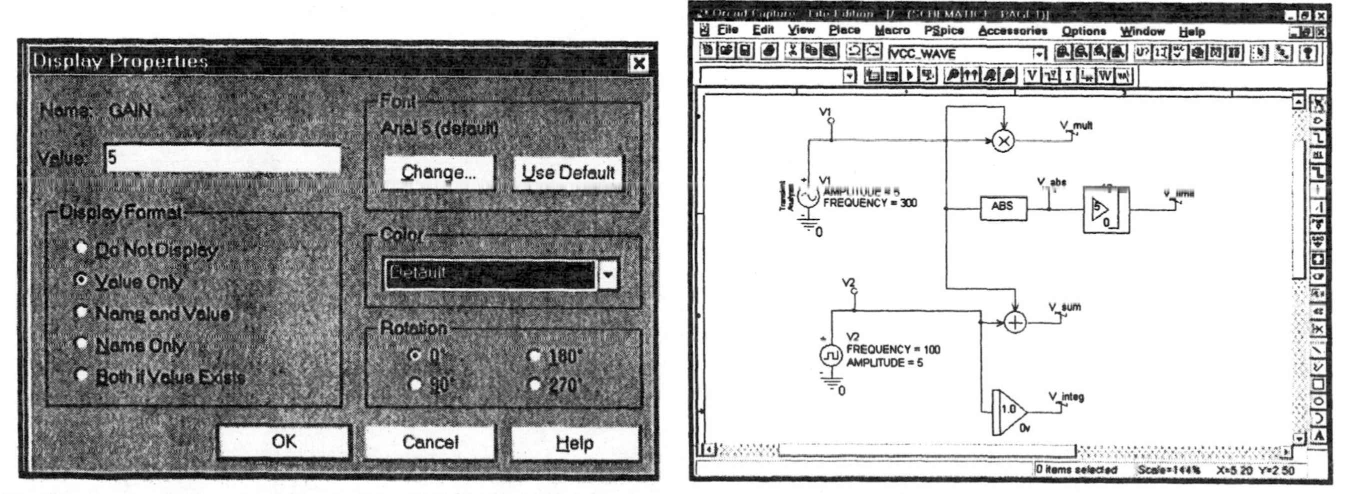


To change the limits, double-click on the ***0*** or the ***10***. Change the limits to –15 and 15:

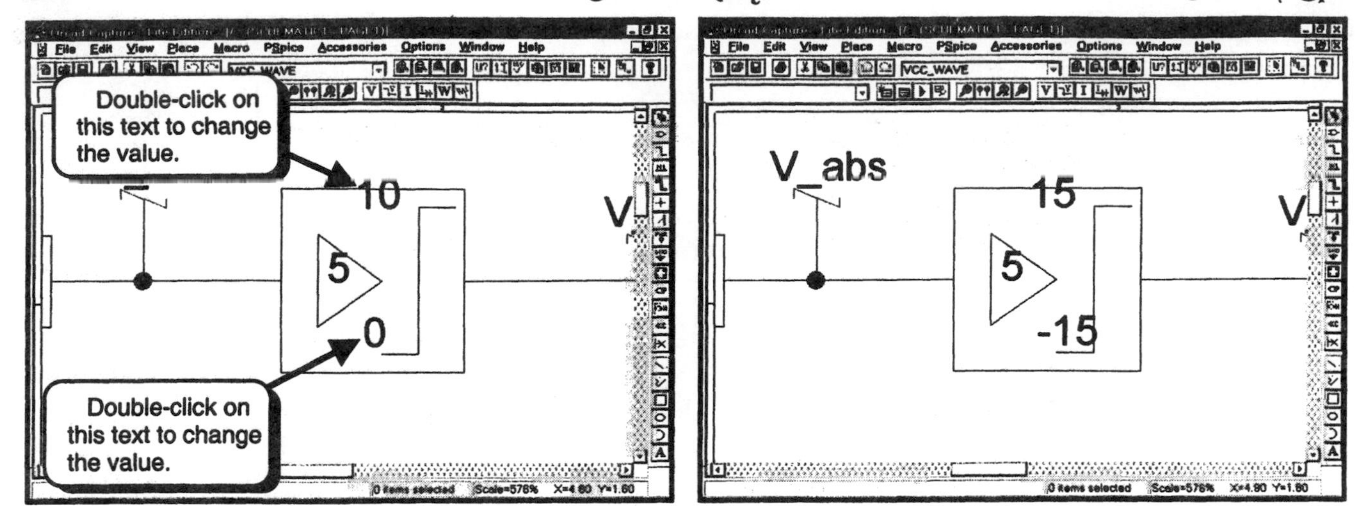


This block can now be viewed as an ideal amplifier with a gain of 5 and ±15 V supplies.

The integrator block 1.0 0v performs the function

$$V_{out}(t) = \left[ A \int_{t=0} V_{in}(t)dt \right] + V_{in}(0)$$

where A is the gain of the integrator and $V_{in}(0)$ is the initial condition. In the block shown on the schematic the gain is ***1.0*** and the initial condition is ***0V***. To change the gain, double-click on the text ***1.0***; to change the initial condition, double-click on the text ***0V***. I will change the gain to 5 and the initial condition to 1 V.

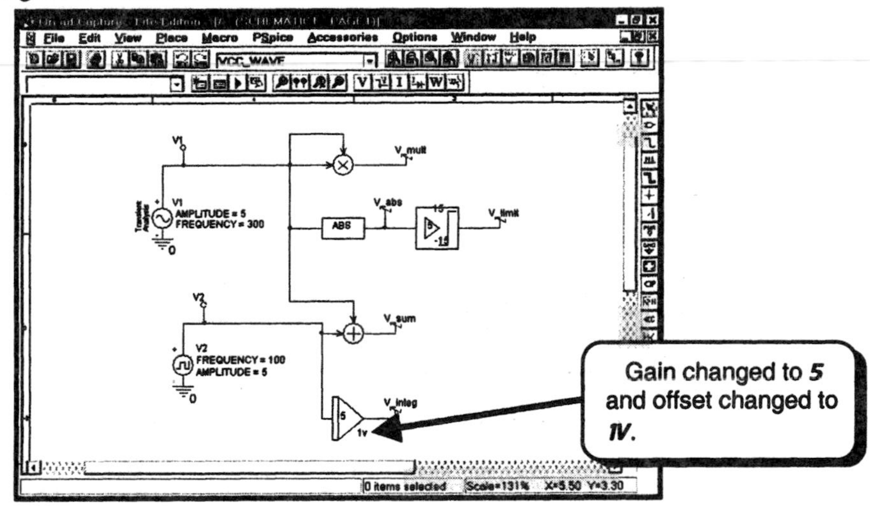


We will now run the circuit with a Transient Analysis for 50 ms:

General | Analysis | Include Files | Libraries | Stimulus | Options | Data Collection | Probe Window

Analysis type: Time Domain (Transient)

Run to time: 50m seconds (TSTOP)

Start saving data after: 0 seconds

Transient options

Maximum step size: 100u seconds

Skip the initial transient bias point calculation (SKIPBP)

Options: Monte Carlo/Worst Case, Parametric Sweep, Temperature (Sweep), Save Bias Point, Load Bias Point

Output File Options...

OK | Cancel | Apply | Help

Run the simulation and plot the sine wave and square wave on separate plots. To add plots to the Probe window, select **Plot** and then **Add Plot to Window** from the Probe menus.

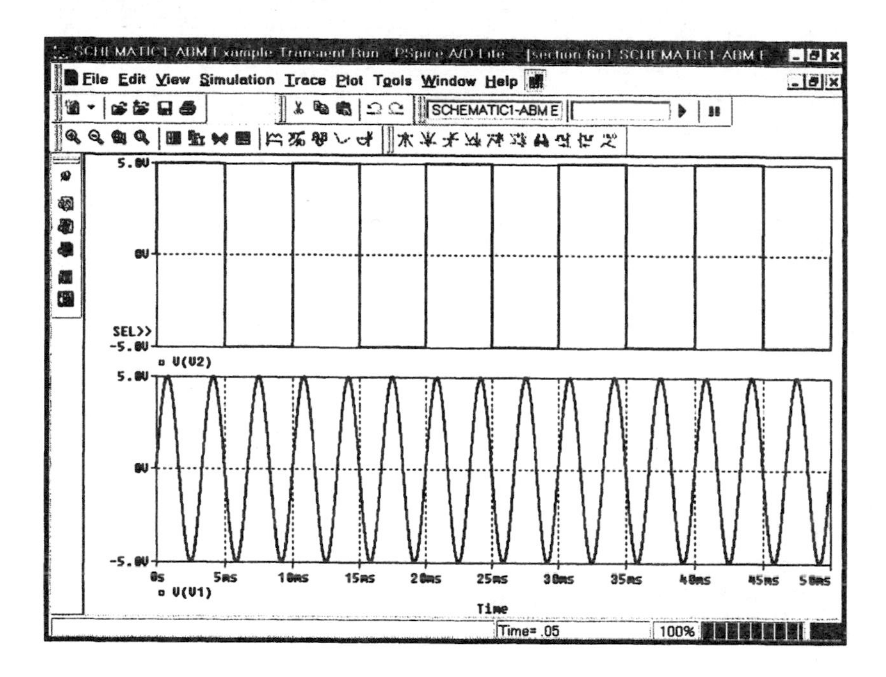


We will now look at each output individually and compare them to the two input waveforms. The output of the multiplier should be the square of a sine wave:

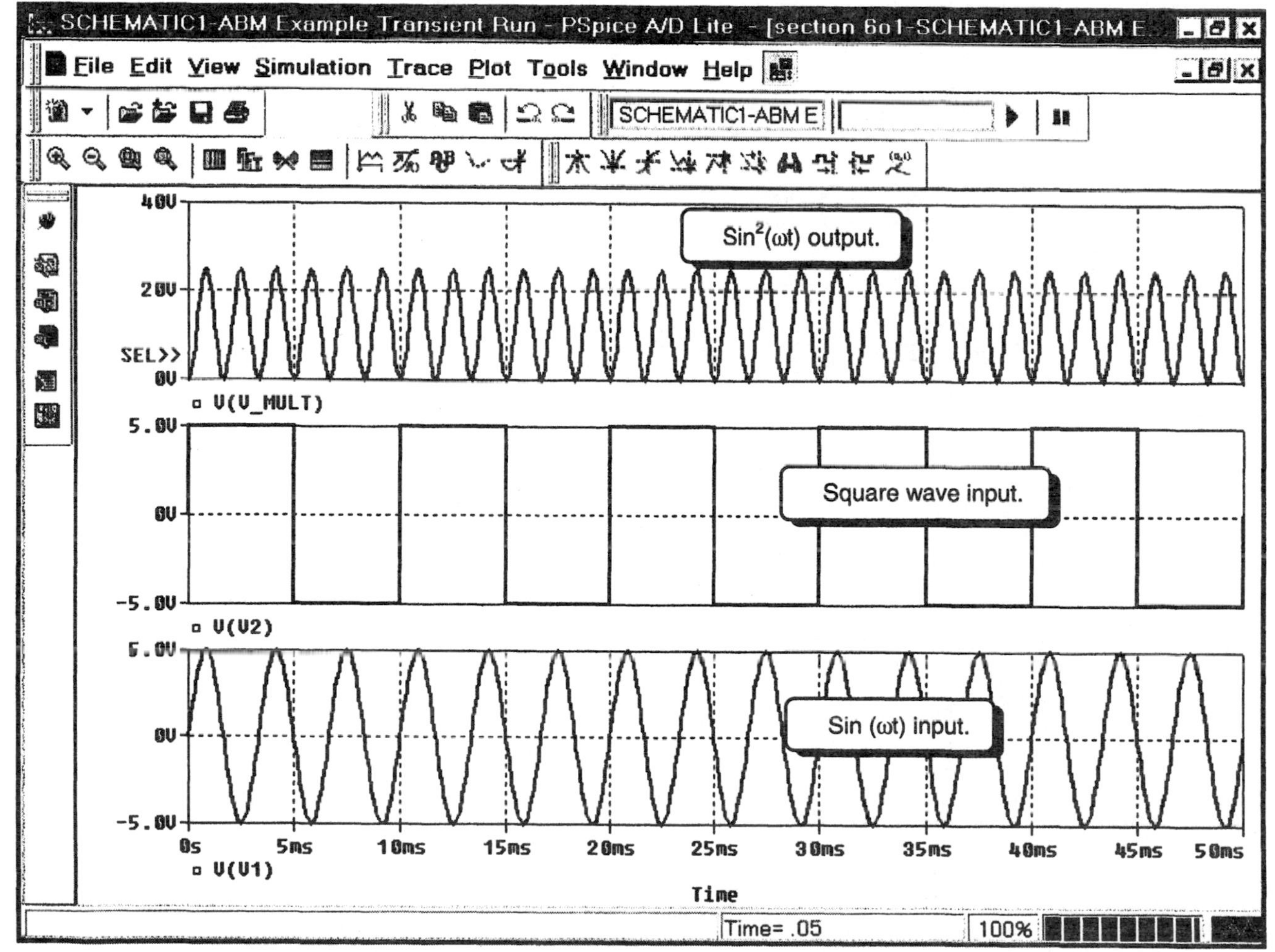


Note that the output of the multiplier is always positive. The output of the ABS function is:

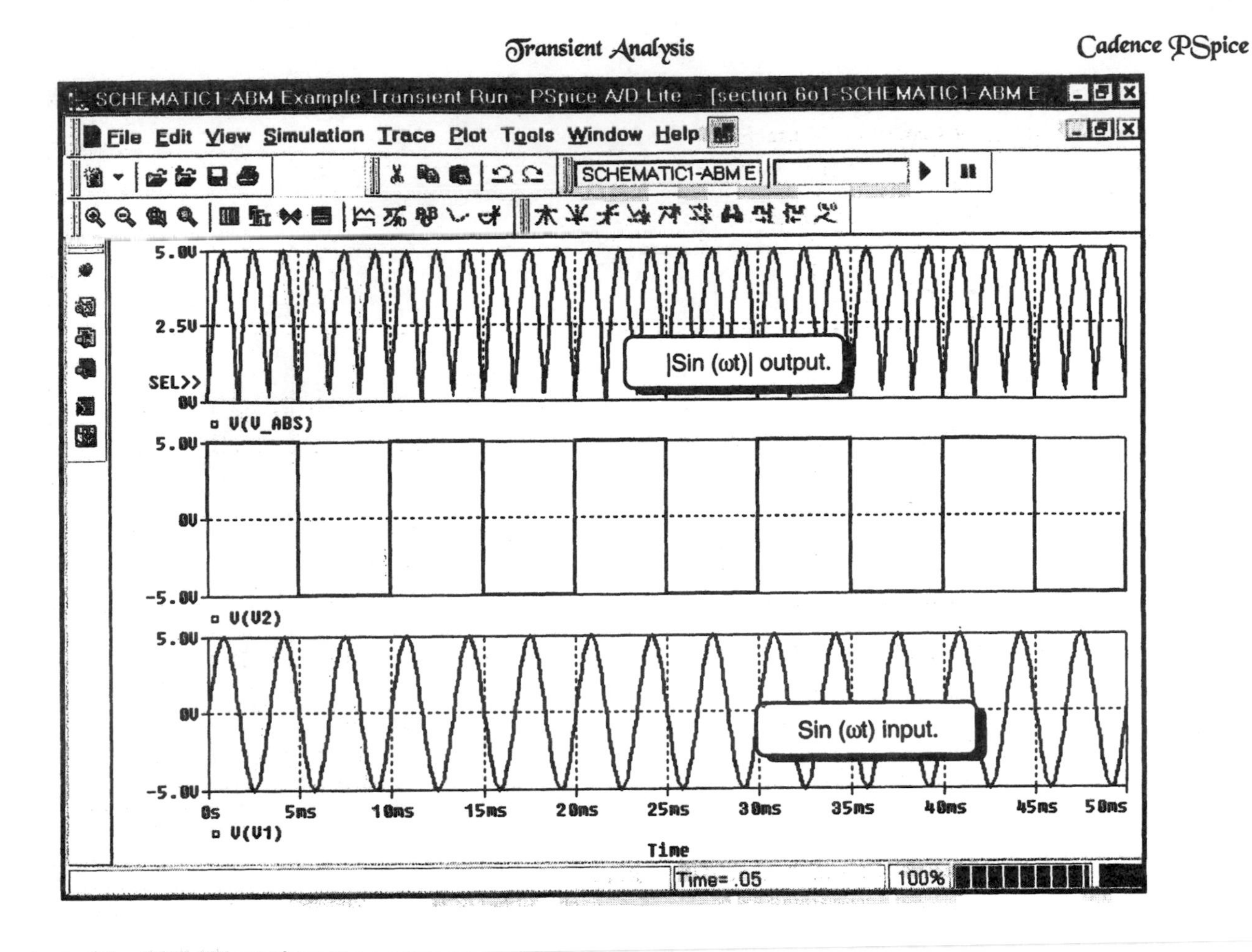

The output of the GLIMIT part is:

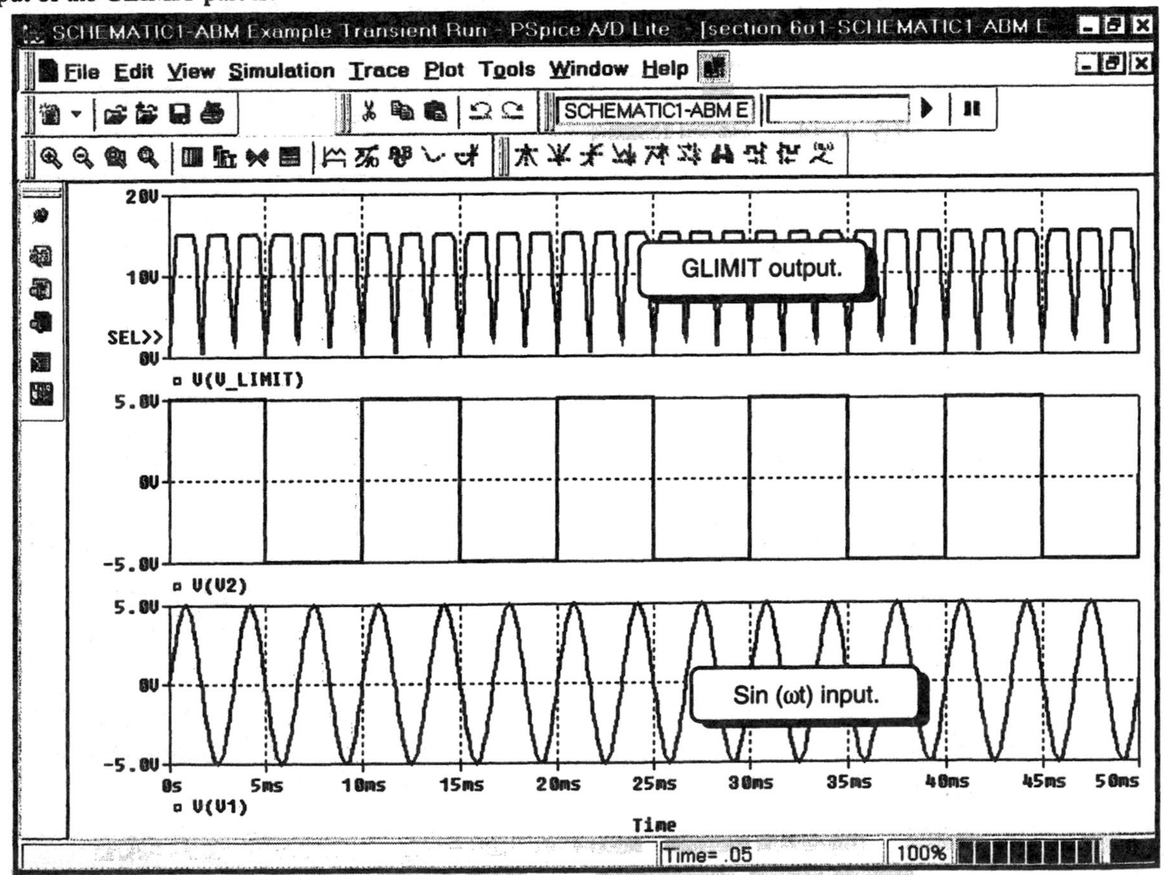


The output of the integrator is:

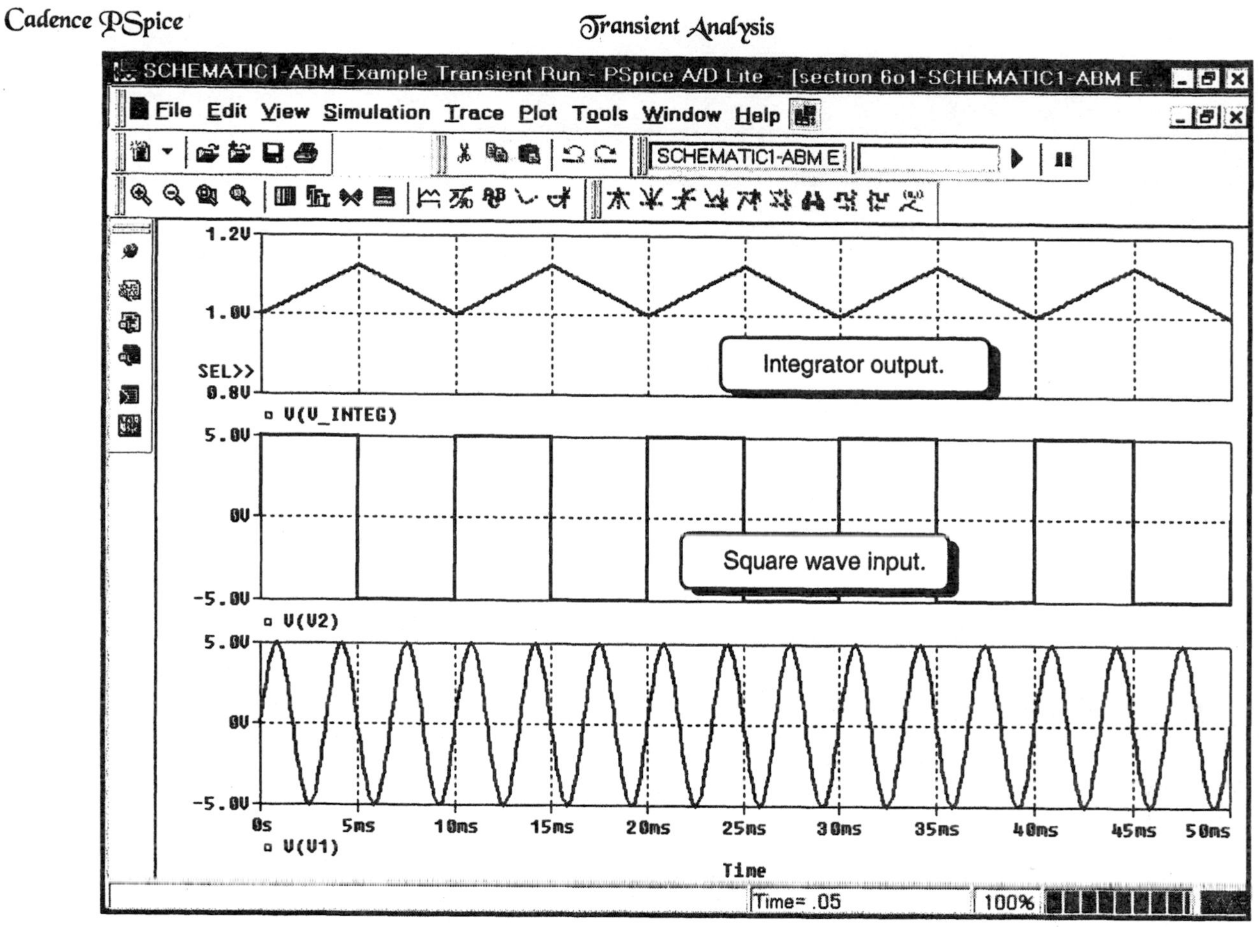


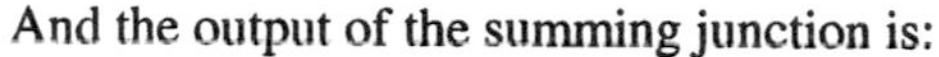
And the output of the summing junction is:

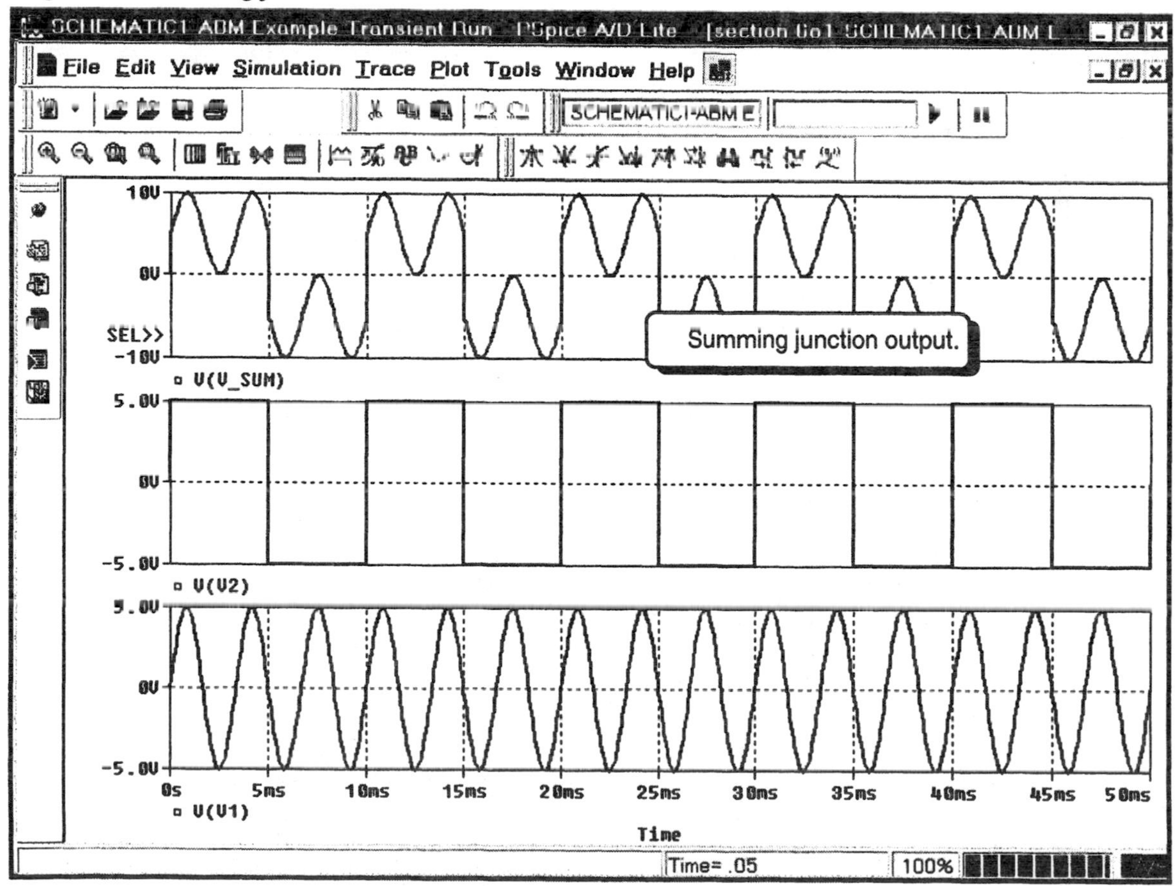

## 6.O.2. Modeling the Step Response of a Feedback System

In this example we will show the performance of a feedback system using only ABM parts. Wire the circuit below:

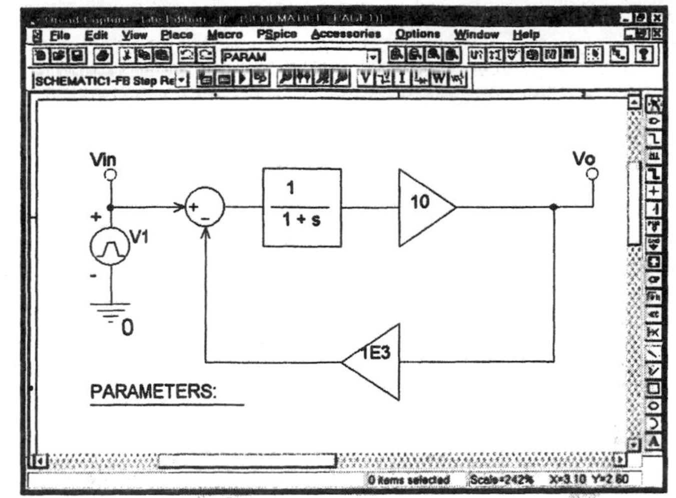


| Diff<br>Difference junction | Laplace<br>Laplace transfer function | 0<br>Ground |
|---|---|---|
| **Vpulse**<br>Pulsed voltage source | **Gain**<br>Gain block | PARAMETERS:<br>**PARAM**<br>Parameters |

We would like to see how the output step response changes for different feedback gains. Presently the feedback gain is ***1E3*** or 1000. We would like to vary the feedback gain to see how the feedback affects the operation of the system. We will use a parameter to change the value of the feedback gain. Double-click on the text ***PARAMETERS:*** to obtain the parts spreadsheet:

File Edit View Place Macro Accessories Options Window Help

New Column... | Apply | Display... | Delete Property | Filter by: < Current properties >

| | | Color | Designator | Graphic | ID | Implementation | Implementation P |
|---|---|---|---|---|---|---|---|
| 1 | SCHEMATIC1 : PAGE1 : 1 | Default | | PARAM.Normal | | | |

Parts / Schematic Nets / Pins / Title Blocks / Globals / Ports

Ready

Click the ***New Column*** button to create a new parameter and fill in the dialog box as shown:

Add New Column

Name:
FB_gain

Value:
1

Enter a name and click Apply or OK to add a column/row to the property editor and optionally the current filter (but not the <Current properties> filter).

No properties will be added to selected objects until you enter a value here or in the newly created cells in the property editor spreadsheet.

Always show this column/row in this filter

Apply | OK | Cancel | Help

Click the ***OK*** button to add the parameter as a new column to the spreadsheet:

Select the ***FB_gain*** column and then click the ***Display*** button:

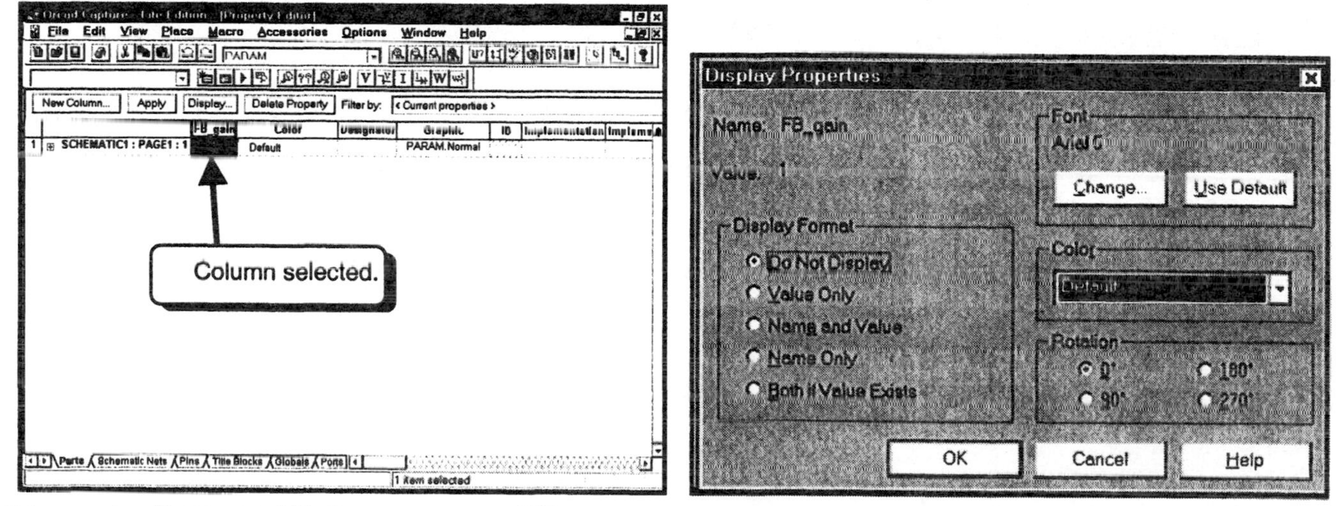


Select option ***Name and Value*** and click the ***OK*** button:

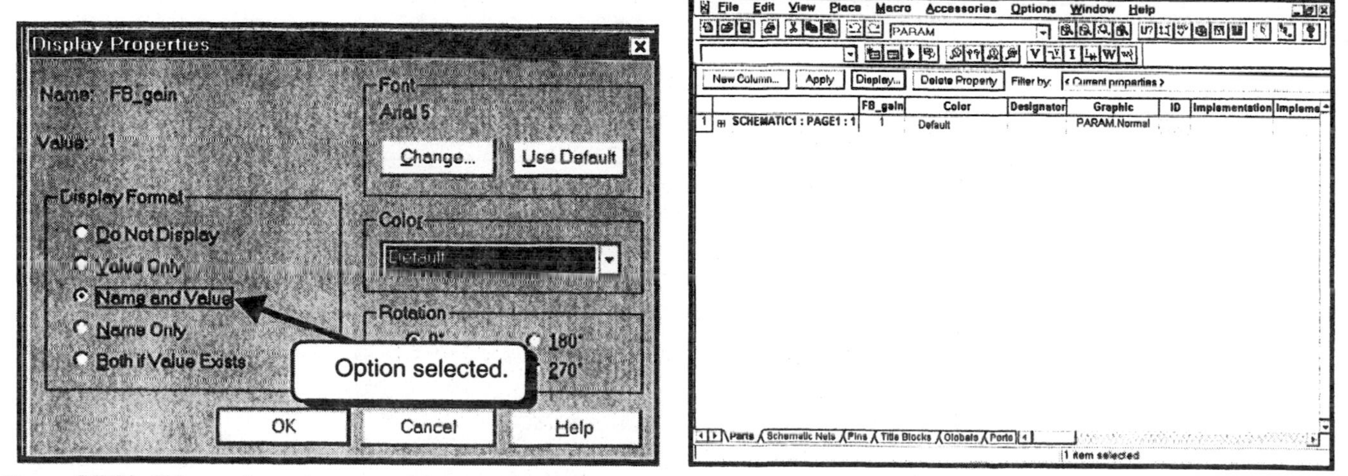


Type **CTRL-F4** to close the spreadsheet and return to the schematic. The new parameter should be displayed on the schematic:

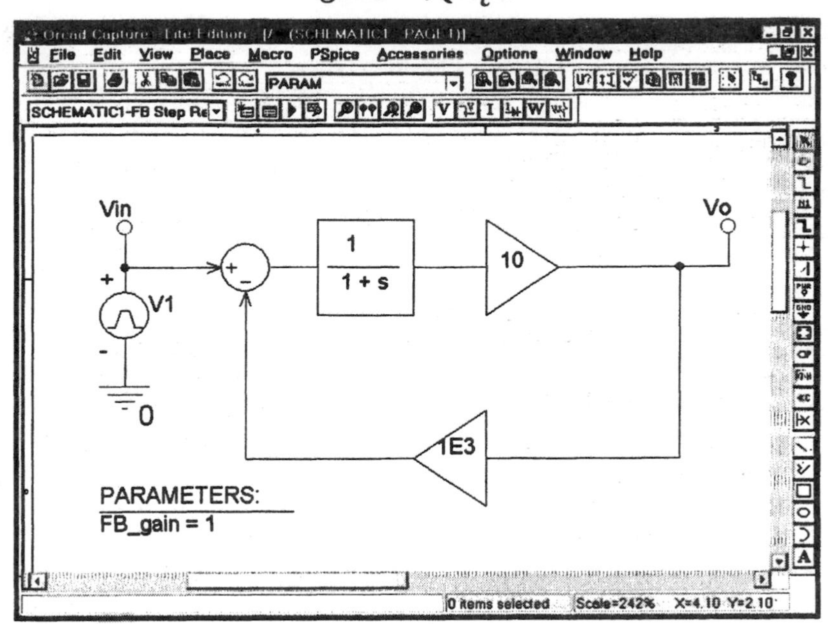


Double-click on the text ***1E3*** of the bottom gain block to change the gain of the feedback block:

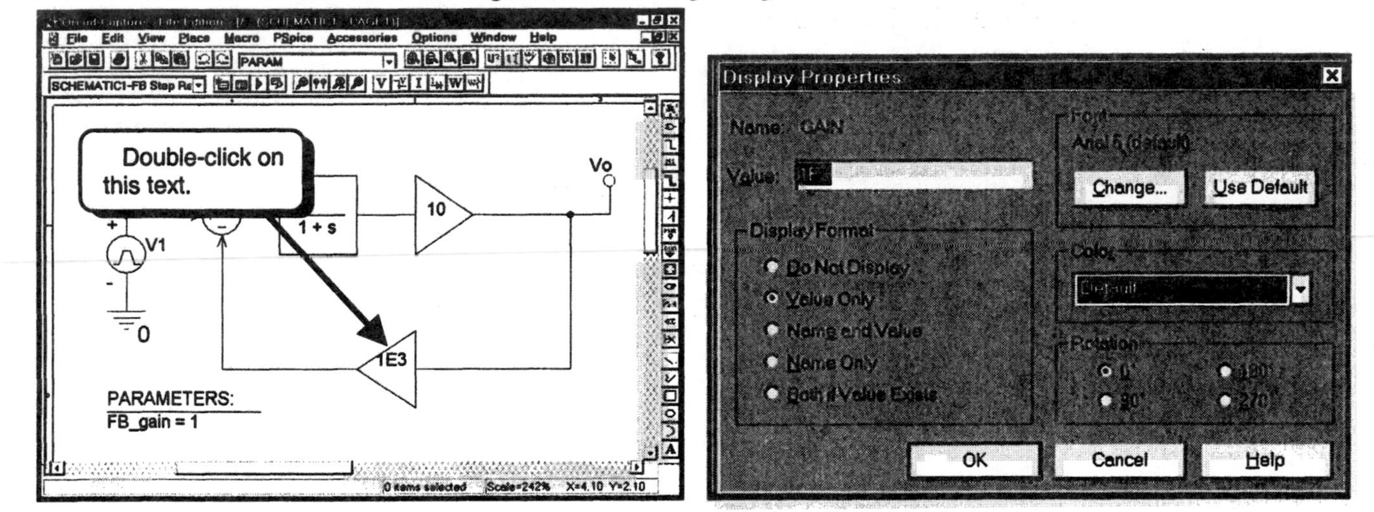


Type the text **{FB_gain}** and click the ***OK*** button to accept the value. In the screen capture below, the text **{FB_gain}** has been moved to a location that makes it easier to read.

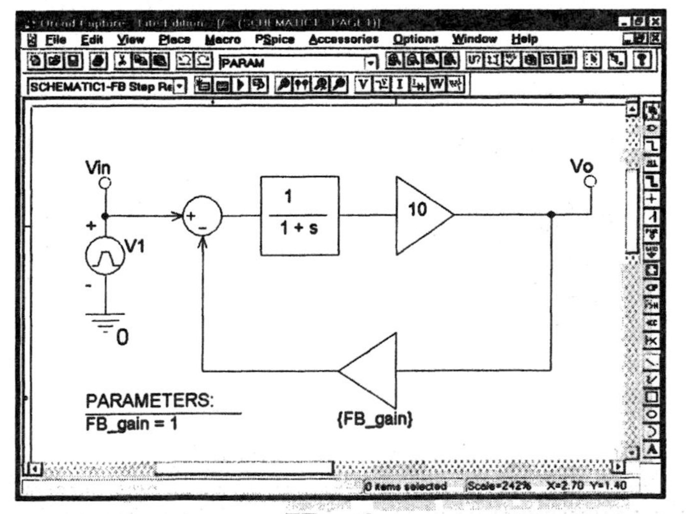


The value of the gain is now the value of the parameter ***{FB_gain}***. We would like the input to be a 0 to 5 V step input. The step will be 0 for the first second and then 5 V for the remainder of the simulation. This waveform can be constructed with a pulsed voltage source. Set the attributes of the VPULSE source as shown. On your screen, the attributes will be arranged in alphabetical order. I have changed the order in my screen so that you can see all of the parameters you need to specify:

Orcad Capture - Lite Edition - [Property Editor]

File Edit View Place Macro Accessories Options Window Help

PARAM

New Column... | Apply | Display... | Delete Property | Filter by: < Current properties > | Help

| | | INITIAL_VOLTAGE | PERIOD | PULSE_WIDTH | PULSED_VOLTAGE | RISE_TIME | FALL_TIME | DELAY_TIME | Designator | Grap |
|---|---|---|---|---|---|---|---|---|---|---|
| 1 | SCHEMATIC1 : PAGE1 : V1 | 0 | 1000 | 999 | 5 | 1u | 1u | 1 | | VPULSE |

Parts / Schematic Nets / Pins / Title Blocks / Globals / Ports / Aliases

Ready    1 item selected

A ***PERIOD*** of ***1000*** seconds and a ***PULSE_WIDTH*** of ***999*** seconds make this a waveform that is at the pulsed voltage for most of the time. The ***DELAY_TIME*** of ***1*** second keeps the voltage at the initial voltage for the first second of the simulation. Thus, when the simulation starts, the voltage will stay at the initial voltage of 0 for the first second. Then the source will change to the pulsed voltage of 5 volts for 999 seconds. The simulation will end well before 999 seconds, so for purposes of this example, the pulsed source is a step from 0 to 5 volts. The waveform of this source is shown below:

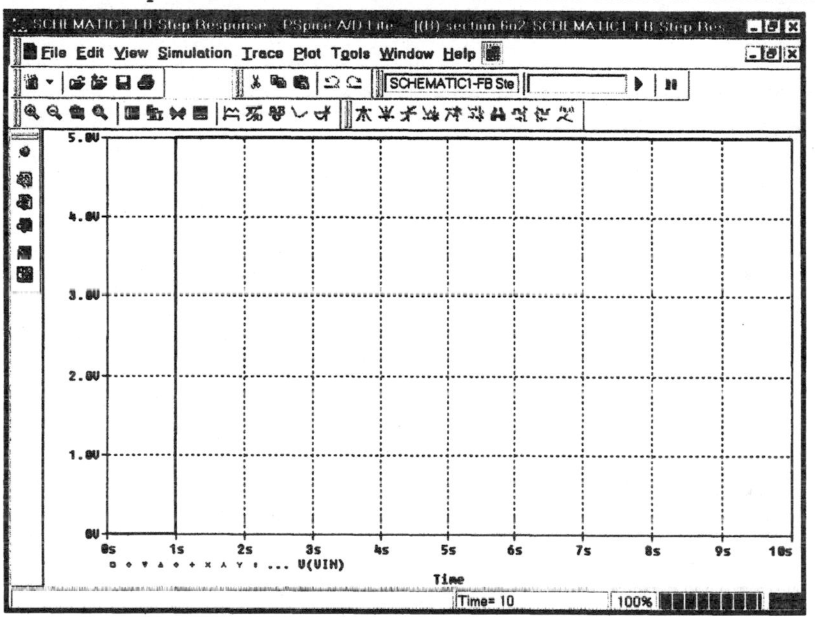


We would like to run the response of the system for 10 seconds. Select **PSpice** and then **New Simulation Profile** from the Capture menus, enter a name for the profile, and then click the ***Create*** button. By default the ***Time Domain (Transient) Analysis type*** is selected. Set up a Transient Analysis with the following parameters:

Simulation Settings - FB Step Response

General | Analysis | Include Files | Libraries | Stimulus | Options | Data Collection | Probe Window

Analysis type: Time Domain (Transient)

Run to time: 10 seconds (TSTOP)

Start saving data after: 0 seconds

Options:
- Monte Carlo/Worst Case
- Parametric Sweep
- Temperature (Sweep)
- Save Bias Point
- Load Bias Point

Transient options
- Maximum step size: seconds
- Skip the initial transient bias point calculation (SKIPBP)

Output File Options...

OK | Cancel | Apply | Help

We would like to observe the step response for several different values of the feedback gain. We will use a Parametric Sweep to vary the value of the parameter FB_gain. Click the *LEFT* mouse button on the square □ next to the text ***Parametric Sweep*** to enable the sweep and display its options. Fill in the Parametric dialog box as shown:

With this ***Parametric*** setup, the ***Transient*** analyses will run eleven times.

We will be running the Transient Analysis 11 times. The first time the feedback gain will be zero, the second time it will be 0.1, the third time it will be 0.2, and so on. Run the simulation and plot the output:

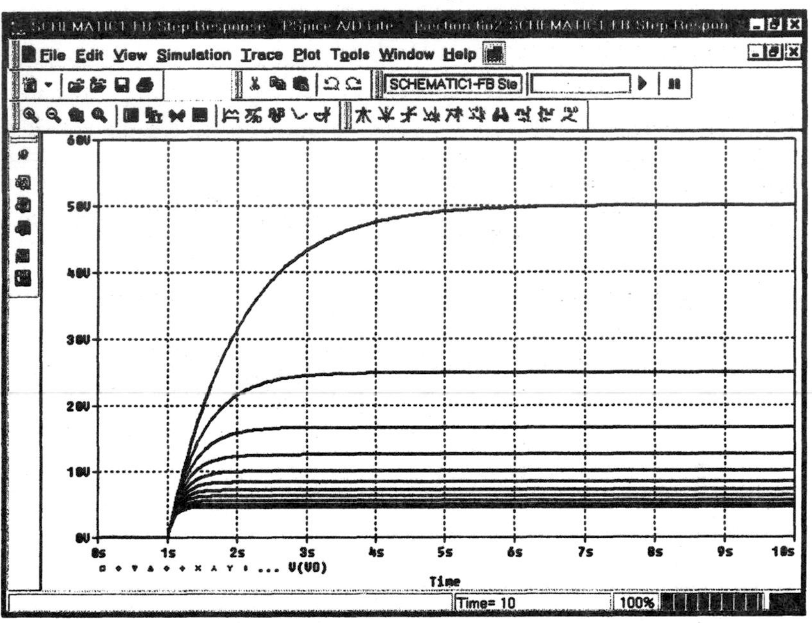


Next, we want to answer the question of which trace belongs to what value of the parameter FB_gain. To answer this question, click the ***RIGHT*** mouse button on one of the traces:

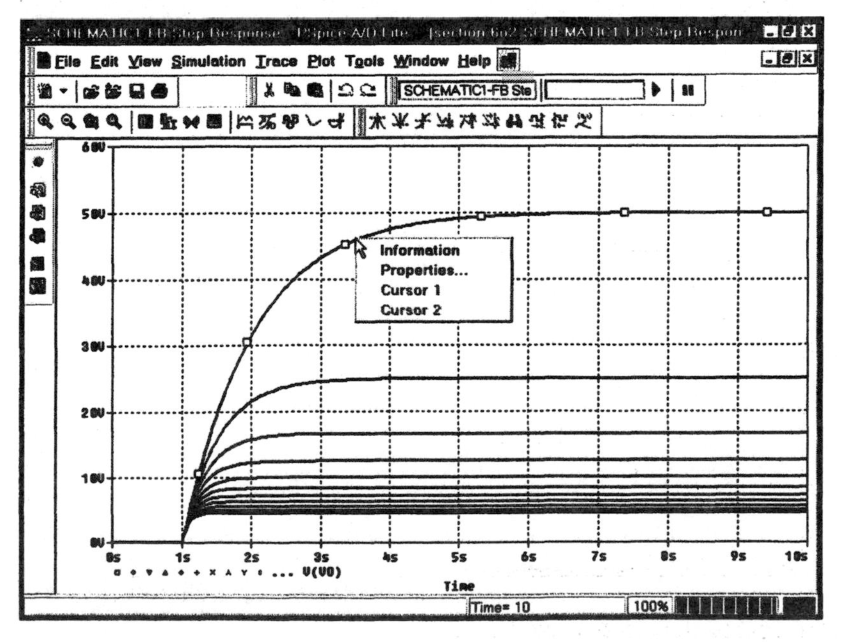

Select **Information**:

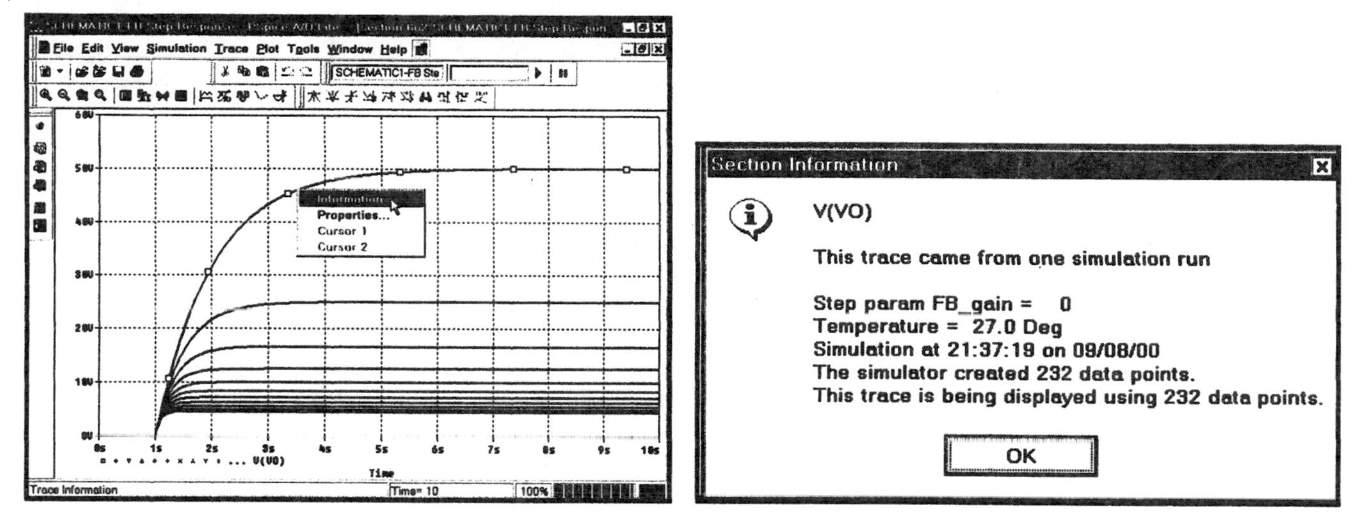


We see that the top trace was generated for a feedback gain of 0.

Just for fun, we can run more complicated transfer functions. We will change the transfer function 1/(s + 1) to (s – 1)/($s^2$ + 2s + 1). To change the numerator, double-click the ***LEFT*** mouse button on the text ***1***:

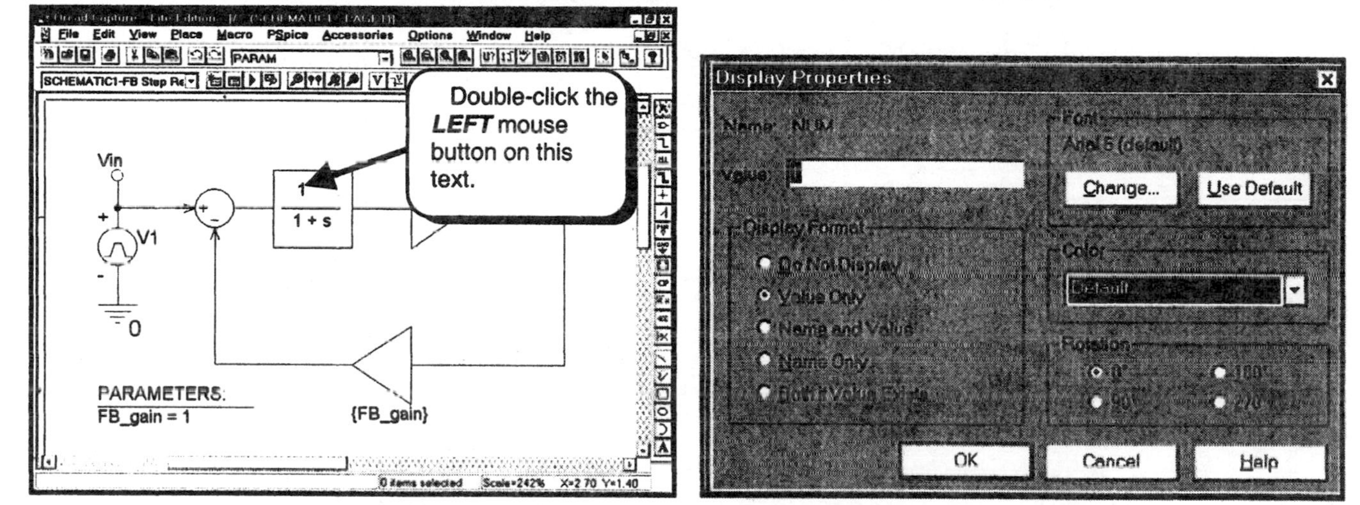


Type the text **s-1** and press the **ENTER** key:

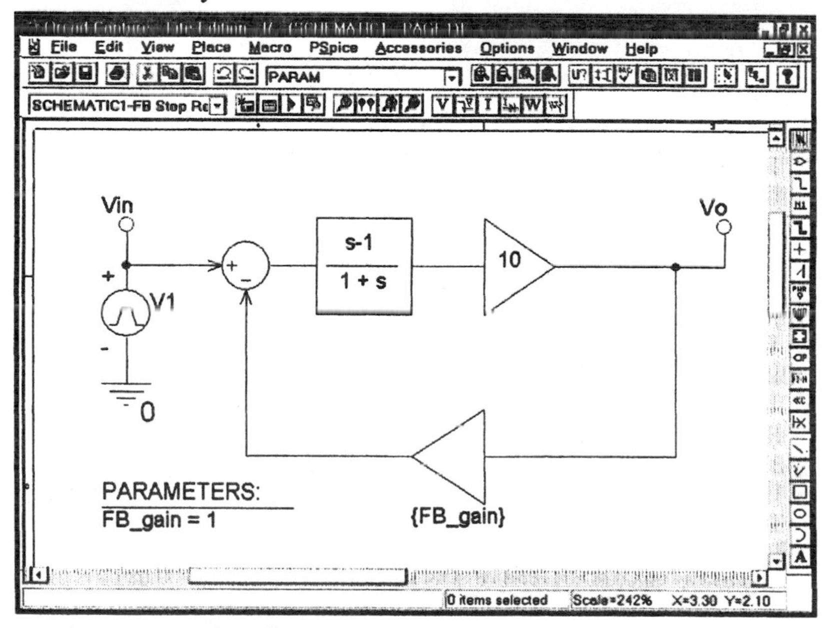


To change the denominator, double-click on the text ***1 + S***:

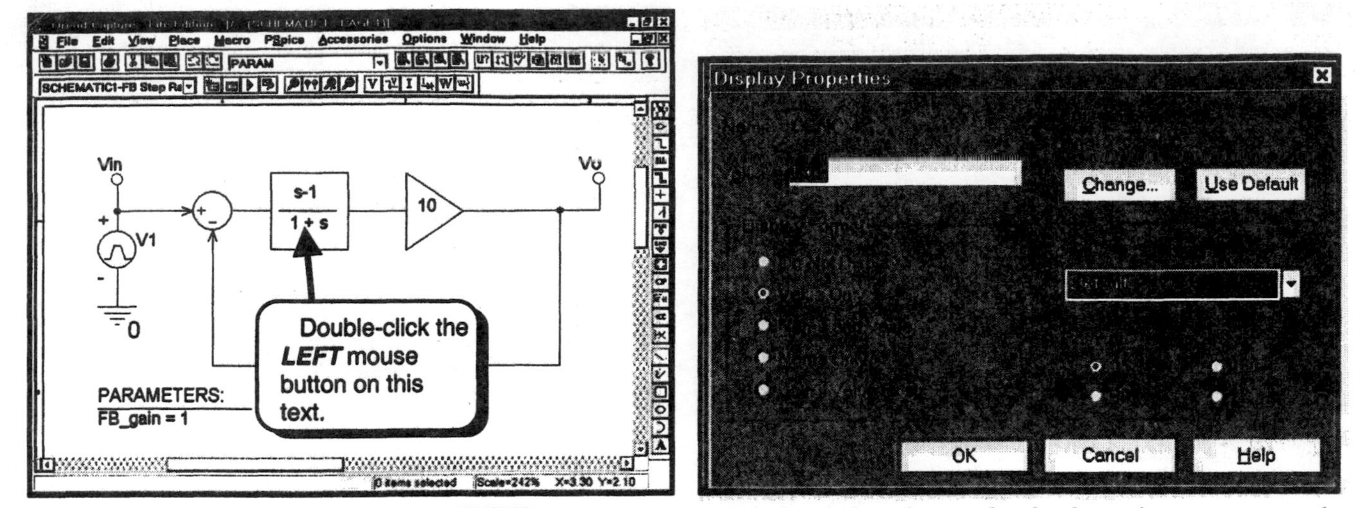


Type the text **s*s+2*s+1** and press the **ENTER** key. In the screen capture below the text for the denominator was moved to make the screen capture more readable:

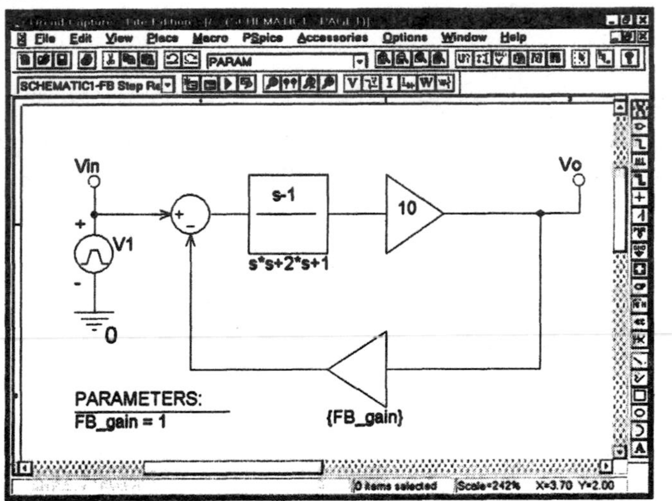


We have now created a system with the open loop transfer function:

$$\frac{Vo}{Vin}(s) = 10\left(\frac{s-1}{s^2+2s+1}\right)$$

Run the circuit and plot the output for the first 5 seconds:

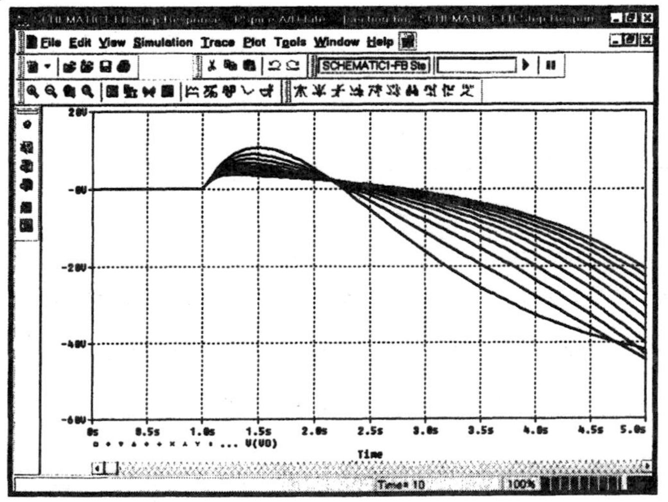

## 6.O.3. Op-Amp Models with ABM Parts

It is easy to create ideal parts using ABM parts. The circuit below has infinite input impedance and a gain of 1,000,000. The output has no supply limits:

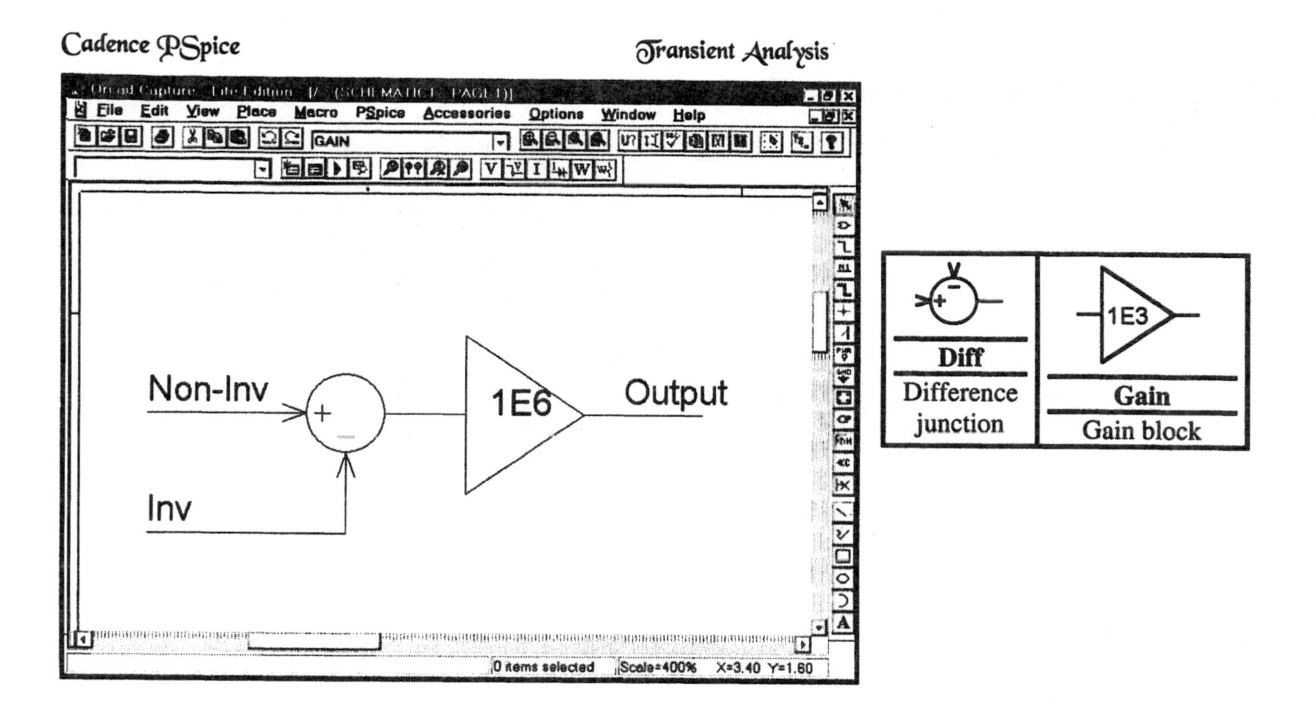


If we make a subcircuit out of this circuit, we could run into problems when the circuit is used. A problem that may result is that, if any of the terminals are not connected to other circuit elements, PSpice may generate an error message that only one element is connected to a certain node. To avoid this problem, we will add resistors to the circuit that do not affect its operation:

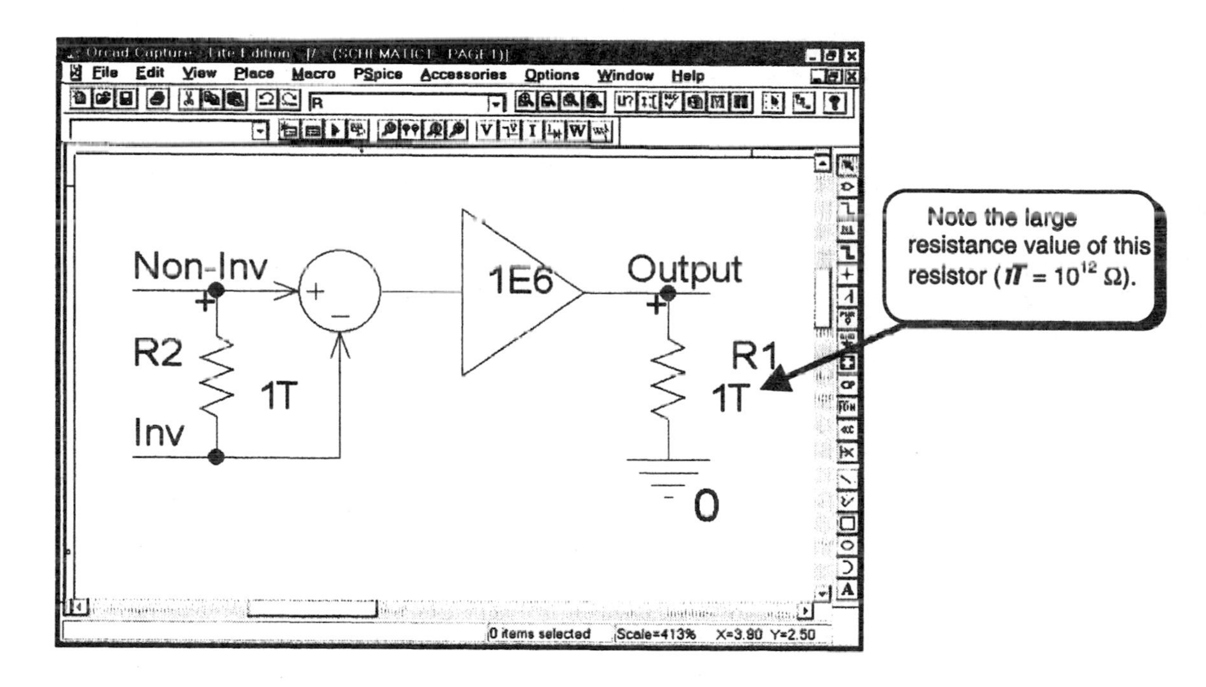


The resistors have a value of 1T or $10^{12}$ Ω. These resistors are so large that they will not affect the operation of the circuit in most cases, and they avoid the problem of only one circuit element being connected to each node.

The circuit below is an improved model for an op-amp. It has a low frequency gain of $10^6/30$ and infinite input impedance, but we have added frequency dependence and supply limits of ±15 V:

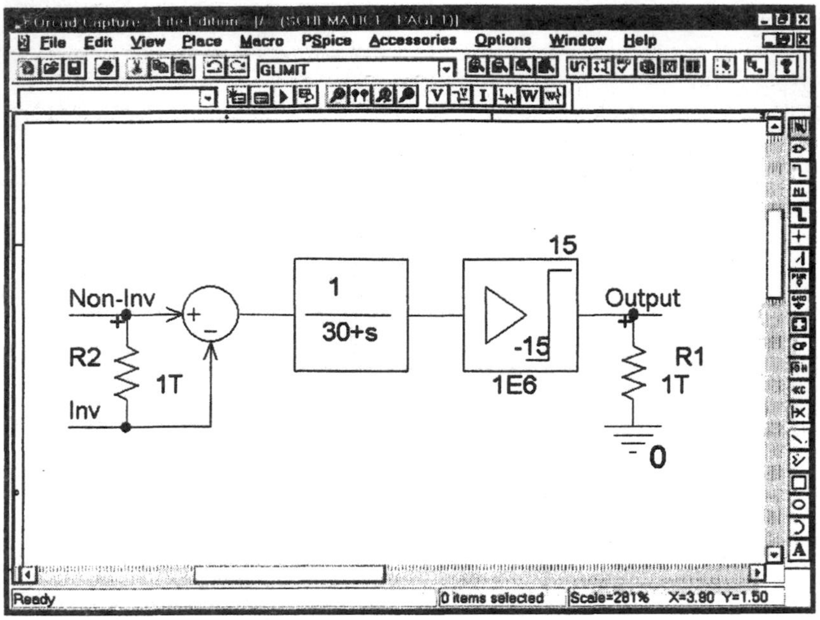

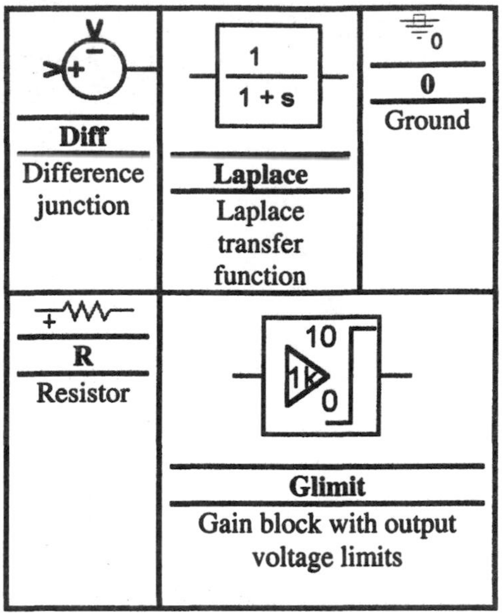

| Diff | Laplace | Ground |
|---|---|---|
| Difference junction | Laplace transfer function | 0 |
| R | Glimit | |
| Resistor | Gain block with output voltage limits | |

The output voltage swing is limited to ±15 V and we have added a pole at 30 rad/sec. This is approximately the frequency response of a 741 op-amp. Note that this circuit is still ideal because it does not include many of the other non-ideal characteristics of a 741 op-amp such as bias currents, offset voltages, and slew rate.

## 6.O.4. AC Sweep with ABM Parts

Next, we will use the op-amp circuit created in the previous section to demonstrate an AC Sweep. We created an op-amp with frequency dependence in the previous section. We will now show how the frequency response varies with feedback. In the circuit below, the op-amp model is used as a non-inverting amplifier with gain 1 + (Rf/R4):

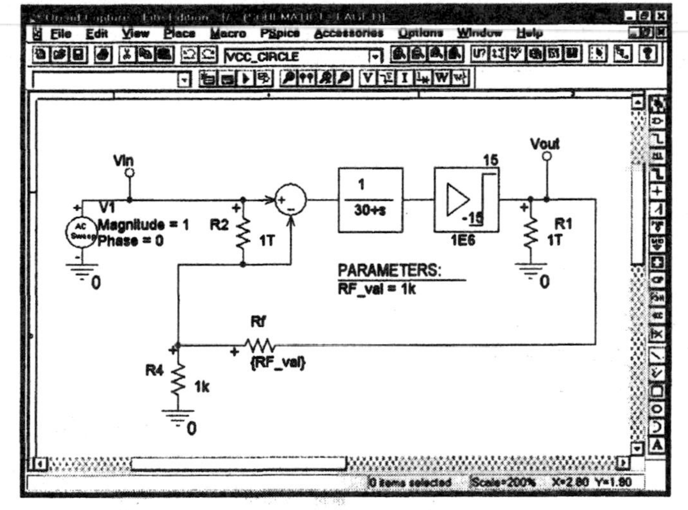


Set up the AC and Parametric sweeps as shown:

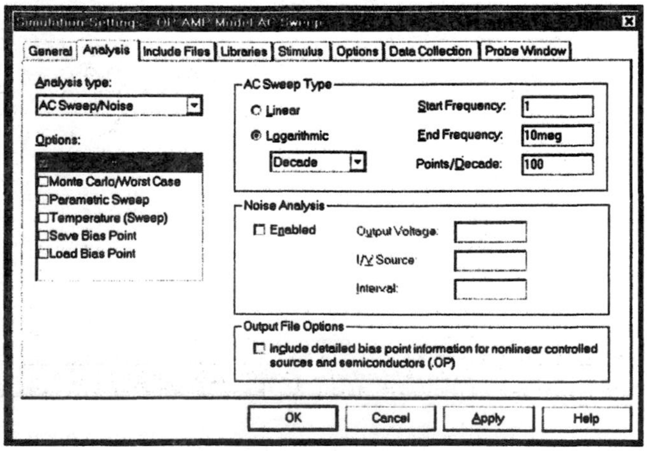

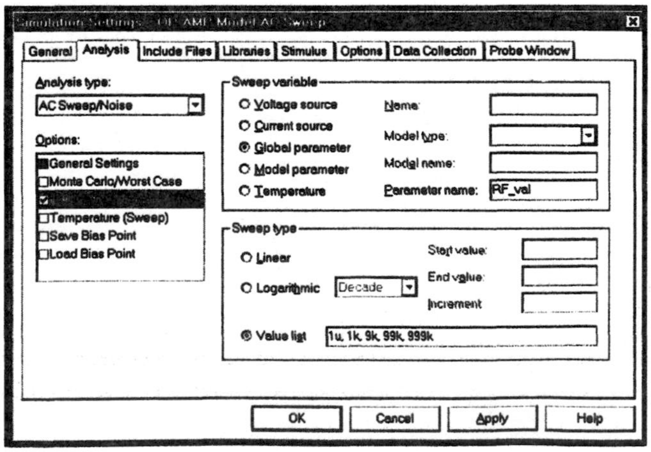

For the specified values of RF_val, the gain of the amplifier should be 1, 2, 10, 100, and 1000. Run the simulation and plot the gain in decibels:

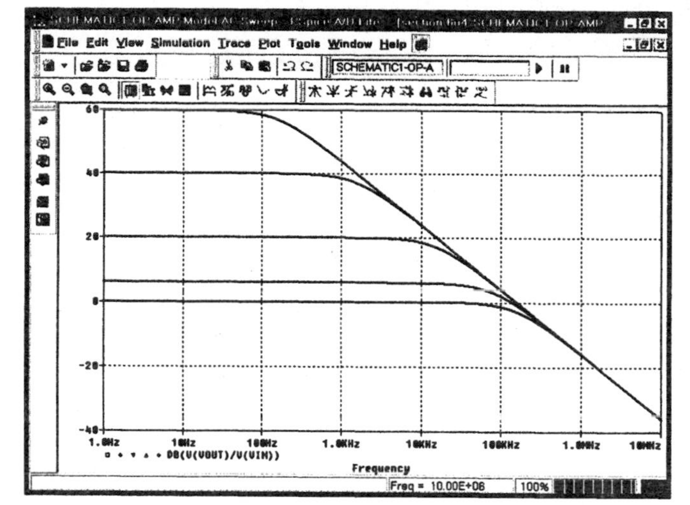

The plot shows the expected result that, as we reduce the gain, the bandwidth increases.

## 6.O.5. Switching Power Supply

One of the advantages of the ABM parts is that we can create circuits with complicated functions easily and with fewer parts than creating the functions with circuit components. Reducing the complexity of a circuit by using the ABM parts gives us the added benefit of being able to run circuits that would otherwise exceed the limitations of the Lite version. An example would be the switching power supply below:

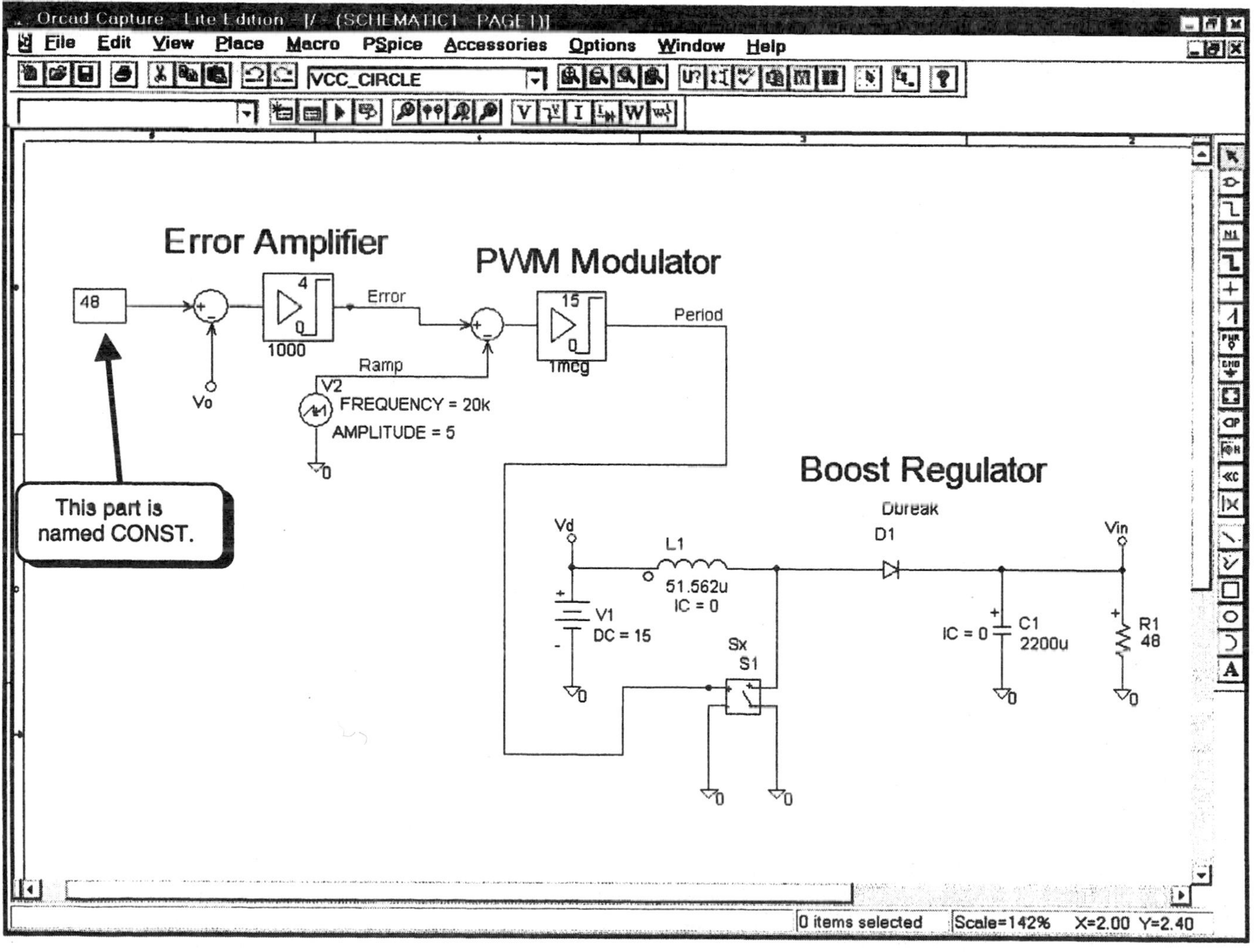

To run this circuit with real components we would need to create an error amp with output limits and then a ramp generator with output limits. Notice that the limits are not traditional supply limits and would require limiting circuits. This circuit would take at least two op-amps plus some limiting circuits, which could be Zeners or active clamps that also use op-amps. A real circuit of this complexity could easily exceed the limits of the Lite version. The purpose of this simulation was not to accurately run the control circuit, but to create an "ideal controller" so that we could test the theory behind the switching circuit. The circuit also uses the PSpice part Vswitch rather than a MOSFET or IGBT. Vswitch is an ideal switch. Using this switch rather than a MOSFET or IGBT eliminates the circuitry needed for driving the MOSFET.

We will not run the circuit here because it would take more than an hour to complete. The circuit is given as an example of how the ABM parts can be used to simplify a complicated control circuit.

# 6.P. Summary

- The Transient Analysis is used to look at plots of voltages and currents versus time. This simulation displays waveforms as you would see them on an oscilloscope screen.
- Use Probe to view the results graphically.
- Use the Transient Analysis with sources such as Vsin, Isin, Vpulse, Ipulse, VPWL, IPWL, etc.
- The AC sources VAC and IAC are set to zero.
- DC sources keep their specified value.
- Use the Transient Analysis to view inductor and capacitor transient responses.
- Use the Transient Analysis to observe an amplifier's voltage swing.
- The Fourier components of a time signal can be viewed graphically with Probe by selecting **Axis** and then **Fourier** from the Probe menu.
- The Fourier components of a time signal can be displayed in the output file by enabling the Fourier option in the Transient Analysis dialog box.
- The Parametric Sweep can be used to see how values of devices affect the performance of a circuit.
- Goal functions can be used to obtain numerical data from Probe graphs. The result of evaluating a goal function is a single numerical value.
- The Performance Analysis is the use of a goal function in conjunction with a Parametric Sweep. The result of the goal function is plotted versus the swept parameter.
- The user can create complex waveforms with the Stimulus Editor without knowing all of the parameters of a particular source.

# PART 7
# Creating and Modifying Models Using Capture

In this part we will demonstrate how to modify existing PSpice models and how to create new models. We will assume that the user is familiar with PSpice models and knows how he or she would like to modify the models. A discussion of the various models requires too much detail to be given here. The user is referred to the PSpice Reference Manual available from Orcad Corporation for model details. This manual is contained on the CD-ROM that accompanies this text. You will probably need to review the many references that Orcad gives to understand the model parameters. Here, we will show how to make changes to existing models or create simple new models. Section 7.E contains simplified models for some of the commonly used parts. The model parameters given are for first-time users. For more accurate models, you will need to refer to more detailed texts covering SPICE models. If you are more familiar with the models, you can use these procedures to modify all parameters in a model.

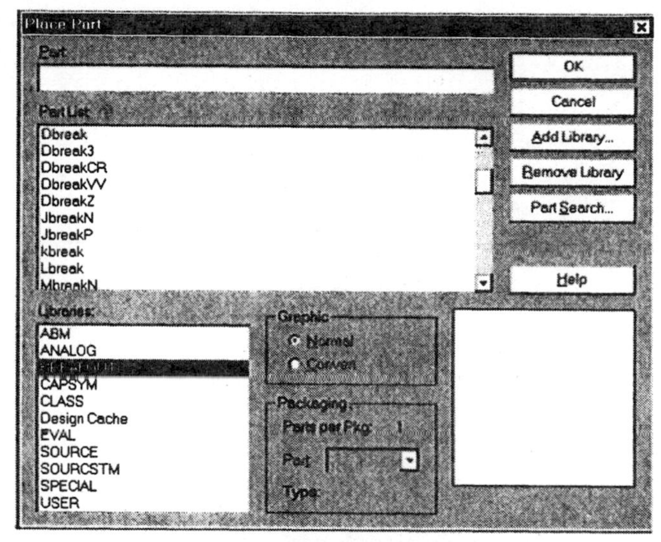


There are three ways to create new models in PSpice. One way is to modify an existing model and give it a new name. The second way is to get a "breakout" part and create a new model. The third way is to create a new model using Orcad's Parts program. Since the Lite version of Parts creates only diode models, we will not discuss it here. The breakout parts are contained in the library called ***BREAKOUT***. This library contains graphic symbols for all parts available in Capture. As you scroll down through the ***Part List*** pane you will notice many different parts:

**Table 7-1: Standard Breakout Models**

| Model | Description |
|---|---|
| Cbreak | Capacitor model. |
| Dbreak | Generic diode. |
| DbreakZ | Zener diode model. The PSpice models used to create a diode and a Zener diode are the same. The only difference is in the graphic symbols of the parts. |
| JbreakN | N-type junction FET. |
| JbreakP | P-type junction FET. |
| Lbreak | Inductor model. |
| MbreakN<br>MbreakN4 | N-channel MOSFET. Both graphic symbols have four terminals. The fourth terminal is the substrate. |
| MbreakN3 | N-channel MOSFET. The graphic symbol has only three terminals available. The substrate terminal is tied to the source in the graphic. |
| MbreakP<br>MbreakP4 | P-channel MOSFET. Both graphic symbols have four terminals. The fourth terminal is the substrate. |
| MbreakP3 | P-channel MOSFET. The graphic symbol has only three terminals available. The substrate terminal is tied to the source in the graphic. |
| QbreakN | NPN bipolar junction transistor. |
| QbreakP | PNP bipolar transistor. |
| Rbreak | Resistor model. |
| Sbreak | Voltage controlled switch. |
| ZbreakN | IGBT model. |

To demonstrate changing models, we will use the circuit components below. We will not bother wiring the parts since we are interested only in creating models. The only purpose of this example is to demonstrate how to change models.

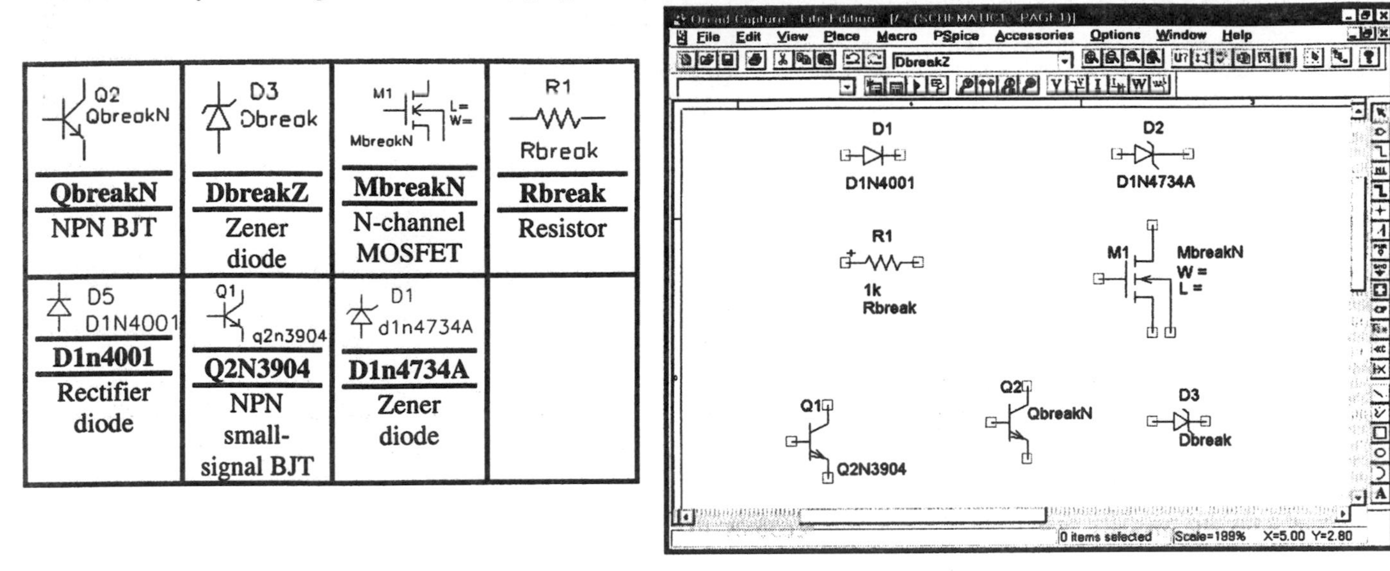

| QbreakN | DbreakZ | MbreakN | Rbreak |
|---|---|---|---|
| NPN BJT | Zener diode | N-channel MOSFET | Resistor |
| **D1n4001** | **Q2N3904** | **D1n4734A** | |
| Rectifier diode | NPN small-signal BJT | Zener diode | |

This project is named "Section 7A.opj."

# 7.A. Changing the Model Reference

If you have a PSpice library with models you would like to use, an easy way to use them is to change the model reference of a graphic symbol to use the name of the model in your library. In this section, we show how to change the model reference of a graphic symbol. We assume that you know the name of the PSpice model that you would like to use.

Double-click the ***LEFT*** mouse button on the graphic symbol for the breakout Zener diode ***D3***, , to obtain the properties spreadsheet for the part. Two screen captures are shown to display the properties we need for this discussion.

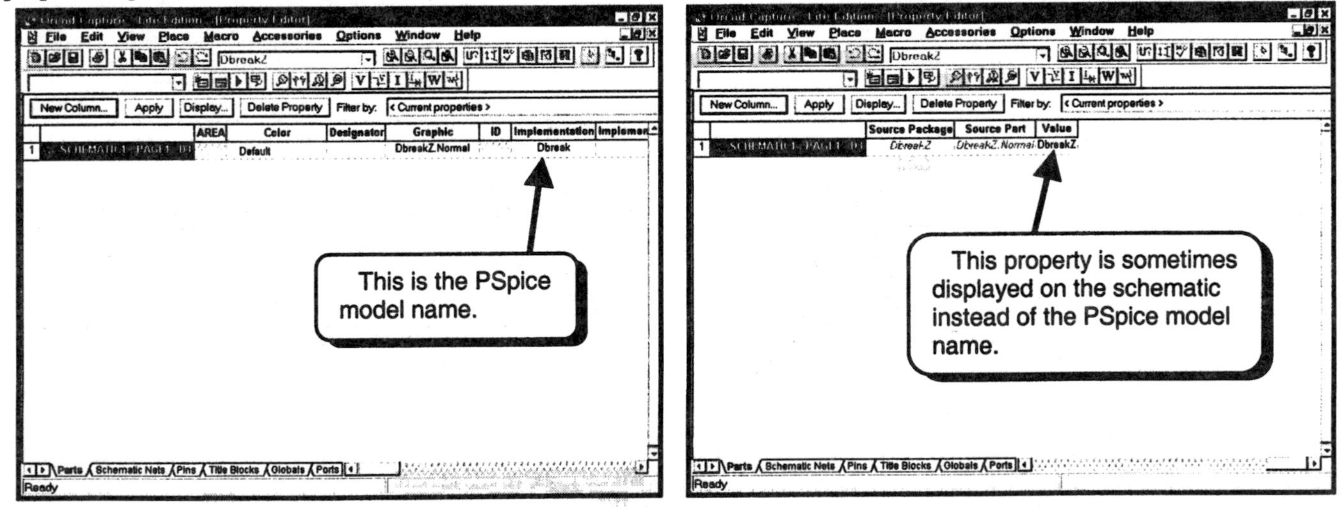


There are two properties that display the name of the model, the ***Value*** property and the ***Implementation*** property. The ***Value*** property is not used by PSpice, but this property is sometimes displayed on the schematic instead of the PSpice model name. The ***Implementation*** property is the name of the PSpice model and it is the property used when creating the netlist. If you change the ***Value*** property to the name of your new model, nothing will change in the simulation. If you change the ***Implementation*** property to the name of your new model, the new model will be used. Always change the ***Implementation*** property.

The ***Implementation*** property shows that the model for this diode is ***Dbreak***. Suppose we want both Zener diodes in the circuit to use the model D1n4734A. Change the ***Implementation*** property to **D1n4734A** and type **CTRL-F4** to close the spreadsheet and return to the schematic. In the schematic, the text Dbreak will change to ***D1n4734A***:

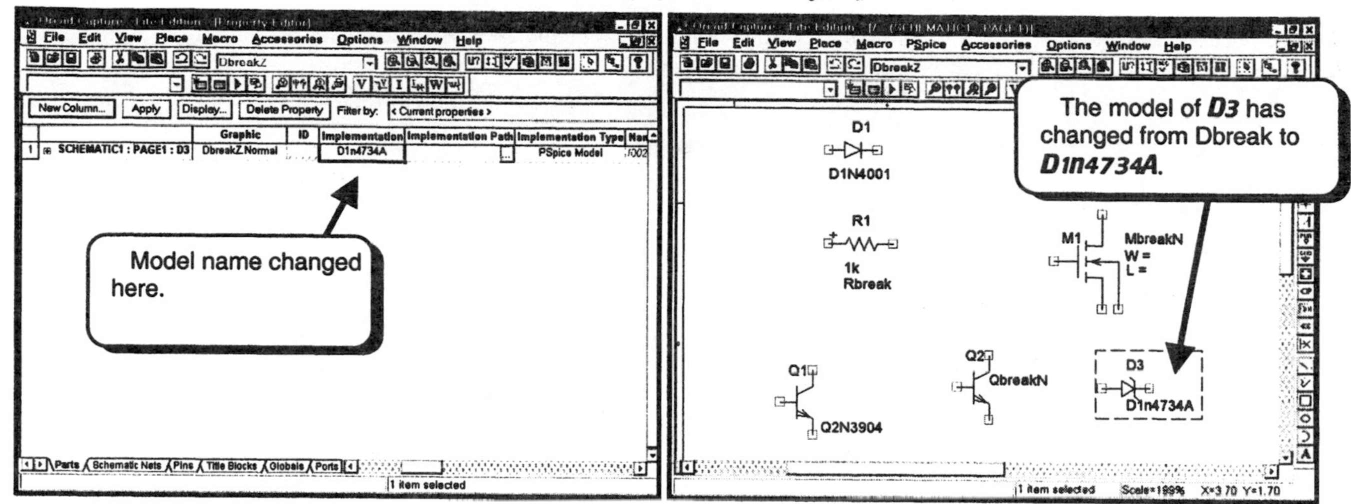


The model ***D1n4734A*** must be defined in one of the PSpice libraries. If you look in Appendix E on page 619, you will see the text below in the file class.lib:

```
.model D1N4734A D(Is=1.085f Rs=.7945 Ikf=0 N=1 Xti=3 Eg=1.11 Cjo=157p M=.2966
+           Vj=.75 Fc=.5 Isr=2.811n Nr=2 Bv=5.6 Ibv=.37157 Nbv=.64726
+           Ibvl=1m Nbvl=6.5761 Tbv1=267.86u)
*           Motorola     pid=1N4734     case=DO-41
*           89-9-19 gjg
*           Vz = 5.6 @ 45mA, Zz = 40 @ 1mA, Zz = 4.5 @ 5mA, Zz = 1.9 @ 20mA
```

Lines that begin with a + sign are continuation lines and are part of the model definition. Lines that begin with an * are comment lines and are ignored by PSpice. This text defines model D1N4734A. The graphic symbol does not define the model. The graphic symbol refers to a model named D1N4734A and this model must be defined in a PSpice library (files with a .lib extension). In this case, the library is file class.lib.

If you add new model libraries you can change model references, rather than define new graphic symbols for each part in the library. Tens of thousands of models ship with the professional version of PSpice and are available on the WWW. Instead of creating a new graphic symbol for each part, you should get a breakout part and then change the model reference to a model in the ".lib" file. You are encouraged to look through all files with the suffix ".lib." These files define the models available to you. Not all of these models have predefined graphic symbols that refer to them. You may use the models in the ".lib" files by changing the model reference of a breakout part.

As a second example of changing a model reference, we will change diode ***D1***. Note that ***D1*** is a part with a predefined model, ***D1N4001***. Double-click the ***LEFT*** mouse button on the graphic for diode ***D1***, ─▷|─, to obtain the spreadsheet for the part. Two screen captures are shown to display the properties of interest.

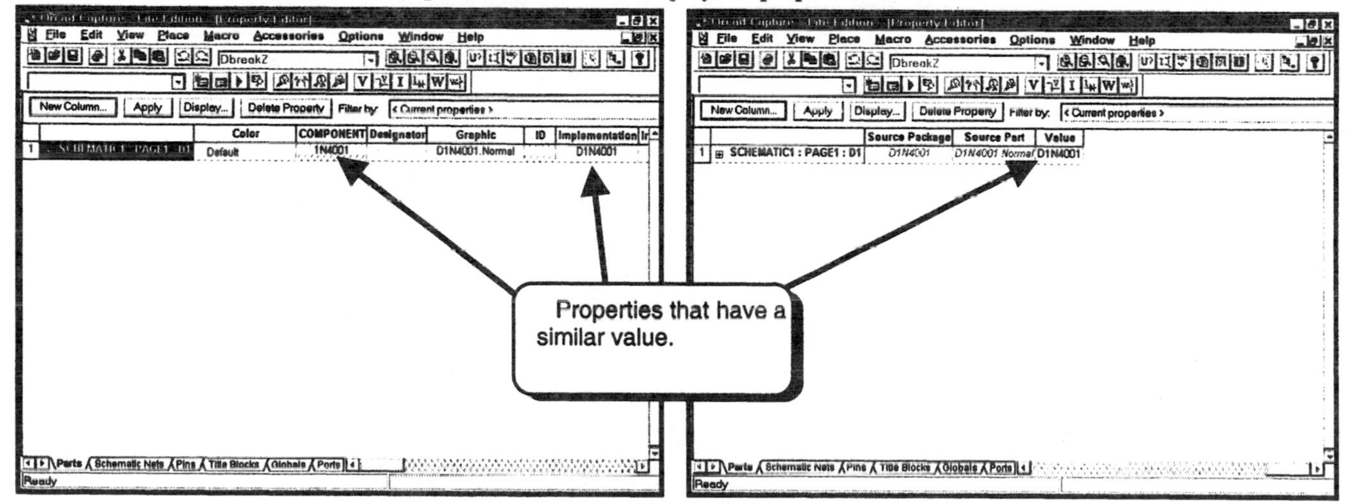


We see that the ***Implementation*** is ***D1N4001***, the ***COMPONENT*** is ***1N4001***, and the ***Value*** is ***D1N4001***. If you select a property and then click the ***Display*** button, you will find that the ***Value*** property is displayed and not the ***Implementation*** attribute.

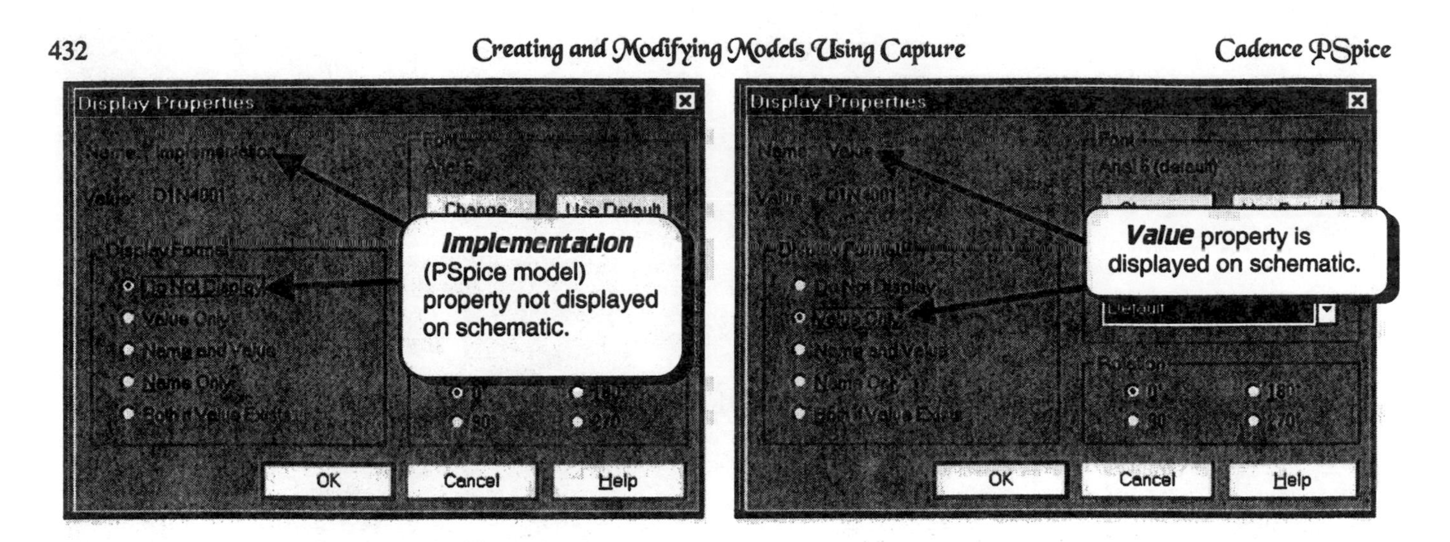


Thus, when we change the model (***Implementation*** property), the change will not be displayed on the screen. Click the ***Cancel*** button to return to the spreadsheet.

Change the ***Implementation*** property to **Dx** and then type **CTRL-F4** to close the spreadsheet to return to the schematic:

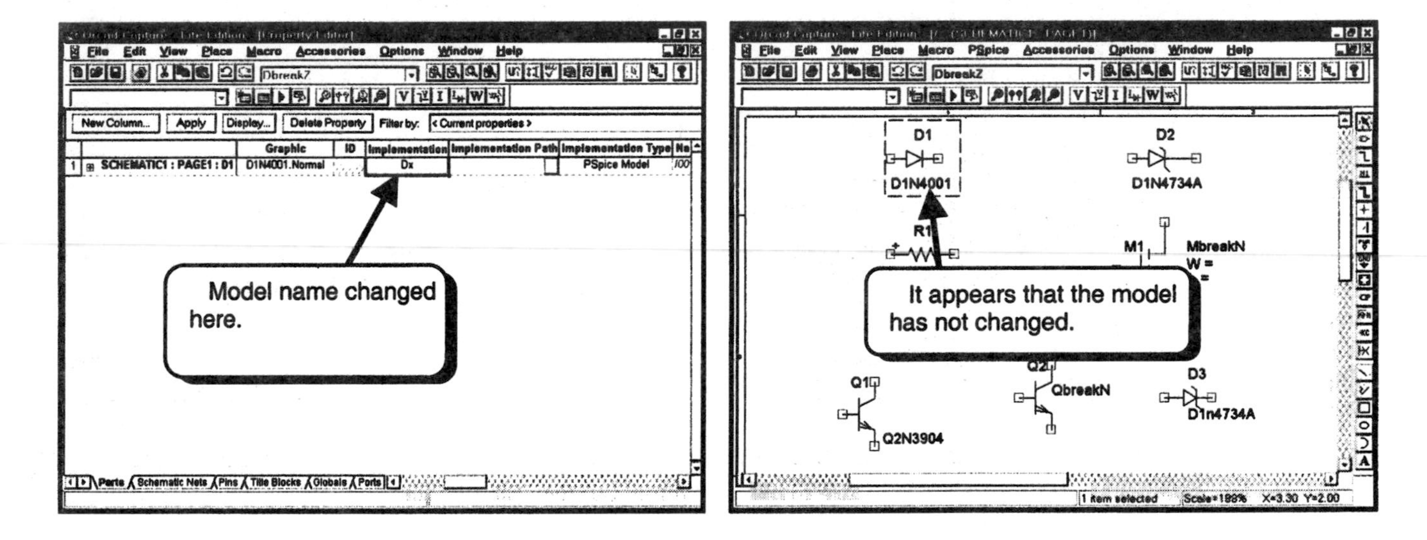


Note that the model Dx does not appear in the schematic. This is because, with predefined models like the ***D1N4001***, the text ***D1N4001*** displayed on the schematic is the ***Value*** property and not the ***Implementation*** property. Since the ***Implementation*** attribute has been changed to ***DX*** in the property spreadsheet, PSpice will use model Dx in the simulation.

The PSpice model name (Implementation property) is displayed on the screen with breakout parts. When the model is changed, the changed model name appears on the schematic. This is because Capture assumes that you wish to change the breakout model to a different model, and that you wish to identify that model on the screen. Capture assumes that when you place a breakout part, you will change it. If you know you will be changing the model reference of a part, you should use the breakout parts.

We will change the PSpice model of ***D3*** back to Dbreak using a different method. Since, for breakout parts, the text shown on the screen is the PSpice model, we can double-click on the text to change the model. Double-click on the text ***D1n4734A*** for D3:

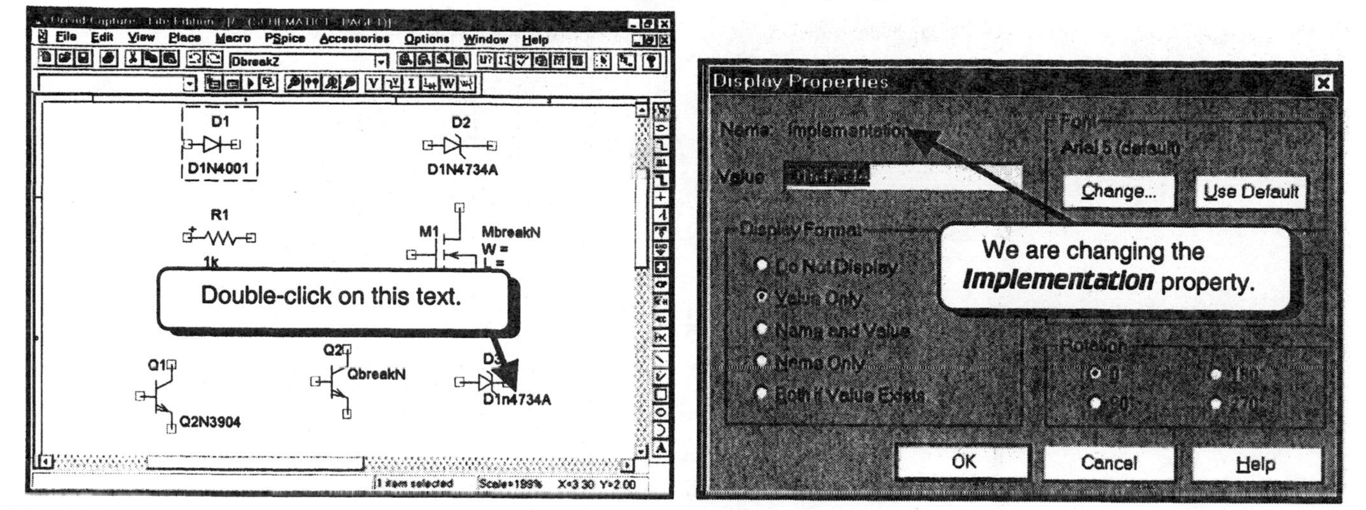

The dialog box shows that we are changing the ***Implementation*** property, which is the PSpice model. Change the value to Dbreak and then click the ***OK*** button:

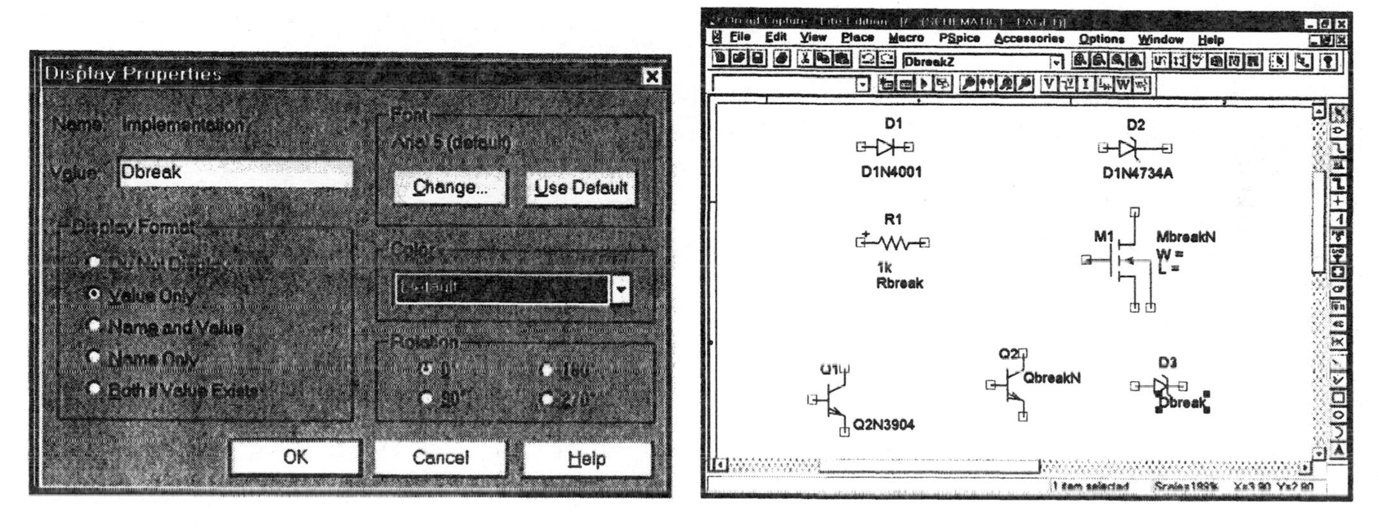


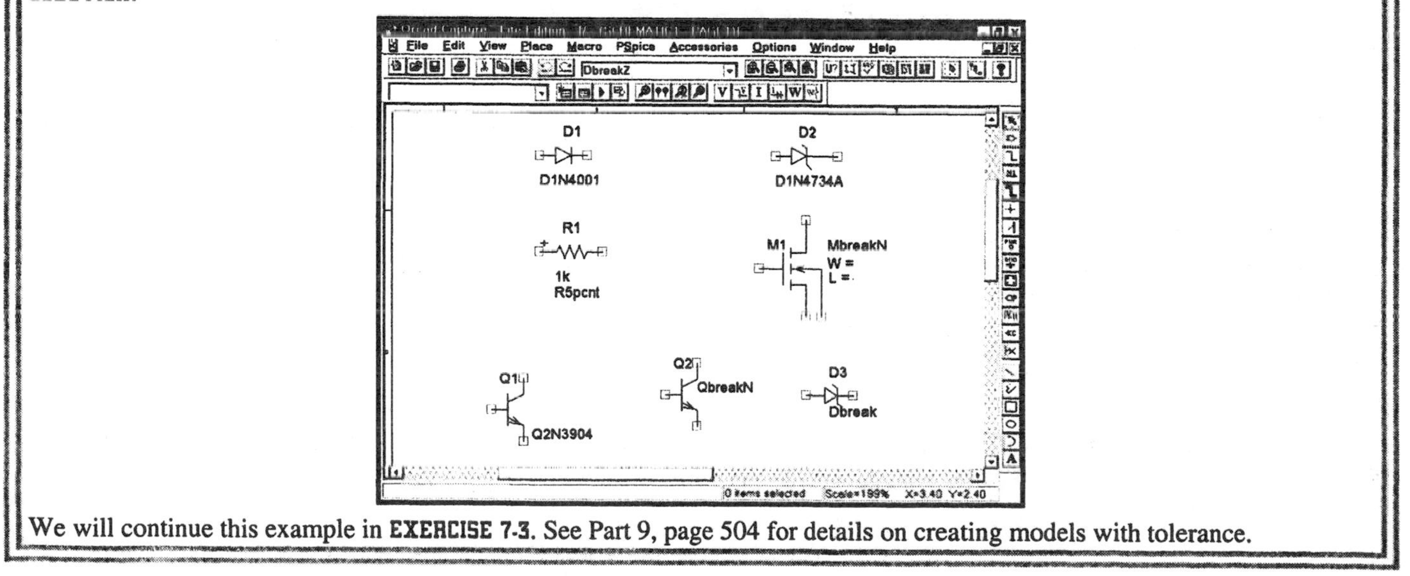

**EXERCISE 7-1**: Change the model reference of R1 to the 5 percent resistor model R5pcnt.

**SOLUTION:**



We will continue this example in **EXERCISE 7-3**. See Part 9, page 504 for details on creating models with tolerance.

# 7.B. Creating New Models Using the Breakout Parts

We will continue with the schematic from the previous section. The schematic was saved as Section 7A. Open project Section 7A.opj and open the schematic.

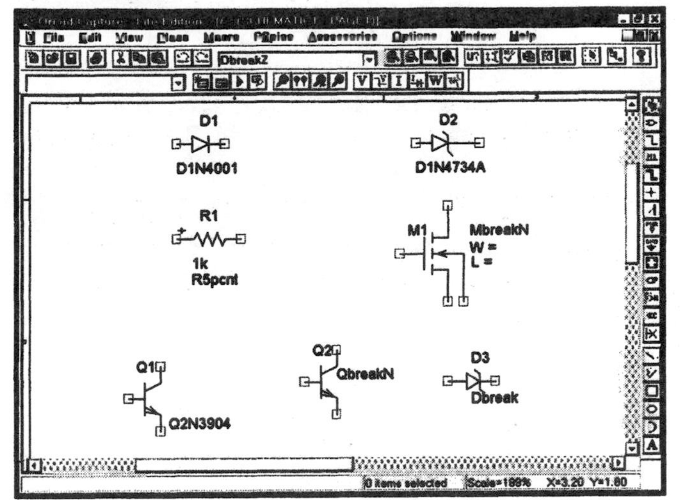


Click the ***LEFT*** mouse button on the graphic symbol for the breakout Zener diode ***D3***, . It should turn pink, indicating that it has been selected. Select **Edit** and then **PSpice Model** from the Capture menus:*

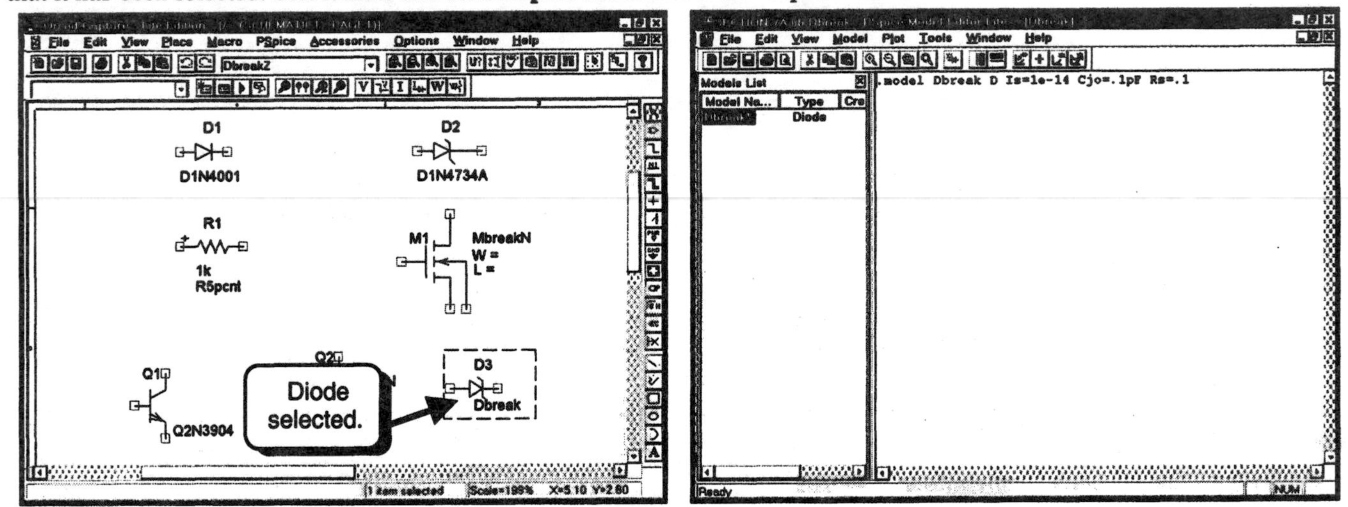


We can now modify the model. Some points are highlighted below:

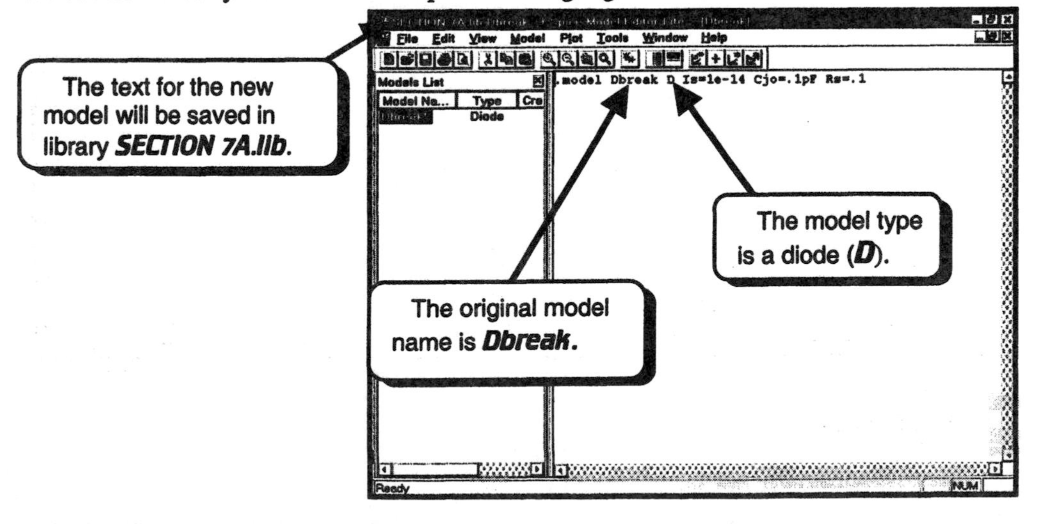


---

*If the text for **PSpice Model** appears grayed out in the menu (**PSpice Model**), you have either not selected a part or the part you have selected does not have a model. Attempt to select the part again or select a different part.

This window is a text editor and allows us to edit this model. Notice in the window title bar that the model we create will be saved in library ***SECTION 7A.lib***. This is a new library created for the current project. When we create a new model, we are also creating a new PSpice library file for the schematic. The project is named Section 7A.opj, so the library file will be named ***SECTION 7A.lib***. When you create new models for a project, a new library file is created with the name of the schematic and the extension ".lib."

Let us create a new Zener diode with a breakdown voltage of 3 volts. We will name this new Zener model Dz3V. The model parameter in PSpice that controls the breakdown voltage is called BV. Change the model as shown:

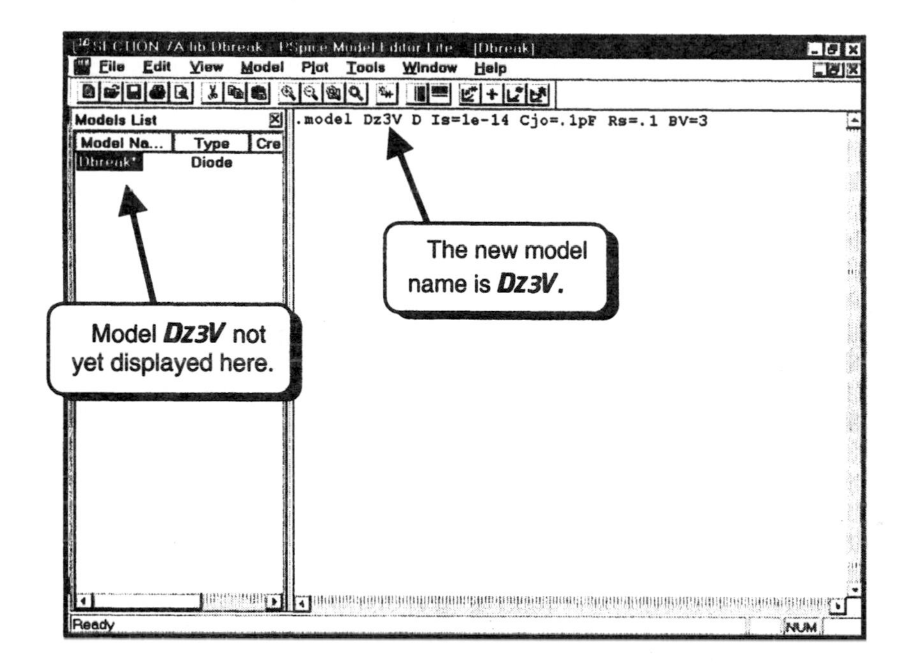


The only model parameter we are defining is BV. All other parameters will be left at the PSpice defaults. Notice that the new model we created is not yet displayed in the left window pane. Select **File** and then **Save** from the menus. The file will be saved and the model will be added to the left pane of the window:

Select **File** and then **Exit** to return to the schematic:

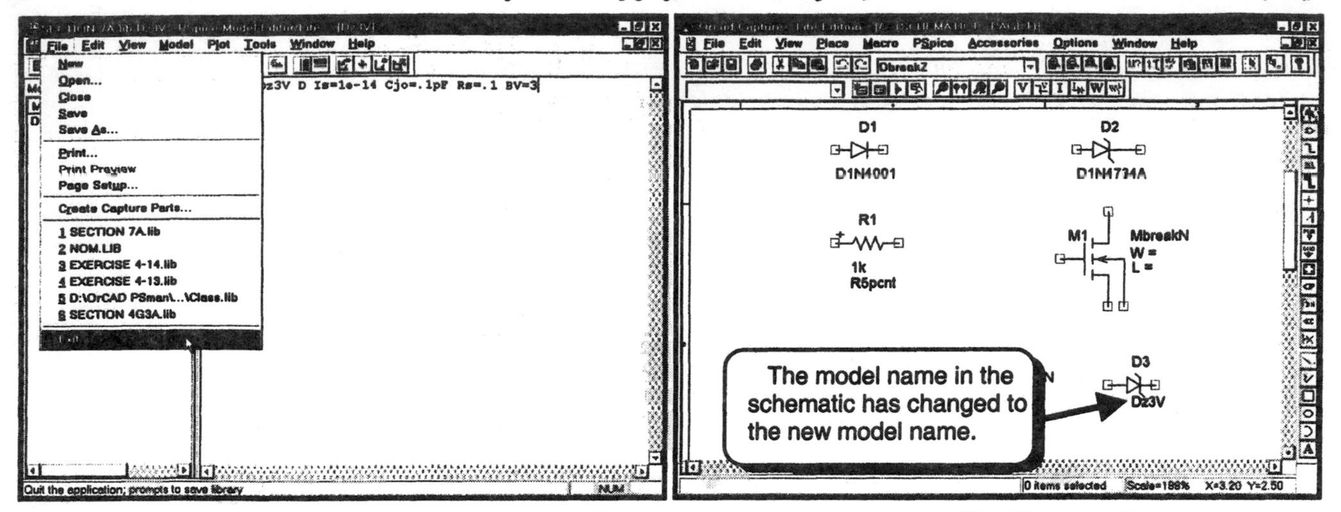


Notice in the schematic that the model reference for ***D3*** has changed from Dbreak to ***Dz3V***. Thus, ***D3*** will now use the newly defined model ***Dz3V***.

Next, suppose we wish part ***Q2*** to have an $H_{FE}$ of 33. Chances are, no model in the library has this parameter. We will define a new model for this part. Click the ***LEFT*** mouse button on the NPN graphic for ***Q2***, . It should turn red, indicating that it has been selected. Select **Edit** and then **PSpice Model** from the Capture menus:

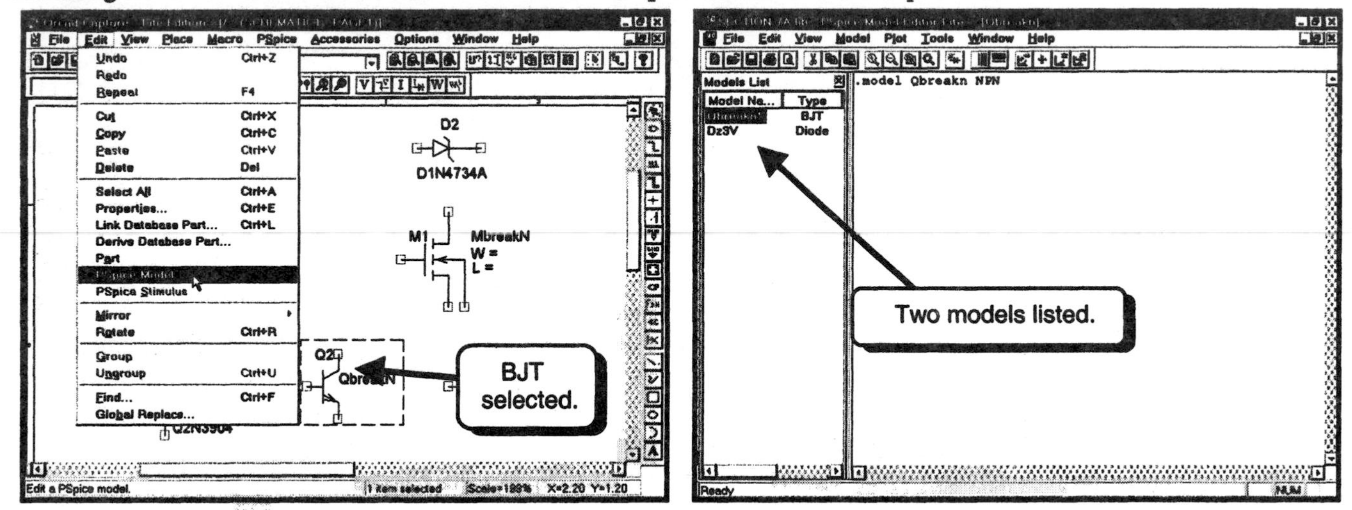


Notice now that there are two models listed in the left window pane. This window displays all of the models contained in the library we are editing, ***Section 7A.lib.*** The name of the model for the part we selected is ***Qbreakn***. This model uses all default model parameters since no model parameters are set in the model. The PSpice parameter that represents $H_{FE}$ is the parameter BF. Change the model as shown. We have renamed the model Qbf33:

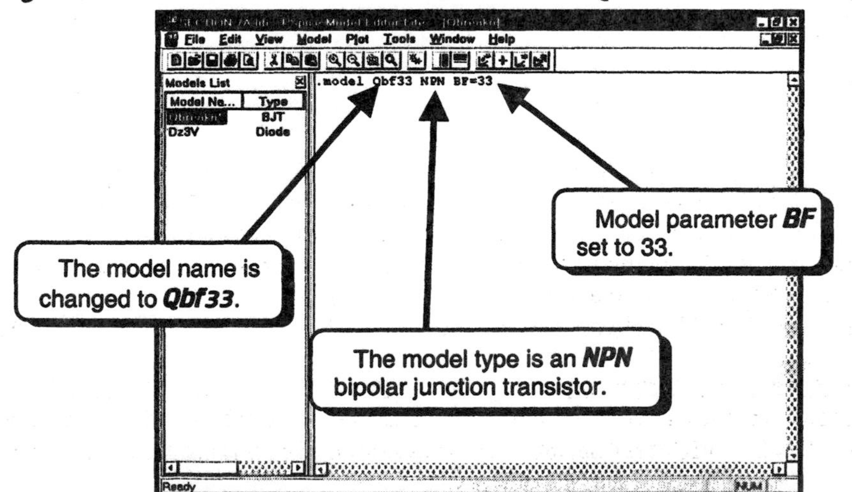


Select **File** and then **Save** to save the changes and add the new model to the list in the left pane of the window:

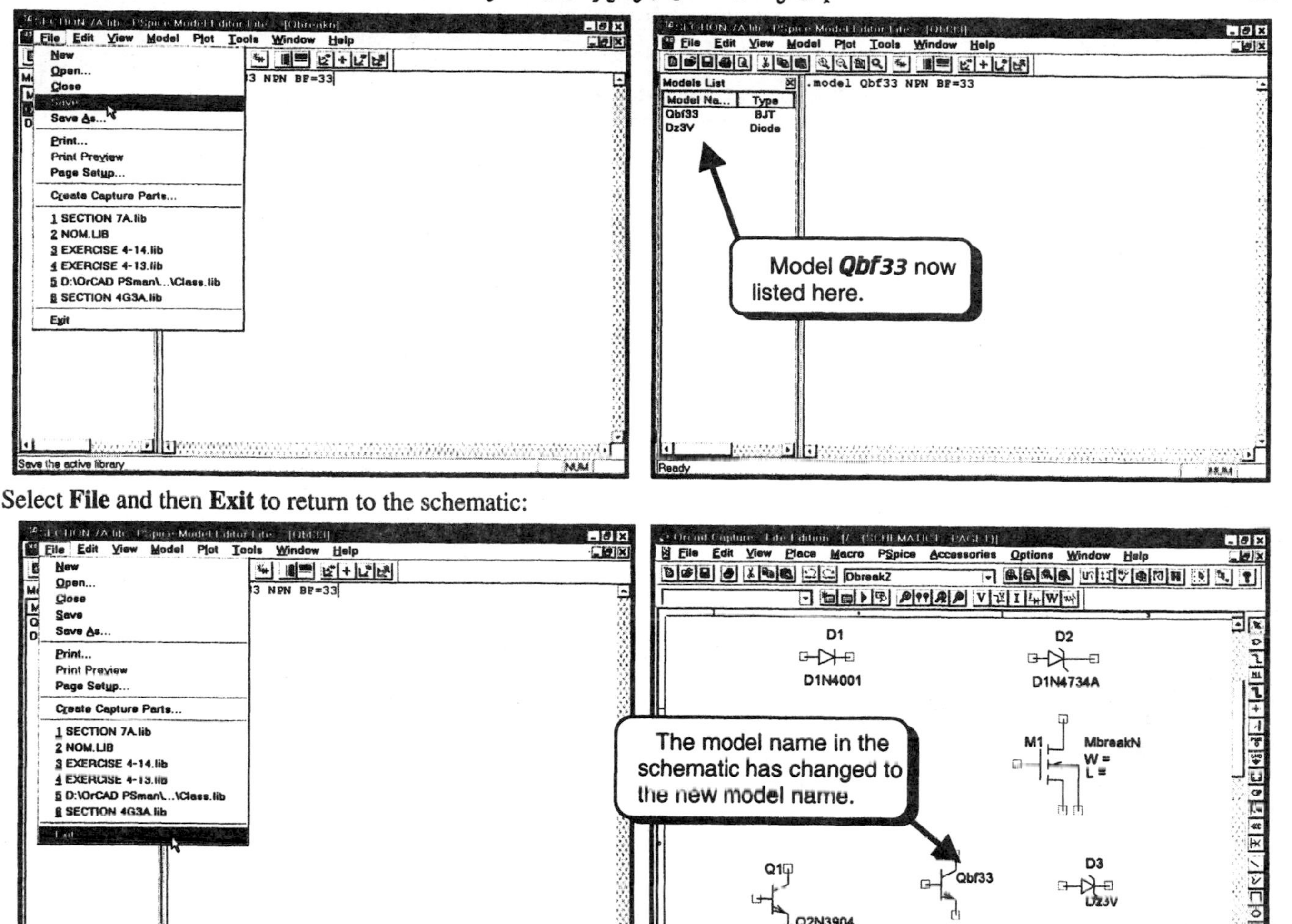


Select **File** and then **Exit** to return to the schematic:

Notice that the model reference has changed from QbreakN to **Qbf33**. If you use the Windows Notepad to edit the file Section 7A.lib, you will see the complete contents of the file and the models you have just created:

```
SECTION 7A.lib - Notepad
File  Edit  Format  Help  Send
* PSpice Model Editor - Version 9.2

*$
.model Qbf33 NPN BF=33
*$
.model Dz3V D Is=1e-14 Cjo=.1pF Rs=.1 BV=3
*$
```

The procedures we have covered can be used to create a model using any of the breakout parts. Before you create any new parts, you must know the definition of the model parameters. The user is strongly urged to refer to the Orcad PSpice A/D Reference Manual for the definitions of the model parameters. This manual is contained on the CD-ROM that accompanies this text and is located in file D:\Document\pspcref.pdf. This file can be viewed with the Adobe Acrobat Reader.

**EXERCISE 7-2:** Change the model parameters of M1 in the circuit above to define an enhancement MOSFET with the parameters $K = 20\ \mu A/V^2$ and $V_T = 3$ V. MOSFET operation in the saturation region is governed by the equation:

$$I_D = K(V_{GS} - V_T)^2$$

In PSpice, the equation that defines the MOSFET operation in the saturation region is:

$$I_D = \frac{K_P}{2}(V_{GS} - V_{TO})^2$$

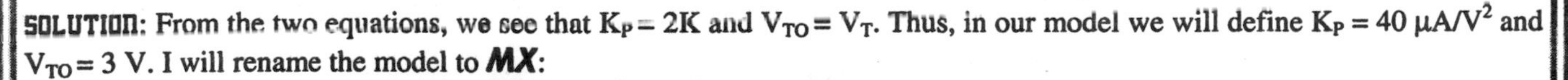

**SOLUTION**: From the two equations, we see that $K_P = 2K$ and $V_{TO} = V_T$. Thus, in our model we will define $K_P = 40\ \mu A/V^2$ and $V_{TO} = 3$ V. I will rename the model to ***MX***:

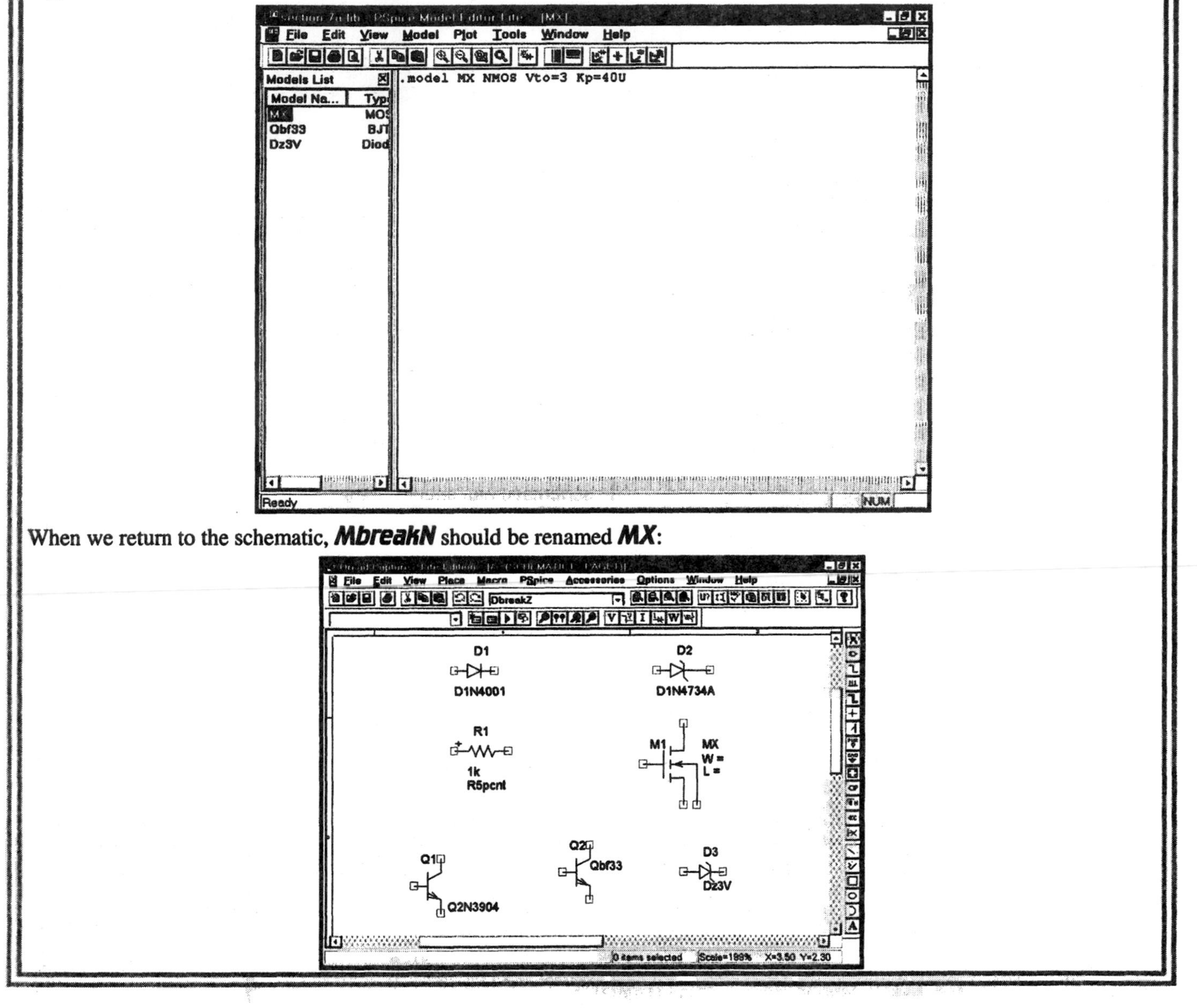


When we return to the schematic, ***MbreakN*** should be renamed ***MX***:

# 7.C. Modifying Existing Models

When you edit an existing model, Capture makes a copy of the original model and saves it in the library for the project, leaving the original model unchanged. This feature allows you to edit entire existing models and modify the parameters you wish.

We would like to use the Q2N3904 model, but we would like to set $H_{FE}$ to its minimum value of 50. We will continue with the circuit named Section 7A.opj created in the previous section. Click the ***LEFT*** mouse button on the ***Q2*** graphic, . It should turn red, indicating that it has been selected. Select **Edit** and then **PSpice Model** from the Capture menus.*

---

*If the text **PSpice Model** appears grayed out in the menu (**PSpice Model**), you have either not selected a part or the part you have selected does not have a model. Attempt to select the part again or select a different part.

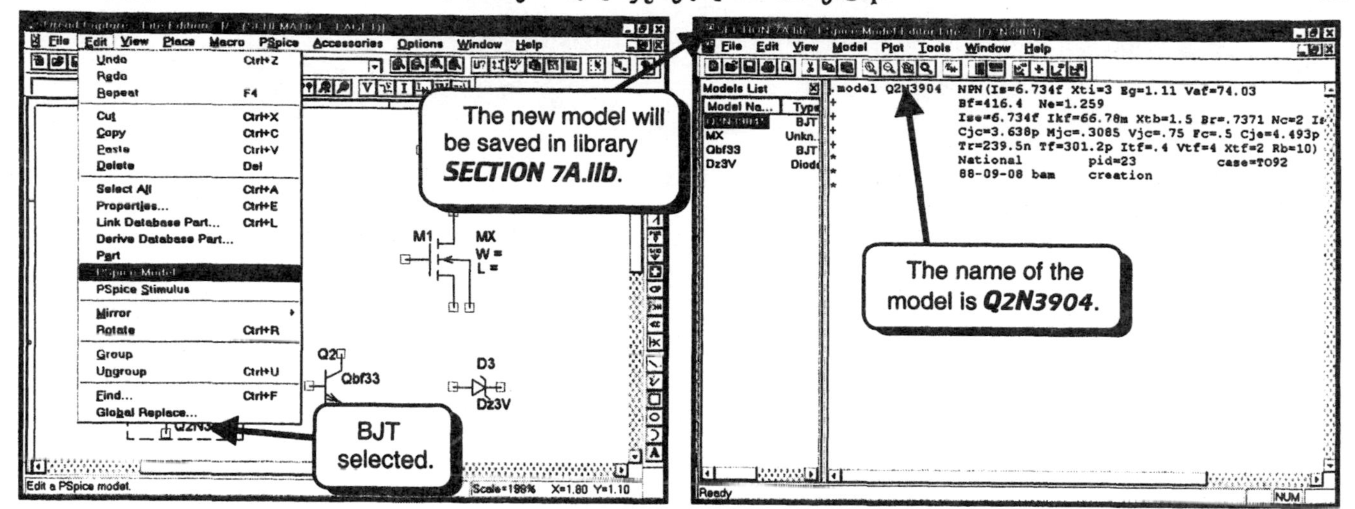


This screen gives all of the model parameters for the 2n3904 model. To distinguish the new model we are about to create from the original model, we will rename the model to ***Q2N3904-X***. This new model will be saved in file ***SECTION 7A.lib***.

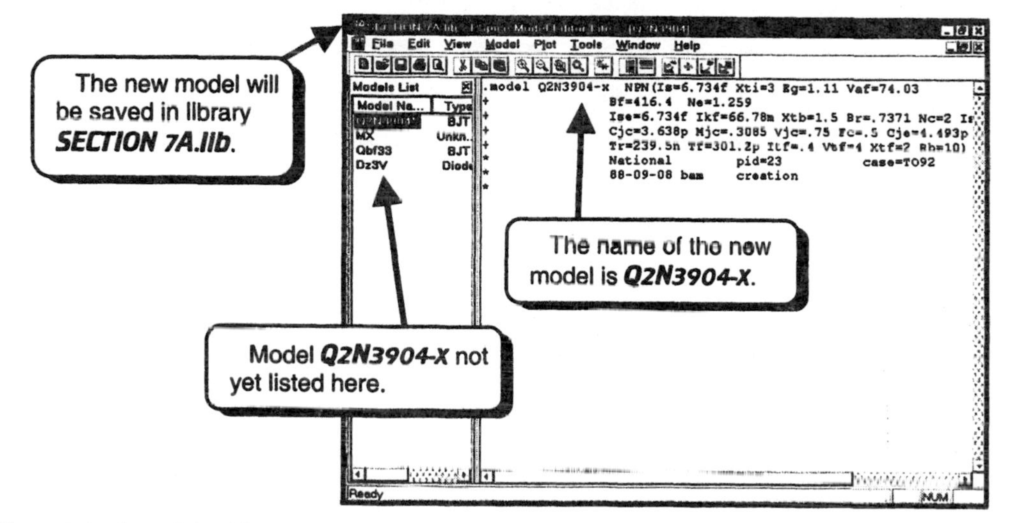


The original model will remain unchanged. Change the model parameters that you want and then select **File** and then **Save** from the menus:

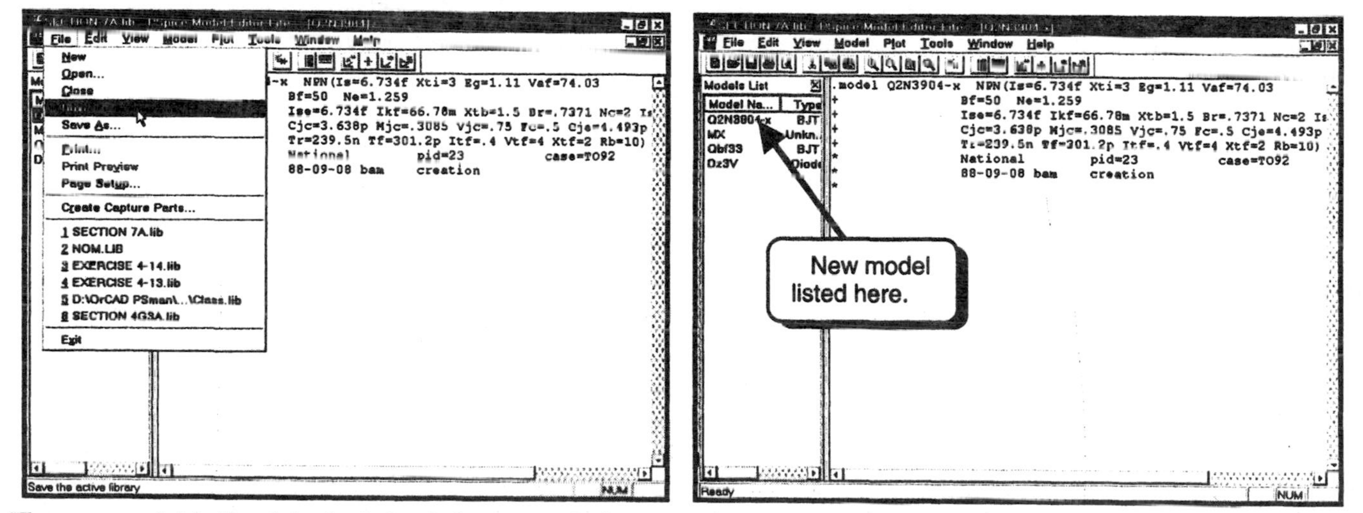


The new model is listed in the left window pane, indicating that it has been added to the library for the project, Section 7A.lib. Select **File** and then **Exit** to close the Model Editor window and return to the schematic:

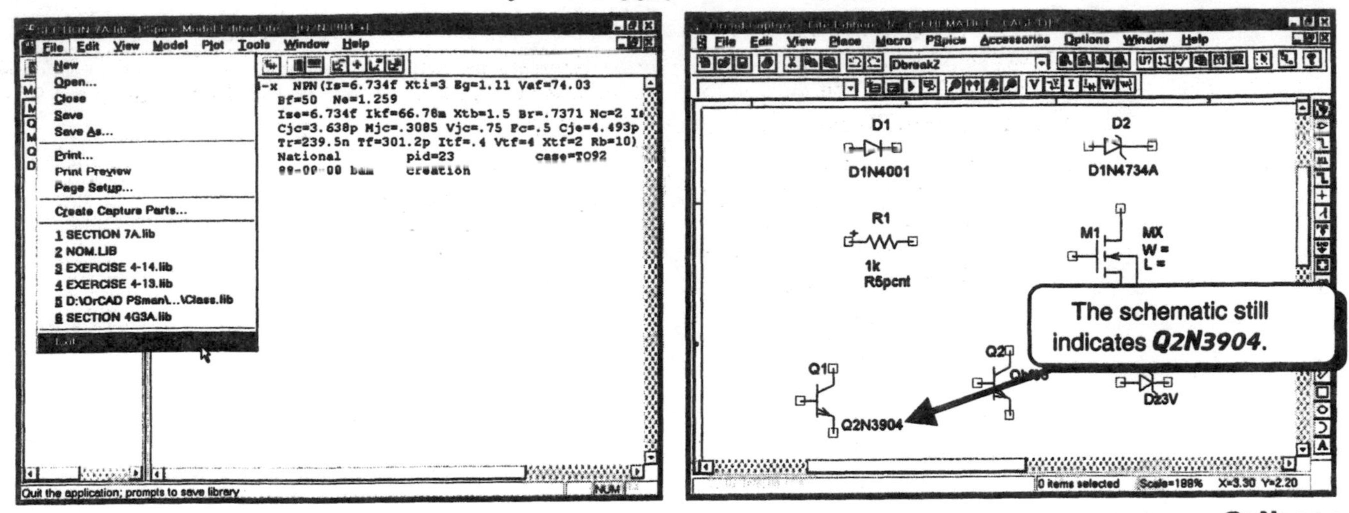


We see that the text ***Q2N3904*** has not changed to Q2N3904-x. This is because, with predefined models like the ***Q2N3904***, the text ***Q2N3904*** is the ***Value*** property and not the model name. If you double-click the *LEFT* mouse button on the ***Q2*** graphic, , you will see the Properties for ***Q2***:

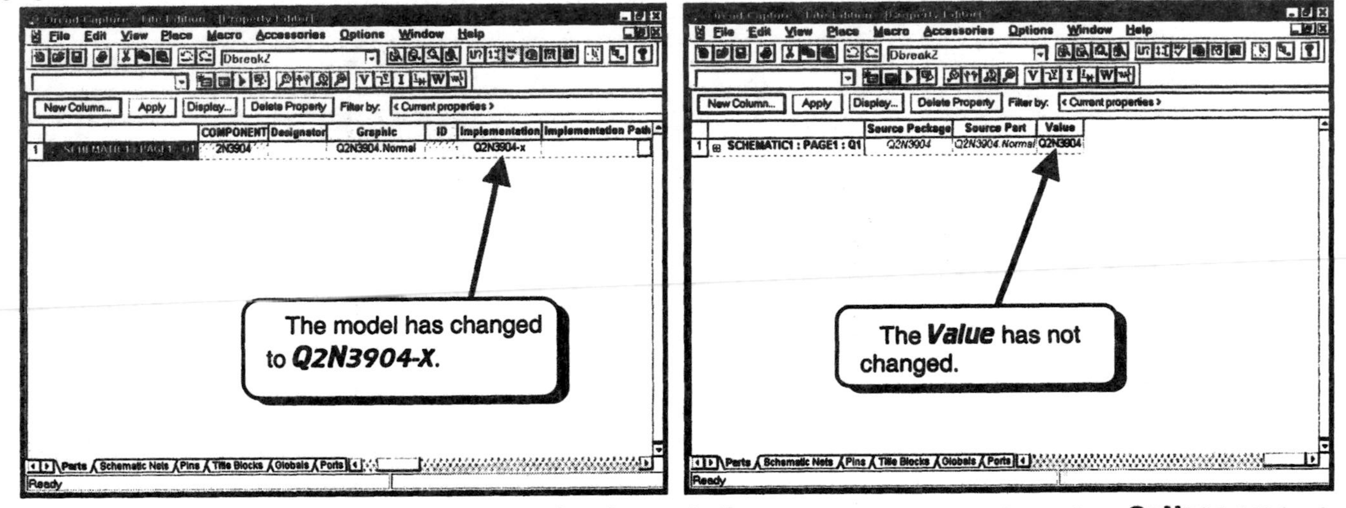


We see that the PSpice model indicated by the ***Implementation*** property has been changed to ***Q2N3904-X*** in the attribute list. When PSpice runs it will use model ***Q2N3904-X***. Note that the ***Value*** has not changed, and it is this property that is displayed in the schematic. This is different from the breakout parts. With breakout parts, the new model name is displayed because the breakout part assumes that you wish to create a new part and to identify that part on the schematic. When you modify an existing model, the original model name is displayed because Capture assumes that you are making only small modifications to the original model. The modified model is still basically the original model.

If you use the Windows Notepad program to display the contents of the file Section 7A.lib, you will see the new model ***Q2N3904-X***:

```
* PSpice Model Editor - Version 9.2

*$
.model Q2N3904-x  NPN(Is=6.734f Xti=3 Eg=1.11 Vaf=74.03
+               Bf=50  Ne=1.259
+               Ise=6.734f Ikf=66.78m Xtb=1.5 Br=.7371 Nc=2 Isc=0 Ikr=0 Rc=1
+               Cjc=3.638p Mjc=.3085 Vjc=.75 Fc=.5 Cje=4.493p Mje=.2593 Vje=.7
+               Tr=239.5n Tf=301.2p Itf=.4 Vtf=4 Xtf=2 Rb=10)
*               National        pid=23          case=TO92
*               88-09-08 bam    creation
*
*$
.model MX reakn NMOS Vto=3 Kp=40U
*$
.model Qbf33 NPN BF=33
*$
.model Dz3V D Is=1e-14 Cjo=.1pF Rs=.1 BV=3
*$
```

We see that the original model for the Q2n3904 has been copied, with the exception of the changes we made to the model.

**EXERCISE 7-3**: Continuing with **EXERCISE 7-1**, edit the 5 percent model of ***R1*** and change the model to a 10 percent model. The model reference of R1 was changed from Rbreak to ***R5pcnt*** in **EXERCISE 7-1**.

**SOLUTION**: Select the graphic and then select **Edit** and **PSpice Model** to edit the resistor model ***R5pcnt***:

```
.MODEL R5pcnt RES(R=1 DEV/UNIFORM 5%)
* The models below are 1% and 5% resistors with Gaussian
* probability distributions
```

Change the name of the model to `R10pcnt` and change the ***5%*** tolerance to `10%`:

```
.MODEL R10pcnt RES(R=1 DEV/UNIFORM 10%)
* The models below are 1% and 5% resistors with Gaussian
* probability distributions
```

Save the changes and return to the schematic. The new model name should change on the schematic since the part was originally a breakout part:

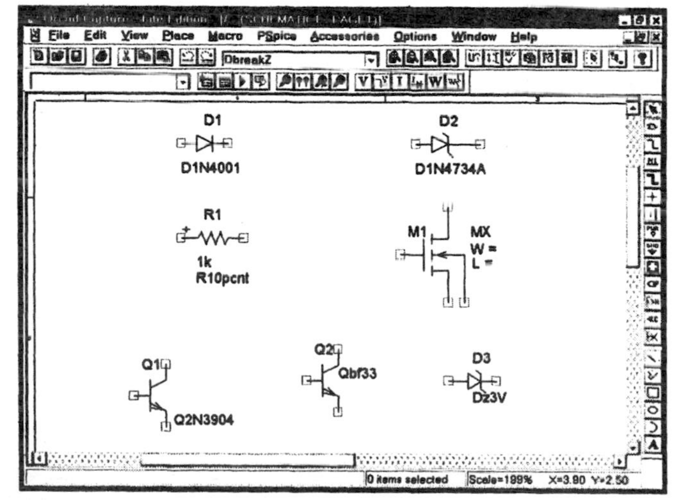


If you look in the file Section 7A.lib, you will see the newly defined model:

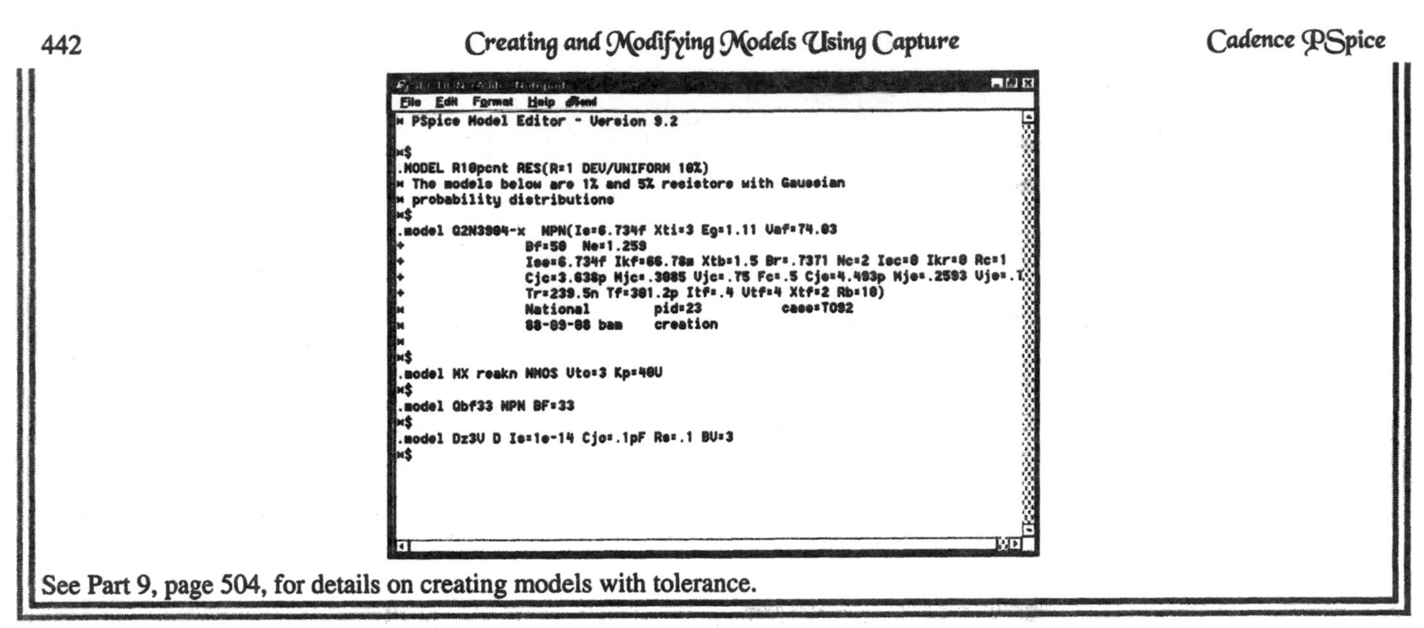

```
File  Edit  Format  Help  Send
* PSpice Model Editor - Version 9.2

*$
.MODEL R10pcnt RES(R=1 DEV/UNIFORM 10%)
* The models below are 1% and 5% resistors with Gaussian
* probability distributions
*$
.model Q2N3904-x  NPN(Is=6.734f Xti=3 Eg=1.11 Vaf=74.03
+               Bf=50  Ne=1.259
+               Ise=6.734f Ikf=66.78m Xtb=1.5 Br=.7371 Nc=2 Isc=0 Ikr=0 Rc=1
+               Cjc=3.638p Mjc=.3085 Vjc=.75 Fc=.5 Cje=4.493p Mje=.2593 Vje=.7
+               Tr=239.5n Tf=301.2p Itf=.4 Vtf=4 Xtf=2 Rb=10)
*               National       pid=23         case=TO92
*               88-09-08 bam   creation
*
*$
.model MX reakn NMOS Vto=3 Kp=40U
*$
.model Qbf33 NPN BF=33
*$
.model Dz3V D Is=1e-14 Cjo=.1pF Rs=.1 BV=3
*$
```

See Part 9, page 504, for details on creating models with tolerance.

# 7.D. Changing the Library Path

Suppose you have created a number of new models and you would like to use them in another circuit. This can be done by adding the library name to the library path. The current project we are using is SECTION 7A.opj. The new models created in this circuit are stored in the file called SECTION 7A.lib. Use the Windows Notepad program to edit the file ***SECTION 7A.lib***.

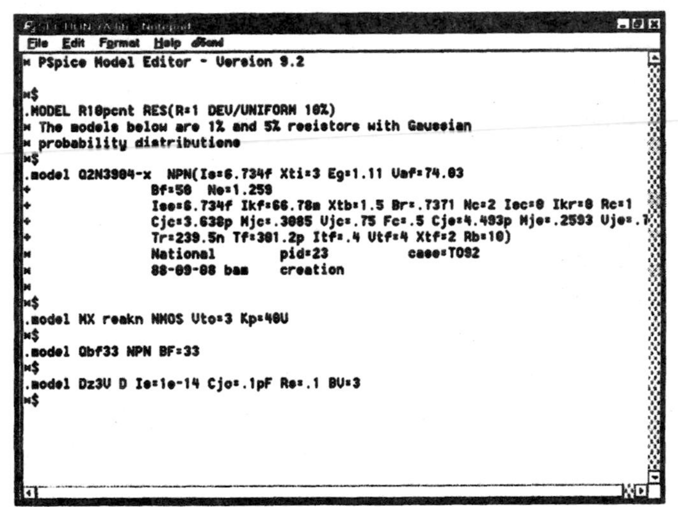


```
File  Edit  Format  Help  Send
* PSpice Model Editor - Version 9.2

*$
.MODEL R10pcnt RES(R=1 DEV/UNIFORM 10%)
* The models below are 1% and 5% resistors with Gaussian
* probability distributions
*$
.model Q2N3904-x  NPN(Is=6.734f Xti=3 Eg=1.11 Vaf=74.03
+               Bf=50  Ne=1.259
+               Ise=6.734f Ikf=66.78m Xtb=1.5 Br=.7371 Nc=2 Isc=0 Ikr=0 Rc=1
+               Cjc=3.638p Mjc=.3085 Vjc=.75 Fc=.5 Cje=4.493p Mje=.2593 Vje=.7
+               Tr=239.5n Tf=301.2p Itf=.4 Vtf=4 Xtf=2 Rb=10)
*               National       pid=23         case=TO92
*               88-09-08 bam   creation
*
*$
.model MX reakn NMOS Vto=3 Kp=40U
*$
.model Qbf33 NPN BF=33
*$
.model Dz3V D Is=1e-14 Cjo=.1pF Rs=.1 BV=3
*$
```


This file contains all of the models we created for the current schematic. Close the Notepad and return to the schematic. We will now look at the current library path. To view the libraries, we must set up a simulation profile. Select **PSpice** and then **New Simulation Profile** from the Capture menus:

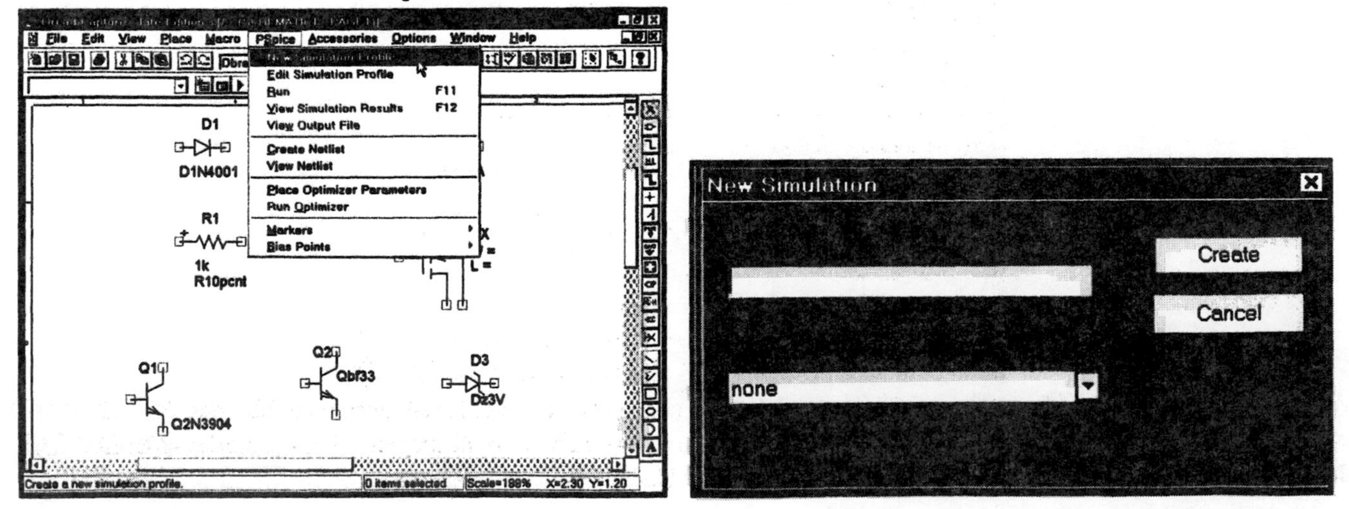

Enter a name for the profile and then select the ***Create*** button:

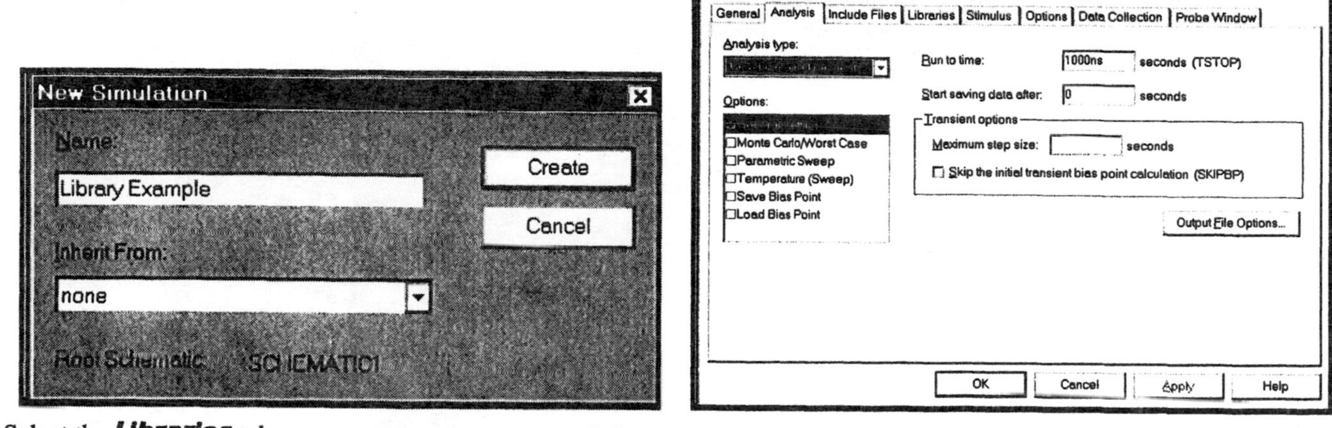


Select the ***Libraries*** tab:

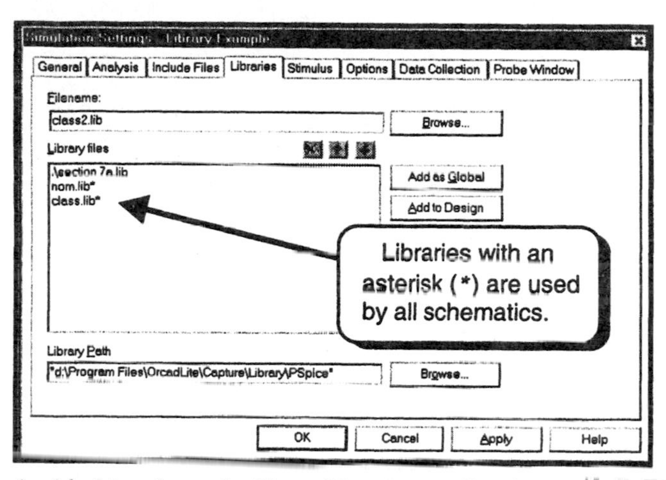


This screen lists all libraries used with this schematic. Three libraries are listed, ***section 7a.lib***, ***nom.lib***, and ***class.lib***. Libraries that have an asterisk (*) are global libraries that are available to all projects. Libraries that do not have an asterisk are available only to the present project. Thus, library ***section 7a.lib*** is available only to the present project and ***nom.lib*** and ***class.lib*** are available to all schematics.

Library ***nom.lib*** is actually a short file that lists all of the available PSpice libraries. If you installed the software in the default directory, the library files are located in directory C:\Program Files\OrcadLite\Capture\library\pspice. Open the ***nom.lib*** file with the Notepad:

```
* Sample standard device library
*
*  Copyright OrCAD, Inc. 1998 All Rights Reserved.
* This is a reduced version of OrCAD's standard parts libraries. Some
* components from several types of component libraries have been included
* here.  You are welcome to make as many copies of it as you find convenient.
*
* $Revision:   1.5  $
* $Author:    RPEREZ  $
* $Date:    20 Apr 1998 09:36:36  $
*
*
* ---------------------------------------------------------------------------
*
* The OrCAD library included with the production version of PSpice
* includes over 10,000 analog devices, and over 1,800 digital devices.
*
*
* It takes time for PSpice to scan a library file.  To speed this up, PSpice
* creates an index file, called <filename>.IND. The index file is re-created
* whenever PSpice senses that the library file has changed.
*
.lib "breakout.lib"              : generic devices for OrCAD Capture

* "regular" device libraries

.lib "eval.lib"                  : reduced version of OrCAD's standard
```

We see that the file ***nom.lib*** contains no PSpice models. Instead, all this file does is reference other libraries. Since ***nom.lib*** is available to all projects, the libraries listed in ***nom.lib*** are also available to all projects. Thus, libraries ***breakout.lib***, ***eval.lib***, and ***class.lib*** are available to all projects. Close the Notepad when you are done.

The next thing we will look at is how to make the models in library Section 7A.lib available to other schematics. We have two choices: (1) we can make the library available to a specific project, or (2) we can make the library available to all projects.

Close the Simulation Settings dialog box, close the current project, create a new project called Section 7D, and then display the schematic for the project:

Select **PSpice** and then **New Simulation Profile** from the Capture menus, enter a name for the profile, and then click the ***Create*** button. Select the ***Libraries*** tab:

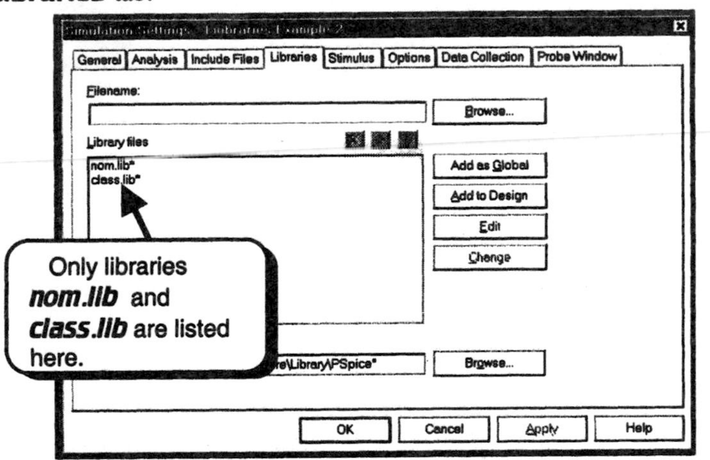


Only libraries ***nom.lib*** and ***class.lib*** are listed. These are the libraries available to all projects. If we want the library file Section 7A.lib to be available, we will need to add it to this list. Click the ***Browse*** button to locate and select the file Section 7A.lib:

Select the ***Open*** button:

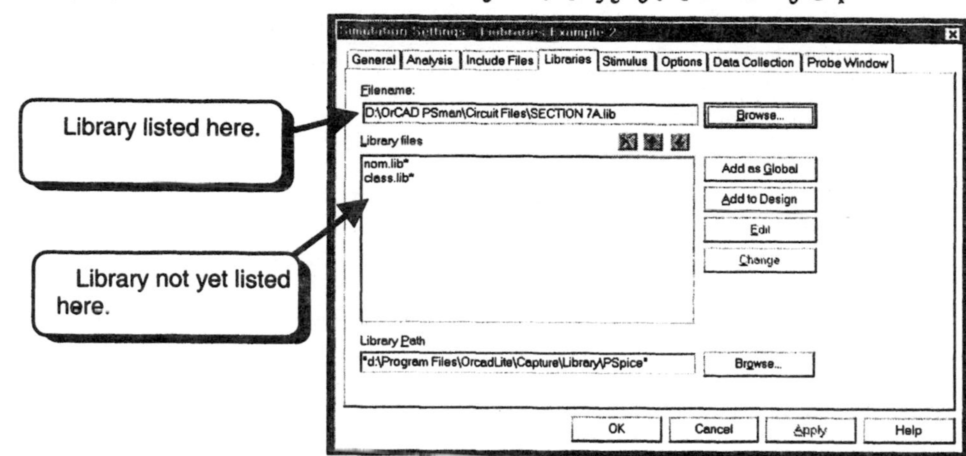


The library is not yet configured. If you want the library to be available to all schematics, click the ***Add as Global*** button. With this selection, the library will be added and displayed with an * indicating that it is a global library available to all projects. If you want the library to be available only to the current project, select the ***Add to Design*** button. The library will be added to the list without an asterisk, indicating that it is only available to the present project. Select either of the ***Add*** buttons:

Click the ***OK*** button. The models in the library can now be used by this project, and since I added the library as global, the models can be used by all projects.

# 7.E. Model Parameters for Commonly Used Parts

This section gives brief descriptions of model parameters for most of the parts mentioned in this manual. Only a few of the parameters for each model are given to allow the user to create simple models. To create more detailed models, the reader is referred to the *PSpice Reference Manual*, which is contained on the CD-ROM that accompanies this manual. The semiconductor models presented here include only the simple DC parameters because the models get complicated very quickly when dynamics are included. Temperature dependence parameters for some models are discussed.

**CAPACITOR**

The capacitor model is used to specify capacitor voltage dependence and tolerance. The main use of the capacitor model in this manual is to specify capacitor tolerance. See Part 9, page 505, for an example of using this model to specify tolerance.

The value of the capacitance used in the simulation is:

| Model Parameter | Description | Units | Default Value If Not Specified |
|---|---|---|---|
| **C** | Capacitance multiplier | none | 1 |
| **VC1** | Linear voltage coefficient | $volt^{-1}$ | 0 |
| **VC2** | Quadratic voltage coefficient | $volt^{-2}$ | 0 |
| **TC1** | Linear temperature coefficient | $°C^{-1}$ | 0 |
| **TC2** | Quadratic temperature coefficient | $°C^{-2}$ | 0 |

$$\text{Capacitance Value} = (\text{SCH\_VAL}) \cdot \mathbf{C} \cdot \left[1 + (V_C \cdot \mathbf{VC1}) + (V_C^2 \cdot \mathbf{VC2})\right]\left[1 + \mathbf{TC1}(T - Tnom) + \mathbf{TC2}(T - Tnom)^2\right]$$

The variables in this equation are:

- Capacitance Value – the value of capacitance used in the simulation
- SCH_VAL – the value of capacitance specified in the schematic
- $V_C$ – the voltage across the capacitor, determined during the simulation
- T – temperature at which the simulation is run
- Tnom – the nominal temperature, 27°C

Note that if a capacitor model is not specified, all model parameters are set at their default values and the value in the simulation is the value specified in the schematic.

<u>EXAMPLE CAPACITOR MODELS:</u>

```
.Model Cideal CAP ( C=1 )
.Model Clin CAP ( VC1=5 )
.Model Cdouble CAP ( C=2 )
```

- .Model Cideal CAP ( C=1 ) – equivalent to not using a model
- .Model Clin CAP ( VC1=5 ) – capacitor with voltage dependence
- .Model Cdouble CAP ( C=2 ) – The capacitance used in the simulation is twice the value specified in the schematic.

**DIODE**

The diode model is used to create rectifier, signal, and Zener diodes.

| Model Parameter | Description | Units | Default Value If Not Specified |
|---|---|---|---|
| **IS** | Saturation current | amp | $1 \times 10^{-15}$ |
| **N** | Emission coefficient | none | 1 |
| **BV** | Diode reverse breakdown voltage. Used to set the diode breakdown voltage in rectifier diodes, and the Zener breakdown voltage in Zener diodes. | volt | ∞ |
| **RS** | Diode parasitic resistance. Can be viewed as the series resistance of the diode. | ohm | 0 |

The diode equation using these parameters is:

$$\text{Diode Current} = \mathbf{IS}\left[\exp\left(V_D / \mathbf{N} \cdot V_T\right) - 1\right]$$

EXAMPLE DIODE MODELS:

.Model Dx D ( IS=1.5e–15 RS=3 BV=100 )

.Model Dideal D ( IS=1e–15 )

.Model Dzener D (BV=4.7)

**JUNCTION FETs**

Two different models are given to specify n- and p-type jFETs. Both models use the same parameters.

| Model Parameter | Description | Units | Default Value If Not Specified |
|---|---|---|---|
| VTO | Threshold voltage | volt | 0 |
| BETA | Transconductance coefficient | amp/volt$^2$ | $1\times10^{-4}$ |
| LAMBDA | Channel-length modulation | volt$^{-1}$ | 0 |

The equation for the drain current using these parameters is:

$$\text{DrainCurrent} = \begin{cases} \textbf{BETA}\cdot(1+\textbf{LAMBDA}\cdot V_{DS})\cdot\left[2\cdot(V_{GS}-\textbf{VTO})V_{DS}-V_{DS}^2\right] & -\textit{Linear Region} \\ \textbf{BETA}\cdot(1+\textbf{LAMBDA}\cdot V_{DS})\cdot(V_{GS}-\textbf{VTO})^2 & -\textit{Saturation Region} \end{cases}$$

Some readers may be familiar with a jFET described by the equation:

$$\text{DrainCurrent} = I_{DSS}\cdot\left(1-\frac{V_{GS}}{V_P}\right)^2 \qquad -\textit{Saturation Region}$$

This is an equivalent description with LAMBDA equal to zero, VTO =$V_P$, and BETA = $I_{DSS}/V_P^2$.

EXAMPLE jFET MODELS:

.Model jxx NJF ( BETA=250u VTO=–3 ) – n-type jFET model

.Model Jy PFJ ( BETA=100u VTO=2 ) – p-type jFET model

**INDUCTOR**

The inductor model is used to specify inductor voltage dependence. The value of the inductance used in the simulation is:

| Model Parameter | Description | Units | Default Value If Not Specified |
|---|---|---|---|
| L | Inductance multiplier | none | 1 |
| IL1 | Linear current coefficient | amp$^{-1}$ | 0 |
| IL2 | Quadratic current coefficient | amp$^{-2}$ | 0 |
| TC1 | Linear temperature coefficient | °C$^{-1}$ | 0 |
| TC2 | Quadratic temperature coefficient | °C$^{-2}$ | 0 |

$$\text{InductanceValue} = (\text{SCH\_VAL})\cdot\textbf{L}\cdot\left[1+(I_L\cdot\textbf{IL1})+(I_L^2\cdot\textbf{IL2})\right]\left[1+\textbf{TC1}(T-Tnom)+\textbf{TC2}(T-Tnom)^2\right]$$

The variables in this equation are:

- Inductance Value – the value of inductance used in the simulation
- SCH_VAL – the value of inductance specified in the schematic
- $I_L$ – the current through the inductor, determined during the simulation
- T – temperature at which the simulation is run

- Tnom – the nominal temperature, 27°C

Note that if an inductor model is not specified, all model parameters are set at their default values and the value in the simulation is the value specified in the schematic.

EXAMPLE INDUCTOR MODELS:

.Model Lideal IND ( L=1 ) – equivalent to not using a model
.Model L_lin IND ( IL1=5 ) – inductor with current dependence
.Model Ldouble IND ( L=2 ) – The inductance used in the simulation is twice the value specified in the schematic.

**MOSFET**

Two different models are given to specify n- and p-type MOSFETs. Both models use the same parameters. These models are used to specify both enhancement- and depletion-type MOSFETs.

| Model Parameter | Description | Units | Default Value If Not Specified |
|---|---|---|---|
| VTO | Threshold voltage | volt | 0 |
| KP | Transconductance coefficient | amp/volt$^2$ | $2 \times 10^{-5}$ |
| L | Channel length | meter | $100 \times 10^{-6}$ |
| W | Channel width | meter | $100 \times 10^{-6}$ |
| LAMBDA | Channel-length modulation | volt$^{-1}$ | 0 |

The equation for the drain current using these parameters is:

$$\text{DrainCurrent} = \begin{cases} \left(\frac{\mathbf{W}}{\mathbf{L}}\right)\frac{\mathbf{KP}}{2}\cdot(1+\mathbf{LAMBDA}\cdot V_{DS})\cdot\left[2\cdot(V_{GS}-\mathbf{VTO})V_{DS}-V_{DS}^2\right] & -\textit{Linear Region} \\ \left(\frac{\mathbf{W}}{\mathbf{L}}\right)\frac{\mathbf{KP}}{2}\cdot(1+\mathbf{LAMBDA}\cdot V_{DS})\cdot(V_{GS}-\mathbf{VTO})^2 & -\textit{Saturation Region} \end{cases}$$

EXAMPLE MOSFET MODELS:

.Model M1 NMOS ( KP=25u LAMBDA=0.01 VTO=3 ) – enhancement-type NMOS
.Model M2 NMOS ( KP=20u LAMBDA=0.1 VTO=–2 W=10u L=5u ) – depletion-type NMOS
.Model M3 PMOS ( KP=8u LAMBDA=0.01 VTO=–2 ) – enhancement-type PMOS
.Model M4 PMOS ( KP=6u LAMBDA=0.05 VTO=1.5 W=10u L=2u ) – depletion-type PMOS

**BIPOLAR JUNCTION TRANSISTORS**

Two different models are given to specify n- and p-type BJTs. Both models use the same parameters.

| Model Parameter | Description | Units | Default Value If Not Specified |
|---|---|---|---|
| BF | Ideal maximum forward current gain | none | 100 |
| BR | Ideal maximum reverse current gain | none | 1 |
| RE | Emitter ohmic resistance | ohm | 0 |
| RC | Collector ohmic resistance | ohm | 0 |
| RB | Base ohmic resistance | ohm | 0 |
| VAF | Forward Early voltage | volt | ∞ |

EXAMPLE BJT MODELS:

.Model Qnpn1 NPN ( BF=50 ) – NPN model
.Model Qpnp PNP ( BF=25) – PNP model

**RESISTOR**

The resistor model is used to specify resistor tolerance and temperature dependence. We will not discuss temperature dependence here. The main use of the resistor model in this manual is to specify resistor tolerance. See Part 9, page 504, for examples of using this model to specify tolerance.

| Model Parameter | Description | Units | Default Value If Not Specified |
|---|---|---|---|
| R | Resistance multiplier | none | 1 |
| TC1 | Linear temperature coefficient | $°C^{-1}$ | 0 |
| TC2 | Quadratic temperature coefficient | $°C^{-2}$ | 0 |
| TCE | Exponential temperature coefficient | %/°C | 0 |

The value of the resistor used in the simulation is:

$$\text{Resistance Value} = \begin{cases} (\text{SCH\_VAL})\cdot \mathbf{R}\left[1+\mathbf{TC1}(T-Tnom)+\mathbf{TC2}(T-Tnom)^2\right] & \textit{if } \mathbf{TCE} \textit{ is not defined} \\ (\text{SCH\_VAL})\cdot \mathbf{R}\left[1.01^{\mathbf{TCE}(T-Tnom)}\right] & \textit{if } \mathbf{TCE} \textit{ is defined} \end{cases}$$

The variables in this equation are:

- Resistance Value – the value of resistance used in the simulation
- SCH_VAL – the value of resistance specified in the schematic
- T – temperature at which the simulation is run
- Tnom – the nominal temperature, 27°C

Note that if a resistor model is not specified, all model parameters are set at their default values and the value in the simulation is the value specified in the schematic.

<u>EXAMPLE RESISTOR MODELS:</u>

.Model Rideal RES ( R=1 ) – equivalent to not using a model

.Model Rdouble RES ( R=2 ) – The resistance used in the simulation is twice the value specified in the schematic.

# 7.F. Downloading Models Using the World Wide Web

Many semiconductor manufacturers have data sheets and PSpice models available online for easy access by engineers. In this section we will show how to obtain those models so that we can use them in a simulation. The types of models we will show can be split into two types. The first type of models are primitives that use only a .model statement. Examples of these are diodes, bipolar junction transistors (BJT), and MOSFETs. The second type of models we will download are subcircuit models such as op-amps, IGBT's, Darlington transistors, and MOSFET subcircuit models.

The libraries that come with the Lite version of Capture contain a few parts for creating your own models out of subcircuits. However, several models have been added for this text so that you can download the most common types of subcircuit models. A list of the models is given in **Table 7-2:**

**Table 7-2: Subcircuit Breakout Models**

| Model | Description |
|---|---|
| Darlington_NPN_Subckt | Subcircuit model for a Darlington NPN transistor |
| Darlington_PNP_Subckt | Subcircuit model for a Darlington NPN transistor |
| IGBT_Subckt | IGBT subcircuit model without an anti-parallel diode |
| IGBT_Subckt_D | IGBT subcircuit model with an anti-parallel diode |
| Mosfet_N_Subckt_D | N-type MOSFET subcircuit model with an anti-parallel diode |
| Mosfet_P_Subckt_D | P-type MOSFET subcircuit model with an anti-parallel diode |
| Op-Amp_breakout | Operational Amplifier. The default model is equivalent to a 741 op-amp. |

Note that if you wish to use a MOSFET model that does not contain an anti-parallel diode, you can use the standard MOSFET breakout parts such as MbreakN and MbreakP.

For all models in this section, we will use the circuit below:

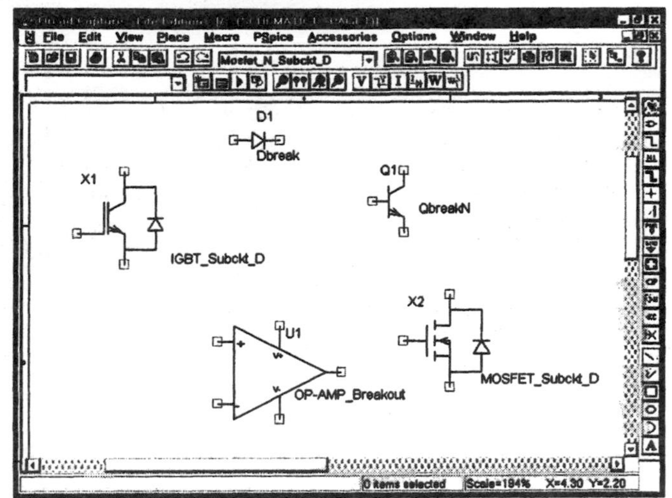


The parts used in this circuit are Dbreak, QbreakN, IGBT_Subckt_D, MOSFET_N_Subckt_D, and OP-AMP_Breakout.

In these examples, we will show web sites for two manufacturers. These sites are always under construction so you should expect that the web pages shown here will be different from what you see when you explore these sites. Our goal here is not to show how to use these specific sites, but how to obtain models in general. We will use the Microsoft Internet Explorer web browser for these examples. Before proceeding, save the schematic.

## 7.F.1. Obtaining a .model from the Web: Diode Model

As a first example, we will show how to download a model for a Schottky diode. This example will be the same for any device that uses a .model statement. Thus, you can use this same procedure for BJT's, MOSFETs, and jFETs. For this example, we will obtain the model from International Rectifier. Run your browser and open site www.irf.com:

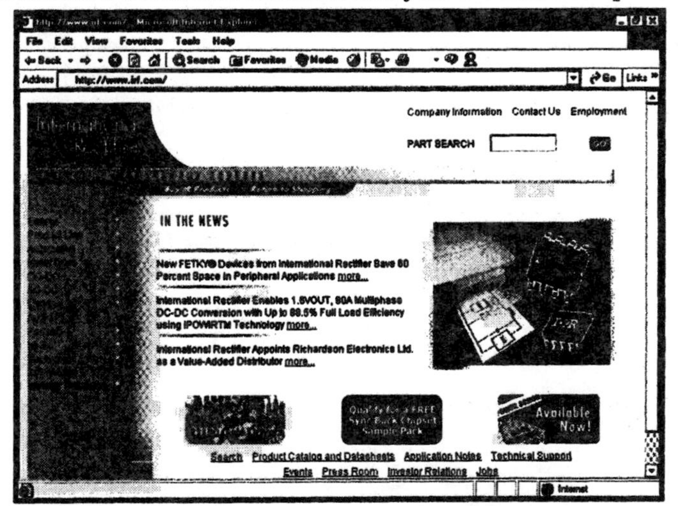


Enter the name of the device for which you need a model in the ***PART SEARCH*** field and click the ***GO*** button. I will look for a model of the Schottky diode 10BQ040:

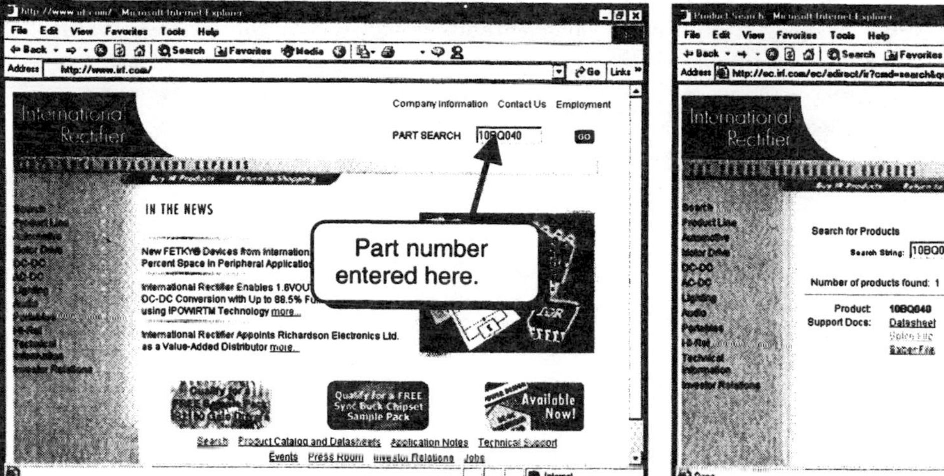


Select the link ***Spice File***:

Product Search - Microsoft Internet Explorer
File Edit View Favorites Tools Help
Address http://ec.irf.com/ec/adirect/ir?cmd=search&query=10BQ040
Company Information Contact Us Employment
International Rectifier
PART SEARCH GO
Search for Products
Search String: 10BQ040 Search
Number of products found: 1
Product 10BQ040
Support Docs: Datasheet
Spice File
Saber File
Click on this link.
Order Parts
ALLIED ELECTRONICS, INC. 42 10/03/2001
DIGI-KEY CORPORATION 15509 10/03/2001
ACI ELECTRONICS 487 10/03/2001
NEWARK ELECTRONICS 4925 09/27/2001
SOLID STATE INC. 7388 09/20/2001
Not Given
Done
Internet

This link is to a text file that contains this model. When you select this link with the Microsoft Internet Explorer, the text file is displayed in the browser window:

http://www.irf.com/product-info/models/spice/10bq040.spi - Microsoft Internet Explorer
File Edit View Favorites Tools Help Send
Back Search Favorites History
Address http://www.irf.com/product-info/models/spice/10bq040.spi Go

```
.MODEL 10bq040 d
+IS=2.13126e-05 RS=0.123203 N=1.53952 EG=0.600841
+XTI=3.78803 BV=40 IBV=0.0001 CJO=1.53747e-10
+VJ=1.5 M=0.476132 FC=0.5 TT=0
+KF=0 AF=1
* Model generated on May 28, 96
* Model format: SPICE3
```

Done
Internet

Use the mouse to highlight all the text in this window:

Select **Edit** and then **Copy** from the Internet Explorer menus:

Now that we have copied the text, we can paste it into a model in Capture. Switch back to Capture and select the Dbreak part. It should be highlighted in pink when selected:

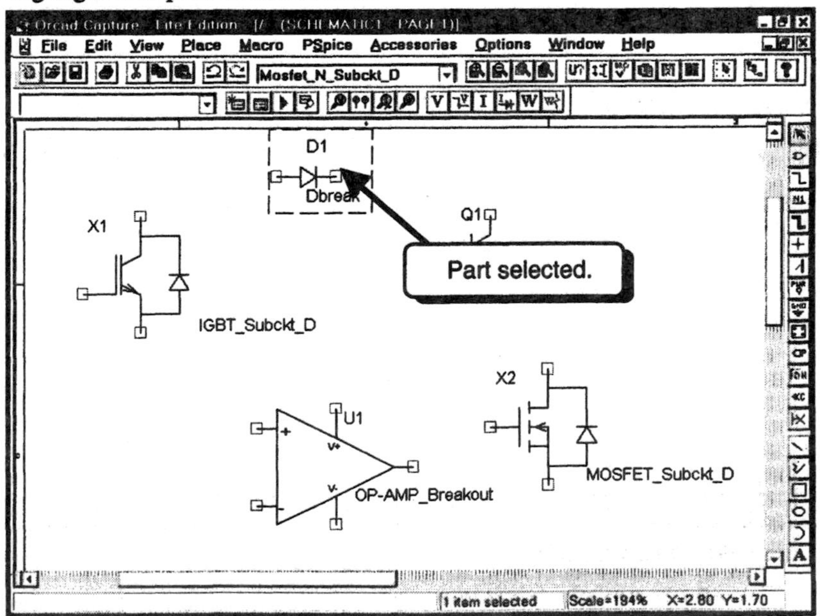


With the Dbreak part selected, select **Edit** and then **PSpice Model** from the Capture menus:

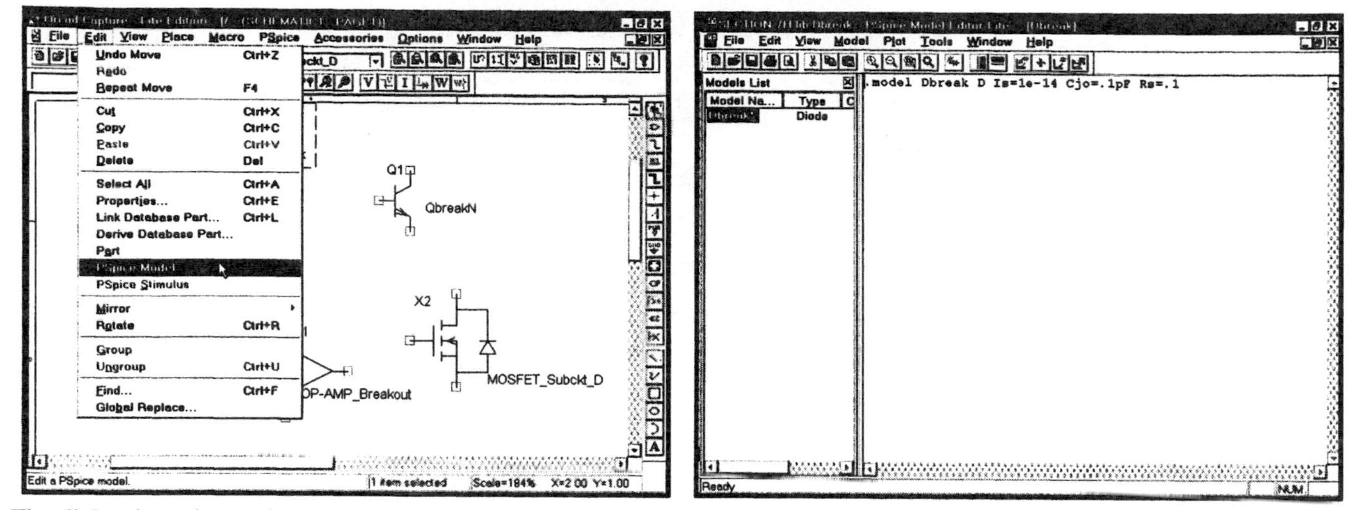


The dialog box shows the present model for Dbreak. We wish to replace the entire model with the one we just copied. Use the mouse to highlight all of the text in the dialog box:

We would like to paste the text we copied into this window. Select **Edit** and then **Paste** from the Model Editor menus to replace the selected text with the model copied from International Rectifier:

Select **File** and then **Save** to create the new model:

Select **File** and then **Exit** to close the model editor and return to Capture. The new model name will be displayed on the schematic:

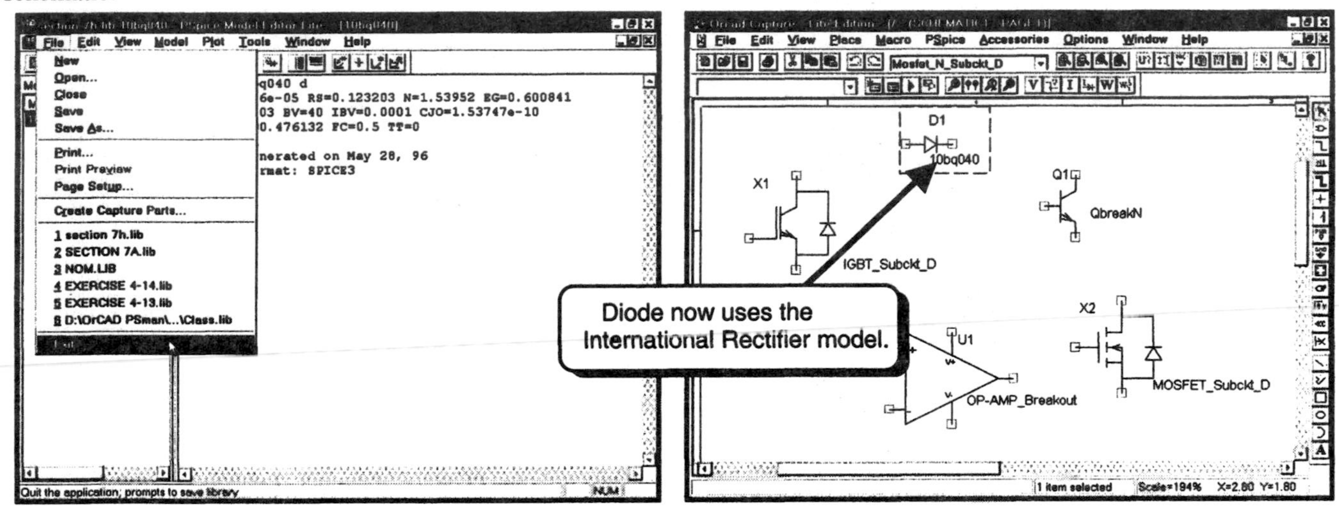


## 7.F.2. Obtaining Subcircuit Models from the Web: Op-Amp

For our second example, we will copy an op-amp subcircuit model from the Maxim Integrated Products web site. This example will be the same for any device that uses a .subckt model. Thus, you can use this same procedure for IGBT subcircuits, MOSFET subcircuits, and Darlington subcircuits. Run your web browser and open site www.maxim-ic.com:

Scroll down the window until you see the ***SPICE Models*** link:

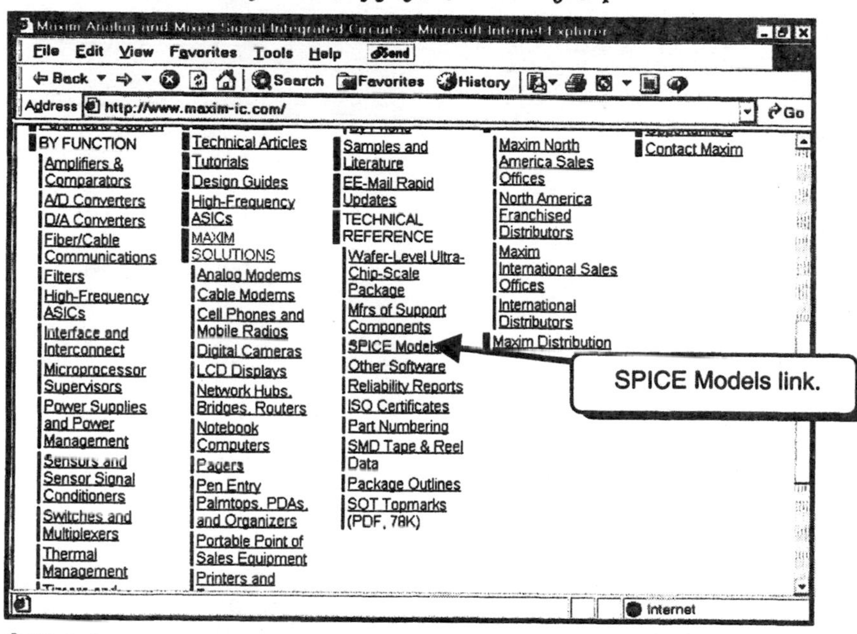


Select the ***SPICE Models*** link:

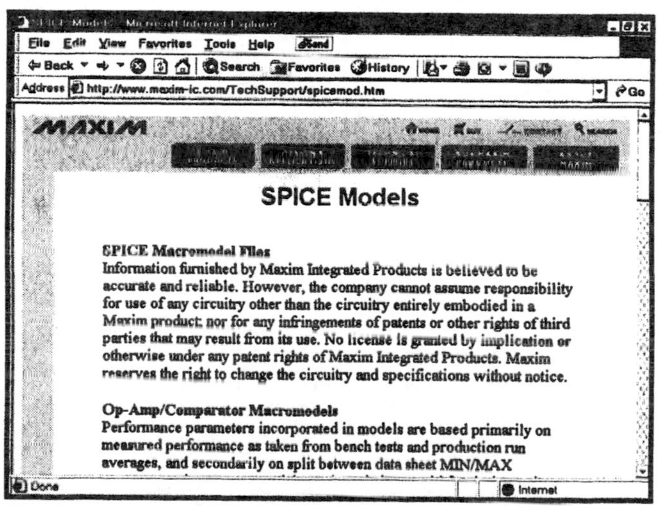


If you scroll down the page, you will see the listing of models you can download:

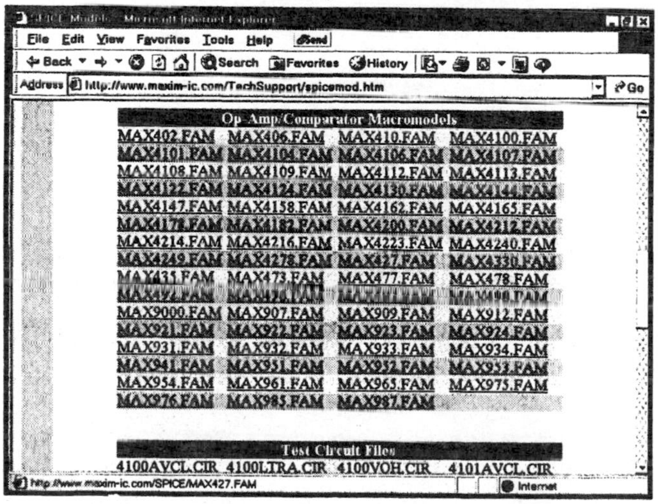

| Op-Amp/Comparator Macromodels | | | |
|---|---|---|---|
| MAX402.FAM | MAX406.FAM | MAX410.FAM | MAX4100.FAM |
| MAX4101.FAM | MAX4104.FAM | MAX4106.FAM | MAX4107.FAM |
| MAX4108.FAM | MAX4109.FAM | MAX4112.FAM | MAX4113.FAM |
| MAX4122.FAM | MAX4124.FAM | MAX4130.FAM | MAX4144.FAM |
| MAX4147.FAM | MAX4158.FAM | MAX4162.FAM | MAX4165.FAM |
| MAX4178.FAM | MAX4182.FAM | MAX4200.FAM | MAX4212.FAM |
| MAX4214.FAM | MAX4216.FAM | MAX4223.FAM | MAX4240.FAM |
| MAX4249.FAM | MAX4278.FAM | MAX427.FAM | MAX4330.FAM |
| MAX435.FAM | MAX473.FAM | MAX477.FAM | MAX478.FAM |
| [illegible] | [illegible] | [illegible] | MAX498.FAM |
| MAX9000.FAM | MAX907.FAM | MAX909.FAM | MAX912.FAM |
| MAX921.FAM | MAX922.FAM | MAX923.FAM | MAX924.FAM |
| MAX931.FAM | MAX932.FAM | MAX933.FAM | MAX934.FAM |
| MAX941.FAM | MAX951.FAM | MAX952.FAM | MAX953.FAM |
| MAX954.FAM | MAX961.FAM | MAX965.FAM | MAX975.FAM |
| MAX976.FAM | MAX985.FAM | MAX987.FAM | |

| Test Circuit Files | | | |
|---|---|---|---|
| 4100AVCL.CIR | 4100LTRA.CIR | 4100VOH.CIR | 4101AVCL.CIR |

I would like a model in the MAX 402 family of op-amps. Select link MAX402.FAM. The text for models in this family will be displayed in the browser window:

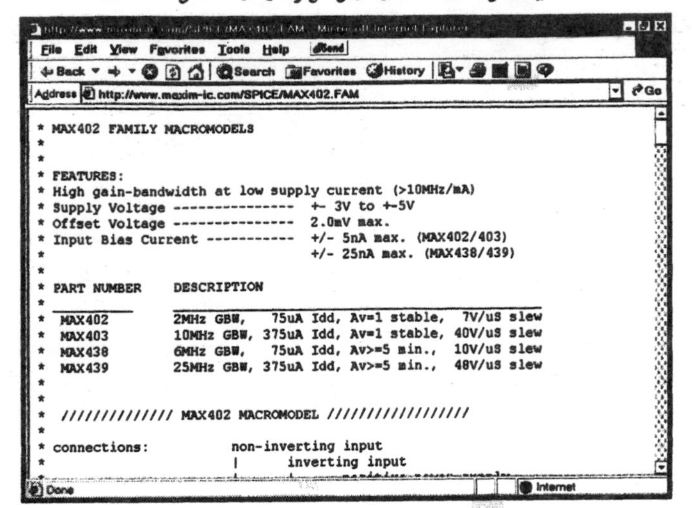


We can now copy the text from the file and paste it into Capture. First locate the model you wish to use and highlight the text using the mouse. I will copy model MAX439:

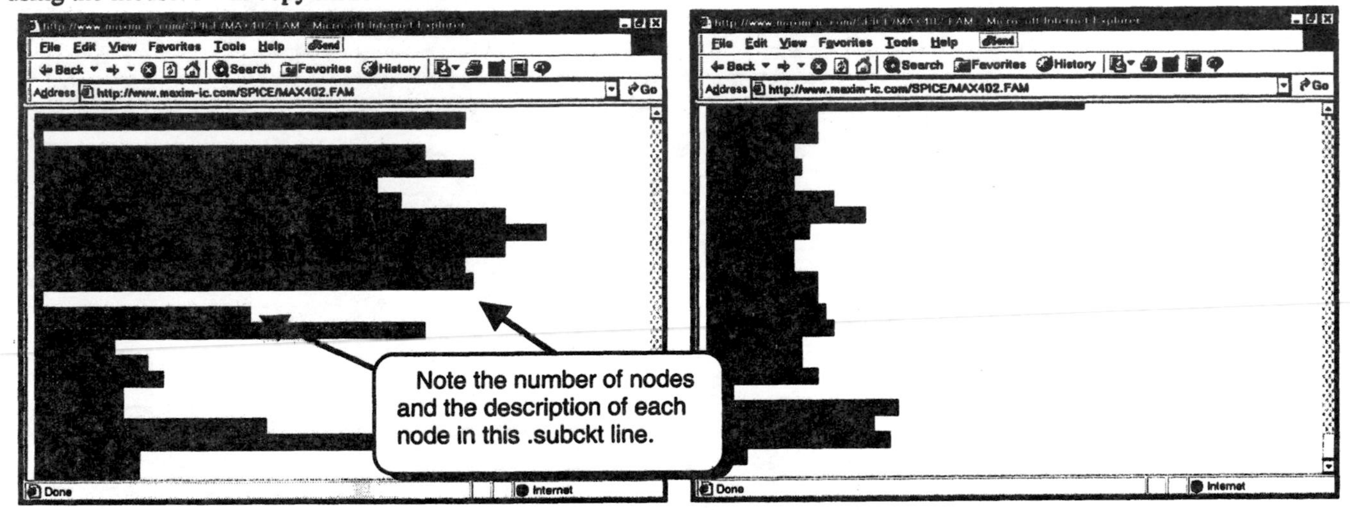


To copy the text, select **Edit** and then **Copy** from the Internet Explorer menus. Next, switch back to Capture and select the op-amp part:

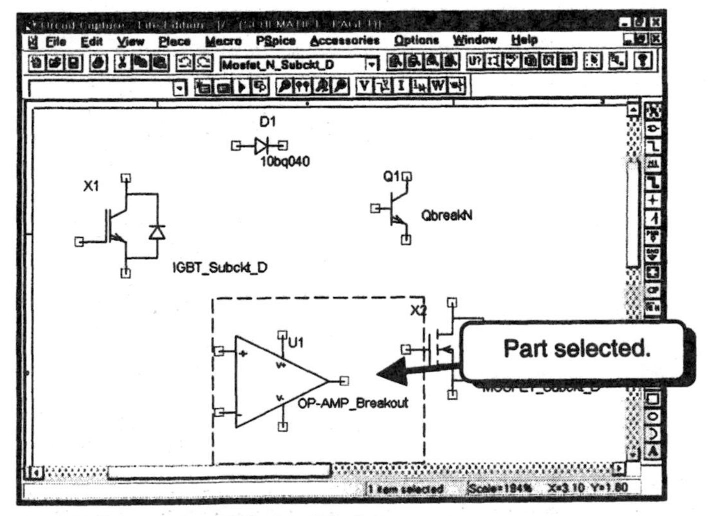


With the op-amp part selected, select **Edit** and then **PSpice Model**:

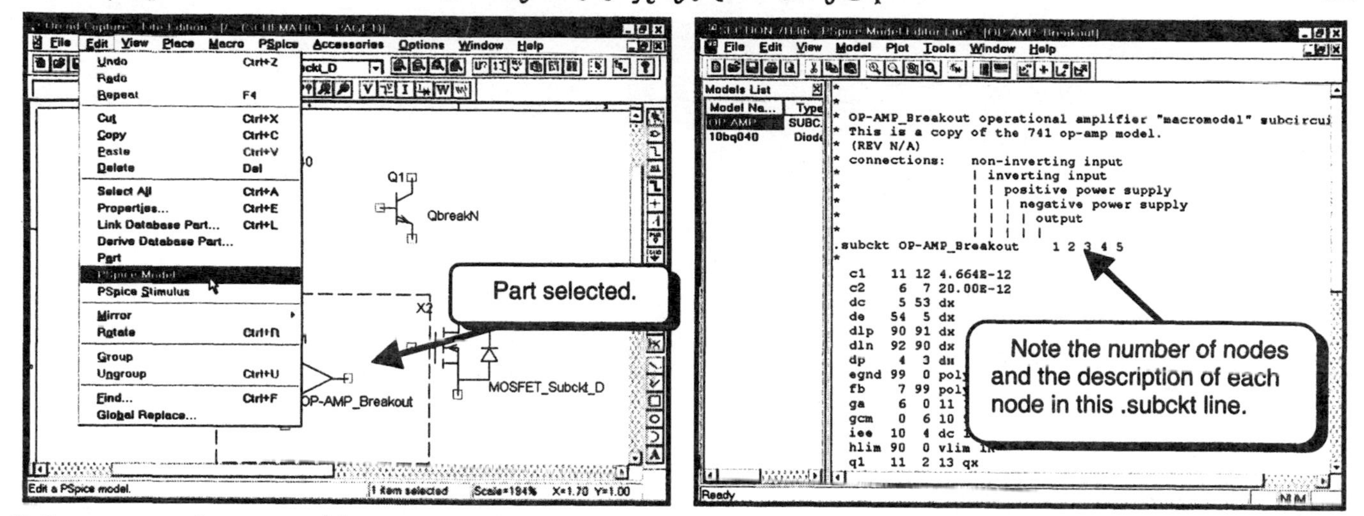
Part selected.
Note the number of nodes and the description of each node in this .subckt line.

Before we copy the new model, we must make sure that each .subckt line has the same number of calling nodes and that the description of the nodes and their order are the same. The numbering of the nodes is not important, only the number of nodes, their description, and their order. If these three items do not match up, we cannot use the model. If the .subckt lines match up (they do in this case), we can proceed. Delete all of the text in the window:

Next we need to paste the text copied from the browser into this window. Select **Edit** and then **Paste** from the menus:

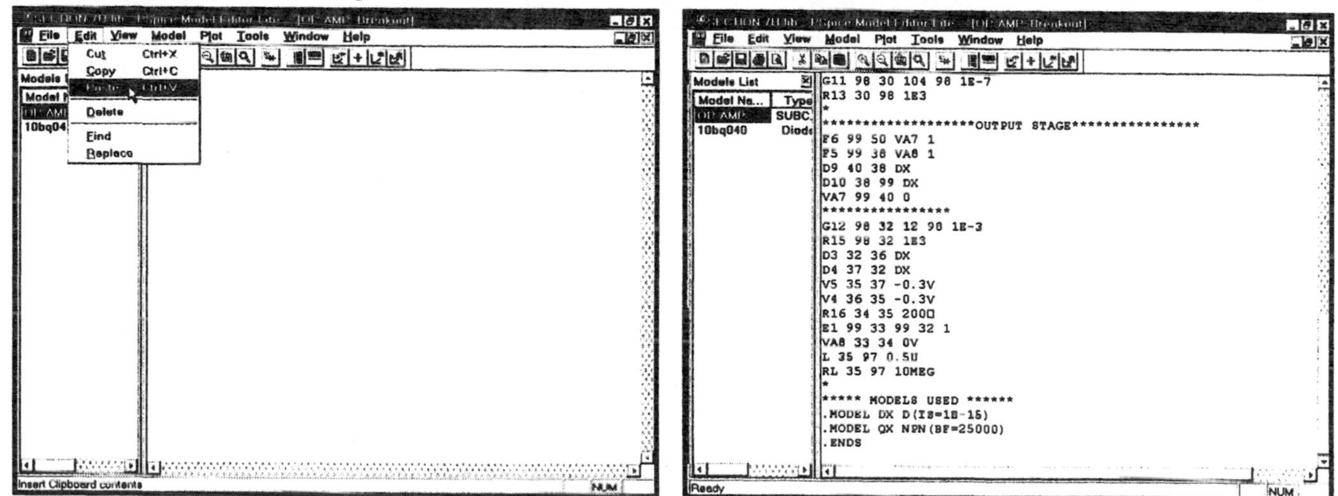

Select **File** and then **Save** to save the model, and then select **File** and then **Exit** to return to Capture:

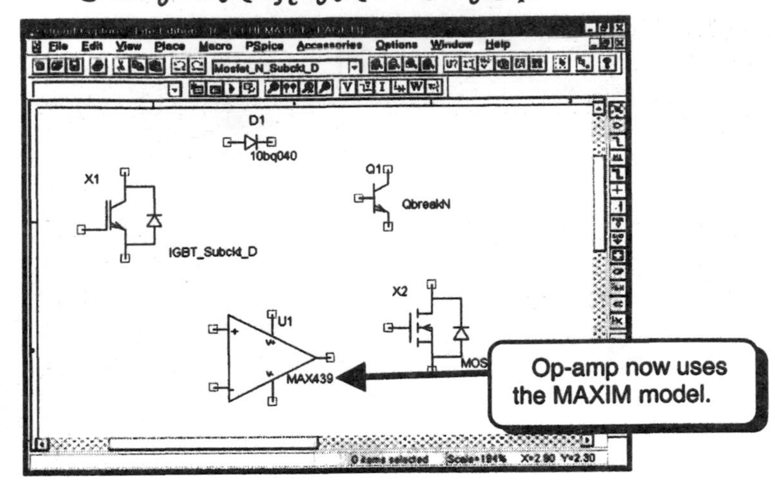


We can now use the new model in our simulations.

# 7.G. Creating and Using Subcircuits

In this section we will show how to take a schematic and change it into a subcircuit so that you can use it with other circuits. In order to follow this section you must have access to the libraries and configuration files for Capture. If you are running Capture on a network, you may not have access to these files. If you are running this example on your own computer, you should be able to follow the example with no problems.

## 7.G.1. Creating the Subcircuit

We will start with the op-amp circuit created from ABM parts in Section 6.O.3. Create a new blank project and draw the circuit below:

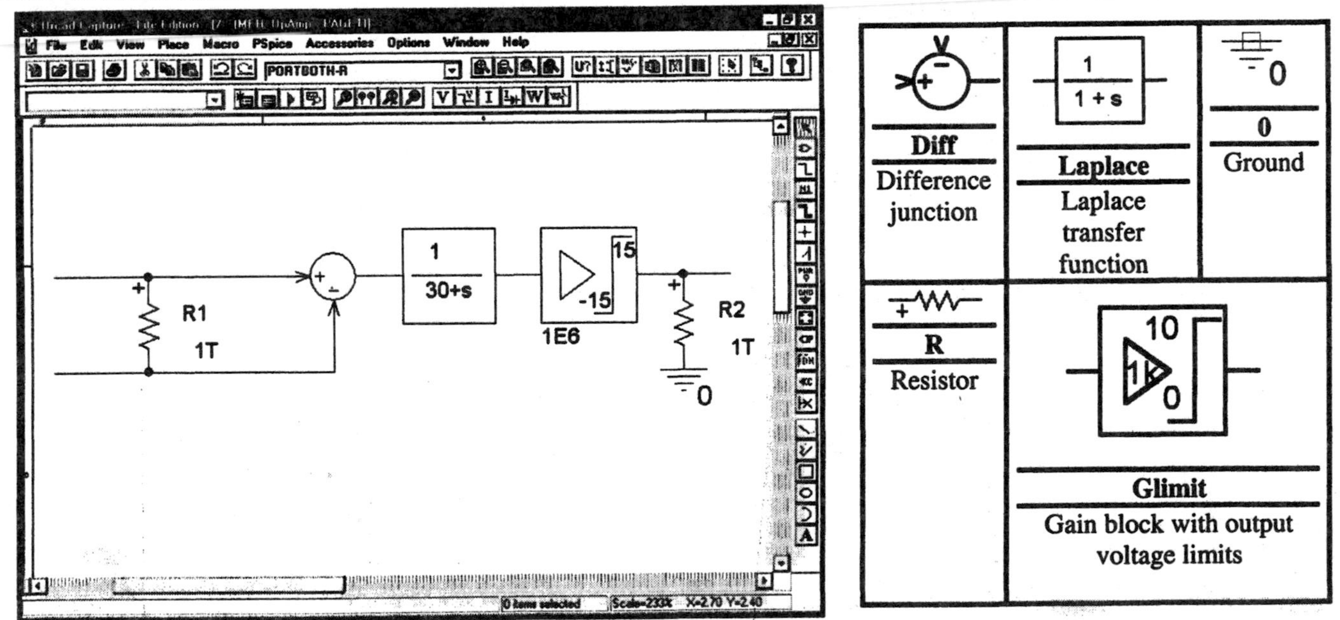


Note that none of the wires are labeled in this example. The text labels in the circuit of Section 6.O.3 were removed and the names of the resistors are not important.

We must now add interface ports to the circuit. Select **Place** and then **Hierarchical Port** from the Capture menus:

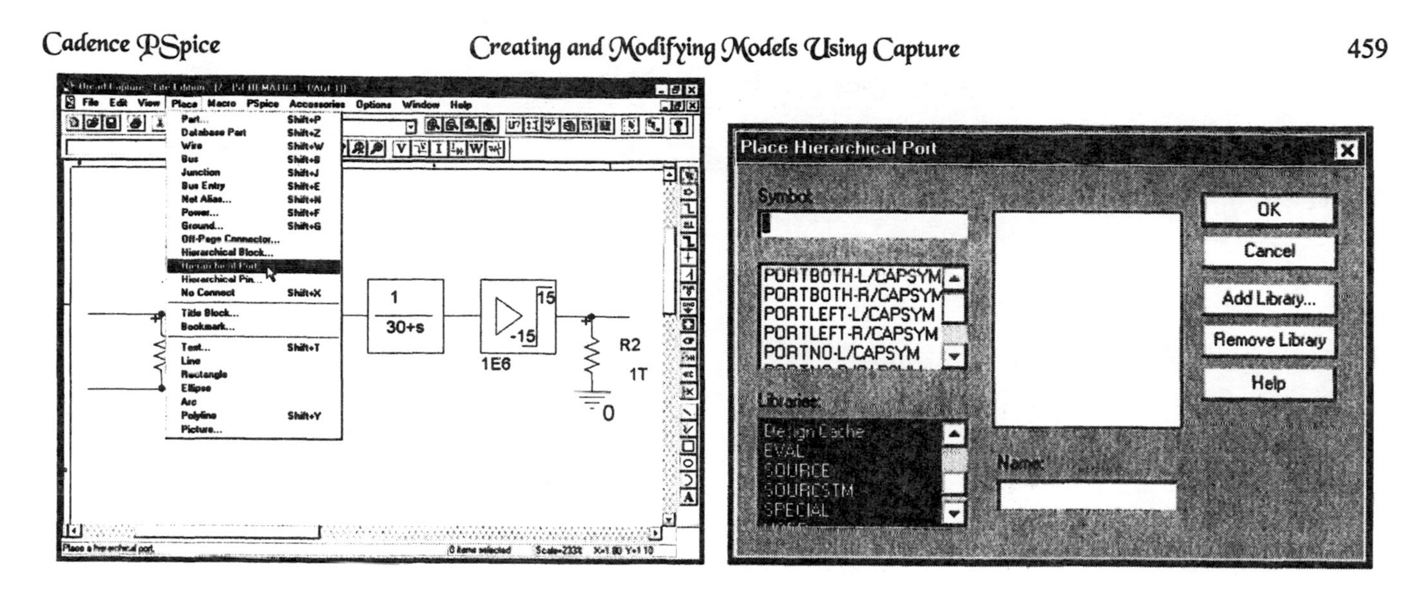


We can use any of the ports in this list. Some of the graphics may suggest an input or an output and you are free to use those if you wish. I will select the ***PORTBOTH-R*** and use it for all of my ports:

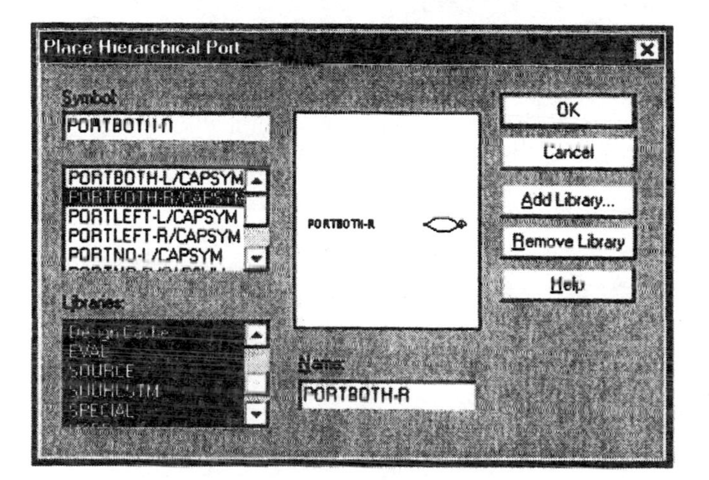


Place ports on the three terminals and label the ports as **Vp**, **Vm**, and **Vo**:

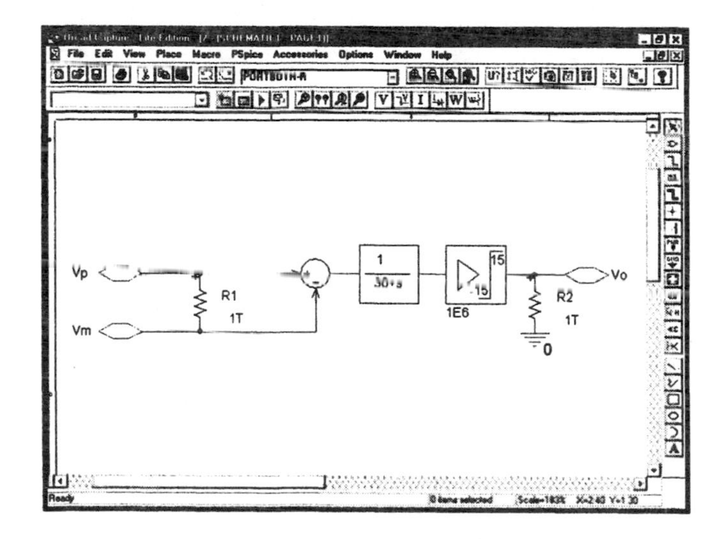


We are now ready to create the subcircuit netlist for this circuit. Select **File** and then **Save** to save the circuit, and then select **File** and then **Close** to close the window. You should return to the project tree for this circuit. Expand all the branches of the tree:

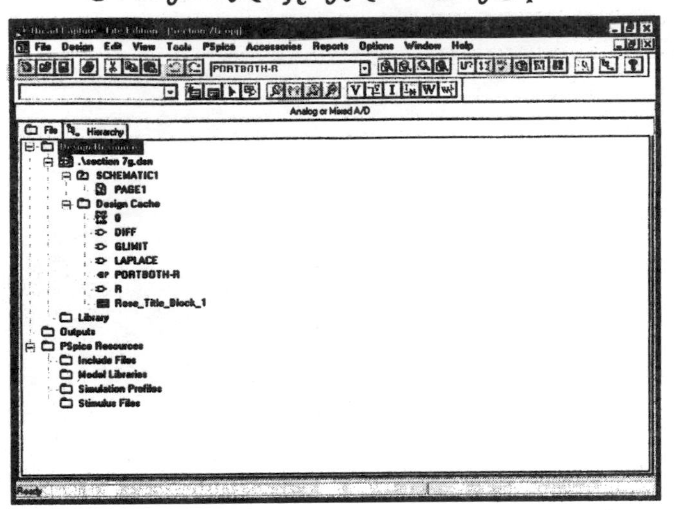

Click the ***LEFT*** mouse button on the text ***SCHEMATIC1*** to select it:

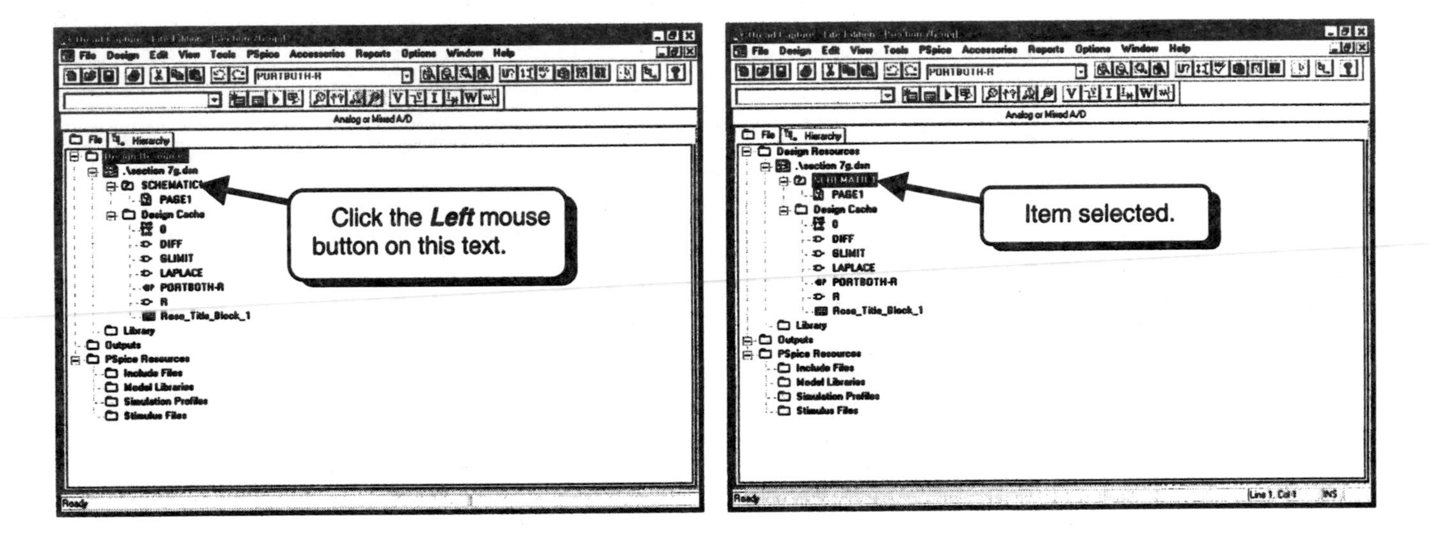


Click the ***RIGHT*** mouse button on the highlighted text to obtain the pull-down menu:

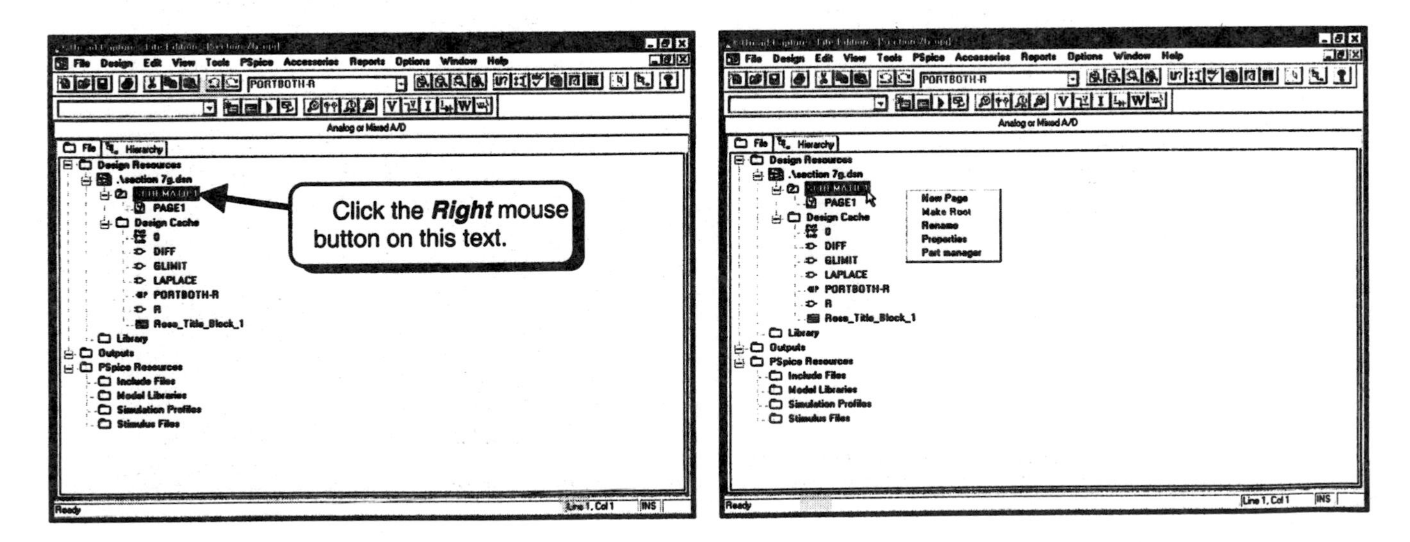


Select **Rename** to rename this item:

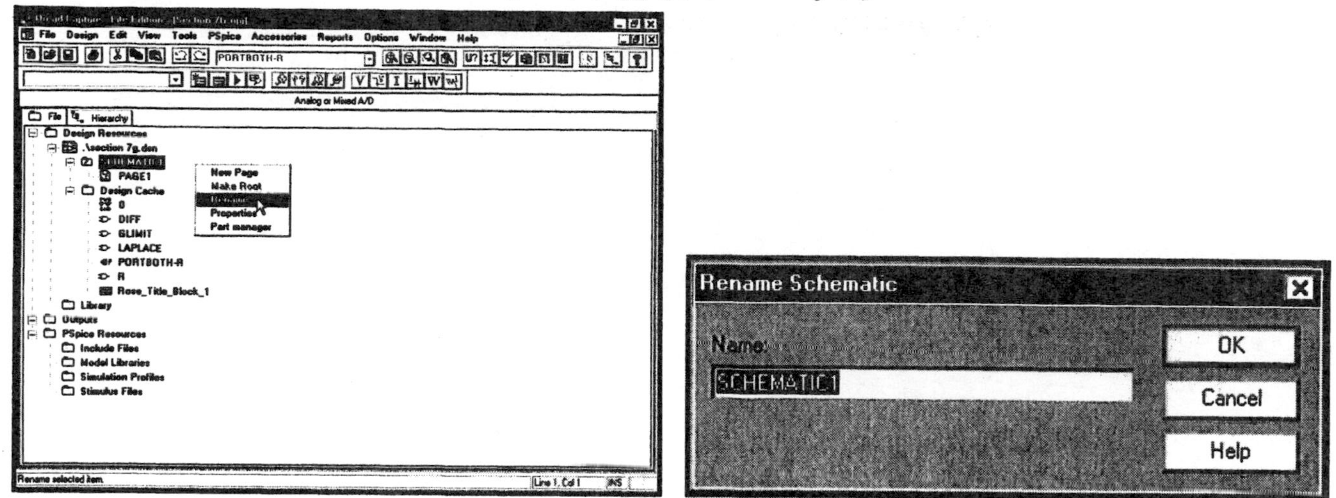


The name you specify here will be the name of the subcircuit and the name you use when you place this subcircuit part in a schematic. I will name my subcircuit MEH_OpAmp since there are already several op-amps in Orcad's libraries. Specify a new name and click the ***OK*** button:

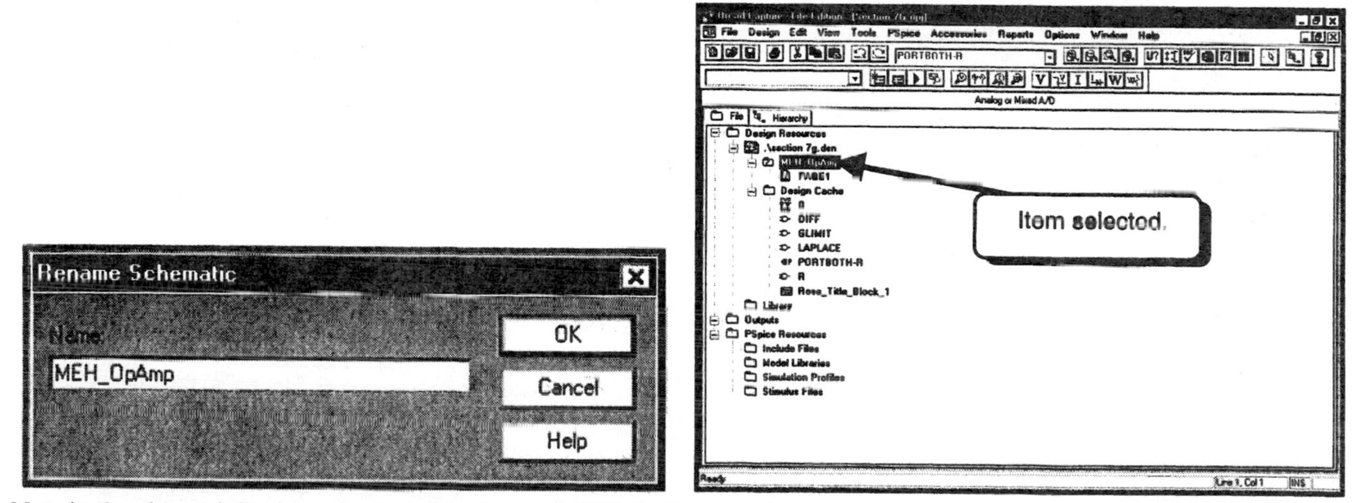


Note in the above right screen capture that the text MEH_OpAmp is selected. If the subcircuit name is not selected, click the ***LEFT*** mouse button on the text to select it. We can now create the subcircuit netlist. Select **Tools** and **Create Netlist** from the Capture menus:

Select the ***PSpice*** tab:

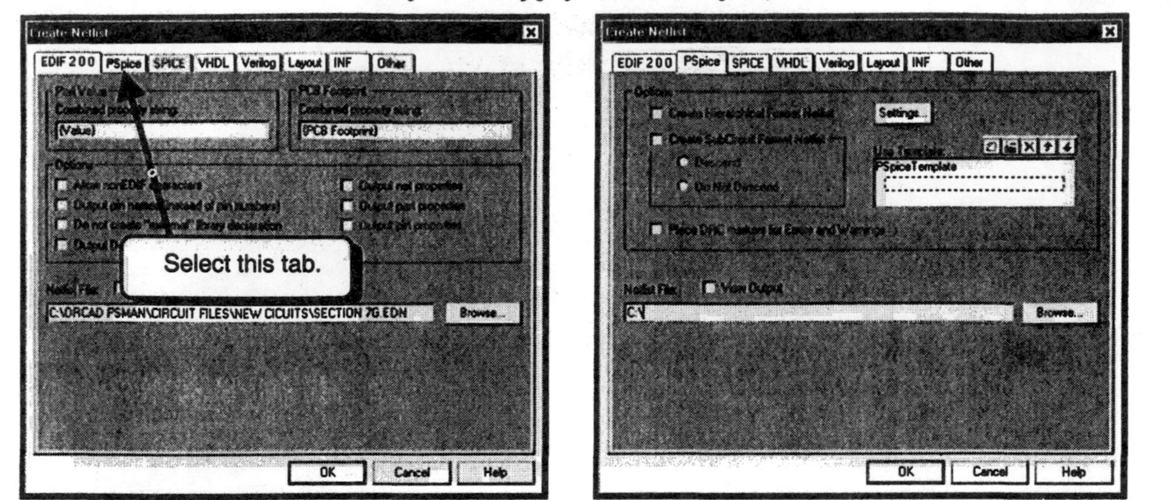


Select the options ***Create a SubCircuit Format Netlist*** and ***View Output***, and then click the ***Browse*** button:

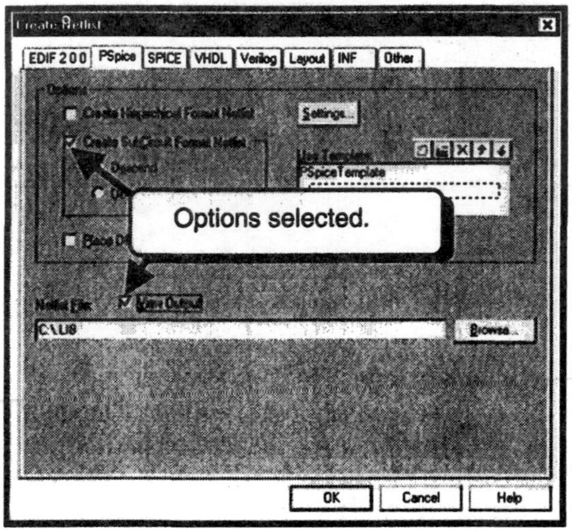

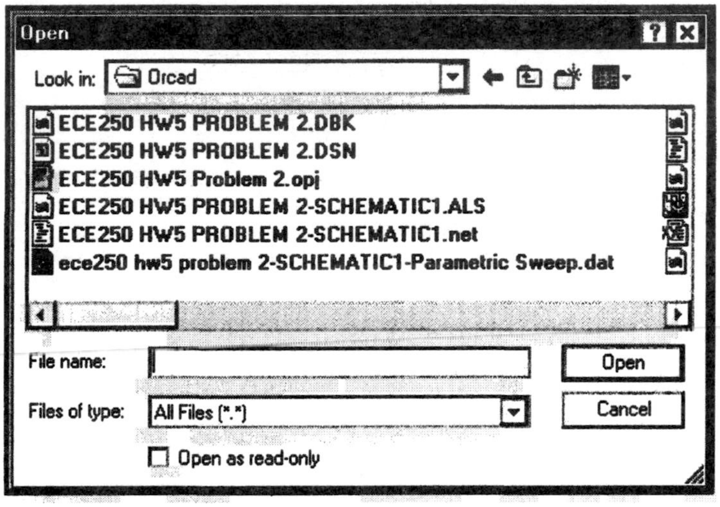

We need to specify the name of the netlist file and its location with this dialog box. This is the name of the file where the netlist will be stored. We will name the file User_subcircuits.lib and place it in the directory c:\Program Files\OrcadLite\Capture\Library\PSpice:

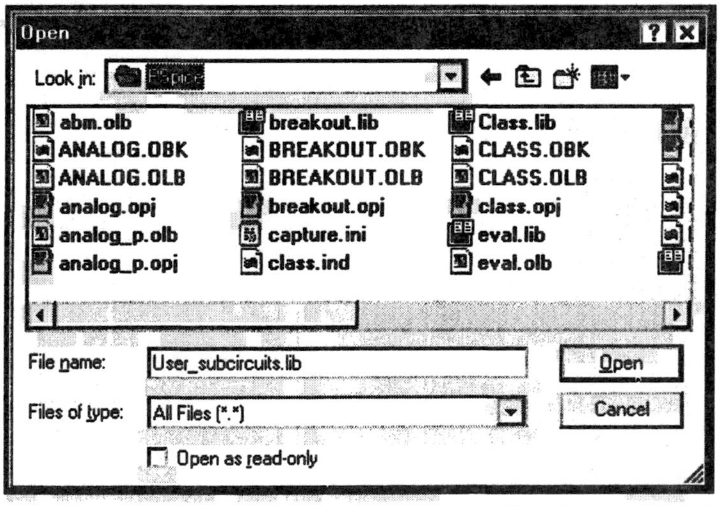

We have chosen this directory because this is the location of the library files that come with Orcad Lite. If we place the library here, we can easily make the library available to other projects. Click the ***Open*** button when you have finished modifying this dialog box:

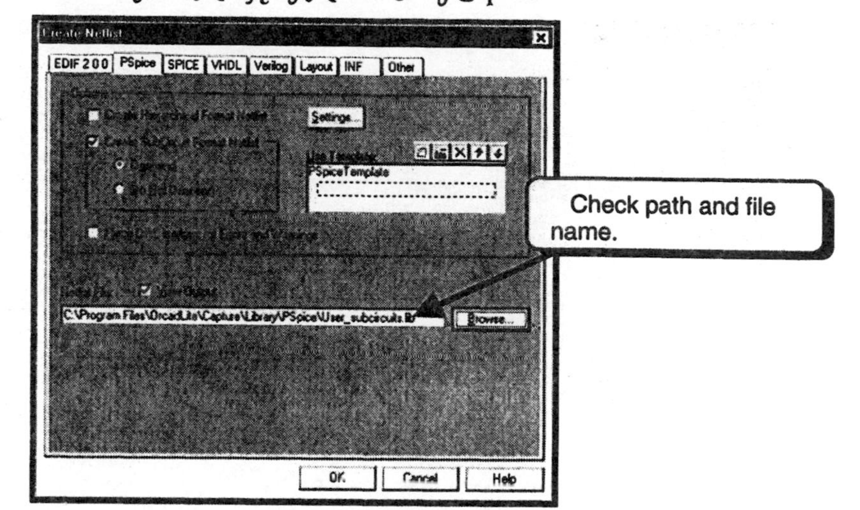


Make sure that the path and filename are entered correctly in the box. When the settings are correct, click the ***OK*** button to create the subcircuit netlist. After a few moments, the netlist will be created and a window will open and display the netlist:

```
1: * source SECTION 7G
2: .SUBCKT MEH_OpAmp Vm Vo Vp
3: R_R1          VP VM 1T
4: E_DIFF1          N000690 0 VALUE {V(VP,VM)}
5: E_LAPLACE1          N000831 0 LAPLACE {V(N000690)} {(1)/(30+s)}
6: E_GLIMIT1          VO 0 VALUE {LIMIT(V(N000831)*1E6,-15,15)}
7: R_R2          VO 0 1T
8: .ENDS
9:
```

This netlist describes the circuit we created in Capture. Since we chose to create a subcircuit netlist, the second line of the file declares the netlist as a subcircuit. The name of the subcircuit is ***MEH_OpAmp***, or whatever you named your subcircuit, and the calling nodes are ***Vm***, ***Vo***, and ***Vp***. These node names were derived from the ports we added to the circuit. The order is not important because we will have Capture create the symbol for this subcircuit, and it will use the same nodes in the correct order.

Select **File** and then **Close** to close the netlist window. You will be returned to the project tree. Notice that the netlist file we created is now listed in the tree:

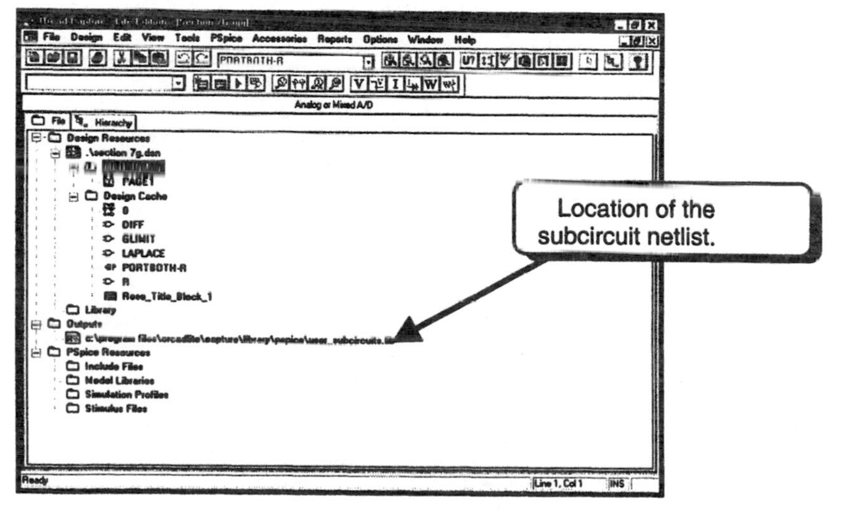

## 7.G.2. Creating and Modifying the Symbol

In the previous section, we created the netlist for our subcircuit. Next, we have to create a graphic symbol to place in our schematic. We will continue with the previous example where we are displaying the project tree. Note that the subcircuit name is still selected:

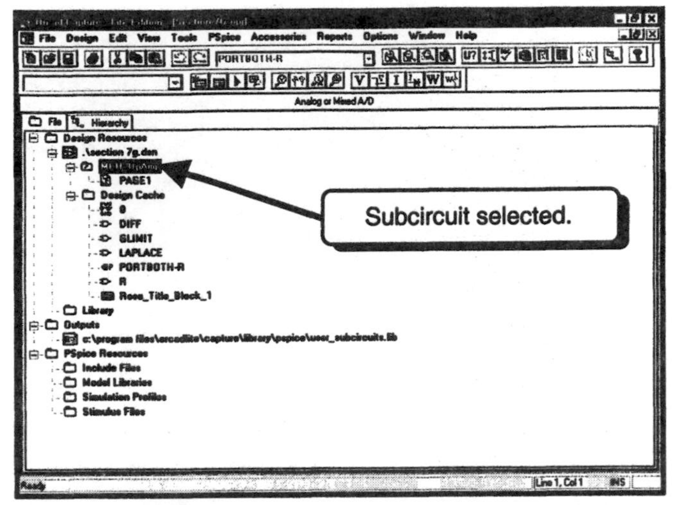


If your circuit is not highlighted as in the screen capture above, click the *LEFT* mouse button on the text to select it. We will now create the symbol for this subcircuit. Select **Tools** and then **Generate Part**:

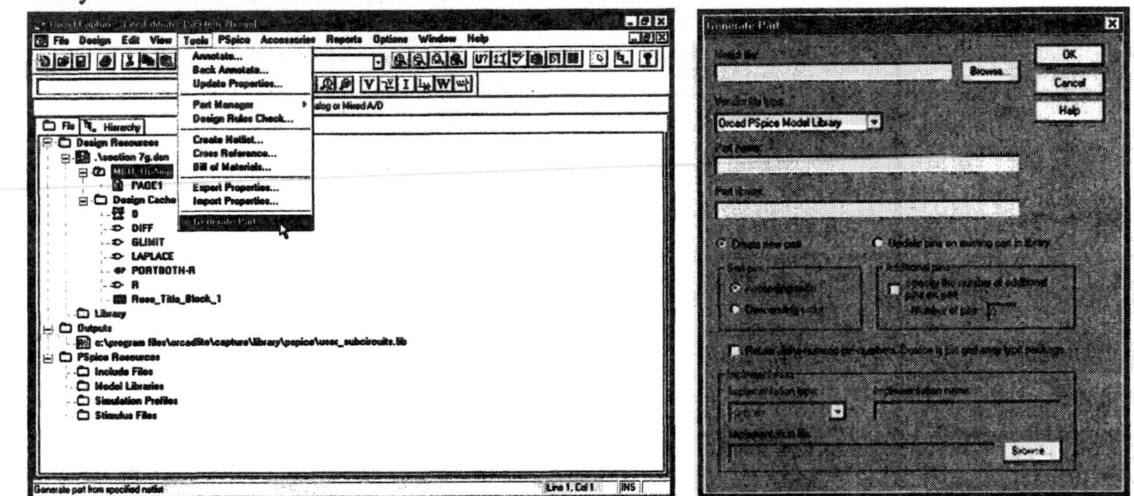

In the ***Netlist file*** field, enter the name of the netlist we created in the previous section. You can use the ***Browse*** button to select the file we created. I named my netlist file C:\Program Files\OrcadLite\Capture\Library\PSpice\User_subcircuits.lib.

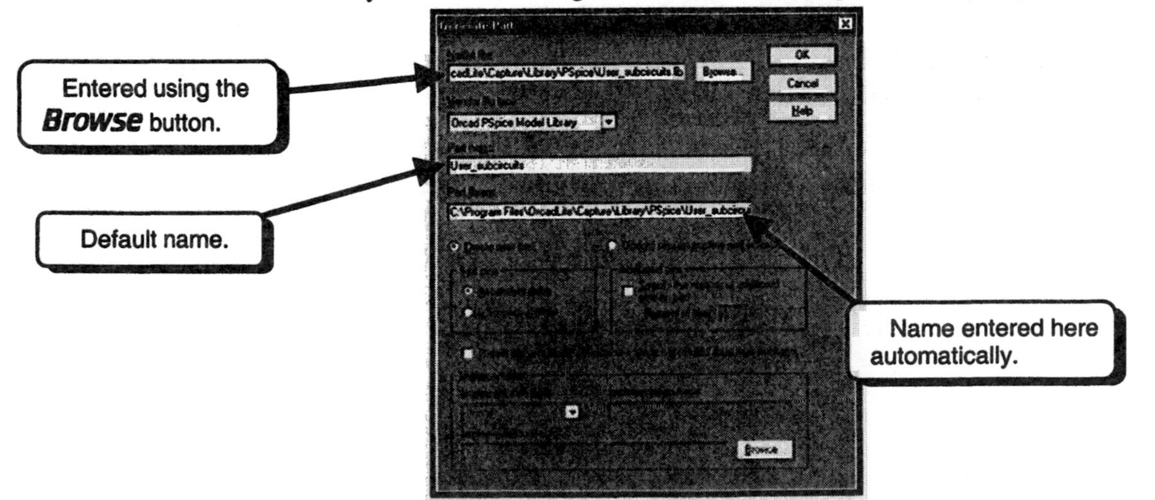


When you use the ***Browse*** button, the ***Part library*** field is filled in automatically. A part library is also referred to as a symbol library and has an .olb extension. By default, Capture uses the same name you specified for the ***Netlist file*** as for the ***Part library*** file except it changes the extension to .olb. In my case the library is named C:\Program Files\OrcadLite\Capture\Library\PSpice\User_subcircuits.OLB.

We see in the box above that Capture names the part the same as the netlist file name, or ***User_subcircuits***. I will change the ***Part name*** to the name of this subcircuit, **MEH_OpAmp**:

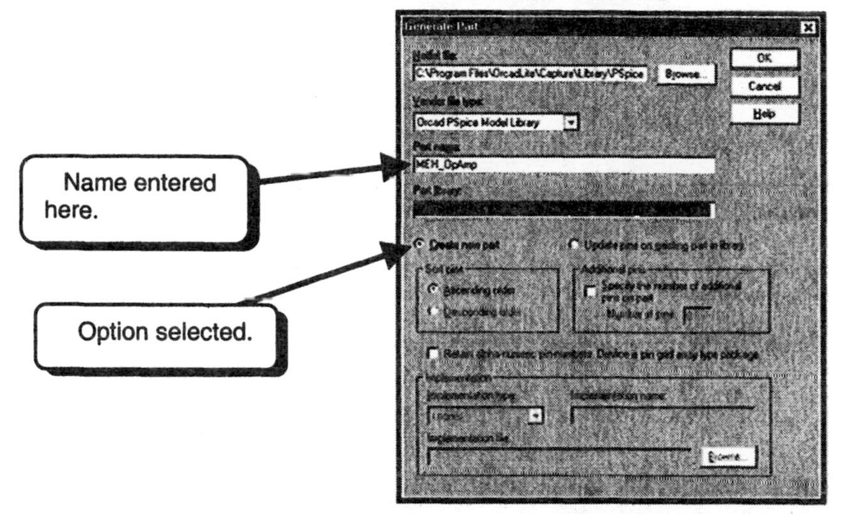


Make sure that the ***Create new part*** option is selected and then click the ***OK*** button to create the symbol. When the symbol is complete, you will return to the project tree. Note that a new line has been added to the tree in the ***Library*** folder. Expand the entire tree:

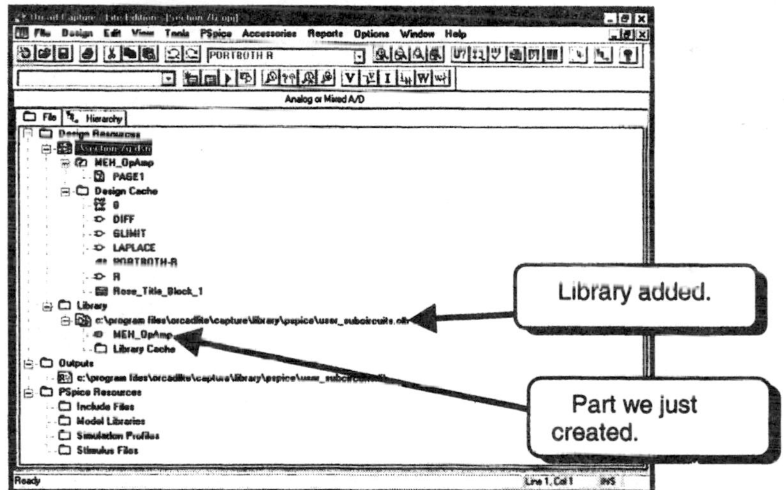


The new .olb file is added to our project tree. We see that this library only contains one graphic symbol, the one we just created for our subcircuit.

The last thing we need to do for our new part is modify the graphic symbol. Double-click the *Left* mouse button on the text ***MEH_OpAmp*** to edit the symbol:

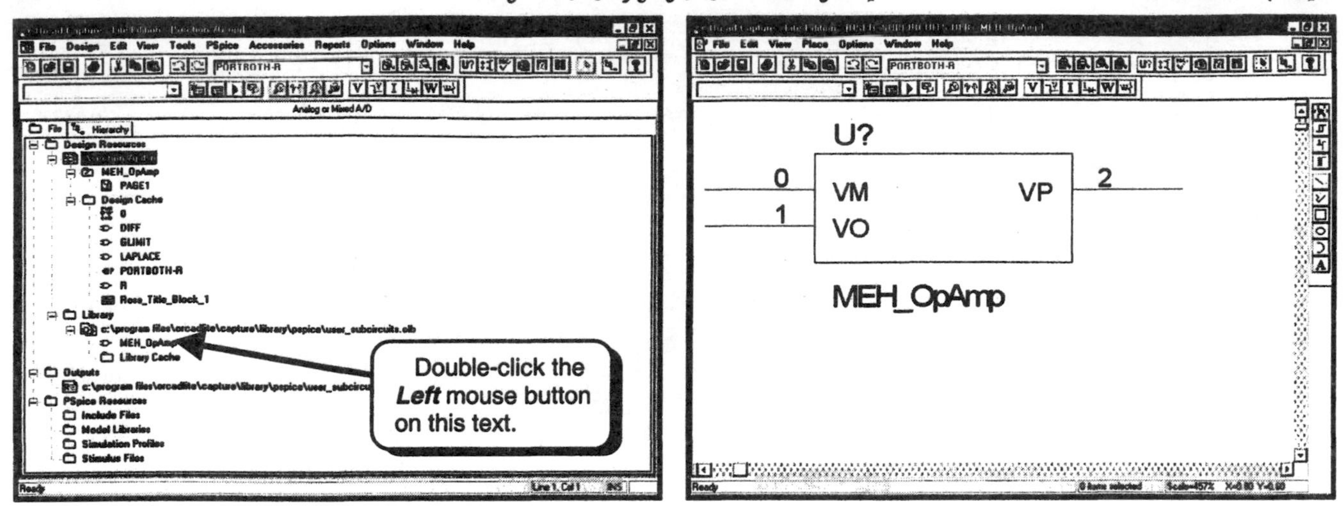


The first thing we need to do is hide the pin numbers ***0***, ***1***, and ***2***. Select **Options** and then **Part Properties**:

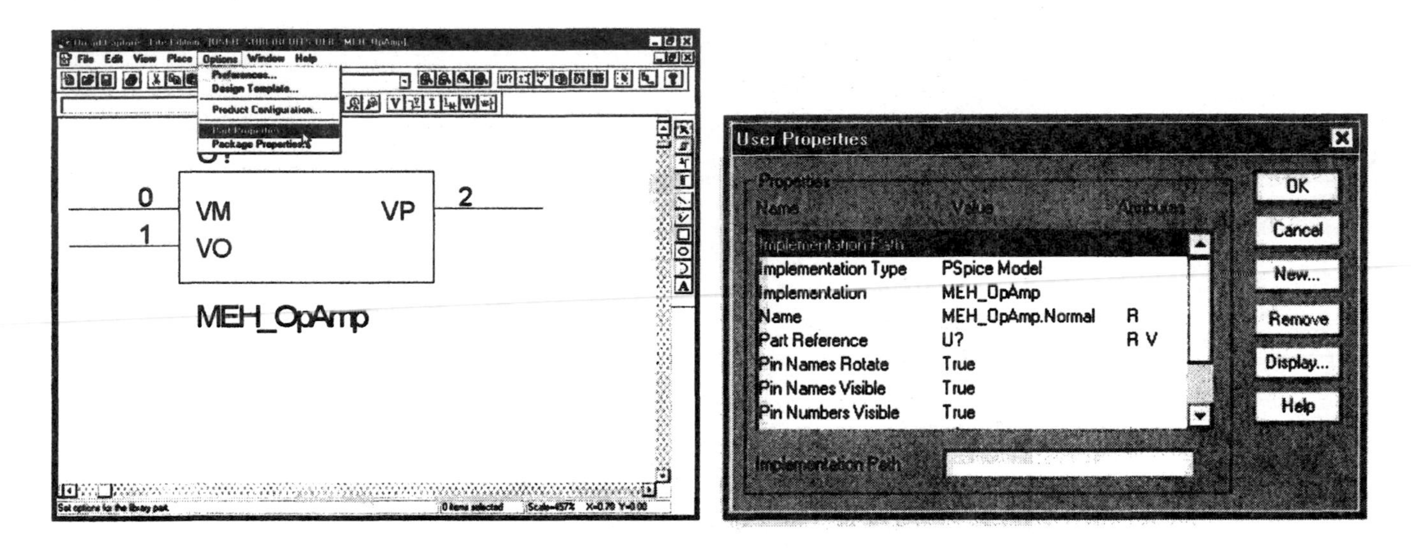


Select the ***Pin Numbers Visible*** property and change the value to ***False***:

User Properties

| Name | Value | Attributes |
| --- | --- | --- |
| Implementation Path | | |
| Implementation Type | PSpice Model | |
| Implementation | MEH_OpAmp | |
| Name | MEH_OpAmp.Normal | R |
| Part Reference | U? | R V |
| Pin Names Rotate | True | |
| Pin Names Visible | True | |
| Pin Numbers Visible | False | |

Property selected.

***False*** selected.

Click the ***OK*** button. The pin numbers should no longer be displayed:

Next, we need to modify the graphic symbol. You can move the pins, but do not change them or rename them. You can hide the pin names (Vp, Vm, or Vo) if you wish, but you cannot change their names because they correspond to the calling nodes used in the subcircuit netlist. You can delete the box and then add any graphics you wish using the place menu. My modified symbol is shown below:

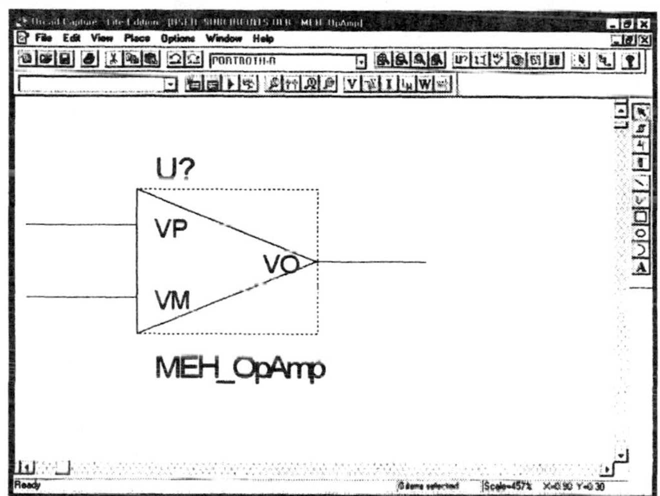


The dashed box around the graphics is a bounding box that tells Capture how large your part is. Make the bounding box as small as possible while still enclosing all the graphics for your symbol. Note that the bounding box does not enclose the pins. You can resize the bounding box by dragging corners or sides.

The symbol is now useable. However, I do not like the location of the pin names so I will turn off their display. This is done using the same method we used to hide the pin numbers. (Select **Options** and then **Part Properties** from the menu and change property ***Pin Names Visible*** to ***False***.)The result is shown below:

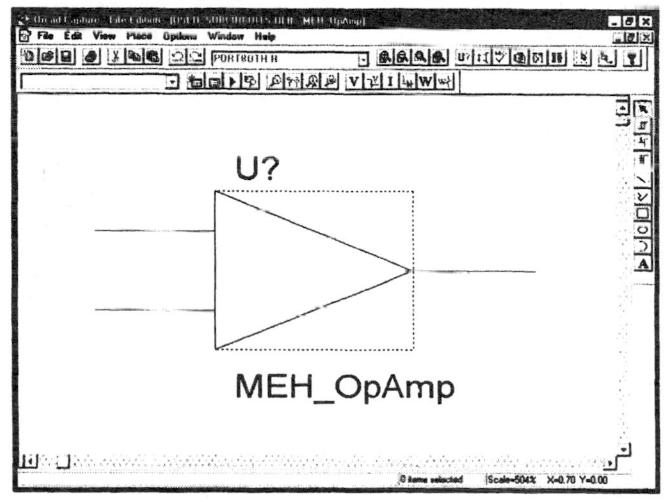

The output pin is obvious to a user of this part but the input pins are not. I will add + and – symbols to the symbol. I know which is which because a few moments ago, the pin names where displayed on the symbol. You can place text by selecting **Place** and then **Text** from the menus:

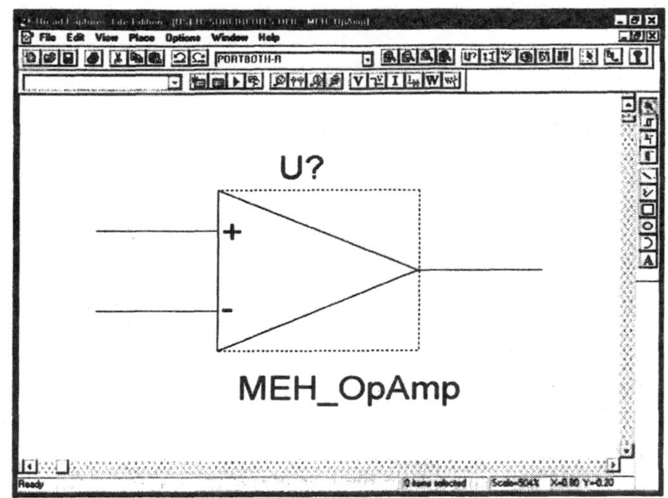


We are now finished with the symbol and the creation of the subcircuit. Save the changes to the symbol and then close the project. Answer yes to any question asking if you want to save something.

## 7.G.3. Making the Subcircuit Available to Future Designs

The last thing we need to do is configure Capture and PSpice to use the subcircuit and symbol libraries we created. Create a new blank project:

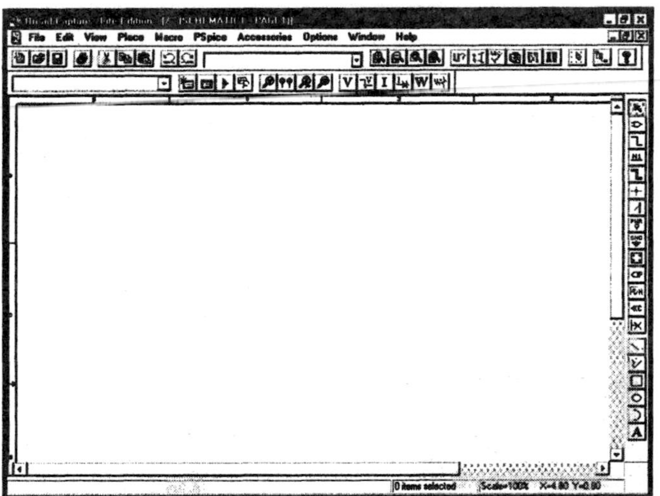


Select **Place** and then **Part** to obtain the ***Place Part*** dialog box:

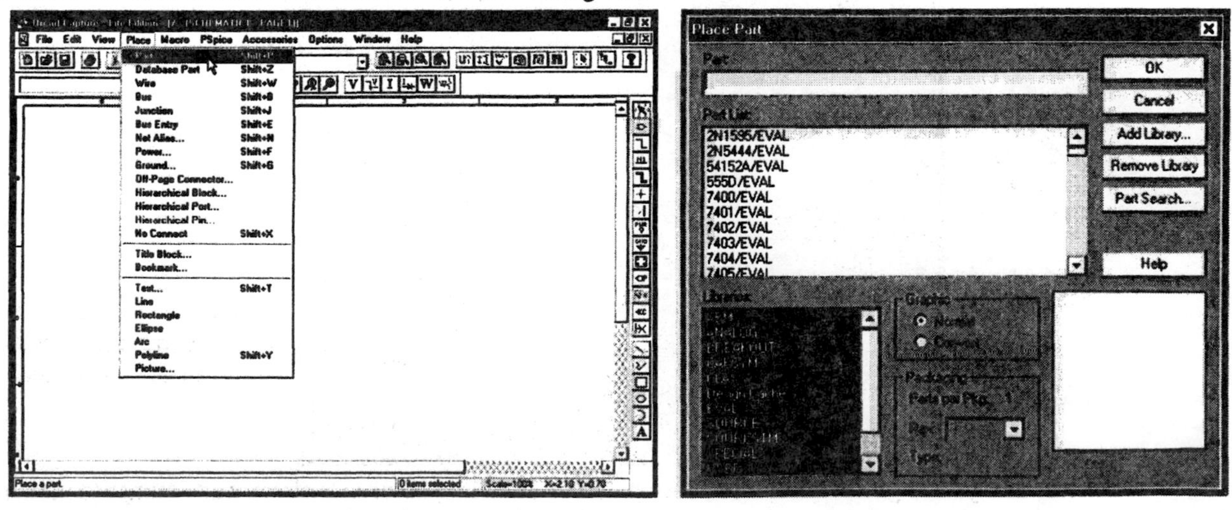

Click the ***Add Library*** button:

Since we placed the new symbol library in the default library directory, our new library is easily found. Select the ***USER_SUBCIRCUITS.OLB*** library and click the ***Open*** button:

The library is added to the list of libraries. All projects can now use this symbol library. If we select a part in this library, it should be displayed in the dialog box:

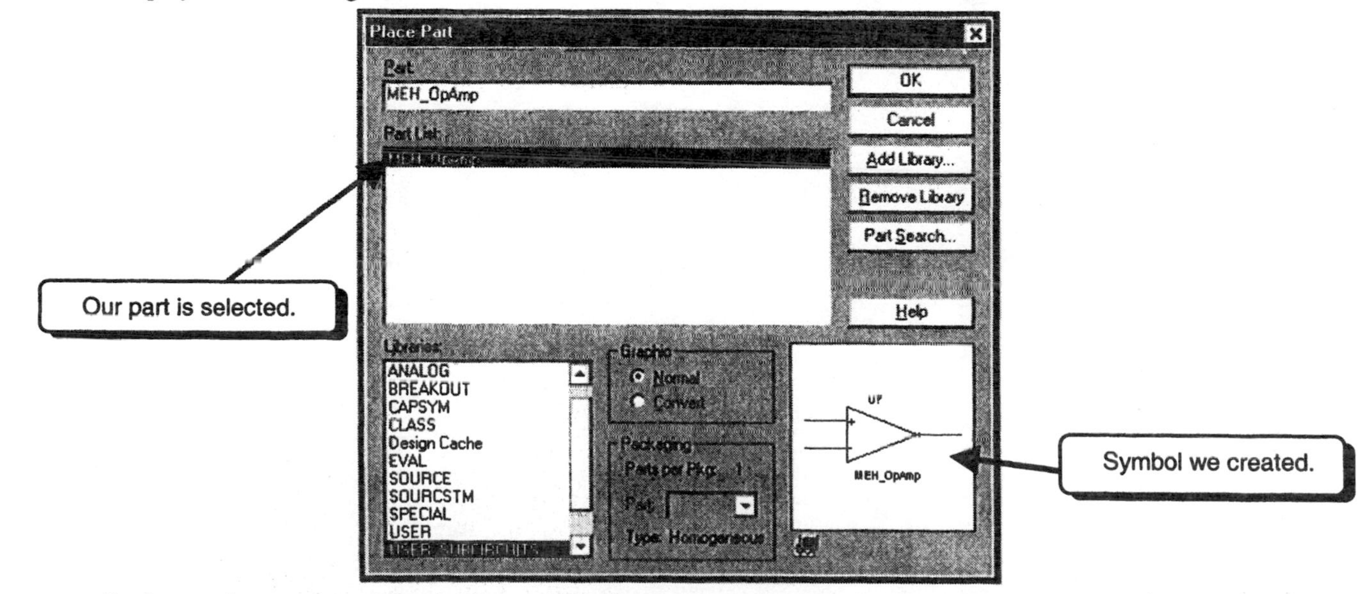


So far, we have only configured the symbol library. If we use this symbol in a circuit and attempt to simulate it, PSpice will generate an error stating that it cannot find the model for this part. Thus, the next thing we will do is configure the PSpice libraries. Click the ***Cancel*** button to close the ***Place Part*** dialog box. To configure the PSpice libraries, we must set up a simulation. Select **PSpice** and then **New Simulation Profile**, give the profile a name, and click the ***Create*** button:

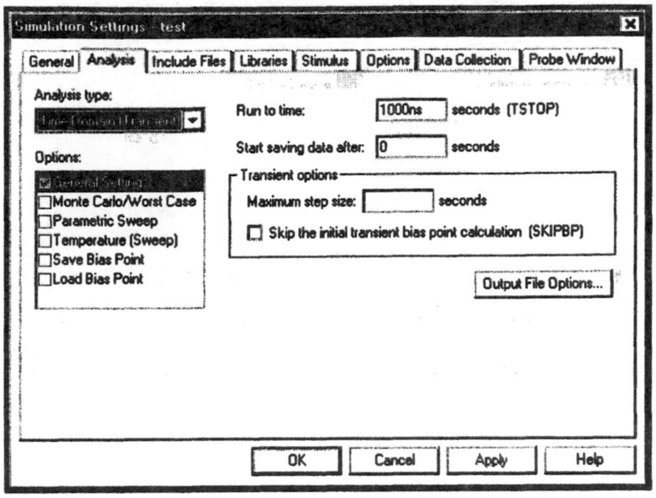

Select the ***Libraries*** tab:

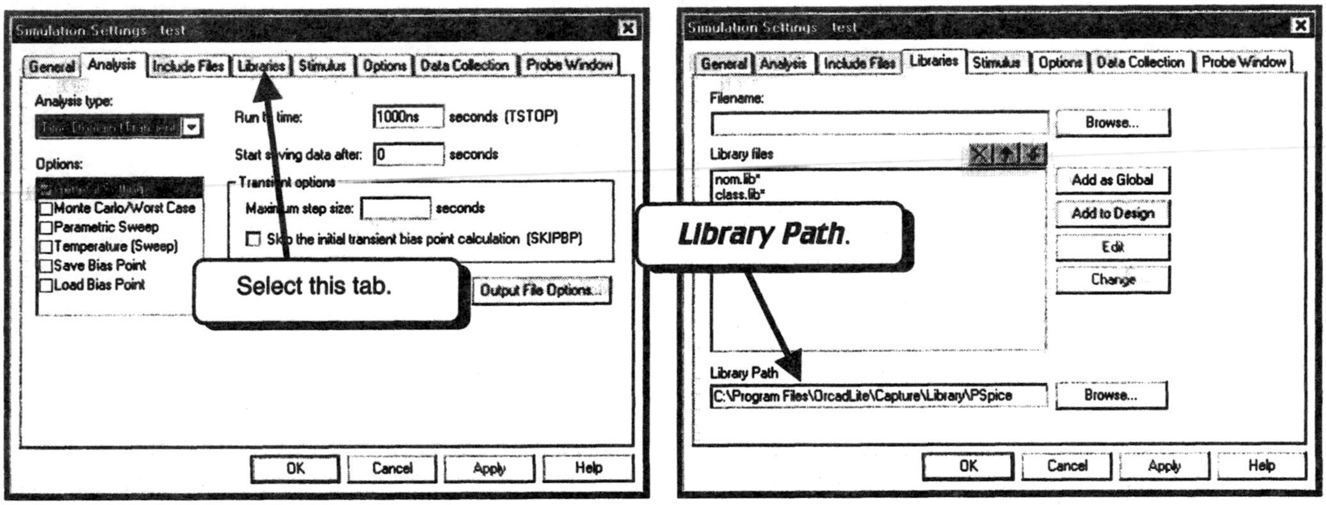

Notice that the ***Library Path*** is the same directory in which we placed the .lib file for our subcircuit. Select the ***Browse*** button. If it is not already selected, change the directory to C:\Program Files\OrcadLite\Capture\Library\PSpice:

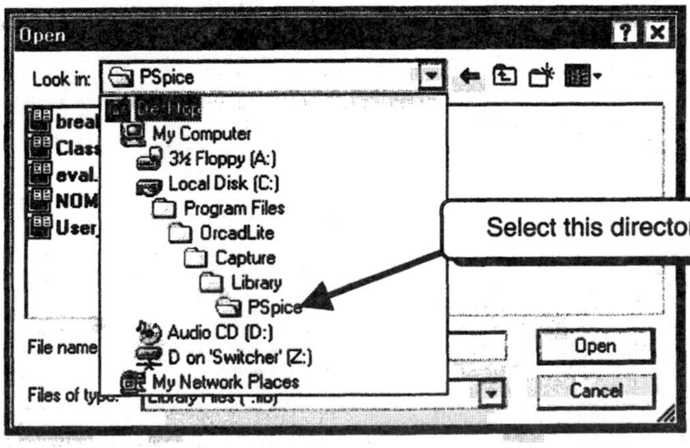

The ***User_subcircuits.lib*** file should be in this directory. Select the file and click the ***Open*** button:

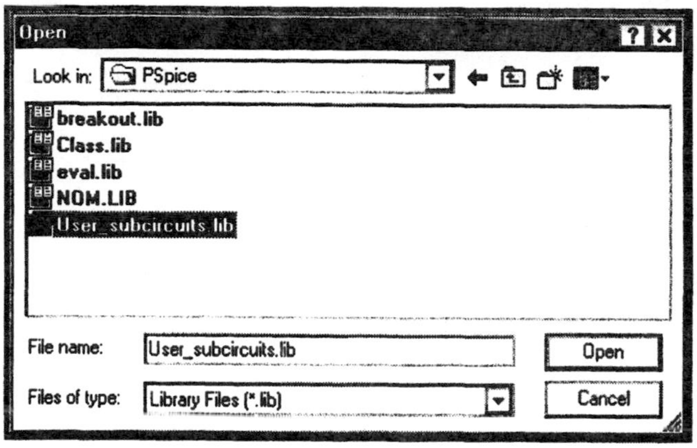

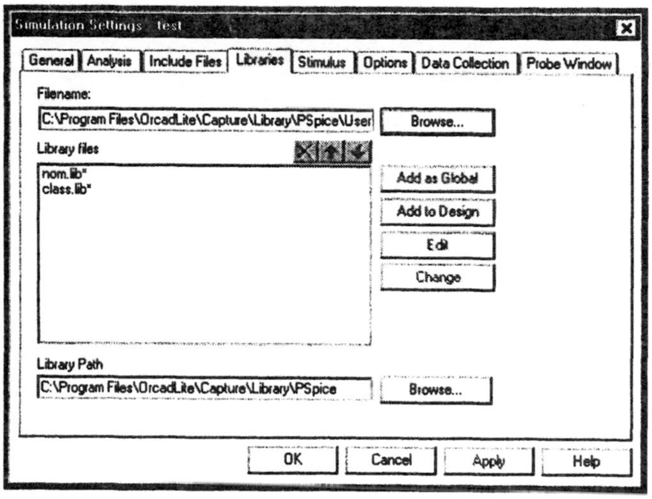

Select the ***Add as Global*** button.

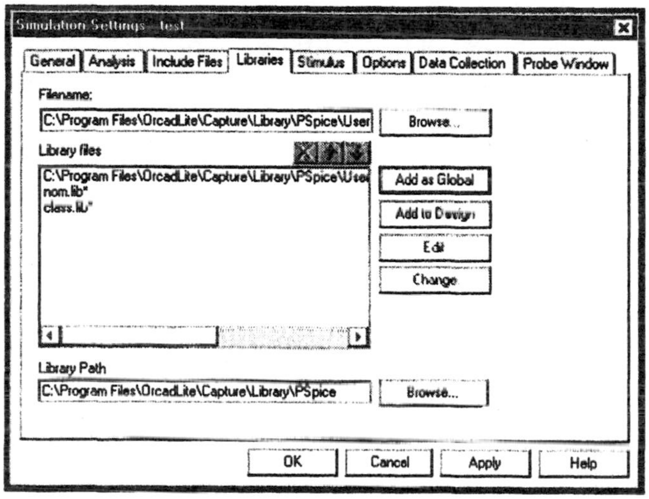

The ***Add as Global*** button adds a library so that all projects can use the library. The ***Add to Design*** button adds the library for the current project only. If you look in the list of libraries, all libraries that have an asterisk added to the name are global libraries and available to all projects. Libraries without an asterisk are only available to the current project. If we scroll the window to the right, we can see that the library we added does have an asterisk added to the name:

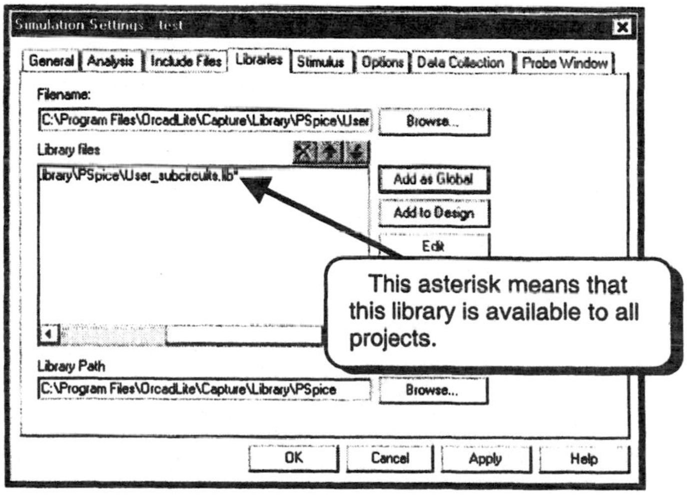

Click the ***OK*** button. You will return to the empty project:

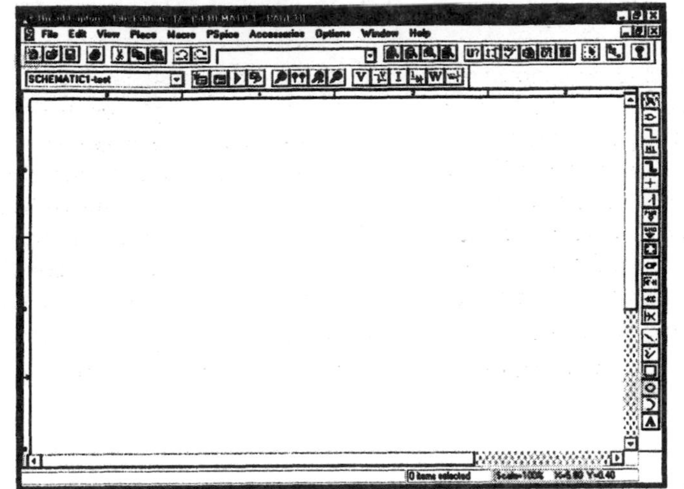


You can close and delete this empty project if you wish. The libraries are now configured to use the subcircuit.

# 7.H. Summary

- Changing the model reference does not create a new part. It tells PSpice to use a model that is already located in one of the library files (all files ending with the .lib suffix). You must know the name of the model and the name of the library file.
- Editing an instance of a model creates a new model by changing the contents of an older model. The original model is left unchanged. This method is used to create new models with slightly different characteristics from those of the original model.
- The breakout models are used to create completely new models.
- If Capture or PSpice has trouble finding a model, check the library path to see if the files listed in the library exist.

# PART 8
# Digital Simulations

PSpice can simulate pure analog, mixed analog/digital, and pure digital circuits. This part describes how to run pure digital and mixed analog/digital simulations. The circuits given are fairly simple, but the examples can be applied to larger systems. Although the digital components can be used with any of the previously described simulations, digital circuits are usually simulated with the Transient Analysis (Time Domain) because we are interested in a gate's output at a particular time. Only Transient simulations will be demonstrated here. If you are not familiar with the Transient Analysis, review Part 6 before proceeding. This part also assumes that you are familiar with displaying traces using Probe. If this is your first time using Capture and PSpice, you should review Part 1 for instructions on drawing a circuit, Part 2 for instructions on using Probe, and run a few of the examples in Part 6 to become familiar with using the Transient Analysis.

The library "eval.olb" shipped with the Lite version contains 134 logic circuits ranging from a 7400 to a 74490. This library is rich enough for students to simulate most circuits found in first- and second-semester logic classes.

**NOTE ON CIRCUIT LIMITATIONS**

The limitations of the Lite version of Capture and PSpice on digital simulations are: (1) The number of parts you can draw on a page is limited to 60 parts, and (2) the number of parts contained in a netlist that can be read by PSpice is limited to 65 digital primitive devices. Some of the digital circuits are fairly complicated and contain a large number of primitives. Even though a circuit may appear small on the screen, when it is compiled into its subcircuits, it may have a large number of internal parts and may reach the limit of 60 parts. The limitations on the mixed analog/digital simulations are the same as on pure analog simulations. However, when simulating mixed analog/digital circuits, PSpice must add interface circuits between the analog and digital components. The interface circuits increase the part count and can cause a circuit to reach the part limitation sooner than if the circuit contained only digital parts. These limitations are fairly generous and allow for the simulation of fairly large circuits.

## 8.A. Digital Signal Sources

Although a digital circuit could have only analog inputs, digital signal sources are available to provide a digital signal to the digital components. If an analog voltage source is used to provide a signal to a digital gate, PSpice will insert analog-to-digital conversion circuits between the source and the gate. These conversion circuits could result in longer simulation times. For example, you could use the analog source Vpulse to generate a 0 to 5 volt square wave for use as a clock. This would require more circuit elements than if you used the digital sources described below.

## 8.A.1. Digital Signal

The "Digital_signal" source allows us to set a single wire to any bit sequence in time. To get this part, type **P**, and then enter the name **`Digital_signal`**:

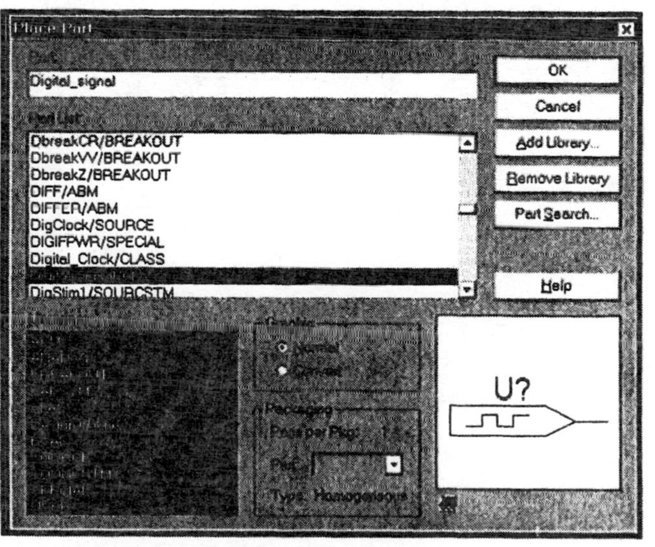


Click the ***OK*** button. The graphic for the ***Digital_signal*** part, , will become attached to the mouse pointer. Place the part in your schematic by clicking the ***LEFT*** mouse button. To stop placing parts, press the **ESC** key.

Double-click the ***LEFT*** mouse button on the ***Digital_signal*** graphic, , to obtain the property spreadsheet for the part:

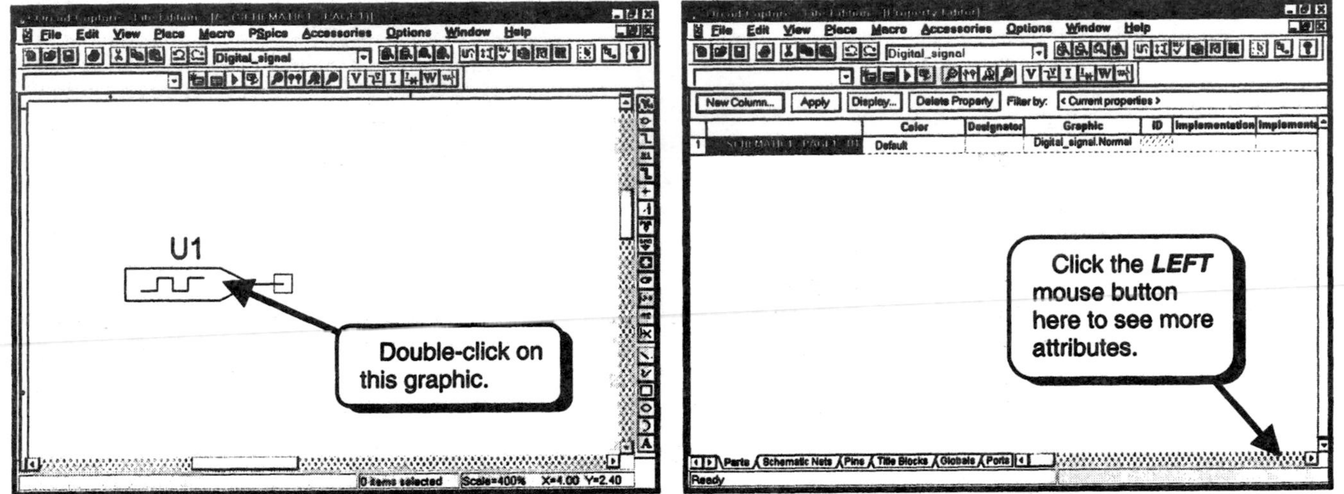


The properties for specifying the bit sequence of this source are not displayed. Scroll the spreadsheet to the right until you see the properties ***LINE1*** to ***LINE13***.

Not all of the lines must contain data, but they must be filled in sequentially, ***LINE1***, ***LINE2***, then ***LINE3***, and so on. At a minimum, you must define the attribute ***LINE1***. These lines may be defined in several different ways, but only three will be presented here.

## ABSOLUTE TIME

Table 8-1 and the screen capture below display the properties of the digital source using absolute time:

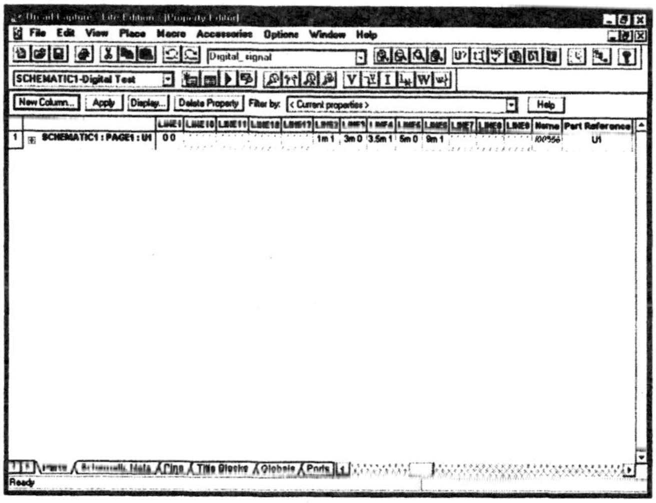

**Table 8-1**

| Property | Value |
|---|---|
| LINE1 | 0 0 |
| LINE2 | 1m 1 |
| LINE3 | 3m 0 |
| LINE4 | 3.5m 1 |
| LINE5 | 5m 0 |
| LINE6 | 9m 1 |

These attributes define the source as the following bit stream: At time zero the source has a value of logic zero. The source remains zero until time equals 1 ms. At 1 ms, the source changes to a logical one. The source remains a one until time equals 3 ms. At 3 ms, the source changes to a zero. It remains at zero until time equals 3.5 ms. At 3.5 ms, it switches back to a one. It remains a one until time equals 5 ms, when it changes back to a zero. The source remains at zero until time equals 9 ms, when it changes back to a logical one. For time greater than 9 ms, the source will remain at logical one.

From this example we see that ***LINE1*** through ***LINE13*** consist of time and transition pairs. The output of the source remains constant until time reaches one of the times specified in the attributes. At that time, the source makes the transition from its present value to the value specified by the attribute. You need to define only as many attributes as necessary. In this example only six of the thirteen "Line" attributes were needed. The waveform this source describes is:

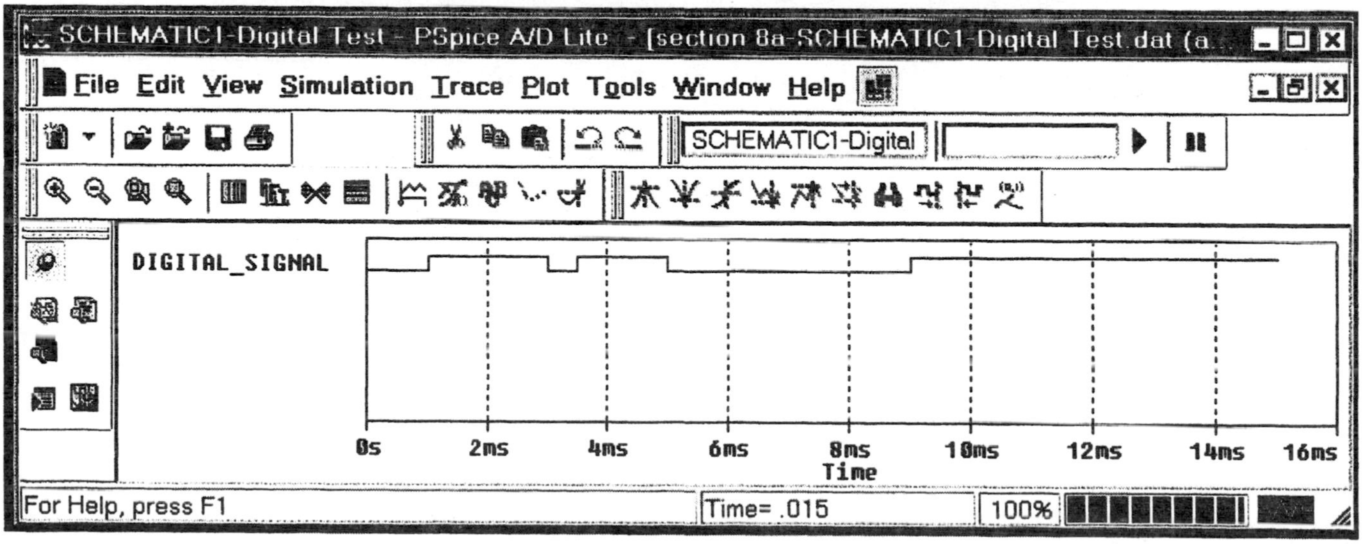

## RELATIVE TIME

We will now define a source that produces the same waveform as above, but we will use relative timing. Relative timing is similar to absolute time, except that the time-transition pairs specify the amount of time from the last transition. The attributes in Table 8-2 specify the waveform specified previously using relative time. Note the plus (+) signs before the time values. The attributes are displayed in the screen capture below and in Table 8-2:

Table 8-2

| Property | Value |
|---|---|
| LINE1 | 0 0 |
| LINE2 | +1m 1 |
| LINE3 | +2m 0 |
| LINE4 | +0.5m 1 |
| LINE5 | +1.5m 0 |
| LINE6 | +4m 1 |

At time equals zero the source is a logic zero. It remains at zero for 1 ms and then changes to a one. It remains a one for 2 ms and then changes to a zero. It remains a zero for 0.5 ms and then changes to a one. It remains a one for 1.5 ms and then changes to a zero. It remains a zero for 4 ms and then changes to a one. It remains at one for the remainder of the simulation.

**REPEATED LOOPS**

Since many digital waveforms are periodic, a GOTO statement is provided for looping. The attributes in Table 8-3 and shown in the screen capture produce a clock frequency of 1 kHz for four cycles and then 500 Hz for an infinite number of cycles:

Table 8-3

| Property | Value |
|---|---|
| LINE1 | 0 0 |
| LINE2 | LABEL=LOOP1 |
| LINE3 | +0.5m 1 |
| LINE4 | +0.5m 0 |
| LINE5 | +0.5m GOTO LOOP1 3 TIMES |
| LINE6 | +0.5m 1 |
| LINE7 | LABEL=LOOP2 |
| LINE8 | +1m 0 |
| LINE9 | +1m 1 |
| LINE10 | +1m GOTO LOOP2 –1 TIMES |

Note that not all of the attributes are shown in the screen capture above. The line ***GOTO LOOP1 3 TIMES*** means execute the ***GOTO*** statement 3 times. The line ***GOTO LOOP2 -1 TIMES*** means execute the ***GOTO*** statement an infinite number of times. These attributes describe the waveform:

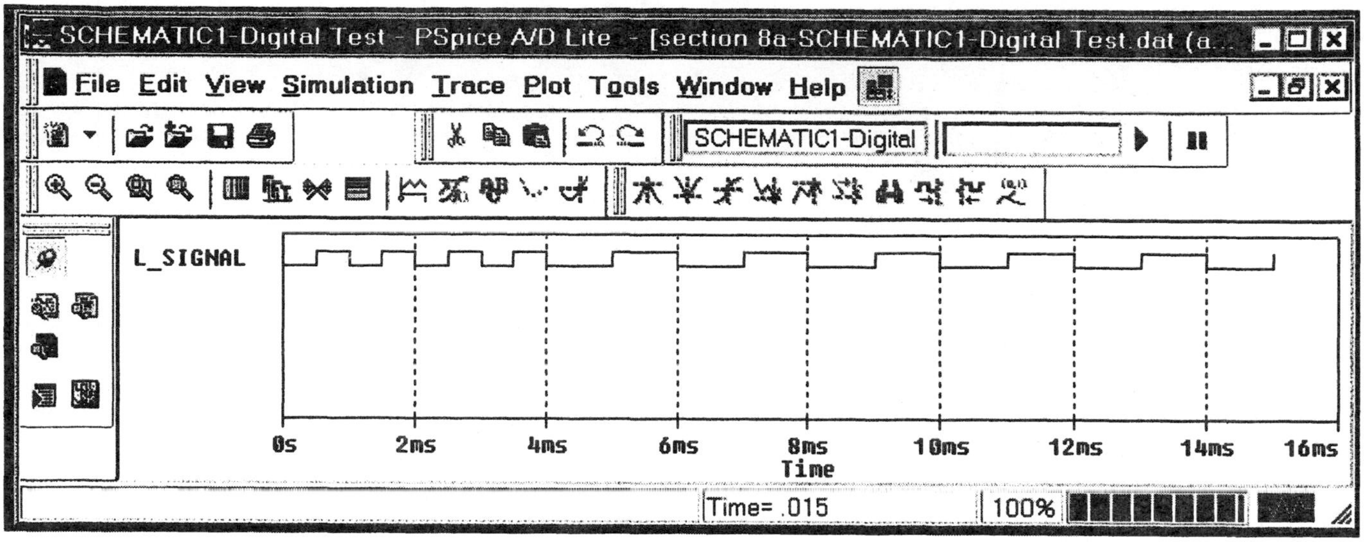

## 8.A.2. Digital Clock

Since most digital systems use a clock with a constant frequency, a special part called "Digital_Clock" is provided to generate clock waveforms. This part is nothing more than a special case of the "Digital_signal" part using the GOTO statement. To get this part, type **P** and then enter the name **`Digital_Clock`**:

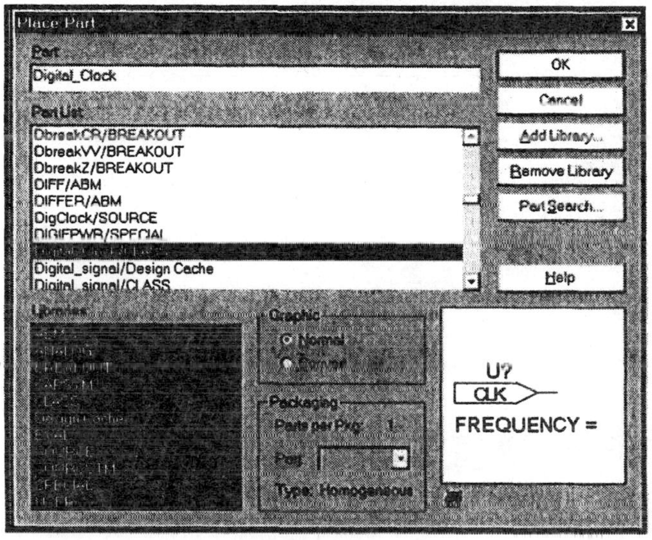

Click the ***OK*** button. The mouse pointer will become attached to the graphic for the ***Digital_Clock*** part. Place the part in your schematic by clicking the ***LEFT*** mouse button. To stop placing parts, press the **ESC** key. Notice that the frequency is indicated on the schematic:

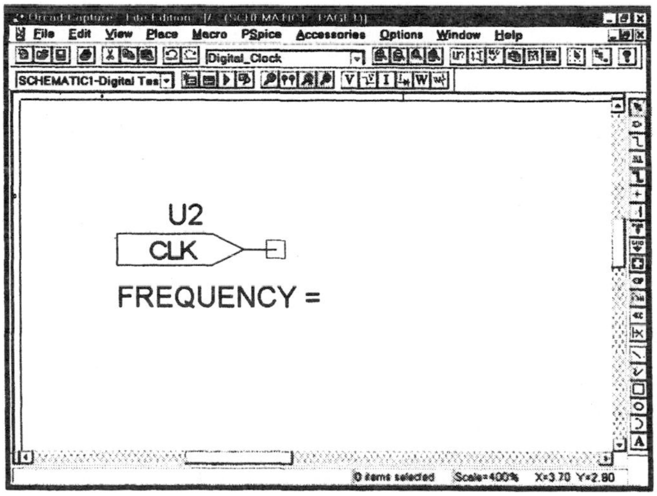

To set the frequency, double-click on the text ***FREQUENCY=***

Type in the value for the frequency and then click the ***OK*** button:

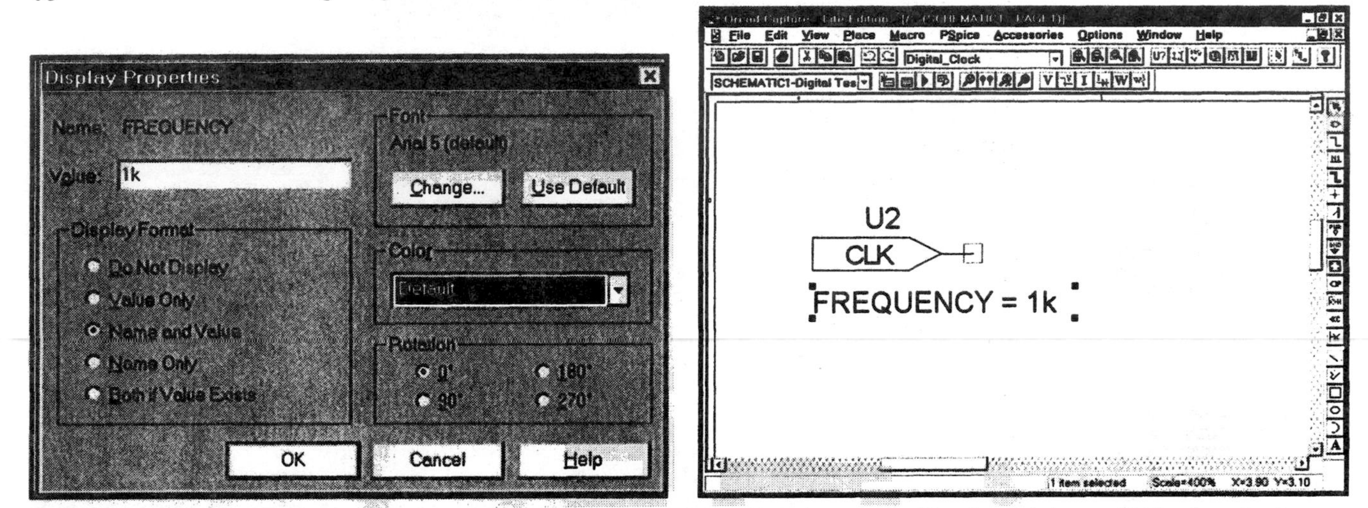


The Digital_Clock part will produce the specified frequency with a 50% duty cycle (the time high equals the time low).

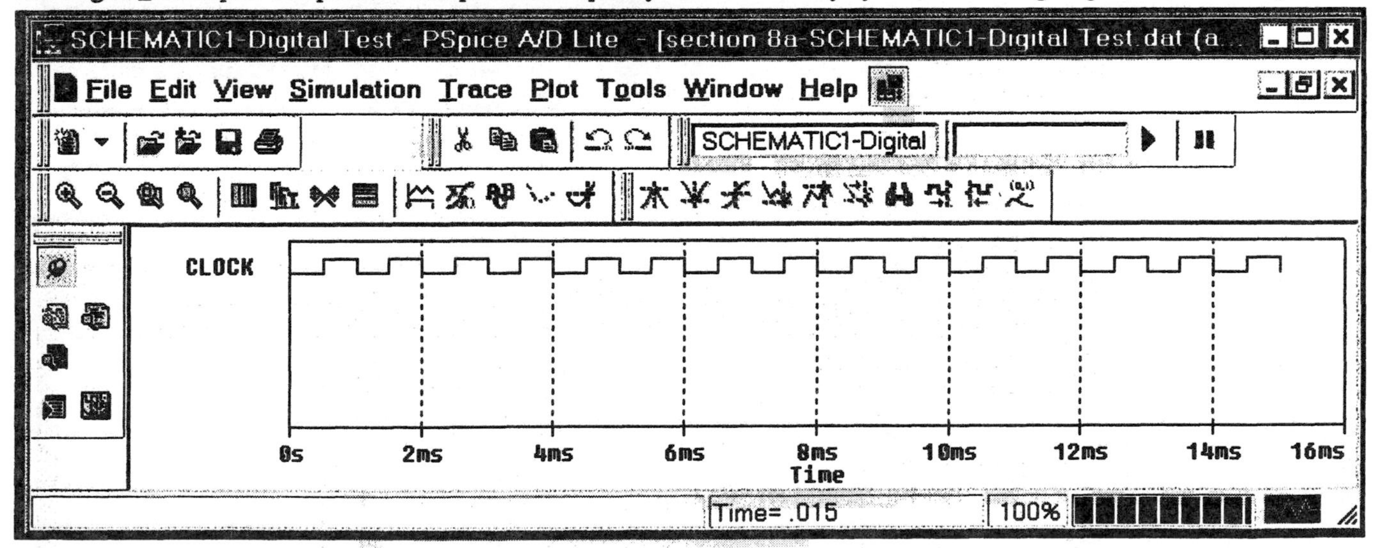


## 8.A.3. Digital Stimulus Part (DigStim)

The Stimulus Editor is a tool that allows us to create and view a signal source before we run a simulation. It can be used with analog voltage and current sources, and digital sources. Here, we will demonstrate its use with a digital signal. The Stimulus Editor is a very useful tool for creating waveforms if you are not familiar with the various sources available with PSpice. Unfortunately, for digital signals the Lite version is limited to clock signals. To get this part, type **P** and then enter the name **DigStim1**. This is a 1-bit digital signal. Parts DigStim2, 4, 8, 16, and 32 provide digital signals with 2, 4, 8, 16, and 32-bit signals:

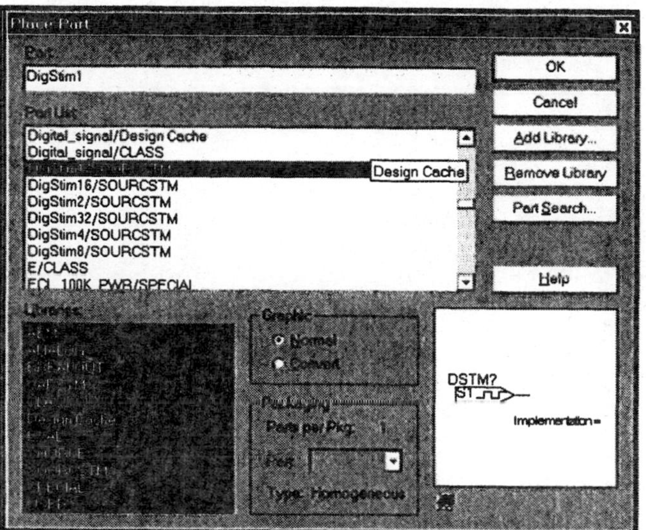


Place the DigStim1 part in your schematic:

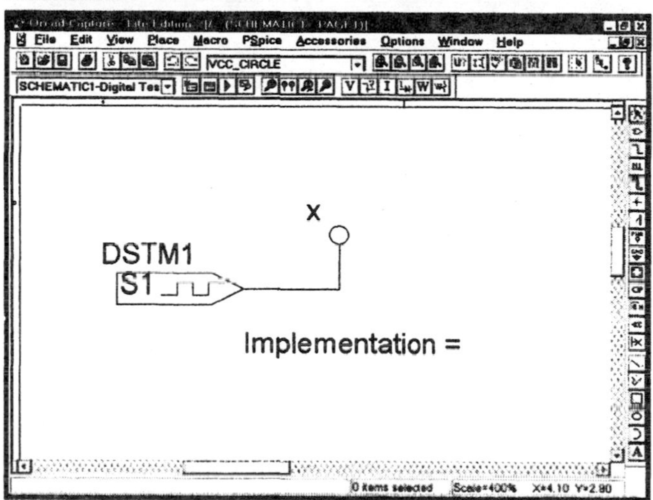


Notice that the graphic has the attribute ***Implementation=***. When you create a waveform with the Stimulus Editor, you will give the waveform a name. The name specified with the Stimulus Editor will be specified in the schematic using the ***Implementation=*** line. A part named VCC_Circle was added to the circuit with the label ***X*** so that we can view the waveform after we run the simulation.

To create the waveform for the part, click the ***LEFT*** mouse button on the DigStim1 graphic, S1 to select the part and then select **Edit** and **PSpice Stimulus** from the menus:

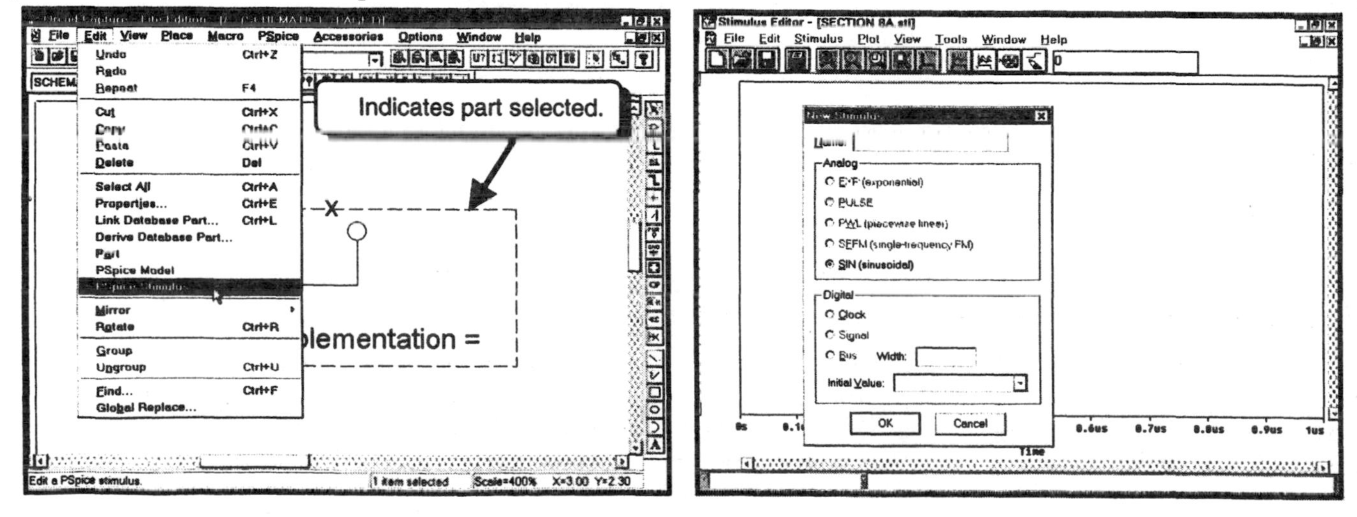

We must now specify a name for the stimulus and select the waveform. For the Lite version only a sinusoid is available for analog sources and only a 1-bit clock waveform is available for digital sources. Name the source **CLK** and specify a ***Digital Clock*** as the waveform:

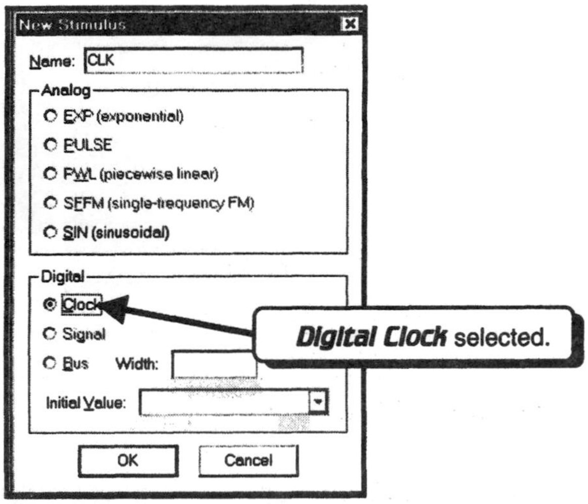

New Stimulus
Name: CLK
Analog
EXP (exponential)
PULSE
PWL (piecewise linear)
SFFM (single-frequency FM)
SIN (sinusoidal)
Digital
Clock
Signal
Bus Width:
Initial Value:
OK Cancel

***Digital Clock*** selected.

Click the ***OK*** button:

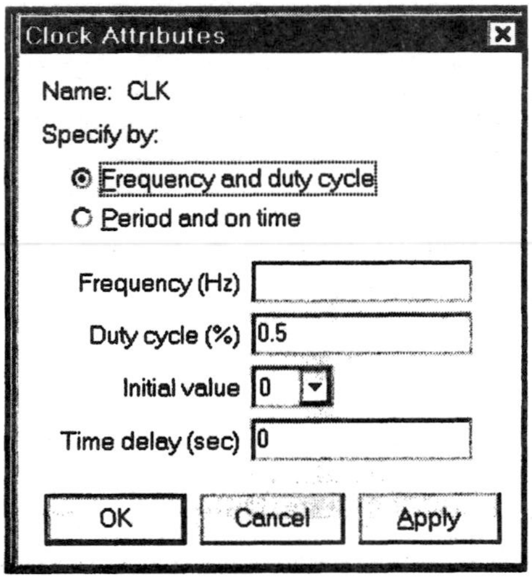

Clock Attributes
Name: CLK
Specify by:
Frequency and duty cycle
Period and on time
Frequency (Hz)
Duty cycle (%) 0.5
Initial value 0
Time delay (sec) 0
OK Cancel Apply

The ***Clock Attributes*** dialog box allows us to change the properties of the waveform. For this type of waveform we can specify the clock in terms of a frequency or a period. When we specify the attributes in terms of ***Frequency***, we are allowed to specify a ***Duty cycle***, ***Initial value***, and ***Time delay***. The dialog box below specifies a 1 kHz clock with a 50% duty cycle:

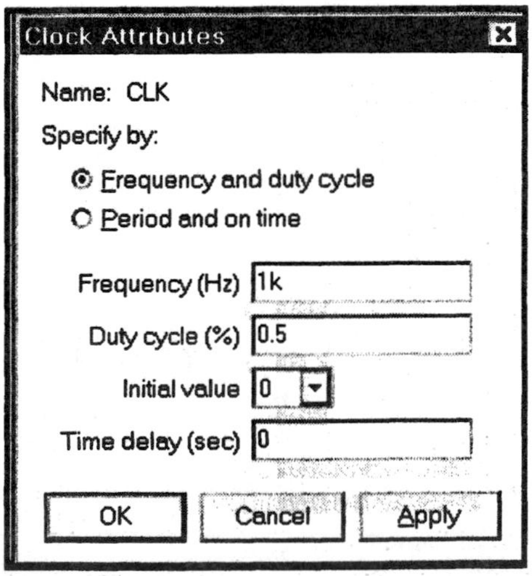

Clock Attributes
Name: CLK
Specify by:
Frequency and duty cycle
Period and on time
Frequency (Hz) 1k
Duty cycle (%) 0.5
Initial value 0
Time delay (sec) 0
OK Cancel Apply

The dialog box below shows settings that specify the same waveform using a ***Period*** and an ***On time***:

Clock Attributes

Name: CLK

Specify by:

Frequency and duty cycle

Period and on time

Period (sec) 1ms

On time (sec) 0.5m

Initial value 0

Time delay (sec) 0

OK Cancel Apply

To view the waveform with the Stimulus Editor, click the ***OK*** button:

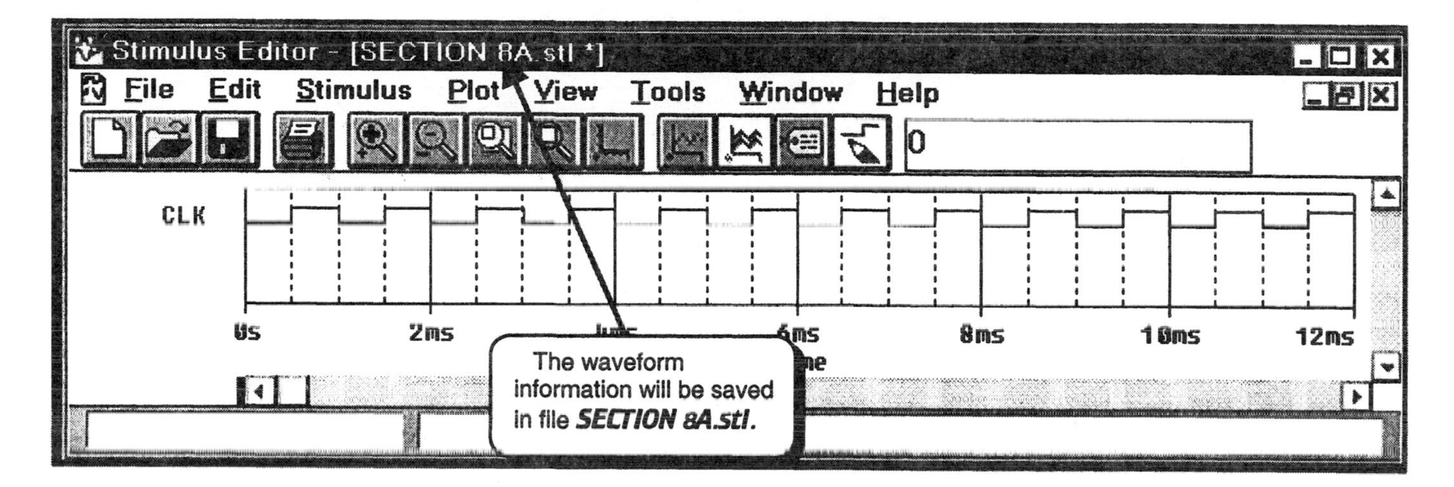


The waveform information will be saved in file ***SECTION 8A.stl***. If you copy this circuit to a floppy, make sure you copy the files sec_8a3.opj and ***SECTION 8A.stl***. If you do not copy the file ***SECTION 8A.stl***, you will lose the waveform information.

To apply the waveform to the S1 part in the schematic and return to the schematic, select **File** and then **Exit** from the Stimulus Editor menu bar:

Stimulus Editor

Save changes to "D:\ORCAD PSMAN\CIRCUIT FILES\SECTION 8A.stl"?

Yes No Cancel

Click the ***Yes*** button to save the changes:

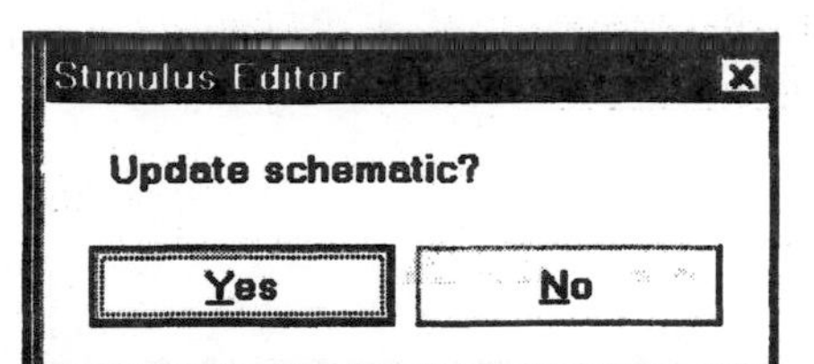


Click the ***Yes*** button to update the changes to the schematic:

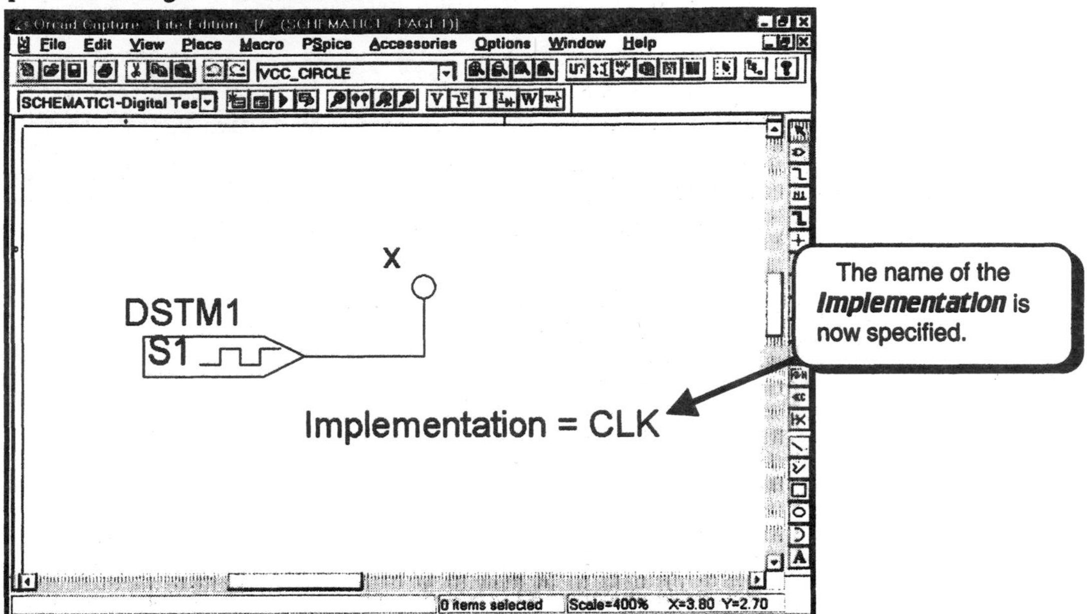


Notice that the name of the stimulus is now specified in the schematic, ***Implementation = CLK***. If you run a Transient Analysis, the waveform at node ***X*** will have the waveform created with the Stimulus Editor.

As a second example, we will create a two-phase non-overlapping clock. Add two more DigStim1 parts to the circuit as shown:

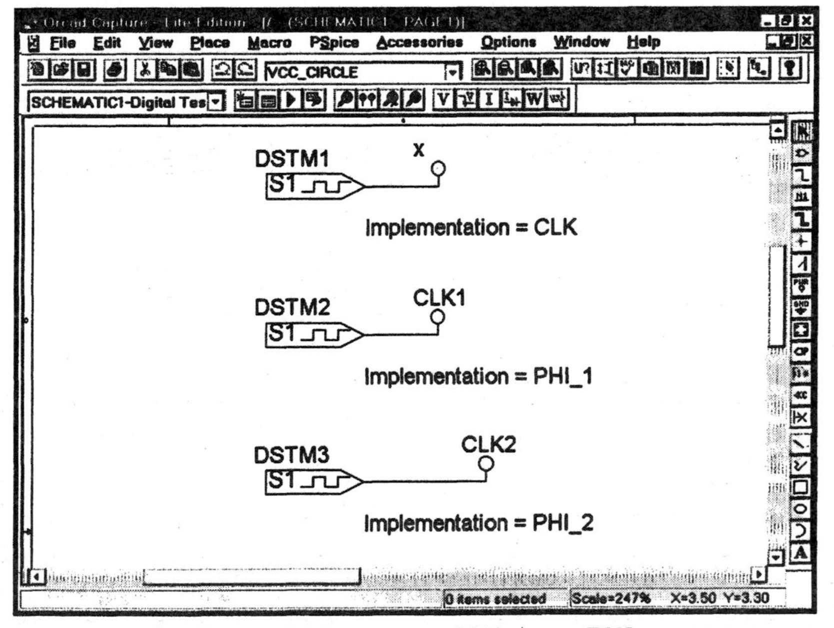


Notice that we have labeled the two ***Implementation*** signals ***PHI_1*** and ***PHI_2***. Click the ***LEFT*** mouse button on the graphic for ***DSTM2***, S1 to select the graphic and then select **Edit** and then **PSpice Stimulus** from the menu. The Stimulus Editor will run and bring up the dialog box for waveform ***PHI_1***:

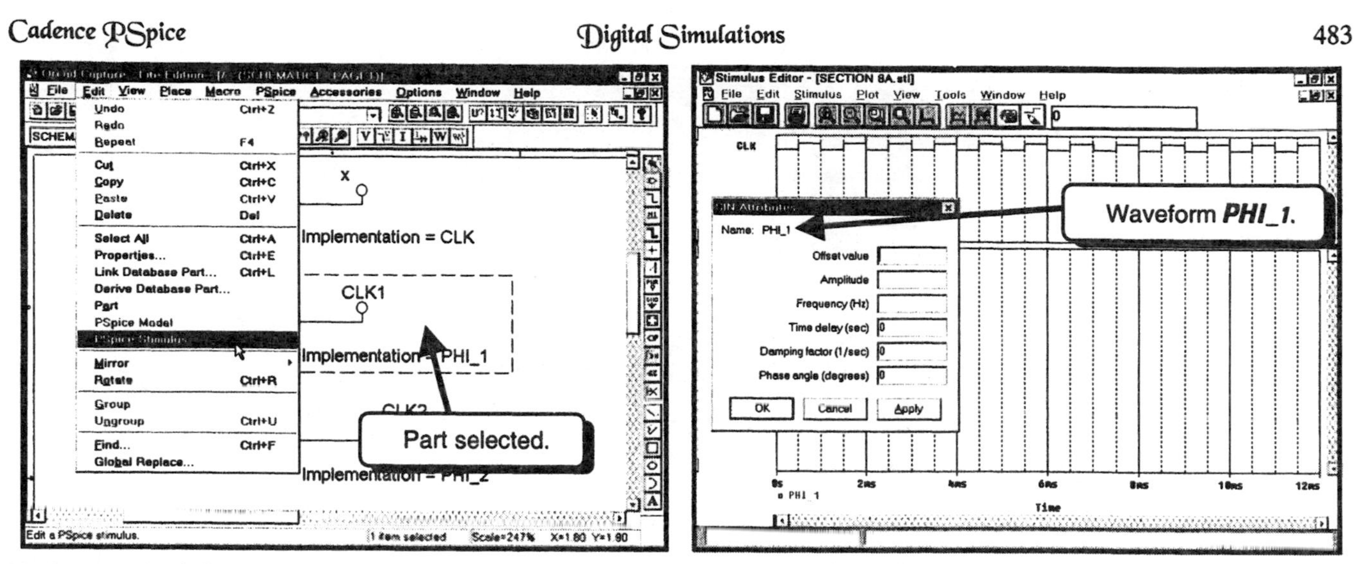


Notice that the default waveform is for a sine wave. To change it, click the ***Cancel*** button and then select **Stimulus** and then **New** from the menus:

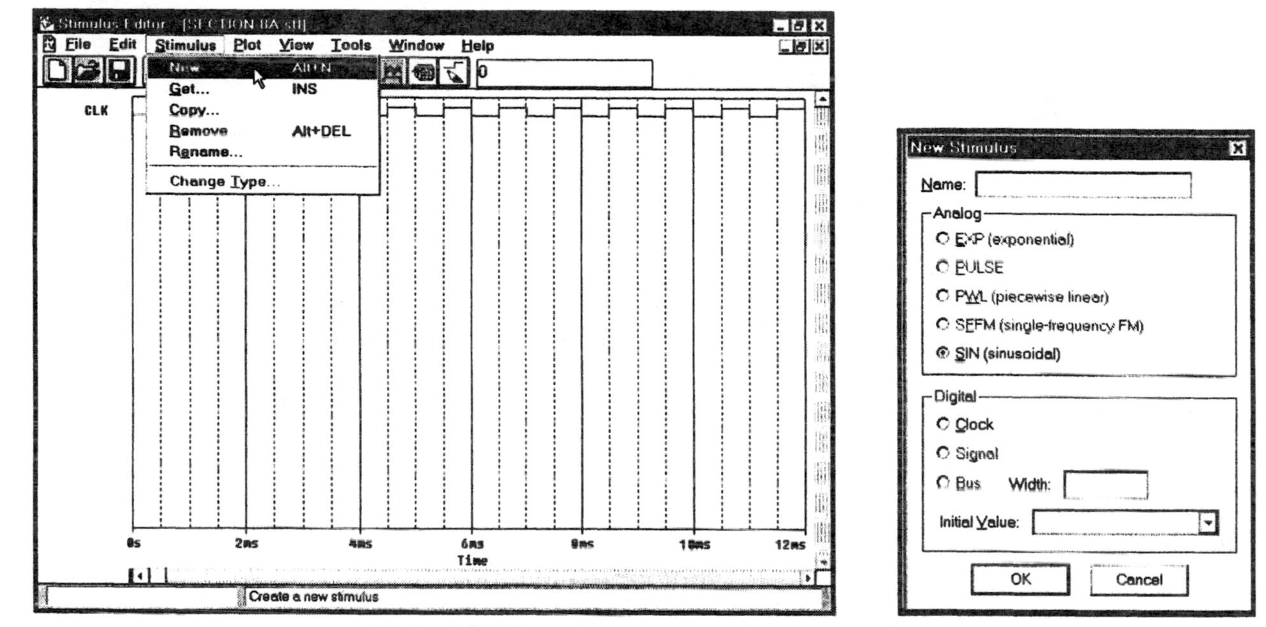


Specify the name as `PHI_1`,select a ***Digital Clock*** waveform, and then click the ***OK*** button:

We will create a 1 kHz clock with a duty cycle of 25%. Fill in the dialog box as shown:

Clock Attributes

Name: PHI_1

Specify by:

Frequency and duty cycle

Period and on time

Frequency (Hz) 1k

Duty cycle (%) 25

Initial value 0

Time delay (sec) 0

OK Cancel Apply

Click the ***OK*** button to view the waveform:

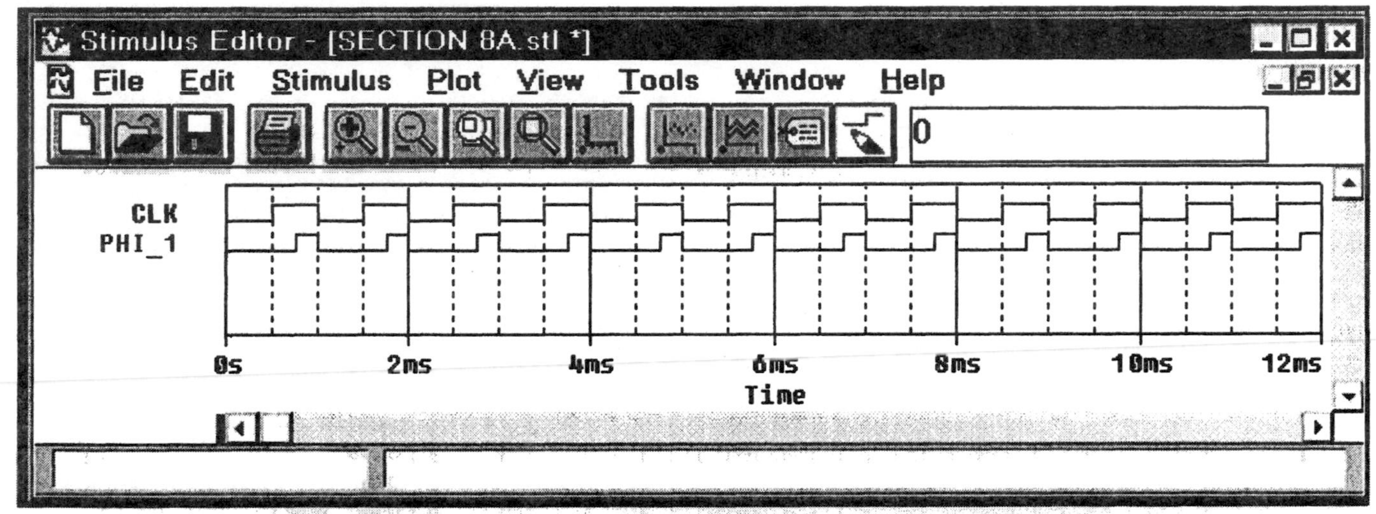


Next we need to create the waveform for PHI_2. Select **Stimulus** and then **New** from the Stimulus Editor menu bar:

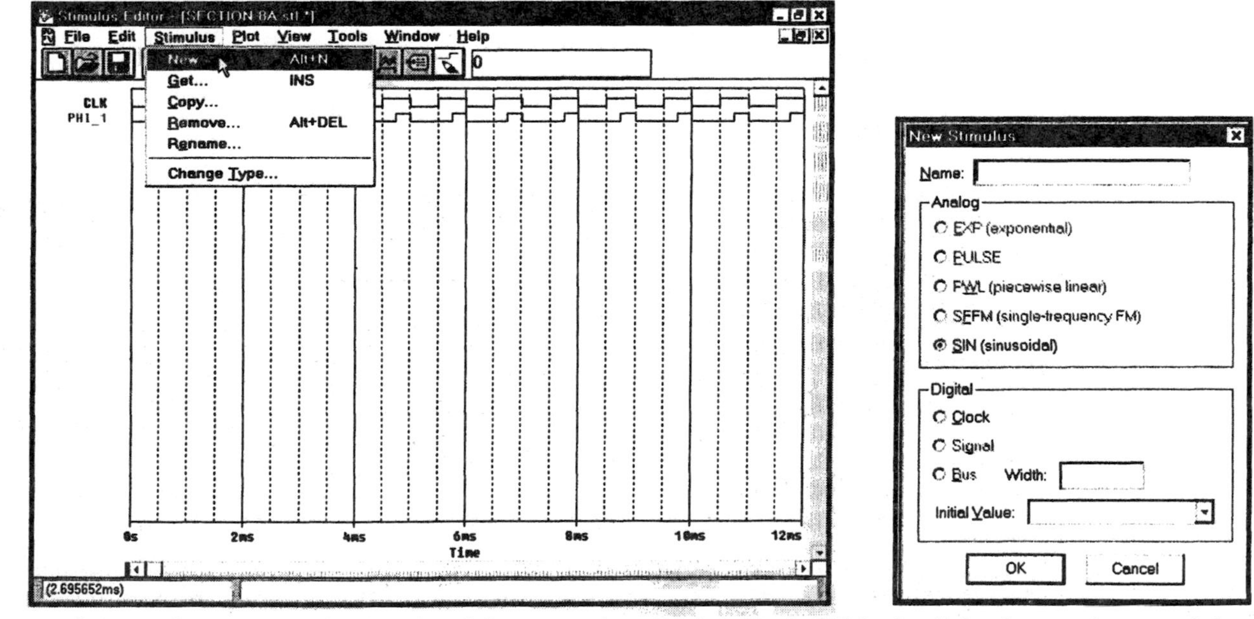


We want the waveform name to be PHI_2 and the type to be a Digital Clock. Fill in the dialog box as shown and then click the ***OK*** button:

New Stimulus

Name: PHI_2

Analog
- EXP (exponential)
- PULSE
- PWL (piecewise linear)
- SFFM (single-frequency FM)
- SIN (sinusoidal)

Digital
- Clock
- Signal
- Bus Width:

Initial Value:

OK Cancel

Clock Attributes

Name: PHI_2

Specify by:
- Frequency and duty cycle
- Period and on time

Frequency (Hz)

Duty cycle (%) 0.5

Initial value 0

Time delay (sec) 0

OK Cancel Apply

We want ***PHI_2*** to have the same frequency and duty cycle as PHI_1, but the pulse should be delayed by 0.25 ms. Fill in the dialog box as shown:

Clock Attributes

Name: PHI_2

Specify by:
- Frequency and duty cycle
- Period and on time

Frequency (Hz) 1k

Duty cycle (%) 25

Initial value 0

Time delay (sec) 0.25m

OK Cancel Apply

Click the ***OK*** button to view the waveforms:

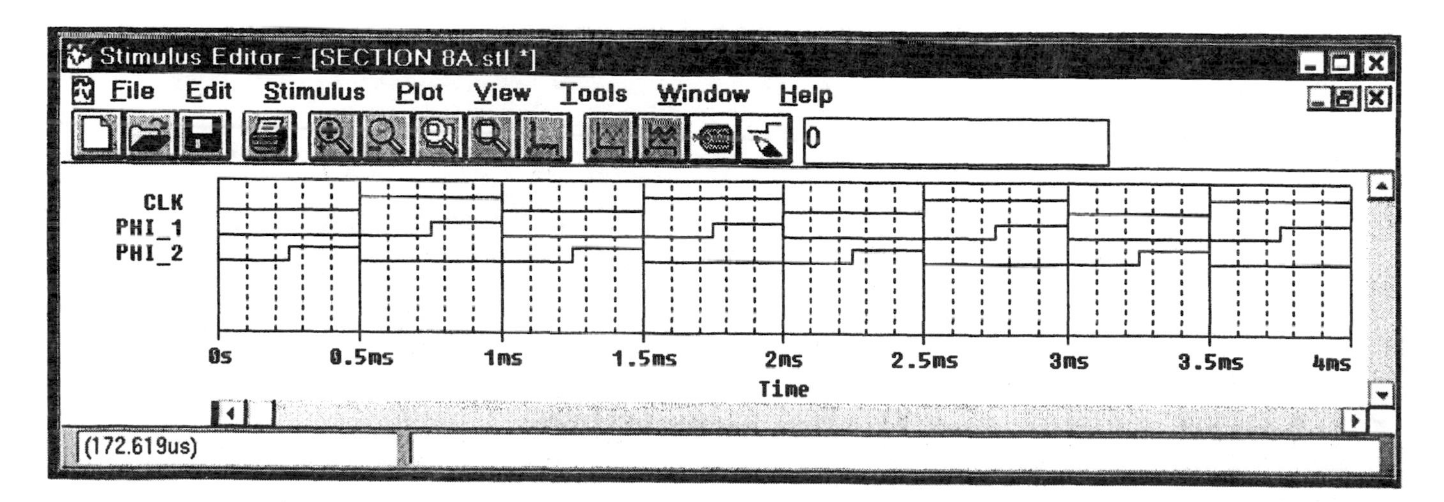


To save the waveform information and return to the schematic, select **File** and then **Exit** from the Stimulus Editor menu bar:

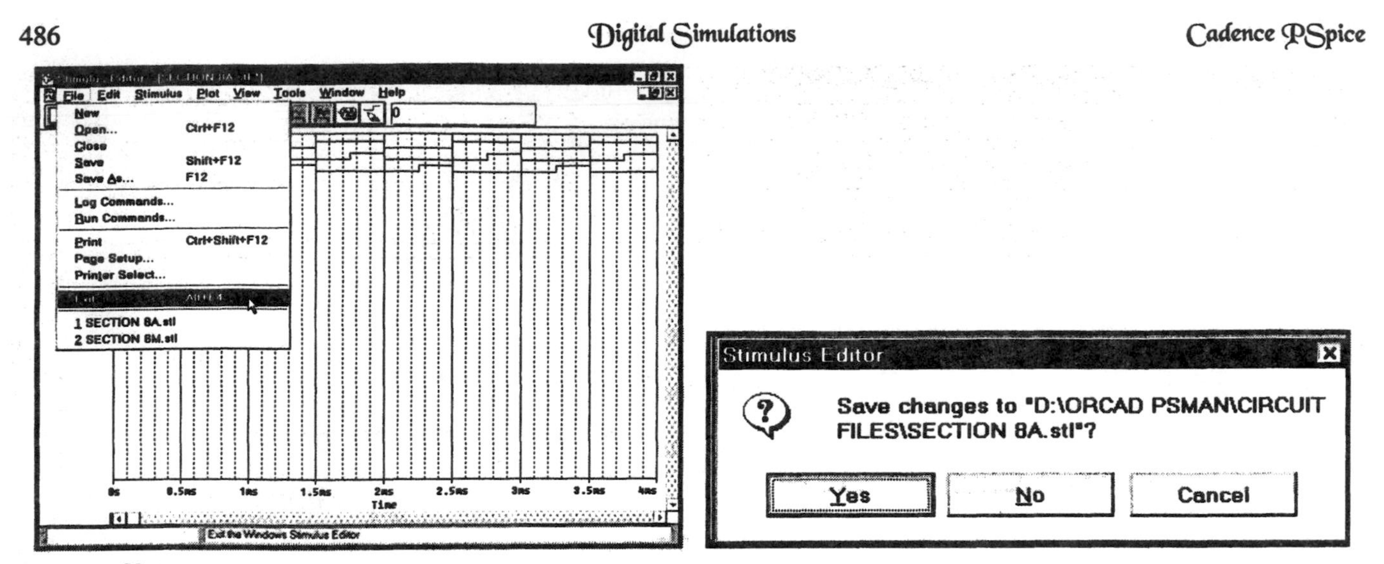


Click the ***Yes*** button to save the changes.

Stimulus Editor

Update schematic?

Yes No

Click the ***Yes*** button to update the schematic:

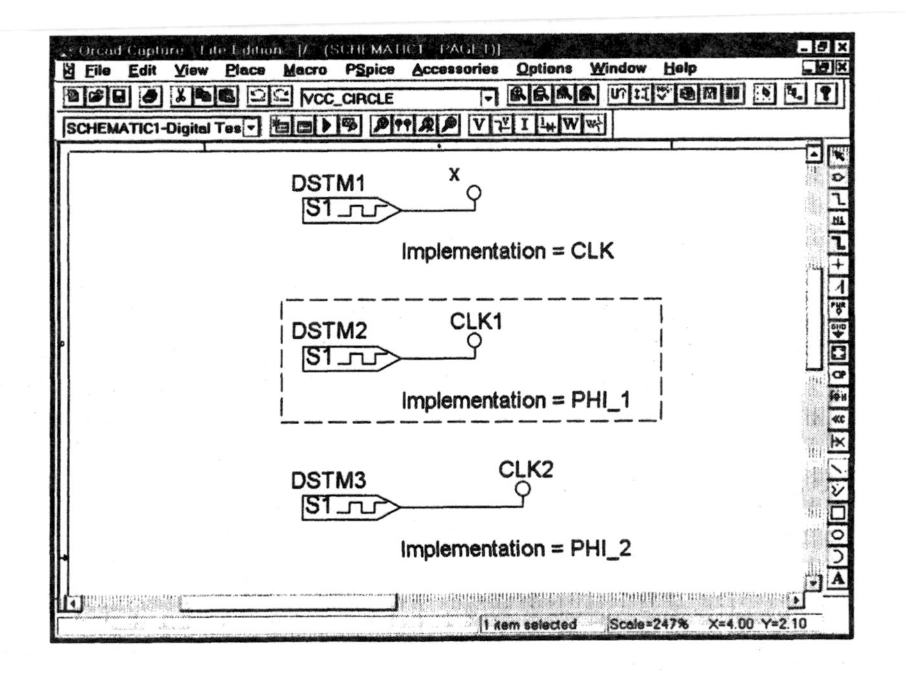


If you run a Transient Analysis, the waveforms at ***X***, ***CLK1***, and ***CLK2*** will be the same as those seen on the Stimulus Editor window.

# 8.B. Mixed Analog and Digital Simulations

The first circuit we will look at is an op-amp circuit that drives the clock of a J-K flip-flop. Wire the circuit shown below:

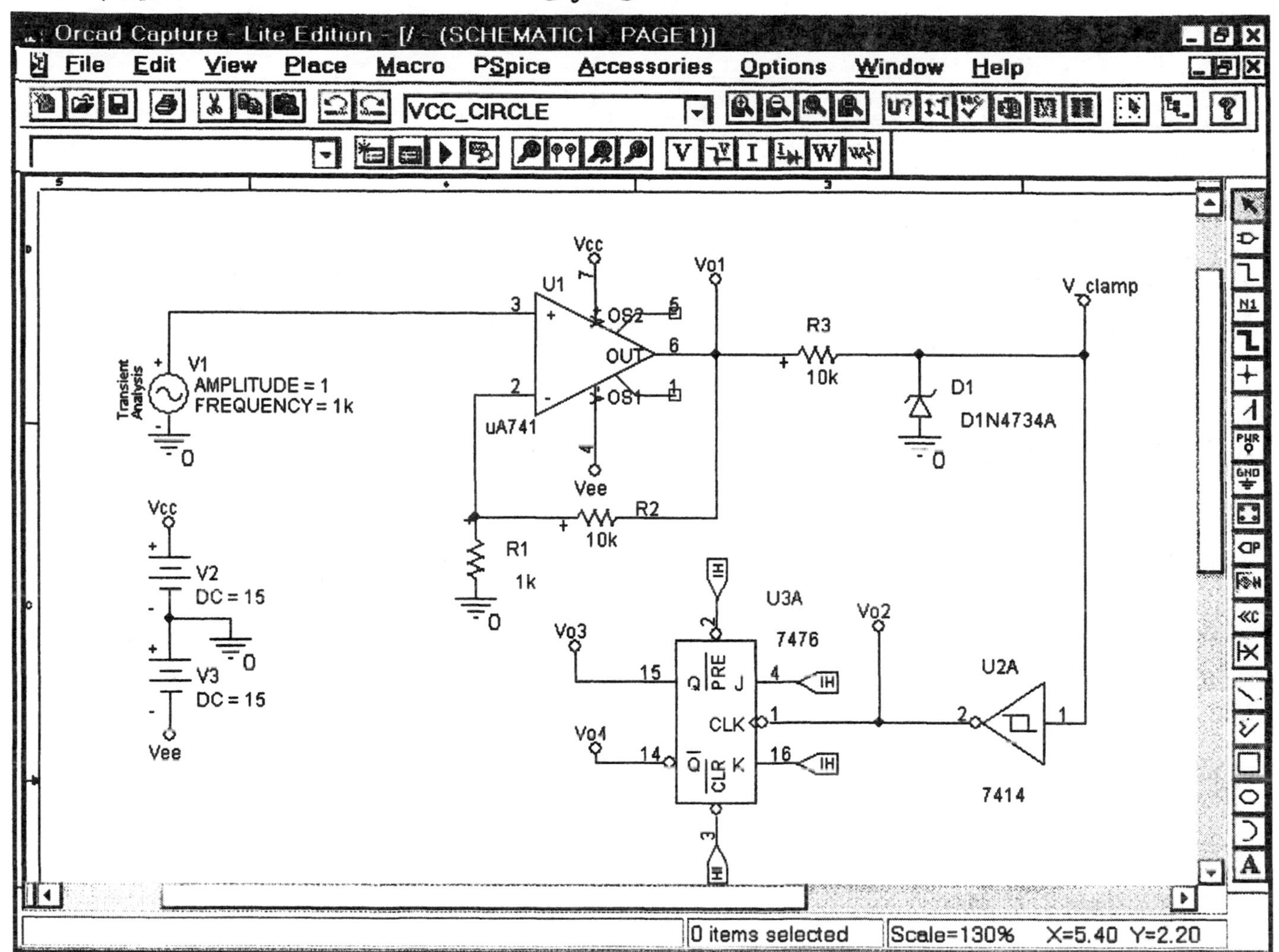


| Part | Description |
|---|---|
| R | Resistor |
| d1n4734A | 5.6 volt Zener |
| Vcc_Circle | Node label |
| 0 | Ground |
| UA741 | Operational amplifier |
| 7476 | J-K flip-flop |
| $D_HI | Constant logic 1 – select **Place Power** from the menus to place this part. |
| 7414 | Inverting Schmitt Trigger |
| VSIN | Sinusoidal voltage source |

The circuit is drawn as though we were going to wire the circuit in the lab. No special circuits are required between the analog circuitry and the digital logic gates. Note that the J and K inputs of the flip-flop are held high so that the flip-flop toggles at each negative clock edge.

The sinusoidal voltage waveform produces a 1 kHz sine wave with voltages between –1 and 1 volts. The ***UA741*** op-amp circuit has a gain of 11 and produces a ±11 volt sine wave of 1 kHz at node ***Vo1***. This waveform goes into a Zener clipping circuit that limits the voltage to approximately +5.6 and –0.7 volts at node ***V_clamp***. This voltage is TTL compatible and can be connected to the Schmitt Trigger input. The output of the Schmitt Trigger should be a 0 to 5 volt square wave at 1 kHz. The J-K flip-flop is wired as a divide-by-two counter, so Q and $\overline{Q}$ should be 0 to 5 volt square waves at 500 Hz. They should also be 180 degrees out of phase.

Since the frequency of the pulse generator is 1 kHz, we will run a Transient simulation for 15 ms to allow 15 cycles. To run a digital simulation we must set up the Transient Analysis, as well as specify parameters for the digital simulation. We will first set up the Transient Analysis. Select **PSpice**, and then **New Simulation Profile** from the Capture menus:

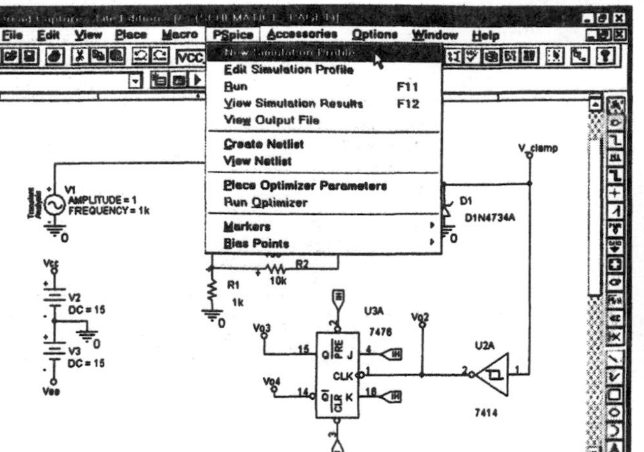


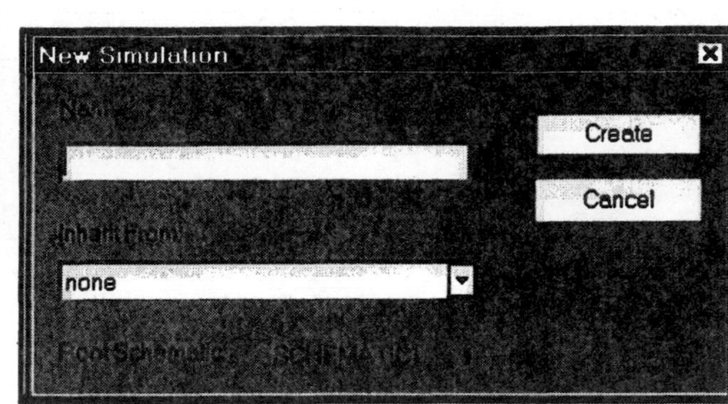


Specify a name for the profile and click the ***Create*** button:

New Simulation
Name:
Digital Simulation Example
Create
Cancel
Inherit From:
none
Root Schematic: SCHEMATIC1

General | Analysis | Include Files | Libraries | Stimulus | Options | Data Collection | Probe Window
Analysis type:
Options:
Monte Carlo/Worst Case
Parametric Sweep
Temperature (Sweep)
Save Bias Point
Load Bias Point
Run to time: 1000ns seconds (TSTOP)
Start saving data after: 0 seconds
Transient options
Maximum step size: seconds
Skip the initial transient bias point calculation (SKIPBP)
Output File Options...
OK Cancel Apply Help

Fill in the dialog box as shown to run the simulation for 15 ms:

Simulation Settings - Digital Simulation Example
General | Analysis | Include Files | Libraries | Stimulus | Options | Data Collection | Probe Window
Analysis type:
Time Domain (Transient)
Options:
Monte Carlo/Worst Case
Parametric Sweep
Temperature (Sweep)
Save Bias Point
Load Bias Point
Run to time: 15m seconds (TSTOP)
Start saving data after: 0 seconds
Transient options
Maximum step size: seconds
Skip the initial transient bias point calculation (SKIPBP)
Output File Options...
OK Cancel Apply Help

Next, select the ***Options*** tab:

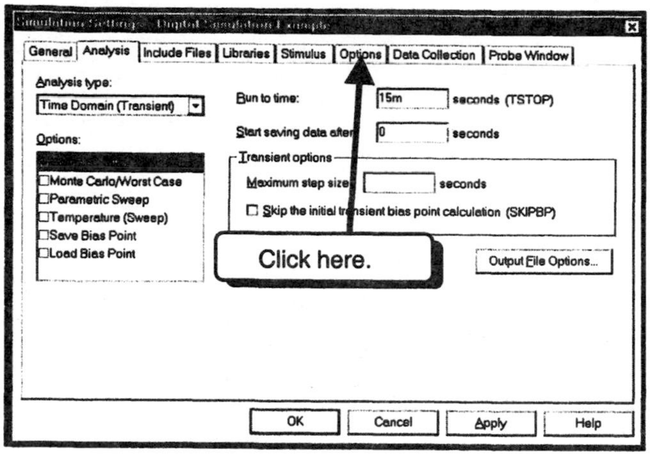

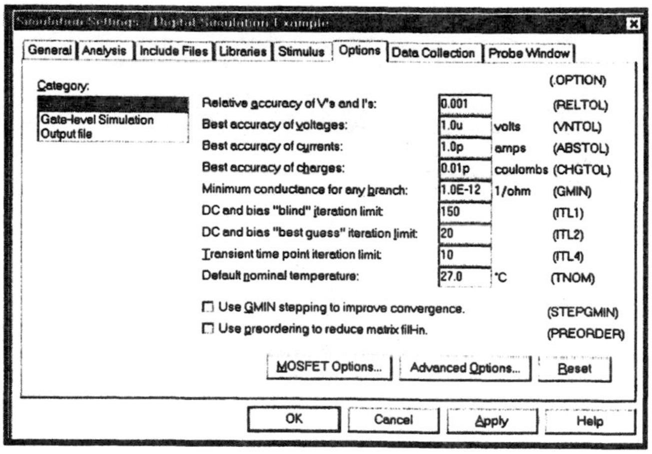

Select the ***Gate-level Simulation*** menu selection:

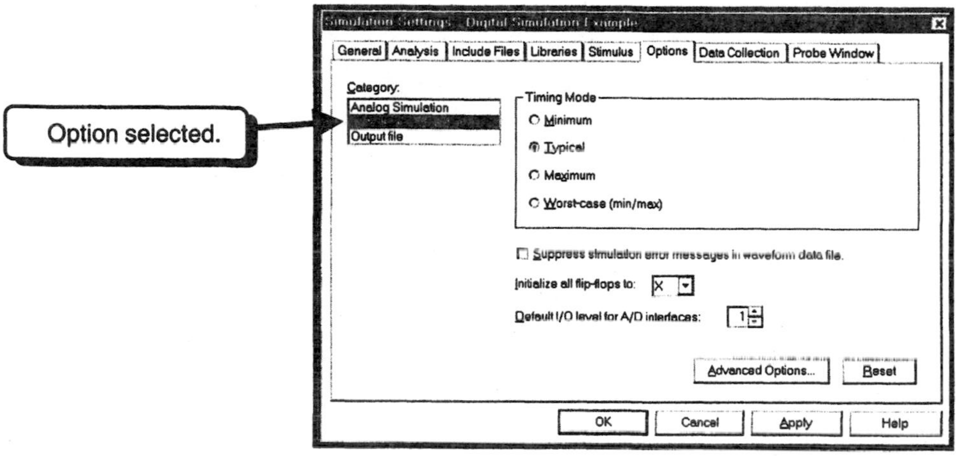

This dialog box allows us to specify timing and initial conditions for the digital circuits. We are interested in the initial state of the flip-flops. The default initial state for all flip-flops is ***X*** (unknown). We wish to set the initial state of all flip-flops to zero. Select an initial state of 0:

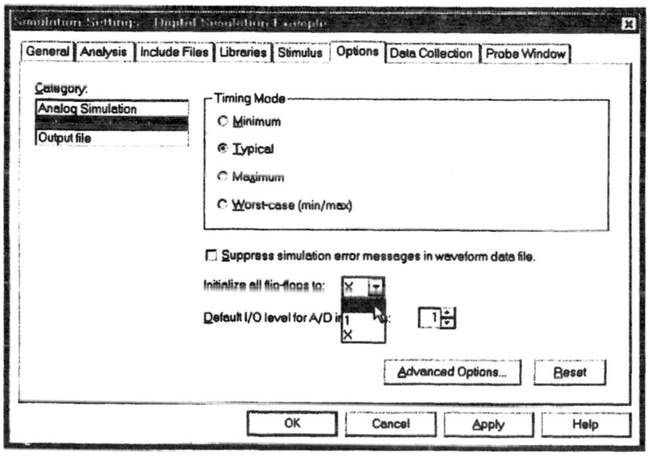

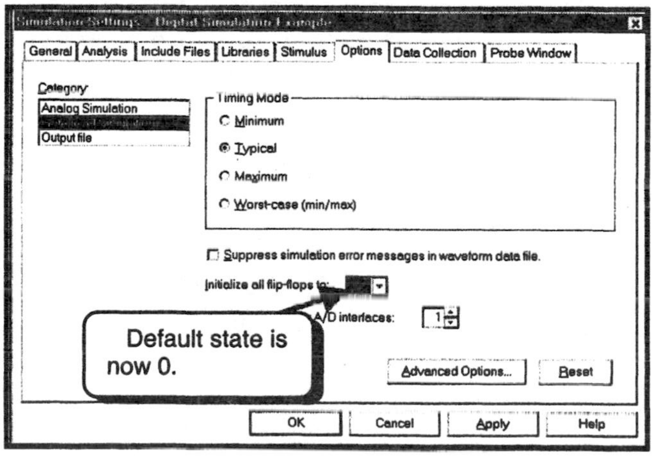

Click the ***OK*** button to return to the schematic.

Run the simulation (**PSpice, Run**). Display the traces ***V(VO1)***, ***V(V_CLAMP)***, ***VO2***, ***VO3***, and ***VO4*** with Probe. To display traces, select **Trace** and then **Add Trace** from the Probe menu bar or press the **INSERT** key:

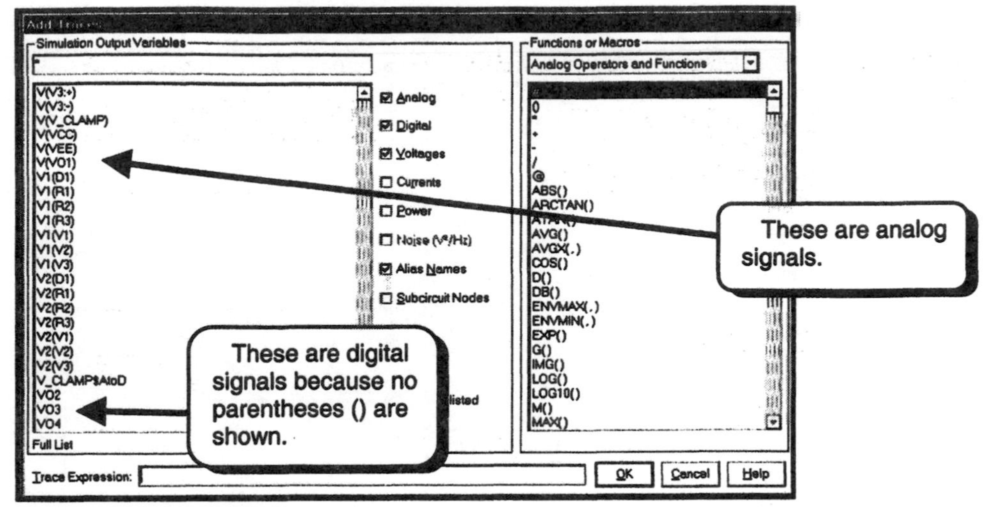


Some of the traces are displayed a little differently in this dialog box. At analog nodes we see that traces are displayed as ***V(VO1)*** or ***V(V_CLAMP)***. The currents through analog components are shown as ***I(D1)*** or ***I(V1)***. The waveforms at digital nodes are shown as ***VO3*** or ***VO4***. This is how Probe allows you to distinguish between digital and analog nodes. Display the traces at all nodes of interest.

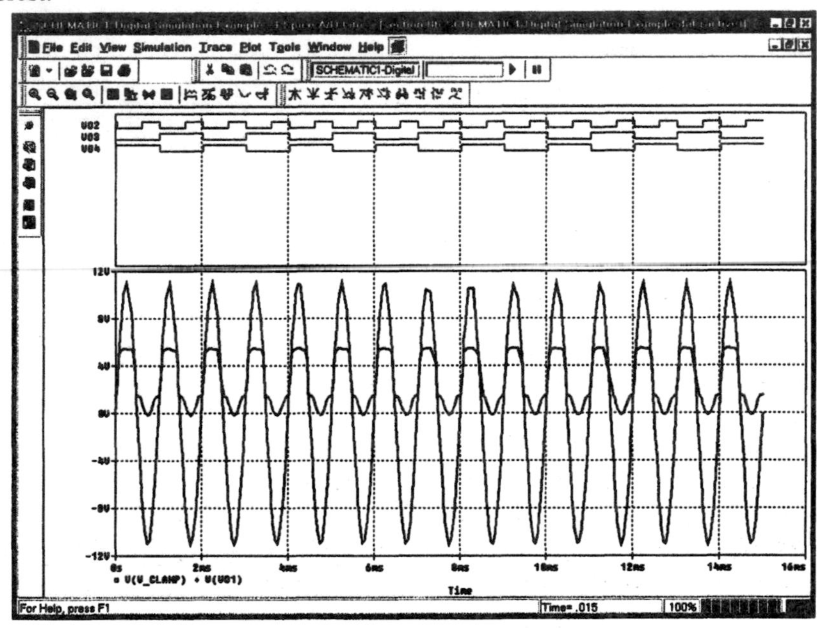


We see that the digital traces are shown on a different graph than the analog traces, but they share the same time axis. These traces show results similar to the expected results.

| R<br>Resistor | D1 d1n4734A<br>**d1n4734A**<br>5.6 volt Zener | **Vcc_Circle**<br>Node label | 0<br>**0**<br>Ground | U4 7414<br>**7414**<br>Inverting Schmitt Trigger | **$D_HI**<br>Constant logic 1 – select **Place Power** from the menus to place this part. | V3 AMPLITUDE = FREQUENCY = Transient Analysis<br>**VSIN**<br>Sinusoidal voltage source |
|---|---|---|---|---|---|---|
| U2 V+ V− ua741<br>**UA741**<br>Operational amplifier | | **7476**<br>J-K flip-flop | | **74161**<br>Binary counter | | **VDC**<br>Independent DC voltage source |

For a second example, we will simulate the circuit below:

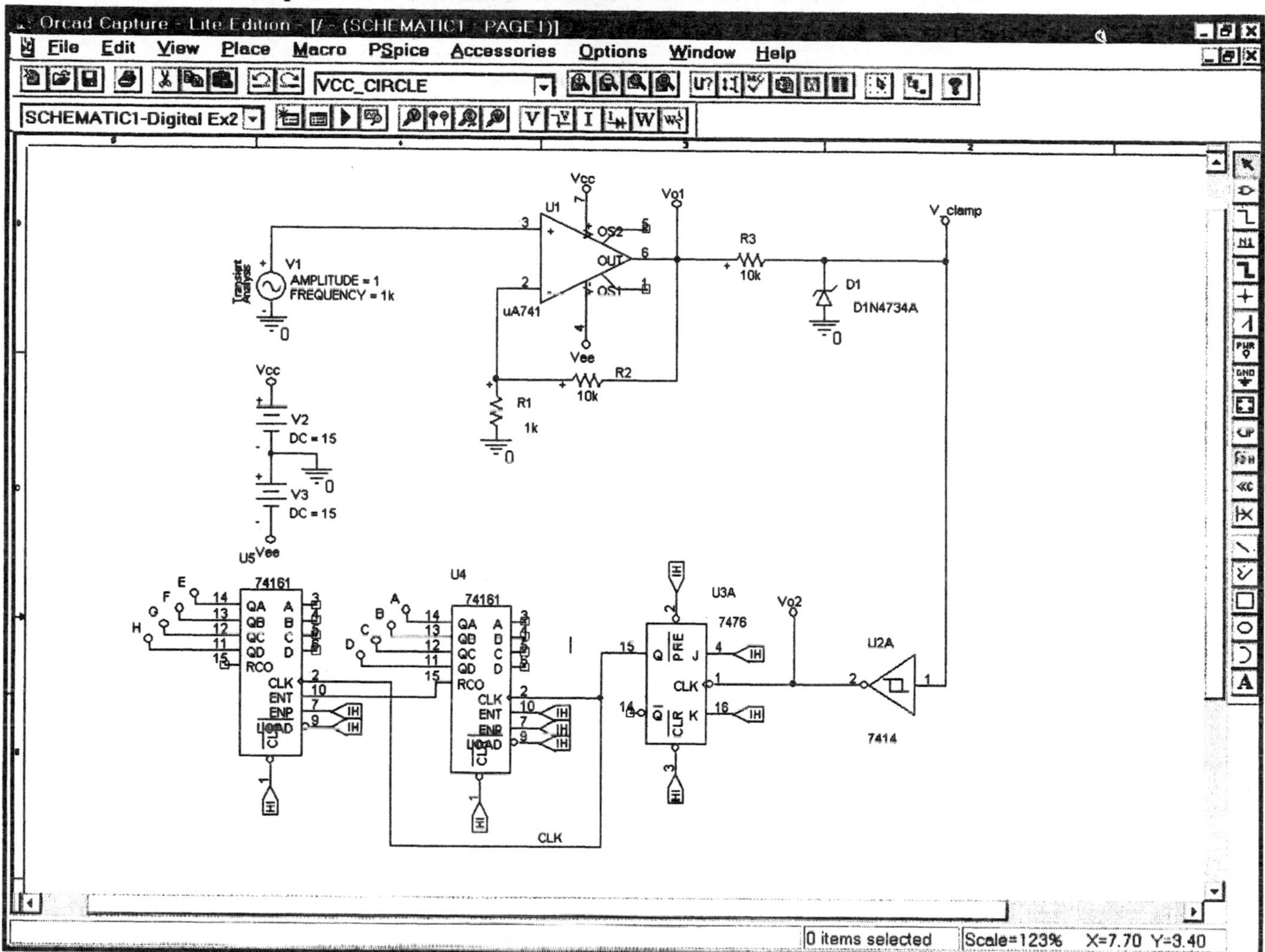


As in the previous example, the flip-flops in the 74161 must be initialized at the start of the simulation. Select the ***Gate-level Simulation Options*** as shown in the right screen capture below. We will run a Transient Analysis for 130 ms. Fill in the ***Time Domain (Transient)*** options as shown in the left screen capture below.

Click the ***OK*** button to return to the schematic. Run the simulation by selecting **PSpice** and then **Run** from the menus. The results of the simulation are displayed with Probe. To add a trace, select **Trace** and then **Add Trace** from the Probe menus or press the **INSERT** key.

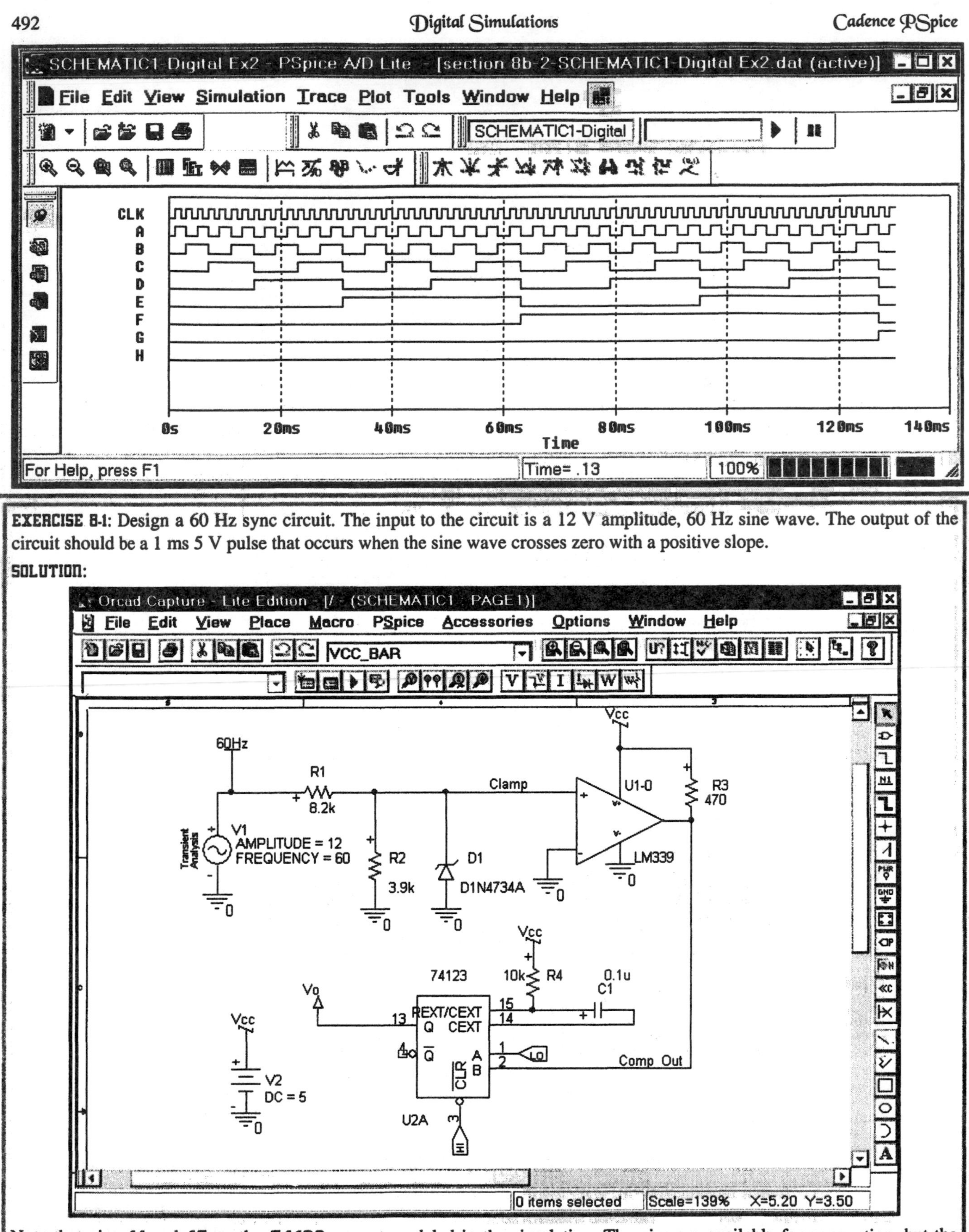


**EXERCISE 8-1:** Design a 60 Hz sync circuit. The input to the circuit is a 12 V amplitude, 60 Hz sine wave. The output of the circuit should be a 1 ms 5 V pulse that occurs when the sine wave crosses zero with a positive slope.

**SOLUTION:**

Note that pins ***14*** and ***15*** on the ***74123*** are not modeled in the simulation. The pins are available for connection, but the circuit connected to the pins is not used in the simulation. Thus, ***C1*** and ***R4*** do not determine the pulse timing in the simulation. They are included for completeness and could be omitted if only a simulation is required. If you were constructing a PC board, you would need to include them. The pulse width is determined by one of the attributes of the ***74123***:

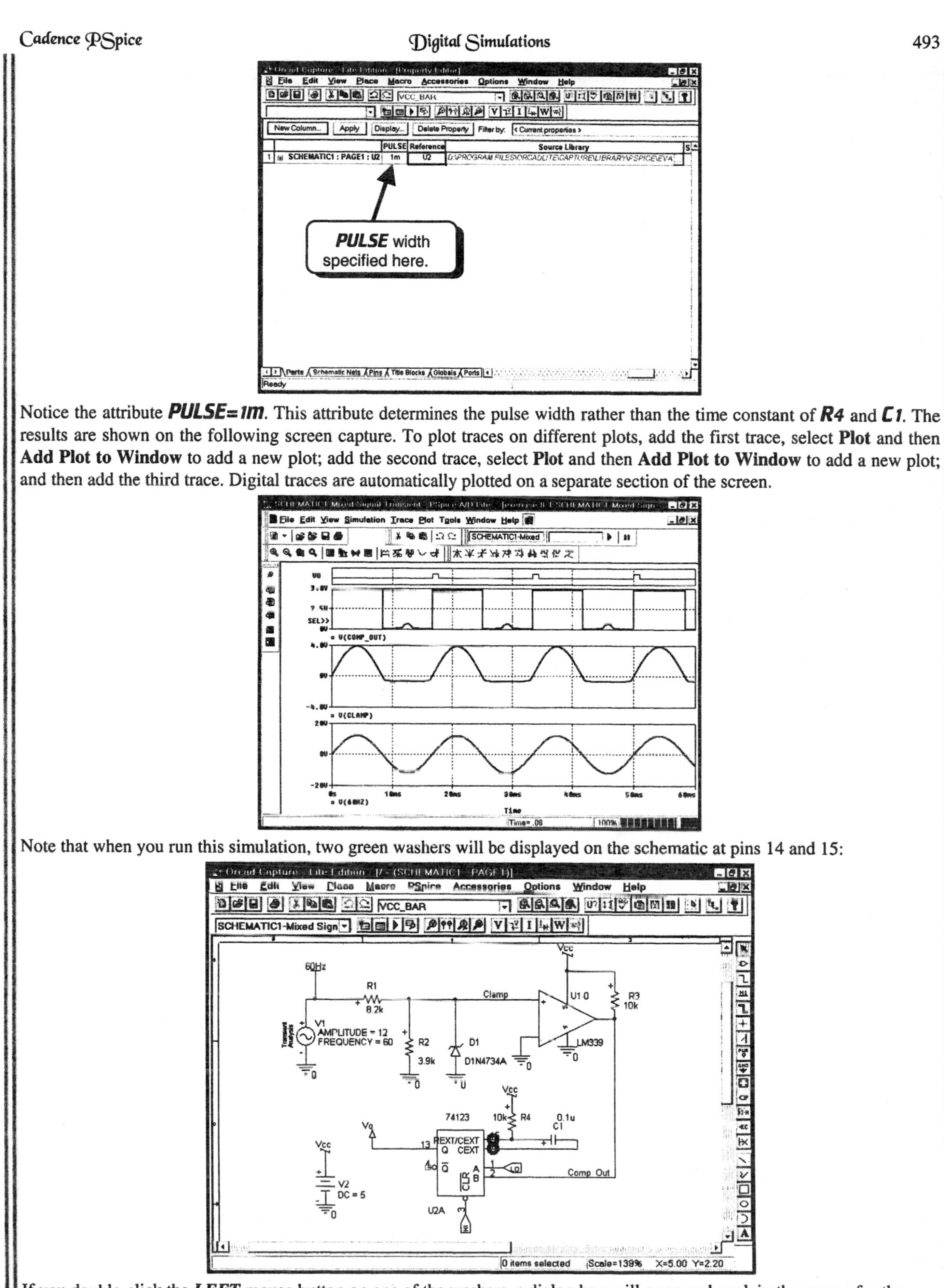


Notice the attribute ***PULSE=1m***. This attribute determines the pulse width rather than the time constant of ***R4*** and ***C1***. The results are shown on the following screen capture. To plot traces on different plots, add the first trace, select **Plot** and then **Add Plot to Window** to add a new plot; add the second trace, select **Plot** and then **Add Plot to Window** to add a new plot; and then add the third trace. Digital traces are automatically plotted on a separate section of the screen.

Note that when you run this simulation, two green washers will be displayed on the schematic at pins 14 and 15:

If you double-click the ***LEFT*** mouse button on one of the washers, a dialog box will open and explain the reason for the washer:

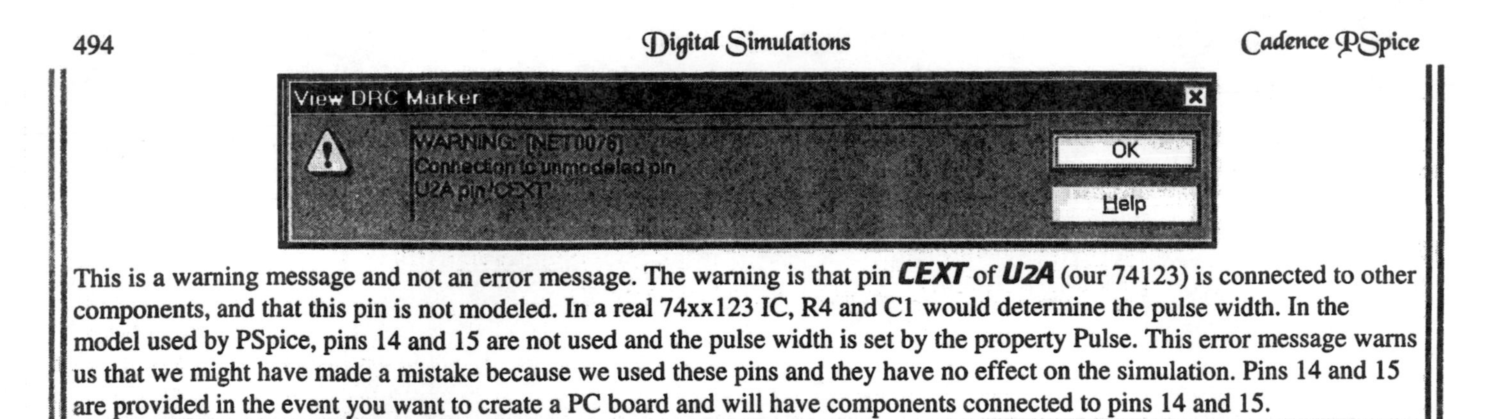


This is a warning message and not an error message. The warning is that pin ***CEXT*** of ***U2A*** (our 74123) is connected to other components, and that this pin is not modeled. In a real 74xx123 IC, R4 and C1 would determine the pulse width. In the model used by PSpice, pins 14 and 15 are not used and the pulse width is set by the property Pulse. This error message warns us that we might have made a mistake because we used these pins and they have no effect on the simulation. Pins 14 and 15 are provided in the event you want to create a PC board and will have components connected to pins 14 and 15.

# 8.C. Effect of Not Initializing Flip-Flops

In the previous section we used the Digital Setup to initialize all flip-flops to the zero state. Suppose that instead of clearing the flip-flops, we specify the initial states as unknown (***X***). We will be using the circuit shown on page 487. Follow the procedure for running the analysis, except fill in the ***Gate-level Simulation Options*** as shown below:

Initial state of the flip-flop is unknown.

Note that the initial state of the flip-flop is specified as unknown. Run the simulation and then display the traces ***V(VO1)***, ***V(V_CLAMP)***, ***VO2***, ***VO3***, and ***VO4***. To display traces, select **Trace** and then **Add Trace** from the Probe menu bar or press the **INSERT** key:

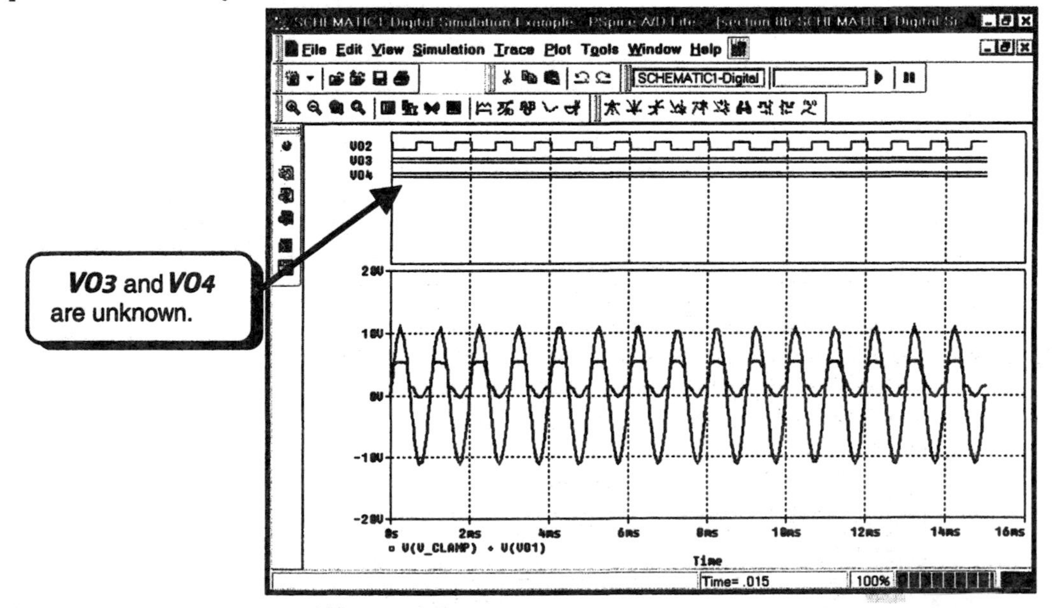


Note that the flip-flop outputs (***VO3*** and ***VO4***) appear as double lines. The double lines indicate that PSpice does not know if the output we are plotting is a logic one or zero. This is because if PSpice does not know the initial state, it cannot determine any of the following states. To run digital simulations you need to specify the initial states using the 1 or 0 parameter in the Gate-level Simulation Options dialog box, or create a circuit that initializes flip-flops. The following section gives a circuit for initializing digital circuits.

# 8.C.1. Start-Up Clear Circuit

Suppose we want to initialize all flip-flops to an initial state but do not want to use the initial condition of 0 or 1. The circuit below clears the flip-flop at the beginning of the simulation:

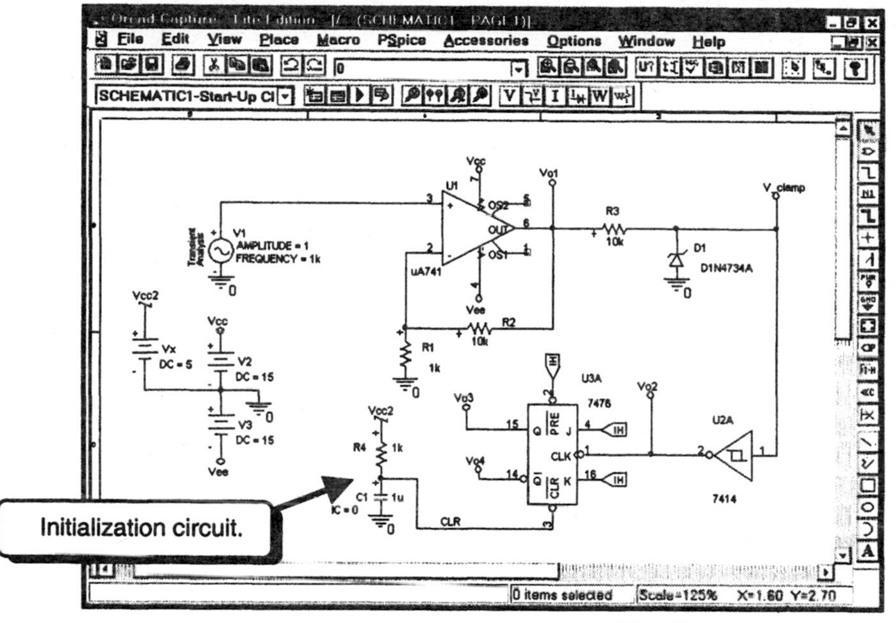


This is the same circuit as that shown on page 487 with the addition of the ***R4***, ***C1***, and a 5-volt DC source. The resistor and capacitor create the initialization circuit. The initial condition of the capacitor is specified as 0 volts. When the simulation runs, the capacitor starts at 0 volts (logic zero) and charges to 5 volts. While the capacitor is close to 0 volts, it provides a logic zero input to the flip-flop, which clears the flip-flop. For most of the simulation the capacitor is charged to 5 volts, providing a logic 1 to the flip-flop clear input, which has no effect. See Section 6.B for specifying capacitor initial conditions.

The ***Gate-level Simulation Options*** should be specified as ***X*** as shown below:

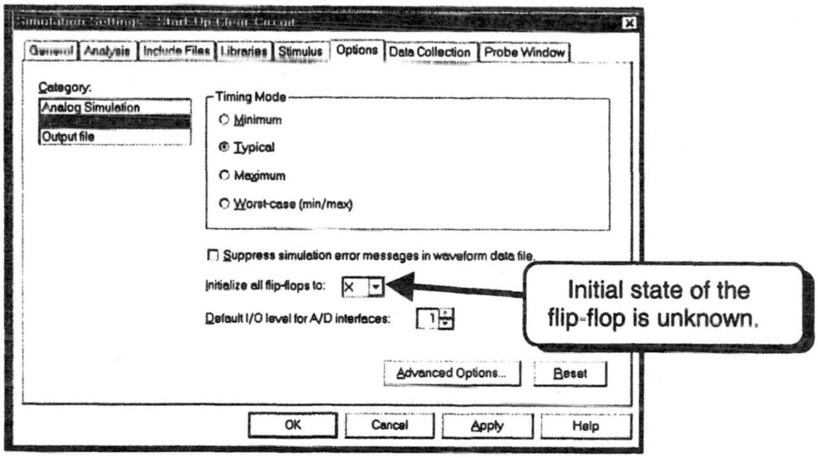

Initial state of the flip-flop is unknown.

The Transient Analysis setup is the same as in the example without the start-up clear circuit:

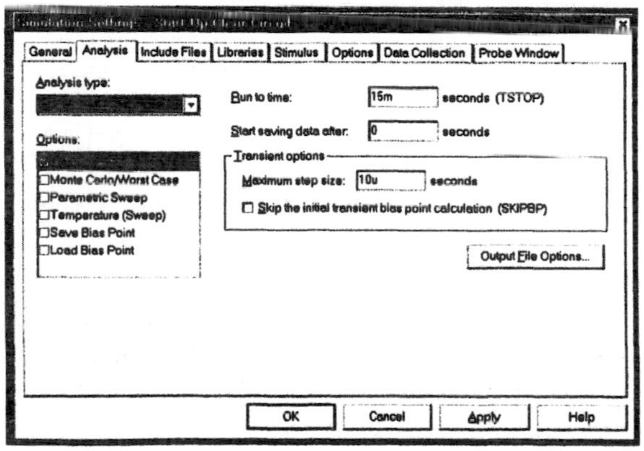

Run the simulation and then run Probe. Add the traces **VO3**, **VO4**, and **V(CLR)**:

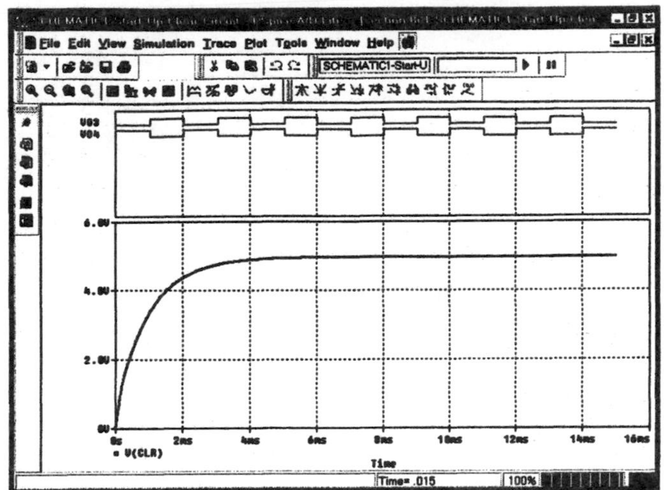


We see that the flip-flop is initialized to an initial state of zero.

# 8.D. Pure Digital Simulations

PSpice can be used as a logic simulator as well as a mixed analog/digital simulator. We will now simulate a switch-tail counter. Wire the circuit as shown:

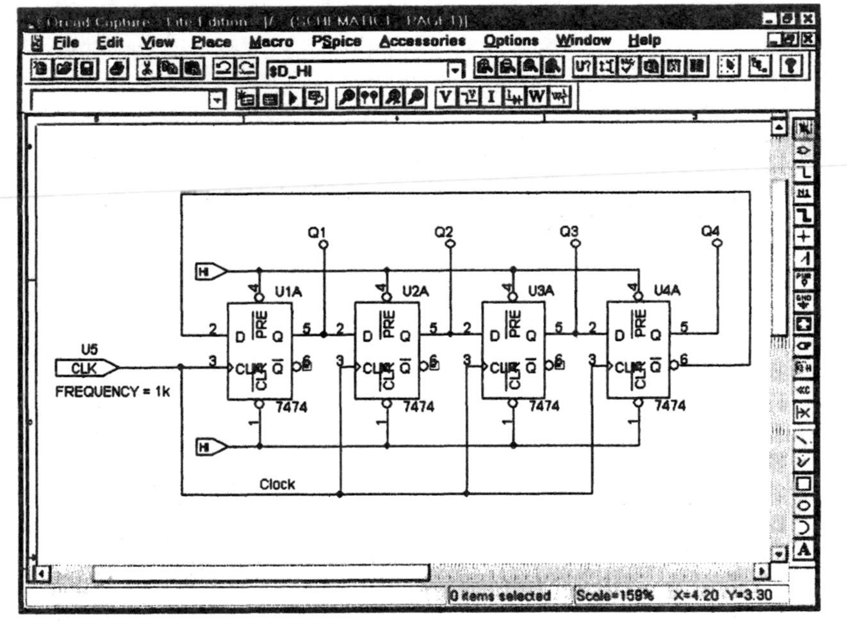


| Digital_Clock | HI | Vcc_Circle | 7474 |
|---|---|---|---|
| 1-bit digital bit stream | Digital logic one | Node label | D flip-flop |

There are two things that are different about this circuit. The first is that there are no ground or positive supply connections in the circuit. When Orcad made the graphics for the digital parts, they included the ground and power connections in the graphics but hid them to keep Capture from getting cluttered. The power and ground connections are in the circuit but are not shown.

Instead of a node label, such as a bubble, having been placed on the ***Clock*** wire, the wire has been labeled. To label a wire, select **Place** and then **Net Alias** from the menus. A dialog box will appear, asking you for a label. Enter a name for the label and then place the label next to a wire. It is sometimes better to label a wire rather than add a bubble, because a bubble adds more clutter to a circuit drawing.

When we run a digital simulation we must set up the Transient Analysis, as well as specify parameters for the digital simulation. We will first set up the Transient analysis. Select **PSpice** and then **New Simulation Profile** from the Capture menus, enter a name for the profile, and then click the ***Create*** button. By default the ***Time Domain (Transient) Analysis type*** is selected. Fill in the dialog box as shown to run the simulation for 20 ms:

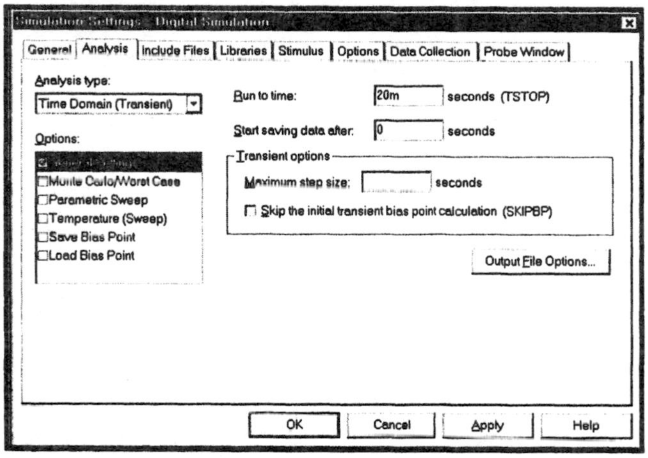

We must also set up the parameters for the digital simulation. Click the *LEFT* mouse button on the ***Options*** tab and select the ***Gate-level Simulation Category***:

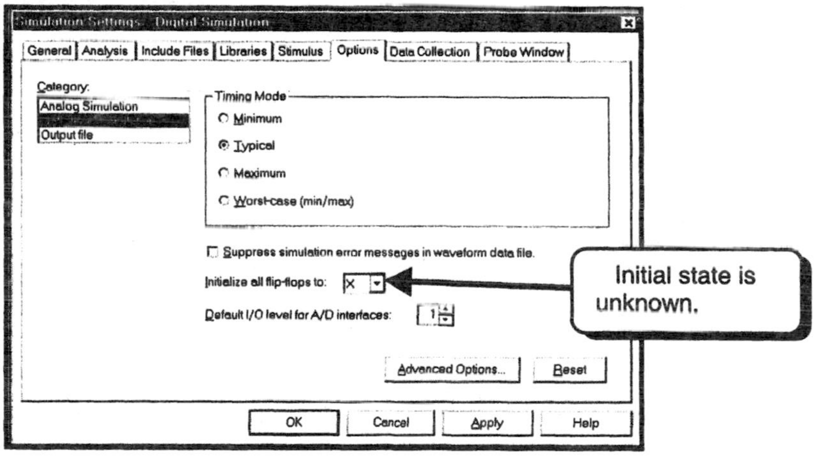

This dialog box allows us to specify timing and initial conditions for the digital circuits. We are interested in the initial state of the flip-flops. The default initial state for all flip-flops is ***X*** (unknown). We wish to set the initial state to all zeros:

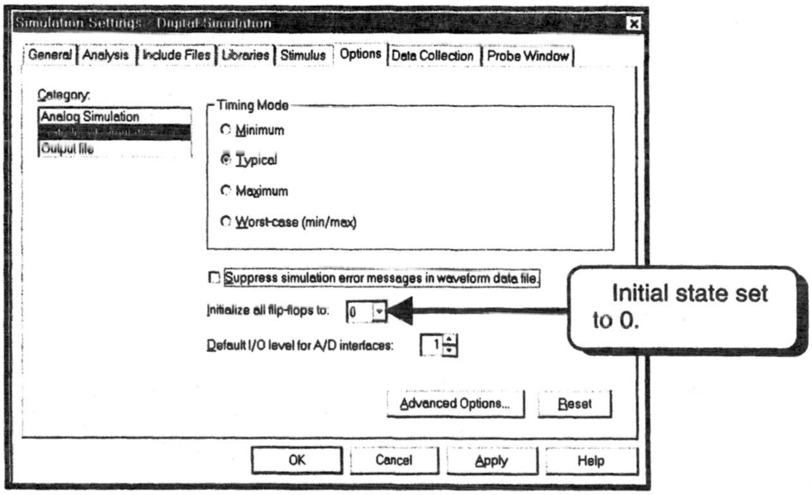

Click the ***OK*** button to accept the settings and to return to the schematic.

Run the simulation (**PSpice, Run**) and then run Probe. Add the traces ***Q1***, ***Q2***, ***Q3***, and ***Q4***:

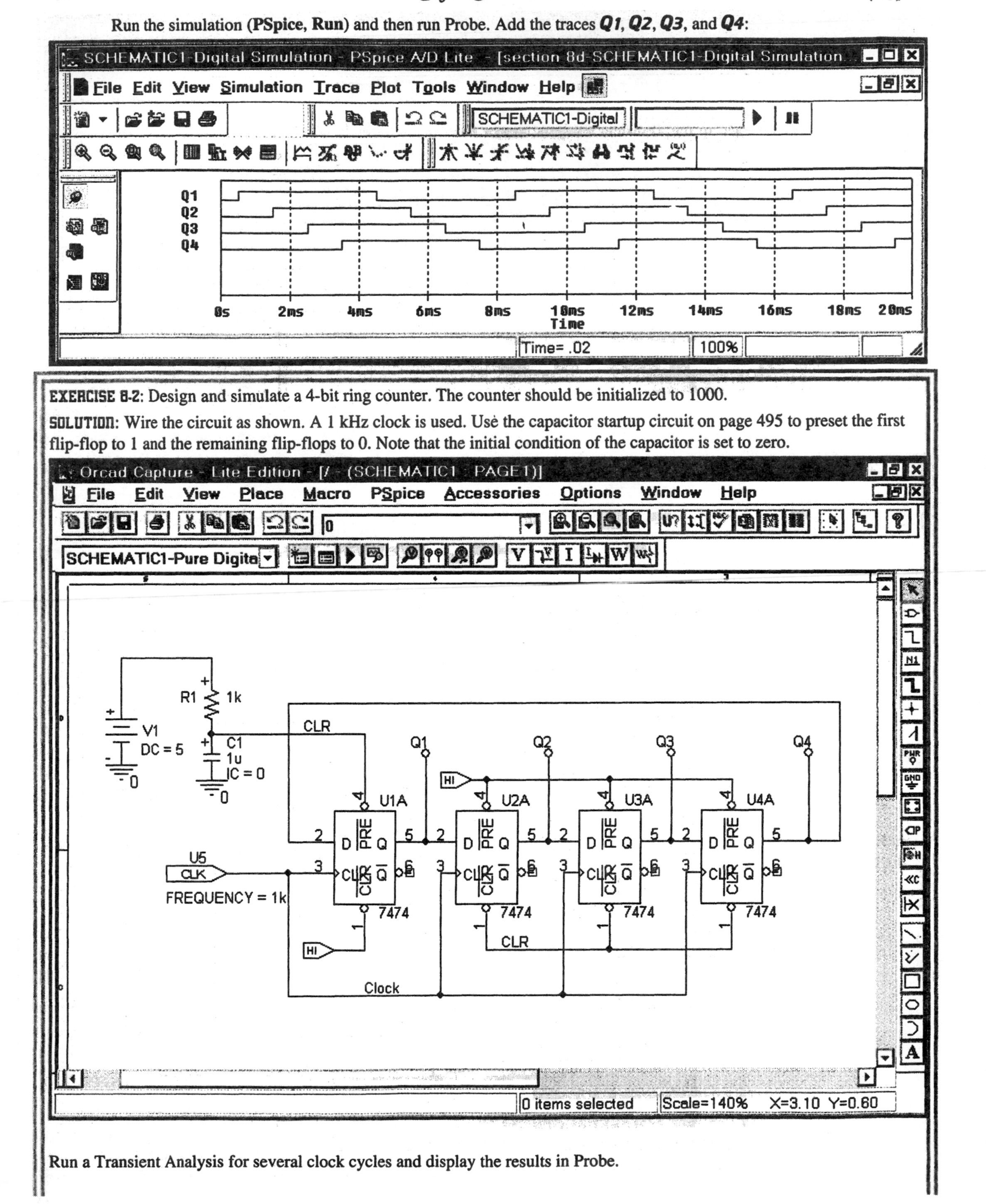


**EXERCISE 8-2**: Design and simulate a 4-bit ring counter. The counter should be initialized to 1000.

**SOLUTION**: Wire the circuit as shown. A 1 kHz clock is used. Use the capacitor startup circuit on page 495 to preset the first flip-flop to 1 and the remaining flip-flops to 0. Note that the initial condition of the capacitor is set to zero.

Run a Transient Analysis for several clock cycles and display the results in Probe.

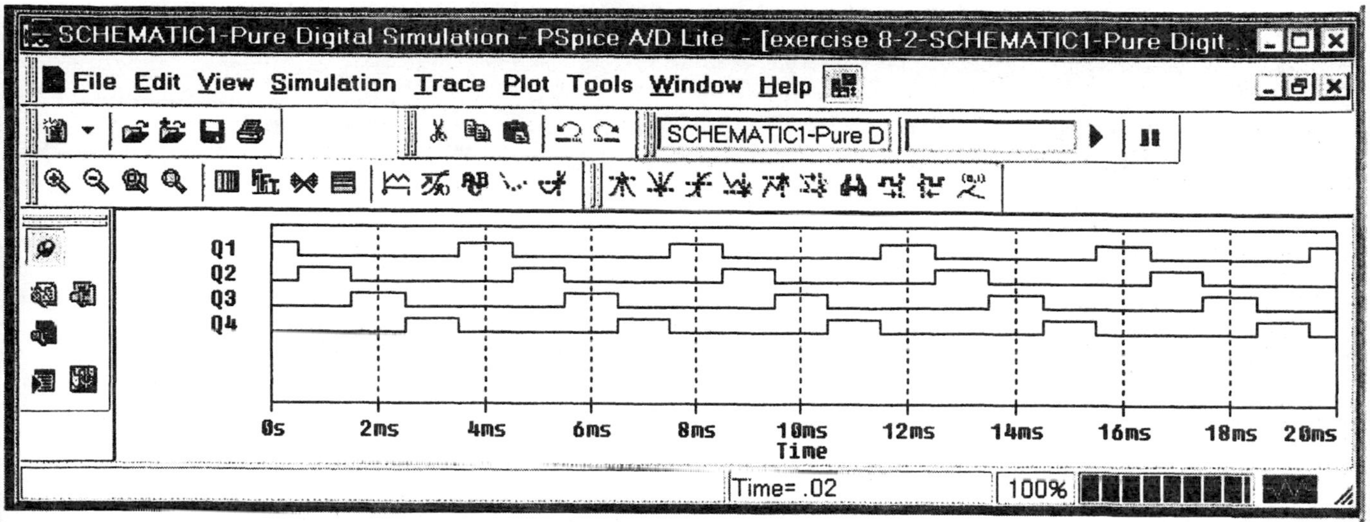

**EXERCISE 8-3**: Design and simulate a 4-bit decade counter. The counter should be initialized to 0000 at the start of the simulation.

**SOLUTION**: Wire the circuit as shown. A 1 kHz clock is used. Use the Gate-level Simulation Options dialog box to set all flip-flops to an initial state of 0.

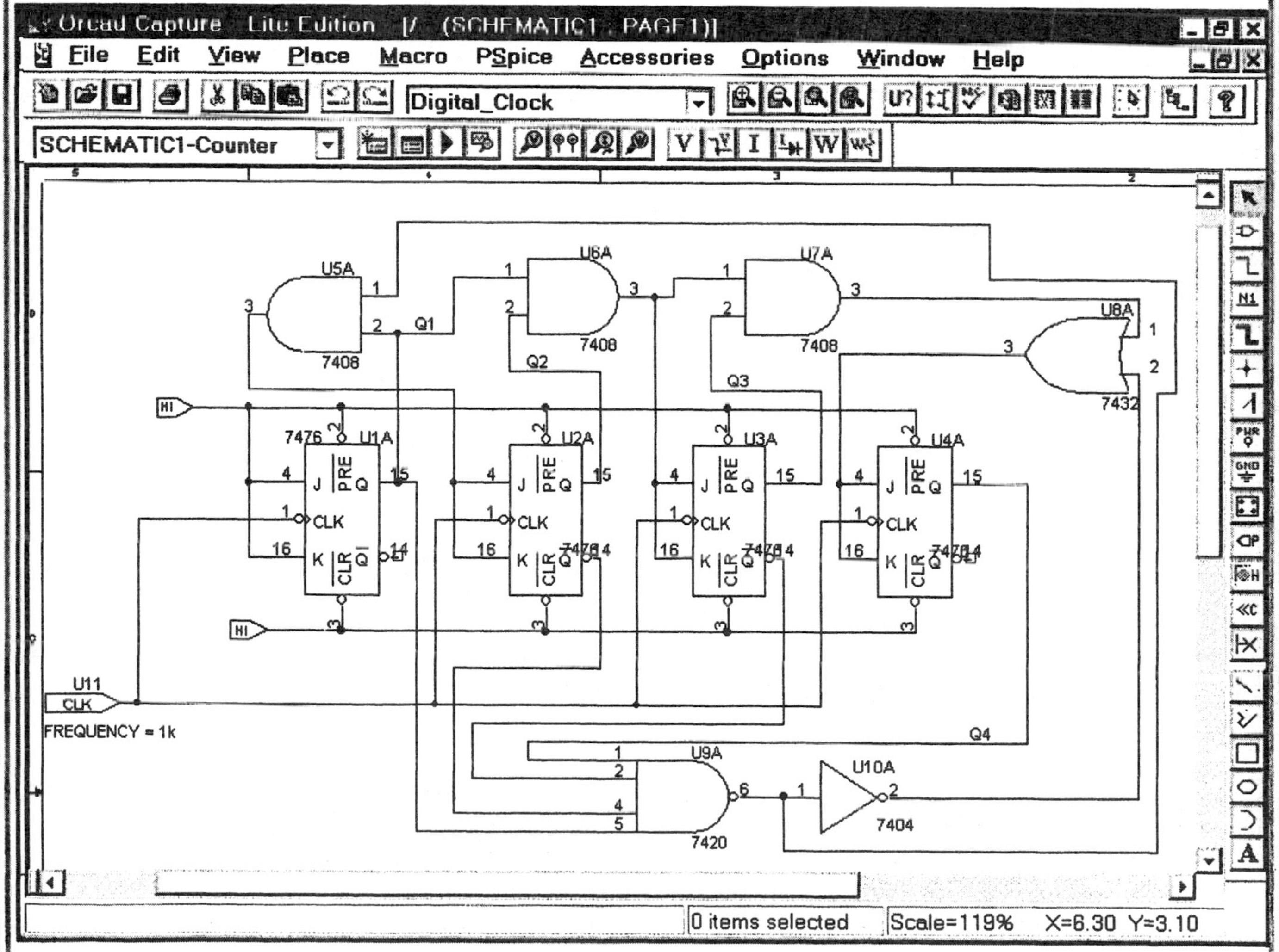

The circuit is a bit large to fit on a screen capture, but it does show the capabilities of the digital simulations. Run a Transient Analysis for several clock cycles and display the results in Probe.

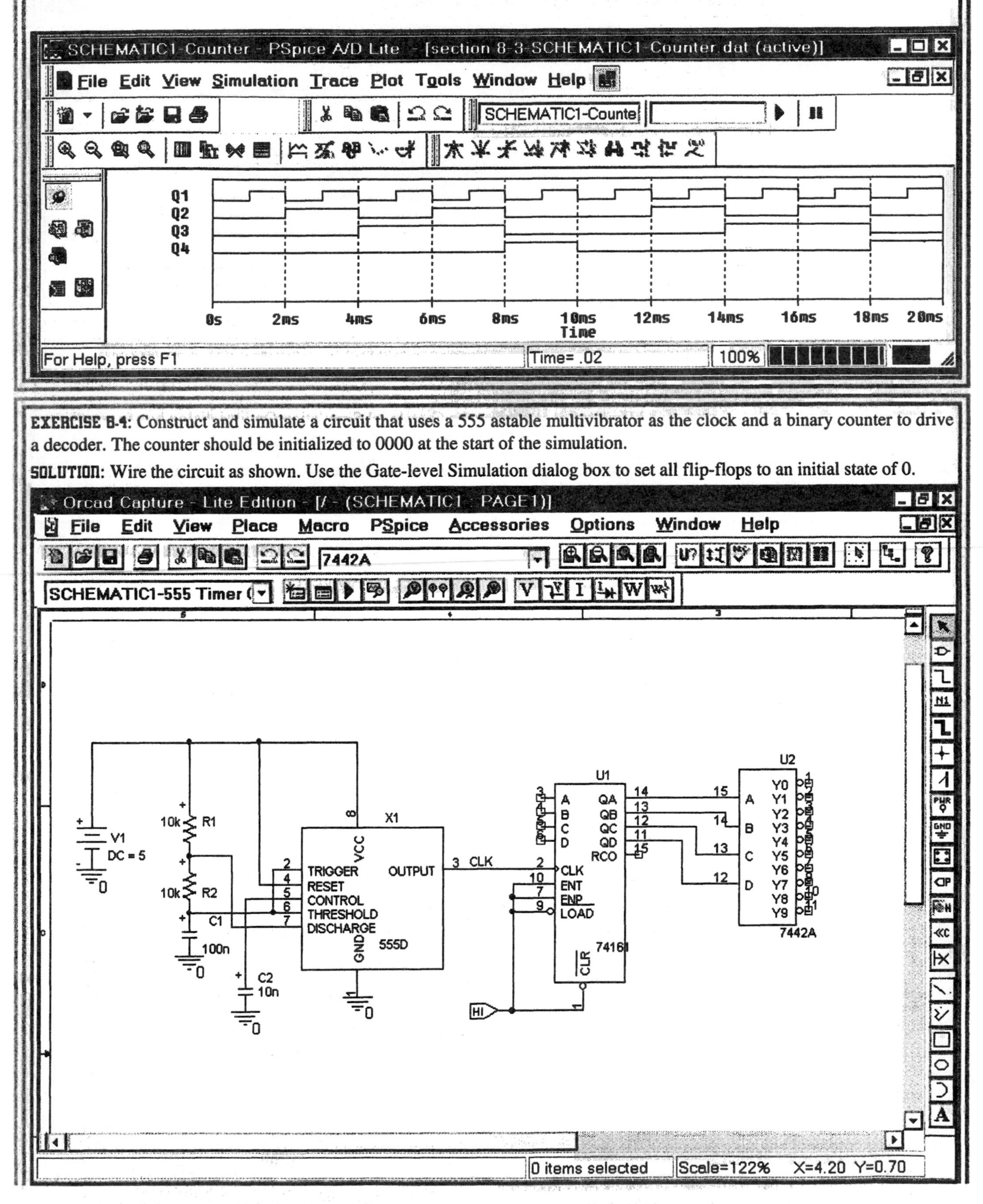


**EXERCISE 8-4**: Construct and simulate a circuit that uses a 555 astable multivibrator as the clock and a binary counter to drive a decoder. The counter should be initialized to 0000 at the start of the simulation.

**SOLUTION**: Wire the circuit as shown. Use the Gate-level Simulation dialog box to set all flip-flops to an initial state of 0.

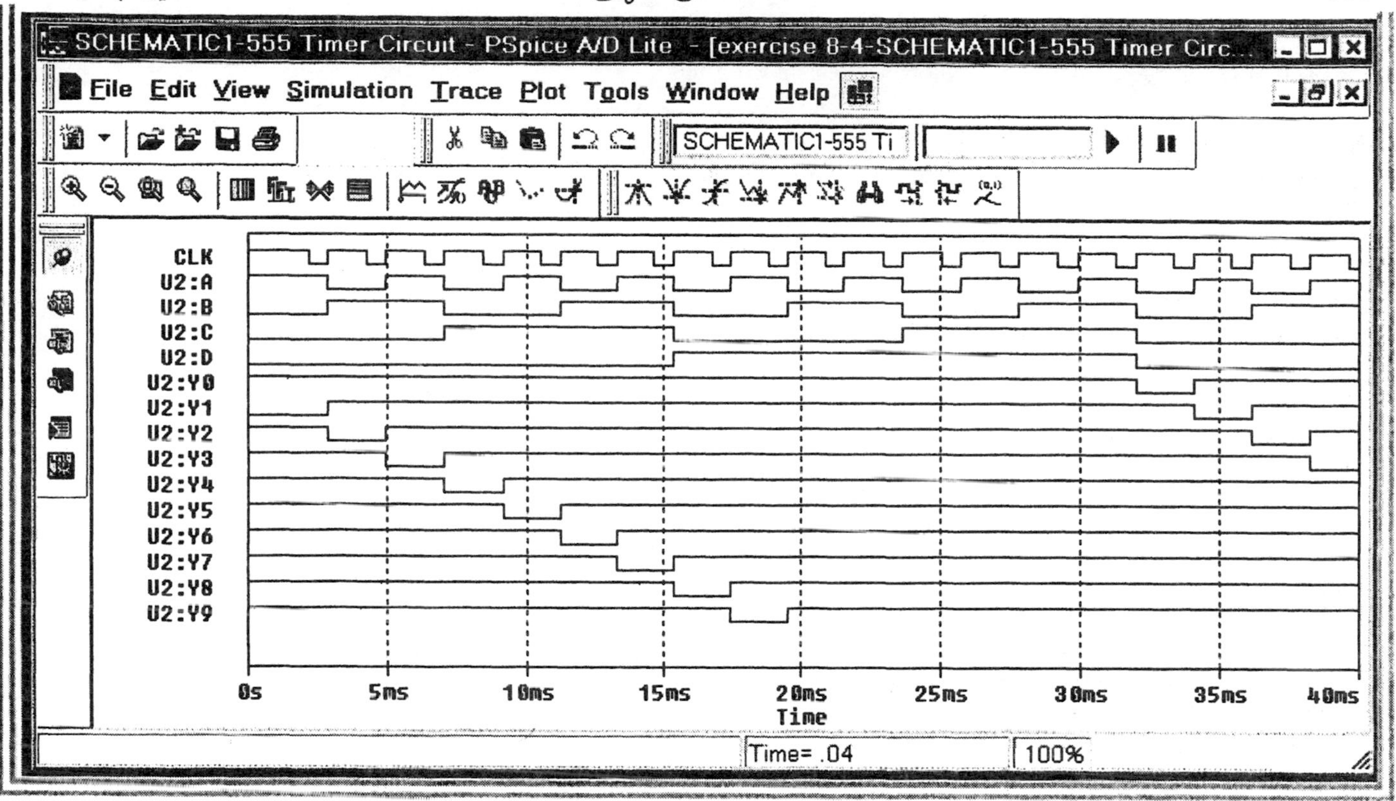


# 8.E. Gate Delays

PSpice includes gate delays in its digital simulations. To illustrate gate delays, we will create a two-phase, non-overlapping clock using gate delays. Wire the circuit below:

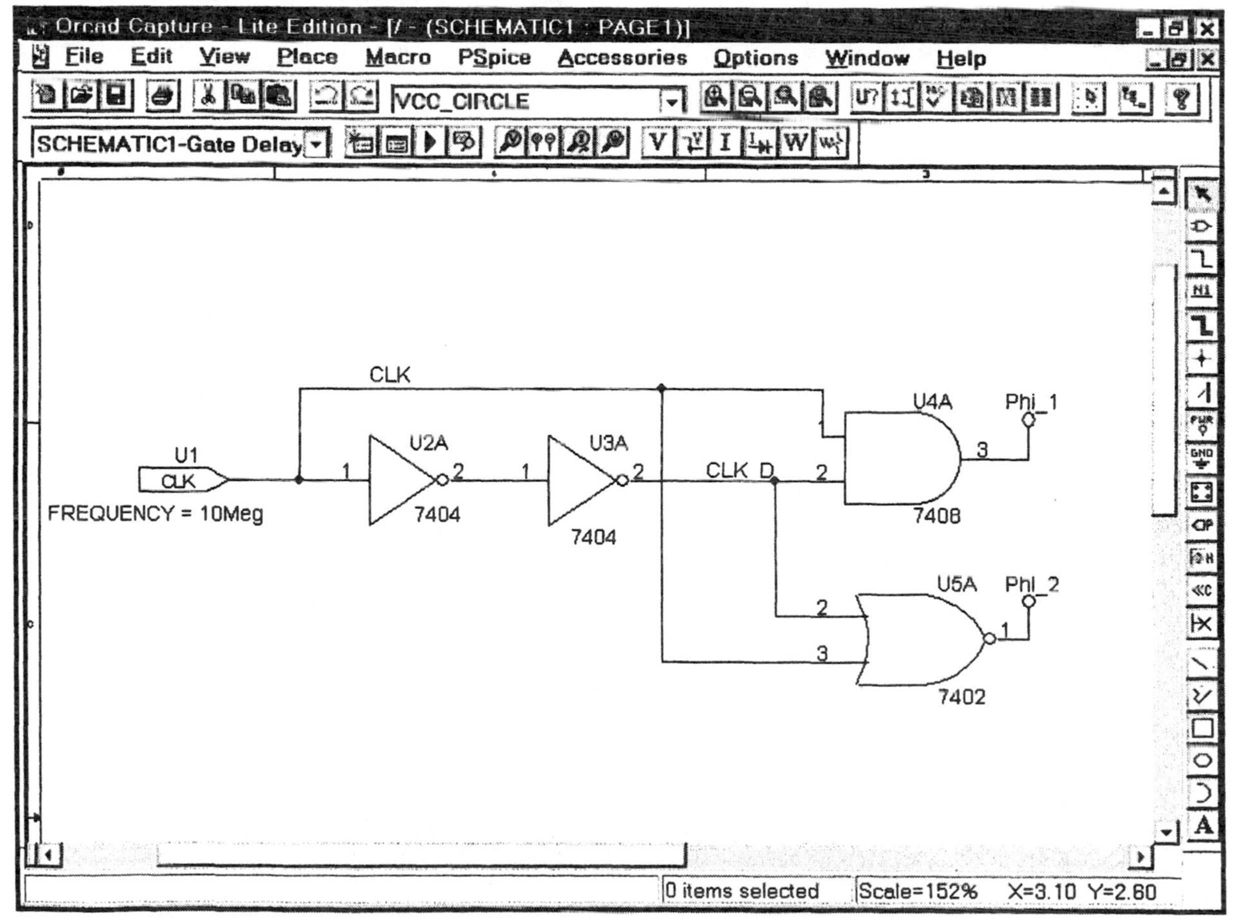

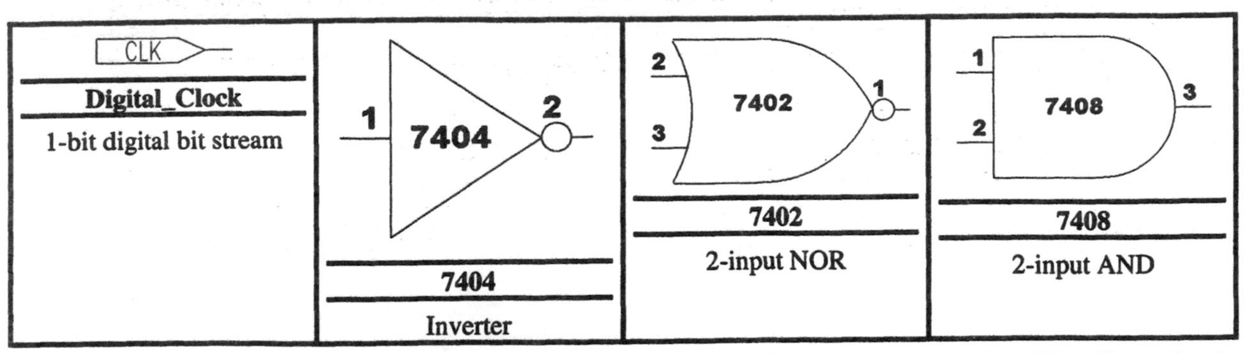


Since gate delays are on the order of nanoseconds, we will set the clock frequency to 10 MHz. We will run the simulation for 500 ns. Fill in the ***Time Domain (Transient)*** dialog box as shown:

General | Analysis | Include Files | Libraries | Stimulus | Options | Data Collection | Probe Window

Analysis type:

Options:

Monte Carlo/Worst Case
Parametric Sweep
Temperature (Sweep)
Save Bias Point
Load Bias Point

Run to time: 500n seconds (TSTOP)

Start saving data after: 0 seconds

Transient options

Maximum step size: seconds

Skip the initial transient bias point calculation (SKIPBP)

Output File Options...

OK Cancel Apply Help

Run the simulation and then run Probe. Display the traces ***CLK***, ***CLK_D***, ***PHI_1***, and ***PHI_2***:

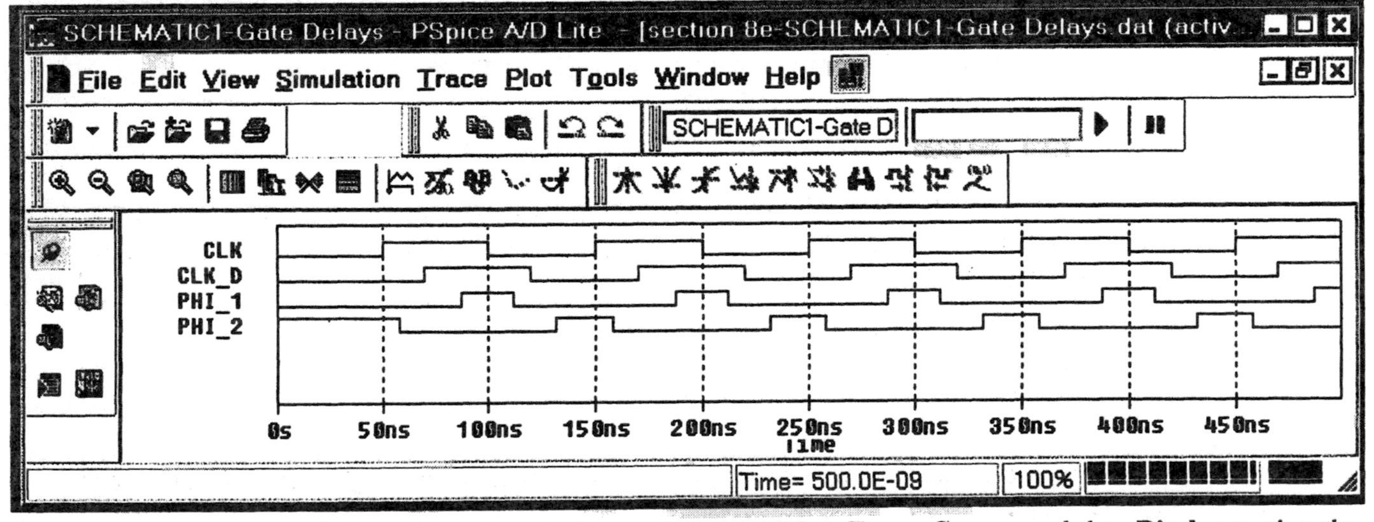


We can use the cursors to find the delay through the two inverters. Select **Trace, Cursor,** and then **Display** to view the cursors:

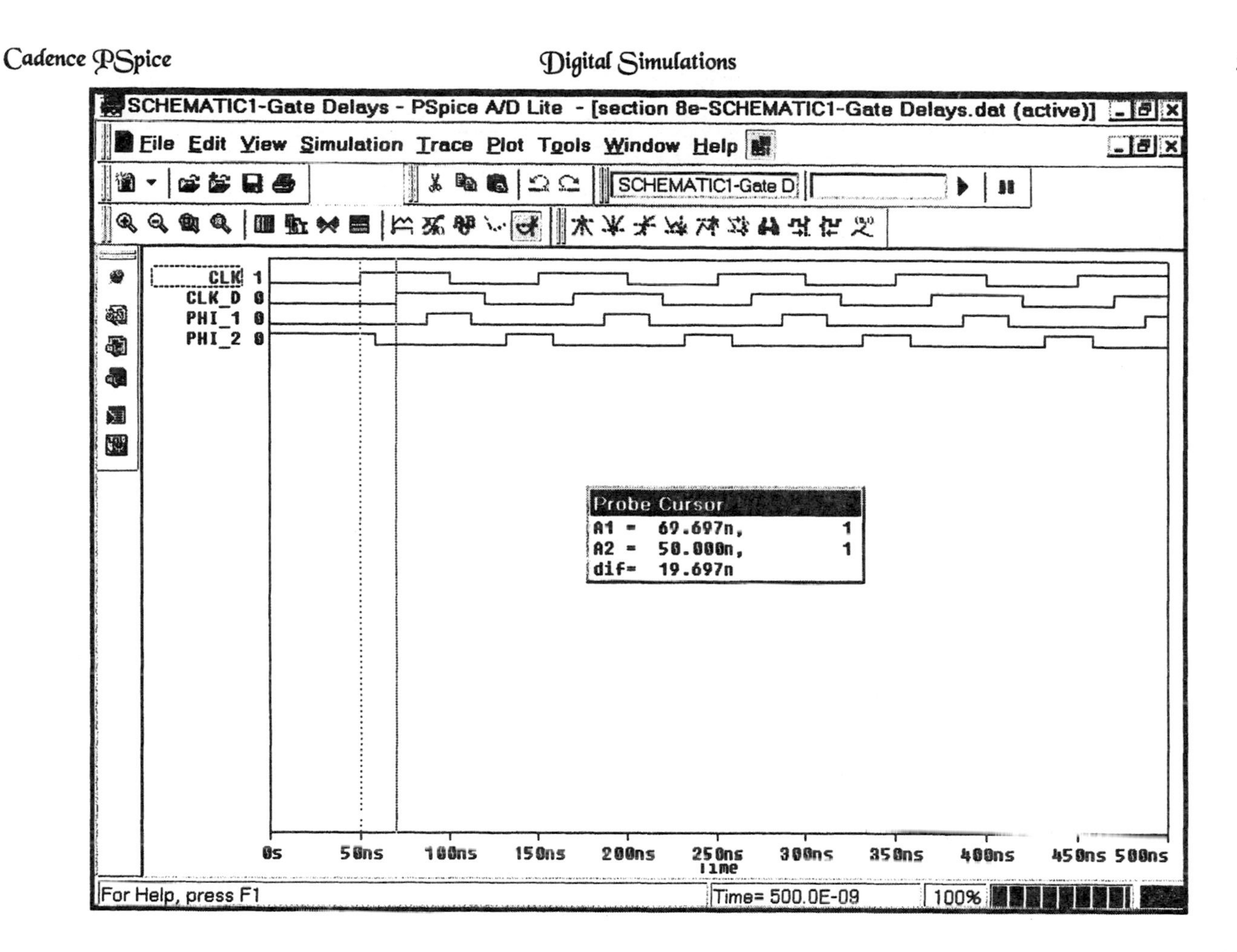


The ***LEFT*** mouse button controls cursor 1 and the ***RIGHT*** mouse button controls cursor 2. The cursor coordinates are displayed in the dialog box as ***A1*** and ***A2***. The difference in time between the two cursors is given as ***dif=19.697*** nanoseconds. Thus, the total delay through the two inverters is approximately 20 ns.

# 8.F. Summary

- PSpice can be used for pure digital simulations, pure analog simulations, or mixed analog/digital simulations.
- Always preset or clear all flip-flops before running a simulation. Use the digital setup dialog box or a start-up clear circuit to set the initial state of the flip-flops.
- PSpice simulations include gate delays.

# 8.G. Bibliography

[1] MicroSim Corporation. *MicroSim PSpice A/D- Circuit Analysis Software - Reference Manual*, Version 7.1, Irvine, CA, October, 1996, p. 3-78

# PART 9
# Monte Carlo Analyses

The Monte Carlo analyses are used to observe how device tolerances can affect a design. There are two analyses that can be performed. The Worst Case analysis is used to find the maximum or minimum value of a parameter given device tolerances. Device tolerances are varied to their maximum or minimum limits such that the maximum or minimum of the specified parameter is found. The Monte Carlo analysis is used to find production yield. If the Worst Case analysis shows that not all designs will pass a specific criterion, the Monte Carlo analysis can be used to estimate what percentage will pass. The Monte Carlo analysis varies device parameters within the specified tolerance. The analysis randomly picks a value for each device that has tolerance and simulates the circuit using the random values. A specified output can be observed.

This part discusses the syntax of models with device tolerance and how to use the models in a simulation. This part does not discuss how to easily create models within Capture. After you have understood the syntax for the models and how to run the simulations, you may wish to look at Part 7 to see how to create new models to specify the tolerance you need.

## 9.A. Device Models

The first thing we need to know is how to create devices with tolerance. The libraries already have a number of parts with tolerance, but these are usually not enough. PSpice allows you to create several types of distributions.

### 9.A.1. Uniform Distribution

The uniform distribution specifies that the part is equally likely to have a value anywhere in the specified tolerance range. The first example we will look at is the 5% resistor model included in class.lib. If you look at Appendix E on page 620, you will see the following line in the file class.lib:

```
.MODEL R5pcnt RES(R=1 DEV/UNIFORM 5%)
```

The model name is R5pcnt. It is a resistor model because the model type is RES. The nominal value of the model is R=1 and it has a 5% tolerance. The distribution is uniform. A uniform distribution means that the model parameter R is equally likely to have a value of 1.05, 1, 0.95, or any other value between 1.05 and 0.95. When you use this model, the actual value of the resistor is the value specified in the schematic times the parameter R. Thus, if the value of a 5% resistor is specified as 1k in the schematic, it may have a value anywhere between 950 and 1050 when used with the Monte Carlo analyses. An equivalent model is:

```
.MODEL R5pcnt RES(R=1 DEV/UNIFORM 0.05)
```

This model has an absolute tolerance rather than a percentage tolerance. The parameter R can have a value in the range $1 \pm 0.05$.

The next model we will look at is an NPN bipolar junction transistor. The model below is a transistor with a value of $\beta_F$ that may vary between 50 and 350:

```
.MODEL QBf NPN( Bf=200 DEV/UNIFORM 150)
```

The nominal value of $\beta_F$ is 200. The range of $\beta_F$ is $200 \pm 150$. The transistor is equally likely to have a value anywhere in this range. The name of the model is QBf. The text NPN specifies the model as an NPN bipolar transistor. This model is included in class.lib. A limitation of this model is that none of the other transistor parameters have been specified. This model is almost ideal. Its limitations will become apparent when you observe that the high-frequency response does not roll off, even at frequencies beyond 1 GHz.

Another transistor model that includes tolerance is the Q2N3904B model shown below:

```
.MODEL Q2N3904B    NPN(Is=6.734f Xti=3 Eg=1.11 Vaf=74.03
+       Bf=416.4 DEV/UNIFORM 80% Ne=1.259
+       Ise=6.734f Ikf=66.78m Xtb=1.5 Br=.7371 Nc=2 Isc=0 Ikr=0 Rc=1
+       Cjc=3.638p Mjc=.3085 Vjc=.75 Fc=.5 Cje=4.493p Mje=.2593 Vje=.75
+       Tr=239.5n Tf=301.2p Itf=.4 Vtf=4 Xtf=2 Rb=10)
```

This is a copy of the standard Q2N3904 model, except that tolerance has been added to the $\beta_F$ parameter. This model will have the same properties as the Q2N3904 but will allow variations in $\beta_F$.

We are not limited to tolerance in one parameter. The above model has tolerance only in $\beta_F$. The model shown below specifies tolerance in $I_S$ as well as $\beta_F$:

```
.MODEL Q2N3904B NPN(Is=6.734f DEV/UNIFORM 10% Xti=3 Eg=1.11
```

```
+        Vaf=74.03
+        Bf=416.4 DEV/UNIFORM 80% Ne=1.259
+        Ise=6.734f Ikf=66.78m Xtb=1.5 Br=.7371 Nc=2 Isc=0 Ikr=0 Rc=1
+        Cjc=3.638p Mjc=.3085 Vjc=.75 Fc=.5 Cje=4.493p Mje=.2593 Vje=.75
+        Tr=239.5n Tf=301.2p Itf=.4 Vtf=4 Xtf=2 Rb=10)
```

The last model we will look at is a capacitor model with +80% and –20% tolerance. This model is called CAP20_80 and is included in class.lib:

```
.MODEL CAP20_80 CAP(C=1.3 DEV/UNIFORM 38.461538%)
```

Tolerances in PSpice are set up to have equal plus and minus ranges. With a +80% and –20% capacitor we have to fudge things a little. A capacitor with a +80% and –20% tolerance has a maximum value 1.8 times the nominal value, and has a minimum value 0.8 times the nominal value. This range is equivalent to a capacitor with a nominal value of (1.8 + 0.8)/2 = 1.3 and a ±38.461538% tolerance. Note that 1.3*(1 + 0.38461538) = 1.8, and 1.3*(1 – 0.38461538) = 0.8. Thus, a capacitor with a nominal value of 1 and +80% and –20% tolerance has the same capacitance range as a capacitor with a nominal value of 1.3 and ±38.461538% tolerance. The only difference is the nominal value.

When you use the CAP20_80 model in a schematic, the actual value of the capacitor is the value specified in the schematic times the parameter C. If the value of a +80% and –20% capacitor is specified as 470 μF in the schematic, its nominal value is 470 μF*1.3 = 611 μF. Its maximum value is 611 μF*(1 + 0.38461538) = 846 μF. Its minimum value is 611 μF*(1 – 0.38461538) = 376 μF.

## 9.A.2. Gaussian Distribution

The Gaussian tolerance distribution generates the distribution shown below. The graph shows a 1 kΩ resistor with a ±5% tolerance.

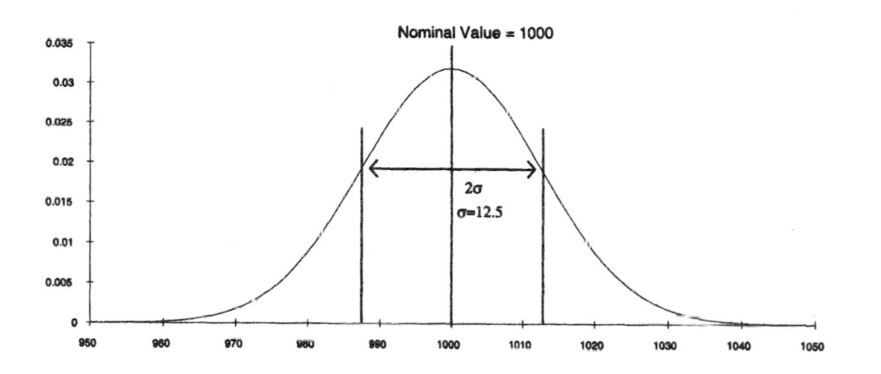


The mean value of the distribution is 1000 and the standard deviation, σ, is 12.5. A 1 kΩ, 5% resistor will have values from 950 Ω to 1050 Ω. Deviations of ±4σ achieve this spread. This distribution shows us that almost all of the resistors are within plus or minus three standard deviations from the nominal value, ±3σ = ±37.5. It is very rare that we see a resistor value above 1040 Ω or below 960 Ω.

The PSpice Gaussian distribution specifies the nominal value and the standard deviation. All part distributions are limited to ±4σ since the probability of finding a resistor outside this range is extremely small. The model below describes a 5% resistor with a Gaussian distribution:

```
MODEL R5gauss RES(R=1 DEV/GAUSS 1.25%)
```

For this model, the nominal value is 1 Ω and the standard deviation is σ = 1.25%. The distribution has a maximum distribution of ±4σ = ±5%.

# 9.B. Voltage Divider Analysis

To illustrate the basic operation of the Monte Carlo and Worst Case analyses, we will simulate a voltage divider. Create a voltage divider using 5% resistors with Gaussian distributions as follows.

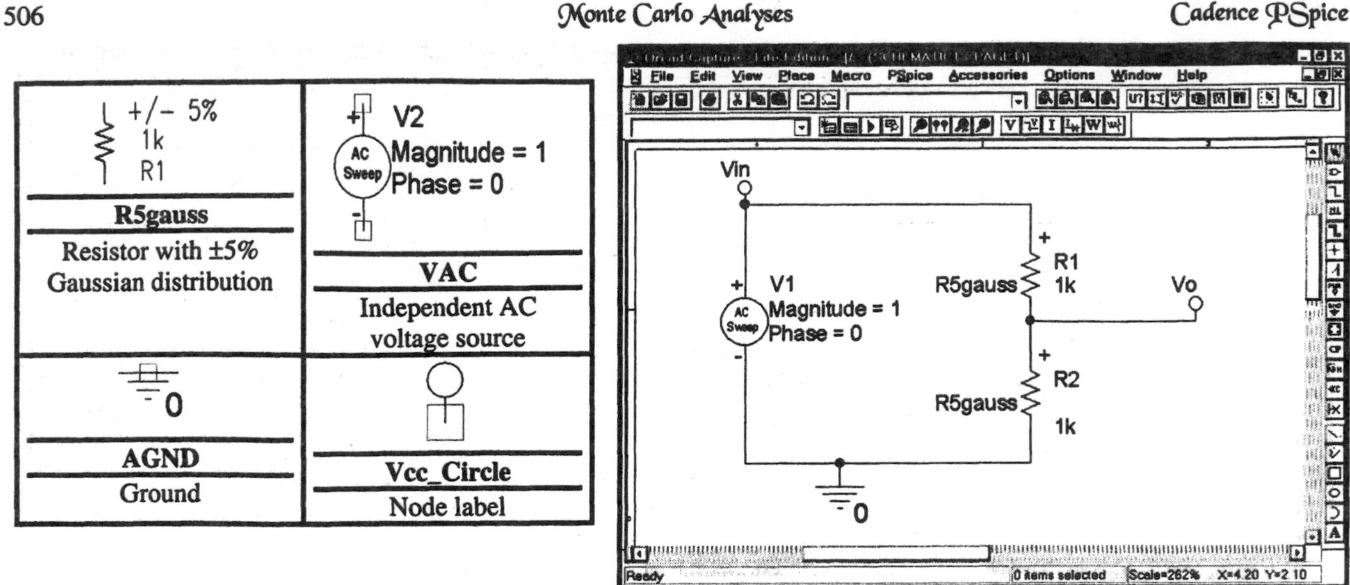


## 9.B.1. Voltage Divider Minimum and Maximum Voltage Gain

We will now use PSpice to find the worst case minimum and maximum voltage gain of this circuit. Since this circuit is very simple, we will do the calculations by hand and compare the results to PSpice.

The nominal voltage gain of this divider network is:

$$\frac{V_o}{V_{in}} = \frac{R_2}{R_1 + R_2} = \frac{1000}{1000 + 1000} = 0.5$$

The worst case maximum gain of the network is:

$$\frac{V_o}{V_{in}} = \frac{R_{2_{max}}}{R_{1_{min}} + R_{2_{max}}} = \frac{1050}{950 + 1050} = 0.525$$

The worst case minimum gain of the network is:

$$\frac{V_o}{V_{in}} = \frac{R_{2_{min}}}{R_{1_{max}} + R_{2_{min}}} = \frac{950}{1050 + 950} = 0.475$$

Now that we know what limits to expect we can set up the analysis. Since we want to find gain, we need to set up an AC Sweep. Let us find the gain of the circuit at 1 kHz. Select **PSpice** and then **New Simulation Profile** from the Capture menus, enter a name for the profile, and then click the ***Create*** button. Select the ***AC Sweep/Noise Analysis type***. Fill in the parameters as shown. The AC Sweep is set up to simulate one point at 1 kHz.

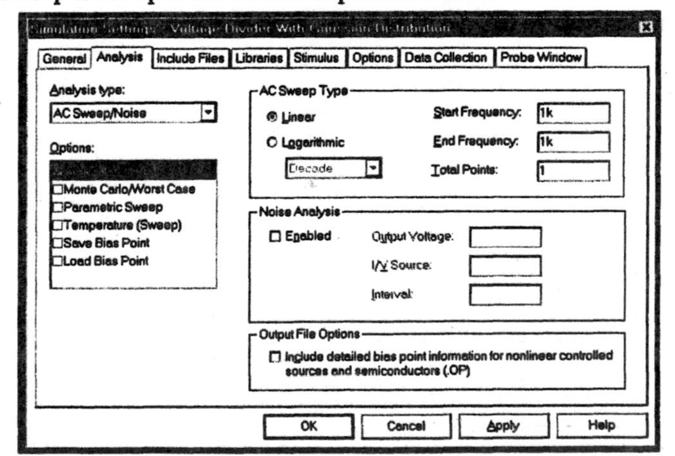


We must also set up the Worst Case analysis. Click the ***LEFT*** mouse button on the □ next to the text ***Monte Carlo/Worst Case*** button. This will enable the analysis and display its settings:

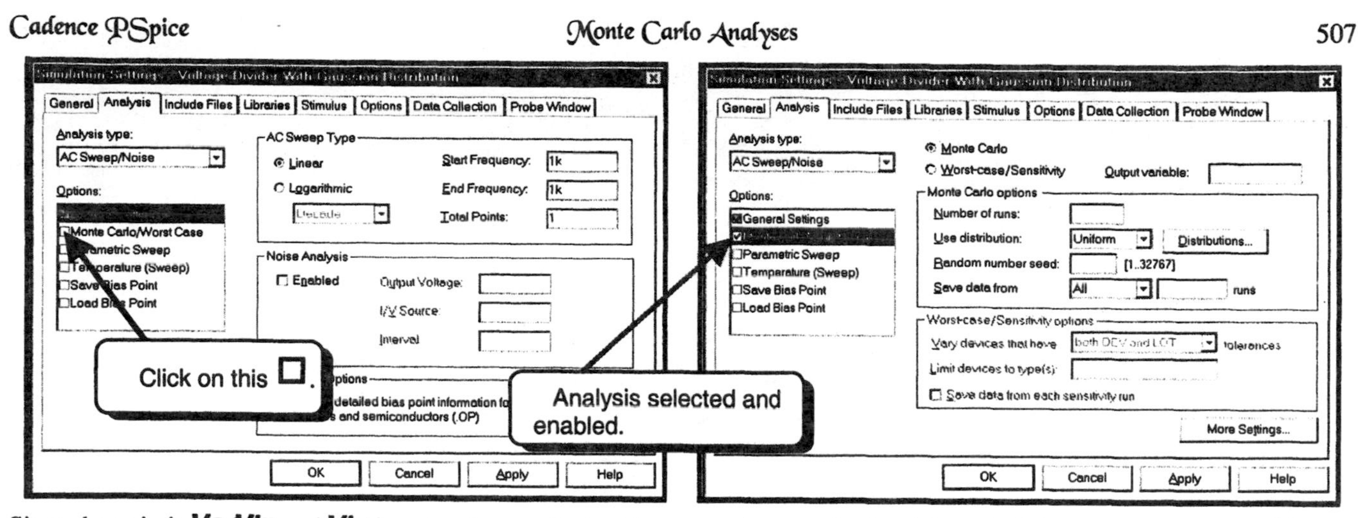


Since the gain is ***Vo/Vin*** and ***Vin*** is a 1-volt magnitude AC source, the magnitude of the gain is just the magnitude of ***Vo***:

Simulation Settings - Voltage Divider With Gaussian Distribution
General | Analysis | Include Files | Libraries | Stimulus | Options | Data Collection | Probe Window
Analysis type: AC Sweep/Noise
Options: General Settings, Monte Carlo/Worst Case, Parametric Sweep, Temperature (Sweep), Save Bias Point, Load Bias Point
Monte Carlo
Worst-case/Sensitivity
Output variable: V(Vo)
Monte Carlo options
Number of runs:
Use distribution: Uniform
Distributions...
Random number seed: [1..32767]
Save data from: All runs
Worst-case/Sensitivity options
Vary devices that have: both DEV and LOT tolerances
Limit devices to type(s):
Save data from each sensitivity run
More Settings...
OK | Cancel | Apply | Help

The analysis is set to ***Worst-case/Sensitivity***. The ***Output variable*** is ***V(Vo)*** because we are interested in the maximum voltage at this node. Click the ***More Settings*** button:

Monte Carlo/Worst Case Output File Options
Function
The function is performed on an output variable (for example, V(1)).
The result is listed in the output (.OUT) file only.
Find:
Threshold value:
Evaluate only when the sweep variable is in the range:
to
Worst-Case direction
Hi
Low
List model parameter values in the output file for each run
OK
Cancel

Specify the ***Find*** field as ***the maximum value(MAX)*** and make sure that the ***Worst-Case direction*** is set to ***HI*** :

Monte Carlo/Worst Case Output File Options
Function
The function is performed on an output variable (for example, V(1)).
The result is listed in the output (.OUT) file only.
Find: the maximum value (MAX)
Threshold value:
Evaluate only when the sweep variable is in the range:
to
Worst-Case direction
Hi
Low
List model parameter values in the output file for each run
OK
Cancel

***MAX*** option selected.

***HI*** option selected.

***Find*** is set to ***MAX*** since we wish to find the worst case maximum value of Vo. Click the ***OK*** button to return to the ***Simulation Settings*** dialog box:

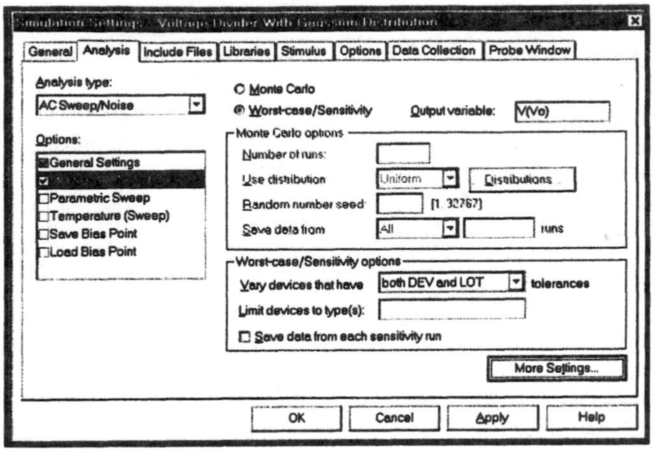

The results of this simulation will be contained in the output file and we do not need Probe to run. Select the ***Probe Window*** tab and deselect the options that specify that Probe should display a window:

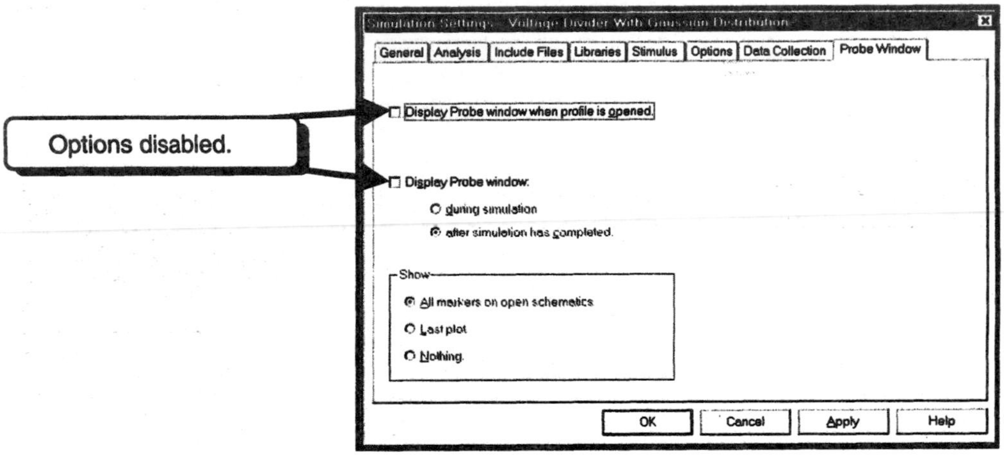

Click the ***OK*** button to return to the schematic. Run PSpice by selecting **PSpice** and then **Run** from the Capture menu bar:

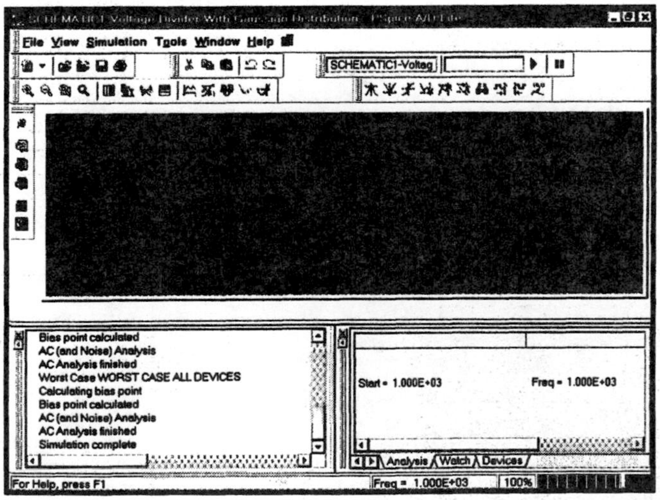

The results of the Worst Case analysis are saved in the output file. Select **PSpice** and then **View Output File** from the Capture menu bar. The results are given at the bottom of the output file.

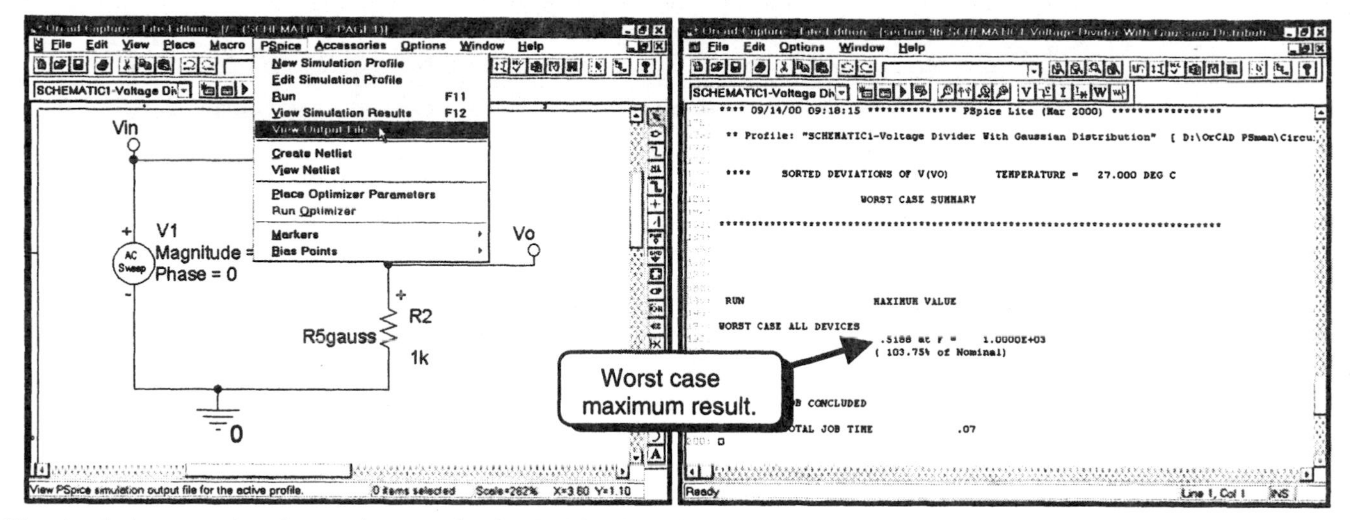


The simulation says that the maximum value is ***.5188***, which is less than the expected value. Remember that for the resistor with the 5% Gaussian distribution, the standard deviation was 1.25%, and the absolute limits on the distribution were $\pm 4\sigma = \pm 5\%$. **In the Worst Case analysis, a device with a Gaussian distribution is varied by only $\pm 3\sigma$.** Had we calculated the maximum value with a 3.75% resistor variation, we would have come up with a maximum gain of 0.51875, which agrees with the PSpice result. **To obtain the worst case limits, I prefer to use the uniform distribution.** Type **CTRL-F4** to close the output file and display the schematic.

We will now change the voltage divider circuit to use the resistor model ***R5pcnt***. To change the model reference of a resistor, double-click the ***LEFT*** mouse button on the text ***R5gauss***:

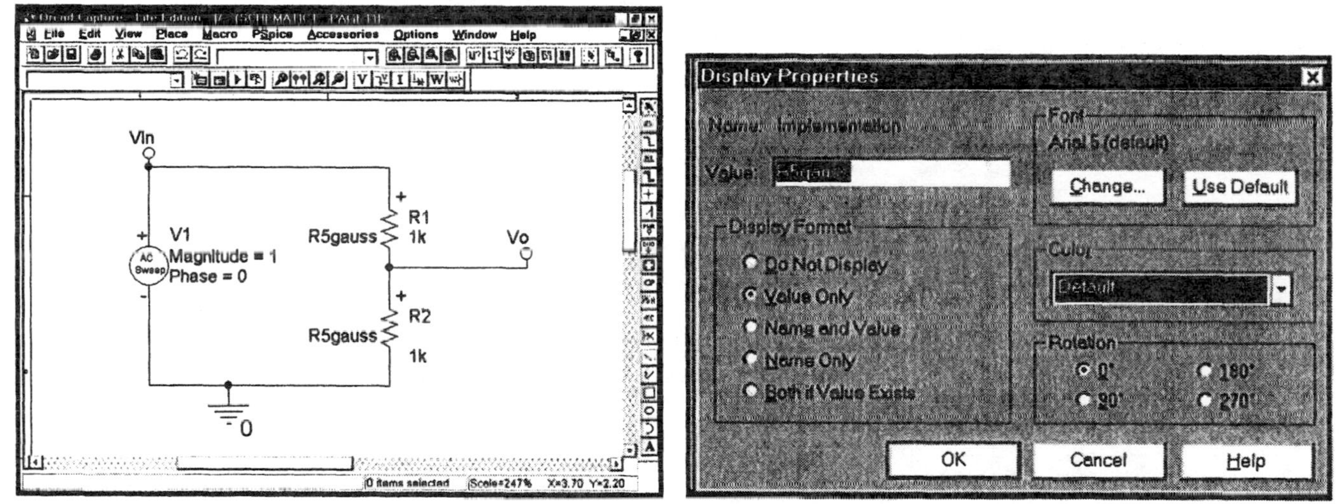

Type **R5pcnt** and click the ***OK*** button

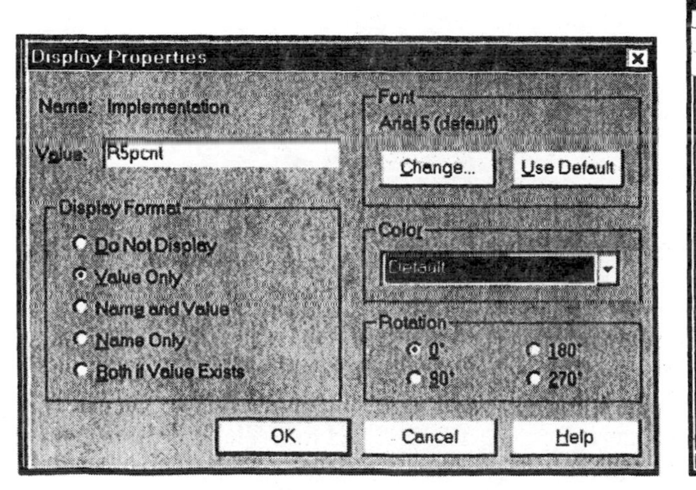

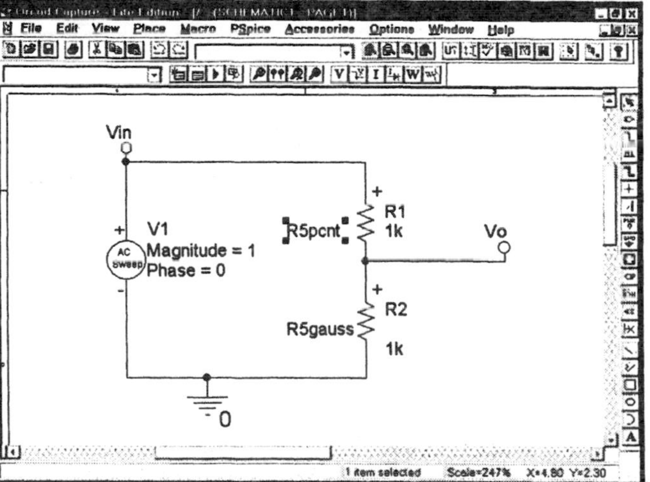

Change both resistors to R5pcnt and then run PSpice. The results will again be stored in the output file. At the end of the output file you will see the results of the Worst Case analysis:

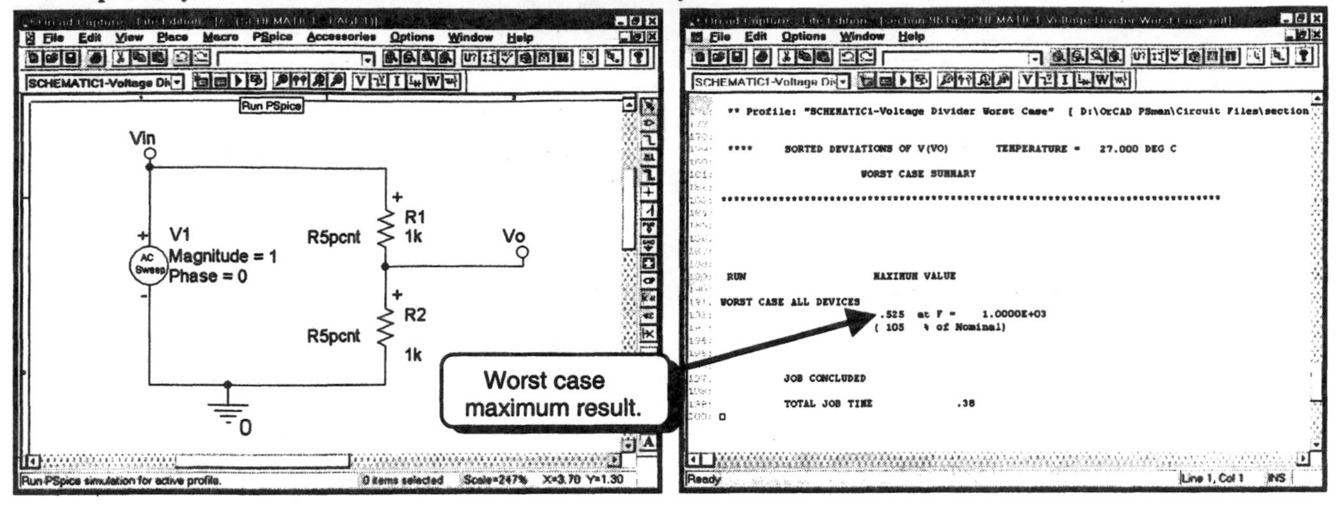


The results show the expected maximum value of ***.525***. Type **CTRL-F4** to close the output file and display the schematic.

We would now like PSpice to find the minimum value. Obtain the Monte Carlo setup dialog box. Select **PSpice** and then **Edit Simulation Profile** from the menus, and then select ***Monte Carlo/Worst Case***:

Simulation Settings - Voltage Divider Worst Case

General | Analysis | Include Files | Libraries | Stimulus | Options | Data Collection | Probe Window

Analysis type: AC Sweep/Noise

Options: General Settings, Monte Carlo/Worst Case, Parametric Sweep, Temperature (Sweep), Save Bias Point, Load Bias Point

○ Monte Carlo
◉ Worst-case/Sensitivity Output variable: V(Vo)

Monte Carlo options
Number of runs:
Use distribution: Uniform Distributions...
Random number seed: [1..32767]
Save data from: All runs

Worst-case/Sensitivity options
Vary devices that have both DEV and LOT tolerances
Limit devices to type(s):
☐ Save data from each sensitivity run

More Settings...

OK Cancel Apply Help

Select the ***More Settings*** button and fill in the dialog box as shown:

Monte Carlo/Worst-Case Output File Options

Function
The function is performed on an output variable (for example, V(1)).
The result is listed in the output (.OUT) file only.

Find: the minimum value (MIN)
Threshold value:
Evaluate only when the sweep variable is in the range: to

Worst-Case direction
○ Hi ◉ Low

☐ List model parameter values in the output file for each run

OK Cancel

***MIN*** selected.

***Low*** selected.

Click the ***OK*** button twice to return to the schematic.

Run PSpice again to find the minimum value. The results will again be stored in the output file. At the end of the output file you will see the results of the Worst Case analysis:

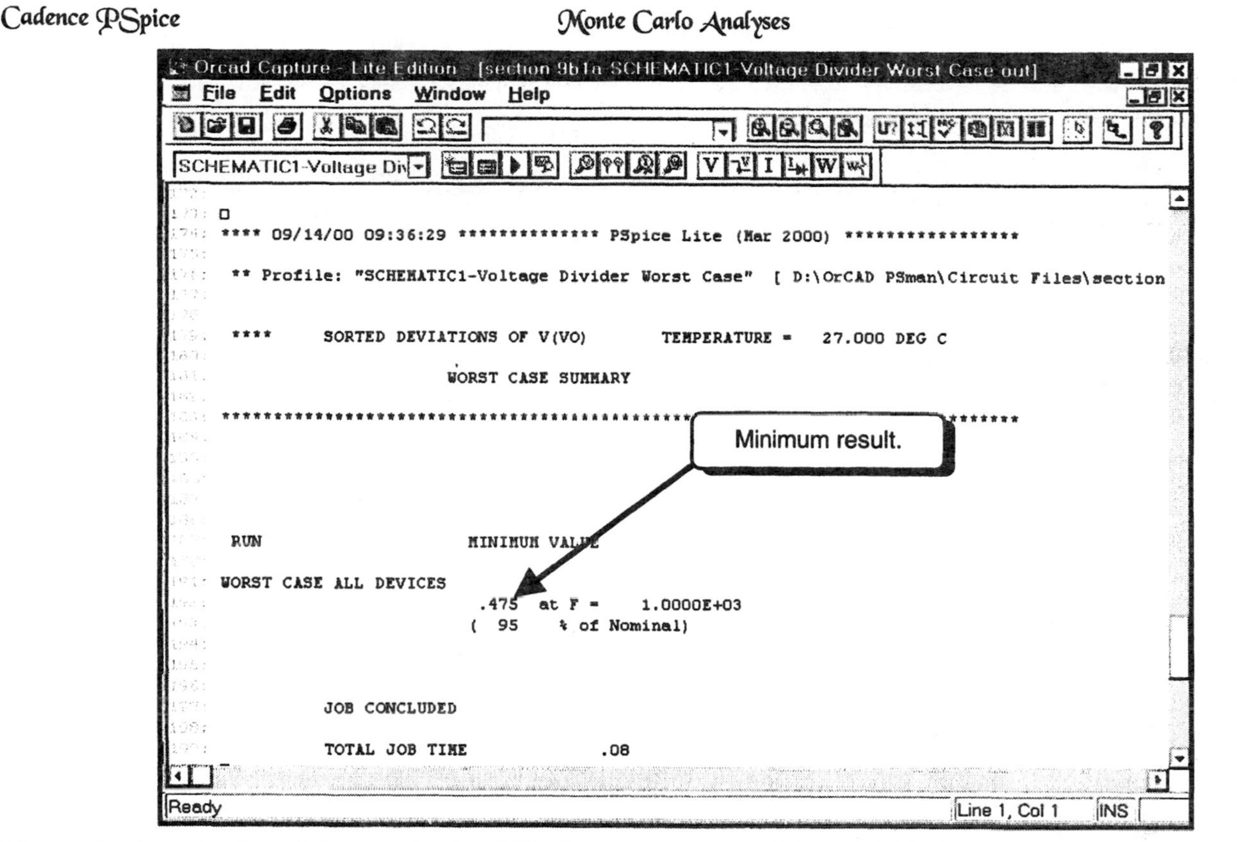


The results show that the minimum value is ***.475***. These values agree with our calculations.

**EXERCISE 9-1**: Find the worst case minimum and maximum voltage gain of the voltage divider if 10% resistors are used instead of 5%. See Section 7.C for instructions on making modifications to existing models, and see **EXERCISE 7-3** on page 441 to change the models of the resistors to 10%.

**SOLUTION**: From calculations, the maximum gain should be 0.55 and the minimum gain should be 0.45.

## 9.B.2. Voltage Divider Monte Carlo Analysis

The Monte Carlo analysis is used to answer the question, "What percentage of my circuits will achieve or exceed my specifications?" Usually you would run a Worst Case analysis to see if all of the circuits pass the specifications. If they all pass, there is no need to run the Monte Carlo analysis. If they do not all pass, the Monte Carlo analysis is used to estimate what percentage of the circuits will pass.

An example would be the gain of our voltage divider. We may ask the question, "A minimum gain of 0.4 is required; what percentage of the circuits will have a gain of 0.4 or higher?" From the Worst Case analysis we know that the worst case minimum gain is 0.475, so we know that all of the circuits will achieve the specification. There is no need to run a Monte Carlo analysis to see whether the circuit passes this specification since the Worst Case analysis told us that all of the circuits will achieve the specification. However, we may ask the question, "A minimum gain of 0.49 is required; what percentage of the circuits will achieve the specification?" The Worst Case analysis told us that some of the circuits may have a gain as low as 0.475. Not all of the circuits may achieve the specification, and we need to run a Monte Carlo analysis to answer the question.

The accuracy of the Monte Carlo analysis depends greatly on knowing the tolerance distributions of your parts. The Gaussian distribution is considered a better model of a part's distribution than the uniform distribution [1]. To make an accurate simulation you should find out the distributions of your parts. If the Gaussian or uniform distributions are not good models for your parts, you may have to make up your own distributions using the PSpice ".Distribution" statement [2]. See the PSpice Circuit Analysis manual for more information on this statement. To illustrate the effect of different tolerance distributions on the Monte Carlo analysis, we will simulate the voltage divider using resistors with Gaussian and uniform distributions.

### 9.B.2.a. Voltage Divider Gain Analysis with Uniform Tolerance Distribution

We will run a Monte Carlo analysis on the voltage divider circuit previously described. The circuit and parts are repeated below:

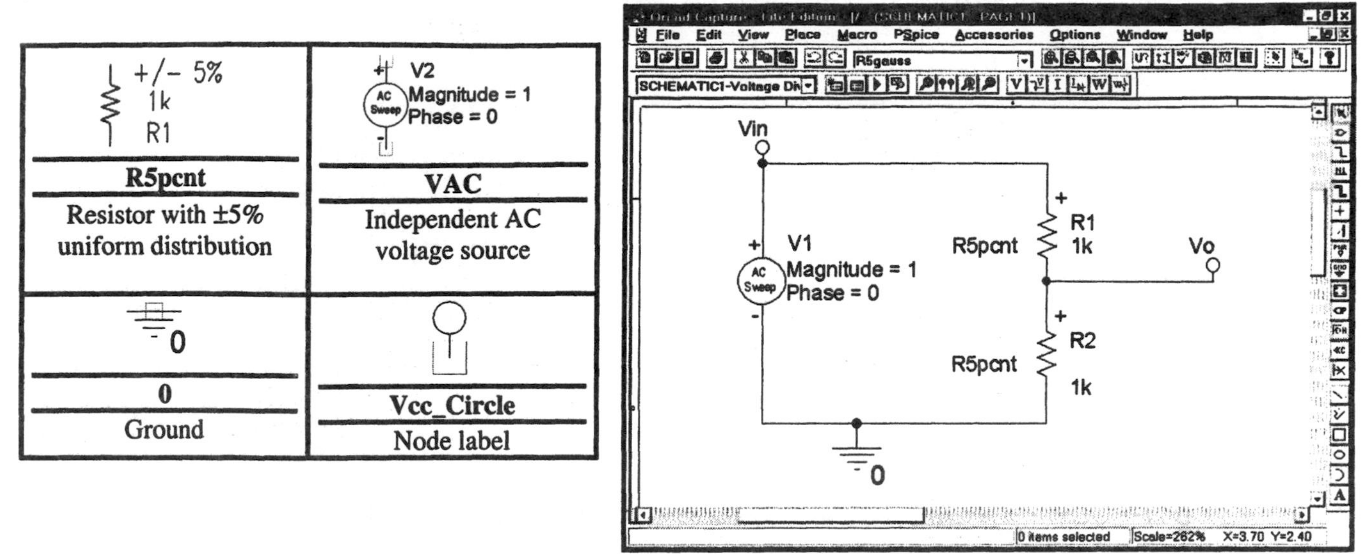


It is important to note that the resistor part used is ***R5pcnt*** and not R5gauss. The model ***R5pcnt*** has a 5% uniform distribution. We would like to find the gain of this circuit, so we must set up an AC Sweep. Select **PSpice** and then **New Simulation Profile** from the Capture menus, enter a name for the profile, and then click the ***Create*** button. Select the ***AC Sweep/Noise Analysis type***. Fill in the parameters as shown:

The AC Sweep is set up to simulate the circuit at a single frequency of 1 Hz.

Next, we need to set up the Monte Carlo analysis. Click the ***LEFT*** mouse button on the □ next to the text ***Monte Carlo/Worst Case*** button. This will enable the analysis and display its settings:

Click on this □.

Analysis selected and enabled.

Fill in the options as shown below:

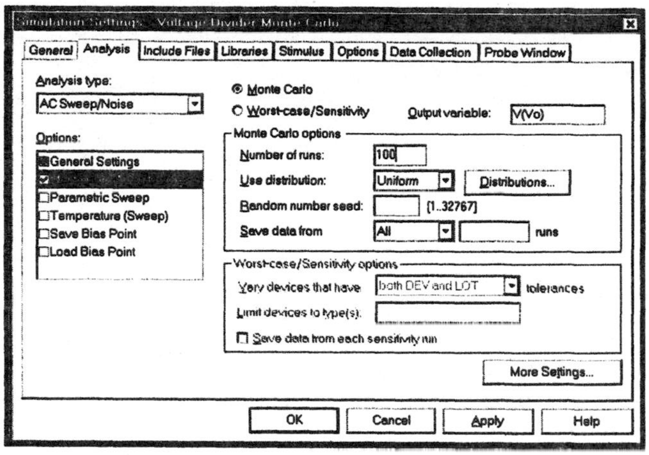

The ***Analysis*** is set up for ***Monte Carlo***. The number of runs is ***100***. The AC Sweep will run 100 times. For each run, each part that has tolerance will have a value randomly chosen within its tolerance range. The ***Output variable*** is ***V(Vo)***.

Click the ***More Settings*** button:

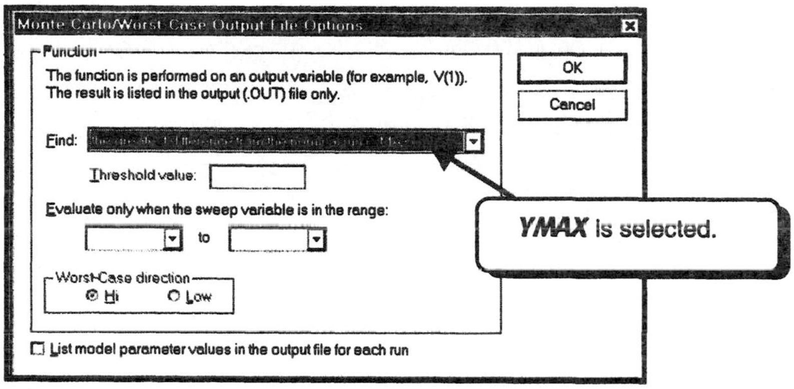

YMAX is selected.

Make sure that ***YMAX*** is selected as in the dialog box above. This function instructs PSpice to sort the output according to the maximum difference from the nominal value. We found that the nominal value of the gain was 0.5. ***YMAX*** specifies the output function:

$$f = |V(Vo) - 0.5|$$

The output from PSpice is the function $f$ sorted in descending order. Since we know the nominal value, we can obtain ***V(Vo)*** from the nominal value and $f$. Click the ***OK*** button to return to the ***Simulation Settings*** dialog box:

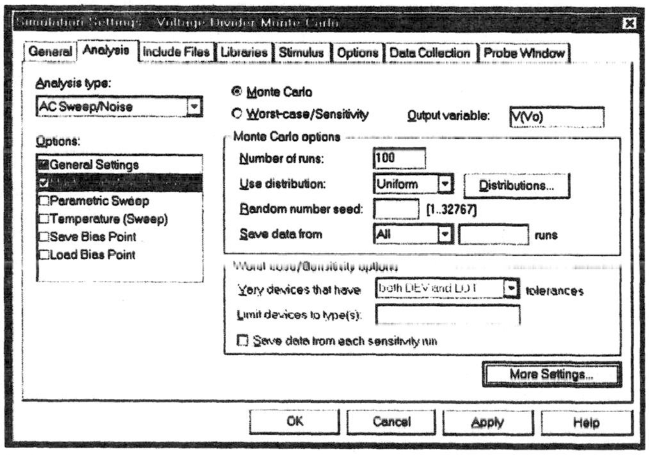

The results of this simulation will be contained in the output file and we do not need Probe to run. Select the ***Probe Window*** tab and deselect the options that specify that Probe should display a window:

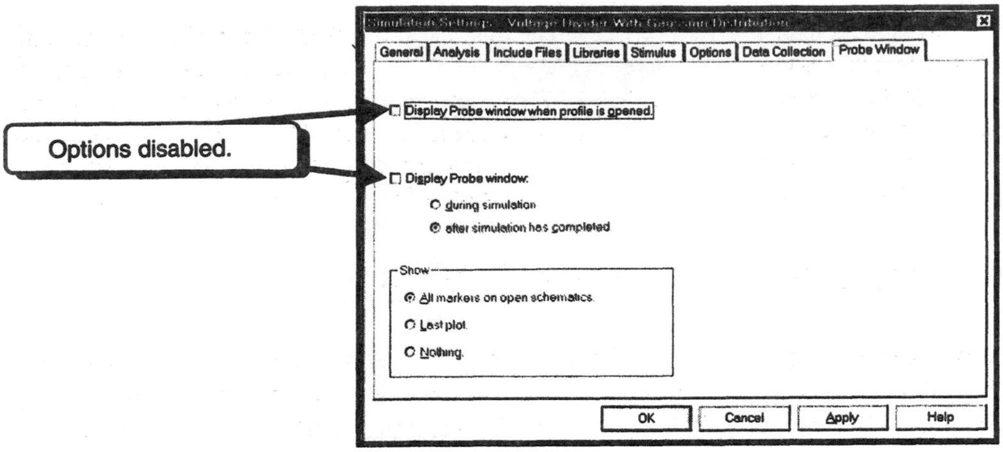

Options disabled.

Click the ***OK*** button to return to the schematic.

Run PSpice. Note that the message window tells you which pass is being simulated.

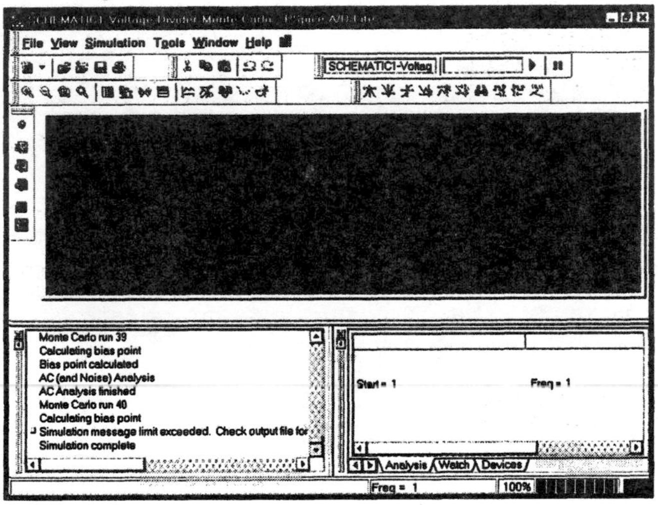

The results of the Monte Carlo analysis are stored in the output file. Examine the output file: Select **PSpice** and then **View Output File** from the Capture menus. As you page down through the output file you will see a screen similar to the one shown below:

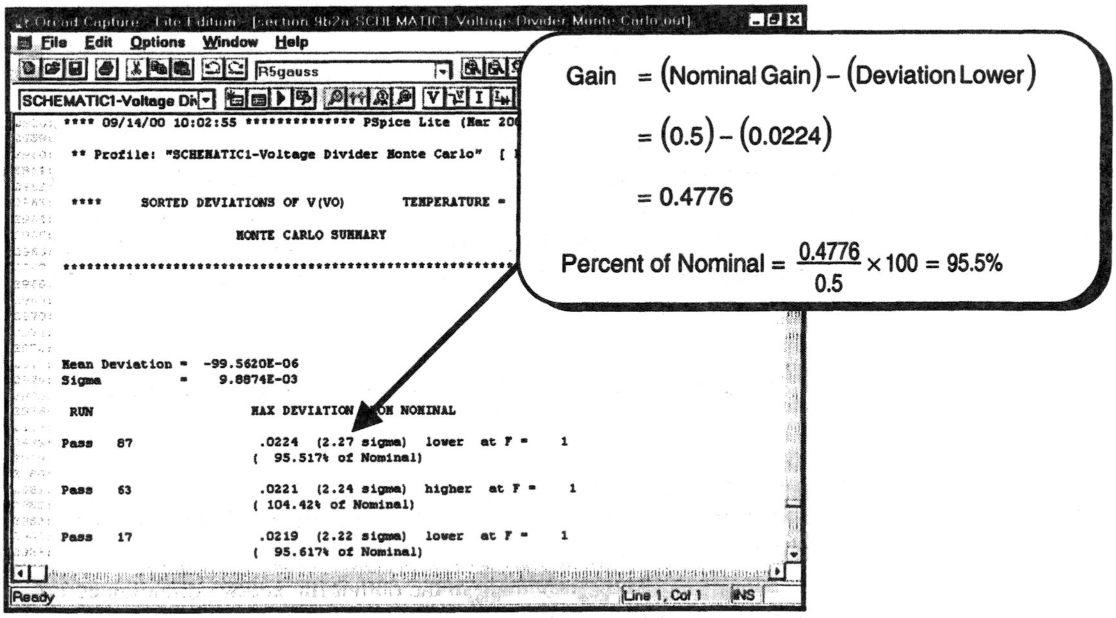

$$\text{Gain} = (\text{Nominal Gain}) - (\text{Deviation Lower})$$
$$= (0.5) - (0.0224)$$
$$= 0.4776$$
$$\text{Percent of Nominal} = \frac{0.4776}{0.5} \times 100 = 95.5\%$$

```
**** 09/14/00 10:02:55 ************** PSpice Lite (Mar 20
** Profile: "SCHEMATIC1-Voltage Divider Monte Carlo"
****     SORTED DEVIATIONS OF V(VO)      TEMPERATURE =
                  MONTE CARLO SUMMARY

Mean Deviation =  -99.5620E-06
Sigma          =    9.8874E-03

RUN                    MAX DEVIATION FROM NOMINAL

Pass   87              .0224  (2.27 sigma)  lower  at F =   1
                       (  95.517% of Nominal)

Pass   63              .0221  (2.24 sigma)  higher  at F =   1
                       ( 104.42% of Nominal)

Pass   17              .0219  (2.22 sigma)  lower  at F =   1
                       (  95.617% of Nominal)
```

The results are given as deviations from the nominal. From calculations we know that the nominal value is ***0.5***. We wanted to know how many circuits would have a gain of 0.49 or greater. This means that all circuits with a ***lower*** deviation of more than 0.01 from the nominal will not pass. Circuits with a ***higher*** deviation have a gain greater than the nominal and pass the specification. All we need to do is count the number of ***lower*** deviations larger than 0.01. These will be the circuits that do not pass the specification. Since this is a random distribution, your simulation will give slightly different results. In my simulation I found that 17 runs had ***lower*** deviations greater than 0.01. This means that 17% of my circuits will have a gain less than 0.49. It is important to note that the more runs you do, the more accurate the simulation will be. I chose 100 runs to reduce the simulation time. To get a more accurate estimate, you may wish to perform more than 100 runs. The more runs, the more accurate your results. The maximum number of runs you can do in any Monte Carlo simulation is 9999.

### 9.B.2.b. Voltage Divider Gain Analysis with Gaussian Tolerance Distribution

We will now simulate the voltage divider using resistors with a Gaussian distribution. Change the model reference of the two resistors from ***R5pcnt*** to **R5gauss**. See Section 7.A for instructions on changing the model reference. The R5gauss resistors have a Gaussian tolerance distribution of ±5% with a standard deviation of 1.25%. The setup is the same as in the previous simulation, so all we have to do is run the simulation. Run PSpice and then examine the output file.

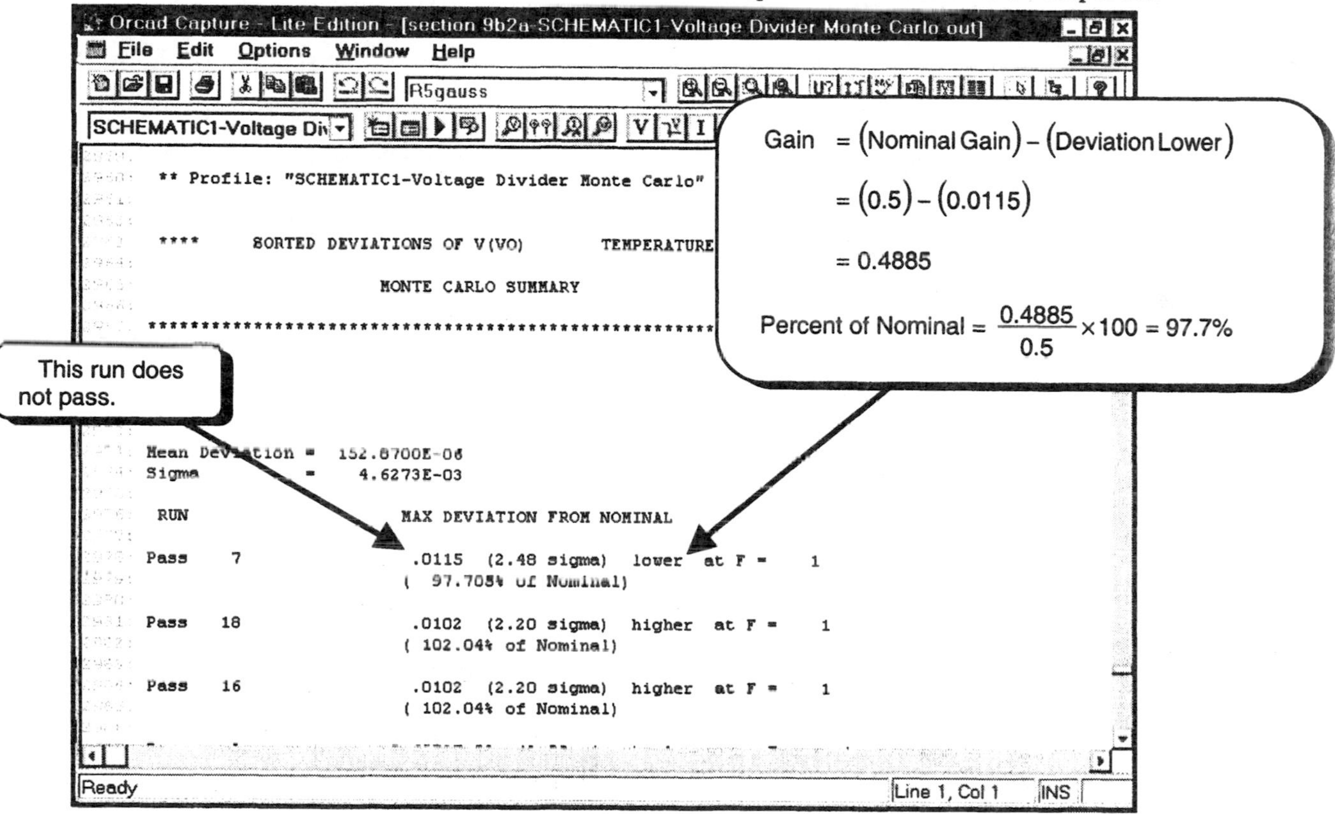


The output is similar to the previous simulation except that the deviations from the nominal value are smaller. Only one run did not pass the specification. The run had a lower deviation of 0.0115, which corresponds to a gain of 0.5 – 0.0115 = 0.4885. Smaller deviations should be expected since, in the Gaussian distribution, the bulk of the resistors were within plus or minus one standard deviation. From the results of the two previous simulations, we conclude that the tolerance distribution has a large impact on the Monte Carlo results.

To get more accurate results, the above simulation was run again with 9999 runs, showing that 113 of the runs did not pass the gain specification, or 1.13% failed.

**EXERCISE 9-2**: What percentage of the voltage divider circuits will have a gain of 0.49 or greater if 10% resistors with a Gaussian distribution are used?

**SOLUTION**: First, create a model with a 10% Gaussian distribution. See **EXERCISE 7-3** on page 441 to change the models of the resistors to 10%. The model for a 10% resistor with a Gaussian distribution is:

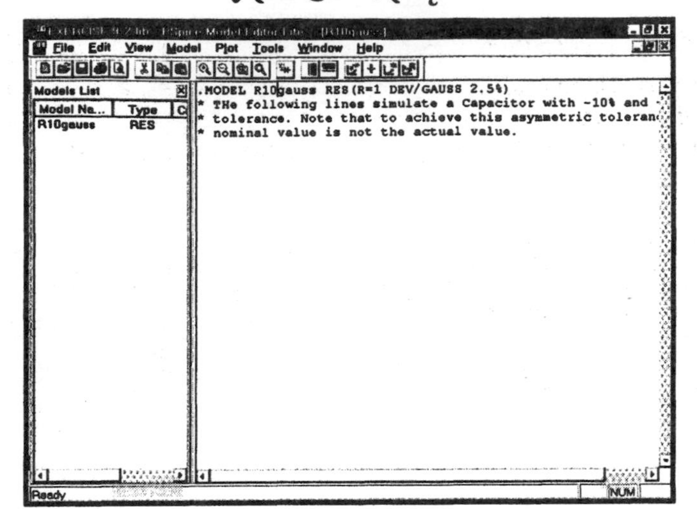


Remember that in a Gaussian distribution, the tolerance specified is the standard deviation σ, and the full distribution extends ±4σ. Change both resistors to R10gauss:

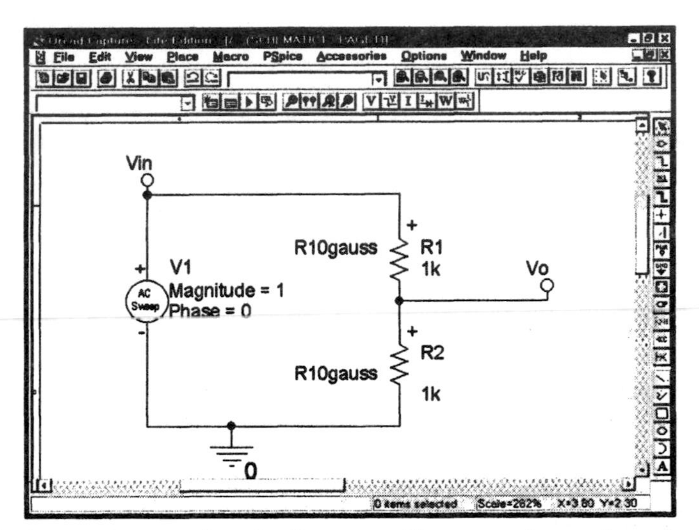


Run the simulation as shown in the previous section. In the output file, count the number of runs that have a lower deviation greater than 0.01. The results below are for 1000 Monte Carlo runs:

```
File  Edit  Options  Window  Help
SCHEMATIC1-Monte Carl
**** 09/14/00 10:23:44 ************** PSpice Li

 ** Profile: "SCHEMATIC1-Monte Carlo Ex 9-2"  [

****      SORTED DEVIATIONS OF V(VO)        TEMP

                     MONTE CARLO SUMMARY

*************************************************

Mean Deviation =   56.7300E-06
Sigma          =    8.8644E-03

 RUN                      MAX DEVIATION FROM NOMINAL

Pass 9800                  .0359  (4.05 sigma)  lower  at F =    1
                          (  92.815% of Nominal)

Pass 4832                  .0359  (4.05 sigma)  lower  at F =    1
                          (  92.823% of Nominal)

Pass 7335                  .0356  (4.02 sigma)  higher  at F =    1
                          ( 107.12% of Nominal)
Ready                                                Line 1, Col 1   INS
```

$$\text{Gain} = (\text{Nominal Gain}) - (\text{Deviation Lower})$$
$$= (0.5) + (0.0359)$$
$$= 0.4641$$

$$\text{Percent of Nominal} = \frac{0.4641}{0.5} \times 100 = 92.82\%$$

With 1000 runs, I found that 129 out of 1000 runs had a gain lower than 0.49, or 12.9% failed.

## 9.B.3. Performance Analysis — Voltage Divider Gain Spread

The Performance Analysis can be used in conjunction with the Monte Carlo analysis to view the distribution of a parameter as a function of device tolerances. For this example, we will display how the spread of the gain V(Vo)/V(V1) varies with resistor tolerances. We will use the voltage divider of the previous section and 5% resistors with a uniform distribution:

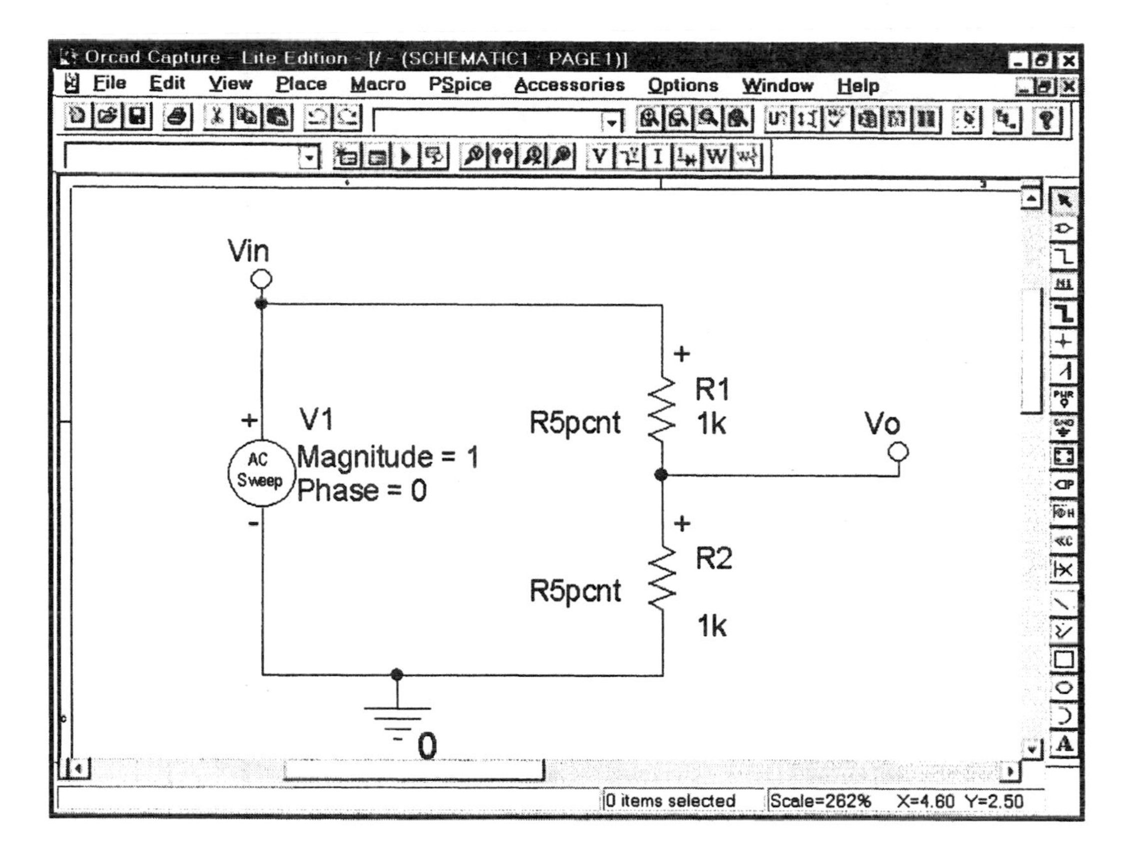


Set up the AC Sweep and Monte Carlo analysis as in the previous section. For this simulation, set the number of Monte Carlo runs to 9999. This is the largest number of runs we can view with Probe. (You may want to use fewer runs to reduce the run time of this example.)

***All*** must be selected.

Note that in the ***Monte Carlo options*** portion of the dialog box, ***All*** is selected. If this option is not selected, you will not see any results in the performance analysis.

We will be viewing the results with Probe. Select the ***Probe Window*** tab. If you are continuing this example for the previous sections, your options may be different than those shown below. Select the options as shown:

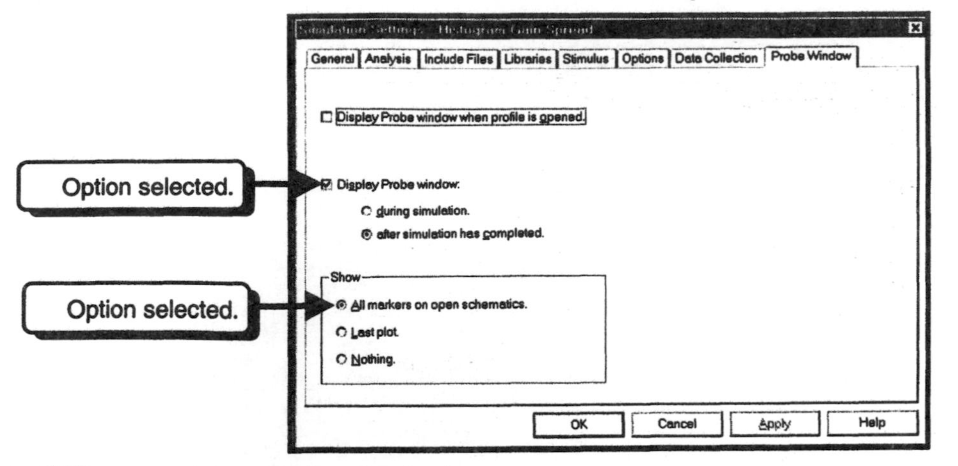


Click the ***OK*** button and then simulate the circuit:

By default, Probe selects all of the runs. There are a total of 9999 Monte Carlo runs. We would like to compose a histogram of all of the results, so click the ***OK*** button to enter Probe and use the 9999 selected traces.

Since we ran a Monte Carlo analysis and each run contains only a single point (the gain at 1 kHz), the only graph we can create is a histogram. Probe recognizes this and automatically comes up with an empty histogram. Add the trace **V(Vo)**:

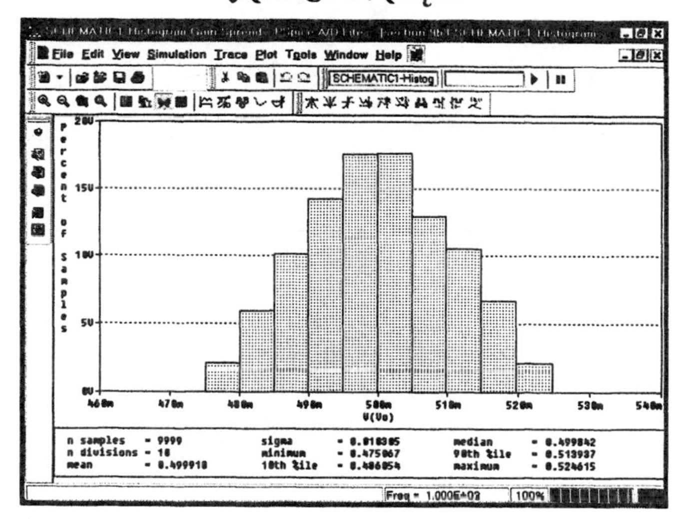


This plot shows us the distributions of gains for the specified tolerances in the circuit. The plot gives us a great deal of information: typical maximum and minimum gain to expect (not necessarily worst case maximum and minimum), mean and median gains, as well as the standard deviation (sigma).

This histogram has 10 bars (***n divisions = 10***). If you would like a histogram with more bars, select ***Tools*** and then ***Options*** from the Probe menu:

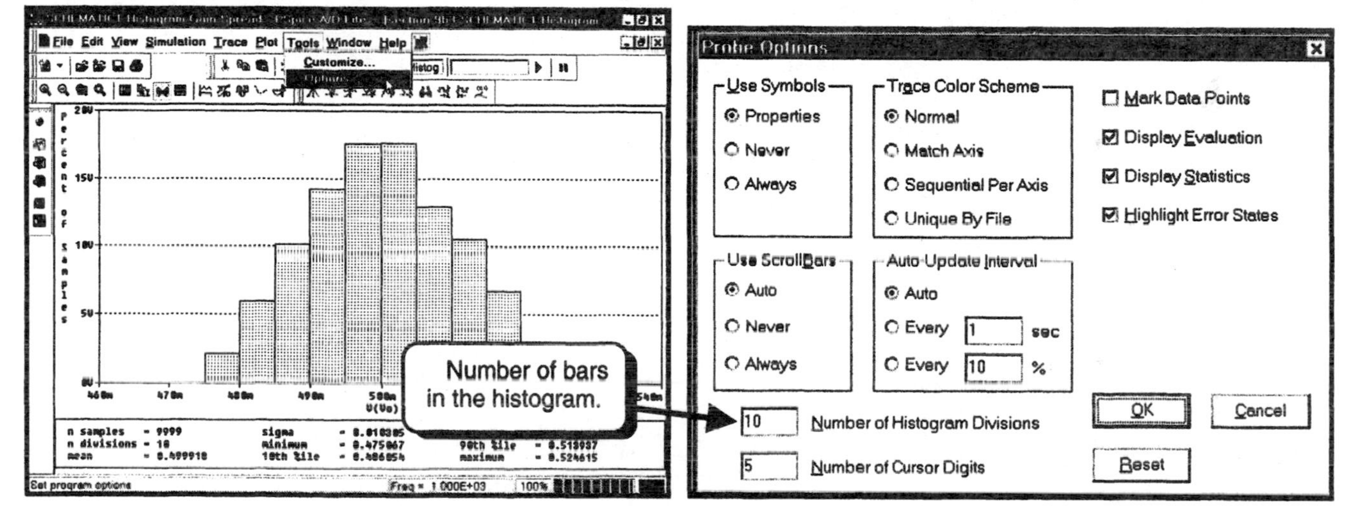


The number of bars in the histogram is set to ***10***. We would like to see more divisions, so set the ***Number of Histogram Divisions*** to `100`:

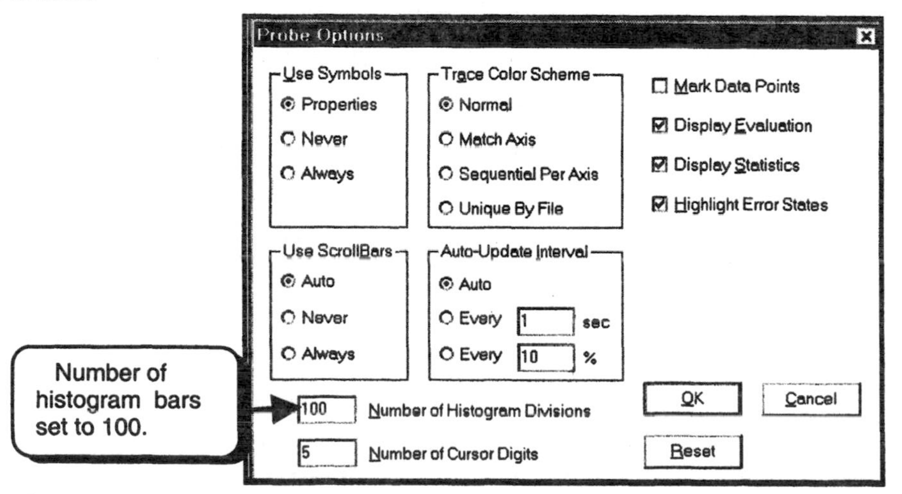


Click the ***OK*** button to view the graph:

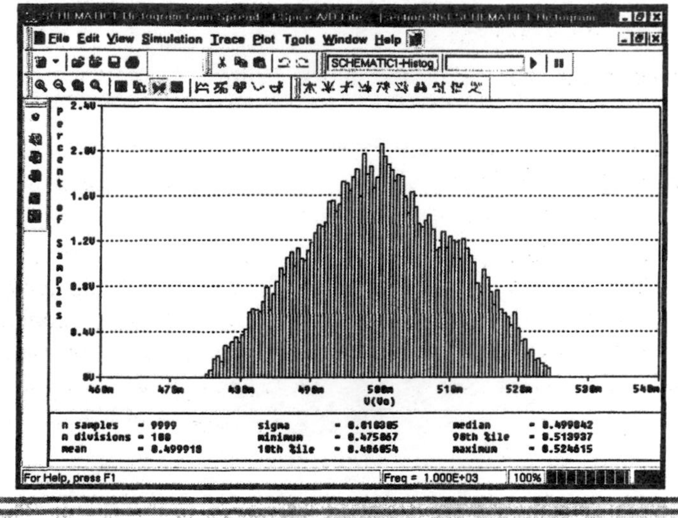

**EXERCISE 9-3**: Find the histogram for the gain of the voltage divider if 5% resistors with a Gaussian distribution instead of a uniform distribution are used. Compare the standard deviation, and minimum and maximum values of the histogram to the histogram using a uniform distribution.

**SOLUTION**: Change the model of the resistors to use the model ***R5gauss***:

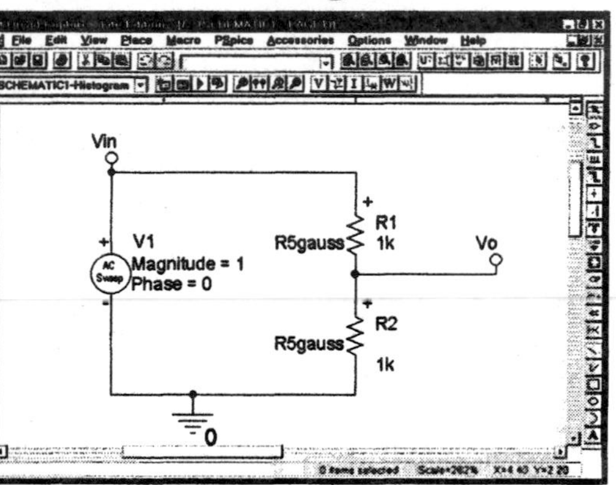


Run the simulation. When Probe runs, select all of the traces and then add the trace V(Vo):

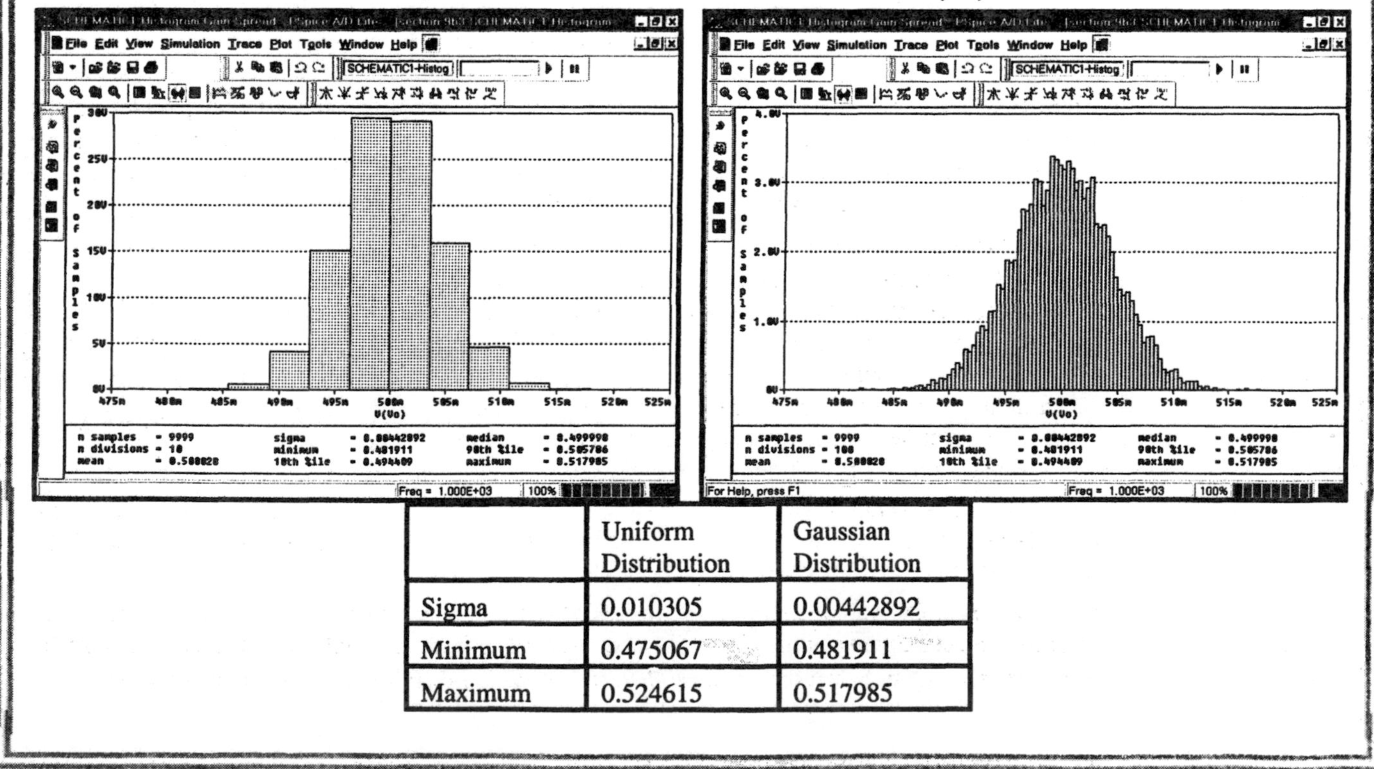

| | Uniform Distribution | Gaussian Distribution |
|---|---|---|
| Sigma | 0.010305 | 0.00442892 |
| Minimum | 0.475067 | 0.481911 |
| Maximum | 0.524615 | 0.517985 |

## 9.B.4. Voltage Divider Summary

We can draw a few conclusions about using the Monte Carlo and Worst Case analyses from the results of the voltage divider circuit.

- For Worst Case analyses, use uniform distributions. In a uniform distribution, PSpice uses the absolute limits of the distribution to find the worst case limits. In a Gaussian distribution, PSpice uses $\pm 3\sigma$ to calculate the worst case limits, even though the part can have a maximum deviation of $\pm 4\sigma$.
- In the Monte Carlo analysis the tolerance distribution has a major effect on the results. It is best to know the distribution of your parts before you trust the results of the Monte Carlo analysis. You can use the PSpice ".Distribution" statement to define your own probability distributions. If you do not know the distribution of your parts, the Gaussian distribution is a better representation of parts than the uniform distribution.
- When using a Gaussian distribution in a model, the specified deviation is one standard deviation, $\sigma$. The limits of the distribution are $\pm 4\sigma$. If you have $\sigma > 25\%$, the specified parameter could become negative.

# 9.C. BJT Bias Analysis

## 9.C.1. BJT Maximum and Minimum Collector Current

| Part | Description |
|---|---|
| **C** | Capacitor |
| **R5pcnt** | Resistor with 5% uniform tolerance distribution |
| **QBF** | Small-signal NPN BJT with $350 \geq \beta_F \geq 50$ |
| **VAC** | AC voltage source |
| **VDC** | DC source |
| **0** | Ground |
| **Vcc_Circle** | Node label |

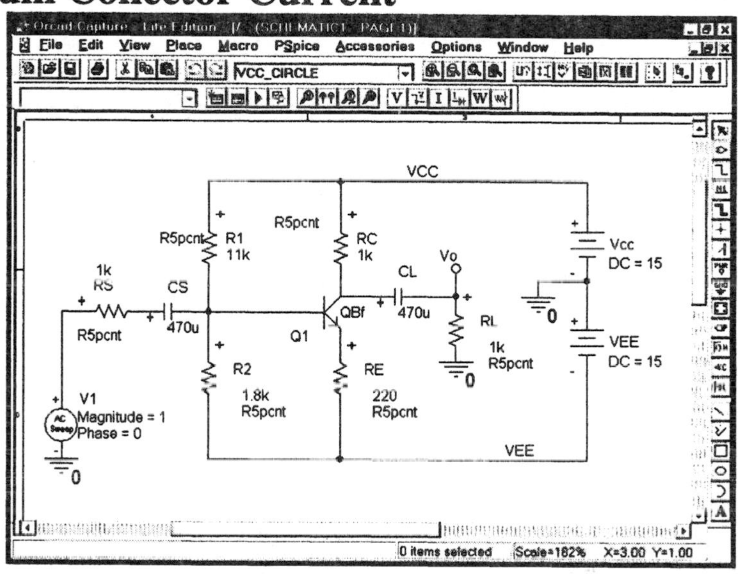


In amplifier design it is important to know how your bias will change with device tolerances. In this section we will find the minimum and maximum collector current of a BJT when we include variations in the transistor current gain, $\beta_F$, and resistor tolerances. The circuit above was previously simulated in the Transient Analysis and AC Sweep parts. We will use the same resistor values as before, but we will change the resistor models to include tolerance. The BJT is also changed to the model QBf. This model allows $\beta_F$ to have a uniform distribution between 50 and 350.

Note that all resistors have ±5% tolerance. All tolerance distributions are uniform. In addition, note that all resistors in the schematic have a plus sign at one of the terminals. This plus sign indicates PSpice's designation of the positive current reference of the device. This reference becomes important if we wish to know the current through a two-terminal device in PSpice. If we wish to know the current through a resistor, Rc, for example, we would specify I(Rc). This text string does not specify a direction for the current, so we need to indicate a positive direction of current in the schematic. The current direction is specified by the plus sign. Current is positive when it enters the positive terminal of the device, as shown in Figure 9-1.

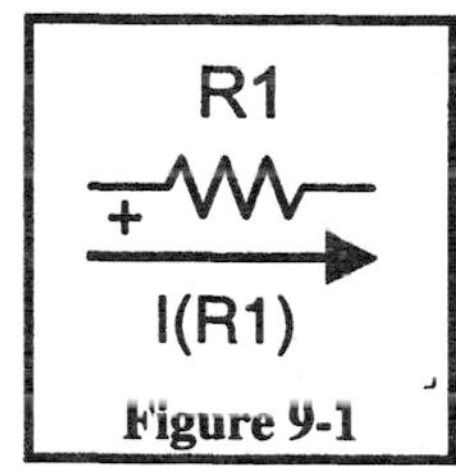


Figure 9-1

In the previous schematic, the plus sign on Rc is at the top of the resistor. This indicates that I(Rc) will be positive if it flows down. Remember, by definition, the collector current Ic is positive when it enters the collector terminal. Thus, I(Rc) = Ic. If the plus sign were at the bottom of Rc, then I(Rc) would equal –Ic.

The bias collector current is a DC value, so we must run a DC Sweep to find its minimum or maximum value. Obtain the DC Sweep dialog box: Select **PSpice** and then **New Simulation Profile** from the Capture menus, enter a name for the profile, and then click the ***Create*** button. Select the ***DC Sweep Analysis type***. Fill in the parameters as shown:

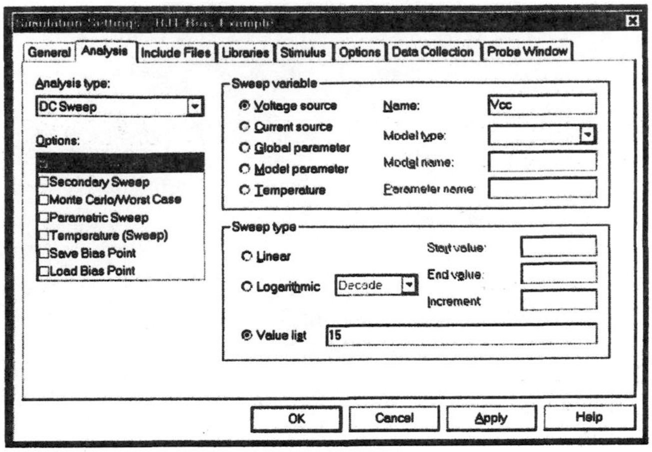

The dialog box is set up to simulate the circuit at Vcc = 15 V. This was already specified in the circuit and appears to be redundant. It must be specified because the Worst Case analysis must run in conjunction with an AC Sweep, a DC Sweep, or a Transient Analysis. Since we are interested in the DC collector current we must set up a DC Sweep.

Next we need to set up the Worst Case analysis. Click the ***LEFT*** mouse button on the □ next to the text ***Monte Carlo/Worst Case*** to select the analysis. Fill in the dialog box as shown:

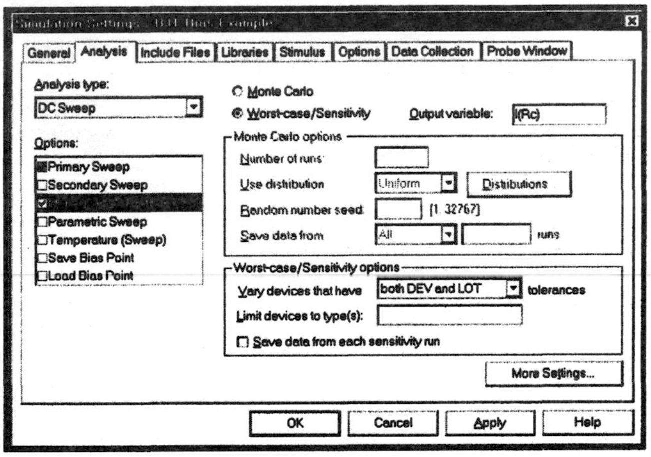

The dialog box is set up to find the worst case value of ***I(Rc)***. Since the current through Rc is the collector current, we are asking for the worst case value of the collector current.* Click the ***More Settings*** button and select ***MAX*** and ***HI*** as shown:

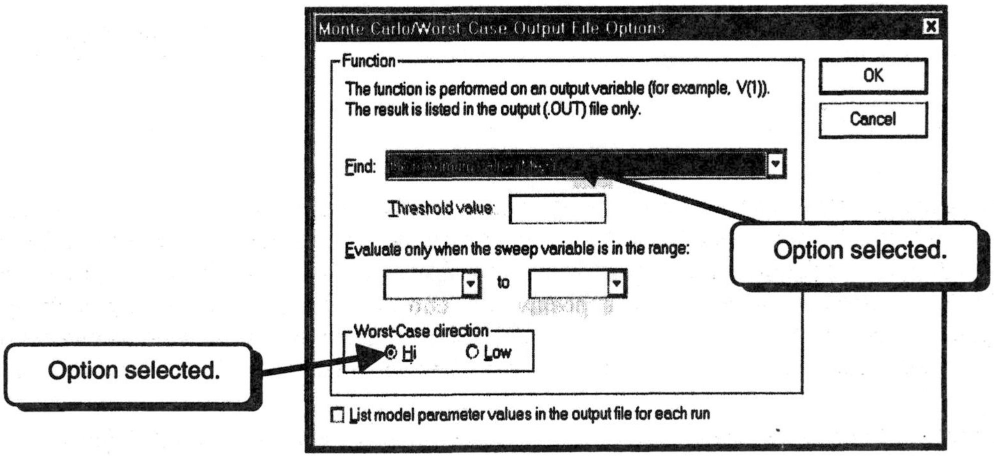

These settings specify that we will look for the worst case maximum value of the output variable I(Rc). Click the ***OK*** button to return to the ***Simulation Settings*** dialog box.

*We could also have asked for Ic(Q1). However, I(Rc) does not yield the same answer as Ic(Q1). I(Rc) agrees with the calculated results.

The results are stored in the output file so we do not need to run Probe. Select the ***Probe Window*** tab and fill in the options as shown to keep Probe from running after the simulation is complete:

General | Analysis | Include Files | Libraries | Stimulus | Options | Data Collection | Probe Window

☐ Display Probe window when profile is opened.

☐ Display Probe window:

○ during simulation

◉ after simulation has completed

Show

◉ All markers on open schematics

○ Last plot

○ Nothing.

OK | Cancel | Apply | Help

Click the ***OK*** button to return to the schematic.

Run PSpice and then open the output file by selecting **PSpice**, and then **View Output File** from the Capture menus. At the bottom of the output file you will see the results of the Worst Case analysis:

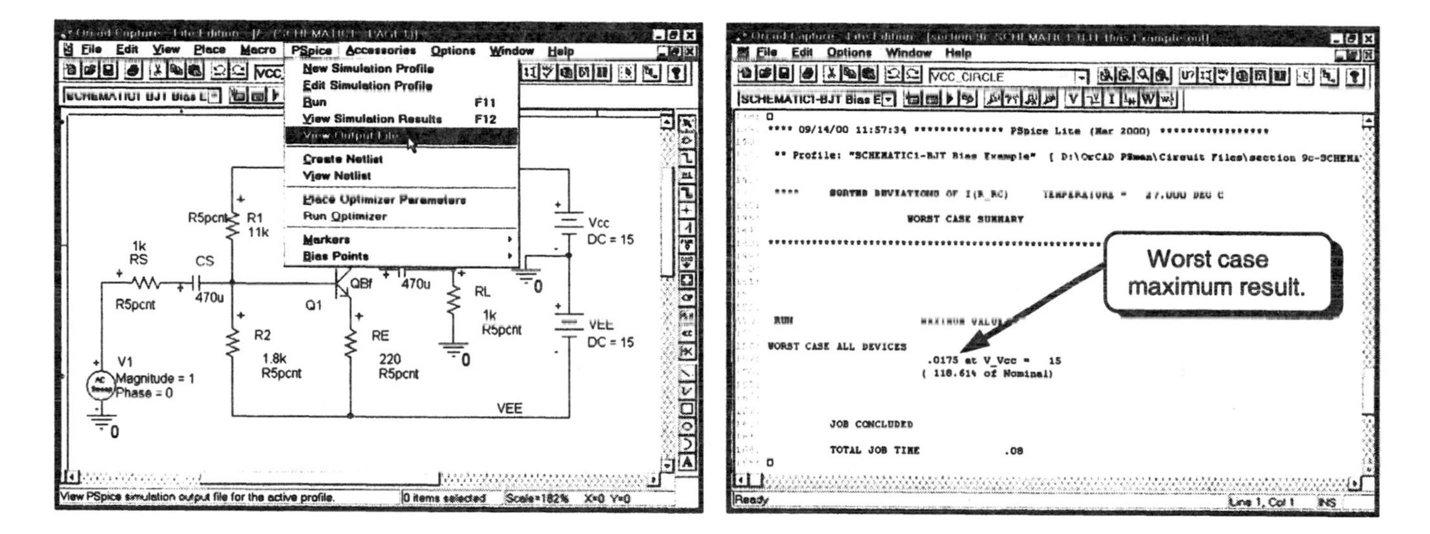


The results show us that the maximum value is ***17.5*** mA.

**Note: When you run the simulation you may come up with negative values for the current through Rc.** This result is not wrong and is due to the polarity reference of the resistor. The resistor graphic has a positive reference, indicated by the plus sign. If the plus sign for Rc is at the bottom of the graphic, you will get negative values for I(Rc) because current is flowing down into the collector terminal. If you do not like negative values for the current, rotate the Rc graphic by 180 degrees, until the plus sign appears at the top of the resistor graphic. Also note that PSpice finds the numerical minimum and maximum values of the specified parameter. Negative currents for Rc switch the interpretation of minimum and maximum as well.

The output file also contains information on how each element was changed to achieve the result. If you page up in the output file, you will see the text shown below:

```
Orcad Capture - Lite Edition - [section 9c SCHEMATIC1-BJT Bias Example.out]
File  Edit  Options  Window  Help
VCC_CIRCLE
SCHEMATIC1-BJT Bias E

 ** Profile: "SCHEMATIC1-BJT Bias Example"  [ D:\OrCAD PSman\Circuit Files\section 9c-SCHEMA

 ****     UPDATED MODEL PARAMETERS          TEMPERATURE =   27.000 DEG C

                    WORST CASE ALL DEVICES

 ******************************************************************************

 Device       MODEL          PARAMETER    NEW VALUE
 Q_Q1         QBf            BF            350              (Increased)
 R_RS         R5pcnt         R               1              (Unchanged)
 R_R1         R5pcnt         R                .95           (Decreased)
 R_R2         R5pcnt         R              1.05            (Increased)
 R_RC         R5pcnt         R               1              (Unchanged)
 R_RE         R5pcnt         R                .95           (Decreased)
 R_RL         R5pcnt         R               1              (Unchanged)

 **** 09/14/00 11:57:34 ************** PSpice Lite (Mar 2000) ****************

 ** Profile: "SCHEMATIC1-BJT Bias Example"  [ D:\OrCAD PSman\Circuit Files\section 9c-SCHEMA

 ****     SORTED DEVIATIONS OF I(R_RC)      TEMPERATURE =   27.000 DEG C
Ready                                                   Line 1, Col 1    INS
```

This information tells us that to achieve maximum collector current, $\beta_F$ was increased to 350, ***RE*** was decreased by 5%, ***R1*** was decreased by 5%, and ***R2*** was increased by 5%. The other resistors had tolerance, but their values had no effect on the collector current.

To find the minimum collector current, all we have to do is modify the ***Monte Carlo/Worst-Case Output File Options*** dialog box to find the minimum value. Modify the dialog box as shown:

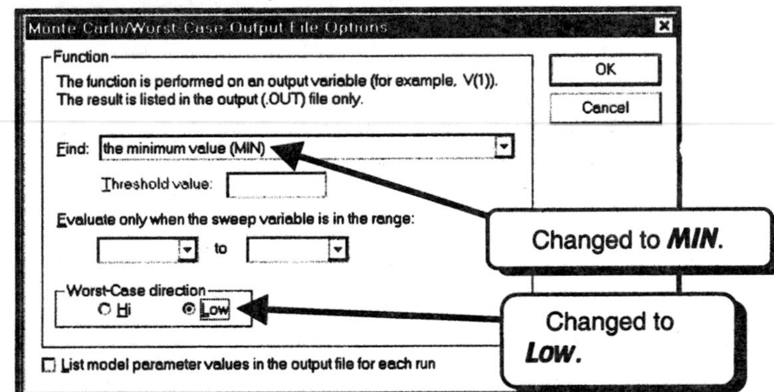


Run PSpice and then examine the output file. At the bottom of the output file you will see the results:

```
Orcad Capture - Lite Edition - [section 9c SCHEMATIC1-BJT Bias Example.out]
File  Edit  Options  Window  Help
VCC_CIRCLE
SCHEMATIC1-BJT Bias E

 **** 09/14/00 12:10:23 ************** PSpice Lite (Mar 2000) ****************

 ** Profile: "SCHEMATIC1-BJT Bias Example"  [ D:\OrCAD PSman\Circuit Files\section 9c-SCHEMA

 ****     SORTED DEVIATIONS OF I(R_RC)      TEMPERATURE =   27.000 DEG C

                    WORST CASE SUMMARY

 ******************************************************************************

 RUN                              MINIMUM VALUE

 WORST CASE ALL DEVICES
                                   .0114 at V_Vcc =   15
                                 (  77.447% of Nominal)

           JOB CONCLUDED

           TOTAL JOB TIME            .08

Ready                                                   Line 1, Col 1    INS
```

Worst case minimum result.

The results show that the minimum collector current is ***11.4*** mA.

**EXERCISE 9-4**: Find the maximum and minimum collector current of the circuit in this section if the base resistors have 1% tolerance rather than 5% tolerance.

**SOLUTION**:

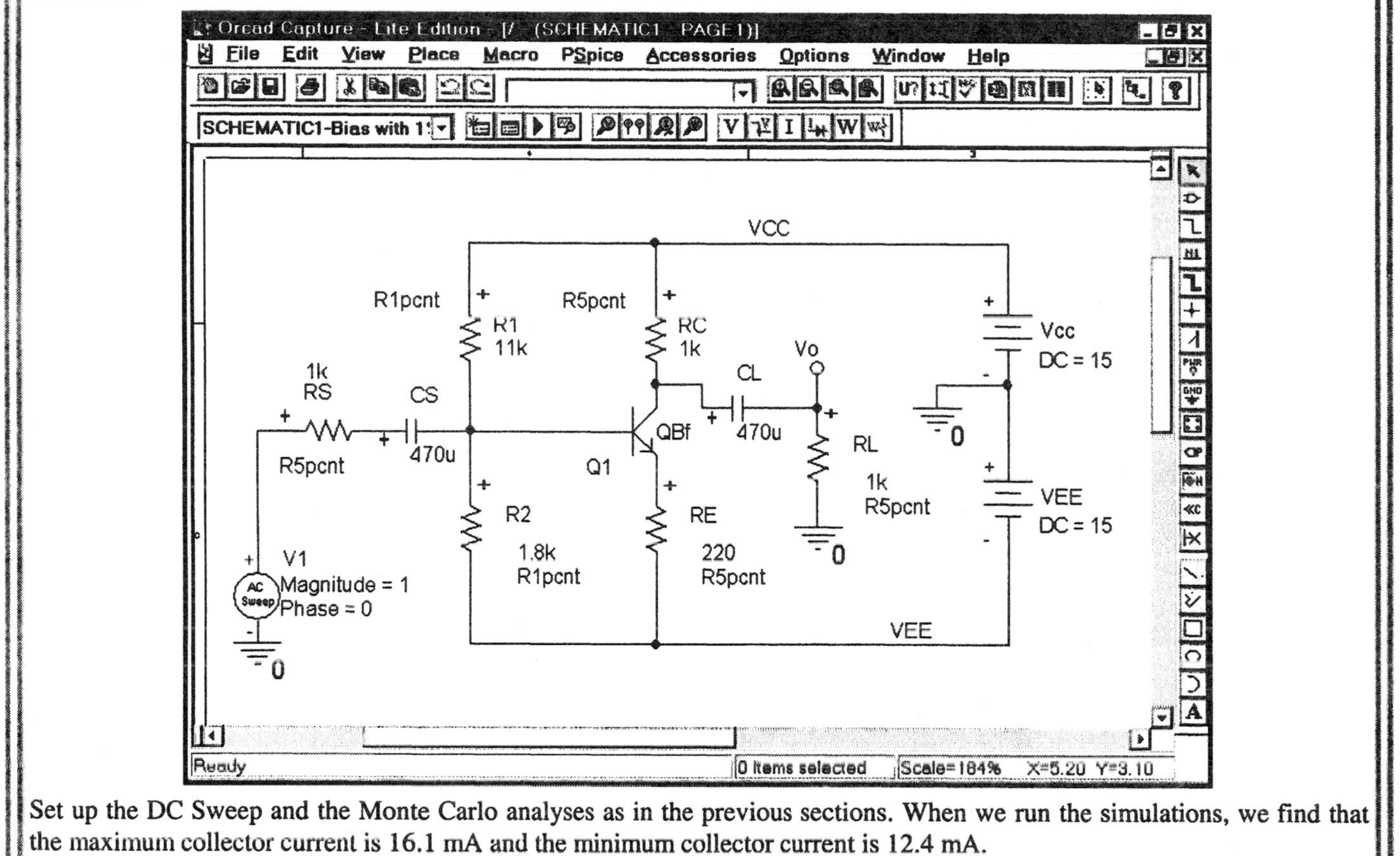


Set up the DC Sweep and the Monte Carlo analyses as in the previous sections. When we run the simulations, we find that the maximum collector current is 16.1 mA and the minimum collector current is 12.4 mA.

## 9.C.2. BJT Minimum $V_{CE}$

When biasing a BJT, we are also interested in the collector to emitter voltage, $V_{CE}$. The minimum or maximum value of $V_{CE}$ can also be easily found using the Worst Case analysis. We can use the same setup that was used to find the collector current. All we have to do is modify the ***Monte Carlo/Worst Case*** settings. Fill in the dialog boxes as shown below:

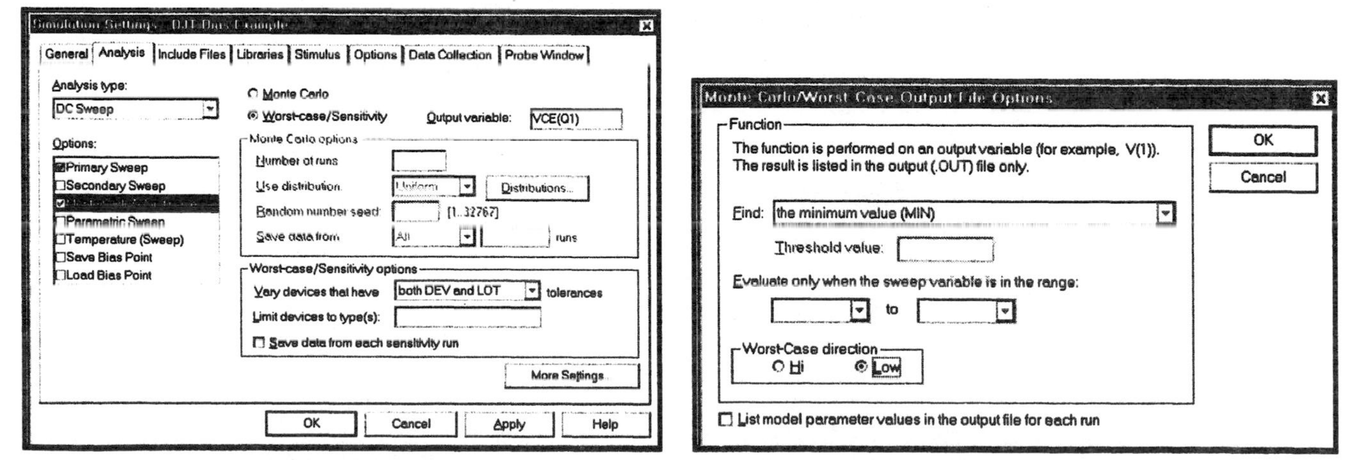


Note that the ***Output variable*** is ***VCE(Q1)***, the collector-emitter voltage of ***Q1***. ***MIN*** and ***Low*** are selected, so we are asking for the minimum value of $V_{CE}$. Run the analysis. The results are given at the end of the output file:

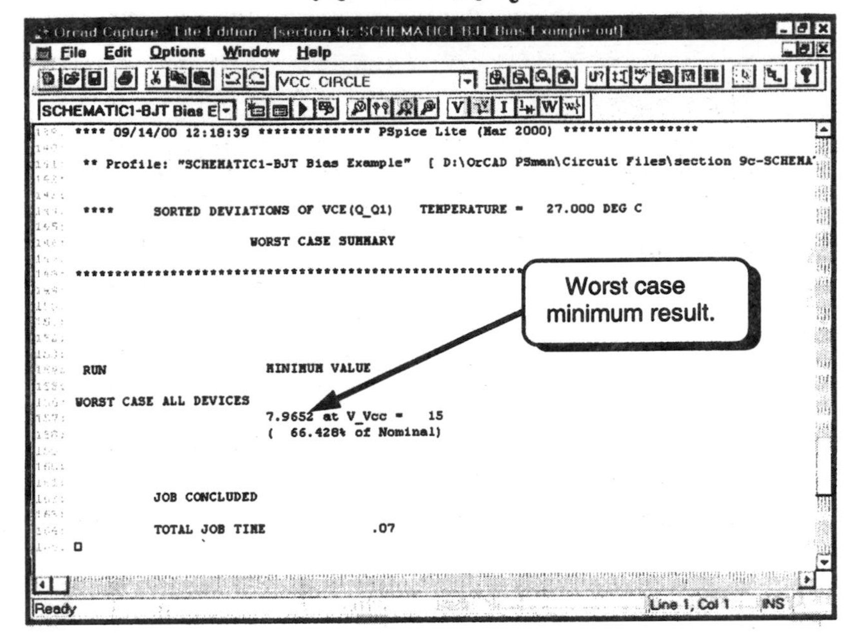


The results show that the minimum value of $V_{CE}$ is ***7.9652*** V.

**EXERCISE 9-5**: Find the minimum value of $V_{CE}$ of the circuit in this section if the base resistors have a 1% tolerance rather than a 5% tolerance.

**SOLUTION**: Use the circuit of **EXERCISE 9-4**. The result of the simulation shows that the minimum value of $V_{CE}$ is 9.7233 V.

# 9.D. BJT Amplifier Minimum and Maximum Gain

We will now find the minimum and maximum gain of the amplifier shown on page 298. The amplifier circuit is repeated below:

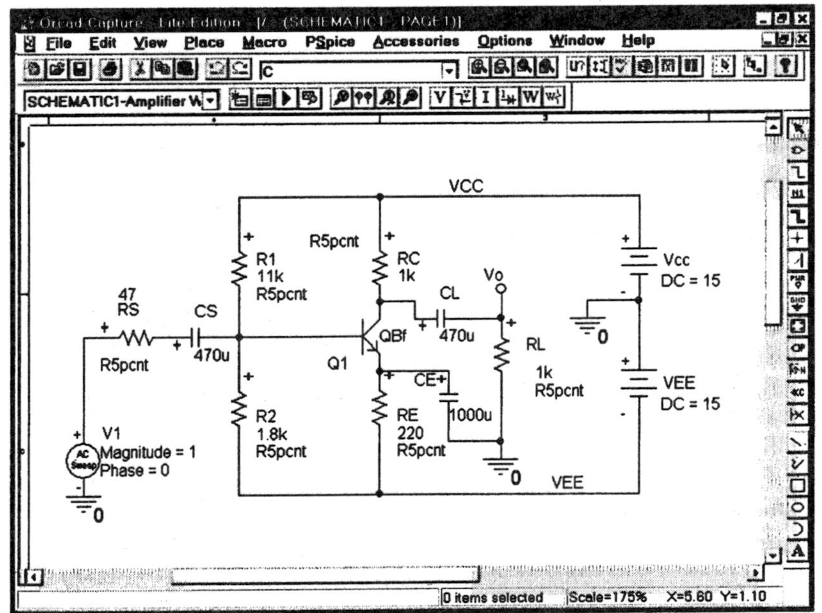


Note that ***RS*** has been changed to ***47*** Ω and an emitter bypass capacitor (***CE***) has been added. The frequency response of this amplifier was found in Section 5.D on page 298. We found the mid-band gain to be 45.7 dB and the upper and lower 3 dB frequencies to be 66 Hz and 6.3 MHz, respectively. The amplifier above is the same as that in Section 5.D except that the BJT is described by the model ***QBf*** rather than Q2N3904. The model ***QBf*** has $350 \geq \beta_F \geq 50$ and a nominal value of $\beta_F = 200$. The Q2N3904 model has no tolerance and a nominal value of $\beta_F = 416$. Thus, we should expect the nominal value of the gain of the amplifier above to be slightly different from the nominal gain of the amplifier from Section 5.D.

We would like to find the minimum and maximum gain at mid-band. We will set up an AC Sweep at 1 kHz. Fill in the AC Sweep dialog box as shown:

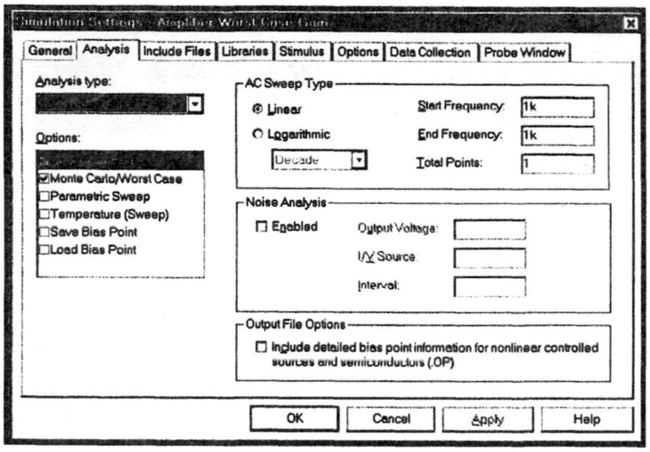

The dialog box is set to run an AC Sweep at a single frequency of 1 kHz.

Next we need to set up the Worst Case analysis. Fill in the ***Monte Carlo/Worst Case*** dialog boxes as shown below.

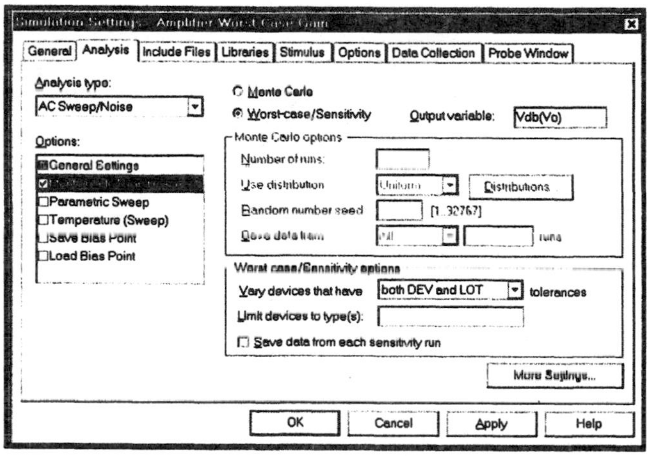

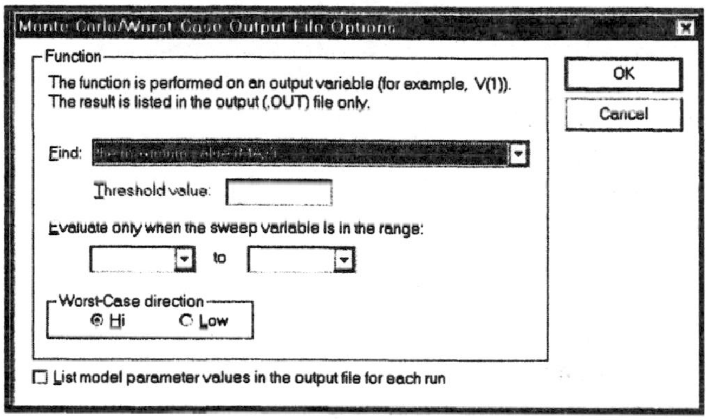

The dialog box is set up to find the Worst Case maximum value of the output variable ***Vdb(Vo)***. VdB is the voltage at the specified node in decibels, $VdB(V_o) = 20 \log_{10}(V_o)$. Since our only input is ***V1*** and the magnitude of ***V1*** is 1,

$$VdB(V_o) = 20 \log_{10}(V_o) = 20 \log_{10}\left(\frac{V_o}{V_1}\right)$$

Thus, ***Vdb(Vo)*** gives us the gain of our amplifier in decibels.

Run PSpice and then view the output file. The results of the Worst Case analysis are shown at the end of the file:

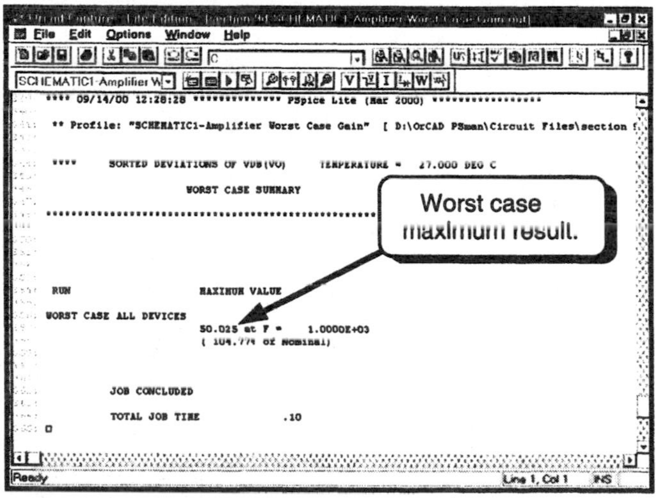

The results show a maximum gain of ***50.025*** dB.

To find the minimum gain, change the ***Monte Carlo/Worst-Case Output File Options*** dialog box as shown:

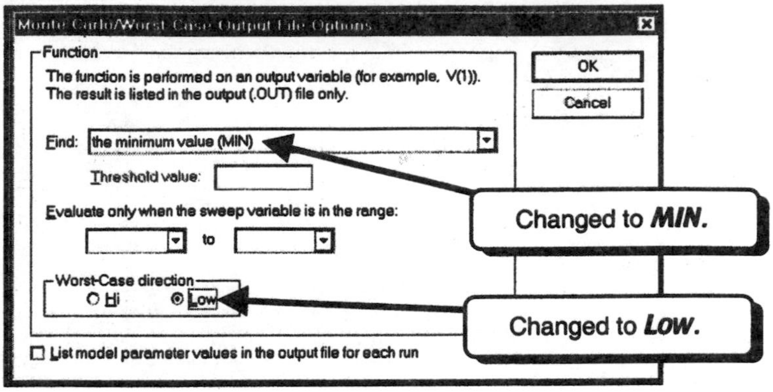

Monte Carlo/Worst Case Output File Options

Function — The function is performed on an output variable (for example, V(1)). The result is listed in the output (.OUT) file only.

Find: the minimum value (MIN)

Threshold value:

Evaluate only when the sweep variable is in the range: to

Worst-Case direction: Hi / Low

List model parameter values in the output file for each run

OK / Cancel

Changed to ***MIN.***

Changed to ***Low.***

Run PSpice and then view the output file. The results of the analysis are given at the bottom of the output file:

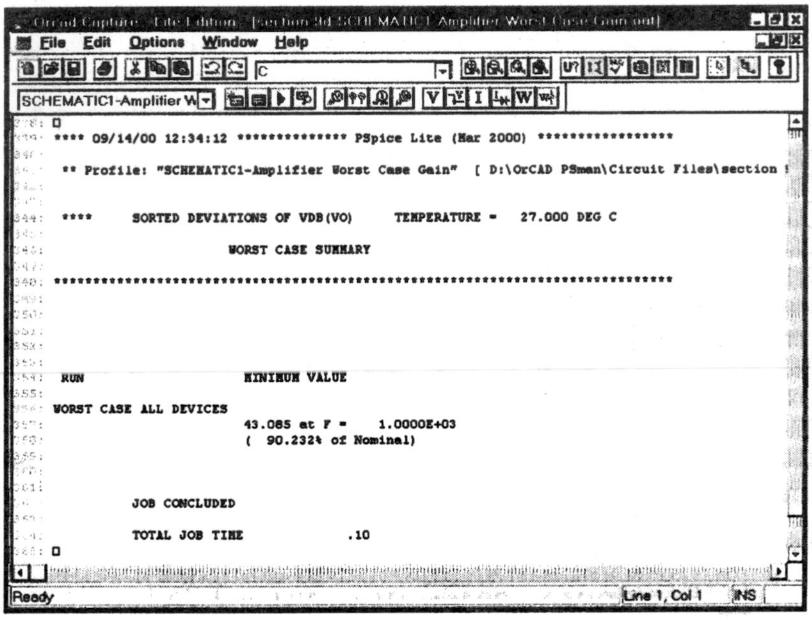

```
**** 09/14/00 12:34:12 ************* PSpice Lite (Mar 2000) ****************

 ** Profile: "SCHEMATIC1-Amplifier Worst Case Gain"  [ D:\OrCAD PSman\Circuit Files\section

 ****     SORTED DEVIATIONS OF VDB(VO)     TEMPERATURE =   27.000 DEG C

                    WORST CASE SUMMARY

 ******************************************************************************

 RUN                          MINIMUM VALUE

WORST CASE ALL DEVICES
                              43.085 at F =     1.0000E+03
                              (  90.232% of Nominal)

          JOB CONCLUDED

          TOTAL JOB TIME            .10
```

The results show a minimum gain of ***43.085*** dB.

**EXERCISE 9-6:** Find the minimum and maximum gain of the circuit in this section if the base resistors have 1% tolerance rather than 5% tolerance.

**SOLUTION:** The results of the simulation show that the minimum and maximum values of the gain are 43.591 dB and 49.356 dB.

# 9.E. Performance Analysis — Amplifier Frequency Response

When we design a circuit with tolerance, we may sometimes want to find the worst case upper or lower 3 dB frequency with component tolerances. Unfortunately, calculating a 3 dB frequency requires that we find the mid-band gain and then find the frequency where the gain is 3 dB less than the mid-band. This type of calculation cannot be specified in the Monte Carlo/Worst Case dialog box. However, we can run a Monte Carlo analysis and then determine the 3 dB frequency using the Performance Analysis capabilities available in Probe. In this example, we will illustrate finding the maximum lower 3 dB frequency ($F_L$), minimum upper 3 dB frequency ($F_H$), and maximum and minimum bandwidth ($F_H - F_L$) for a common-emitter amplifier. Wire the circuit below:

| Symbol | Part | Description |
|---|---|---|
| +80% – 20% 1U C1 | **Cap20_80** | Capacitor with –20%, +80% tolerance |
| +/- 5% 1k R1 | **R5pcnt** | Resistor with 1% uniform tolerance distribution |
| Q3 q2n3904 | **Q2N3904** | Small-signal NPN BJT with $350 \geq \beta_F \geq 50$ |
| V2 Magnitude = 1 Phase = 0 | **VAC** | AC voltage source |
| V | **Voltage Level Marker** | In Capture type **CTRL-M** to place marker |
| | **VDC** | DC source |
| 0 | **0** | Ground |
| | **Vcc_Circle** | Node label |

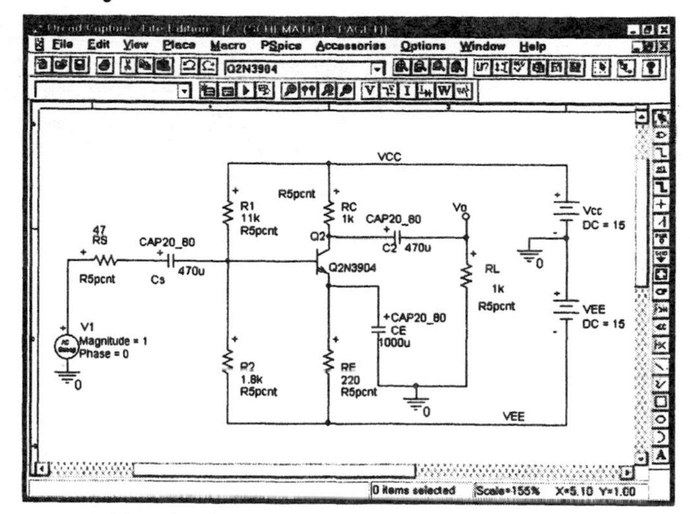


This circuit contains 5% resistors, capacitors with +80%, –20% tolerance, and a 2N3904 transistor that has been modified to include tolerance in β. The modified model is shown below. See Section 7.C for instructions on how to modify a BJT model.

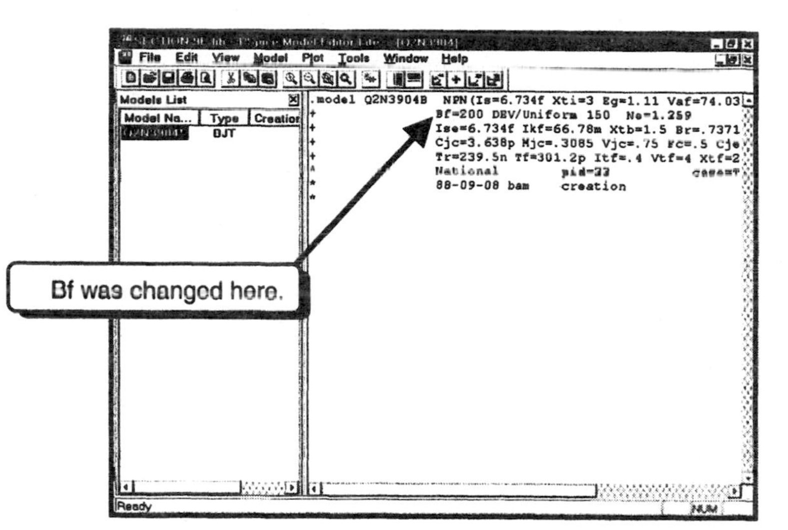


All distributions are uniform. We will run an AC Sweep in conjunction with the Monte Carlo analysis. Fill in the AC Sweep and Monte Carlo dialog boxes as shown below:

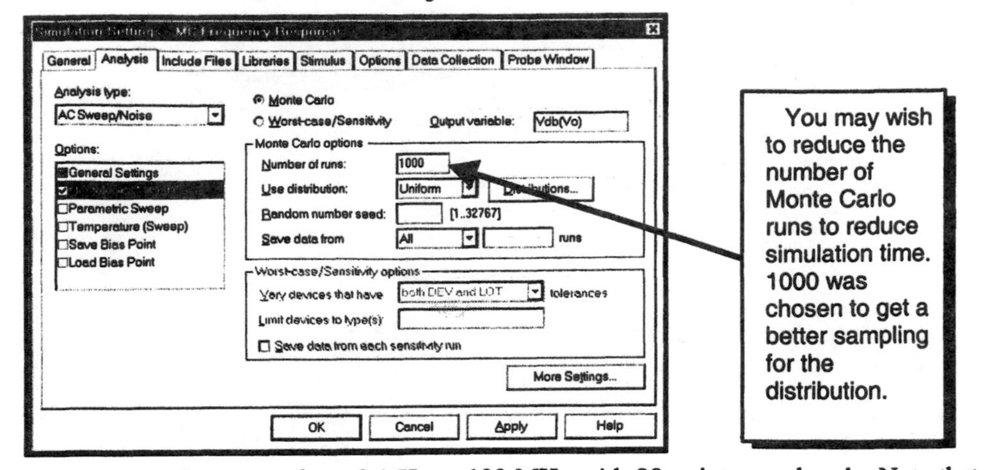


The AC Sweep dialog box is set up to sweep frequency from 0.1 Hz to 100 MHz with 20 points per decade. Note that the number of Monte Carlo runs is set to 1000. This will take a large amount of simulation time but will give us more accurate results. You may wish to reduce the number of runs in order to reduce the simulation time. Click the ***OK*** button to return to the schematic.

Since we are running 1000 simulations and the circuit is fairly large, the data file created by PSpice could be huge if we collect voltage and current data for all circuit elements. To reduce the size of the data file we will collect data only for the output voltage. This can be done by placing a marker at the output. To place a marker, select **PSpice**, **Markers**, and then **Voltage Level** from the Capture menus. Place the marker at the output as shown:

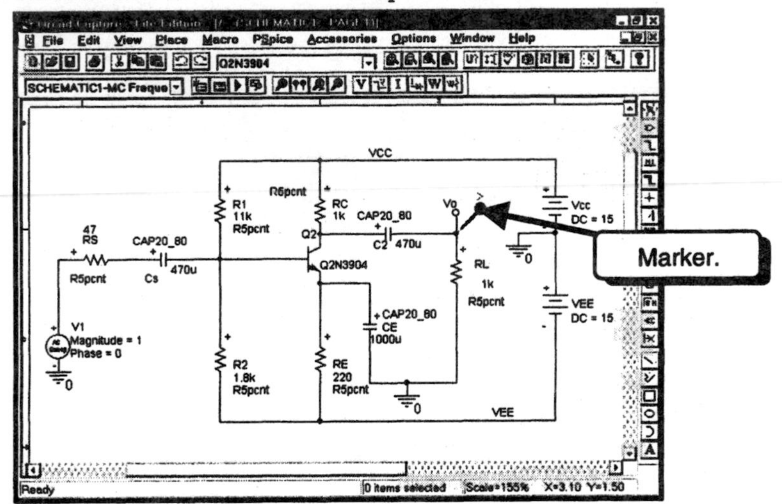


After the marker is placed, we must tell Probe to collect data only at the markers. Select **PSpice** and then **Edit Simulation Profile** from the menus, and then select the ***Data Collection*** tab. Select the options as shown:

Next, select the ***Probe Window*** tab and specify the following settings:

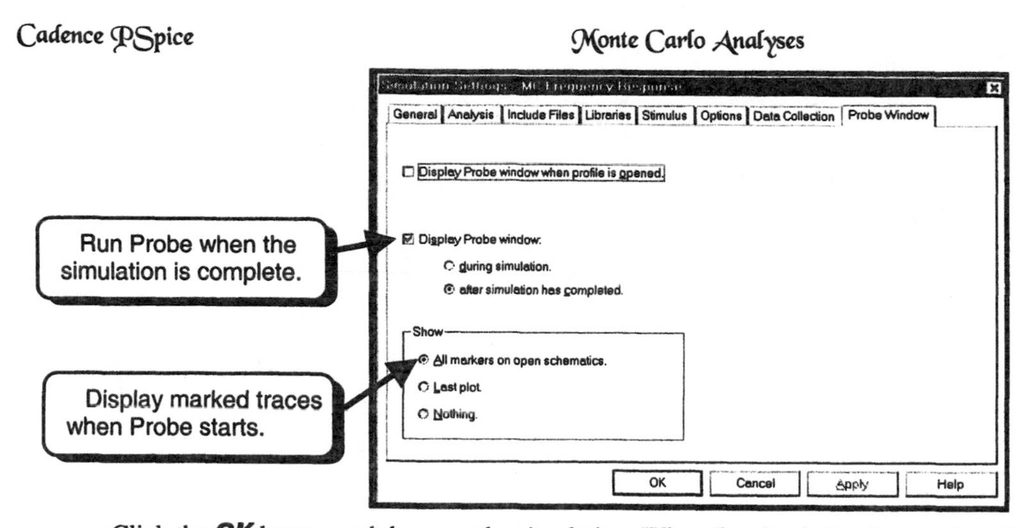


Click the ***OK*** button and then run the simulation. When the simulation is complete, Probe will automatically run:

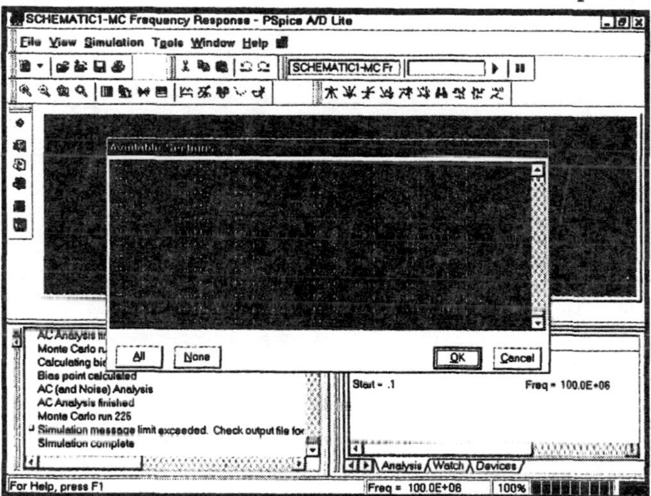


Click the ***OK*** button to select all of the traces. The gain plot for all traces will be displayed by Probe:

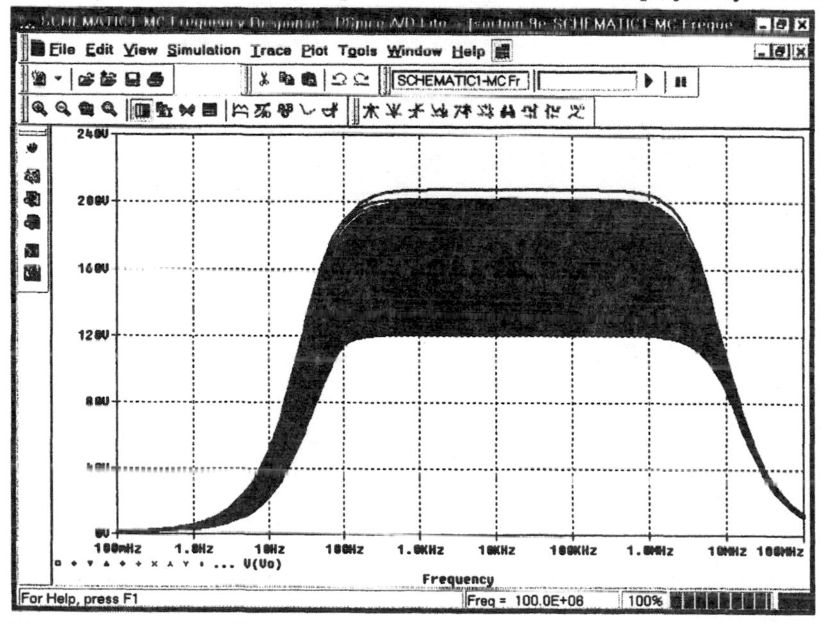


Next, we would like to display a histogram of the lower 3 dB frequency. We will first delete the displayed trace. Click the *LEFT* mouse button on the text ***V(Vo)*** to select the trace. When the trace is selected, the text ***V(Vo)*** will be highlighted in red. Press the **DELETE** key to delete the trace. (Your trace may be labeled differently than ***V(Vo)***.) A blank Probe screen will result.

To plot a function like bandwidth, we must use the Performance Analysis. Select ***Plot*** and then ***Axis Settings***:

Under ***Processing Options*** we notice that ***Performance Analysis*** is not selected. Click the ***LEFT*** mouse button on the square ☐ next to ***Performance Analysis***. It should fill with a checkmark, ☑:

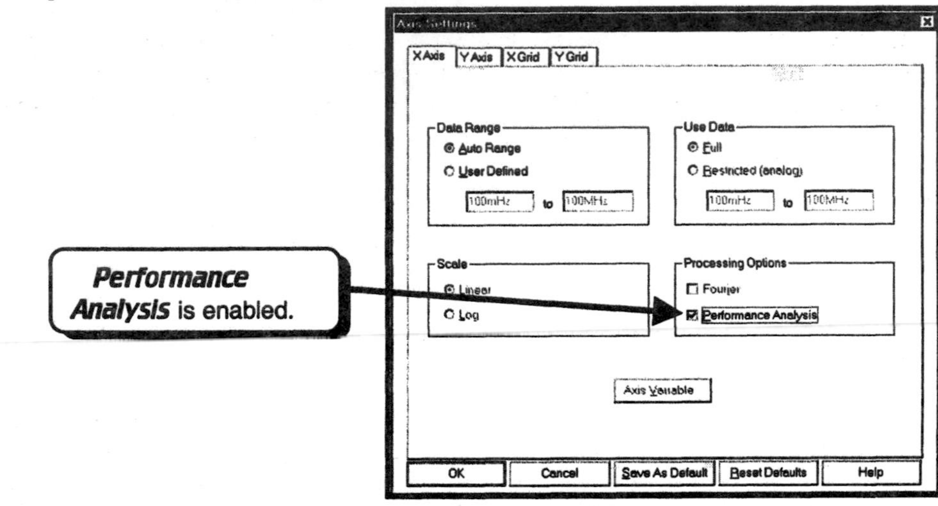


Click the ***OK*** button. You will return to Probe with a blank histogram displayed:

To add a trace, select **Trace** and then **Add Trace**:

Add Traces

Simulation Output Variables

Frequency
V(C2:2)
V(RL:1)
V(VO)
V1(RL)
V2(C2)

Full List

☑ Analog
☐ Digital
☑ Voltages
☐ Currents
☐ Power
☐ Noise (V²/Hz)
☑ Alias Names
☐ Subcircuit Nodes

6 variables listed

Functions or Macros

Goal Functions

Bandwidth(1,db_level)
BPBW(1,db_level)
BW(1)
CenterFreq(1,db_level)
Falltime(1)
GainMargin(1,2)
GenFall(1)
GenRise(1)
HPBW(1,db_level)
lower3dB(1)
LPBW(1,db_level)
Max(1)
MAXr(1,begin_x,end_x)
Min(1)
MINr(1,begin_x,end_x)
NMh(1,2)
NMl(1,2)
Overshoot(1)
Peak(1,n_occur)
Period(1)
PhaseMargin(1,2)
Pulsewidth(1)
Risetime(1)
SWINGr(1,begin_x,end_x)

Trace Expression: OK Cancel Help

The left pane shows the normal voltage and current traces that we are familiar with. The right pane displays the goal functions. These functions are available using the Performance Analysis. The functions are defined in a file called C:\Program Files\OrcadLite\PSpice\Common\pspice.prb. If you view this file using the Windows Notepad, you will see the function below near the end of the file:

```
upper3dB(1) = x1
* Find the upper 3 dB frequency
{
1|sfle(max-3dB,n) !1;
}
```

The name of the function is **upper3dB**. The text **(1)** indicates that it has **1** input argument. **1|sfle** means search the first input forward and find a level. The level we are looking for is 3 dB less than the maximum **(max-3)**. The **n** means find the specified level when the trace has a negative-going slope. When the point is found, the text **!1** designates its coordinates as x1 and y1. The function returns the x-coordinate of the point (**upper3dB(1) = x1**). The x-axis of a frequency trace is frequency, so this function returns the frequency of the upper 3 dB point.

A second function is :

```
lower3dB(1) = x1
* Find the lower 3 dB frequency
{
1|sfle(max-3dB,p) !1;
}
```

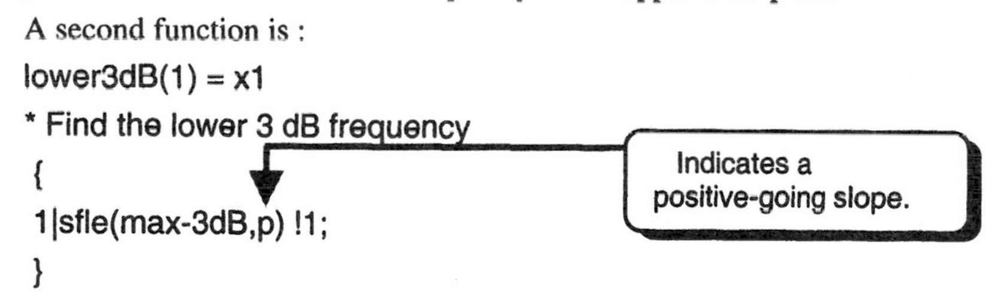
Indicates a positive-going slope.

This function is similar to the upper3dB function except that it finds the 3 dB point when the trace has a positive-going slope. This will mark the coordinates of the lower 3 dB point. A third function is:

```
BW(1) = x2-x1
* Find the 3 dB bandwidth of a signal
{
1|Search forward level(max-3dB,p) !1
Search forward level(max-3dB,n) !2;
}
```

This function finds the lower 3 dB frequency and marks the coordinates of the point x1 and y1. It then finds the upper 3 dB frequency and marks the coordinates of the point x2 and y2. The function returns the bandwidth, **x2 - x1**.

We will first plot a histogram of the lower 3 dB frequency. Enter the trace **`lower3dB(V(VO))`**:

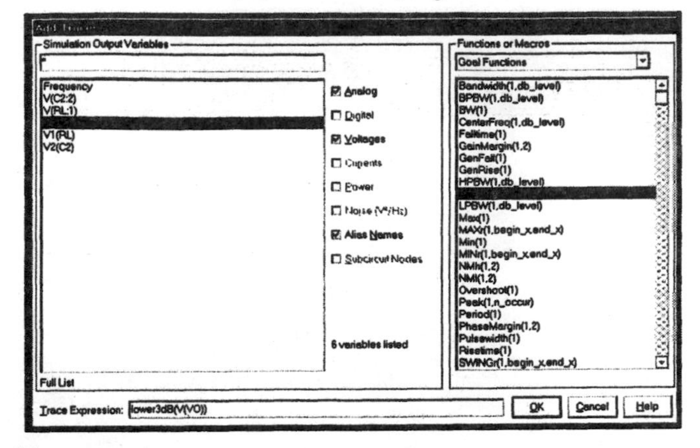


Click the ***OK*** button to plot the histogram:

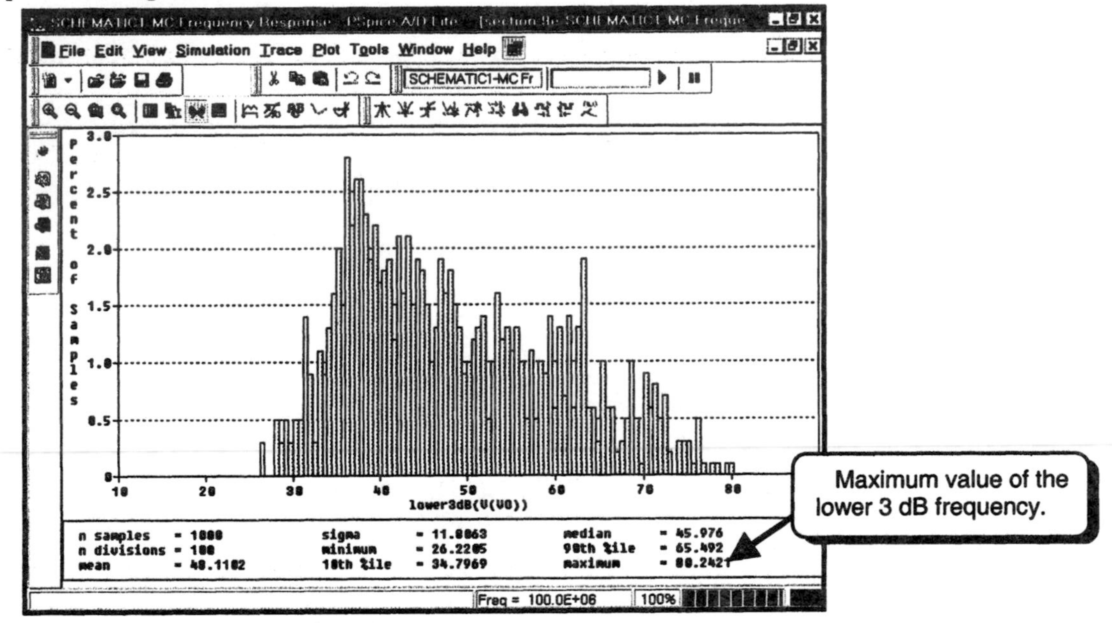


We see that the maximum lower 3 dB frequency is at ***80.2421*** Hz. In the screen capture above, the number of histogram divisions has been set to 100. See page 519 for instructions on how to change this setting. To view the upper 3 dB frequencies, add the trace **`upper3dB(V(VO))`**:

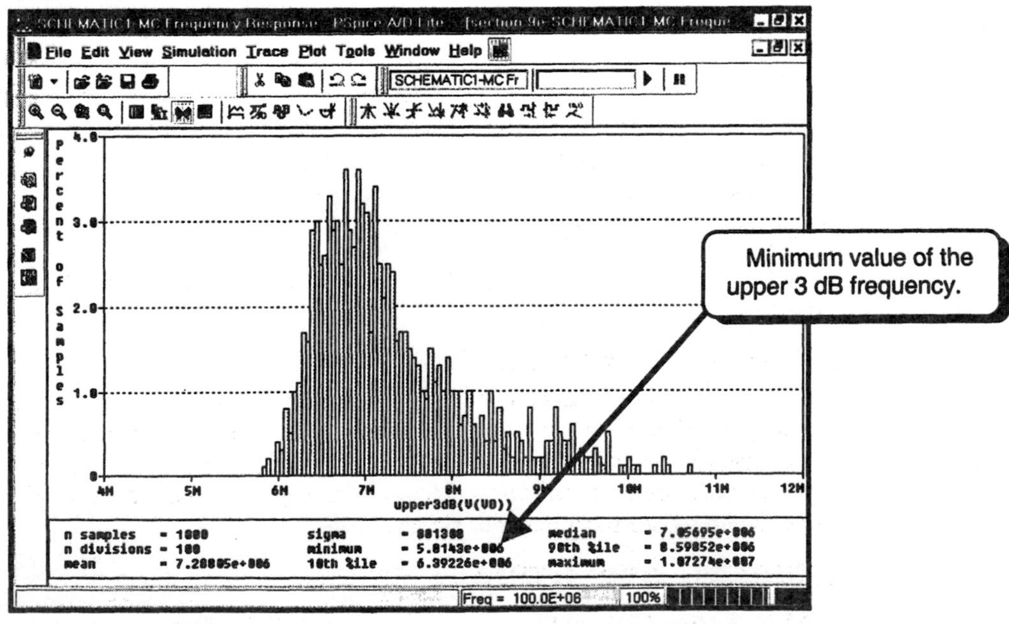


We see that the minimum upper 3 dB frequency is ***5.8143*** MHz. To view the bandwidth of the amplifier, add the trace **`BW(V(VO))`**:

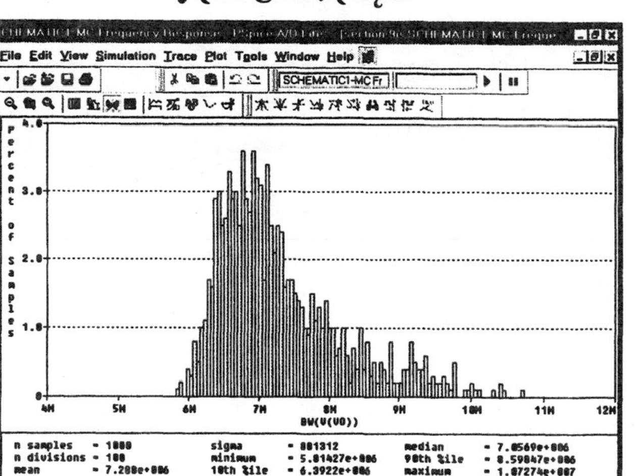


Although the BW function is not too helpful for a circuit of this type, it is very useful for bandpass filters.

# 9.F. jFET Minimum and Maximum Drain Current

The jFET provided in the libraries is the 2N5951. This jFET has the characteristics:

$$7\ \text{mA} \le I_{DSS} \le 13\ \text{mA}$$

$$-5\ \text{V} \le V_P \le -2\ \text{V}$$

The equation that governs the jFET's operation in the saturation region is:

$$I_D = I_{DSS}\left(1 - \frac{V_{GS}}{V_P}\right)^2$$

$I_D$ is maximum when $I_{DSS}$ = 13 mA and $V_P$ = –5 V. $I_D$ is minimum when $I_{DSS}$ = 7 mA and $V_P$ = –2 V. We shall let $I_{DSS(max)}$ = 13 mA, $I_{DSS(min)}$ = 7 mA, $V_{P(max)}$ = –5 V, and $V_{P(min)}$ = –2 V.

The equation used by PSpice to describe the jFET is:

$$I_D = \beta\left(V_{GS} - V_{TO}\right)^2$$

where $V_{TO} = V_P$ and $\beta = I_{DSS}/(V_P)^2$. When we are running a Worst Case analysis on a jFET circuit, we would like to let both $I_{DSS}$ and $V_P$ vary at the same time. However, because of the non-linear relationship of $\beta$ to $I_{DSS}$ and $V_P$, a spread in the values of $I_{DSS}$ and $V_P$ does not correspond to an equivalent spread in $\beta$ and $V_{TO}$. To model the worst case limits of the jFET we will have to make two models: one that corresponds to $I_{DSS(max)}$ and $V_{P(max)}$, and a second that corresponds to $I_{DSS(min)}$ and $V_{P(min)}$.

In the library class.lib there are two jFET models called jMAX and jMIN. These models correspond to the minimum and maximum limits of the model J2n5951. The model jMAX is shown below:

```
.model jMAX NJF(Vto=–5 Beta=.52m)
```

In the jMAX model $V_{TO} = V_{P(max)}$ = –5 V, and $\beta = I_{DSS(max)}/(V_{P(max)})^2 = 0.00052$. The model jMIN is:

```
.model jMIN NJF(Vto=–2 Beta=1.75m)
```

In the jMIN model $V_{TO} = V_{P(min)}$ = –2 V, and $\beta = I_{DSS(min)}/(V_{P(min)})^2 = 0.00175$.

To find the minimum and maximum of a quantity, you will have to run the Worst Case analysis four times. To find the maximum of the quantity, you will have to run the simulation once with the jMAX model and once with the jMIN model. To find the minimum of the quantity, you will also have to run the simulation once with the jMAX model and once with the jMIN model. If you happen to know which of the two models will give you the maximum or minimum of the quantity you are looking for, you may be able to reduce the number of simulations.

## 9.F.1. jFET Minimum Bias Drain Current

We would like to find the minimum drain current of the circuit shown below:

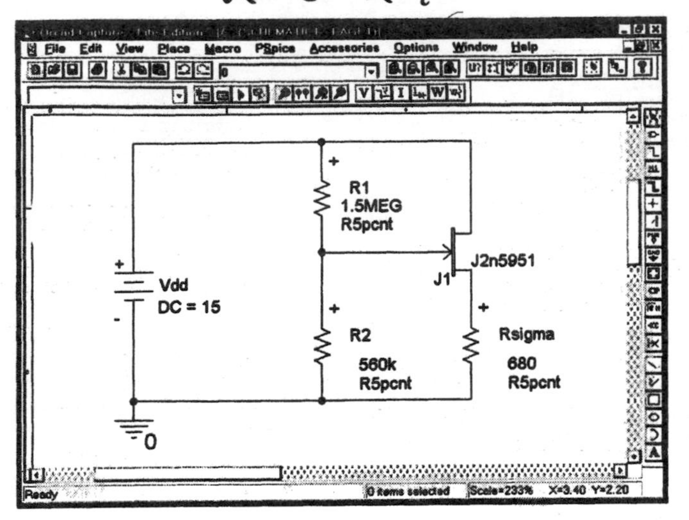


In order to include the tolerance of the jFET we must use the models jMAX and jMIN. Replace the J2n5951 jFET with the jMAX part:

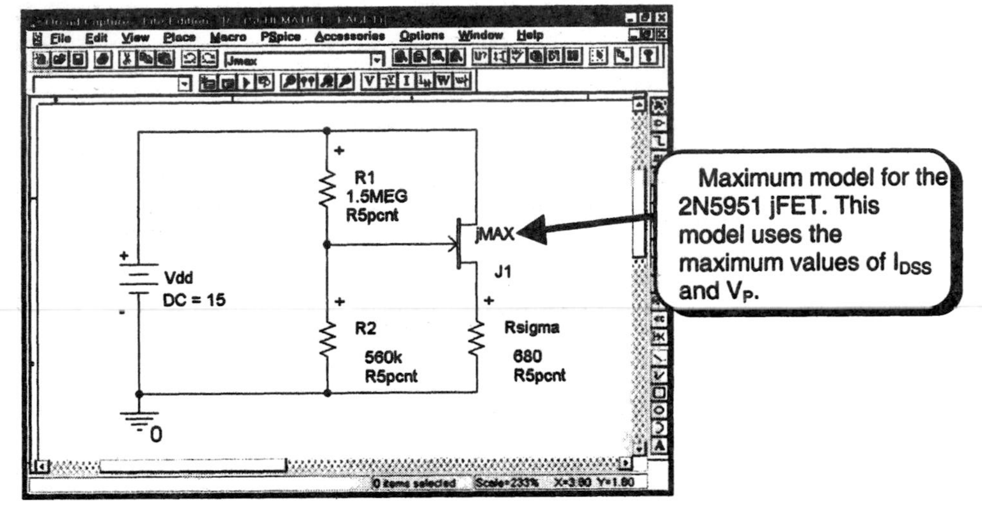


We must now set up a DC Sweep since the bias drain current is a DC quantity. Fill in the DC Sweep dialog box as shown:

General | Analysis | Include Files | Libraries | Stimulus | Options | Data Collection | Probe Window

Analysis type: DC Sweep

Options: Secondary Sweep, Monte Carlo/Worst Case, Parametric Sweep, Temperature (Sweep), Save Bias Point, Load Bias Point

Sweep variable: Voltage source (selected), Current source, Global parameter, Model parameter, Temperature — Name: Vdd; Model type; Model name; Parameter name

Sweep type: Linear, Logarithmic (Decade), Value list (selected): 15 — Start value; End value; Increment

OK | Cancel | Apply | Help

The dialog box is set up to run a DC Sweep with Vdd equal to 15 V. Setting Vdd to 15 V is redundant, but a DC Sweep is required for the Worst Case analysis and Vdd is the only DC source in the circuit.

Next, we must set up the Worst Case analysis. Fill in the **Monte Carlo/Worst Case** dialog boxes as shown:

The dialog box is set up to find the worst case minimum value of the quantity ***I(Rsigma)***. This is the current through resistor Rsigma as well as the drain current of device ***J1***.*

Run PSpice and then examine the output file. The results are stored at the bottom of the output file. The screen below tells us that the minimum drain current with the model jMAX is ***7.1235*** mA.

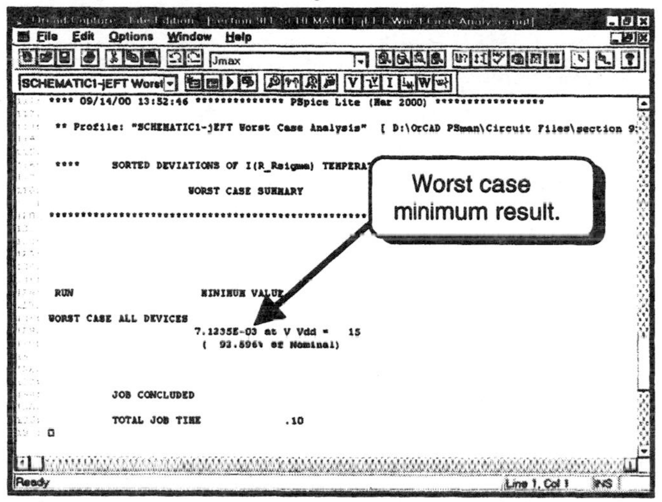


We must now run the simulation again with the model jMIN. Delete the jFET and place a part named jMIN in your circuit:

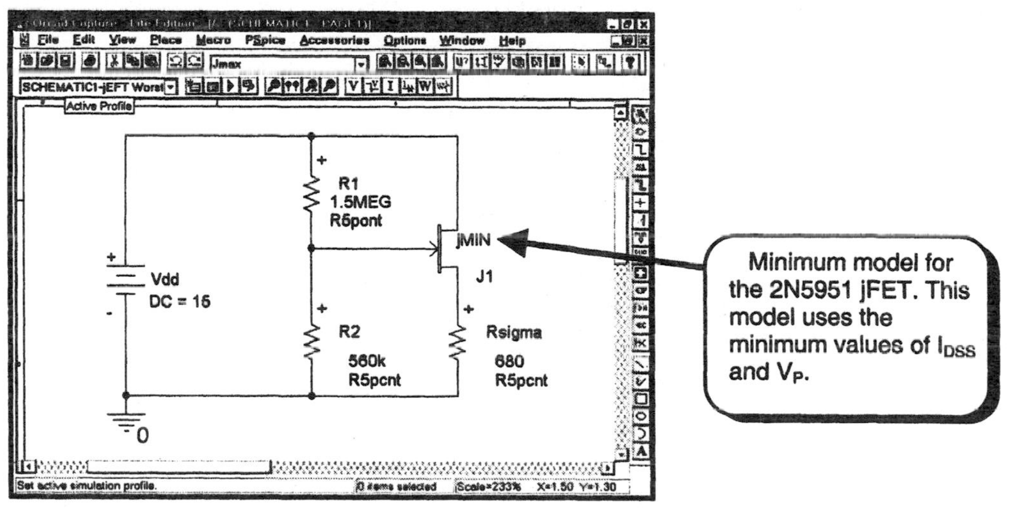


Run PSpice and then examine the output file:

* We could also have asked for ID(J1). However, I(R7) does not yield the same answer as ID(J1). I(R7) appears to agree with the calculated results.

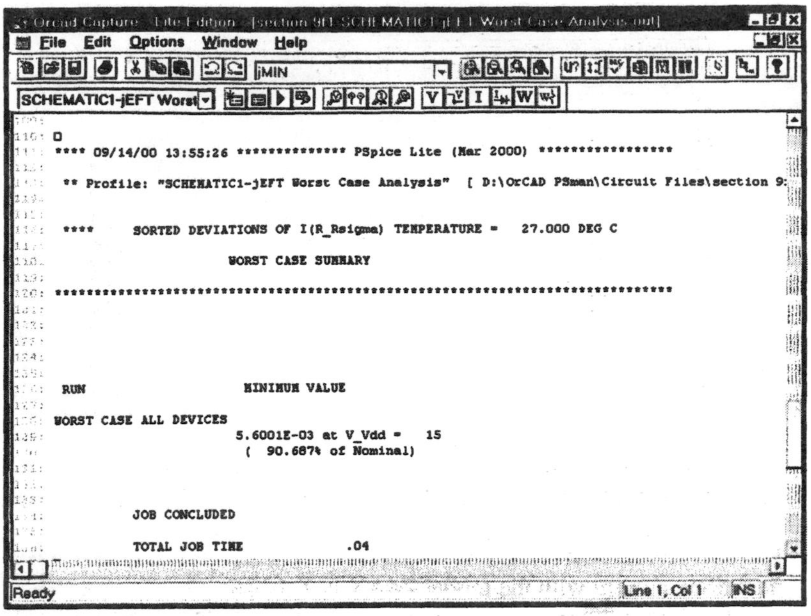

```
OrCAD Capture - Lite Edition - [section 9F-SCHEMATIC1-jEFT Worst Case Analysis.out]
File  Edit  Options  Window  Help
jMIN
SCHEMATIC1-jEFT Worst

**** 09/14/00 13:55:26 ************* PSpice Lite (Mar 2000) ****************

 ** Profile: "SCHEMATIC1-jEFT Worst Case Analysis"  [ D:\OrCAD PSman\Circuit Files\section 9:

 ****      SORTED DEVIATIONS OF I(R_Rsigma) TEMPERATURE =   27.000 DEG C

                      WORST CASE SUMMARY

 ******************************************************************************

 RUN                       MINIMUM VALUE

 WORST CASE ALL DEVICES
                           5.6001E-03 at V_Vdd =   15
                           ( 90.6874% of Nominal)

          JOB CONCLUDED

          TOTAL JOB TIME           .04

Ready                                             Line 1, Col 1   INS
```

The results show that the minimum drain current is ***5.6001*** mA.

From these two runs we determined that the minimum drain current occurs with the model jMIN and its value is ***5.6001*** mA. There is no general rule to determine whether the maximum or minimum model of the jFET will give you the maximum or minimum of the output variable. In general, you will have to make multiple runs to find the minimum and multiple runs to find the maximum.

## 9.F.2. jFET Maximum Bias Drain Current

We will now find the maximum drain current of the circuit in Section 9.F.1. The procedure is exactly the same as in Section 9.F.1, except that we must specify PSpice to find the maximum value. Follow the procedure given in Section 9.F.1, but use the ***Monte Carlo/Worst-Case Output File Options*** dialog box settings shown below:

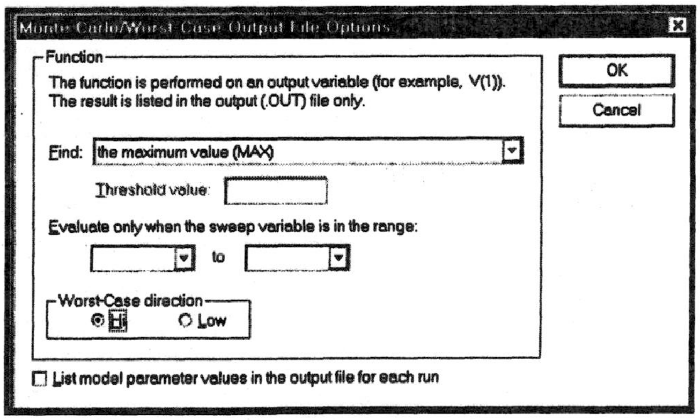

```
Monte Carlo/Worst Case Output File Options
Function
The function is performed on an output variable (for example, V(1)).
The result is listed in the output (.OUT) file only.
Find: the maximum value (MAX)
Threshold value:
Evaluate only when the sweep variable is in the range:
to
Worst-Case direction
(•) Hi   ( ) Low
[ ] List model parameter values in the output file for each run
OK
Cancel
```

If you run the simulation with the jMAX and jMIN models as shown in Section 9.F.1, you should get the results shown below:

| Maximum Drain Current | |
|---|---|
| Model | Drain Current (mA) |
| jMAX | 8.3277 |
| jMIN | 6.8221 |

**EXERCISE 9-7**: In the circuit below, the jFET has the following parameters: 6 mA ≤ IDSS ≤ 14 mA, and –6 V ≤ VP ≤ –1 V. Find the minimum and maximum drain current.

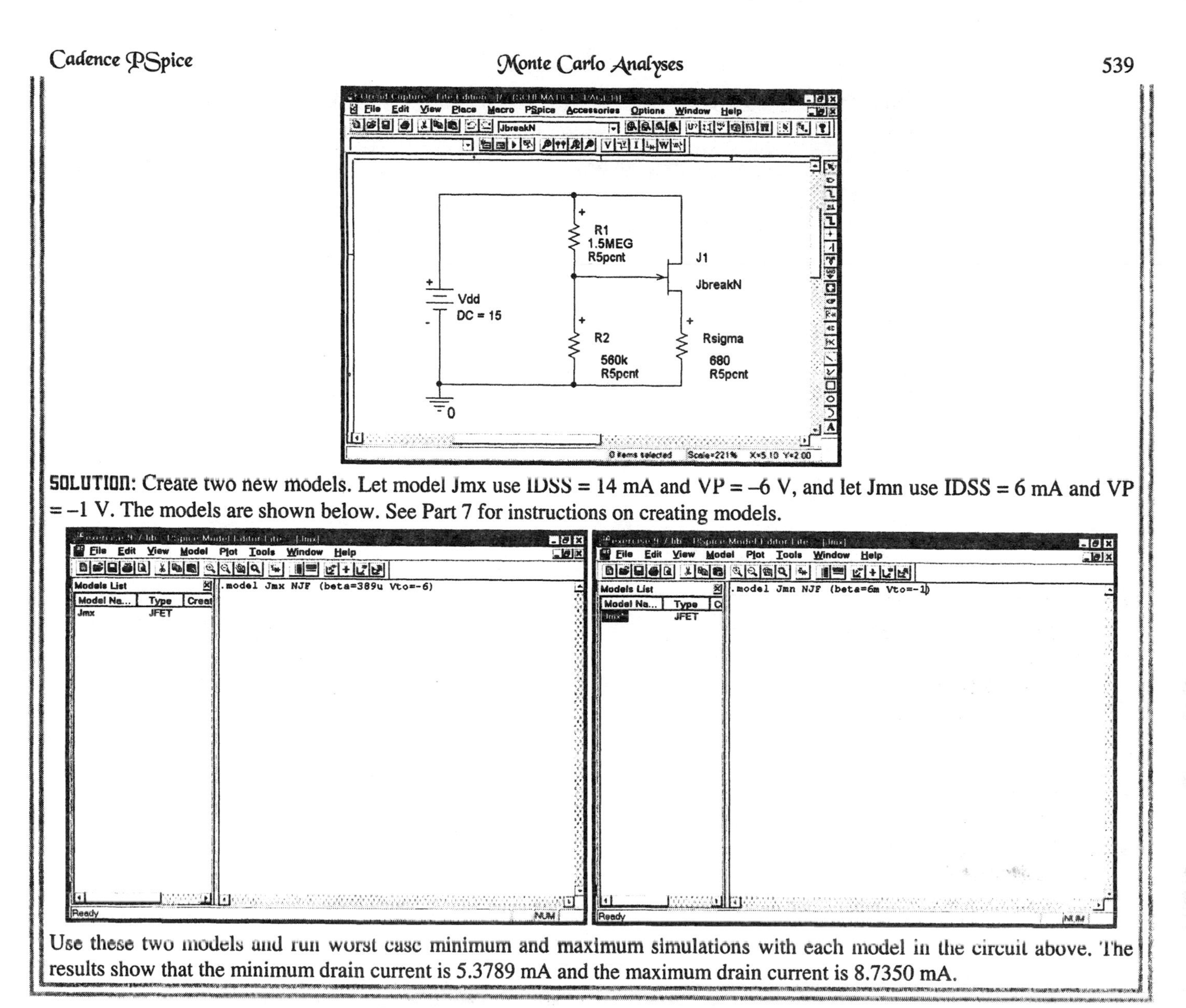


**SOLUTION**: Create two new models. Let model Jmx use IDSS = 14 mA and VP = –6 V, and let Jmn use IDSS = 6 mA and VP = –1 V. The models are shown below. See Part 7 for instructions on creating models.

Use these two models and run worst case minimum and maximum simulations with each model in the circuit above. The results show that the minimum drain current is 5.3789 mA and the maximum drain current is 8.7350 mA.

# 9.G. Performance Analysis — Inverter Switching Speed

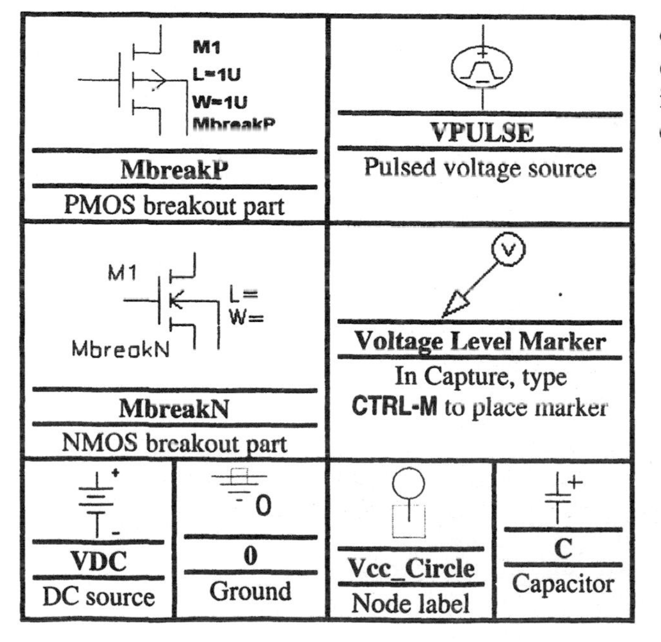


When designing digital circuits we are usually concerned with the rise and fall times of the design, given device tolerances. The example given here is for a CMOS inverter, but the procedure used can be applied to any switching circuit with device tolerances. Wire the circuit below:

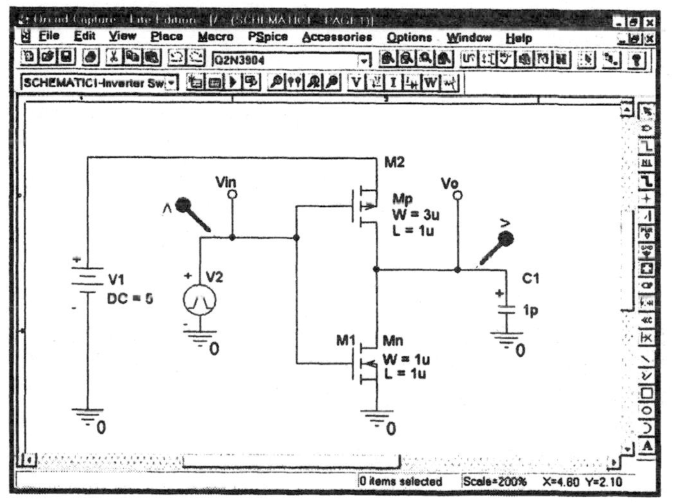

The models for the MOSFETs have tolerances in their threshold voltages and transconductances. The models for the MOSFETs are given below:

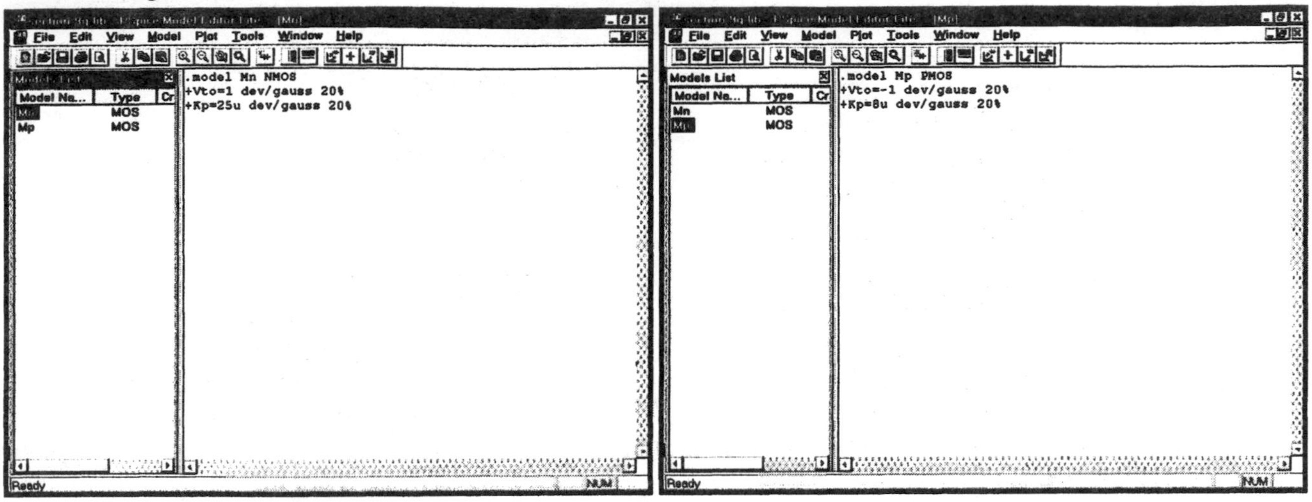

```
.model Mn NMOS
+Vto=1 dev/gauss 20%
+Kp=25u dev/gauss 20%
```

```
.model Mp PMOS
+Vto=-1 dev/gauss 20%
+Kp=8u dev/gauss 20%
```

See Part 7 for instructions on creating models.

We would like to see how the rise and fall times vary with random device tolerances. We must set up the Transient Analysis to view waveforms versus time, and the Monte Carlo analysis to allow for device variations. First we will look at the input pulsed waveform. The property spreadsheet for ***V1*** is:

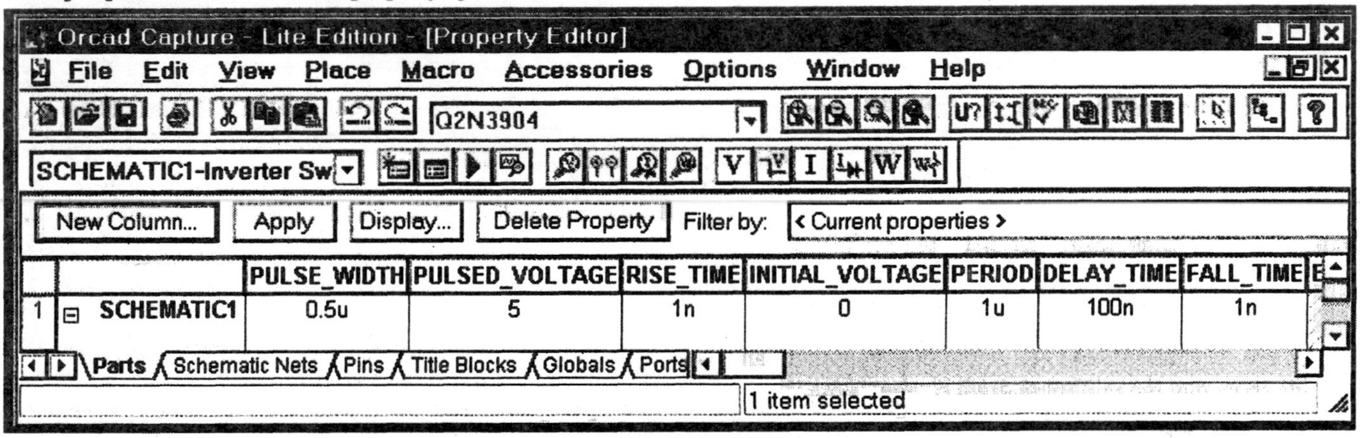

| | | PULSE_WIDTH | PULSED_VOLTAGE | RISE_TIME | INITIAL_VOLTAGE | PERIOD | DELAY_TIME | FALL_TIME |
|---|---|---|---|---|---|---|---|---|
| 1 | ⊟ SCHEMATIC1 | 0.5u | 5 | 1n | 0 | 1u | 100n | 1n |

The attributes specify a 0 to 5 V pulse with a 0.5 μs pulse width and 1 μs period. A delay time of 100 ns is specified so that the pulse does not start until 100 ns after the beginning of the simulation. The rise and fall times are 1 ns. We would like to set up a Transient analysis to simulate one cycle of the input. Select **PSpice** and then **New Simulation Profile** from the Capture menus, enter a name for the profile, and then click the ***Create*** button. By default the ***Time Domain (Transient) Analysis type*** is selected. Fill in the parameters as shown in the ***Time Domain*** dialog box below:

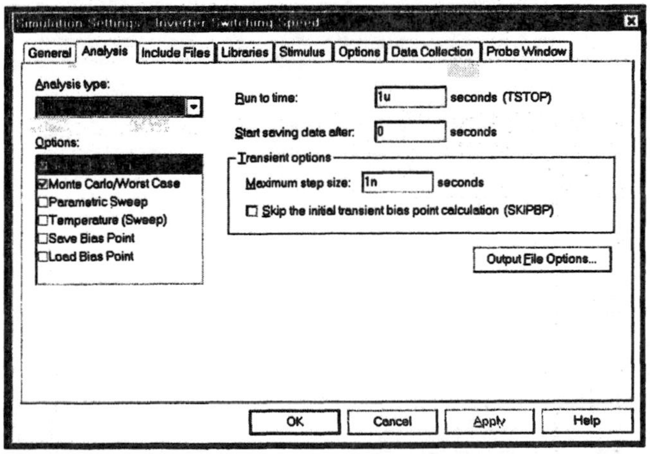

Next, we need to set up the Monte Carlo analysis. We will run the simulation 1000 times to get a good sampling. The dialog boxes below select the Monte Carlo analysis and specify 1000 Transient Analysis runs. This simulation will take a very long time to run. You may want to reduce the number of runs to reduce the simulation time.

Simulation Settings - Inverter Switching Speed

General | Analysis | Include Files | Libraries | Stimulus | Options | Data Collection | Probe Window

Analysis type: Time Domain (Transient)
Options: General Settings, Parametric Sweep, Temperature (Sweep), Save Bias Point, Load Bias Point
Monte Carlo / Worst-case/Sensitivity — Output variable: V(Vo)
Monte Carlo options — Number of runs: 1000; Use distribution: Uniform; Distributions...; Random number seed: [1..32767]; Save data from: All runs
Worst-case/Sensitivity options — Vary devices that have both DEV and LOT tolerances; Limit devices to type(s); Save data from each sensitivity run
More Settings...
OK | Cancel | Apply | Help

Monte Carlo/Worst-Case Output File Options

Function — The function is performed on an output variable (for example, V(1)). The result is listed in the output (.OUT) file only.
Find:
Threshold value:
Evaluate only when the sweep variable is in the range: to
Worst-Case direction — Hi / Low
List model parameter values in the output file for each run
OK | Cancel

Transient Analyses create a large amount of data. Even though the circuit is fairly small, running the analysis 1000 times will generate a large amount of data. To reduce the size of the data file, markers are added to the circuit and the Probe ***Data Collection*** options are set to collect voltage data at markers only. Since 1000 traces will take a long time to plot, we will specify that the Probe window not display anything when Probe starts. The ***Probe Window*** and ***Data Collection*** tabs are filled out as shown:

Simulation Settings - Inverter Switching Speed

General | Analysis | Include Files | Libraries | Stimulus | Options | Data Collection | Probe Window

Display Probe window when profile is opened.
Display Probe window: during simulation. / after simulation has completed.
Show — All markers on open schematics. / Last plot / Nothing
Do not display any traces when Probe starts.
OK | Cancel | Apply | Help

Simulation Settings - Inverter Switching Speed

General | Analysis | Include Files | Libraries | Stimulus | Options | Data Collection | Probe Window

Data Collection Options — Voltages: At Markers Only; Currents: None; Power: None; Digital: At Markers Only; Noise: None
Collect voltage data at markers only.
Save data in the CSDF format (.CSD)
OK | Cancel | Apply | Help

Click the ***OK*** button to return to the schematic and then run PSpice. When Probe runs, all of the runs will be selected by default:

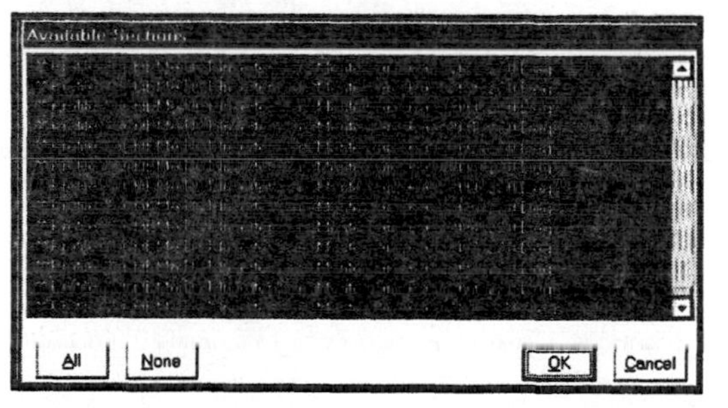


Click the ***OK*** button to enter Probe:

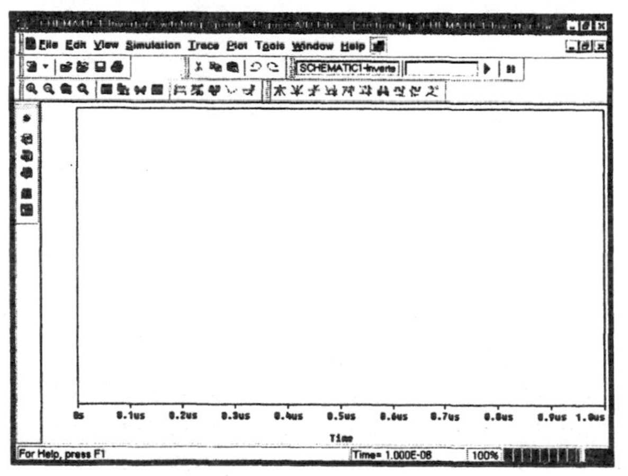

Since we have so many traces, we will instruct Probe not to use symbols to mark the traces. Select **Tools** and then **Options** from the Probe menus. Fill in the dialog box as shown:

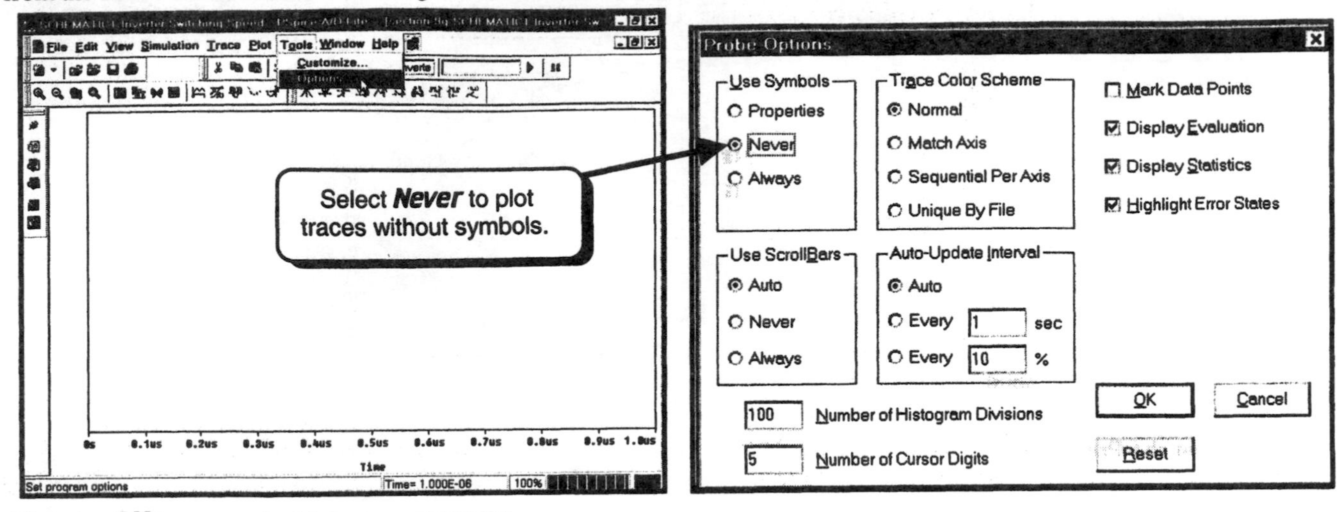


Click the ***OK*** button and add the trace `V(VIN)`:

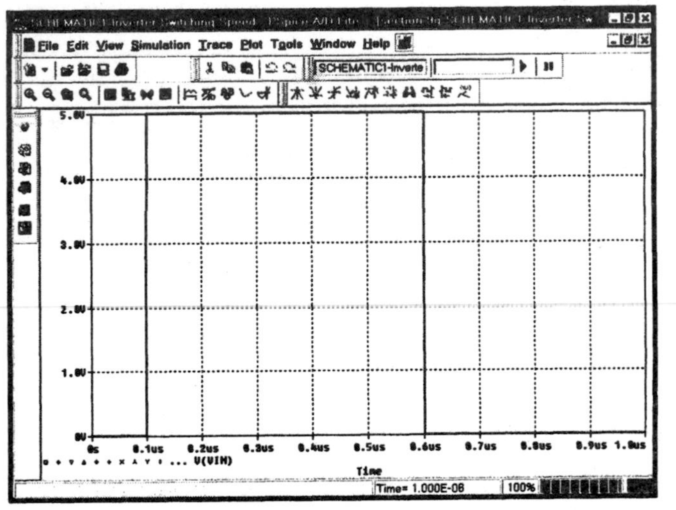

This is our input trace. We see that it has no variation with device tolerances, as should be expected. Delete the trace ***V(VIn)*** and then add the trace `V(VO)`:

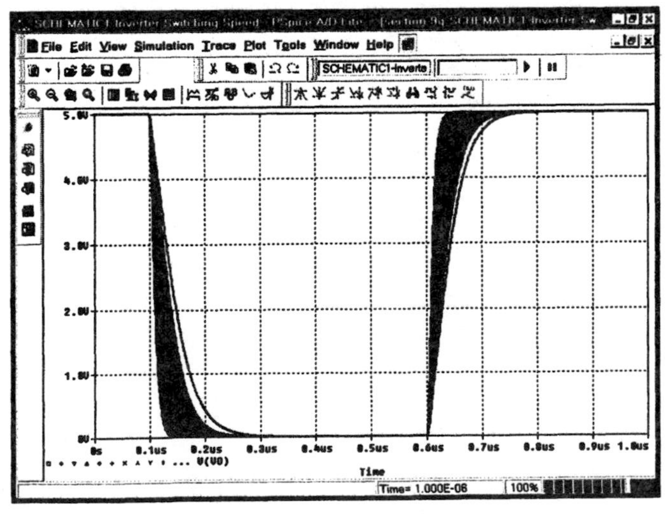

This is how the output pulse looks with time.

We would like to find out what the minimum and maximum rise and fall times are from this data. We can view the results as a histogram. First, delete the trace ***V(VO)*** to obtain an empty window:

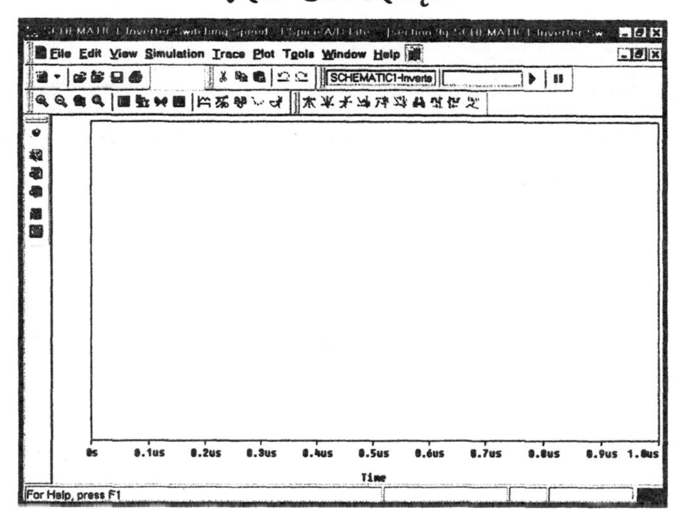

We must use the Performance Analysis to plot the information in which we are interested. Select **Plot** and then **Axis Settings** to obtain the ***Axis Settings*** dialog box. By default the properties for the x-axis are displayed. Specify ***Performance Analysis*** as follows:

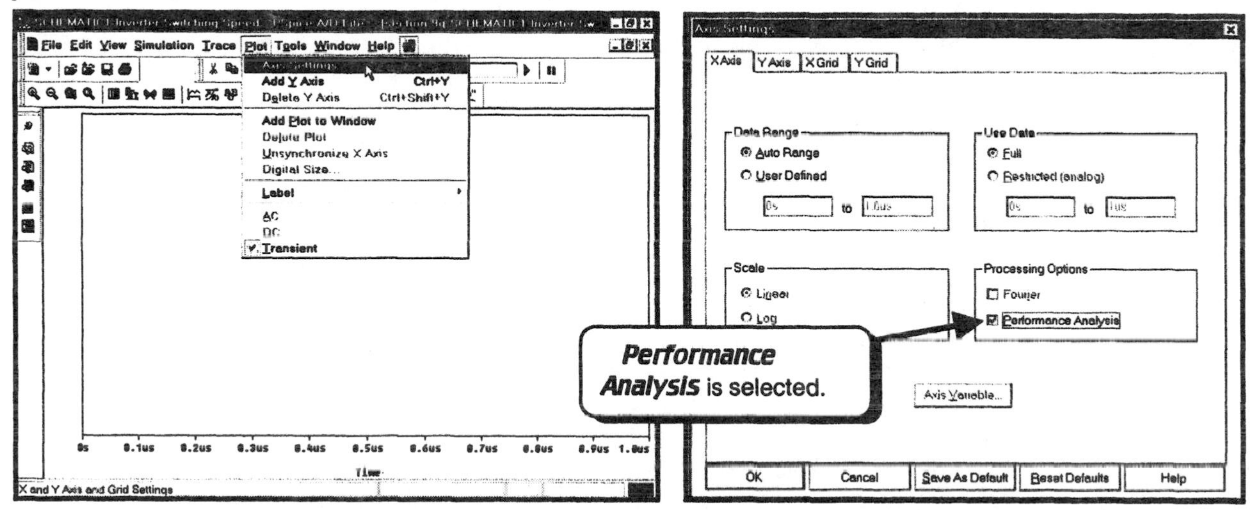


Click the ***OK*** button to return to Probe. The plot window will display a blank histogram plot:

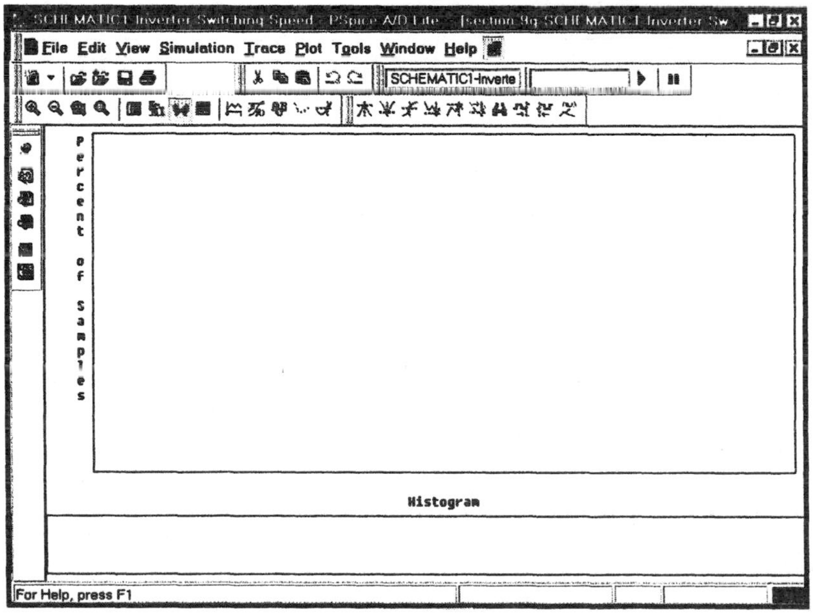

Select **Trace** and then **Add Trace** to add a trace:

The screen shows all of the Performance Analysis traces available to us. Enter the trace **Risetime(V(Vo))**:

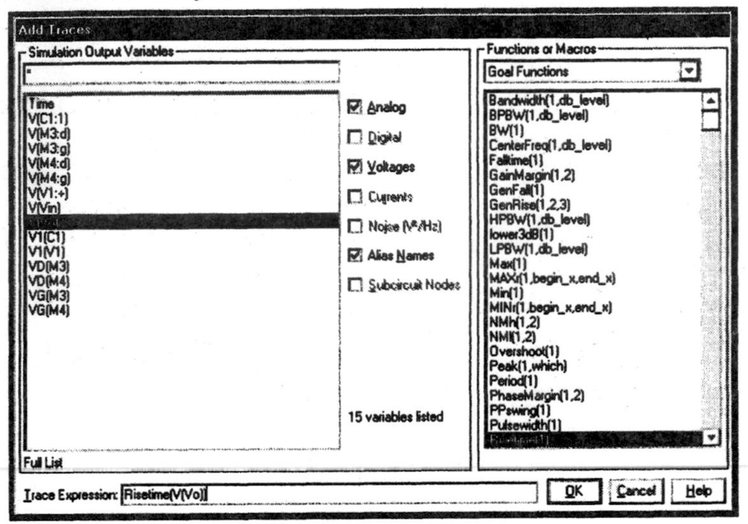


Click the ***OK*** button to plot the histogram:

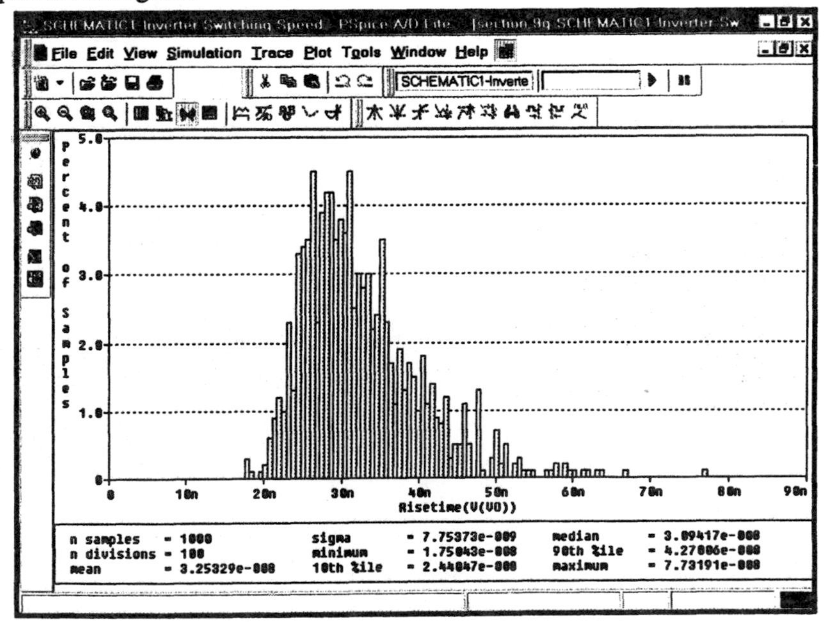


We see that the minimum and maximum rise times are 17.5 ns and 77.3 ns. In the screen capture above, the number of histogram divisions has been set to 100. See page 519 for instructions on how to change this setting.

To find the fall times, add the trace **Falltime(V(Vo))**:

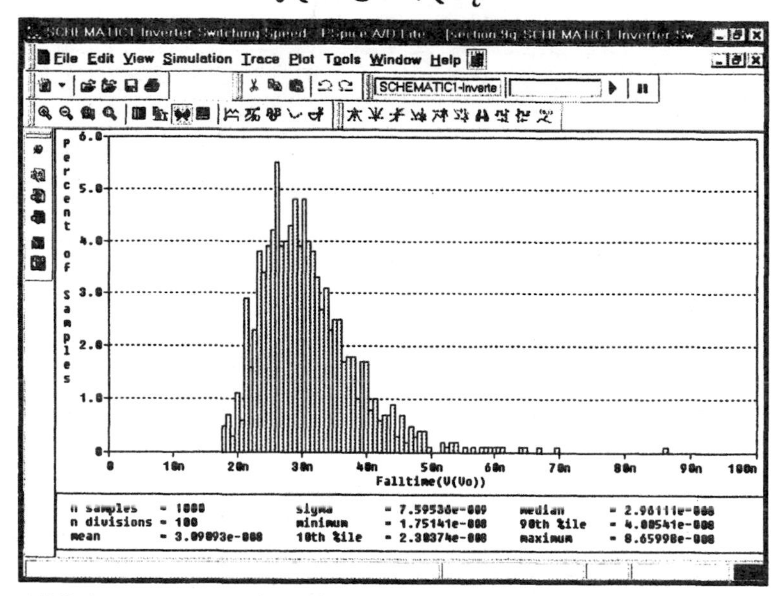


The minimum and maximum fall times are approximately 17.5 ns and 86.6 ns, respectively.

**EXERCISE 9-8**: Find the minimum and maximum rise and fall times for the BJT inverter studied in Section 6.K. Let the resistors have 20% Gaussian distributions and let $\beta_F$ have a uniform distribution from 50 to 350. Start with the circuit from Section 6.K, but use the 5% resistor model and the 2n3904 BJT model (R5pcnt and Q2n3904):

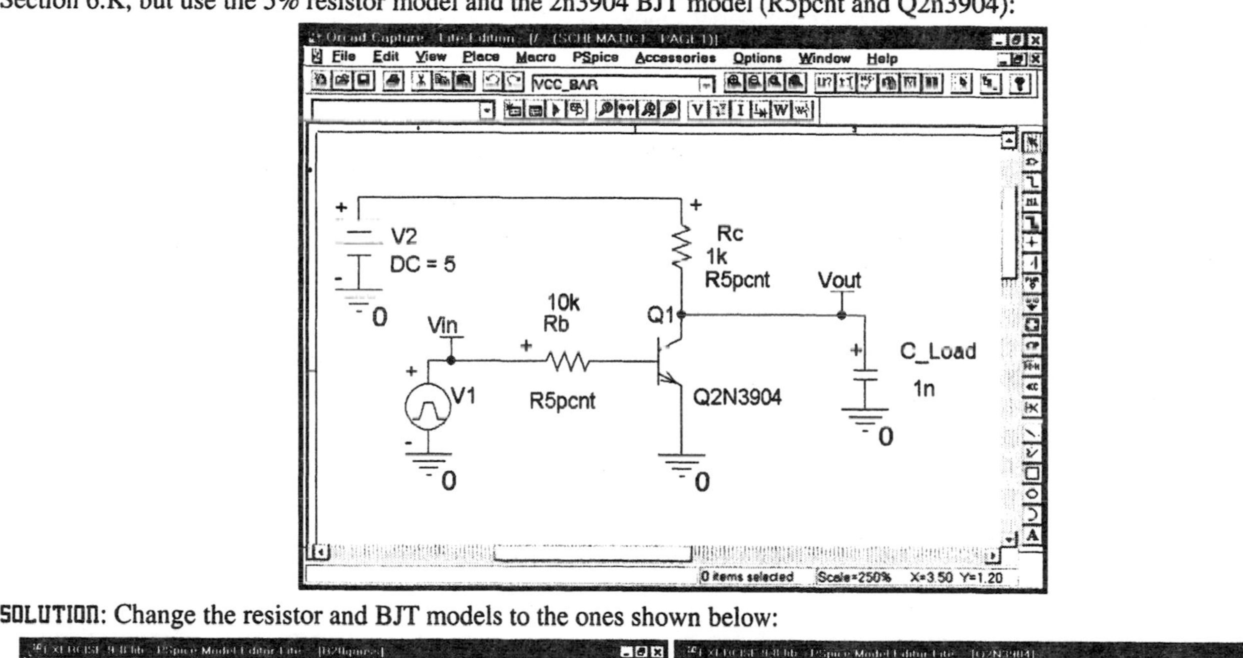


**SOLUTION**: Change the resistor and BJT models to the ones shown below:

```
.MODEL R20gauss RES(R=1 DEV/GAUSS 5%)
* The models below are 1% and 5% resistors with Gaussian
* probability distributions
```

```
.model Q2N3904   NPN(Is=6.734f Xti=3 Eg=1.11 Vaf=74.03
+                Bf=200 DEV/UNIFORM 150
+                  =1.259
+                 e=6.734f Ikf=66.78m Xtb=1.5 Br=.7371 Nc=2 Is
+                Cjc=3.638p Mjc=.3085 Vjc=.75 Fc=.5 Cje=4.493p
+                Tr=239.5n Tf=301.2p Itf=.4 Vtf=4 Xtf=2 Rb=10)
*                National          pid=23          case=TO92
*                88-09-08 bam      creation
*
```

This line was changed.

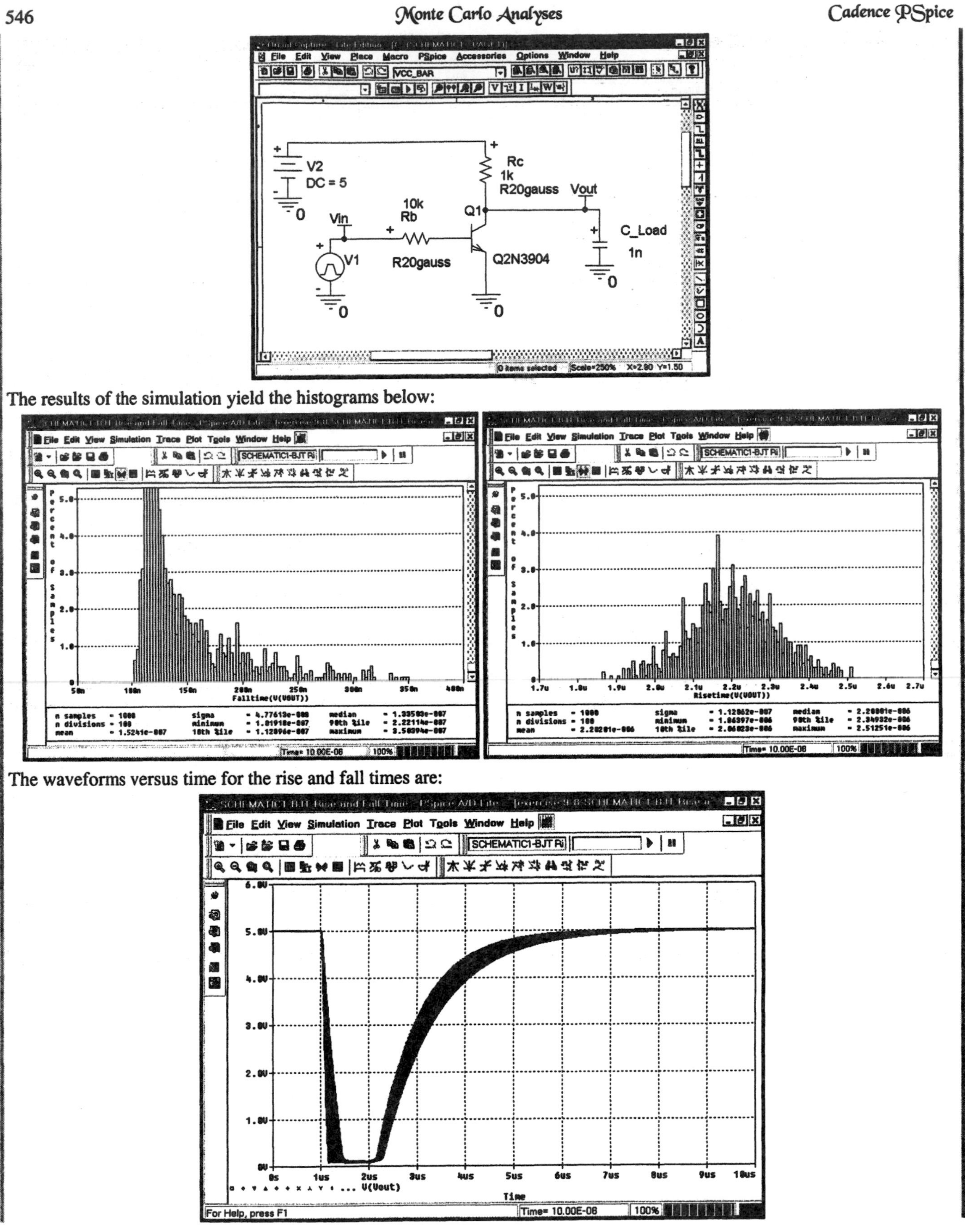


The results of the simulation yield the histograms below:

The waveforms versus time for the rise and fall times are:

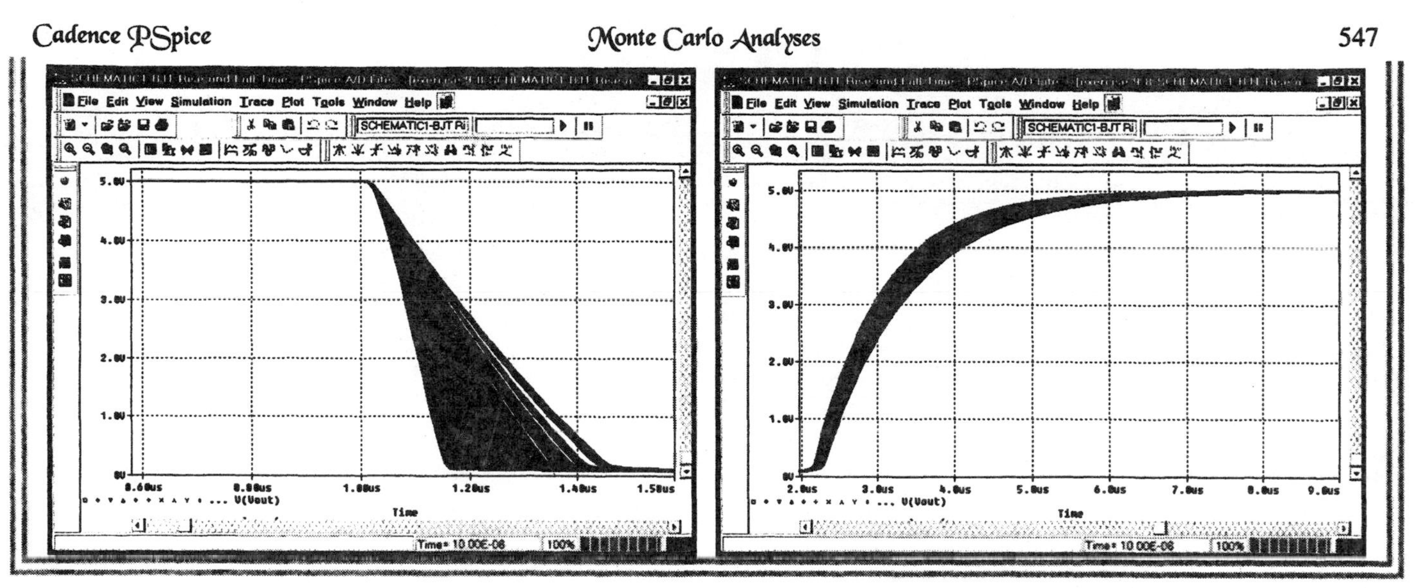

# 9.H. Summary

- The Monte Carlo analyses (Monte Carlo and Worst Case) are used to determine how tolerance in component values will affect circuit performance.
- The Worst Case analysis determines the absolute maximum or minimum value of a parameter for given component tolerances.
- If the Worst Case analysis determines that not all circuits will pass a specified performance parameter, the Monte Carlo analysis may be used to estimate what percentage of the circuits will pass.
- The uniform distribution is the easiest to use for the Worst Case analysis since this distribution specifies the upper and lower limits of a component's tolerance.
- The Gaussian distribution is the best to use with the Monte Carlo analysis since it more closely matches the actual distribution of components.
- For accurate results in a Monte Carlo simulation, the actual distribution of the components should be used rather than the Gaussian or uniform distributions. Use the PSpice ".Distribution" command to specify a nonstandard distribution.
- The Performance Analysis can be used together with the Monte Carlo analysis to display a histogram. The histogram will display properties such as minimum and maximum values of an output versus random variations. The Performance Analysis can find the minimum and maximum values of quantities not available with the Worst Case analysis, such as bandwidth and rise time.

# 9.I. Bibliography

[1] P. R. Gray and R. G. Meyer. *Analysis and Design of Analog Integrated Circuits*, 2nd ed. New York: Wiley, 1984, p. 225.

[2] MicroSim Corporation. *MicroSim PSpice A/D- Circuit Analysis Software - Reference Manual*, Version 7.1, Irvine, CA, October, 1996, p. 1-10.

# PART 10
# Project Management with Orcad Capture CIS

In this chapter we will cover creating and maintaining a project using Orcad Capture CIS. Topics include: (1) using parts contained in the Digikey Parts Database, (2) downloading parts from Orcad's online database called Activeparts and saving those parts in your local database, (3) adding your own parts to your database, and (4) creating a bill of materials (BOM) from the parts in your circuit using the data stored in the database. Since the BOM is created from information in your own database, the data can include price, supplier names, part numbers, supplier part numbers, your own part number, part details such as capacitance, working voltage, tolerance, and any other data you want to use. Every time you place a part from your database, the schematic has a link to the part database. When you generate a BOM, any information contained in your database can be displayed on the BOM. Thus, with the click of a mouse to place a part, all vital information about the part is included. BOM's created with the schematic are accurate and complete. If the information in the database is updated, the updated information is included when you generate a new BOM from the schematic. This eliminates the time needed and mistakes created by creating a BOM manually from the parts contained in a schematic.

The database used here is a modified version of the Digikey starter database available on Orcad's web site. The modified database contains an added table that can be used to store information for your own parts. This allows you to use the Digikey database and also have your own company database in the same file. This way, if the Digikey database is updated, you can easily identify your company's parts and copy the parts to the updated Digikey database when available.

The Digikey database is desirable because if you place a part from the database in your schematic and generate a BOM from that schematic, the BOM will contain the Digikey order numbers for all parts available from Digikey. You can then easily order the needed parts. If you do not wish to use this feature, you can choose to use only the parts in your own portion of the database.

The parts contained in the Digikey database and the parts you can download from Activeparts do not have PSpice models, although this feature may be added in the future. Thus, the projects we create can only be used for documentation purposes or for PC board layouts. No simulations can be performed on a circuit we create from these parts. Here, we will only show how to add these parts and create bills of materials.

## 10.A. Setting Up Orcad CIS to Use the Digikey Database

To use the database functions of Orcad Capture you need the five components listed below. Most users will only need to follow step 5, which shows how to install and set up the data base. Make sure that you follow these instructions completely.

1. You must have Microsoft Access installed on your computer if you wish to modify the contents of the database. This is a necessary tool for maintaining a company database. If you do not have Microsoft Access, you can still use the database with Capture CIS and you can still download parts from the Internet. However, you will not be able to modify the downloaded data unless you have Microsoft Access.
2. You must have the Microsoft Internet Explorer version 5 or higher installed on your system. The Explorer does not need to be your default browser, but it must be installed because it is used as a plug-in by Orcad Capture CIS. If you do not already have it installed, you can either download and install the most recent version from Microsoft's web site at www.microsoft.com, or install version 5 from the CD-ROM that accompanies this text. On the CD-ROM, the installation files are located in the IE5 directory. You will need to select the proper subdirectory for the language you wish to use.
3. You must have the Microsoft ODBC management tool installed. If this tool was not on your machine when you installed Orcad Capture CIS, the installation should have asked if you wanted to install it. If you selected no, you will need to install it now. This tool is located at different locations depending on your operating system. In Windows NT, it is located in the Control Panel. In Windows 2000, it is located in the Administrative Tools directory inside the Control Panel. If you cannot find this tool, run the Orcad Lite setup program again and be sure to install Capture CIS. If you are asked to install the Microsoft ODBC utility select yes. If you are not asked, then the tool is already installed on your system and you will need to locate it. If you have to reinstall the Orcad Lite software, you will also need to reinstall the libraries for this text as shown in Appendix A.2.
4. You must have Orcad Capture **CIS** installed, not Orcad Capture. Capture CIS has the extensions that allow you to download parts from the Internet and use parts from a database. When you installed the software for this text, you had the choice of installing Orcad Capture or Orcad Capture CIS. If you did not install Capture CIS, you must reinstall the software.

To check which version of Capture you are using, run Capture and then select **Help** from the menus:

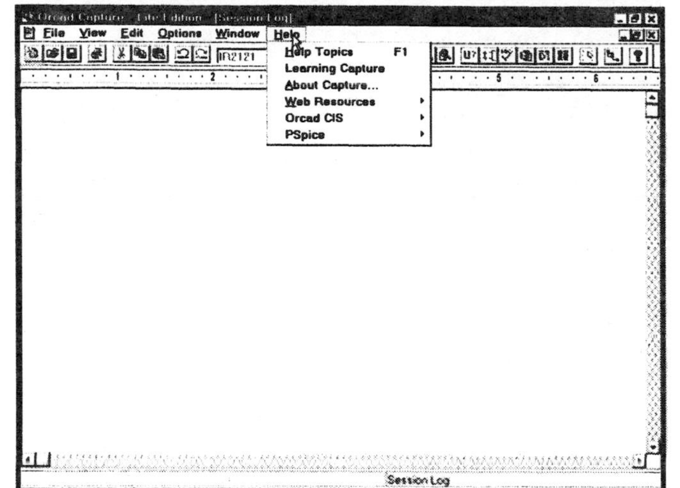


If your menu says **Orcad** rather than **Orcad CIS**, you do not have the proper software installed and will need to reinstall the software. If your software says **Orcad CIS**, then you have the proper version installed. If you need to reinstall the software, reinstall both the Orcad Lite software as shown in Appendix A.1 and the libraries for this text as shown in Appendix A.2. **Do not forget to reinstall the libraries.** When you install the Orcad software, make sure you select Capture CIS as shown:

5. The last thing we need to do is install the Digikey database components. The setup program is located on the CD-ROM that accompanies this text. Run the program D:\Database\setup.exe, where D: is the drive letter for your CD-ROM drive. After a few moments, you will see the screen below:

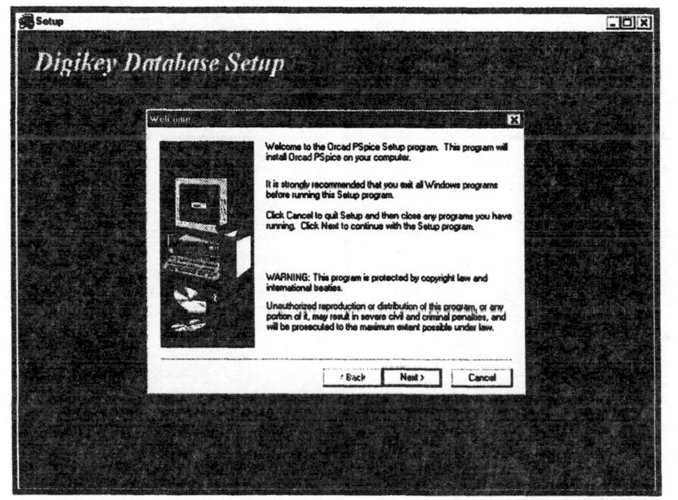


Click the ***Next*** button:

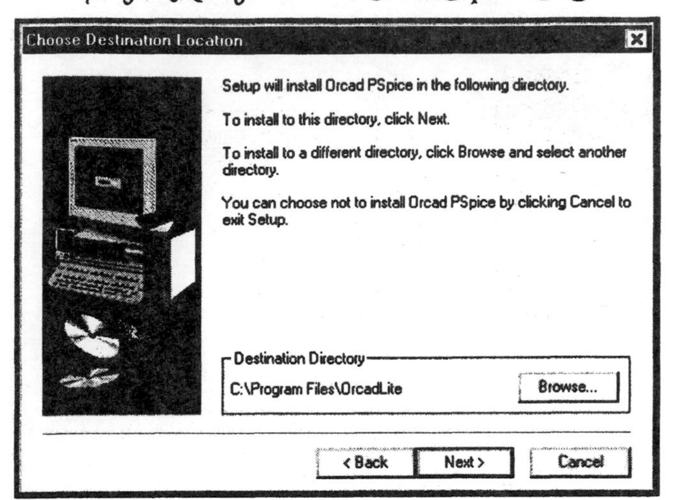


You must choose the same directory and drive here as you chose to install the Orcad software. If you chose the default settings when you installed the Orcad software, the directory shown here is correct and you can click the ***Next*** button. If you chose a different directory when installing the Orcad software, click the ***Browse*** button and select the directory where you installed the Orcad software. When you return to the screen above, click the ***Next*** button. The software will be installed:

Click the ***Finish*** button to complete the installation.

When you have completed the setup, there are a few steps you must follow to set up the database for use by Capture CIS. First, you need to run the Microsoft ODBC tool. If you are using Windows 98, the ODBC tool is located in the Control Panel folder. I am using Windows 2000, so this tool is located in the **Administrative Tools** folder of the Control Panel:

Select the ***Add*** button:

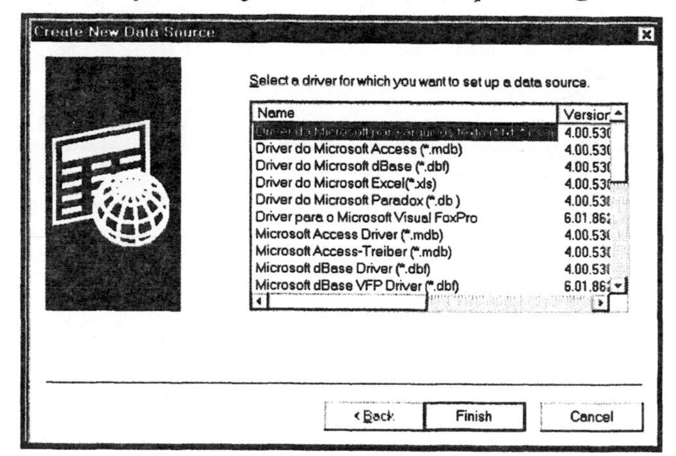


Select ***Microsoft Access Driver(*.mdb)*** and then click the ***Finish*** button:

Click the ***Select*** button:

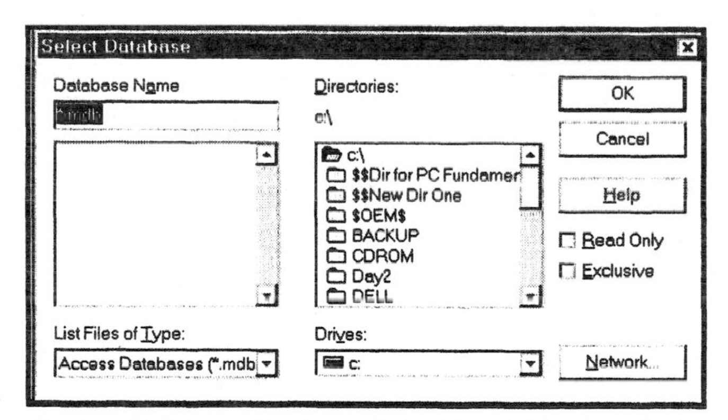


The database is located in the Capture/Database directory of the Orcad Lite installation directory. If you installed the database in the default directory but selected a different drive, the file is located in directory D:\Program Files\OrcadLite\Capture\Database, where D: is the name of the drive:

Select file ***DIGIKEY_SC.mdb*** and then click the ***OK*** button:

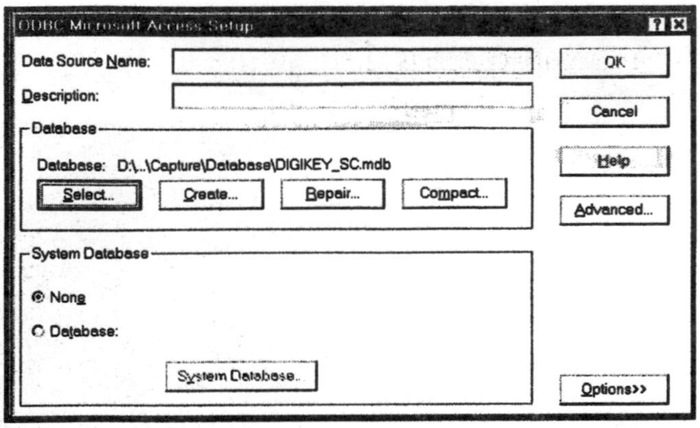

Fill in the ***Data Source Name*** and the ***Description*** fields, and then click the ***OK*** button. The ***Data Source Name*** must be ***Digikey_SC***

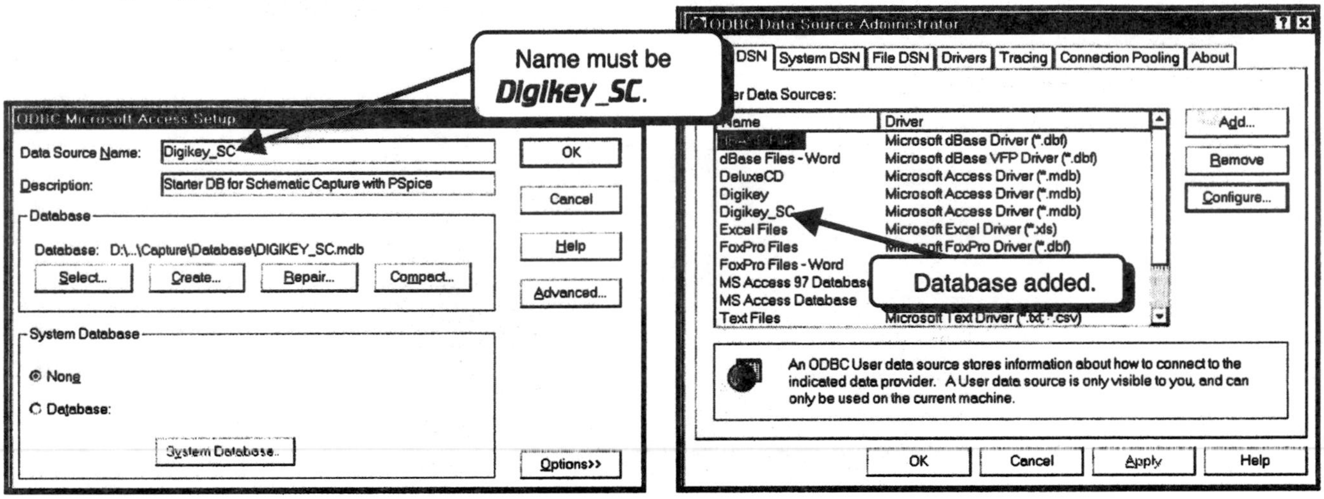

If you were successful, the database should be added to the list as shown above. Click the ***OK*** button to close the ODBC tool.

Next, we need to run Orcad Capture CIS and select the database:

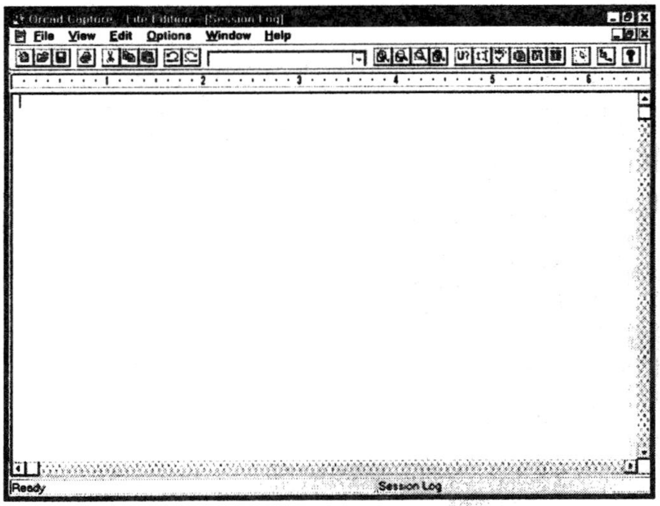

Select **File**, **New**, and then **Project** to create a new project:

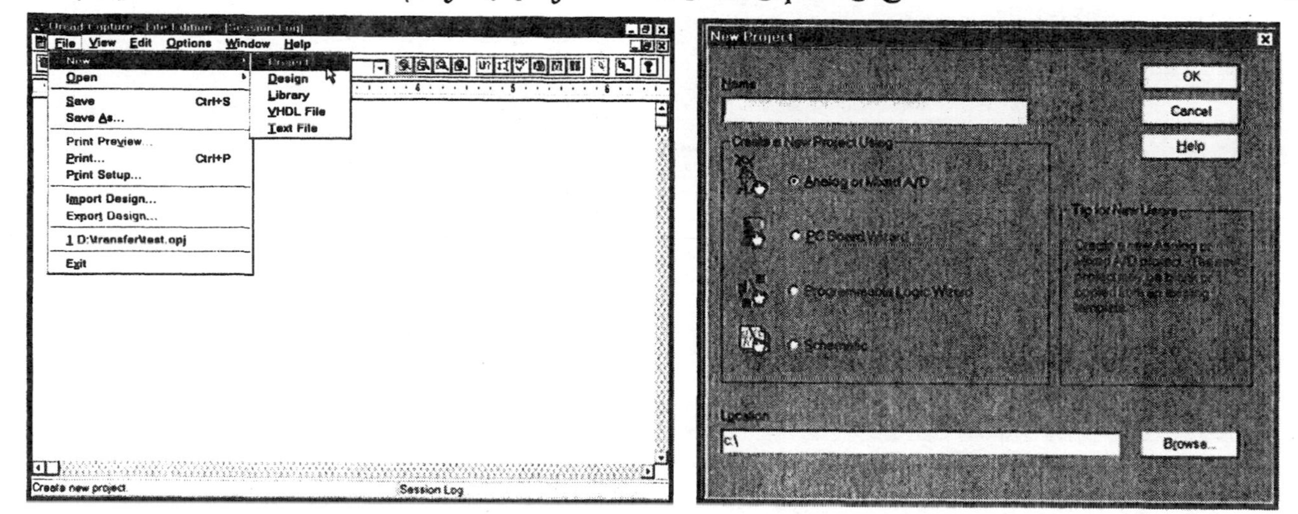


We cannot run a simulation with the parts in the database or the parts we download from the Internet, so select either the ***PC Board Wizard*** or the ***Schematic*** options:

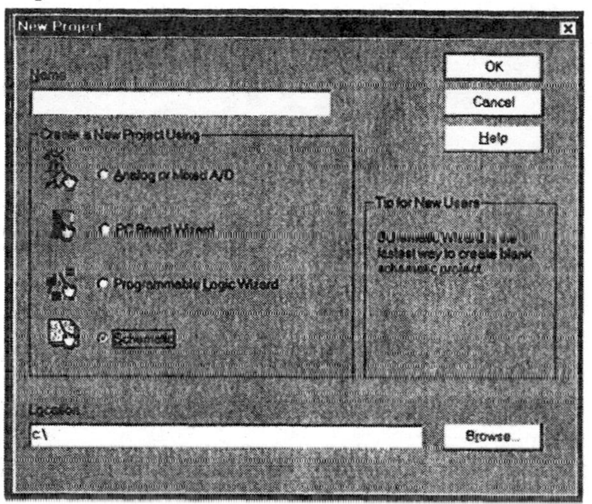


Select a ***Name*** and a ***Location*** for the project and then click the ***OK*** button:

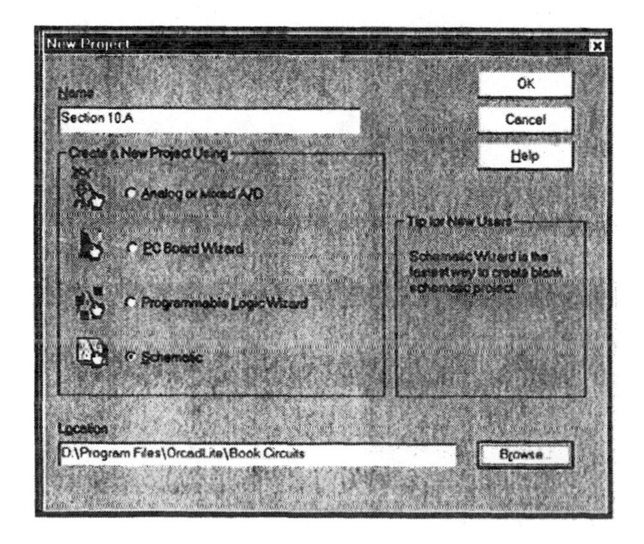


Next we need to close the schematic window and then select the database. Type **CTRL-F4** to close the schematic window. Select **Options** and then **CIS Configuration** from the menus:

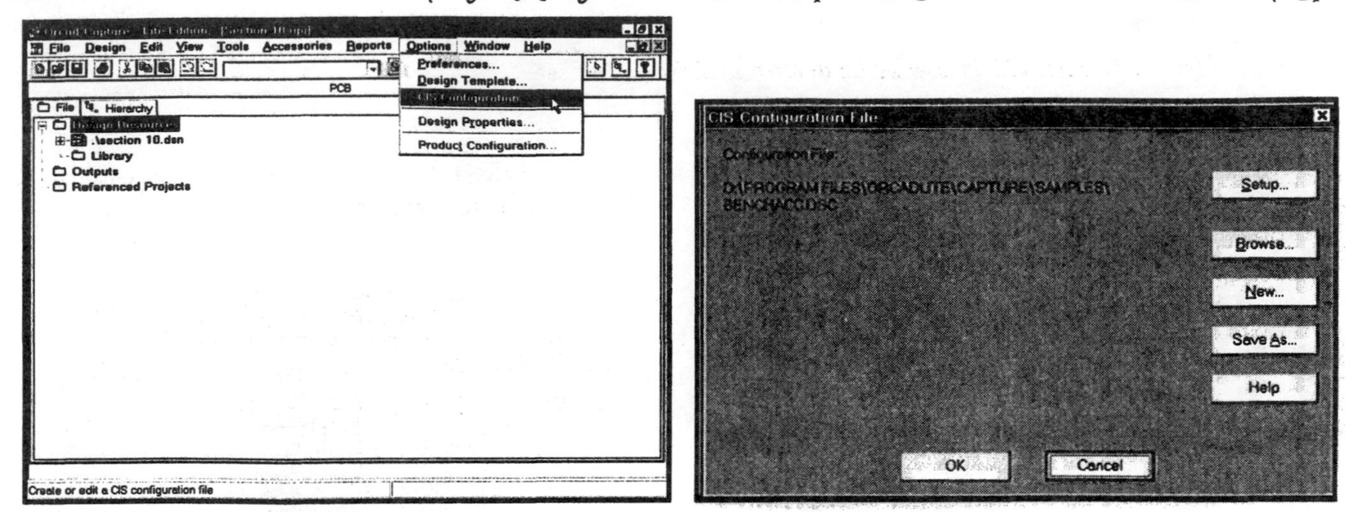


The default installation of Capture CIS uses a small database named ***BENCHACC.DBC***. We need to change the database to our Digikey database. Select the ***Browse*** button:

We are looking for the file named Digikey_SC.dbc. This file is located in the Capture\Database directory of the Orcad Lite installation directory. Locate and select this file:

Click the ***Open*** button to select the file:

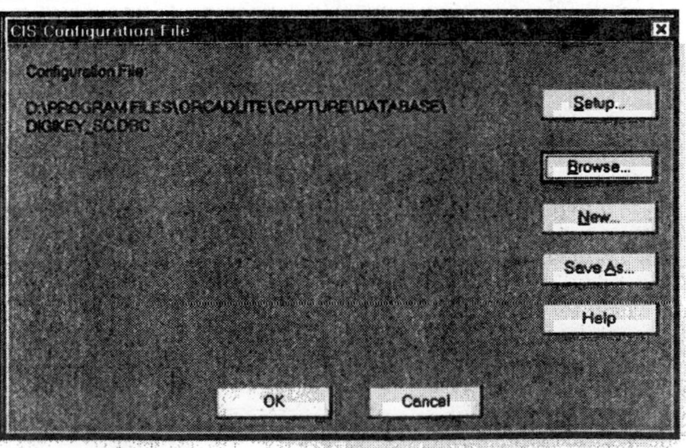

Click the ***OK*** button. Capture CIS is now set up to use our Digikey database.

# 10.B. Starting a Project

The parts available in the database and the parts that can be downloaded from the Activeparts web site do not contain PSpice models, so we cannot use these parts to run a simulation. When you create a project, you should select either the Schematic or PC Board Wizard option. Run Capture CIS and then select **File**, **New**, and then **Project** to create a new project:

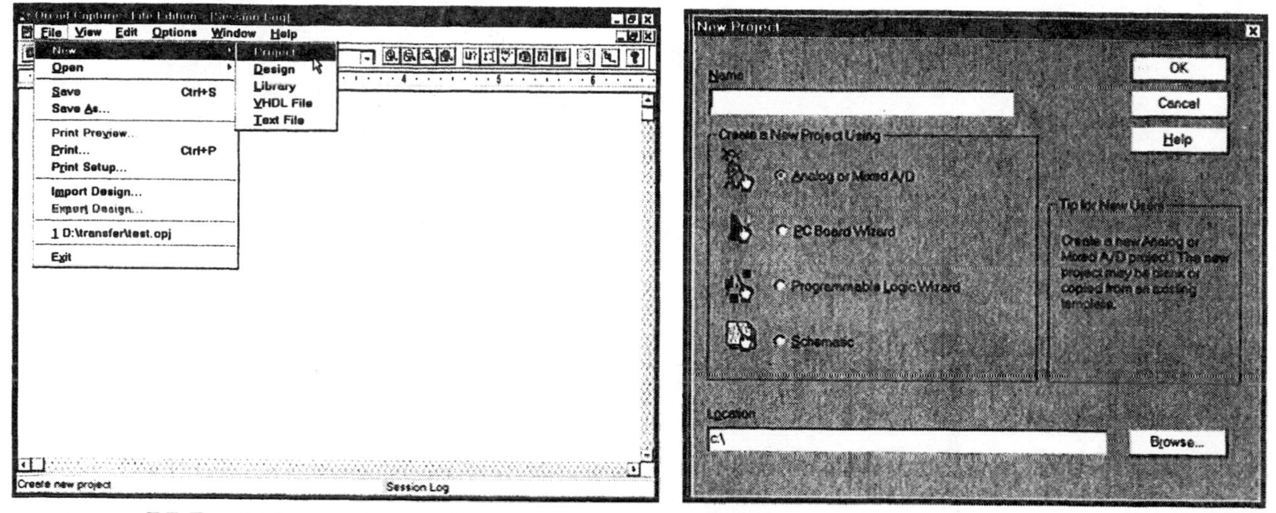


Select either the ***PC Board Wizard*** or the ***Schematic*** option

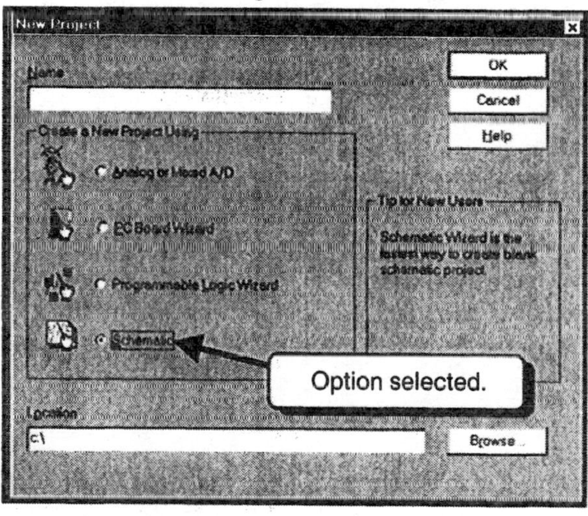


Select a ***Name*** and a ***Location*** for the project and then click the ***OK*** button:

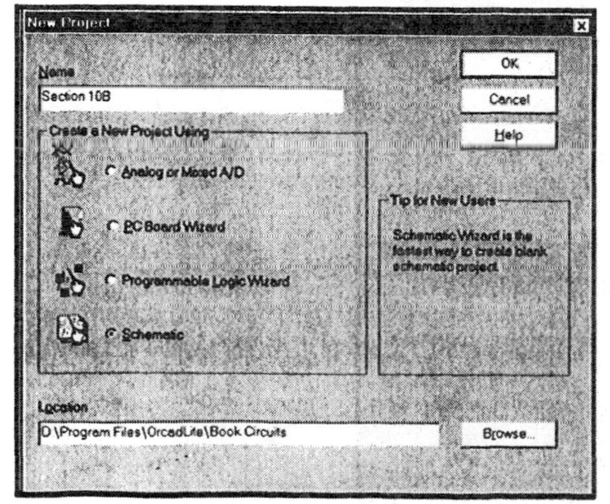


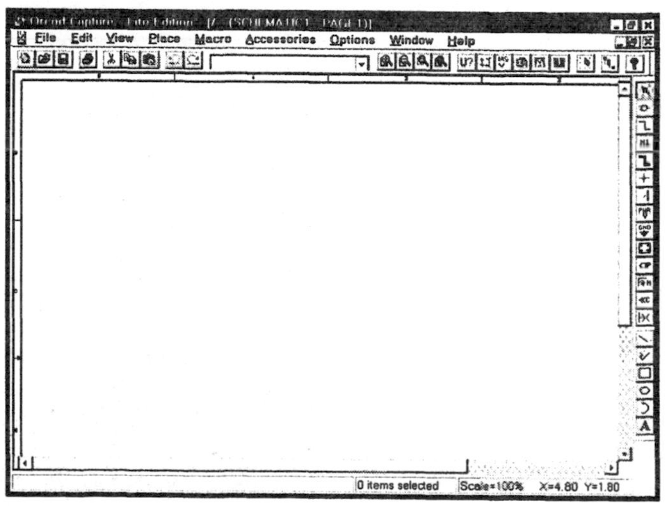

We can now add parts to this schematic using the database or Activeparts.

# 10.C. Using Parts in the Database

With over 70,000 parts in the Digikey database, finding the part you need can be a challenge. Fortunately, the Capture CIS provides tools to help you search for parts. To place a part on your schematic, select **Place** and then **Database Part** from the Capture menus:

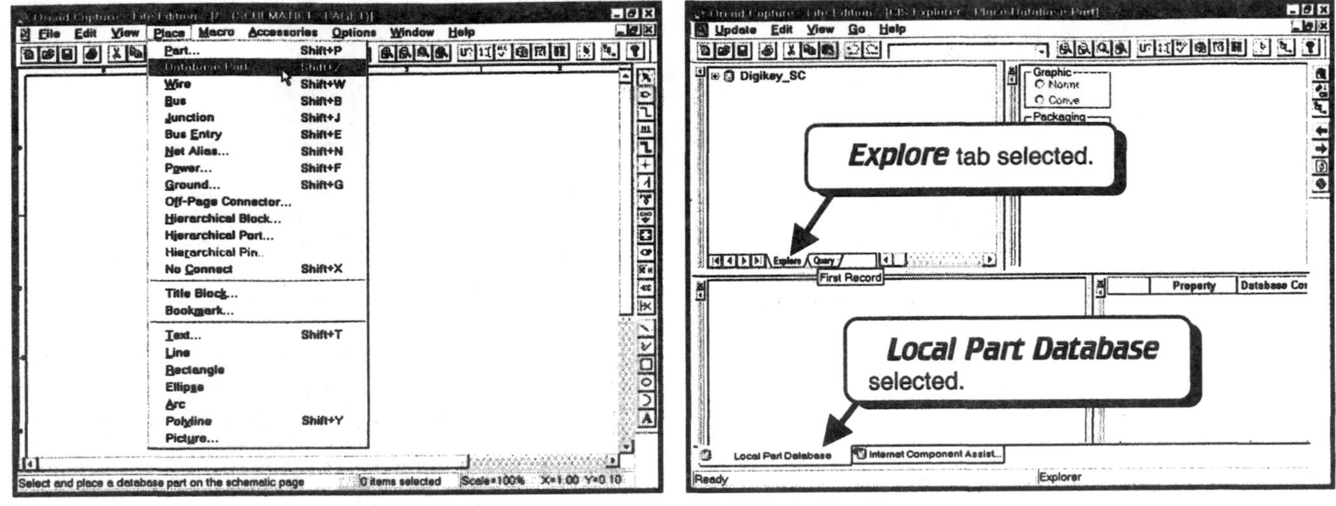


## 10.C.1. Exploring the Database

In the above left screen capture, the ***Explore*** tab is selected. In this mode, we can explore the database like we would a hard disk using the Windows Explorer. To expand the tree, click the ***LEFT*** mouse button on the ⊞ as shown:

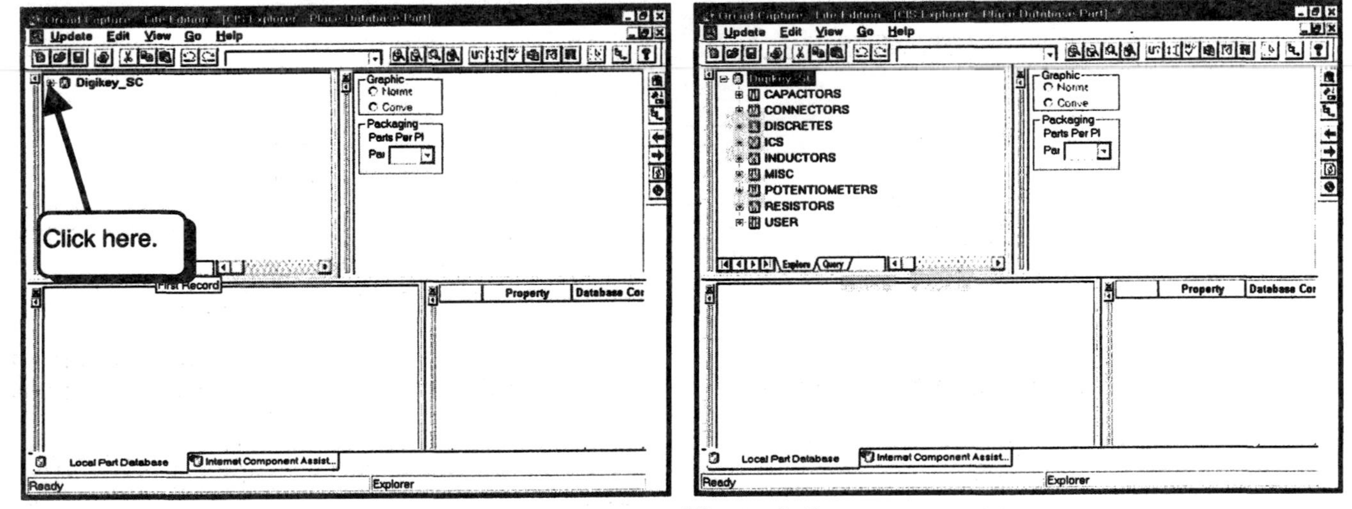

Expand the ***CAPACITORS*** branch or the tree until you see the ***Electrolytic*** section:

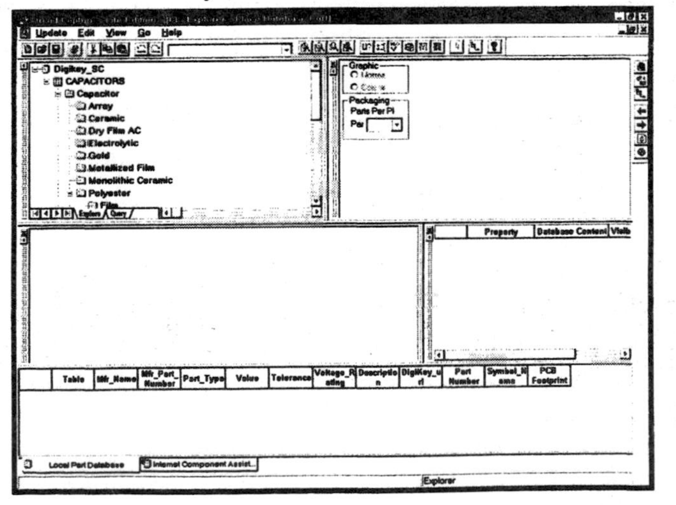

Click on the text ***Electrolytic*** to display the available electrolytic capacitors:

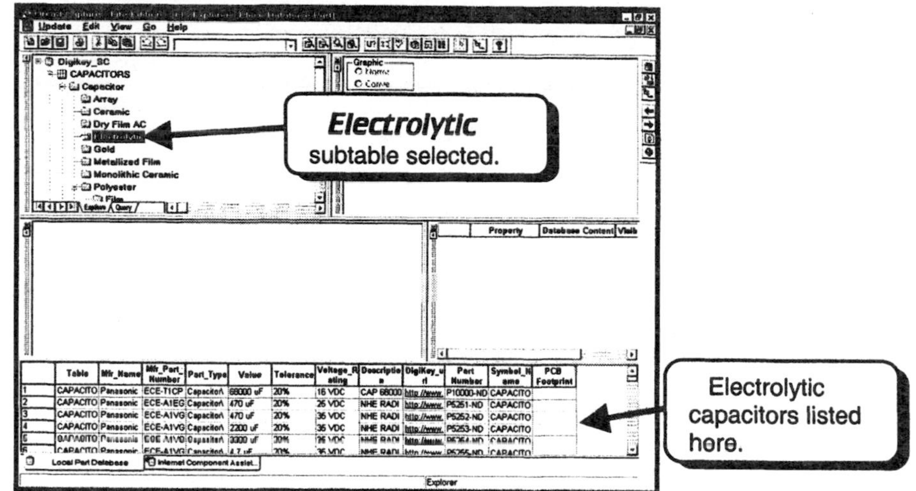


You can scroll through the list if you like. Hundreds of capacitors are contained in this list. To select a part, click the ***LEFT*** mouse button on the line in the database you want to select. If the line is highlighted in green, then the part is correctly linked to a graphic symbol in one of the symbol libraries. If a symbol and PCB layout footprint are found, they will be displayed in separate windows:

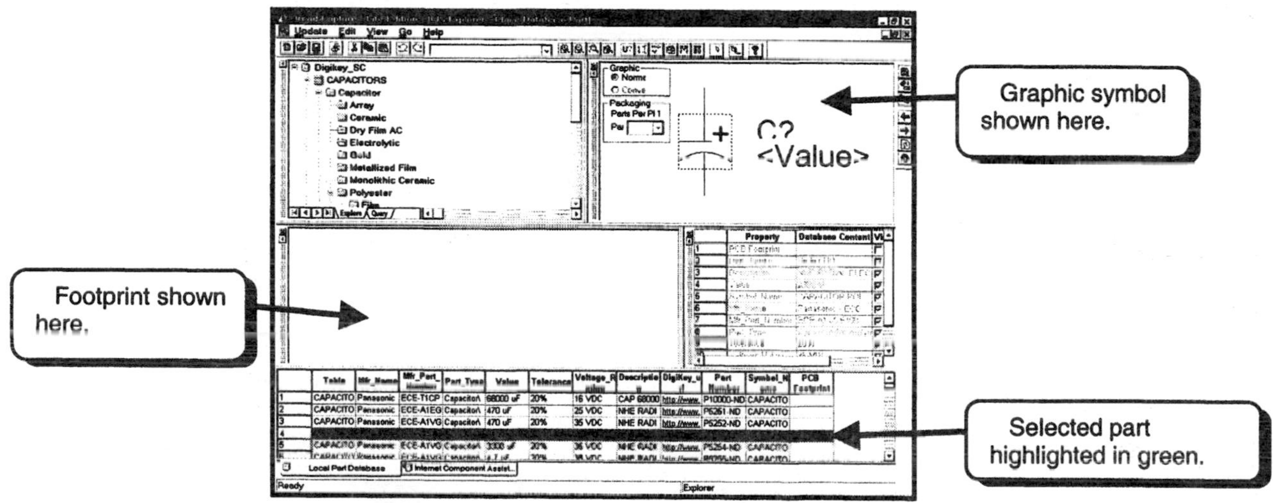


A footprint is not displayed because the parts in the Digikey database do not have footprints associated with them. To place the selected part in your schematic, click the ***RIGHT*** mouse button on the highlighted part and then select **Place Database Part**:

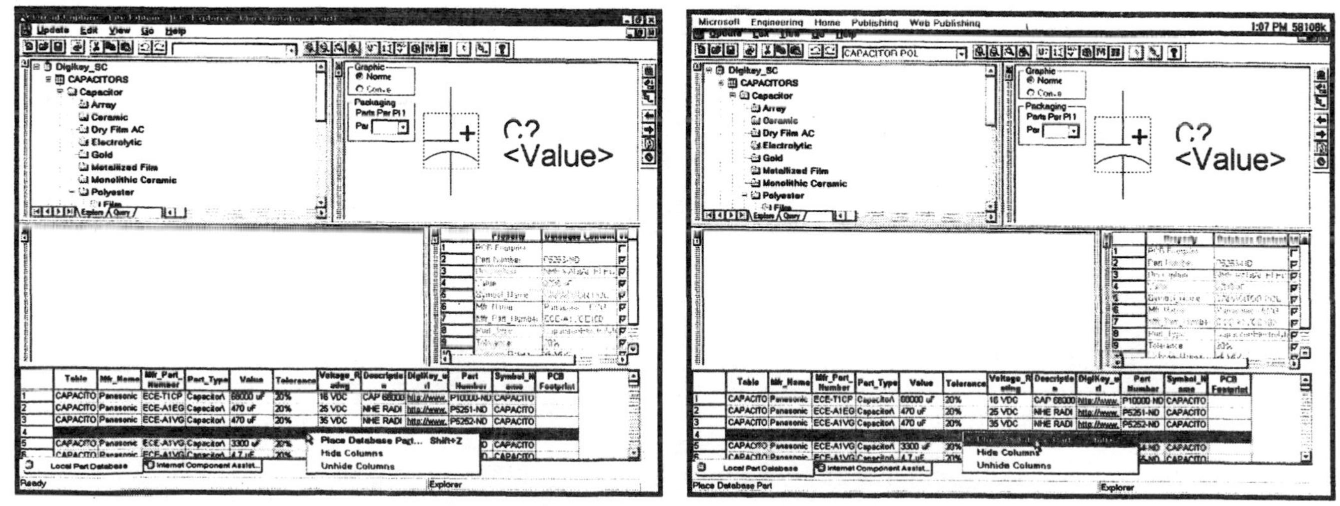

You will return to the schematic with the part symbol attached to the mouse pointer:

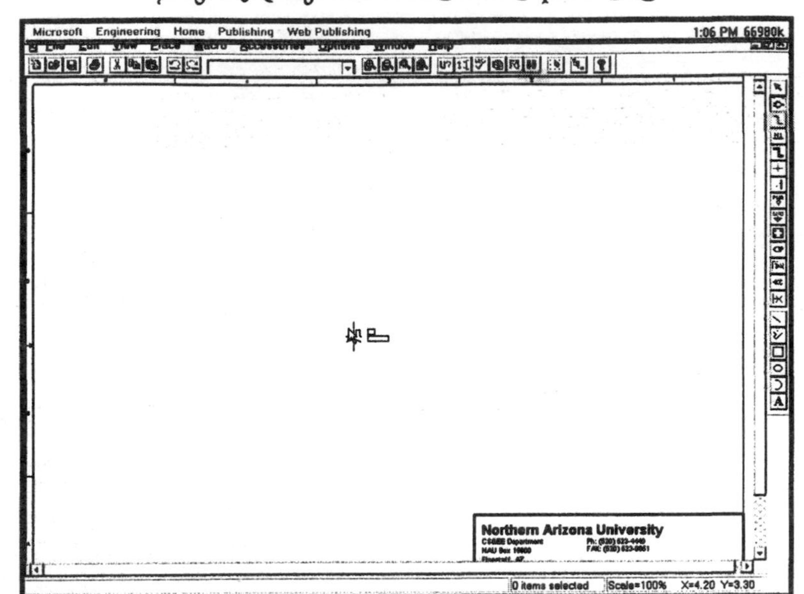


Place the part as you would any part and then press the **ESC** key to stop placing parts:

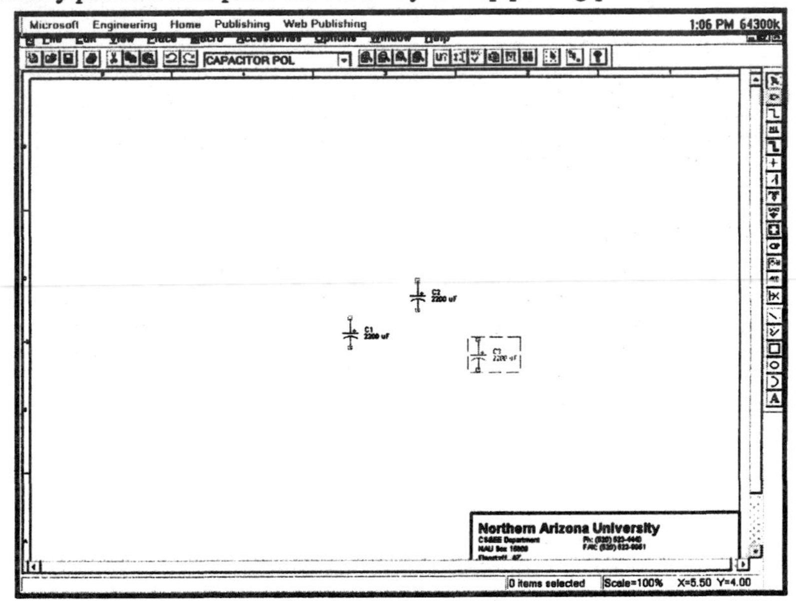


We would like to further explore the database, so press the **z** key to place a database part:

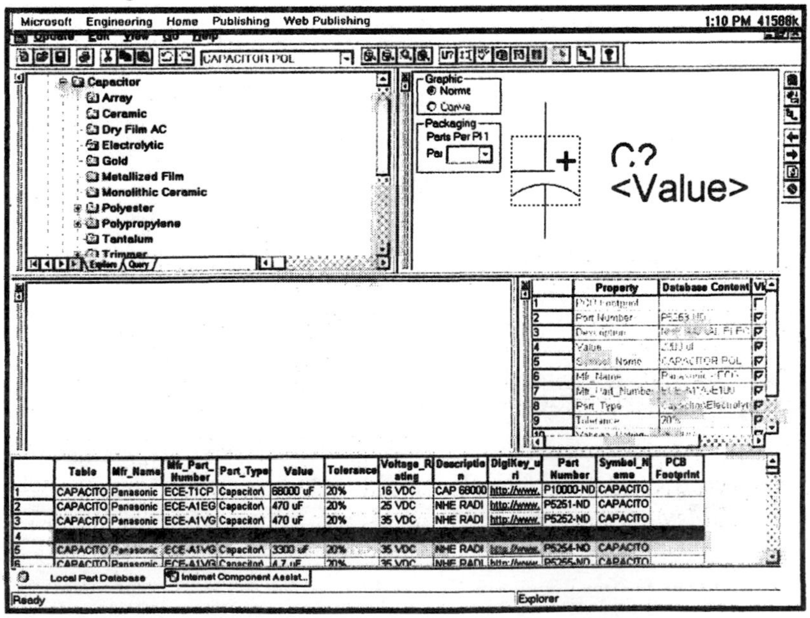

Microsoft Engineering Home Publishing Web Publishing 1:10 PM 41588k

CAPACITOR POL

Capacitor
- Array
- Ceramic
- Dry Film AC
- Electrolytic
- Gold
- Metallized Film
- Monolithic Ceramic
- Polyester
- Polypropylene
- Tantalum
- Trimmer

Explore Query

Graphic: Norme / Conve

Packaging: Parts Per Pl 1

C?
<Value>

| | Property | Database Content |
|---|---|---|

| | Table | Mfr_Name | Mfr_Part_Number | Part_Type | Value | Tolerance | Voltage_Rating | Description | DigiKey_url | Part Number | Symbol_Name | PCB Footprint |
|---|---|---|---|---|---|---|---|---|---|---|---|---|
| 1 | CAPACITO | Panasonic | ECE-T1CP | Capacitor\ | 68000 uF | 20% | 16 VDC | CAP 68000 | http://www. | P10000-ND | CAPACITO | |
| 2 | CAPACITO | Panasonic | ECE-A1EG | Capacitor\ | 470 uF | 20% | 25 VDC | NHE RADI | http://www. | P5251-ND | CAPACITO | |
| 3 | CAPACITO | Panasonic | ECE-A1VG | Capacitor\ | 470 uF | 20% | 35 VDC | NHE RADI | http://www. | P5252-ND | CAPACITO | |
| 4 | | | | | | | | | | | | |
| 5 | CAPACITO | Panasonic | ECE-A1VG | Capacitor\ | 3300 uF | 20% | 35 VDC | NHE RADI | http://www. | P5254-ND | CAPACITO | |
| 6 | CAPACITO | Panasonic | ECE-A1VG | Capacitor\ | 4.7 uF | 20% | 35 VDC | NHE RADI | http://www. | P5255-ND | CAPACITO | |

Local Part Database | Internet Component Assist...

Ready | Explorer

Scroll down the tree view of the database and locate and expand the ***USER*** section:

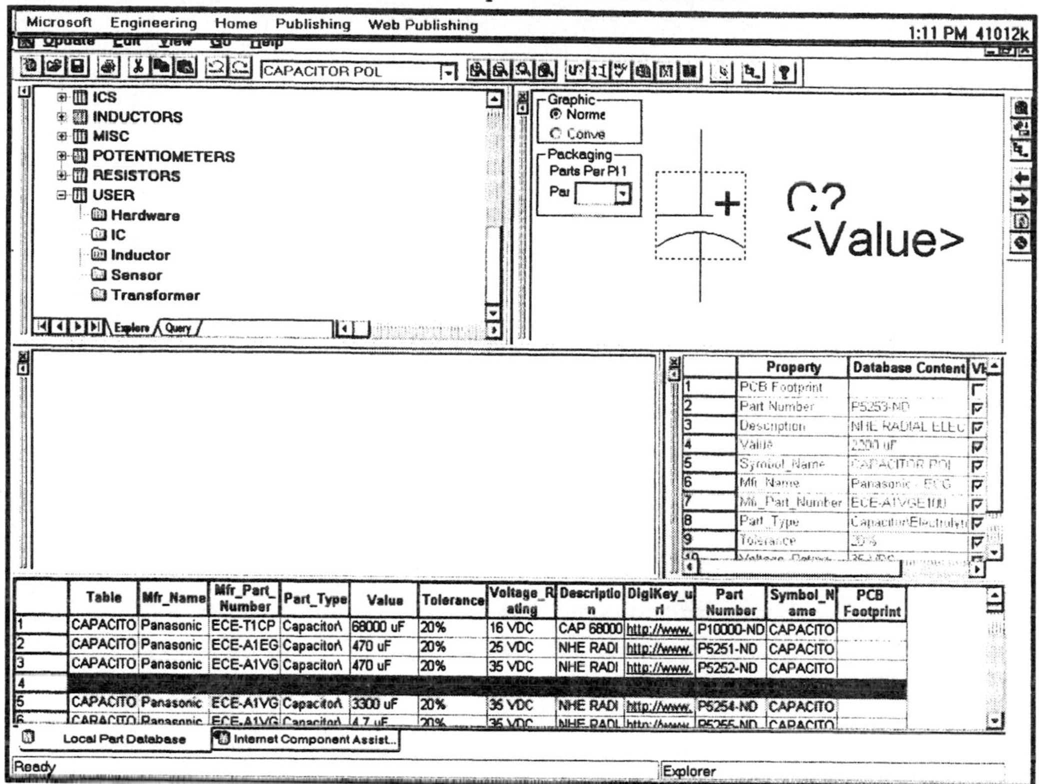


The user portion of the database is where you will save parts for your company. Notice that the ***USER*** portion is split into subsections. When you download parts from Activeparts, they will be added to the ***USER*** section. Using Microsoft Access, you can place them in one of the existing subsections, or create new subsections. Select the ***Hardware*** subsection:

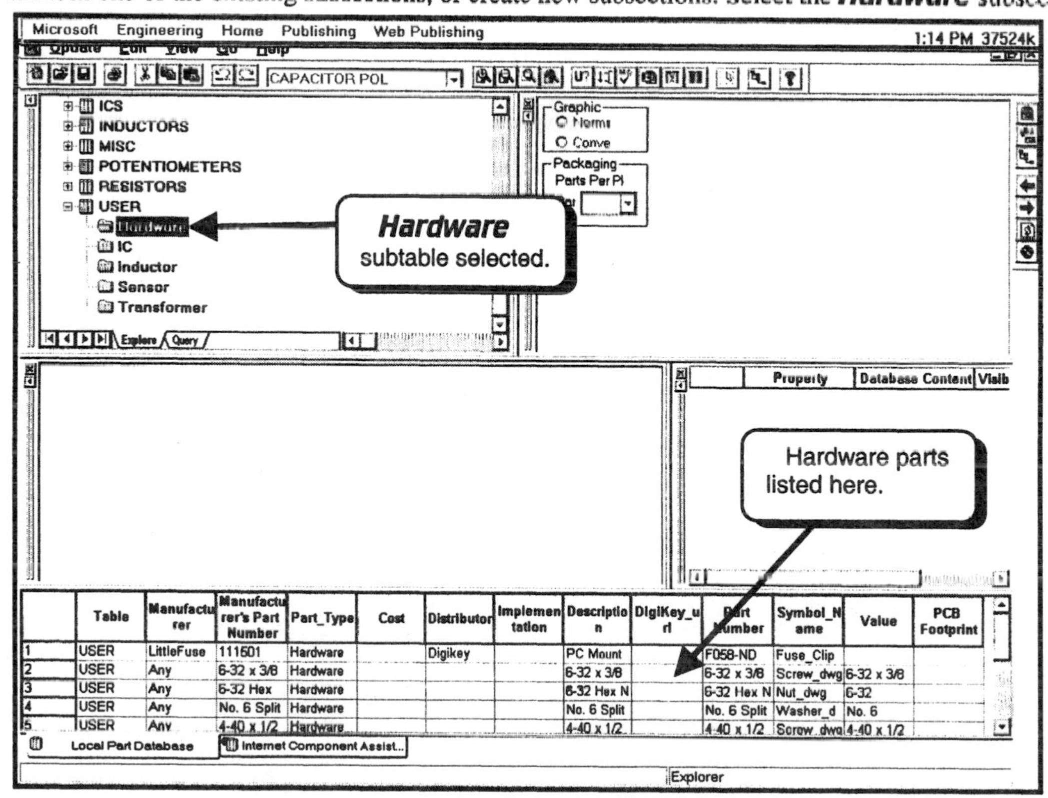


Hardware is provided for projects that need hardware for assembly. If you place hardware items in your schematic, the hardware will be listed on the BOM when you generate the BOM from Capture CIS. Select a 6-32 screw and place it in your schematic:

The Explore view is useful to see some of the major categories available in your database, but with over 70,000 available parts, we need to use the search tools available in Capture CIS.

## 10.C.2. Searching for Parts in the Database

We will show two methods for finding parts in the database. In the first method we use the Query tab and search for specific criteria. In the second method, we will find a part number in the Digikey catalog and then immediately find the part since we know the part number.

### 10.C.2.a. Searching with the Query Tab

While viewing the schematic, press the **z** key to place a database part:

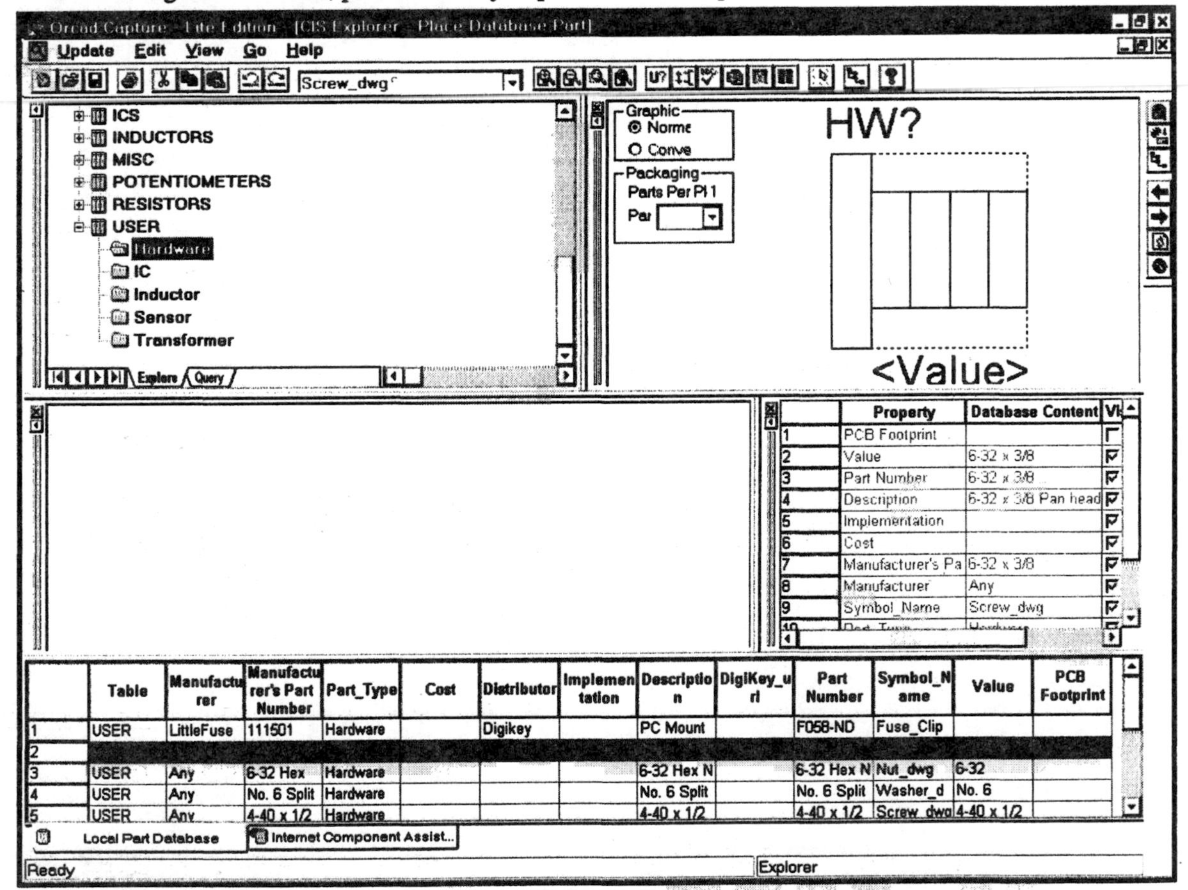


We would like to find a 0.1 μF ceramic capacitor. Select the ***Ceramic*** subtable in the ***Explore*** view:

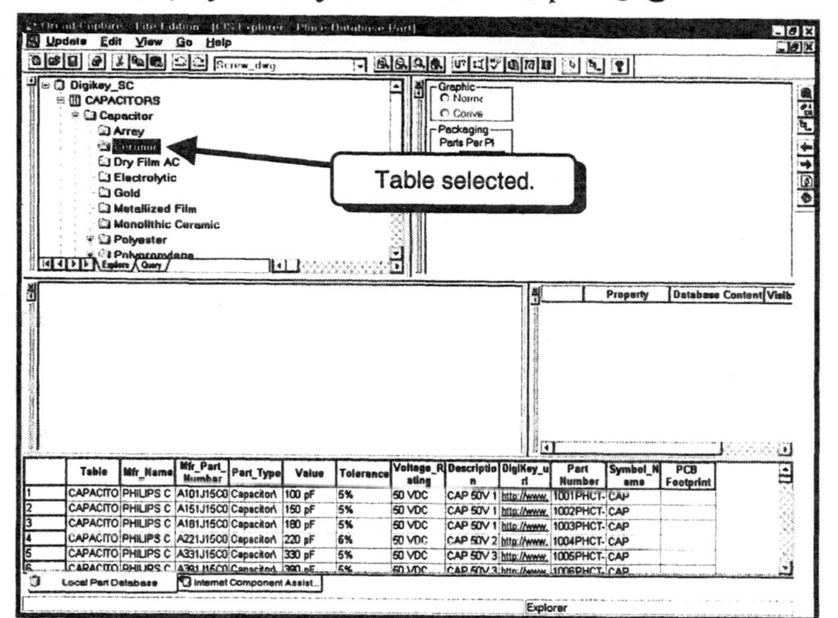


Select the ***Query*** tab:

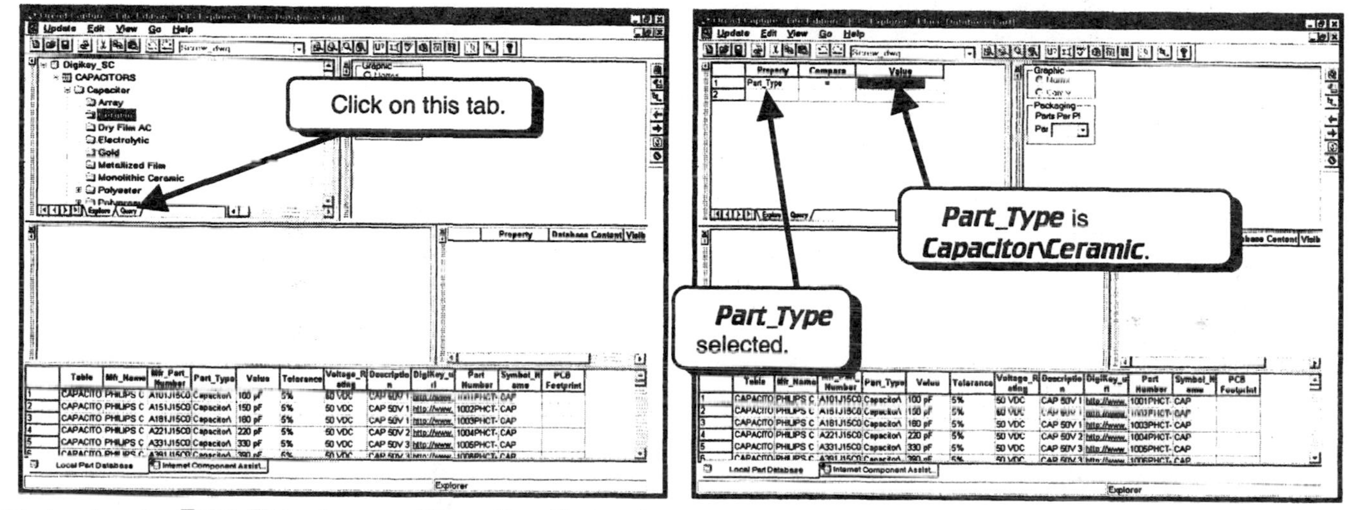


Notice that the ***Part_Type*** is set to ***CapacitorCeramic***. This automatically narrows down our search to that portion of the database.

Next, we need to add another restriction to our search. Click the ***LEFT*** mouse button as shown to select another property:

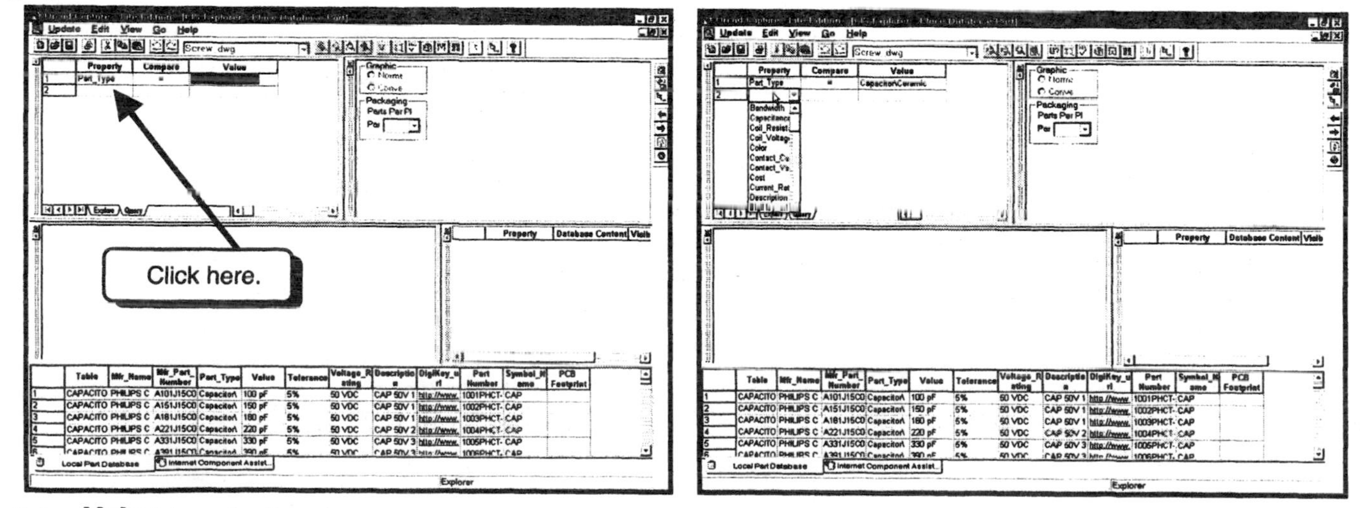


Select ***Value*** from the list:

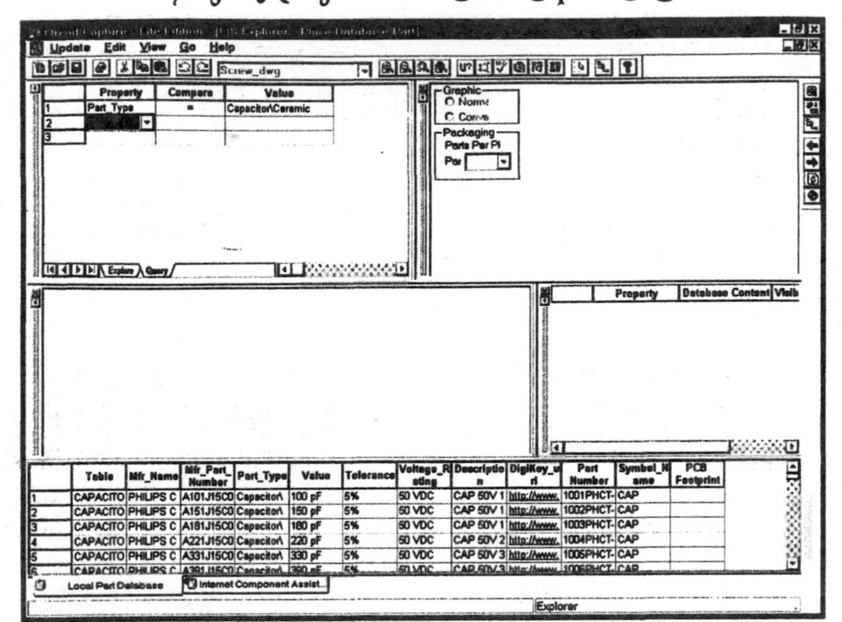

Next, click the ***LEFT*** mouse button as shown below:

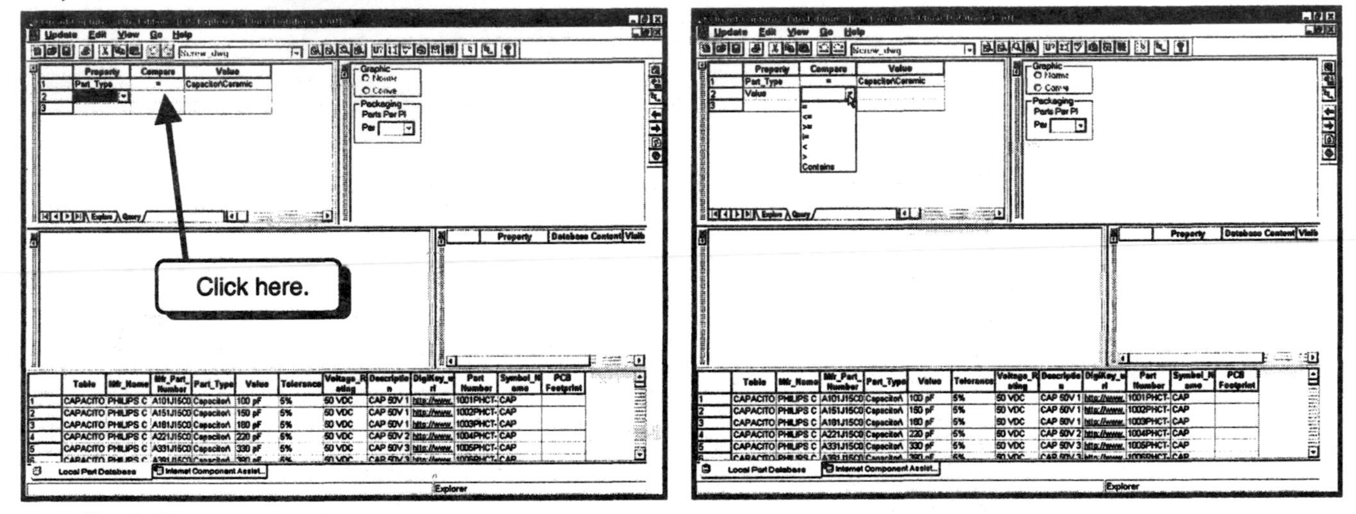


Select ***Contains***:

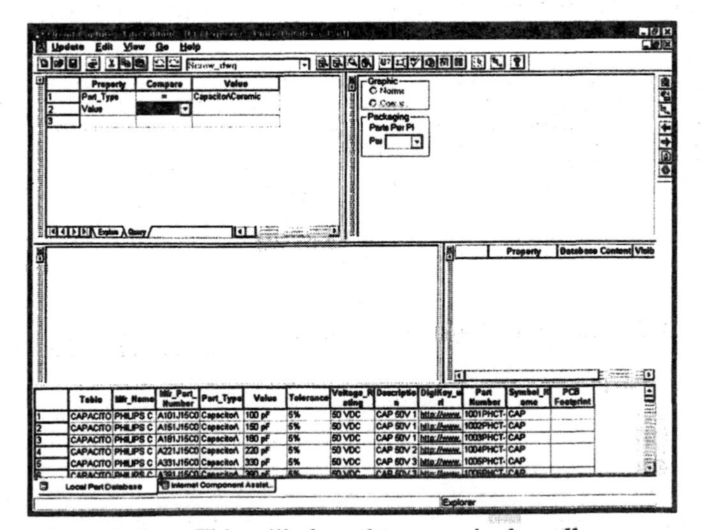

Click the ***LEFT*** mouse button as shown below. This will place the cursor in the cell:

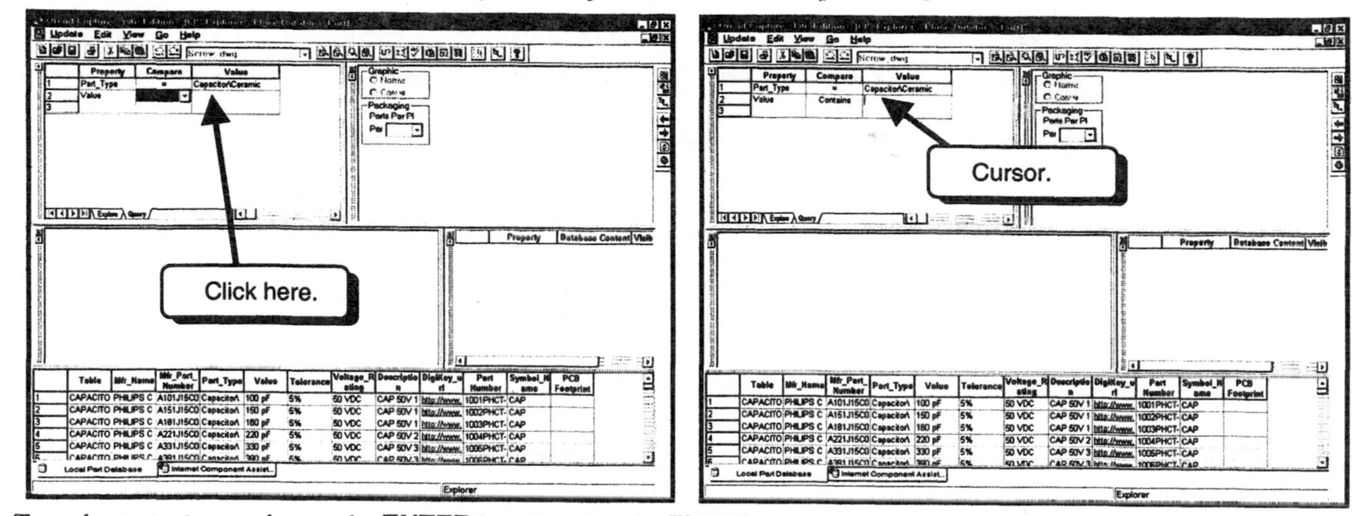


Type the text **.1 u** and press the **ENTER** key. Pressing the **ENTER** key will start the search. Make sure that there is a space between the 1 and the u:



There are too many items in the list. We will narrow down the search to include parts with 5% tolerance. Fill in the Query spreadsheet as shown:

When you press the **ENTER** key, the search results will be updated to contain the new criteria:

Orcad Capture - Lite Edition - [CIS Explorer - Place Database Part]

Update Edit View Go Help

Screw_dwg

| | Property | Compare | Value |
|---|---|---|---|
| 1 | Part_Type | = | Capacitor\Ceramic |
| 2 | Value | Contains | .1 u |
| 3 | Tolerance | = | 5% |
| 4 | | | |

Explorer / Query

Graphic
Normal
Convert
Packaging
Parts Per Pkg

| | Property | Database Content | Visib |
|---|---|---|---|

Ceramic 0.1 μF capacitors with 5% tolerance listed here.

| | Table | Mfr_Name | Mfr_Part_Number | Part_Type | Value | Tolerance | Voltage_Rating | Description | DigiKey_url | Part Number | Symbol_Name | PCB Footprint |
|---|---|---|---|---|---|---|---|---|---|---|---|---|
| 1 | CAPACITO | Panasonic | ECU-S1H1 | Capacitor\ | 0.1 uF | 5% | 50 VDC | CAP CERA | http://www. | P4923-ND | CAP | |
| 2 | CAPACITO | Panasonic | ECU-S1H1 | Capacitor\ | 0.1 uF | 5% | 50 VDC | CAP CERA | http://www. | P4924-ND | CAP | |

Local Part Database | Internet Component Assist...

Explorer

The updated list now contains only two parts. To place one of the parts in your schematic, double-click the ***LEFT*** mouse button on one of the items in the list. You will return to the schematic with the part attached to the mouse pointer:

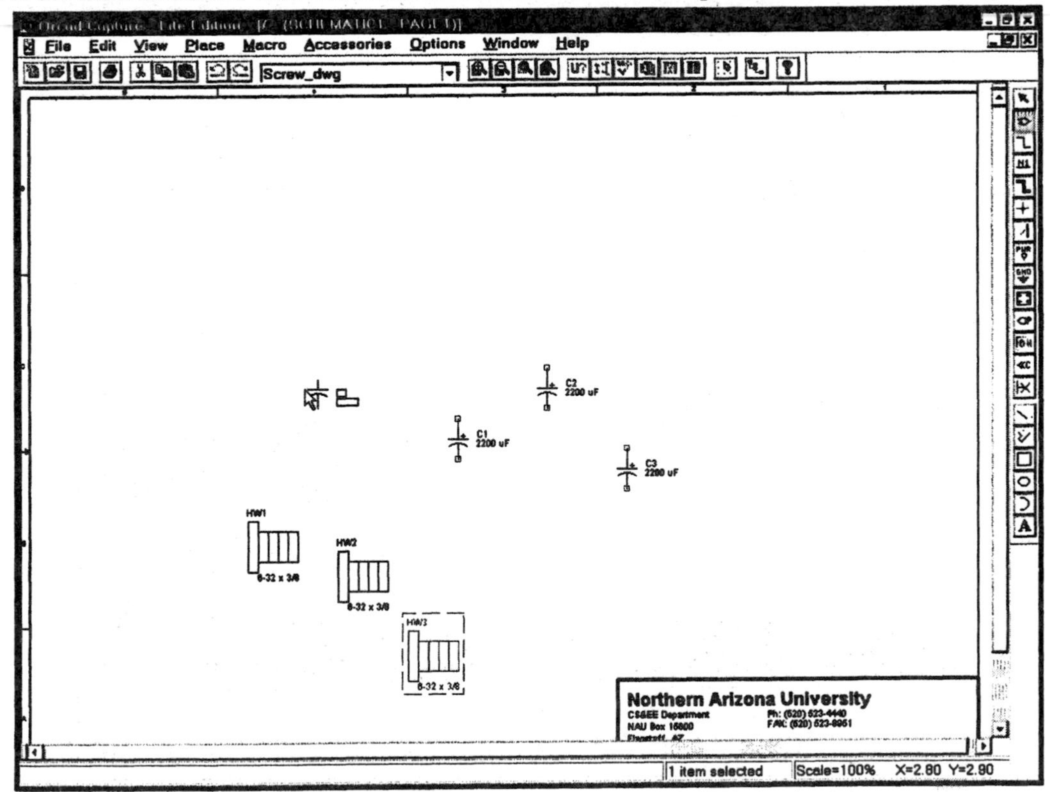


Place the part as you would a standard part. Press the **ESC** key to stop placing parts.

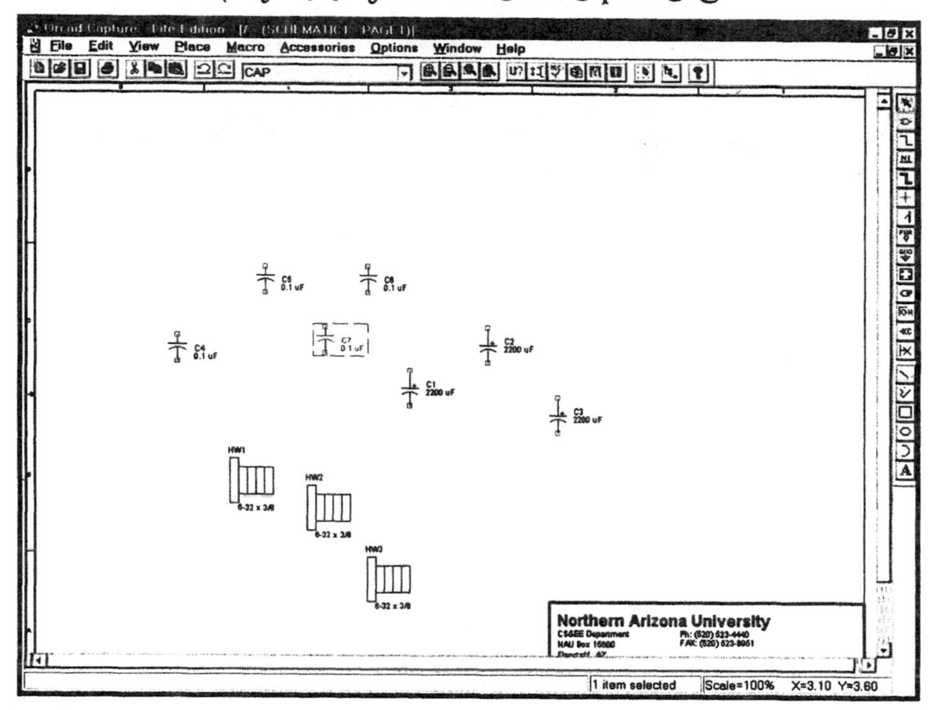


## 10.C.2.b. Searching for Parts Using the Digikey Catalog

The Digikey database contains a huge number of parts and a lot of part information about each part. Unfortunately, it does not contain all of the information needed to select a part. I have found that the easiest way to use the Digikey portion of the database is to find the part you need using a hard copy of the Digikey catalog (which you can request from the Digikey web site) and then use the Digikey part number to locate your part in the database.

I wish to place a 1.5 kΩ, 1/8 watt resistor in my circuit. It should be a surface mount part with a 1206 footprint and come packaged on a tape and reel. It would be difficult to find this part using the query tab of Capture CIS. After browsing the Digikey catalog, we find that the Digikey part number for this part is p1.5ketr-nd. Assuming that you have the schematic displayed on your computer, type **z** to place a database part:

The Query tab has our last search on it. To delete a row of the search, click the ***LEFT*** mouse button on the number 3 as shown. This will select row 3:

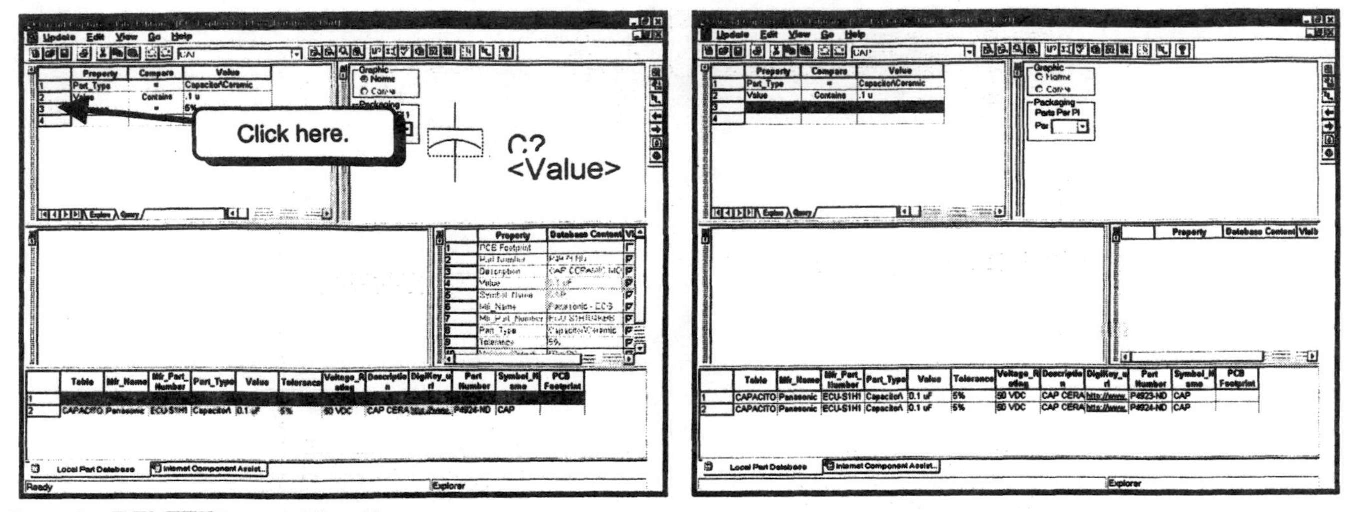


Press the **DELETE** key to delete the row:

Update Edit View Go Help

CAP

| | Property | Compare | Value |
|---|---|---|---|
| 1 | Part_Type | = | Capacitor\Ceramic |
| 2 | Value | Contains | .1 u |
| 3 | | | |

Graphic: Nome, Conve

Packaging: Parts Per Pl, Per

Explore Query

| | Property | Database Content | Visib |
|---|---|---|---|

| | Table | Mfr_Name | Mfr_Part_Number | Part_Type | Value | Tolerance | Voltage_Rating | Description | DigiKey_url | Part Number | Symbol_Name | PCB Footprint |
|---|---|---|---|---|---|---|---|---|---|---|---|---|
| 1 | CAPACITO | Panasonic | ECU-S1H1 | Capacitor\ | 0.1 uF | 5% | 50 VDC | CAP CERA | http://www. | P4923-ND | CAP | |
| 2 | CAPACITO | Panasonic | ECU-S1H1 | Capacitor\ | 0.1 uF | 5% | 50 VDC | CAP CERA | http://www. | P4924-ND | CAP | |

Local Part Database | Internet Component Assist...

Explorer

Use the same method to delete the second row:

Update Edit View Go Help

CAP

| | Property | Compare | Value |
|---|---|---|---|
| 1 | Part_Type | = | Capacitor\Ceramic |
| 2 | | | |

Graphic: Nome, Conve

Packaging: Parts Per Pl, Per

Explore Query

| | Property | Database Content | Visib |
|---|---|---|---|

| | Table | Mfr_Name | Mfr_Part_Number | Part_Type | Value | Tolerance | Voltage_Rating | Description | DigiKey_url | Part Number | Symbol_Name | PCB Footprint |
|---|---|---|---|---|---|---|---|---|---|---|---|---|
| 1 | CAPACITO | Panasonic | ECU-S1H1 | Capacitor\ | 0.1 uF | 5% | 50 VDC | CAP CERA | http://www. | P4923-ND | CAP | |
| 2 | CAPACITO | Panasonic | ECU-S1H1 | Capacitor\ | 0.1 uF | 5% | 50 VDC | CAP CERA | http://www. | P4924-ND | CAP | |

Local Part Database | Internet Component Assist...

Explorer

To find a part with a specific Digikey part number, set the ***Property*** column to ***Part Number***, the ***Compare*** column to =, and the ***Value*** column to the part number:

Update Edit View Go Help

CAP

| | Property | Compare | Value |
|---|---|---|---|
| 1 | Part Number | = | p1.5ketr-nd |
| 2 | | | |

Graphic
Norma
Conve
Packaging
Parts Per Pl
Par

Explore Query

| | Property | Database Content | Visib |
|---|---|---|---|

| | Table | Mfr_Name | Mfr_Part_Number | Part_Type | Value | Tolerance | Voltage_Rating | Description | DigiKey_url | Part Number | Symbol_Name | PCB Footprint |
|---|---|---|---|---|---|---|---|---|---|---|---|---|
| 1 | CAPACITO | Panasonic | ECU-S1H1 | Capacitor\ | 0.1 uF | 5% | 50 VDC | CAP CERA | http://www. | P4923-ND | CAP | |
| 2 | CAPACITO | Panasonic | ECU-S1H1 | Capacitor\ | 0.1 uF | 5% | 50 VDC | CAP CERA | http://www. | P4924-ND | CAP | |

Local Part Database
Internet Component Assist...
Explorer

When you press the **ENTER** key, only one part should be displayed in the part list:

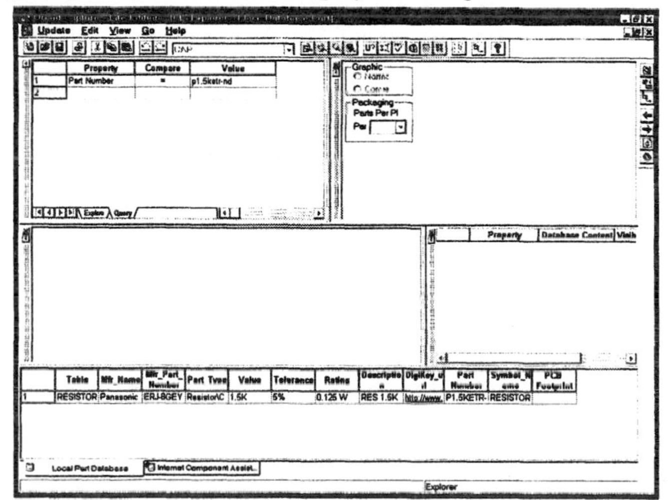


Each part in the database has a unique Digikey part number, so the search should find only the part for which you are looking. Double-click on the part to place it in your schematic:

File Edit View Place Macro Accessories Options Window Help

R

C5 0.1 uF
C6 0.1 uF
C4 0.1 uF
C7 0.1 uF
C2 2200 uF
C1 2200 uF
C3 2200 uF
R1 1.5K
HW1 6-32 x 3/8
HW2 6-32 x 3/8
HW3 6-32 x 3/8

Northern Arizona University
CS&EE Department
NAU Box 15600
Ph: (520) 523-4440
FAX: (520) 523-9961

0 items selected Scale=100% X=7.00 Y=4.00

## 10.C.3. Using the Part Manager

The Part Manager tells us whether the part information in our schematic is consistent with the part information in the database. As an example, we will change the value of the 1.5k resistor to 1.8k:

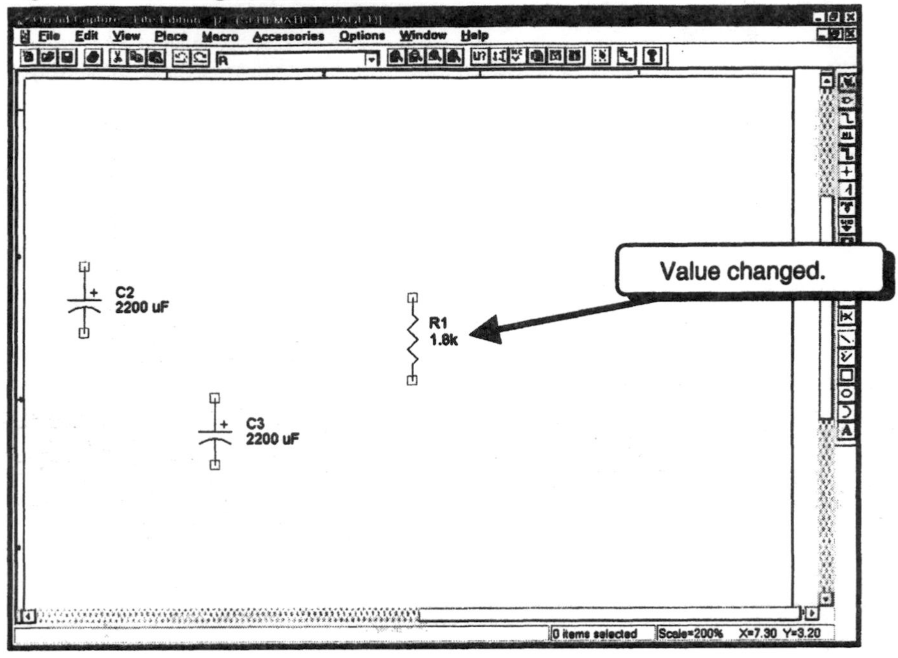


The part listed in the database is part number p1.5ketr-nd, which is a 1.5k resistor. In the schematic, the value is 1.8k, so the schematic information is inconsistent with the database information. If we create a BOM with this schematic, it will be incorrect. Capture CIS provides a facility for detecting these inconsistencies.

We must return to the tree view of the project to check for inconsistencies. Type **CTRL-F4** to close the schematic:

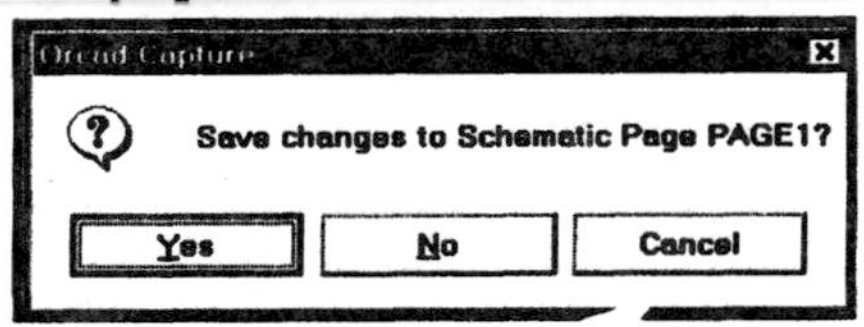


Select the ***Yes*** button to save the changes:

If you see the window above, click the ***OK*** button and then type **CTRL-F4** to close the ***Place Database Part*** window. You should be at the tree view of the project:

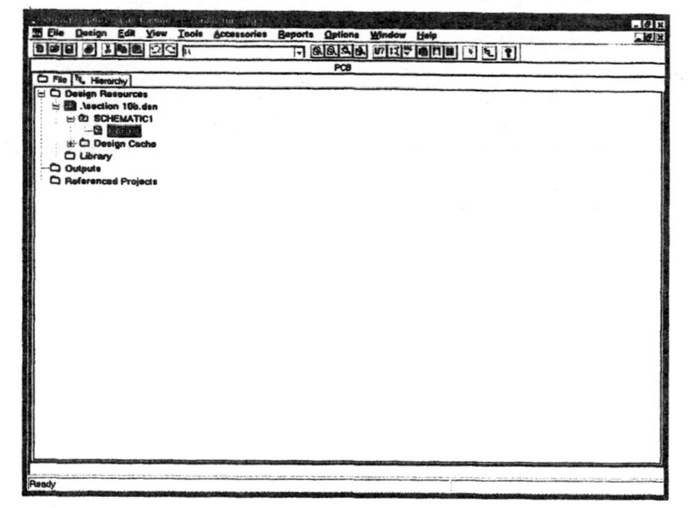


Select **Tools**, **Part Manager**, and then **Open**:

File Design Edit View Tools Accessories Reports Options Window Help

PCB

File Hierarchy

Design Resources
.\section 10b.dsn
SCHEMATIC1
Design Cache
Library
Outputs
Referenced Projects

Annotate...
Back Annotate...
Update Properties...
Part Manager
Design Rules Check...
Create Netlist...
Cross Reference...
Bill of Materials...
Export Properties...
Import Properties...
Generate Part...

Open
Update

Open the CIS part manager

File Edit View Tools Reports Options Window Help

| # | Schema... | Part Re... | Value | Part Nu... | Part Sta... | Databa... | Source ... | Source ... |
|---|---|---|---|---|---|---|---|---|
| 1 | SCHEMATIC... | C1 | 2200 uF | P5253-ND | O Approved... | | D:\PROGRA... | CAPACITOR... |
| 2 | SCHEMATIC... | HW1 | 6-32 x 3/8 | 6-32 x 3/8 | O Approved... | | D:\PROGRA... | Screw_dwg |
| 3 | SCHEMATIC... | C7 | 0.1 uF | P4923-ND | O Approved... | | D:\PROGRA... | CAP |
| 4 | SCHEMATIC... | HW2 | 6-32 x 3/8 | 6-32 x 3/8 | O Approved... | | D:\PROGRA... | Screw_dwg |
| 5 | SCHEMATIC... | HW3 | 6-32 x 3/8 | 6-32 x 3/8 | O Approved... | | D:\PROGRA... | Screw_dwg |
| 6 | SCHEMATIC... | C8 | 0.1 uF | P4923-ND | O Approved... | | D:\PROGRA... | CAP |
| 7 | SCHEMATIC... | C5 | 0.1 uF | P4923-ND | O Approved... | | D:\PROGRA... | CAP |
| 8 | SCHEMATIC... | C3 | 2200 uF | P5253-ND | O Approved... | | D:\PROGRA... | CAPACITOR... |
| 9 | SCHEMATIC... | C4 | 0.1 uF | P4923-ND | O Approved... | | D:\PROGRA... | CAP |
| 10 | SCHEMATIC... | C2 | 2200 uF | P5253-ND | O Approved... | | D:\PROGRA... | CAPACITOR... |
| 11 | SCHEMATIC... | R1 | 1.8k | P1.5KETR-ND | O Approved... | | D:\PROGRA... | R |

Part status is yellow.

Ready Part Manager

All items are shown with yellow part status to indicate that we need to update the list to see the actual status.

To check the schematic and database for consistency, select **Tools** and then **Update Part Status**:

File Edit View Tools Reports Options Window Help

Link Database Part ... Ctrl+L
Update Part Status

| # | Schema... | Part Re... | Value | Part Nu... | Part Sta... | Databa... | Source ... | Source ... |
|---|---|---|---|---|---|---|---|---|
| 1 | SCHEMATIC... | C1 | 2200 uF | P5253-ND | O Approved... | | D:\PROGRA... | CAPACITOR... |
| 2 | SCHEMATIC... | HW1 | 6-32 x 3/8 | 6-32 x 3/8 | O Approved... | | D:\PROGRA... | Screw_dwg |
| 3 | SCHEMATIC... | C7 | 0.1 uF | P4923-ND | O Approved... | | D:\PROGRA... | CAP |
| 4 | SCHEMATIC... | HW2 | 6-32 x 3/8 | 6-32 x 3/8 | O Approved... | | D:\PROGRA... | Screw_dwg |
| 5 | SCHEMATIC... | HW3 | 6-32 x 3/8 | 6-32 x 3/8 | O Approved... | | D:\PROGRA... | Screw_dwg |
| 6 | SCHEMATIC... | C8 | 0.1 uF | P4923-ND | O Approved... | | D:\PROGRA... | CAP |
| 7 | SCHEMATIC... | C5 | 0.1 uF | P4923-ND | O Approved... | | D:\PROGRA... | CAP |
| 8 | SCHEMATIC... | C3 | 2200 uF | P5253-ND | O Approved... | | D:\PROGRA... | CAPACITOR... |
| 9 | SCHEMATIC... | C4 | 0.1 uF | P4923-ND | O Approved... | | D:\PROGRA... | CAP |
| 10 | SCHEMATIC... | C2 | 2200 uF | P5253-ND | O Approved... | | D:\PROGRA... | CAPACITOR... |
| 11 | SCHEMATIC... | R1 | 1.8k | P1.5KETR-ND | O Approved... | | D:\PROGRA... | R |

Check schematic parts against the part database Part Manager

Orcad Capture - Lite Edition - [Part Manager - SECTION 10B.DSN]

File Edit View Tools Reports Options Window Help

| # | Schema... | Part Re... | Value | Part Nu... | Part Sta... | Databa... | Source ... | Source ... |
|---|---|---|---|---|---|---|---|---|
| 1 | SCHEMATIC... | C1 | 2200 uF | P5253-ND | O Approved... | | D:\PROGRA... | CAPACITOR... |
| 2 | SCHEMATIC... | HW1 | 6-32 x 3/8 | 6-32 x 3/8 | O Approved... | | D:\PROGRA... | Screw_dwg |
| 3 | SCHEMATIC... | C7 | 0.1 uF | P4923-ND | O Approved... | | D:\PROGRA... | CAP |
| 4 | SCHEMATIC... | HW2 | 6-32 x 3/8 | 6-32 x 3/8 | O Approved... | | D:\PROGRA... | Screw_dwg |
| 5 | SCHEMATIC... | HW3 | 6-32 x 3/8 | 6-32 x 3/8 | O Approved... | | D:\PROGRA... | Screw_dwg |
| 6 | SCHEMATIC... | C8 | 0.1 uF | P4923-ND | O Approved... | | D:\PROGRA... | CAP |
| 7 | SCHEMATIC... | C5 | 0.1 uF | P4923-ND | O Approved... | | D:\PROGRA... | CAP |
| 8 | SCHEMATIC... | C3 | 2200 uF | P5253-ND | O Approved... | | D:\PROGRA... | CAPACITOR... |
| 9 | SCHEMATIC... | C4 | 0.1 uF | P4923-ND | O Approved... | | D:\PROGRA... | CAP |
| 10 | SCHEMATIC... | C2 | 2200 uF | P5253-ND | ● Approved... | CAPACITORS | D:\PROGRA... | CAPACITOR... |
| 11 | SCHEMATIC... | R1 | 1.8k | P1.5KETR-ND | O Approved... | | D:\PROGRA... | R |

| | Value | Mfr_Name | Mfr_Part_Number | Part_Type | Tolerance | Rating |
|---|---|---|---|---|---|---|
| DB Props | | Panasonic - EC | ERJ-8GEYJ152 | Resistor\Carbon | 5% | 0.125 W |
| Sch Props R1 | | | | | 5% | 0.125 W |

Yes Yes All No No All Cancel Help

Checking part status, processing R1... (Press [Esc] to cancel)

A table pops up showing us that the resistor has inconsistent information. In the schematic, the value is 1.8k while in the database, the value is 1.5k. We will show how to correct this problem in the next section. Click the ***No*** button to continue:

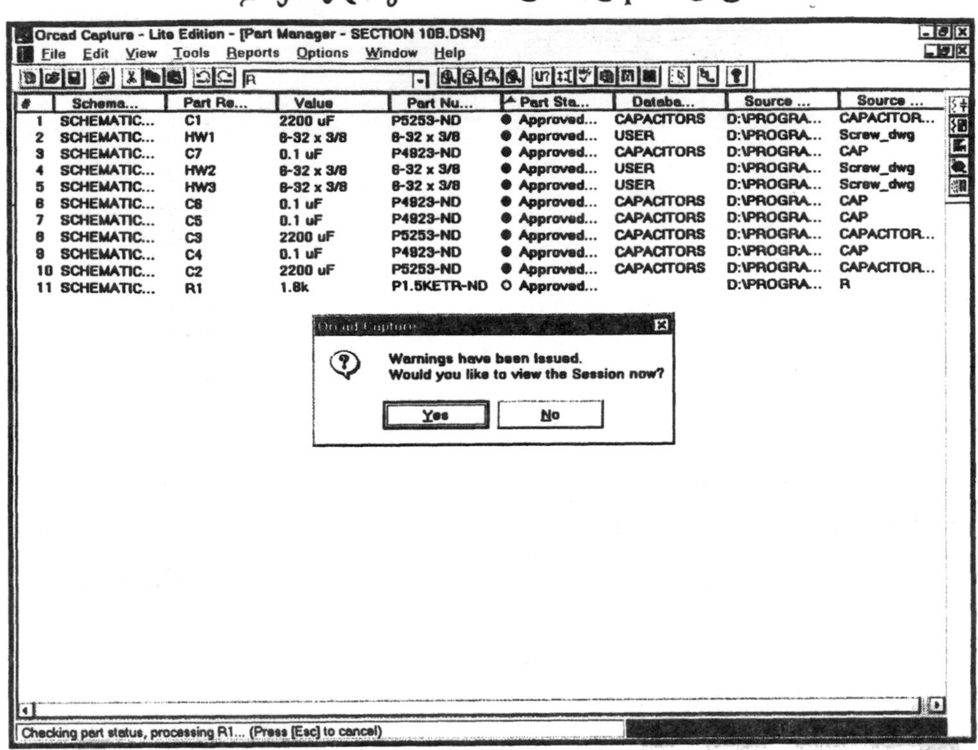


Select the ***NO*** button:

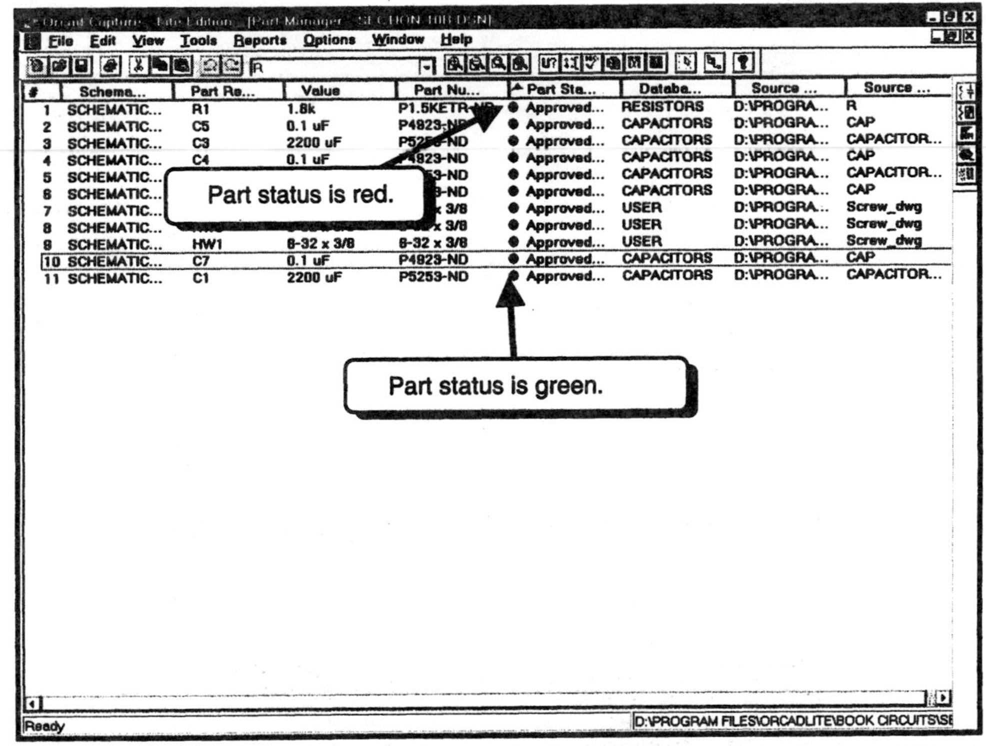


We see that all parts have a green part status except for the resistor, which has a red part status. This shows that all parts except for the resistors have consistent information in the schematic and the database. In the next section we will show how to correct this problem with the resistor. Type **CTRL-F4** to close the ***Part Manager*** window.

## 10.C.4. Linking to Parts in the Database

One method of fixing inconsistencies between the schematic and database is to delete the offending part from the schematic, find the correct part in the database, and then place that part in your schematic. A second method is to link the part to the database. Here, we will show this second method.

Open the schematic and zoom in around the resistor. Click the ***LEFT*** mouse button on the resistor graphic to select it:

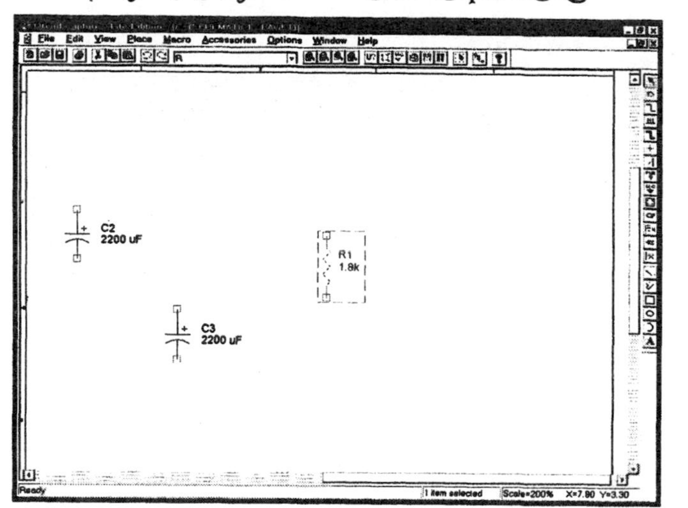


With the resistor selected, click the ***RIGHT*** mouse button on the graphic and then select **Link Database Part**:

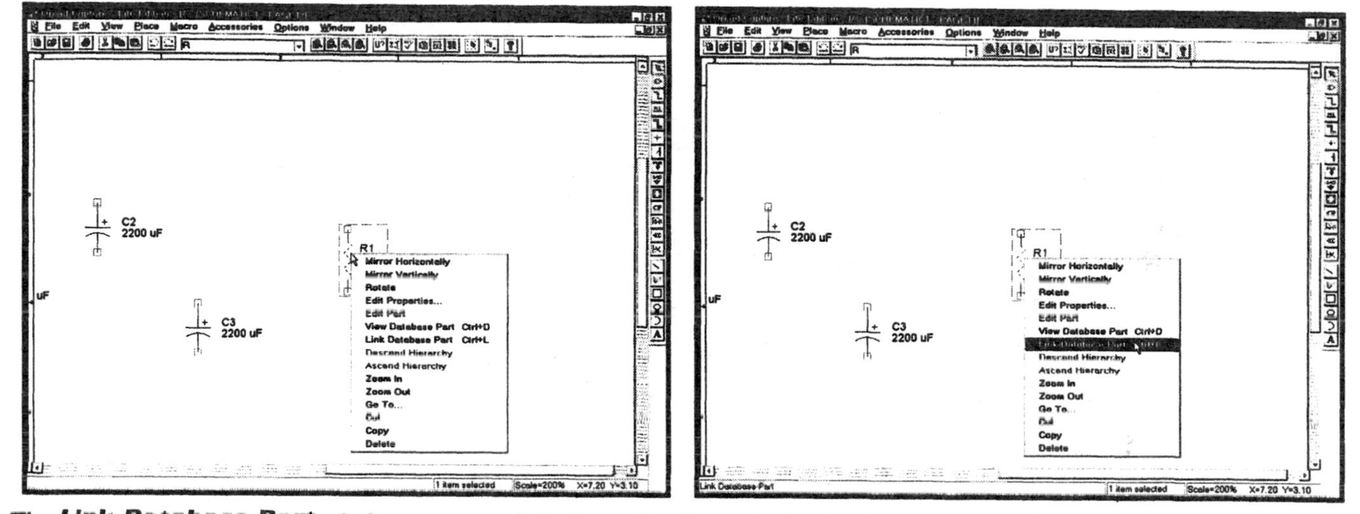


The ***Link Database Part*** window opens and displays all resistors in the database with a value of 1.8k:

A large number of resistors are available because the database is so large. You may wish to narrow the list by adding more criteria. The list is so large that you may just want to delete the part in the schematic, find a new Digikey part number using the Digikey catalog, and then place the new correct part in your schematic. Here, we will select a part from the database window.

When you find the part you want to use, click the ***LEFT*** mouse button on the part to select it:

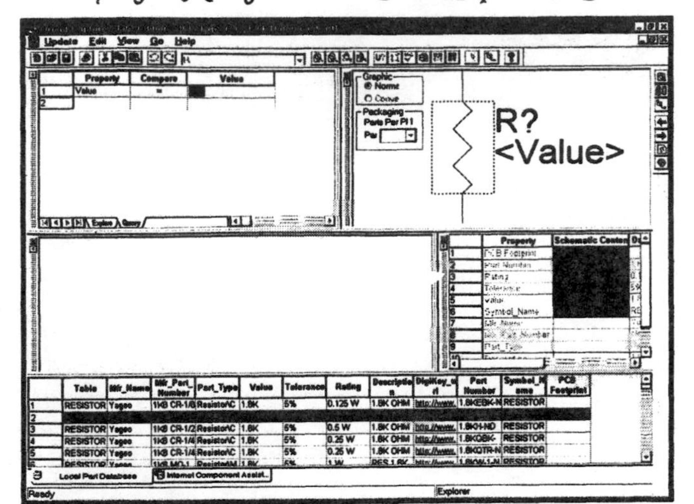


Select **Update** and then **Link Database Part** from the menus:

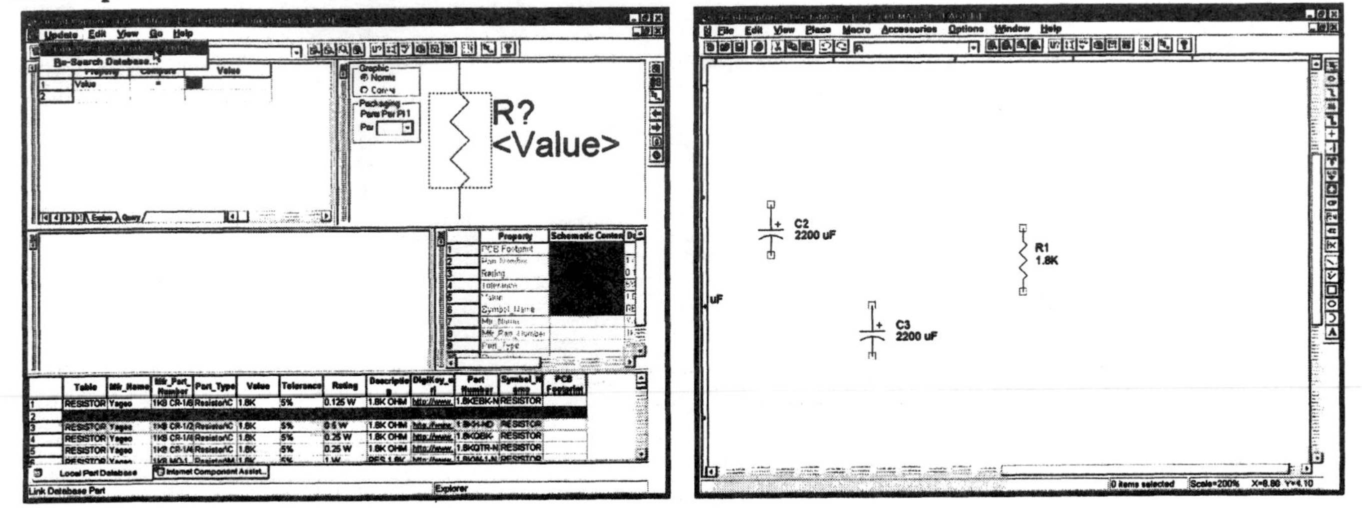


If you now close the schematic, open the Part Manager, and then update the status as shown in section 10.C.3, you will see that all parts are displayed with green part status, indicating that the schematic is consistent with the database:

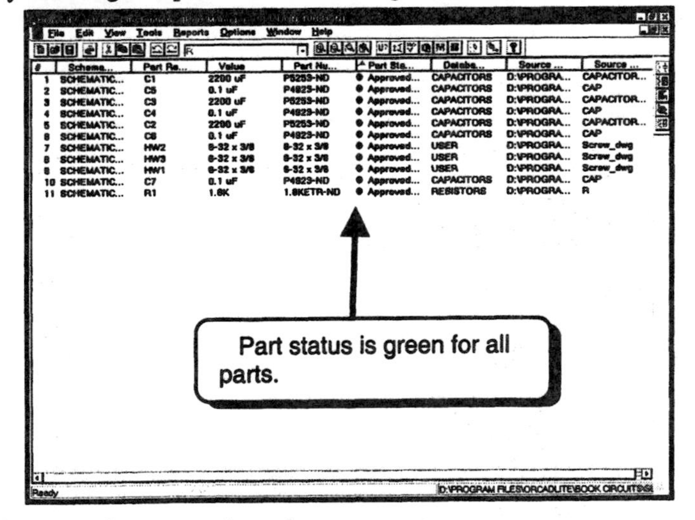


# 10.D. Placing a Part from Activeparts

Activeparts is a web site that contains a database of over 1,000,000 parts. Orcad Capture CIS can link to this site and download part information to your schematic. This information can include vendor information, a graphic symbol for the part, and a footprint for PC board layout. To access the database with the schematic displayed, type **z** to place a database part:

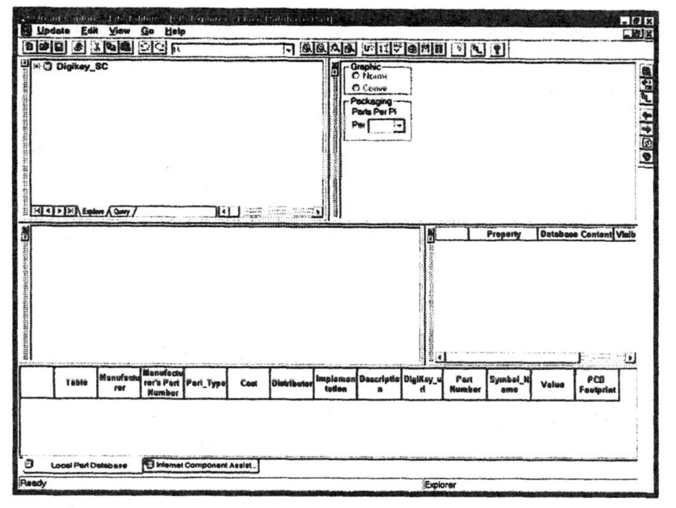

Select the ***Internet Component Assistant*** tab:

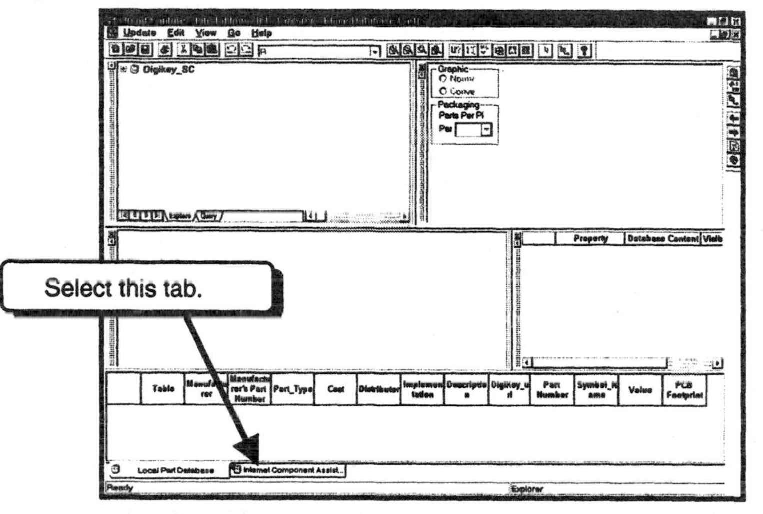


Please note that web sites are continually updated, and the screens you see may be slightly different than those shown here. The screens you see should be similar to the ones shown. If this is the first time you have used this web site, you will see the screen below and you will be required to create an account. If you already have an account, log in and skip to page 575 of this manual where it says "Select the ***Component Search*** button" and continue. Please note that if you make a mistake with the browser, you can press the **Backspace** key to go back to the previous web page.

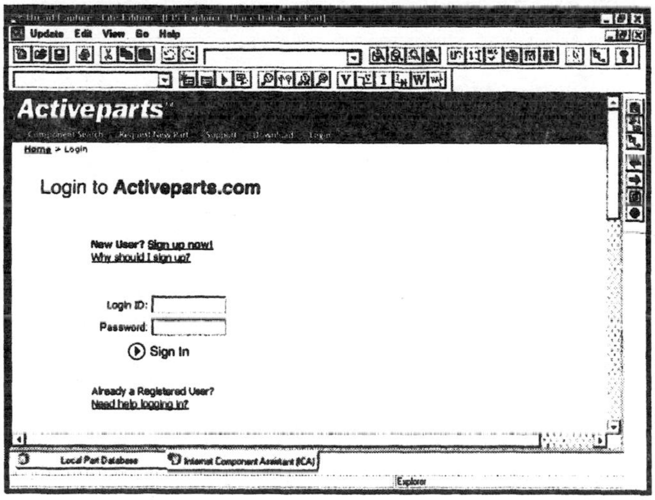

Select the ***Sign up now*** link:

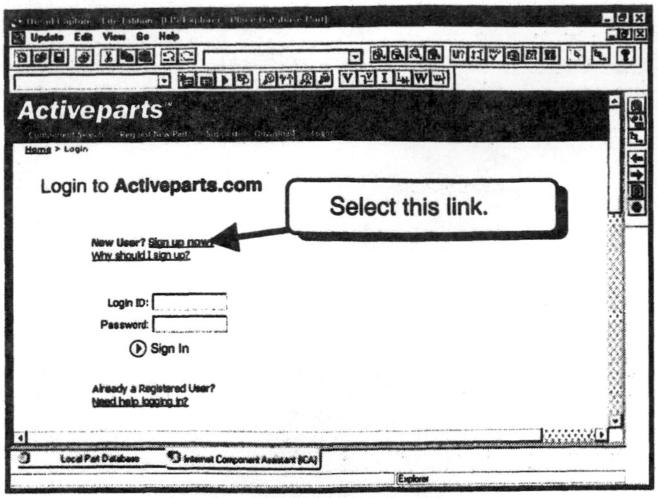

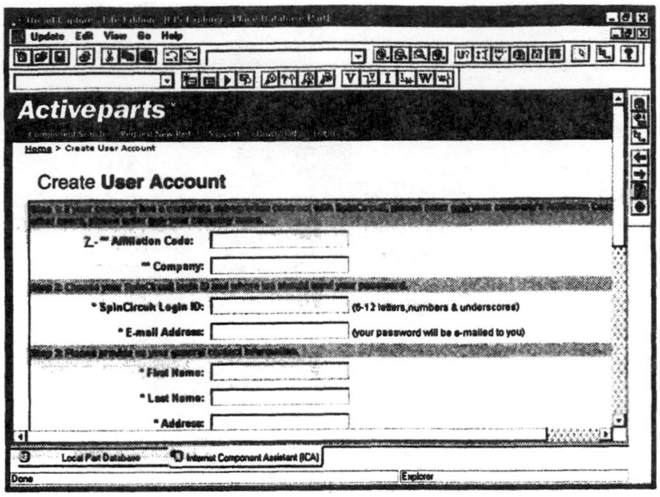

Update Edit View Go Help

Activeparts

Home > Login

Login to Activeparts.com

Select this link.

New User? Sign up now!
Why should I sign up?

Login ID:
Password:
Sign In

Already a Registered User?
Need help logging in?

Local Part Database | Internet Component Assistant (ICA)

Explorer

Update Edit View Go Help

Activeparts

Home > Create User Account

Create User Account

2.-** Affiliation Code:
** Company:
* SpinCircuit Login ID: (6-12 letters, numbers & underscores)
* E-mail Address: (your password will be e-mailed to you)
* First Name:
* Last Name:
* Address:

Local Part Database | Internet Component Assistant (ICA)

Done | Explorer

Fill in the form. Fields with an asterisk are required:

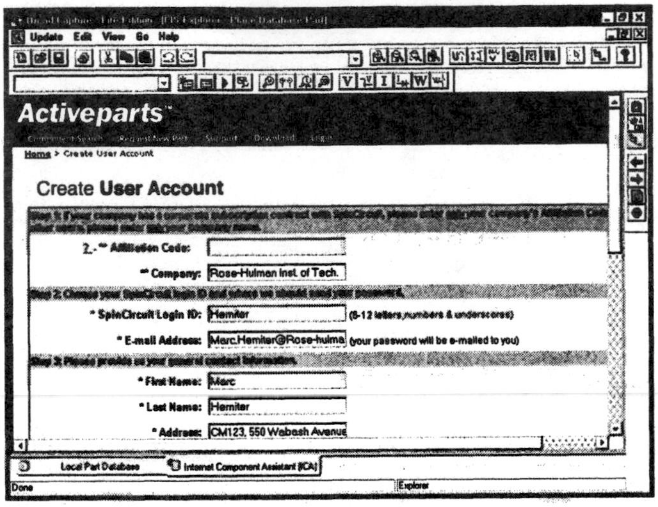

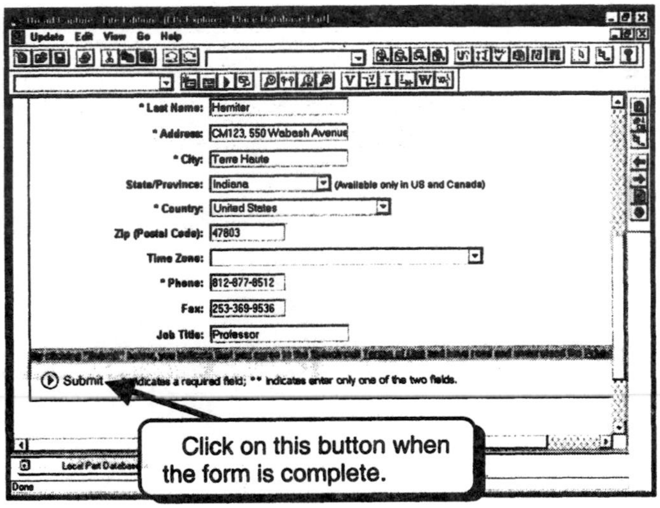

Update Edit View Go Help

Activeparts

Home > Create User Account

Create User Account

2.-** Affiliation Code:
** Company: Rose-Hulman Inst. of Tech.
* SpinCircuit Login ID: Hemiter (6-12 letters, numbers & underscores)
* E-mail Address: Marc.Hemiter@Rose-hulma (your password will be e-mailed to you)
* First Name: Marc
* Last Name: Hemiter
* Address: CM123, 550 Wabash Avenue

Local Part Database | Internet Component Assistant (ICA)

Done | Explorer

Update Edit View Go Help

* Last Name: Hemiter
* Address: CM123, 550 Wabash Avenue
* City: Terre Haute
State/Province: Indiana (Available only in US and Canada)
* Country: United States
Zip (Postal Code): 47803
Time Zone:
* Phone: 812-877-8512
Fax: 253-369-9536
Job Title: Professor

Submit * indicates a required field; ** indicates enter only one of the two fields.

Click on this button when the form is complete.

Local Part Database

Done

Click the ***Submit*** button when your form is complete. If you entered all of the required fields, you will receive the screen below:

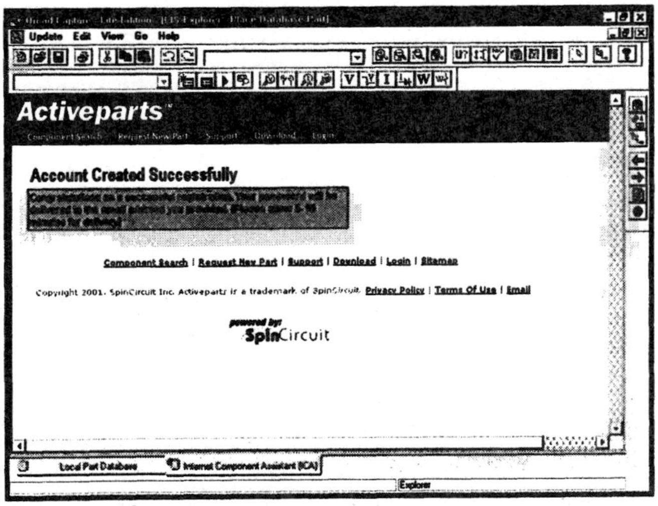

Update Edit View Go Help

Activeparts

Account Created Successfully

Component Search | Request New Part | Support | Download | Login | Sitemap

Copyright 2001. SpinCircuit Inc. Activeparts is a trademark of SpinCircuit. Privacy Policy | Terms Of Use | Email

powered by:
SpinCircuit

Local Part Database | Internet Component Assistant (ICA)

Explorer

Click the ***Login*** button and wait for your password to arrive through email.

Update Edit View Go Help
Activeparts
Account Created Successfully
Component Search | Request New Part | Support | Do
Select this link.
Copyright 2001. SpinCircuit Inc. Activeparts is a trademark of SpinCircuit. Privacy Policy | Terms Of Use | Email
powered by:
SpinCircuit
Local Part Database
Internet Component Assistant (ICA)
Login
Explorer

Update Edit View Go Help
Activeparts
Home > Login
Login to Activeparts.com
New User? Sign up now!
Why should I sign up?
Login ID:
Password:
Sign In
Already a Registered User?
Need help logging in?
Local Part Database
Internet Component Assistant (ICA)
Done
Explorer

When your password arrives, enter your ID and password and click the ***Sign In*** button:

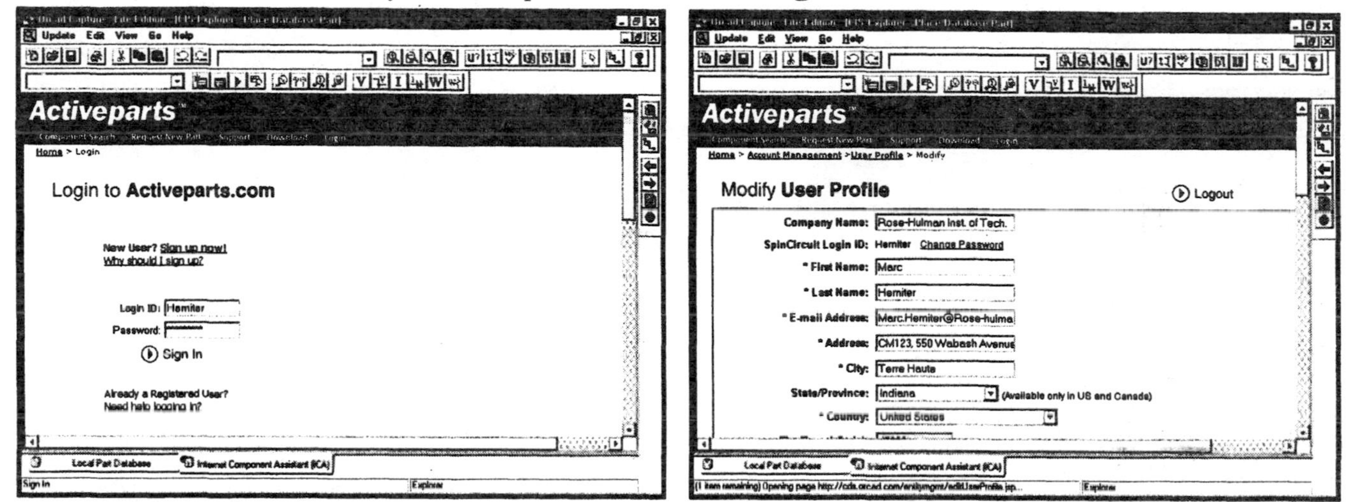


Select the ***Component Search*** button:

Update Edit View Go Help
Activeparts
Home > Account Management > User Profile > Modify
Modify User Profile
Logout
Click on this button.
* Last Name: Hemiter
* E-mail Address: Marc.Hemiter@Rose-hulme
* Address: CM123, 550 Wabash Avenue
* City: Terre Haute
State/Province: Indiana (Available only in US and Canada)
* Country: United States
Local Part Database
Internet Component Assistant (ICA)
(1 item remaining) Opening page http://cds.orcad.com/entitymgmt/editUserProfile.jsp...
Explorer

Update Edit View Go Help
Activeparts
Component Search
Search Parameters (Maximum 250 Search Results)
Back Forward Help Logout
Select from following Component
HEAT SINK
CONNECTOR
TRANSFORMER
CLOCK SOURCE
R/F MICROWAVE
FASTENER
BATTERY
VIRTUAL COMPONENT
PASSIVE
IC
ELECTRO-MECHANICAL
OPTO-DEVICE
DISCRETES
WIRE/CABLE
Manufacturer Name: ANY
MFR's Part Number:
Starting With Exact *Contains
Part Description:
Search Example: CAP TANT 10UF
Go
* Using the "contains" operator could result in longer search times.
Local Part Database
Internet Component Assistant (ICA)
Component Search
Explorer

We will look for a MAX492 op-amp available from Maxim. Fill in the parameters as shown:

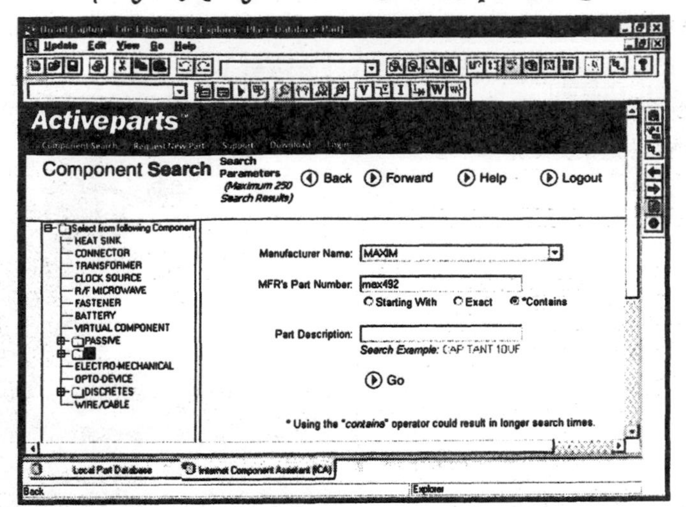


You can use the tree in the left pane to narrow the search to the portion of the database in which the part is likely to be located. In the screen capture above, I have selected the ***IC*** portion of the database. I have selected the ***Contains*** option because it is usually very hard to pick a part number exactly. Like any search engine, it will take some experience with the engine before you can search for parts efficiently. Click the ***Go*** button when you have entered the information above:

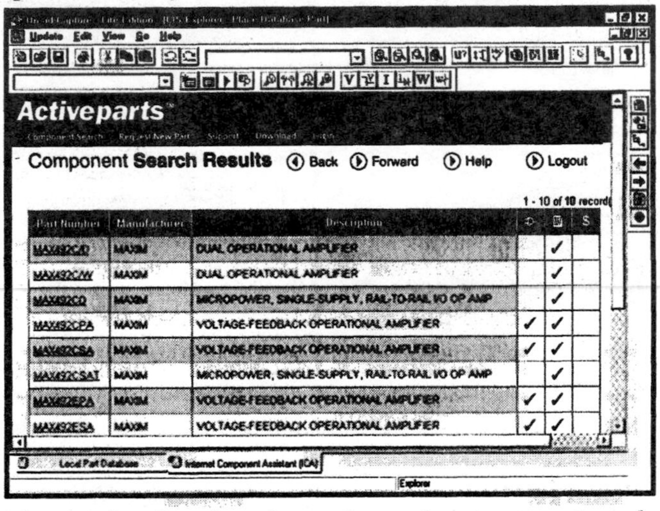

At this point, you may need to refer to a datasheet in order to choose the correct part number for your design. I want to use the MAX492CPA part, so I will select that link. The columns on the right indicate that this part has both a datasheet and a symbol available for download.

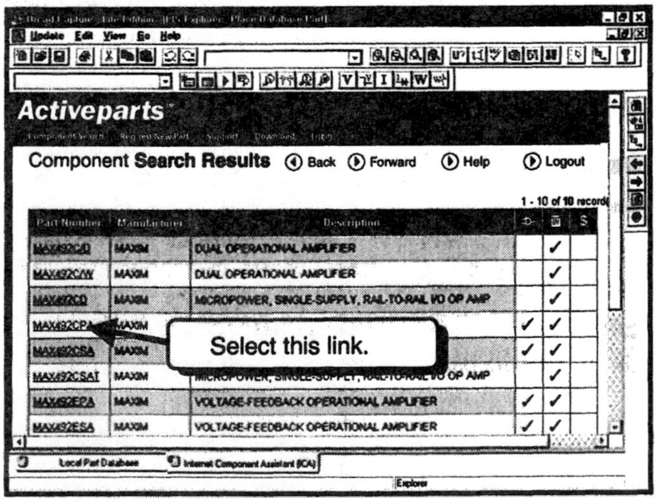

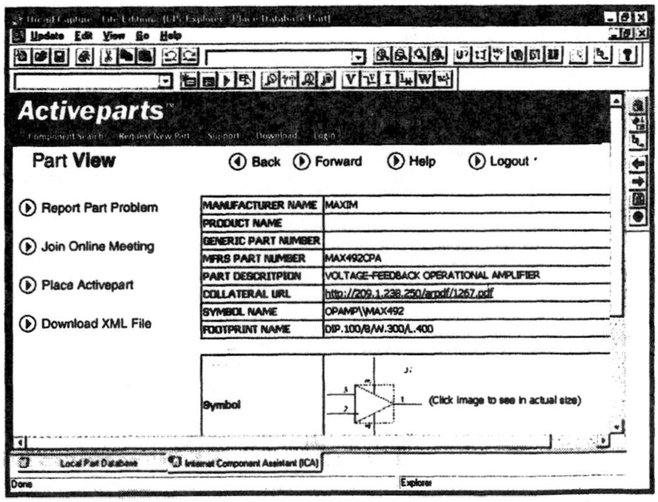

The information includes a symbol, a link to a datasheet we can download, and some technical information.

If this is the part you want to use, click the ***Place Activepart*** button:

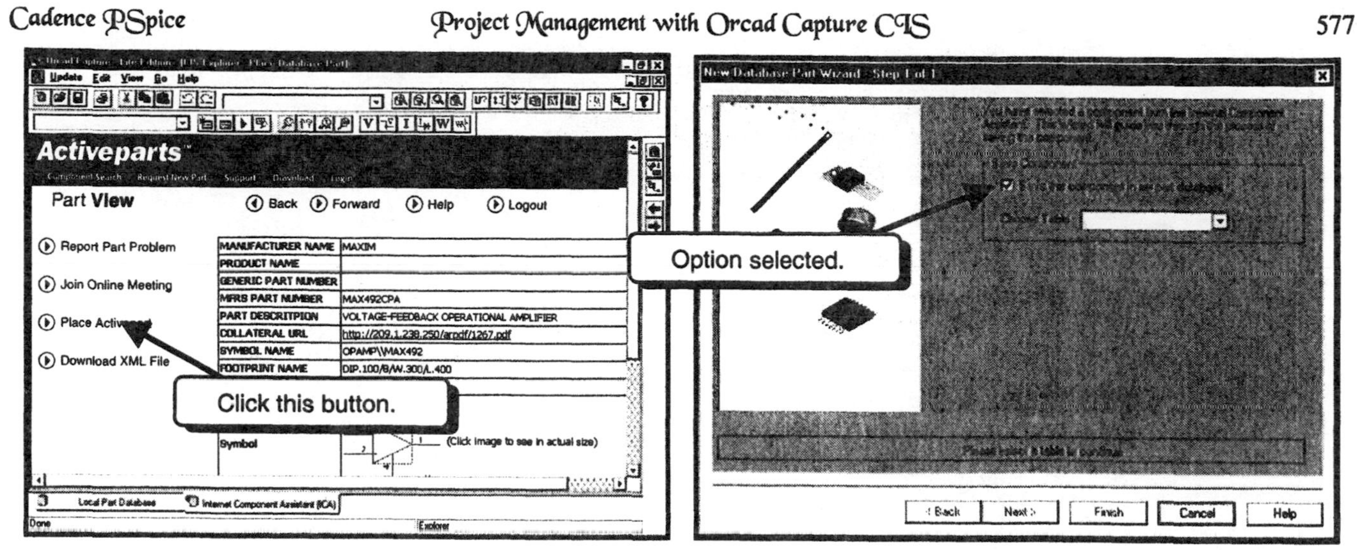


The option to save the part information to your database is selected. Saving the part in your database will allow you to place the part from your database the next time you need it rather than find the part on the Internet. Saving the part to your database also allows you to easily build up a company database of parts.

The ***USER*** table in the Digikey database was added for creating your own parts and to save parts downloaded from the Internet. If you save the part in one of the Digikey tables, it will be hard to find among the 70,000 parts. Also, if the Digikey database is upgraded in the future, all of your parts will be located in one easy-to-copy table. Select the ***USER*** table and click the ***Finish*** button:

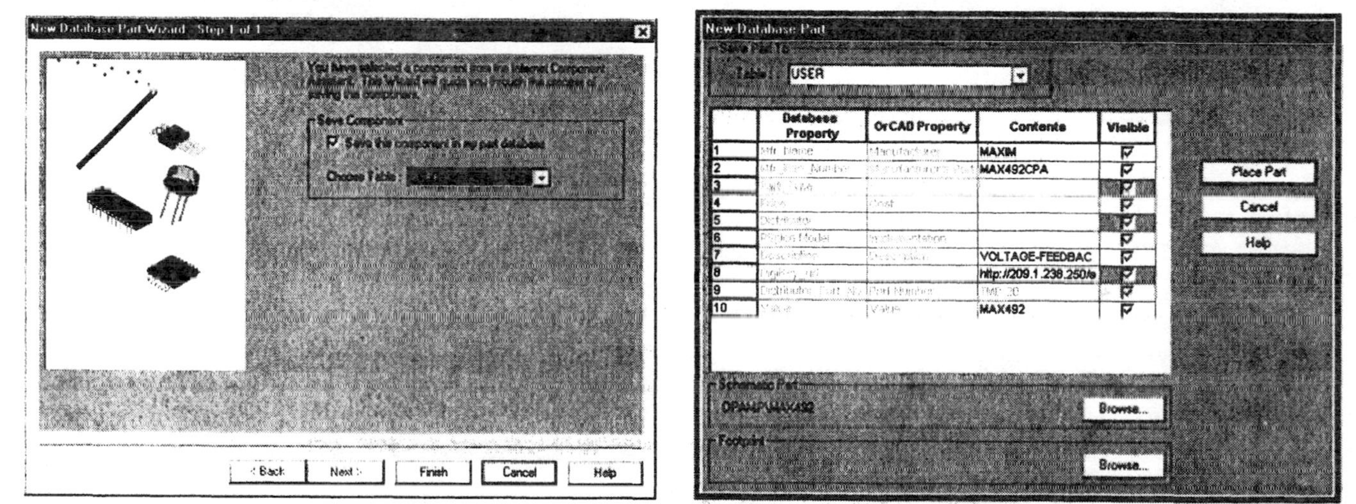


The above right dialog box tells you what information will be saved in your database. Click the ***Place Part*** button to place the part in your schematic. You will return to the schematic with the symbol attached to the mouse pointer:

Place the part in your schematic:

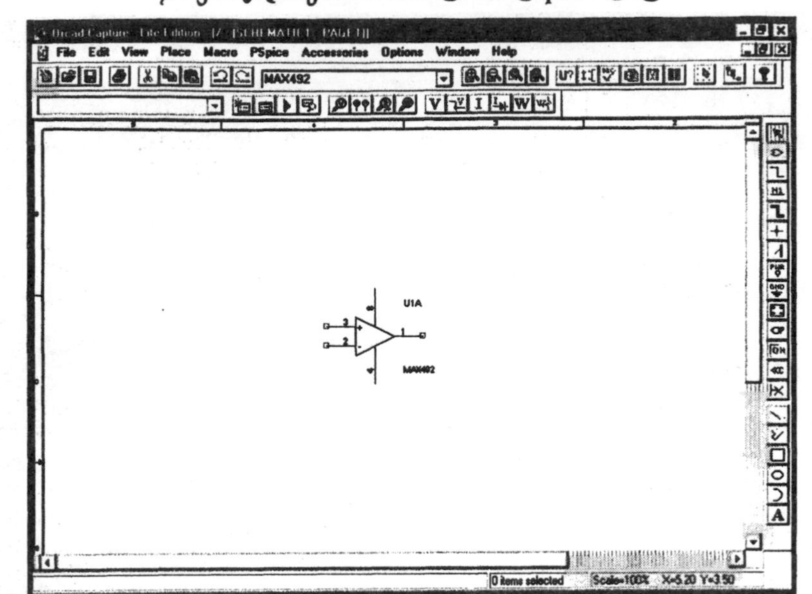

The next thing that we would like to do is view our database and see where the part was added. Save the project and then run Microsoft Access. Open file DIGIKEY_SC.mdb. It should be located in the Capture\Database directory in the OrcadLite directory:

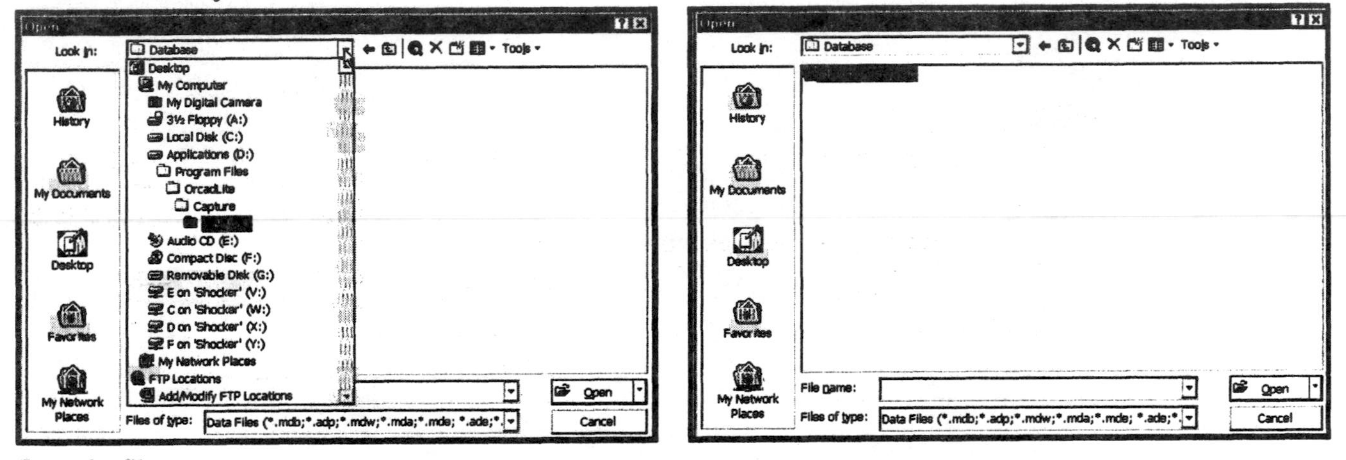

Open the file:

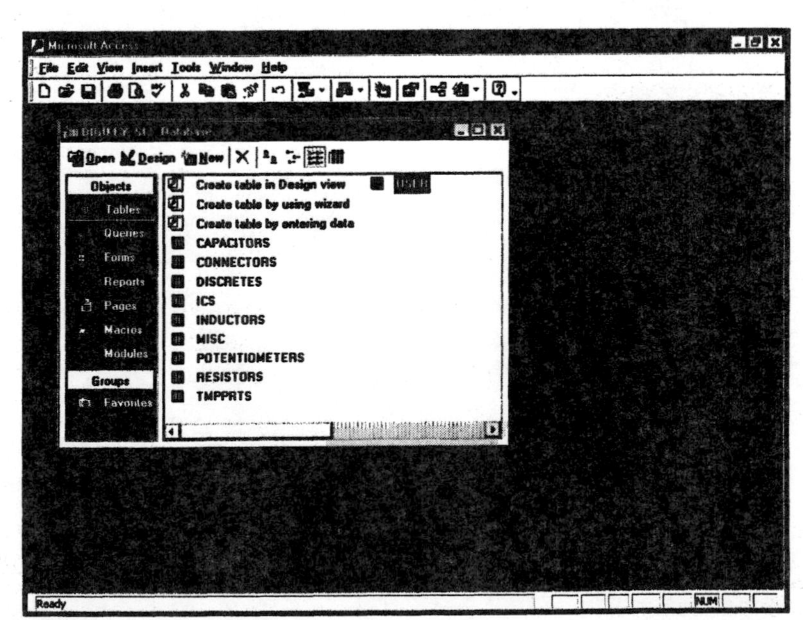

Double-click the ***LEFT*** mouse button on the ***USER*** table to open the table:

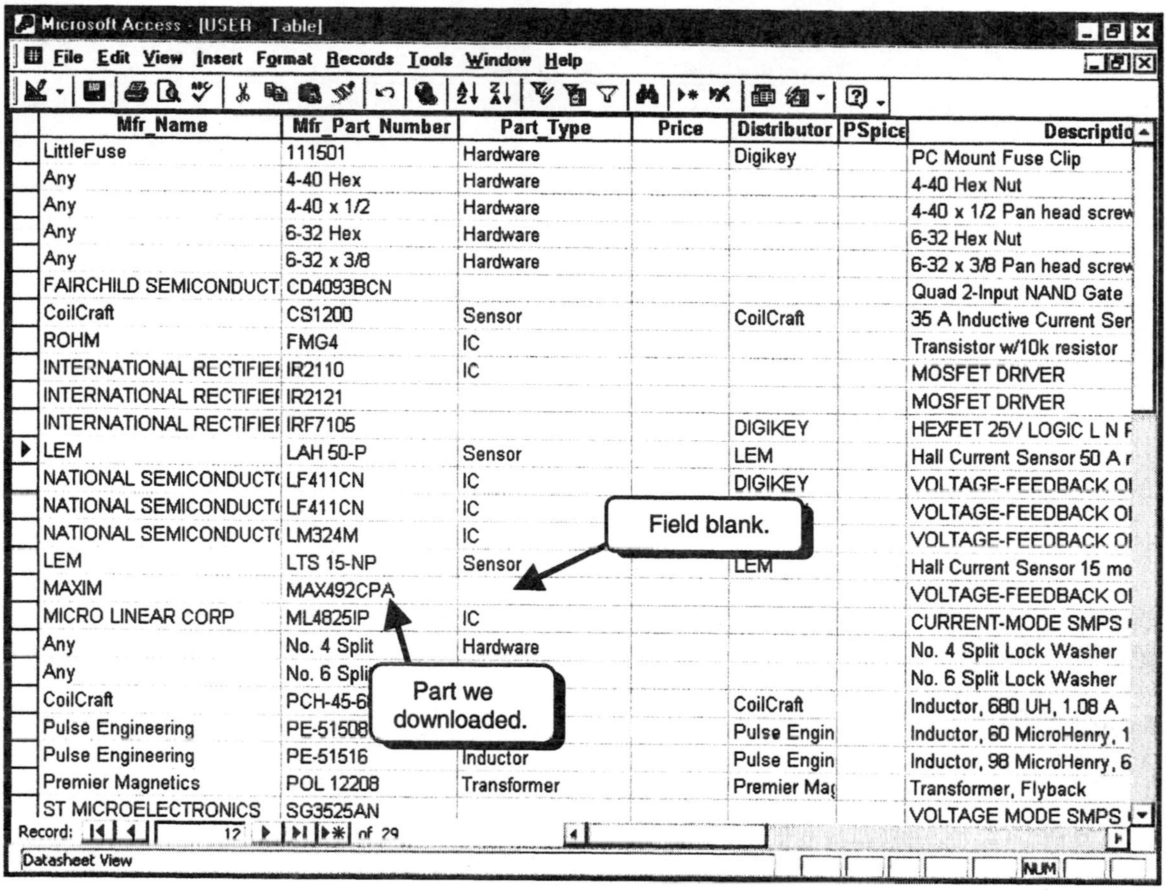

Microsoft Access - [USER : Table]

File Edit View Insert Format Records Tools Window Help

| Mfr_Name | Mfr_Part_Number | Part_Type | Price | Distributor | PSpice | Descriptio |
|---|---|---|---|---|---|---|
| LittleFuse | 111501 | Hardware | | Digikey | | PC Mount Fuse Clip |
| Any | 4-40 Hex | Hardware | | | | 4-40 Hex Nut |
| Any | 4-40 x 1/2 | Hardware | | | | 4-40 x 1/2 Pan head screw |
| Any | 6-32 Hex | Hardware | | | | 6-32 Hex Nut |
| Any | 6-32 x 3/8 | Hardware | | | | 6-32 x 3/8 Pan head screw |
| FAIRCHILD SEMICONDUCT | CD4093BCN | | | | | Quad 2-Input NAND Gate |
| CoilCraft | CS1200 | Sensor | | CoilCraft | | 35 A Inductive Current Ser |
| ROHM | FMG4 | IC | | | | Transistor w/10k resistor |
| INTERNATIONAL RECTIFIEI | IR2110 | IC | | | | MOSFET DRIVER |
| INTERNATIONAL RECTIFIEI | IR2121 | | | | | MOSFET DRIVER |
| INTERNATIONAL RECTIFIEI | IRF7105 | | | DIGIKEY | | HEXFET 25V LOGIC L N F |
| LEM | LAH 50-P | Sensor | | LEM | | Hall Current Sensor 50 A r |
| NATIONAL SEMICONDUCT( | LF411CN | IC | | DIGIKEY | | VOLTAGE-FEEDBACK OI |
| NATIONAL SEMICONDUCT( | LF411CN | IC | | | | VOLTAGE-FEEDBACK OI |
| NATIONAL SEMICONDUCT( | LM324M | IC | | | | VOLTAGE-FEEDBACK OI |
| LEM | LTS 15-NP | Sensor | | LEM | | Hall Current Sensor 15 mo |
| MAXIM | MAX492CPA | | | | | VOLTAGE-FEEDBACK OI |
| MICRO LINEAR CORP | ML4825IP | IC | | | | CURRENT-MODE SMPS |
| Any | No. 4 Split | Hardware | | | | No. 4 Split Lock Washer |
| Any | No. 6 Spli | | | | | No. 6 Split Lock Washer |
| CoilCraft | PCH-45-6 | | | CoilCraft | | Inductor, 680 UH, 1.08 A |
| Pulse Engineering | PE-51508 | | | Pulse Engin | | Inductor, 60 MicroHenry, 1 |
| Pulse Engineering | PE-51516 | Inductor | | Pulse Engin | | Inductor, 98 MicroHenry, 6 |
| Premier Magnetics | POL 12208 | Transformer | | Premier Ma( | | Transformer, Flyback |
| ST MICROELECTRONICS | SG3525AN | | | | | VOLTAGE MODE SMPS |

Record: 17 of 29

Datasheet View NUM

Field blank.

Part we downloaded.

We can see the part we just downloaded. Also note that the ***Part_Type*** field for the part is blank. The ***Part_Type*** is the field that determines the subtable in which the part will be placed. I will change this field to `IC`:

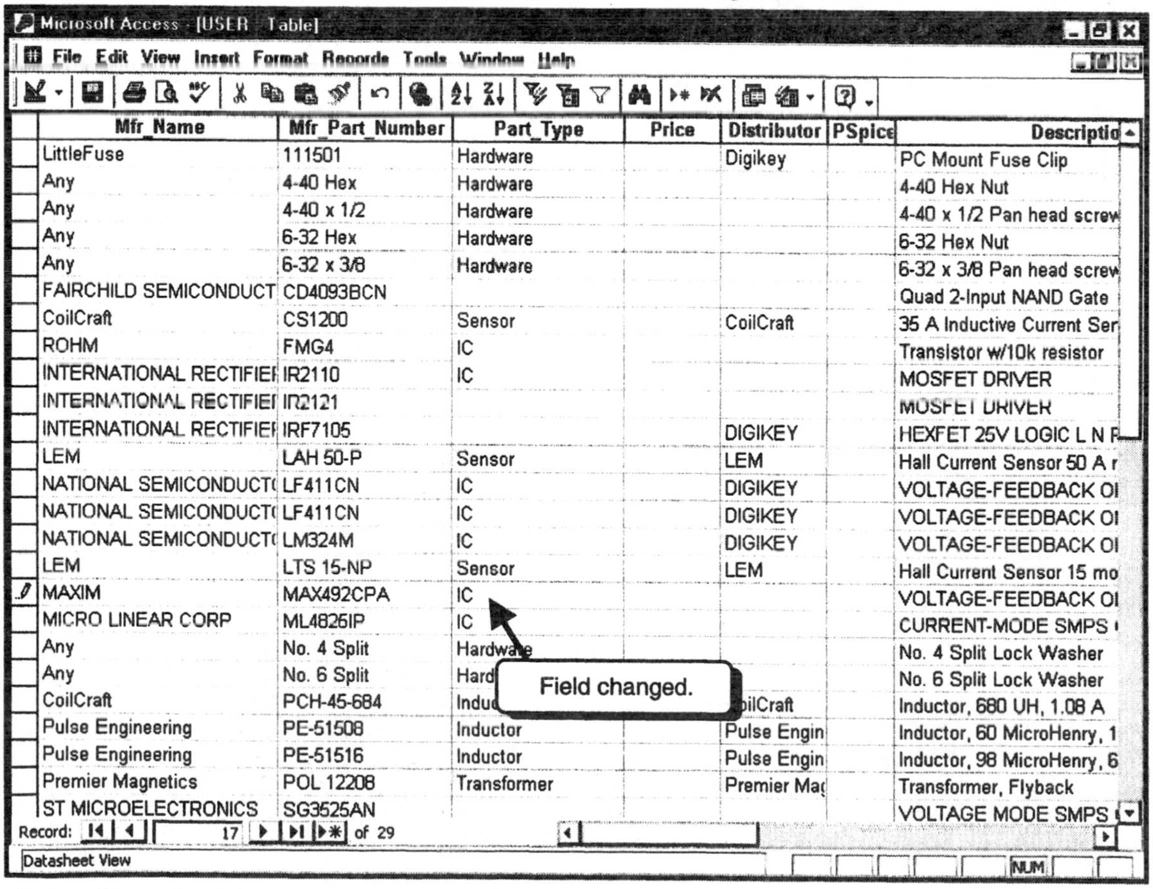

Microsoft Access - [USER : Table]

File Edit View Insert Format Records Tools Window Help

| Mfr_Name | Mfr_Part_Number | Part_Type | Price | Distributor | PSpice | Descriptio |
|---|---|---|---|---|---|---|
| LittleFuse | 111501 | Hardware | | Digikey | | PC Mount Fuse Clip |
| Any | 4-40 Hex | Hardware | | | | 4-40 Hex Nut |
| Any | 4-40 x 1/2 | Hardware | | | | 4-40 x 1/2 Pan head screw |
| Any | 6-32 Hex | Hardware | | | | 6-32 Hex Nut |
| Any | 6-32 x 3/8 | Hardware | | | | 6-32 x 3/8 Pan head screw |
| FAIRCHILD SEMICONDUCT | CD4093BCN | | | | | Quad 2-Input NAND Gate |
| CoilCraft | CS1200 | Sensor | | CoilCraft | | 35 A Inductive Current Ser |
| ROHM | FMG4 | IC | | | | Transistor w/10k resistor |
| INTERNATIONAL RECTIFIEI | IR2110 | IC | | | | MOSFET DRIVER |
| INTERNATIONAL RECTIFIEI | IR2121 | | | | | MOSFET DRIVER |
| INTERNATIONAL RECTIFIEI | IRF7105 | | | DIGIKEY | | HEXFET 25V LOGIC L N F |
| LEM | LAH 50-P | Sensor | | LEM | | Hall Current Sensor 50 A r |
| NATIONAL SEMICONDUCT( | LF411CN | IC | | DIGIKEY | | VOLTAGE-FEEDBACK OI |
| NATIONAL SEMICONDUCT( | LF411CN | IC | | DIGIKEY | | VOLTAGE-FEEDBACK OI |
| NATIONAL SEMICONDUCT( | LM324M | IC | | DIGIKEY | | VOLTAGE-FEEDBACK OI |
| LEM | LTS 15-NP | Sensor | | LEM | | Hall Current Sensor 15 mo |
| MAXIM | MAX492CPA | IC | | | | VOLTAGE-FEEDBACK OI |
| MICRO LINEAR CORP | ML4825IP | IC | | | | CURRENT-MODE SMPS |
| Any | No. 4 Split | Hardwa | | | | No. 4 Split Lock Washer |
| Any | No. 6 Split | Hard | | | | No. 6 Split Lock Washer |
| CoilCraft | PCH-45-684 | Indu | | ilCraft | | Inductor, 680 UH, 1.08 A |
| Pulse Engineering | PE-51508 | Inductor | | Pulse Engin | | Inductor, 60 MicroHenry, 1 |
| Pulse Engineering | PE-51516 | Inductor | | Pulse Engin | | Inductor, 98 MicroHenry, 6 |
| Premier Magnetics | POL 12208 | Transformer | | Premier Ma( | | Transformer, Flyback |
| ST MICROELECTRONICS | SG3525AN | | | | | VOLTAGE MODE SMPS |

Record: 17 of 29

Datasheet View NUM

Field changed.

If you scroll the table to the right, you can see more fields:

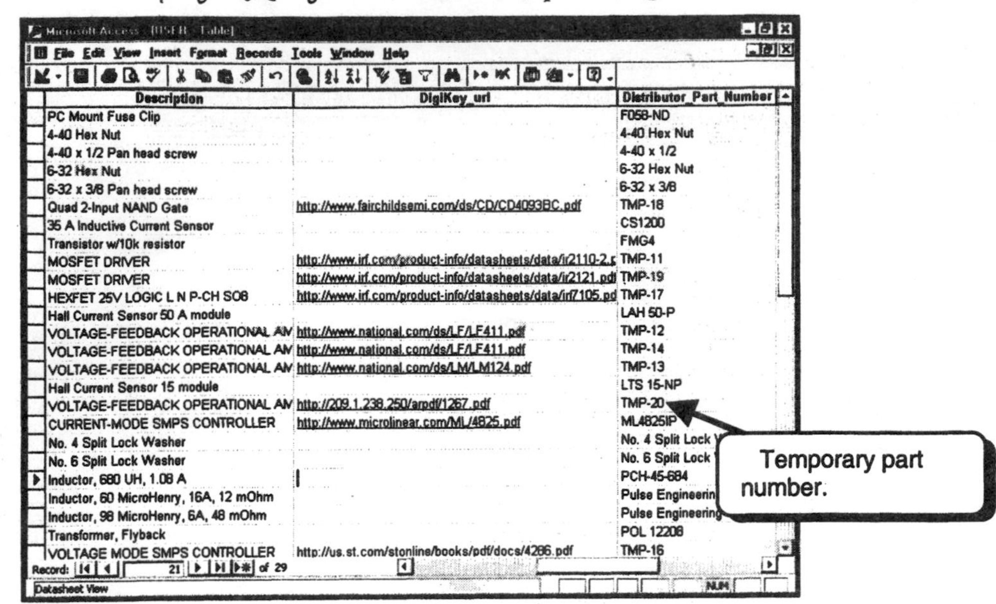

| Description | DigiKey_url | Distributor_Part_Number |
|---|---|---|
| PC Mount Fuse Clip | | F058-ND |
| 4-40 Hex Nut | | 4-40 Hex Nut |
| 4-40 x 1/2 Pan head screw | | 4-40 x 1/2 |
| 6-32 Hex Nut | | 6-32 Hex Nut |
| 6-32 x 3/8 Pan head screw | | 6-32 x 3/8 |
| Quad 2-Input NAND Gate | http://www.fairchildsemi.com/ds/CD/CD4093BC.pdf | TMP-18 |
| 35 A Inductive Current Sensor | | CS1200 |
| Transistor w/10k resistor | | FMG4 |
| MOSFET DRIVER | http://www.irf.com/product-info/datasheets/data/ir2110-2.p | TMP-11 |
| MOSFET DRIVER | http://www.irf.com/product-info/datasheets/data/ir2121.pdf | TMP-19 |
| HEXFET 25V LOGIC L N P-CH SO8 | http://www.irf.com/product-info/datasheets/data/irf7105.pd | TMP-17 |
| Hall Current Sensor 50 A module | | LAH 50-P |
| VOLTAGE-FEEDBACK OPERATIONAL AM | http://www.national.com/ds/LF/LF411.pdf | TMP-12 |
| VOLTAGE-FEEDBACK OPERATIONAL AM | http://www.national.com/ds/LF/LF411.pdf | TMP-14 |
| VOLTAGE-FEEDBACK OPERATIONAL AM | http://www.national.com/ds/LM/LM124.pdf | TMP-13 |
| Hall Current Sensor 15 module | | LTS 15-NP |
| VOLTAGE-FEEDBACK OPERATIONAL AM | http://209.1.238.250/arpdf/1267.pdf | TMP-20 |
| CURRENT-MODE SMPS CONTROLLER | http://www.microlinear.com/ML/4825.pdf | ML4825IP |
| No. 4 Split Lock Washer | | No. 4 Split Lock |
| No. 6 Split Lock Washer | | No. 6 Split Lock |
| Inductor, 680 UH, 1.08 A | | PCH-45-684 |
| Inductor, 60 MicroHenry, 16A, 12 mOhm | | Pulse Engineerin |
| Inductor, 98 MicroHenry, 6A, 48 mOhm | | Pulse Engineering |
| Transformer, Flyback | | POL 12208 |
| VOLTAGE MODE SMPS CONTROLLER | http://us.st.com/stonline/books/pdf/docs/4286.pdf | TMP-16 |



Note that all downloaded parts are given a temporary distributor part number. This is how Capture CIS indicates a downloaded part that is not yet officially accepted by your company. You can change this field when you find the actual distributor information.

Close the table and then close Microsoft Access. Switch to the schematic view of Orcad Capture CIS and then type **z** to place a database part:

| | |
|---|---|
| MANUFACTURER NAME | MAXIM |
| PRODUCT NAME | |
| GENERIC PART NUMBER | |
| MFRS PART NUMBER | MAX492CPA |
| PART DESCRIPTION | VOLTAGE-FEEDBACK OPERATIONAL AMPLIFIER |
| COLLATERAL URL | http://209.1.238.250/arpdf/1267.pdf |
| SYMBOL NAME | OPAMP\MAX492 |
| FOOTPRINT NAME | DIP.100/8/W.300/L.400 |

Select the ***Local Part Database*** tab and then display the ***IC*** subtable of the ***USER*** table with the ***Explore*** tab:

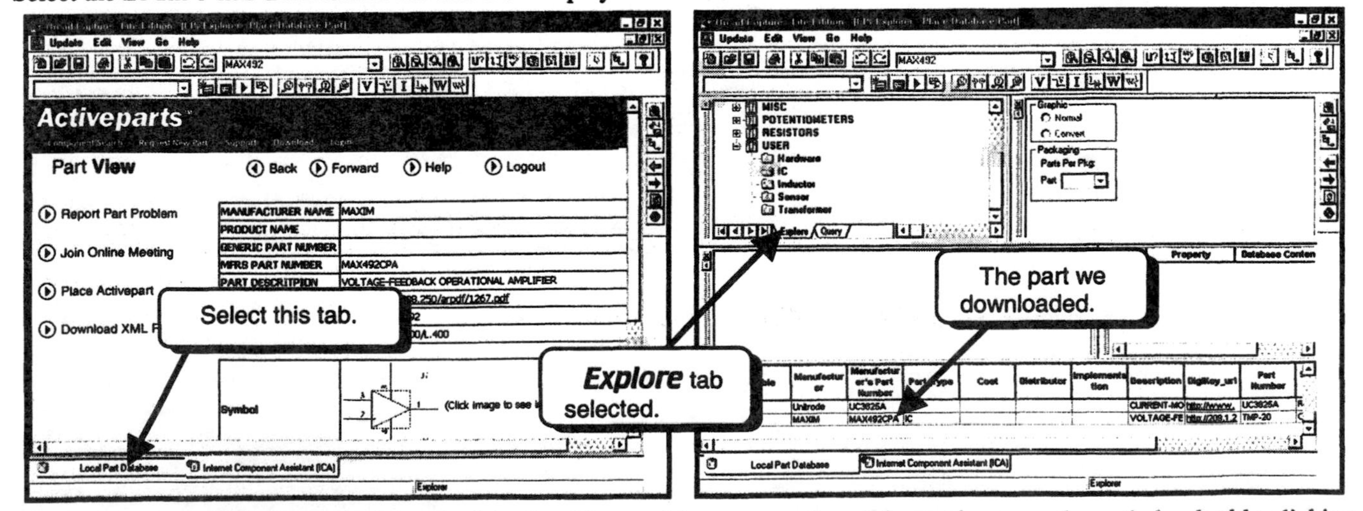



The part we downloaded is at the bottom of the list. If you wish, you can place this part in your schematic by double-clicking the ***LEFT*** mouse button on the part in the list:

The part is now available offline in your local database.

# 10.E. Creating your Own Part and Adding It to Your Database

To add a part to the database we must create a symbol for the part using Orcad capture, unless it is a standard symbol that already exists, and then we must add the part information to the database using Microsoft Access. We will show how to create a new library, add a new symbol to that library, and then add the part information to the Digikey database.

## 10.E.1. Creating a Symbol Using Orcad Capture

Here we will show how to create a new library and then add a part to that library. Since Orcad Lite only allows us to modify a library with fifteen parts or less, you will need to know how to create new libraries and add parts to those libraries. Run Orcad Capture CIS and then select **File, New**, and then **Library** from the menus:



Click the ***RIGHT*** mouse button the text ***library1.olb*** to obtain a pull-down menu:

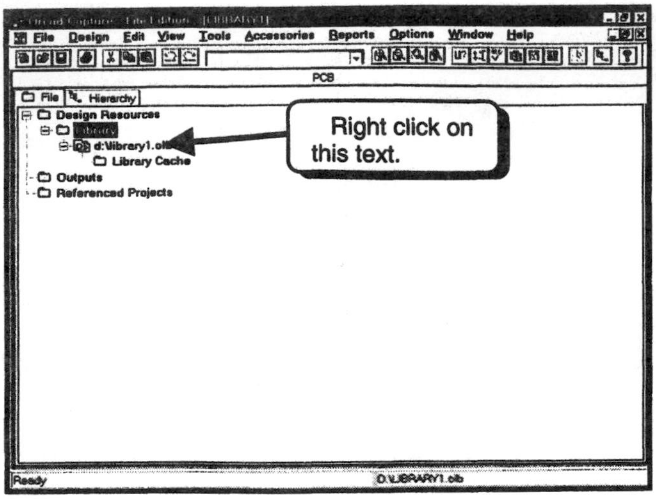

File Design Edit View Tools Accessories Reports Options Window Help

PCB

File Hierarchy

Design Resources
Library
d:\library1.olb
Library Cache
Outputs
Referenced Projects

Right click on this text.

Ready D:\LIBRARY1.olb

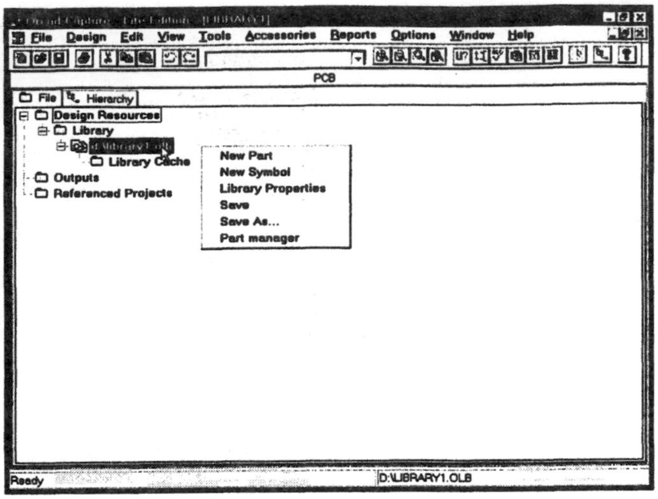

File Design Edit View Tools Accessories Reports Options Window Help

PCB

File Hierarchy

Design Resources
Library
Library Cache
Outputs
Referenced Projects

New Part
New Symbol
Library Properties
Save
Save As...
Part manager

Ready D:\LIBRARY1.OLB

Select **Save As**:

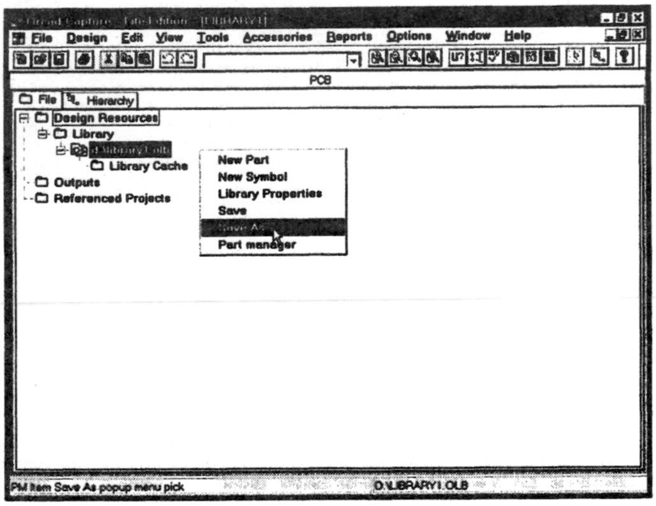

File Design Edit View Tools Accessories Reports Options Window Help

PCB

File Hierarchy

Design Resources
Library
Library Cache
Outputs
Referenced Projects

New Part
New Symbol
Library Properties
Save
Save As
Part manager

PM Item Save As popup menu pick D:\LIBRARY1.OLB

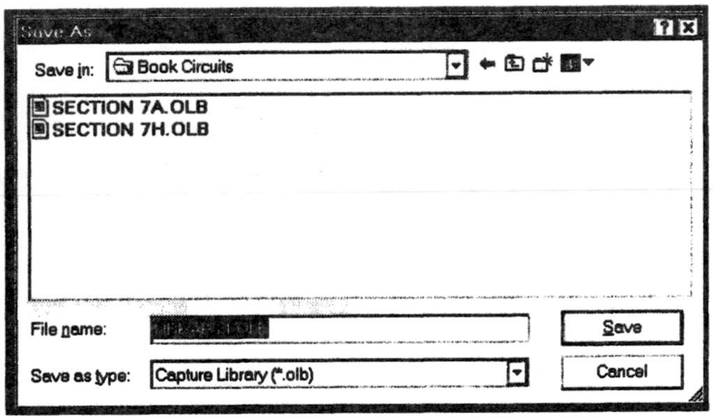

Save As

Save in: Book Circuits

SECTION 7A.OLB
SECTION 7H.OLB

File name:
Save as type: Capture Library (*.olb)
Save
Cancel

We need to rename the library and then save it in directory Capture\Library of the Orcad Lite installation directory. I will name the library meh1.olb and save it in directory d:\Program Files\OrcadLite\Capture\Library:

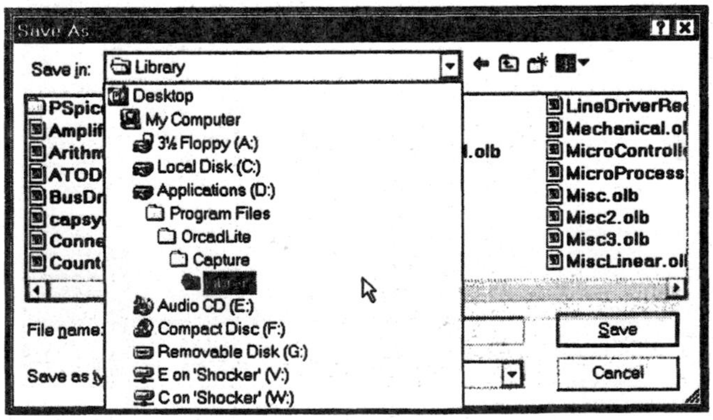

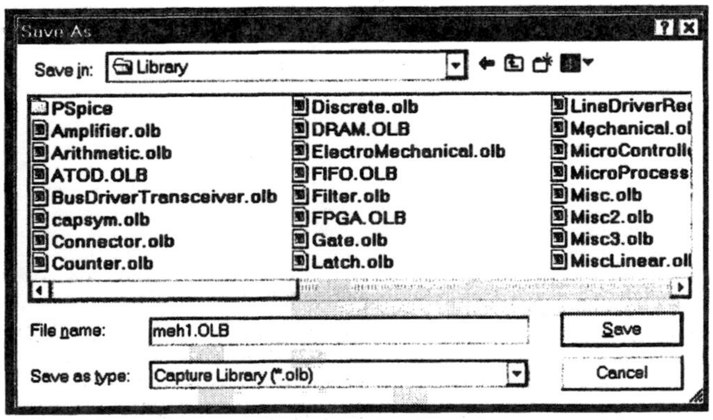

Save As

Save in: Library

Desktop
My Computer
3½ Floppy (A:)
Local Disk (C:)
Applications (D:)
Program Files
OrcadLite
Capture
Audio CD (E:)
Compact Disc (F:)
Removable Disk (G:)
E on 'Shocker' (V:)
C on 'Shocker' (W:)

PSpic
Amplif
Arithm
ATOD
BusDr
capsy
Conne
Count
.olb
LineDriverRe
Mechanical.ol
MicroControll
MicroProcess
Misc.olb
Misc2.olb
Misc3.olb
MiscLinear.ol

File name:
Save as ty
Save
Cancel

Save As

Save in: Library

PSpice
Amplifier.olb
Arithmetic.olb
ATOD.OLB
BusDriverTransceiver.olb
capsym.olb
Connector.olb
Counter.olb
Discrete.olb
DRAM.OLB
ElectroMechanical.olb
FIFO.OLB
Filter.olb
FPGA.OLB
Gate.olb
Latch.olb
LineDriverRe
Mechanical.ol
MicroControll
MicroProcess
Misc.olb
Misc2.olb
Misc3.olb
MiscLinear.ol

File name: meh1.OLB
Save as type: Capture Library (*.olb)
Save
Cancel

After clicking the ***Save*** button, we see the screen below:

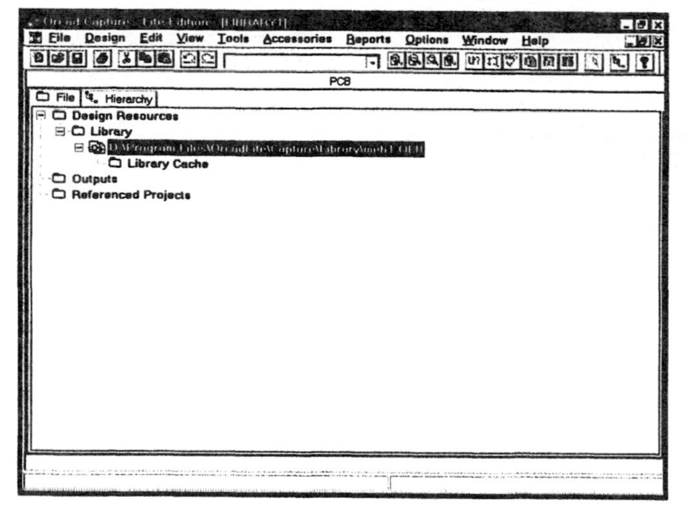


We have created a new library in the default directory. We can now add a part to that library.

Click the ***RIGHT*** mouse button on the text ***meh1.OLB***:

Select **New Part:**

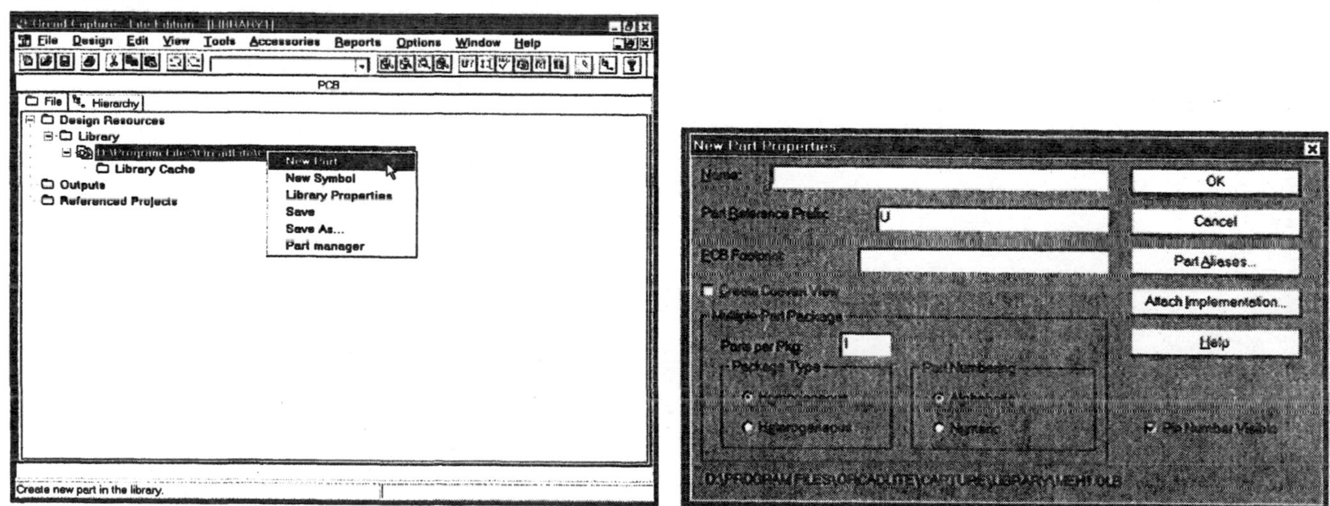


I will create a symbol for a power integrated circuit made by Power Integrations. The part number is TOP104YAI and it has three pins. I will name the part TOP104YAI. The ***Part Reference Prefix*** is the prefix given to the part when it is placed in the schematic. The default prefix is U, which we will not change. Thus this part will be numbered U1, U2, U3, etc. when multiple parts are placed in the schematic. You can change the prefix if you want. You can also assign a PCB Footprint if you wish. Since we have yet to cover the layout tool, we will leave this field unassigned. We can always add the footprint later. Fill in the dialog box as shown below and click the ***OK*** button:

We will draw a rectangle with three pins. First, draw a rectangle by selecting **Place** and then **Rectangle** from the menus. The cursor will be replaced by crosshairs:

Draw a rectangle as shown below:

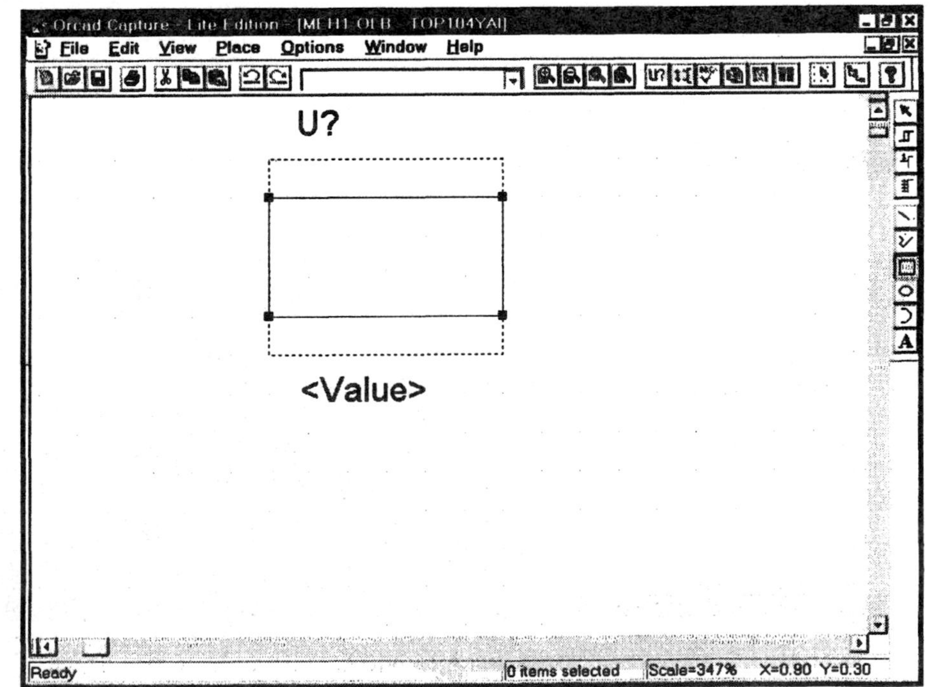


Next, resize the dashed box so that it is the same size as the rectangle. You can resize the dashed box by clicking the ***LEFT*** mouse button on it to select it. Once selected, you can drag the handles to resize it:

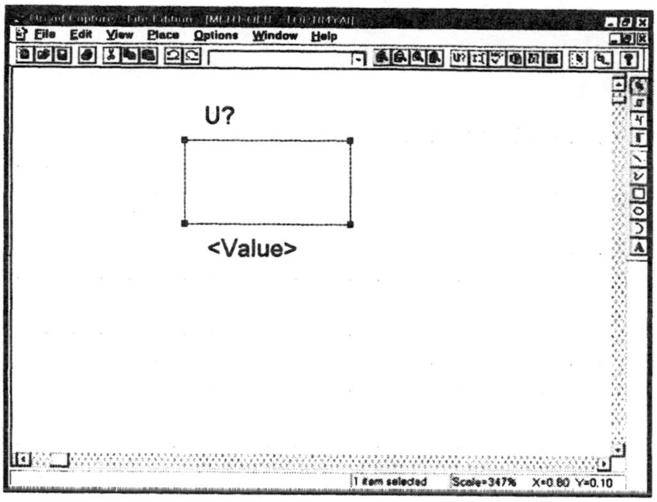

Next, drag the ***<Value>*** property inside the rectangle. This property will display the part name, in this case TOP104YAI, when displayed on the schematic. Place the property as shown:

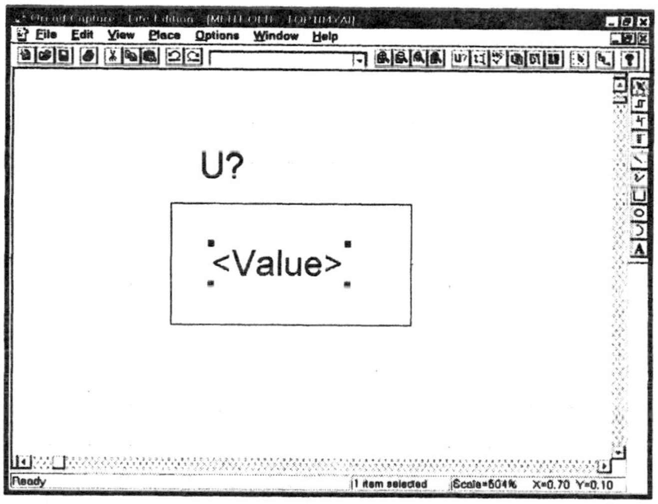

Next, we need to add three pins to this part. Select **Place** and then **Pin**:

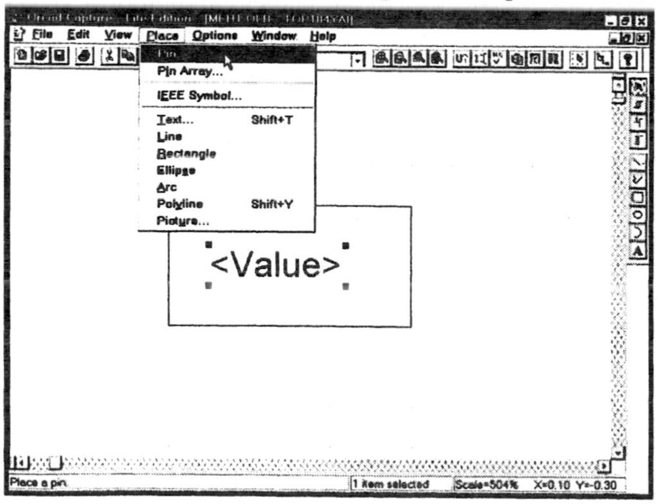

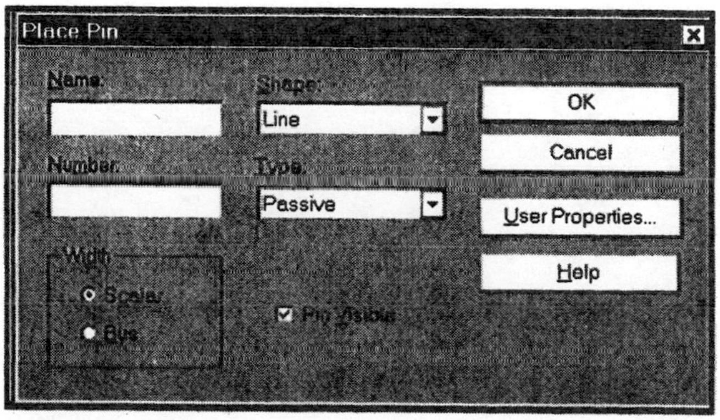

From the datasheet for the TOP104 we see that pin 1 is called "Control", which we will abbreviate as "C." Fill in the dialog box as shown:

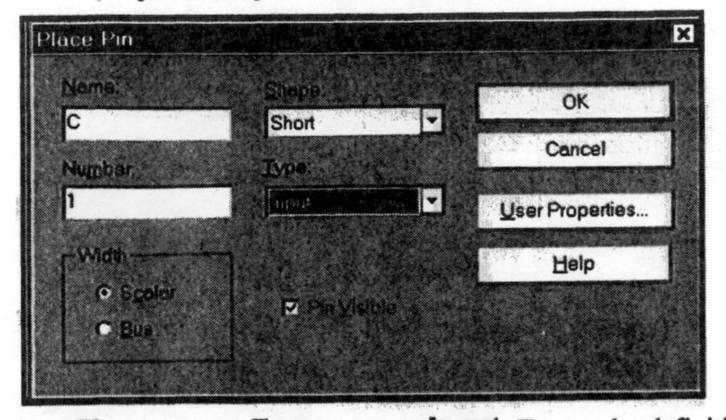


Notice that the pin ***Shape*** is set to ***Short*** and the ***Type*** is set to ***Input***. To see the definition of the pin ***Type***, select the ***Help*** button. The ***Type*** property can also be used for error checking and for hiding pins. For example, two outputs that are connected together will generate an error message, and power pins can be hidden. From the datasheet, this pin is used to monitor a signal, so it does indeed expect an input signal. Click the ***OK*** button when done:

The pin can be placed only on the dashed box, which is the outline of the part. As you move the mouse, you will notice that the pin only moves along the outline defined by the dashed box, which also happens to be the same dimensions as the rectangle we placed. Place the pin as shown below and then press the **ESC** key to stop placing pins:

Place two more pins using the same procedure. Pin 2 should be named "S" and pin 3 should be named "D." Place the pins as shown below:

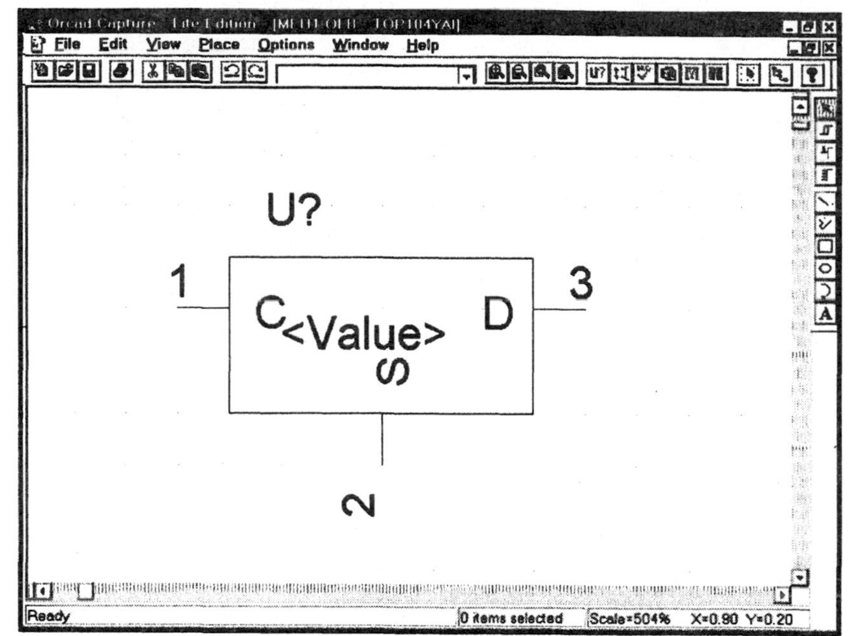


If your pins do not look as shown, you can always double-click on a pin to edit its properties. The properties I chose for pins 2 and 3 are:

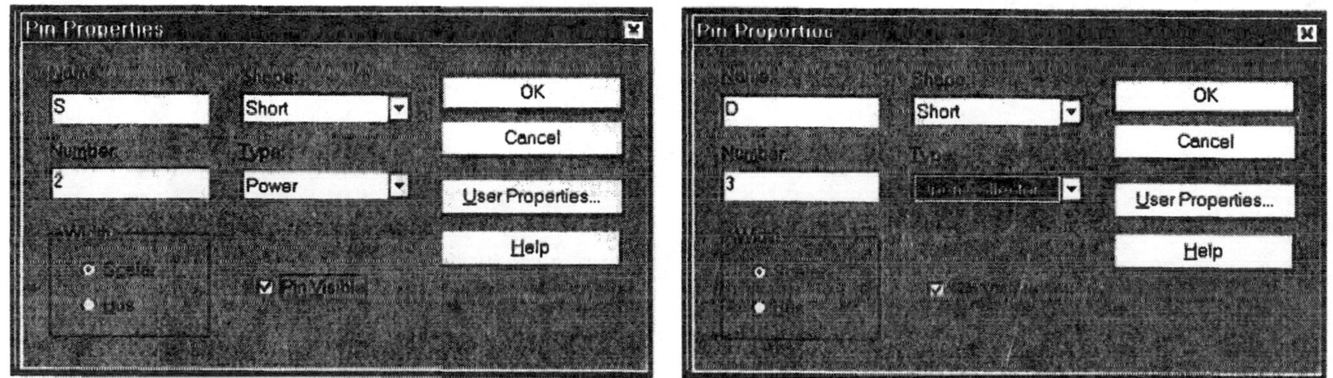


The ***<Value>*** property no longer appears to be in a good spot, so I will rearrange the text in the part as shown:

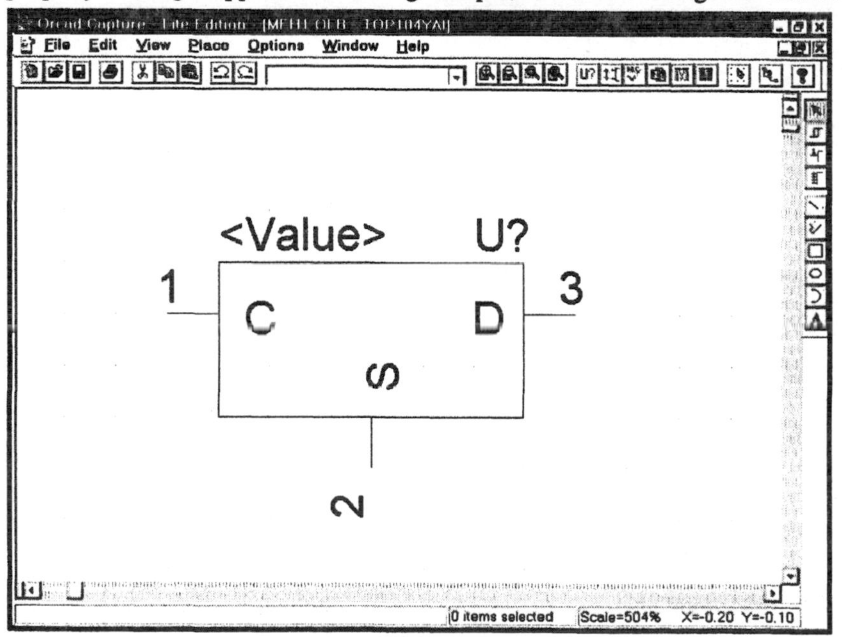

I want the pin names to be displayed, but I do not want to display the pin numbers. To change the display, select **Options** and then **Part Properties**:

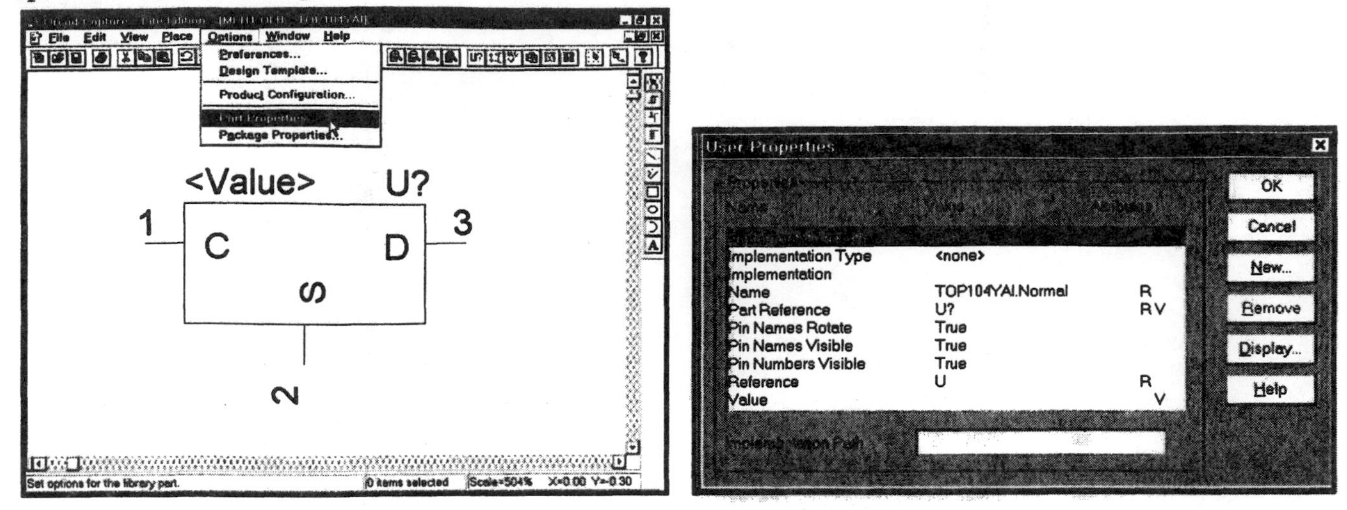


Notice that the ***Pin Numbers Visible*** property has a value of ***True***. To hide the pin numbers, we must select False. Click the ***LEFT*** mouse button on the text ***Pin Numbers Visible*** to select it:

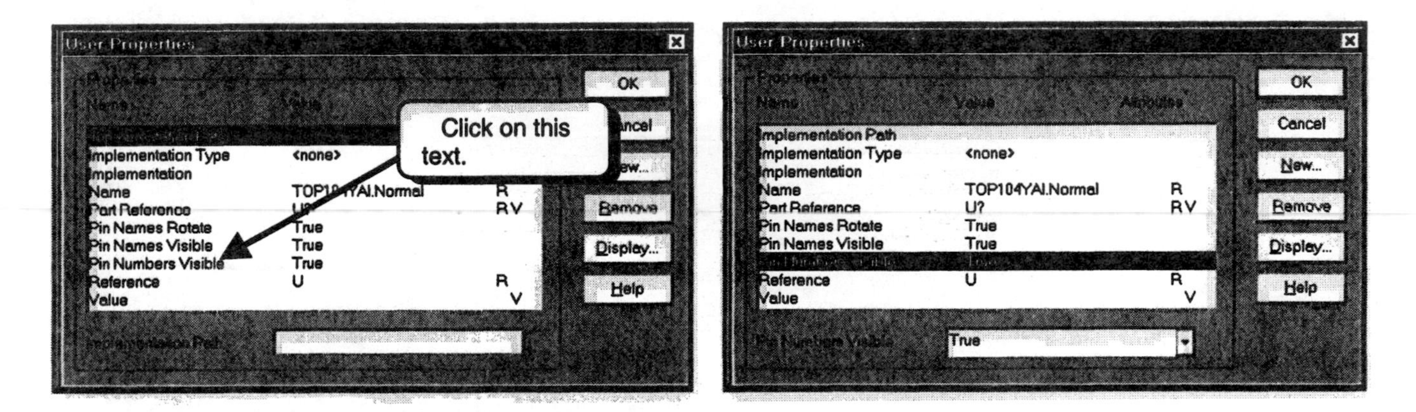


To change the value, click the ***LEFT*** mouse button on the down triangle ▼ and then select ***False***:

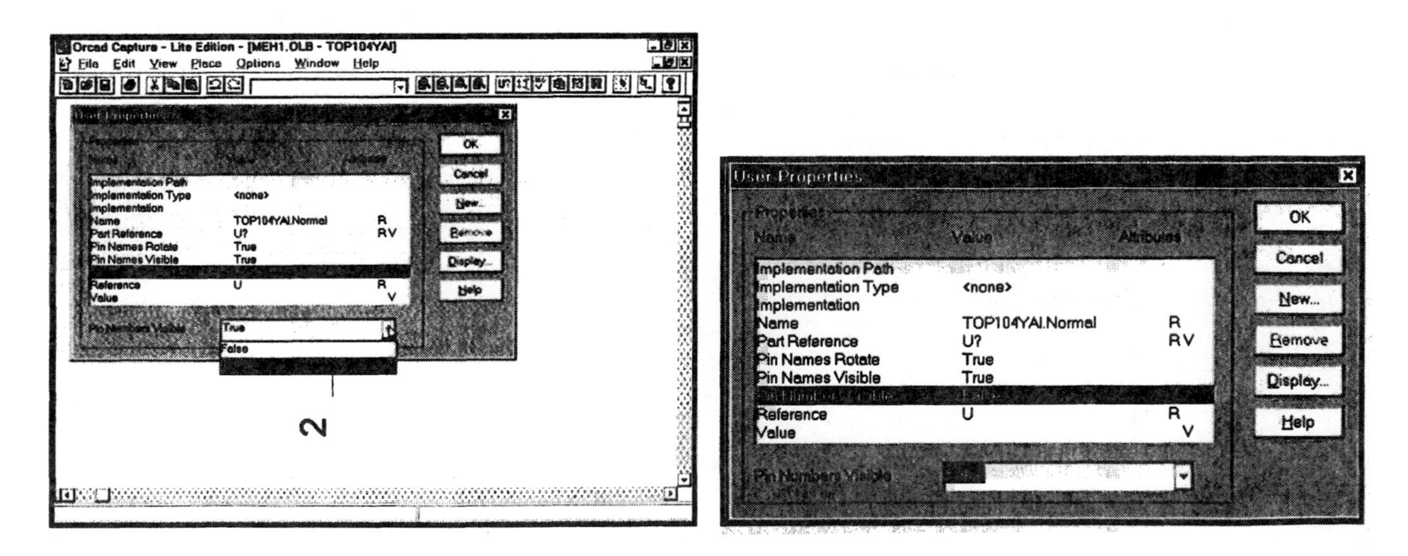


Click the ***OK*** button to accept the changes:

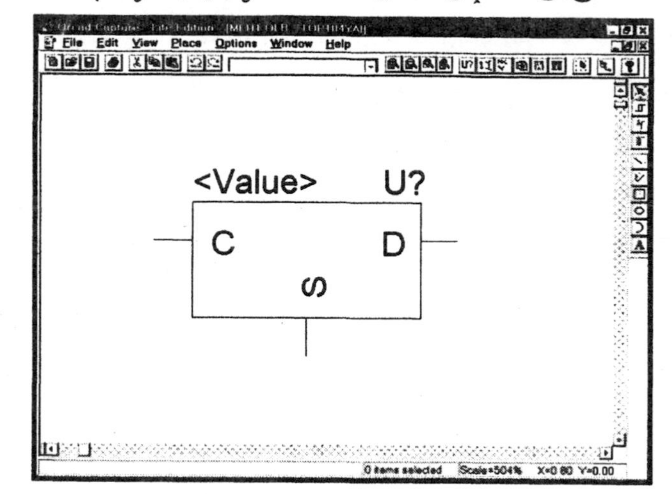


The part is now acceptable.

Type **CTRL-F4** to close the symbol drawing:

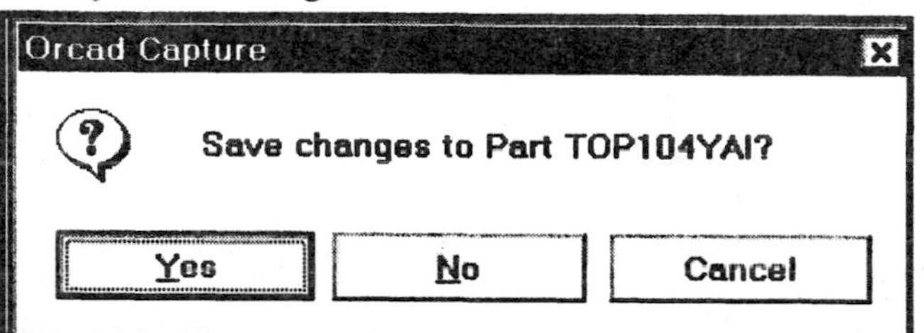


Select ***Yes***:

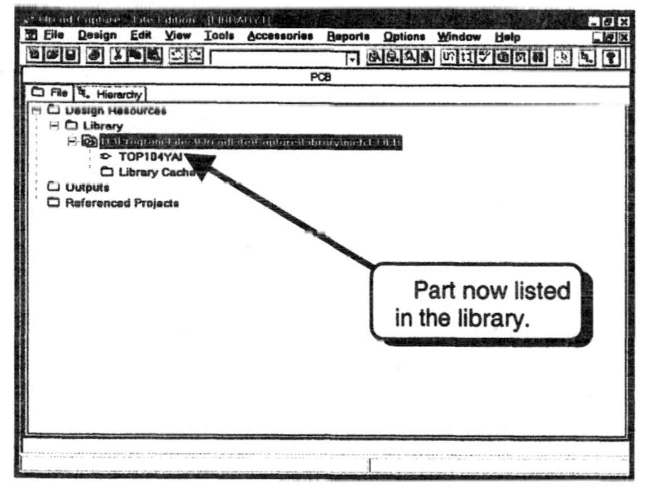


Select **File** and then **Close Project** to close the library:

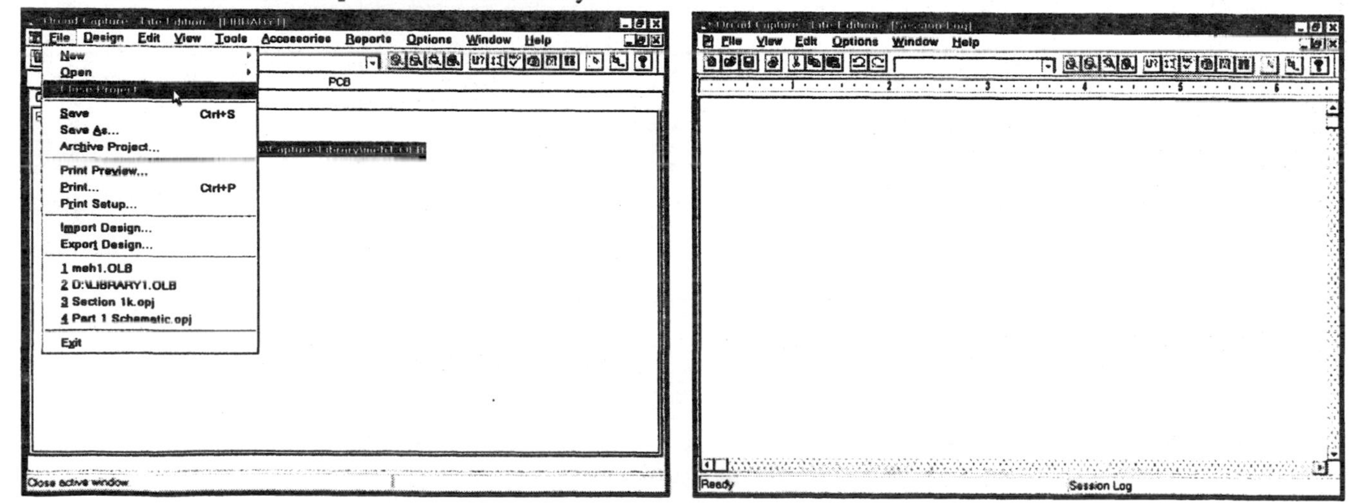

We are now ready to add the part to the database.

## 10.E.2. Adding a Part to the Database Using Microsoft Access

Run Microsoft Access, open the file DIGIKEY_SC.mdb, and then open the USER table as shown on pages 578-579:

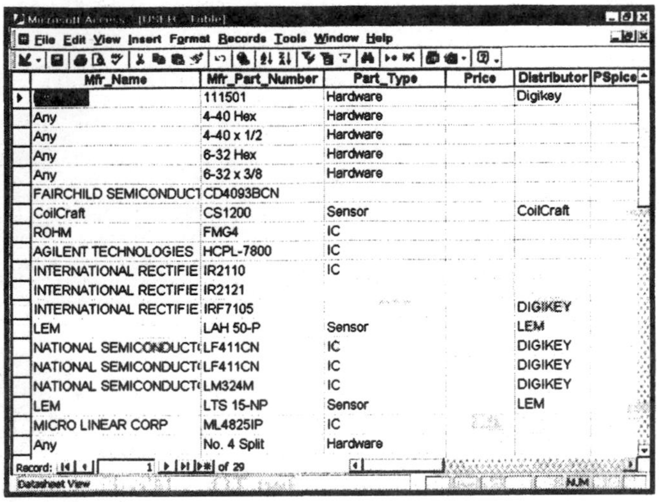

| Mfr_Name | Mfr_Part_Number | Part_Type | Price | Distributor | PSpice |
|---|---|---|---|---|---|
| [illegible] | 111501 | Hardware | | Digikey | |
| Any | 4-40 Hex | Hardware | | | |
| Any | 4-40 x 1/2 | Hardware | | | |
| Any | 6-32 Hex | Hardware | | | |
| Any | 6-32 x 3/8 | Hardware | | | |
| FAIRCHILD SEMICONDUCT | CD4093BCN | | | | |
| CoilCraft | CS1200 | Sensor | | CoilCraft | |
| ROHM | FMG4 | IC | | | |
| AGILENT TECHNOLOGIES | HCPL-7800 | IC | | | |
| INTERNATIONAL RECTIFIE | IR2110 | IC | | | |
| INTERNATIONAL RECTIFIE | IR2121 | | | | |
| INTERNATIONAL RECTIFIE | IRF7105 | | | DIGIKEY | |
| LEM | LAH 50-P | Sensor | | LEM | |
| NATIONAL SEMICONDUCT( | LF411CN | IC | | DIGIKEY | |
| NATIONAL SEMICONDUCT( | LF411CN | IC | | DIGIKEY | |
| NATIONAL SEMICONDUCT( | LM324M | IC | | DIGIKEY | |
| LEM | LTS 15-NP | Sensor | | LEM | |
| MICRO LINEAR CORP | ML4825IP | IC | | | |
| Any | No. 4 Split | Hardware | | | |

Record: 1 of 29

The easiest way to enter the information for a new part is to copy a part with nearly the same properties as the one you are adding. Since we are adding an integrated circuit, I will select the LM324 part. Click the ***LEFT*** mouse button as shown below to highlight the entire line for the part:

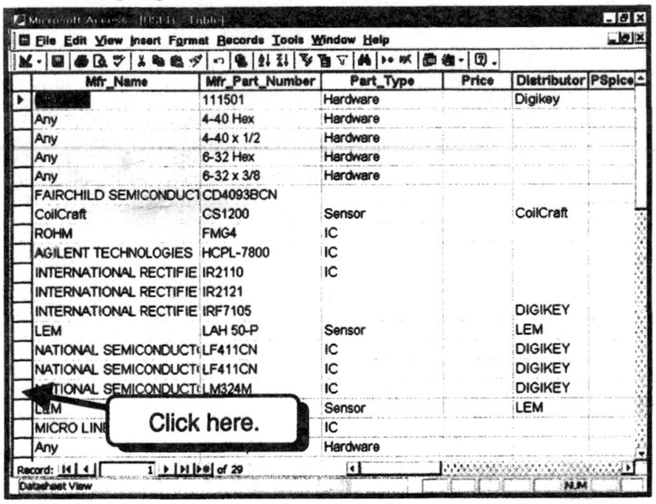

| Mfr_Name | Mfr_Part_Number | Part_Type | Price | Distributor | PSpice |
|---|---|---|---|---|---|
| [illegible] | 111501 | Hardware | | Digikey | |
| Any | 4-40 Hex | Hardware | | | |
| Any | 4-40 x 1/2 | Hardware | | | |
| Any | 6-32 Hex | Hardware | | | |
| Any | 6-32 x 3/8 | Hardware | | | |
| FAIRCHILD SEMICONDUCT | CD4093BCN | | | | |
| CoilCraft | CS1200 | Sensor | | CoilCraft | |
| ROHM | FMG4 | IC | | | |
| AGILENT TECHNOLOGIES | HCPL-7800 | IC | | | |
| INTERNATIONAL RECTIFIE | IR2110 | IC | | | |
| INTERNATIONAL RECTIFIE | IR2121 | | | | |
| INTERNATIONAL RECTIFIE | IRF7105 | | | DIGIKEY | |
| LEM | LAH 50-P | Sensor | | LEM | |
| NATIONAL SEMICONDUCT( | LF411CN | IC | | DIGIKEY | |
| NATIONAL SEMICONDUCT( | LF411CN | IC | | DIGIKEY | |
| …TIONAL SEMICONDUCT( | LM324M | IC | | DIGIKEY | |
| …M | | Sensor | | LEM | |
| MICRO LIN… | | IC | | | |
| Any | | Hardware | | | |

Click here.

Record: 1 of 29

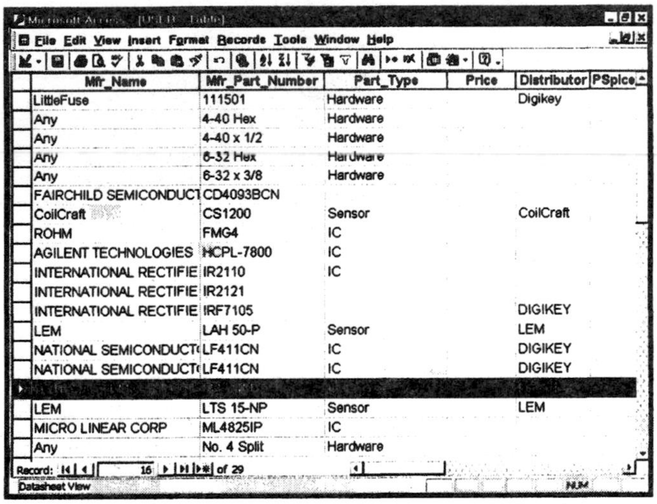

| Mfr_Name | Mfr_Part_Number | Part_Type | Price | Distributor | PSpice |
|---|---|---|---|---|---|
| LittleFuse | 111501 | Hardware | | Digikey | |
| Any | 4-40 Hex | Hardware | | | |
| Any | 4-40 x 1/2 | Hardware | | | |
| Any | 6-32 Hex | Hardware | | | |
| Any | 6-32 x 3/8 | Hardware | | | |
| FAIRCHILD SEMICONDUCT | CD4093BCN | | | | |
| CoilCraft | CS1200 | Sensor | | CoilCraft | |
| ROHM | FMG4 | IC | | | |
| AGILENT TECHNOLOGIES | HCPL-7800 | IC | | | |
| INTERNATIONAL RECTIFIE | IR2110 | IC | | | |
| INTERNATIONAL RECTIFIE | IR2121 | | | | |
| INTERNATIONAL RECTIFIE | IRF7105 | | | DIGIKEY | |
| LEM | LAH 50-P | Sensor | | LEM | |
| NATIONAL SEMICONDUCT( | LF411CN | IC | | DIGIKEY | |
| NATIONAL SEMICONDUCT( | LF411CN | IC | | DIGIKEY | |
| [illegible] | [illegible] | [illegible] | | [illegible] | |
| LEM | LTS 15-NP | Sensor | | LEM | |
| MICRO LINEAR CORP | ML4825IP | IC | | | |
| Any | No. 4 Split | Hardware | | | |

Record: 16 of 29

Select **Edit** and then **Copy** from the Access menus to copy the data for the LM324. Next, scroll the table down until you see the last line of the table. It will be empty:

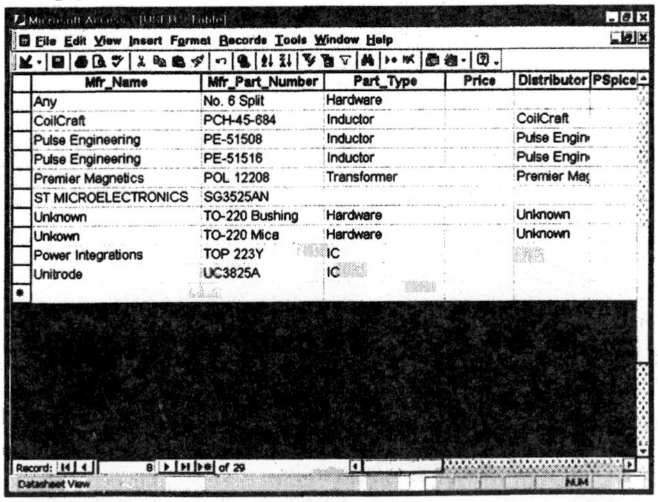

| Mfr_Name | Mfr_Part_Number | Part_Type | Price | Distributor | PSpice |
|---|---|---|---|---|---|
| Any | No. 6 Split | Hardware | | | |
| CoilCraft | PCH-45-684 | Inductor | | CoilCraft | |
| Pulse Engineering | PE-51508 | Inductor | | Pulse Engin | |
| Pulse Engineering | PE-51516 | Inductor | | Pulse Engin | |
| Premier Magnetics | POL 12208 | Transformer | | Premier Mag | |
| ST MICROELECTRONICS | SG3525AN | | | | |
| Unknown | TO-220 Bushing | Hardware | | Unknown | |
| Unkown | TO-220 Mica | Hardware | | Unknown | |
| Power Integrations | TOP 223Y | IC | | | |
| Unitrode | UC3825A | IC | | | |
| * | | | | | |

Record: 8 of 29

Click the ***LEFT*** mouse button as shown below to select the entire line:

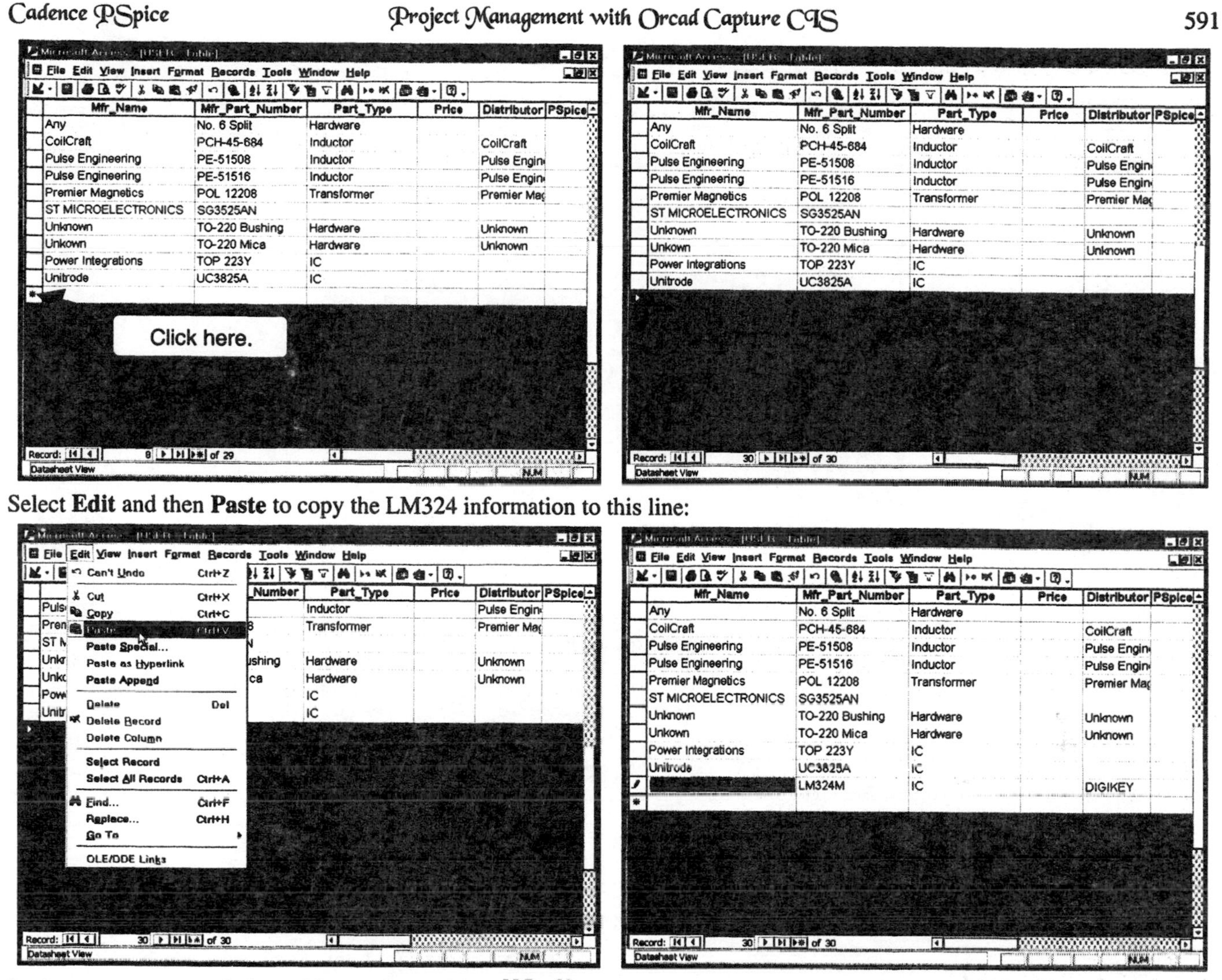


Select **Edit** and then **Paste** to copy the LM324 information to this line:



In the screen capture below, I have changed the ***Mfr_Name*** to `Power Integrations`, the ***Mfr_Part_Number*** to `TOP104YAI`, and the ***Distributor*** to `Unknown`. To change a value, click the *LEFT* mouse button on a cell and then type in the new value:

Microsoft Access - [USER : Table]

File Edit View Insert Format Records Tools Window Help

| Mfr_Name | Mfr_Part_Number | Part_Type | Price | Distributor | PSpice |
|---|---|---|---|---|---|
| Pulse Engineering | PE-51516 | Inductor | | Pulse Engin | |
| Premier Magnetics | POL 12208 | Transformer | | Premier Mag | |
| ST MICROELECTRONICS | SG3525AN | | | | |
| Unknown | TO-220 Bushing | Hardware | | Unknown | |
| Unkown | TO-220 Mica | Hardware | | Unknown | |
| Power Integrations | TOP 223Y | IC | | | |
| Unitrode | UC3825A | IC | | | |
| Power Integrations | TOP104YAI | IC | | Unknown | |

Record: 30 of 30

Datasheet View NUM

Scroll the table to the right until you see the ***Description*** column:

| Description | DigiKey_url |
|---|---|
| Inductor, 98 MicroHenry, 6A, 48 mOhm | |
| Transformer, Flyback | |
| VOLTAGE MODE SMPS CONTROLLER | http://us.st.com/stonline/books/pdf/docs/4286.pdf |
| TO-220 Insulating Bushing | |
| TO-220 Mica Insulator | |
| IC, Power Supply - Flyback | |
| CURRENT-MODE SMPS CONTROLLER | http://www.microlinear.com/ML/4825.pdf |
| VOLTAGE-FEEDBACK OPERATIONAL AM | http://www.national.com/ds/LM/LM124.pdf |

Record: 30 of 30

Change the description to one appropriate for the part:

| Description | DigiKey_url |
|---|---|
| Inductor, 98 MicroHenry, 6A, 48 mOhm | |
| Transformer, Flyback | |
| VOLTAGE MODE SMPS CONTROLLER | http://us.st.com/stonline/books/pdf/docs/4286.pdf |
| TO-220 Insulating Bushing | |
| TO-220 Mica Insulator | |
| IC, Power Supply - Flyback | |
| CURRENT-MODE SMPS CONTROLLER | http://www.microlinear.com/ML/4825.pdf |
| IC, Power Supply, Flywack | http://www.national.com/ds/LM/LM124.pdf |

Record: 30 of 30

Scroll the table to the right again until you see the columns below:

| Distributor_Part_Number | Symbol_Name | Value | Footprint_Nam |
|---|---|---|---|
| Pulse Engineering | L | 98UH | |
| POL 12208 | POL 12208 | POL 12208 | |
| TMP-16 | MISCLINEAR\IP15 | SG3525A | |
| TO-220 Bushing | TO-220 Bushing | TO-220 Bushing | |
| TO-220 Mica | TO-220 Mica | TO-220 Mica | |
| TOP 223Y | TOP 223Y | TOP 223Y | |
| UC3825A | REGULATOR\ML48 | UC3825A | |
| TMP-13 | MHV\124 | | |

Record: 30 of 30

The important information here is the ***Symbol_Name*** and the ***Value*** columns. The ***Symbol_Name*** must be the same name you gave the symbol. In our case, we named the symbol TOP104YAI. The ***Value*** column is the information that will be used by the <Value> property of the symbol. For the symbol we created, the value property is displayed on the schematic. Thus, the value we place in the ***Value*** column here will be the information displayed on the schematic. If this was a capacitor, we would use C for the ***Symbol_Name*** column and the value of the capacitance, 47 UF for example, for the

***Value*** column. For the capacitor, a capacitor symbol and a value of 47 UF would be displayed on the schematic. For the part we created, we would like the part number displayed on the schematic, so we will set the ***Value*** column to TOP104YAI:

File Edit View Insert Format Records Tools Window Help

| Distributor_Part_Number | Symbol_Name | Value | Footprint_Nam |
|---|---|---|---|
| Pulse Engineering | L | 98UH | |
| POL 12208 | POL 12208 | POL 12208 | |
| TMP-16 | MISCLINEAR\IP15 | SG3525A | |
| TO-220 Bushing | TO-220 Bushing | TO-220 Bushinc | |
| TO-220 Mica | TO-220 Mica | TO-220 Mica | |
| TOP 223Y | TOP 223Y | TOP 223Y | |
| UC3825A | REGULATOR\ML48 | UC3825A | |
| TOP104YAI | TOP104YAI | TOP104YAI | |

Record: 30 of 30
Datasheet View NUM

Save and close the database when you have made the changes.

## 10.E.3. Using the Symbol in a Schematic

To test the part we just made, we will open the schematic we created earlier and place the part in it:

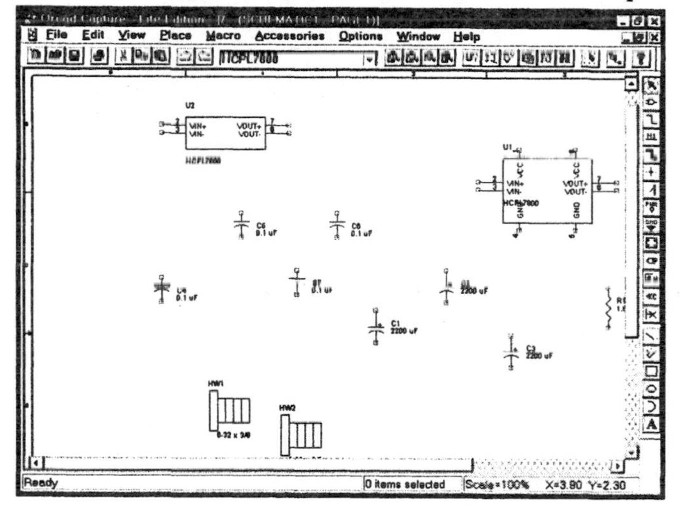


Type **z** to place a database part:

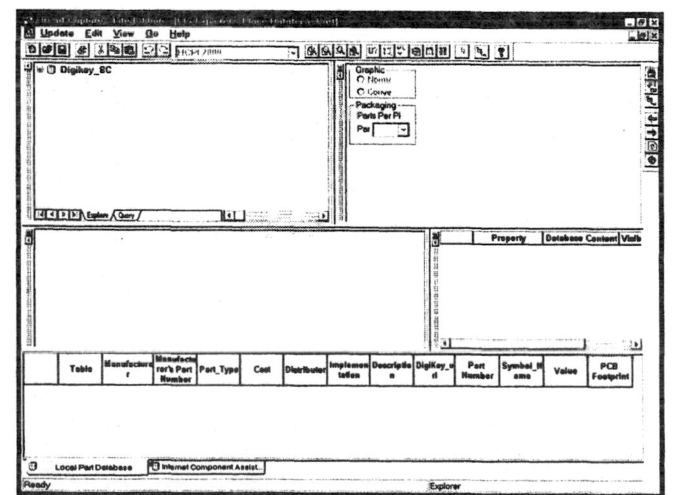


Expand the tree and select the ***IC*** section of the ***USER*** table. At the bottom of my window, we see the part we just added:

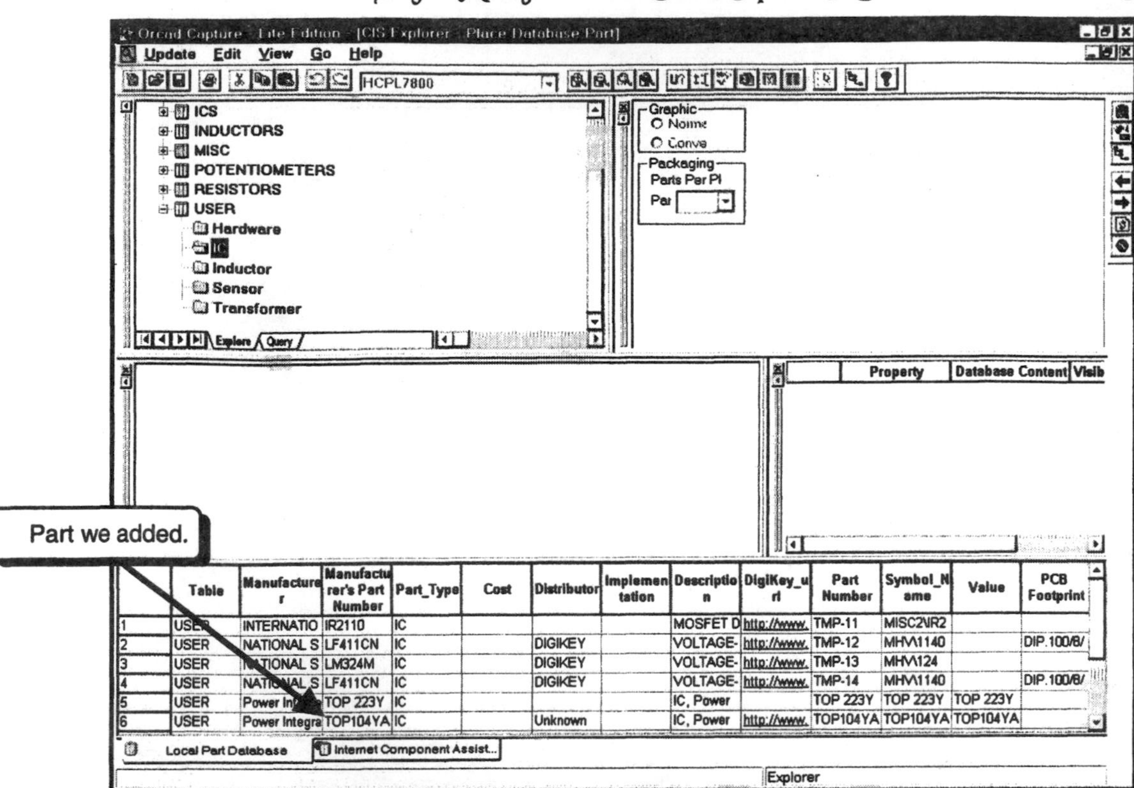


If the part is set up correctly, when we click the ***LEFT*** mouse button on the part, it should be highlighted in green, and the symbol should be displayed on the screen:

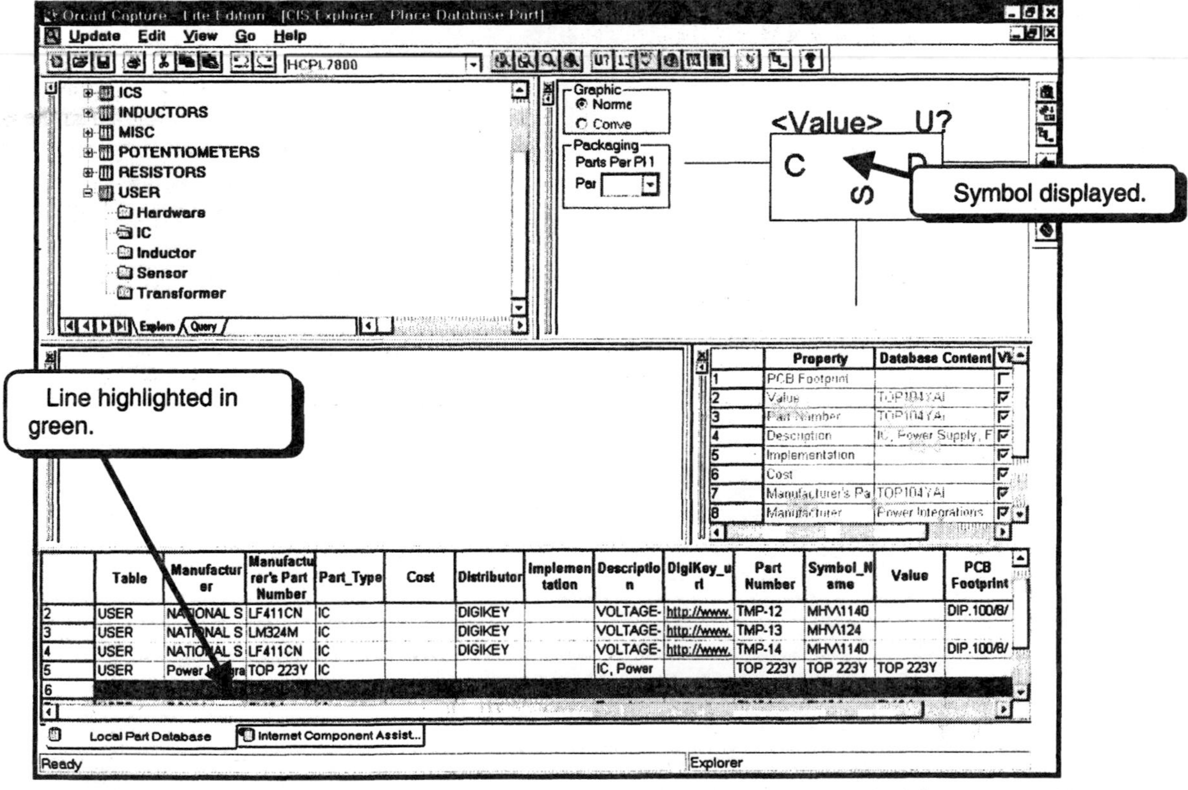


If the line was highlighted in red or yellow, chances are that Capture cannot locate the symbol for the part. There are three possibilities: (1) You did not create the part. (2) The library that contains the part is not in the directory OrcadLite\Capture\Library. (3) You specified the wrong name in the Symbol_Name column of the database. Since my line is highlighted in green, I can use the part. To place the highlighted part, click the ***RIGHT*** mouse button on the highlighted line and then select **Place Database Part** from the menu that appears. Place a few copies of the part in your schematic:

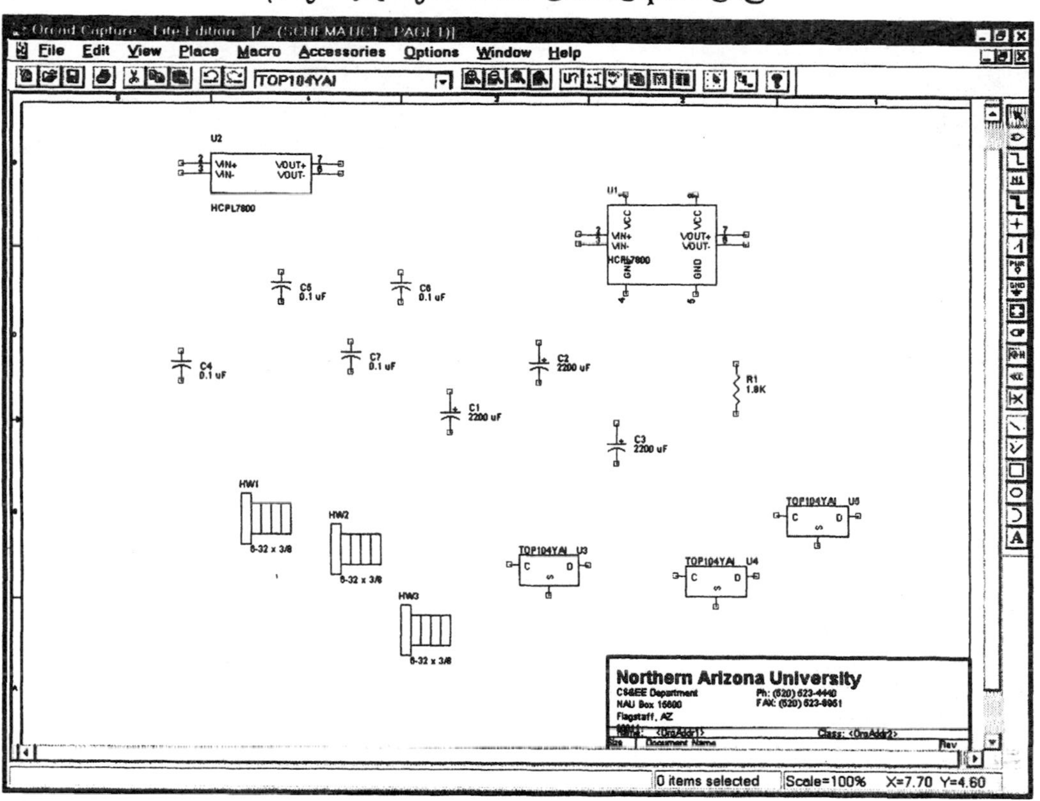


In my schematic, the part number TOP104YAI and the reference names (U3, U4, or U5) were too close together, so I moved the part references to separate the part number from the reference names.

# 10.F. Creating a Bill of Materials

Since the schematic we created in previous sections only contains parts in our Digikey database, we can easily create a bill of materials (BOM) that contains information from the database. The information contained in the BOM can be easily configured, and the BOM can be viewed with a text editor or exported to Microsoft Excel. Here, we will show how to create a custom BOM and export the information to Excel, where it can be easily manipulated. We will start with the project created in the earlier sections of this manual. If you have the screen in the previous window, type **CTRL-F4** to close the schematic and return to the tree view of the project:

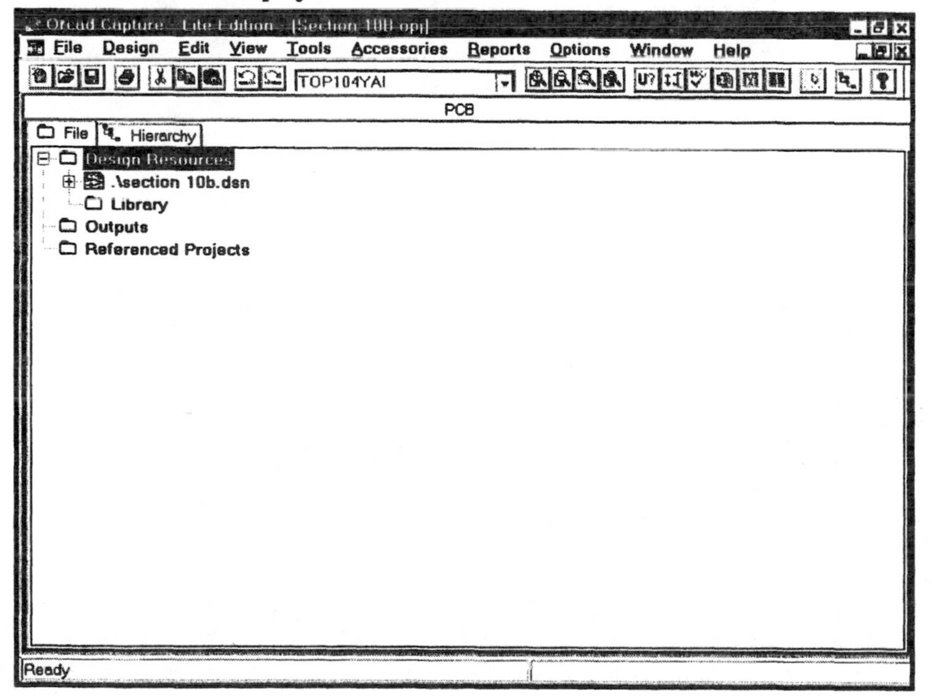


Select **Reports**, **CIS Bill of Materials**, and then **Standard** from the menus:

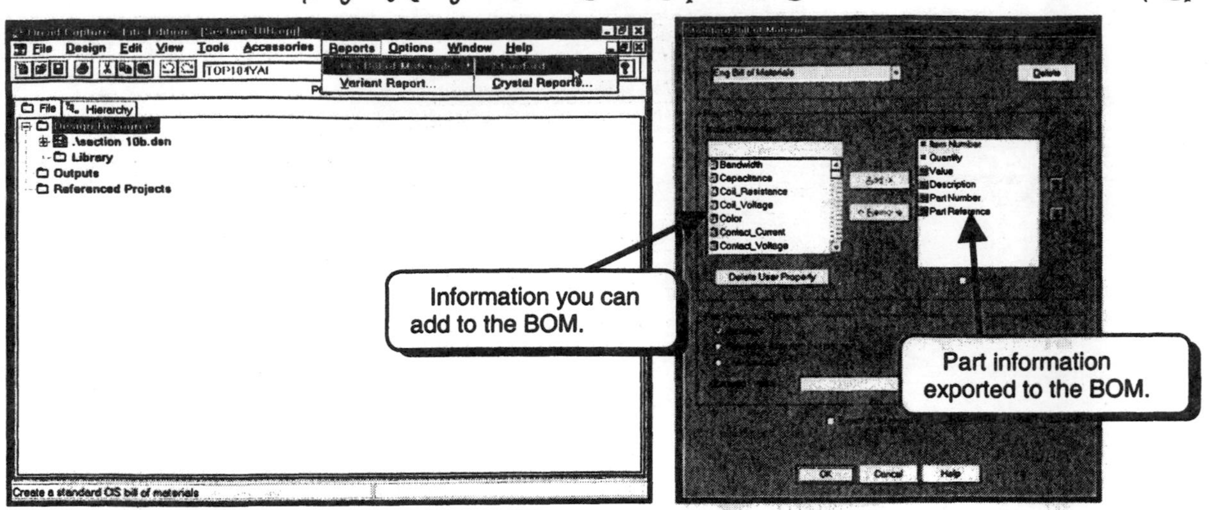


The left pane of the ***Report Properties*** of the portion of the dialog box contains the list of available information that can be added to the BOM. This information can come from several sources such as:

- information transferred from the schematic

- information transferred from your database

- information transferred over the Internet through the Internet Component Assistant

The ***Output Format*** pane contains the information that will be placed in the BOM. You can add information by selecting an item in the ***Select Properties*** pane and then clicking the ***Add*** button. You can delete properties from the BOM by selecting an item in the ***Output Format*** pane and then clicking the ***Remove*** button. I will change the BOM to use the information from our database:

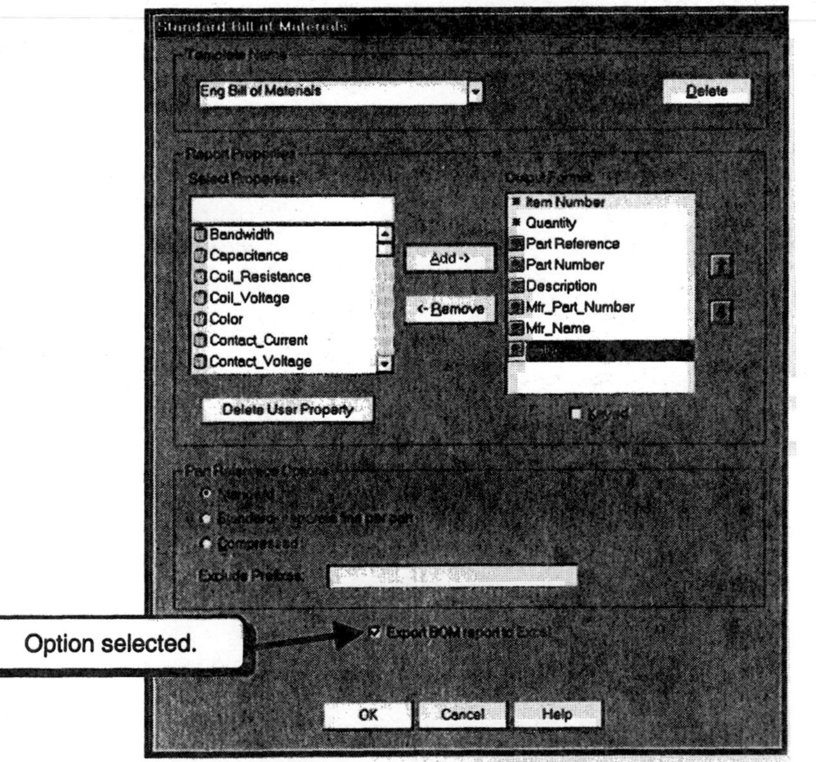


You will need to experiment with the different items to generate the information in which you are interested. The last thing we will do is choose the option to export the information to Microsoft Excel. Select the option as shown above.

When you click the ***OK*** button, Excel will run, and the information for the BOM will be exported to the spreadsheet:

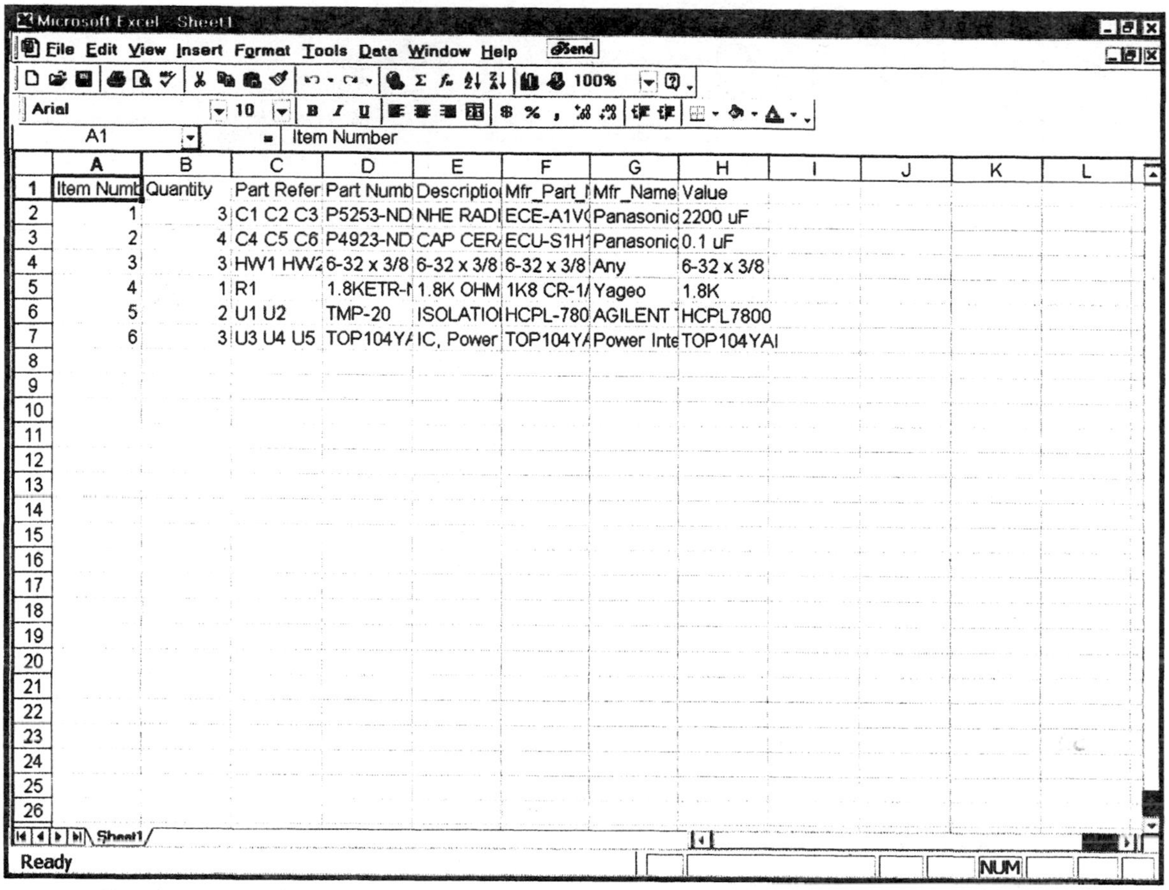

Microsoft Excel - Sheet1

File Edit View Insert Format Tools Data Window Help Send

100%

Arial 10

A1 = Item Number

| | A | B | C | D | E | F | G | H | I | J | K | L |
|---|---|---|---|---|---|---|---|---|---|---|---|---|
| 1 | Item Numb | Quantity | Part Refer | Part Numb | Descriptio | Mfr_Part_I | Mfr_Name | Value | | | | |
| 2 | 1 | 3 | C1 C2 C3 | P5253-ND | NHE RADI | ECE-A1V( | Panasonic | 2200 uF | | | | |
| 3 | 2 | 4 | C4 C5 C6 | P4923-ND | CAP CER/ | ECU-S1H1 | Panasonic | 0.1 uF | | | | |
| 4 | 3 | 3 | HW1 HW2 | 6-32 x 3/8 | 6-32 x 3/8 | 6-32 x 3/8 | Any | 6-32 x 3/8 | | | | |
| 5 | 4 | 1 | R1 | 1.8KETR-I | 1.8K OHM | 1K8 CR-1/ | Yageo | 1.8K | | | | |
| 6 | 5 | 2 | U1 U2 | TMP-20 | ISOLATIO | HCPL-780 | AGILENT T | HCPL7800 | | | | |
| 7 | 6 | 3 | U3 U4 U5 | TOP104Y/ | IC, Power | TOP104Y/ | Power Inte | TOP104YAI | | | | |
| 8 | | | | | | | | | | | | |
| 9 | | | | | | | | | | | | |
| 10 | | | | | | | | | | | | |
| 11 | | | | | | | | | | | | |
| 12 | | | | | | | | | | | | |
| 13 | | | | | | | | | | | | |
| 14 | | | | | | | | | | | | |
| 15 | | | | | | | | | | | | |
| 16 | | | | | | | | | | | | |
| 17 | | | | | | | | | | | | |
| 18 | | | | | | | | | | | | |
| 19 | | | | | | | | | | | | |
| 20 | | | | | | | | | | | | |
| 21 | | | | | | | | | | | | |
| 22 | | | | | | | | | | | | |
| 23 | | | | | | | | | | | | |
| 24 | | | | | | | | | | | | |
| 25 | | | | | | | | | | | | |
| 26 | | | | | | | | | | | | |

Sheet1

Ready NUM

You can now use Excel to manipulate the data or make the information look more presentable:

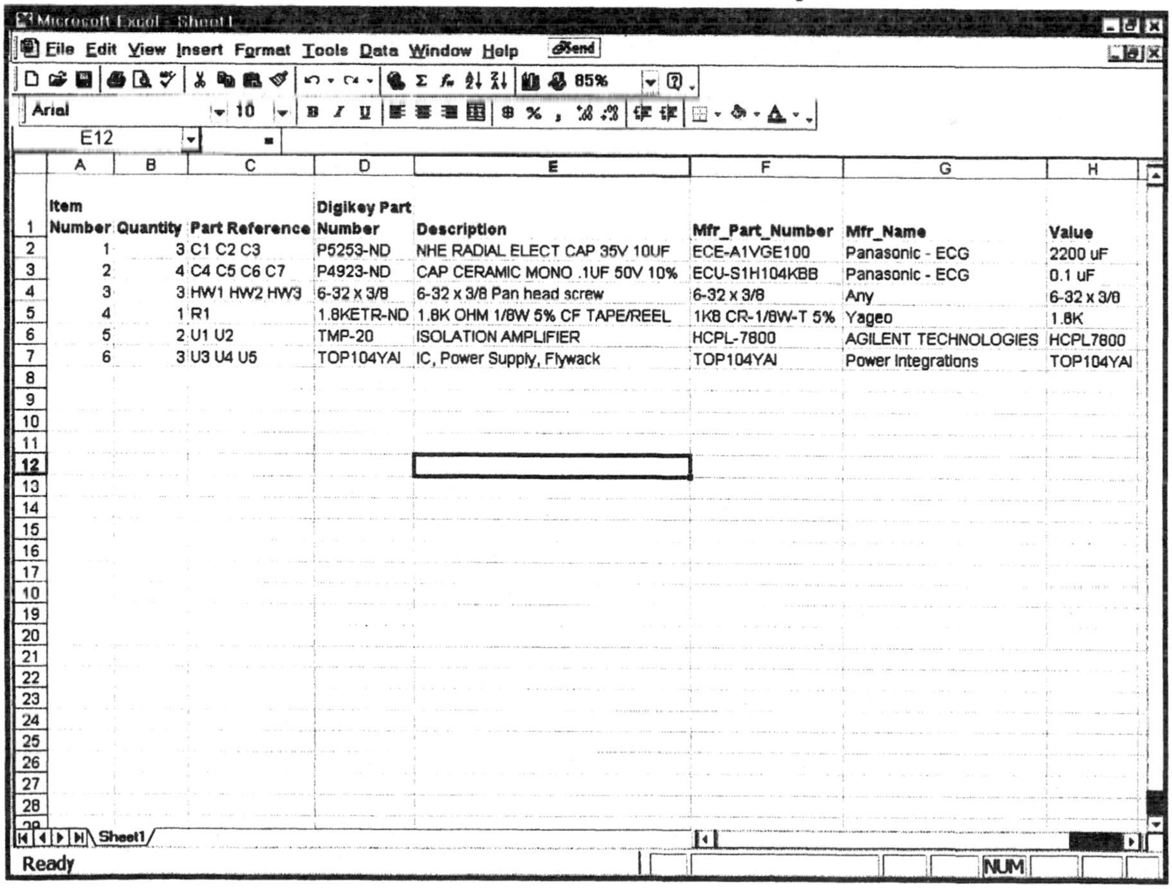

Microsoft Excel - Sheet1

File Edit View Insert Format Tools Data Window Help Send

85%

Arial 10

E12 =

| | A | B | C | D | E | F | G | H |
|---|---|---|---|---|---|---|---|---|
| 1 | **Item Number** | **Quantity** | **Part Reference** | **Digikey Part Number** | **Description** | **Mfr_Part_Number** | **Mfr_Name** | **Value** |
| 2 | 1 | 3 | C1 C2 C3 | P5253-ND | NHE RADIAL ELECT CAP 35V 10UF | ECE-A1VGE100 | Panasonic - ECG | 2200 uF |
| 3 | 2 | 4 | C4 C5 C6 C7 | P4923-ND | CAP CERAMIC MONO .1UF 50V 10% | ECU-S1H104KBB | Panasonic - ECG | 0.1 uF |
| 4 | 3 | 3 | HW1 HW2 HW3 | 6-32 x 3/8 | 6-32 x 3/8 Pan head screw | 6-32 x 3/8 | Any | 6-32 x 3/8 |
| 5 | 4 | 1 | R1 | 1.8KETR-ND | 1.8K OHM 1/8W 5% CF TAPE/REEL | 1K8 CR-1/8W-T 5% | Yageo | 1.8K |
| 6 | 5 | 2 | U1 U2 | TMP-20 | ISOLATION AMPLIFIER | HCPL-7800 | AGILENT TECHNOLOGIES | HCPL7800 |
| 7 | 6 | 3 | U3 U4 U5 | TOP104YAI | IC, Power Supply, Flywack | TOP104YAI | Power Integrations | TOP104YAI |
| 8 | | | | | | | | |
| 9 | | | | | | | | |
| 10 | | | | | | | | |
| 11 | | | | | | | | |
| 12 | | | | | | | | |
| 13 | | | | | | | | |
| 14 | | | | | | | | |
| 15 | | | | | | | | |
| 16 | | | | | | | | |
| 17 | | | | | | | | |
| 18 | | | | | | | | |
| 19 | | | | | | | | |
| 20 | | | | | | | | |
| 21 | | | | | | | | |
| 22 | | | | | | | | |
| 23 | | | | | | | | |
| 24 | | | | | | | | |
| 25 | | | | | | | | |
| 26 | | | | | | | | |
| 27 | | | | | | | | |
| 28 | | | | | | | | |

Sheet1

Ready NUM

# APPENDIX A
# Installing Orcad Lite Version 9.2

Orcad Lite version 9.2 can be installed on any system that runs Windows 95, 98, or NT. To install Orcad Lite and use the parts shown in this manual, you must run two installation programs. The first installation is for the Orcad Lite software. The second installation is for the part libraries that accompany this text. **If you do not follow the installation instructions in this appendix, the libraries for this manual will not be installed correctly.**

In the installation instructions we assume that your CD-ROM drive is designated as drive D:. If your CD-ROM drive has a different drive label, substitute your drive label where D: is specified.

When you insert the CD-ROM in your computer, the following screen will appear if the auto-run feature is enabled for your CD-ROM:

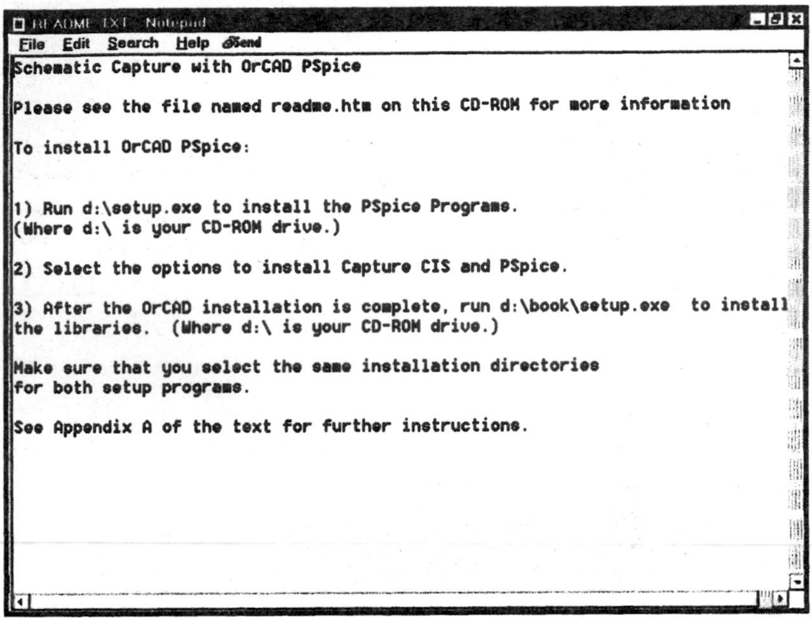

```
Schematic Capture with OrCAD PSpice

Please see the file named readme.htm on this CD-ROM for more information

To install OrCAD PSpice:

1) Run d:\setup.exe to install the PSpice Programs.
(Where d:\ is your CD-ROM drive.)

2) Select the options to install Capture CIS and PSpice.

3) After the OrCAD installation is complete, run d:\book\setup.exe  to install
the libraries.  (Where d:\ is your CD-ROM drive.)

Make sure that you select the same installation directories
for both setup programs.

See Appendix A of the text for further instructions.
```

This screen reminds you that you must run two installation programs to properly install the software for this text. To exit the program, select **File** and then **Close** from the menus.

## A.1. Installing the Orcad Lite Software

To run the installation program, click on the **Start** button in the Windows Start menu, and then select **Run**:

Enter the text **`d:\setup.exe`** where d: is the drive letter for your CD-ROM:

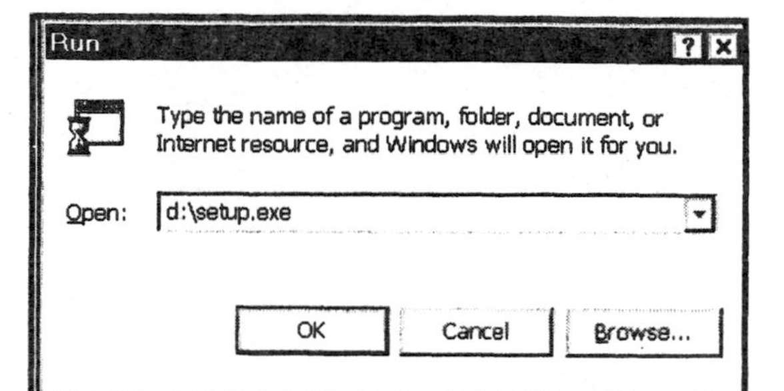


Click the ***OK*** button to run the installation program:

Disable any virus checking programs you have running on your computer and then click the ***OK*** button:

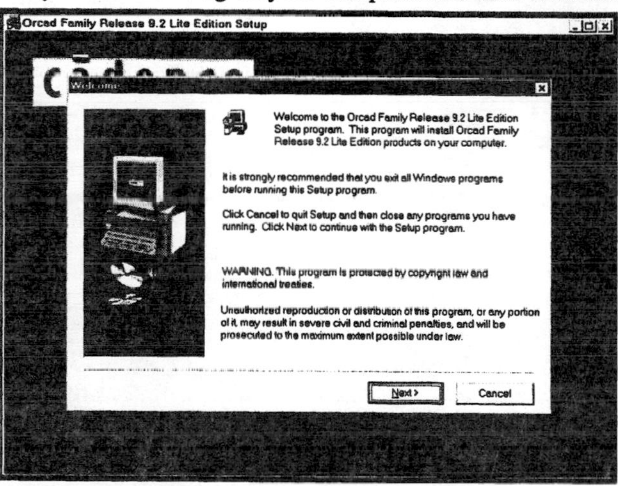


Click the ***Next*** button:

Capture CIS includes many extra features, such as downloading parts from the Internet and choosing parts from a large company-created database. For this text, you should check the Capture CIS and the PSpice options. If you would like to use the Orcad layout tool, you may also select the Layout option. However, the layout program is not discussed in this text. The minimum selections you need to choose are shown below:

Select Products

Choose the products listed below which you would like to install.

☐ Capture
☑ Capture CIS
☑ PSpice
☐ Layout

Once you have selected the appropriate installation option(s), click Next to continue. Click Cancel to exit.

< Back | Next > | Cancel

Click the ***Next*** button:

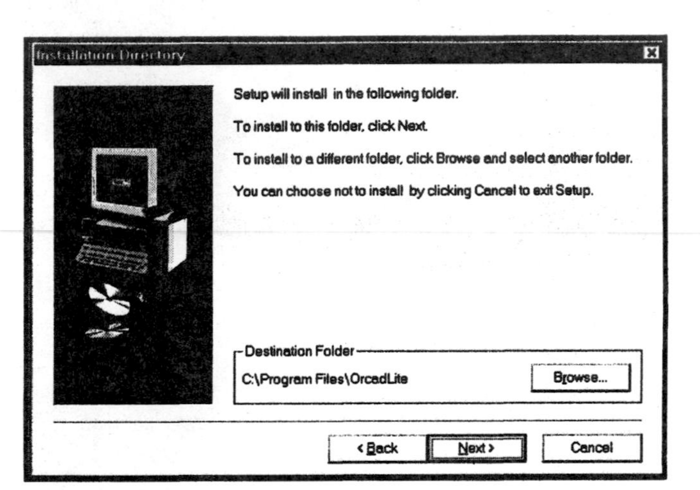


If you wish to change the ***Destination Folder***, click the ***Browse*** button and specify another drive. **Important note: If you change the destination drive and directory here, you must use the same drive and directory when installing the libraries. Make sure you write down the installation drive and directory and use the same drive and directory when you specify the *Destination Folder* for the library installation on page 603.** I will use the default directory, so I will click the ***Next*** button:

Select Program Folder

Setup will add program icons to the Program Folder listed below. You may type a new folder name, or select one from the existing Folders list. Click Next to continue.

Program Folders:

Existing Folders:

Accessories
Administrative Tools
Communications
Engineering
Enroute Imaging
Husqvarna Professional Embroidery System
Iomega
Microsoft
Microsoft Office Server Extensions

< Back | Next > | Cancel

Here you can specify the group name in which the icons will be placed. Select the ***Next*** button to accept the name:

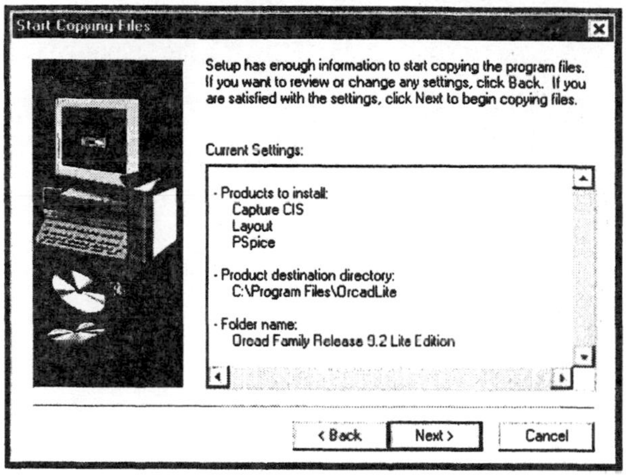

Start Copying Files

Setup has enough information to start copying the program files. If you want to review or change any settings, click Back. If you are satisfied with the settings, click Next to begin copying files.

Current Settings:

- Products to install:
  Capture CIS
  Layout
  PSpice
- Product destination directory:
  C:\Program Files\OrcadLite
- Folder name:
  Orcad Family Release 9.2 Lite Edition

< Back | Next > | Cancel

This screen is a summary of our selections so far. If you wish to change any of the listed items, click the ***Back*** button. Otherwise, click the ***Next*** button to begin the installation. **Important Note: If you are asked to install the Microsoft ODBC Data Source Administrator, select yes to install the facility. You will need it if you want to use the database capabilities of Orcad Capture. The ODBC installation is not shown here because it is already installed on my computer.**

cadence

Orcad Capture

9.2

orcad

c:\program files\orcadlite\capture\mfc40.dll

5 %

Cancel

When the files have been copied to your hard drive, you will be reminded that you need the Acrobat Reader to view the Orcad manuals that are installed with the software:

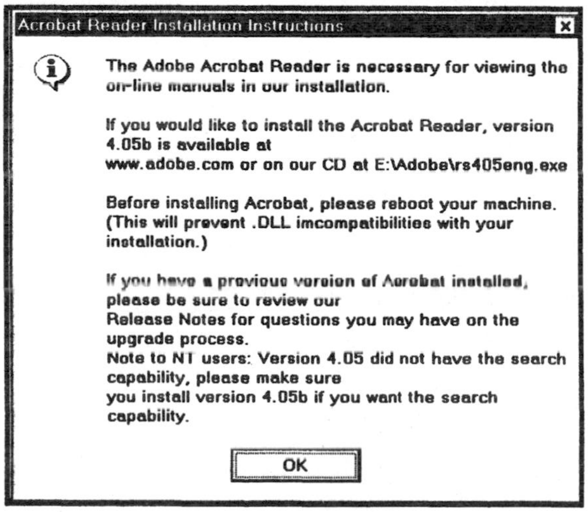

Acrobat Reader Installation Instructions

The Adobe Acrobat Reader is necessary for viewing the on-line manuals in our installation.

If you would like to install the Acrobat Reader, version 4.05b is available at www.adobe.com or on our CD at E:\Adobe\rs405eng.exe

Before installing Acrobat, please reboot your machine. (This will prevent .DLL imcompatibilities with your installation.)

If you have a previous version of Acrobat installed, please be sure to review our Release Notes for questions you may have on the upgrade process.
Note to NT users: Version 4.05 did not have the search capability, please make sure you install version 4.05b if you want the search capability.

OK

Read the information and then click the ***OK*** button to continue:

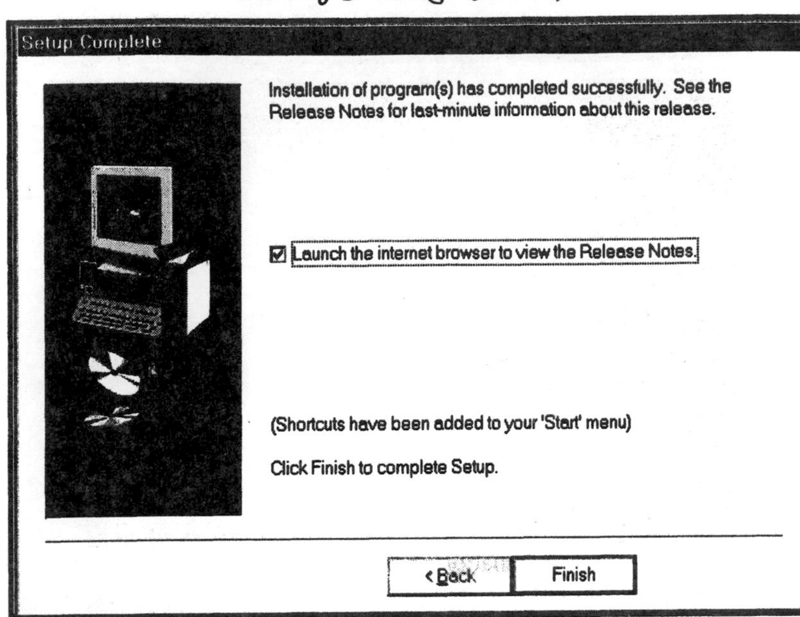


If you want to read the Release Notes, leave the option checked and then click the **Finish** button to complete the installation.

# A.2. Installing the Libraries

Several parts have been created for Orcad Capture to make this text easier for students to use. In order to have your circuits look like those shown in this manual, you must install the libraries specific to this text. This section shows how to install those libraries. This installation will do two things: (1) It will install the parts libraries used in this text. (2) It will copy all circuit files used as examples in this text onto your hard disk so that you can look at and run the example files. The libraries will be copied to the standard library subdirectory. The circuit files will be copied to a directory named "Book Circuits" in the installation directory.

To run the installation program, click on the **Start** button in the Windows Start menu, and then select **Run**:

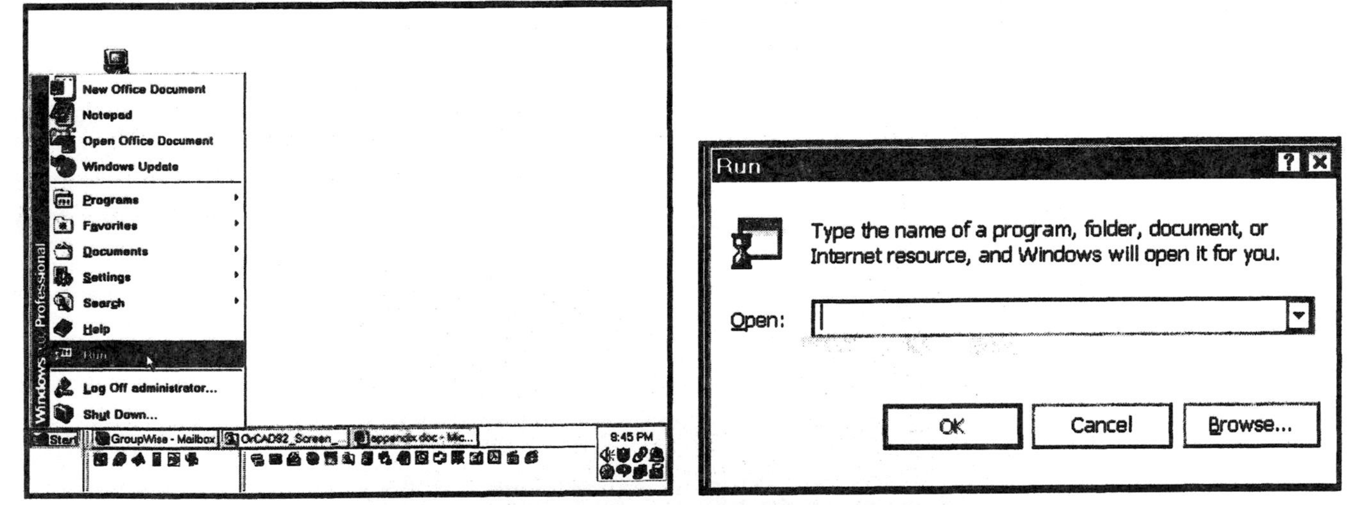


Enter the text D:\Book\setup.exe where D: is the drive letter for your CD-ROM:

Run

Type the name of a program, folder, document, or Internet resource, and Windows will open it for you.

Open: D:\Book\setup.exe

OK Cancel Browse...

Click the ***OK*** button to begin the installation:

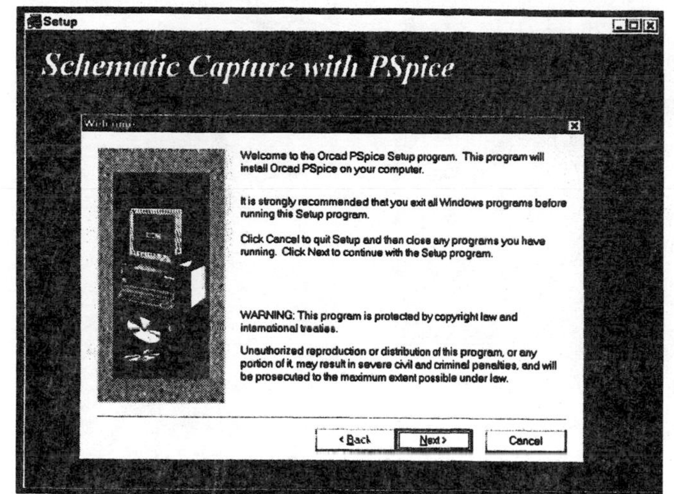


Click the ***Next*** button:

The only important parameter in the installation is the destination directory. The destination directory must be the same as the one you chose for the Orcad Lite program on page 600. Choose the same drive and same directory as was chosen in the Capture setup shown on page 600. I installed the Orcad Lite software in the default directory, so I will click the ***Next*** button:

Library files and files for all of the examples in this book will be copied to your system, so the above screen may be displayed for a while.

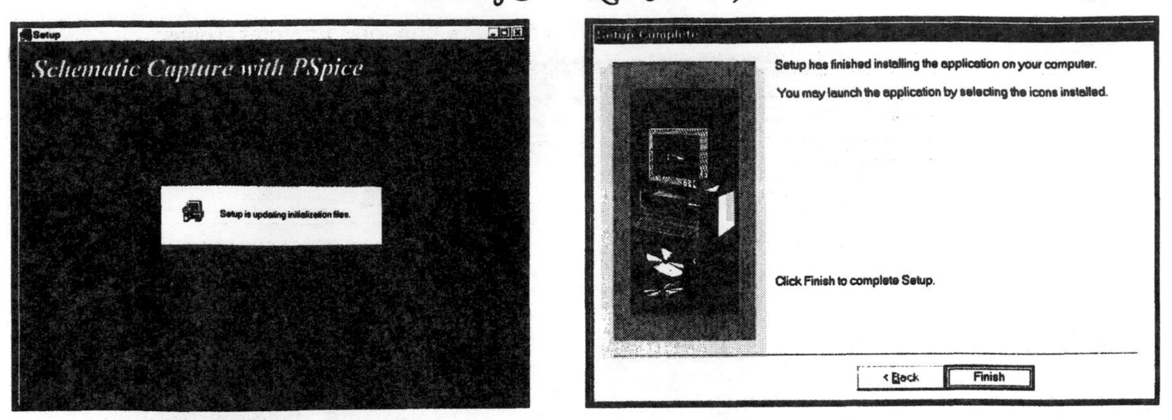


The installation is complete. Click the **Finish** button to exit the installation program.

**Note that you may need to restart your computer in order to use the circuit files installed for this book.**

# A.3. Circuit Files Used in the Text

The library installation program copies all of the circuit files used as examples in the book to your hard drive. The files are placed in a directory named "Book Circuits" in the directory you specified for installation.

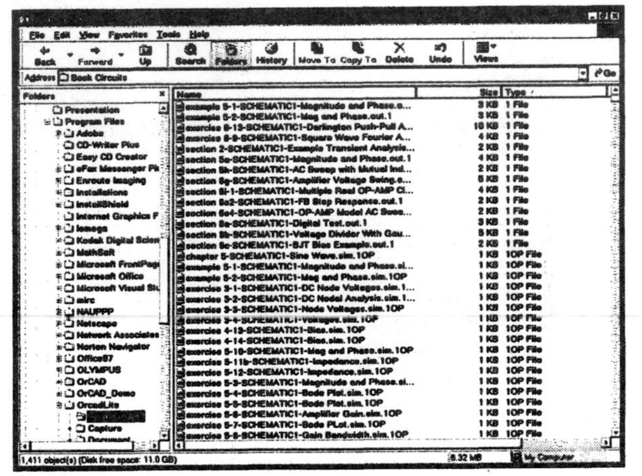

You may run these files and modify them.

For some of the examples, you may wish to simulate a circuit provided with the software installation. The easiest way to use the example files is to open the files installed on the hard drive. However, in some cases, these files may not be available and you must use the files from the CD-ROM that accompany this text. To use these files you must take two steps: (1) use the Windows Explorer to copy the files of interest to a directory on your hard drive, and (2) use the Windows Explorer to change the properties of those files to not read-only.

The following screen capture shows a directory named **ckt_files**. We used the Windows Explorer to copy three files from the CD-ROM to this directory. We will now look at the properties of one of these files. Select one of the files:

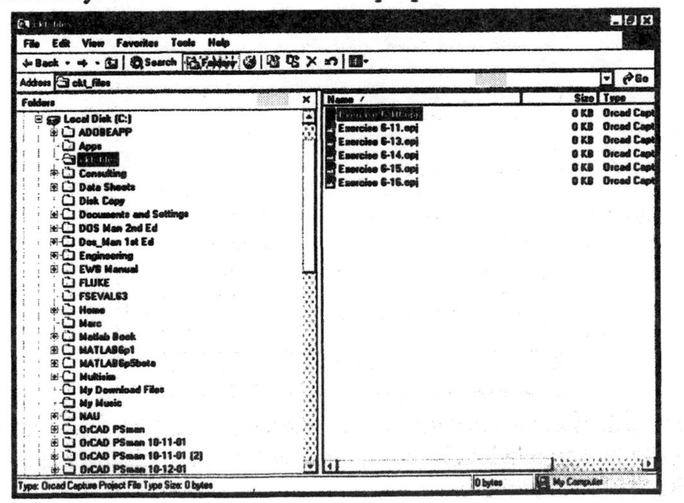

To view the properties of this file, select **File** and then **Properties** from the Windows Explorer menus:

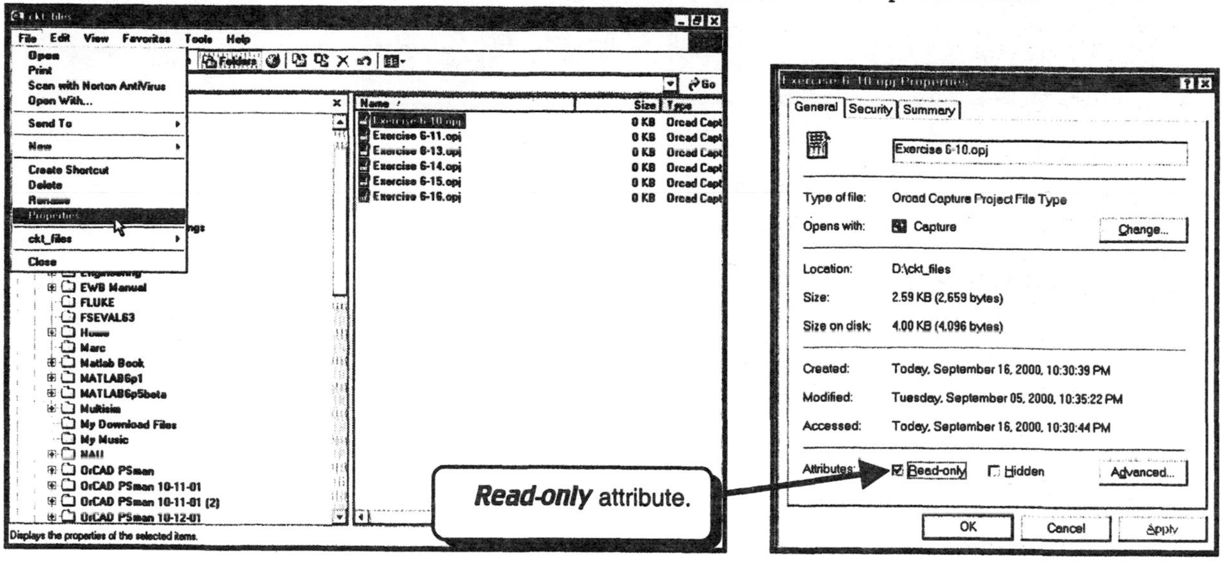

File Edit View Favorites Tools Help
Open
Print
Scan with Norton AntiVirus
Open With...
Send To
New
Create Shortcut
Delete
Rename
Properties
ckt_files
Close

EWB Manual
FLUKE
FSEVAL63
Home
Marc
Matlab Book
MATLAB6p1
MATLAB6p5beta
Multisim
My Download Files
My Music
NAII
OrCAD PSman
OrCAD PSman 10-11-01
OrCAD PSman 10-11-01 (2)
OrCAD PSman 10-12-01
Displays the properties of the selected items.

| Name | Size | Type |
|---|---|---|
| Exercise 6-10.opj | 0 KB | Orcad Capt |
| Exercise 6-11.opj | 0 KB | Orcad Capt |
| Exercise 6-13.opj | 0 KB | Orcad Capt |
| Exercise 6-14.opj | 0 KB | Orcad Capt |
| Exercise 6-15.opj | 0 KB | Orcad Capt |
| Exercise 6-16.opj | 0 KB | Orcad Capt |

***Read-only*** attribute.

Exercise 6-10.opj Properties
General | Security | Summary
Exercise 6-10.opj
Type of file: Orcad Capture Project File Type
Opens with: Capture Change...
Location: D:\ckt_files
Size: 2.59 KB (2,659 bytes)
Size on disk: 4.00 KB (4,096 bytes)
Created: Today, September 16, 2000, 10:30:39 PM
Modified: Tuesday, September 05, 2000, 10:35:22 PM
Accessed: Today, September 16, 2000, 10:30:44 PM
Attributes: ☑ Read-only ☐ Hidden Advanced...
OK Cancel Apply

We see that the ***Read-only*** attribute for this file is set. This will cause problems when we simulate the file. We need to disable the read-only attribute. Click the ***LEFT*** mouse button on the ☑ to toggle the checkmark off:

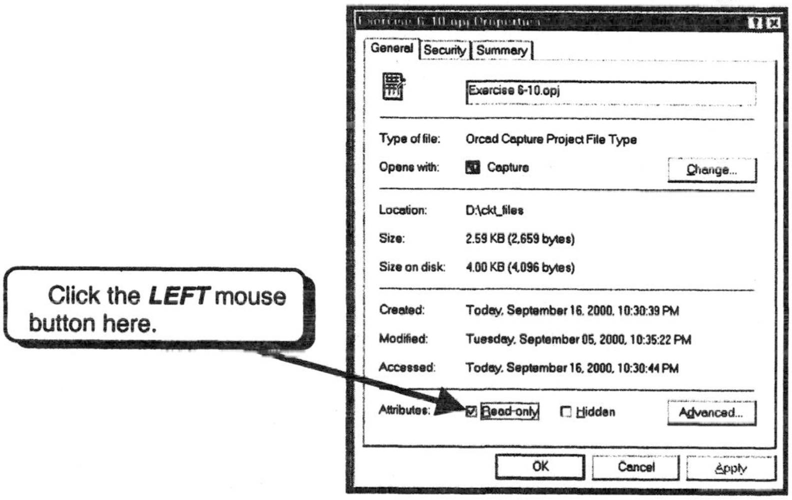

Click the ***LEFT*** mouse button here.

Exercise 6-10.opj Properties
General | Security | Summary
Exercise 6-10.opj
Type of file: Orcad Capture Project File Type
Opens with: Capture Change...
Location: D:\ckt_files
Size: 2.59 KB (2,659 bytes)
Size on disk: 4.00 KB (4,096 bytes)
Created: Today, September 16, 2000, 10:30:39 PM
Modified: Tuesday, September 05, 2000, 10:35:22 PM
Accessed: Today, September 16, 2000, 10:30:44 PM
Attributes: ☑ Read-only ☐ Hidden Advanced...
OK Cancel Apply

After clicking on the square, the checkmark should be removed:

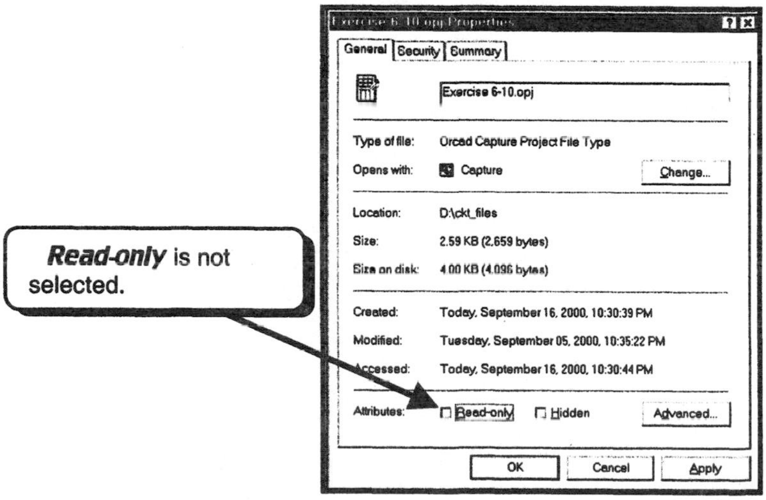

***Read-only*** is not selected.

Exercise 6-10.opj Properties
General | Security | Summary
Exercise 6-10.opj
Type of file: Orcad Capture Project File Type
Opens with: Capture Change...
Location: D:\ckt_files
Size: 2.59 KB (2,659 bytes)
Size on disk: 4.00 KB (4,096 bytes)
Created: Today, September 16, 2000, 10:30:39 PM
Modified: Tuesday, September 05, 2000, 10:35:22 PM
Accessed: Today, September 16, 2000, 10:30:44 PM
Attributes: ☐ Read-only ☐ Hidden Advanced...
OK Cancel Apply

In the screen capture above, the ***Read-only*** attribute is not selected. Click the ***OK*** button to accept the changes to the attributes. We can now use the file in a simulation. Note that you will need to disable the read-only attribute of all files you directly copy from the CD-ROM to your hard drive.

# APPENDIX B
# Scale Multipliers for PSpice and Capture

| SYMBOL [1] | SCALE | NAME |
|---|---|---|
| F | $10^{-15}$ | femto- |
| P | $10^{-12}$ | pico- |
| N | $10^{-9}$ | nano- |
| U | $10^{-6}$ | micro- |
| MIL | $25.4\times10^{-6}$ | |
| M | $10^{-3}$ | milli- |
| K | $10^{+3}$ | kilo- |
| MEG | $10^{+6}$ | mega- |
| G | $10^{+9}$ | giga- |
| T | $10^{+12}$ | tera- |
| C | | Clock cycle |

## Bibliography


[1] MicroSim Corporation. *The Design Center - Circuit Analysis - Reference Manual*, Version 6.1. Irvine, CA, July, 1994, p. 3-1.

# APPENDIX C
# Functions Available with Probe

| Function [1] | Meaning | Comments |
|---|---|---|
| ABS(x) | $\|x\|$ | |
| SGN(x) | +1 (if x>0), 0 (if x=0) –1 (if x<0) | |
| SQRT(x) | $x^{1/2}$ | |
| EXP(x) | $e^x$ | |
| LOG(x) | ln(x) | log base e |
| LOG10(x) | log(x) | log base 10 |
| M(x) | magnitude of x | |
| P(x) | phase of x | result in degrees |
| R(x) | real part of x | |
| IMG(x) | imaginary part of x | |
| G(x) | group delay of x | result in seconds |
| PWR(x,y) | $\|x\|^y$ | |
| SIN(x) | sin(x) | x in radians |
| COS(x) | cos(x) | x in radians |
| TAN(x) | tan(x) | x in radians |
| ATAN(x) | $\tan^{-1}(x)$ | result in radians |
| ARCTAN(x) | $\tan^{-1}(x)$ | result in radians |
| d(x) | derivative of x with respect to the x-axis variable | |
| s(x) | integral of x over the range of the x-axis variable | |
| AVG(x) | running average of x over the range of the x-axis variable | |
| AVGX(x,d) | running average of x (from x-d to x) over the range of the x-axis variable | |
| RMS(x) | running RMS average of x over the range of the x-axis variable | |
| DB(x) | magnitude of x in decibels | |
| MIN(x) | minimum of the real part of x | |
| MAX(x) | maximum of the real part of x | |

## Bibliography


[1] MicroSim Corporation. *The Design Center - Circuit Analysis - Reference Manual*, Version 6.1. Irvine, CA, July, 1994, pp. 8-17–8-18.

# APPENDIX D
# Schematic Errors

This appendix contains a brief discussion of the common errors encountered in drawing schematics. There are two types of errors you may encounter when you use Orcad Capture with PSpice. The first type are drawing errors, which are detected by the Capture program. When you create a netlist, any drawing errors will be detected by Capture and the circuit will not be simulated. Thus, all drawing errors must be corrected before the circuit can be simulated.

The second type are run-time errors, which the PSpice program detects. Once you have a circuit free of drawing errors, the PSpice program will run. Errors generated with PSpice are identified in the output file. The simulation will stop and you will be returned to the Capture program. To see the run-time errors you must look at the output file (select **PSpice** and then **View Output File** from the Capture menu bar).

## D.1. Drawing Errors

We will first look at drawing errors. We will start with the circuit below. This circuit has some obvious errors and a few you might never spot.

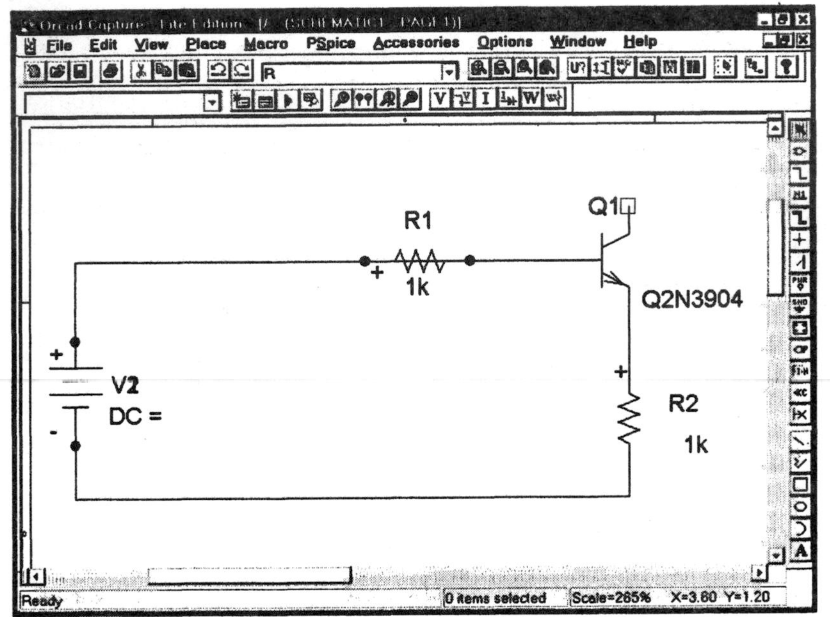


To check for errors, select **PSpice** and then **Create Netlist**:

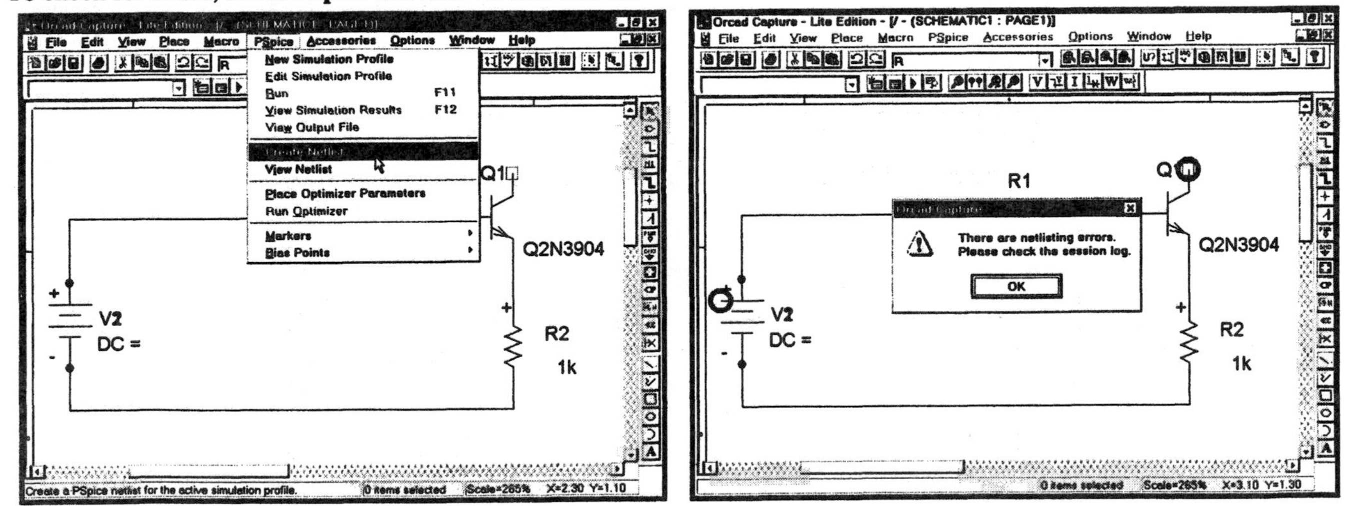


Creating a netlist performs an electrical rule check, so we shall attempt to create a netlist. Errors are indicated by the dialog box and on the circuit by the presence of the green washers. Click the ***OK*** button to close the dialog box.

To view the error indicated by a green washer, double-click the ***LEFT*** mouse button on the washer:

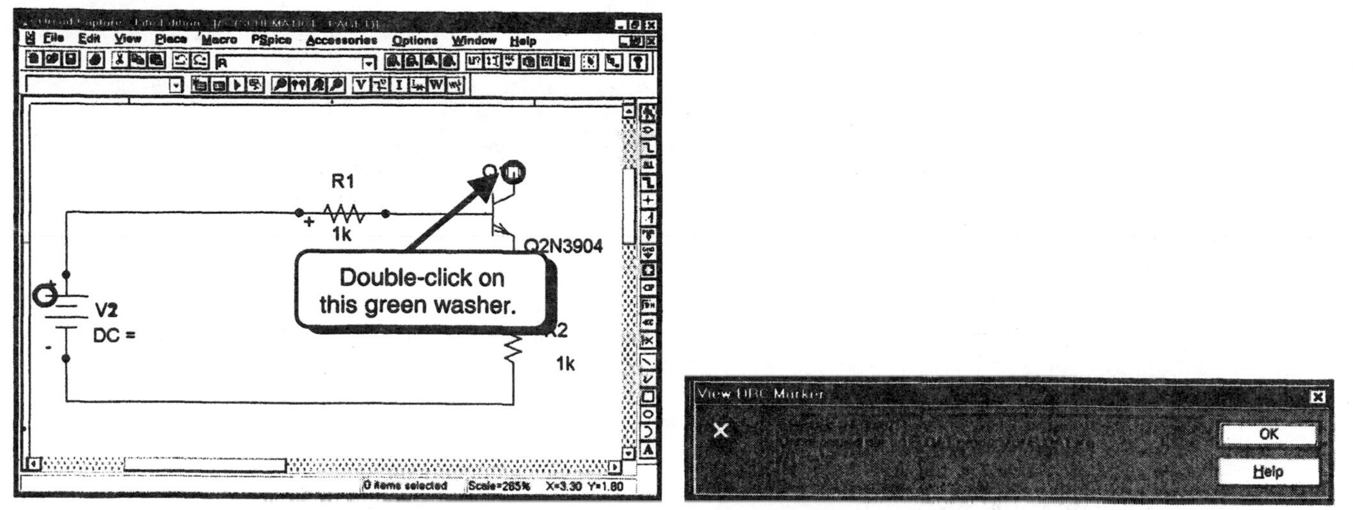


The dialog box states that pin ***C*** of ***Q1*** is not connected. This is an error we will correct. Click the ***OK*** button to close the dialog box:

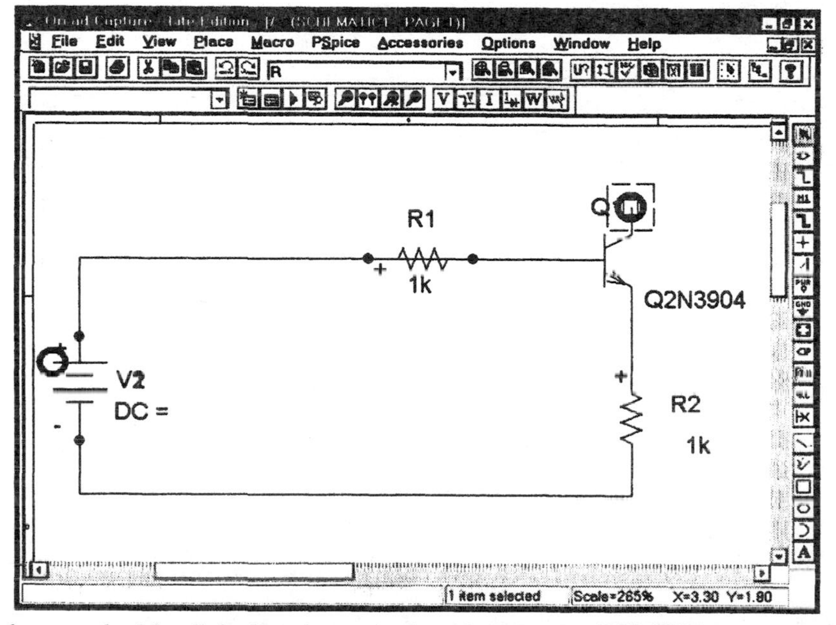


Note that the green washer we double-clicked on is now selected. Press the **DELETE** key to delete the washer from the schematic:

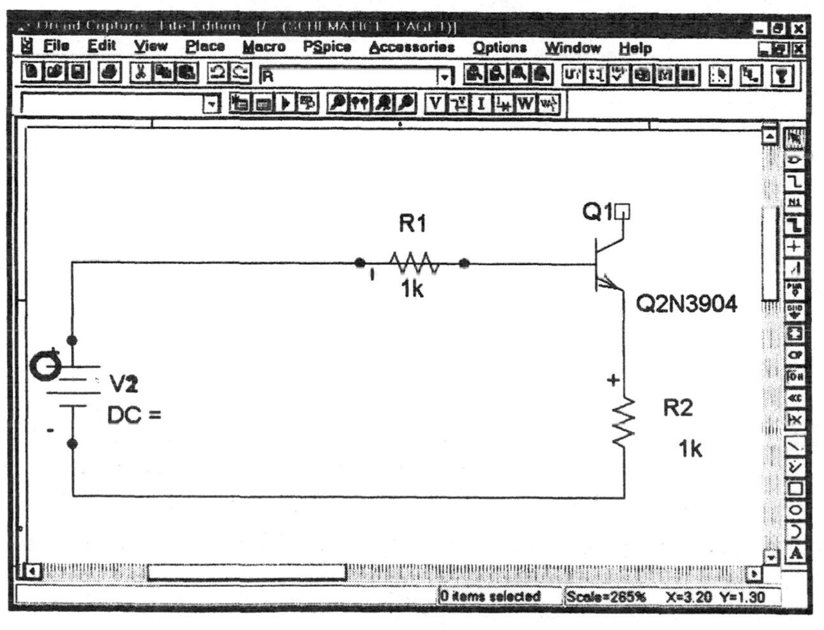

To fix the problem I will make a connection as shown below. Although this may not result in a useful circuit, it does fix the problem.

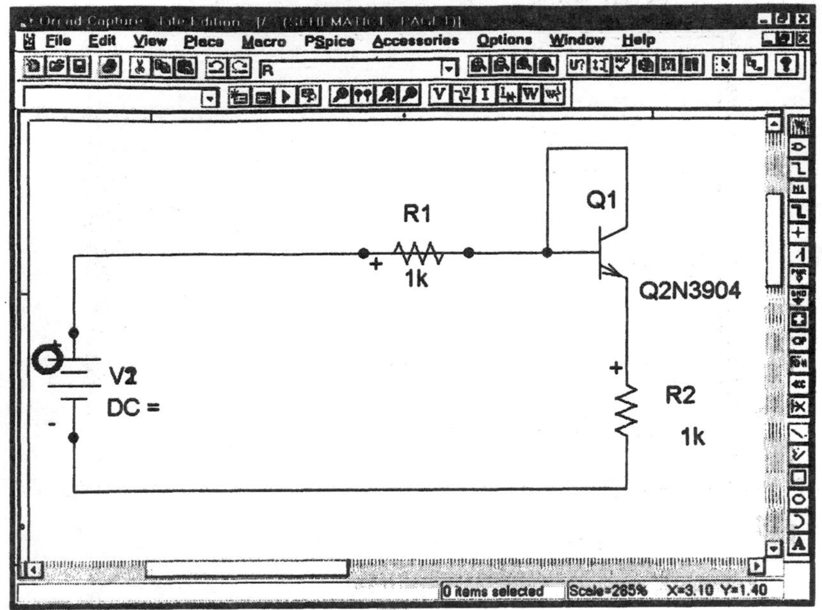


Next, double-click the ***LEFT*** mouse button on the green washer near the DC source:

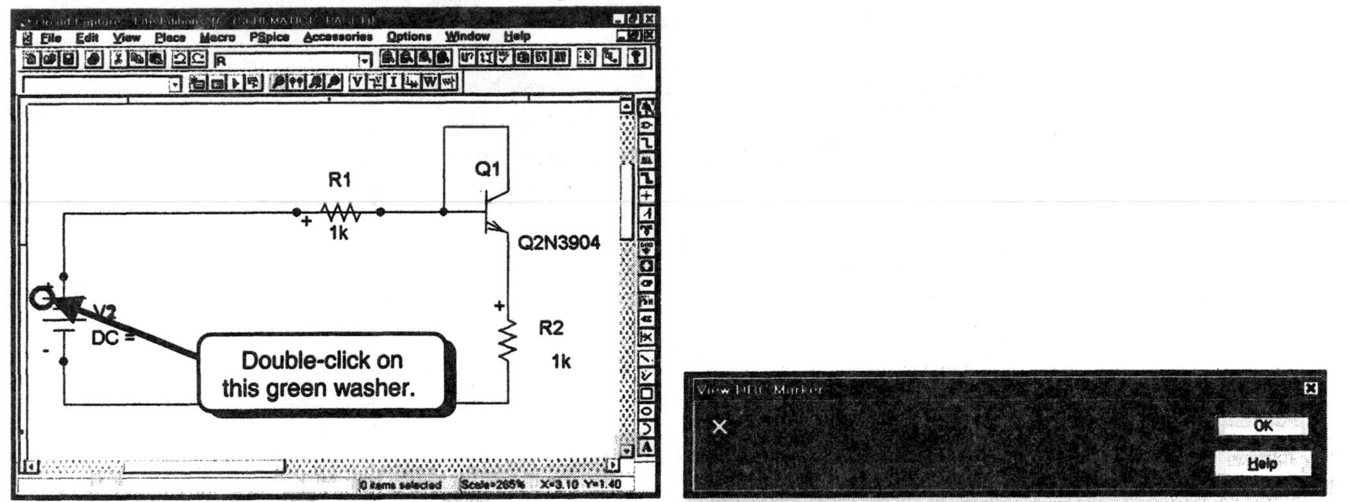


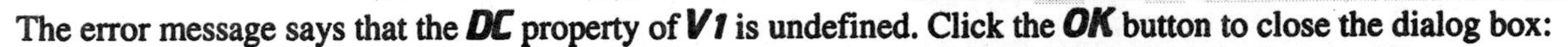

The error message says that the ***DC*** property of ***V1*** is undefined. Click the ***OK*** button to close the dialog box:

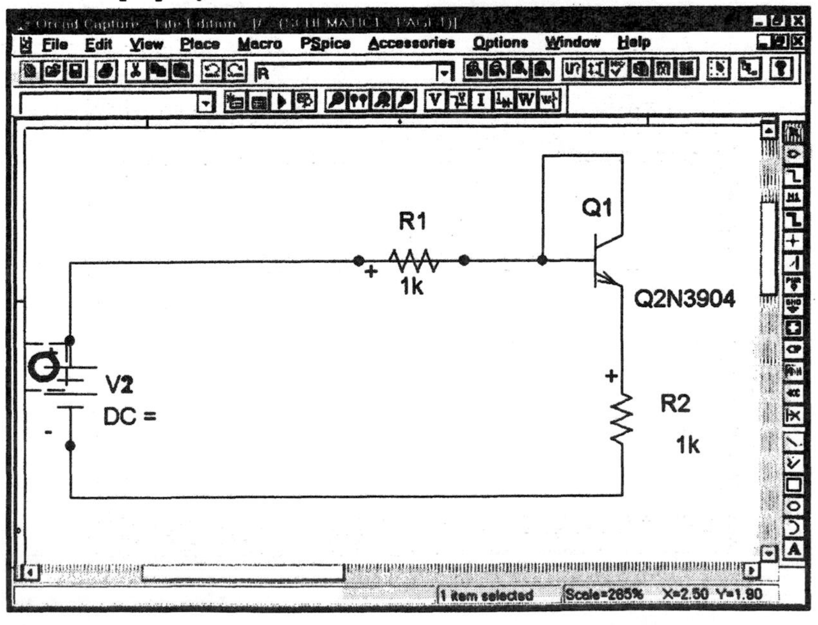

Note that the green washer is selected. Press the delete key to **DELETE** the washer:

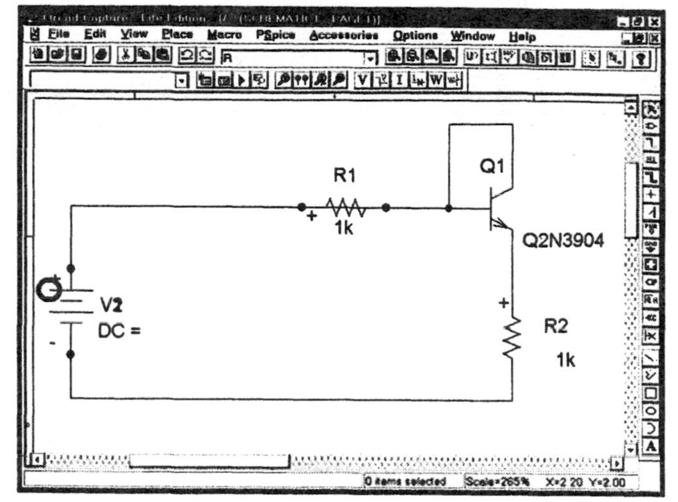


For some reason, the washer is still displayed on the schematic. Just for fun, double-click the ***LEFT*** mouse button on the green washer again:

The error message says that the ***DC*** property of ***V2*** is undefined. Click the ***OK*** button to close the dialog box:

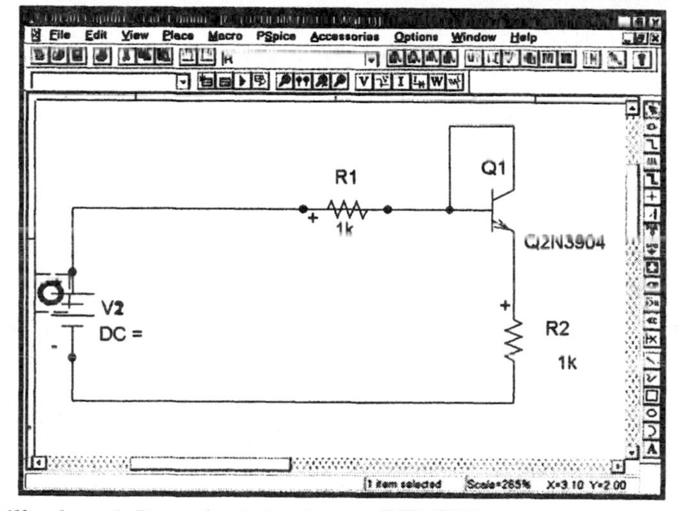


Note that the green washer is still selected. Press the delete key to **DELETE** the washer. This time the washer disappears:

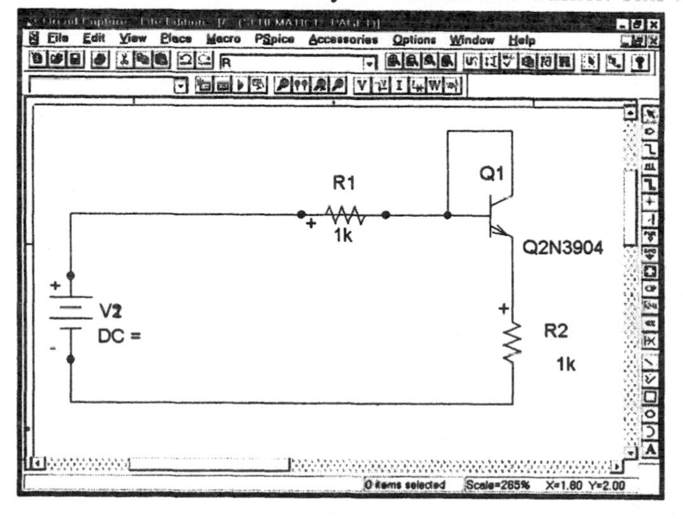

The DC attribute of this source must be specified before the circuit can be simulated. I will edit the attributes of the DC source and change the attribute to ***DC=12***, as shown below:

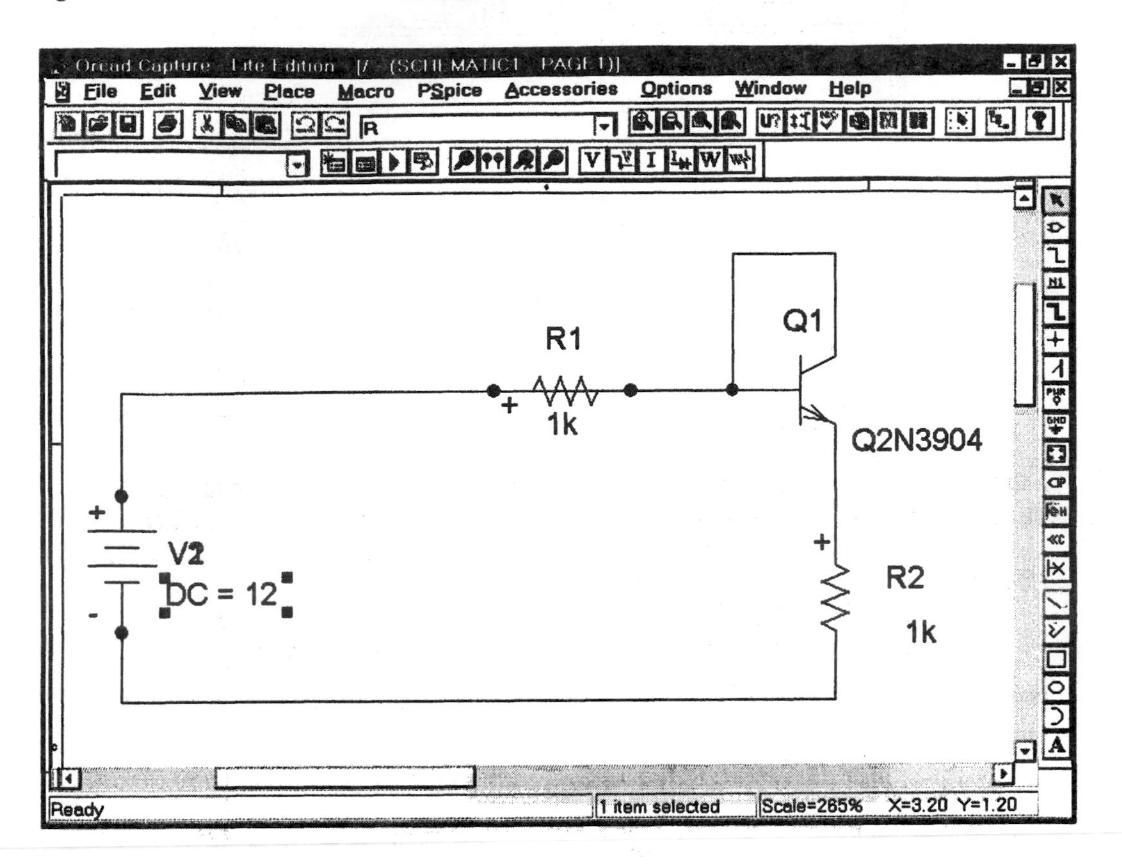


We now check the circuit again by creating a netlist. Once again there are errors:

When we double-click on the green washer we get the error message:

The error message says that the ***DC*** property of ***V2*** is still undefined. Click the ***OK*** button and then press the **DELETE** key to delete the washer.

The text of the DC source looks a little garbled. We will zoom in around the source:

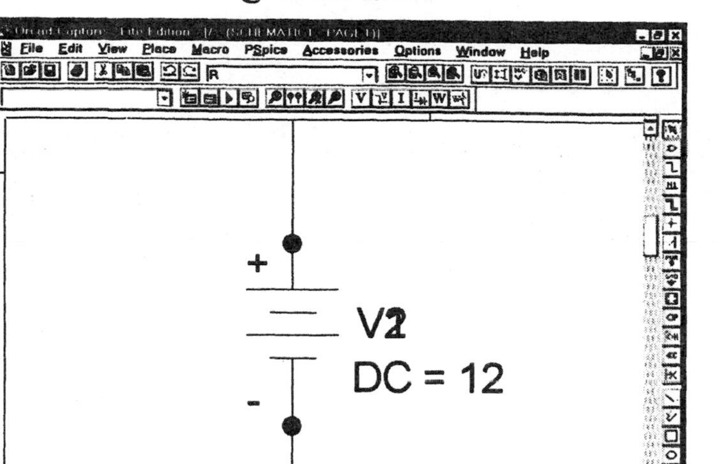


In the text V2, the 2 actually looks like a 2 written over a 1. If you look closely at the error dialog boxes on pages 610 and 611, you will notice that the messages are slightly different. One message says that V1 is the problem and the other message says that V2 is the problem. It turns out that there are actually two sources placed on top of one another. Click the ***LEFT*** mouse button on the dc source and drag it to a new location:

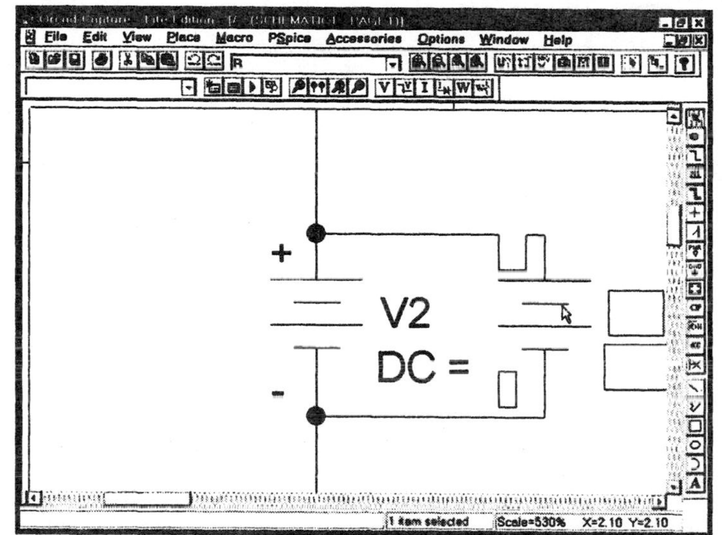


When we release the mouse button, it becomes clear that we had two sources placed on top of one another:

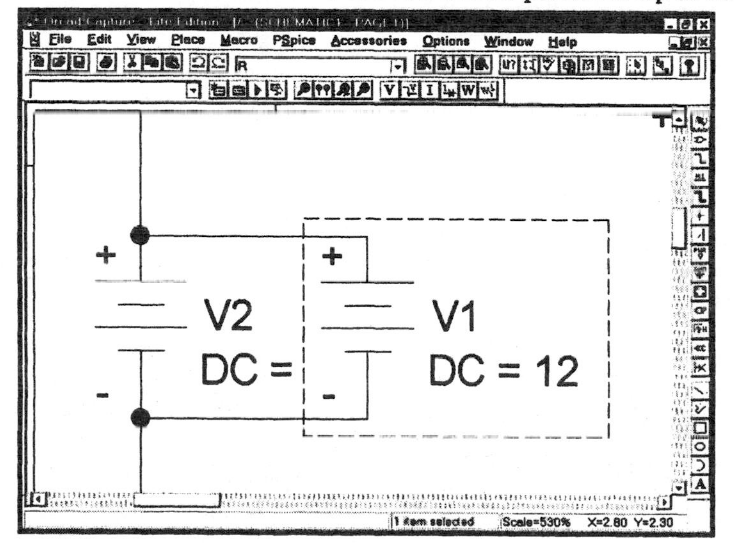


I will delete one of the sources to fix the problem:

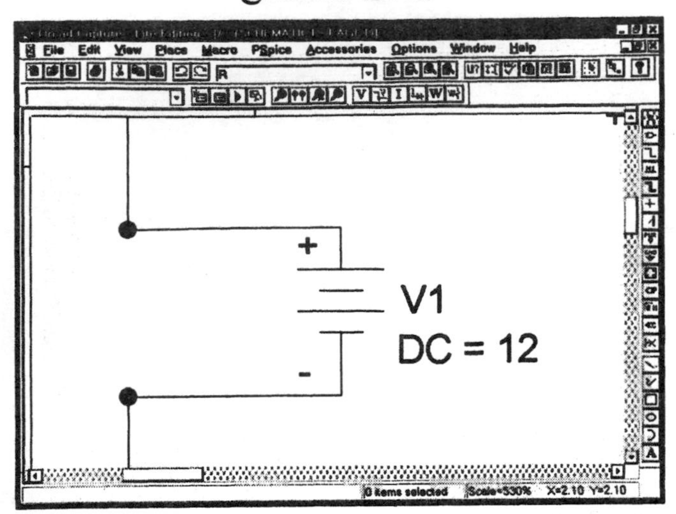


**This error was caused by two identical parts being placed directly on top of each other. It was impossible to see the bottom source because the two graphics were the same. This is a very hard problem to spot and can happen with any part, not just a voltage source.**

After we delete the duplicate DC source, the circuit is free of drawing errors as far as Capture is concerned:

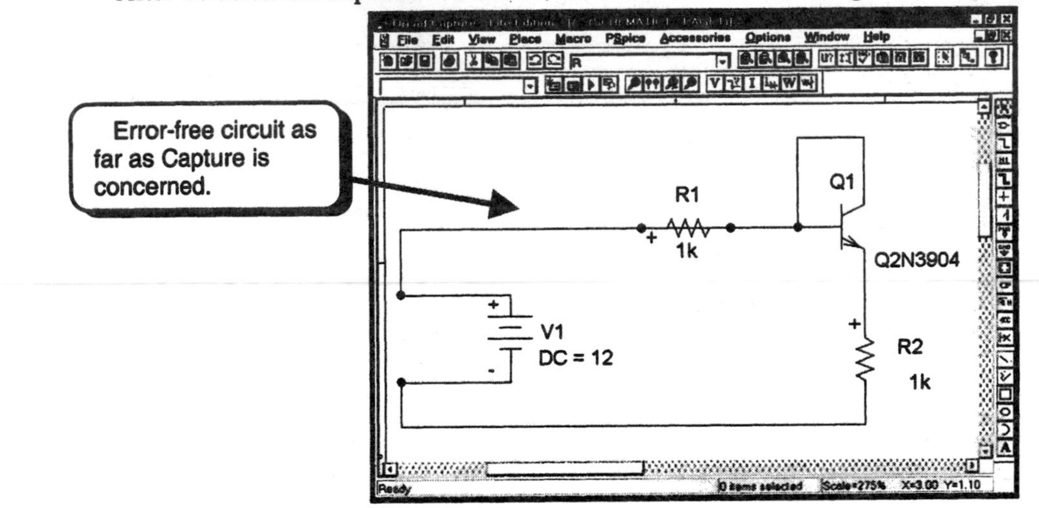


No more errors will be indicated when we create a netlist. However, there are still two errors in the drawing. The first is that the circuit is not grounded. The entire circuit is floating. This error would have been caught when we ran the simulation. You must add a part called 0 to your circuit. Select **Place** and then **Power** from the Capture menus to place the ground.

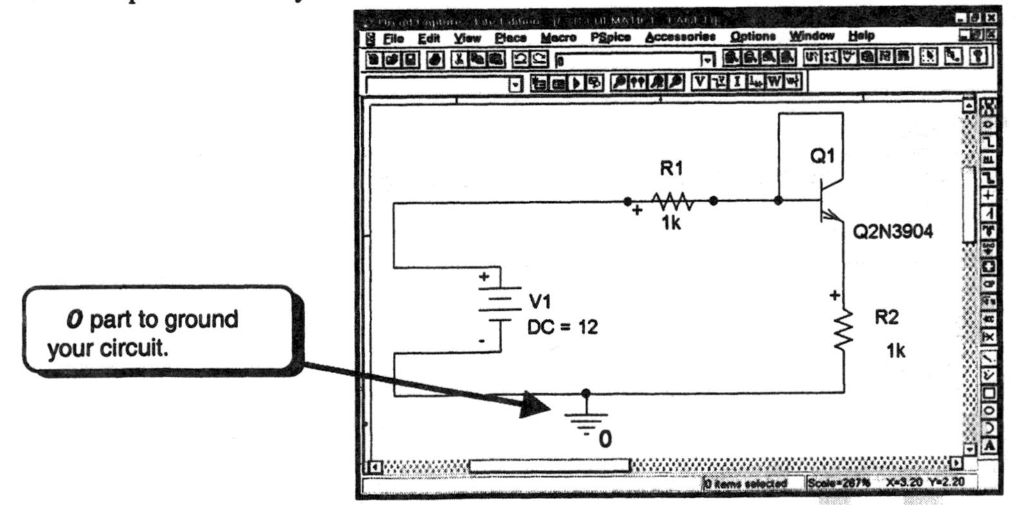


The last error is a little more subtle. If we zoom in around ***R1***, the error becomes more obvious:

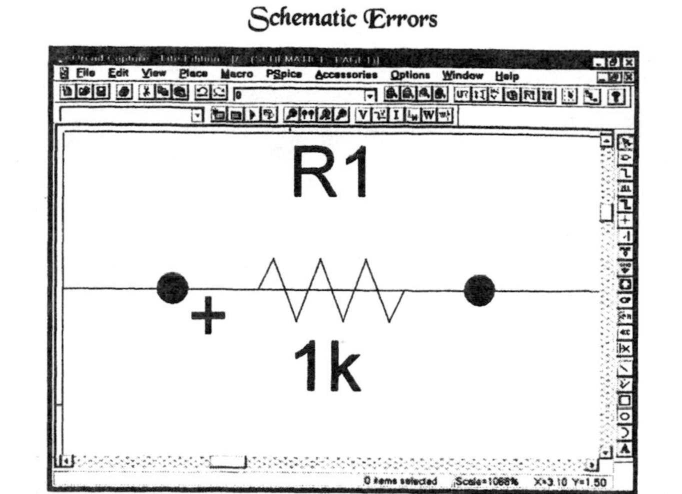


The wire through ***R1*** is shorting out the resistor. This wire was probably not intentional since ***R1*** will have no effect. This error would not be detected by either Capture or PSpice. This is a valid connection as far as the simulator is concerned. The simulation would run, but the results would be incorrect. The wire through ***R1*** should be deleted. To correct the problem you will have to delete the part, delete the wire, and then place the part in your circuit again. The correct appearance of the resistor is shown below:

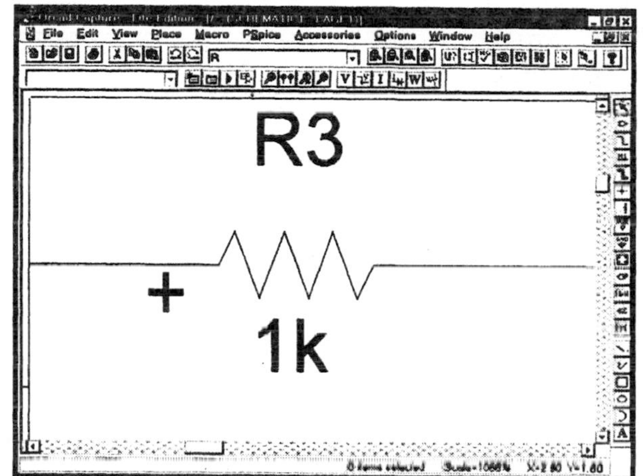


The resistor is shown as ***R3*** because I had to delete the original part and then place a new resistor in the circuit. The new resistor was automatically numbered ***R3***.

# D.2. PSpice Errors

## D.2.1. Floating Nodes

The circuit below has an error:

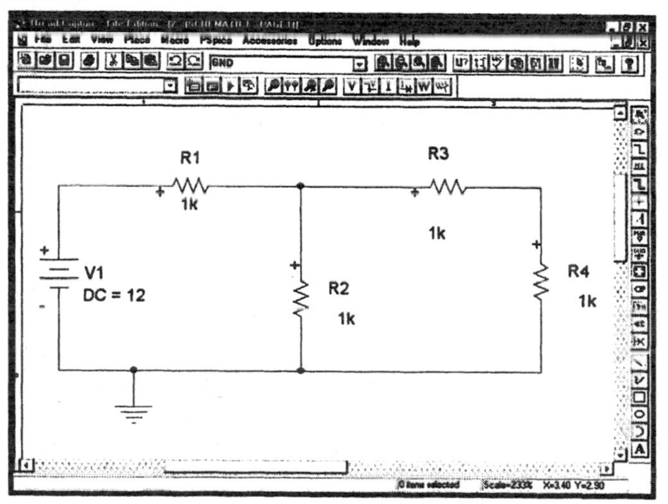

The circuit looks fine, so we will run a bias point analysis similar to the one covered in section 3.A. The error is displayed in the simulation window when we attempt to run the simulation:

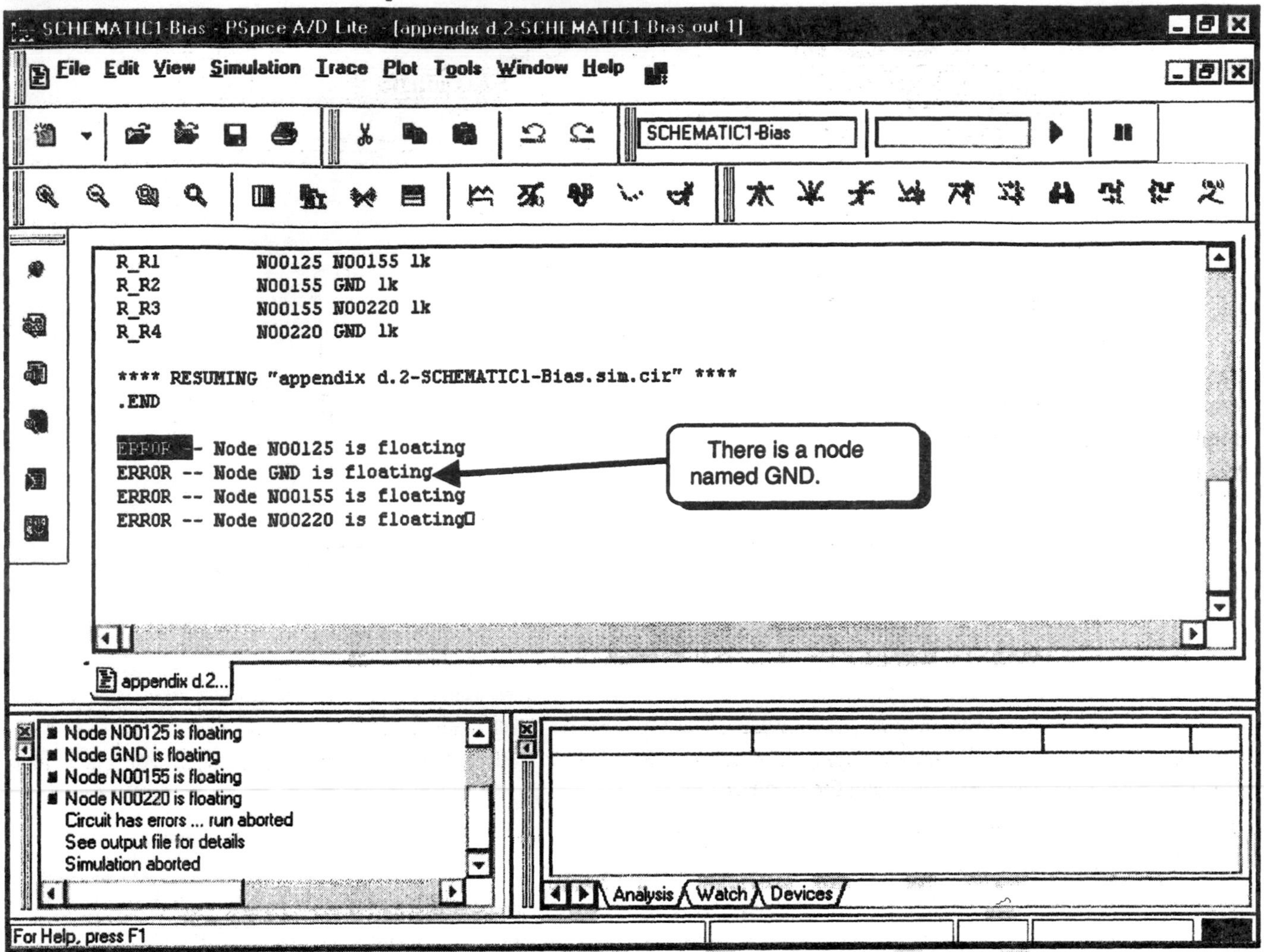

```
R_R1        N00125 N00155 1k
R_R2        N00155 GND 1k
R_R3        N00155 N00220 1k
R_R4        N00220 GND 1k

**** RESUMING "appendix d.2-SCHEMATIC1-Bias.sim.cir" ****
.END

ERROR -- Node N00125 is floating
ERROR -- Node GND is floating
ERROR -- Node N00155 is floating
ERROR -- Node N00220 is floating
```

There is a node named GND.

The errors indicate that all nodes are floating. When a node is floating, there is no DC path to ground. How can this be? Our circuit has a ground symbol in it. Remember that in PSpice, ground must be numbered as node 0 (the number zero). If we look in the error listing above, we see that there is a node called ***GND***. Our error is that we chose the wrong ground symbol to ground the circuit. You must always use a part called 0. In the ***Place Ground*** dialog box, the correct ground to use is displayed as ***0/SOURCE*** as shown below:

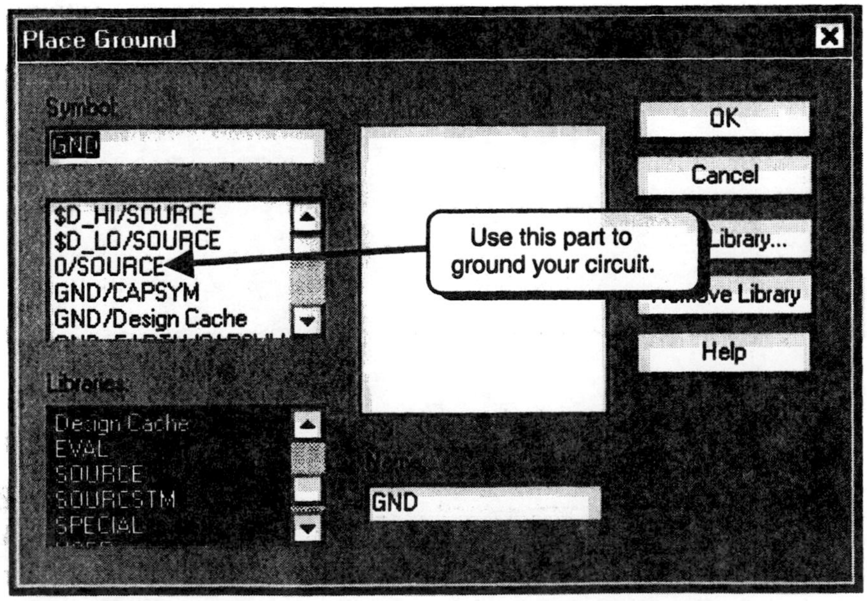

Use this part to ground your circuit.

To fix this problem, delete the ground symbol in your schematic and place the ***0/SOURCE*** part in your schematic to ground it. The solution is shown below:

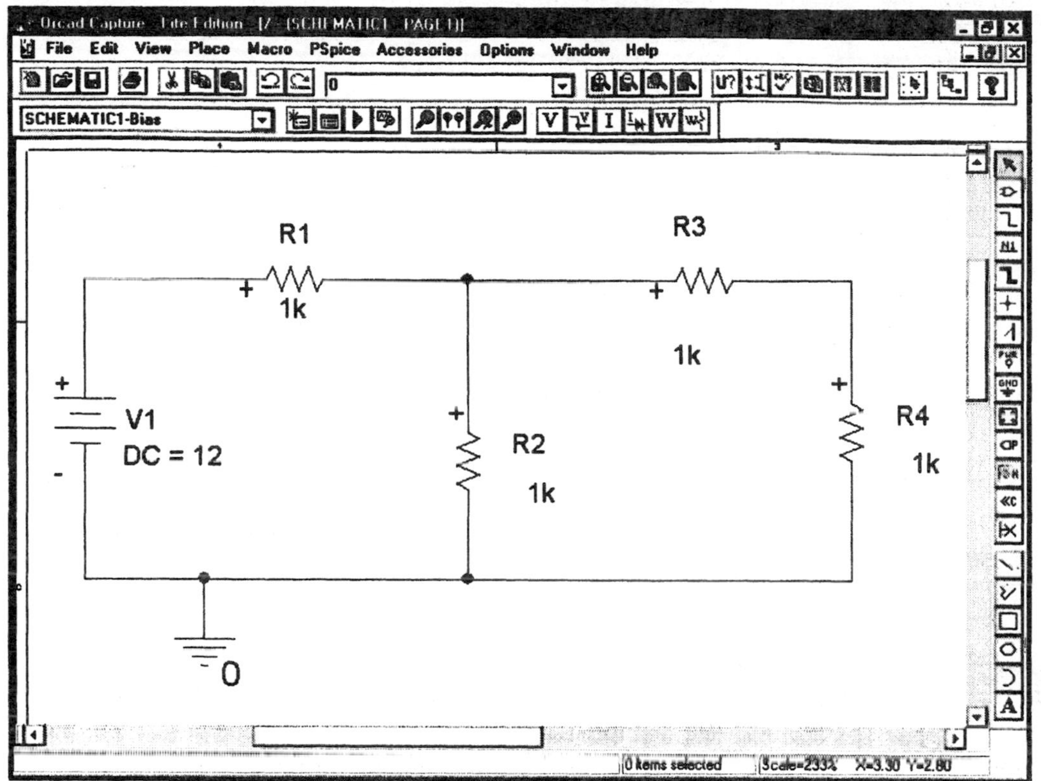


Note that the correct ground symbol displays the number ***0*** on the schematic. To place this ground, select **Place** and then **Ground** from the menus, and then select the 0/source part.

## D.2.2. Space Between a Number and Its Multiplier

The schematic below has an error:

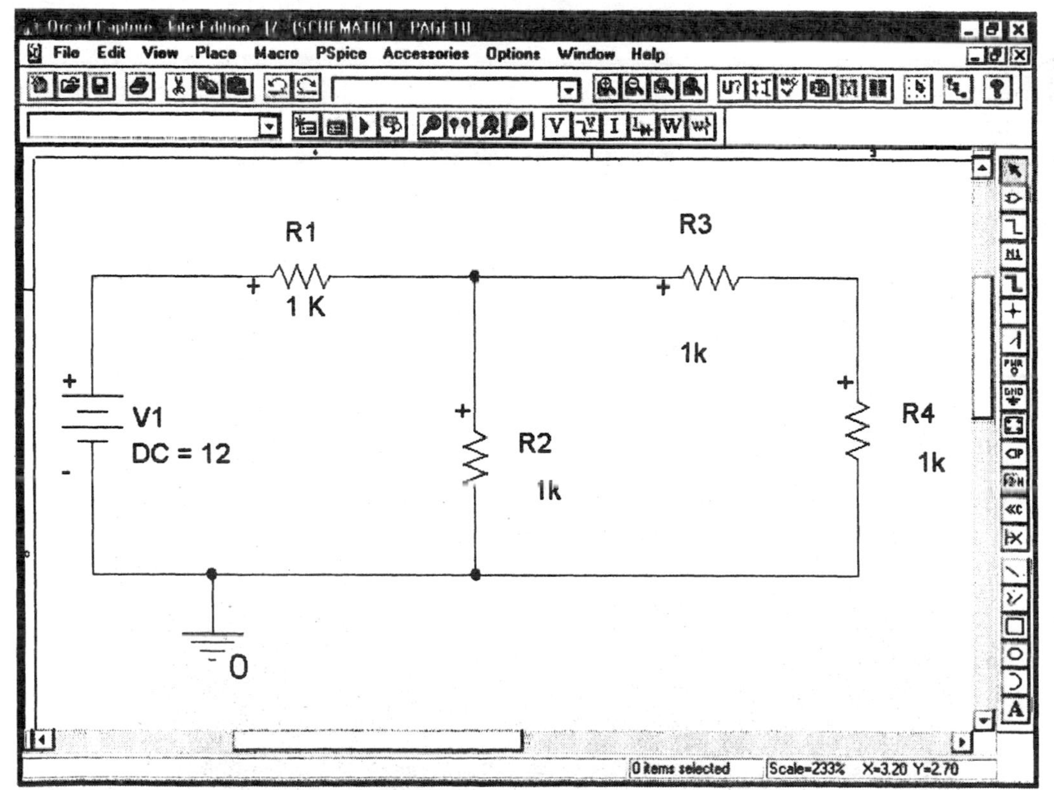


When we run the simulation we get the error below:

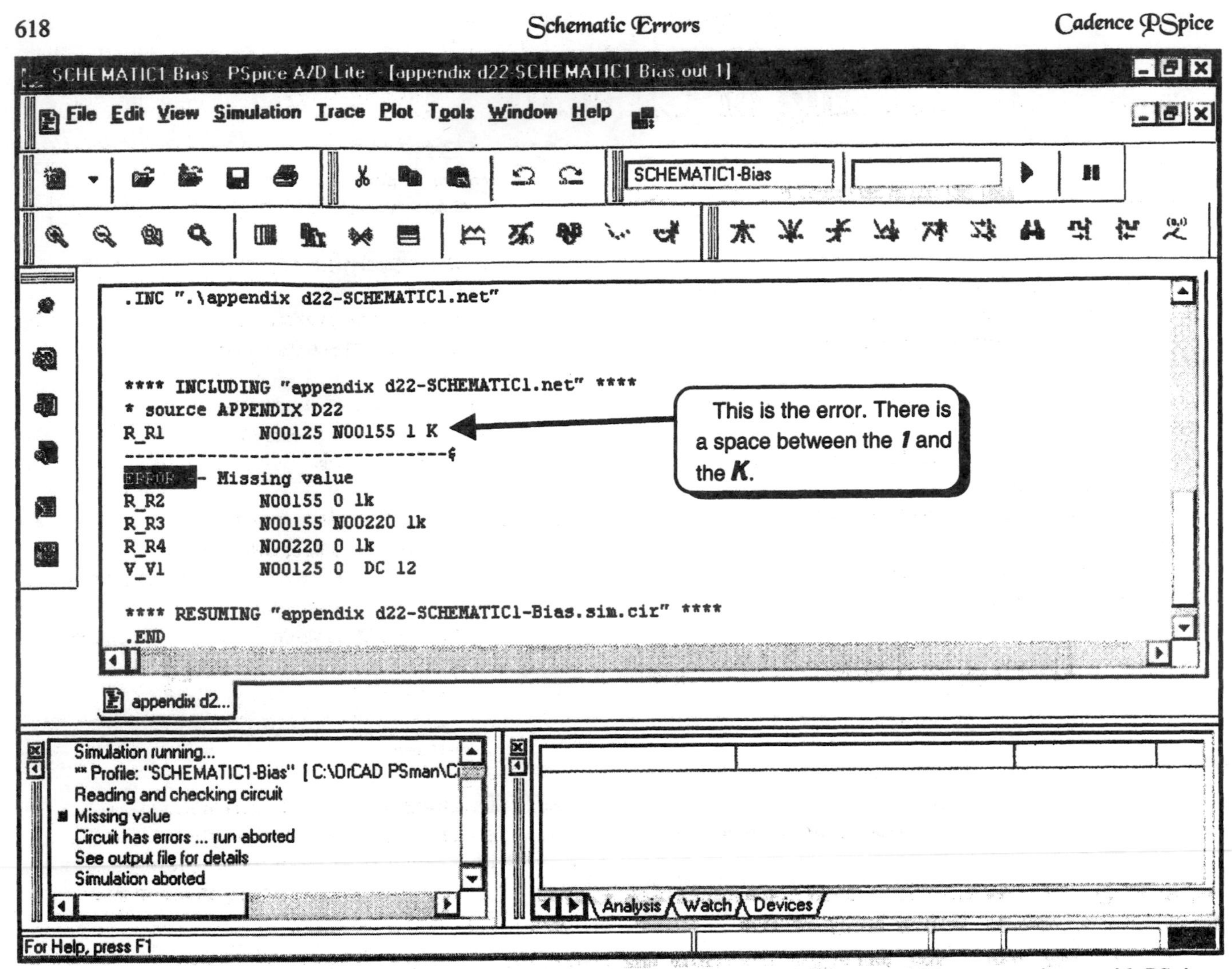


PSpice indicates that the error is for resistor R1. Since PSpice points directly to the error, and I have experience with PSpice, I see that there is a space between the ***1*** and the ***K***. There are not supposed to be any spaces between a number and its multiplier. The correct way to enter the number is 1K, not 1 K. Go back to the schematic, double-click the ***LEFT*** mouse button on the text to edit it, and change it to 1K. The correct value is shown below:

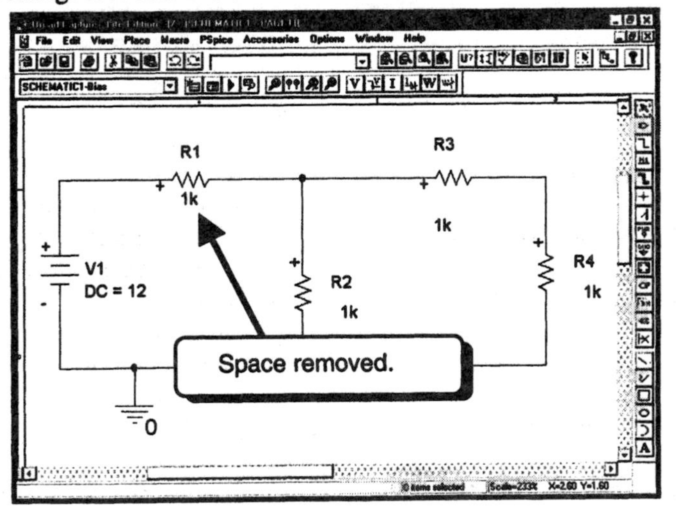


PSpice can now simulate the circuit.

# APPENDIX E
# Listing of Class.lib Library

```
*
.model D1N4734A D(Is=1.085f Rs=.7945 Ikf=0 N=1 Xti=3 Eg=1.11
Cjo=157p M=.2966
+          Vj=.75 Fc=.5 Isr=2.811n Nr=2 Bv=5.6 Ibv=.37157 Nbv=.64726
+          Ibvl=1m Nbvl=6.5761 Tbv1=267.86u)
*          Motorola     pid=1N4734     case=DO-41
*          89-9-19 gjg
*          Vz = 5.6 @ 45mA, Zz = 40 @ 1mA, Zz = 4.5 @ 5mA, Zz = 1.9
@ 20mA
.model D1N4148  D(Is=0.1p Rs=16 CJO=2p Tt=12n Bv=100 Ibv=0.1p)
*          85-??-??     Original library
.model D1N5401  D(Is=11.5f Rs=8.254m Ikf=3.87 N=1 Xti=3 Eg=1.11
Cjo=130.4p
+          M=.3758 Vj=.75 Fc=.5 Isr=79.29u Nr=2)
*          Motorola     pid=1N5400     case=267-01
*          88-09-12 rmn
.model D1N914   D(Is=0.1p Rs=16 CJO=2p Tt=12n Bv=100 Ibv=0.1p)
*          85-??-??     Original library
.model D1N4007  D(Is=0.1p Rs=4 CJO=2p Tt=3n Bv=1000 Ibv=0.1p)
*          85-??-??     Original library
.model D1N4001  D(Is=0.1p Rs=4 CJO=2p Tt=3n Bv=50 Ibv=0.1p)
*          85-??-??     Original library
.model D1N4004  D(Is=0.1p Rs=4 CJO=2p Tt=3n Bv=400 Ibv=0.1p)
*          85-??-??     Original library
*---------------------------------------------------------------------------
.MODEL Q2N2222    NPN(IS=3.108E-15 XTI=3 EG=1.11 VAF=131.5
+    BF=200 DEV/UNIFORM 150 NE=1.541
+     ISE=190.7E-15 IKF=1.296 XTB=1.5 BR=6.18 NC=2 ISC=0 IKR=0
RC=1
+     CJC=14.57E-12 VJC=.75 MJC=.3333 FC=.5 CJE=26.08E-12
VJE=.75
+     MJE=.3333 TR=51.35E-9 TF=451E-12 ITF=.1 VTF=10 XTF=2)
*.MODEL Q2N2907 PNP(IS=9.913E-15 XTI=3 EG=1.11 VAF=90.7
BF=197.8 NE=2.264
.MODEL Q2N2907  PNP(IS=9.913E-15 XTI=3 EG=1.11 VAF=90.7
+          BF=200
+          ISE=6.191E-12 IKF=.7322 XTB=1.5 BR=3.369 NC=2 ISC=0
IKR=0 RC=1
+          CJC=14.57E-12 VJC=.75 MJC=.3333 FC=.5 CJE=20.16E-12
VJE=.75
+          MJE=.3333 TR=29.17E-9 TF=405.7E-12 ITF=.4 VTF=10
XTF=2)
*          88-09-09 bam    creation
      pwt    change Rb
.model Q2N3904  NPN(Is=6.734f Xti=3 Eg=1.11 Vaf=74.03
+          Bf=416.4  Ne=1.259
+          Ise=6.734f Ikf=66.78m Xtb=1.5 Br=.7371 Nc=2 Isc=0 Ikr=0
Rc=1
+          Cjc=3.638p Mjc=.3085 Vjc=.75 Fc=.5 Cje=4.493p Mje=.2593
Vje=.75
+          Tr=239.5n Tf=301.2p Itf=.4 Vtf=4 Xtf=2 Rb=10)
*          National     pid=23      case=TO92
*          88-09-08 bam    creation
*
.model Q2N3906  PNP(Is=1.41f Xti=3 Eg=1.11 Vaf=18.7
+          Bf=180.7 Ne=1.5 Ise=0
+          Ikf=80m Xtb=1.5 Br=4.977 Nc=2 Isc=0 Ikr=0 Rc=2.5 Cjc=9.728p
+          Mjc=.5776 Vjc=.75 Fc=.5 Cje=8.063p Mje=.3677 Vje=.75
Tr=33.42n
+          Tf=179.3p Itf=.4 Vtf=4 Xtf=6 Rb=10)
*          National     pid=66      case=TO92
*          88-09-09 bam    creation
*
.model Q2N3904B NPN(Is=6.734f Xti=3 Eg=1.11 Vaf=74.03
+          Bf=416.4 DEV/UNIFORM 80% Ne=1.259
+          Ise=6.734f Ikf=66.78m Xtb=1.5 Br=.7371 Nc=2 Isc=0 Ikr=0
Rc=1
+          Cjc=3.638p Mjc=.3085 Vjc=.75 Fc=.5 Cje=4.493p Mje=.2593
Vje=.75
+          Tr=239.5n Tf=301.2p Itf=.4 Vtf=4 Xtf=2 Rb=10)
*          National     pid=23      case=TO92
*          88-09-08 bam    creation
*
.model Q2N3906B PNP(Is=1.41f Xti=3 Eg=1.11 Vaf=18.7
+          Bf=180.7 DEV/UNIFORM 73% Ne=1.5 Ise=0
+          Ikf=80m Xtb=1.5 Br=4.977 Nc=2 Isc=0 Ikr=0 Rc=2.5 Cjc=9.728p
+          Mjc=.5776 Vjc=.75 Fc=.5 Cje=8.063p Mje=.3677 Vje=.75
Tr=33.42n
+          Tf=179.3p Itf=.4 Vtf=4 Xtf=6 Rb=10)
*          National     pid=66      case=TO92
*          88-09-09 bam    creation
*
.model MPF102   NJF( Betatce=-.5 Rd=1 Rs=1 Lambda=2m
+     Vto=-3.41 Beta=1.04m
+          Vtotc=-2.5m Is=33.57f Isr=322.4f N=1 Nr=2 Xti=3 Alpha=311.7
+          Vk=243.6 Cgd=1.6p M=.3622 Pb=1 Fc=.5 Cgs=2.414p
Kf=11.73E-18
+          Af=1)
.model J2N5951  NJF( Betatce=-.5 Rd=1 Rs=1 Lambda=7.25m
+     Vto=-3.427 Beta=736.9u
+          Vtotc=-2.5m Is=33.57f Isr=322.4f N=1 Nr=2 Xti=3 Alpha=311.7
+          Vk=243.6 Cgd=1.6p M=.3622 Pb=1 Fc=.5 Cgs=2.414p
Kf=5.642E-18
+          Af=1)
*          National     pid=50      case=TO92
*          88-08-02 rmn    BVmin=30
*
*

* Create a transistor with Beta variations of 50 to 350.
.model QBf NPN( Bf=200 DEV/UNIFORM 150)

* QLM3046 model created using Parts version 4.04 on 01/01/80 at 02:25
*
```

```
.model QLM3046   NPN(Is=2p Xti=3 Eg=1.11 Vaf=16.64 Bf=128.3
Ise=17.62p Ne=1.667
+           Ikf=31.29m Nk=.4878 Xtb=1.5 Br=1 Isc=2p Nc=2 Ikr=0 Rc=0
+           Cjc=991.8f Mjc=.3333 Vjc=.75 Fc=.5 Cje=1.026p Mje=.3333
Vje=.75
+           Tr=10n Tf=274.7p Itf=.4434 Xtf=31.73 Vtf=10)

.subckt diff_pair c1 c2 b1 b2 e
Q1 c1 b1 e QLM3046
Q2 C2 B2 e QLM3046
.ends
*
*
*
*---------------------------------------------------------------------------
*---------------------------------------------------------------------------
.SUBCKT Ideal_OPAMP    v_plus v_minus v_out
*
* This is an ideal OpAmp model with the output voltage
* limited to +/- 15V.
*
Rin         v_plus  v_minus 10MEG
Eamp   v_out  0     value={LIMIT(1MEG*v(v_plus,v_minus),-15,15)}
R0          v_out  0      10MEG
.ENDS
*
*
*---------------------------------------------------------------------------
.SUBCKT Ideal_OPAMP_2   v_plus v_minus v_out
*
* This is an ideal OpAmp model with the output voltage
* limited to +/- 15V.
*
Rin         v_plus  v_minus 10MEG
Eamp   v_out  0     TABLE {1MEG*(v(v_plus,v_minus))}=(-11.25,-11.25)
(11.25,11.25)
R0          v_out  0      10MEG
.ENDS
*
*
*---------------------------------------------------------------------------
* The following models simulate the max and min transistors for
* a 2n5951 N-jFET
.model Jmax NJF(Vto=-5 Beta=.52m)
.model Jmin NJF(Vto=-2 Beta=1.75m)
*
* The following models simulate resistors with tolerances
*
* The two models below are 1% and 5% resistors with uniform
* probability distributions
.MODEL R1pcnt RES(R=1 DEV/UNIFORM 1%)
.MODEL R5pcnt RES(R=1 DEV/UNIFORM 5%)
* The models below are 1% and 5% resistors with Gaussian
* probability distributions
.MODEL R1gauss RES(R=1 DEV/gauss 0.25%)
.MODEL R5gauss RES(R=1 DEV/gauss 1.25%)
* THe following lines simulate a Capacitor with -10% and +50 %
* tolerance. Note that to achieve this asymmetric tolerance the
* nominal value is not the actual value.
.model CAP10_50 CAP(C=1.2 DEV/UNIFORM 25%)
* THe following lines define a -20 + 80% tolerance Capacitor.
.model CAP20_80 CAP(C=1.3 DEV/UNIFORM 38.461538%)
*
* UA741 operational amplifier "macromodel" subcircuit
* created using Parts release 4.01 on 07/05/89 at 09:09
* (REV N/A)
* connections:   non-inverting input
*                | inverting input
*                | | positive power supply
*                | | | negative power supply
*                | | | | output
*                | | | | |
.subckt UA741    1 2 3 4 5
*
  c1   11 12 4.664E-12
  c2    6  7 20.00E-12
  dc    5 53 dx
  de   54  5 dx
  dlp  90 91 dx
  dln  92 90 dx
  dp    4  3 dx
  egnd 99  0 poly(2) (3,0) (4,0) 0 .5 .5
  fb    7 99 poly(5) vb vc ve vlp vln 0 10.61E6 -10E6 10E6 10E6 -10E6
  ga    6  0 11 12 137.7E-6
  gcm   0  6 10 99 2.574E-9
  iee  10  4 dc 10.16E-6
  hlim 90  0 vlim 1K
  q1   11  2 13 qx
  q2   12  1 14 qx
  r2    6  9 100.0E3
  rc1   3 11 7.957E3
  rc2   3 12 7.957E3
  re1  13 10 2.740E3
  re2  14 10 2.740E3
  ree  10 99 19.69E6
  ro1   8  5 150
  ro2   7 99 150
  rp    3  4 18.11E3
  vb    9  0 dc 0
  vc    3 53 dc 2.600
  ve   54  4 dc 2.600
  vlim  7  8 dc 0
  vlp  91  0 dc 25
  vln   0 92 dc 25
.model dx D(Is=800.0E-18)
.model qx NPN(Is=800.0E-18 Bf=62.50)
.ends
* LF411C operational amplifier "macromodel" subcircuit
* created using Parts release 4.01 on 06/27/89 at 08:19
* (REV N/A)
* connections:   non-inverting input
*                | inverting input
*                | | positive power supply
*                | | | negative power supply
*                | | | | output
*                | | | | |
```

```
.subckt LF411C  1 2 3 4 5
*
 c1   11 12 3.498E-12
 c2   6  7 15.00E-12
 dc   5 53 dx
 de   54  5 dx
 dlp  90 91 dx
 dln  92 90 dx
 dp   4  3 dx
 egnd 99  0 poly(2) (3,0) (4,0) 0 .5 .5
 fb   7 99 poly(5) vb vc ve vlp vln 0 28.29E6 -30E6 30E6 30E6 -30E6
 ga   6  0 11 12 282.8E-6
 gcm  0  6 10 99 1.590E-9
 iss  3 10 dc 195.0E-6
 hlim 90  0 vlim 1K
 j1   11  2 10 jx
 j2   12  1 10 jx
 r2   6  9 100.0E3
 rd1  4 11 3.536E3
 rd2  4 12 3.536E3
 ro1  8  5 50
 ro2  7 99 25
 rp   3  4 15.00E3
 rss  10 99 1.026E6
 vb   9  0 dc 0
 vc   3 53 dc 2.200
 ve   54  4 dc 2.200
 vlim 7  8 dc 0
 vlp  91  0 dc 30
 vln  0 92 dc 30
.model dx D(Is=800.0E-18)
.model jx PJF(Is=12.50E-12 Beta=250.1E-6 Vto=-1)
.ends
* LM301A operational amplifier "macromodel" subcircuit
* created using Parts release 4.01 on 09/01/89 at 13:14
* (REV N/A)
* connections:   non-inverting input
*                | inverting input
*                | | positive power supply
*                | | | negative power supply
*                | | | | output
*                | | | | |  compensation
*                | | | | | /\
.subckt LM301A  1 2 3 4 5 6 7
*
 c1   11 12 7.977E-12
 dc   5 53 dx
 de   54  5 dx
 dlp  90 91 dx
 dln  92 90 dx
 dp   4  3 dx
 egnd 99  0 poly(2) (3,0) (4,0) 0 .5 .5
 fb   7 99 poly(5) vb vc ve vlp vln 0 42.44E6 -40E6 40E6 40E6 -40E6
 ga   6  0 11 12 188.5E-6
 gcm  0  6 10 99 3.352E-9
 iee  10  4 dc 15.14E-6
 hlim 90  0 vlim 1K
 q1   11  2 13 qx
 q2   12  1 14 qx
 r2   6  9 100.0E3
 rc1  3 11 5.305E3
 rc2  3 12 5.305E3
 re1  13 10 1.839E3
 re2  14 10 1.839E3
 ree  10 99 13.21E6
 ro1  8  5 50
 ro2  7 99 25
 rp   3  4 16.81E3
 vb   9  0 dc 0
 vc   3 53 dc 2.600
 ve   54  4 dc 2.600
 vlim 7  8 dc 0
 vlp  91  0 dc 25
 vln  0 92 dc 25
.model dx D(Is=800.0E-18)
.model qx NPN(Is=800.0E-18 Bf=107.1)
.ends
* LM324 operational amplifier "macromodel" subcircuit
* created using Parts release 4.01 on 09/08/89 at 10:54
* (REV N/A)
* connections:   non-inverting input
*                | inverting input
*                | | positive power supply
*                | | | negative power supply
*                | | | | output
*                | | | | |
.subckt LM324   1 2 3 4 5
*
 c1   11 12 5.544E-12
 c2   6  7 20.00E-12
 dc   5 53 dx
 de   54  5 dx
 dlp  90 91 dx
 dln  92 90 dx
 dp   4  3 dx
 egnd 99  0 poly(2) (3,0) (4,0) 0 .5 .5
 fb   7 99 poly(5) vb vc ve vlp vln 0 15.91E6 -20E6 20E6 20E6 -20E6
 ga   6  0 11 12 125.7E-6
 gcm  0  6 10 99 7.067E-9
 iee  3 10 dc 10.04E-6
 hlim 90  0 vlim 1K
 q1   11  2 13 qx
 q2   12  1 14 qx
 r2   6  9 100.0E3
 rc1  4 11 7.957E3
 rc2  4 12 7.957E3
 re1  13 10 2.773E3
 re2  14 10 2.773E3
 ree  10 99 19.92E6
 ro1  8  5 50
 ro2  7 99 50
 rp   3  4 30.31E3
 vb   9  0 dc 0
 vc   3 53 dc 2.100
 ve   54  4 dc .6
 vlim 7  8 dc 0
```

```
 vlp 91 0 dc 40
 vln 0 92 dc 40
.model dx D(Is=800.0E-18)
.model qx PNP(Is=800.0E-18 Bf=250)
.ends
*-------------------------------------------------------------LM7915C
.SUBCKT LM7915C Input Output Ground
  x1 Input Output Ground x_LM79XX PARAMS:
+   Av_feedback=555, R1_Value=13845,
+   Rbg_Tc1=-9.50E-7, Rbg_Tc2=-6.53E-7,
+   Rout_Value=0.01, Rreg_Value=11.3k
.ENDS
*-------------------------------------------------------------LM7918C
.SUBCKT LM7815C Input Output Ground
  x1 Input Output Ground x_LM78XX PARAMS:
+   Av_feedback=550, R1_Value=3060
.ENDS
*-------------------------------------------------------------LM7805C
.SUBCKT LM7805C Input Output Ground
  x1 Input Output Ground x_LM78XX PARAMS:
+   Av_feedback=1665, R1_Value=1020
.ENDS
*
*
*
*
*** Voltage regulators (positive)

.SUBCKT x_LM78XX Input Output Ground PARAMS:
+    Av_feedback=1665, R1_Value=1020
*
* SERIES 3-TERMINAL POSITIVE REGULATOR
*
* Note: This regulator is based on the LM78XX series of
*    regulators (also the LM140 and LM340). The model
*    will cause some current to flow to Node 0 which
*    is not part of the actual voltage regulator circuit.
*
* Band-gap voltage source:
*
*    The source is off when Vin<3V and fully on when Vin>3.7V.
*    Line regulation and ripple rejection) are set with
*    Rreg= 0.5 * dVin/dVbg. The temperature dependence of this
*    circuit is a quadratic fit to the following points:
*
*            T       Vbg(T)/Vbg(nom)
*           ---      ---------------
*            0         .999
*           37.5        1
*           125        .990
*
*    The temperature coefficient of Rbg is set to 2 * the band gap
*    temperature coefficient. Tnom is assumed to be 27 deg. C and
*    Vnom is 3.7V
*
Vbg 100 0 DC 7.4V
Sbg (100,101) (Input,Ground) Sbg1
Rbg 101 0 1 TC=1.612E-5,-2.255E-6
Ebg (102,0) (Input,Ground) 1
Rreg 102 101 7k
.MODEL Sbg1 VSWITCH (Ron=1 Roff=1MEG Von=3.7 Voff=3)
*
* Feedback stage
*
*    Diodes D1,D2 limit the excursion of the amplifier
*    outputs to being near the rails. Rfb, Cfb Set the
*    corner frequency for roll-off of ripple rejection.

*
*    The opamp gain is given by:  Av = (Fores/Freg) * (Vout/Vbg)
*    where Fores = output impedance corner frequency
*              with Cl=0 (typical value about 1MHz)
*       Freg  = corner frequency in ripple rejection
*              (typical value about 600 Hz)
*       Vout  = regulator output voltage (5,12,15V)
*       Vbg   = bandgap voltage (3.7V)
*
*    Note: Av is constant for all output voltages, but the
*    feedback factor changes. If Av=2250, then the
*    Av*Feedback factor is as given below:
*
*              Vout    Av*Feedback factor
*              ----    ------------------
*               5        1665
*               12        694
*               15        550
*
Rfb 9 8 1MEG
Cfb 8 Ground 265PF
* Eopamp 105 0 VALUE={2250*v(101,0)+Av_feedback*v(Ground,8)}
Vgainf 200 0 {Av_feedback}
Rgainf 200 0 1
Eopamp 105 0 POLY(3) (101,0) (Ground,8) (200,0) 0 2250 0 0 0 0 0 0 1
Ro 105 106 1k
D1 106 108 Dlim
D2 107 106 Dlim
.MODEL Dlim D (Vj=0.7)
Vl1 102 108 DC 1
Vl2 107 0 DC 1
*
* Quiescent current modelling
*
*    Quiescent current is set by Gq, which draws a current
*    proportional to the voltage drop across the regulator and
*    R1 (temperature coefficient .1%/deg C). R1 must change
*    with output voltage as follows:  R1 = R1(5v) * Vout/5v.
*
Gq (Input,Ground) (Input,9) 2.0E-5
R1 9 Ground {R1_Value} TC=0.001
*
* Output Stage
*
*    Rout is used to set both the low frequency output impedence
*    and the load regulation.
*
```

```
Q1 Input 5 6 Npn1
Q2 Input 6 7 Npn1 10
.MODEL Npn1 NPN (Bf=50 Is=1E-14)
* Efb Input 4 VALUE={v(Input,Ground)+v(0,106)}
Efb Input 4 POLY(2) (Input,Ground) (0,106) 0 1 1
Rb 4 5 1k TC=0.003
Re 6 7 2k
Rsc 7 9 0.275 TC=1.136E-3,-7.806E-6
Rout 9 Output 0.008
*
* Current Limit
*
Rbcl 7 55 290
Qcl 5 55 9 Npn1
Rcldz 56 55 10k
Dz1 56 Input Dz
.MODEL Dz D (Is=0.05p Rs=3 Bv=7.11 Ibv=0.05u)
.ENDS
*
*----------------------------------------------------------------LM7815C
*** Voltage regulators (negative)

.SUBCKT x_LM79XX Input Output Ground PARAMS:
+     Av_feedback=1660, R1_Value=4615,
+     Rbg_Tc1=8.13E-5, Rbg_Tc2=0.0,
+     Rout_Value=0.01, Rreg_Value=1.2k
*
* SERIES 3-TERMINAL NEGATIVE REGULATOR
*
* Note: This regulator is based on the LM79XX series of
*     regulators (also the LM120 and LM320). The
*     LM79XX regulators are unstable and will
*     oscillate unless a 1 uFarad solid tantalum
*     capacitor is placed on the output with an ESR
*     betweed .5 and 1.5. This model is stable without
*     a capacitor on the output. When performing
*     simulations a 1 uFarad capacitor should still be
*     placed on the output. However, it it not necessary
*     to include a resistor in series with this capacitor
*     to model the ESR of the capacitor. See the
*     comments and circuit description of the x_LM78XX
*     regulator for more information on this model.
*
* Band-gap voltage source:
*
Vbg 100 0 DC -7.4V
Sbg (100,101) (Ground,Input) Sbg1
Rbg 101 0 Rbg1 1
.MODEL Rbg1 RES (Tc1={Rbg_Tc1},Tc2={Rbg_Tc2})
Ebg (102,0) (Input,Ground) 1
Rreg 102 101 {Rreg_Value}
.MODEL Sbg1 VSWITCH (Ron=1 Roff=1MEG Von=3.7 Voff=3)
*
* Feedback stage
*
Rfb 9 8 1MEG
Cfb 8 Ground 265PF
* Eopamp 105 0 VALUE={2250*v(101,0)+Av_feedback*v(Ground,8)}
Vgainf 200 0 {Av_feedback}
Rgainf 200 0 1
Eopamp 105 0 POLY(3) (101,0) (Ground,8) (200,0) 0 2250 0 0 0 0 0 0 1
Ro 105 106 1k
D1 108 106 Dlim
D2 106 107 Dlim
.MODEL Dlim D (Vj=0.7)
Vl1 108 102 DC 1
Vl2 0 107 DC 1
*
* Quiescent current modelling
*
Gq (Ground,Input) (9,Input) 9.0E-7
R1 9 Ground {R1_Value} TC=0.001
Fl (Ground,0) Vmon 3.0E-4
*
* Output Stage
*
Q1 9 5 6 Npn1
Q2 9 6 7 Npn1 10
.MODEL Npn1 NPN (Bf=50 Is=1E-14)
* Efb 4 Ground VALUE={v(Input,Ground)+v(0,106)}
Efb 4 Ground POLY(2) (Input,Ground) (0,106) 0 1 1
Rb 4 5 1k TC=0.003
Re 6 7 2k
Rsc 7 Input 0.13 TC=1.136E-3,-7.806E-6
Rout 9 Imon {Rout_Value}
Vmon Imon Output DC 0.0
*
* Current Limit
*
Qcl1 54 52 53 Npn1
Qcl3 Input 54 5 Pnp1
.MODEL Pnp1 PNP (Bf=250 Is=1E-14)
Rcl3 5 54 1.8k
Qcl2 52 52 51 Npn1
Veset 53 Input DC 0.3v
Ibias Input 52 DC 300u
Rcl1 50 51 20k
Rcl2 51 7 115
Dz1 50 9 Dz
.MODEL Dz D (Is=0.05p Rs=3 Bv=7.11 Ibv=0.05u)
.ENDS
*


*
* Digital Circuits for EGR482
* TTL Circuit Simulation
*
.subckt ttl_gate in out VCC
Q1   4   2   in   Q2n2222
Q2   6   4   5    Q2n2222
Q3   out 5   0    Q2n2222
Q4   7   6   8    Q2n2222
Q5   8   8   out  Q2n2222
Rb   Vcc 2   4k
Re   5   0   1k
Rc4  Vcc 6   1.4k
```

```
Rc2  Vcc  7   100
.ends
*
* DTL Circuit
.subckt dtl_gate in out Vcc
Q1    2    1    3    Q2n2222
Q2    out  4    0    Q2n2222
Q3    3    3    4    Q2n2222
Q4    1    1    IN   Q2n2222
R1    Vcc  2    1.6K
R2    2    1    2.15K
Rb    4    0    5k
Rc    Vcc  out  2K
.ends
*
* RTL Circuit
*
.subckt rtl_gate in out Vcc
Rb    in   1    10K
Rc    Vcc  out  1K
Q1    out  1    0    Q2n2222
.ends
*
* Power Mosfets
*
.model IRFD1Z3  NMOS(Level=3 Gamma=0 Delta=0 Eta=0 Theta=0
Kappa=0 Vmax=0 Xj=0
+        Tox=100n Uo=600 Phi=.6 Rs=2.063 Kp=21.17u W=.2 L=2u
Vto=3.936
+        Rd=.6512 Rds=266.7K Cbd=84.39p Pb=.8 Mj=.5 Fc=.5
Cgso=932.5p
+        Cgdo=116.3p Rg=26.48 Is=1.135p N=1 Tt=470n)
*        Int'l Rectifier  pid=IRFC1Z0    case=4 Pin DIP
*        88-08-26 bam    creation
.model IRFD9113 PMOS(Level=3 Gamma=0 Delta=0 Eta=0 Theta=0
Kappa=0 Vmax=0 Xj=0
+        Tox=100n Uo=300 Phi=.6 Rs=.5286 Kp=10.29u W=.25 L=2u
Vto=-3.909
+        Rd=.54 Rds=266.7K Cbd=309.5p Pb=.8 Mj=.5 Fc=.5
Cgso=3.761n
+        Cgdo=498.1p Rg=3.99 Is=8.282f N=4 Tt=8800n)
*        Int'l Rectifier  pid=IRFC9110    case=4 Pin DIP
*        88-08-26 bam    creation
.model IRF252   NMOS(Level=3 Gamma=0 Delta=0 Eta=0 Theta=0
Kappa=0 Vmax=0 Xj=0
+        Tox=100n Uo=600 Phi=.6 Rs=2.081m Kp=20.86u W=1 L=2u
Vto=3.794
+        Rd=67.06m Rds=888.9K Cbd=3.481n Pb=.8 Mj=.5 Fc=.5
Cgso=1.585n
+        Cgdo=442.1p Rg=5.549 Is=168.3p N=1 Tt=340n)
*        Int'l Rectifier pid=IRFC250     case=TO3
*        88-08-25 bam    creation
.model MJE3055  NPN(Is=974.4f Xti=3 Eg=1.11 Vaf=50
+        Bf=99.49 DEV/UNIFORM 80% Ne=1.941
+        Ise=902.5p Ikf=4.029 Xtb=1.5 Br=2.949 Nc=2 Isc=0 Ikr=0 Rc=.1
+        Cjc=276p Vjc=.75 Mjc=.3333 Fc=.5 Cje=569.1p Vje=.75
Mje=.3333
+        Tr=971.7n Tf=39.11n Itf=20 Vtf=10 Xtf=2 Rb=.1)
*        Texas Inst.   pid=2N3055    case=TO3
*        Original Library
*        02 Jan 91     pwt    change Rb

.model Q2N3055  NPN(Is=974.4f Xti=3 Eg=1.11 Vaf=50
+        Bf=99.49 DEV/UNIFORM 80% Ne=1.941
+        Ise=902.5p Ikf=4.029 Xtb=1.5 Br=2.949 Nc=2 Isc=0 Ikr=0 Rc=.1
+        Cjc=276p Vjc=.75 Mjc=.3333 Fc=.5 Cje=569.1p Vje=.75
Mje=.3333
+        Tr=971.7n Tf=39.11n Itf=20 Vtf=10 Xtf=2 Rb=.1)
*        Texas Inst.   pid=2N3055    case=TO3
*        Original Library
*        02 Jan 91     pwt    change Rb
.model MJE2955  PNP(Is=66.19f Xti=3 Eg=1.11 Vaf=100
+        Bf=137.6 DEV/UNIFORM 85% Ise=862.2f
+        Ne=1.481 Ikf=1.642 Nk=.5695 Xtb=2 Br=5.88 Isc=273.5f
Nc=1.24
+        Ikr=3.555 Rc=79.39m Cjc=870.4p Mjc=.6481 Vjc=.75 Fc=.5
+        Cje=390.1p Mje=.4343 Vje=.75 Tr=235.4n Tf=23.21n Itf=71.33
+        Xtf=5.982 Vtf=10 Rb=.1)
*        National Semiconductor
*        Transistor Databook, 1982, process 5A, pg 9-30
*        30 Nov 90     pwt    creation
.model TIP31   ako:NSC_4F    NPN()  ; case TO-220

.model TIP31A  ako:NSC_4F    NPN()  ; case TO-220

.model TIP31B  ako:NSC_4F    NPN()  ; case TO-220

.model TIP31C  ako:NSC_4F    NPN()  ; case TO-220

.model TIP32   ako:NSC_5F    PNP()  ; case TO-220

.model TIP32A  ako:NSC_5F    PNP()  ; case TO-220

.model TIP32B  ako:NSC_5F    PNP()  ; case TO-220

.model NSC_5F  PNP(Is=51.23f Xti=3 Eg=1.11 Vaf=100 Bf=434.1
Ise=51.23f Ne=1.22
+        Ikf=.3883 Nk=.5544 Xtb=2.2 Br=55.47 Isc=51.23f Nc=1.205
+        Ikr=10.87 Rc=.3443 Cjc=136.9p Mjc=.3155 Vjc=.75 Fc=.5
+        Cje=179.9p Mje=.4294 Vje=.75 Tr=20.25n Tf=13.05n Itf=6.85
+        Xtf=1.573 Vtf=10 Rb=.1)
*        National Semiconductor
*        Transistor Databook, 1982, process 5F, pg 9-36
*        30 Nov 90     pwt    creation

.model NSC_4F  NPN(Is=2.447p Xti=3 Eg=1.11 Vaf=100 Bf=208.2
Ise=70.69p
+        Ne=1.565 Ikf=.9743 Nk=.6134 Xtb=1.5 Br=12.59 Isc=11.68n
+        Nc=1.835 Ikr=3.86 Rc=.4685 Cjc=142p Mjc=.4353 Vjc=.75
Fc=.5
+        Cje=188.5p Mje=.4878 Vje=.75 Tr=194.2n Tf=19.85n Itf=164.1
+        Xtf=5.945 Vtf=10 Rb=.1)
*        National Semiconductor
*        Transistor Databook, 1982, process 4F, pg 9-13
*        30 Nov 90     pwt    creation
* TL064 operational amplifier "macromodel" subcircuit
```

```
* created using Parts release 4.01 on 06/28/89 at 10:42
* (REV N/A)
* connections:   non-inverting input
*                | inverting input
*                | | positive power supply
*                | | | negative power supply
*                | | | | output
*                | | | | |
.subckt TL064/TI 1 2 3 4 5
*
 c1   11 12 3.498E-12
 c2    6  7 15.00E-12
 dc    5 53 dx
 de   54  5 dx
 dlp  90 91 dx
 dln  92 90 dx
 dp    4  3 dx
 egnd 99  0 poly(2) (3,0) (4,0) 0 .5 .5
 fb    7 99 poly(5) vb vc ve vlp vln 0 318.3E3 -300E3 300E3 300E3 -300E3
 ga    6  0 11 12 94.26E-6
 gcm   0  6 10 99 1.607E-9
 iss   3 10 dc 52.50E-6
 hlim 90  0 vlim 1K
 j1   11  2 10 jx
 j2   12  1 10 jx
 r2    6  9 100.0E3
 rd1   4 11 10.61E3
 rd2   4 12 10.61E3
 ro1   8  5 200
 ro2   7 99 200
 rp    3  4 150.0E3
 rss  10 99 3.810E6
 vb    9  0 dc 0
 vc    3 53 dc 2.200
 ve   54  4 dc 2.200
 vlim  7  8 dc 0
 vlp  91  0 dc 15
 vln   0 92 dc 15
.model dx D(Is=800.0E-18)
.model jx PJF(Is=15.00E-12 Beta=100.5E-6 Vto=-1)
.ends
.subckt LM339    1 2 3 4 5
*
 x_lm339 1 2 3 4 5 LM139
*
* the LM339 is identical to the LM139, but has a more limited temp. range
*
.ends
* connections:   non-inverting input
*                | inverting input
*                | | positive power supply
*                | | | negative power supply
*                | | | | open collector output
*                | | | | |
.subckt LM139    1 2 3 4 5
*
 f1    9  3 v1 1
 iee   3  7 dc 100.0E-6
 vi1  21  1 dc .75
 vi2  22  2 dc .75
 q1    9 21  7 qin
 q2    8 22  7 qin
 q3    9  8  4 qmo
 q4    8  8  4 qmi
.model qin PNP(Is=800.0E-18 Bf=2.000E3)
.model qmi NPN(Is=800.0E-18 Bf=1002)
.model qmo NPN(Is=800.0E-18 Bf=1000 Cjc=1E-15 Tr=475.4E-9)
 e1   10  4  9  4  1
 v1   10 11 dc 0
 q5    5 11  4 qoc
.model qoc NPN(Is=800.0E-18 Bf=20.69E3 Cjc=1E-15 Tf=3.540E-9
Tr=472.8E-9)
 dp    4  3 dx
 rp    3  4 37.50E3
.model dx  D(Is=800.0E-18 Rs=1)
*
.ends


.subckt N/O_switch 1 2 PARAMS: to=0
s1 1 2 3 0 sx
V_s1 3 0 pulse(0 2 {to} 1n 1n 999 9999)
.model sx Vswitch( Ron=.001 Roff=1g Von=1 Voff=0.5)
.ends


.subckt N/C_switch 1 2 PARAMS: to=0
s1 1 2 3 0 sx
V_s1 3 0 pulse(2 0 {to} 1n 1n 999 9999)
.model sx Vswitch( Ron=.001 Roff=1g Von=1 Voff=0.5)
.ends
*
* Ideal Transformer with V2 = aV1.
*
.subckt Ideal_XFMR_1/a 1 2 3 4 PARAMS: a=1
Rs1 1 a 1U
Rs2 3 b 1U
Rp1 1 2 1G
Rp2 3 4 1T
L1 a 2 1
L2 b 4 {a*a}
K1 L1 L2 .99999999
R1 2 4 1T
.ends
*
*
* Ideal Transformer Sprcifying Vout and Vin
*
.subckt Ideal_XFMR_Vo/Vin 1 2 3 4 PARAMS: Vin=115 Vo=12
Rs1 1 a 1U
Rs2 3 b 1U
Rp1 1 2 1G
Rp2 3 4 1T
L1 a 2 1
L2 b 4 {(Vo*Vo)/(Vin*Vin)}
K1 L1 L2 .99999999
R1 2 4 1T
.ends
```

```
*
*Switched Models
.model Dswitch  D(Is=0.1p Rs=1 CJO=2p Tt=12n Bv=1000 Ibv=0.1p)
.model Sx vswitch (Voff=0.5 Von=1.5)


* Laplace Transform Block Library
*
* Vo=K/(s+a)
*
.subckt Xform1 1 2 PARAMS: K=1, a=1
E1 1 0 LAPLACE {V(2)} = { K /(s+ a )}
Ro 1 0 1k
Rin 2 0 1T
.ends
*
* Vo=K(s+a)/s
*
.subckt Xform2 1 2 PARAMS: K=1, a=1
E1 1 0 LAPLACE {V(2)} = { K *(s+ a )}
Ro 1 0 1k
Rin 2 0 1T
.ends
*
.subckt Xform3 1 2 PARAMS: K=1,
a0=1,a1=0,a2=0,a3=0,a4=0,b0=1,b1=0,b2=0,b3=0,b4=0
E1 1 0 LAPLACE {V(2)} =
+ {(K
*(a4*s*s*s*s+a3*s*s*s+a2*s*s+a1*s+a0)/(b4*s*s*s*s+b3*s*s*s+b2*s*s+b1*
s+b0))}
Ro 1 0 1k
Rin 2 0 1T
.ends
*
*
*  Two input Summer
.subckt two_sum 1 2 3
E1 1 0 POLY(2) 2 0  3 0 0 1 1
Ro 1 0 1K
R1 2 0 1T
R2 3 0 1T
.ends
*
*  Two input Difference amp
.subckt two_diff 1 2 3
E1 1 0 POLY(2) 2 0  3 0 0 1 -1
Ro 1 0 1K
R1 2 0 1T
R2 3 0 1T
.ends
*
*
* Gain block
*
.subckt gain_block 1 2 PARAMS: K=1
Eo 1 0 VALUE={v(2,0)*K}
Ro 1 0 1K
Ri 2 0 1T
.ends
*

* Schematics Subcircuit *

.SUBCKT lm555 Vcc Threshold Control Trigger Ground Output Discharge
R_R1    Vcc Control 5K
R_R2    Control $N_0001 5K
R_R3    $N_0001 Ground 5K
X_U1A   R $N_0002 Output $G_DPWR $G_DGND 7402 PARAMS:
+ IO_LEVEL=0 MNTYMXDLY=0
X_U2A   Output S $N_0002 $G_DPWR $G_DGND 7402 PARAMS:
+ IO_LEVEL=0 MNTYMXDLY=0
R_R4    $N_0002 $N_0003 10K
Q_Q1    Discharge $N_0003 Ground Q2N3904
X_U19   Threshold Control  R 555_comp
X_U20   $N_0001 Trigger  S 555_comp
.ENDS   lm555



*----------------------------------------------------------------------------
.SUBCKT 555_COMP        v_plus v_minus v_out
*
* This is an ideal OpAmp model with the output voltage
* limited to 0 to +5 Volts+/- 15V.
*
Rin         v_plus  v_minus 10MEG
Eamp    v_out   0       TABLE {100000*(v(v_plus,v_minus))}=(0,0) (5,5)
R0          v_out   0       10MEG
.ENDS
*
*
* Op-Amp_Breakout operational amplifier "macromodel" subcircuit
* This is a copy of the 741 op-amp model.
* (REV N/A)
* connections:   non-inverting input
*                | inverting input
*                | | positive power supply
*                | | | negative power supply
*                | | | | output
*                | | | | |
.subckt Op-Amp_Breakout    1 2 3 4 5
*
  c1   11 12 4.664E-12
  c2    6  7 20.00E-12
  dc    5 53 dx
  de   54  5 dx
  dlp  90 91 dx
  dln  92 90 dx
  dp    4  3 dx
  egnd 99  0 poly(2) (3,0) (4,0) 0 .5 .5
  fb    7 99 poly(5) vb vc ve vlp vln 0 10.61E6 -10E6 10E6 10E6 -10E6
  ga    6  0 11 12 137.7E-6
  gcm   0  6 10 99 2.574E-9
  iee  10  4 dc 10.16E-6
  hlim 90  0 vlim 1K
  q1   11  2 13 qx
```

```
 q2   12  1 14 qx
 r2    6  9 100.0E3
 rc1   3 11 7.957E3
 rc2   3 12 7.957E3
 re1  13 10 2.740E3
 re2  14 10 2.740E3
 ree  10 99 19.69E6
 ro1   8  5 150
 ro2   7 99 150
 rp    3  4 18.11E3
 vb    9  0 dc 0
 vc    3 53 dc 2.600
 ve   54  4 dc 2.600
 vlim  7  8 dc 0
 vlp  91  0 dc 25
 vln   0 92 dc 25
.model dx D(Is=800.0E-18)
.model qx NPN(Is=800.0E-18 Bf=62.50)
.ends

.SUBCKT MbreakN_Sub 1 2 3
M1 1 2 3 3 MbreakN
.ends

.SUBCKT MbreakP_Sub 1 2 3
M1 1 2 3 3 MbreakP
.ends *        85-??-??     Original library
```

# INDEX

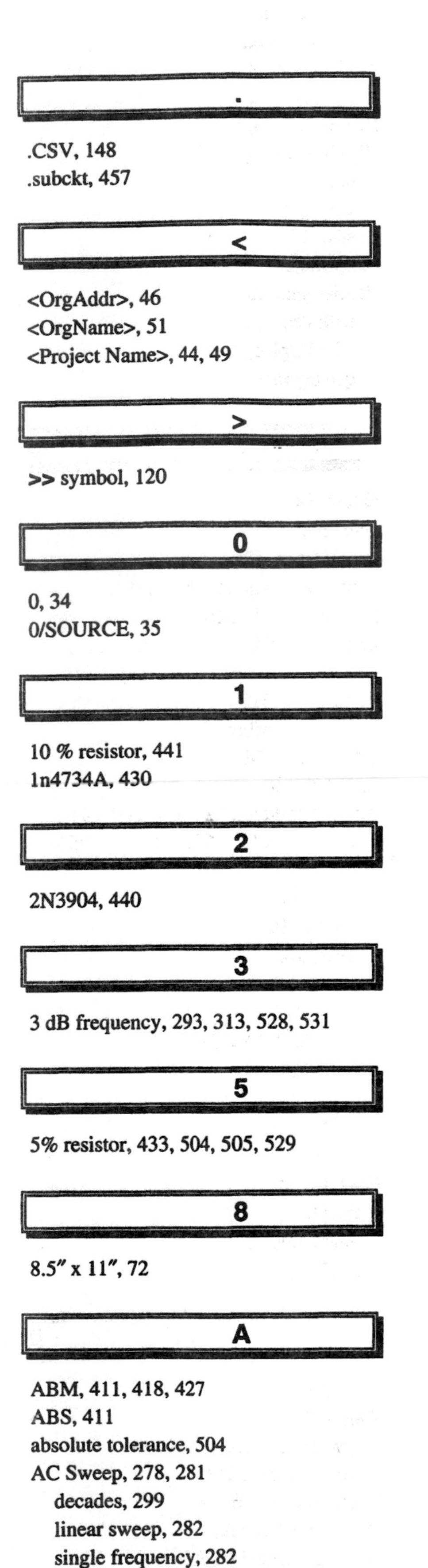

## O



## P

## Q



## R



## S



## T



## U



## V